D1591595

Sensory Mechanisms of the Spinal Cord

SECOND EDITION

Sensory Mechanisms of the Spinal Cord

SECOND EDITION

W.D. Willis, Jr.
and
R.E. Coggeshall

The University of Texas Medical Branch at Galveston
Galveston, Texas

Plenum Press • New York and London

Library of Congress Cataloging-in-Publication Data

Willis, William D., 1934-
Sensory mechanisms of the spinal cord / W.D. Willis and R.E. Coggeshall. -- 2nd ed.
p. cm.
Includes bibliographical references and index.
ISBN 0-306-43781-3
1. Spinal cord. 2. Senses and sensation. 3. Afferent pathways. 4. Spinal cord--Localization of function. I. Coggeshall, Richard E., 1932- . II. Title.
[DNLM: 1. Receptors, Sensory. 2. Spinal Cord. WL 400 W735s]
QP374.W54 1991
612.8'3--dc20
DNLM/DLC
for Library of Congress

91-21065
CIP

QP
374
.W54
1991

ISBN 0-306-43781-3

© 1991, 1978 Plenum Press, New York
A Division of Plenum Publishing Corporation
233 Spring Street, New York, N.Y. 10013

All rights reserved

No part of this book may be reproduced, stored in a retrieval system, or transmitted in any form or by any means, electronic, mechanical, photocopying, microfilming, recording, or otherwise, without written permission from the Publisher

Printed in the United States of America

Preface

As stated in the preface to the first edition, the goal of this monograph is to provide an overview of current thought about the spinal cord mechanisms responsible for sensory processing. We hope that the book will be of value to both basic neuroscientists and clinicians.

The organization of the monograph has followed the original plan in most respects, although the emphasis has changed with respect to many topics because of recent advances. In particular, a substantial increase in the number of investigations of the dorsal root ganglion has led us to devote a chapter to this topic. The treatment of chemical neuroanatomy in the dorsal horn, as well as the relevant neuropharmacology, has also been expanded considerably. Another major emphasis is on the results of experiments employing microneurography in human subjects.

We thank Margie Watson and Lyn Schilling for their assistance with the typing and Griselda Gonzales for preparing the illustrations.

Contents

Chapter 3

Dorsal Root Ganglion Cells and Their Processes ... **47**

Chapter 4

Structure of the Dorsal Horn ... **79**

1 Introduction

SPECIFICITY VERSUS PATTERN THEORIES OF SENSATION

Historical Perspective

Specificity Theory. The way in which the nervous system differentiates among the various forms of sensory experience has been a central issue since the beginning of sensory physiology. A brief history of the major theories is given by Sinclair (1981). The notion of specificity of cutaneous sensation is attributed by Sinclair to Bell (1811), although it was actually Magendie (1822) who demonstrated the sensory function of the spinal cord dorsal roots. The idea of specificity of cutaneous sensation was forwarded by Müller's doctrine of specific nerve energies (Müller, 1840–1842). Müller had in mind the Aristotelean five senses and lumped together the sensations derived from the body surface under the category of "touch." According to Sinclair (1981), Volkmann, Natanson and others in the 1840s extended the specificity concept to include the postulate of separate nerve endings for each variety of sensation arising from cutaneous stimulation. However, Sinclair emphasizes the distinction between "specific nerve energies" (i.e., particular nerves evoke particular sensations) and "specific irritability" (i.e., particular stimuli activate particular sense organs). It is one thing to show that specific stimuli preferentially activate certain sensory receptors; it is another to show that the same sensory receptors are responsible for a particular quality of sensory experience.

Evidence supporting the notion that the doctrine of specific nerve energies applied to the different cutaneous senses came from the observations of Blix (1884), who discovered that stimulation of separate localized points on the skin gave rise to distinct sensations of pressure, warmth, cold, or pain. This observation was confirmed by numerous other investigators, including Goldscheider (1884) and Donaldson (1885). Specificity was shown by the observation that a cold spot, for example, could be stimulated by cold, but not by heat, whereas a warm spot responds to heat but not cold (Blix, 1884). Even if a cold (or warm) spot is stimulated by an electric current, the sensation that results is still of cold (or warm). Mapping of marked areas of skin proved that the cold and warm spots are in fixed positions that can be identified on subsequent days with a high degree of accuracy (Fig. 1.1A), provided that care is taken to avoid a number of sources of technical error (Dallenbach, 1927).

Cold and warm spots were shown by careful mapping to be distinct from pressure and pain spots (Blix, 1884; Goldscheider, 1884; von Frey, 1896). The density of the different kinds of sensory spots was found to vary from one part of the body surface to another. For example, von Frey (1896) estimated that there are about 100–200 pain spots/cm^2 on the back of the hand, about 8 times the density of pressure points in the same region (Fig. 1.1B). However, there are 100–200 pressure spots/cm^2 on the palm. In an area of 29 mm^2 on the conjunctiva, von Frey (1896) found

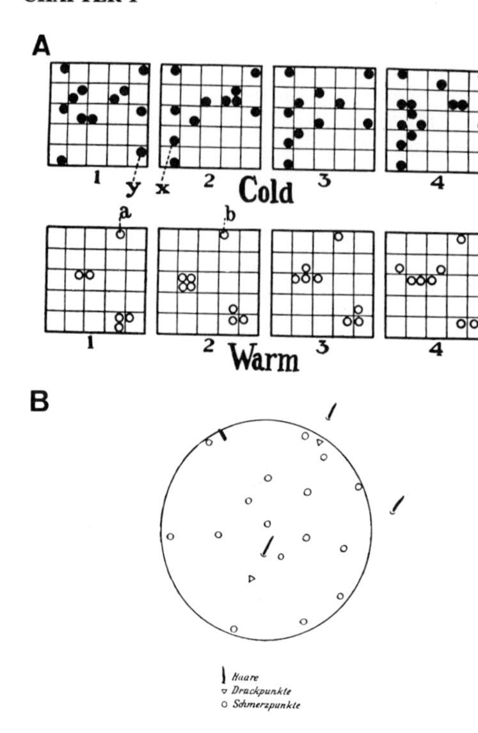

Fig. 1.1. (A) Maps of the distribution of cold and warm spots within an area of 1 cm^2 in a single subject. The maps were made on four different days (1–4). The cold spots were mapped before the warm spots during a given session. (From Dallenbach, 1927.) (B) Map of pressure (Druckpunkte) and pain (Schmerzpuncte) spots in an area of 12.5 mm^2 on the back of the hand. The locations of hairs (Haare) are also indicated. (From von Frey, 1896.)

35 pain spots and 10 cold spots. In general, pain spots are the most common. Cold spots are somewhat more frequent than warm spots (Donaldson, 1885).

The spotlike distribution of mechanically and thermally sensitive areas suggested that each spot was associated with a specific sensory receptor organ. If so, there would be a causal linkage between individual sensory receptors and specific sensations. Histologists had begun to describe a variety of types of cutaneous sense organs (Krause, 1859; Meissner, 1859; Ruffini, 1894), and so it was reasonable for von Frey (1906, 1910) to suggest that each kind of sensory spot was associated with a particular type of sense organ. On the basis of his understanding of the distribution of sense organs, von Frey proposed the following relations between sensation and activation of particular sense organs: touch, hair follicle endings in hairy skin and Meissner's corpuscles in glabrous skin; cold, Krause's end bulbs; warmth, Ruffini endings; pain, free endings.

The match suggested by von Frey between particular sensory receptors and sensory spots was found to be incorrect with respect to Ruffini endings and Krause's end bulbs and the warm and cold spots (Donaldson, 1885; Dallenbach, 1927; Weddell, 1955; Weddell and Sinclair, 1953; Sinclair, 1955; Lele and Weddell, 1956, 1959; Weddell and Miller, 1962). Ruffini endings have now been identified with SA II slowly adapting mechanoreceptors (M. R. Chambers *et al.*, 1972), and Krause's end bulbs are a variety of rapidly adapting mechanoreceptor (Iggo and Ogawa, 1977). Presumably, these have tactile sensory functions, although, at least for SA II receptors, this has not been directly demonstrated. Thermal sensibility, like pain, depends upon the activation of "free" (i.e., unencapsulated) endings, although the nature of the receptor membrane is different for different types of sensory endings (Andres and von Düring, 1973; Hensel *et al.*, 1974; Kruger *et al.*, 1981, 1985). It seems likely that molecular differences, such as peptide content and intracellular or surface antigens, will eventually provide the means for distinguishing morphologically between different functional types of "free endings" (Jessell and

Dodd, 1985; Lawson *et al.*, 1985; O. Johansson and Vaalasti, 1987; Kruger, 1987; Alvarez *et al.*, 1988; Kruger *et al.*, 1989).

The criticisms by Weddell and colleagues of von Frey's theory were directed not so much at the notion of sensory spots as at the histological correlates of the spots. However, even the existence of sensory spots was called into question (Guilford and Lovewell, 1936; Jenkins, 1941a,b; see Sinclair, 1955). Furthermore, Weddell and Miller (1962) point out that it is unlikely that a naturally occurring stimulus applied to a sensory spot would activate only a single afferent fiber, since a stimulus applied to just 1 mm^2 of skin could activate more than 100 underlying endings. This criticism loses some force if the differential sensitivity of the endings is taken into account. Nevertheless, Burgess *et al.* (1974) estimate that the application of a weak stimulus (less than 100 mg) by a probe with a diameter of 0.5–1 mm will excite 10–15 rapidly adapting mechanoreceptive afferent fibers from several types of receptors found in cat hairy skin. Stronger stimuli, if favorably placed, would also activate slowly adapting mechanoreceptors. Similarly, R. S. Johansson and Vallbo (1976) provide evidence "that practically all naturally occurring tactile stimuli to the human hand excite a large number of sensory units, setting up a pattern of neural activity from the population of mechanosensitive units" and that "even the smallest variation of a suprathreshold stimulus . . . will appreciably change this pattern." Thus, pressure spots are likely to be complex entities involving several afferent fibers innervating more than one receptor type.

Pattern Theory. Because of these difficulties with the specificity theory, another theory of cutaneous sensation seemed to be required. An alternative was a pattern theory. This was based on the observation that sensory stimuli are encoded by trains of nerve impulses in the nerve fibers that supply sensory receptor organs (Adrian, 1946). Nafe (1927, 1929) suggested that a sensation results from a patterned input from sense organs of the skin that is usually, but not necessarily always, associated with a particular kind of stimulus. Learning provides the name for the sensation associated with the stimulus. Specific sensory channels are not needed, just particular spatial and temporal patterns of nerve impulses in the central nervous system. This theory was supported by Sinclair (1955) and by Weddell (1955).

However, a difficulty for the pattern theory was the finding that large myelinated afferent fibers play a special role in touch and small myelinated and unmyelinated fibers in pain and temperature sensations (Fig. 1.2) (Heinbecker *et al.*, 1933, 1934; Lewis and Pochin, 1938a,b; Torebjörk and Hallin, 1973). Furthermore, single-unit recording studies began to demonstrate that mechanoreceptors are often supplied by large myelinated fibers and nociceptors and thermoreceptors by small myelinated and unmyelinated fibers (Hensel and Zotterman, 1951a; Dodt, 1952; Dodt and Zotterman, 1952a; Iggo 1959, 1960; Hensel *et al.*, 1960; C. C. Hunt and McIntyre, 1960c; Iriuchijima and Zotterman, 1960), although some mechanoreceptors are also innervated by unmyelinated fibers (W.W. Douglas and Ritchie, 1957; Iggo, 1960).

Gate Theory. With the recent strengthening of the evidence that sensory receptors play specific sensory roles, a compromise theory was proposed in which specific information from the sense organs generates patterned activity centrally. The model of Melzack and Wall (1965) is known as the "gate theory" of pain. However, this theory does not attempt to account for modalities of cutaneous sensation other than pain. The details of the original neural circuit suggested by Melzack and Wall have been challenged, as discussed in Chapter 5.

Clinical evidence suggests that sensory pathways in the central nervous system are specific, since dissociated sensory loss can be produced in such disorders as the Wallenberg syndrome and after therapeutic interventions such as anterolateral cordotomy for the relief of pain (Willis, 1985). Melzack and Wall (1962) caution that a lesion placed in the spinal cord white matter can affect sensory transmission in a number of ways in addition to the possible interruption of a specific sensory pathway. For example, a lesion will reduce the total number of tract neurons capable of responding to sensory input; furthermore, the pattern of activity in the ascending

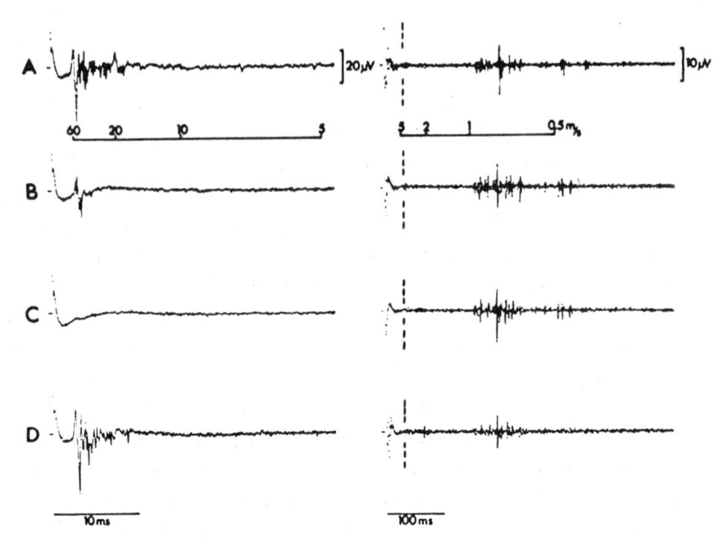

Fig. 1.2. Recordings of the activity of A and C fibers of the radial nerve at the wrist in a human subject during pressure block of the nerve. The records in the left column were taken on a faster time base than those in the right column (see conduction velocity scales in panel A). Note the reduction in the A-fiber response in panel B and its elimination in panel C as the block progressed and the recovery in panel D. The C-fiber response was not blocked, as shown in the records at the right. Sensation associated with the A-fiber volley was tactile, whereas that associated with the C fiber volley was a prolonged, burning pain. (From Torebjörk and Hallin, 1973.)

pathways will be altered. Another important consideration is that pathways descending from the brain that control the activity in sensory pathways may be interrupted, with a consequent change in the operation of the sensory pathways. However, a case report by Noordenbos and Wall (1976) provides additional evidence that sensory information concerning pain, temperature, and certain aspects of tactile sense is conveyed to the brain by pathways ascending in the anterolateral quadrant of the spinal cord, whereas other mechanical senses depend upon pathways in other parts of the spinal cord. These observations are consistent with the notion that there are specific sensory pathways in the central nervous system.

Recent Evidence from Microneurography

Sinclair's Challenge. As mentioned above, specificity of responsiveness of sense organs must be distinguished from specificity of the sensations that their activity elicits. Response specificity does not necessarily lead to sensory specificity. For example, baroreceptor afferents respond in a rather specific way to changes in blood pressure, but there is no corresponding sensory experience. Another question is whether activation of a cold receptor would necessarily lead to a sensation of cold. Is the receptor both a cold detector and a provoker of cold sensation? Sinclair (1981) stated that "There is as yet no unassailable evidence that stimulation of a single fibre in isolation can give rise to a sensation of any kind. To stimulate a single fibre in an intact human subject, to prove satisfactorily that only that fibre and no other has been stimulated, and to record a simultaneous meaningful sensory judgement is an almost incredibly difficult technical and psychophysiological feat, and it may be a long time before unequivocal evidence can be obtained."

Sensory Role of Single Tactile Afferents. Recently, experiments that appear to meet the stringent criteria described above by Sinclair have been done by using intraneural microneurography in human subjects (Hagbarth and Vallbo, 1967; Torebjörk and Ochoa, 1980, 1990;

Konietzny *et al.*, 1981; Vallbo, 1981; Ochoa and Torebjörk, 1983, 1989; Schady and Torebjörk, 1983; Schady *et al.*, 1983; Torebjörk *et al.*, 1984a,b; 1987; Macefield *et al.*, 1990). A metal microelectrode is introduced into a peripheral nerve. A search is conducted while either recording or stimulating with the electrode. When the search is conducted by using stimulation, the experience of a unitary sensation is the end point. The sensation is investigated psychophysically, and then the electrode is switched to a recording mode; there is consistently a sensory unit within recording range whose receptive field corresponds to the area of sensory referral (Fig. 1.3A) (Torebjörk and Ochoa, 1980; Schady and Torebjörk, 1983). The response properties of the unit are examined (Fig. 1.3B), and the conduction velocity determined (Fig. 1.3C). Verification that the same nerve fiber is responsible for both the unitary sensation and the recorded response is obtained by application of a high-frequency train of stimuli. This causes the nerve fiber to develop a state of hyperexcitability, as shown by increases in its responses to stimulation of the receptive field, either mechanically or electrically (Fig. 1.3D and E). Fine myelinated and unmyelinated afferent fibers do not become hyperexcitable, but instead their conduction velocity shows when they are "marked" by repetitive stimulation (Torebjörk and Ochoa, 1990).

The correspondence between the projected and receptive fields is shown for three different units in Fig. 1.3F (Ochoa and Torebjörk, 1983). If the stimulus intensity is raised, activation of an adjacent sensory unit is signaled by the development of an additional area of sensory referral. For example, in recordings from the median nerve, the receptive field might be on a digit. Raising the stimulus intensity causes no change at first, but another sensation is detected eventually. This is typically referred to a different receptive field, for example on another digit (Torebjörk and Ochoa, 1980; Ochoa and Torebjörk, 1983), although sometimes the original area of referral enlarges (Vallbo, 1981; Schady *et al.*, 1983). In some experiments, a second electrode was inserted into the nerve at a more proximal location to show that the unitary sensation corresponds in an all-or-none fashion to the activation of a single sensory unit (Ochoa and Torebjörk, 1983). These observations are evidence that only a single afferent fiber is being stimulated when a unitary sensation is experienced.

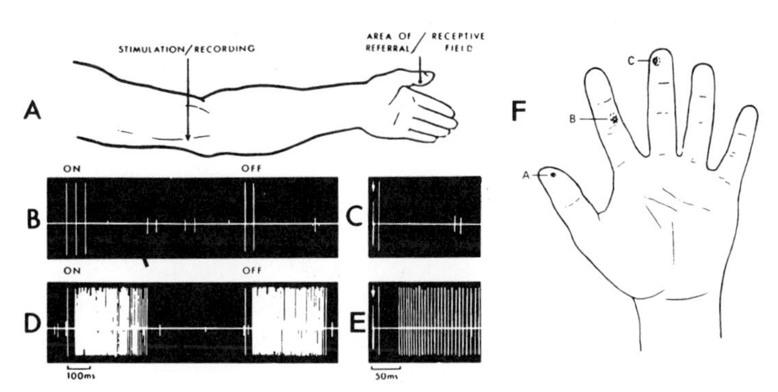

Fig. 1.3. (A) Method used for stimulating single afferent nerve fibers during microneurography. Recordings are made from a nerve fiber by using a metal microelectrode inserted percutaneously into a peripheral nerve. The receptive field is mapped, and then the nerve fiber is stimulated through the recording electrode. The area to which sensation is referred is compared with the receptive field of the receptor supplied by the fiber. The recordings in panels B–E were made from the axon from a rapidly adapting receptor, whose stimulation evoked a sensation of flutter-vibration. (B) On- and off-responses of the unit produced by brief indentation of the skin. (C) Conduction delay following electrical stimulation of the skin; the conduction velocity was 45 m/s. (D and E) Responses to mechanical and electrical stimulation after the receptor was sensitized by high-frequency stimulation. (From Torebjörk and Ochoa, 1980.) (F) Correspondence between the receptive field and the area of sensory projection for three units in the median nerve. One unit was a rapidly adapting receptor (A), and the other two were Merkel cell receptors (B and C). (From Ochoa and Torebjörk, 1983).

A single electrical stimulus applied to a single nerve fiber usually does not evoke any sensation, although for FA I receptors it may. Generally, repetitive stimulation is required. However, when a sensation is produced (by using a suprathreshold stimulus intensity lower than the intensity required to cause recognition of a second sensory experience), it is always of the same modality, whatever the frequency of stimulation. The modality appears to be more "pure" than that experienced following a natural stimulus. As mentioned above, a natural stimulus would activate several kinds of sensory receptors, and the mixed input would produce mixed sensation. The particular modality of sensation elicited by stimulation of a single sensory axon depends upon the receptor type, and the intensity of the sensation depends upon the frequency of stimulation (for rapidly adapting receptors, increases in frequency at first affect just the quality of the sensation, but above some frequency the intensity increases [Ochoa and Torebjörk, 1983]). Stimulation of afferents from most of the commonly observed receptor types produces the sensations listed in Table 1.1. Of particular note is that pain is produced only when nociceptors are stimulated; no pain results from high-frequency stimulation of mechanoreceptors. It is possible that the negative results from stimulation of SA II afferents from Ruffini endings reflect the need for activation of more than one fiber to meet the requirements for spatial summation in the central pathways.

The general conclusion that can be reached from these experiments is that there is specificity of primary afferent fibers supplying particular cutaneous sense organs with respect both to the adequate stimulus and to the sensation evoked. Although it can be questioned whether only single afferent fibers were excited in all of these experiments, particularly those involving small fibers, the conclusion about specificity appears to hold even if a small number of afferent fibers were sometimes being stimulated.

Criticism of Microneurography Results. Wall and McMahon (1985) have challenged the interpretation of the microneurography experiments. They suggest that the large microelectrodes used in such studies (Fig. 1.4) are unlikely to be able to record and stimulate single axons selectively unless adjacent fibers are blocked by the mechanical trauma produced by the search procedure. If a large number of axons are blocked, the sensory experience reported might be abnormal. Furthermore, a number of the observations from the microneurography experiments suggest that the sensations reported might depend upon central processing of input over a number of afferent fibers rather than a single fiber, and so the identification of sensory modalities may depend upon a patterned input to the central nervous system in these experiments.

Rebuttal of Criticism. Although Wall and McMahon (1985) present a reasonable case for caution in interpreting the microneurography experiments, it is important to recognize that if a

TABLE 1.1. Sensations Evoked by Stimulation of Single Identified Cutaneous Afferent Fibers during Intraneural Microneurography in Human Subjects[a]

Receptor	Sensation
FA I (RA)	Tapping at 1 Hz; flutter at 10 Hz; vibration at 50 Hz
FA II (PC)	Tickling or vibration over 20–50 Hz
SA I	Sustained pressure over 5–10 Hz
SA II	No sensation
A delta mechanical nociceptor	Sharp pain
C polymodal nociceptors	Dull pain, burning pain or itch

[a]From Torebjörk and Ochoa (1980); Ochoa and Torebjörk (1983); Schady *et al.* (1983); Torebjörk *et al.* (1984a,b); Ochoa and Torebjörk (1989); Torebjörk and Ochoa (1990); and Macefield *et al.* (1990). For C fibers, small groups of axons rather than individual fibers may have been stimulated.

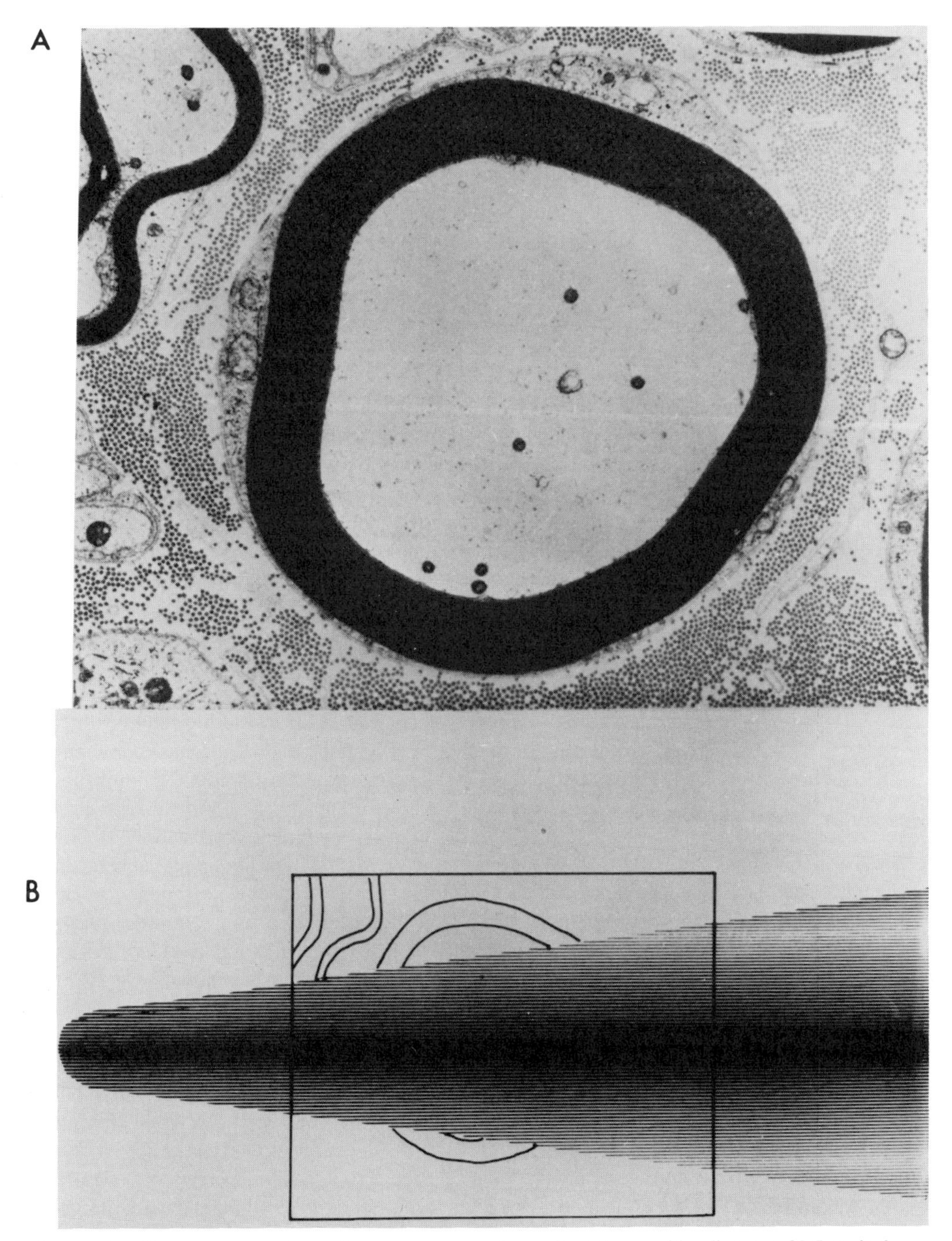

Fig. 1.4. (A) Electron micrograph showing a cross-section of a myelinated axon with a diameter of 9.5 μm in the human sural nerve. (B) Drawing of the same axon, with a representation of the tip of a microneurography recording electrode superimposed. (From Wall and McMahon, 1985.)

single afferent fiber of a particular type can be activated and can be shown to evoke a specific sensation, the evidence in favor of the specificity theory is strengthened. Although one can never be certain that only one axon is excited in experiments on humans, it is still hard to dismiss many of the observations described in the microneurography experiments. Furthermore, in animal experiments, in which it is possible to obtain better stimulus control, activation of an individual sensory receptor can result in the activation of neurons in the spinal cord dorsal horn and the dorsal column nuclei, cortical evoked potentials, and behavioral responses (McIntyre *et al.*, 1967; Tapper and Mann, 1968; Tapper, 1970; Mann *et al.*, 1972; P. B. Brown *et al.*, 1973; Ferrington *et al.*, 1987a). Therefore, it certainly seems plausible that stimulation through a microneurography electrode can activate a single afferent fiber, which in turn evokes a unitary sensation.

Torebjörk *et al.* (1987) deal carefully with the issues raised by Wall and McMahon (1985). The microneurography technique is reviewed, and evidence is presented that there is not severe damage to the nerve with a consequent block of conduction of many afferent fibers. Torebjörk *et al.* (1987) state that there is no reason to believe that conduction is blocked in most of the fibers being recorded from, although there is a change in the action potential configuration as a result of the presence of the recording electrode. The sensory experience is argued not to depend upon changes due to a loss of afferent barrage from the damage, since damage was limited and since the only cutaneous mechanoreceptive afferent units that are normally active are from SA II receptors, which make up only 19% of the mechanoreceptor population. The current intensities required to produce threshold sensations averaged only 0.81 μA, which agrees with the current needed to activate single axons in animal experiments. The currents used were anodal. Since the microelectrode tip was thought to be within the myelin sheath or partially intracellular in the axon, as indicated by the configuration of the action potential, adjacent fibers to the one stimulated should not be excited. Torebjörk *et al.* conclude that "specificity is a key principle for tactile systems."

Torebjörk *et al.* (1987) list a series of proposals that were made by Melzack and Wall (1962) and by Wall and McMahon (1985) and for which there is contrary evidence from microneurography. These proposals and the contrary findings from microneurography include the following. (1) "More than one nerve impulse from a single afferent fibre, or more than one fibre carrying single nerve impulses, is essential for central cells to detect the characteristics of a sensory stimulus" (Melzack and Wall, 1962). This does not hold for FA I receptors in the finger, since a single impulse in the axon of at least some of these receptors can evoke a specific sensation. (2) "Every discriminably different somaesthetic perception is produced by a unique pattern of nerve impulses" (Melzack and Wall, 1962). Changing the temporal pattern of impulses in an SA I unit does not change the quality of pressure sensation that is experienced, and stimulating an FA I unit or an SA I unit with the same pattern evokes quite different sensations. (3) "Perceived sensation is not determined in quality and location and time course by the presence of action potentials in uniquely specialized fibres" (Wall and McMahon, 1985). This does not hold for FA I and SA I units. (4) "Threshold of perceived sensations depends not on the presence or absence of impulses in a particular type of fibre but on their achieving certain levels of firing" (Wall and McMahon, 1985). This does not hold for FA I units in the fingers, but does apply to SA I units. (5) "The time course of the afferent barrage is poorly related to the time course of perception" (Wall and McMahon, 1985). This does not apply to FA I units whose activity reproduces the time course of the sensation. (6) "Wide spatial gradients of stimulus and of afferent activity are abstracted by the brain to achieve a perception of apparent elementary sensations with a spot location" (Wall and McMahon, 1985). This holds in part, since surround inhibition may sharpen receptive fields, but spatial summation is not needed for accurate localization of the fields of FA I and SA I units. However, spatial summation may be required for SA II receptors to evoke sensation.

Two "essential conclusions" reached by Torebjörk *et al.* (1987) are that "physiological specificity of low-threshold mechanoreceptors in the hand is linked to distinct attributes of very simple tactile sensations" and that "afferent inputs are processed differently in tactile subsystems."

Microneurography and Pain. Ochoa and Torebjörk (1989) performed an extensive investigation of the sensations associated with the activation of C polymodal nociceptors during microneurography. They used microstimulation in a peripheral nerve as a search stimulus. The sensory quality and localization of the subjective responses to the stimuli were then correlated with the ability to record from one or more single C units at the same electrode position. They found that stimulation of C fibers supplying glabrous skin of the hand more often provoked a sensation of dull pain than of burning pain, but that stimulation of C fibers from hairy skin always caused a burning pain sensation.

Commonly, the activity of more than one C fiber could be recorded from a stimulation site that evoked pain. Furthermore, the projected receptive fields for pain were often larger than the receptive fields of single C fibers, and the areas of the receptive fields grew larger as the stimulus intensity was increased. These observations indicate that several C fibers were probably activated simultaneously by the stimuli. Nevertheless, a pure sensation of dull or burning pain resulted (although, in some instances, stimulation of C fibers produced itch). It was suggested that the difference between the glabrous and hairy skin with respect to the sensations of dull and burning pain might relate to the number of C fibers that were simultaneously activated, since it appeared that fewer glabrous- than hairy-skin C fibers were stimulated at a given site. That is, burning pain might require spatial facilitation.

An important observation was that the quality of the pain was unchanged when the myelinated fibers in the nerve were blocked by pressure, indicating that the sensory quality depended strictly on the input rather than on modulation by activity in more rapidly conducting afferent fibers.

Signals from the C-fiber volleys were relatively well localized if the projected receptive fields were compared with those mapped. The correspondence was generally within 10 mm. This can be compared with the psychophysical finding that a hot stimulus applied to glabrous skin on the hand is localized within 7–10 mm (tactile stimuli are localized within 4–7 mm).

EPICRITIC VERSUS PROTOPATHIC SENSATIONS

Brief consideration should be given to the proposal of Head (1920) that the cutaneous senses can be divided into two broad categories—protopathic and epicritic. The protopathic system, according to this view, mediates pain and the extremes of temperature sensation, while the epicritic system is responsible for touch, size and two-point discrimination, and detection of small thermal gradients. The experimental basis for the proposal was the introspective analysis of sensory changes in an area of skin after denervation and during regeneration of the cut nerve. Head himself was the experimental subject. Many of the observations and most of the interpretations of this experiment have been contested (Trotter and Davis, 1909; Boring, 1916; Walshe, 1942; Weddell *et al.*, 1948), and the hypothesis should be laid to rest. However, perhaps because of the appeal of the terminology (cf. Walshe, 1942), the protopathic–epicritic dichotomy lingers. In fact, the terminology is sometimes applied to central nervous system pathways, such as the spinothalamic tract and the dorsal column pathway, even though Head clearly stated that the information conveyed by the two proposed systems in the periphery became intermingled centrally.

The notion of Head (1920) that somatic sensory functions can be described in terms of a dual system of epicritic and protopathic sensory mechanisms has been superseded by two other dualities: lemniscal versus nonlemniscal systems and large- versus small-fiber systems.

LEMNISCAL VERSUS NONLEMNISCAL SYSTEMS

The lemniscal versus nonlemniscal nomenclature was developed by Poggio and Mountcastle (1960, 1963) when describing the responses of neurons in the thalamus to cutaneous stimulation. Lemniscal responses were defined as those with the following attributes: the thalamic neurons have small, contralateral receptive fields; the kinds of effective stimuli are restricted, indicating that only one or a few kinds of sensory receptors are involved in activating a given thalamic neuron; and synaptic transmission is secure, so the cells can follow relatively high stimulus rates. Nonlemniscal neurons have the following traits: the thalamic cells have large, often bilateral receptive fields; they receive convergent input from different kinds of receptors; and they are unable to follow repetitive stimulation well. The assumption was made that the lemniscal neurons of the thalamus, in recordings from the ventrobasal complex, were activated by the dorsal column-medial lemniscus pathway, whereas the nonlemniscal neurons, whose activity was recorded from other parts of the thalamus, such as the posterior nuclear complex, were excited by the spinothalamic tract (Poggio and Mountcastle, 1960). This nomenclature has been criticized (Boivie and Perl, 1975), since both the medial lemniscus and the spinothalamic tract each project to both the ventrobasal and posterior complexes. Furthermore, neurons of both the dorsal column-medial lemniscus system and the spinothalamic tract can be found that have properties that are consistent with either category of thalamic neuron (as do neurons in other sensory pathways accessing the thalamus). Boivie and Perl (1975) suggest that the terms "lemniscal" and "nonlemniscal" be replaced by "specified" and "unspecified," although they warn that "unspecified" may simply reflect ignorance of the functional role of the neuron.

LARGE- VERSUS SMALL-FIBER SYSTEMS

The other duality often cited in the current literature is the notion of large-fiber versus small-fiber systems (Noordenbos, 1959; Melzack and Wall, 1965). The implication of this terminology is that the large-fiber system has to do with innocuous forms of mechanoreception, whereas activation of the small-fiber system is required for pain. Interactions occur between the inputs conveyed by the large- and small-fiber systems to central neurons, and the central nervous system determines from the outcome of these interactions whether or not a stimulus is painful. The difficulty with this nomenclature is that it tends to obscure the fact that both large (Aα and Aβ) and small (Aδ and C) fiber groups in cutaneous nerves include axons that supply a variety of receptor types. Although most nociceptors have small axons (except for a few that conduct at Aβ velocities), there are a number of different kinds of cutaneous receptors that have small axons but that are not nociceptors. These include sensitive mechanoreceptors and specific thermoreceptors (see previous discussion). Consequently, it is dangerous to assume that the results of stimulation of small fibers (for example, by electrical stimulation) should necessarily be attributed to the activation of nociceptors. Furthermore, mixed nerves contain muscle and joint afferents as well as cutaneous afferents, and so there are more receptor types whose axons are represented in mixed nerves than in purely cutaneous nerves. Therefore, the interpretation of the results of stimulating mixed nerves is even more problematic.

SENSORY MODALITIES

According to Boring (1942), the term "sensory modality" was introduced by Helmholtz. A modality is a class of sensations connected by qualitative continua. That is, sensations of two different modalities are qualitatively different (e.g., hearing versus vision), whereas sensations within a single modality are only quantitatively different (e.g., different intensities of two sounds). Although Müller (1840–1842) regarded the general sense of "feeling" or "touch" derived from skin stimulation as an entity, following Aristotle's classification of the five senses, most later investigators have considered touch (or pressure), cold, warmth, and pain to be discrete modalities. A given sensory modality would have a number of characteristics or attributes, such as quality, intensity, duration, and extension. On the basis of subjective awareness and also of clinical and psychophysical testing, a number of sensory modalities pertaining to the sensory experiences resulting from stimulation of the skin or of subcutaneous tissue may be recognized. These include touch-pressure, flutter-vibration, tickle, warmth, cold, pain, itch, position sense, and kinesthesia. Some of these may be further subdivided. For example, pain may be considered to have several submodalities, including pricking pain, burning pain, and aching pain. More complicated somatic sensations, perhaps involving combinations of modalities, are also of interest, especially in clinical testing. These include two-point discrimination, stereognosis, graphesthesia, and the abnormal sensations called paresthesias. Visceral sensations include the awareness of distention, hunger, nausea, and visceral pain. [See Geldard (1953) for a more extensive list of somatovisceral modalities.]

SENSORY CHANNELS

Each modality of sensation depends upon information transmitted along one or more sensory pathways. Transection of the spinal cord completely abolishes any awareness of sensation from regions of the body below the transection. Interruption of one or more of the sensory pathways may partially or totally eliminate a particular kind of sensory experience.

The mechanism for transmission of information concerning a sensory modality can be defined as a "sensory channel." A sensory channel would include one or more sets of sensory receptors, one or more sensory pathways, and particular regions of the thalamus and cerebral cortex that are involved in receiving and processing the information. Activity in a sensory channel would be under centrifugal control by way of descending pathways.

* * *

The first part of this book will be concerned with the organization of the peripheral nervous system and spinal cord. Then the sensory pathways in the spinal cord will be discussed, along with the kinds of information they carry. Their descending control will also be considered. Portions of several of the sensory channels will be discussed in the last chapter.

CONCLUSIONS

1. The specificity theory of cutaneous sensation was originally based on the discovery of localized sensory "spots" that respond specifically to tactile, cold, warm, or painful stimuli.
2. The correlations of cold spots with Krause's end bulbs and of warm spots with Ruffini endings proved to be incorrect. Krause's end bulbs and Ruffini endings are mechanoreceptors. Specific thermoreceptors appear to have free nerve endings with specialized membranes, as do nociceptors.
3. Stimulation of a sensory spot is likely to activate a number of sensory receptors, although activation of single afferent nerve fibers can evoke behavioral or sensory events (see below).
4. A pattern theory of cutaneous sensation does not require a high degree of specificity of

sensory receptors or of central nervous system pathways to elicit a particular sensation. It is also possible to develop theories of sensation that accept an admixture of specific and nonspecific elements.

5. Lesions that affect sensory pathways can do so by reducing the total number of neurons conveying information to the brain, by altering the pattern of activity in ascending pathways, and by changing the operation of control systems that originate in the brain and that regulate activity in sensory pathways.

6. Evidence from microneurography experiments in humans indicates that stimulation of the individual axons of particular types of sensory fibers can lead to a specific sensory experience. For FA I units in the fingers, a single stimulus can evoke a sensation. The quality of sensation depends upon the type of receptor stimulated. Changing the frequency of stimulation may increase the intensity of the sensation for one type of receptor and the frequency of the sensation for another type of receptor. The evidence from microneurography experiments has been criticized, but counterarguments to the criticisms have been made.

7. Head's theory of the epicritic and protopathic divisions of the peripheral nervous system should be discarded.

8. It is preferable not to use the terms "lemniscal" and "nonlemniscal" to distinguish between activity evoked via the dorsal column pathway and that evoked via the spinothalamic tract. Better terms are needed to classify somatosensory neurons that have contrasting response properties (e.g., neurons with small, contralateral receptive fields, restricted convergence, and secure synaptic coupling versus neurons with large, bilateral receptive fields, wide convergence, and weak synaptic coupling).

9. The subdivision of afferent fibers into large- and small-fiber systems is misleading, since sensory receptors of a variety of types fall into each of these subdivisions. Small fibers cannot be equated with nociceptive afferents.

10. Sensory modalities are classes of sensation. A number of cutaneous, subcutaneous, and visceral sensory modalities can be recognized.

11. A sensory channel is the sensory mechanism responsible for conveying the information needed for recognition of a sensory modality. A sensory channel would include one or more sets of sensory receptors, one or more ascending pathways, regions of the thalamus and cerebral cortex, and also the descending pathways that can modify the ascending activity.

2 Peripheral Nerves and Sensory Receptors

COMPOSITION OF PERIPHERAL NERVES

Peripheral nerves are composed of the axons of sensory neurons and somatic and autonomic motor neurons, along with the investing connective-tissue sheaths (endoneurium, perineurium, and epineurium) (see Landon, 1976). The axons may be myelinated or unmyelinated. Since the emphasis will be on sensory mechanisms, the characteristics of motor fibers will not be considered here.

In cutaneous nerves, the largest sensory axons belong to the Aαβ class (Erlanger and Gasser, 1937), whereas the small myelinated fibers belong to the Aδ group. Aαβ fibers conduct at 30–100 m/s, and Aδ fibers conduct at 4–30 m/s (Boivie and Perl, 1975). Unmyelinated fibers are often designated as C fibers; they conduct at less than 2.5 m/s (Gasser, 1950).

A different terminology is used for muscle and joint nerves. The myelinated afferent fibers of muscle nerves are subdivided into groups I, II and III (Lloyd and Chang, 1948; Rexed and Therman, 1948); in cats, these groups of axons conduct at 72–120, 24–71, and 6–23 m/s, respectively (C. C. Hunt, 1954). Muscle nerves also contain numerous unmyelinated, or group IV, afferent fibers (Stacey, 1969), conducting at less than 2.5 m/s.

Joint nerves resemble muscle nerves, except that they have only a few group I fibers, which appear to arise from Golgi tendon organs in the ligaments around the joint or from muscle spindles of nearby muscles (Gardner, 1944, 1948; Skoglund, 1956; H. T. Andersen *et al.*, 1967; Burgess and Clark, 1969b).

Visceral nerves share the terminology of cutaneous nerve axons.

For at least the large myelinated fibers, it is possible to predict the conduction velocity in meters per second by multiplying the axon diameter (in micrometers), including the myelin sheath, by a factor of 6 (Hursh, 1939) or 5.7 (Arbuthnott *et al.*, 1975). A factor of 4.5 is more accurate for group II and III myelinated fibers (Arbuthnott *et al.*, 1975). The conduction velocities of Aδ and C fibers, especially of nociceptors, decrease during repetitive firing (Raymond *et al.*, 1990).

ELECTRICAL STIMULATION OF PERIPHERAL NERVES

The threshold for exciting axons with an electrical stimulus is an inverse function of axonal diameter. The larger the fiber, the lower is the threshold. In many experimental situations, advantage is taken of this property of axons to stimulate a selected population of axons by graded electrical pulses. The strength of stimulation is commonly expressed as a multiple of the threshold of the largest axons of the nerve (e.g., 1.5T means 1.5 times the threshold of the most excitable axon in the nerve).

In some nerves, an electrical stimulus that is near threshold can evoke a volley essentially confined to one class of afferent fibers. For instance, stimuli below 1.3T applied to the sural nerve of the rabbit activate an almost pure population of axons belonging to SA I receptors (see below) and stimulation of the posterior tibial nerve of the same animal at strengths up to 1.5T will activate just guard hair afferent fibers (A. G. Brown and Hayden, 1971). Group Ia afferent fibers from muscle spindle primary endings (see below) can often be activated almost exclusively by weak stimuli applied to the hamstring or quadriceps muscle nerves in cats (Bradley and Eccles, 1953; J. C. Eccles *et al.*, 1957). However, in general, electrical stimuli will coactivate afferent fibers from more than one class of receptor. This is certainly the case when the flexor reflex afferents (FRA) are stimulated. The FRA include cutaneous afferent fibers of all sizes, plus high-threshold muscle and joint afferent fibers; stimulation of the FRA tends to cause the excitation of flexor motoneurons and the inhibition of extensor motoneurons, the pattern of the flexion reflex (R. M. Eccles and Lundberg, 1959a).

Another approach to the use of electrical stimulation of cutaneous nerves is to excite either the Aαβ fibers alone, these plus the Aδ fibers, or all of the A fibers plus the C fibers. The assumption often seems to be that the Aαβ sensory fibers all belong to mechanoreceptors, whereas many of the Aδ and C fibers are connected with nociceptors, and that any central effects may reflect activity in either the mechanoreceptive or the nociceptive afferent fibers. However, several reservations must be kept in mind about this approach. For example, some nociceptors have axons that conduct at velocities greater than 30 m/s (Burgess and Perl, 1967; Georgopoulos, 1976), and these may be included in an Aαβ volley. Small-fiber groups include afferent fibers from mechanoreceptors and from thermoreceptors, as well as from nociceptors (Burgess and Perl, 1967), and so caution must be exercised before a central action is attributed just to nociceptors.

Activation of the whole spectrum of afferent fibers is complicated by the fact that whereas fibers of all sizes are stimulated simultaneously in a peripheral nerve, the volley in the largest fibers will reach the spinal cord far in advance of the volley in the smallest fibers (Erlanger and Gasser, 1937). The central neural circuits activated by the largest fibers may cause an alteration in the effects that would otherwise have been produced by the smallest afferent fibers. This problem can be circumvented in animal experiments by use of anodal blockade of the large-fiber volley (see, e.g., A. G. Brown *et al.*, 1975) or, in humans, by an ischemic block with a blood pressure cuff (LaMotte *et al.*, 1982). However, the problem of the admixture of impulses in afferent fibers having different functions remains. This problem is usually dealt with by the use of more natural forms of stimulation to determine the central effects of specific classes of sensory receptors. Another approach is to stimulate single afferent fibers through a microneurography electrode (see Chapters 1 and 10). Since this can be done in humans, it is possible with this technique to relate the properties of individual sensory receptors to the sensations they elicit.

SENSORY RECEPTORS

Although the characteristics of the somatic and visceral sensory receptors have been amply reviewed elsewhere (P. B. C. Matthews, 1964, 1972, 1981; Burgess and Perl, 1973; Hensel, 1973a, 1974; Boivie and Perl, 1975; D. D. Price and Dubner, 1977; Burgess *et al.*, 1982; Willis, 1985), it is necessary to provide a brief description of them here as background for later descriptions of the effects evoked by the different receptors in central neurons.

The somatovisceral sensory receptors can be grouped by their location in the skin, skeletal muscles, joints, or viscera. The receptor types will be discussed in this order. A summary of the main types and their response properties in mammals is given in Table 2.1.

TAXONOMIC DISTRIBUTION

Somatosensory receptors of several types have been found in poikilothermic vertebrates. These include rapidly and slowly adapting cutaneous mechanoreceptors in amphibia (Maruhashi *et al.*, 1952; Lindblom, 1962) and reptiles (Siminoff, 1968; Siminoff and Kruger, 1968; Kenton and Kruger, 1971; Kenton *et al.*, 1971). Nociceptors with myelinated and unmyelinated axons in amphibia have also been investigated (Maruhashi *et al.*, 1952).

CUTANEOUS RECEPTORS

Receptors in the skin (and the adjacent subcutaneous connective tissue) include mechanoreceptors, nociceptors, and thermoreceptors. There may also be a chemoreceptor class.

Mechanoreceptors

Cutaneous mechanoreceptors are most readily activated by mechanical forces in the skin. They are sometimes also affected by thermal changes (C. C. Hunt and McIntyre, 1960b; Iggo, 1963), but the thermal responses are thought to be too weak to be significant in thermoreception (Werner and Mountcastle, 1965; Duclaux and Kenshalo, 1972; however, cf. Burton *et al.*, 1972). Noxious stimuli produce no greater excitation than do innocuous ones (Burgess and Perl, 1973).

The cutaneous mechanoreceptors can be classified in several ways. The adaptation rate to a maintained stimulus is an important parameter (Adrian and Zotterman, 1926a,b; Adrian, 1928). Slowly adapting receptors tend to continue to discharge repetitively so long as a stimulus is maintained (Fig. 2.1A, top traces; Fig. 2.1B, top). On the other hand, rapidly adapting receptors discharge only near the time that a stimulus is applied (and sometimes again when it is removed) (Fig. 2.1A, second two traces; Fig. 2.1B, bottom).

Another approach to the classification of mechanoreceptors is based on the most likely property of the stimulus for which they encode information (Burgess and Perl, 1973). For slowly adapting cutaneous mechanoreceptors, this may be the amount of indentation of the skin. The discharge rate of such receptors would then be a function of skin displacement or position (Fig. 2.2A). Rapidly adapting receptors might signal the rate at which the skin is displaced (stimulus velocity) (Fig. 2.2B, top) or some higher derivative of skin position (acceleration, jerk) (Fig. 2.2B, bottom). Thus, the number of discharges that occur when such receptors are activated is a function of stimulus velocity, acceleration, or jerk. Receptors signaling velocity can be termed velocity detectors, and those signaling acceleration or jerk can be grouped together as transient detectors. Sometimes a given receptor will behave as if it signaled two properties of the stimulus. For instance, the slowly adapting (or static) response of a displacement detector might be preceded by a velocity-sensitive (dynamic) response (Fig. 2.3B). Such a receptor would signal both velocity and displacement.

A key that is useful for the classification of cutaneous mechanoreceptors in mammals (Fig. 2.4) has been published by Horch *et al.* (1977). The proportions of different classes of myelinated sensory fibers supplying the hairy skin innervated by a branch of the sural nerve in cats have been estimated by Whitehorn *et al.* (1974). A mathematical model has been developed by A. W. Freeman and Johnson (1982a,b) that accounts for the responses to vibratory stimuli of slowly adapting, rapidly adapting, and Pacinian corpuscle receptors in the monkey.

Cutaneous Displacement and Velocity Detectors. Two kinds of slowly adapting cutaneous mechanoreceptors have been described. Each signals both displacement and velocity. They are often referred to as SA I and SA II (SA for slowly adapting) mechanoreceptors, but each has synonyms.

Table 2.1. Characteristics of Sensory Receptors

Receptor type	Best stimulus	Signal	Background activity	Conduction velocity (m/s)
Cutaneous mechanoreceptors				
Type I	Indentation of dome	Displacement and velocity	None	A. G. Brown and Iggo, 1967 Cat: 57.2 ± 0.99 (33–95) Rabbit: 47.3 ± 1.77 (16–96) Burgess *et al.*, 1968 Cat: 65 (47–84) Tapper *et al.*, 1973 Cat: 63.6 ± 2.3 Perl, 1968 Monkey: 46 ± 4, 51 ± 2 Knibestöl, 1975 Human: 58.7 ± 2.3
Type II	Skin deformation	Displacement and velocity	Sometimes	A. G. Brown and Iggo, 1967 Cat: 53.6 ± 2.21 (20–100) Rabbit: 31.4 ± 2.40 (24–45) Burgess *et al.*, 1968 Cat: 54 (39–68) Perl, 1968 Monkey: 34 ± 12, 40 ± 10 Knibestöl, 1975 Human: 45.3 ± 3.6
G_2 (and T) hair	Movement of guard (or tylotrich) hair or of skin	Velocity	None	A. G. Brown and Iggo, 1967 Cat T: 68 ± 2.72 (44–72) Rabbit T: 35.6 ± 1.7 (8–53) Burgess *et al.*, 1968 Cat G_2: 53 (39–73) Perl, 1968 Monkey G_2: 36 ± 14, 42 ± 16

D hair	Movement of down or guard hairs or skin	Velocity; very low threshold	None	A. G. Brown and Iggo, 1967 Cat: 17.9 ± 0.23 (15–24) Rabbit: 9 ± 0.2 (5–16) Burgess and Perl, 1967 Cat: 18.3 (11–30) Burgess *et al.*, 1968 Cat: 21 (15–32) Perl, 1968 Monkey: 16 ± 6, 19 ± 6
Field	Skin indentation	Velocity; to some extent displacement	None	Burgess *et al.*, 1968 Cat: 55 (36–72) Perl, 1968 Monkey: 39 ± 13, 39 ± 8
C mechanoreceptor	Skin indentation	Velocity (slow); to some extent displacement	None	Iggo, 1960 Cat: (0.55–1.25) Bessou *et al.*, 1971 Cat: (0.5–1.0)
Meissner's corpuscle and Krause's end bulb	Skin indentation	Velocity	None	Jänig *et al.*, 1968a Cat: 50% >60 Iggo and Ogawa, 1977 Cat: 36.1 ± 6.2 W. H. Talbot *et al.*, 1968 Monkey: (40–80) Knibestöl, 1973 Human: 55.3 ± 3.4 (26–91)
G_1	Rapid movement of guard hair or of skin	Transients	None	Burgess *et al.*, 1968 Cat: 75 (56–85) Perl, 1968 Monkey: 47 ± 12, 49 ± 13

(continued)

Table 2.1. *(Continued)*

Receptor type	Best stimulus	Signal	Background activity	Conduction velocity (m/s)
Pacinian corpuscle	Vibration	Transients	None	C. C. Hunt and McIntyre, 1960b Cat: (54–84) Burgess *et al.*, 1968 Cat: 65 (54–82) Jänig *et al.*, 1968a Cat: 50% > 57 Perl, 1968 Monkey: 41 ± 8, 56 ± 11 Knibestöl, 1973 Human 46.9 ± 3.6 (34–61)
Cutaneous nociceptors				
Aδ mechanical	Damage to skin	Threat or damage	None	Fitzgerald and Lynn, 1977 Rabbit: 15 (5–32.5) Cat: 27 (5.5–49) Burgess and Perl, 1967 Cat: (6–51) Burgess *et al.*, 1968 Cat: (6–65) Perl, 1968 Monkey: 25 ± 11, 21 ± 13 Georgopoulos, 1976 Monkey: (4–44)
C mechanical	Damage to skin	Threat or damage	None	Bessou and Perl, 1969 Cat (0.6–1.4) Georgopoulos, 1976 Monkey: (most 0.8–1)
Aδ heat	Noxious heat or mechanical damage	Threat or damage	None	Iggo and Ogawa, 1971 Monkey: (3.9–6.8) Georgopoulos, 1976 Monkey: (4–40)
Aδ and C cold	Severe cold; mechanical damage	Threat or damage	None	Georgopoulos, 1976 Monkey: (4–40, 0.8–1)

C-polymodal	Noxious heat; sometimes severe cold; mechanical damage; algesic chemicals	Threat or damage	None	Bessou and Perl, 1969 Cat: (0.4–1.1) Beck *et al.*, 1974 Cat: (0.4–1.8) Croze *et al.*, 1976 Monkey: (0.6–1.1) Van Hees and Gybels, 1972 Human: 0.89 (0.66–1.1)
Cutaneous thermoreceptors				
Cold	Reduced temperature	Cooling	Yes	Iriuchijima and Zotterman, 1960 Rat: (0.5–1.0) Hensel *et al.*, 1960 Cat: (0.6–1.5) Perl, 1968 Monkey: 8 (4–14) Iggo, 1969 Monkey, baboon: (<1.5–15.3) Dog (lip): (9–18) Rat: (<1.5) Hensel and Iggo, 1971 Monkey: 5.8 ± 2.5 (2.2–9.5), 0.7 ± 0.3 (0.3–1.3) Darian-Smith *et al.*, 1973 Monkey: 14.5 (5–31) R. R. Long, 1977 Monkey Glabrous: 13.3 ± 2.8 Hairy: 7.0 ± 2.8 Junction: 10.8 ± 3.1

(*continued*)

Table 2.1. *(Continued)*

Receptor type	Best stimulus	Signal	Background activity	Conduction velocity (m/s)
Warm	Increased temperature	Warming	Yes	Iriuchijima and Zotterman, 1960 Rat: (0.7–1.5) Hensel *et al.*, 1960 Cat: (0.84 for 1 fiber) Hensel and Iggo, 1971 Monkey: 0.7 ± 0.2 (0.4–0.9) Duclaux and Kenshalo, 1980 Monkey: (0.8 ± 0.09) Konietzny and Hensel, 1975 Human: (0.5–0.75)
Muscle mechanoreceptors				
Muscle spindle				
Primary ending	Change of spindle length and rate of change	Length and velocity	Sometimes	C. C. Hunt, 1954 Cat: (72–120)
Secondary ending	Change of spindle length	Length	Sometimes	C. C. Hunt, 1954 Cat: (24–72)
Golgi tendon organ	Change of muscle length or contraction	Tension	Sometimes	C. C. Hunt, 1954 Cat: (72–120)
Pacinian corpuscle (see above)				
Muscle nociceptors				
Group III	Pressure; damage	Pressure; threat or damage	None or slight	Paintal, 1960 Cat: (6–91, mostly 10–15)
Group IV	Pressure; damage	Threat or damage	Little	Mense and Schmidt, 1974 Cat: <2.5 Franz and Mense, 1975 Cat: <2.5
Joint mechanoreceptors				
Ruffini ending	Flexion or extension (to extremes)	Joint pressure	Only at extreme positions	Burgess and Clark, 1969b Cat: (20–70) F. J. Clark, 1975 Cat: (9–77)

Golgi tendon organ (see above)				
Paciniform ending (see Cutaneous mechanoreceptors, Pacinian corpuscle)				Burgess and Clark, 1969b Cat: 12–33
Joint nociceptors				
Aδ	Extreme bending, probing	Threat or damage	Little or none	F. J. Clark and Burgess, 1975 Cat: <30
C	Extreme bending, probing	Threat or damage	Little or none	Schaible and Schmidt, 1983a,b Cat: 2.5–20 0.3–1.3
Visceral mechanoreceptors				
Intestine	Distension, tension on mesentery or blood vessels (intestine)	Movement	May have respiratory, cardiovascular, or gastrointestinal rhythm	Bessou and Perl, 1966 Cat: (2–21) Ranieri *et al.*, 1973 Cat: (0.6–5) Clifton *et al.*, 1976 Cat: Aδ or <2.5 Morrison, 1977 Cat: (0.6–30) Dog: (0.5–36)
Bladder	Distension or contraction	Tension	Maybe	Winter, 1971 Cat: (<2–>22) Clifton *et al.*, 1976 Cat: Aδ
	Distention	Volume	Maybe	Clifton *et al.*, 1976 Cat: <2.5
	Mesenteric Pacinian corpuscles	Vibration	May have cardiac rhythm	Ranieri *et al.*, 1973 Cat: (61–70)
Visceral nociceptors	Intense mechanical, thermal, and chemical stimuli	Threat or damage	Little or none	Clifton *et al.*, 1976 Cat: <2.5

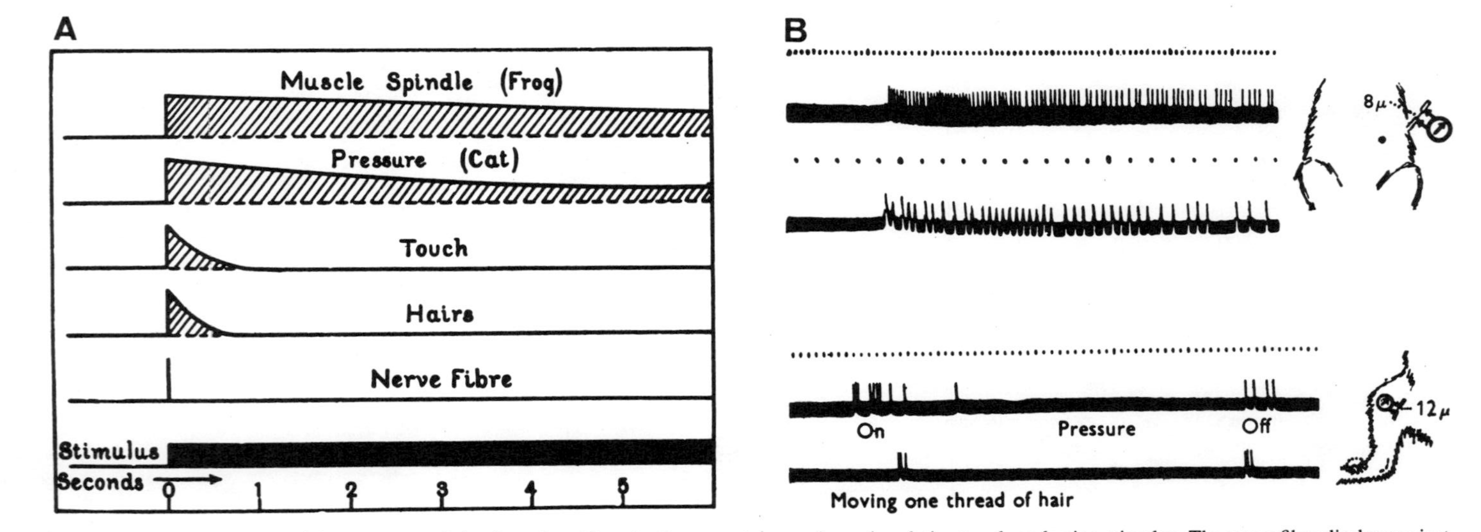

Fig. 2.1. (A) The durations of the responses of slowly and rapidly adapting receptors are shown in relation to a long-lasting stimulus. The nerve fiber discharges just once at the start of the stimulus. (From Adrian, 1928, with permission.) (B) Illustration of the discharges of an afferent fiber from a slowly adapting tactile receptor (above) and of an afferent fiber from a rapidly adapting hair follicle receptor (below). Time marks 10 ms. (From Maruhashi *et al.*, 1952.)

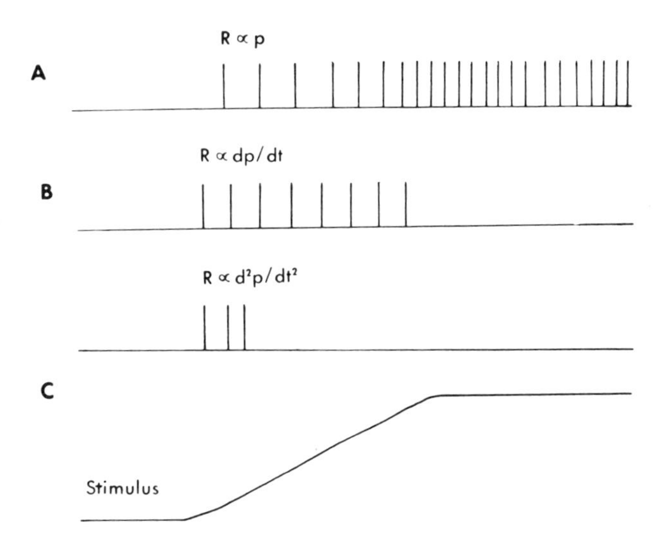

Fig. 2.2. Responses of different types of cutaneous mechanoreceptors to a ramp displacement of the skin. The receptor whose discharge is shown in the trace in panel A signals skin displacement or position (p). The first response in panel B is in proportion to the first derivative of position (or velocity). The second response in panel B is proportional to the second derivative of position (acceleration). Panel C shows the stimulus. (From Willis and Grossman, 1981.)

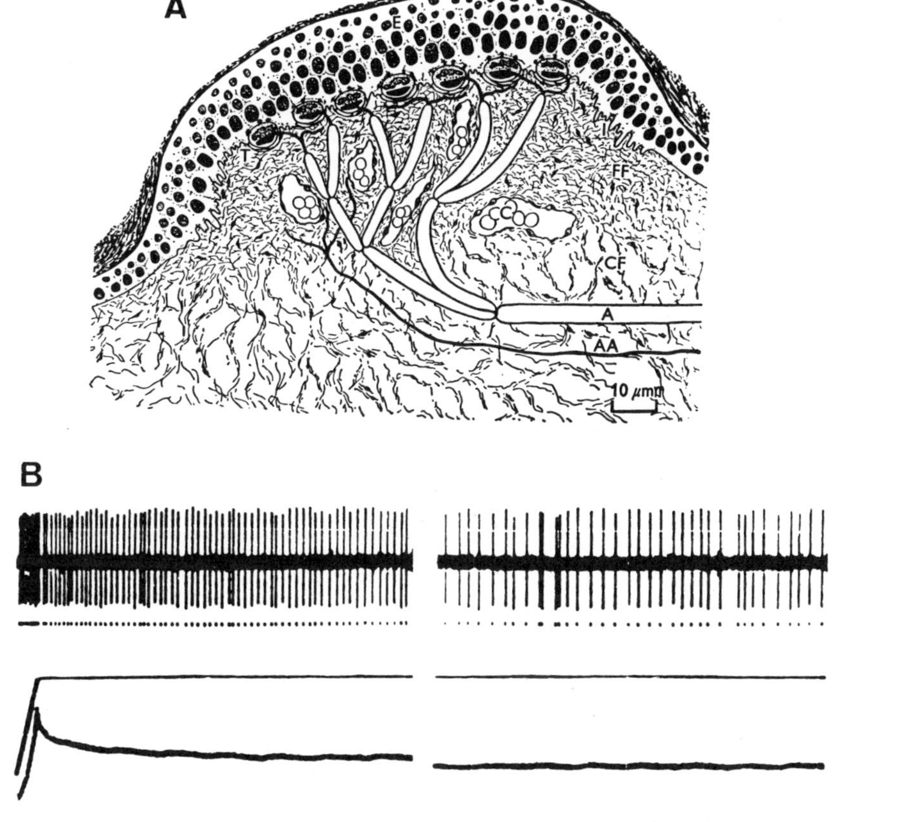

Fig. 2.3. Structure and function of an SA I receptor. (A) Tactile dome. The afferent fiber divides and ends in relation to a set of Merkel cells in the basal layer of the epidermis. (From Iggo and Muir, 1969.) (B) Discharge of an SA I receptor located in the glabrous skin of the hand of a squirrel monkey in response to a constant displacement stimulus (force is also shown in the lower trace; note the decrease in force as a result of skin compliance). The left traces are for the first 2 s of stimulation. Note the dynamic and static responses. The traces on the right are for the last 2 s of a 10-s stimulus. Note the irregularity of the discharges. (From B. H. Pubols and Benkich, 1986.)

SA I slowly adapting mechanoreceptors are associated with Merkel cell complexes. Merkel cells are specialized cells in the epidermis adjacent to the basement membrane. Afferent nerve terminals contact the Merkel cells. The sense organs take the form of domelike structures (Fig. 2.3A) in some animals, including cats, rabbits, mice, rats, and guinea pigs. However, demarcation of tactile domes in the skin of primates (including humans) is difficult. Often, a tylotrich hair follicle is associated with a tactile dome (K. R. Smith, 1968). These morphological features give rise to alternative names for this receptor type: Merkel cell ending, tactile dome, touch corpuscle, and Haarscheibe. (For details about the structural features of SA I receptors, see Merkel, 1875; Iggo, 1963; Pinkus, 1964; Munger, 1965; K. R. Smith, 1968, 1970; Iggo and Muir, 1969; Straile, 1969; Jänig, 1971a,b.)

In the absence of a stimulus, SA I endings are normally silent. However, they discharge following indentation of the skin directly over the ending and have a low threshold (displacement of less than 15 μm). They are relatively insensitive to displacement of the skin immediately adjacent to the dome and to skin stretch. A step-displacement stimulus applied to the dome produces a dynamic response followed by a static response (Fig. 2.3B). The discharges during the static response tend to occur at irregular intervals. A given afferent fiber can supply as many as seven separate domes, although more commonly there are only two or three branches. (For more information about the functional properties of SA I endings, see Frankenhaeuser, 1949; Maruhashi *et al.*, 1952; Witt and Hensel, 1959; C. C. Hunt and McIntyre, 1960a; Tapper, 1965; Werner and Mountcastle, 1965; Burgess *et al.*, 1968; Perl, 1968; Iggo and Muir, 1969; Harrington and Merzenich, 1970; Hagbarth *et al.*, 1970; Kenton and Kruger, 1971; Pubols *et al.*, 1971; Horch *et al.*, 1974; Whitehorn *et al.*, 1974; Horch and Burgess, 1975, 1976; B. H. Pubols and L. M. Pubols, 1976; Aitken and Lal, 1982a; Lynn and Carpenter, 1982; B. H. Pubols and Benkich, 1986; Handwerker *et al.*, 1987; B. H. Pubols, 1990; see also Figs. 10.1–10.10).

SA II slowly adapting mechanoreceptors have been identified with Ruffini endings (Fig. 2.5A) (Ruffini, 1894; M. R. Chambers *et al.*, 1972). These endings are located in the dermis. There is generally one low-threshold spot for each SA II fiber. These endings may show a background discharge in the absence of an overt stimulus, and they respond to small displacements of the skin either directly over the receptor (threshold of about 15 μm) or as a result of stretch of the skin from a distance. The discharge has both a dynamic and a static component following a step displacement of the skin, but the dynamic responses are smaller than is typical of the SA I receptor (Fig. 2.5B). The discharges of SA II endings are quite regular during the static response, in contrast to those of the SA I receptor. (See Burgess *et al.*, 1968; Perl, 1968; Harrington and Merzenich, 1970; Hagbarth *et al.*, 1970; Kenton and Kruger, 1971; M. R. Chambers *et al.*, 1972; Horch and Burgess, 1975, 1976; Aitken and Lal, 1982b; Lynn and Carpenter, 1982; Handwerker *et al.*, 1987; see also Figs. 10.2, 10.5, and 10.7).

Cutaneous Velocity Detectors. There are at least five kinds of cutaneous sensory receptors that detect stimulus velocity. Four of these are found only in hairy skin: G_2 hair follicle receptors (and T hair follicle receptors), D hair follicle receptors, field receptors, and C mechanoreceptors. The velocity detector in the glabrous (nonhairy) skin is Meissner's corpuscle or the FA I receptor in primates (including humans) and Krause's end bulb in cats.

The vast majority of hair follicles in many mammals can be subdivided into three types: tylotrichs, guard hairs, and down hairs (Noback, 1951; Straile, 1960, 1961; A. G. Brown and Iggo, 1967; A. G. Brown *et al.*, 1967). A fourth type of hair, the sinus hair, is found only in certain locations; sinus hairs will be discussed separately below. Tylotrichs are the largest and least numerous of the hair types. Both tylotrichs and guard hairs arise individually from follicles. Down hairs are the smallest type of hair; they are wavy, and they arise in groups from single follicles.

Many of the myelinated fibers supplying the skin of the cat can be activated by hair movement (Hunt and McIntyre, 1960c; Ray *et al.*, 1985). The afferent fibers supplying guard

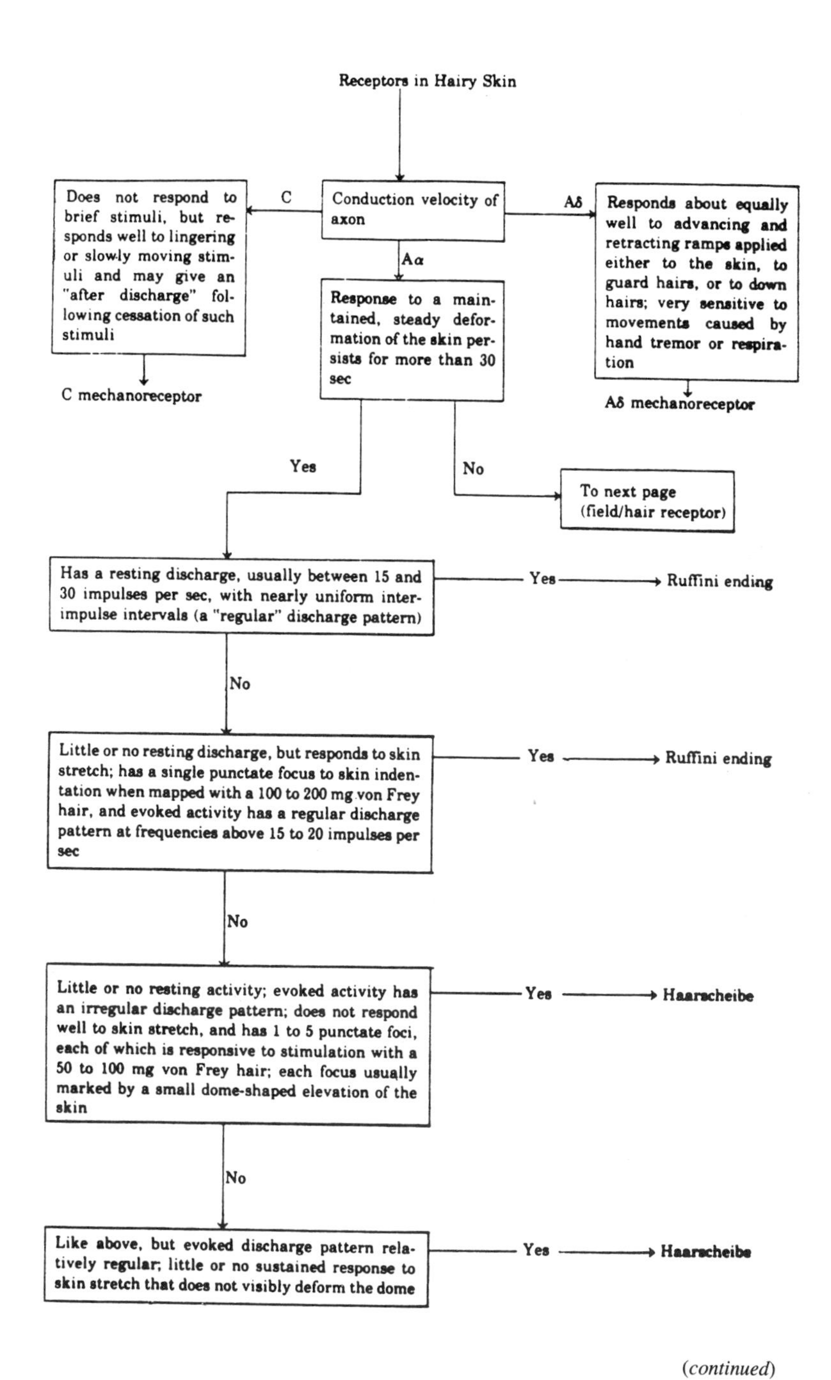

(*continued*)

Fig. 2.4. Key for classification of cutaneous receptors in mammals. (From Horch, *et al.*, reprinted with permission.)

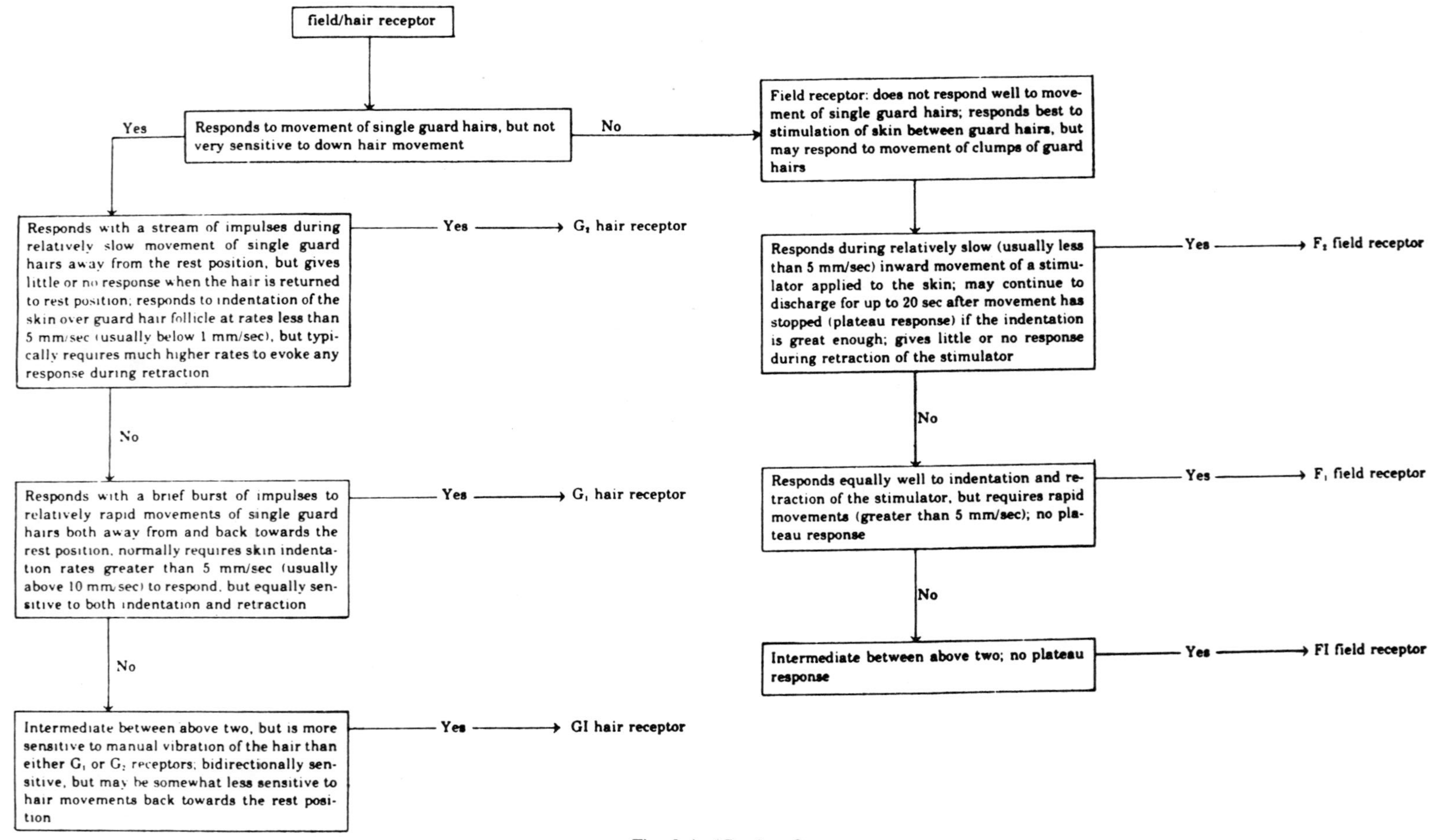

Fig. 2.4. (*Continued*)

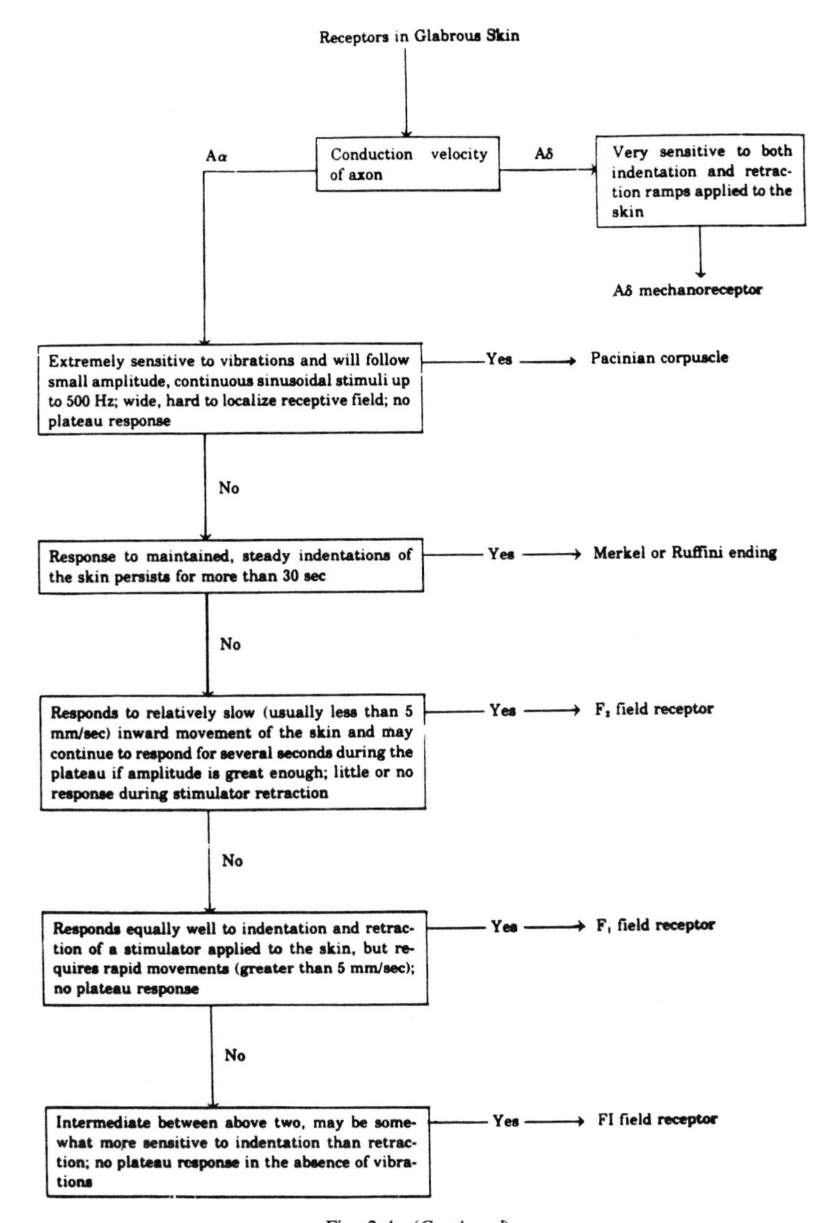

Fig. 2.4. (*Continued*)

hair follicles can be subdivided into two groups based on their responses to mechanical stimulation. G_1 hair follicle receptors are transient detectors and will be discussed below; G_2 hair follicle receptors are velocity detectors (however, Hahn and Wall [1975] failed to find a clear separation between G_1 and G_2 receptors). Some T (for tylotrich) hair follicle receptors appear to behave like G_1 receptors and others like G_2 receptors, and so the T hair follicle receptors will be grouped here with the appropriate type of G hair follicle receptor (cf. A. G. Brown and Iggo, 1967; A. G. Brown *et al.*, 1967; Burgess *et al.*, 1968; Perl, 1968; Tuckett *et al.*, 1978). The association between different response types and different types of hairs has also been examined by Tuckett (1982).

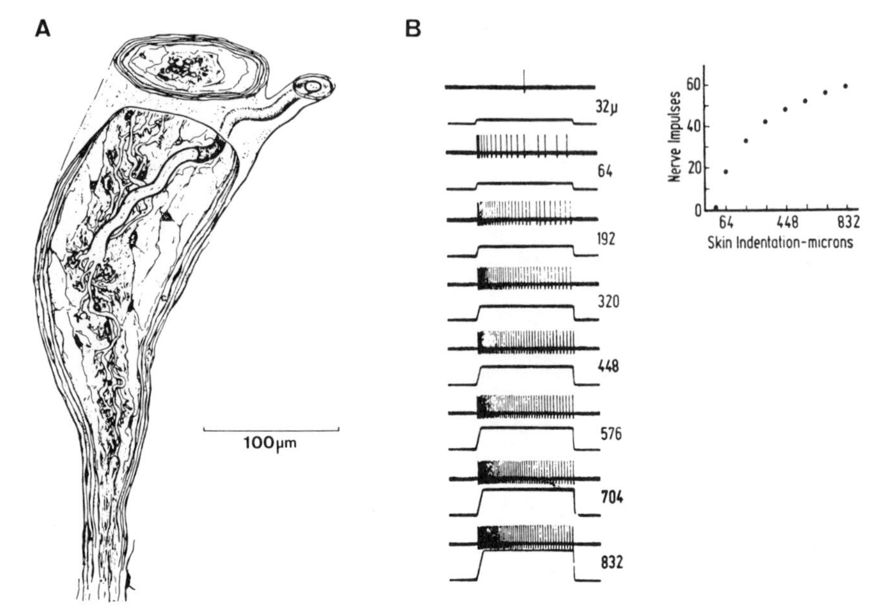

Fig. 2.5. Structure and function of an SA II mechanoreceptor. (A) Ruffini or SA II ending. (From M. R. Chambers *et al.*, 1972.) (B) Series of responses of an SA II ending to different amounts of displacement of the skin and a graph of the stimulus–response function. (From Harrington and Merzenich, 1970.)

G_2 hair follicle receptors respond to the velocity of hair movement (Fig. 2.6). Thresholds to ramp movements of individual hairs in cats are 0.5–1.5 μm/ms. When the stimulus is applied to the skin, the threshold is less than 0.05 μm/ms. While the hair is moved gently, there is a steady discharge. However, no discharge occurs during a maintained stimulus. A given afferent fiber supplies about 10 guard hairs (Burgess *et al.*, 1968).

D hair follicle receptors also signal stimulus velocity, but their thresholds are lower than those of G_2 receptors. For hair movement, threshold responses are seen with ramp stimuli of 0.1–1 μm/ms. These receptors are extremely sensitive and can be activated by movement of either down or guard hairs, the latter causing indirect activation of the D hair follicle receptor. There is no discharge in the absence of an overt stimulus, but very small movements of the skin will activate these receptors (Maruhashi *et al.*, 1952; A. G. Brown and Iggo, 1967; Burgess *et al.*, 1968; Perl, 1968; Merzenich and Harrington, 1969). They may appear to be slowly adapting when activated by a hand-held stimulator, perhaps because of repeated stimulation by physiological tremor. D hair mechanoreceptors respond vigorously to air jet stimuli (Ray *et al.*, 1985).

Field receptors can be distinguished from hair follicle receptors because they cannot be activated by the movements of single hairs. They are excited by brushing large numbers of hairs, probably by spread of the mechanical stimulus to the underlying skin. Direct stimulation of the skin, along with hair movement, is the best stimulus. The morphology of the field receptor sense organ is unknown. Field receptors signal stimulus velocity and have low thresholds to ramp stimuli (0.1–1 μm/ms). Some field receptors have a tendency to continue to discharge during a maintained stimulus. Several subcategories of field receptors (F1, intermediate, F2) can be distinguished on the basis of gradations between rapidly and slowly adapting responses (Burgess *et al.*, 1968; Perl, 1968; Burgess and Perl, 1973; Tuckett *et al.*, 1978).

Primary afferent fibers of the guard hair, down hair and field receptor types similar to those observed in cats have also been found in the hairy skin of rats (Lynn and Carpenter, 1982).

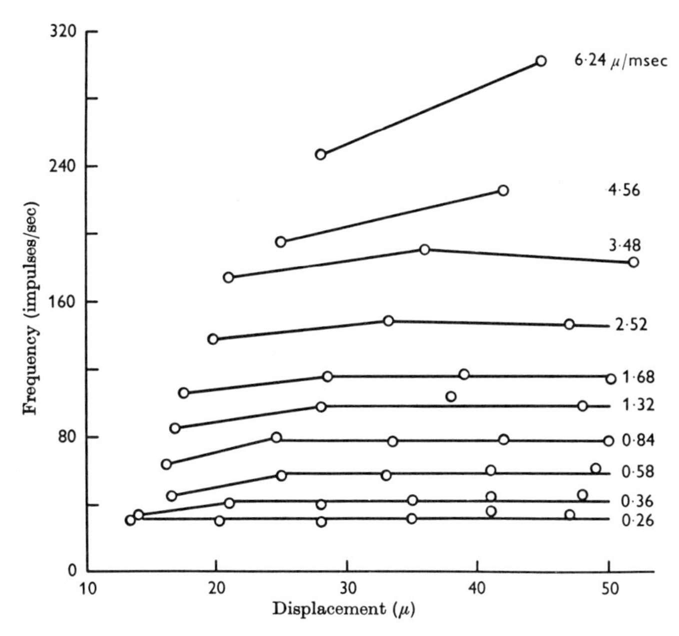

Fig. 2.6. Input–output curves for a rapidly adapting receptor (T hair follicle unit) that signaled stimulus velocity. The stimuli were constant-velocity displacements of the hair. The ordinate shows the instantaneous frequency of the discharge, and the abscissa shows the displacement. The stimuli were applied over a range of velocities. The family of curves indicates that the discharge rate was constant for a given velocity and was a function of stimulus velocity. (From A. G. Brown and Iggo, 1967.)

C mechanoreceptors are sensitive mechanoreceptors that have unmyelinated afferent fibers. They respond to skin indentation, but they are especially sensitive to stimuli that move slowly across the receptive field. They signal chiefly velocity, although there is a tendency for their discharge to continue during a maintained stimulus (Fig. 2.7) (Iggo, 1960; Bessou *et al*, 1971). Following cessation of stimulation, there may be an after discharge. (For additional details, see Zotterman, 1939; W. W. Douglas and Ritchie, 1957; Iggo, 1960; Iriuchijima and Zotterman, 1960; Bessou and Perl, 1969; Bessou *et al.*, 1971; Iggo and Kornhuber, 1977; Lynn and Carpenter, 1982.) C mechanoreceptors are not distributed universally; they are found in the proximal hairy skin, but not in the pad of the cat's foot (Bessou *et al.*, 1971; Beck *et al.*, 1974), nor in the glabrous skin of the monkey hand (Georgopoulos, 1976). They have not yet been identified in recordings from limbic nerves in human subjects (Van Hees and Gybels, 1972; Torebjörk, 1974; see Chapter 10), although they have recently been described in nerves to the face (Nordin, 1990). They occur, but are uncommon, in the hairy skin of monkeys (Kumazawa and Perl, 1977).

In the glabrous skin, there appears to be just a single type of velocity-sensitive receptor. The morphology of this receptor varies, even within a single species (Málinovský, 1966). However, it appears that the receptor, which is often called the rapidly adapting or FA I receptor (FA for fast adapting), can be identified with Meissner's corpuscle (Fig. 2.8A) in the primate (including human) glabrous skin (see Figs. 10.5–10.7, 10.14, and 10.16–10.18) (Meissner, 1859; Cauna, 1956; Cauna and Ross, 1960; Munger, 1971; Quilliam, 1975) and with Krause's end bulb in the cat foot and toe pads (Krause, 1859; Lynn, 1969; Jänig, 1971a,b; Iggo and Ogawa, 1977). Meissner's corpuscles and Krause's end bulbs may be evolutionarily derived from a common type of ending, and both may be morphological equivalents for glabrous skin of hair follicle endings in hairy skin (Málinovský, 1966; Munger, 1971). Meissner's corpuscles in rat skin

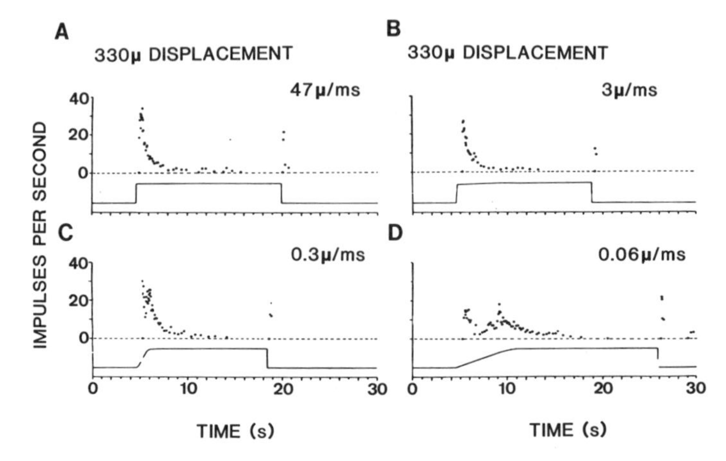

Fig. 2.7. Responses of a C mechanoreceptor to identation of the skin at different velocities. (From Bessou *et al.*, 1971.)

receive accessory innervation from unmyelinated fibers that contain immunoreactivity for calcitonin gene-related peptide and substance P (Ishida-Yamamoto *et al.*, 1988). The function of these accessory fibers is unclear.

The FA I receptor has a localized receptive field, and its threshold in the skin of the cat is low (often less than 10 μm of indentation; ramp thresholds average 2 μm/ms). There is no discharge during a maintained stimulus (Jänig *et al.*, 1968a; Jänig, 1971b; Iggo and Ogawa, 1977). The FA I receptors in primate glabrous skin have similar properties, although their thresholds seem to be higher than in the cat. They are activated best when repetitive stimuli at rates of 5–40 Hz are used (Fig. 2.8A) (for further details, see Lindblom, 1965; W. H. Talbot *et al.*, 1968; L. M. Pubols *et al.*, 1971; Knibestöl, 1973; K. O. Johnson, 1974; B. H. Pubols and L. M. Pubols, 1976; Iggo and Ogawa, 1977).

Cutaneous Transient Detectors. There are two types of cutaneous receptors that are designed to detect transients: G_1 hair follicle receptors and Pacinian corpuscles.

G_1 hair follicle receptors are activated by rapid movements of guard hairs, especially the longest ones. The thresholds to ramp stimuli exceed 80 μm/ms when single hairs are moved and 5–20 μm/ms when the skin is stimulated (Burgess *et al.*, 1968; cf. Tuckett *et al.*, 1978).

Pacinian corpuscles are subcutaneous receptors that are very sensitive to mechanical transients resulting from cutaneous stimulation. The capsule of the Pacinian corpuscle (Fig. 2.8B) serves as a mechanical filter, allowing mechanical transients to affect the terminal but preventing lower-frequency mechanical events from doing so. (For details about the structure, distribution, and mechanical properties of Pacinian corpuscles, see Vater, 1741; Pacini, 1840; Quilliam and Sato, 1955; Pease and Quilliam, 1957; Cauna and Mannan, 1958; Hubbard, 1958; Loewenstein and Skalak, 1966; Málinovský, 1966; Lynn, 1969; see also Figs. 10.5, 10.7, and 10.14–10.16.)

Pacinian corpuscles adjacent to the interosseous membrane or beneath the pads of the cat's foot are extraordinarily sensitive. They can be activated by hair movement, as well as by stimuli applied to the skin or bony prominences; in cats, they can be excited by movements of the floor at some distance from the experimental preparation, by dropping a paper match folder on the experimental table, or even by a loud sound (C. C. Hunt and McIntyre, 1960a). The threshold to indentation of the skin is often less than 1 μm (C. C. Hunt, 1961; Jänig *et al.*, 1968a; Lynn, 1969, 1971). Pacinian corpuscles are activated best by frequencies of vibration in the range of 60–300

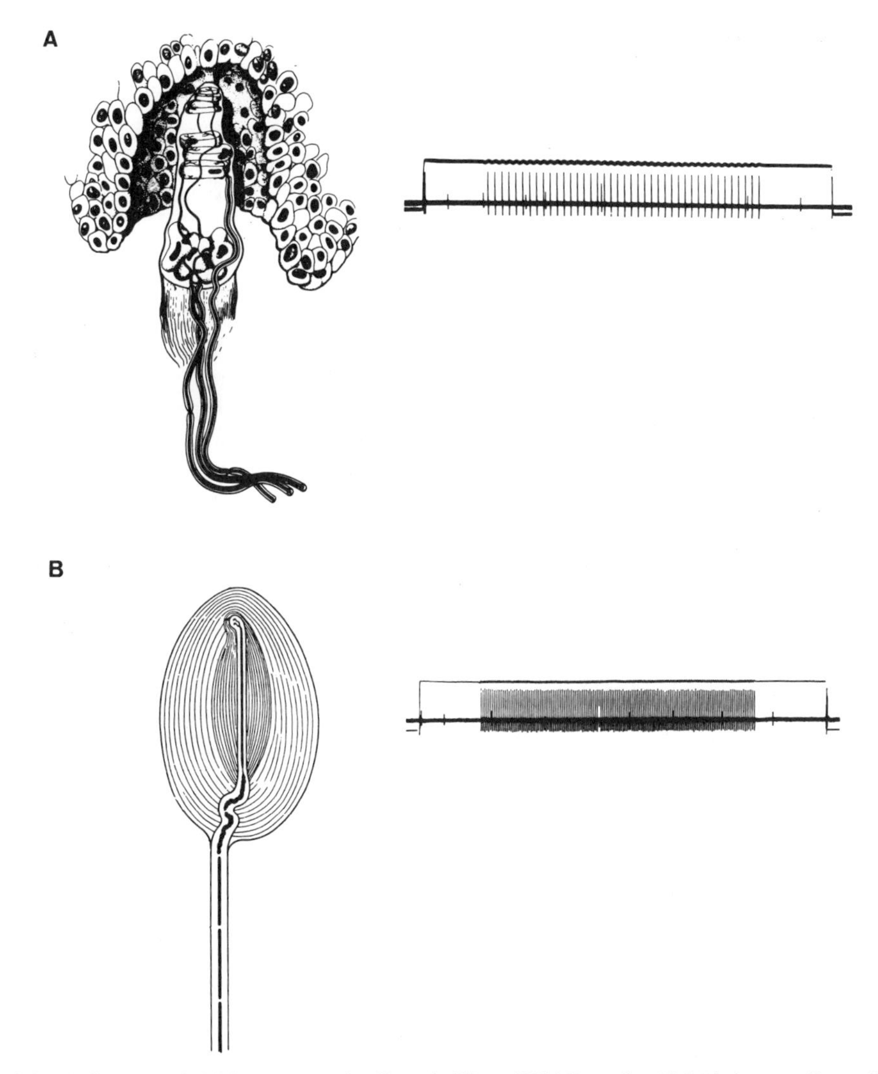

Fig. 2.8. (A) Structure of a Meissner corpuscle. (From Quilliam, 1975.) Recording (right) is from an afferent fiber supplying a rapidly adapting (FA I) receptor in the glabrous skin of a monkey's finger, presumably a Meissner corpuscle. The stimulus was a sinusoidal 18-μm indentation of the skin at 40 Hz. (From W. H. Talbot *et al.*, 1968.) (B) Structure of a Pacinian corpuscle. (From Quilliam and Sato, 1955.) Recording (right) is from an afferent fiber supplying an FA II receptor, presumably a Pacinian corpuscle. The stimulus was a sinusoidal 19-μm indentation of the palmar surface of a monkey's hand at 150 Hz. (From W. H. Talbot *et al.*, 1968.)

Hz (Fig. 2.8B) (W. H. Talbot *et al.*, 1968; Iggo and Ogawa, 1977), but they can follow frequencies of 500 Hz or even higher (C. C. Hunt, 1961; Burgess *et al.*, 1968). Pacinian corpuscles do not appear to project by way of cutaneous nerves in monkeys, although they do in cats (Merzenich and Harrington, 1969).

Sinus Hairs. A special sensory apparatus is associated with sinus hairs. These hairs form the mystacial vibrissae on the snout and are also present in other locations, such as the carpal hairs in the region of the wrist. The innervation of sinus hairs is complex and includes several types of endings (Vincent, 1913; Nilsson and Skoglund, 1965; Andres, 1966; Nilsson, 1969a,b).

Recordings from the afferent fibers innervating the vibrissae show that these sinus hairs are supplied by two kinds of slowly adapting receptors that behave like SA I and SA II endings and also by two types of rapidly adapting receptors, one with a high threshold and one with a low threshold (Gottschaldt *et al.*, 1973; cf. Nilsson and Skoglund, 1965). Carpal hairs also have the two kinds of slowly adapting receptors, but the rapidly adapting receptors are missing. Instead, movements of carpal hairs excite adjacent Pacinian corpuscles (Gottschaldt *et al.*, 1973).

Cutaneous Nociceptors

Nociceptors can be defined as sensory endings that respond to stimuli that threaten or that actually damage tissue (Sherrington, 1906). Receptors that respond at threshold to moderate pressure are included in the nociceptor category, since they respond progressively more as the stimulus intensity is increased, with the greatest response occurring when the stimulus becomes damaging. Some nociceptors in the skin respond selectively to very intense stimuli. Others are silent, being unresponsive to any kind of mechanical or thermal stimulus under normal conditions (Lynn and Carpenter, 1982). However, these receptors can become sensitized during inflammation (Reeh *et al.*, 1987). At least some silent nociceptors may be chemoreceptive (Baumann *et al.*, 1991; cf. Lang *et al.*, 1990). Special techniques need to be used to locate and identify such receptors (Meyer and Campbell, 1988).

There are two main categories of cutaneous nociceptor: Aδ mechanical nociceptors and C-polymodal nociceptors. These are named according to the sizes of the afferent fibers that innervate them and to the types of adequate stimuli. Other cutaneous nociceptors respond well to various combinations of intense mechanical, thermal, and chemical stimuli. These receptors include Aδ mechanoheat nociceptors, Aδ and C cold nociceptors, and C mechanical nociceptors.

Aδ Mechanical Nociceptors. Aδ mechanical nociceptors are excited best by mechanical stimuli that damage the skin (Fig. 2.9) (Hunt and McIntyre, 1960c; Burgess and Perl, 1967; Perl, 1968; Handwerker *et al.*, 1987). Thresholds for these receptors vary. The lowest thresholds are in the innocuous range. The receptive field of a single afferent fiber consists of a set of small spots (Fig. 2.9D) distributed over an area of about 2 cm^2. There is no response to noxious heat or to intense cold, nor is there one to algesic chemicals (Burgess and Perl, 1967; Burgess *et al.*, 1968; Perl, 1968; Beck *et al.*, 1974; Georgopoulos, 1976; Lynn and Carpenter, 1982; see also Fig. 10.27). However, repeated applications of noxious heat stimuli may sensitize these receptors to heat (Fitzgerald and Lynn, 1977; see also Fig. 10.28). The morphology of Aδ mechanical nociceptors has recently been studied (Kruger *et al.*, 1981). The terminals invade the epidermis and are largely ensheathed by Schwann cells. This feature suggests that the term "free nerve endings" is inappropriate (Kruger *et al.*, 1981).

C Polymodal Nociceptors. C polymodal nociceptors are an abundant receptor type, especially in primates (including humans). They respond well to noxious mechanical, thermal, and chemical stimuli (Fig. 2.10) (Bessou and Perl, 1969). However, their chemical sensitivity is often not tested, and in such cases these receptors may be called C mechanothermal nociceptors. As for many other nociceptors, threshold stimuli can be in the innocuous range. The effective thermal stimuli are noxious heat (greater than 45°C) and sometimes intense cold. Effective chemical stimuli include topical application of acid or injections of algesic chemicals. (For more detail, see Iggo, 1959; Iriuchijima and Zotterman, 1960; Fjällbrant and Iggo, 1961; Bessou and Perl, 1969; Iggo and Ogawa, 1977; Van Hees and Gybels, 1972; Beck and Handwerker, 1974; Beck *et al.*, 1974; Torebjörk, 1974; Georgopoulos, 1976, 1977; Handwerker and Neher, 1976; Croze *et al.*, 1976; Kumazawa and Perl, 1977; R. H. LaMotte and Campbell, 1978; Lynn and Carpenter, 1982; Fleischer *et al.*, 1983; R. H. LaMotte *et al.*, 1983; Handwerker *et al.*, 1987; Kenins, 1988; Lang *et al.*, 1990; see also Figs. 10.23, 10.26, 10.27 and 10.29). Some C

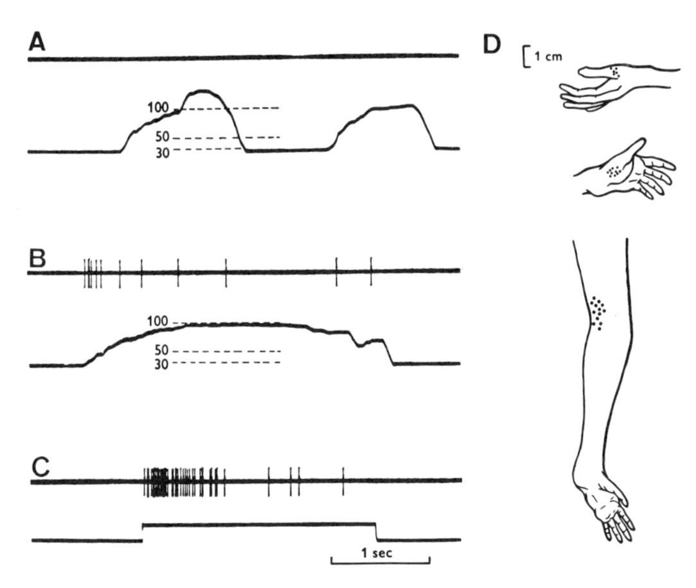

Fig. 2.9. (A–C) Responses of an Aδ mechanical nociceptor to pressure or to damaging stimuli in the glabrous skin of a monkey. In panel A there was no response to pressure by a probe with a 2.2-mm tip diameter (numbers indicate force applied), whereas in panel B there was a response when pressure was applied with a needle tip. In panel C the stimulus was provided by pinching with serrated forceps. (D) Receptive fields of three different Aδ mechanical nociceptors in monkey skin. The receptive fields consisted of sets of punctate spots separated by insensitive zones. (From Perl, 1968.)

polymodal nociceptors respond to cowhage, an itch-provoking substance (Tuckett and Wei, 1987b); myelinated afferent fibers do not respond selectively to cowhage (Tuckett and Wei, 1987a).

Other Cutaneous Nociceptors. Other types of cutaneous nociceptors have been described, including C mechanical nociceptors (Iggo, 1960; Bessou and Perl, 1969; Beck *et al.*, 1974; Georgopoulos, 1976; Kumazawa and Perl, 1977; R. H. LaMotte and Campbell, 1978; Fleischer *et al.*, 1983). It is possible that some of these receptors are the same as the C cold nociceptors, since adequate cold stimuli may not always have been used in characterizing them (Bessou and Perl, 1969).

Aδ mechanoheat nociceptors respond well to both noxious mechanical and heat stimuli (see Iggo and Ogawa, 1971; Beck *et al.*, 1974; Georgopoulos, 1976, 1977; Kumazawa and Perl, 1977; R.H. LaMotte *et al.*, 1982), although the heat threshold is higher than that of C polymodal nociceptors (LaMotte *et al.*, 1983). Some also respond to intense cold. Aδ and C cold nociceptors respond both to extreme cold and to intense mechanical stimuli (Iggo, 1959; Bessou and Perl, 1969; Georgopoulos, 1976, 1977).

Peptide Content of Putative Nociceptors. Some cutaneous nerve terminals that are distributed in the dermis and epidermis and that may belong to nociceptors are immunoreactive to antibodies against calcitonin gene-related peptide, substance P and/or other peptides (Dalsgaard *et al.*, 1983; O'Shaughnessy *et al.*, 1983; Gibbins *et al.*, 1985, 1987a; Alvarez *et al.*, 1988; Kruger *et al.*, 1989; Ishida-Yamamoto *et al.*, 1989). The presence of these peptides may reflect the synthesis of these substances by the sensory neurons for release centrally as neurotransmitters or modulators. However, it is likely that sensory neurons also have an effector role in the periphery involving the release of peptides; such a peripheral release of peptides plays an important role in inflammation (Lembeck and Holzer, 1979; Kenins, 1981; J. C. Foreman *et*

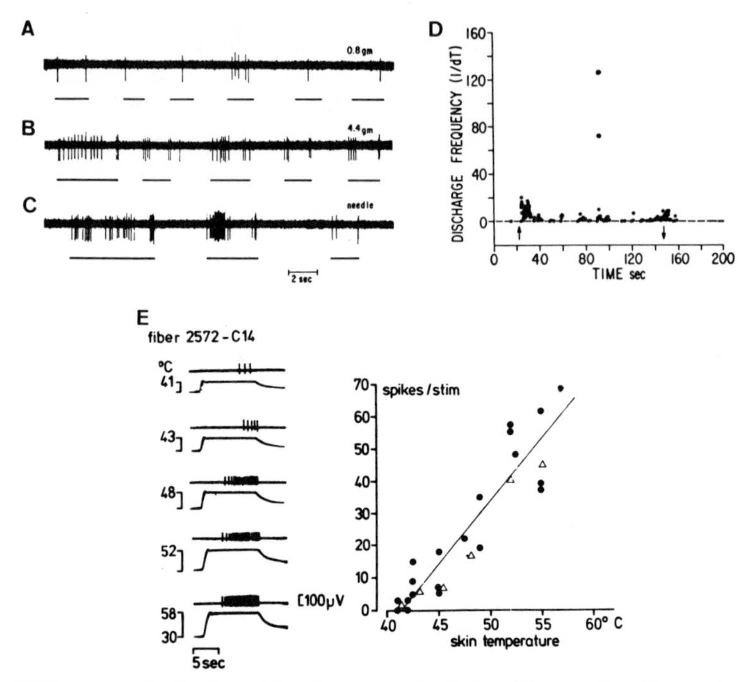

Fig. 2.10. (A) Responses of a C-polymodal nociceptor to stimulation with a von Frey filament that bent with a force of 0.8 g. (B) Response when the stimulus was a 4.4 g von Frey filament. This response is greater than that in panel A. (C) Response to stimulation with a needle that penetrated the skin. (D) Discharge produced in a C-polymodal nociceptor by the application of dilute acid to the skin. (From Bessou and Perl, 1969.) (E) Responses of a C-polymodal nociceptor to graded heat stimuli. The relationship between the number of discharges evoked by each stimulus and skin temperature is shown by the graph at the right. (From Beck *et al.*, 1974.)

al., 1983; Kenins *et al.*, 1984; Brain *et al.*, 1985; Gamse and Saria, 1985; Uddman *et al.*, 1986; Magerl *et al.*, 1987; Szolcsanyi, 1988; see reviews by Yaksh and Hammond, 1982; Fitzgerald, 1983a).

Cutaneous Thermoreceptors

Two kinds of receptors in the skin signal innocuous changes in temperature. These specific thermoreceptors include cold receptors and warm receptors. Thermoreceptors respond very poorly or not at all to mechanical stimuli.

Cold Receptors. The morphology of cold receptors has been described (Hensel, 1973a; Hensel *et al.*, 1974). The afferent fibers from cold receptors may be myelinated or unmyelinated; they are chiefly myelinated in humans (Fruhstorfer *et al.*, 1974). Cold receptors are activated by changes in skin temperature as small as 0.1°C. They show a background discharge at normal skin temperatures. With cooling, there is an increase in the discharge rate, with both a dynamic and a static component (Fig. 2.11). The dynamic response signals the rate at which the temperature is changed, while the static response signals the temperature level. Static discharges are seen at temperatures from 5 to 43°C (Fig. 2.12). The maximum static discharge occurs between about 18 and 34°C. (For further information about cold receptors, see Hensel and Boman, 1960; Hensel *et al.*, 1960; Iriuchijima and Zotterman, 1960; Perl, 1968; Iggo, 1969; Poulos and Lende, 1970a,b; Hensel and Iggo, 1971; Darian-Smith *et al.*, 1973; Hellon *et al.*, 1975; Pierau *et al.*, 1975; Dykes,

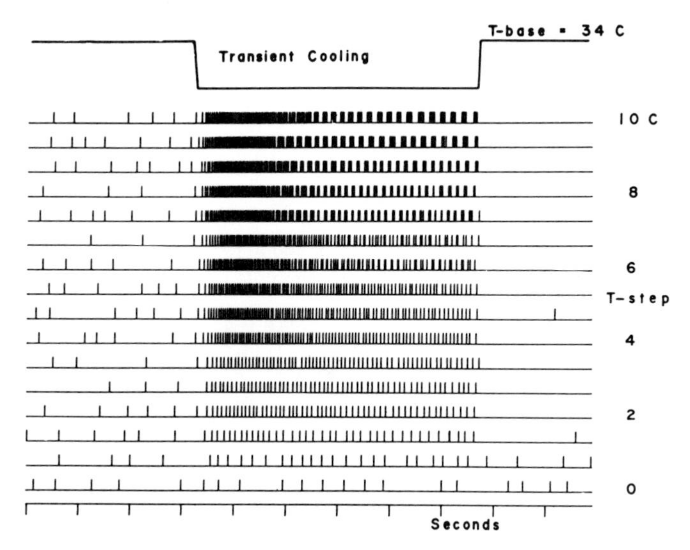

Fig. 2.11. Responses of an afferent fiber that supplied a cold receptor to graded cooling pulses. (From Darian-Smith *et al.*, 1973.)

1975; Kenshalo and Duclaux, 1977; Sumino and Dubner, 1981; Lynn and Carpenter, 1982; Fleischer *et al.*, 1983; see also Fig. 10.36A and B).

It is of interest that the static discharge rate is the same for pairs of temperature readings above and below the optimum temperature (Fig. 2.12). Cold receptors tend to have a bursting pattern of discharge in the lower part of their temperature range (Fig. 2.11) (Dodt, 1952), and it has been suggested that this bursting activity may permit the central nervous system to distinguish between temperatures above and below the optimum (Iggo, 1969; Poulos, 1971;

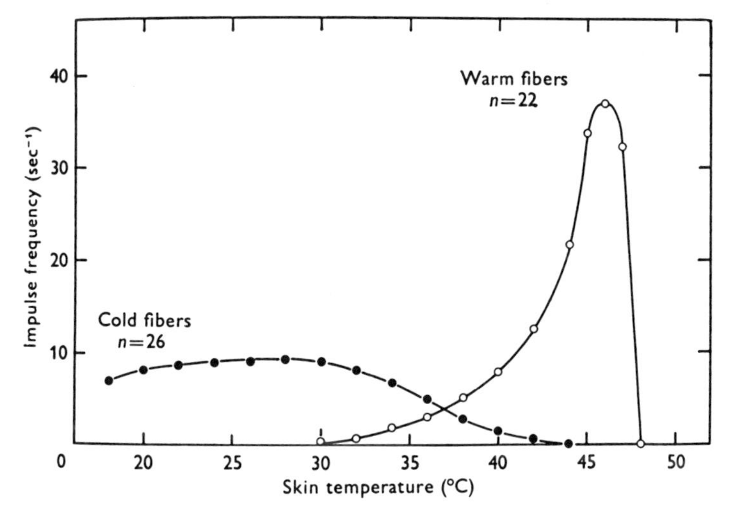

Fig. 2.12. Average static discharge rates for populations of cold and warm receptors. (From Hensel and Kenshalo, 1969.)

Dykes, 1975). However, the coding for static temperature by bursts of discharges in cold fibers may be more important for thermoregulation than for thermal perception (Dykes, 1975). Noxious heat may also evoke a discharge in cold receptors, the "paradoxical cold response" (Dodt and Zotterman, 1952b); this depends in part upon the core body temperature (R. R. Long, 1977).

R. H. LaMotte and Thalhammer (1982) described a class of high-threshold cold receptors in the skin of monkeys with Aδ- and C-sized axons. The receptors responded to colder, but innocuous, cold stimuli below the range for maximal responses of low threshold cold receptors.

Warm Receptors. The morphology of warm receptors is unknown, but they are presumed to be free endings. Warm receptors are a distinctly separate group from cold receptors. They are active at normal body temperature. Their discharge slows with cooling of the skin, and it accelerates with warming, reaching a maximum at temperatures of about 45°C (Fig. 2.12). However, warm receptors are silenced by noxious levels of heat (above 48°C). They have unmyelinated axons. (For additional detail, see Dodt and Zotterman, 1952a; Hensel *et al.*, 1960; Iriuchijima and Zotterman, 1960; Martin and Manning, 1969, 1972; Iggo, 1969; Hensel and Iggo, 1971; Stolwijk and Wexler, 1971; Pierau *et al.*, 1975; Hellon *et al.*, 1975; Konietzny and Hensel, 1975; Handwerker and Neher, 1976; Duclaux and Kenshalo, 1980; R. H. LaMotte and Campbell, 1978; Sumino and Dubner, 1981; see also Fig. 10.36C and D.)

MUSCLE RECEPTORS

Receptors in and around skeletal muscles include the stretch receptors (muscle spindles, Golgi tendon organs) and pressure-pain endings (D. Barker, 1962) with either myelinated or unmyelinated afferent fibers. There are also Pacinian corpuscles in fascial planes; these respond like the ones found in subcutaneous tissue (see above).

Stretch Receptors

The structure and function of the stretch receptors have been reviewed thoroughly (P. B. C. Matthews, 1964, 1972, 1981; D. Barker, 1962; D. Barker *et al.*, 1974). No attempt will be made here to provide an inclusive reference list for these important receptors. (For a bibliography of the literature up to the mid-1970s, see Eldred *et al.*, 1977.)

Muscle Spindles. Muscle spindles have two types of sensory endings: a primary ending and one or more secondary endings. The primary ending of a muscle spindle is supplied by a large myelinated fiber classified as group Ia (Fig. 2.13A). The primary ending is responsive both to the rate of stretch of the spindle (dynamic response) and to the new length (static response) (Fig. 2.13B). The sensitivity of the dynamic response is set, in part, by the activity of dynamic fusimotor fibers (Bessou *et al.*, 1968). The secondary endings are supplied by afferent fibers of intermediate size, the group II fibers (Fig. 2.13A). Secondary endings show little dynamic responsiveness; instead, they signal chiefly muscle length (Fig. 2.13B). Static responsiveness is controlled, in part, by static fusimotor fibers (Appelberg *et al.*, 1966). Some muscle spindles receive a motor supply from branches of α motor axons (β innervation). When a muscle nerve is stimulated at a strength that activates α motor axons, but not γ motor axons, muscle spindle afferent fibers stop discharging during the resultant twitch, in contrast to Golgi tendon organs, which discharge during the twitch (Fig. 2.13C).

Golgi Tendon Organs. Golgi tendon organs are located in the dense connective tissue of muscle tendons and aponeuroses, and their terminals intertwine among bundles of collagen (Fig. 2.14A; Schoultz and Swett, 1972). Their afferent fibers are large myelinated axons classified as group Ib. They respond at a high threshold to passive muscle stretch, but at a low threshold to the contraction of muscle fibers ending on the tendon slip containing the receptor (Fig. 2.14B) (Houk

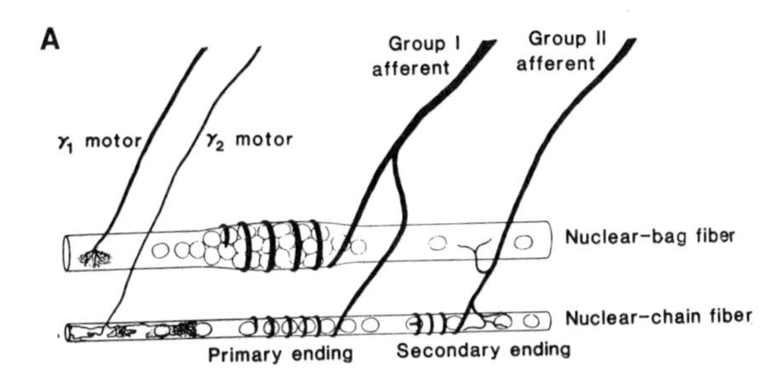

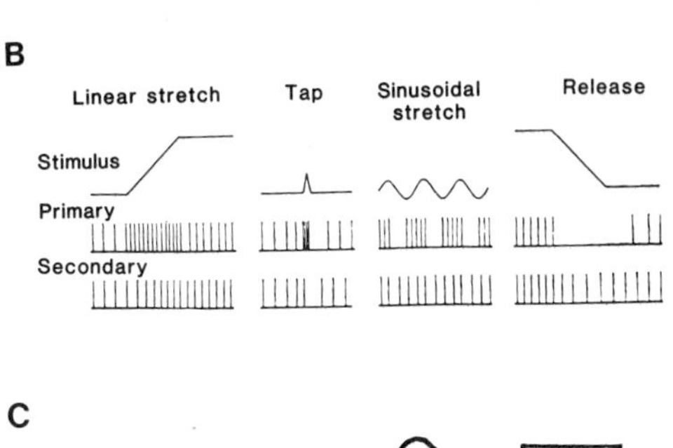

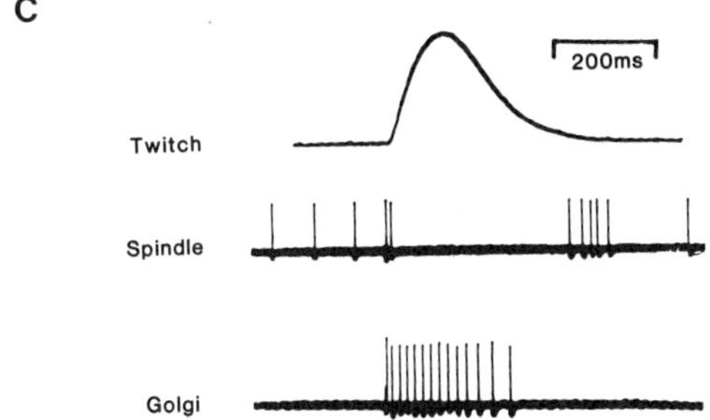

Fig. 2.13. Structure and function of muscle spindle afferent fibers. (A) Simplified diagram of the equatorial region of a muscle spindle. One of the several nuclear bag fibers is shown receiving a primary ending from a group Ia afferent fiber and a plate ending from a dynamic fusimotor neuron. One of the four to six nuclear chain fibers has both a primary ending from the same group Ia fiber and a secondary ending from a group II fiber. It also receives a trail ending from a static fusimotor neuron. (From P. B. C. Matthews, 1964.) (B) Responses of typical primary and secondary endings to linear stretch, tap, sinusoidal stretch and release. (C) Unloading effect of a twitch on the discharges of a muscle spindle afferent fiber and the activation of a Golgi tendon organ afferent fiber. (Panels B and C from Matthews, P. B. C., 1972, copyright Edward Arnold, Publishers.)

and Henneman, 1967). Golgi tendon organs signal tension. There is only a small dynamic response (Fig. 2.14B).

Pressure-Pain Endings

There are numerous unencapsulated endings in the fascia and in the adventitia of blood vessels of muscles. These form as much as 75% of the innervation of skeletal muscle (Stacey, 1969). Some endings are supplied by small myelinated group III afferent fibers, whereas others are supplied by group IV or unmyelinated fibers. A few of the endings are connected with larger afferent fibers of group II size (Paintal, 1960; Stacey, 1969).

Group III afferent fibers can be excited by mechanical stimulation of muscle. Several types of pressure receptors can be recognized. Some respond with a slowly adapting discharge to pressure applied at the junction of muscle with tendon; others with large receptive fields are excited by pressure on the belly of the muscle; another group of rapidly adapting receptors have receptive fields in a small region of the muscle; one group is activated by manipulations affecting the space between muscles or along the surface of a muscle; the final group is discharged by muscle stretch (Bessou and Laporte, 1961).

Group III afferent fibers can also be activated or sensitized by intra-arterial injections of algesic chemicals (Mense, 1977; see also Hník *et al.*, 1969) and by thermal stimuli applied to the

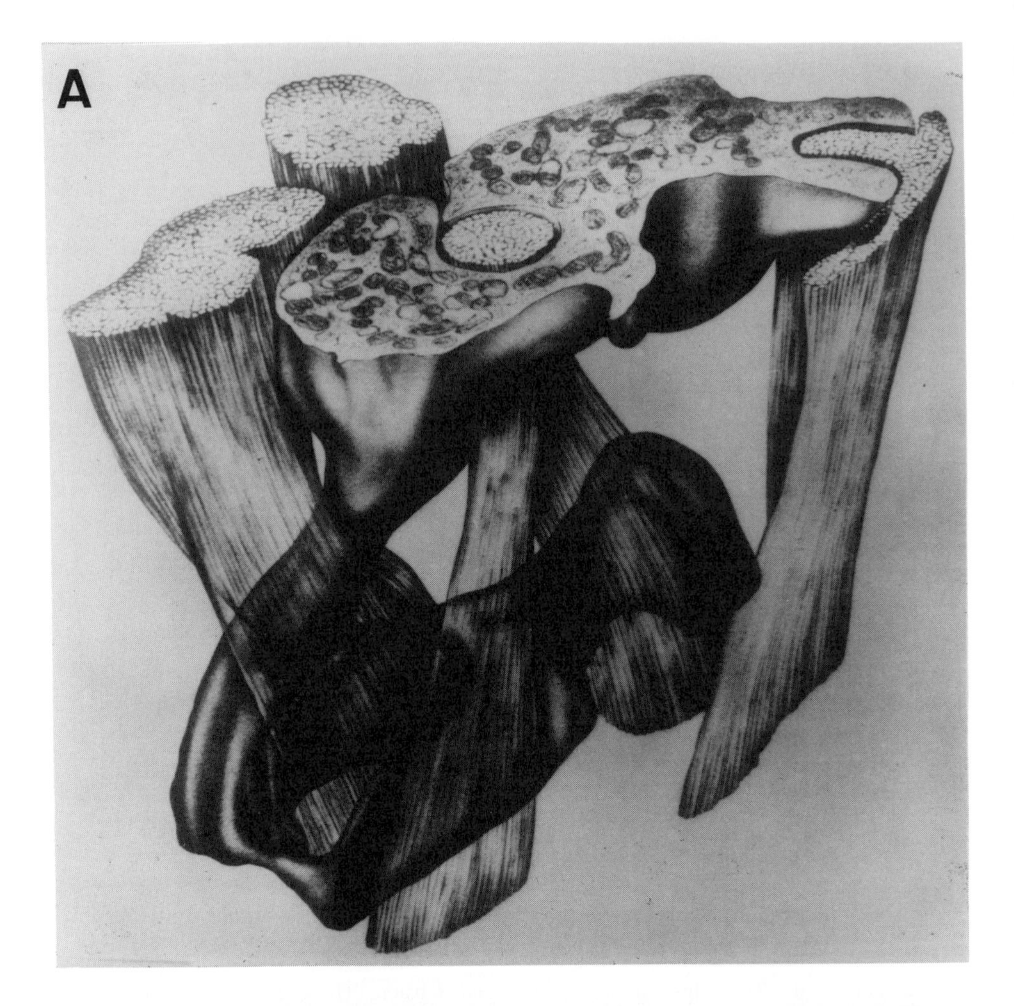

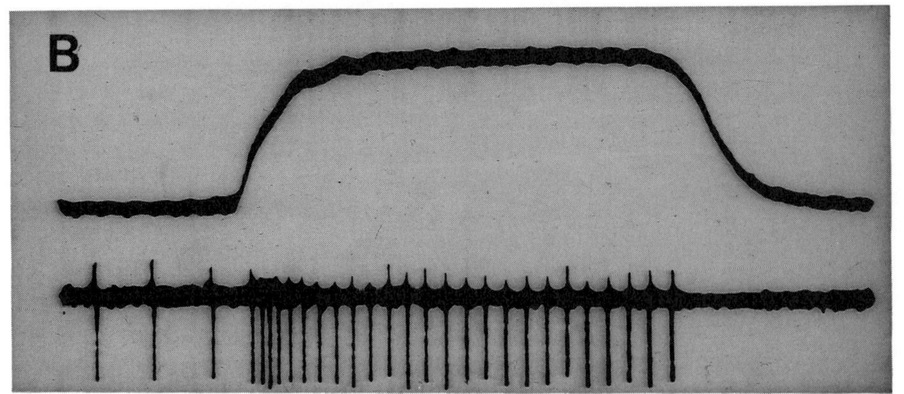

Fig. 2.14. Structure and function of Golgi tendon organs. (A) Drawing of a three-dimensional reconstruction of a Golgi tendon organ. The nerve terminals are shown intertwined with collagen bundles. (From Schoultz and Swett, 1972, reprinted with permission.) (B) Response of a group Ib afferent fiber to contraction of a motor unit. (From Houk and Henneman, 1967.)

muscle belly (Hertel *et al.*, 1976; Kumazawa and Mizumura, 1976, 1977b; Mense and Meyer, 1988). Another effective stimulus is the injection of hypertonic sodium chloride solution into the muscle (Paintal, 1960; Abrahams *et al.*, 1984a).

Mense and Stahnke (1983) confirmed that many group III muscle afferent fibers are activated by muscle stretch. The same fibers often respond to muscle contractions as well. However, the responses to muscle contraction are reduced following a period of ischemia. The suggestion was made that these muscle afferent fibers would be activated during exercise and that they were more likely to function as ergoreceptors than as nociceptors.

Mense and Meyer (1985) found that the most common type of group III muscle afferent fiber in their sample (44%) responded to innocuous mechanical stimuli. However, numerous group III fibers were classified as nociceptive (33%) or contraction-sensitive units (23%).

Group IV muscle afferent fibers are similar in some of their response properties to group III afferent fibers, but their thresholds to mechanical stimuli tend to be higher (Bessou and Laporte, 1961; Iggo, 1961). Many are readily activated or are sensitized by algesic chemicals (Fig. 2.15) (Mense and Schmidt, 1974; Franz and Mense, 1975; Fock and Mense, 1976; Hiss and Mense, 1976; Mense, 1977, 1981; Kumazawa and Mizumura, 1976, 1977b; Kniffki *et al.*, 1978; Mense and Meyer, 1988), and they may also discharge in response to thermal stimuli (Hertel *et al.*, 1976).

Kniffki *et al.* (1978) proposed that group IV muscle afferent fibers that respond to algesic

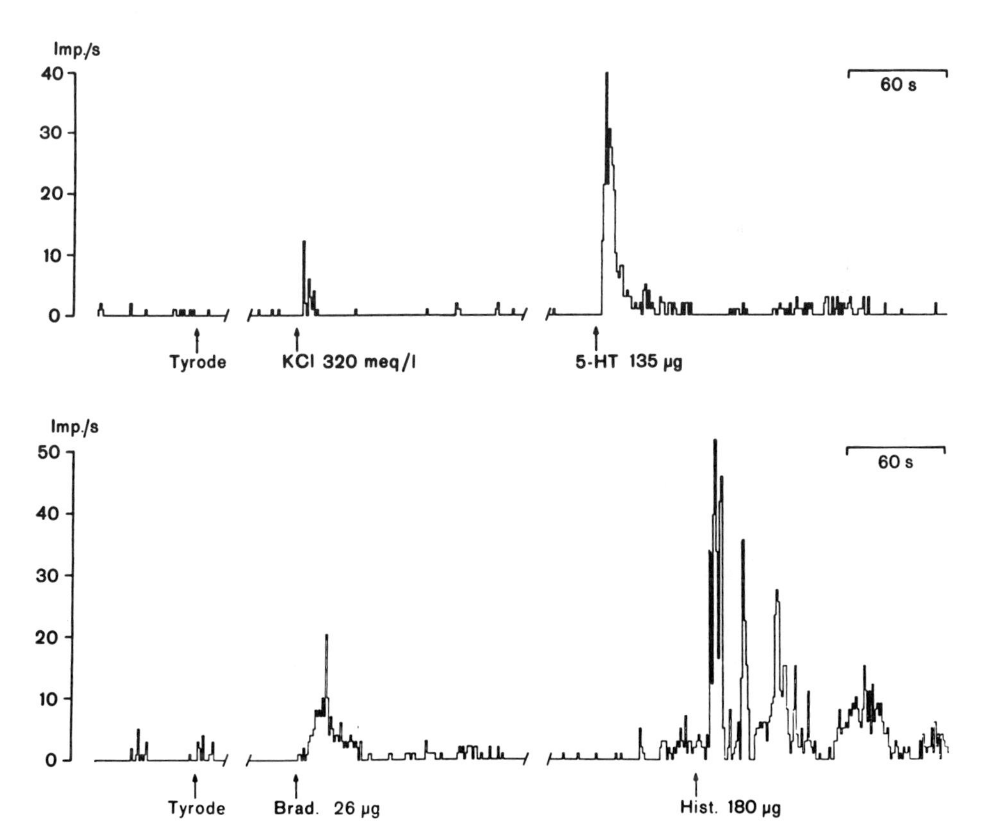

Fig. 2.15. Responses of a group IV muscle afferent fiber to the intra-arterial injection of algesic chemicals (5-HT, 5-hydroxytryptamine; Brad., bradykinin; Hist., histamine). (From Mense and Schmidt, 1974.)

chemicals but not muscular activity are nociceptors, whereas those that respond to contractions but not to algesic chemicals are ergoreceptors. However, some group IV fibers have both response properties.

Mense and Stahnke (1983) found that few group IV muscle afferent fibers respond to muscle stretch or contraction. However, some discharge vigorously when muscle contractions occur during ischemia. They suggested that such afferent fibers might be nociceptors contributing to the pain of muscle ischemia.

Mense and Meyer (1985) observed that the most common type of group IV unit in their sample (43%) was nociceptive. However, there were also group IV units that responded to innocuous mechanical stimuli or to muscle contractions. A few group IV units behaved like specific thermoreceptors. It was suggested that these might participate in thermoregulation, rather than in thermal sensation.

Muscle afferent fibers have less effect than cutaneous ones in evoking plasma extravasation; this observation correlates with a smaller content of substance P in muscle nerves than in cutaneous nerves (S. B. McMahon *et al.*, 1984).

JOINT RECEPTORS

A number of investigations have been done on the innervation of joints (Gardner, 1944; Samuel, 1952; Andrew, 1954; I. A. Boyd, 1954; Skoglund, 1956; Polacek, 1961; M. A. R. Freeman and Wyke, 1967; O'Conner and McConnaughey, 1978; Tracey, 1979; O'Conner and Woodbury, 1982; Langford and Schmidt, 1983; Langford, 1983; Schultz *et al.*, 1984; Zimny *et al.*, 1986; Heppelmann *et al.*, 1988, 1990; Sjölander *et al.*, 1989). The receptor types that have been described anatomically include Ruffini endings, Pacinian corpuscles, Golgi tendon organs, and "free nerve endings." There are also reports of Golgi-Mazzoni endings in joint capsules (Polacek, 1961; M. A. R. Freeman and Wyke, 1967; Grigg *et al.*, 1982).

Joint Mechanoreceptors

The receptor types associated with joints based on the discharge properties of the afferent fibers include both slowly adapting (Golgi tendon organs; Ruffini endings) and rapidly adapting (Paciniform) mechanoreceptors (Gardner, 1944; I. A. Boyd, 1954; Skoglund, 1956; Halata, 1975, 1977; Halata *et al.*, 1984, 1985). The Golgi tendon organs and Paciniform endings respond like comparable endings found elsewhere.

Slowly Adapting Joint Receptors. Slowly adapting discharges have been recorded from joint nerves in response to joint movements (Gardner, 1948; Andrew and Dodt, 1953; Andrew, 1954; I. A. Boyd and Roberts, 1953; I. A. Boyd, 1954; Cohen, 1955; Skoglund, 1956; Eklund and Skoglund, 1960). Golgi tendon organs seem to respond maximally when the joint is moved to one end of its range and are relatively insensitive; Ruffini endings may respond over a narrow range of joint angles and are very sensitive (Skoglund, 1973). For these reasons, it has been proposed that Ruffini endings signal joint position (I. A. Boyd and Roberts, 1953; I. A. Boyd, 1954; Cohen, 1955; Skoglund, 1956, 1973; Carli *et al.*, 1975, 1979, 1981).

However, it has been shown that Ruffini endings are seldom active at intermediate positions of joints (at least of the knee and hip joints of cats), and so it is unlikely that they signal joint position in proximal joints (Fig. 2.16). Instead, they appear to signal the torque that develops as the joint is extended, flexed, or rotated to the extremes of its range (Burgess and Clark, 1969b; F. J. Clark and Burgess, 1975; F. J. Clark, 1975; Grigg, 1975, 1976; Grigg and Greenspan, 1977; Grigg and Hoffman, 1982, 1984; A. Rossi and Grigg, 1982; Aloisi *et al.*, 1988; see also the review by Burgess *et al.*, 1982).

The slowly adapting responses seen in early work when a joint was at midrange have been

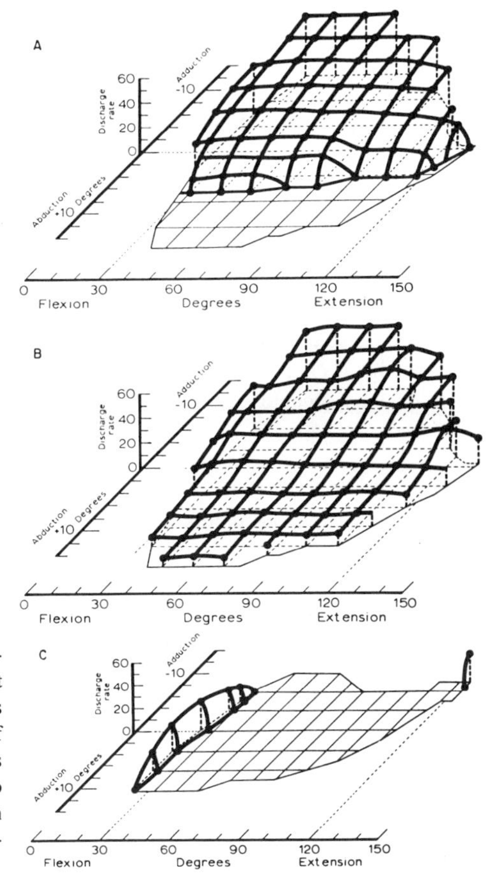

Fig. 2.16. Responses of slowly adapting joint mechanoreceptors of the cat knee joint to different positions of the joint. The activity was measured 15 s after a new position was established. The type of response shown in panel A occurred four times as frequently as that shown in panel B. Responses to extremes of both flexion and extension, as shown in panel C, were common. (From F. J. Clark and Burgess, 1975.)

attributed to muscle afferent fibers that occasionally distribute through some joint nerves (Burgess and Clark, 1969b; McIntyre *et al.*, 1978; Aloisi *et al.*, 1988). However, there are reports of activity in slowly adapting joint afferent fibers when the joint is held at midrange (Ferrell, 1980; Ferrell *et al.*, 1986). Of particular interest are recordings from the nerve supplying the elbow of the cat (Millar, 1973a, 1975; Baxendale and Ferrell, 1983), since this nerve does not appear to be contaminated by muscle afferent fibers (Baxendale and Ferrell, 1983). Several reports have analyzed the dynamic as well as the static reponses of slowly adapting joint afferent fibers (McCall *et al.*, 1974; Nade *et al.*, 1987)

Grigg *et al.* (1982) provided evidence for another type of slowly adapting joint afferent fiber that responds to compression of the joint capsule but not to stretch. They argue that these receptors are Golgi-Mazzoni corpuscles that might function to signal deep pressure.

Rapidly Adapting Joint Receptors. The Pacinian corpuscles in joint nerves respond to vibration, like Pacinian corpuscles elsewhere (Aloisi *et al.*, 1988). These receptors presumably signal mechanical transients in the joint (Skoglund, 1956, 1973).

Joint Nociceptors

Nociceptors in joints have now been described in some detail (Burgess and Clark, 1969b; F. J. Clark and Burgess, 1975; F. J. Clark, 1975; Schaible and Schmidt, 1983a,b). Many group III and IV afferent fibers supplying the joint capsule respond to innocuous movements of the joint.

However, others respond best or only when the joint is moved beyond its normal range, a noxious stimulus (Fig. 2.17). Some fine afferent fibers supplying the joint are not activated at all by joint movements unless the joint is inflamed (Coggeshall *et al.*, 1983; Schaible and Schmidt, 1983a,b, 1985; Grigg *et al.*, 1986). Joint nociceptors appear to be selectively sensitive to capsaicin (He *et al.*, 1990).

VISCERAL RECEPTORS

A variety of receptor types have been identified in viscera. Only those that enter the spinal cord will be considered here. These gain access to the cord either by way of the splanchnic nerves and sympathetic chain or by way of the pelvic parasympathetic nerves. The receptor types include mechanoreceptors and nociceptors.

Visceral Mechanoreceptors

Visceral mechanoreceptors include Pacinian corpuscles located in the mesentery and in connective tissue around visceral organs (Sheehan, 1932). They behave like Pacinian corpuscles elsewhere, except that they sometimes discharge in phase with the heartbeat; it is not known whether signals from these receptors are of consequence in cardiovascular regulation (Gammon and Bronk, 1935; Gernandt and Zotterman, 1946).

Other visceral mechanoreceptors have been found associated with the mesentery, with the serosal surface of the gut or of other organs, and along blood vessels (Bessou and Perl, 1966; Ranieri *et al.*, 1973; Paintal, 1957; J. K. Todd, 1964; Morrison, 1973, 1977; Floyd and Morrison, 1974; Floyd *et al.*, 1976, 1977; Clifton *et al.*, 1976; Cervero and Sharkey, 1988). They appear to respond to movements or to distension of the viscera. However, the discharges do not signal intraluminal pressure or volume with precision, since the receptors are at a distance from the contractile region (Morrison, 1977). Another type of visceral mechanoreceptor seems to be

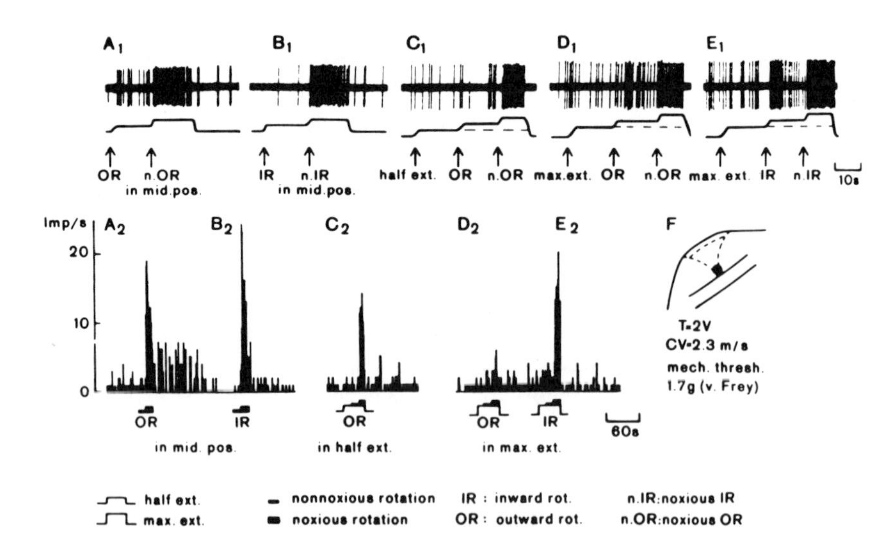

Fig. 2.17. Responses of a group IV primary afferent fiber that innervated the knee joint of a cat. The fiber responded only weakly to innocuous movements of the joint, but strongly to noxious movements. The location of the receptive field is shown in panel F. (From Schaible and Schmidt, 1983b.)

inserted in series with smooth muscle and responds either to distention or to contraction of the muscle; such receptors are typical of the bladder (Talaat, 1937; Iggo, 1955; Winter, 1971; cf. Floyd *et al.*, 1976).

Berkley *et al.* (1990) have examined the responses of afferent fibers supplying the reproductive organs, as well as other pelvic organs, in female rats. Both mechanoreceptive and chemoreceptive responses were observed. The responses of some afferent fibers supplying reproductive organs vary during the estrous cycle.

Visceral Nociceptors

Nociceptors have been found in visceral organs, including the heart, gastrointestinal tract, and reproductive organs (Gernandt and Zotterman, 1946; Lim *et al.*, 1962; A. M. Brown, 1967; Peterson and Brown, 1973; Uchida and Murao, 1974; D. G. Baker *et al.*, 1980; Cervero, 1982b; Cervero and Sharkey, 1988; Sengupta *et al.*, 1990).

Kumazawa and Mizumura (1977a, 1979, 1980a,b, 1983), Kumazawa *et al.* (1987) and Sato *et al.* (1989) have described in some detail a class of polymodal nociceptors in the testis of the dog. Most have conduction velocities in the Aδ range, and they respond to mechanical, thermal, and chemical stimuli. The ultrastructure of some of these endings has been described (Kruger *et al.*, 1988a).

Another type of visceral nociceptor has multiple spotlike receptive fields in the mucosal lining of the rectal canal; this receptor has an unmyelinated axon, and it responds to noxious mechanical, thermal, and chemical stimuli, thus resembling the C-polymodal nociceptors of the skin (Clifton *et al.*, 1976).

Some afferent fibers innervating the reproductive organs of female rats respond to both innocuous and noxious mechanical stimuli, as well as to noxious chemical stimuli (Berkley *et al.*, 1990).

CONCLUSIONS

1. The sensory axons in peripheral nerves can be subdivided into myelinated and unmyelinated classes. There are two broad classes of myelinated cutaneous and visceral afferent fibers, the Aαβ fibers, which conduct at 30–100 m/s, and the Aδ fibers, which conduct at 4–30 m/s. The unmyelinated or C fibers have conduction velocities less than 2.5 m/s. Muscle and joint nerves have three classes of myelinated axons: group I, which conduct at 72–120 m/s; group II, which conduct at 24–71 m/s, and group III, which conduct at 6–23 m/s. There are also unmyelinated group IV axons that have conduction velocities less than 2.5 m/s.
2. The conduction velocity of a myelinated axon in meters per second can be predicted by multiplying its total diameter in micrometers by 6 if it is a large fiber or by 4.5 if it is medium sized or small. The conduction velocities of fine afferent fibers slows during repetitive discharge.
3. The thresholds of axons for electrical stimulation are inversely related to diameter. A weak stimulus can activate the largest axons of a nerve selectively. In some nerves, this permits activation of the axons of a particular type of sensory receptor. However, electrical stimuli usually excite a mixture of axons from different classes of receptors.
4. Strong electrical stimuli can be used to activate unmyelinated, as well as myelinated axons. However, the more rapidly conducting myelinated axons can produce central effects that may influence the responses to a volley in the slow afferent fibers. Volleys restricted to unmyelinated fibers can sometimes be produced by anodal or ischemic blockade of conduction in myelinated axons.

5. Somatosensory receptors in amphibia and reptiles have been investigated. Apart from hair follicle receptors, most of the cutaneous receptor types that are seen in mammals have been recognized in poikilothermic vertebrates.

6. Cutaneous sensory receptors can be classified as mechanoreceptors, nociceptors, thermoreceptors, and perhaps chemoreceptors. They can also be subdivided into slowly adapting and rapidly adapting receptors.

7. Another way of classifying cutaneous receptors is by the most likely information that they encode. For example, cutaneous mechanoreceptors may encode skin indentation (position of the skin), velocity of skin indentation, or a higher derivative of indentation (acceleration, jerk).

8. Cutaneous slowly adapting mechanoreceptors (position detectors) include SA I and SA II receptors. The SA I receptors are associated with Merkel cells in the basal layer of the epidermis. In many species, these endings can be visualized as tactile domes. They respond with a dynamic and then an irregular static discharge when the dome is indented, but they are insensitive to stimuli applied to adjacent skin or to skin stretch. SA II receptors are associated with Ruffini endings. They may have a background discharge, and they respond with a dynamic and a regular static discharge to skin indentation or stretch over a broad area.

9. Cutaneous velocity detectors include G_2 hair follicle receptors, D hair follicle receptors, field receptors, and C mechanoreceptors in hairy skin and the FA I receptor (Meissner's corpuscle in primates and Krause's end bulb in cats) in the glabrous skin. G_2 hair follicle receptors supply guard hairs and are readily activated by movements of these hairs. D hair follicle receptors innervate down hairs; they are exquisitely sensitive to movements of down hairs by stimuli applied anywhere over a broad area. Field receptors are in the skin; they can be excited by movement of many hairs, but not of single ones, or by contact with the skin. C mechanoreceptors are especially responsive to stimuli that move slowly across the hairy skin; they are much less common in primates than in cats. FA I receptors are found in the glabrous (nonhairy) skin; they are very sensitive, responding best to repetitive stimulation at 5–40 Hz.

10. Cutaneous transient detectors include G_1 hair follicle receptors and Pacinian corpuscles. G_1 hair follicle receptors innervate guard hairs and respond only to rapid movement of these hairs. Pacinian corpuscles are found in subcutaneous tissue (as well as in fascial planes, along the interosseous membranes, adjacent to joints, and in the mesentery). They are extraordinarily sensitive to mechanical transients, sometimes having a threshold of less than 1 μm. They respond best to vibratory stimuli at 60–300 Hz, but can often follow at 500 Hz or more.

11. Sinus hairs are a special type of hair follicle with a rich innervation. Examples include the mystacial vibrissae and the carpal hairs. Vibrissae are innervated by two types of slowly adapting and two types of rapidly adapting receptors. Carpal hairs have two types of slowly adapting receptors, but no rapidly adapting receptors. Instead, movement of carpal hairs excites a cluster of adjacent Pacinian corpuscles.

12. Cutaneous nociceptors of several types have been described. Aδ mechanical nociceptors are activated selectively by noxious mechanical stimuli. If sensitized, they may become responsive to noxious heat stimuli. Their terminals invade the epidermis but remain ensheathed by Schwann cells; therefore, they should not be considered free nerve endings. C-polymodal nociceptors respond to noxious mechanical, thermal, and chemical stimuli. Some can be activated by the itch-provoking substance cowhage. Other nociceptors in the skin include C mechanical nociceptors, Aδ mechanoheat nociceptors, and Aδ and C cold nociceptors.

13. Many of the fine afferent fibers that innervate the skin contain peptides, such as calcitonin gene-related peptide, substance P, and others. The sensory neurons may release these peptides centrally, but in addition they may release them peripherally as part of the process of inflammation.

14. Cutaneous thermoreceptors include cold and warm receptors. Cold receptors may be innervated by Aδ fibers or by C fibers. They respond with a dynamic and a static discharge to

sudden cooling of the skin. The maximum static discharge occurs at 18–34°C. At low temperatures, cold receptors show a bursting discharge which may help distinguish between temperatures above and below the level of maximum responsiveness. Cold receptors may show a paradoxical cold response when exposed to noxious heat. Warm receptors are supplied by unmyelinated fibers. They discharge when the skin is warmed, but they stop firing when noxious temperatures are reached.

15. Muscle contains stretch receptors and pressure-pain endings. The stretch receptors include muscle spindles and Golgi tendon organs. Muscle spindles are complex sense organs with two types of sensory endings and at least two kinds of motor endings. The primary endings are supplied by the largest type of sensory axons, the group Ia afferent fiber. They are sensitive to the rate of stretch of the muscle and to the length of the muscle. The sensitivity of the primary ending is set by activity in dynamic fusimotor fibers. Secondary endings are innervated by intermediate-sized or group II axons and they signal muscle length. Their sensitivity is controlled by static fusimotor axons. Golgi tendon organs are innervated by group Ib axons and are located in tendons and aponeuroses. They respond to both muscle stretch and contraction.

16. Group III muscle afferent fibers have been called pressure-pain endings. Some are activated by pressure on the junction between the muscle and its tendon or on the belly of the muscle; others are in fascial planes; some can be excited by muscle stretch. Intra-arterial injection of algesic chemicals or intramuscular injection of hypertonic saline will activate many of these afferent fibers. Although many are excited by muscular contractions, their responses are decreased following ischemia. Therefore, it has been proposed that these fibers are more likely to be ergoreceptors than nociceptors.

17. Group IV muscle afferent fibers often have high thresholds for mechanical stimuli. They can be excited or sensitized by intra-arterial injection of algesic chemicals and often by thermal stimuli. It has been suggested that those which respond preferentially to chemical stimuli may be nociceptors and those which are activated by muscle contractions but not by algesic chemicals may be ergoreceptors. Some of these afferent fibers discharge vigorously during muscle ischemia and are suited for a role in ischemic muscle pain.

18. Joints are innervated by mechanoreceptors and nociceptors. The mechanoreceptors include Ruffini endings, Golgi tendon organs, Pacinian corpuscles, and Golgi-Mazzoni endings. The Ruffini endings appear to discharge best when the joint is brought to the limit of its range, and so they may signal joint pressure. Golgi tendon organs respond like those in muscle and Pacinian corpuscles like those in subcutaneous tissue. Golgi-Mazzoni endings appear to respond with a slowly adapting discharge when they are compressed, and so they may provide deep pressure information.

19. There is a controversy about the presence or absence of slowly adapting discharges when the joint is at mid-range. One opinion is that when such discharges are seen, they reflect the activity of muscle stretch receptors that have entered the joint nerve; joint position is signaled by muscle receptors rather than by joint receptors. The opposing view is that there are some slowly adapting joint afferent fibers that discharge best when the joint is at mid-range and that these can provide information to the central nervous system about joint position.

20. Joint nociceptors have been well characterized. They may have group III or IV afferent fibers. They respond best or only to noxious movements of the joint. Some do not respond normally to any joint movement, but become very sensitive during inflammation.

21. Viscera are supplied by mechanoreceptors and, at least in some organs, by nociceptors. Mechanoreceptors associated with the gastrointestinal tract respond to movements of the bowel or of the mesentery. Some in the wall of the urinary bladder respond either to distension or to contraction of the bladder. Visceral nociceptors have been found in the heart, esophagus, gallbladder, and testis. Those in the testis have been well characterized. Most are supplied by Aδ axons, and they are responsive to noxious mechanical, thermal, and chemical stimuli.

3 Dorsal Root Ganglion Cells and Their Processes

General somatic primary afferent neurons have an especially important place in considerations of spinal sensory mechanisms. These neurons are housed in dorsal root ganglia and their cranial nerve equivalents. At spinal levels they are referred to as dorsal root ganglion cells (DRG cells). These cells are unipolar neurons (sometimes called pseudounipolar since they originated as bipolar cells, [Dogiel, 1908; Ranson, 1912]) with a peripheral process or processes that receive information from sensory receptor organs and a central process or processes that transmit this information centrally (Fig. 3.1). Thus, we can divide the primary afferent neuron into a cell body, a peripheral process or processes, and a central process or processes. The central processes are often referred to as dorsal root axons because they travel in the dorsal root on their way to the cord.

DORSAL ROOT GANGLION CELL BODIES

DRG cells are often described as a homogeneous population because their cell bodies are located in similar peripheral ganglia, their processes are organized in a similar way, and they have a common function, to transduce environmental stimuli and transmit the resulting information centrally (Scharf, 1958; Lieberman, 1976). A major advance since the previous edition of this book, however, is the perception that despite the similarities, these cells are remarkably diverse. At present this diversity is manifested primarily by differences in the cell body. This section of the chapter will be concerned with documenting these diversities. The categories we will consider are size, cytology, development, morphologic markers, physiologic characteristics, and organizational features.

Size

The most widely accepted subdivision of DRG cell bodies is by size. Indeed, it would be hard to find a paper concerned with DRG cells that did not relate the size of the cell to the topic under discussion. The usual statement is that dorsal root ganglion cell bodies can be divided into one of two categories: large or small. The actual evidence that this is the case, however, should come from histograms that relate cell size to cell numbers.

Histograms. Probably the first histogram that related cell size to cell numbers was that of Warrington and Griffith (1904), who showed that the diameters of C2 and T11 cat primary afferent neurons formed a unimodal histogram with a peak at approximately 40 μm whereas

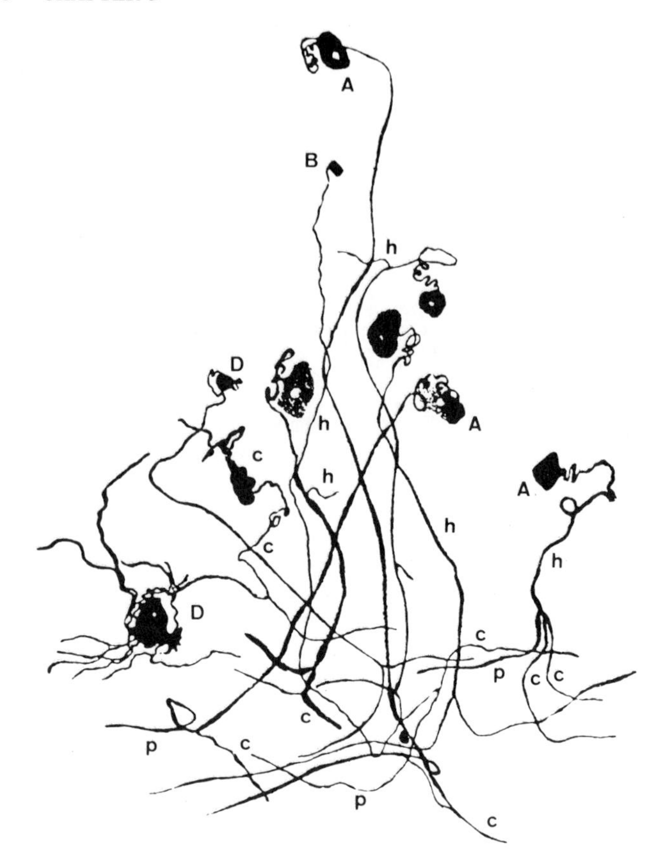

Fig. 3.1. A drawing from a methylene blue stained dorsal root ganglion (DRG) cell. A and B are large and small unipolar ganglion cells respectively. D is a multipolar ganglion cell. The division of the axons (h) of the ganglion cells into peripheral (p) and central (c) processes is shown for several neurons. Note the trifurcation of the axon of cell A at the right. (From Dogiel, 1896.)

those at L7 were bimodal with peaks at 40 and 70 μm. More recently, bimodal histograms (Fig. 3.2) have been described for mouse, rat, cat, and human DRGs (as a partial list see Lawson, 1979; Lawson and Biscoe, 1979; Szarijanni and Réthelyi, 1979; Lawson *et al.*, 1984, 1985; McDougal *et al.*, 1985; K. H. Lee *et al.*, 1986; Ohnishi and Ogawa, 1986; Schmalbruch, 1987). Inspection of the histograms shows that there is considerable overlap of the two peaks, however. Also, not all studies agree that there are two peaks. There is the single peak for the C2 and T11 ganglia in cats (Warrington and Griffiths, 1904), and several studies report three peaks in certain human ganglia (Fig. 3.3) (Dyck *et al.*, 1979; Kawamura and Dyck, 1978; Kawamura *et al.*, 1981) and certain trigeminal cell populations (Sugimoto *et al.*, 1988). Furthermore, a number of studies describe intermediate-sized cells, albeit without histograms (Matsuura, 1967; Matsuura *et al.*, 1969; Honda *et al.*, 1983; Lindh *et al.*, 1983; Otten and Lorez, 1983; Kuo *et al.*, 1985; Tuchscherer and Seybold, 1985; Hanko *et al.*, 1986; Kuwayama *et al.*, 1987). Notwithstanding, the notion that there are two populations by size of DRG cells is generally accepted, and correlations of cell size with other features of ganglion cell organization are a major goal of almost all studies on DRG cell bodies.

Correlations. A major thrust in somatic sensory physiology has been to correlate the type of information transmitted with the conduction velocity of the axon (see Chapter 2). Conduction velocity is in turn thought to be correlated with axon size, whether or not the axon is myelinated, and with cell body size. Accordingly, an examination of the precise relationships between these variables bears on sensory mechanisms.

Axon Size. A general relation between cell body and axon size (large cells give rise to large

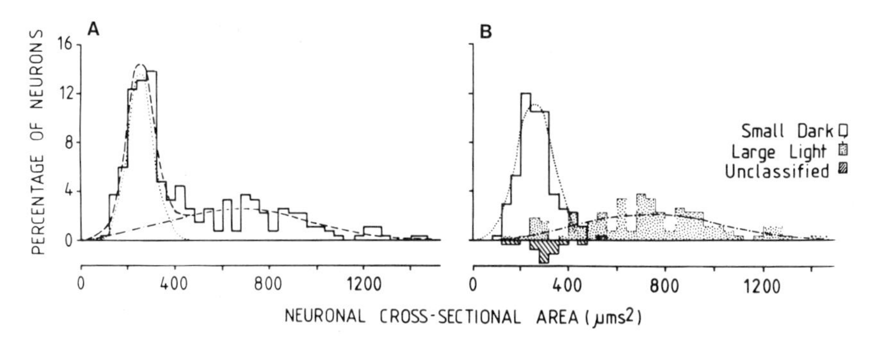

Fig. 3.2. Histograms of dorsal root ganglion cell body sizes. A indicates all cells and B shows the cells split into large light and small dark categories. Note that the histograms are essentially the same. Also note that there are some small "large, light" cells. If the large light cells are labeled by the neurofilament antibody RT-97, the histograms are essentially identical. (From Lawson *et al.*, 1984; reprinted by permission of Wiley-Liss.)

axons and small cells to small axons) became apparent when whole-neuron staining techniques came into widespread use (Dogiel, 1908; Ramón y Cajal, 1909; Marinesco, 1909). For DRG cells, this correlation is substantiated by many indirect observations. For example, there is a loss of small DRG cells and unmyelinated axons after neonatal capsaicin treatment (Fitzgerald, 1983a) and cell and axon sizes are correlated in the loss that occurs in certain diseases (Ohnishi and Dyck, 1974; Ohnishi and Ogawa, 1986). Such observations, although suggestive, are not direct proof, nor do they provide mathematical relations between axon and cell body size. To obtain such relations, it would be desirable to measure individual cell bodies and their processes. Yoshida and Matsuda (1979) did this with mouse DRG cells filled with horseradish peroxidase (HRP) and described a linear relationship between cell and axon diameters, but a formula was not provided. A linear correlation with an appropriate formula was reported for HRP-filled cat

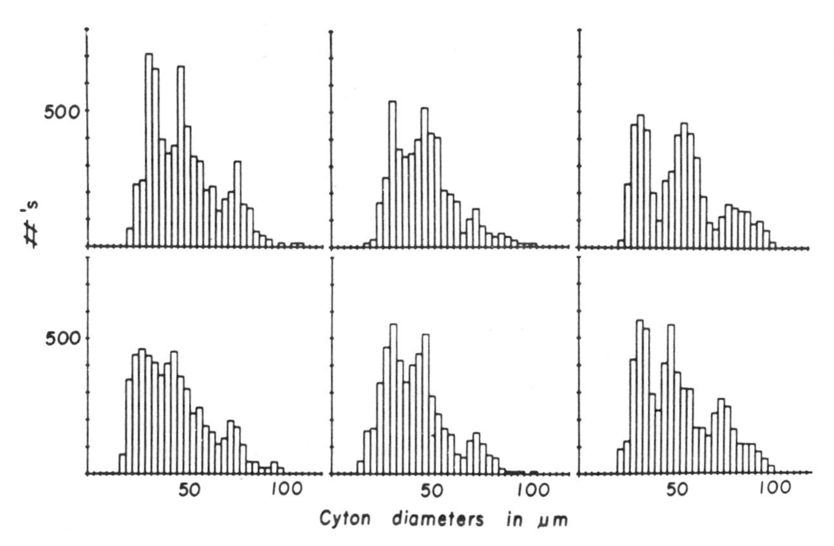

Fig. 3.3. Histograms of DRG cell diameters from the L5 spinal ganglion in man. Note that in some of the pictures there are what appear to be 3 peaks of cell sizes. This is not clear in all of the histograms, however. (From Kawamura and Dyck, 1978; reprinted with permission.)

L5–S3 dorsal root ganglion cells whose processes conducted impulses more rapidly than 2.5 m/s, but no correlation could be noted for cells whose processes conducted at less than 2.5 m/s (K. H. Lee *et al.*, 1986). Furthermore, the data show considerable scatter, and a power function would have fitted the data as well as a linear function, the linear function being chosen because of simplicity.

Conduction Velocity. Conduction velocity is more frequently used as an estimate of axon size than morphologic measurements of axon diameters. Two studies report a linear relationship between DRG cell body size and conduction velocities of the processes of the cells (Yoshida and Matsuda, 1979; Cameron *et al.*, 1986). In two other studies the relation is more complicated. Harper and Lawson (1985a) report a loose correlation between cell size and conduction velocity for rapidly conducting fibers and a much tighter correlation between cell body size and conduction velocity for slowly conducting axons. Furthermore, there is a different relation for fibers that conduct at 2.2–8 m/s (δ fibers) than for those that conduct at 0.8–1.4 m/s (C fibers). By contrast K. H. Lee *et al.* (1986) report a good correlation between cell size and conduction velocity for rapidly conducting axons (mean conduction velocity, 26.6 m/s), but no correlation for slowly conducting cells (mean conduction velocity, 0.55 m/s). Finally, two studies find no correlations at all. Thus sensory cells in the nodose ganglia whose processes conduct action potentials in the A range (21 m/s) are the same size as those whose processes conduct in the C range (1 m/s) (Gallego and Eyzaguirre, 1978). Furthermore, all sizes of dorsal root ganglion cells gave rise to group IV (unmyelinated) afferents from joints (Hoheisel and Mense, 1986).

Myelination. The information carried by myelinated sensory axons is different from that carried by unmyelinated sensory axons. It is also generally accepted that large axons are myelinated and small axons are unmyelinated, although the actual formulae relating axon size to myelination are still a subject of study. Thus it is reasonable to assume that large cells give rise to myelinated axons and small cells to unmyelinated axons (Ranson, 1912). Much indirect evidence supports this idea, but direct evidence in which sensory cell bodies are measured and the processes of the same cells are examined by appropriate morphologic techniques, is not to our knowledge available. Axonal conduction velocity is often taken as an indicator of myelination, however, and when conduction velocity is compared with cell size, the results are variable (see above).

In summary, there is a general perception that cell body size, axon size, axon conduction velocity, the presence or absence of a myelin sheath and the type of information carried by the axon are closely interlinked. Much indirect evidence supports this. Quantitation of these relationships is a relatively recent development, however, and when this is done some of the correlations are not very tight and a few seem to be nonexistent. The quantitative work is just beginning, however, and needs to be continued. It is particularly important to measure all parameters for the same cell.

Cytology

The cytology of the DRG cell at the light microscopic level has been well reviewed (Scharf, 1958; Lieberman, 1976). Early cytologic subdivisions were on the basis of whether the cells appeared "light" or "dark" with the usual basic stains and on the organization of the Nissl material. More complicated subdivisions have recently been proposed.

Light (Clear) and Dark Cells. A common early theme was that there are two basic types of DRG cells, lightly staining cells that tend to be large (large, clear cells) and intensely staining cells that tend to be small (small, dark cells) (Dogiel, 1908; Ramón y Cajal, 1909; Scharf, 1958; Lieberman, 1976). Some did not accept this classification as representing the living state, either because they did not observe distinctive cytologic differences or because poor fixation often produces small dark cells (summarized by Lieberman [1976]). However, there are other studies that show what appear to be distinctive differences in material that is adequately fixed, (Lawson *et al.*, 1974, Lieberman, 1976). Also the last birthdays of the large cells are different from those of the small cells (Lawson and Biscoe 1979), and the clear cells but not the dark cells are immunostained by RT97 antibodies which recognize neurofilaments (Lawson *et al.*, 1984). Thus the distinction between large, clear cells and small, dark cells is generally accepted (Lieberman, 1976).

Are All Clear Cells Large? One problem with the above subdivision is that not all the clear cells seem to be large. This is seen best with the use of an antibody (RT97) that recognizes neurofilaments and thus labels the clear cells. In this case, cells of all sizes, not just the large cells, are labeled (Lawson *et al.*, 1984). Reconstructions of histograms also reveal two bell-shaped populations of DRG cells, the clear cells with a diameter or cross-sectional area range that spans the whole population and the dark cells, which are small (Fig. 3.2). Thus, the dark cells are always small, but the clear cells are of all sizes. This is presumably one of the reasons why the separation between large and small cells is sometimes indistinct in histograms.

Nissl Substance. Early cytologic subdivisions of DRG cell bodies beyond the large clear and small dark categories were on the basis of the arrangement of the Nissl substance (summarized by Orr and Rows [1901], and S. L. Clark [1926]). These can serve as descriptive subdivisions, but they do not correlate with the function of the sensory cell or its responses to various types of stimuli (S. L. Clark, 1926; Duce and Keen, 1977). In agreement with Lieberman (1976), therefore, these classifications should be put aside until increasing knowledge clarifies the meaning of differences in the arrangement of Nissl substance.

Modern Classifications. More recently, DRG cells have been subdivided on multiple cytologic criteria. Andres (1961), for example, using phase and electron microscopy, divided the cells in the large light category (A cells) into three subgroups (A1, A2, and A3) and the cells in the small, dark category (B cells) into three subgroups (B1, B2, and B3) and related these subgroups to those in earlier studies. J. Jacobs *et al.* (1975) proposed a somewhat simpler grouping with two types of A cells and one type of B cell. Duce and Keen (1977) divided the A cells into three groups, and the B cells into four groups. Rambourg *et al.* (1983) divided DRG cells into three types (A, B, and C), with the A type being subdivided into three subgroups and the B type into two subgroups (Fig. 3.4). Finally, Sommer *et al.* (1985), although the criteria are different, also divided the cells into three A types, two B types and a single C type. As with the earlier classifications, an obvious goal is to correlate the various cytologic subdivisions with functional or biochemical variables. Beginnings in this regard are correlations with concanavalin A binding (Rambourgh *et al.*, 1983; Malchiodi *et al.*, 1986), histochemical correlations (Sommer *et al.*, 1985), and electron microscopic correlations with single functionally identified cells (Sugiura *et al.*, 1988).

A Note on Terminology. The A, B, C terminology that is often applied to DRG cell bodies is ambiguous. For example, DRG cells are often labeled in reference to the size and conduction velocity of their processes. This is essentially applying the peripheral-nerve compound action potential terminology of Erlanger and Gasser (1937) to DRG cells. Thus A cells are cells whose peripheral processes conduct at large myelinated-fiber velocities, B cells at small myelinated-fiber velocities and C cells at unmyelinated-fiber velocities. Subdivisions of this terminology would, for example, be Aβ or Aδ cells. Cytologic subdivisions also use an A, B, C terminology, and none of these categories are completely equivalent to each other or to the above categories (Andres 1961; J. Jacobs *et al.*, 1975; Duce and Keen, 1977; Rambourg *et al.*, 1983; Sommer *et al.*, 1985). In addition, Yoshida *et al.* (1978) use the term "A cells" for cells with brief sodium spikes insensitive to tetrodotoxin. Thus for the present one must remember which type of terminology one is using. It is hoped that advancing knowledge will result in a unification.

In summary, DRG cell bodies have classically been subdivided into large light and small dark categories, but recent work indicates that the clear (relatively organelle-free) cells are of all

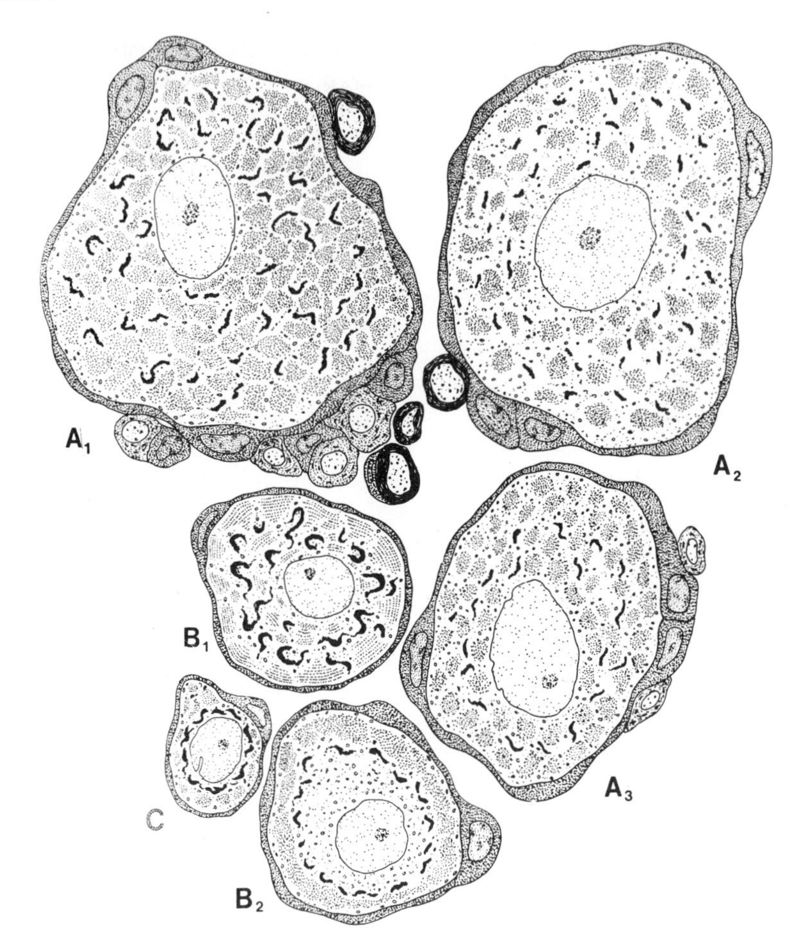

Fig. 3.4. A diagram of the various cell types proposed by Rambourgh *et al.* (1983). A1 is a large neuron with evenly distributed Nissl bodies and Golgi saccules. A2 is a large neuron with widely spaced Nissl bodies and Golgi stacks. A3 is a smaller neuron with closely packed Nissl bodies and the stacks of Golgi saccules forming a perinuclear ring. B1 is a small neuron whose Nissl bodies have parallel cisternae and with the stacks of Golgi saccules forming a pericellular net. B2 is a small cell with a cortical zone consisting of Nissl substance and the Golgi apparatus forming a ring separated form the nucleus by mitochondria and agranular reticulum containing cytoplasm. C is a very small cell with a poorly demarcated Nissl substance and Golgi apparatus. (From Rambourgh *et al.*, 1983; reprinted with permission.)

sizes whereas the dark (relatively organelle-concentrated) cells are small. Further interesting cytologic subdivisions have been proposed, and correlations of these with functional variables are desirable.

Development

Two points are relevant to DRG cell diversity. First, large cells have earlier final birthdays than small cells (Lawson *et al.*, 1974; Sims and Vaughan, 1979; Altman and Bayer, 1984). This may be related to the development of reflex arcs (Sims and Vaughan, 1979). Second, in embryonic birds there is a clear separation of large ventrolateral and small dorsomedial dorsal root ganglion cell bodies (Hamburger *et al.*, 1981). Furthermore, the small cells seem to contain substance P and are derived from the neural crest, whereas the large cells do not contain substance P and are probably derived from neural placodes (Fontaine-Perus *et al.*, 1985). Such clear subdivisions are not seen in adult birds nor in mammals at any age (Altman and Bayer, 1984), although Tennyson (1964) did report such a subdivision in cranial nerve ganglia of the rabbit.

Morphologic Markers

The term "morphologic marker" indicates a compound that can be found by anatomic means in a subpopulation of DRG cells. The two largest groups in this category are peptide and enzyme markers. A third group of more heterogeneous markers will also be considered.

Peptides. The demonstration that a population of DRG cells was selectively immunostained for substance P (Hökfelt *et al.*, 1975b,c) began an extensive series of studies localizing peptides in primary afferent neurons (Fig. 3.5). Before proceeding, however, there are some caveats. First, because of uncertainties about exactly what amino acid sequences the antibodies are recognizing, the label is often referred to as "peptidelike" (substance P-like or calcitonin gene-related peptide-like, for example). This terminology is sometimes not used, as here, because it is cumbersome, but the uncertainty should be remembered. Second, the immunolabels do not measure absolute amounts of an antigen, nor do they determine whether the cell is actively making the compound or just storing it. Thus, it cannot be assumed that all cells that contain or can make a peptide are immunolabeled. As a complicating factor, it is often necessary to apply a microtubule inhibitor such as colchicine or to tie the roots to build up the antigen in the cell body if cells are to be labeled. Notwithstanding these difficulties, the development of immunocytochemical labeling will undoubtedly lead to major advances in our understanding of spinal sensory mechanisms.

Localizations. The studies summarized in Table 3.1 have demonstrated subpopulations of DRG cells immunostained for various peptides. Not surprisingly, the number of studies roughly parallels the length of time the antibodies have been available. These studies are restricted to those that demonstrate immunolabeled cells morphologically. Other evidence that these peptides are localized in sensory neurons (e.g., demonstrations of peptide localizations in primary afferent receiving areas of the dorsal horn or chemical demonstrations of a peptide in the DRG) will be presented in Chapter 5.

From Table 3.1, it can be seen that many peptides have been localized in subpopulations of DRG cells. This number will undoubtedly increase. An early hope was that specific sensory categories (see Chapter 2) might be correlated with specific single peptides. These hopes have not been sustained (S. P. Hunt and Rossi, 1985; Leah *et al.*, 1985a). Many peptides are colocalized, however, and some of the patterns of colocalization may have functional or topographic significance (see below).

Sizes of Labeled Cells. Almost all early studies note that the peptides are found in the small DRG cell population. This does not imply that these populations are indistinguishable, however, for J. Price (1985) noted significant differences in the means and shapes of the histograms of the cell diameters. Also, various investigators noted that some larger cells were labeled. These are often referred to as intermediate-sized cells (Honda *et al.*, 1983; Lindh *et al.*, 1983; Otten and Lorez, 1983; Kuo *et al.*, 1985; K. H. Lee *et al.*, 1986; Tuchscherer and Seybold, 1985; Hanko *et al.*, 1986; Kuwayama *et al.*, 1987) because they are larger than the small cells but not the largest in the ganglion. The observation is correct, but the term "intermediate-sized cells" implies a separate category. This category is not found (except possibly in humans [Kawamura and Dyck, 1978]), however, in terms of cell sizes, cell birthdays, or general cell morphology. Thus, more evidence is needed if intermediate-sized cells are to be regarded as a meaningful separate

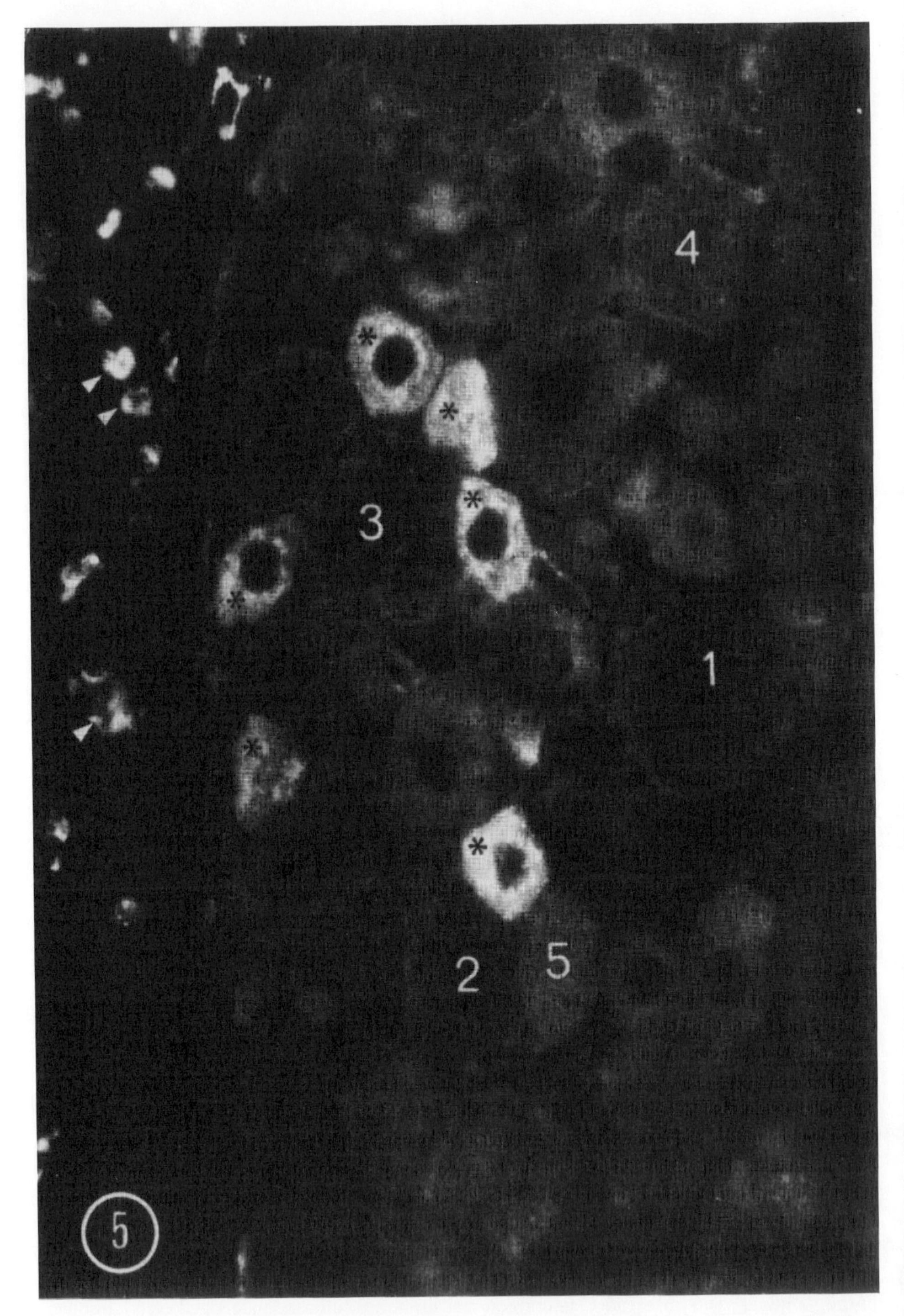

Fig. 3.5. An example of immunoreactive DRG cell bodies. In this case the reaction is for somatostatin. (From Hökfelt *et al.*, 1976; reprinted with permission.)

TABLE 3.1. Immunocytochemical Peptides Localizations in DRG Cells

	Component	References
A.	*Substance P (SP)*	Hökfelt *et al.*, 1975b,c; 1976; Chan-Palay and Palay, 1977a,b; Cuello and Kanazawa, 1978; Lundberg *et al.*, 1978; DiTirro *et al.*, 1981; Janscó *et al.*, 1981; Tervo *et al.*, 1981; Dalsgaard *et al.*,1982b, 1984; Nagy and Hunt, 1982; Charnay *et al.*, 1983; Fuxe *et al.*, 1983; DeGroat *et al.*, 1983; Lindh *et al.*, 1983; Panula *et al.*, 1983; Kuo *et al.*, 1984; Lehtosalo 1984; Sharkey *et al.*, 1984; Wiesenfeld-Hallin *et al.*, 1984; Leah *et al.*, 1985a,b; K. H. Lee *et al.*, 1985; Ninkovic and Hunt, 1985; J. J. Price, 1985; Skofitsch and Jacobwitz, 1985c; Sommer *et al.*, 1985; Tuchscher and Seybold, 1985; Chiba *et al.*, 1986; Dockray and Sharkey, 1986; Hanko *et al.*, 1986; Kai-Kai *et al.*, 1986; Wanaka *et al.*, 1986; Gibbins *et al.*, 1987b; Green and Dockray, 1987; Kuwayama *et al.*, 1987; D. L. McNeill and Burden, 1987; Molander *et al.*, 1987; Marti *et al.*, 1987; Su *et al.*, 1987; Cameron *et al.*, 1988; reviews by Hökfelt *et al.*, 1980; Buck *et al.*, 1982; S. P. Hunt *et al.*, 1982; Salt and Hill, 1983; Ruda *et al.*, 1986; Jessell and Dodd, 1986; Blumenkopf, 1988.
B.	*Somatostatin (SOM)*	Hökfelt *et al.*, 1975a; 1976; O. Johansson, 1978; Lundberg *et al.*, 1978; Dalsgaard *et al.*, 1984; Lehtosalo, 1984; Leah *et al.*, 1985a,b; Price, 1985; Skofitsch *et al.*, 1985a; Tuchscherer and Seybold, 1985; Kummer and Heym, 1986; Shehab *et al.*, 1986a,b; Ju *et al.*, 1987b; Marti *et al.*, 1987; Molander *et al.*, 1987; Ositelu *et al.*, 1987; Cameron *et al.*, 1988.
C.	*Cholecystokinin (CCK)*	Lundberg *et al.* 1978; Dalsgaard *et al.*, 1982b, 1984; Otten and Lorez, 1983; Leah *et al.*, 1985b; Tuchscherer and Seybold, 1985; Kuwayama and Stone, 1986a; Kuwayama *et al.*, 1987; D. L. McNeill and Burden, 1987; Gibbins *et al.*, 1987b; Ju *et al.*, 1987b.
D.	*Calcitonin gene-related peptide (CGRP)*	Rosenfeld *et al.* ,1983; Gibson *et al.*, 1984b; Wiesenfeld-Hallin *et al.*, 1984; Gibbins *et al.*, 1985; 1987b; Lee *et al.*, 1985; Skofitsch and Jacobwitz 1985a,b; Terenghi *et al.*, 1985; Uddman *et al.*, 1985a,b; Franco-Cereceda *et al.*, 1986; Hanko *et al.*, 1986; Inyama *et al.*, 1986; Matsuyama *et al.*, 1986; Terenghi *et al.*, 1986; Wanaka *et al.*, 1986; Yokokawa *et al.*, 1986; Green and Dockray, 1987; Ju *et al.*, 1987b; Kuwayma *et al.*, 1987; Marti *et al.*, 1987; Su *et al.*, 1987; Cameron *et al.*, 1988; Merighi *et al.*, 1988, Tsai *et al.* 1988.
E.	*Bombesin (BOMB)*	Lundberg *et al.*, 1978; Fuxe *et al.*, 1983; Panula *et al.*, 1983; Cameron *et al.*, 1988.
F.	*Vasoactive intestinal polypeptide (VIP)*	DeGroat *et al.*, 1983; Honda *et al.*, 1983; Kawatani *et al.*, 1983; Kuo *et al.*, 1985; Leah *et al.*, 1985a,b; DeGroat, 1987.
G.	*Galanin (GAL)*	Ch'ng *et al.*, 1985; Skofitsch and Jacobwitz, 1985d; Ju *et al.*, 1987b; Hökfelt *et al.*, 1987.
H.	*Vasopressin and Oxytocin (AVP, OXY)*	Kai-Kai *et al.*, 1985, 1986.
I.	*Dynorphin (DYN)*	Kummer and Heym, 1986; Gibbins *et al.*, 1987b
J.	*Enkephalin (ENK)*	Skofitsch *et al.*, 1985a; Kummer and Heym, 1986; Senba *et al.*, 1989.
K.	*α-neo-Endorphin (END)*	Kummer and Heym, 1986.
L.	*Corticotropin-releasing factor (CRF)*	Skofitsch *et al.*, 1985.
M.	*Neurokinin- A (NKA)*	Dalsgaard *et al.*, 1985.

population. One type of evidence that would have considerable bearing on spinal sensory mechanisms is if it could be shown that small cells give rise to unmyelinated axons and intermediate cells to small myelinated axons.

Percentages of Labeled Cells. Most studies provide percentages of labeled cells, and these are summarized in Table 3.2. Note that there is considerable variability in the ranges of the

TABLE 3.2. Percentages of Labeled DRG Cells

Peptide	Mean	Range
SP	19	8–33
SOM	10	5.3–20
CGRP	38	10–63
CCK	8	2–10
BOMB	6.3	5–7.6
VIP	8.1	5.3–11
GAL	7	2–10
AVP	55	50–60
OXY	45	35–60
DYN	3	3

[a]Abbreviations are defined in Table 3.1.

percentages of labeled cells. This is partially due to such factors as species, particular antibody used, or whether or not colchicine is given. Despite this, there seems to be an overall pattern that undoubtedly has functional significance. A hypothesis is that compounds that are found in large numbers of cells are more "important" in sensory mechanisms than those found in only a few cells. Proof of this hypothesis obviously rests on determining the function of these compounds. An arbitrary but potentially useful ranking is into many labeled cells, moderate numbers of labeled cells, and few labeled cells (Table 3.3). It should be noted that many of these peptides can be colocalized (Table 3.4).

Colocalizations. It was generally assumed that each neuron contains one transmitter, which is released at all terminals of that cell. Thus, the discovery that many neurons, among them DRG cells, contain more than one neuroactive substance caused a considerable stir (e.g. Lundberg and

TABLE 3.3. Classification of Labeled Cells by Percent Labeling of DRG Cells[a,b]

A. *Many labeled cells* (more than 30% of the total DRG population)
   1. CGRP
   2. AVP
   3. OXY
B. *Moderate numbers of labeled cells* (10–30% of the total DRG population)
   1. SP
C. *Infrequently labeled cells* (clear labeling but 10% or less of the total DRG population)
   1. SOM
   2. CCK
   3. BOMB
   4. VIP
   5. GAL
   6. DYN
   7. END
   8. ENK
   9. CRF

[a]A caveat is that some of these localizations have only been reported in one or two studies, and for these further work is obviously necessary. This is particularly true for arginine-vasopressin, oxytocin and items 5–9 in category C. Somewhat higher percentages for some of these compounds are given in a very recent review (Kai-Kai, 1989).

[b]Abbreviations are defined in Table 3.1.

TABLE 3.4. DRG Peptide Colocalizations[a]

A. SP colocalizes with:

CGRP

Wiesenfeld-Hallin *et al.*, 1984; Gibbins *et al.*, 1985; Y. Lee *et al.*, 1985; Skofitsch and Jacobwitz, 1985c; Terenghi *et al.*, 1985; Hanko *et al.*, 1986; Franco-Cereceda *et al.*, 1986; Kuwayama and Stone, 1986a; Kuwayama *et al.*, 1987; Wanaka *et al.*, 1986; Yokokawa *et al.*, 1986; Gazelius *et al.*, 1987; Gibbins *et al.*, 1987b; Ju *et al.*, 1987b; Cameron *et al.*, 1988

There is no doubt that SP and CGRP colocalize. It is generally believed that all or almost all SP cells contain CGRP, but that many CGRP neurons do not contain SP.

CCK

Dalsgaard *et al.*, 1982b; Tuchscherer and Seybold, 1985; Gibbins *et al.*, 1987b

SOM

Hökfelt *et al.*, 1976 (not colocalized); Leah *et al.*, 1985b; Ju *et al.*, 1987b

SP and SOM were first reported not to colocalize in the rat (Hökfelt *et al.*, 1976). Leah *et al.* (1985b) then reported extensive colocalization in the cat, and Ju *et al.* (1987b) reported an occasional cell with both compounds in the rat.

BOMB

Fuxe *et al.*, 1983; Cameron *et al.*, 1988

VIP

Leah *et al.*, 1985b; Kummer and Heym, 1986 (rare)

DYN or *END*

Kummer and Heym, 1986; Weike *et al.*, 1986; Gibbins *et al.*, 1987b

B. SOM colocalizes with:

SP

See above under SP

CGRP

Cameron *et al.*, 1988

CCK

Leah *et al.*, 1985b

VIP

Leah *et al.*, 1985b

SOM usually does not colocalize with other peptides or other nonpeptide markers such as FRAP (see, e.g., Ruda *et al.*, 1986), except in the cat (Leah *et al.*, 1985b; Cameron *et al.*, 1988).

C. CGRP colocalizes with:

SP

See above under SP

CCK

Gibbins *et al.*, 1987b

DYN

Gibbins *et al.*, 1987b

D. CCK colocalizes with:

SP

See above under SP

SOM

Leah *et al.*, 1985b

CGRP

Gibbins *et al.*, 1987b; Ju *et al.*, 1987b

VIP

Leah *et al.*, 1985b

DYN

Gibbins *et al.* 1987b

(continued)

TABLE 3.4. *(Continued)*

E. BOMB colocalizes with:	
	CGRP, SP, and SO
Cameron *et al.*, 1988	
F. VIP colocalizes with:	
	SP
Leah *et al.*, 1985b; Ju *et al.*, 1987b	
	CCK
Leah *et al.*, 1985b	
	SOM
Leah *et al.*, 1985b	
	END
Kummel and Hayes, 1986	
G. GAL colocalizes with:	
	SP *CGRP*
Ju *et al.*, 1987b	
H. DYN colocalizes with:	
	ENK and END
Kummer and Hayes, 1986	
	SP, CGRP, and DYN
Gibbins *et al.*, 1987	
I. NKA colocalizes with:	
	SP
Dalsgaard *et al.*, 1985	
J. AVP and OXY colocalize with:	
	SP (and FRAP but not RT97)
Kai Kai *et al.*, 1986	

[a]Abbreviations are defined in Table 3.1.

Hökfelt, 1986). This finding considerably complicates the classification of DRG cells. For example, if only five compounds are considered, and if they occur in all combinations, there would be 26 different categories. Admittedly, not all combinations occur, but many do (see, e.g., Leah *et al.*, 1985b; Cameron *et al.*, 1988). Furthermore, since at least 13 peptides have been identified in DRG cells, and since peptides can colocalize with other compounds, the possibilities increase greatly. At present the findings are so new that the main task is simply to determine which DRG cells contain various combinations of peptides or other substances. The meaning of these combinations will be a most important topic in considerations of sensory mechanisms.

The most common colocalization studies are what we can call "single colocalizations", which are demonstrations that two peptides (or a peptide and another compound) are found in the same DRG cell. The single colocalizations can now be supplemented by multiple colocalizations that have been reported in recent studies (Leah *et al.*, 1985b; Gibbins *et al.* 1987b; Gibbons and Morris, 1987; Ju *et al.*, 1987b; Cameron *et al.*, 1988). It seems apparent that as these studies increase, the data on localizations of peptides and other morphologically demonstrable neuroactive substances will be best presented as matrices. At present, however, this is premature. The colocalizations reported to date are presented in Table 3.4. An obvious question is whether the colocalizations have functional implications. No simple patterns have emerged, but there is a suggestion that certain combinations of peptides may be associated with the innervation patterns of different parts of the body (see below).

Peripheral Correlations. One way to get insight into the function of peptide containing primary afferent cells would be to correlate peptide immunolocalization with patterns of peripheral innervation. This has recently become feasible because retrograde labeling of a cell from a particular part of the body can be combined with immunocytochemical labeling. These studies can be divided into five categories by peripheral regions; (1) large viscera (Lindh *et al.*, 1983; Kuo *et al.*, 1984; Sharkey *et al.*, 1984; Uddman *et al.*, 1985b; Inyama *et al.*, 1986; Terenghi *et al.*, 1986; Yokokawa *et al.*, 1986; Gibbins *et al.*, 1987b; Green and Dockray, 1987; D. L. McNeill and Burden, 1987; Molander *et al.*, 1987); (2) blood vessels (Uddman *et al.*, 1985a; Hanko *et al.*, 1986; Matsuyama *et al.*, 1986; Wanaka *et al.*, 1986; Gazelius *et al.*, 1987; Gibbins *et al.*, 1987b; Tsai *et al.*, 1988); (3) eye (Lehtosalo 1984; Terenghi *et al.*, 1985; Kuwayama and Stone, 1986b, Kuwayama *et al.*, 1987); (4) skin (Leah *et al.*, 1985b; Gibbins *et al.*, 1987b; Molander *et al.*, 1987; Ositelu *et al.*, 1987; O'Brien *et al.*, 1989; Gibbins and Morris, 1987); and (5) skeletal muscles (Molander *et al.*, 1987; Ositelu *et al.*, 1987; O'Brien *et al.*, 1989). A great deal of valuable information about the innervation of a particular organ, a particular vascular tree, etc., has emerged from these studies. In addition, there are some suggestive patterns. For example, in one study axons of SOM cells do not project into a visceral nerve but do project into somatic nerves (Molander *et al.*, 1987). This is in contrast to axons labeled for CGRP, SP, or FRAP, which are seen in both kinds of nerves. In another study, SOM cells innervated facial skin but not masticatory muscles or tongue (Ositelu *et al.*, 1987). Another example is that the DRG cells whose axons project in the greater splanchnic nerve or innervate the stomach almost all contain CGRP (Molander *et al.*, 1987; Green and Dockray, 1987). By contrast, almost no gastric afferent cells in the nodose ganglion contain CGRP (Green and Dockray, 1987). Furthermore, it may be that the search should not be for correlations with single peptides. Gibbins *et al.* (1987b) found relatively precise correlations of the part of the body innervated with different colocalizations of four different peptides, and O'Brien *et al.* (1989) found that the chemical expression of primary afferents is characteristic of a peripheral target. It is obviously too early to draw definitive conclusions, but it would appear that certain patterns of peptide localization might be correlated with the area of peripheral innervation. Interestingly, peptide expression can be changed when nerves are attached to different peripheral targets (McMahon and Gibson, 1987).

Function. The function of the neuroactive peptides is an extremely active area of research. Here we make only a few points related to their possible role in sensory transmission. The most commonly discussed possibility is that the peptides are used as sensory neurotransmitters or neuromodulators. The evidence is suggestive. The peptides are found in subpopulations of DRG cells, particularly small cells, and in fine fibers which end in appropriate laminae of the dorsal horn. These compounds can be released by various types of stimuli, and they usually affect postsynaptic neurons directly or enhance the action of other presumed transmitters. What, then, is the difficulty in regarding these compounds as transmitters? One difficulty is the lack of precise quantitative information relating the amount of peptide release to any particular activity (Wall and Fitzgerald, 1982). Another problem is more philosophical. It is difficult to see why so many different peptides are needed as neurotransmitters unless each sensory category uses a different peptide, but evidence does not support this idea (Lynn and Hunt, 1984; Leah *et al.*, 1985a; Cameron *et al.*, 1986). Thus, the evidence is suggestive but not definitive that these compounds are transmitters (see Chapter 5 for further discussion).

Noxious input is carried by fine fibers which presumably arise from small cells. Since the peptides are found primarily in small cells, noxious-information transfer is frequently discussed. As before, the evidence is suggestive but not definitive. In addition, particularly for substance P, the time course of the changes in noxious-information transmission does not correlate exactly with the time course of changes in peptide concentrations in the dorsal horn (Wall and Fitzgerald, 1982).

Two other suggestions are as follows. First, there may be two distinct fine-fiber pathways innervating similar peripheral structures and conveying similar information, but to different areas within the dorsal horn. The peptidergic neurons would form one of these pathways, and the FRAP system the other (S. P. Hunt and Rossi, 1985). There is also the idea that the peptides may be more involved in regulatory or developmental events than in impulse transmission (Wall and Fitzgerald, 1982). These ideas are not mutually exclusive.

In summary, much of the interest in the peptides found in DRG cells is based on the fact that these compounds are not distributed randomly. They tend, for example, to be located in cells of similar size, the percentages of labeled cells for any peptide are roughly similar from study to study, there are suggestions that they may be organized in relation to specific peripheral areas, and many of them have interesting physiologic effects. Nevertheless, it must be admitted that we really do not know exactly what these compounds do. It seems clear that insights into the function of the peptides will be of major import in considerations of spinal sensory mechanisms.

Enzymes. It has been known for some time that DRG cells can be subdivided into separate populations on the basis of the presence or absence of a histochemical reaction for a particular enzyme. This is due, to a certain extent, to the development of the technology for anatomically demonstrating the activity of a large number of enzymes. The surprising thing is that subpopulations of DRG cells are labeled, implying that a number of cells are intensely stained and that the others are essentially unstained. These are striking findings, and when the functional implications are understood, they will undoubtedly be important considerations in sensory mechanisms. As with the peptides, however, the reactions are not particularly quantitative, so a negative reaction does not mean that the cell contains no enzyme, nor do these reactions tell whether a cell has the genetic apparatus to make the enzyme that is demonstrated.

Acetylcholinesterase and Choline Acetyltransferase. Acetylcholinesterase (ACHE) breaks down acetylcholine (ACH), and so the presence of this enzyme is often taken to suggest cholinergic transmission. ACH is not thought to be a sensory transmitter, however (see, e.g., Gwyn and Flummerfelt, 1971), and early histochemical localizations in DRG cells showed little esterase in ganglion cells (Hard and Peterson, 1950; Koelle, 1951). Thus it was a considerable surprise when Sauer (1954), Csillik and Savay (1954), and Koelle (1955), who modified his stain, noted that some DRG cells were intensely labeled (Fig. 3.6). Giacobini (1956, 1958, 1959a, b, 1960), using both histochemistry and the Cartesian diver technique, then showed that 85–90% of the DRG cells had no demonstrable ACHE activity but that 10–15% had a high activity. Since then, all investigators agree that some DRG cells are intensely labeled after the histochemical procedures for ACHE (Coupland and Holmes, 1957; Gerebtzoff, 1959; Tewari and Bourne, 1962b, 1963a; Cauna and Naik, 1963; Kokko, 1965; Nandy and Bourne, 1964; Novikoff *et al.*, 1966; Matsuura, 1967; Shantha *et al.*, 1967; Dixon, 1968; Schlaepfer, 1968; Kalina and Bubis, 1969; Matsuura *et al.*, 1969; 1970; Kalina and Wolman, 1970; Lukas *et al.*, 1970; Sarrat, 1970; Gwyn and Flummerfelt, 1971; Robain and Jardin, 1972; Manocha, 1973; Mazza *et al.*, 1973; Hanker and Peach, 1976; Thomas, 1977; Ambrose and McNeill, 1978; Homor and Kasa, 1978; M. E. McNeill and Norvell, 1978; Schoenen, 1978; Malatova *et al.*, 1985; Vega *et al.*, 1989) (Fig. 3.6). It is usually the small cells that are the most intensely labeled (Giacobini, 1959a,b; Tewari and Bourne, 1962b, 1963a; Kokko, 1965; Nandy and Bourne, 1964; Schlaepfer, 1968; Kalina and Bubis, 1969; Kalina and Wolman, 1970; Robain and Jardin, 1972; Ambrose and M. E. McNeill, 1978; McNeill and Norvell, 1978), but some studies do not show such a correlation (Cauna and Naik, 1963; Novikoff *et al.*, 1966; Matsuura, 1967; Matsuura *et al.*, 1969, 1970; Lukas *et al.*, 1970; Gwyn and Flummerfelt, 1971; Homor and Kasa, 1978; Malatova *et al.*, 1985; Vega *et al.*, 1989). Reviews usually emphasize that the small cells are labeled (see, e.g., Thomas, 1977).

The usual speculation about ACHE is that it indicates a site of cholinergic transmission (see, e.g., Giacobini, 1959a,b). If this is so, there would seem to be two possibilities. First, the

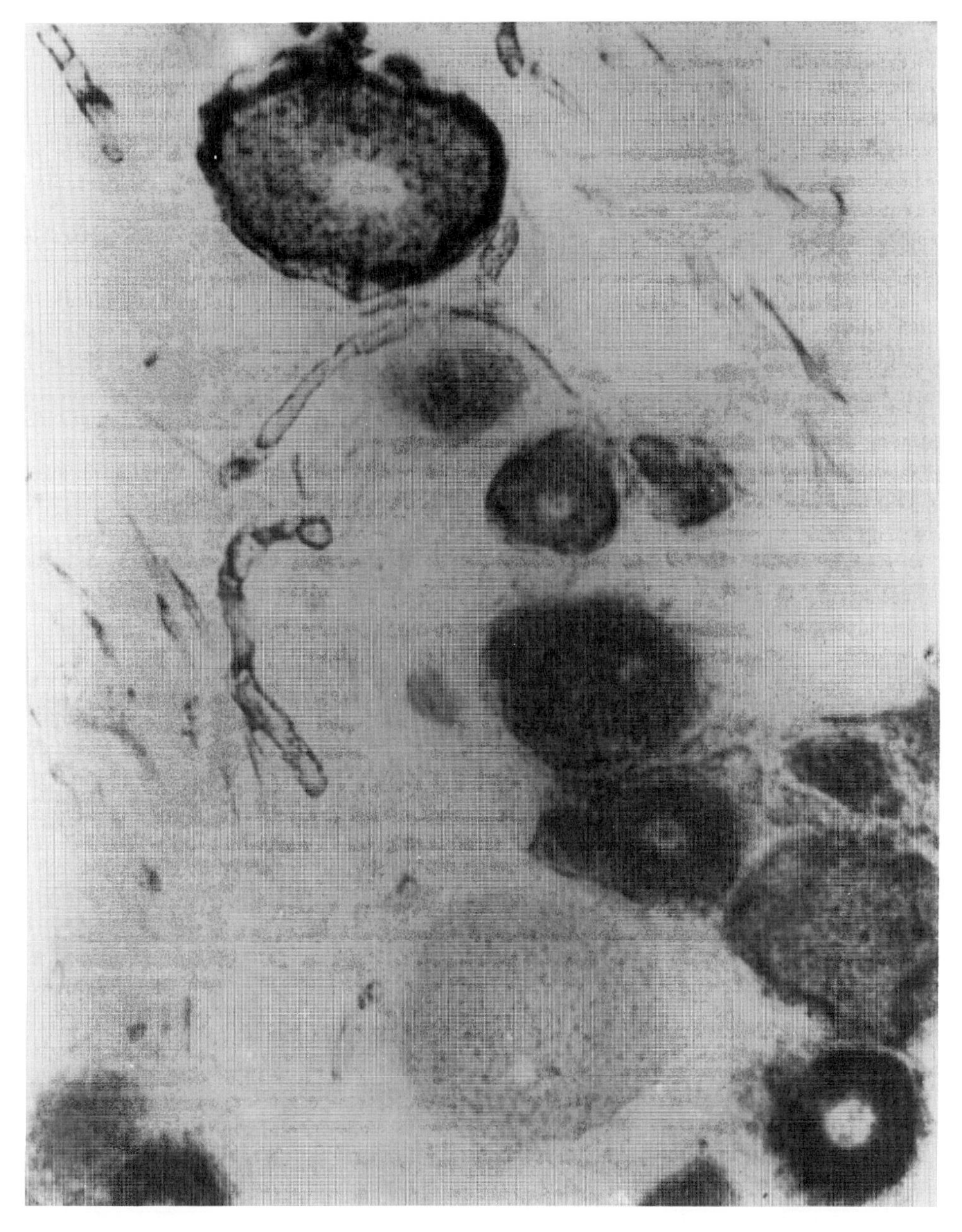

Fig. 3.6. Acetylcholinesterase staining of DRG cell bodies. Note that many cells are heavily labeled but that one cell is essentially unstained. (From Thomas, 1977.)

labeled cells use ACH as a transmitter and are manufacturing ACHE to be distributed to their terminals. In accord with this, ACHE or choline acetyltransferase (CHAT)-positive synaptic fields are seen in appropriate laminae of the dorsal horn (Navaratham and Lewis, 1970; H. Kimura *et al.*, 1981; Barber *et al.*, 1984). The other possibility is that there are cholinergic endings on DRG cells which need the ACHE to break down ACH after it is released. In this regard, there are α-bungarotoxin and nicotinic receptors on DRG cell bodies (see below). In addition, synapses that are stated to arise from motoneurons have been found on cat DRG cells and the published pictures clearly fulfill the criteria for synapses (see below). There is little other evidence for cholinergic primary afferent transmission or cholinergic synapses on cell bodies, however, and the possibilities that ACHE indicates cholinergic transmission or that DRG cells have cholinergic terminals on them are usually not considered. Clearly, however, the function of the ACHE and cholinergic receptors must be determined, and more data on the numbers of the presumed cholinergic terminals must be obtained.

If the presence of ACHE-labeled cells does not indicate primary afferent cholinergic transmission, as would undoubtedly be the majority opinion, why do so many DRG cells make ACHE? Some suggestions are that ACHE may influence membrane permeabilities of DRG cells or their processes in the ganglion (Koelle, 1955) or that ACHE may be used to break down ACH and thus protect the perikarya from ACH released from nearby fibers (Mazza *et al.*, 1973). These remain speculations, however, and we really do not know the function of ACHE in DRG cells. The possibility of cholinergic primary afferent transmission should be kept open, but again there is little evidence for this beyond the demonstration of ACHE in a population of primary afferent cells.

During development, ACHE appears before birth and rapidly becomes prominent in all DRG cells (Kalina and Wolman, 1970; Sarrat, 1970; Tennyson and Brzin, 1970; Schoenen, 1978). Kalina and Wolman (1970) note that the stain becomes less intense in large cells as development proceeds. Thus, in their study only the small cells are labeled as the animal approaches adulthood.

ChAT is involved in the manufacture of ACH and most investigators would accept the demonstration of ChAT in a population of neurons as a better indicator of putative cholinergic function than the demonstration of ACHE. Karczmar *et al.* (1980) showed that both ChAT and ACH could be found in the prenatal rabbit DRGs but that the amounts of these compounds waned to very low levels as development proceeded. His conclusion was that adult DRG cells do not use ACH as a transmitter. Many physiologic data are consistent with this conclusion. Some studies report ChAT immunolabeled DRG cells (Burt, 1971; Motavkin and Okhotin, 1985), but Burt (1971) indicates that the label is much lighter than in known cholinergic cells and the pictures shown by Motavkin and Okhotin (1985) are not impressive. In addition, ChAT labeling in DRG cells is denied by Barber *et al.* (1984) and Borges and Iversen (1986). ChAT is immunologically demonstrable in areas of the spinal cord where many afferent fibers terminate, but it is thought that this substance comes from interneurons (see Chapter 4).

In summary, ACHE labels many DRG cells intensely. The labeled cells are predominantly from the small DRG cell population. No functional correlations for the labeled cells are known. ChAT-positive DRG cells have been reported, but the findings must be repeated. Obvious questions are as follows. Does a population of DRG cells use ACH as a neurotransmitter? If not, as seems probable, what is the function of ACHE in DRG cells?

Intermediary Metabolism Enzymes. Intermediary metabolism enzymes which have been studied histochemically in DRG cells are the aerobic enzymes succinic, maleic, and isocitrate dehydrogenase, cytochrome oxidase, and NADH and NADPH reductase; the anaerobic enzymes lactate and α-glycerophosphate dehydrogenase; and the pentose shunt enzymes glucose-6-phosphate and 6-phosphoglucuronic dehydrogenase. The first major study indicating a differential activity in DRG cells showed that some DRG cells contained large amounts of succinic dehydrogenase and cytochrome oxidase and that others did not (Hyden *et al.*, 1958). Similar findings have been reported for essentially all the reactions in the present section, namely that some cells stain intensely by appropriate histochemical methods whereas others do not (Scharf and Rowe, 1957; Potanos *et al.*, 1959; Klein, 1960; Samorajski, 1960; Thomas and Pearse, 1961;

Tewari and Bourne, 1962b,c; Kumamoto and Bourne, 1963a,b; Thomas 1963, 1972, 1977; Rudolph and Klein, 1964; Carpenter, 1965; Gerebtzoff, 1966; Matsuura, 1967; Novikoff, 1967b; Shantha *et al.*, 1967; Diculesceu *et al.*, 1968; Matsuura *et al.*, 1969; Sarrat, 1970; Sethi and Tewari, 1971; Robain and Jardin, 1972; Manocha, 1973; Hanker *et al.*, 1973; Thomas, 1977) (Fig. 3.7). Most studies report that the small cells are labeled (Klein, 1960; Thomas and Pearse, 1961; Tewari and Bourne, 1962a,b,c; Thomas, 1963, 1977), although not always (Robain and Jardin, 1972; Thomas, 1977; Matsuura, 1967; Matsuura *et al.*, 1970; Shantha *et al.*, 1967; Manocha, 1973). No measurements of cell sizes are reported in any of these studies, however.

An obvious question is whether DRG cells can be grouped into enzyme-containing and enzyme-lacking categories? There have been no suggestions that a population of DRG cells lack any of these enzymes, however. In this regard it should be remembered that DRG cells are subdivided into large light and small dark categories, with the small dark category reflecting a relatively high concentration of metabolically active organelles. Thus it might be expected that small cells would have a more intense reaction but that the reactions would be somewhat variable.

There are suggestions that the differences have functional significance (Robain and Jardin, 1972), but as yet there is no suggestion about what the functional differences are. One interesting possibility is that activity might be correlated with the intensity of these reactions.

Monoamine Oxidase. Because of the role of monoamine oxidase in breaking down monoamine neurotransmitters, a subpopulation of DRG cells positive for this enzyme would be of considerable interest. In the first histochemical study for this compound, Koelle and Valk (1954) noted that all DRG cells were moderately stained in the cat and faintly stained in the rabbit. Others also report that DRG cells stain, usually faintly, but do not report differential staining (Nandy and Bourne, 1964; Shantha *et al.*, 1967; Matsuura, 1967; Matsuura *et al.*, 1969; Giacobini and Kerpel-Fronius, 1970; Kalina and Wolman, 1970; Hanker *et al.*, 1973). Thus, DRG cells react for monoamine oxidase, but monoamine oxidase-positive and monoamine oxidase-negative categories do not seem to exist. In this regard, the finding of a small population of catecholamine-containing DRG cells in the L4 ganglion of the rat and the transient appearance of a large population of such cells during development is of interest (see below).

Carbonic Anhydrase. Carbonic anhydrase hydrates carbon dioxide. Korhonen and Hyyppa (1967) claimed that some DRG cells were labeled for this compound, but Parthe (1981) could not duplicate the finding. Therefore, it was a surprise to find an intense reaction for carbonic anhydrase in a population of large and some intermediate-sized DRG cells in the rat (Wong *et al.*, 1983) (Fig. 3.8). This finding has been confirmed (Riley *et al.*, 1984; Sommer *et al.*, 1985; Kazimierczak *et al.*, 1984; 1986; Peyronnard *et al.*, 1986, 1988a, b; Droz and Kazimierczak *et al.*, 1987; Robertson and Grant, 1989), and it is agreed that the label is found mostly in large cells. Interestingly the carbonic anhydrase reactivity was closely correlated with reactivity to an antibody (RT-97) that labels neurofilaments (Robertson and Grant, 1989). A hypothesis is that the high carbonic anhydrase activity serves to maintain the acid-base balance in tonically active cells (Riley *et al.*, 1984; Peyronnard *et al.*, 1988a). Functional correlates are not known.

Cytochrome oxidase. This enzyme has been used as a marker for neuronal activity. A strong reaction is found in some DRG cells of all sizes (Wong-Riley and Kageyama, 1986).

Glycogen Phosphorylase. Takeuchi (1958) noted that some trigeminal ganglion cells react for glycogen phosphorylase (Fig. 3.9). Robain and Jardin (1972) and Thomas (1977) pointed out that the reaction was more intense in large cells. This reaction was then used following nerve stimulation, turpentine injection, or cyclic AMP administration (Woolf *et al.*, 1985) and administration of excitatory amino acids (Woolf, 1987b). Only a small proportion of the neurons were labeled under normal conditions, but exposure to cyclic AMP resulted in an increase in cell labeling (Woolf *et al.*, 1985) (Fig. 3.9). The numbers of labeled cells were also increased after the other procedures, and the increase was greatest in the small-cell population when the animal

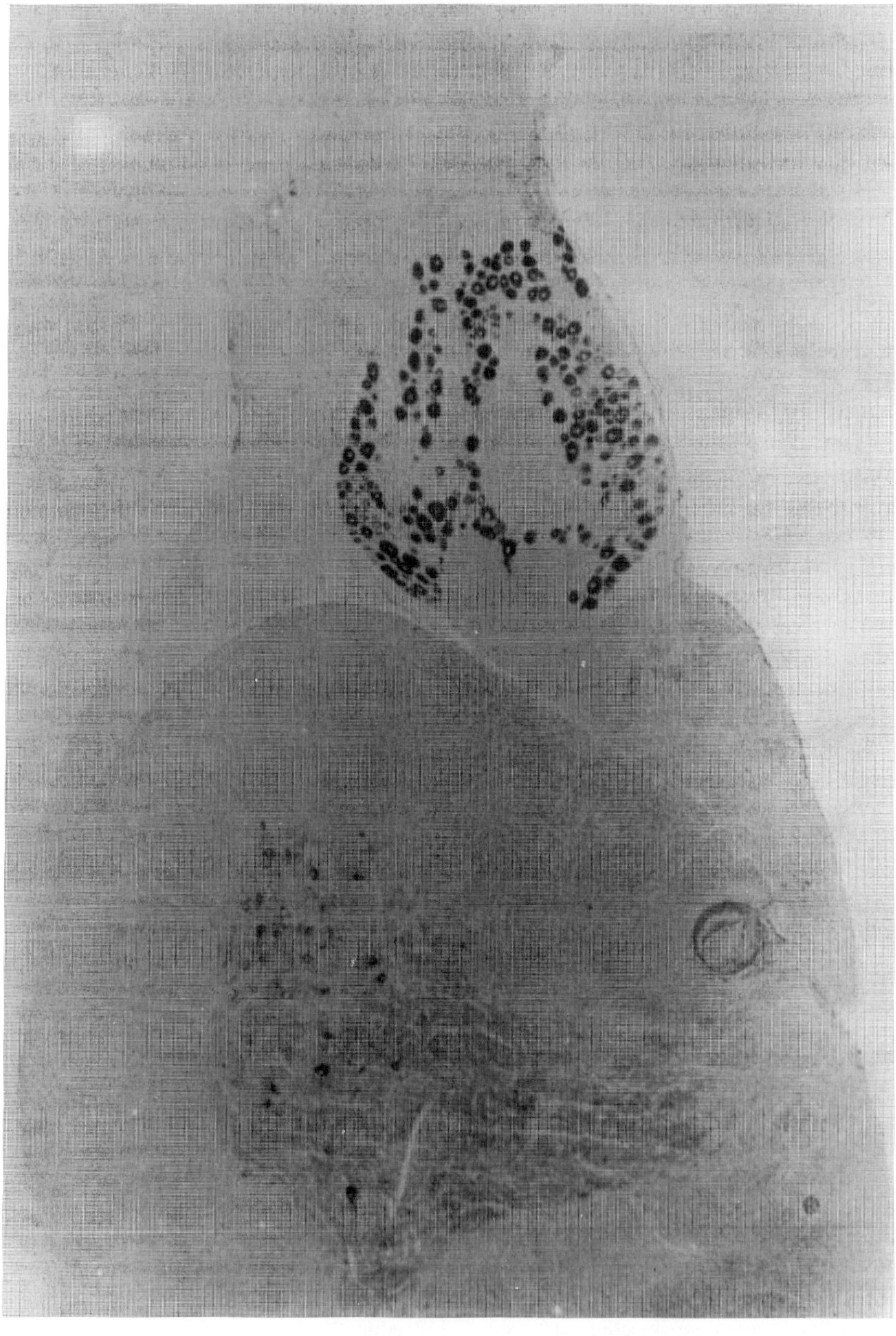

Fig. 3.7. A section of dorsal root ganglion and spinal cord reacted for succinate dehydrogenase. Note that many DRG cells are intensely labeled, as well as some cells in the ventral horn. (From Thomas, 1977.)

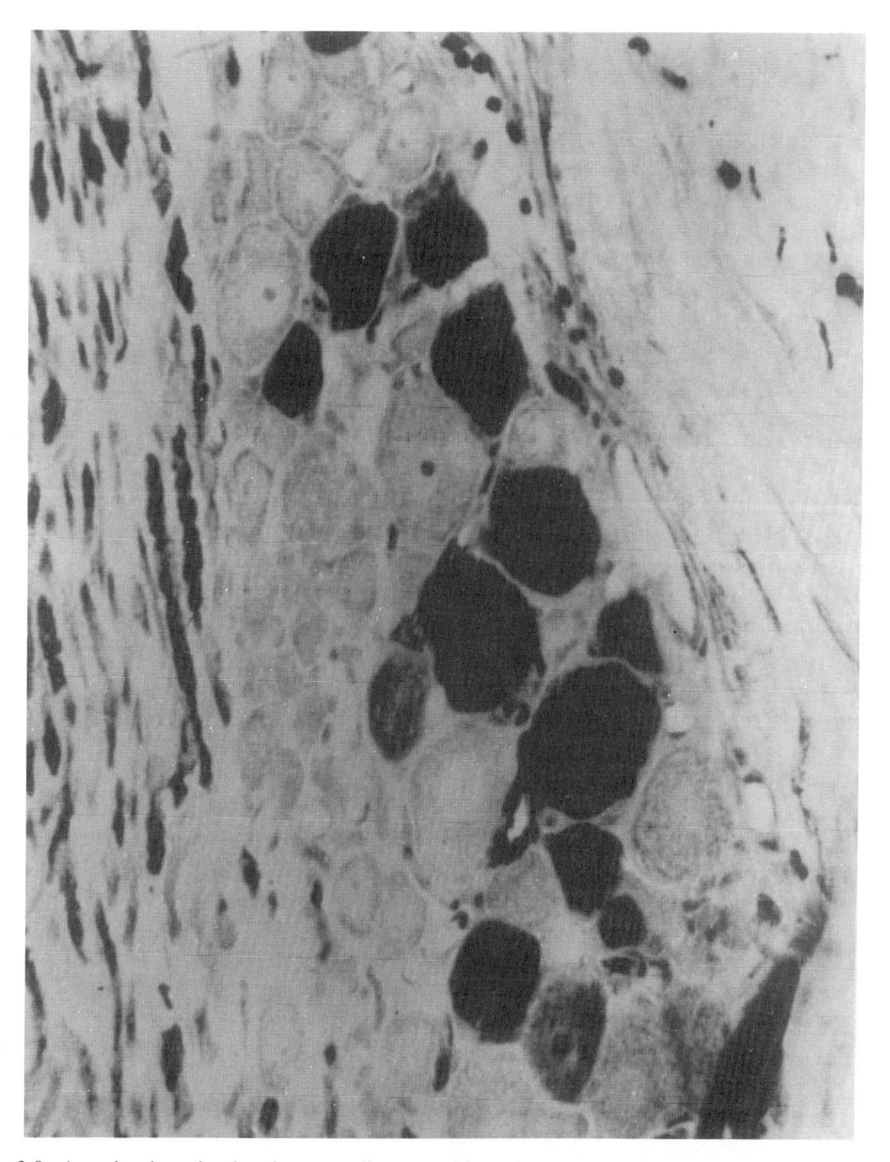

Fig. 3.8. A section through a dorsal root ganglion reacted for carbonic anhydrase. Note that some cells, many of which are large, are intensely labeled. (From Wong *et al.*, 1983; reprinted with permission.)

was stimulated in such a way as to activate C fibers. This technique gives promise that populations of ganglion cells can be marked by relatively precise stimuli. From such markings, greater insights into primary sensory cell organization will presumably be forthcoming.

Adenosine Deaminase. Adenosine deaminase was found in a subpopulation of small (type B) DRG neurons (Nagy *et al.*, 1984; Nagy and DaDonna, 1985). The suggestion is that this reaction distinguishes a population of sensory neurons that use purines as transmitters or cotransmitters.

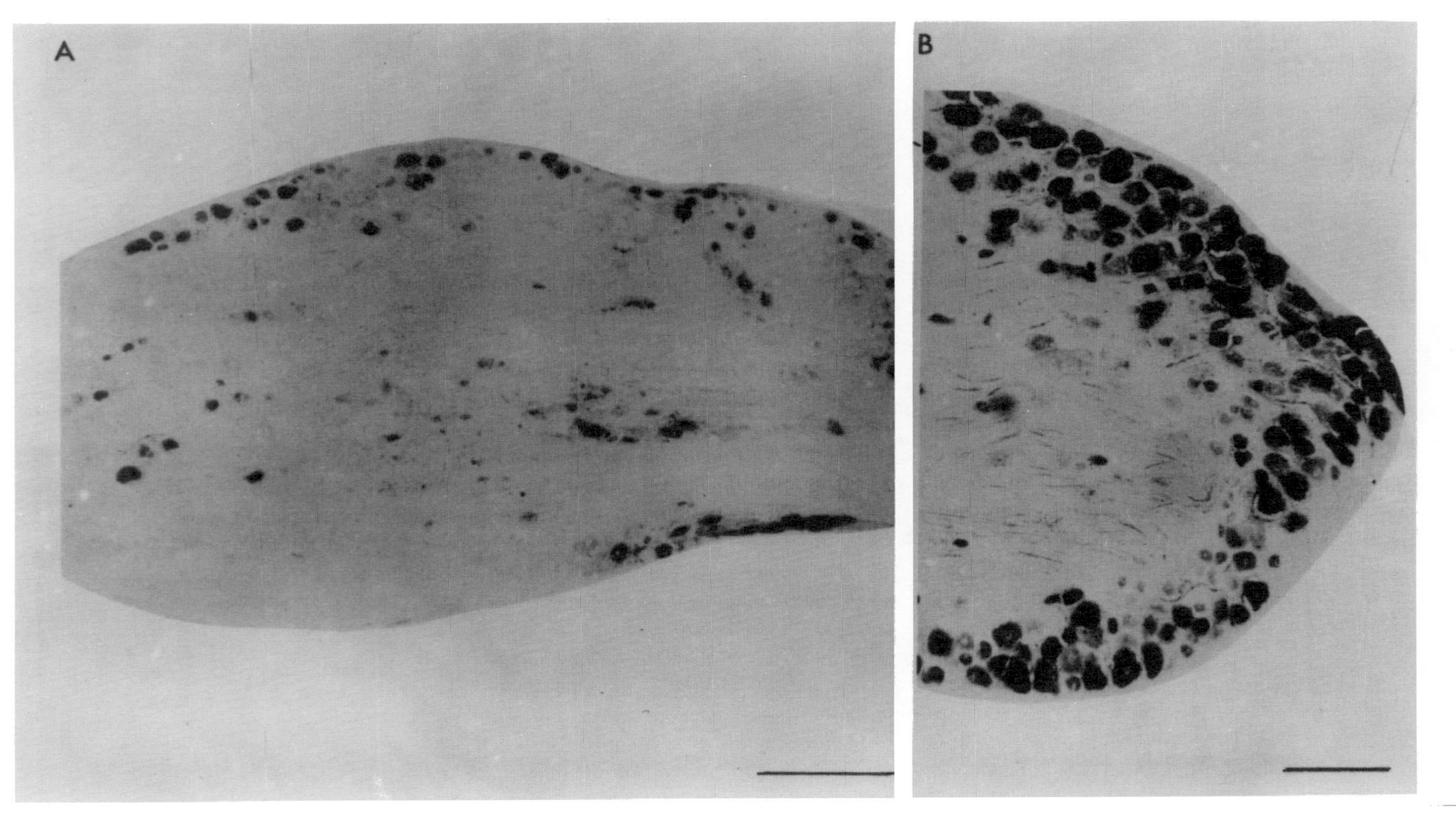

Fig. 3.9. Glycogen phosphorylase activity in a dorsal root ganglion from a normal animal in A and from after AMP incubation in B. Note the increase in the number of labeled cells in B. A similar increase is seen after intense peripheral stimuli. (From Woolf *et al.*, 1985; reprinted with permission of Wiley-Liss.)

Hydrolases: Alkaline Phosphatase. A number of studies report alkaline phosphatase labeling of DRG cells (Tewari and Bourne, 1962a,c; E. Thomas, 1972, 1977; Shantha *et al.*, 1967; Thakar and Tewari, 1967; Sethi *et al.*, 1969; Sarrat, 1970; Tewari *et al.*, 1970; Sethi and Tewari, 1971; Manocha, 1973; Kopaczyk *et al.*, 1974). The label is light and found at the periphery of the cell. Matsuura *et al.* (1969) used the electron microscope to show that the reaction is confined to the capsular cells and the membrane of the neuron where it faces the capsular cells. If this is correct, it suggests that some of the light microscopic localizations might be the result of diffusion from the satellite cells, which are always intensely labeled. Because of the superficial location of the enzyme, several investigators suggest that it may be involved in surface activity such as ion transfer (Matsuura *et al.*, 1970). There has been no suggestion that this enzyme delineates different sensory cell populations, however. Thus, alkaline phosphatase, whatever its role in the metabolism of the DRG cell, does not seem to be key in the neural organization of the primary sensory cells.

Hydrolases: Acid Phosphatase. Acid phosphatase (AP) is a hydrolase found in almost all cells as an important constituent of lysosomes, and its cytochemical localization in DRG cells has been important in cell biology (Novikoff, 1967a,b). Interest in AP in the context of sensory systems first arose when Colmant (1959) observed that many small sensory neurons stained intensely for AP, whereas large cells were much more lightly labeled (Fig. 3.10). Subsequent investigators confirmed this finding (Kokko, 1965; Kalina and Bubis, 1968; Kalina and Wolman, 1970; Sarrat, 1970; Novikoff *et al.*, 1971; Peach, 1972; Robain and Jardin, 1972; Kopuczyuk, 1974).

Hydrolases: Fluoride-Resistant Acid Phosphatase (FRAP). A major finding was that the acid phosphatase (AP) reaction that intensely labeled a population of small sensory cells was not due to regular AP but to a nonlysosomal AP that was resistant to fluoride ions (FRAP) (Schoenen *et al.*, 1968; Gerebtzoff and Maeda, 1968; Knyihár, 1971) (Fig. 3.10). FRAP labeling is also found in axons at the periphery and in the dorsal horn. The generally accepted hypothesis to explain these observations is that a population of small DRG cells make FRAP which is then transported proximally and distally in their processes (Knyihár, 1971; Knyihár-Csillik and Csillik, 1981). As a terminologic note, lysosomal AP is referred to in this literature as "regular" or "GERL" or "trivial" AP whereas the AP resistant to fluoride ions in sensory cells and processes is referred to as "FRAP" or as "extralysosomal" or "diffuse", or "free" AP (Csillik and Knyihár-Csillik, 1986).

One problem has been that the FRAP localizations are most easy to demonstrate in rodents, so much so that the phenomenon was thought to be species specific. More recently, FRAP localizations have been obtained from a variety of mammals (Knyihár-Csillik and Csillik, 1981; Silverman and Krueger, 1988b), but they are most easily demonstrated in rodents.

The FRAP-labeled cells are of two types: (1) cells which have a diffuse label in the light microscope and (2) cells which show striations in the light microscope, the striations being caused by heavily labeled cisternae of the granular endoplasmic reticulum (Fig. 3.10) (Knyihár-Csillik and Csillik, 1981; Csillik and Knyihár-Csillik, 1986). Functional differences between the two cell types are not known.

A common point of discussion concerns the function of FRAP in sensory transmission. FRAP itself is not thought to be a transmitter, and synaptic transmission is not obviously affected when FRAP is markedly depleted centrally by crush of a peripheral nerve distally (Wall and Devor, 1981). It is frequently suggested, however, that FRAP might be involved in the metabolism of a transmitter (Knyihár-Csillik and Csillik, 1981; Dodd *et al.*, 1983; Jahr and Jessell, 1983; Csillik and Knyihár-Csillik, 1986), but FRAP does not coexist, or coexists only to a small extent, with other well known compounds found in sensory cells (substance P, somatostatin, cholecystokinin, adenosine deaminase [Nagy and Hunt, 1982; Dalsgaard *et al.*, 1984; Nagy and DaDonna, 1985]). Thus, FRAP is presumably not involved with the metabolism of these compounds. It has also been suggested that FRAP might be involved with membrane

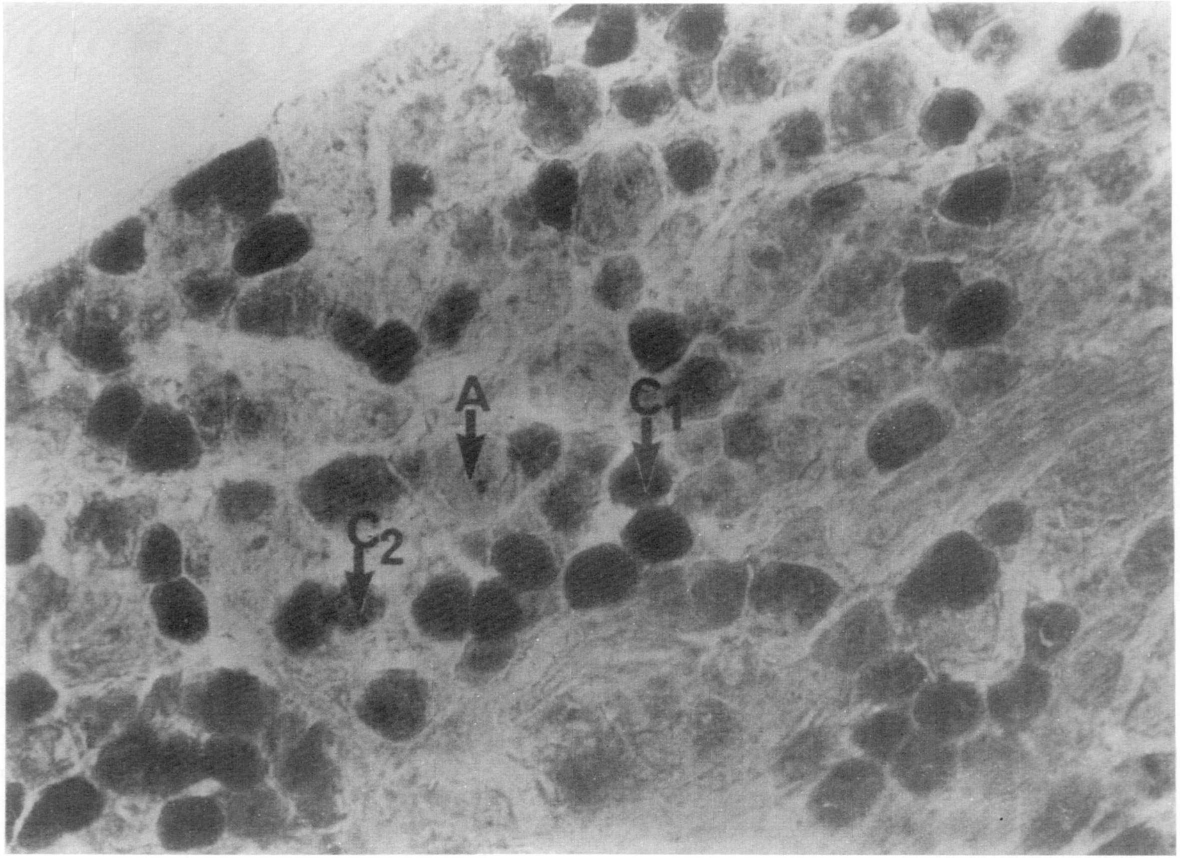

Fig. 3.10. FRAP activity in dorsal root ganglion cells. The cell at A is a large essentially unreactive cell. Cell C1 shows an intensely labeled small cell with the enzyme reaction being dispersed homogeneously in the cytoplasm. Cell C2 is equally reactive but the reaction is more punctate. (From Knyihár-Csillik and Csillik, 1981.)

components or turnover (Dodd *et al.*, 1983; Silverman and Krueger, 1988b), but these remain speculations. Probably the most interest in FRAP is because it might be involved in nociceptive transmission (Knyihár, 1971; Jancsò and Knyihár, 1975). Evidence for this, outside of the observations that it is found in small DRG cells and in appropriate laminae of the dorsal horn, is that there is an extensive plexus of AP-containing axons in the cornea and that these axons are lost when the trigeminal ganglion is removed (Szonyi *et al.*, 1979; Knyihár-Csillik and Csillik, 1981). Since many corneal sensory axons are nociceptive, the findings are suggestive, but other axon types are present in the cornea (see Chapter 1).

One interesting finding is that FRAP can be demonstrated in cutaneous and visceral nerves but not in muscle nerves (S. B. McMahon *et al.*, 1984; S. B. McMahon, 1986) and that this pattern changes after cross anastomosis of different nerves (S. B. McMahon and Moore, 1988). Another finding illustrating the plasticity of FRAP in DRG cells is the loss of the reaction product following nerve transection and its subsequent return (Devor and Claman, 1980; Tenser, 1985; Shehab and Atkinson, 1986a,b).

Hydrolases: Thiamine Pyrophosphatase. Shanthaveerappa and Bourne (1965) and Shantha *et al.* (1967) showed thiamine pyrophosphatase-stained elements in the cytoplasm of DRG cells. Novikoff and collaborators then showed that thiamine pyrophosphatase and nucleoside diphosphatase specifically labeled the inner one or two lamellae in the Golgi complex in rat DRG cells (Novikoff *et al.*, 1971; Boutry and Novikoff, 1975), but there is no clear relation of these reactions to the sensory functions of DRG cells, except that small cells are labeled in the adult (Kalina and Wolman, 1970).

Hydrolases: Thiamine Monophosphatase. Knyihár-Csillik *et al.* (1986, 1989) noted that thiamine monophosphatase had the same localizations and responses to dorsal rhizotomy or nerve crush as did FRAP. They further noted that thiamine monophosphatase reactivity was restricted to a population of small DRG cells, whereas thiamine pyrophosphatase and thiamine triphosphatase were ubiquitously distributed among neurons. This relatively specific distribution is consistent with a role for this compound in sensory function. They believe that thiamine monophosphatase is a relatively specific marker.

In summary, histochemical reactions for a number of enzymes delineate subpopulations of DRG cells. Functional correlations with these morphologic demonstrations are not known, although monoamine oxidase, adenosine deaminase, and possibly the fluoride-resistant phosphatases are probably involved in transmitter metabolism. Carbonic anhydrase is one of the few morphologic markers that delineates a population of large DRG cells, and the number of cells reactive for glycogen phosphorylase greatly increases after particular types of stimuli.

Heterogeneous Markers. Morphologic techniques for various compounds that are neither peptides nor enzymes delimit subpopulations of DRG cells. The most widely studied will be summarized here. This list will undoubtedly increase.

Monoamines. The usual perception is that there are no monoamine-containing DRG cells. However, there have been apparently unequivocal demonstrations of tyrosine hydroxylase-containing small sensory cells (Katz *et al.*, 1983; J. Price and Mudge, 1983; J. Price, 1985), Jonakait *et al.* (1984) emphasize that this staining appears and then disappears in early development. Serotonin (5-HT) containing primary afferent neurons have also been described (Calas *et al.*, 1981; Kai-Kai and Keen, 1985), although the latter pointed out that these are not always distinguishable from mast cells. This morphologic evidence suggests that there are monoamine-containing afferent neurons. Further work must establish the functions, however, and the 5-HT work needs to be expanded.

Glutamate and Aspartate. Glutamic acid (glutamate) is a candidate primary afferent transmitter. Morphologically, (1) glutamate is localized immunologically in a population of DRG cells, with one group noting that small cells labeled (Battaglia *et al.*, 1987; Battaglia and Rustioni, 1988) and another implicating large cells (Wanaka *et al.*, 1987); (2) cytosolic and mitochondrial glutamic oxaloacetic transaminase is localized in DRG cells of all sizes (N. Inagaki *et al.*, 1987); (3) small, dark DRG cells accumulate much more radioactive glutamine than do large cells, and a significant fraction of this glutamine is converted to glutamate (Duce and Keen, 1983); (4) glutaminase is localized in a population of small DRG cells (Cangro *et al.*, 1985); (5) *N*-acetylaspartylglutamate is localized in small DRG cells (Cangro *et al.*, 1987), and a subpopulation of DRG cells retrogradely transports peripherally injected radioactive aspartate (Barbaresi *et al.*, 1985; Jessell *et al.*, 1986). These findings presumably indicate that a subpopulation of DRG cells is especially involved in glutamate storage and metabolism. The discrepancy in the size of the cells implies that more work must be done, but since fine sensory axons (Westlund *et al.*, 1989a) and terminals in the superficial laminae of the dorsal horn where fine primary afferents end (De Biasi and Rustioni, 1988) are selectively labeled, and since glutamate and substance P colocalize (Battaglia and Rustioni, 1988), the small-cell population must be especially considered.

There is much less evidence that aspartate is a primary afferent transmitter. One finding of note, however, is that there are many aspartate labeled dorsal root axons (Westlund *et al.*, 1989b).

Carbohydrate Differentiation Antigens. In the last few years, a series of monoclonal antibodies have been developed that recognize lactoseries and globoseries carbohydrate differentiation antigens (Dodd *et al.*, 1983, 1984; Dodd and Jessell, 1985, 1986; Chou *et al.*, 1988; Hynes et al., 1989; Kusunoki *et al.*, 1989). The lactoseries antibodies label small and some intermediate-sized DRG cells and terminals in lamina I and II of the dorsal horn (Jessell and Dodd, 1985; Dodd and Jessell, 1985, 1986). By contrast, the globoseries antibodies recognize larger cells and terminals primarily in laminae III and IV of the cord. Also, the lactoseries oligosaccharides seem to be implicated in certain aspects of cell adhesion (Dodd and Jessell, 1986). In addition another antibody to presumably a glycoprotein labeled only a subclass of small DRG neurons in culture (Rougon *et al.*, 1983). These results are extremely interesting, and the carbohydrate differentiation antigens have been used in combination with other markers to delineate probable primary afferent neurons in the enteric plexus (Kirchgessner *et al.*, 1988). In addition, these antigens may be recognition signals in the development of sensory systems (Jessell and Dodd, 1985). Further work on colocalizations and precise anatomic arrangements will presumably lead to insights into sensory function.

Lectins. Lectins have been used to indicate subpopulations of DRG cells. For example, wheat germ agglutinin-HRP is transported to the marginal zone and substantia gelatinosa (laminae I and II), whereas choleragenoid-HRP is transported to the deeper dorsal horn (Robertson and Grant, 1985). This work was extended to the DRG cell by showing that large (and some intermediate) neurons were labeled for the receptor for cholera toxin, and choleragenoid, which is the toxin binding subunit. The large cells were also carbonic anhydrase and RT97 neurofilament antibody positive but peptide and FRAP negative (Robertson and Grant, 1989). Localization to small cells and laminae I and IIo of the dorsal horn was shown with soybean agglutinin lectin (Plenderleith *et al.*, 1988, 1989; Peyronnard *et al.*, 1989). An interesting combination of lectin and peptide labeling has recently been done (Silverman and Kruger, 1988a). Lectin-HRP conjugates were used to show localizations of a glycoconjugate with terminal α-galactose residues in small cells and that a characteristic galactosylconjugate was found in small cells colocalized with substance P (Streit *et al.*, 1985, 1986). It was suggested that these localizations might have significance particularly in the area of pain and temperature transmission. Some glycoconjugates and endogenous lectins may be involved in cellular interactions that are important in development (Regan *et al.*, 1986). Finally, lectin binding changes after nerve and spinal cord injuries (Peyronnard *et al.*, 1989). Although it is early, all of these general areas would seem to have considerable potential for providing insights into sensory function.

Growth Hormone Releasing Factor. Growth hormone releasing factor was found in 1% of all sensory cells in rat spinal and trigeminal ganglia (Józsa *et al.*, 1987). The function of this compound in sensory cells is not known.

Parvalbumin and Calbindin. Parvalbumin and calbindin are calcium-binding proteins. They label 14 and 22% of rat DRG cells, respectively, and these cells are predominantly of the large type (Carr *et al.*, 1989).

Nerve Growth Factor Receptors. Many DRG cells are selectively labeled for nerve growth factor receptors (Richardson and Riopelle, 1984; Yip and Johnson, 1984; Richardson *et al.*, 1986; Sobue *et al.*, 1989). From the point of view of DRG organization, the interesting thing is that some cells are labeled and some are not, a phenomenon that is also seen with the retrograde transport of NGF (Stoeckel *et al.*, 1975). There is a tendency for small cells to be labeled (Richardson *et al.*, 1986). It would be most interesting if a particular sensory class of dorsal root ganglion cells was NGF positive. Clues might be the localization of NGF receptors on afferent endings or to lamina I and II of the spinal cord (Eckenstein, 1988; Yip and Johnson, 1987).

Opiate and Histamine Receptors. Opiate and histamine receptors are localized to a subpopulation of DRG cells (Ninkovic *et al.*, 1982; Gundlach *et al.*, 1986). The small ganglion cells are preferentially labeled (Ninkovic *et al.*, 1982).

Muscarinic Cholinergic and α-Bungarotoxin Receptors. Muscarinic cholinergic receptors and nicotinic α-bungarotoxin sites were localized to a population of DRG cells (Wamsley *et al.*, 1981a,b; Ninkovic and Hunt, 1983). The bungarotoxin receptors were localized in large DRG cells (Ninkovic and Hunt, 1983). These findings raise the question of cholinergic transmission, but more work remains to be done before this can be proved (see above). It would be of considerable interest to determine whether the cells that are characterized by the above receptors also make ACHE.

In Situ Hybridization. In situ hybridization will locate the genetic apparatus for making peptides or other compounds. CGRP- mRNA was found in DRG cells and motoneurons (Gibson *et al.*, 1988; Réthelyi *et al.*,1989; Noguchi *et al.*, 1990a.b). Noxious stimulation of the foot caused an increase in preprotachykinin and preprosomatostatin in DRG cells and their mRNAs (Henken *et al.*, 1988; Noguchi *et al.*, 1988). A similar increase in the gene encoding prodynorphin was seen after an experimental arthritis (Weihe *et al.*, 1989). The fact that the numbers of labeled cells are changing after these behavioral stimuli has considerable import for understanding sensory mechanisms. This useful technique will undoubtedly be rapidly and widely applied.

Physiologic Characteristics

DRG cells are not a uniform population in relation to their electrophysiologic properties. An early finding was that cells with rapidly conducting axons (A cells) had a shorter action potential with less overshoot, a shorter time constant and a smaller input capacitance than did cells with slowly conducting axons (C cells) (Gallego and Ezyaguirre, 1978). In an analysis of cells from isolated DRGs of the mouse, three types of neuron were described: F neurons with a tetrodotoxin-sensitive (TTX) Na^+ spike, a large cell body, and a myelinated axon; A neurons with a TTX-resistant Na^+ spike, a small cell body, and an unmyelinated axon (this is a reversal of the usual meaning of an A cell); and H neuron with a TTX-resistant Na^+-Ca^{++} spike and a hump on the falling phase of the action potential (Yoshida *et al.*, 1978; Yoshida and Matsuda, 1979). That some of these membrane properties might be related to functional categories was emphasized by finding that chemoreceptors (conduction velocity = 11 m/s) and slow baroreceptors (conduction velocity = 10 m/s) from the nodose ganglion had long action potentials with a hump, whereas fast baroreceptors (conduction velocity, 33 m/s) had short action potentials with no hump (Belmonte and Gallego, 1983). For rat DRG cells, Aα, Aδ, and some Aβ cells had fast action potentials and some Aβ and C cells had longer action potentials, primarily owing to a hump on the falling limb (Harper and Lawson, 1985b). These last findings are interesting in that the C cells, which are presumably small, and some Aβ cells, which are presumably large, have one kind of action potential and Aα cells, which are presumably large, and Aδ cells, which are presumably small or intermediate in size, have another. In another type of correlation, cells giving rise to group I or II afferents had brief action potentials and group III and IV cells had long action potentials because of the hump (Cameron *et al.*, 1986). Finally, in important recent studies, different types of mechanoreceptors were identified and correlated with action potential duration, rate of rise, amplitude, after-hyperpolarization, and inward rectification. Interestingly, the physiologic parameters correlated better with the type of information transmitted than with the conduction velocity of the axon (Rose *et al.*, 1986; Koerber *et al.*, 1988; Koerber and Mendell, 1988).

In summary, DRG cells can be separated into separate categories on the basis of various membrane characteristics of the cell body. These subdivisions are, at least to some extent, correlated with functional parameters. Further work in this area will obviously be fruitful.

Organizational Features

Topographic Organization. One way to subdivide a population of primary sensory cells would be to show that certain categories of neurons were located in particular places in the ganglion. The usual finding, however, is that the cells seem to be distributed randomly (see, e.g., Norcio and DeSantis, 1976; McLachlan and Jänig, 1983; Peyronnard *et al.*, 1986). This seems to be particularly true for visceral neurons (McLachlan and Jänig, 1983; Cervero *et al.*, 1984a,b). Some organization is evident, however, in that the position of the perikaryon in the L7 ganglion of the cat predicts which dorsal root filament will carry the central axon to the spinal-cord (Burton and McFarlane, 1973). Similarly, a lamellar arrangement of cell bodies was demonstrated following restricted injections of HRP into different parts of the periphery or the spinal cord (Kausz and Réthelyi, 1985), and Szarijanni and Réthelyi (1979) noted that the large neurons in cat sacrococcygeal ganglia were distributed laterally in the ganglion. It should be understood that these patterns are not as precise as, for example, the arrangements of the motoneurons in the spinal cord. Nevertheless, there is some indication of topographic organization within the ganglion. Whether these patterns have functional meaning is not known.

Synapses on Dorsal Root Ganglion Cell Bodies. There is an extensive early literature on synaptic relations in the DRG itself. This is well summarized by Scharf (1958), who published diagrams of the possible arrangements, and Lieberman (1976), who provided a critique. The original claims are that there are processes, which may arise from DRG cells or from outside the ganglion, that form synaptic endings on DRG cells (summarized by Scharf [1958]). It is obvious that if a population of DRG cells is under direct synaptic control at the level of the cell body, this would have important consequences for sensory function. Lieberman (1976), however, ascribes the observations primarily to difficulties with fixation and states that, "in the 20 years or so during which sensory ganglia have been subjected to electron microscopic study, no multipolar neurons, no convincing afferent or efferent nerve endings, and no synaptic contacts between an axon terminal and a ganglion cell body have been convincingly demonstrated". We strongly supported this view. More recently, however, two sets of studies deserve consideration. First, synaptic terminals that end on cat cervical DRG cell bodies have been shown electron microscopically (Kayahara *et al.*, 1981). These synaptic terminals seem to arise from motor neurons (Kayahara *et al.* 1984; Kayahara, 1986). These are important claims, and the pictorial documentation is adequate. We examined 10,000 lumbar DRG cells in the rat and could find no such terminals. Second, catecholaminergic or peptidergic pericellular baskets that surround a few DRG cell bodies have been observed at the light-microscopic level (Owman and Santini, 1966; Lukas *et al.*, 1970; R. T. Stevens *et al.*, 1983; Kuwayama and Stone, 1986b). Further work obviously must be done, but if there is a population of DRG cell bodies that are under direct synaptic control, this would have obvious import for synaptic mechanisms.

PROCESSES OF DORSAL ROOT GANGLION CELLS

Four issues related to sensory spinal mechanisms will be considered here: dermatomes, branching, ventral root afferents, and segregation of fibers in the dorsal root.

Dermatomes

A dermatome is the area of skin innervated by the sensory axons from a single dorsal root. The dermatomes are obviously important in neurologic diagnosis. They have been delineated in several ways: (1) surgical isolation of a root and mapping of the resulting area of projected sensation, (2) electrical stimulation of the root and determination of the area of sensation, (3) reflex vasodilation, (4) the distribution of blisters after herpes zoster infection, (5) dorsal root potentials, (6) sensory levels after spinal cord transection, and (7) sensory examination following disk and other diseases that impinge on the dorsal roots (Kirk and Denny-Brown, 1970; Keegan and Garrett, 1948). The usually accepted maps are those of Sherrington (1893, 1898).

If a dermatome simply represents the distribution of the peripheral sensory axons from a root, which was thought to be the case, it should be relatively unchanging. Therefore, it was a surprise when Denny-Brown *et al.* (1973) showed that a dermatome could be enlarged by giving strychnine, by cutting the lateral part of Lissauer's tract, or by sectioning the proximal peripheral nerves rather than the dorsal roots to isolate the ganglion. By contrast, it could be decreased by cutting more dorsal roots cranially and caudally. Thus, the dermatome, as measured behaviorally or clinically, does not result only from the anatomic distribution of the peripheral sensory axons from a root. Denny-Brown *et al.* (1973) suggested that the above size changes resulted from an interaction of primary afferent axons with intrinsic axons in Lissauer's tract. There is also the possibility that interactions of primary afferent fibers with each other are important, and primary afferent fibers synapsing upon each other have been seen (D. L. McNeill *et al.*, 1988a). These findings caution against taking too simplistic a view of a dermatome and emphasize that the way a dermatome is outlined may have a considerable bearing on its size.

Branching

Primary afferent neurons are classically described as having a single central process that passes to the spinal cord through the dorsal root and a single peripheral process that passes to a distal sense organ. The major pieces of evidence supporting this formulation were the depictions of primary afferent neurons as seen by single-cell staining techniques (see, e.g., Ramón y Cajal, 1909) and counts showing equal numbers of DRG cells and dorsal root axons (Duncan and Keyser, 1936). However, Dogiel (1908) showed DRG cells with branched processes (Fig. 3.1), and the early counts were hampered by the inability of the light microscope to visualize unmyelinated axons. When the counts were repeated by using the electron microscope, a considerable surplus of dorsal root and peripheral sensory axons, as compared with the numbers of DRG cells, was seen (Langford and Coggeshall, 1979, 1981a; Aldskogius and Risling, 1981; K. Chung and Coggeshall, 1984). These data are consistent with the supposition that sensory axons branch, and this is occasionally seen (Hoheisel and Mense, 1985). The major interest in this phenomenon is that it might provide a peripheral explanation for referred pain, as postulated by Sinclair *et al.* (1948). Confirmation was obtained when it was shown that double-labeled DRG cells could be found when distinguishable markers were placed in different peripheral nerves or organs (Pierau *et al.*, 1982, 1984; Taylor and Pierau, 1982; Taylor *et al.*, 1983; Borges and Moskowitz, 1983; Alles and Dom, 1985; Laurberg and Sorenson, 1985; D. L. McNeill and Burden, 1986), although the percentages of such cells varied widely, and some studies using similar paradigms could find no double-labeled cells (Katan *et al.*, 1982; S. B. McMahon and Wall, 1987). Similarly, DRG cells could be fired by stimulation of different peripheral nerves or areas (Mense *et al.*, 1980; Bahr *et al.*, 1981; Pierau *et al.*, 1982; Devor *et al.*, 1984; McMahon and Wall, 1987), but again the percentages of cells of this type varied widely. McMahon and Wall (1987) suggest that the high percentages of double-labeled or stimulated cells reported in some of the above studies might be due to artifact. They did find that more than half of their single units sent more than one branch into a single peripheral nerve, but they believed that most of the "second branches" probably ended without making functional connections.

In summary, it seems clear that primary afferent axons branch. The exact percentages of branching are still a subject of study, and the functional significance of the branching is not known. The most frequently discussed possibility is that this branching might be related to referred pain, but more evidence must be obtained before this is accepted.

Ventral Root Afferent Fibers

It has been known since the time of Sherrington that a few myelinated fibers survive on the distal side and die on the proximal side of a ventral rhizotomy (Sherrington, 1894; Windle, 1931). That these recurrent fibers are sensory is shown by physiologic recordings (Dimsdale and Kemp, 1966; Kato and Hirata, 1968; Ryall and Piercey, 1970; Kato and Tanji, 1971; Clifton *et al.*, 1976; Coggeshall and Ito, 1977). More recently, unmyelinated ventral root sensory fibers have been discovered (Coggeshall *et al.*, 1974; Clifton *et al.*, 1976; Coggeshall and Ito, 1977; Wee *et al.*, 1985). These exist in considerably greater numbers than the myelinated sensory axons (Coggeshall *et al.*, 1974; M. L. Applebaum *et al.*, 1976; Emery *et al.*, 1977; Coggeshall, 1980; Holmes, 1982; Hiura, 1982). The two major questions about these fibers are how are they organized and what do they do?

Organization of Ventral Root Afferents (Fig. 3.11). *Direct Entry into the Spinal Cord.* The evidence that some ventral root afferent fibers enter the spinal cord directly through the ventral root is that, first, DRG cells can be labeled by placing HRP in the spinal cord when the dorsal roots are cut (Maynard *et al.*, 1977; T. Yamamoto *et al.*, 1977). Although there was disagreement about the proportion of aberrant cells that were labeled, it was agreed that fibers in the ventral root were transporting HRP to the labeled DRG cells. Second, afferent fibers are seen coming from the ventral root and going to the superficial dorsal horn or intermediolateral cell column when HRP is placed within the ventral roots (Light and Metz, 1978, Mawe *et al.*, 1984, 1986; Beattie *et al.*, 1987). Thus, it seems clear that some ventral root afferents enter the spinal cord directly through the ventral roots. If these fibers can be shown to have a definite function, they may necessitate a revision of our ideas about segregation of function in the spinal roots. Relatively few fibers seem to enter the cord this way, however.

Blind Fibers. Evidence regarding latencies shows that many primary afferent fibers enter the ventral root and end without entering the cord or looping (Kim *et al.*, 1987). In terms of percentages, Häbler *et al.*, (1990) suggest that 85% of the ventral root fibers that arise from DRG cells fall into this category.

Looping Fibers. Looping fibers have been postulated ever since Magendie (1822) discovered that stimulation of the ventral root caused pain which was relieved by section of the dorsal root (Bernard, 1858; Frykholm, 1951; Frykholm *et al.*, 1953). There is evidence that the number of ventral root unmyelinated axons decreases as one approaches the cord (Risling and Hildebrand, 1982; Risling *et al.*, 1984a,c), and that "U" turns of fine axons can be seen morphologically or demonstrated physiologically in the ventral roots (Bostok, 1981; Risling *et*

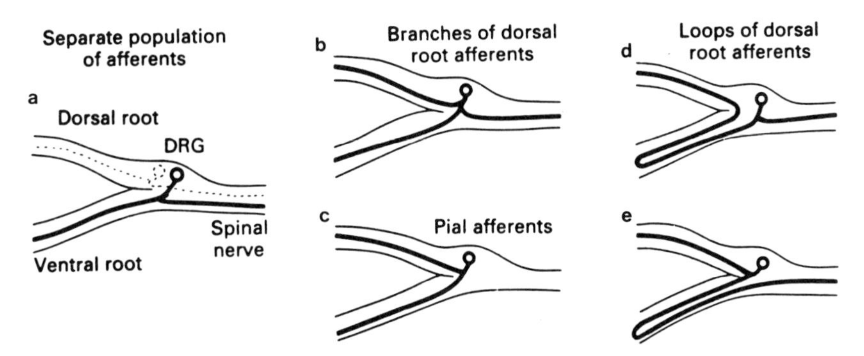

Fig. 3.11. Diagrams of various arrangements of ventral root afferents. In a the ventral root afferent will either end blindly in the ventral root or enter the spinal cord directly. b is similar to a except that the DRG cell has a third branch in the dorsal root. The pial afferents are shown in c. The two possibilities of looping afferents are shown in d and e. (From Häbler *et al.*, 1990.)

al., 1984b; Azerad *et al.*, 1986; Vergara *et al.*, 1986; Baik-Han *et al.*, 1989). It is not clear whether the central or the peripheral process of the DRG cell is looping, but Veraga *et al.*, (1986) provide suggestive cytologic evidence that it is the central process.

Pial Fibers. Small numbers of fibers coming from the pia and entering the root have been seen (S. L. Clark, 1931), particularly following immunocytochemistry for substance P (Dalsgaard *et al.*, 1982a).

Third Branches. Following the original observation of Loeb (1976), Kim *et al.*, (1987) show that many lumbar DRG cells that give rise to apparently blind ventral root afferents also send branches into the dorsal root. By contrast, however, Häbler *et al.*, (1990), working on sacral segments, found that 85% of the ventral root afferents come straight into the ventral root with no evidence of looping or of a third branch in the dorsal root. Both studies appear well done, so it is probable that this represents a difference in the segmental organization of the cord.

Proportions of Fiber Types. Apparently all of the above fiber types exist, with the possible exception of looping peripheral primary afferents. As regards proportions, the number of pial fibers seems to be relatively small, but it must be understood that these fibers are identified by being labeled with SP and if other pial afferents do not contain SP, as is certainly possible, the numbers might be larger. It also seems clear that the number of fibers that enter the cord directly is relatively small, as evidenced by the small numbers of cells that are retrogradely labeled (Maynard *et al.*, 1977; T. Yamamoto *et al.*, 1977), the small numbers of unmyelinated axons (actually none were found) that enter the cord from the ventral root (Risling and Hildebrand, 1982) and the lack of effect of stimulating the central stump of a transected ventral root (see below). The looping fibers (Fig. 3.11B and C) were thought to be dominant, but more recent evidence indicates that fibers entering directly into the root, with or without a third branch in the dorsal root, are more common (Kim *et al.*, 1987; Häbler *et al.*, 1990). If this is accepted, this would imply that the majority of ventral root afferent fibers end blindly in the ventral roots.

Function of the Ventral Root Afferents. It has been known from the time of the formulation of the law of separation of the spinal roots that stimulation of the ventral root causes pain (Magendie, 1822). Furthermore, interruption of the dorsal roots averts this pain (Bernard, 1858; Frykholm, 1951; Frykholm *et al.*, 1953). Similarly, sciatic nerve stimulation with the appropriate dorsal roots cut leads to a small rise in blood pressure (Longhurst *et al.*, 1980), whereas if the dorsal roots are intact the rise is much more dramatic (J. M. Chung *et al.*, 1986c). It has also been reported that intra-arterial injection of bradykinin following dorsal rhizotomies led to an increase in discharges of flexor motoneurons and dorsal horn cells (Voerhoove and

Zwaagstra, 1984). The difficulty is that bradykinin seems to have a direct effect on these cells (Shin *et al.*, 1985), and so this activity cannot be ascribed to ventral root afferents. Finally, when the proximal stump of a cut ventral root is stimulated, no change in the activity of dorsal horn or motoneurons is observed, whereas stimulation of the distal stump markedly excites these cells (J. M. Chung *et al.*, 1983b, 1985; Endo *et al.*, 1985). Thus, if the dorsal root is intact, strong stimulation of the ventral root acts as if the dorsal root is being stimulated, whereas if the dorsal root is interrupted, the only known result of stimulation of the ventral root afferents is a slight rise in blood pressure.

One possible way to obtain insight into the functions, if any, of ventral root fibers that enter the cord directly would be to increase the numbers of these axons, and hence presumably the measurable effects of stimulating these fibers. In this regard, the ventral root afferent population seems to be relatively plastic. There is, for example, a considerable increase in the numbers of unmyelinated fibers postnatally in cat ventral roots (Risling *et al.*, 1981; D. D. Heath *et al.*, 1986). More interestingly, neonatal sciatic nerve interruption or amputation of the neonatal limb leads to a considerable increase in numbers of unmyelinated ventral root axons (Risling *et al.*, 1984a,c; D. D. Heath *et al.*, 1986; Nam *et al.*, 1989; Oh *et al.*, 1989). If the increased number of fibers in the ventral root implies an increased number that enter the cord, it might be possible to obtain some insight into the function of ventral root afferent fibers that enter the cord directly in these animals.

In summary, the mammalian ventral root contains a considerable number of primary afferent fibers, most of which are unmyelinated. It would seem that the majority of these fibers enter the ventral root and end blindly. Several other axonal arrangements can be found, but these seem to be present in smaller numbers. Only minimal effects can be demonstrated by activating the ventral root afferents if dorsal root transmission is blocked. What, then, could be the function of the blind ventral root afferents? They would not seem to be synapsing, so perhaps they could have some kind of "trophic" or maintenance function similar to that discussed for other unmyelinated afferents (Wall and Fitzgerald, 1982). Obviously, insight into any such function would have important consequences for sensory mechanisms.

Segregation of Dorsal Root Fibers

Dorsal roots break up into a series of rootlets as they approach the cord. Many studies report that the large and small fibers in rootlet segregate as they enter the cord (Lissauer, 1886; Ranson, 1914; Ingvar, 1927; O'Leary *et al.*, 1932; Szentágothai, 1964; Sindou *et al.*, 1974; Kerr, 1975a,b). There were some disagreements, however, particularly about whether these subdivisions applied only to primary afferent fibers or might be more properly applied to fibers in Lissauer's tract (Earle, 1952; Wall, 1962). A major problem was that the unmyelinated fibers could not be well visualized with the light microscope. A careful electron-microscopic examination showed that there was a parcelling of coarse and fine axons in the monkey proximal rootlets but not in the cat (Snyder, 1977). In our experience, segregation is also not prominent in the rat. From a functional point of view, the interest in such a segregation would be that fine afferent fibers alone could be cut, thus hopefully relieving intractable pain without making the segment completely anesthetic. This was attempted in the cat, and there seemed to be pain relief with intact touch following damage to the lateral part of each rootlet and subjacent cord (Ranson and Billingsley, 1916). These effects are now ascribed to interruption of the blood supply to the dorsal horn (Wall, 1962; Sindou *et al.*, 1974), however. Nevertheless, such lesions have been used to interrupt fine sensory fibers selectively for experimental purposes, (C. LaMotte, 1977; Light and Perl, 1977), so the procedure would seem feasible. Electron-microscopic examination to see exactly what fiber systems are being removed would obviously be desirable before accepting the idea that fine afferents can be removed surgically while the coarse afferents are left intact.

The question of the arrangements of different sizes of primary afferent fibers as they enter the cord has moved to the clinic in recent years. Many patients get intractable pain with nerve and root avulsions, and sometimes the only effective method of treatment is a selective posterior rhizotomy or a dorsal root entry zone lesion (Sindou *et al.*, 1987; Nashold, 1987). In addition, spasticity is a difficult clinical problem that is sometimes amenable to selective primary afferent removal. This is too large a subject for this book, but it is clear that if only fine fibers in the dorsal root could be removed, pain might be relieved with other important sensory modalities left intact. This is the goal of the selective rhizotomy. The dorsal root entry zone operation is more extensive and essentially removes laminae I–V over a relatively short distance. We are unable to evaluate the clinical results precisely, but it would seem that relief is often obtained. The point here, however, is that seemingly esoteric questions about the arrangements of the primary afferent fibers as they enter the cord and their arrangements in the dorsal horn have a considerable bearing on the relief of human suffering. As with the experimental work, it would be desirable to examine the normal human dorsal root entry zone with the electron microscope (to see how the fine fibers are arranged) and then to examine the same areas after surgery to see exactly what has been removed. The technical difficulties are formidable, however, given the exigencies of fixation in humans and the large size of the areas in question.

CONCLUSIONS

1. DRG cells are unipolar neurons that convey information from the periphery to the spinal cord.
2. DRG cells are usually divided into two categories, small and large. These categories correlate with function, cell birthdays, and histochemistry and are almost universally accepted.
3. The large DRG cells classically contain relatively few organelles and stain lightly or weakly with the usual stains, whereas the small cells contain many organelles and stain intensely (large light and small dark cells).
4. A relatively recent finding is that the light cells are of all sizes, not just large, whereas the dark cells are always small.
5. Many investigators mention the possibility of intermediate-sized cells as a separate category, but as yet there is relatively little reason to accept this category as a separate group of DRG cells.
6. Many further subdivisions of DRG cells have been proposed on the basis of different arrangements of the organelles, but as yet there are few functional correlations with these subdivisions.
7. A major development in recent years is the discovery that DRG cells can be subdivided on the basis of their content of morphologically visible markers. These markers are most commonly peptides or enzymes, but many other compounds are found in subpopulations of DRG cells as well.
8. To date 13 peptides can be visualized in subpopulations of DRG cells. This number will undoubtedly increase.
9. The cells that contain the peptides are almost always in the small-cell category, but for some compounds, a few larger cells are also labeled.
10. The presence of a peptide does not yet have a clear functional correlate. Since the small-cell population is primarily involved, however, the relation of the peptides to nociceptive transmission is frequently discussed.
11. Many of the peptides are colocalized in DRG cells, and there are some preliminary indications that patterns of colocalization may be correlated with area of peripheral innervation.
12. Many different classes of enzymes delineate subpopulations of DRG cells. Interesting

categories here are ACHE because of its possible relation to cholinergic function, carbonic anhydrase because it labels primarily the large cells, glycogen phosphorylase because the amount of labeling increases after functionally relevant stimuli, and FRAP because of its possible relation to nociceptive transmission.

13. DRG cells can also be subdivided on the basis of the physiologic characteristics of their membranes. Some of these characteristics correlate with functional variables.

14. Peripheral and central processes of DRG cells seem to branch in the general vicinity of the ganglion. This branching is usually discussed as a possible mechanism for referred pain, but further evidence must be obtained before this can be accepted.

15. Ventral root afferent fibers, which were classically not thought to exist, have been intensively studied recently. Of the several possible arrangements of these fibers, the commonest seems to be that they end blindly in the ventral root. The only functional correlate that has clearly been identified as coming from ventral root afferent fibers that enter the cord through the ventral root is a slight rise in blood pressure. Stimulation of the ventral root with the dorsal root intact, however, results in marked responses.

16. The old question of the segregation of dorsal root fibers as they enter the cord has clinical relevance.

4 Structure of the Dorsal Horn

This chapter will take as its primary mission the description of laminae I–VI (Fig. 4.1). The relations of these laminae to earlier organizational schemes of the dorsal horn were presented by Rexed (1952, 1954). The earlier edition of this book was concerned primarily with describing classic work, particularly studies using the Golgi method, and correlating this material with advances made by electron-microscopic examination of the dorsal horn. Since that time there has been an explosion of structural studies, many using completely new techniques, that are providing further insight into the organization of the dorsal horn. To keep the size of this chapter in reasonable bounds, we will focus on the primary afferent input (and not consider the equally important descending input) and restrict the cited studies essentially to those concerned with mammals.

LAMINA I

Lamina I is the classical marginal zone. Rexed (1952) described it as "a thin veil of gray substance, forming the dorsal-most part of the spinal gray matter." Bundles of myelinated axons penetrate this lamina, giving it a reticular appearance and sometimes obscuring the boundary between this lamina and the overlying white matter. Some early investigators regarded this lamina as the superficial part of the substantia gelatinosa (Ramón y Cajal, 1909; Ranson, 1913b), but it is now clear that lamina I is a separate structural and functional entity (Cervero and Iggo, 1980). The classic features of this region are large, horizontal neurons (the marginal cells of Waldeyer) and a plexus of numerous horizontally arranged fine axons. These contrast with the tightly packed, radially arranged, small neurons and longitudinally oriented axons in the underlying gelatinosa.

Cell Types

Classic Types. *Marginal and Smaller Cells.* The neuronal cell bodies in lamina I have a wide range of sizes. Interest has focused on the largest cells, the marginal cells of Waldeyer (1888) (Fig. 4.2). These cells, although prominent, are relatively sparse. They overlie the gelatinosa, and their major dendrites are arranged horizontally (Clarke, 1859; Waldeyer, 1888; Ramón y Cajal, 1909; Szentágothai, 1964; Scheibel and Scheibel, 1968) (Figs. 4.2–4.4). Thus, the dendritic fields of these cells were reported to have a discoid shape (Figs. 4.5–4.8). However, reconstructions show that the dendritic trees of many lamina I cells are markedly elongated in the rostrocaudal direction (Light *et al.*, 1979; Woolf and Fitzgerald, 1983; Lima and Coimbra, 1986, 1989), so a better depiction would be as a flattened oval with the long axis in the craniocaudal direction.

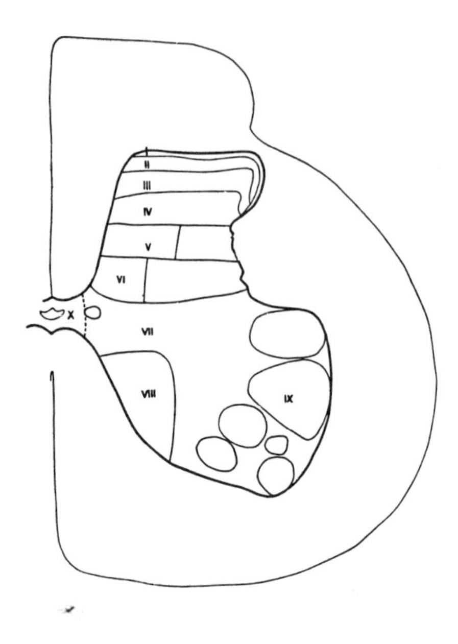

Fig. 4.1. Schematic indicating the laminae in the L7 segment of cat spinal cord. Laminae I–VI make up the dorsal horn. (From Rexed, 1952; reprinted with permission.)

The marginal cells, which are large lamina I neurons with horizontal dendrites, make up a minority of the neurons in lamina I. The more numerous smaller cells were not extensively studied classically. When physiologic work demonstrated diverse lamina I neuronal populations, however, it was asked whether more neuronal subtypes than "marginal neurons" and "smaller lamina I neurons" could be shown.

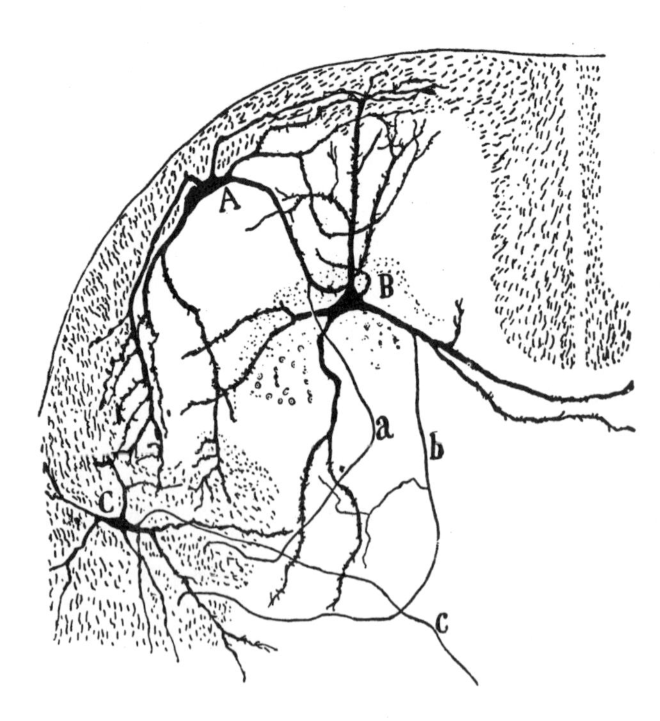

Fig. 4.2. Drawing of a Golgi-stained section of spinal cord. Cell A is a large marginal cell. Note that majority of the dendrites of this cell pass over the surface of the dorsal horn, but that some pass ventrally into the grey matter. The axon (a) of cell A arises from a ventral dendrite and passes into the lateral funiculus. Cell B is a large cell in the center of the dorsal horn (approximately lamina IV). The dendrites pass medially, laterally, and dorsally. The dorsal dendrites are characteristic of antenna-type neurons. The axon of this cell goes to the lateral funiculus. Cell C does not belong to the dorsal horn but appears to be in the lateral spinal nucleus. (From Ramón y Cajal, 1909.)

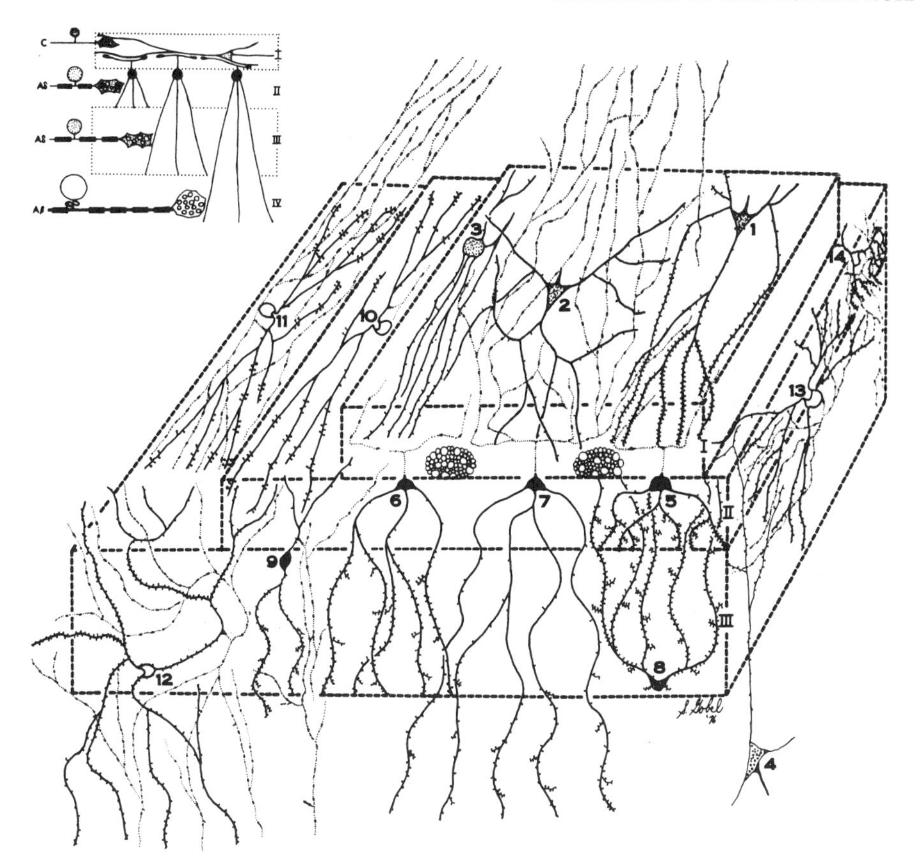

Fig. 4.3. Schematic diagram of the major cell types of laminae I–IV of the medullary dorsal horn. Lamina I neurons are spiny (1) and smooth (2) pyramids and multipolar (3) neurons. Cell 4 is a lamina IV neuron which sends dendrites dorsally into laminae I–III (antenna type neuron.) Cells 5–9 are stalked cells, cells 10 and 11 islet cells, cell 12 a spiny cell, cell 13 a lamina II/III border cell, and cell 14 an arboreal cell. (From Gobel, 1978; reprinted with permission.)

Modern Classifications. Narotzky and Kerr (1978) subdivided lamina I cells into large marginal neurons and smaller neurons of unknown type. They then divided the marginal neurons into type P (projection) and type A (association) neurons. This subdivision reflects primarily the destinations of the axons of the marginal cells (see below). In the marginal layer of the nucleus caudalis of the medullary dorsal horn, Gobel (1978a) described two types of pyramidal cells, one with numerous dendritic spines and one without, and two types of multipolar neurons, one with a compact dendritic arbor and one with a loose dendritic arbor (Fig. 4.3). The axonal destinations of these cells were not determined. Gobel (1978a) emphasized that the dendrites of these cells remain in lamina I. Beal *et al*. (1981) proposed a more complicated subdivision with four groups of neurons, three of which are split into A and B subtypes. Some of these correspond to the subtypes of Gobel (1978a), but others do not. Lima and Coimbra (1983), on the basis of reconstructions of cell bodies and proximal dendrites, described four lamina I cell types: fusiform, flattened, multipolar and encased. More complete reconstructions of the dendritic trees of these cells led to subdivisions of the fusiform and multipolar categories, largely on the basis of ventral spread of the dendritic arbors (Lima and Coimbra, 1986) (Fig. 4.9). Marginal cells fall primarily into the flattened and encased categories (Lima and Coimbra, 1986). These

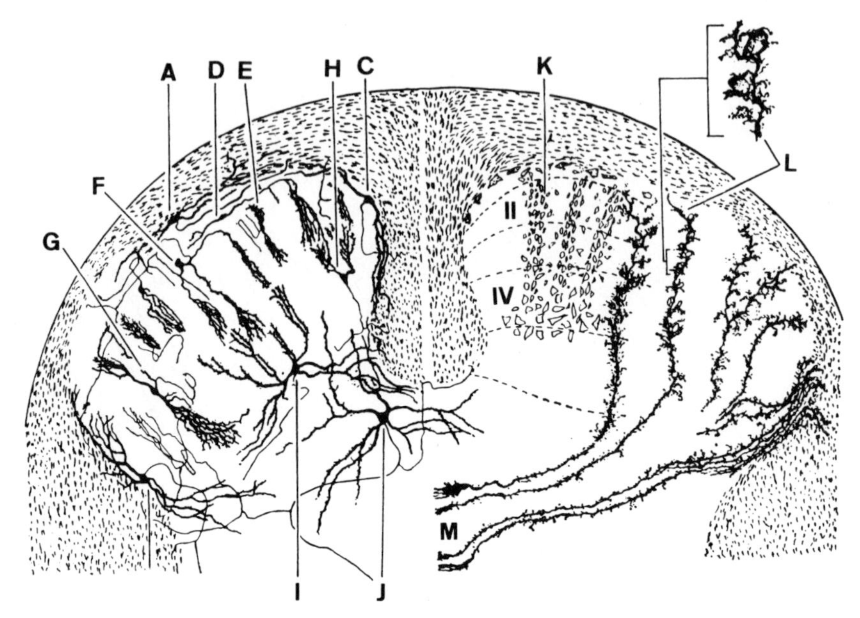

Fig. 4.4. Schematic diagram of the organization of the dorsal horn. Cells A and C are marginal cells of Waldeyer. Cell D is a cell that has characteristics of both marginal and central cells and would thus be a limiting cell of Cajal. Cells E–G are central cells of the substantia gelatinosa. Cell H is probably deep in lamina III, and cell I is in lamina IV. In the right-hand side of the diagram are cell bodies of dorsal horn neurons as seen in Nissl stains and glial processes. (From Scheibel and Scheibel, 1968.)

various classifications show some similarities to one another, but there are many differences. As yet, none of the various cell types fit clearly with functional categories (Light *et al.*, 1979; Woolf and Fitzgerald, 1983; Hylden *et al.*, 1986a; Rethelyi *et al.*, 1989).

The above classifications result from reconstructions following Golgi and other classic techniques. Retrograde or direct labeling techniques also provide insight into the shape of the cells and the arrangements of their dendritic trees. In this work, receptive field type was correlated with the laminar location of the dendrites but not with the location of the cell body or general shape of the dendritic tree (Light *et al.*, 1979). Bennett *et al.* (1981) reconstructed 14 lamina I neurons with an emphasis on axonal collaterals that overlapped the dendritic fields of these cells. Molony *et al.* (1981) reconstructed a lamina I nociceptive cell and found it to be a typical marginal cell. Woolf and Fitzgerald (1983) emphasized that cell and dendritic morphology did not correlate with function and pointed out that some neurons, which are presumably part of the lamina I system, are found in the overlying white matter. Steedman *et al.* (1985) also could not correlate function with dendritic morphology. Hylden *et al.* (1986b), in a study concerned with lamina I cells that project to the midbrain, found a diversity of morphologic types but could not correlate structure with function.

It is an important issue as to which laminae the dendrites of lamina I cells enter. All investigators agree that the horizontal dendrites are wide spread. Most investigators also point out that some dendrites of marginal and other lamina I cells dip into deeper laminae (Ramón y Cajal, 1909; A. A. Pearson, 1952; Scheibel and Scheibel, 1968; Beal, 1979b) (Fig. 4.2), but Gobel (1978a) denied this for adult animals and Hylden *et al.* (1986b) showed that the dendritic trees of lamina I cells that projected to the mesencephalon were largely restricted to lamina I. On the other hand, when adult lamina I cells were filled with markers from microelectrodes or

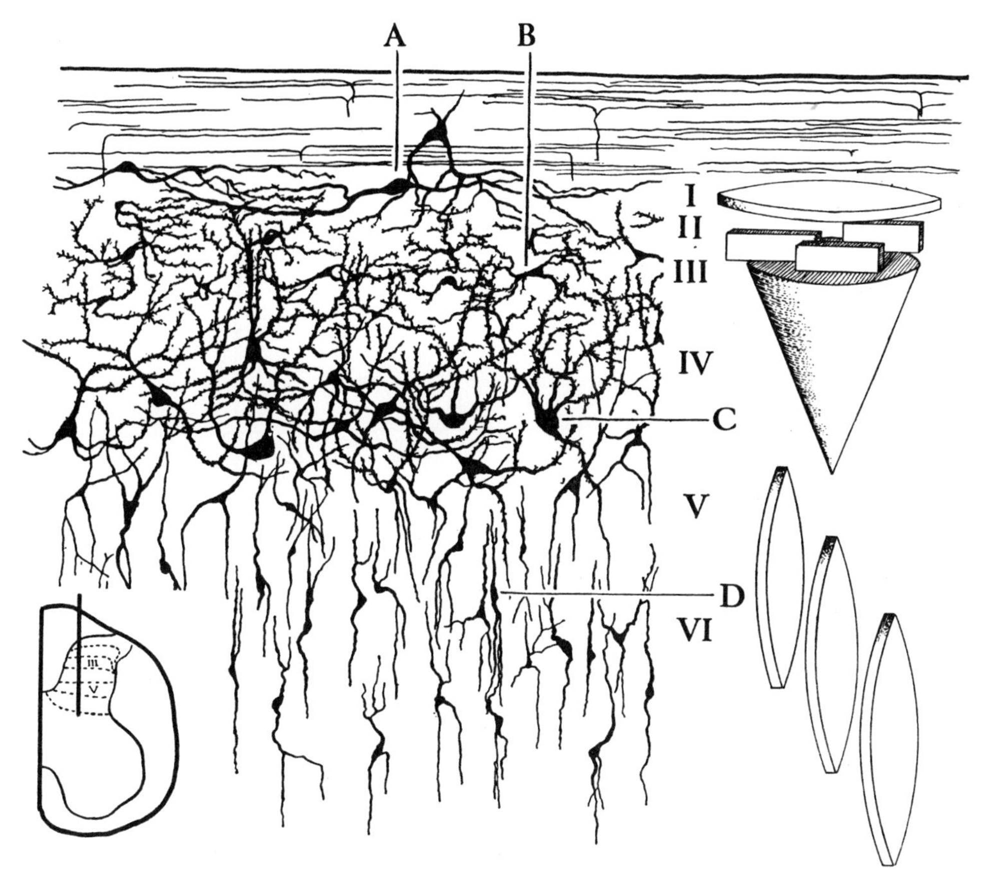

Fig. 4.5. Shapes of the dendritic patterns of cells in laminae I–VI. The geometric patterns of the dendritic fields are shown in the right-hand side of the figure. (From Scheibel and Scheibel, 1968.)

reconstructed in several planes, it was clear that dendrites of some cells did pass to deeper laminae (Light *et al.*, 1979; Woolf and Fitzgerald, 1983; Hylden *et al.*, 1985; Lima and Coimbra, 1986). It is not yet clear whether the presence of dendrites that pass ventrally characterizes only certain lamina I cell types. The types of afferents that characterize a lamina correlate with the functional responses of the cells whose dendrites enter that lamina (Light *et al.*, 1979).

In summary, the neurons of lamina I have classically been divided into large marginal cells characterized by wide-ranging horizontal dendrites and smaller neurons. In recent years, however, these categories are seen as being too simplistic, and more detailed categories have been proposed. As yet there is not complete agreement on the newer categories, and precise function cannot be correlated with dendritic morphology, although the types of afferents that enter a lamina where the dendrites are located have a bearing on the responses of the cell.

Chemical Localizations. The term "chemical neuroanatomy" is used for the structural findings that result from the visualization of potential transmitters (usually peptides or amino acids), the enzymes that degrade or manufacture these compounds or other compounds that can be localized in spinal neurons. An overall review of this subject is given at the end of this chapter. Here we will simply list the compounds found in lamina I cells and any anatomic insights that

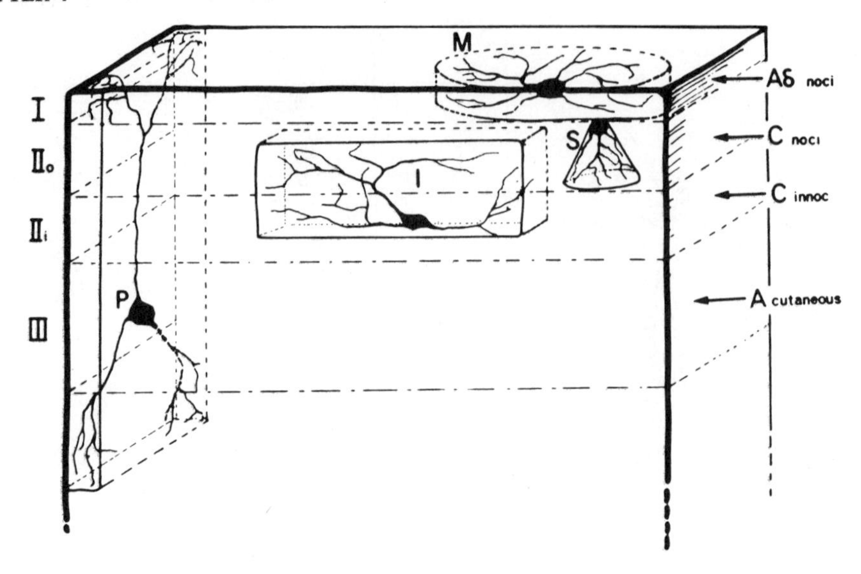

Fig. 4.6. Diagrammatic representation of the dendritic trees of neurons in laminae I–III as viewed in a parasagittal view of the cord. Cell M is a marginal cell in lamina I, cell I is an islet cell and cell S a stalked cell, both in lamina II, and cell P is a large pyramidal cell in lamina III. (From A. G. Brown, 1981.)

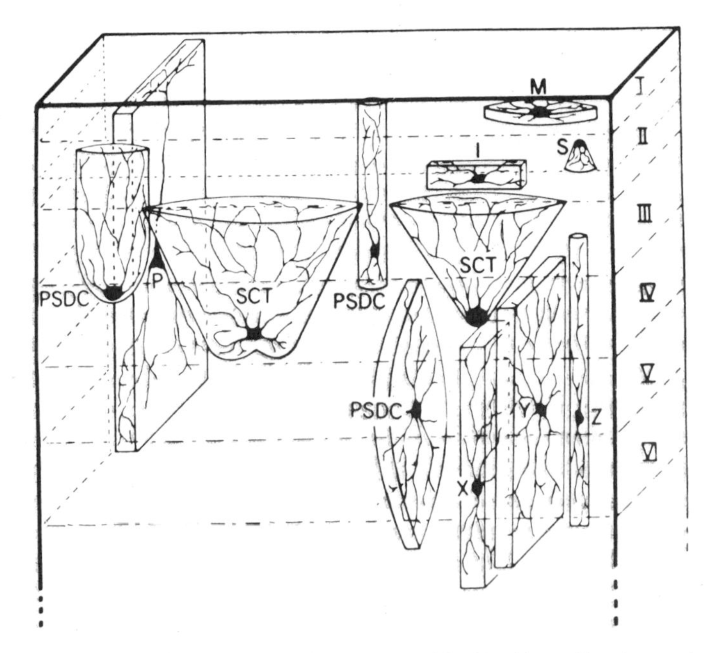

Fig. 4.7. Depiction of the dendritic trees of the major cell types of the dorsal horn. Note the marginal (M), stalked (S), islet (I) and pyramidal cells as in Fig. 4.6. In addition, there are postsynaptic dorsal column (PSDC) and spinocervical tract (SCT) cells in laminae III–V and interneurons (X, Y, and Z) in laminae V and VI. (From A. G. Brown, 1981.)

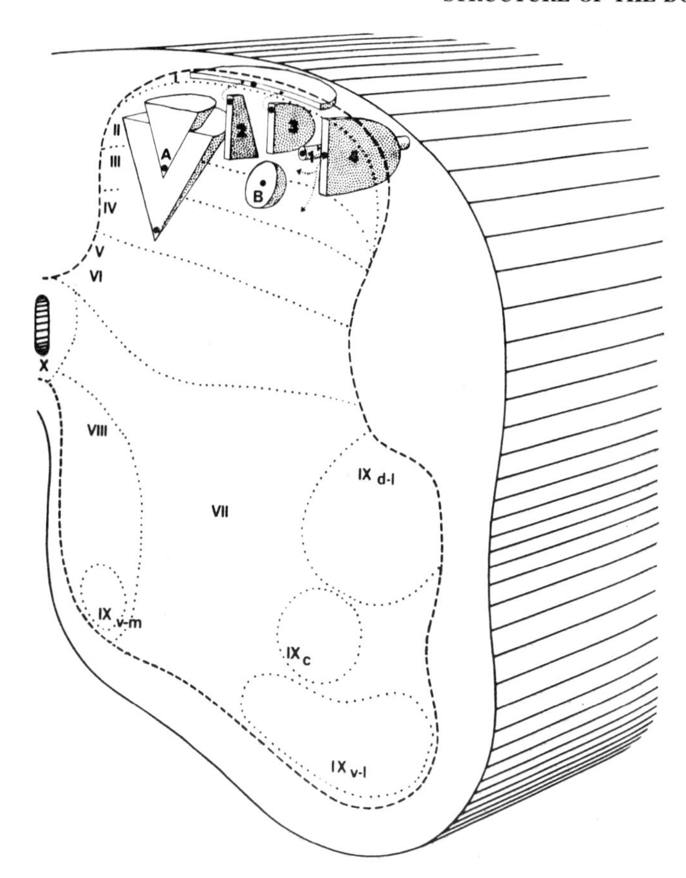

Fig. 4.8. Schematic depiction of the shape of the dendritic fields of the major cell types of the dorsal horn. The cell in lamina I is a marginal cell. Cells 1–4 are islet cells, filamentous cells, curly cells, and stellate cells, respectively. The cells at A are antenna-type neurons, and cell B is a radial neuron of lamina III. (From Schoenen, 1982; reprinted with permission of Pergamon Press.)

come from this work. Before beginning, one caveat is that many of the localization studies do not describe the criteria for laminar boundaries. Thus, particularly for narrow laminae such as lamina I, the laminar localizations may not be as precise as desirable.

The compounds that are found in lamina I cells are as follows.

1. Acetylcholinesterase (ACHE) and cholineacetyltransferase (ChAT). The interest in these compounds is that they may indicate a cholinergic system. ACHE labeled cells have been in lamina I (Odutola, 1972; Marchand and Barbeau, 1982). No ChAT cells have been found specifically in lamina I to our knowledge.

2. Corticotrophin releasing factor (CRF). A few immunolabeled cells are seen in lamina I (Merchanthaler, 1984).

3. Dynorphin. Dynorphin-labeled cells are seen in lamina I (Khachaturian *et al.*, 1982; Vincent *et al.*, 1982; L. Cruz and Basbaum, 1985; Basbaum *et al.*, 1986b; Cho and Basbaum, 1988, 1989; Ruda *et al.*, 1988; Weihe *et al.*, 1989). K. E. Miller and Seybold (1989) and Standaert *et al.* (1986) report that some dynorphin cells in lamina I project to the thalamus.

4. Enkephalin. A large number of studies report enkephalin labeled lamina I neurons (Hökfelt *et al.*, 1977a; Aronin *et al.*, 1981; Finley *et al.*, 1981a; Glazer and Basbaum, 1981, 1982;

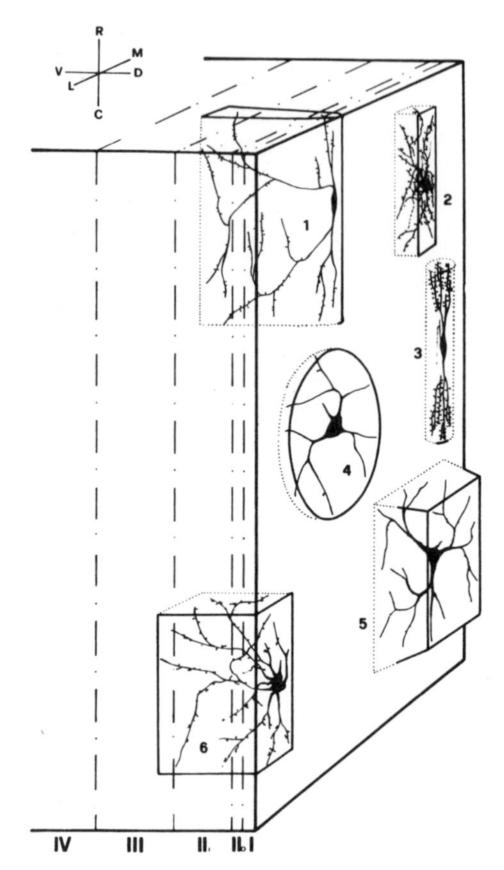

Fig. 4.9. Diagrammatic representation of the dendritic territories of cell types in the superficial dorsal horn. Type 1 is a fusiform neuron with longitudinal and ventral dendrites, type 2 is a multipolar spiny neuron with many dendritic branches, type 3 is a fusiform neuron with longitudinal arbors, type 4 is a flattened spiny neuron, type 5 is a pyramidal neuron, and type 6 a multipolar neuron with few dendritic branches. (From Lima and Coimbra, 1986; reprinted with permission of Wiley-Liss.)

S. P. Hunt *et al.*, 1981a; Bennett *et al.*, 1982; Sumal *et al.*, 1982; Arluison *et al.*, 1983a; Conrath-Verrier *et al.*, 1983; Petrusz *et al.*, 1985; Ruda *et al.*, 1986; Standaert *et al.*, 1986; K. E. Miller and Seybold 1987, 1989; Weihe *et al.*, 1988b; Nahin, 1988). There are slight disagreements about the relative numbers of these cells. Several investigators note that some of the labeled cells are large and have a horizontal orientation and are thus thought to be Waldeyer or marginal neurons (Glazer and Basbaum, 1981, 1982; Hunt *et al.*, 1981a; Sumal *et al.*, 1982; Ruda *et al.*, 1986; K. E. Miller and Seybold, 1987). Double labeling studies show that some lamina I enkephalin neurons project to the parabrachial area of the midbrain (Standaert *et al.*, 1986) or the thalamus (Nahin, 1988). The projecting cells are both marginal and smaller neurons. Enkephalin and substance P coexist in a significant number of lamina I neurons (Senba *et al.*, 1988).

5. Galanin. Some small cells containing galanin are seen in lamina I (Rökaeus *et al.* 1984; Ch'ng *et al.*, 1985; Melander *et al.*, 1986).

6. γ-aminobutyric acid (GABA). Glutamic acid decarboxylase (GAD)-containing cells are seen in lamina I (Ribeiro-DaSilva and Coimbra, 1980; S. P. Hunt *et al.*, 1981; Barber *et al.*, 1982; Mugnaini and Oertel, 1985). Ribeiro-DaSilva and Coimbra (1980) did not think that these were marginal cells. Barber *et al.* (1982) believed that most cells of the dorsal horn were immunoreactive. Kosaka *et al.* (1988) found that some of the dorsal horn GAD cells contained ChAT, but no laminar localizations were reported.

7. Glutamate. Glutamate-labeled cells are found in lamina I (Greenamyre *et al.*, 1984; Weinberg *et al.*, 1987; De Biasi and Rustioni, 1988; K. E. Miller *et al.*, 1988). Some glutamate neurons in lamina I project to the thalamus (K. R. Magnusson *et al.*, 1987).

8. Glycine. In a study of both the glycine receptor and glycine itself, label was seen in lamina I but individually labeled cells were not reported (van den Pol and Gorcs, 1988).

9. Neurotensin. The greatest concentration of neurotensin cells was seen in lumbosacral segments, but even here these cells were fewer than those labeled for dynorphin or enkephalin (K. E. Miller and Seybold, 1989).

10. Serotonin (5-HT). A few serotonin immunolabeled neurons have been reported in lamina I (C. C. LaMotte *et al.*, 1982; Bowker, 1986; Newton *et al.*, 1986; Newton and Hammell, 1988).

11. Somatostatin. A few somatostatin-labeled neurons are seen in lamina I (Schroder, 1984; Ruda *et al.*, 1986; Mizukawa, 1988), although most studies on adults do not mention such cells. Some of the labeled cells seem to be large and horizontally oriented (S. P. Hunt *et al.*, 1981a). More cells label in young animals (Ho, 1983, 1988; Ho and Berlowitz, 1984; Charnay *et al.*, 1987).

12. SP. Some SP cells are seen in lamina I, although many more have been found in lamina II. References will be given in the section on lamina II.

13. Thyrotropin-releasing hormone (TRH). TRH has been found in cells in laminae I–III (Coffield *et al.*, 1986; Tsuruo *et al.*, 1987).

Axonal Projections

Axons of cells in lamina I project to the thalamus (see Chapter 9) or to other parts of the spinal cord. Other important destinations are the midbrain and brain stem reticular formation (see Chapter 9). In addition, connections to the cerebellum are reported (Snyder *et al.*, 1978) but must be confirmed. Many of the lamina I cell axons branch. For example, some lamina I neurons project to both mesencephalon and thalamus (see Chapter 9). Presumably, knowledge about the precise projections of lamina I neurons will rapidly increase.

One question is whether the distribution of the dendritic tree can be correlated with the destination of the axon. Most studies do not report such correlations, but Lima and Coimbra noted that flattened and pyramidal cells project to the thalamus (Lima and Coimbra, (1988) and that pyramidal cells project to the periaqueductal grey and fusiform cells to the parabrachial nuclei of the midbrain (Lima and Coimbra, 1989).

Primary Afferent Input into Lamina I

The morphology of the primary afferent input into the various laminae of the dorsal horn is a key issue in understanding spinal sensory mechanisms. Classic work used primarily Golgi and degeneration techniques. More recently the intra-axonal filling techniques that allow single fibers to be visualized along with a simultaneous determination of their function have had a major impact in this area. Excellent reviews are available (A. G. Brown, 1981; Fyffe, 1984; Réthelyi, 1984a; Fitzgerald, 1989).

The classic view is that large primary afferent fibers segregate by size either in the proximal dorsal root or as they enter the cord, with the large fibers entering the dorsal columns and fine fibers the tract of Lissauer (Ramón y Cajal, 1909). Both types then bifurcate into a cranial and a caudal branch, and the input to the gray matter of the cord comes from collaterals from these branches (Ramón y Cajal, 1909). This has been a useful construct, but certain aspects of it do not seem to be correct. For example, large numbers of unmyelinated primary afferents are found in the dorsal columns (Langford and Coggeshall, 1981b; K. Chung and Coggeshall, 1983; K. Chung *et al.*, 1985, 1987; Patterson *et al.*, 1989, 1990), and there seem to be more primary afferent fibers, at least in sacral segments of the rat, in the dorsolateral funiculus than in the tract of Lissauer and dorsal columns combined (K. Chung *et al.*, 1987). In addition, many hair follicle afferents do not bifurcate (A. G. Brown *et al.*, 1977b; A. G. Brown, 1981) (Fig. 10.20). Accordingly, the classic picture must be modified.

The Marginal Plexus. Large numbers of fine fibers enter this lamina from the overlying tract of Lissauer and surrounding white matter. These fibers form a horizontal plexus, the marginal plexus (Ramón y Cajal, 1909); (Fig. 4.10). Many fibers in this plexus are primary afferents, but most studies do not provide quantitative data on the proportions of primary afferents and propriospinal fibers, which are the other major component of this plexus (Szentágothai, 1964; Scheibel and Scheibel, 1968). Recently, however, quantitative evaluation shows that approximately 50% of the synapses in lamina I of the rat dorsal horn are lost following dorsal rhizotomy and thus presumably arise from primary afferents (D. L. McNeill, *et al.*, 1988c; K. Chung *et al.*, 1989b). The remaining synapses presumably arise from propriospinal and descending neurons. Subdivisions of the longitudinal plexus have been proposed. Beal and Bicknell (1981) describe a transverse plexus superficial to the well-studied longitudinal plexus. The transverse plexus is most apparent laterally. They suggested that the primary afferent terminals in both plexuses arise from Aδ: fibers. A somewhat different arrangement was proposed by F. Cruz *et al.* (1987). These investigators state that the lateral third of laminae I and II of the rat cord has a common longitudinal plexus characterized by parallel 0.8-μm fibers that give rise to longitudinal side branches with many *en passant* endings. In the middle third of laminae I and IIo are dichotomizing 1-μm fibers with elongated boutons, and in the medial third is a dense plexus arising from 1.3-μm fibers that give rise to large boutons. In the dorsal parts of medial and middle lamina IIi are terminal bouquets arising from 1-μm fibers, and the whole middle and medial parts of lamina IIi contain clusters of ultrafine boutons arising from very fine fibers. Functional correlates of these plexuses are not known, but it was suggested that fibers less than 1-μm in diameter are C fibers, whereas larger fibers are mostly Aδ fibers (F. Cruz *et al.*, 1987). (Useful summary diagrams of the types of primary afferent input into various laminae of the dorsal horn are given in Fig. 4.11–4.13; see also Figure 10 in Chapter 5).

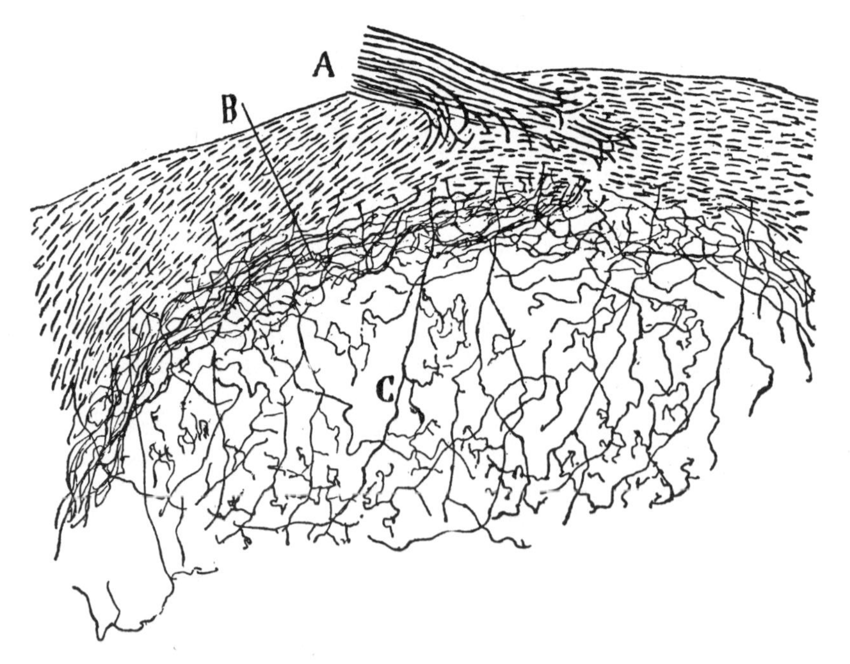

Fig. 4.10. Depiction of the fine fibers that permeate the superficial dorsal horn. A indicates dorsal root fibers, B the fine fibers of the marginal plexus, and C the fine fibers that enter the substantia gelatinosa (From Ramón y Cajal, 1909.)

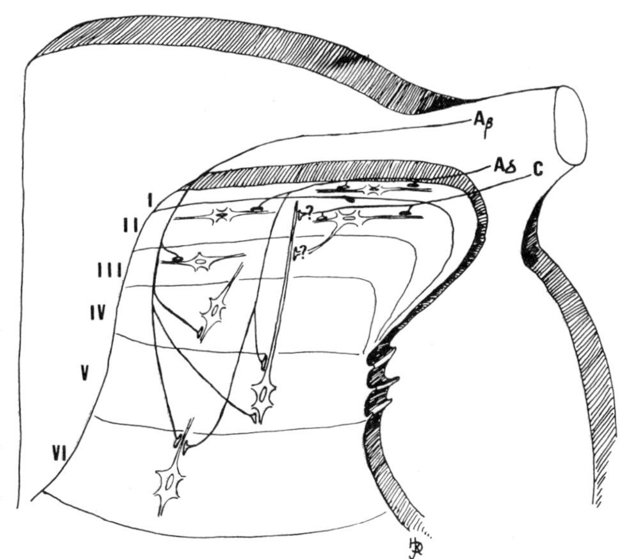

Fig. 4.11. Schematic diagram showing the distribution of Aβ (large myelinated), Aδ (fine myelinated), and C (unmyelinated) primary afferent fibers in the dorsal horn. Note in particular the extensive distribution of the large myelinated afferents in laminae III–VI and the fine myelinated fibers in laminae I and II. (From Ralston and Ralston, 1982; reprinted by permission of Wiley-Liss.)

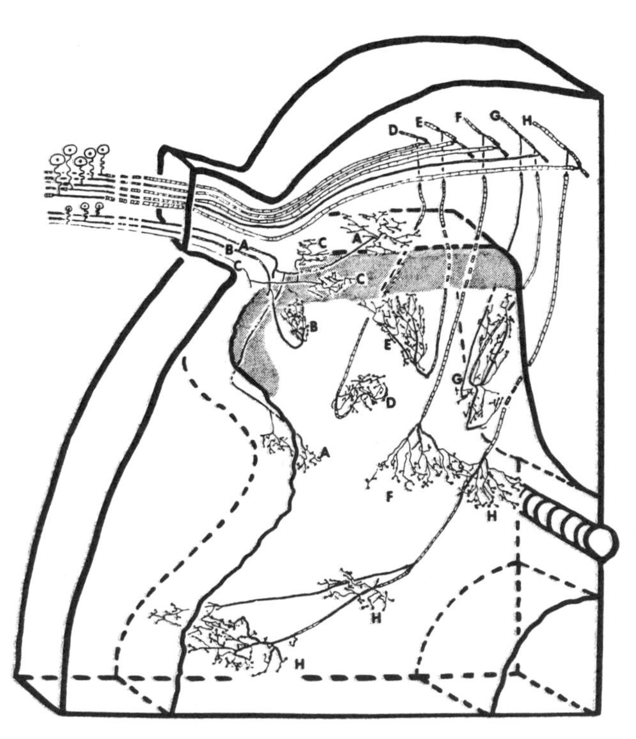

Fig. 4.12. Summary diagram of the spinal terminations of various types of primary afferent fibers in a transverse slab of the spinal cord. The shaded area indicates the superficial dorsal horn. Note that the fine fibers (A–C) end in or near the superficial dorsal horn. The main point of the diagram is to depict the specificity of the termination patterns of different types of primary afferent fibers. (Reprinted from Fyffe, 1984.)

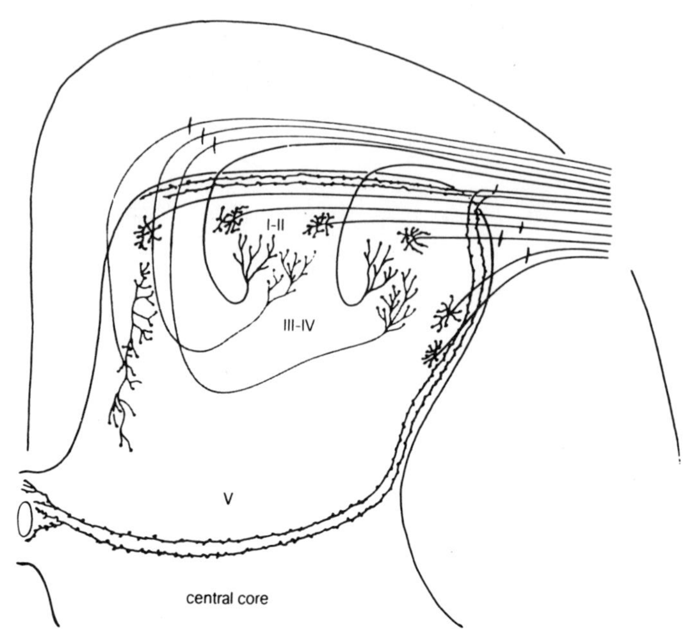

Fig. 4.13. Another summary diagram of the specific patterns of termination of different types of primary afferent fibers. Since the original is in color, the original text must be consulted for details. (From Maxwell and Réthelyi, 1987.)

Large *vs.* Small Fibers. Ramón y Cajal (1909) noted that fine fibers entering primarily from the tract of Lissauer and the surrounding white matter make up most of the input into the marginal plexus (Fig. 4.10). This has been confirmed by retrograde transport studies after selective surgery or by reconstructions of individually labeled axons whose conduction velocities and receptive fields have been determined (C. C. LaMotte, 1977; Light and Perl, 1977, 1979a) (Figs. 4.11–4.13; see also fig. 5.10). The recent successful labeling of many lamina I fibers and terminals by wheatgerm agglutinin conjugated horseradish peroxidase (WGA-HRP), (Robertson and Grant, 1985; Plenderleith *et al.*, 1988) is of great interest because WGA-HRP seems to be a marker associated with fine primary afferent axons. There is no evidence that collaterals from large primary afferent fibers enter lamina I (A. G. Brown, 1981).

Aδ *vs.* C Fibers. Fine primary afferent fibers enter lamina I. A recent question is whether these fine fibers are collaterals from Aδ (fine myelinated) or C (unmyelinated) primary afferent fibers. One suggestion, on the basis of a differential primary afferent degeneration in lamina I as opposed to lamina II and the finding of many ultrafine primary afferent fibers in lamina I, is that the main primary afferent input into lamina I is from C fibers (Gobel and Binck, 1977; Gobel *et al.*, 1981). By contrast, however, Beal and Fox (1976) and Beal and Bicknell (1981) found relatively large axons giving rise to axonal terminals in lamina I, and C. C. LaMotte (1977) found that the afferent endings in lamina I degenerated more slowly than those in lamina II. Accordingly, these investigators suggested that Aδ primary afferents gave rise to the lamina I afferent input. This was confirmed in functional studies (Kumazawa and Perl, 1977). The most definitive work comes from individual filling of high-threshold cutaneous mechanoreceptors conducting in the Aδ range, and these end primarily in lamina I and less frequently in lamina V (Light and Perl, 1979b) (Fig. 10.30). Thus, most investigators regard lamina I as a major Aδ primary afferent termination site (Fyffe, 1984) (Figs. 4.11–4.13; fig. 5.10).

The above conclusion does not imply that the only types of afferent fibers in lamina I are Aδ fibers. Gobel *et al.* (1981) show, for example, that there are two types of fibers in lamina I, one group arising from "ultrafine", presumably unmyelinated, axons and the other arising from larger-caliber axons. In addition, some of the individually labeled C fibers from the guinea pig had terminations in lamina I (Sugiura *et al.*, 1986, 1989); (Figs. 10.34 and 10.38). Also, many terminals immunolabeled for compounds that are found only in small dorsal root ganglion (DRG) cells or unmyelinated primary afferent fibers (Chapter 3) are found in lamina I. Thus, it is clear that there is unmyelinated primary afferent input into lamina I. It is still fair to say, however, that lamina I is a major nociceptive Aδ primary afferent termination area (Light and Perl, 1979b; Fyffe, 1984; Mense and Prabhakar, 1986) and that the unmyelinated afferents congregate more in lamina II. What is obviously needed is a quantitative evaluation of the numbers of terminals that arise from unmyelinated as opposed to fine myelinated fibers in lamina I.

Propriospinal *vs.* Primary Afferent Fibers. Pioneers in the study of the tract of Lissauer and the marginal layer of the dorsal horn regarded most of the fine fibers as primary afferents (Lissauer, 1885, 1886; Ramón y Cajal, 1909; Ranson, 1913b). Later studies, based primarily on the seeming lack of change following dorsal rhizotomy or spinal isolation, suggested that most of the fibers in the marginal plexus were propriospinal (Earle, 1952; Szentágothai, 1964). Earle (1952) and Szentágothai (1964) emphasized that the lateral part of the tract of Lissauer consisted almost exclusively of propriospinal fibers. Therefore, this part of the tract was cut to remove propriospinal fibers (Kirk and Denny-Brown, 1970; Narotzky and Kerr, 1978). More recent evidence indicates that fine primary afferent fibers are distributed throughout the tract of Lissauer, however (K. Chung *et al.*, 1979; K. Chung and Coggeshall, 1979; Light and Perl, 1979a; Rethelyi *et al.*, 1979; Coggeshall *et al.*, 1981), so that a clean separation between propriospinal and primary afferents cannot be achieved by cutting or stimulating a specific part of the tract of Lissauer. Quantitative work shows large numbers of terminals from both primary afferents and other fibers (presumably propriospinal or descending) in lamina I of the rat (K. Chung *et al.*, 1989b).

Cutaneous, Muscular, and Visceral Afferent Input. There are laminar differences in the termination patterns of the primary afferents in different types of peripheral nerves (Fyffe, 1984; McMahon and Wall, 1985b; Fitzgerald, 1989). Visceral nerve projections are reported to occur particularly to laminae I and V (DeGroat *et al.*, 1978; Morgan *et al.*, 1981; Nadelhaft *et al.*, 1983; Cervero and Connell, 1984a,b; Kuo and DeGroat, 1985). By contrast, terminals of cutaneous afferent fibers, although also in lamina I, are prominent in laminae II–IV (G. Grant and Ygge, 1981; Abrahams *et al.*, 1984b; Cervero and Connell, 1984b; Molander and Grant, 1986). For primary afferents in muscle nerves, the story is more complicated. Early studies in which muscles were injected with tracer show terminals in laminae I–V (Brushart and Mesulam, 1980; Kalia *et al.*, 1981); others report terminals mainly in lamina I (Craig and Mense, 1983; Nyberg and Blomqvist, 1984; Abrahams and Swett, 1986; Mense and Prabhakar, 1986; Mense and Craig, 1988). Others find essentially no or very sparse labeling in lamina I (Ammann *et al.*, 1983; Abrahams *et al.*, 1984b, Bakker *et al.*, 1984; Molander and Grant, 1987). Most investigators now believe that the intramuscular injection results are the result of spread of label, but the issue of whether primary afferents in muscle nerves terminate in lamina I is not settled. The most precise information comes from Mense and Prabhakar (1986) who found lamina I terminals after injecting individual fibers. Thus, it seems likely that in some instances muscle afferents terminate in lamina I, but whether this occurs for all muscle nerves in all species is unclear. Such things as technique and species differences may be important.

In summary, the primary afferent input into lamina I is classically described as deriving from the marginal plexus which is in turn derived from fine fibers from the tract of Lissauer and other parts of the surrounding spinal white matter. Aδ fibers are now singled out as providing the main primary afferent input to lamina I, although there is also input from unmyelinated fibers.

Propriospinal fibers also make a major contribution to this lamina. There are laminar differences in the distribution of terminals from cutaneous as opposed to muscle as opposed to visceral afferent fibers.

Chemically Identified Fiber Systems in Lamina I. Several compounds are found in plexuses of axons and terminals in lamina I (Fig. 4.14). Although there are variations in the amount of labeling, the labeled plexuses are not otherwise distinctive. References are provided in the chemical neuroanatomy section at the end of the description of lamina II. In the future, these markers will undoubtedly be key to a better understanding of the structural organization of lamina I. The compounds that can be localized in axons and terminals in lamina I are ACHE, bombesin, cholecystokinin, calcitonin gene-related peptide (CGRP), CRF, dynorphin, enkephalin, fluoride- resistant acid phosphatase (FRAP) and thiamine monophosphatase, galanin, GABA, glycine, glutamate, neurotensin, neuropeptide Y, noradrenaline and dopamine, serotonin (5-HT), somatostatin, substance P (SP), thyroid-releasing hormone (TRH), and vasoactive intestinal polypeptide (VIP).

Neuropil Organization

Various investigators note that the majority of axons in lamina I are horizontal, parallel to the surface of the cord (Ralston, 1965, 1968a, 1979; Gobel *et al.*, 1977; Kerr, 1966, 1970a,b; Narotzky and Kerr, 1978). These fibers give rise to numerous presynaptic terminals of various types. Primary afferent terminals can be categorized as containing (1) round clear vesicles intermixed with only a few large granular vesicles, (2) many large granular vesicles intermixed with round clear vesicles, (3) flattened vesicles, and (4) very large numbers of small clear vesicles, which are the central elements of glomeruli (Kerr, 1970a,b; Narotzky and Kerr, 1978; Ralston, 1979). Round vesicle terminals (Fig. 4.14) predominate in lamina I, but large granular vesicle terminals are also common (Ralston, 1979; Snyder, 1982). Flat vesicle terminals are relatively rare except in the peripheral parts of glomeruli. Ralston (1979) and Snyder (1982) find that the round vesicle terminals which predominate in lamina I give way to flat vesicle terminals in deeper laminae. Ralston (1979) makes the reasonable hypothesis that this indicates a predominance of excitatory processes in lamina I, which changes to a predominance of inhibitory processes in deeper laminae. The central terminals of glomeruli, although very important, are relatively rare compared to the other types.

Kerr's group (Kerr, 1970a,b, 1975a,b; Narotzky and Kerr, 1978) reported that 40% of the boutons on distal dendrites, 20% on proximal dendrites, and 10% of those on the cell bodies of lamina I cells arise from primary afferent fibers. By contrast, 20% of the boutons on distal dendrites, 60% on proximal dendrites, and 50% on the soma were thought to arise from propriospinal cells (Narotzky and Kerr, 1978). If this were correct, the primary afferent input would be distal and the propriospinal input would be proximal on cells which presumably transmit nociceptive information centrally. Ralston (1979) and Gobel *et al.* (1981) could not confirm the large numbers of primary afferent terminals on the cell bodies, however. Furthermore, one part of Kerr's strategy was to remove propriospinal terminals by cutting the lateral part of the tract of Lissauer, but recent work demonstrates many primary afferent fibers here as well (K. Chung et al., 1979; Light and Perl, 1979b). Accordingly, the conclusions of Narotzky and Kerr (1978) are not accepted (Ralston, 1979; Gobel *et al.*, 1981).

Glomeruli (complex synaptic arrays, large synaptic complexes) are striking synaptic arrangements that characterize laminae I–III, particularly lamina II. They consist of a prominent central terminal surrounded by dendrites and peripheral terminals, the whole being somewhat isolated by glial cell processes. The glomeruli will be described and illustrated in detail in the section on lamina II. For lamina I, there has been some debate about the presence of these structures. For example, glomeruli are not reported in lamina I of rat dorsal horn (Ribeiro-

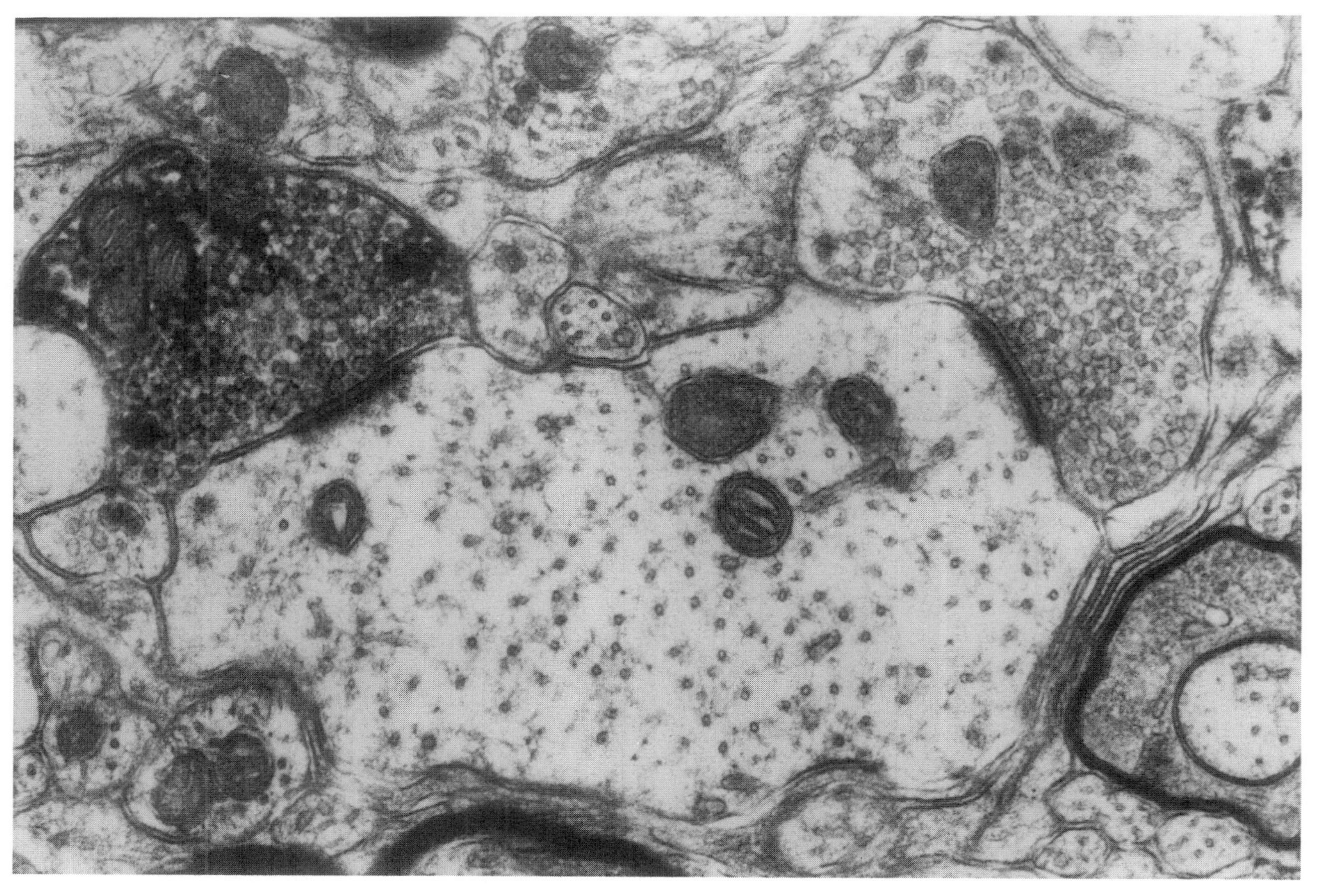

Fig. 4.14. Electron micrograph showing a large, relatively clear dendrite being contacted by two synaptic terminals in rat lamina II. Both synaptic terminals contain predominantly small round clear vesicles and make asymmetric synapses on the dendrite. The darker terminal is immunolabeled for CGRP. (From D. L. McNeill *et al.*, 1988a.)

DaSilva and Coimbra, 1982) or cat trigeminal nucleus caudalis (Kerr, 1970a), but they are found in monkey (Ralston, 1979), and cat (Ralston, 1968a,b; Gobel *et al.*, 1981) lamina I. In the monkey many central terminals are characterized by large dense core vesicles, but otherwise they are relatively simple in that they do not seem to have axoaxonic endings on them (Ralston, 1979). By contrast, the glomeruli are very complex in lamina I of the cat, with ample possibilities for presynaptic interaction (Gobel, 1974, 1976). Central terminals of glomeruli are immunolabeled by various compounds found in primary afferents. This will be discussed in more detail in the section on lamina II.

Nonglomerular primary afferent presynaptic terminals in lamina I are primarily axodendritic (Fig. 4.14), but occasional axosomatic and axoaxonal synapses are seen (Ralston, 1968a, 1979; Ribeiro-DaSilva and Coimbra, 1982). The reports of large numbers of primary afferent synapses on lamina I somata (Kerr, 1970b; Narotzky and Kerr, 1978) are not accepted (Ralston and Ralston, 1979; Gobel *et al.*, 1981).

In summary, subpopulations of lamina I neurons contain certain compounds, the majority of which are transmitter candidates. Labeled cells include both marginal and "other" cells. Axonal projections of lamina I cells are to the thalamus, the midbrain, possibly the cerebellum, and other areas of the spinal cord. Primary afferent input consists of fine fibers coming through the marginal plexus. A major finding is that this lamina is an important area for fine myelinated fiber terminations, although many unmyelinated fibers end here as well. Laminar differences are seen in cutaneous, muscle, and visceral input. Many different compounds can be localized to axons and synaptic terminals in this lamina. The usual speculation is that these are neurotransmitters or neurotransmitter related, but further work must be done before this can be accepted. Axonal terminals in lamina I are characterized by small, round, clear vesicles, in contrast to deeper laminae, but large dense-core vesicles are also found in many terminals.

LAMINA II (THE SUBSTANTIA GELATINOSA)

The "substantia gelatinosa" is a term coined by Rolando (Cervero and Iggo, 1980) for a grossly distinguishable layer of the superficial dorsal horn. The gelatinous appearance of this area is due to the concentration of small neurons and their processes plus a striking absence of myelinated axons. Rexed (1952, 1954) equated the substantia gelatinosa with lamina II. Szentágothai (1964) pointed out, however, that the cells in lamina III had similar dendritic patterns and axonal projections, and physiologic studies did not show a sharp boundary between laminae II and III (Wall, 1967). Accordingly, these authors suggested that although lamina II and III could be distinguished architectonically, they are a functional unit. Although these observations are still valid, it has become clear that there are important differences between lamina II and III (e.g., different types of primary afferent input and striking differences in the distribution of neuronal markers). Accordingly, the earlier admonitions that there are certain common organizational features of the two laminae should not be forgotten, but for now the differences between the laminae are emphasized and most investigators use the term "substantia gelatinosa" as a synonym for lamina II.

Cell Types

The terminology for lamina II cells is complex enough that it is helpful to summarize it in tabular form (Table 4.1).

Before considering these various cell types, some general comments are in order. First, an important distinction is between projection neurons (relay neurons, Golgi type I neurons, etc.) which convey information to distant regions, and local interneurons (intrinsic neurons, local

TABLE 4.1. Summary of Nomenclature of Cell Types in Lamina II

Name	Reference
A. *Classical*	
1. Limiting cells	Ramón y Cajal, 1909; Scheibel and Scheibel, 1968
2. Central cells	
B. *Presently accepted terminology—major cell types*	
1. Stalked cells	Gobel, 1975, 1978b.
2. Islet cells	
C. *Other "Gobel" interneurons*	
1. Arboreal cells Gobel, 1975, 1978b	
2. II–III border cells	
3. Spiny cells	
D. *Cells categorized by dendritic geometry*	
1. Stalked cells	Schoenen, 1982
2. Filamentous cells	
3. Curly cells	
4. Stellate cells	
1–4. Cell types II (1–4) (these types are not named)	Abdel-Maguid and Bowsher, 1984; Bowsher and Abdel-Maguid, 1984
E. *Cells categorized by both dendritic morphology and development*	
1. Projection-propriospinal neurons	Bicknell and Beal, 1984
a. Limiting cells	
b. Large central cells	
c. Small central cells	
d. Transverse cells	
2. Intrinsic neurons	
a. Stalked cells	
b. Islet cells	
c. Inverted stalk-like cells	
d. Vertical cells	

circuit neurons, Golgi type II neurons, etc.), which are enclosed within a structure and have a local integrative function (Rakić, 1975). In many areas of the nervous system this is a clear dichotomy, but not in the dorsal horn. One problem is that most lamina II neurons seem to have extensive local axonal arbors. Thus, no matter how far the axon projects, there is some local integration. Another problem is that the axonal trees of "short-axoned cells" do not seem particularly short. The axons usually branch several times, and even for the smallest cells, some branches travel a considerable distance. Thus, the old distinction between Golgi type I projection cells and Golgi type II local integrative cells is blurred because most cells can presumably do both. Second, there is an increasing appreciation of the multiple projection targets of lamina II neurons. For example, some lamina II cells project out of the spinal cord. These would be "true" projection neurons. Other lamina II cells project to distant segments. These are propriospinal cells. Third, many lamina II cells project to other laminae in the same or nearby segments. These are not generally called projection cells, but they are carrying information a considerable distance from the cell body. Finally, there are cells whose axons are restricted to a single lamina. The axons of both these last types can be restricted to the dendritic domains of these cells. Thus, both might be called Golgi type II cells even though functionally they are different. We suggest that the terms "intralaminar interneurons" and "interlaminar interneurons" be used to express this distinction. It should also be understood that, a cell can fit into several or all of these categories. These considerations may seem esoteric, but questions of this type, particularly pertaining to Golgi type II cells, permeate structural studies of the dorsal horn, particularly lamina II, where the smallest cells are located. Thus, it is important to keep the distinctions in mind as the various cell types are discussed.

Classic Cell Types. The neurons of the gelatinosa were divided into limiting (limitrophe) and central cells by Ramón y Cajal (1909) (Fig. 4.15). This terminology was widely accepted until the middle 1970's.

Limiting (Limitrophe, Border) Cells. Limiting cells are larger than central cells, and their cell bodies are located in the outer parts of lamina II, particularly at the lamina I–II border. It was on this basis that Rexed (1952) divided lamina II into an outer part with large (limiting) cell bodies and an inner part with smaller (central) cell bodies. The dendrites of these cells pass anteriorly and laterally as well as tangentially in the upper part of the dorsal horn (Ramón y Cajal, 1909; Scheibel and Scheibel, 1968) (Fig. 4.15). Thus, these cells have characteristics of dendritic organization that seem to be a mixture of marginal (lamina I) and central cell organizations (Scheibel and Scheibel, 1968). The axonal destinations of these cells are not clearly known. Axons of stalked cell, which are the modern equivalent of limiting cells, end in lamina I (Gobel, 1975, 1978b).

Central Cells. The name "central cell" is a slight misnomer because these cells are found throughout the gelatinosa and not just in central areas. These are the common cells of the gelatinosa. The cell bodies are small and have relatively little cytoplasm, and the proximal dendrites usually arise from the dorsal and ventral poles of the cell (Fig. 4.15). This gives a radially organized appearance to lamina II in transverse sections following cytoarchitectonic stains.

The central cells are among the smallest in the spinal cord. Despite this, the dendrites of

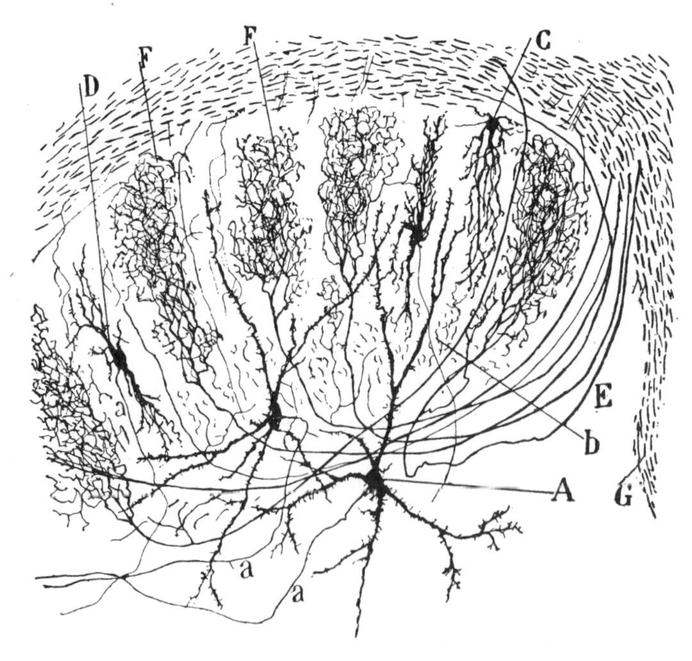

Fig. 4.15. Diagram of the coarse primary afferent collaterals (E) that end in bushy arborizations (F) in the substantia gelatinosa. The endings depicted in F have been termed the flame-shaped arbors. Cell D is a central cell, and cell C is a limiting (limitrophe) cell. (From Ramón y Cajal, 1909.)

these cells branch and rebranch, and the richness of these dendritic branches is an important contributor to the gelatinous appearance of this region (Ramón y Cajal, 1909). Ramón y Cajal (1909) emphasized that the dendritic branches interdigitate between the flame-shaped arbors of the large primary afferent axons (Fig. 4.15). These arbors are now known to be largely confined to lamina III, however (A. G. Brown, 1981). Probably the most striking feature of the dendritic trees of the central cells in early studies is that they are flattened mediolaterally and spread widely in a craniocaudal direction (Ramón y Cajal, 1909; Pearson, 1952; Szentágothai, 1964; Scheibel and Scheibel, 1968). Thus, these dendritic domains are depicted as thin rectangular sheets (Figs. 4.6 to 4.9).

Sugiura (1975) reconstructed several central cells in lamina II from serial Golgi sections (Fig. 4.16). Although not directly disagreeing with the earlier work, which was done primarily with single sections of Golgi-impregnated material, Sugiura (1975) noted that the dendritic trees were much more extensive than previously depicted. The dendritic domains were still somewhat flattened in the mediolateral dimension, but not as prominently as would have been thought from earlier work. As a parenthetical note, it has been a general pattern to find that dendritic and axonal trees are more extensive when reconstructed in serial Golgi sections or by intracellular injection of such markers as HRP than were indicated in earlier Golgi studies which used single histologic sections from young animals.

The central cells were generally split into two types on the basis of the distribution of their axons: funicular cells and short-axoned cells.

Funicular Cells. Funicular cells are central cells whose axons travel in the spinal white matter. Such cells were assumed to be propriospinal. Szentágothai (1964) stated that the axons of these cells reentered the gelatinosa after traveling a variable distance in the white matter, and thus he emphasized that the transmitted information remained within the gelatinosa. This is an

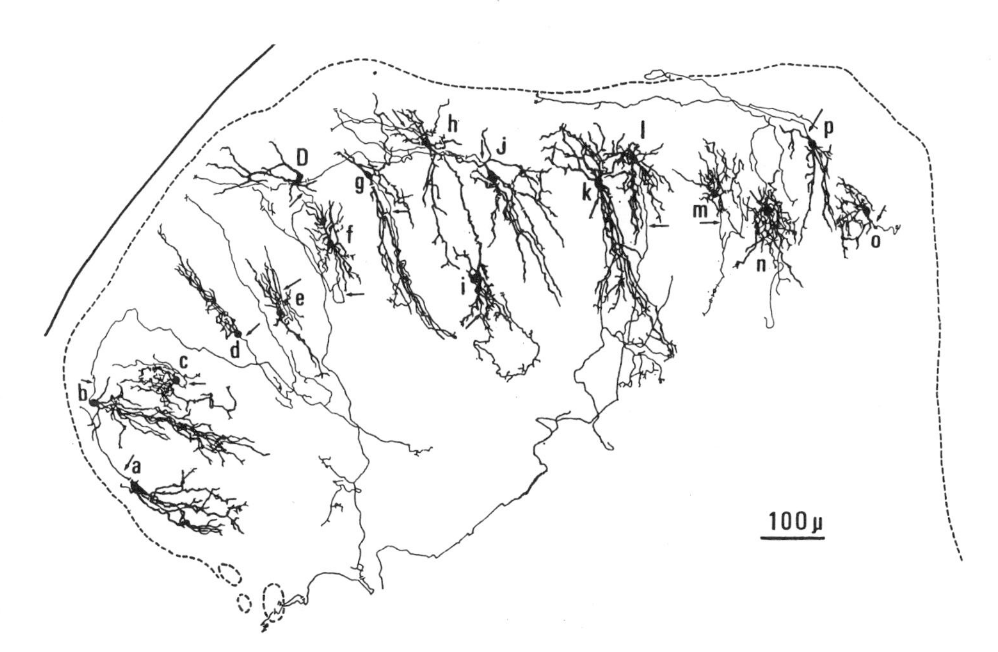

Fig. 4.16. Diagram of cells of lamina II. The cells were reconstructed form serial sections of Golgi stained spinal cord. The letters represent individual cells. Many of the cells have characteristics of both central and limiting cells. (From Sugiura, 1975, reprinted with permission.)

important part of the consideration that the gelatinosa is a closed system (see below). The percentage of funicular as opposed to short-axoned cells is not known, but many central cells from young animals were shown to have axons that travel in the white matter (Lenhossék, 1895; Ramón y Cajal, 1909; Sugiura, 1975). According to Gobel (1979), this is not seen so extensively in adult animals.

Short-Axoned Cells. The axon of a short-axoned central cell ramifies within the dendritic tree and ends near the cell body. Such cells would be Golgi type II cells. There was disagreement about the numbers of such cells in the substantia gelatinosa. Golgi, for example, thought that most of the neurons in the gelatinosa fell into this category. Ramón y Cajal (1909) disagreed because he could find few central cells that did not send their axons into the spinal white matter. This debate over Golgi type II cells is still continuing and will be reconsidered in discussions of the modern classifications of lamina II cell types.

Is the Gelatinosa a Closed System? The idea that the gelatinosa may be a closed system is an important concept that we owe to Szentágothai (1964). Most early investigators found a few gelatinosa neurons whose axons ended in the gelatinosa, but the axons of most gelatinosa cells entered the white matter of the cord (see, e.g., Ramón y Cajal, 1909). From this point their axonal destinations were not known. Szentágothai (1964) then claimed that the axons of these neurons, as well as the axons of the short-axoned cells, ended in the gelatinosa (which was for him both laminae II and III). He concluded that the "substantia gelatinosa proper, i.e., laminae II and III, form a self-contained neuron system, all axons of which either remain in the (gelatinous) substance or return to it . . ." The axons that never enter the white matter form a "short-range intrinsic system" and the axons that enter the white matter and then return form a "longer range longitudinal system." If this were true, axons of the gelatinosa cells would not transmit information out of the gelatinosa, resulting in a lack of "forward transmission," and the gelatinosa would be a "closed" system (Szentágothai, 1964). An important corollary was that the central cells could function only by influencing dorsal dendrites that arise from neurons in deeper laminae (antenna-type neurons). The conclusion was that the substantia gelatinosa is a closed system with its main output through the dorsal dendrites of antenna-type cells in laminae IV and V.

The above idea, although seminal in the development of ideas about the organization of the gelatinosa, needs to be modified. For example, some cells in lamina II have been demonstrated to project to the thalamus or brainstem (Willis *et al.*, 1978; Giesler *et al.*, 1978). Furthermore, stalked cells project to lamina I (see below), and many individually reconstructed or microelectrode-filled cells are seen to project to deeper laminae (Sugiura, 1975; Light and Kavookjian, 1988). Thus, it can no longer be claimed that the gelatinosa is a closed system. Lamina II cells do make extensive synapses on dorsal dendrites from neurons in deeper laminae, however.

Modern Cell Types. The classification of gelatinosa cells into central and limiting categories was generally accepted until the middle 1970s. At that time it became apparent that these categories were too simple. Gobel (1975, 1978a, 1978b) made an extensive study of the gelatinosa of the cat trigeminal nucleus. Unfortunately, he subdivided what we would call lamina II into lamina II and III, (Gobel, 1975, 1978b), rather than IIo and IIi, as most others have done. This causes confusion (A. G. Brown, 1981), and in later papers the terminology was changed (Gobel, 1979; Gobel *et al.*, 1980). Here we will use the IIo and IIi terminology. The cell types proposed by Gobel represent an important advance, and one should not be confused by the laminar terminology. He proposed that the two predominant cell types in lamina II are the stalked and the islet cells. Three other types of interneurons were also identified (see below).

Stalked Cells. (Figs. 4.3, 4.7–4.9, 4.17, and 4.18). Stalked cells are the limiting cells of Ramón y Cajal (1909). They are named because of their short stalklike spines. The cell bodies are located in IIo, especially at the I–IIo junction. The main characteristic of the dendrites of the

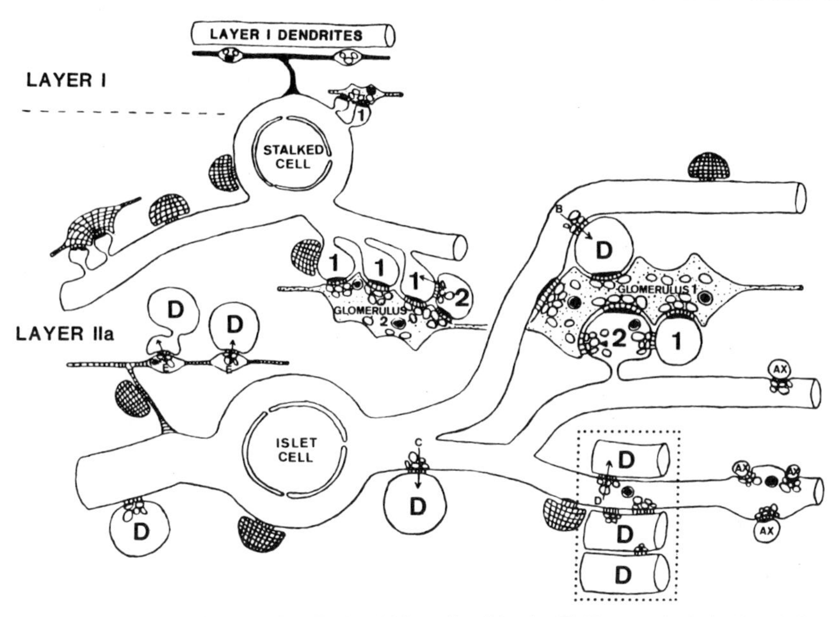

Fig. 4.17. Gobel's summary diagram of stalked and islet cells of lamina II. Note particularly the relation of the dendrites of these cells to the glomeruli. Also note that the dendrites of the islet cells from synapses on other neural processes. Other dendrites participating in these synapses are labeled (1). (From Gobel *et al.*, 1980; reprinted with permission.)

stalked cells is that they form a cone emanating from the cell body and passing ventrally through laminae II-IV (Figs. 4.3, 4.6, 4.7), although Bennett *et al.* (1981) describe these cells as having a fanlike shape. Gobel emphasizes that the major axonal destination of the stalked cells is into laminae I, but axonal ramifications into deeper laminae have also been shown (Schoenen, 1982; Light and Kavookjian, 1988). Thus, these cells are interlaminar interneurons. Gobel (1979) states that the axons of these cells arborize outside their dendritic arbors; he therefore believes that they are not Golgi type II neurons. Gobel (1978b) also emphasizes that lamina I is the site of many long transmission neurons involved in nociception (see Chapter 9) and that these are probably synaptic targets for stalked cell axons. He also suggests that the stalked cells are excitatory interneurons that receive input from primary afferent fibers that end in the central terminals of the glomeruli (see below). This idea needs further confirmation, because some stalked cells contain enkephalin (Bennett *et al.*, 1982), which is a presumed inhibitory transmitter. Recordings from these cells indicate that they are either nociceptive specific or wide-dynamic-range cells (Bennett *et al.*, 1980).

Stalked cells do not possess presynaptic dendrites (Fig. 4.17), which is an important difference from islet cells (Gobel *et al.*, 1980).

Islet Cells. (Figs. 4.3, 4.17, and 4.18). The name "islet cells" was chosen by Gobel (1979) because Ramón y Cajal (1909) described the small central cells in the medullary trigeminal nucleus as being located in clusters or islands. The term "islet cells" is now used for both the medullary trigeminal nucleus and dorsal horn of the spinal cord. Presumably, these would be equivalent to the central cells of the gelatinosa, although other interneurons described by Gobel (1979) are probably also central cells. The dendritic trees of these cells extend for the full thickness of lamina II and are oriented in the rostrocaudal plane (i.e., flattened mediolaterally) (Gobel, 1978b; Schoenen, 1982) (Figs. 4.7, and 4.8). Islet cells whose dendrites are located

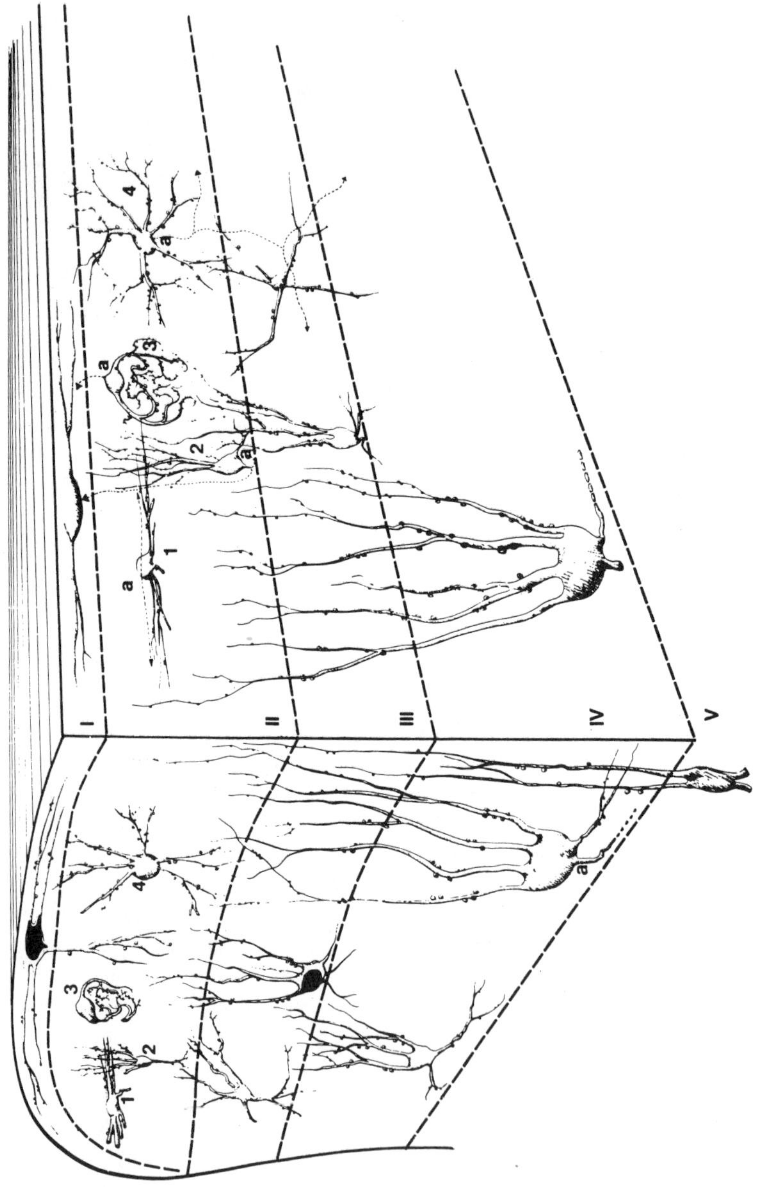

Fig. 4.18. Depiction of the dendritic fields of the dorsal horn neurons in the human. Roman numerals indicate laminae, axons are indicated (a) and the numbers indicate specific cell types. Note particularly the antenna-type neurons in laminae III and IV with dorsal dendrites. (From Schoenen, 1982.)

largely in lamina IIo are reported to be nociceptive specific, and those whose dendrites are in lamina IIi to be mechanoreceptive (Bennett *et al.*, 1980). Other investigators, however, are unable to make such clear correlations of structure with functional type (Light *et al.*, 1979; Woolf and Fitzgerald, 1983).

The axons of islet cells are reported by Gobel (1978b) to end largely in lamina II, and these cells are thought to be Golgi type II cells (Gobel, 1979; Schoenen, 1982; Bicknell and Beal, 1984). Gobel (1978b) suggests that the islet cells are inhibitory interneurons. One extremely interesting feature of these cells is that their dendrites form synapses on nearby dendrites and axonal terminals (Fig. 4.17). These dendrodendritic and dendroaxonic terminals will be considered in the section on the organization of the neuropil of lamina II.

Other "Gobel" Cell Types. Gobel (1978b, 1979) described three other types of interneurons in the gelatinosa (Fig. 4.3). These have not been as widely mentioned as islet and stalked cells in subsequent studies.

Arboreal Cells. Arboreal cells are interneurons that are found predominantly in lamina IIo. They are characterized by extensive dendritic branching in lamina IIo but with branches extending to laminae I and IIi (Gobel, 1978b). The axons of these cells arborize primarily in the vicinity of the dendrites, but some axonal branches also extend into laminae I and IIi.

II–III Border Cells. Border cells are the rarest of the lamina II interneurons. They give rise to dendrites from their medial and lateral poles, and these dendrites arborize extensively in laminae IIo and IIi. The axon is extensively branched and, on the basis of frequency of spines, is regarded as having a maximum effect in IIo and the outer half of IIi.

Spiny Cells. Spiny cells have a spherical or slightly elliptical dendritic domain (Gobel, 1975). The dendrites of spiny cells are reported to be the only ones to enter the magnocellular part of the nucleus caudalis (Gobel, 1975). The axons of spiny cells distribute to laminae I–III.

Gobel (1975, 1978b) suggests that islet cells, arboreal cells, II–III border cells and spiny cells are Golgi type II interneurons. By this he means that the axonal trees are confined to the region of the dendritic arborizations of these same cells (Gobel, 1979). There is, however, an important difference between islet cells on the one hand and the arboreal, II–III border, and spiny cells on the other in that the islet cell axon is confined to the lamina in which the cell body is located whereas the axons of the other cells span laminar boundaries (Gobel, 1979).

One other point in relation to Gobel's findings is that the stalked cells are essentially the same as Cajal's limiting cells. This presumably leaves the other types of interneurons in lamina II, particularly the islet cells, as equivalent to Cajal's central cells. A major difference between the central cells, as described classically, and the more modern subdivisions by Gobel and others, however, is that the axons of most central cells were depicted as entering the spinal white matter, whereas the modern studies do not find this. According to Gobel (1979), this is because Cajal and other early investigators used tissue from embryonic or very young animals.

Cells Characterized by Dendritic Geometry. Schoenen (1982) describes different cell types in the human lamina II from those above (Figs. 4.8 and 4.18). He agrees that there are islet cells, which he regards as Golgi type II cells, but he labeled the other types filamentous cells, curly cells, and stellate cells. These cells are described primarily on the basis of their dendritic domains, which are all longitudinally oriented (Figs. 4.8 and 4.18).

Filamentous Cells. Filamentous cells have a number of fine expansions on their terminal dendrites, thus giving a filamentous appearance to the dendritic trees. The dendritic trees are oriented vertically with cells in IIo having a ventral orientation and in IIi having a dorsal orientation. These cells represent about 20% of the total in lamina II.

Curly Cells. The dendrites of curly cells curve toward each other, giving a curly appearance to the dendritic trees. These are the rarest cells of lamina II, and their axonal trees give relatively few branches in the vicinity of the dendrites.

Stellate Cells. Stellate cells have a simple dendritic tree that spreads in all directions from the cell body, thus leading to a stellate appearance of the cells.

Maguid (1984) and Abdel-Maguid and Bowsher (1984). These investigators subdivide the cells in laminae I–III into 10 types. The four types in the gelatinosa are labeled types II1–II4. These types do not have specific names. Types II3 and II4 are thought to be Golgi type II cells. In addition, some II3 cells are thought to be short propriospinal neurons. Type II1 cells resemble spinothalamic cells, and type II2 cells are of unknown type. Correlating the axonal trees with these dendritic patterns would be desirable.

Cells Characterized by Both Dendritic Geometry and Developmental Age. Bicknell and Beal (1981, 1984) show developmental differences between lamina II cells whose axons enter the white matter (projection/propriospinal neurons) and those whose axons stay within the gray matter (intrinsic/nonprojection neurons). The dendrites of projection/propriospinal cells develop prenatally, and their dendritic arbors increase by elongation and branching. Their axons can be traced (Beal, 1983). These cells are subdivided into limiting, large and small central, and transverse categories. By contrast, the dendrites of the intrinsic cells arise postnatally as short, beaded processes that arise from the cell body. These cells are divided into islet, stalked, inverted stalklike, and vertical categories. There is some difficulty in terminology in that terms "stalked," "limiting," "central," and "islet" are not used exactly as in earlier studies, but the finding of different developmental patterns for these two general cell groups is of considerable interest.

Adequacy of Dendritic Classifications for Lamina II Cells. A number of investigators noted that many of the cells they study do not fit into any category of published cell types (Beal and Cooper, 1978; Light *et al.*, 1979; Woolf and Fitzgerald, 1983; A. J. Todd and Lewis, 1986; A. J. Todd, 1988). In particular, several investigators point out that the dendritic morphology of lamina II cells is a spectrum and suggest that any subdivision into a limited number of specific types may be misleading (Beal and Cooper, 1978; Cervero and Iggo, 1980; Molony *et al.*, 1981). Furthermore, the shape of the dendritic tree and the synaptic ultrastructure around the individual cell does not always seem to correlate with cell function (Light *et al.*, 1979; Woolf and Fitzgerald, 1983; Light and Kavookjian, 1988; Rethelyi *et al.*, 1989), except that the dendrites are found primarily in the laminae of the particular afferent input that excites the cell (Light *et al.*, 1979; Rethelyi *et al.*, 1989). We cannot resolve this lack of structure–function correlation at present and can only suggest that the classifications of lamina II cells into specific categories on the basis of dendritic morphology be treated with caution until clearer functional correlates can be established.

Chemical Localizations

The compounds that are found in lamina II cells will be listed below. Emphasis will be on anatomic insights gained from these studies.

1. Acetylcholinesterase (ACHE) and choline acetyltransferase (ChAT). ACHE- and ChAT-reactive cells have been found in the substantia gelatinosa (Roessmann and Friede, 1967; H. Kimura *et al.*, 1981; Marchand and Barbeau, 1982).

2. Dynorphin. Immunolabeled cells are seen in lamina II, primarily in IIo, as well as laminae I and V (Khachaturian *et al.*, 1982; Vincent *et al.*, 1982; Cruz and Basbaum, 1985; Basbaum *et al.*, 1986b; Cho and Basbaum, 1988, 1989; Ruda *et al.*, 1988; Weihe *et al.*, 1989). The number of these cells increases after experimental inflammation (Ruda *et al.*, 1988) or peripheral nerve and dorsal root section (Cruz and Basbaum, 1985).

3. Enkephalin. Enkephalin-positive cells are seen in lamina II (Hökfelt *et al.*, 1977a,b; Uhl, 1979a; Del Fiacco and Cuello, 1980; S. P. Hunt *et al.*, 1980; Aronin *et al.*, 1981; Finley *et al.*, 1981b; Glazer and Basbaum, 1981, 1982; C. C. LaMotte and DeLanerolle, 1981; Bennett *et al.*, 1982; DeLanerolle and LaMotte, 1982; Ruda, 1982; Senba *et al.*, 1982; Sumal *et al.*, 1982; Arluison *et al.*, 1983a; Conrath-Verrier *et al.*, 1983; Bresnahan *et al.*, 1984; Charnay *et al.*, 1984;

Cruz and Basbaum, 1985; Schoenen *et al.*, 1985a; Standaert *et al.*, 1986; K. E. Miller and Seybold, 1987; Katoh *et al.*, 1988a; Nahin, 1988). Bennett *et al.* (1982) find that both stalked and islet cells in lamina II contain enkephalin.

4. GABA. GABA-ergic cells have been localized, primarily by antisera directed against GAD, in lamina II cells (Ribeiro-DaSilva and Coimbra, 1980; S. P. Hunt *et al.*, 1981a; Barber *et al.*, 1982; Mugnaini and Oertel, 1985). Barber *et al.* (1982) believed that the labeled cells were both stalked and islet cells. Kosaka *et al.* (1988) point out that some GABA cells contain ChAT.

5. Galanin. Galanin has been found in small neurons in laminae II (Rökaeus *et al.*, 1984; Ch'ng *et al.*, 1985; Melander *et al.*, 1986).

6. Glutamate. Glutamate is found in cells in lamina II of the spinal cord (Greenamyre *et al.*, 1984; Weinberg *et al.*, 1987; De Biasi and Rustioni, 1988; K. E. Miller *et al.*, 1988).

7. Neurotensin. Neurotensin cells are found in lamina II (Uhl *et al.*, 1979b; DiFiglia *et al.*, 1982a, 1984; Jennes *et al.*, 1982; Seybold and Elde, 1982; Yaksh *et al.*, 1982b; Vincent *et al.*, 1982; Seybold and Maley, 1984; Cruz and Basbaum, 1985). On the basis of admittedly incomplete staining of the dendrites, the labeled cells have been suggested to be spiny cells (Uhl *et al.*, 1979b); islet cells (Seybold and Elde, 1982; DiFiglia *et al.*, 1984), stalked cells (DiFiglia *et al.*, 1984), and II–III border cells (Seybold and Elde, 1982). Seybold and Maley (1984) emphasize that the labeled cells are at the I–II and II–III laminar borders.

8. Neuropeptide Y, peptide YY, and the pancreatic polypeptides. Neuropeptide Y, peptide YY, and the pancreatic polypeptides are seen in lamina II cells (J. M. Lundberg *et al.*, 1980; S. P. Hunt *et al.*, 1981a,b; Gibson *et al.*, 1984c; Sasek and Elde, 1985; Krukoff, 1987).

9. Serotonin (5-HT). Serotonin in spinal cord axons and terminals is generally attributed to supraspinal sources (Ruda *et al.*, 1986). Nevertheless, some 5-HT-containing cells have been found in the gelatinosa (C. C. LaMotte *et al.*, 1982; Bowker, 1986; Newton and Hamill, 1988).

10. Somatostatin. Somatostatin neurons are found in lamina II (Seybold and Elde, 1980; Dalsgaard *et al.*, 1981; S. P. Hunt *et al.*, 1981a; Senba *et al.*, 1982; Ho, 1983; O. Johansson *et al.*, 1984; Schroder, 1984; Schoenen *et al.*, 1985a; Vincent *et al.*, 1985; Krukoff *et al.*, 1986; Ruda *et al.*, 1986; Papadopoulos *et al.*, 1986b; Charnay *et al.* 1987; Mizukawa *et al.*, 1988). The labeled cells are small and spindle- shaped (Dalsgaard *et al.*, 1981; S. P. Hunt *et al.*, 1981a; O. Johansson *et al.*, 1984; Schoenen *et al.*, 1985; Vincent *et al.*, 1985; Ruda *et al.*, 1986; Mizukawa *et al.*, 1988). Some investigators state that the cells are concentrated in lamina IIi (Dalsgaard *et al.*, 1981; S. P. Hunt *et al.*, 1981a; Ho, 1983). Schoenen *et al.* (1985) state that the somatostatin cells are like islet cells.

11. Substance P (SP). There are many SP neurons in the dorsal horn, lamina II in particular (Hökfelt *et al.*, 1977b; Del Fiacco and Cuello, 1980; Seybold and Elde, 1980; Gibson *et al.*, 1981; S. P. Hunt *et al.*, 1981a; Nagy *et al.*, 1981; Tessler *et al.*, 1981; Senba *et al.*, 1982; Ho, 1983; Ruda *et al.*, 1986; Tohyama and Shiotani, 1986; Katoh *et al.*, 1988a,b). The cells are reported to be relatively small (S. P. Hunt *et al.*, 1981a; Ruda *et al.*, 1986). Cells that contain the mRNA for encoding SP are found in the same locations (Warden and Young, 1988). Enkephalin and SP are colocalized in neurons in lamina II (Katoh *et al.*, 1988a; Senba *et al.*, 1988).

12. TRH. TRH cell bodies are found in laminae I-III (Coffield *et al.*, 1986; Tsuruo *et al.*, 1987). Coffield *et al.* (1986) note that some of the labeled cells in IIi resemble stalked cells and others resemble spiny cells.

13. Vasoactive intestinal polypeptide (VIP). Some VIP neurons are seen in the lateral part of laminae II and III (Gibson *et al.*, 1984a).

In summary, many compounds have been localized to cells in lamina II. These compounds have provided some insights into the anatomic arrangements of these cells, but as yet the label is generally restricted to the cell body, with little visualization of the dendritic and axonal trees. Advances in this area will lead to major increases in our understanding of the organization of the dorsal horn.

Axonal Projections

Early studies were concerned primarily with the question of whether or not cells of the substantia gelatinosa sent their axons into the spinal white matter. This was the basic distinction between short axon and funicular cells. The assumption was that the cells that sent their axons into the white matter were propriospinal and those that did not were local interneurons. Szentágothai (1964) denied that these were two cell types, however, because the axons of both ended in the gelatinosa, although the ones that sent their axons into the white matter projected for longer distances. Then Gobel (1978b, 1979) added detail by renaming the lamina II cells and pointing out that stalked cells project to lamina I, islet cells to lamina II, arboreal cells to laminae I and II, II–III border cells to laminae IIo and IIi, and spiny cells to laminae I–III. Gobel went on to state that the stalked cells were not Golgi type II cells because their axonal trees extended beyond their dendritic arbors, whereas the other four types were Golgi type II cells. He then separated islet cells from the arboreal, II–III border, and spiny cells because the axons of the former were restricted to lamina II whereas the axons of the latter spanned laminae (laminae IIo and IIi being regarded as separate). Later studies by Schoenen (1982), Abdel-Maguid and Bowsher (1984), and Bicknell and Beal (1984) agree with the basic thesis of local interneurons (Golgi type II cells) as contrasted to longer-projecting cells, although these studies did not provide exact axonal terminations.

In summary, gelatinosa cells were classically divided into limiting and central categories. The central cells were further divided into funicular and short-axoned cells. The dendritic trees of limiting cells seemed to be a mixture of marginal and central cell organization. The dendritic trees of the central cells were thought to be flattened, but more complete reconstructions reveal that this may be an oversimplification. Axons of funicular cells entered the white matter, and axons of the short-axoned cells were restricted to the area of the cell body and dendritic tree.

Szentágothai (1964) then suggested that the axonal terminals of both types of central cells ended only in the gelatinosa, which for him was both laminae II and III. This is an important part of the idea that the gelatinosa is a closed system. This idea is not presently accepted, however, because many lamina II cells have been shown to project out of these areas.

In recent years, more detailed subdivisions based on dendritic geometry have been proposed. The most widely accepted is to split the majority of lamina II cells into stalked and islet cell categories, and the remainder into several other subtypes. There is not complete agreement over these types, however. None of the proposed dendritic organization patterns have had physiologic correlates established, except that the laminae into which the dendrites project seem to determine the types of primary afferent input that excites the cell. The axonal projections of lamina II cells have turned out to be more diverse than was classically thought. A useful classification based on axonal destinations would seem to be: (1) cells whose axons are restricted to lamina II (intralaminar interneurons), (2) cells whose axons project to different laminae (interlaminar interneurons), (3) cells whose axons project to other segments (propriospinal neurons), and (4) cells whose axons project out of the spinal cord (projection neurons). The relative numbers of these types are not known, except that category 4 is relatively sparse and some investigators believe that there are also few cells in category 1. These categories are not exclusive. For example, Golgi type II neurons would presumably be concentrated in category 1, but many category 2 cells would also qualify in the sense that their axonal trees are restricted to the region of the dendritic field of that same cell. Furthermore, axons in all four categories of cells have some arborizations in the region of the cell body, and many papers emphasize that these could perform the local integrative functions usually thought to be performed only by Golgi type II neurons. One technical point is that cells reconstructed from serial Golgi sections or after a intracellular filling have more extensive axonal and dendritic trees than was seen after study of single Golgi sections from young animals, which is how most of the classic work was done.

Primary Afferent Input

Primary afferent input into lamina II comes from collaterals that arise from coarse sensory fibers, which are large and myelinated, or fine sensory fibers, which may be either myelinated (Aδ) or unmyelinated (C).

Coarse Primary Afferent Collaterals. Collaterals from large primary afferent fibers are shown in classic studies curving ventrally through the medial part of the dorsal horn and then recurving to end in large flame-shaped arbors in the substantia gelatinosa (Ramón y Cajal, 1909; Szentágothai, 1964; Scheibel and Scheibel, 1968) (Fig. 4.19). These studies were done before the laminae were known or when the gelatinosa was regarded as both lamina II and III. Thus, although the flame -shaped arbors clearly exist (Beal and Fox, 1976; A. G. Brown *et al.*, 1977b; Hamano *et al.*, 1978; Beal, 1979a; A. G. Brown, 1981; Fyffe, 1984; Ralston *et al.*, 1984; Woolf, 1987a), it has been questioned whether they enter lamina II (see, e.g., Hamano *et al.*, 1978; Fyffe, 1984). Several studies report, however, that the distal parts of the flame-shaped arbors enter lamina IIi (Proshansky and Egger, 1977; Light and Perl, 1979b; Beal, 1979a; Woolf, 1987a, Shortland *et al.*, 1989). This is somewhat species dependent in that it is clear in rats and monkeys but not seen or very limited in cats (Beal and Fox, 1976; A. G. Brown *et al.*, 1977b; Proshansky and Egger, 1977; Réthelyi, 1977; Hamano *et al.*, 1978; Ralston *et al.*, 1984; Woolf, 1987a). Work on individual fibers shows that the flame-shaped arbors arise from hair follicle afferents (A. G. Brown, 1981). Thus, information from hair follicle afferents seems to enter IIi as well as deeper laminae in monkeys and rats but possibly not in cats, and in all animals the number of terminals is much smaller than in lamina III. No other types of large primary afferent collaterals have been found entering lamina II.

Fine Primary Afferent Collaterals. Early studies showed large numbers of fine fibers entering the gelatinosa from the surrounding white matter, particularly the tract of Lissauer, and from the marginal plexus (Ramón y Cajal, 1909; Pearson, 1952; Szentágothai, 1964; Scheibel and Scheibel, 1968) (Fig. 4.10). There was debate about whether primary afferent or proprio-

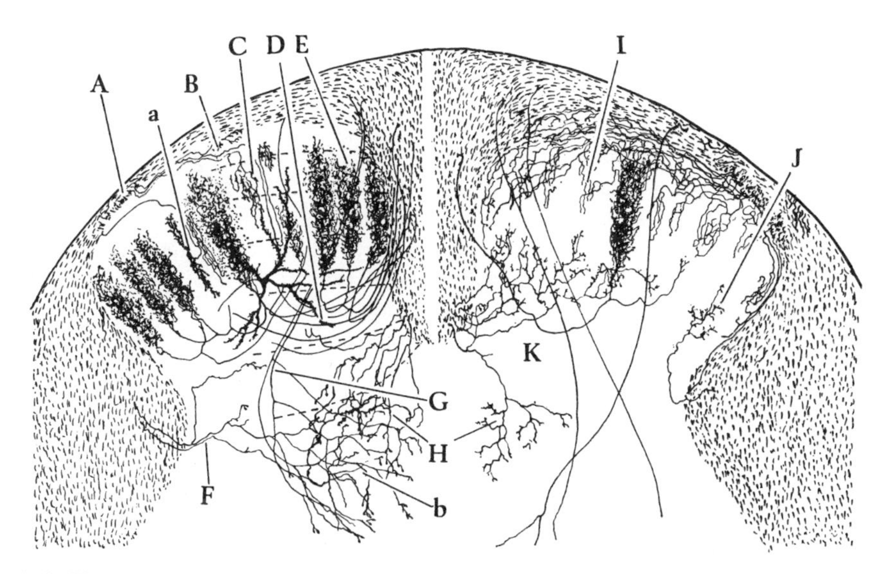

Fig. 4.19. The apical (C) and basal (K) plexuses in the substantia gelatinosa. Note particularly that the fibers that make up K appear to be in lamina III or perhaps IV. This is one of the few depictions indicating a significant fine primary afferent input into deeper laminae of the dorsal horn. (From Scheibel and Scheibel, 1968.)

spinal fibers predominated in this input (see, e.g., Lissauer, 1885, 1886; Ramón y Cajal, 1909; Ranson, 1913b; Earle, 1952; Szentágothai, 1964; Scheibel and Scheibel, 1968). Early opinion was that primary afferent input dominated, particularly with the flame-shaped arbors being so prominent in the gelatinosa. Then the realization that the arbors were restricted largely to lamina III (A. G. Brown, 1981) and the apparent lack of degeneration following dorsal rhizotomy in lamina II (Ralston, 1965, 1968b; Sterling and Kuypers, 1967; Carpenter *et al.*, 1968; Shriver *et al.*, 1968) led to the impression that propriospinal fibers dominated. It has become clear, however, that there is a considerable fine primary afferent input into lamina II (Sprague and Ha, 1964; Szentágothai, 1964; Heimer and Wall, 1968; Réthelyi, 1977; Gobel and Binck, 1977; C. C. LaMotte, 1977; Proshansky and Egger, 1977; Light and Perl, 1977, 1979a; Gobel and Falls, 1979; Ralston and Ralston, 1979; Gobel *et al.*, 1981; K. Chung *et al.*, 1989b), with the single afferent fiber-labeling studies providing probably the most direct evidence (Light and Perl, 1979b; Réthelyi *et al.*, 1982; Ralston *et al.*, 1984; Mense and Prabhakar, 1986; Sugiura *et al.*, 1986, 1989). It is probable that differences in degeneration time of different types of primary afferents were responsible for some of the earlier disagreements (Heimer and Wall, 1968).

Although it is now clear that many fine primary afferent fibers enter lamina II, this does not imply a paucity of propriospinal terminals. What are obviously needed are quantitative estimates of the proportion of propriospinal and primary afferent innervation to lamina II. In this regard, K. Chung *et al.* (1989b) showed that approximately 50% of the synapses in lamina II were lost following dorsal rhizotomy, these presumably being the primary afferents, with the remaining terminals presumably being a mixture of propriospinal and descending input. However, dorsal rhizotomy in the cat is reported to lead to the loss of essentially no synapses, even though primary afferents are removed, presumably because of sprouting (Murray and Goldberger, 1986). Further quantitative work is obviously desirable.

A key issue in recent considerations of the primary afferent input into lamina II is whether the primary afferent collaterals are from fine myelinated or unmyelinated fibers.

Fine Myelinated (Aδ) Primary Afferent Input into Lamina II. Light and Perl (1979b) and Réthelyi *et al.* (1982) labeled fine myelinated (as defined by conduction velocity) mechanical nociceptive fibers and found that they ended primarily in the marginal zone (lamina I) and lamina V. Mense and Prabhakar (1986) also found that fine myelinated nociceptors from deep tissues ended in the marginal layer and lamina V. The conclusion that fine myelinated primary afferents essentially do not end in lamina II is in accord with other morphologic work (C. C. LaMotte, 1977; Ralston and Ralston, 1979) and physiological studies (Kumazawa and Perl, 1977). The arguments that significant numbers of fine myelinated primary afferent fibers end in lamina II depend on differential degeneration times (Gobel and Binck, 1977) or on diameters of parent axons that give rise to terminal branches in lamina II (Gobel and Falls, 1979). The latter observations, although undoubtedly correct, are more indirect than the single-fiber work. Thus, the conclusion would seem to be that fine myelinated (Aδ) primary afferent input into lamina II is relatively sparse.

Unmyelinated (C) Primary Afferent Input into Lamina II. It is clear that numerous fine axons enter the gelatinosa from the tract of Lissauer and other parts of the spinal white matter. There was, as stated above, argument about how many of these fibers were primary afferents, with some studies indicating a relative lack of primary afferent input. It is now accepted, however, that large numbers of unmyelinated primary afferent fibers end in lamina II. Organizational insights are that presumed propriospinal fibers form synapses "en passant" whereas presumed primary afferents form "boutons terminaux" and that fine primary afferent unmyelinated fibers end in a shower of boutons that form sagittally oriented sheets (Scheibel and Scheibel, 1968; Réthelyi, 1977; Réthelyi and Capowski, 1977). Of particular interest are studies based on individual fibers. In the first study of unmyelinated axons as defined by conduction velocity, one high-threshold mechanoreceptor, two polymodal nociceptors, two mechanical and cold nociceptors and two low-threshold mechanoreceptors were shown with endings in IIo (Sugiura *et al.*,

1986). Many of these fibers also had endings in IIi, several had endings in lamina I and a polymodal nociceptor ended in laminae III and IV as well as I and IIo (see Chapter 5). The authors concluded that the terminals of functionally defined sensory fibers appear in a specific region of the spinal cord and that the termination patterns of a functional type of fiber are characteristic (Sugiura *et al.*, 1986), although more data would be desirable for this conclusion. The authors stressed that lamina II appears to be the main projection area for unmyelinated primary afferent fibers from the skin, but it seems clear from their work that other laminae are also the targets of these axons. In the second study, unmyelinated visceral afferents were labeled (Sugiura *et al.*, 1989). The fibers terminated in laminae I and II, as well as deeper laminae. Thus, lamina II also receives unmyelinated afferents from noncutaneous areas. The numbers of terminals for each fiber are much sparser for visceral than for cutaneous fibers, but the visceral fibers distribute to a wider area of the cord. The suggestion is that this may be related to the poor localization of visceral sensation. Clearly, more work of this type is needed, but the technical obstacles are formidable.

Neuropil Organization

The most prominent structures in the neuropil of lamina II are glomeruli, so these will be considered first. This will be followed by a consideration of the "nonglomerular" neuropil.

Glomeruli. A glomerulus consists of a central terminal, which is a primary afferent ending, in synaptic contact with several peripheral dendrites and axonal terminals (Figs. 4.20–4.28). Because the peripheral structures (mostly dendrites) indent the central terminal, the name "scalloped ending" is sometimes used synonymously with central ending. The whole complex is partially separated from the surrounding neuropil by glial processes (Fig. 4.21). These synaptic arrangements were previously given other names ("complex synaptic arrays", [Ralston, 1965, 1968a, 1971]; "central synaptic complexes" or "large synaptic complexes", [Réthelyi and Szentágothai 1969]), but the term now in general use is "glomeruli". The glomeruli are regarded as key structures of the dorsal horn because they offer a morphologic basis for a more complex modulation of information transfer than do the more common axodendritic synapses.

Although the history of glomerular architecture is not a primary goal of this chapter, understanding is facilitated if the early arrangements are presented as a basis for the more complicated picture that is presently emerging. A relatively simple glomerular arrangement (Fig. 4.21) shows a central terminal ending on several surrounding dendrites, with a peripheral terminal ending on the central terminal. Note that the vesicles in the central terminal are round whereas those in the peripheral terminal are flattened. This organization was delineated in early studies (Ralston, 1965, 1968a, 1971; Kerr, 1966, 1970a, 1975a,b). The work of Coimbra *et al.* (1974) is essentially compatible, except that these investigators point out that the peripheral terminals may be of different kinds (Fig. 4.23). A dissenting view was that the synapses between the central and peripheral terminals went the other way, that is the central terminal synapsed on the peripheral terminals (Réthelyi and Szentágothai, 1969, 1973; Knyihár and Gerebtzoff, 1973; Knyihár *et al.*, 1974) (Fig. 4.22). These latter formulations are not presently accepted, but they serve as forerunners of recent findings that the relation between the peripheral and central terminals are more complicated than previously envisaged.

More recent studies widen our understanding in three main areas: (1) the central terminals, which have been shown to be of several types; (2) the surrounding dendrites, some of which make synaptic endings on nearby processes; and (3) the peripheral terminals, which can be either presynaptic or postsynaptic to central terminals and surrounding dendrites.

Central Terminals. Central terminals of glomeruli are "boutons en passage" (Ralston, 1971; Knyihár and Gerebtzoff, 1973). The axons that give rise to the central process are usually thought to be unmyelinated (Ralston, 1971), but Knyihár and Gerebtzoff (1973) mention that the

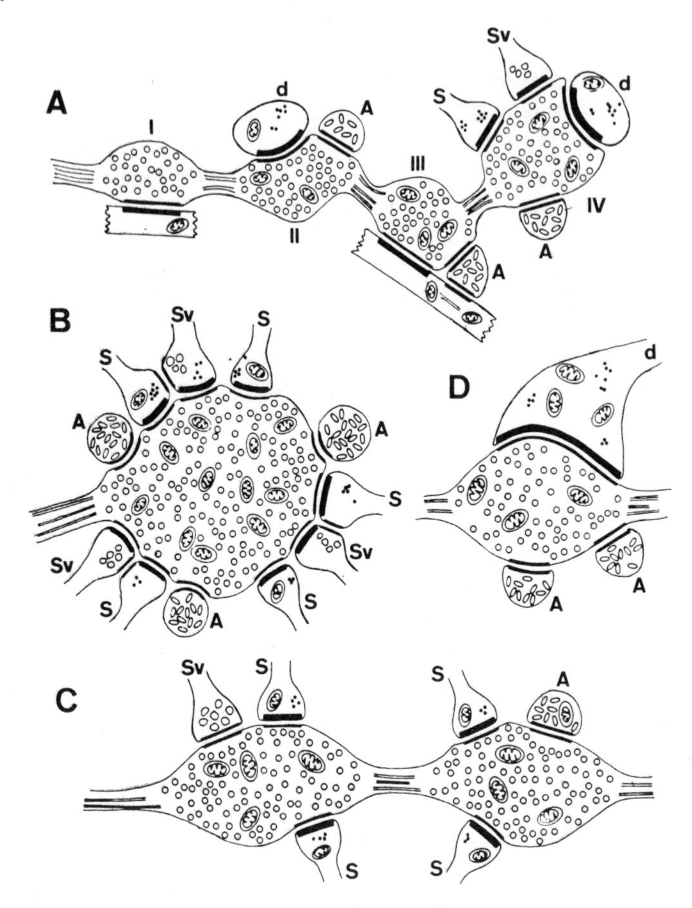

Fig. 4.20. Schematic drawings of the different synaptic organizations that surround primary afferent fibers. Note in particular the numerous axoaxonic, dendrodendritic and dendroaxonic terminals. A = Axon; d = dendrite; and S = dendritic spine. The complex in panel B is a typical glomerulus. (From Maxwell and Réthelyi, 1987.)

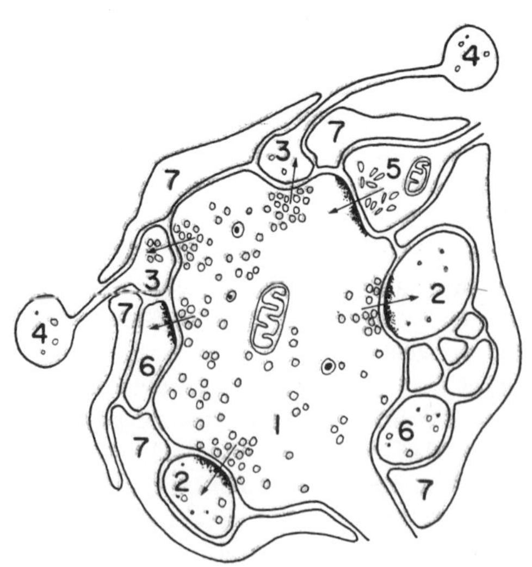

Fig. 4.21. Diagram of a glomerulus. The components are as follows (1) the central terminal; (2) dendrites, probably from antenna-type neurons; (3 and 4) dendritic spine heads presumably from substantia gelatinosa neurons; (5) a peripheral terminal filled with flattened vesicles; (6) an unknown dendrite; and (7) glial processes. (From Kerr, 1975a.)

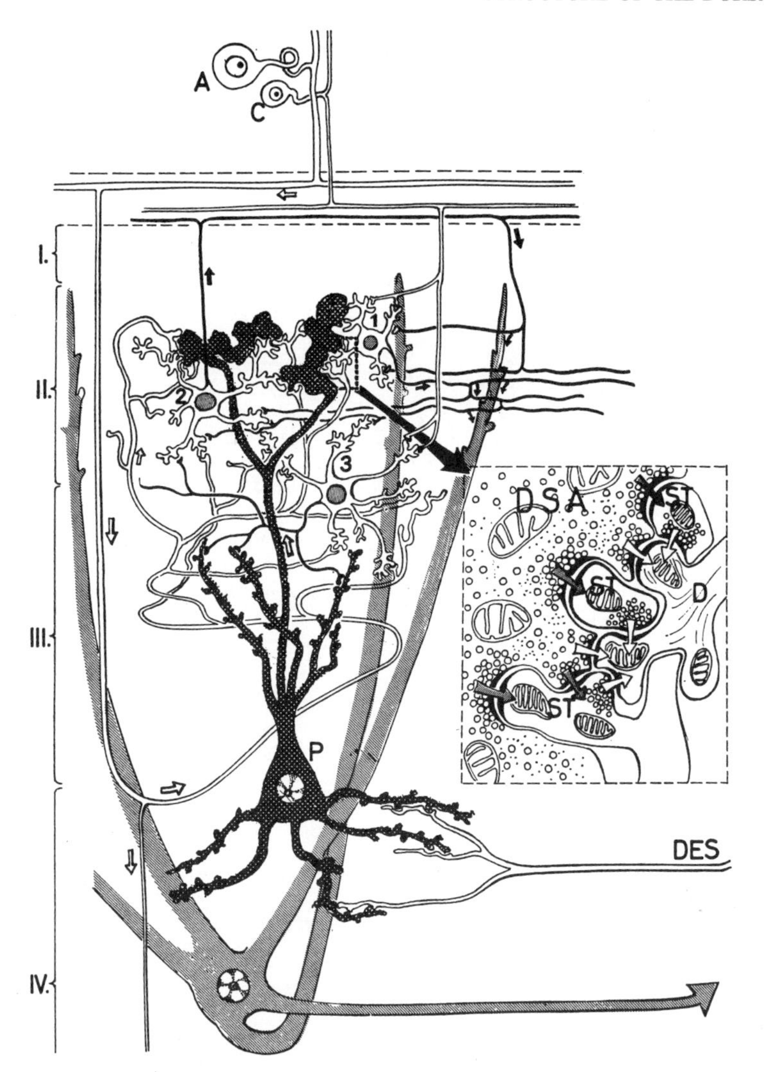

Fig. 4.22. Another schematic diagram of a glomerulus in the substantia gelatinosa. In the large diagram, 1, 2, and 3 are neurons of the gelatinosa and P is a large pyramidal neuron at the junction of laminae III and IV. In the inset, DSA is a terminal arising from neuron P, ST is a small-axon terminal, and D is a dendrite of a substantia gelatinosa neuron. The main thrust of this drawing is that the dense terminal arises from an intrinsic spinal neuron, which is no longer accepted, but this depiction is a forerunner of the more complex glomerular patterns that are illustrated below. (From Réthelyi and Szentágothai, 1973.)

preterminal fibers are sometimes covered with a thin myelin sheath. Early work did not subdivide the central terminals, but more recently several morphologic types have been described. We will try and provide a synthesis, but some of the descriptions are slightly different and there are species differences, so this work will be presented individually where necessary.

Dense Terminals. Dense terminals are the common central terminals that have been described ever since lamina II was examined in the electron microscope (Ralston, 1965, 1968a, 1971, 1979; Kerr, 1966, 1970a, 1975a; Coimbra *et al.*, 1974) (Figs. 4.17, 4.22, and 4.25). The axoplasm is relatively electron dense, hence the name. They were called dense sinusoidal

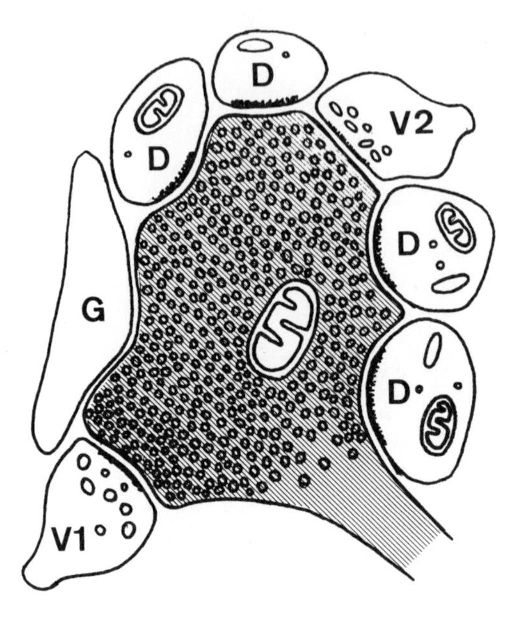

Fig. 4.23. Diagram of glomerular organization according to the study of Coimbra *et al.* (1976). The central terminal is darkened, processes D are "ordinary dendrites," V1 is a peripheral terminal that contains clear vesicles of various sizes, V2 is a peripheral terminal with mostly flattened vesicles, and G is a glial process. (From Coimbra *et al.*, 1974.)

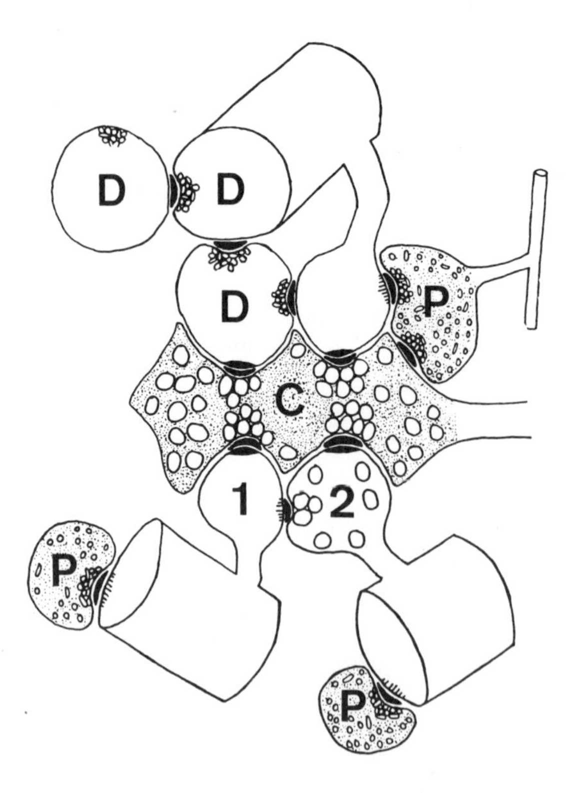

Fig. 4.24. Gobel's schematic diagram indicating that dendrites forming synapses on each other are an important part of glomerular organization. Abbreviations: C, central terminal; 1 and 2, dendritic spines contacted by the central terminal; D, dendritic shafts (dendrites), some of which are also contacted by the central element (note that some of the dendritic spines and shafts make synaptic contacts with each other); P, peripheral endings. (From Gobel, 1974; reprinted with permission.)

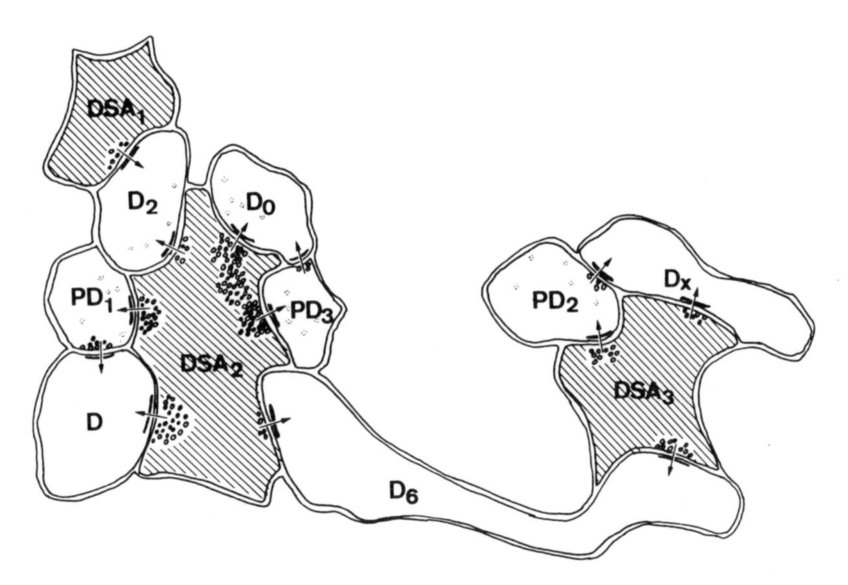

Fig. 4.25. Schematic drawing of the synaptic arrangements around two primary sensory terminals in lamina II of the monkey. In particular, note the "triads" where two dendrites in synaptic contact with one another are contacted by the central terminal. (From Knyihár-Csillik *et al.*, 1982; reprinted with permission.)

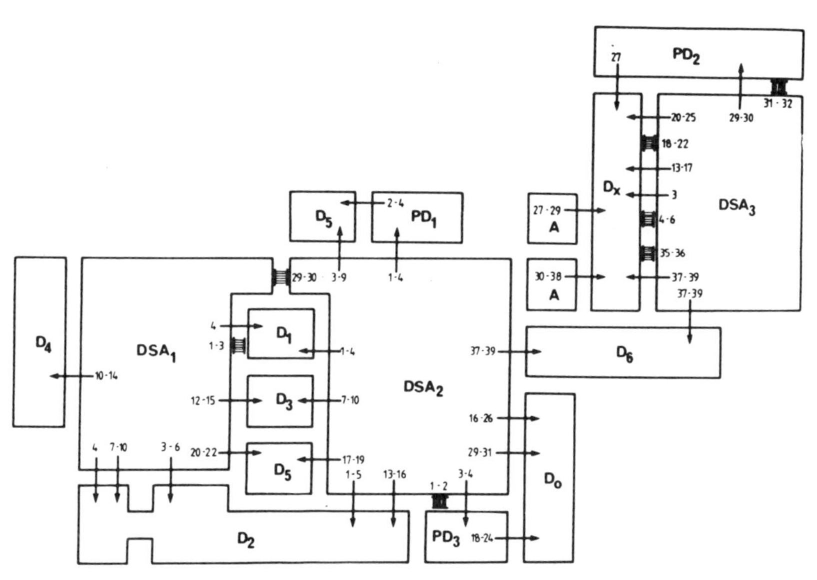

Fig. 4.26. Block diagram constructed from serial sections of glomerular arrangements around primary afferent terminals in the monkey dorsal horn. The major point here is to indicate the complexity of the synaptic arrangements revealed in serial sections. Abbreviations: D = dendrites; PD = presynaptic dendrites; A = axons; DSA = electron-dense sinusoid axon terminals that arise as superficial collaterals from primary sensory fibers. (From Knihár-Csillik *et al.*, 1982; reprinted with permission.)

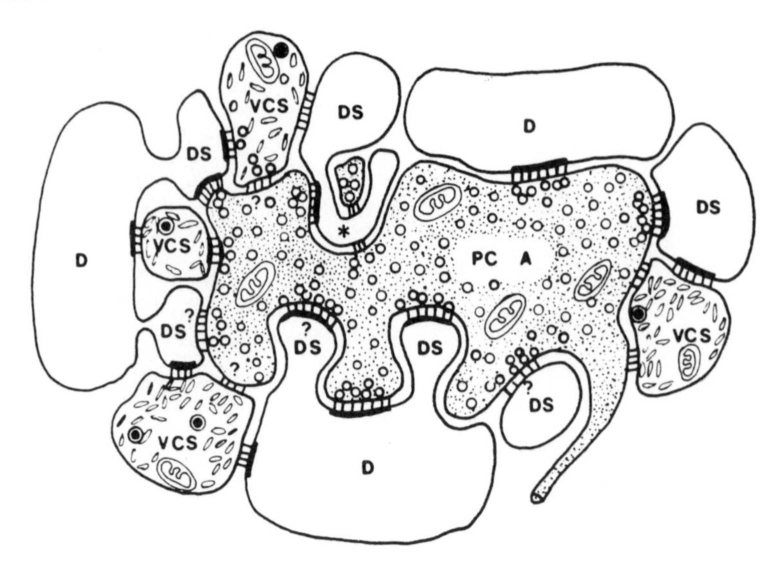

Fig. 4.27. Glomeruli were previously thought to occur primarily in relation to fine sensory axons. It has become clear, however, that typical glomerular arrangements are found around some large sensory fibers. This figure is a schematic diagram of the terminals that surround an individually filled Pacinian corpuscle afferent (PC A) in lamina IV. Abbreviations: D = dendrites, DS = dendritic spine, VCS = vesicle-containing structures. (From Semba *et al.*, 1984.)

terminals by Knyihár-Csillik *et al.* (1982a). They contain predominantly round, clear vesicles of varying sizes but whose average diameter is approximately 50 nm. Numerous mitochondria and some large dense-core vesicles are also seen. The electron density of the axoplasm appears to be intrinsic to the axoplasm, but some investigators also note fine filaments in the axoplasm (Knyihár and Gerebtzoff, 1973).

Light Terminals. Ralston (1979) noted that central terminals could be subdivided on the basis of the electron density of the axoplasm into those with a light axoplasmic matrix, and those with a dark or dense axoplasmic matrix. He did not distinguish between the light and dark central terminals on the basis of differences in vesicle population or synaptic connections, but he did note that light central terminals were found deeper in the dorsal horn. A distinction between light and dark central terminals was also seen in the rat by Ribeiro-DaSilva and Coimbra (1982),

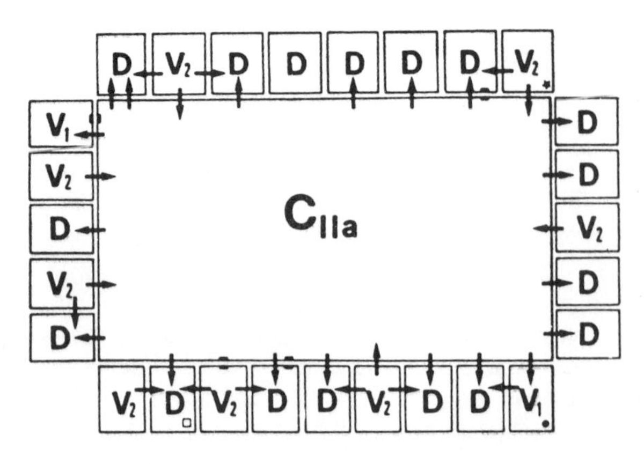

Fig. 4.28. Glomeruli in rats are as complex as those in higher animals. This figure is a schematic depiction of glomerular organization in the rat superficial dorsal horn. Dyads and triads, as indicated by double and triple arrows are emphasized. Abbreviations: CIIa = a IIa type of central terminal; D = dendrite; V = a peripheral terminal, which has two types. (From Ribeiro-DaSilva *et al.*, 1985; reprinted with permission.)

who called them electron-dark and electron-lucent terminals, and in the monkey by Knyihár-Csillik *et al.* (1982a,b), who called them regular synaptic vesicle terminals and dense sinusoidal terminals. Both of these investigators noted that the vesicles in the light terminals were round, clear, relatively uniform in size, and present in smaller relative numbers than in the dark terminals. Both of these studies confirm that the light terminals were more prevalent in deeper laminae. In the rat, the light terminals were subdivided into those that contained numerous fine filaments (type IIa) and those that did not (type IIb) (Ribeiro-DaSilva and Coimbra, 1982; Ribeiro-DaSilva *et al.*, 1985). Type IIa terminals were found primarily in lamina IIi and type IIb in lamina III.

Large Dense-Core Vesicle Terminals. The superficial laminae of the dorsal horn, particularly in the monkey, are characterized by presynaptic terminals that contain numerous large dense-core vesicles. In the monkey many of these are central terminals in glomeruli (Ralston, 1979; Knyihár-Csillik *et al.*, 1982a).

Are All Central Terminals Primary Afferents? Although there was some early disagreement (Ralston, 1968b; Réthelyi and Szentágothai, 1969), most investigators found that central terminals degenerated following dorsal rhizotomy (Kerr, 1970b; Coimbra *et al.*, 1974; Gobel, 1974; Knyihár *et al.*, 1974; Knyihár and Csillik, 1976; Ralston and Ralston, 1979; Knyihár-Csillik *et al.*, 1982a). One question, however, is whether all such terminals degenerate. The importance of this question is that if some terminals do not degenerate, the survivors might arise from "nonprimary afferents", which would be intrinsic or descending fibers. That all eventually degenerated was claimed by Knyihár-Csillik *et al.* (1982a) and Coimbra *et al.* (1984), but others argued that some survive (Knyihár *et al.*, 1974; Ralston and Ralston, 1979). Coimbra *et al.* (1984) suggested that the surviving central terminals may come from distant uncut roots, which would imply that all central terminals arise from primary afferents. In addition, there are different times and types of degeneration for the different types of terminals (Heimer and Wall, 1968; Ralston and Ralston, 1979; Knyihár-Csillik *et al.*, 1982b; Coimbra *et al.*, 1984), which may have given the impression that some survived if they were examined early. It must be remembered, however, that 5HT-containing central terminals have been reported (Ruda and Gobel, 1980), and 5-HT is thought not to be in primary afferents (but see Chapter 3). Thus, the preponderance of evidence indicates that all central terminals arise from primary afferents, but the issue is not completely settled.

Morphologic Markers in Central Terminals. The following substances have been found in central terminals of glomeruli: CGRP, cholecystokinin, glutamate, serotonin, SP, and FRAP. (references are given in the chemical neuroanatomy section). One point of interest is that serotonin, which is not supposed to be found in primary afferents, labels some central terminals (see above). This may, as previously stated, imply that not all central terminals are of primary afferent origin.

Laminar Localization of Central Terminals. In the monkey, the large dense-core vesicle central terminals are found predominantly in lamina I, dense terminals in lamina II, and light terminals in lamina III (Knyihár-Csillik *et al.*, 1982b). In the rat, dense terminals are found in the ventral part of IIo, light terminals without filaments in IIi, and light terminals with filaments in lamina III (Coimbra *et al.*, 1984).

Axonal Types That Give Rise to the Central Terminals. In relation to the above laminar localizations, Knyihár-Csillik *et al.* (1982b) suggest that the dense and large dense-core vesicle terminals arise from fine primary afferent axons (the superficial and marginal collaterals of Cajal, respectively). Coimbra *et al.* (1984) suggest that the dense terminals arise from unmyelinated fibers, the light terminals without filaments from fine myelinated fibers, and the light terminals with filaments from large recurrent myelinated fibers. These correlations are of considerable interest, but more evidence is necessary for confirmation.

Peripheral Dendrites. All investigators agree that most of the processes that surround the central terminal are dendrites from cells in the spinal cord (Figs. 4.20–4.28). Speculation about which particular cells give rise to the dendrites centers on large cells from laminae IV and V (the antenna neurons of Réthelyi and Szentágothai [1969, 1973]) (Fig. 4.22) and neurons of the gelatinosa. An advance in our understanding of this aspect of glomerular organization came with the observation that there were two types of dendrites: (1) a typical dendrite that received only presynaptic terminals, and (2) an atypical dendrite that, in addition to receiving presynaptic terminals, had synaptic vesicles and formed synapses with surrounding dendrites and the central terminal (Figs. 4.17, and 4.24) (Gobel, 1974, 1976; Gobel *et al.*, 1980). The typical dendrites were reported to arise from stalked cells and the atypical dendrites from islet cells (Fig. 4.17), but the fact that some may arise from antenna type neurons in laminae IV and V must not be forgotten. Further studies with individually labeled spinal interneurons will be necessary to determine this.

Peripheral Terminals. Peripheral terminals are axonal endings that are in synaptic contact with central terminals in glomeruli. Thus, these terminals form axoaxonic complexes with the central terminal. Axoaxonic terminals are also found in the general neuropil (Zhu *et al.*, 1981), although they are more frequently found in glomeruli. The typical peripheral terminal contains flat vesicles and makes a symmetric synapse with the central terminal (Figs. 4.19–4.27). Thus, these terminals have the morphologic characteristics of inhibitory synapses. These are referred to as V2 endings by Coimbra *et al.* (1974), P endings by Gobel (1974), and F endings by Knyihár-Csillik *et al.* (1982a). The cells of origin for these terminals are unknown, but the terminals do not seem to be affected by dorsal rhizotomy. Some of these terminals contain GABA or GAD (Barber *et al.*, 1978; Ribeiro-DaSilva and Coimbra, 1980; Carlton and Hayes, 1990; Todd and Lochhead, 1990). There is a suggestion that there are different types of peripheral terminals (Coimbra *et al.*, 1974; Ribeiro-DaSilva *et al.*, 1985). Care must be taken not to confuse peripheral terminals with presynaptic dendrites

Glomerular Circuitry. Further insight into glomerular architecture has been obtained with serial electron-microscopic sections. Knyihár-Csillik *et al.* (1982b) (Figs. 4.25, 4.26), found that it was often difficult to make definitive identifications of presynaptic dendrites in the monkey without serial reconstructions Figs. 4.25 and 4.26. These investigators emphasized "triads" where a central terminal makes contact with two dendrites, one of which is presynaptic to the other (Fig. 4.25). These investigators also noted that the organization of the processes around each type of presynaptic terminal, although having many common features, is distinctive. They did not find direct synaptic interrelationships of central terminals with one another, but they did note that presynaptic dendritic interactions seem to link primary afferent central terminals. A similar analysis in the rat (Ribeiro-DaSilva *et al.*, 1985) shows equal complexity (Fig. 4.27). The studies emphasize the possibilities for modulation of primary afferent information transfer in the glomeruli. Further work of this type is much needed.

Functional Speculations. An early speculation was that primary afferent impulses arriving in a central ending would depolarize the surrounding dendrites, some of which are from antenna type projection neurons in laminae IV and V and some of which are from central neurons of the gelatinosa (Kerr, 1975a). The gelatinosa neurons would then discharge back onto the central terminal, presumably to produce inhibition of primary afferent excitation, thus shutting off and thereby sharpening the primary afferent information flow to the projection neurons. The discovery of glomerular presynaptic dendrites (Gobel, 1974) (Figs. 4.17, and 4.24) then led to a much greater appreciation of the complexity of the glomerular circuit. Gobel (1976) suggested that those dendritic terminals that synapsed back onto the central terminal (dendroaxonic synapses) might be excitatory. In later work, however, Gobel *et al.* (1980) found that the dendritic synapses were in islet cell dendrites, and islet cells were thought to be inhibitory. If this is true, and if the peripheral terminals are also inhibitory, as is usually postulated, excitation of a primary afferent central terminal would activate a complicated inhibitory system. Physiologic confirmation of these speculations is needed. Insights into glomerular function are important steps in understanding spinal sensory information transfer.

Several more recent publications note that a basic arrangement in the glomerulus can be described as a triad (Knyihár-Csillik *et al.*, 1982a,b; Ribeiro-DaSilva *et al.*, 1985). A triad occurs when a central ending makes synaptic contact with two other processes which are in contact with each other. Such triads have been seen in other parts of the nervous system (Knyihár-Csillik *et al.*, 1982b). From several of these studies comes the notion that triads might be a system which serves to block or shorten transmission at the first primary afferent terminal in the spinal cord (Knyihár-Csillik *et al.*,1982b). Whether or not these speculations turn out to be true, the glomerular apparatus that surrounds the primary afferent terminal in the superficial laminae of the dorsal horn would seem to allow abundant opportunity for both presynaptic and postsynaptic modulation of the primary afferent input.

Nonglomerular Synapses. Although the glomeruli are the most prominent and extensively studied elements of the lamina II neuropil, they are in the minority in terms of total numbers of terminals. Several studies mention, for example, that glomerular terminals make up less than 5% of the total synapses in lamina II (Ralston, 1971; Duncan and Morales, 1973; 1978; Ralston and Ralston, 1979). The large majority of synapses in the neuropil of lamina II are simple and axodendritic in nature. They are not particularly remarkable. The majority of the presynaptic terminals contain round, clear, similar-sized vesicles approximately 50 nm in diameter clustered against the presynaptic membrane (Fig. 4.14). A few large dense-core vesicles are also seen. The cytoplasmic thickenings in these synapses are asymmetric. Such synapses are usually thought to be excitatory. Other axodendritic synapses contain flattened or pleomorphic vesicles and have symmetric cytoplasmic thickenings. Such synapses are usually thought to be inhibitory. It has been pointed out that the round-vesicle terminals predominate in more superficial laminae and the flat-vesicle terminals predominate in deeper laminae (Ralston, 1971). Axosomatic, axoaxonic, and dendrodendritic terminals are also seen (Zhu *et al.*, 1981), but the nonglomerular axodendritic terminals far outnumber these other categories.

In summary, large myelinated primary afferents do not enter lamina II except for distal parts of hair follicle afferents in some species. Fine myelinated afferents also seem to be largely excluded from lamina II. There is an extensive unmyelinated primary afferent input into lamina II. The primary afferent endings seem to fall into two general categories: central terminals in glomeruli and simple axodendritic endings. Most attention has been paid to glomeruli because they seem to offer the morphologic basis for a modification of primary afferent input to the cord. Glomeruli consist of a central ending, which is apparently an *en passant* primary afferent terminal, that makes synaptic contact with a group (usually four to eight) of surrounding dendrites and other (peripheral) axonal terminals, the whole being at least partially set off from the rest of the neuropil by glial processes. A complication is that many of the dendrites, presumably those of islet cells, make synapses either on some of the neighboring dendrites or on the central terminal. The peripheral terminals have the morphologic characteristics of inhibitory synapses and some contain GABA or enkephalin. The central terminals have the morphologic characteristics of excitatory terminals, and some contain CGRP, cholecystokinin, glutamate, 5-HT, SP, or FRAP. The simplest hypothesis would seem to be that the glomerulus is a complicated inhibitory synaptic arrangement that surrounds an excitatory primary afferent ending, but much further work will be necessary to establish this. The more numerous primary afferent axodendritic endings are not particularly remarkable, except that they contain a large number of compounds that are also found in DRG cells.

LAMINA III

Laminae III–VI make up the dorsal horn deep to the substantia gelatinosa. Many different cytoarchitectonic subdivisions of this part of the dorsal horn have been proposed, but with the exception of the Rexed laminae, none has been accepted. These have been summarized by Rexed in his excellent historical reviews (Rexed, 1952, 1954). A major contributory factor to the general acceptance of these laminae was the study of Wall (1967), who noted that the receptive fields of dorsal horn neurons changed in a predictable way from lamina to lamina.

Lamina III forms a broad band across the dorsal horn (Fig. 4.1). In cytoarchitectonic studies, lamina III is distinguished from lamina II by having slightly larger and more widely spaced cells (Rexed, 1952). This boundary is imprecise. In myelin-stained or plastic-embedded material, however, the boundary is sharp because lamina II contains relatively few myelinated fibers and lamina III contains many.

Cell Types

In Nissl-stained sections, the cells of lamina III are spindle-shaped and have relatively little cytoplasm (Ramón y Cajal, 1909; Rexed, 1952). Thus, in this regard they resemble, but are somewhat larger than, the central cells of lamina II. Dendritic patterns of lamina III cells were not distinguished from lamina II cells in early studies, except that the extension of the dendrites of lamina III cells was somewhat greater (Ramón y Cajal, 1909; Pearson, 1952; Szentágothai, 1964; Scheibel and Scheibel, 1968). The main feature of the dendritic trees in these studies is that they were flattened mediolaterally and extended craniocaudally (Figs. 4.4–4.7; (Scheibel and Scheibel, 1968). The first study to consider the dendritic fields of neurons whose cell bodies were clearly located in lamina III was that of Mannen and Sugiura (1976), who reconstructed three such cells in serial sections of Golgi material (Fig. 4.29). The dendritic fields of the identified cells were more complicated than indicated earlier in terms both of dendritic branching and of spanning more laminae, and they seemed only to be slightly flattened in a mediolateral direction. A similar complexity of dendritic organization was reported by Beal and Cooper (1978), although they confirmed the general craniocaudal orientation reported by earlier investigators. They stated that the dendrites of these cells formed three longitudinally oriented plexuses, one in IIo, one in IIi and the outer part of III, and one in the inner part of III. The suggestion was that

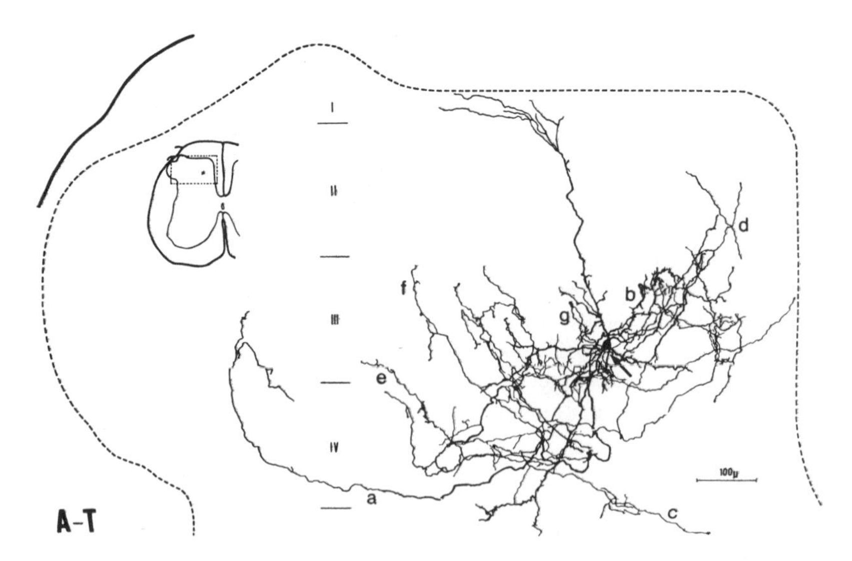

Fig. 4.29. Reconstruction from serial sections of a lamina III cell. The dendritic and axonal arborizations are more complex than the depictions by earlier investigators. (From Mannen and Sugiura, 1976; reproduced with permission.)

these different strata might be related to the different types of primary afferent input. Finally, they showed that some of the cells in ventral lamina III sent dendrites dorsally so that these cells, as well as cells in deeper laminae, could be classed as antenna-type neurons.

A major advance in our understanding of the dendritic and cellular architecture of lamina III came when two particular cell types, spinocervical tract cells and postsynaptic dorsal column cells, were demonstrated in this lamina (A. G. Brown *et al.*, 1977b; A. G. Brown and Fyffe, 1981; A. G. Brown, 1981) (Fig. 4.7). Although it is agreed that lamina III contains these two cell types, as well as a large population of as yet unidentified cells, there has been disagreement about the dendritic organization of the identified cells. The work of the Brown group indicates that the spinocervical and postsynaptic dorsal column cells are quite distinctive (A. G. Brown *et al.*, 1977b; A. G. Brown and Fyffe, 1981; A. G. Brown, 1981; Maxwell *et al.*, 1982b). The spinocervical tract cells have their dendrites oriented more in the longitudinal than in the transverse plane (i.e., flattened mediolaterally), in agreement with the depictions of Scheibel and Scheibel (1968), and although they have many dorsal dendrites, these dendrites do not travel into lamina I or IIo (A. G. Brown, 1981) (Fig. 4.7). A. G. Brown (1981) points out that it almost appears as if there was a barrier to the dorsal spread of these dendrites. An important consequence is that these cells would not get direct input from many fine afferent fibers (A. G. Brown, 1981). By contrast, dendrites of the postsynaptic dorsal column cells do travel dorsally into laminae II and I, and their dendritic trees are not restricted mediolaterally. These cells are depicted as dorsoventrally oriented cylinders or cones (Fig. 4.7), and because of their dorsal dendrites they would be antenna type neurons. Thus, these cells could have extensive monosynaptic contacts from fine afferent fibers.

The above patterns were not confirmed by Bennett *et al.* (1984), who reported that spinocervical tract and postsynaptic dorsal column neurons had a similar dendritic organization and suggested that a single cell might give rise to both a dorsal column and a spinocervical tract axon. A. G. Brown *et al.* (1986b) disagreed but showed that spinocervical tract cells made excitatory connections onto postsynaptic dorsal column cells. It is difficult for us to make a judgment. Both studies show high technical skill. Further work is necessary to resolve the issue.

Several studies have dealt with the dendritic fields of lamina III cells that are not identified, but are clearly not spinocervical or postsynaptic dorsal column neurons. A. G. Brown (1981) points out, for example, that there are pyramidal neurons in lamina III whose dendrites extend dorsally through laminae I–II and ventrally through laminae IV–V (Fig. 4.7). These cells are flattened in the transverse plane and in this regard are the opposite of the longitudinally oriented cells of Scheibel and Scheibel (1968). Such a cell could receive direct synaptic input from essentially all types of primary afferent fibers because its dendrites span all the laminae of the dorsal horn. In Golgi material, Schoenen (1982) found essentially two types of cells, one with dorsal dendrites (antenna type cells) and one with radial dendrites, but both of these types are simpler than similar cells in lamina II. Similarly, Bowsher and Abdel-Maguid (1984) found a Golgi type II neuron that resembles Schoenen's radial cells and one with dorsal dendrites that resembles the candelabra cells of Schoenen (1982). Maxwell (1985) also found two cell types, one whose dendrites arborizes dorsoventrally with the dorsal dendrites penetrating lamina II and one whose dendrites are confined to lamina III. Finally, Beal *et al.* (1988) note that there are two cell types, one that develops before birth and sends axons into the white matter (these are thought to correspond to the types described by A. G. Brown, [1981]) and one that develops considerably after birth and is regarded as a local circuit neuron.

Some studies have been done correlating dendritic morphology with function. For example, there are the reports of the dendritic organization of spinocervical and postsynaptic dorsal column neurons (A. G. Brown *et al.*, 1977b; A. G. Brown and Fyffe, 1981; A. G. Brown, 1981; Bennett *et al.*, 1984) and of the synaptic types that end on these cells (Maxwell *et al.*, 1984b). For cells that are neither of the above types, however, the shape of dendritic trees did not seem to correlate in any obvious way with the function of the cell (Maxwell *et al.*, 1983a).

Axonal Projections

In early studies, the projections of lamina III cells were thought to be essentially the same as those of lamina II cells. Thus, lamina III cells were thought to send axons either to the substantia gelatinosa or to the white matter that surrounds the dorsal horn (Ramón y Cajal, 1909; Szentágothai, 1964; Scheibel and Scheibel, 1968). Szentágothai (1964) added two caveats. First, some axons were shown to cross the midline and terminate in the opposite gelatinosa. Second, axons that entered the white matter were thought to reenter the gray matter and again terminate in the gelatinosa (it should be remembered that the gelatinosa for Szentágothai was both laminae II and III). The key point here was that axons of the laminae II and III cells ended only in these laminae.

Evidence against the idea that lamina III cells projected only to the gelatinosa was quickly obtained. For example, Matsushita (1969, 1970) showed that many lamina III cells project to deeper laminae (Fig. 4.30). Furthermore, reconstructions from serial sections of Golgi material showed a much wider and more detailed axonal arborization than was previously appreciated (Fig. 4.30). In particular, branches of some lamina III cells clearly go to lamina I and IV (Mannen and Sugiura, 1976), (Fig. 4.29). Intracellular injections and retrograde filling studies clearly show that significant populations of lamina III neurons project to the spinocervical nucleus and into the dorsal columns (A. G. Brown *et al.*, 1977b; A. G. Brown and Fyffe, 1981; A. G. Brown, 1981; Bennett *et al.*, 1984; [see above]). In addition, other cells in lamina III also have extensive axonal arborizations. For example, an individually injected cell in lamina III of the cat had axonal ramifications in laminae IV–VI plus branches in the ipsilateral dorsolateral funiculus and the contralateral ventral funiculus (Light and Kavookjian, 1988). It is probable that some of the Golgi type II cells identified by Schoenen (1982), Bowsher and Abdel-Maguid (1984), and Maxwell *et al.* (1983a) have axons restricted to lamina III, but it seems clear that

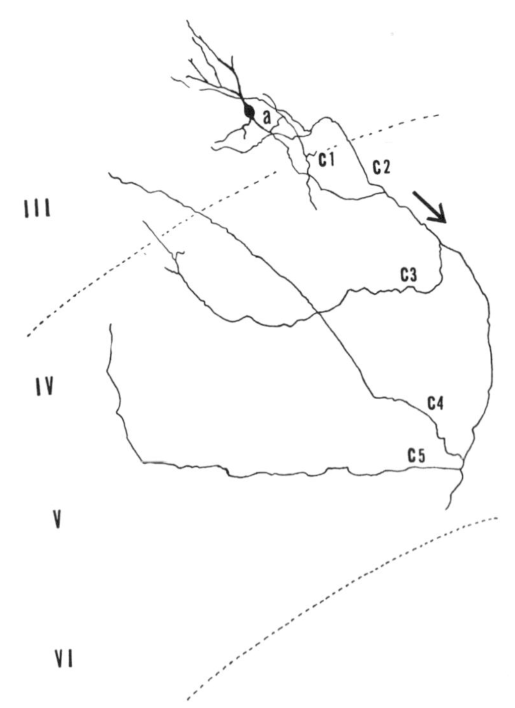

Fig. 4.30. Axonal tree of a lamina III cell. Note that some branches ramify in the general region of the cell body. Also note that the axonal tree of this cell passes to laminae IV and V. (From Matsushita, M., 1969; reprinted with permission.)

many, if not most, lamina III cells have axonal trees that arborize at least partially outside lamina III. It should be emphasized, however, that as for the other laminae, almost all cells have local branches that could serve to mediate the activity previously ascribed to Golgi type II cells.

Primary Afferent Input

General Studies. Early studies on primary afferent input into lamina III were concerned with the flame-shaped arbors (see below) and the question of coarse *vs.* fine primary afferent input into this area. Since the late 1970s, the development of intra-axonal labeling has markedly advanced our understanding of the details of this input. The early material is important as a basis for the later ideas that have emerged, but the single-axon work is where our focus will be directed (A. G. Brown, 1981; Fyffe, 1984; Maxwell and Réthelyi, 1987).

It has always been agreed that lamina III has a major primary afferent input. The shape of the primary afferent endings was shown to be distinctive in early studies. Thus, Ramón y Cajal (1909) described large primary afferent fibers that entered from the medial division of the dorsal root, curved ventrally through the medial part of the dorsal horn, and then recurved to end in a characteristic pattern in the substantia gelatinosa (Fig. 4.15). He called these the "collaterales grosses et profondes de la substance de Rolando." Scheibel and Scheibel (1968) called these endings the "flame-shaped arbors." This work has been abundantly confirmed (Szentágothai, 1964; Beal and Fox, 1976; Hamano *et al.*, 1978), except that the arbors are now known to be extended in the craniocaudal direction and so are flame shaped only on transverse section (Scheibel and Scheibel, 1968). It must be admitted that the early work either did not recognize laminae or regarded the gelatinosa as both laminae II and III, so the question of whether the flame-shaped arbors were found in both laminae II and III or only in lamina III was not considered. Recent work suggests that the flame-shaped arbors are located primarily in lamina III (Fyffe, 1984), although in the rat and monkey they enter lamina IIi (see above).

The flame-shaped arbors arise from coarse primary afferents (A. G. Brown, 1981), whereas fine afferents are thought to be restricted largely to laminae I and II (Fyffe, 1984). There are two reports of fine primary afferents in lamina III, however. First, Scheibel and Scheibel (1968) report a plexus of fine axons that they regarded as primary afferent in origin at the base of the flame-shaped arbors (the basal capping plexus) (Fig. 4.19). This plexus is not often mentioned today. Second, C. C. LaMotte (1977) reported that following selective surgery, fine-fiber degeneration is found in laminae I–III and coarse-fiber degeneration in laminae IV–VI. Another degeneration study found, however, that removing fine primary afferent fibers caused degeneration mainly in laminae I and II and that coarse-fiber degeneration was occurred in lamina III and below (Light and Perl, 1977). Some single-fiber work shows that unmyelinated axons synapse in lamina III (Sugiura *et al.*, 1986, 1989), even though the large majority of endings from these fibers are more superficial. Thus, there is fine primary afferent input into lamina III, but it seems to be much less important than the coarse-fiber input.

Single-Fiber Studies. With the advent of intracellular marking techniques, the distribution of single, functionally identified axons could be studied (see, e.g., A. G. Brown, 1981). One early finding was that the flame-shaped arbors (Fig. 10.15) arise from hair follicle afferents (A. G. Brown, 1981). It also became apparent that other types of coarse primary afferent fibers were ending in lamina III. A major conclusion from this work is that each functional type of primary afferent has its own distinctive termination pattern. Many of these fibers end in deeper laminae as well. Much material relevant to this section is discussed in Chapter 10. The somatotopic patterns of these terminations are considered in Chapters 5 and 10.

Hair Follicle Afferents. The terminology for these fiber types is not uniform (see Chapter 2). From the point of view of termination in lamina III, however, there are large hair follicle afferents that respond to guard and tylotrich hair movements and smaller fibers that respond to movement of down hairs. *Large Hair Follicle Afferents.* Classic studies reported that all primary afferent fibers that entered the spinal cord from the dorsal roots bifurcated into an ascending and a descending branch and that collaterals from these branches then passed to appropriate parts of the dorsal horn and deeper laminae (Ramón y Cajal, 1909; Réthelyi and Szentágothai, 1973). By contrast, 75% of the large myelinated hair follicle afferents do not so divide but simply turn dorsally and enter the dorsal columns (Fig. 10.19). These fibers, either divided or undivided, then give off numerous collaterals (Fig. 10.20), which terminate in the classic flame-shaped arbors emphasized by Ramón y Cajal (1909) and Scheibel and Scheibel (1968). These arborizations are much elongated, with axons from a single collateral extending as much as 1800 μm (compared with a width of no more than 400 μm), and the endings from sequential collaterals merging (A. G. Brown *et al.*, 1977c; A. G. Brown, 1981). Thus, an extraordinarily long shelf is formed that extends for many millimeters longitudinally (Fig. 10.20). These terminal fields are essentially restricted to lamina III in cats but extend to lamina IIi in rats and monkeys (Woolf 1987a; Shortland *et al.*, 1989). *Small Hair Follicle Afferents.* Four fibers of this type were filled by Light and Perl (1979a) (Fig. 10.33). These fibers bifurcate as they enter the cord. They end in the dorsal part of the nucleus proprius (lamina III) and ventral part of the substantia gelatinosa (lamina IIi). Their conduction velocities indicate that they are Aδ fibers. It should be remembered that Aδ nociceptors terminate primarily in lamina I (Light and Perl, 1979a).

Pacinian Corpuscle (FA II) Afferents. These axons bifurcate on entering the cord and travel cranially and caudally in the spinal white matter (A. G. Brown *et al.*, 1980c). Collaterals from these fibers are emitted which travel ventrally through laminae I and II. The fibers then arborize extensively in laminae III and IV, with a lesser arborization in laminae V and VI (Fig. 10.21) (A. G. Brown, 1981). The two arborizations tend to be separate from one another, with the III–IV arborization predominantly in the longitudinal plane and the V–VI arborization in clusters that have a dorsoventral orientation.

Rapidly Adapting (FA I) Mechanoreceptive Afferents in Glabrous Skin. These fibers bifurcate upon entering the cord, and collaterals are distributed to the medial dorsal horn (A. G. Brown *et al.*, 1980c). There, each collateral passes to lamina III (not in the recurrent fashion of

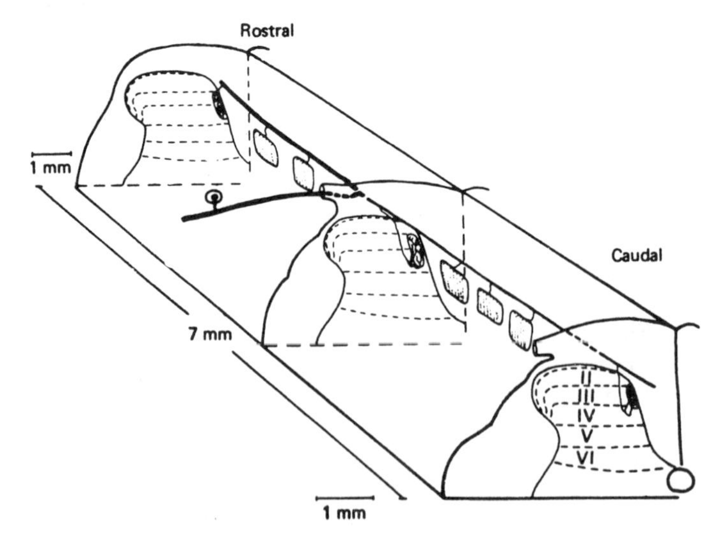

Fig. 4.31. Schematic depiction of the spadelike arborizations of rapidly adapting primary afferent fibers. (From A. G. Brown *et al.*, 1980c.)

the hair follicle afferents) and breaks up into a spadelike axonal arborization (Fig. 4.31). The arborization is basically confined to lamina III and is constricted mediolaterally. Note that this arborization is quite similar to that of the hair follicle afferents, except that the rapidly adapting mechanoreceptors form separate individual arborizations whereas the hair follicle afferents form a continuous shelf or strip.

Type I Slowly Adapting (SA I) Mechanoreceptive Afferents. These fibers bifurcate on entering the cord. Collaterals pass laterally and curve back to form globular termination regions that extend through laminae III–V but are centered in lamina IV (A. G. Brown *et al.*, 1978) (Figs. 10.2 and 10.11).

Type II Slowly Adapting (SA II) Mechanoreceptive Afferents. These fibers bifurcate, and the collaterals spread through laminae III–VI in the shape of transversely oriented plates (A. G. Brown *et al.*, 1981) (Fig. 4.32).

These descriptions of the geometry of the primary afferents that end in lamina III and deeper laminae should be supplemented by reading the physiologic characteristics of these fibers in Chapter 10 and the original literature. The dramatic correlations of terminal geometry with function for these fiber types represent a major advance in the understanding of somatic sensation.

Neuropil Organization

The neuropil of lamina III is described as being similar to that of lamina II (see, e.g., Kerr, 1975a; Ralston, 1979). There is a predominance of axodendritic synapses, a few axoaxonic and axosomatic synapses, and many glomeruli that seem to be similar to those described for the neuropil of lamina II (see above). There are some differences, however. For example, the proportion of flat vesicle terminals with symmetric synaptic thickenings is higher in lamina III than in laminae I and II (Ralston, 1979). Furthermore, in rats the central terminals in the glomeruli have a much higher proportion of filaments than in more superficial laminae (Ribeiro-DaSilva and Coimbra, 1982).

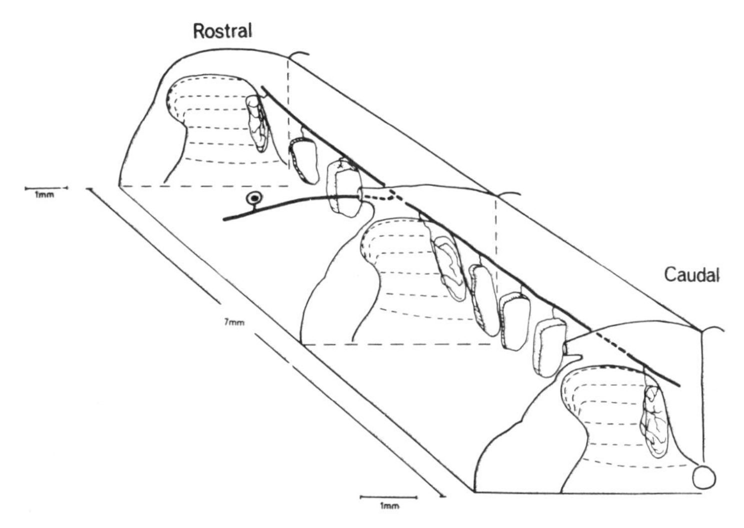

Fig. 4.32. Schematic depiction of the platelike arborizations of slowly adapting type II fibers. (From A. G. Brown *et al.*, 1981.)

Fine Structure of Terminals of Identified Axons

Terminals that arise from the above types of axons (hair follicle, Pacinian corpuscle, rapidly adapting type I's, slowly adapting type Is, slowly adapting type IIs) have been described. It is claimed that the terminal arrangements, like the axonal arborizations, are distinctive (Ralston *et al.*, 1984). Similarities are also impressive, however (Fig. 4.33). For example, all presynaptic terminals that arise from these axons contain spherical granular vesicles and make asymmetric synapses primarily with surrounding dendrites (Maxwell and Réthelyi, 1987) (Fig. 4.20). Thus, these terminals are thought to have an excitatory function. In addition, the terminals that arise from any of the above axon types are contacted by other, as yet unidentified, terminals that almost always contain flattened or pleomorphic vesicles and form symmetric junctions onto the labeled terminals (Maxwell and Réthelyi, 1987) (Figs. 4.20 and 4.32). It is true that the complexities of the above arrangements vary (see below), but it is not yet clear that these variations are related to the primary afferent axon type. Obviously more work in this important area is necessary.

Hair Follicle Afferents. In an electron-microscopic study, hair follicle afferents were seen to synapse on unknown dendrites, and other terminals ended on these endings (Maxwell *et al.*, 1982a) (Fig. 4.33). Similar findings were reported in a later study, but dendritic synapses back onto the labeled primary afferent were also seen (Ralston *et al.*, 1984). These arrangements were seen to be parts of glomerular complexes in the cat and monkey (Ralston *et al.*, 1984). Some of the axonal terminals onto the primary afferent terminals contain glutamic acid decarboxylase, an enzyme which synthesizes GABA (Maxwell and Noble, 1987).

Pacinian Corpuscle Afferents. The terminals of Pacinian corpuscle afferents are similar at the ultrastructural level to the terminals of hair follicle afferents. In particular, the labeled axons form synapses on dendritic shafts and spines, and there are a significant number of

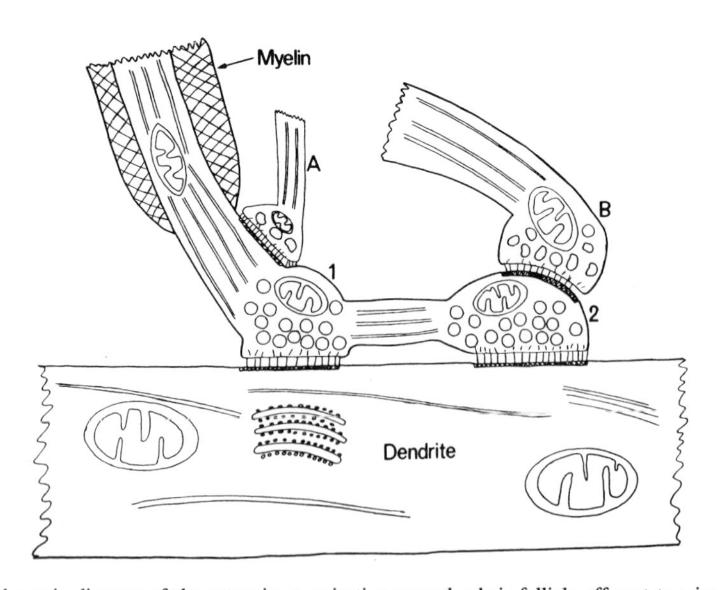

Fig. 4.33. Schematic diagram of the synaptic organization around a hair follicle afferent terminal. Note the axoaxonic synapses at A and B. The endings of the presynaptic fiber may be en passant (1) or terminal (2.) (From Maxwell *et al.*, 1982a.)

terminals from unknown neurons back onto the primary afferent terminals (Semba *et al.*, 1984; Maxwell *et al.*, 1984a; Ralston *et al.*, 1984). The secondary terminals usually contain flattened or pleomorphic vesicles and are considered to be of both axonal and dendritic origin.

Rapidly Adapting Afferents. Rapidly adapting afferents are similar at the fine-structural level to the Pacinian corpuscle afferents (Maxwell *et al.*, 1984a). Sometimes the arrangements resemble glomeruli (Semba *et al.*, 1985) (Fig. 4.33).

Slowly Adapting Afferents. The study by Semba *et al.* (1983) illustrates a synaptic organization similar to the above, but Bannatyne *et al.* (1984) proposed a slightly simpler organization and stated that they did not find glomeruli.

In summary, lamina III is distinguished from lamina II in having slightly larger cells and a neuropil that contains myelinated axons. Two types of cells have been identified in lamina III, postsynaptic dorsal column cells and spinocervical tract cells. Although there is not complete agreement, the postsynaptic dorsal column cells are characterized by dorsal dendrites that penetrate laminae I and II, whereas the dendrites of spinocervical tract cells do not. The unidentified cells in this lamina are of various types. One type is a pyramidal neuron that is flattened in the frontal plane and has dendrites passing through essentially all the laminae of the dorsal horn. Local interneurons (Golgi type II cells) and antenna-type neurons (neurons with prominent dorsal dendrites) are also seen in abundance.

The classic primary afferent input into this lamina comes from the flame-shaped arbors which have recently been shown to carry information from hair follicles. Other types of coarse primary afferents that enter lamina III arise from Pacinian corpuscles and rapidly and slowly adapting fibers. The axonal arborizations of each of these fiber types are distinctive. This is a major finding in somatic sensory mechanisms. Some fine-fiber primary afferent input has been demonstrated in lamina III, but it is relatively insignificant compared to the coarse fiber input.

The combination of electron microscopy and intracellular labeling is a major advance in our attempts to analyze the spinal neuropil. For the large-diameter primary afferents in the dorsal horn, use of this technique reveals what appear to be similar organizational patterns. All agree that the primary afferent endings contain round, clear vesicles and that they make asymmetric synapses on surrounding dendritic profiles. There are also terminals, which may be axonal or dendritic in origin, that end back on the primary afferent terminal, which in one study are shown to be GABA-ergic. Furthermore, in some cases glomerular arrangements are shown. Thus, there are similarities with the fine primary afferent synaptic organization. Further work to elicit more precisely some of the differences in these basic organizational patterns that underlie different functional categories would be desirable.

LAMINA IV

Lamina IV is a relatively thick layer that extends across the dorsal horn (Fig. 4.1). Its medial border is the white matter of the dorsal column, and its lateral border is the ventral bend of laminae I–III (Fig. 4.1). The neurons in this layer are of various sizes, ranging from small (approximately 8 by 11 μm) to large (35 by 45 μm). The largest cells are relatively infrequent, but they are so prominent that there is a general impression that this is a layer with large cells. Rexed (1952) was struck, however, by the heterogeneity of cell sizes rather than the large cells as being most characteristic of this layer. Lamina IV can be distinguished from lamina III by the heterogeneity of neuronal sizes and the presence of some very large cells compared with more homogeneous, smaller cells that characterize lamina III. Lamina IV can be distinguished from lamina V because the neurons of lamina V are even more heterogeneous than those of lamina IV and there are many longitudinally oriented myelinated axons in lamina V, particularly in its lateral part.

Cell Types

It is difficult to be certain of the exact laminar location of the cells that Ramón y Cajal (1909) illustrated in the basal part of the dorsal horn, but cell B in Fig. 4.2 and cell A in Fig. 4.15 are representative of cells of the nucleus proprius, and they are found in the general region of lamina IV. These cells have long spiny dendrites that pass dorsally, laterally, and ventrally and an axon that passes into the white matter. This is a general picture that would probably be valid for most of the large neurons in laminae IV and V.

Szentágothai (1964) emphasized particularly the dorsal dendrites of the large cells in laminae IV and V (Figs. 4.6, 4.15, and 4.18). The reason for this is that the dorsal dendrites penetrate the substantia gelatinosa, where they are contacted by the axons from central cells. It should be remembered that Szentágothai (1964) claimed that the substantia gelatinosa is a system of cells that receive primary afferent input and have rich synaptic interconnections but possess no obvious pathways for direct forward conduction. The importance of the large cells in laminae IV and V with dorsal dendrites that penetrate the gelatinosa is that they were thought to represent the output of this enigmatic region. Although this view is now known to be simplistic (see above), there can be no question that large numbers of dorsal dendrites from cells in laminae III–V enter the gelatinosa and are undoubtedly strongly influenced by synaptic input from this region.

Scheibel and Scheibel (1968) state that lamina IV consists of a relatively small number of moderate to large cells (20–40 μm) whose dendrites radiate medially, laterally, ventrally, and dorsally from the soma. They point out that these dendritic fields are cone shaped, with the apex of the cone directed ventrally (Fig. 4.5). This is because the ventral dendrite is usually single, the longitudinal dendrites do not extend far from the cell body, and the dorsal dendrites ramify extensively.

Réthelyi and Szentágothai (1973) point out that "it is difficult to translate into modern architectonic terms (lamination of Rexed, 1954) the classical expression head (or center) of the dorsal horn (nucleus proprius cornuus posterioris). The difficulty is caused mainly because the cytoarchitectonic borders do not match with those of dendroarchitectonics and neuropil architectonics." These authors point out that laminae III, IV, and the upper part of V are dominated by a longitudinal axonal plexus that had previously been described by Sterling and Kuypers (1967). This plexus should not be confused with the intrinsic longitudinal axonal plexus of laminae II and III that Szentágothai (1964) had described earlier. Réthelyi and Szentágothai (1973) described three types of neurons within the deeper plexus, on the basis of their dendritic projections: (1) antenna-type neurons, smaller in lamina III and larger in laminae IV and V, whose dendrites project dorsally into the substantia gelatinosa (these they thought were the output neurons for the substantia gelatinosa); (2) central cells with longitudinally oriented dendrites; and (3) cells, primarily in lamina V but also in IV, with transverse dendrites. Réthelyi and Szentágothai (1973) mentioned, however, that many cells in these laminae have all three types of dendrites.

Proshansky and Egger (1977) noted that neurons in laminae IV, V, and VI could be distinguished on the basis of differences in dendritic spread. These investigators point out that most neurons in lamina IV have dense, bushy dendritic fields. By contrast, lamina V has some cells with bushy dendrites and other cells with radiating fields, and lamina VI has cells with primarily radiating fields (the term "radiating" implying that the dendrites are relatively straight and have few branches). The dendritic spread, according to Proshansky and Egger (1977), is least for the lamina IV cells, intermediate for lamina V cells, and greatest for lamina VI cells. These investigators correlate this increase with the increasing size of the receptive fields from lamina to lamina as measured physiologically (Wall, 1967).

A possible problem is that many processes of Golgi-stained neurons are lost if the cells are not examined in serial sections (Mannen, 1975). For example, one cell located at the boundary of laminae IV and V had dorsal dendrites that spread through laminae I–V of the ipsilateral side and laminae III–IV of the opposite side of the spinal cord and ventral dendrites that spread down into laminae VII (Fig. 4.34). The side-to-side spread of these dendrites is greater than noted by previous investigators, and the shape of a dendritic field does not resemble a cone, which is the shape suggested by Scheibel and Scheibel (1968). Mannen studied only one lamina IV cell by his technique, but the results of this technique on cells from many laminae in the spinal cord (Mannen, 1975; Mannen and Sugiura, 1976) seem to imply that studies on single sections of Golgi-stained neurons may result in an underestimate of cell size because axonal and dendritic processes are lost unless serial sections are done.

The above studies are concerned with cells that are unidentified as to function or axonal termination. There are two types of identified cells in lamina IV, however, the spinocervical cells and the postsynaptic dorsal column cells (A. G. Brown, 1981; Bennett *et al.*, 1984). These cells have been described in the section on lamina III, and, except for the location of the cell body, the cells in lamina IV are the same. Actually, these two cell types are found in laminae III, IV, and upper V, with the greatest concentration in lamina IV (A. G. Brown, 1981). A. G. Brown (1981) also notes that there are two types of cells in lamina IV that do not fit into the above categories. The first have dendritic fields that resemble spinocervical tract cells or postsynaptic dorsal column neurons, but the axons of these cells are local. Thus, they might be local interneurons with extensive dendritic trees. The other cells are large and have strongly directed dorsal dendrites with an axon that passes cranially in the cord to an unknown destination. Schoenen (1982) emphasizes that most cells in lamina IV have dorsally directed dendrites. Finally, spinothalamic cells are also found in lamina IV, although more are found in laminae I and V (see Chapter 9). In the older literature, spinothalamic cells were thought to be characterized by axons that passed to the anterior white commissure (see below). As yet, however, we do not have complete descriptions of the organization of the dendritic trees of spinothalamic cells.

Axonal Projections

The classic view was provided by Ramón y Cajal (1909), who illustrated several cells deep in the dorsal horn (lamina IV or V) that sent their axons into the white matter of the lateral white

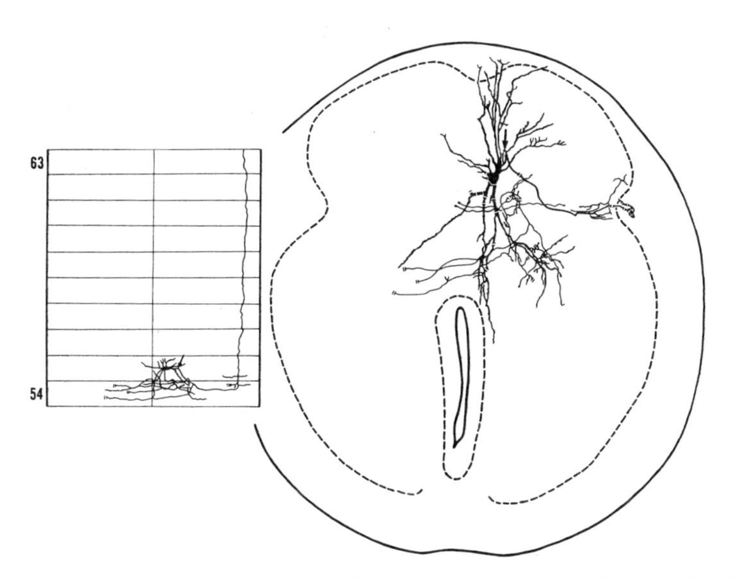

Fig. 4.34. Large neuron reconstructed from serial sections located at the boundary of laminae IV and V. Note the dorsal dendrites and how widespread the whole dendritic tree is. (From Mannen, 1975.)

column (Fig. 4.2 and 4.15). Ramón y Cajal (1909) also illustrated neurons in this area that sent their axons across the midline. Szentágothai (1964) noted that most textbooks state that large dorsal horn neurons send their axons to the opposite spinothalamic tract through the anterior white commissure in the cat. Szentágothai (1964) further noted, however, that he never observed a lamina IV or V neuron sending its axon into the anterior white commissure. It is not clear whether this implies that the axons do not cross in the anterior white commissure or whether they ascend for a segment or so before crossing, a fact that could not be ascertained in his Golgi studies on single sections.

Scheibel and Scheibel (1968) note that the axons of lamina IV cells "enter propriospinal bundles either in the lateral or dorsolateral white matter, most often in the central area of Marie." They emphasize the tortuous course of these axons before they get to the white matter. They also point out that the axons often bifurcate, a fact which was also noted by some earlier investigators, and that collaterals of these axons penetrate into deeper laminae.

Matsushita (1969, 1970) notes that "the main axons of lamina IV cells enter the dorsal part of the lateral funiculus," which agrees with earlier work. Before the axons enter the lateral white matter, however, they give off collaterals in the region of the cell body that ramify in laminae III, IV, and V. Matsushita (1970) points out that the axons of some lamina IV neurons travel ventrally into lamina VII before passing dorsolaterally to enter the lateral white matter. The question of Golgi type II cells is also considered. Matsushita noted the absence of short-axoned cells in this part of the spinal cord and then pointed out that the axonal collaterals of the lamina IV cells given off before the main axon or axons enter the white matter must be responsible for the inhibitory phenomena that are interpreted by physiologists as resulting from Golgi type II interneurons. By contrast to Szentágothai (1964), Matsushita found cells in laminae IV and V whose axons could be followed to the anterior white commissure. These may be spinothalamic cells.

Mannen (1975) considers the axonal distributions of Golgi-impregnated lamina IV cells in serial sections. He confirms that "Neurons found in the dorsal horn (laminae IV and V) send their axons either to the ipsilateral lateral funiculus . . . or to the ipsilateral lateral funiculus as well as to the contralateral anterior funiculus," and also says that "neurons found in the dorsal horn send their axonal branches into the dorsal half of the intermediate region, including laminae VII and X of the ipsilateral side . . . or contralateral side . . . or both." Thus, the axon collaterals of lamina IV cells (as well as lamina V cells) send numerous branches into deeper laminae of the spinal cord. Most lamina IV cells, including those that send their axons to the white matter or across the midline or to deeper laminae, usually have a "local" axonal arborization in the region of the dendritic fields.

Primary Afferent Input

Before discussing the input from primary afferent fibers into lamina IV, it should be remembered that many of the lamina IV neurons send their dorsal dendrites superficially into laminae I–III. Thus, many cells in lamina IV can receive a direct primary afferent input from fibers that are entering more superficial laminae.

Many primary afferent axons directly enter lamina IV. In describing these axons, earlier work focused on plexuses, but single-fiber studies now predominate. An understanding of the plexuses is important, however, as a basis for the single fiber studies. It was earlier noted that numerous axons come from the dorsal parts of the posterior columns to ramify in the head ("tête") of the dorsal horn (Ramón y Cajal, 1909) (Fig. 7.3), which is located approximately in lamina IV. These axons form a longitudinal plexus (the longitudinal plexus of the dorsal horn), and synaptic contacts are made on the cell bodies of the large neurons of the dorsal horn. Most of these fibers are thought to be terminal ramifications of large myelinated primary afferent fibers, and although they come through the gelatinosa, they give off no collaterals to this substance (Ramón y Cajal, 1909). They were regarded as separate from the coarse primary afferent fibers that break up into the flame-shaped arbors (Fig. 4.4). Scheibel and Scheibel (1968) agreed, but Szentágothai (1964) pictured them as coming from the same population of fibers. Single-fiber studies show that the flame-shaped arbors are endings from hair follicle afferents. The longitudinal orientation of the plexus in lamina IV and vertical orientation in deeper laminae resembles the organization of Pacinian corpuscle input to some extent, but further work is needed to confirm this.

There is not complete agreement about the laminar location of Ramón y Cajal's (1909) longitudinal plexus to the head of the dorsal horn. Sprague and Ha (1964) stated that this plexus is found in lamina IV. By contrast, Réthelyi and Szentágothai (1973) located the plexus in laminae III, IV, and the upper part of V, and Sterling and Kuypers (1967) showed a longitudinal primary afferent axonal plexus in laminae III and IV, but did not relate it to the plexus of Ramón y Cajal (1909). It should be remembered that there are an enormous number of fibers passing in various directions in these laminae, so that plexuses with an overall fiber orientation may be difficult to discern.

Scheibel and Scheibel (1968) provide a relatively complex picture of the afferent innervation to lamina IV. These investigators believe that the main primary sensory input to these cells is to their dorsal dendrites that are ascending through the substantia gelatinosa, whereas the corticospinal fibers end on the lateral dendrites of these cells. The primary afferent input is perhaps more important because the dorsal dendrites have longer and more numerous spines than the lateral dendrites. Fibers from the cornucommissural bundle of Marie intertwine in lamina IV, and Scheibel and Scheibel (1968) suggest that they form a "wealth of axoaxonal contacts with resulting possibilities for presynaptic modulation." To the medial dendrites of the lamina IV cells come a heterogeneous input: (1) terminal collaterals emerging from the cornucommissural bundle and adjacent white matter, (2) fibers from the contralateral dorsal horn crossing in the posterior commissure, and (3) branches of contralateral primary afferent collaterals (Scheibel and Scheibel, 1968). They did not relate any of these systems to Ramón y Cajal's longitudinal plexus.

The dense longitudinal axonal plexus of the head of the dorsal horn (Ramón y Cajal, 1909) has been named by Réthelyi and Szentágothai (1973) "the central longitudinally orientated axonal plexus of the dorsal horn." They believe that this plexus defines a nucleus of the dorsal horn. Thus, this part of the dorsal horn (laminae III, IV, and upper V), which corresponds to most of the nucleus proprius, is defined by neuropil architectonics rather than cytoarchitectonics. Kerr (1975a) also provided a special name for this longitudinal axonal plexus. He called it the "dorsal intracornual tract" and noted that it exists primarily in laminae IV, V, and VI. Kerr (1975a) noted that Ramón y Cajal (1909) thought that these fibers were collaterals of primary afferents. Kerr found, however, that a significant number of these axons remain after dorsal rhizotomy, and thus many of the fibers must arise from interneurons. He speculated that this pathway might be involved in nociceptive sensation, perhaps as "a supplementary pathway in instances of long-term interruptions of nociceptive pathways in the human."

Experimental studies do not show a fine primary afferent input into lamina IV (C. C. LaMotte, 1977; Light and Perl, 1977). At the electron-microscopic level, the degeneration of primary afferent fibers and terminals in lamina IV is characterized by neurofilamentous and electron-lucent degeneration, in contrast to the electron-dense degeneration that characterizes laminae I and II (Ralston and Ralston, 1982). The neurofilamentous degeneration is thought to characterize large myelinated afferents, the electronlucent degeneration fine myelinated afferents, and the electron-dense degeneration unmyelinated afferent fibers (Ralston and Ralston, 1982). If these findings can be generalized, they will allow an efficient search of the neuropil at appropriate times after dorsal rhizotomy. In addition, the central terminals of glomeruli and many of the round, clear vesicle-containing presynaptic endings that make simple axodendritic synapses are of primary afferent origin. It is of interest that flat or pleomorphic vesicle endings do not seem to arise from primary afferent fibers (Ralston and Ralston, 1982).

The above studies are concerned with primary afferent fibers in bulk. In recent years, single-fiber studies have provided more specific data. This material was given in the section on lamina III.

Neuropil Organization

The synapses in lamina IV are axodendritic, axosomatic, and axoaxonal (Ralston, 1968a, 1979; Kerr, 1970a). This pattern was thought to differ from that of lamina III, in which axosomatic synapses were rare and glomeruli were common (Ralston, 1968a; Kerr, 1970a), but typical glomeruli have now been seen in lamina IV and deeper laminae (Ralston, 1982), although they are not as prominent as in the superficial laminae. Two other characteristics of the synapses of lamina IV are that large dense core vesicles are much less prominent than in more superficial laminae, particularly laminae I and IIo (Ralston, 1979), and that the dense-core vesicle terminals that do exist in lamina IV do not arise from primary afferents, whereas many of them do in superficial laminae (Ralston and Ralston, 1982). The synaptic terminals in the upper three laminae are characterized by an abundance of round clear vesicles superficially and flattened vesicles more deeply (Ralston, 1979). In lamina IV, about two-thirds of the terminals contain round clear vesicles, and this proportion decreases as laminae V and VI are reached. The suggestion is that inhibitory mechanisms are more prominent in lamina IV than in V or VI. Axoaxonic terminals and presynaptic dendrites are seen, but they are not common.

The above findings are supplemented by the fine-structural analysis of the synaptic organization around individually labeled fibers (Maxwell and Réthelyi, 1987). These have been described in the section on the neuropil of lamina III. Two points should be made here, however. First, typical glomeruli can be found in lamina IV, which is contrary to early reports (e.g., Kerr, 1970a). Second, axoaxonal synapses and presynaptic dendrites are of special interest because they are thought to be the morphologic basis of presynaptic inhibition. These structures are relatively rare in the general neuropil of lamina IV (Ralston, 1982). By contrast, they are common in the synaptic organization that surrounds primary afferent terminals. A reasonable conclusion would seem to be that the majority of axoaxonal synapses and presynaptic dendrites in lamina IV (and laminae V and VI) are associated with primary afferent terminals.

In summary, lamina IV is a thick layer consisting of nerve cells of various sizes, ranging from very small to quite large. The large cells have dendrites that pass laterally, dorsally, and medially. Szentágothai (1964) gave special emphasis to the cells with prominent dorsal dendrites (antenna-type cells). Some of the cells of lamina IV are postsynaptic dorsal column and spinocervical tract cells. Some of the primary afferent input to these cells comes to their dorsal dendrites, which are in superficial laminae. The other major input in classic studies is generally described as being from Ramón y Cajal's (1909) longitudinal plexus of the head of the dorsal horn. Several longitudinal plexuses have been described by different authors, however, and these are not always the same. Single-fiber studies are now providing more precise information on the organization of primary afferent input. The neuropil of lamina IV was thought to differ from the neuropil of lamina III in having frequent axosomatic synapses and no glomeruli. Glomeruli have been found in association with primary afferent terminals from coarse fibers in this lamina, however. Most axoaxonic and dendrodendritic synapses in this lamina are associated with these glomeruli.

LAMINA V

Lamina V extends as a thick band across the narrowest part of the dorsal horn (Fig. 4.1). It occupies the zone that is often called the neck of the dorsal horn. It has a sharp boundary against the dorsal funiculus, but its lateral boundary is indistinct because of the many bundles of myelinated fibers coursing longitudinally through this area. These bundles of myelinated fibers give the lateral part of this lamina a reticulated appearance. The lateral reticulated area or zone occupies about one-third of the lamina. Lamina V can be distinguished cytoarchitectonically from lamina IV, since the cells are even more varied than in lamina IV, and from lamina VI, which consists of a medial zone with small, packed, compact cells and a lateral zone with slightly larger cells that are nevertheless more regular than those in lamina V. A good clue to the location of lamina V is the presence of numerous myelinated fibers (the lateral reticulated area) on its lateral side.

Cell Types

The cell bodies in this region are vary from 8–10 μm to 30–45 μm in diameter. Most of the cells fall in the range of 10-13 by 15-20 μm (Rexed, 1952). The cells are large and have clear, distinct nucleoli. Nissl substance is relatively sparse but can be clearly seen, particularly in the large cells.

Ramón y Cajal (1909) did not separate the cells in this region of the dorsal horn on the basis of their dendritic patterns, so his description of lamina IV cells would apply equally well to the cells of lamina V. Scheibel and Scheibel (1968), however, emphasized that there is a major shift in the axonal and dendritic neuropil from lamina IV to lamina V. The axonal changes will be discussed in a later section of this chapter. The dendrites of the lamina V cells "radiate along the dorsoventral and mediolateral planes with little or no extensions along the longitudinal axis of the cord" (Scheibel and Scheibel, 1968). Thus, the dendritic shapes of these cells were described as flattened disks (Figs. 4.5 and 4.7). This contrasts with the orientation of most of the neurons in lamina IV.

Mannen (1975) reconstructed the dendritic and axonal trees of 16 spinal neurons from serial Golgi sections. Three of these neurons were located in lamina V. The dendrites of these cells spread over large areas in all directions (Fig. 4.34). Nevertheless, although one could not regard these dendritic fields as disk-like, as suggested by Scheibel and Scheibel (1968), the dendritic fields are larger in the mediolateral and dorsoventral directions than in craniocaudal extent. A. G. Brown (1981) studied lamina V neurons by the method of intracellular injection and agreed that the dendritic fields of these cells are restricted longitudinally (Fig. 4.7). Thus, it seems agreed that the dendritic fields of lamina V cells are restricted longitudinally, although there is some disagreement about the extent of this restriction and the shape of the dendritic tree.

Proshansky and Egger (1977) note that the dendritic fields are more widely spread in lamina V neurons than in lamina IV neurons and that some of the lamina V cells have radiating dendrites, in contrast to the bushy dendrites of the lamina IV neurons. It must be remembered that the dendritic field of a neuron in this region of the spinal cord cannot be completely visualized unless serial sections are studied, but physiologic work corroborates the increasing size of the dendritic field from lamina IV to lamina V (Wall, 1967). A. G. Brown (1981) emphasizes the vertical orientation of some of the lamina V cell dendrites (Fig. 4.7).

In a precise correlation of structure and function for lamina V cells, Ritz and Greenspan (1985) correlated dendritic morphology with physiologic type. Multireceptive cells were large and had a dendritic tree that extended in all directions. Thus, the transverse flattening described above is not seen here. The cells that responded only to noxious stimuli were smaller than the cells that responded to both noxious and nonnoxious stimuli, but the dendritic fields were approximately the same. Finally, cells that responded only to innocuous stimuli were the smallest and had the least extensive dendritic trees. Thus, there was a positive correlation between mediolateral dendritic spread and the size of the low-threshold component of the receptive field.

Axonal Projections

Similarly to lamina IV neurons, the three main projection sites that are suggested for lamina V neurons are the thalamus, the lateral cervical nucleus, and dorsal column nuclei. There are relatively more spinothalamic cells and fewer spinocervical tract and postsynaptic dorsal column neurons, however. The spinothalamic cells are described in Chapter 9, and the others are described in Chapters 7 and 8 as well as in previous sections of this chapter.

There is relatively little work on the axonal projections of lamina V cells as distinct from lamina IV cells. Ramón y Cajal (1909) showed that cells in this general region of the spinal cord send their axons into the neighboring white matter that surrounds the dorsal horn or to the opposite side of the spinal cord. Matsushita (1969) noted that it was difficult to distinguish neurons in lamina IV from those in lamina V by using silver stains because the axonal pathways are similar. Matsushita did note, however, that there were a number of lamina V neurons that sent their axons to the ipsilateral anterior funiculi. These may be spinothalamic cells. Matsushita also mentioned that some lamina V cells "give thick myelinated axons to the motoneuron groups." Also, there are large dorsal commissural cells "in the lateral part of lamina V and in the central part of lamina VI of the cervicothoracic and the lumbar cord." These could presumably be the same cells that Ramón y Cajal (1909) described as sending axons into the posterior commissure. In later work (1970), Matsushita some cells in laminae V and VI that send their axons across the spinal cord in the anterior commissure. He (1970) emphasized that the axons of the medium-sized neurons in lamina V give off collaterals around the cell body. These collaterals reach ventrally as far as lamina VII and dorsally into lamina III and IV. He suggested that these collaterals may subserve the function of Golgi type II cells, even though they are widespread.

In summary, it would appear that many lamina IV and V cells have a similar axonal organization, with collaterals ramifying in the gray matter around the cells and into deeper laminae and at least one primary axon passing into the white matter of the lateral funiculus. In addition, some of the neurons, particularly in lamina V, send axons contralaterally, from where they may pass to the thalamus.

Primary Afferent Input

Ramón y Cajal (1909), as usual, made many of the important early observations on the primary afferent input into the base of the dorsal horn. It is difficult to be certain about the laminar location of his primary afferent collaterals, but the axons of his longitudinal plexus to the head of the dorsal horn would ramify in the general region of laminae IV and V (see the section on lamina IV). Sprague and Ha (1964), however, emphasized a dorsoventral orientation of the primary afferent input into lamina V, in contrast to the primarily longitudinal orientation of the primary afferent plexus in lamina IV. They also pointed out that degeneration following dorsal rhizotomy is confined to the central area of lamina V.

Sterling and Kuypers (1967) and Scheibel and Scheibel (1968) agree that there is a marked change in the orientation of the axonal neuropil at the junction of laminae IV and V, the orientation changing from longitudinal to dorsoventral at this point. Sterling and Kuypers (1967) noted the central location of the terminals in lamina V and pointed out that the terminals of the primary afferent axons in laminae V and VI are not somatotopically organized as they seem to be in laminae III and IV. Scheibel and Scheibel (1968) noted that the dorsoventral orientation of these fields corresponds to the dorsoventral orientation of the dendritic system of the neurons in this area.

Szentágothai and Réthelyi (1973) also support the view of a dorsoventral orientation to the plexus in lamina V, but their picture is slightly more complicated. They consider the six groups of primary afferent collaterals to the spinal cord originally described by Ramón y Cajal (1909). The collaterals to the center of the dorsal horn are described as having a horizontal course before breaking up into the dense neuropil in the medial part of the neck and bend of the dorsal horn, which they equate with laminae IV and V. Thus, they identify a prominent horizontal afferent axonal plexus in lamina V as well as in lamina IV. For the dorsal horn proper, they state that "The longitudinally orientated axonal plexus, mainly of primary afferent origin . . . corresponds to lamina III, IV and the upper part of V". If this is true, the dorsoventral plexus must be located below the longitudinal plexus, i.e., in ventral lamina V.

The fine structure of the neuropil is not different from that in lamina IV.

In summary, the cells of lamina V are even more heterogeneous than those of lamina IV. In addition, the lateral third of lamina V is heavily myelinated and thus has a reticulated appearance with most histologic stains. The dendritic organization of lamina V cells is not markedly different from that of lamina IV cells, except that the general orientation of the dendrites changes to vertical, in contrast to the generally longitudinal orientation in lamina IV. The axonal projections of these cells appears to be to the thalamus, the dorsal column nuclei, the lateral cervical nucleus and various local destinations in the cord. There is a major shift in primary afferent organization in this lamina with a dorsoventral orientation in contrast to the longitudinal orientation of the plexus in lamina IV. The neuropil is not greatly different from that of lamina IV.

LAMINA VI

Lamina VI exists only in the cervical and lumbosacral enlargements of the spinal cord (Fig. 4.1). This is a part of the cord where primary afferent fibers, although clearly entering this area, do not predominate as in the rest of the dorsal horn. The cells in this layer are smaller and more regular in their arrangement than those in lamina V. The smallest cells are 8–8 μm and the largest are 30–35 μm. There is more Nissl substance in these cells than in the cells of laminae V and VII. Thus, lamina VI has a darker appearance than laminae V and VII in Nissl-stained sections. Lamina VI is divided into halves, the medial half consisting of a compact group of small or medium-sized heavily staining cells and the lateral half consisting of larger well-stained triangular or star-shaped cells. In this layer are the major parts of the internal and external basal nuclei of Ramón y Cajal (1909).

Cell Types

Relatively few investigators have been particularly concerned with the dendritic pattern of lamina VI neurons. Scheibel and Scheibel (1968) point out that these cells are arranged much like those in lamina V, with dendrites that radiate in dorsoventral and mediolateral planes but with almost no extension in the longitudinal axis of the cord. Thus, they depict the dendritic domains as flattened disks (Fig. 4.5). Proshansky and Egger (1977) note that these neurons were isodendritic, with long, straight, relatively unbranched dendrites. Proshansky and Egger also note that the dendritic spread was wider than the dendritic spread of the laminae IV and V neurons. A. G. Brown (1981) essentially agrees except that he stated that the dorsal dendrites never penetrate lamina I and II. If so, these cells would sample wide areas from the base of the dorsal horn, but they would not be directly exposed to fine primary afferent input. Neurons in the lateral part of lamina VI tend to be somewhat more variable in size than in medial parts (Tredici *et al.*, 1985). Finally, Szentágothai and Réthelyi (1973) and Réthelyi and Szentágothai (1973) note that the general dendritic organization of many cells in the middle parts of laminae V–VIII form a cylinder. The original papers for this last finding should be consulted, because these authors proposed an interesting dendritic and cellular organization of the entire gray matter of the spinal cord (Réthelyi, 1984a,b) that is too extensive to be considered in a chapter devoted to only the dorsal horn.

Axonal Projections

As with the dendritic fields, there has been little work on the axonal projections of these cells as determined by anatomic means. It is clear from the various labeling experiments described in other chapters that a few lamina VI cells project to the thalamus or to the lateral cervical nucleus, but the majority of cells are probably propriospinal. Matsushita (1969) points out that the small cells in the medial part of lamina VI possess the most "abundantly ramifying axon collaterals in the spinal cord." These axon collaterals seem to end primarily in laminae IV, V, VI, and VII, and recurrent collaterals end near the parent cell and in neighboring neurons.

Primary Afferent Input

The primary afferent input to lamina VI is complex. Many collaterals from primary afferent axons destined to reach ventral horn cells end in this area (Ramón y Cajal, 1909; Scheibel and Scheibel, 1968). A particularly prominent group of terminals from Ia fibers is found in the center of lamina VI. The fine structure of the endings is described by Maxwell and Bannatyne (1983). There are also several systems of descending fibers ending in this area (Tredici *et al.*, 1985). The article by Tredici *et al.* (1985) should be consulted for a summary of what is known about the descending input. These authors point out that morphologically this lamina appears to be distinct from both the superficial laminae of the dorsal horn and the motor areas of the ventral horn.

CHEMICAL NEUROANATOMY

The term "chemical neuroanatomy" is often used for the explosively expanding field that has arisen following the development of immunohistochemical and other morphologic techniques that label compounds found in certain populations of neurons. Many excellent reviews exist. For example, there is the *Handbook of Chemical Neuroanatomy* (Björkland and Hökfelt, 1983) for a general review and the articles by Ruda *et al.* (1986) and Tohyama and Shiotani (1986) for the spinal cord. This material, as well as being interesting in its own right, is providing insight into the general structure of the cord. In our judgment, however, there is a need for a correlation of classic spinal anatomy with the newer chemical anatomy. This is the purpose of the present section. Specific parts of this material are presented in various sections of this chapter, but the present section is an overview.

Acetylcholine

Three anatomic techniques yield information on cholinergic spinal systems: ACHE histochemistry, ChAT immunolocalization, and the autoradiographic demonstrations of the distribution of cholinergic receptors.

Acetylcholinesterase (ACHE). Early studies showed ACHE reactivity in cells of the substantia gelatinosa (Roessmann and Friede, 1967), and shortly thereafter the neuropil of the gelatinosa was seen to be stained (Navaratnam and Lewis, 1970). Subsequent studies are similar but differ in details. For neuropil labeling, lamina III was implicated by Silver and Wolstencroft (1971) and laminae I–III by Odutola (1972). Silver and Wolstencroft (1971) found labeled cells in lamina III and Odutola (1972) found them in lamina I. In a later study, with a reduced neuropil staining, cells in laminae I–III are labeled, and vertical areas of alternating heavy and light staining can be seen (Marchand and Barbeau, 1982).

Choline Acetyltransferase (ChAT). It would generally be accepted that ChAT localization is a better marker for cholinergic systems than ACHE (Borges and Iversen, 1986). Early studies with antibodies that recognize ChAT showed considerable neuropil and some cell labeling in the substantia gelatinosa (H. Kimura *et al.*, 1981, Houser *et al.*, 1983). In a later study, lamina III was noted to have the most extensive neuropil staining (Barber *et al.*, 1984). Labeled cells were found in laminae III–V (Phelps *et al.*, 1984, 1988), and the dendrites of these cells formed a plexus in lamina III (Barber *et al.*, 1984; Borges and Iversen, 1986). In addition, Barber *et al.* (1984) show that much of the neuropil labeling is in presynaptic terminals. Since ChAT has not been found in primary afferent cells or fibers (Barber *et al.*, 1984; Borges and Iversen, 1986) (see the discussion in Chapter 3), and since spinal cord transection or capsaicin treatment has no effect on ChAT levels in the dorsal horn (Holtzer-Petsche *et al.*, 1986), the conclusion is that the ChAT system in the dorsal horn arises from intrinsic neurons. A cholinergic specific ganglioside is located on neurons in laminae I and III (Obrocki and Borroni, 1988).

Receptor Studies. Arimatsu *et al.* (1981) and Walmsley (1981a,b) showed that cholinergic muscarinic receptors are found in high concentrations in the substantia gelatinosa. This observation has been repeatedly confirmed, although some investigators note that both lamina II and III or just lamina III are labeled (Ninkovic and Hunt, 1983; Whitehouse *et al.*, 1983; Gillberg *et al.*, 1984, 1988; Scatton *et al.*, 1984; Villiger and Faull, 1985; Gillberg and Aquilonius, 1985; Gillberg and Wiksten, 1986). Many of these receptors are associated with primary afferents in that they accumulate proximal to a ligature of the dorsal root and are considerably reduced following dorsal rhizotomy (Walmsley *et al.*, 1981a; Ninkovic and Hunt, 1983; Gillberg and Wiksten, 1986). This accumulation and loss might indicate a cholinergic influence on primary afferent terminals.

Adenosine

The purine nucleoside adenosine seems to be an important neuromodulator with effects on noxious sensory input (Goodman and Snyder, 1982). In accord with this, adenosine neurons are seen in the substantia gelatinosa (Braas *et al*, 1986). In addition, receptors for this substance and adenosine diaminase are found in particularly high concentrations in the substantia gelatinosa (Goodman and Snyder, 1982; Geiger and Nagy, 1985; Yamamoto *et al.*, 1987; Choca *et al.*, 1988).

Bombesin–Gastrin

No bombesin cells have been found in the spinal cord. There is a relatively extensive plexus of bombesin-containing axons and terminals in lamina I and II (Panula *et al.*, 1982, 1983, 1989; Fuxe *et al.*, 1983; Massari *et al.*, 1983; Panula, 1986, Panula *et al.*, 1989a; K. Chung *et al.*, 1989a) or laminae I–III (O'Donohue *et al.*, 1984). In addition, neuromedin, which is closely related, was seen in laminae I–II (Namba *et al.*, 1985). It is agreed that many but not all of these fibers are lost after dorsal rhizotomy (Panula *et al.*, 1982; Fuxe *et al.*, 1983; O'Donohue *et al.*, 1984). Autoradiographic receptor studies show localization to the same general areas as the immunocytochemical staining (Massari *et al.*, 1985; Zarbin *et al.*, 1985), and dorsal rhizotomy causes a decline in the number of these receptors (Massari *et al.*, 1985).

Brain Natriuretic Peptide

Brain natriuretic peptide is found in fibers and presumed terminals in laminae I and IIo, and in sensory cells of the dorsal root, trigeminal, and nodose ganglia (Saper *et al.*, 1989). A closely related compound, atrial natriuretic peptide, is not found in the same locations. These compounds are implicated primarily in blood pressure regulation, but whether this is the case for this system in the spinal cord is not known.

Calcitonin Gene-Related Peptide (CGRP)

CGRP is noteworthy for labeling a large number of DRG cells (see Chapter 3) and many cells in the ventral horn, but no immunolabeled cells in the dorsal horn have been reported (however, see Chapter 5). Dorsal rhizotomy removes essentially all labeled fibers from the dorsal horn (Gibson *et al.*, 1984b; K. Chung *et al.*, 1988), indicating that CGRP fibers in the dorsal horn are almost entirely of primary afferent origin. Thus, CGRP is regarded as a primary afferent marker in the dorsal horn. CGRP-immunolabeled axons and terminals form an extensive mesh in laminae I, IIo, and the reticular part of lamina V, although some fibers are found in other laminae (Rosenfeld *et al.*, 1983; Gibson *et al.*, 1984b; Skofitsch and Jacobowitz, 1985c; Carlton *et al.*, 1987b, 1988; Marti *et al.*, 1987; Harmann *et al.*, 1988b; D. L. McNeill *et al.*, 1988a; K. Chung *et al.*, 1989a). The label in lamina II appears to be primarily in IIo (Carlton *et al.*, 1987b, 1988; Harmann *et al.*, 1988b; McNeill *et al.*, 1988a). In the rat there is relatively less label in the lateral as opposed to the medial part of lamina II (Carlton *et al.*, 1988) There is an increase in the level of CGRP as development proceeds in the human and rat (Marti *et al.*, 1987).

In the electron microscope, the most common CGRP terminals are presynaptic elements with small clear and large dense-core vesicles which form asymmetric axodendritic synapses (Carlton *et al.*, 1987b, 1988; McNeill *et al.*, 1988a) (Fig. 4.14). In addition, particularly in the monkey, central elements of glomeruli are labeled (Carlton *et al.*, 1987b, 1988; McNeill *et al.*, 1988a). Some of the postsynaptic dendrites in these glomeruli arise from dynorphin neurons (O. Takahashi *et al.*, 1988). In addition, in the monkey there are axoaxonic synapses, with the CGRP-immunolabeled element usually being presynaptic, but in some cases both the pre- and post-synaptic elements are CGRP positive (Carlton *et al.*, 1988) (Fig. 4.35). Since CGRP is a primary afferent marker, this last finding presumably implies monosynaptic control of primary afferents by other primary afferents.

Autoradiographic localizations of CGRP receptors indicate that they are concentrated in the areas of the terminal fields of the CGRP axons (Seifert *et al.*, 1985), and the substantia gelatinosa is particularly labeled in human cord (Tschopp *et al.*, 1985; S. Inagaki *et al.*, 1986).

CGRP and SP are shown to colocalize in primary afferent terminals (Tuchscherer and Seybold, 1989).

Cholecystokinin

Cholecystokinin cells are found in the substantia gelatinosa and in laminae IV and V (Schroder, 1983). Some cells were also seen in lamina IIo, and some of the laminae IV and V cells are described as large and multipolar (Fuji *et al.*, 1985; Tohyama and Shiotani, 1986).

CCK fibers and terminals are seen in laminae I and II (Larsson and Rehfeld, 1979; Loren *et al.*, 1979; Vanderhaeghen *et al.*, 1980; Gibson *et al.*, 1981; Schroder, 1983; Conrath-Verrier *et al.*, 1984; Fuji *et al.*, 1985; Schoenen *et al.*, 1985a; Micevych *et al.*, 1986, Tuchscherer *et al.*, 1987; K. Chung *et al.*, 1989a). Fuji *et al.* (1985) state that the labeling in lamina I is much less intense than in lamina II, which no other investigators noted. Lamina IIo is more intensely labeled than IIi (Maderut *et al.*, 1982; Schroder 1983; Micevych *et al.*, 1986).

On electron-microscopic examination, asymmetric axodendritic synapses with small clear and large dense-core vesicles in the presynaptic element are labeled (Conrath-Verrier *et al.*, 1984). Central elements of glomeruli are also labeled. Finally, vesicle-containing dendrites are labeled, which presumably indicates that at least some of the CCK interneurons are islet cells (Conrath-Verrier *et al.*, 1984).

There is debate about how many CCK fibers arise from primary afferents. On the one hand, some studies show that dorsal rhizotomy (Marley *et al.* 1982; Schultzberg *et al.*, 1982; Schroder, 1983) or capsaicin (Schultzberg *et al.*, [1982] does not affect CCK greatly, but many other

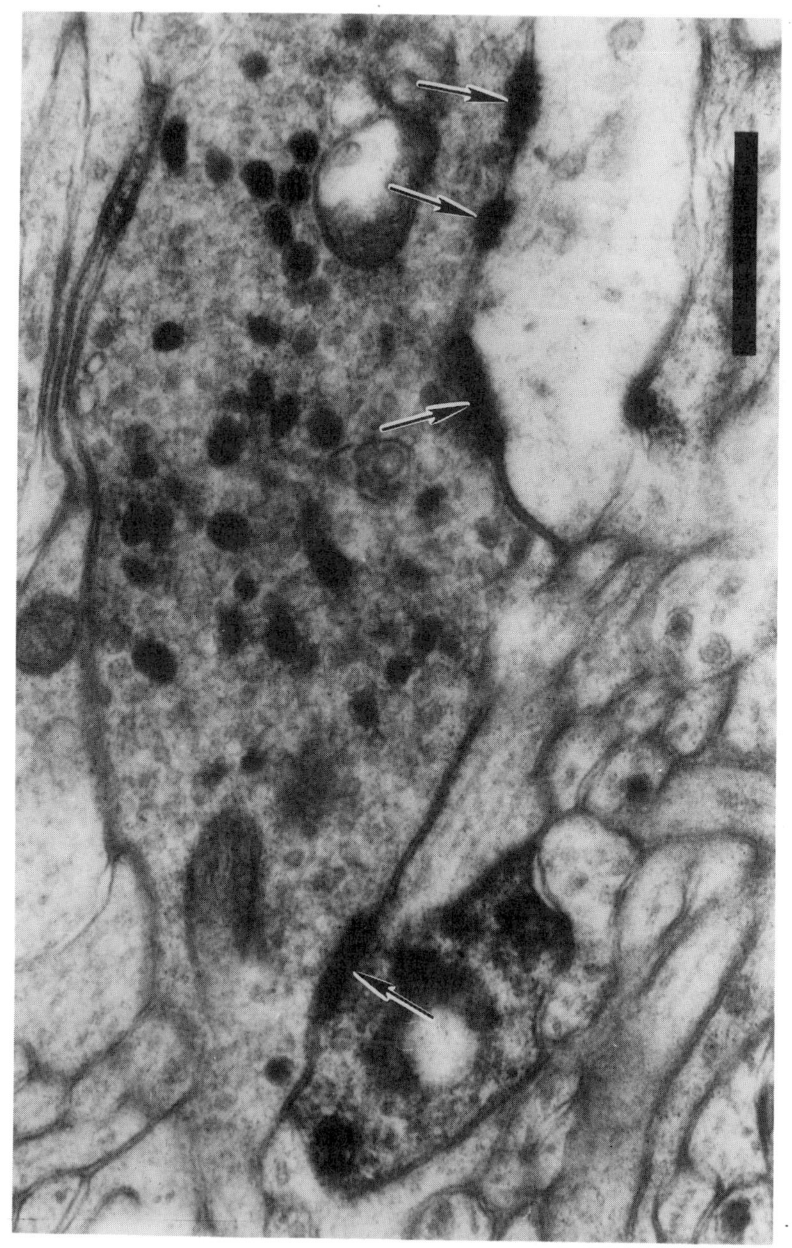

Fig. 4.35. Immunocytochemical stain for CGRP. Note that the immunolabeled terminal at the left-hand arrow makes synaptic contact with a large central immunolabeled terminal, which in turn makes contact with a dendrite. Since CGRP is a primary afferent marker in the dorsal horn, this is evidence that primary afferent fibers are synapsing with each other. (From Carlton *et al.*, 1988.)

studies show a substantial decline (Jancsó *et al.*, 1981; Maderut *et al.*, 1982; Priestly *et al.*, 1982a, Schultzberg *et al.*, 1982; Conrath-Verrier *et al.*, 1984; Micevych *et al.*, 1986), and the latter data seem convincing.

There is colocalization of CCK and SP in terminals of laminae I and II (Tuchscherer *et al.*, 1987). These investigators note that terminals containing SP and CCK seem to be of primary afferent origin, whereas those with only SP are likely to be of intrinsic or descending origin. Furthermore, they find that the CCK terminals that resist capsaicin are in lamina I and they interpret this as indicating that fine myelinated primary afferent fibers terminate here (Tuchscherer *et al.*, 1987).

CCK receptors are concentrated in the substantia gelatinosa of the rat, monkey, and human, but the type of receptor is different in the rat from that in the monkey or human (D. R. Hill *et al.*, 1988).

Corticotropin-Releasing Factor

Spinal CRF cells are not usually reported, but a few labeled cells have been seen in the marginal zone of the dorsal horn (Merchenthaler, 1984). CRF-immunolabeled fibers, terminals and receptors are seen in laminae I and II (Olschowka *et al.*, 1982; Merchenthaler *et al.*, 1983; Schipper *et al.*, 1983; Merchenthaler, 1984; Skofitsch *et al.* 1985a,b; Foote and Cha, 1988), and occasionally in lamina V (Merchenthaler *et al.*, 1983; Schipper *et al.*, 1983). Presumably, some of these fibers come from DRG cells, which recently have been shown to contain CRF (Skofitsch *et al.*, 1985a,b), but another possibility is intrinsic cells. A major CRF system is known at hypothalamic levels, but appropriate lesions show that the spinal CRF system is separate, although hypophysectomy does result in an increase in spinal immunostaining (Schipper *et al.*, 1983).

Cytochrome Oxidase

Cytochrome oxidase is an important energy-deriving enzyme that labels neurons in other parts of the nervous system. It is also found in the neuropil and in marginal and other neurons of laminae I–III (Wong-Riley and Kageyama, 1986). In the same study, DRG cells were labeled. A speculation is that the intensity of the reaction is related to the activity of the cell.

Dynorphin

Dynorphin is an opioid peptide. Immunolabeled cells are seen primarily in laminae I, IIo, and V (Khachaturian *et al.*, 1982; Vincent *et al.*, 1982; L. Cruz and Basbaum, 1985; Basbaum *et al.*, 1986b; Cho and Basbaum, 1988; 1989; Ruda *et al.*, 1988; Weihe *et al.*, 1988a), although some of the above papers emphasize one of these laminae over the others. Standaert *et al.* (1986) show that some lamina I cells that project to the thalamus contain dynorphin (others contain enkephalin). Labeled fibers and presumed terminals are prominent in laminae I, IIo, and V, particularly the lateral part of V (Vincent *et al.*, 1982; L. Cruz and Basbaum, 1985; Basbaum *et al.*, 1986b; Cho and Basbaum, 1988, 1989; Ruda *et al.*, 1988). At the ultrastructural level, dynorphin terminals form predominantly simple axodendritic synapses, but a few axoaxonal synapses are found (Cho and Basbaum, 1989).

A finding of considerable interest is that experimental inflammation (Iadarola *et al.*, 1988; Weihe *et al.*, 1988a) or peripheral nerve and/or root section (Cho and Basbaum, 1988, 1989) causes an increase in the number of immunolabeled dynorphin neurons and in the numbers of neurons that transcribe dynorphin. Both local circuit and projection neurons are affected (Nahin *et al.*, 1989). Furthermore, enkephalin and dynorphin are colocalized in some laminae IV–V cells following experimental arthritis (Weihe *et al.*, 1988b), and genes coding for dynorphin are also induced by this same procedure (Weihe *et al.*, 1989). These morphologic demonstrations follow earlier chemical work, summarized by Iadarola *et al.* (1988), showing an increase in dynorphin synthesis following experimental inflammation. The meaning of these findings is not clear, but the fact that a noxious input into the cord results in morphological changes indicating increases in active substances may well result in insights into somatic sensation in general and noxious information transfer in particular.

Enkephalin (ENK)

Enkephalin-positive cells are seen in the dorsal horn or in the equivalent part of the trigeminal nucleus (Hökfelt *et al.*, 1977a,b; Uhl *et al.*, 1979a; Del Fiacco and Cuello, 1980; S. P. Hunt *et al.*, 1980; Aronin *et al.*, 1981; Finley *et al.*, 1981b; Glazer and Basbaum, 1981, 1982; C. C. LaMotte and DeLanerolle, 1981, 1983b; Bennett *et al.*, 1982; DeLanerolle and LaMotte, 1982; Ruda, 1982; Senba *et al.*, 1982; Sumal *et al.*, 1982; Arluison *et al.*, 1983a,b; Conrath-Verrier *et al.*, 1983; Bresnahan *et al.*, 1984; Charnay *et al.*, 1984; L. Cruz and Basbaum, 1985; Schoenen *et al.*, 1986; Standaert *et al.*, 1986; K. E. Miller and Seybold, 1987; Katoh *et al.*, 1988a; Nahin, 1988) (Fig. 4.30). Early studies (Hökfelt *et al.*, 1977a,b) indicate that the enkephalin cells are found in laminae I–V. Later investigators agree that the cells are most numerous in laminae I and II. Glazer and Basbaum (1981, 1982) and K. E. Miller and Seybold (1987) report more cells in Lamina I, whereas S. P. Hunt *et al.* (1981b) find more in lamina II (Fig. 4.36). Aronin *et al.* (1981) find more in laminae I and IIo, and Finley *et al.* (1981b) a concentration at the II–III border which would probably be considered IIi. Marginal and other neurons are labeled in lamina I, and Bennett *et al.* (1982) found labeled stalked and islet cells. Thus enkephalin cells are of many morphologic types. This presumably indicates that enkephalin-labeled neurons have a wide range of functions in the dorsal horn.

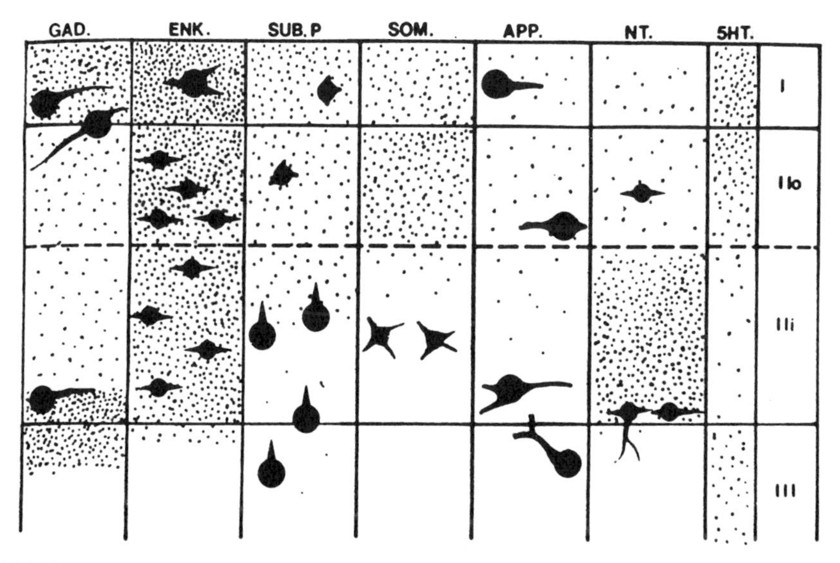

Fig. 4.36. Summary diagram of terminal and perikaryal immunoreactivity in the superficial dorsal horn. Abbreviations: GAD = glutamate decarboxylase, which is important in GABA metabolism; ENK = enkephalin; SUB. P = substance P; APP = avian pancreatic polypeptide; NT = neurotensin; 5-HT = 5-hydroxytryptamine or serotonin. (From Hunt *et al.*, 1981a.)

An extensive plexus of enkephalin fibers and terminals is found in laminae I and II (Hökfelt et el., 1977b; Uhl *et al.*, 1979a; Del Fiacco and Cuello, 1980; S. P. Hunt *et al.*, 1980; Aronin *et al.*, 1981; Finley *et al.*, 1981b; Glazer and Basbaum, 1981; S. P. Hunt *et al.*, 1981b; C. C. LaMotte and DeLanerolle, 1981; DeLanerolle and LaMotte, 1982; Priestley *et al.*, 1982a; Senba *et al.*, 1982; Sumal *et al.*, 1982; Arluison *et al.*, 1983a,b; Conrath-Verrier *et al.*, 1983; Glazer and Basbaum, 1982; Bresnahan *et al.*, 1984; Charnay *et al.*, 1984; L. L. Cruz and Basbaum, 1985; Schoenen *et al.*, 1985a, 1986; Ruda *et al.*, 1986; Tohyama and Shiotani, 1986; Tashiro *et al.*, 1987, 1988; K. E. Miller and Seybold, 1987; Katoh *et al.*, 1988a,b). In cats, enkephalin processes are less apparent in lamina IIo than in lamina I or IIi (Glazer and Basbaum, 1981; Bennett *et al.*, 1982; Conrath-Verrier *et al.*, 1983). More profiles in IIo in the rat are reported by Hunt *et al.* (1981a,b), and fewer by Bresnahan *et al.* (1984). DeLanerolle and LaMotte (1982) found more immunoreactivity in lamina I than in either IIi or IIo. Species findings undoubtedly play a part in these differences, but another possibility is that the results vary because of different physiologic states that are not yet understood.

Laminae I and II do not label for enkephalin until relatively late in development in opossums (DiTirro *et al.*, 1983), but can be found relatively early in humans (Luo *et al.*, 1988). A peptide that is related to enkephalin and modulates morphine activity is found in high concentration in the superficial dorsal horn (Panula *et al.*, 1987).

The precise synaptic localization of enkephalin has been a subject of considerable interest in considerations of pain mechanisms. An early suggestion was that enkephalin terminals might end on fine primary afferent terminals, especially those labeled by SP, and thus provide a morphologic basis for the presynaptic control of nociceptive information (Hökfelt *et al.*, 1977a,b; Del Fiacco and Cuello, 1980). Most immunoreactive enkephalin terminals end on dendrites, however, rather than on axon terminals, and thus they seem to have primarily a postsynaptic action (Fig. 4.37; S. P. Hunt *et al.*, 1980; Aronin *et al.*, 1981; Sumal *et al.*, 1982; Arluison *et al.*, 1983a,b; Glazer and Basbaum, 1982; Bresnahan *et al.*, 1984; Katoh *et al.*, 1988a). It should be noted that some ENK terminals are presynaptic to central terminals in glomeruli (Aronin *et al.*, 1981), but these seem to be rare and are not reported in most studies. In the few axoaxonic synapses involving an enkephalin-labeled element seen by Sumal *et al.* (1982) and Glazer and Basbaum (1982), the enkephalin element was postsynaptic. Sumal *et al.* (1982) emphasize that the enkephalin elements are often located in "triads" where a labeled and an unlabeled terminal contact an unlabeled dendrite (this is a different type of triad from those found in glomeruli), and Bennett *et al.* (1982) show that dendrites from enkephalin cells often contain clusters of vesicles. Glazer and Basbaum (1982) show that there are two types of

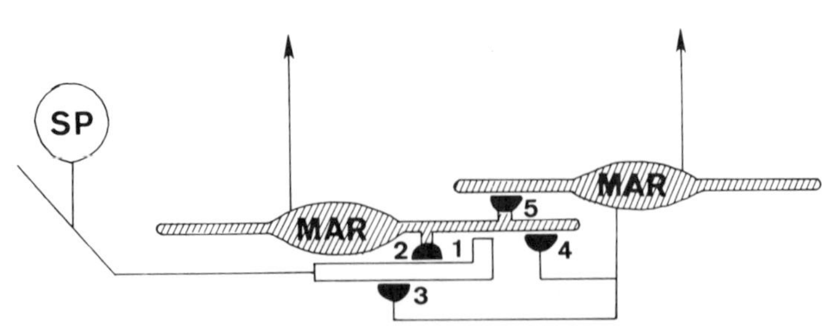

Fig. 4.37. Schematic diagram of some possible relationships in the superficial dorsal horn involving enkephalin-containing (shaded) marginal (MAR) neurons and SP primary afferents. If enkephalin is inhibitory, note that the enkephalin-containing neurons could inhibit both the SP primary afferent input and themselves. (From Glazer and Basbaum, 1981.)

enkephalin terminals in the cat; the first is large and forms the presynaptic part of a typical axodendritic synapse and is located in lamina I, and the second is smaller, is located in lamina II, and is sometimes associated with other axon terminals. A double labeling study showed that 5-HT-containing axons contact enkephalin-containing neurons and dendrites in the superficial dorsal horn (Glazer and Basbaum, 1984). It is interesting that central terminals in glomeruli do not label with enkephalin even though some enkephalin-containing DRG cells have been reported (see Chapter 3). Enkephalin terminals usually contain small clear vesicles and some large dense-core vesicles, and the synaptic thickenings are usually asymmetric, although some symmetric thickenings have been reported (Aronin *et al.*, 1981; Sumal *et al.*, 1982; Glazer and Basbaum, 1982; C. C. LaMotte and DeLanerolle, 1983b; Katoh *et al.*, 1988b). Thus, most enkephalin terminals appear to be of a type considered to be excitatory, which is contrary to the usual suggestion that enkephalin is an inhibitory transmitter.

Enkephalin terminals are shown on spinothalamic neurons (Ruda, 1982; Ruda *et al.*, 1984), and both SP and enkephalin terminals are shown on cells in lamina I and V that are likely to be projection neurons (LaMotte and DeLanerolle, 1981; DeLanerolle and LaMotte, 1982). In addition, enkephalin is localized in neurons that project to the medial thalamus (Nahin, 1988) and in marginal neurons that project to the parabrachial region (Standaert *et al.*, 1986). Thus, this compound is likely to be important in somatic sensory processing.

SP is often suggested as a primary afferent transmitter and enkephalin as an inhibitory substance. It is therefore both interesting and puzzling that these two compounds colocalize in axonal terminals in the gelatinosa (Tashiro *et al.*, 1987) and in cells of the superficial layers of the dorsal horn (Katoh *et al.*, 1988a; Senba *et al.*, 1988). These compounds are found in the same dense core granules (Katoh *et al.*, 1988b). Pro-dynorphin and pro-enkephalin are colocalized in neurons in laminae IV and V of polyarthritic rats (Weihe *et al.*, 1988a,b). Enkephalin and CGRP relationships are markedly changed from normal in rats with a hereditary sensory neuropathy (Kar *et al.*, 1989). Interestingly enkephalin convertase seems to be found in most trigeminal sensory neurons (Lynch *et al.*, 1986).

Endopeptidase-24.11, also called enkephalinase, breaks down enkephalin. This enzyme is especially prominent in lamina II (Matsas *et al.*, 1986; Back and Gorenstein, 1989; Zajac *et al.*, 1989). These authors point out that this localization overlaps the most intense distribution of the peptide. Opiate receptors are also located most strikingly in lamina II, although the resolution of the autographic method sometimes makes it difficult to distinguish laminar boundaries (C. C. LaMotte *et al.*, 1976; Atweh and Kuhar, 1977; Fields *et al.*, 1980; Ninkovic *et al.*, 1981, 1982; Gouarderes *et al.*, 1985; 1986; Gundlach *et al.*, 1986; Faull and Villiger, 1987; Morris and Herz, 1987; Zajac *et al.*, 1989). Gundlach *et al.* (1986) find less labeling in the dorsal horn than in the ventral horn, however. There is a considerable loss of receptor activity but no loss of enkephalinase following dorsal rhizotomy (C. C. LaMotte *et al.*, 1976; Fields *et al.*, 1980; Ninkovic *et al.*, 1981; Zajac *et al.*, 1989). The loss of receptors following dorsal rhizotomy is taken as evidence for presynaptic opiate receptors on sensory fibers, but as stated above, relatively few enkephalin containing terminals are found ending on primary afferent terminals.

FMRF-NH_2, which is antagonistic to ENK, is found in lamina I and may arise from supraspinal cells (Ferrarese *et al.*, 1986).

Fluoride-Resistant Acid Phosphatase

FRAP fibers and terminals are found in great abundance in lamina II (Gerebtzoff and Maeda, 1968; Coimbra *et al.*, 1970, 1974; Knyihár and Gerebtzoff, 1973; Knyihár 1971; Knyihár *et al.*, 1974; Sanyal and Rustioni, 1974; Knyihár-Csillik and Csillik, 1981; McDougal *et al.*, 1985; Csillik and Knyihár-Csillik, 1986). Some investigators emphasize that these fibers are located particularly in lamina IIi (Nagy and Hunt, 1982; Silverman and Krueger, 1988b). At the ultrastructural level, FRAP and a similar but more precise marker, thiamine pyrophosphatase (Csillik *et al.*, 1986; Knyihár-Csillik *et al.*, 1986), are found in the electron-dense central (sinusoid) terminals of glomeruli in lamina II (Coimbra *et al.*, 1970, 1974; Knyihár and Gerebtzoff, 1973; Knyihár *et al.*, 1974; Knyihár and Csillik, 1976; Knyihár-Csillik and Csillik, 1981; Csillik and Knyihár-Csillik, 1986). In addition, Coimbra *et al.* (1970, 1974) note that some of the peripheral elements in the glomeruli are labeled. It is agreed that all or almost all of the FRAP reactivity vanishes with dorsal rhizotomy (Coimbra *et al.*, 1970, 1974; Knyihár, 1971; Knyihár *et al.*, 1974; Csillik and Knyihár Csillik, 1986).

Galanin

Galanin has been found in DRG cells, in small neurons in laminae I and II of the dorsal horn, and in fibers and presumed terminals in these same laminae (Rökaeus *et al.*, 1984; Ch'ng *et al.*, 1985; Melander *et al.*, 1986). Many of the fibers in laminae I and II disappear after dorsal rhizotomy (Ch'ng *et al.*, 1985) or neonatal capsaicin treatment (Skofitsch and Jacobowitz, 1985d). Interestingly, galanin expression is increased in DRG cells and in fibers and terminals in the dorsal horn following a sciatic nerve lesion (Hökfelt *et al.*, 1987; Villar *et al.*, 1989). There are large numbers of galanin labeled axons in the dorsal roots (Klein *et al*, 1990).

γ-Aminobutyric Acid (GABA)

GABA is an important inhibitory transmitter (McLaughlin *et al.*, 1975). It has been localized, by autoradiography (Ljungdahl and Hökfelt, 1973; Kelly *et al.*, 1973), by immunocytochemical reactions for GABA itself or for GAD, in superficial laminae of the dorsal horn (McLaughlin *et al.*, 1975) (Fig. 4.32). Laminae I and IIi were identified as heavily labeled sites (McLaughlin *et al.*, 1975; Barber *et al.*, 1978; S. P. Hunt *et al.*, 1981a; Perez de la Mora *et al.*, 1981; Ruda *et al.*, 1986), and Magoul *et al.*, 1987, and laminae I–II in the trigeminal complex (Basbaum *et al.*, 1986c). GAD cells are located particularly in laminae I and IIo (Ribeiro-DaSilva and Coimbra, 1980; S. P. Hunt *et al.*, 1981a), although Barber *et al.* (1982) found them throughout the dorsal horn. Ribeiro-DaSilva *et al.* (1980) believed that the labeled lamina I cells are not marginal cells, and Barber *et al.* (1982) surmised that the labeled cells in lamina II are both stalked and islet cells. Kosaka *et al.* (1988) point out that some of the GAD cells contain ChAT.

There has been particular interest in the synaptic arrangements of GABA elements because of the pre- and postsynaptic inhibition that this compound mediates. In a study of GAD localization following dorsal rhizotomy (the latter marking primary afferent terminals because of degenerative changes), the GAD terminals were presynaptic to the central terminals in the glomeruli (Barber *et al.*, 1978) (Fig. 4.38). In an autoradiographic study, some of the peripheral terminals in the glomeruli are labeled (Ribeiro-DaSilva and Coimbra, 1980), as well as in immunocytochemical studies using antibodies to GAD and to GABA (Basbaum *et al.*, 1986c; Magoul *et al.*, 1987; M. A. Matthews *et al.*, 1988; Todd and Lochhead, 1990). These data are usually suggested as the morphologic basis for presynaptic inhibition. Simple axodendritic synapses with a labeled presynaptic terminal are also seen and are presumably the basis of postsynaptic inhibition. Such simplicity may be illusory, however, for labeled terminals can have both symmetric and asymmetric thickenings and various types of vesicles. In addition, Basbaum *et al.* (1986c) point out that in a few cases the primary afferent fiber is presynaptic to the GAD terminal. Carlton and Hayes (1990) emphasize that this by showing that the majority of GABA containing structures in the monkeys are post synaptic to primary afferent terminals, and that GABA containing dendrites are an important part of glomerular organization. As a final point, two types of GAD terminals were noted. The first type, which was large and relatively heavily

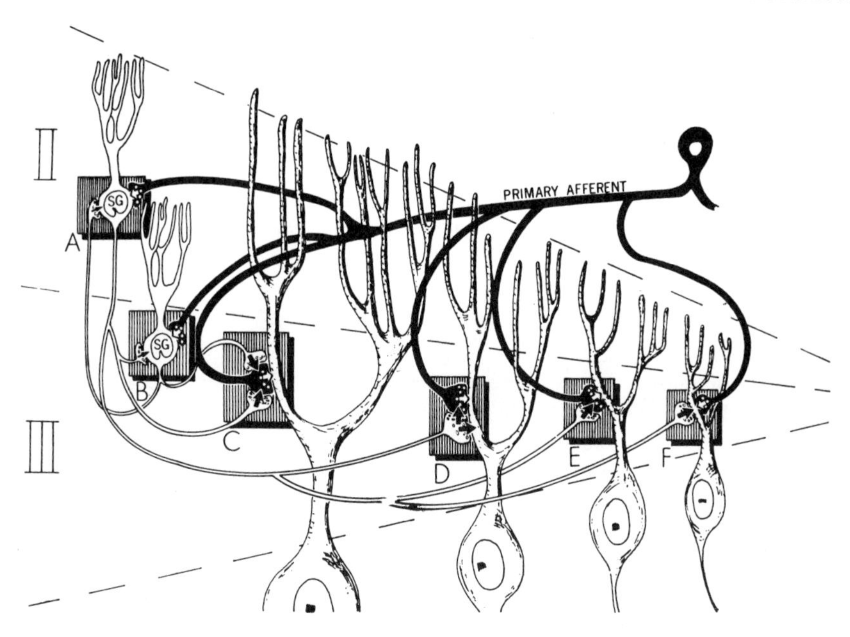

Fig. 4.38. Schematic diagram of the possible synaptic arrangements between primary afferent (Black) fibers and lamina II GABA interneurons (White.) The significance of the synaptic arrangements in A–F is discussed in the text. (From Barber *et al.*, 1978.)

stained, is not found within the cord and is hypothesized to arise from long tract cells. The second, which is smaller and less heavily stained, is the characteristic terminal in the cord, particularly in lamina II, and is hypothesized to arise from Golgi type II neurons (Perez de la Mora *et al.*, 1981).

Receptor binding sites for GABA are of two types: GABAa and GABAb (Bowery *et al.*, 1987). The GABAa sites are numerous but not well localized, at least in the dorsal horn, whereas the bicuculline-insensitive GABAb sites are localized to laminae II and III (Palacios *et al.*, 1981; G. W. Price *et al.*, 1984, 1987; Bowery *et al.*, 1987). The GABAb receptors are reduced following dorsal rhizotomy or neonatal capsaicin administration (G. W. Price *et al.*, 1984, 1987), which would be consistent with the idea that they are on primary afferent axons and terminals, but not with the lack of evidence for GABA containing primary afferent cells (G. W. Price *et al.*, 1987). Benzodiazepine receptors are localized to laminae II (Faull *et al.*, 1986).

Glucocorticoid Receptor

Neurons immunoreactive for glucocorticoid receptors are seen in the substantia gelatinosa of the rat spinal cord (Fuxe *et al.*, 1985). In the brain stem, these neurons are found in the vicinity of the amine containing neurons (Fuxe *et al.*, 1985).

Glutamate

Glutamate, an important excitatory transmitter candidate, is found in many DRG cells (see Chapter 3). Glutamate axons, terminals, cells, and receptors are also found in laminae I–II of the spinal cord (Greenamyre *et al.*, 1984; K. R. Magnusson *et al.*, 1986, Weinberg *et al.*, 1987; De Biasi and Rustioni, 1988; K. E. Miller *et al.*, 1988). Aspartate immunotransferase is reported on the surface of dorsal horn neurons, but the pictures are not completely convincing (Martinez-Rodriguez and Diaz, 1987). Some glutamate-containing neurons project cranially or caudally for three to five segments (Rustioni and Cuénod, 1982), and some of the glutamate-containing neurons in lamina I project to the thalamus (K. R. Magnusson *et al.*, 1987). Since there is some uncertainty about the size of glutamate-containing DRG cells (see Chapter 3), it is interesting that the bulk of the spinal labeling occurs where fine primary afferents terminate. This may implicate glutamate in nociceptive information transfer. Glutamate colocalizes with substance P (De Biasi and Rustioni, 1988).

Glycine

Glycine is an inhibitory transmitter candidate in the mammalian cord (Hökfelt and Ljungdahl, 1971; Ljungdahl and Hökfelt, 1973). The anatomical localization of glycine in the cord is done primarily by autoradiography of radioactive glycine or by localization of glycine receptors, but recently an antibody against glycine has been developed (van den Pol and Gorcs, 1988). There are discrepancies in the laminar localizations. Hökfelt and Ljungdahl (1971) report the heaviest labeling in laminae VI and upper VII, whereas Ribeiro-DaSilva and Coimbra (1980) find it in lamina III. Perhaps these differences reflect different functional states. Receptor localizations show most labeling in laminae II–III and V–VI (Zarbin *et al.*, 1981; Basbaum, 1988), lamina II (Frostholm and Rotter, 1985; A. Probst *et al.*, 1986), and lamina III (Murakami *et al.*, 1988). In a study of both receptors and immunolocalization of glycine itself, the most intense labeling was in deeper laminae but much was also seen in laminae I and II (van den Pol and Gorcs, 1988). Some small glycine-containing cells have been seen in lamina II (Campistron *et al.*, 1986). Glycine is usually found in symmetric flat vesicle terminals (Hökfelt and Ljungdahl, 1971; van den Pol and Gorcs, 1988).

Histamine

Histamine fibers, presumed terminals, and receptors are found in superficial layers of the dorsal horn (Bouthenet *et al.*, 1988; Panula *et al.*, 1989b).

Neurotensin

There is evidence that neurotensin may reduce "pain" (L. Cruz and Basbaum, 1985). Neurotensin is often discussed with enkephalin and dynorphin in this regard. Neurotensin-containing cells are found in lamina II (Uhl *et al.*, 1979b; DiFiglia *et al.*, 1982a, 1984; Jennes *et al.*, 1982; Seybold and Elde, 1982; Yaksh *et al.*, 1982; Vincent *et al.*, 1982; Seybold and Maley, 1984; L. Cruz and Basbaum, 1985; Papadoupolos *et al.*, 1986a) (Fig. 4.36). Labeled cells have been suggested as spiny cells (Uhl *et al.*, 1979b); islet cells (Seybold and Elde, 1982; DiFiglia *et al.*, 1984), stalked cells, (DiFiglia *et al.*, 1984) and II–III border cells (Seybold and Elde, 1982). Seybold and Maley (1984) emphasize that the labeled cells are at the I–II and II–III borders. Labeled cells are also found in laminae I, III, and reticular V.

Fibers and presumed terminals are seen in laminae I–III (Seybold and Elde, 1980; DiFiglia *et al.*, 1982a, 1984; Jennes *et al.*, 1982; Yaksh *et al.*, 1982b; Mai *et al.*, 1987), and IIo is specifically mentioned by DiFiglia *et al.* (1982a, 1984) and Yaksh *et al.* (1982b). At the electron-microscopic level, presynaptic elements in axodendritic synapses are typically labeled, although some axosomatic synapses are seen (DiFiglia *et al.*, 1982a, 1984; Emson *et al.*, 1982; Seybold and Maley, 1984). It is interesting that the labeled terminals contain primarily small, round, clear vesicles and relatively few large dense-core vesicles, although the latter usually were most intensely labeled (DiFiglia *et al.*, 1984; Seybold and Maley, 1984).

Receptors to neurotensin are localized to lamina II or laminae I and II and are not affected by dorsal rhizotomy (Ninkovic *et al.*, 1981; Young and Kuhar, 1981; Emson *et al.*, 1982). A later study found these receptors primarily in laminae II; and III (Faull *et al.*, 1989).

Neuropeptide Y

Neuropeptide Y, peptide YY, and the pancreatic polypeptides are very similar and will be described as for neuropeptide Y. Neuropeptide Y-containing cells are seen in laminae I–III, V, and VI, but especially in lamina II (J. M. Lundberg *et al*; 1980; Hökfelt *et al.*, 1981; S. P. Hunt *et al.*, 1981a,b; Gibson *et al.*, 1984a; Sasek and Elde, 1985; De Quidt and Emson, 1986; Krukoff, 1987) (Fig. 4.36). Earlier papers noted that terminals and presumed endings are found in laminae I and II (S. P. Hunt *et al.*, 1981b; Allen *et al.*, 1984; Chronwall *et al.*, 1985). Gibson *et al.* (1984), however, found fibers in laminae II and III, S. P. Hunt *et al.* (1981a) found fibers primarily in lamina IIi and Krukoff (1987) found terminals in laminae II and III and fibers in laminae I and IV–VI. Neuropeptide Y colocalizes with enkephalin and catecholamines (S. P. Hunt *et al.*, 1981a).

Noradrenaline and Dopamine

Noradrenaline and dopamine arise from supraspinal cells. There is more noradrenaline than dopamine in the spinal cord (Skagerberg *et al.*, 1982). The noradrenaline is found in laminae I, IIo, and V, and the dopamine is in superficial areas of the dorsal horn (Dahlström and Fuxe, 1965; Westlund *et al.*, 1982, 1983; Kondo *et al.*, 1985; Schroder and Skagerberg, 1985). Interestingly, some tyrosine hydroxylase positive and thus presumably dopaminergic neurons are seen in the marginal layer of the S1 segment, and some of these are thought to be marginal cells (Mouchet *et al.*, 1986). A sacral localization of tyrosine hydroxylase-containing DRG cells is also seen (see Chapter 3).

Serotonin

Serotonin in the spinal cord is generally attributed to supraspinal sources (Ruda *et al.*, 1986). Nevertheless, a few 5-HT-containing cells have been found in rat and primate lamina X, around the central canal and in marginal and gelatinosal layers (C. C. LaMotte *et al.*, 1982; Bowker, 1986; Newton and Hamill, 1988). These latter cells are of considerable interest, but their small numbers may imply only a small contribution to the axonal plexuses of the dorsal horn (C. C. LaMotte and DeLanerolle, 1983b; Bowker, 1986; Ruda *et al.*, 1986).

Serotonin-containing axons are seen in the dorsal horn (Dahlström and Fuxe, 1965). The labeling is greatest in laminae I and II (Ruda and Gobel, 1980; Steinbusch, 1981; Ruda *et al.*, 1982; M. Kojima *et al.*, 1982, 1983; C. C. LaMotte and DeLanerolle, 1983b; Light *et al.*, 1983; Maxwell *et al.*, 1983b), although IIo seems to have fewer labeled processes and terminals. 5-HT axons are also seen in the marginal zone (Maxwell *et al.*, 1983b). Ruda and Gobel (1980) marked presynaptic endings in the trigeminal complex by autoradiography or 5,6-dihydroxytryptamine administration and showed seven types of terminals. It is of interest that some scalloped endings are labeled because they are thought to arise from primary afferents, although Ruda and Gobel (1980) think this unlikely because of the absence of 5-HT-containing primary afferent neurons. In the monkey dorsal horn, the presynaptic boutons do not fall into as many categories but are otherwise similar (Ruda *et al.*, 1982). Several studies note that 5-HT synaptic thickenings are symmetric, a morphologic arrangement which is usually correlated with inhibition, although vesicle types vary somewhat (Ruda *et al.*, 1982; C. C. LaMotte and DeLanerolle, 1983b; Light *et al.*, 1983). The processes of several types of spinal cells are contacted by 5-HT endings (Ruda *et al.*, 1982; Light *et al.*, 1983), and C. C. LaMotte and DeLanerolle (1983b) show some axoaxonic synapses with 5-HT terminals as the presynaptic element. They speculate that these may be responsible for the presynaptic effects of 5-HT. Suggested circuitry results in interesting wiring diagrams (Figs. 4.39 and 4.40).

In studies providing further insight into the wiring diagram of the spinal 5-HT system, 5-HT endings on enkephalin cell bodies and dendrites were found (Glazer and Basbaum, 1984). This is an important finding, which correlates with the known effects of 5-HT on opiate-mediated analgesia, but a caution is that actual cytoplasmic thickenings that usually characterize chemical synapses were rare, even though it was common to find 5-HT containing presumed terminals abutting enkephalin processes (Glazer and Basbaum, 1984).

Numerous symmetric 5-HT terminals are also found on spinothalamic cells and other identified dorsal horn cells (Hoffert *et al.*, 1983; Nishikawa *et al.*, 1983; Miletić *et al.*, 1984; Hylden *et al.*, 1986b; Ruda, 1988), emphasizing the possible importance of this substance in noxious and other types of information transfer. 5-HT has been found to coexist with SP (Wessendorf and Elde, 1987; Tashiro and Ruda, 1988), which is consistent with the findings that some brain stem spinally projecting 5-HT neurons contain SP. These double-labeled axons were common in the ventral horn but rare in the dorsal horn. 5-HT also colocalizes with ENK (Tashiro *et al.*, 1988), which complicates considerations of 5-HT opioid connectivity (Tashiro *et al.*, 1988). It has been suggested that there is general release of 5-HT into the cerebrospinal fluid of the cord. In this case the location of the presynaptic terminals would be less important than the locations of the receptors. In an early study, receptors were found in the whole dorsal horn (Segu

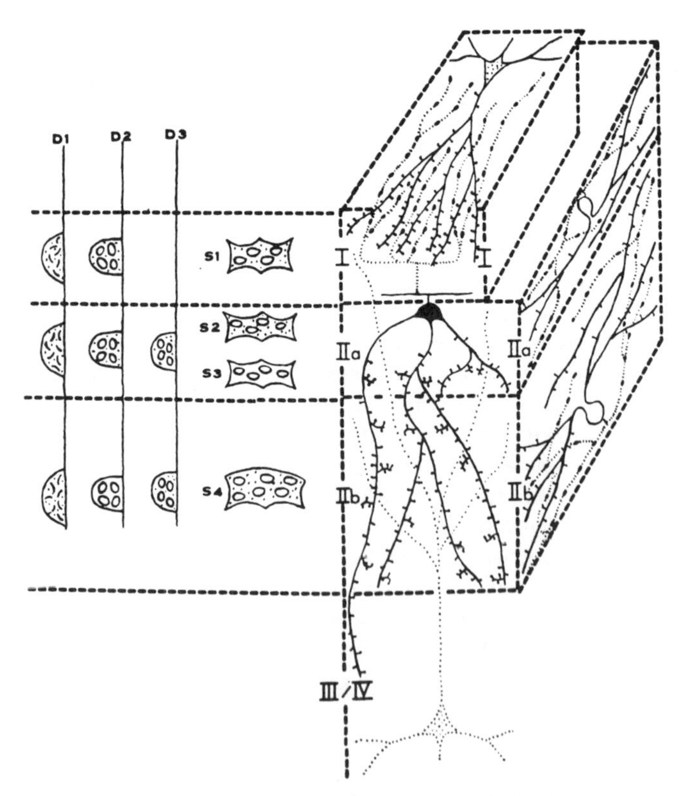

Fig. 4.39. Schematic diagram of the laminar distribution of serotonin terminals on the left and some of the neuronal types in the superficial dorsal horn on the right. The potential synaptic connections of each serotoninergic axon can be determined by reading across each layer from left to right. (From Ruda and Gobel, 1980.)

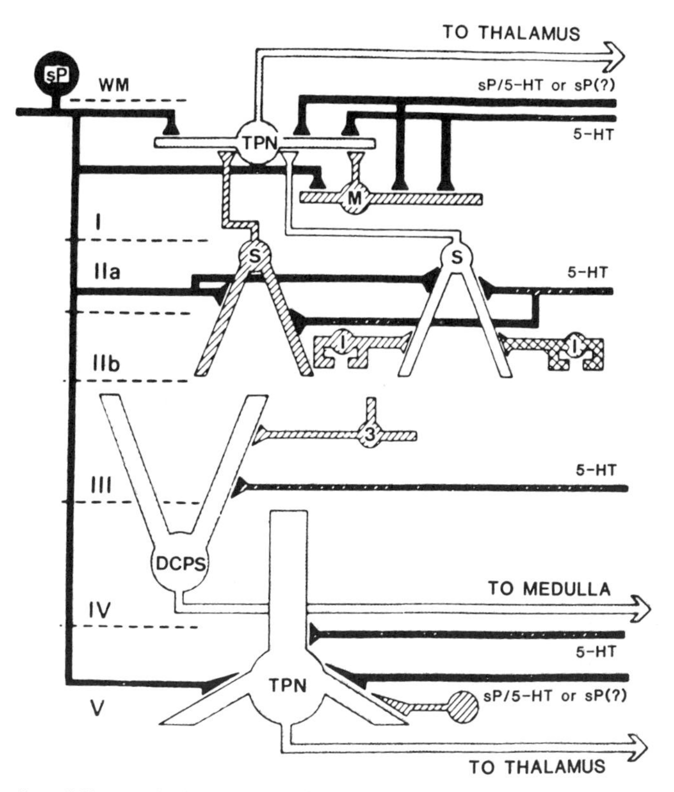

Fig. 4.40. A complicated diagram that is a foretaste of the increase in our knowledge of dorsal horn circuitry that can be expected to follow studies of the various immunolabeled compounds found in this area and in the primary afferents. Labels of immunolabeled compounds are as in Fig. 4.36. Abbreviations: TPN = thalamic projection neuron, M = lamina I local circuit neuron; S = stalked cell, I, = islet cell; DCPS = dorsal column postsynaptic cell. Diagonal striping denotes enkephalin and cross-hatching denote GABA. (From Ruda *et al.*, 1986.)

and Calas, 1978). In more recent work, the localizations were more precisely specified to laminae I and II (Daval *et al.*, 1987) and, particularly to the lamina II–III border (Seybold, 1985).

The possible primary afferent nature of some of the 5-HT terminals in the dorsal horn is an interesting and important question. It is generally believed that there are no 5-HT primary afferents. Kai-Kai and Keen (1985) reported, however, that approximately 9% of DRG cells contain 5-HT, although there was some difficulty in distinguishing DRG cells from mast cells. This observation is hard to reconcile with the low levels of 5-HT in the ganglion (Hadjiconstantinou *et al.*, 1984), and it has been suggested that Kai-Kai and Keen (1985) might have been identifying tyrosine hydroxylase primary afferent cells (Tuchscherer and Seybold, 1989). Nevertheless, the numbers of both 5-HT receptors and 5-HT axons drop in the dorsal horn when the roots are cut or capsaicin is given (Daval *et al.*, 1987; Tuchscherer and Seybold, 1989). Accordingly, the question of whether 5-HT is found in a subpopulation of primary afferent cells deserves further consideration.

Somatostatin

Somatostatin-labeled neurons are found primarily in the substantia gelatinosa or lamina II, with some in lamina I (Seybold and Elde, 1980; Dalsgaard *et al.*, 1981; S. P. Hunt *et al.*, 1981a; Senba *et al.*, 1982; Ho, 1983; O. Johansson *et al.*, 1984; Schroder, 1984; Schoenen *et al.*, 1985a;

Vincent *et al.*, 1985; Krukoff *et al.*, 1986; Ruda *et al.*, 1986; Papadoupolos *et al.*, 1986b; Charnay *et al.* 1987; Mizukawa *et al.*, 1988; Rosenthal and Ho, 1989) (Fig. 4.36). Several of these investigators note that the labeled cells are small and spindle shaped (Dalsgaard *et al.*, 1981; S. P. Hunt *et al.*, 1981a; O. Johansson *et al.*, 1984; Schoenen *et al.*, 1985a; Vincent *et al.*, 1985; Ruda *et al.*, 1986; Mizukawa *et al.*, 1988). Some investigators state that the cells are concentrated in lamina IIi (Dalsgaard *et al.*, 1981; S. P. Hunt *et al.*, 1981a; Ho, 1983). Schoenen *et al.* (1985a) state that the somatostatin cells are like islet cells.

Lamina II contains many somatostatin fibers and presumed terminals (Hökfelt *et al.*, 1975a, 1976; Forssmann, 1978; Burnweit and Forssmann, 1979; Forssmann *et al.*, 1979; Seybold and Elde, 1980; Dalsgaard *et al.*, 1981; S. P. Hunt *et al.*, 1981a; DiTirro *et al.*, 1981, 1983; Senba *et al.*,1982; Ho, 1983; Massari *et al.*, 1983; Ho and Berelowitz, 1984; O. Johansson *et al.*, 1984; Schroder, 1984; Schoenen *et al.*, 1985a; Vincent *et al.*, 1985; Ruda *et al.*, 1986; Charnay *et al.*, 1987; Ho, 1988; Mizukawa *et al.*, 1988; K. Chung *et al.*, 1989a). Several investigators mention that the labeled processes are concentrated in lamina IIo (S. P. Hunt *et al.*, 1981a; Ho, 1983; Ruda *et al.*, 1986), but Schoenen *et al.* (1985a) find more in IIi. Receptors to somatostatin are found in the substantia gelatinosa (Reubi and Maurer, 1985). At the electron-microscopic level, somatostatin was observed mostly in unmyelinated and some fine myelinated axons and in presynaptic nerve terminals in typical asymmetric axodendritic synapses (Mizukawa *et al.*, 1988). The labeled terminals contain many small clear vesicles and some large dense-core vesicles.

Many of the somatostatin-positive fibers in laminae I and II are of primary afferent origin (Hökfelt *et al.*, 1975a, 1976; Ruda *et al.*, 1986), but Schroder (1984) thinks that more arise from intrinsic neurons. The exact proportion of primary afferent *vs.* propriospinal somatostatin terminals in lamina II is not known.

Substance P

There are many (SP) primary afferent neurons (see Chapter 3). There are also many SP neurons in the dorsal horn, lamina II in particular (Hökfelt *et al.*, 1977b; Del Fiacco and Cuello, 1980; Seybold and Elde, 1980; Gibson *et al.*, 1981; S. P. Hunt *et al.*, 1981a; Nagy *et al.*, 1981; Tessler *et al.*, 1981; Senba *et al.*, 1982; Ho, 1983; Ruda *et al.*, 1986; Tohyama and Shiotani, 1986; Katoh *et al.*, 1988a,b) (Fig. 4.36). The cells are reported to be relatively small (S. P. Hunt *et al.*, 1981a; Ruda *et al.*, 1986). Cells that contain the mRNA for encoding SP are found in the same locations (Warden and Young, 1988).

SP fibers and terminals are found in abundance in laminae I and II (Hökfelt *et al.*, 1975b,c, 1976, 1977b; Chan-Palay and Palay, 1977a,b; Pickel *et al.*, 1977; Cuello and Kanazawa, 1978; Cuello *et al.*, 1978; Ljungdahl *et al.*, 1978; Naftchi *et al.*, 1978; Barber *et al.*, 1979; del Fiacco and Cuello, 1980; Seybold and Elde, 1980; Tessler *et al.*, 1980, 1981, 1984; Ainsworth *et al.*, 1981; DiTirro *et al.*, 1981, 1983; Gibson *et al.*, 1981; Hunt *et al.*, 1981a; LaMotte and DeLanerolle, 1981; Nagy *et al.*, 1981; DiFiglia *et al.*, 1982b; Priestley *et al.*, 1982a,b; Senba *et al.*, 1982, 1988; DeGroat *et al.*, 1983; Ho, 1983; Massari *et al.*, 1983; Breshanhan *et al.*, 1984; Kawatani *et al.*, 1985; Schoenen *et al.*, 1985a, 1986; M. S. Davidoff *et al.*, 1986; DeGroat, 1986; Drew *et al.*, 1986; Heii and Emson, 1986; Micevych *et al.*, 1986; Ruda *et al.*, 1986; Tohyama and Shiotani, 1986; Charnay *et al.*, 1987; Sharkey *et al.*, 1987; Tashiro *et al.*, 1987; Tuchscherer *et al.*, 1987; De Biasi and Rustioni, 1988; Katoh *et al.*, 1988a,b; Tashiro and Ruda, 1988; K. Chung *et al.* 1989a; Tuchscherer and Seybold, 1989). Several investigators state that the label is heaviest in laminae I and IIo with less in IIi (S. P. Hunt *et al.*, 1981a; DiFiglia *et al.*, 1982b; DeLanerolle and LaMotte, 1983; Bresnahan *et al.*, 1984; Priestley *et al.*, 1982a,b; Schoenen *et al.*, 1985a; Ruda *et al.*, 1986), although others do not mention differential labeling in the lamina II subdivision. It is probable, however, that the label for SP is somewhat heavier in lamina IIo than IIi.

A major point about SP innervation to laminae I and II of the dorsal horn is that approximately half arises from primary afferent axons and half from intrinsic axons (Tessler *et al.*, 1980, 1981; Mantyh and Hunt, 1985a; Ruda *et al.*, 1986; Howe *et al.*, 1987). Almost none comes from supraspinal sources, as evidenced by lack of change following spinal cord transection (Micevytch *et al.*, 1986). Tessler *et al.* (1980, 1981, 1984) removed the primary afferent input and then showed a return of SP levels in laminae I and II to or toward normal as time progressed. These results were interpreted as sprouting. The fact that somatostatin did not behave in the same way was seen as evidence for the specificity for the phenomenon (Tessler *et al.*, 1984). These results were confirmed by Micevytch *et al.* (1986), who provided further evidence that the sprouting was from intrinsic SP axons. An increase in the level of SP, as well as of FRAP and 5-HT, was seen in the dorsal horn following experimental arthritis (Schoenen *et al.*, 1985b).

Electron-microscopic examination of laminae I and II shows that many unmyelinated and some fine myelinated axons are labeled for SP (see, e.g., DiFiglia *et al.*, 1982b; DeLanerolle and LaMotte, 1983; Bresnahan *et al.*, 1984). The majority of labeled terminals are presynaptic to dendrites and contain many small clear vesicles and some large dense-core vesicles (Chan-Palay and Palay, 1977a,b; Pelletier *et al.*, 1977; Pickel *et al.*, 1977; Barber *et al.*, 1979; DiFiglia *et al.*, 1982b; Priestley *et al.*, 1982a,b; DeLanerolle and LaMotte, 1983; Bresnahan *et al.*, 1984; De Biasi and Rustioni, 1988; Katoh *et al.*, 1988a,b; Ribeiro DaSilva *et al.*, 1989). Axosomatic contacts have been reported but are relatively rare (DeLanerolle and LaMotte, 1983; but see Ribeiro DaSilva, 1989), and some of the axosomatic contacts have symmetric cytoplasmic thickenings, in contrast to the typical terminals, where the thickenings are asymmetric (De-Lanerolle and LaMotte, 1983). A major point of almost all fine-structural studies is that axon terminals ending on SP terminals are nonexistent or rare, and the usual suggestion is that this is compatible with postsynaptic rather than presynaptic mechanisms for inhibition of primary afferent transmission.

Central elements of glomeruli are the other major type of labeled SP terminals in laminae I and II (Barber *et al.*, 1979; DiFiglia *et al.*, 1982b; Priestley *et al.*, 1982a,b; DeLanerolle and LaMotte, 1983; Bresnahan *et al.*, 1984; de Biasi and Rustioni, 1988). In later studies, a distinction is made between light and dark central terminals, and de Biasi and Rustioni (1988) note that both types are labeled. Peripheral terminals in glomeruli do not label.

Receptors for SP have been localized in the dorsal horn (Shults *et al.*, 1984; Charlton and Helke, 1985a,b; Mantyh and Hunt, 1985b; Massari *et al.*, 1985; Helke *et al.*, 1986; Conrath *et al.*, 1988). In normal material, the receptors are concentrated in laminae I and II. Various types of primary afferent denervation caused an increase in the number of SP receptors, and this was regarded as evidence that the SP receptors are located postsynaptically (Massari *et al.*, 1985; Helke *et al.*, 1986). Mantyh and Hunt (1985b) believed that the changes were too small to provide a supersensitivity explanation for such phenomena as phantom limb pain, however. An SP-degrading enzyme has been localized in laminae I and II of the dorsal horn (Probert and Hanley, 1987). In addition, opiate and histamine receptors are seen on a relatively high proportion of DRG cell bodies (Ninkovic and Hunt, 1985).

A major development is the discovery that SP is colocalized with other neuroactive compounds. Both enkephalin and SP are found in neurons in laminae I and II of the dorsal horn (Katoh *et al.* 1988a; Senba *et al.*, 1988). The two groups agree that almost all of the SP neurons contain enkephalin. SP and enkephalin are also colocalized in axons and terminals in laminae I and II (Tashiro *et al.*, 1987; Katoh *et al.*, 1988b). Katoh *et al.* (1988b) note that the two compounds seem to be in the same dense-core vesicles in presynaptic terminals. 5-HT and SP are colocalized in axons and terminals of laminae I and II (Wessendorf and Elde, 1987; Tashiro and Ruda, 1988). The double labeling was relatively rare in the dorsal horn, compared the intermediate and ventral parts of the cord, suggesting that this particular double labeling may be associated more with motor activity. SP and glutamate are colocalized in the dark scalloped central terminals in laminae I and II (De Biasi and Rustioni, 1988). SP and CCK are colocalized in axons and terminals in laminae I and II (Tuchscherer *et al.*, 1987). These investigators note that terminals that contain CCK alone or both CCK and SP seem to be primary afferent fibers, whereas those with only SP probably arise from intrinsic fibers. CGRP, galanin, and dynorphin are localized with SP in the dorsal horn (Tuchscherer and Seybold, 1989). The terminals with SP+CGRP and SP+GAL seem to be from primary afferent fibers.

Thyrotropin-Releasing Hormone

Recently, TRH has been found in the dorsal horn. TRH cell bodies are found in laminae I–III (Coffield *et al.*, 1986; Tsurao *et al.*, 1987). Coffield *et al.* (1986) note that some of the labeled cells in lamina IIi resemble stalked cells and others resemble spiny cells. Fibers and presumed terminals are seen in laminae II and III (Harkness and Brownfield, 1986; Lechan *et al.*, 1987; Ulfhake *et al.*, 1987), and receptors to TRH are seen in the superficial dorsal horn (Mantyh and Hunt, 1985a) or substantia gelatinosa (Manaker *et al.*, 1985a,b; Sharif and Burt, 1985; Winokur *et al.*, 1989), but Mantyh and Hunt (1985a) state that the receptor and TRH itself may not be found together. TRH is located in typical axodendritic synapses (Ulfhake *et al.*, 1987).

Vasoactive Intestinal Polypeptide

VIP cells are present in the spinal cord but not in the dorsal horn (Fuji *et al.*, 1983, 1985; C. C. LaMotte and de Lanerolle, 1986), except that Gibson *et al.* (1984a) note some immunolabeled cells in the lateral part of laminae II and III. VIP fibers and presumed terminals are found in abundance in the tract of Lissauer and lamina I, with labeled elements curving down to the reticular part of lamina V, predominantly in sacral areas of the cord (Gibson *et al.*, 1981; Anand *et al.*, 1983; Basbaum and Glazer, 1983; DeGroat *et al.*, 1983; Honda *et al.*, 1983; Kawatani *et al.*, 1983, 1985; Gibson *et al.*, 1984a; Charnay *et al.*, 1985; C. C. LaMotte and DeLanerolle, 1986; Triepel *et al.*, 1987; see particularly the 1983 articles). There is some debate about whether lamina II, particularly IIo, is labeled. The majority of these fibers and terminals are thought to be from primary afferent fibers.

It is interesting that VIP fibers and terminals seem to coexist with the pelvic visceral afferent distribution (Kawatani *et al.*, 1983, 1985; DeGroat *et al.*, 1983; DeGroat, 1986). Another interesting feature is that the amount of VIP staining increases after various manipulations, in contrast to the behavior of most other peptides that have been studied in this regard (McGregor *et al.*, 1984; Atkinson and Shehab, 1986; Shehab and Atkinson, 1986a,b; Shehab *et al.*, 1986; Crowe and Burnstock, 1988). Obviously, this response to such things as peripheral nerve injury is of interest in terms of monitoring the response of the nervous system to various types of insults.

In summary, cells, axons and terminals in the dorsal horn of the spinal cord contain an extraordinary number of visualizable compounds. Of particular interest for this review are the insights into neural circuitry, particularly the circuitry of the glomeruli, and changes in labeling following various physiologic and traumatic manipulations. This work is just beginning, but as knowledge advances, insights into sensory mechanisms in particular and neural organization in general will undoubtedly emerge. The development of this field is one of the major advances since the previous edition of this book.

CONCLUSIONS

1. The neurons of lamina I are classically divided into large marginal cells and smaller cells on the basis of dendritic organization.

2. The marginal cells have wide-ranging dendrites that travel tangentially over the superficial dorsal horn. The smaller cells are not as well characterized.

3. More recently, the above subdivisions have been shown to be oversimplifications. The new classifications do not yet agree with one another, however, nor do specific dendritic patterns yet correlate with function, but the increase in the power of the technology gives promise of major advances.

4. Some lamina I cells contain certain compounds, among them ACHE, dynorphin, enkephalin, GABA, glutamate, glycine, somatostatin, and SP.

5. Lamina I cells project to the thalamus, midbrain, other parts of the brain stem, possibly the cerebellum, and other parts of the spinal cord.

6. Primary afferent input into lamina I comes through the marginal plexus, which arises from collaterals from fine afferent axons in the tract of Lissauer and surrounding white matter.

7. Both unmyelinated and fine myelinated primary afferents enter lamina I through the marginal plexus, but the fine myelinated input is thought to predominate.

8. Many axons and synaptic terminals contain visualizable compounds. A list of these is provided in the text.

9. Simple axodendritic synapses predominate in the lamina I neuropil, but axosomatic, axoaxonic, dendrodendritic, and dendroaxonic terminals are also seen, albeit in much smaller numbers.

10. Axoaxonic, dendrodendritic and dendroaxonic synapses are usually found in glomeruli (see the section on lamina II).

11. Lamina II is an important region of the dorsal horn that is characterized by large numbers of small cells and few myelinated axons. The term "substantia gelatinosa" is at present equated with lamina II.

12. Lamina II cells were classically divided into central and limiting cells. The limiting cells are found in the outer part of lamina II, and they have a dendritic organization with characteristics of both marginal and central cells. The central cells are the common cells of the gelatinosa, and their dendrites are characterized primarily by being flattened mediolaterally.

13. More recently the above descriptions have been seen to be too simple. A generally accepted classification is into islet and stalked cells, with several other cell types being identified by various investigators.

14. Stalked cells are essentially equivalent to limiting cells.

15. Islet cells make up the bulk of the central cells. Their dendrites are reported to be flattened mediolaterally, but serial-section reconstructions or intracellular filling indicate that this may be an oversimplification. The dendrites of islet cells contain synaptic vesicles and form terminals on surrounding dendrites and axonal terminals.

16. There is not complete agreement about the other cell types in lamina II. Some investigators suggest that the dendritic patterns of lamina II cells do not clearly fall into separate types. Instead, they suggest that dendritic patterns form spectra, with the named types being clear examples. Physiologic correlates with the cell types are not known, except that the laminae into which the dendrites of a cell enter seem to determine the type of primary afferent input that excites the cell.

17. Classically, central cells were thought to have a local axonal tree (Golgi type II cells) or to project into the spinal white matter (propriospinal neurons). Szentágothai (1964) suggested that both ultimately projected only to the gelatinosa, which was an important part of the concept of the "closed" gelatinosa, but this is no longer accepted.

18. More recently, the axonal projections of lamina II cells have been found to be more widespread than previously thought. Some project out of the spinal cord (true projection neurons), some project to different segments (propriospinal neurons), some project to different laminae (interlaminar interneurons), and some have axons confined to a lamina in the region of the dendritic tree of that cell (intralaminar interneurons, Golgi type II cells, local interneurons).

It is generally agreed that there is an important distinction between local interneurons and cells whose axons project to different regions or laminae. It must be remembered, however, that serial-section reconstructions and intracellular fillings indicate that the axonal trees of lamina II neurons are more extensive than was previously thought. In addition, almost all neurons have a local part of the axonal tree that can subserve the functions previously ascribed to Golgi type II cells.

19. Unmyelinated fibers form most of the primary afferent input into lamina II.

20. The neuropil of lamina II consists largely of axodendritic synapses and glomeruli. Axosomatic synapses are rare, and axoaxonic, dendrodendritic, and dendroaxonic synapses are found mostly in glomeruli.

21. The glomeruli are key elements in the synaptic architecture of the dorsal horn because they seem to be the morphologic structures that allow both presynaptic and postsynaptic modification of the primary afferent input at the spinal level.

22. Glomeruli consist of a central terminal, which arises from a primary afferent fiber, surrounded by dendrites and other axonal terminals, the whole being at least partly set off from the general neuropil by glial processes. The central terminal synapses on the peripheral dendrites, and the peripheral terminals synapse back onto the central terminal. The central terminals contain many of the compounds that characterize small dorsal root ganglion cells.

23. Central terminals are divided into light and dark (dense) categories, and in monkeys some are characterized by numerous large dense-core vesicles.

24. Many peripheral dendrites are thought to be dorsal dendrites from neurons in deeper laminae (the antenna-type cells) and others are from intrinsic lamina II neurons. One important point is that the dendrites of one intrinsic cell type, the islet cell, form synapses on other dendrites or onto the central terminal.

25. The origin of the peripheral terminals is not known, but it is presumed that they arise from intrinsic cells of lamina II. Some of these terminals contain GABA.

26. The dendritic patterns of lamina III neurons were classically thought to be similar to lamina II cells except that the dendritic trees were somewhat more extensive.

27. Recently two specific cell types, spinocervical tract cells and postsynaptic dorsal column neurons, have been identified. These cells are also seen in laminae IV and V. Other neurons in lamina III are pyramidal neurons, whose dendrites arborize through laminae I–V, and several investigators have described local circuit neurons.

28. Axons of spinocervical tract cells project to the lateral cervical nucleus and those of the postsynaptic dorsal column neurons project to the dorsal column nuclei. The axonal destinations of the other lamina III neurons are not known, but presumably they are similar to those of lamina II cells.

29. Primary afferent input into lamina III comes largely from collaterals from large sensory axons. Classic work focused on the flame-shaped arbors, but recent single-fiber work has greatly extended our understanding of primary afferent innervation of laminae III–VI. A major generalization is that the different types of myelinated primary afferent fibers have specific patterns of innervation.

30. Large hair follicle afferents do not bifurcate and they form very long, longitudinally oriented shelves located in laminae II/I-IV. These are the axons that form the flame-shaped arbors.

31. Pacinian corpuscle afferents have short, shelflike arborizations in laminae III–IV and clusters in laminae V–VI.

32. Rapidly adapting mechanoreceptors form spade-shaped arbors in lamina III.

33. Slowly adapting type I mechanoreceptive afferents form globular clusters of endings spanning laminae III–V and centering in lamina IV.

34. Slowly adapting type II mechanoreceptive afferents span laminae III–VI in the form of transversely oriented plates.

35. The neuropil of laminae III generally resembles that of lamina II.

36. The synaptic apparatus that surrounds identified endings of the above fiber types has many of the elements of the glomeruli that form around the fine primary afferent endings in lamina II

37. Lamina IV is characterized by a wide range of neuronal sizes from small to very large. Spinocervical tract and postsynaptic dorsal column neurons are found in this lamina, as in lamina III. Some spinothalamic neurons are also found. A major feature of dendritic architecture of many neurons in this lamina is the presence of prominent dorsal dendrites that pass into laminae I and II. Such cells are called "antenna-type neurons."

38. The axons of cells in lamina IV pass to the thalamus, the lateral cervical nucleus the dorsal column nuclei, and other parts of the spinal cord.

39. Primary afferent input to lamina IV is from collaterals from large primary afferent fibers. There is a prominent longitudinal axonal plexus in this lamina that consists primarily of these primary afferent collaterals. The descriptions vary somewhat, but it seems to be agreed that the longitudinal orientation of the fibers in this lamina contrast with the vertical orientation of the plexus in lamina V. The distribution of identified primary afferent fiber types is discussed under lamina III.

40. The neuropil of lamina IV has more axosomatic synapses than that of lamina III, and there are relatively few glomeruli.

41. The neurons in lamina V are as in lamina IV, except that the largest cells are slightly larger than those of lamina IV. Spinothalamic cells are found in relative abundance, along with some spinocervical tract and postsynaptic dorsal column cells. Many cells remain unidentified.

42. The axonal destinations for lamina V cells are similar to those of lamina IV.

43. Primary afferent input to lamina V is from large primary afferent collaterals, and the general orientation of these in the lamina V neuropil is vertical. The morphology of identified primary afferent fiber types is given under lamina III.

44. The neuropil of lamina V resembles that of lamina IV.

45. Lamina VI is a transition between the primary afferent-dominated dorsal horn, and the ventral horn which is more concerned with motor activity. Descending input seems to predominate in this lamina, although there is a prominent group of collaterals from Ia afferent fibers.

46. A major development in structural studies of the dorsal horn is the discovery of a large number of morphologically visualizable compounds that are found in primary afferent, descending, and intrinsic spinal systems. An overview with the specific purpose of relating this chemical anatomy with the classic neuroanatomy is provided.

5 Functional Organization of Dorsal Horn Interneurons

Sensory processing in the spinal cord involves interactions among primary afferent fibers bearing information from sensory receptors, interneurons, ascending-tract cells conveying sensory messages to the brain, and descending-tract cells that modulate spinal cord circuits. Such interactions are complex and still poorly understood. In this chapter, the interactions between primary afferent fibers and spinal interneurons will be discussed, as well as the interactions between descending tracts and interneurons. The organization of the ascending tracts and their descending control form the subject matter of Chapters 6–9.

In this chapter, it is assumed that recordings from neurons not specifically identified as ascending-tract cells are likely to be from propriospinal neurons (see Chapter 3). The justification for this assumption is that by far the majority of spinal cord neurons are interneurons. K. Chung *et al.* (1984) tested this by comparing the number of neurons in the S2 segment of the spinal cord that could be labeled retrogradely with horseradish peroxidase applied to the cut surface of the cervical enlargement in rats with the number of unlabeled neurons. They estimated that only 1% of the neurons at this level contribute to long ascending tracts, 2% of the total are motoneurons, and 97% are propriospinal neurons. Thus, the dominant population of neurons in the spinal cord is the interneuronal component.

POPULATION RESPONSES

Field Potentials

As a first approximation, the distribution of activity evoked by primary afferent fibers in the spinal cord can be gauged by recordings of electrical field potentials. The field potentials reflect in part the excitation of interneurons of the dorsal horn. Field potentials can be recorded either from the surface of the spinal cord with a gross electrode (the "intermediary cord potentials" or "cord dorsum potentials") or from within the cord with a microelectrode.

The potentials that can be recorded from the cord dorsum in response to a volley in myelinated cutaneous afferent fibers following electrical stimulation include (1) an afferent volley, (2) one or more negative waves, and (3) a positive wave (Gasser and Graham, 1933; J. Hughes and Gasser, 1934a,b; Bernhard, 1953; Bernhard *et al.*, 1953; Bernhard and Widén, 1953; Lindblom and Ottoson, 1953a,b; Austin and McCouch, 1955; Willis *et al.*, 1973; Beall *et al.*, 1977).

In monkeys, the cord dorsum potentials include a sequence of three negative waves (Fig.

5.1), the N1, N2, and N3 potentials (Beall *et al.*, 1977). The N1 wave is evoked by activity in Aαβ fibers, the N2 wave is evoked by activity in Aαβ and Aδ fibers, and the N3 wave is produced by Aδ fibers.

N and P waves generated in the human spinal cord can be recorded from the epidural space (Shimoji *et al.*, 1977).

Recordings from within the spinal cord show that the negative waves produced by volleys in A fibers reach a maximum in the middle layers of the dorsal horn and reverse to become positive in the ventral horn (Fig. 5.2) (Howland *et al.*, 1955; Coombs *et al.*, 1956; Fernandez de Molina and Gray, 1957; Willis *et al.*, 1973; Beall *et al.*, 1977). Since the area of peak negativity corresponds to a region that contains many interneurons activated by the cutaneous nerve volley, it can be presumed that the negativity reflects excitatory processes. Negativity would be expected in the extracellular space from the direction of current flow into the sinks produced at the somata and dendrites of interneurons by excitatory postsynaptic potentials and action potentials (see below and Fig. 7.26). The area of positivity in the ventral part of the cord would represent current sources distributed along the axons of interneurons projecting ventrally. It should be pointed out that neurons that are excited by afferent volleys can also be found in the ventral part of the cord. However, the extracellular potentials represent the summed activity of many neurons, and the negativity around excited cells in the ventral part of the cord is evidently swamped by the positivity caused by the many more dorsally situated neurons producing current sources in the ventral region.

In monkeys, a volley in Aαβ fibers of a cutaneous nerve produces a maximum negative field potential in laminae IV and V (Fig. 5.2A), whereas a volley in Aδ fibers has its greatest effects in 2 separate loci (Fig. 5.2C), one along the dorsal border of the dorsal horn and the other in laminae V and VI (Beall *et al.*, 1977; cf. Christensen and Perl, 1970). This implies that interneurons should be identified in laminae IV and V that are strongly excited by Aαβ fibers and that there should be cells in laminae I and V–VI that are powerfully activated by Aδ fibers. This is in fact the case (see below). In rats, the maximum negative field potential produced by volleys in Aβ fibers is in laminae III–IV (Schouenborg, 1984).

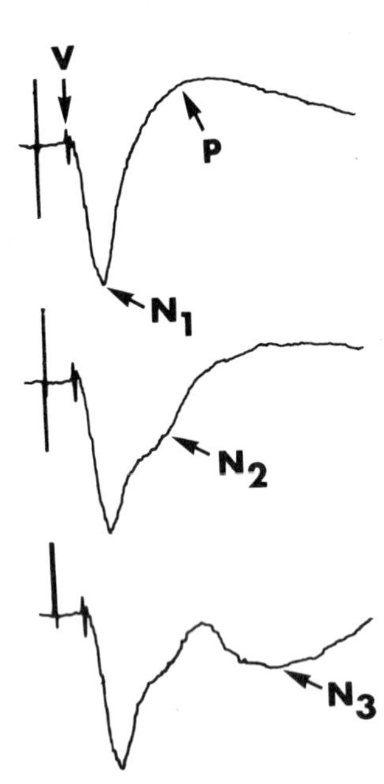

Fig. 5.1. Field potentials recorded from the dorsal surface of the monkey spinal cord in response to stimulation of a cutaneous nerve. The stimulus strength was progressively increased for the three traces. Abbreviations: V, afferent volley; N1, N2, and N3, negative field potentials; P, positive field potential. (From Beall *et al.*, 1977.)

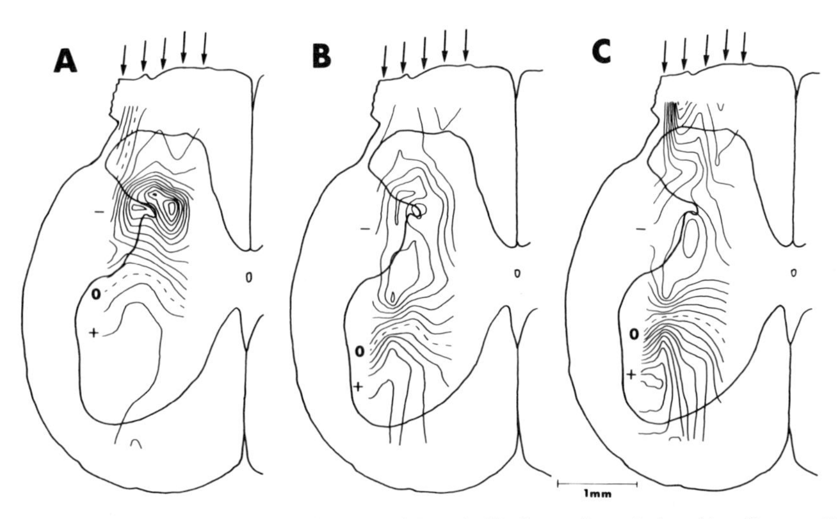

Fig. 5.2. Isopotential contour plots of the field potentials evoked in the monkey spinal cord by afferent volleys in cutaneous fibers of the sural nerve. The stimulus strength was graded, so that the potentials produced by small myelinated fibers could be determined by subtraction of the field potentials produced by large afferents. The measurements were made at fixed time intervals corresponding to the peaks of the N1, N2, and N3 waves recorded at the depth at which the potentials were maximal. The contour plot in panel A shows the spatial distribution of the field potential equivalent to the N1 wave; panel B corresponds to the N2 wave; and panel C corresponds to the N3 wave. (From Beall *et al.*, 1977.)

The positive wave (P wave) that follows the N waves is a reflection of the depolarization of primary afferent fibers and corresponds to the negative dorsal root potential (Barron and Matthews, 1938; Lloyd and McIntyre, 1949; Lloyd, 1952; Koketsu, 1956; J. C. Eccles and Krnjević, 1959; J. C. Eccles *et al.*, 1962b, 1963a). Primary afferent depolarization (PAD) is thought to be responsible for the inhibitory process known as presynaptic inhibition (J. C. Eccles, 1964; R. F. Schmidt, 1971). The P wave evoked by stimulation of cutaneous nerves reverses to negativity in the dorsal horn (Fig. 5.3). It is thought that the depolarization of primary afferent fibers is due to synaptic connections from interneurons onto afferent fiber terminals. Although axoaxonal synapses have been suggested as the morphologic substrate of presynaptic inhibition, dendroaxonal synapses would serve this role as well (see below). The synapses on primary afferent fiber terminals would produce a sink in the dorsal horn, whereas sources would develop along the course of the primary afferent fibers as they enter the spinal cord and cross the dorsal funiculus (Fig. 5.3) (see J. C. Eccles *et al.*, 1962a,b). In addition to being evoked by a specific synaptic system, PAD can be produced by the release of potassium ions into the extracellular space as a consequence of neuronal activity (see below).

Schouenborg (1984) has been able to record negative field potentials in the rat dorsal horn in response to stimulation of C fibers in the sural nerve (Fig. 5.4A). The field potentials are maximal in laminae II and V (Fig. 5.4B). There is a close relationship between the size of the C-fiber volley and the field potential in lamina II, but the threshold for the field potential in lamina V is higher, suggesting that the former is evoked by a monosynaptic connection by C fibers to neurons in lamina II but the latter is caused by a polysynaptic connection to cells in lamina V. The negative field potential in lamina II extends only for about 2 mm longitudinally; beyond this distance, the sign of the potential reverses. The field is undetectable at distances exceeding 6 mm from the central zone of negativity. Anodal blockade of conduction in A fibers and spinal cord transection do not substantially change the potentials. The field potential in

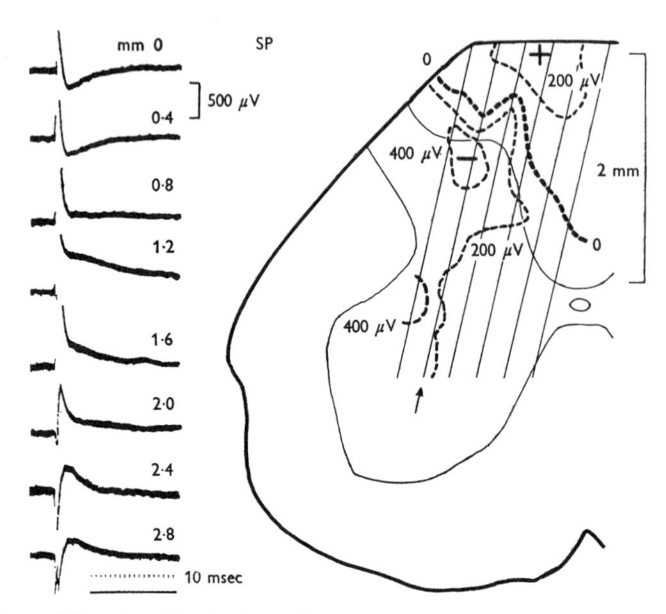

Fig. 5.3. Field potentials produced by stimulation of the superficial peroneal nerve are shown in the column at the left (negativity upward). The P wave is seen to reverse at a depth between 0.4 and 0.8 mm from the surface of the cord. The contour plot at the right was constructed based on measurements of the field potential 40 msec after the afferent volley. The records at the left were from the track indicated by the arrow. Note that the P wave in surface recordings becomes negative in the dorsal horn; the dipoles generating the field potentials are oriented in a dorsomedial to ventrolateral direction. (From J. C. Eccles *et al.*, 1962a.)

lamina II is enhanced when the stimulus frequency is increased from 0.1 to 1 Hz, although the size of the C-fiber volley remains constant (Fig. 5.4B). The increase in the field potential presumably reflects the same process that causes "wind-up" of the discharges of dorsal horn interneurons (see below). The field potential in lamina V is not increased when the stimulus frequency is raised. Noxious heating of the skin also potentiates the C-fiber-evoked field potential in lamina II, whereas innocuous warming of the skin or innocuous mechanical stimulation has no effect. These observations indicate that the field potential produced by C fiber volleys is related largely or entirely to the activation of nociceptors and not thermoreceptors or mechanoreceptors. This finding correlates with the evidence of Lynn and Carpenter (1982) that 78% of C afferent fibers in cutaneous nerves supplying the hairy skin of the rat hindlimb are nociceptors.

Volleys in muscle afferent fibers can also evoke field potentials. In the monkey, muscle afferent volleys produce a series of N waves that can be recorded from the surface of the spinal cord. These have been termed the NI, NII, and NIII waves (Foreman *et al.*, 1979a). The NI wave is evoked by group I fibers, the NII wave by group II fibers, and the NIII wave by group III fibers. Only small N waves are recorded from the surface of the spinal cord of the cat in response to volleys in group I fibers (J. C. Eccles *et al.*, 1954; Coombs *et al.*, 1956; Lucas and Willis, 1974). However, larger N waves are produced in the cat when high-threshold muscle afferent fibers are stimulated (Bernhard, 1953; Mendell, 1972). The maximum negative field due to group I volleys in the cat is located in the intermediate region, and there is a smaller negative peak in the ventral horn centered over parts of laminae VII and IX (Lucas and Willis, 1974). The NII wave in the monkey is largest in laminae IV–VI and the NIII wave in laminae V–VI (Foreman *et al.*, 1979a). An NII wave is also found in laminae IV and V in the cat spinal cord (Fu *et al.*, 1974). The N waves produced by muscle afferent volleys reverse in polarity in the ventral horn. Following the

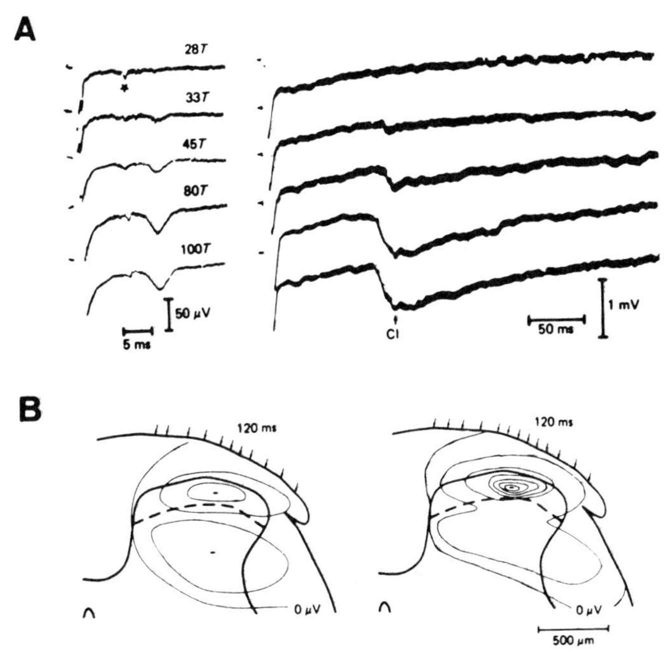

Fig. 5.4. Field potentials evoked by volleys in C fibers. The compound action potentials in A and C fibers produced by different strengths of stimulation of the sural nerve in a rat are shown in the column of records at the left in panel A. In the right column are field potentials recorded in lamina II, using the same stimulus intensities. The contour maps in panel B show the distribution of activity evoked by C fiber volleys in the sural nerve. The map on the left was produced when stimuli were repeated at 0.1 Hz and that on the right at 1 Hz. (From Schouenborg, 1984.)

N waves, there may be a P wave reflecting primary afferent depolarization evoked by the muscle afferent volleys. The P wave is largest when repetitive stimulation is used, especially when nerves to flexor muscles are stimulated (J. C. Eccles *et al.*, 1962b).

Volleys in joint afferent fibers also evoke a series of N waves (Schaible *et al.*, 1986). These have been called the NI, NII, and NIII waves, and they are attributed to volleys in groups of afferent fibers with successively lower conduction velocities. The N waves are maximal in the dorsal horn and reverse sign in the ventral horn.

Similarly, stimulation of visceral afferent fibers evokes N and P waves (see, e.g., Selzer and Spencer, 1969; Yates and Thompson, 1985). The negative field potential that results from stimulation of sympathetic visceral afferent fibers is maximal in lamina V (Selzer and Spencer, 1969).

Changes in Metabolic Activity

2-Deoxyglucose. An increase in the activity of neurons results in a greater utilization of glucose. Excited neurons (or their synapses) also take up 2-deoxyglucose, but this substance is not metabolized. The accumulation of radiolabeled 2-deoxyglucose can be visualized by autoradiography. The technique has been valuable in demonstrating enhanced activity in a number of systems in the brain, but it has not been used as much for demonstrating changes in the spinal cord, perhaps because of the small size of the cord and the limited spatial resolution of the technique. However, several groups have described an increase in glucose metabolism in the dorsal horn in response to electrical stimulation of the sciatic nerve (Piepmeier *et al.*, 1983;

Kadekaro *et al.*, 1985). Since dorsal root ganglia fail to show changes (Kadekaro *et al.*, 1985), it can be concluded that most of the enhanced glucose utilization occurs in afferent terminals.

Proshansky *et al.* (1980) and Crockett *et al.* (1989) were able to demonstrate an enhancement of 2-deoxyglucose uptake in the ipsilateral dorsal horn, as well as in the intermediate region, lamina X, and both ventral horns, in response to innocuous mechanical stimulation of the plantar cushion in the cat's foot. Abram and Kostreva (1986) also found increases in the superficial dorsal horn, the intermediate region, lamina X, and both ventral horns after noxious thermal stimulation of the plantar cushion. The main difference in the distribution of enhanced metabolism in these two studies was that there was increased metabolism in the medial part of the dorsal horn in response to innocuous but not noxious stimulation. Interneurons in the medial dorsal horn are known to participate in the plantar cushion reflex (Egger *et al.*, 1986).

Gonzalez-Lima (1986) found that stimulation in the midbrain reticular formation of rats causes an increase in 2-deoxyglucose activity in laminae II and III.

Glycogen Phosphorylase. The histochemical demonstration of the level of glycogen phosphorylase activity in the spinal cord can be used to demonstrate changes in neural activity (Woolf *et al.*, 1985). The phosphorylated "a" form of this enzyme can break down glycogen, and the nonphosphorylated "b" form is converted to the "a" form when phosphorylase "b" kinase is activated by increases in the levels of intracellular calcium or cyclic AMP. The distribution of the enzyme can be demonstrated immunohistochemically. The active form of glycogen phosphorylase is widely distributed in the spinal cord (Woolf *et al.*, 1985), except for lamina II, and it is particularly concentrated in motoneurons. The inactive form has a similar distribution. Innocuous mechanical and thermal stimuli fail to change the level of glycogen phosphorylase activity. However, noxious mechanical, thermal, or chemical stimuli produce an increase in the level of glycogen phosphorylase activity in the somatotopically appropriate region (see below) of the dorsal horn (Fig. 5.5). The affected region includes laminae I and III-V, but most of lamina II fails to show increased staining. There is also some increase in staining in the contralateral dorsal horn. The increase in staining is first observed in 10 min after noxious chemical stimulation (injection of turpentine), and the maximum change occurs after 20 min. There is less staining when the decerebrate animals are paralyzed.

Electrical stimulation of Aαβ fibers in the sciatic nerve has no effect on staining, beyond that produced by the surgery to expose the nerve (Fig. 5.6A) (Woolf *et al.*, 1985). However, volleys that include Aδ fibers produce dense staining in lamina I and most of the remainder of the dorsal horn, except lamina II (Fig. 5.6B). C-fiber volleys have a similar effect to that of Aδ volleys, except that the region that stains expands and includes part of the contralateral dorsal horn (Fig. 5.6C). Dorsal root ganglion (DRG) cells also show increased staining, especially the smaller-sized ganglion cells.

c-*fos*. The c-*fos* proto-oncogene can be induced to express its protein in neurons of the dorsal horn following stimulation of their receptive fields, especially when noxious stimuli are used (S. P. Hunt *et al.*, 1987; Menétrey *et al.*, 1989; Bullitt, 1989). The gene expression appears to be triggered by calcium influx through voltage-gated channels (J. I. Morgan and Curran, 1986). The c-*fos* protein can be detected immunocytochemically in the superficial dorsal horn within 5–15 min after stimulation; staining in the deeper layers of the dorsal horn increases in the next few hours (S. Williams *et al.*, 1989; see also Mugnaini *et al.*, 1989; Menétrey *et al.*, 1989). The c-*fos* protein enters the nucleus of the cell and interacts with the DNA (Sambucetti and Curran, 1986). It is not clear at present how this affects the later behavior of the cell.

Noxious stimulation of the skin by injection of Freund's adjuvant, application of mustard oil, or exposure to noxious radiant heat results in the appearance of c-*fos* protein in the nuclei of neurons in laminae I, II, IV–VIII, and X (Fig. 5.7A) (S. P. Hunt *et al.*, 1987; Menétrey *et al.*, 1989). A similar pattern of labeling is seen after acute arthritis of the ankle is produced by injection of urate crystals (Fig. 5.7B) (Menétrey *et al.*, 1989). Some neurons in lamina VIII

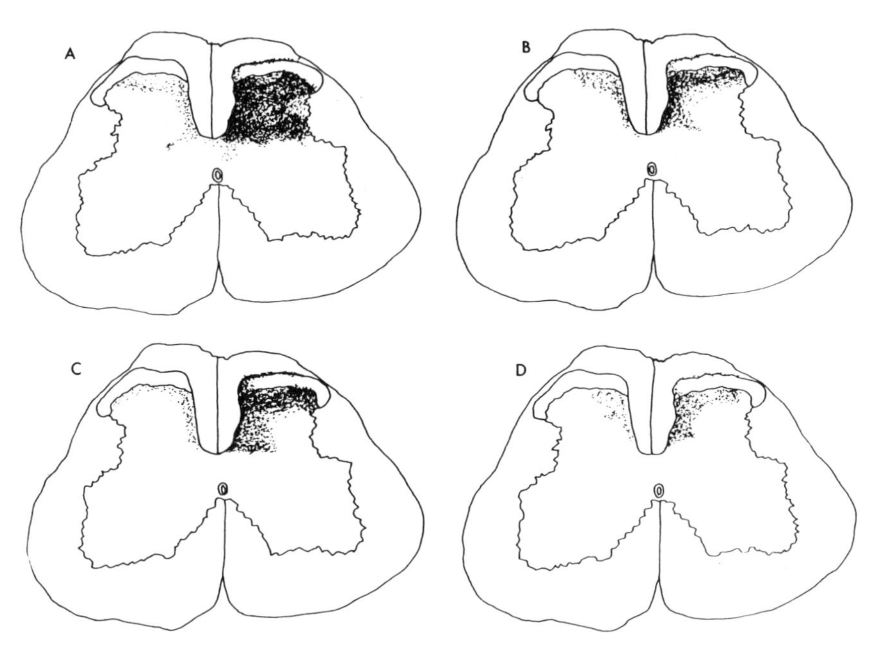

Fig. 5.5. Enhanced histochemical staining for the active form of glycogen phosphorylase in spinal cords of decerebrated rats after each of the following stimuli: (A) subcutaneous injection of turpentine; (B) immersion of the foot in water at 50°C; (C) application of mustard oil to the foot; (D) repeated pinching of the toes. (From Woolf *et al.*. 1985; reproduced with permission.)

contralateral to the noxious stimulus also stain positively. When mustard oil is injected into muscle, neurons stain chiefly in laminae I and IV–VII (S. P. Hunt *et al.*, 1987). Noxious visceral stimulation (intraperitoneal injection of acetic acid) produces labeling in lamina I over a large number of segments and in laminae V and X in a restricted number of segments (Fig. 5.7C) (Menétrey *et al.*, 1989). Interestingly, morphine and a κ receptor agonist produce a naloxone-reversible reduction in the number of stained cells activated by noxious visceral stimulation, especially in the deeper layers of the dorsal horn (Presley *et al.*, 1989). Furthermore, electrical stimulation in the nucleus raphe magnus reduces the number of dorsal horn neurons that show c-*fos* labeling due to noxious heating of the foot (Jones and Light, 1990).

Summary

In summary, activity in populations of neurons in the dorsal horn can be demonstrated in several ways. Field potentials can be recorded electrophysiologically. Synchronous volleys of nerve impulses in primary afferent fibers, evoked by electrical stimulation of peripheral nerves, produce a series of potentials that can be recorded from the surface of the spinal cord or from within the cord. Following the nerve volley are several slow potentials. The first ones are negative in sign when recorded from the dorsal surface of the cord or in the dorsal horn; these reflect excitation of interneurons. A later, positive cord dorsum potential reflects primary afferent depolarization, which is associated with presynaptic inhibition. These slow potentials reverse in sign more ventrally in the cord. Details of the time course and distribution of the slow potentials depend upon the nature of the afferent volley (large or small myelinated fibers versus unmyelinated fibers; cutaneous, muscle, joint, or visceral nerves). Activity in a population of dorsal horn neurons can also be visualized anatomically. The 2-deoxyglucose technique can be used to

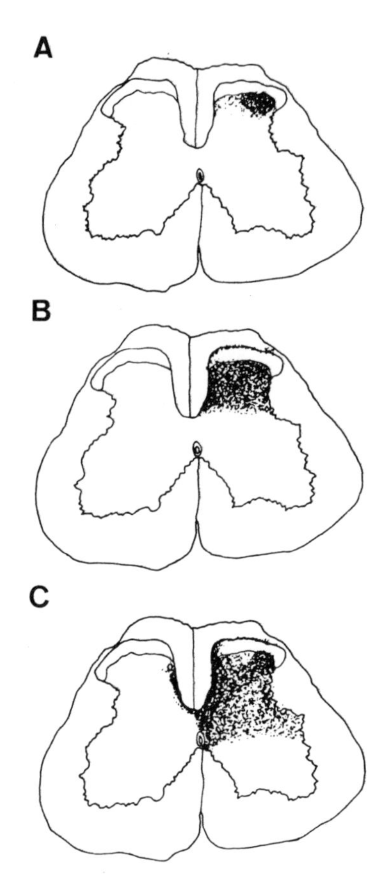

Fig. 5.6. Increases in glycogen phosphorylase activity following stimulation of the sciatic nerve of decerebrated rats at different stimulus intensities for 30 min. The transverse sections are from the L4/L5 junction in the spinal cord. (A) Activity after stimulation at Aβ intensity; however, a similar response is produced by making the incision and not stimulating the nerve. (B) Activity after stimulating at Aδ strength, and (C) Activity after stimulating at C-fiber strength. (From Woolf *et al.*. 1985; reproduced with permission.)

demonstrate increased glucose metabolism, perhaps chiefly in synaptic endings. Glycogen phosphorylase activity is enhanced in most of the dorsal horn except lamina II following noxious stimulation or activation of fine myelinated and unmyelinated afferent fibers. The cellular mechanism involves release of second messengers. Stimulation of nociceptors may also cause the c-*fos* oncogene in dorsal horn neurons to express c-*fos* protein, a nuclear protein that stains immunocytochemically. C-*fos* appears in neurons of laminae I and II, as well as in deeper layers of the dorsal horn and in the ventral horn. C-*fos* expression can be reduced by prior administration of opiate receptor agonists or by stimulation of descending inhibitory projections.

SYNAPTIC EXCITATION AND INHIBITION OF INTERNEURONS

The response properties of individual interneurons in the spinal cord can be investigated by using intracellular or extracellular recording techniques. Interneurons can be distinguished from afferent fibers on the basis of their response properties and spike configuration in extracellular recordings and by the presence of synaptic potentials in intracellular recordings. They can be distinguished from tract cells or motoneurons by their failure to show antidromic activation following stimulation of the white matter of the spinal cord, of projection targets in the brain, or of motor axons (Woodbury and Patton, 1952; Frank and Fuortes, 1956; Wall, 1959; J. C. Eccles *et al.*, 1960; Willis, 1985). Direct evidence concerning the nature of the element from which recordings are made can be obtained by marking the unit intracellularly (see below).

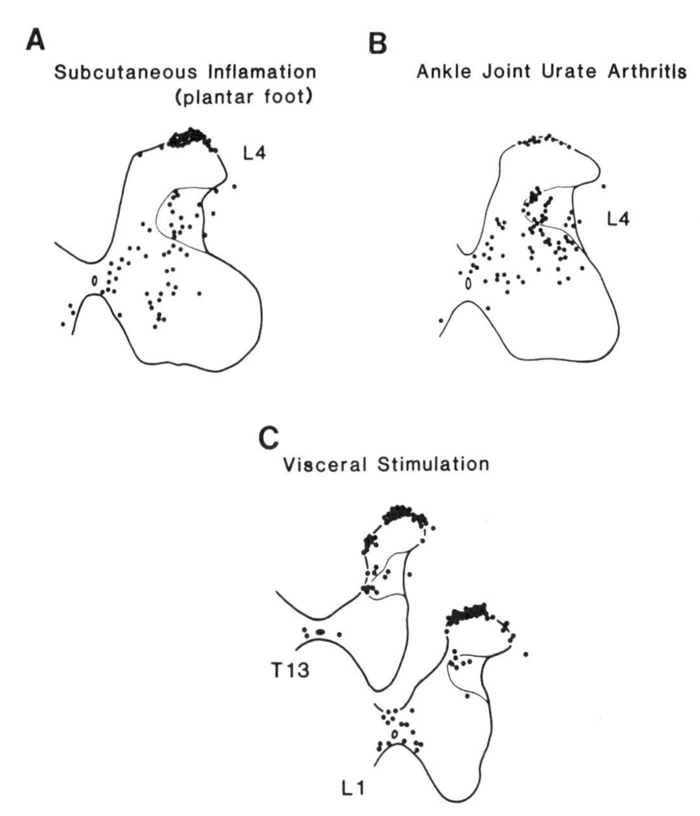

Fig. 5.7. Locations of neurons whose nuclei show c-*fos* immunoreactivity following noxious stimulation in anesthetized rats. Noxious stimuli included injection of Freund's adjuvant into the plantar foot (A), implantation of urate crystals in the ankle (B), and intraperitoneal injection of acetic acid (C). (From Menétrey *et al.*. 1989; reproduced with permission.)

Intracellular recording from interneurons allows the demonstration of their membrane properties, as well as excitatory postsynaptic potentials (EPSPs) and inhibitory postsynaptic potentials (IPSPs) generated in response to activation of pathways impinging synaptically on these cells (Haapanen *et al.*, 1958; Kolmodin and Skoglund, 1958; C. C. Hunt and Kuno, 1959a,b; J. C. Eccles *et al.*, 1960; Kostiuk, 1960; Hongo *et al.*, 1966; D. D. Price *et al.*, 1971; Jankowska *et al.*, 1981; Iggo *et al.*, 1988). When primary afferent fibers are stimulated, the EPSPs in interneurons may be monosynaptic or polysynaptic, whereas the IPSPs are at least disynaptic, indicating an interneuronal relay (Hongo *et al.*, 1966). These observations imply that the action of primary afferent fibers is exclusively excitatory. However, such a judgment depends on measurement of the timing of synaptic events, which is practical only for the large afferent fibers. Therefore, it cannot be said with certainty that small afferent fibers must necessarily all have an excitatory action.

In addition to evoking postsynaptic potentials in interneurons, volleys in primary afferent fibers result in the generation of PAD and presumably presynaptic inhibition. PAD can be demonstrated by intracellular recordings directly from primary afferent fibers (Koketsu, 1956; J. C. Eccles and Krnjević, 1959; J. C. Eccles *et al.*, 1962b, 1963a), by tests for changes in the excitability of primary afferent terminals (Wall, 1958; J. C. Eccles *et al.*, 1962b; 1963a), and by recordings of dorsal root potentials, P waves and dorsal root reflexes (see above) (Toennies, 1938; J. C. Eccles *et al.*, 1961).

A rather specific organization has been found for primary afferent fiber types giving and receiving PAD (see the review by R. F. Schmidt, 1971; see also Whitehorn and Burgess, 1973; Brink *et al.*, 1984). For instance, among cutaneous afferent fibers, the slowly adapting mechanoreceptors preferentially evoke PAD in the terminals of other slowly adapting mechanoreceptors (Fig. 5.8), whereas rapidly adapting receptors tend to produce PAD in the afferent fibers supplying other rapidly adapting receptors (Jänig *et al.*, 1968b). PAD is believed to cause presynaptic inhibition by reducing the amount of excitatory transmitter released by afferent fiber terminals (J. C. Eccles, 1964; Schmidt, 1971).

Excitability testing has also been used to demonstrate changes in the thresholds of the terminals of C fibers in the dorsal horn (Hentall and Fields, 1979; Fitzgerald and Woolf, 1981; Calvillo *et al.*, 1982). Volleys in A fibers have been shown to increase the excitability of C-fiber endings. However, the types of receptors connected with the A and C fibers involved are unknown, and evidence is needed concerning the relationship of presynaptic excitability changes to presynaptic inhibition in C-fiber pathways.

PAD has been attributed to the activity of axoaxonal synapses (E. G. Gray, 1963; Khattab, 1968; Ralston, 1965, 1968a,b, 1979; Conradi, 1969; Réthelyi and Szentágothai, 1969; McLaughlin, 1972; Barber *et al.*, 1978; Ralston and Ralston, 1979, 1982; Zhu *et al.*, 1981; Knyihár-Csillik *et al.*, 1982b), although synaptic contact with primary afferent terminals may actually represent dendroaxonic connections. Functionally, axoaxonic and dendroaxonic synapses would presumably be equivalent.

CLASSIFICATION OF DORSAL HORN INTERNEURONS

Several attempts have been made to develop a system for classifying dorsal horn interneurons, but none of the approaches suggested to date is completely satisfactory. A major difficulty is that the projection path of interneurons is generally unknown, and so the only information available about the cell may be its location in the spinal cord and its response properties.

J. C. Eccles *et al.* (1960) described the responses of interneurons in the region of the intermediate nucleus. The nomenclature introduced by them was based on monosynaptic inputs from group Ia, group Ib, or cutaneous afferent fibers. Interneurons receiving monosynaptic

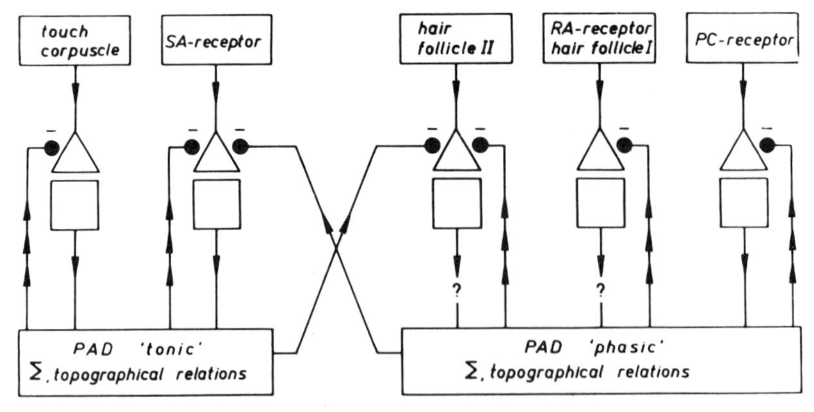

Fig. 5.8. Organization of the cutaneous PAD systems. The diagram indicates that the PAD produced by slowly adapting cutaneous mechanoreceptors acts chiefly on slowly adapting receptors and that the PAD produced by rapidly adapting mechanoreceptors acts chiefly on rapidly adapting receptors, although there is some cross-talk. Abbreviations: SA, slowly adapting; RA, rapidly adapting; PC, Pacinian corpuscle. (From Jänig *et al.*, 1968b.)

inputs from these sources were termed type A, B, and C interneurons, respectively. The inference could be drawn that there are no other significant peripheral inputs to these cells, yet many receive strong polysynaptic effects from afferent fiber types other than those making monosynaptic connections. Furthermore, it is difficult to determine whether inputs from fine afferent fibers are monosynaptic. It is not at all clear that monosynaptic connections reflect the most important connections during the normal operation of interneuronal pathways, although presumably monosynaptic pathways are likely to be subjected to minimal interference. Another objection to this nomenclature is that the C interneurons would include cells receiving an input from a wide variety of cutaneous receptors (Tapper *et al.*, 1973); thus, the use of a categorization of cells based on a single type of cutaneous input would tend to obscure a diversity of functions. J. C. Eccles *et al.* (1960) did subdivide the C class into C and CT, the latter being tract cells (presumably spinocervical tract cells).

The alphabetical list of interneuronal classes was expanded by J. C. Eccles *et al.* (1962a) to include a set of interneurons, labeled D, that had properties suited to the interneurons responsible for evoking PAD. Type D interneurons are not activated monosynaptically by primary afferent fibers, and so the taxonomy of these cells is based on other criteria, including convergence of certain kinds of inputs and the time course of discharge in response to single and repetitive input volleys.

Wall (1967) found that interneurons in laminae IV, V, and VI tend to have different properties. Because of this observation, an informal nomenclature developed in which neurons responding exclusively to tactile stimuli were called "lamina IV" cells and those responding to tactile and noxious stimuli "lamina V" cells (see, e.g., Hillman and Wall, 1969; Besson *et al.*, 1975), irrespective of the actual location of the interneuron. This approach has since been largely abandoned.

Gregor and Zimmermann (1972) determined whether dorsal horn interneurons were excited just by myelinated cutaneous afferent fibers or by both myelinated and unmyelinated fibers. They also determined whether the connections from the A fibers were monosynaptic or polysynaptic. Interneurons could then be classified as activated just by A fibers or by A and C fibers, and then they could be subclassified according to whether the inputs were monosynaptic or polysynaptic. However, it is difficult to be sure whether or not inputs from C fibers are monosynaptic.

A commonly used scheme was introduced by Mendell (1966). This takes into account the intensity "bandwidth" of stimuli that activate interneurons. Cells that respond just to innocuous mechanical stimuli are called "narrow-dynamic-range cells" or "low threshold cells." Interneurons that can be activated by innocuous and by noxious mechanical stimuli are "wide-dynamic-range cells." Although "high-threshold" cells were not generally recognized in 1966, they can be accommodated in this scheme as another form of narrow dynamic range cell.

D. D. Price and Browe (1973) and D. D. Price and Mayer (1974) evaluated the responses of dorsal horn neurons in cats and monkeys to touch, pressure, and pinch. The cells were then subdivided into five classes based on the "bandwidth" of their responses: (1) touch units; (2) touch-pressure units; (3) touch-pressure-pinch units; (4) pressure-pinch units; and (5) pinch units. Class 3 is essentially equivalent to the wide dynamic range category, whereas the other classes represent cells with narrow or intermediate dynamic ranges.

Heavner and DeJong (1973) pursued the classification problem in the direction of identifying the types of stimuli that activate dorsal horn interneurons. Using the responses to a variety of stimuli, including hair movement, prick, pinch, pressure, vibration, and proprioception, they were able to identify 19 classes of response. Tapper *et al.* (1973) examined the patterns of convergence of input from four different cutaneous receptor types that they could activate selectively and found cells representing 16 combinations.

A simplified terminology was suggested by Handwerker *et al.* (1975). They proposed that cells excited just by large myelinated afferent fibers be called class 1 cells and that cells excited by large myelinated and unmyelinated fibers be designated class 2 cells. The effective receptors for class 1 cells could be either sensitive mechanoreceptors of the skin or proprioceptors. Class 2 cells are excited by a variety of sensitive mechanoreceptors, but also by nociceptors. Class 1 cells would be equivalent to "low-threshold cells" and class 2 to "wide-dynamic-range cells."

Cervero *et al.* (1976) expanded the terminology of Handwerker *et al.* (1975) to include another kind of cell, class 3. Class 3 neurons respond only to fine afferent fibers and to noxious stimuli and are equivalent to "high-threshold cells." Two subtypes were recognized: class 3a cells receive their input chiefly from Aδ mechanical nociceptors and generally do not respond to noxious heat stimuli. Class 3b cells are excited by Aδ and C nociceptors, and so they are activated by noxious heat as well as by noxious mechanical stimuli. Some class 3b cells are also excited by group III and IV muscle afferent fibers.

Menétrey *et al.* (1977) elaborated on this scheme. They subdivided dorsal horn interneurons into four classes. Class 1 cells are excited by innocuous stimuli. They may be subdivided into two subclasses, 1A (activated by hair movement and/or touch only) and 1B (hair movement and/or touch and pressure or pressure only). Class 2 cells are excited by innocuous and noxious stimuli. They may also be subdivided into two subclasses, 2A (hair movement and/or touch, pressure, pinch and/or pinprick) and 2B (pressure, pinch and/or pinprick). Class 3 cells are excited just by noxious stimuli (pinch and/or pinprick). Class 4 cells respond to joint movement or deep pressure.

Another approach to the classification of dorsal horn neurons makes use of a multivariate statistical analysis of the responses to several stimuli (J. M. Chung *et al.*, 1986b; Surmeier *et al.*, 1986a,b; 1988; Sorkin *et al.*, 1986). With this approach, the responses of interneurons of the cervical spinal cord of the cat to mechanical stimulation of the skin could be subdivided, on the basis of a k means cluster analysis of the responses to four mechanical stimuli of progressively increasing intensities, into five classes (Sorkin *et al.*, 1986). The response profiles of these are shown in Fig. 5.9 and are compared with the profiles of responses of cells that were classified as low-threshold, wide dynamic range, and high-threshold cells.

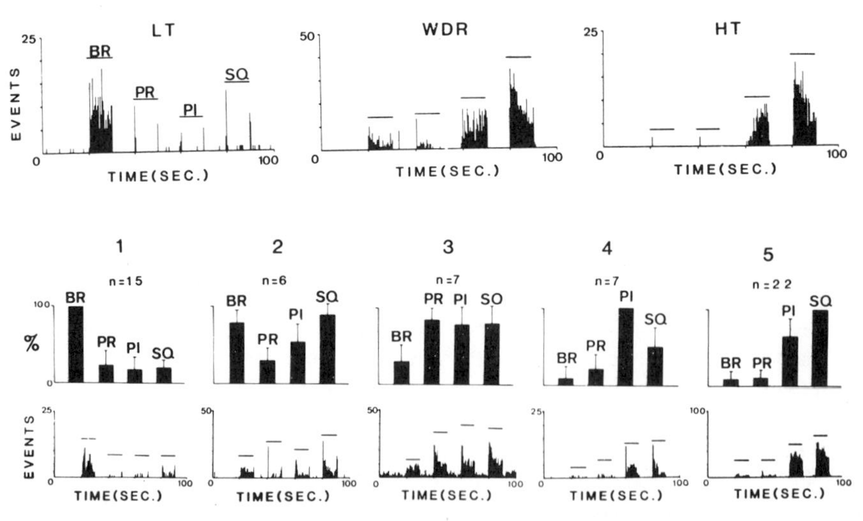

Fig. 5.9. Classification of dorsal horn neurons. The upper row shows peristimulus time histograms of the responses of three neurons classified, respectively, as low-threshold (LT), wide- dynamic-range (WDR), and high-threshold (HT) cells. The responses of these cells to brushing (BR), pressure (PR) from an arterial clip, pinch (PI) from a different arterial clip with a more forceful grip, and squeezing (SQ) with forceps are shown. The lower row of histograms shows the response profiles of five neurons representative of five classes of cells determined by k means cluster analysis, using the same four mechanical stimuli. (From Sorkin *et al.*, 1986.)

AFFERENT PROJECTIONS TO THE DORSAL HORN

Laminae I–IV

Inputs from a variety of receptors having different sizes of primary afferent fibers and the types of neurons in the superficial layers of the dorsal horn are summarized in Fig. 5.10 (Cervero and Iggo, 1980) (see Chapter 4).

Afferent fibers of Aδ size from nociceptors (and presumably thermoreceptors) terminate chiefly in lamina I on neurons that are oriented longitudinally with respect to the spinal cord (cf. Light and Perl, 1979b). Fine muscle, joint, and visceral afferent fibers also project to lamina I (Craig and Mense, 1983; Craig *et al.*, 1988; Sugiura *et al.*, 1989). Cutaneous unmyelinated afferent fibers from nociceptors, thermoreceptors, and sensitive mechanoreceptors project mainly to lamina II, although nociceptors and thermoreceptors also terminate in lamina I. The afferent terminals in lamina II are on limiting cells (stalk cells) at the border between laminae I and II or on central cells (islet cells) within the substantia gelatinosa (cf. Sugiura *et al.*, 1986). The neurons of the substantia gelatinosa, like those in lamina I, are oriented longitudinally. Those in the outer part of the substantia gelatinosa are likely to receive inputs from C nociceptive or thermoreceptive afferent fibers, and those in the inner part receive inputs from C mechanoreceptors.

Aδ fibers from down hair follicles distribute to lamina III (and perhaps the unmyelinated tips of these fibers also end in lamina II). Other sensitive mechanoreceptors supplied by Aαβ fibers end in laminae III and IV. Neurons in lamina III tend to be oriented longitudinally, but those in lamina IV have a more dorsal-ventral orientation (Scheibel and Scheibel, 1968).

Information from lamina II can reach deeper layers of the dorsal horn by ventrally directed axonal projections (Light and Kavookjian, 1988). Neurons in laminae III and IV with dorsally directed dendrites can also receive direct input from primary afferent fibers that project to lamina II (A. J. Todd, 1989; cf. Wall, 1965).

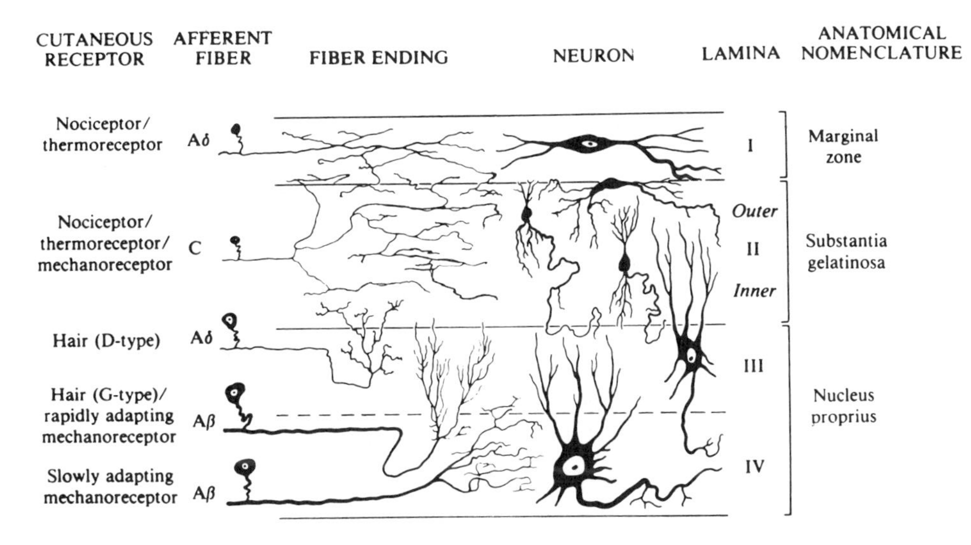

Fig. 5.10. Schematic of afferent input to the superficial layers of the dorsal horn. The types of afferent fibers are indicated at the left, and they are shown projecting onto neuronal types that are typical of laminae I—IV. (From Cervero and Iggo, 1980; reproduced with permission.)

Laminae V, VI, and X

Aδ fibers supplying mechanical nociceptors (Light and Perl, 1979b), as well as sensitive mechanoreceptors, project to lamina V, as do many group III and IV muscle, joint, and visceral afferent fibers (Craig and Mense, 1983; Craig *et al.*, 1988; Sugiura *et al.*, 1989). However, it is likely that cells in lamina V also receive input from other kinds of afferent fibers in the superficial layers of the dorsal horn, since many of these cells have dendrites that extend into more superficial layers (Ritz and Greenspan, 1985).

Proprioceptive afferent fibers project to lamina VI in the dorsal horn and also to the ventral horn (A. G. Brown, 1981).

Aδ mechanical nociceptors and unmyelinated visceral afferent fibers also project to lamina X (Light and Perl, 1979b; Sugiura *et al.*, 1989).

RESPONSES TO ELECTRICAL STIMULATION

Lamina I

Neurons in lamina I are activated by volleys in cutaneous Aδ fibers and sometimes by volleys in C fibers as well (Christensen and Perl, 1970; Kumazawa *et al.*, 1975; Cervero *et al.*, 1976, 1979b; Light *et al.*, 1979; Fitzgerald & Wall, 1980; Bennett *et al.*, 1981; Woolf & Fitzgerald, 1983). There have also been reports of responses by units in lamina I to input from Aαβ fibers (Christensen and Perl, 1970; Woolf and Fitzgerald, 1983). Most neurons in lamina I of the thoracic spinal cord that have been examined respond both to cutaneous input and to volleys in visceral afferent fibers of the greater splanchnic nerve (Cervero and Tattersall, 1987; see also Alarcon and Cervero, 1990) or of the inferior cardiac nerve (M. Takahashi and Yokota, 1983).

Lamina II

Until the past 15 years, there were few studies reporting recordings from neurons of the substantia gelatinosa. However, a number of laboratories have now succeeded in accomplishing this, and some laboratories have been able to mark the recorded cells by intracellular injections of a fluorescent dye or of horseradish peroxidase.

Kumazawa and Perl (1976, 1977, 1978) found that cells in the substantia gelatinosa are activated primarily by C fibers, although there is often some input from Aδ fibers as well. However, other investigators report that fibers of both A and C caliber can often excite neurons of lamina II (Wall *et al.*, 1979a; Bennett *et al.*, 1979, 1980; Fitzgerald, 1981). Similarly to the situation for cells in lamina I, the predominant A-fiber input to lamina II is from Aδ fibers, although there is some Aβ input as well.

Neurons responding to electrical stimulation of fine visceral afferent fibers in the greater splanchnic nerve do not appear to occur in lamina II (Alarcon & Cervero, 1990). However, M. Takahashi & Yokota (1983) found neurons in the outer part of lamina II that respond to stimulation of afferent fibers in the inferior cardiac nerve, as well as to stimulation of the skin.

Since A fibers do not appear to project directly to lamina II (see above and Chapter 3) and the dendrites of substantia gelatinosa neurons do not generally leave lamina II, it can be presumed that the responses to A-fiber input depend upon a polysynaptic pathway.

Laminae III–VI

A population of cells can be found along the medial parts of laminae III–VI of the lumbar spinal cord that are monosynaptically excited by stimulation of Aαβ afferent fibers of the medial plantar nerve (Armett *et al.*, 1961).

Mendell and Wall (1965) and Mendell (1966) found that 60% of the neurons of the nucleus proprius that they investigated could be activated by both A and C fibers, whereas the other 40% responded only to stimulation of A fibers. The cell bodies of such neurons appear to be intermingled throughout the dorsal horn (Gregor and Zimmermann, 1972; Menétrey *et al.*, 1977). Volleys evoked by graded stimulus strengths often demonstrate a convergence of Aαβ, Aδ, and C fibers upon the same neurons (Wagman and Price, 1969).

Repetitive stimulation of C fibers at rates of 1 per 3 sec or greater produces an enhancement of the discharge rate of many dorsal horn neurons, a phenomenon called "windup" (Fig. 5.11) (Mendell and Wall, 1965; Mendell, 1966; Wagman and Price, 1969; Price and Wagman, 1970; Fitzgerald and Wall, 1980; Woolf and King, 1987). Other neurons show a decreasing response to repeated C-fiber volleys, "wind-down" (Fitzgerald and Wall, 1980; Woolf and King, 1987; Alarcon and Cervero, 1990).

The responses to A- and C-fiber volleys in neurons of the rat dorsal horn are triggered by excitatory postsynaptic potentials (Woolf and King, 1987). Either the EPSPs due to C-fiber volleys can consist of a long-lasting depolarization that begins just after the PSPs produced by the myelinated fibers, peaks at 100–200 msec and lasts 300–500 msec or they can occur after a silent period following the response to the A-fiber volley.

In the thoracic and sacral spinal cord, cells in lamina V receive a convergent excitatory input from cutaneous afferent fibers and from visceral afferent fibers of the splanchnic nerve, sympathetic chain, inferior cardiac nerve, or pelvic and pudendal nerves (Pomeranz *et al.*, 1968; Selzer and Spencer, 1969; R. D. Foreman, 1977; R. D. Foreman and Ohata, 1980; S. B. McMahon and Morrison, 1982; M. Takahashi and Yokota, 1983; Alarcon and Cervero, 1990). In the lumbosacral enlargement, comparable cells are activated by cutaneous afferent fibers (Pomeranz *et al.*, 1968; Hancock *et al.*, 1973; Gokin *et al.*, 1977) and by group III muscle afferent fibers (Pomeranz *et al.*, 1968).

Many neurons in the intermediate region (laminae VI and VII) receive monosynaptic

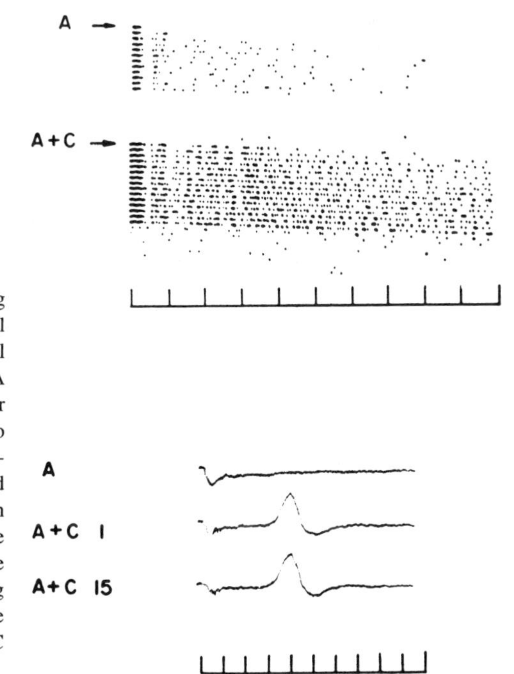

Fig. 5.11. At the top are dot raster records showing the responses of a neuron with its axon in the dorsal lateral funiculus to repeated stimulation of the sural nerve. The stimulus strengths used activated either A fibers or A and C fibers. The responses to A-fiber volleys were consistent, whereas the responses to C-fiber volleys increased with repetition of stimulation (at 1 Hz). This incrementing response is called "wind-up." Below are shown the compound action potentials recorded from the sural nerve in response to A- and C-fiber volleys. The C-fiber volley has the same size for the 1st and 15th repetitions, indicating that wind-up" is a central phenomenon and is not due to a change in the size of the afferent volley in C fibers. (From Mendell, 1966.)

excitatory connections from group I muscle afferent fibers (J. C. Eccles *et al.*, 1956, 1960; R. M. Eccles, 1965; Lucas and Willis, 1974). The same interneurons may also be polysynaptically excited by cutaneous afferent fibers (J. C. Eccles *et al.*, 1960; R. M. Eccles, 1965).

RESPONSES TO NATURAL STIMULATION

Lamina I

Christensen and Perl (1970) found several kinds of responses to natural stimuli in recordings from neurons of lamina I of the cat dorsal horn. One group of cells could be activated only by intense mechanical stimulation (Fig. 5.12). A second group of cells was excited by intense mechanical stimulation of the skin and by noxious heat. A third group of cells could be activated by innocuous temperature changes. Most of these also responded to intense mechanical stimuli. More units were found that responded to cooling rather than to warming.

Kumazawa *et al.* (1975) extended these findings to the monkey dorsal horn and also demonstrated that some lamina I cells projected contralaterally to at least the upper cervical cord; this observation is consistent with a contribution of lamina I to the spinothalamic and spinomesencephalic tracts (see Chapter 9). Many of the cells responded to noxious mechanical stimuli, but some were specifically activated by innocuous cooling or warming. Kumazawa and Perl (1978) subdivided their population of lamina I cells in the monkey into three categories: (1) cells activated only by intense mechanical stimulation, (2) cells excited by innocuous skin cooling, and (3) a few cells that were discharged by noxious mechanical and thermal stimuli.

Light *et al.* (1979) marked several lamina I neurons intracellularly in cats. The cells were nociceptive or thermoreceptive.

Cervero *et al.* (1976) confirmed the finding that a high proportion of cells in lamina I of cats can be excited just by intense stimuli. However, some of the lamina I cells in their sample had a wide dynamic range of responsiveness, indicating that the population of neurons in this lamina is not homogeneous. Wide-dynamic range cells could usually also be activated by group III and IV muscle afferent fibers.

Although there is general agreement that a large proportion of the neurons in lamina I are activated just by noxious intensities of stimulation, several groups besides Cervero *et al.* (1976)

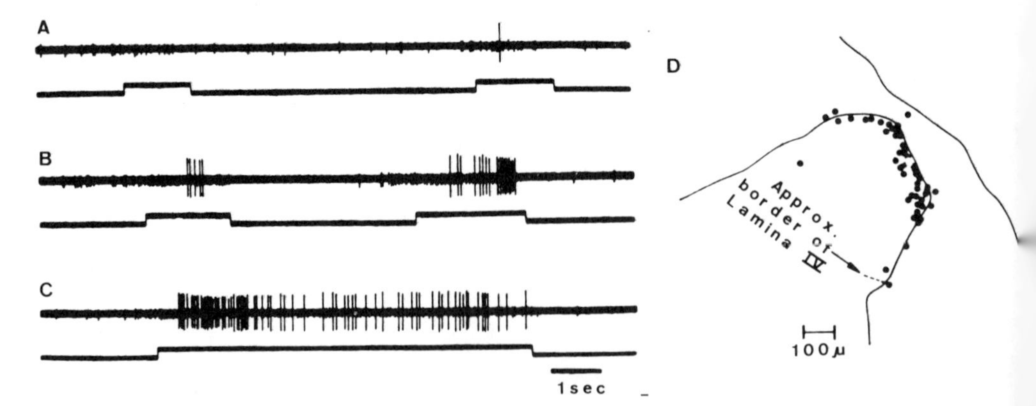

Fig. 5.12. (A) Responses of a lamina I neuron to stroking the skin with a glass rod. (B) Effect of squeezing the skin with smooth-surfaced forceps. (C) Results of squeezing the skin with serrated forceps. (D) Locations of lamina I neurons responsive to strong mechanical stimuli or strong mechanical stimuli and noxious heat. (From Christensen and Perl, 1970.)

have recorded the responses of wide dynamic range neurons in this lamina in rats, cats and monkeys (Menétrey *et al.*, 1977; D. D. Price *et al.*, 1979; Bennett *et al.*, 1981; S. B. McMahon and Wall, 1983c; Woolf and Fitzgerald, 1983; Cervero and Tattersall, 1987). Furthermore, some neurons in the marginal zone of the thoracic spinal cord are mechanoreceptive (Cervero and Tattersall, 1987). Several investigations have confirmed the observation of Christensen and Perl (1970) that there are thermoreceptive neurons in lamina I (Hellon and Misra, 1973a; Iggo and Ramsey, 1976; Kanui, 1988).

D. Mitchell and Hellon (1977) were able to activate several neurons in lamina I of the rat spinal cord by applying noxious heat to the tail. It would be important to know whether these cells play a role in the tail flick reflex.

Lamina I neurons often have input from visceral afferent fibers (M. Takahashi and Yokota, 1983; Cervero and Tattersall, 1987; Ness and Gebhart, 1989; Alarcon and Cervero, 1990). Such cells generally receive a convergent input from cutaneous afferent fibers. Lamina I neurons have also been shown to receive input from joint receptors (Schaible *et al.*, 1986).

Lamina II

Kumazawa and Perl (1976, 1977, 1978) recorded from a number of neurons at recording sites that were shown by dye marks to be in the substantia gelatinosa. None of the cells could be activated antidromically from the cervical cord (cf. Kumazawa *et al.*, 1975). The neurons were excited either by noxious stimuli or by innocuous mechanical stimuli.

Wall *et al.*, (1979a) recorded from neurons in the substantia gelatinosa that were either spontaneously active or could be antidromically or orthodromically activated by stimulation of Lissauer's fasciculus. Many of the cells had small receptive fields, although other cells located deeper in the substantia gelatinosa had somewhat larger fields, which could encompass those of the more dorsal cells. Repeated stimuli often revealed habituation of the responses. Many cells showed a prolonged discharge, lasting seconds to minutes, following a single brush stimulus or, for other cells, following a noxious stimulus. The prolonged discharge was blocked by barbiturate anesthesia. Most cells responded to hair movement or other weak mechanical stimuli, but some cells required intense pressure or pinch before responding.

Cervero *et al.* (1977b, 1979a,c; also see the review by Cervero and Iggo, 1980) also used a volley in the isolated Lissauer tract as a search stimulus for neurons of the substantia gelatinosa in cats. Cells were activated either antidromically or orthodromically by such a volley. One criterion that was used to identify neurons intrinsic to the substantia gelatinosa was that the responses had to differ from those known to characterize cells in laminae I and IV, so that recordings from dendrites of deeper neurons could be ruled out (cf. Wall, 1965); this restriction would, of course, eliminate from the sample any substantia gelatinosa neurons that behaved like the cells of neighboring laminae. Another restriction was that the extracellular spikes should show an inflection on the rising phase. The neurons meeting these criteria had a high background discharge, were usually inhibited by cutaneous stimulation (Fig. 5.13), and sometimes showed prolonged discharges lasting seconds after brief periods of stimulation. The classification system used for the cells was inverse to that applied to most other dorsal horn neurons. Inhibition was produced in some neurons by tactile stimuli (inverse class 1), in others by innocuous and noxious stimuli (inverse class 2), and still others by noxious stimuli (inverse class 3). Inverse classes 1 and 3 are excited by noxious and by innocuous stimuli, respectively. Molony *et al.* (1981) marked several lamina II interneurons intracellularly. The cells also belonged to the "inverse" categories.

Hentall (1977) found units in lamina II that were inhibited and then activated by weak mechanical stimuli. Habituation was often seen. Noxious stimuli could excite the cells, provided that background activity was present.

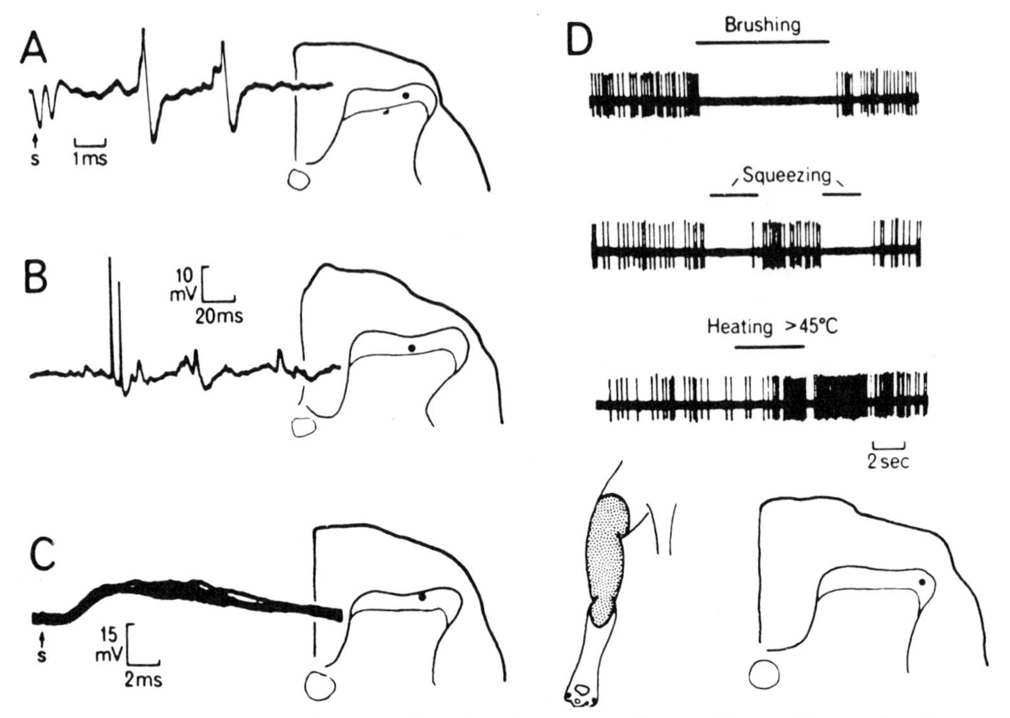

Fig. 5.13. Response properties of neurons of the substantia gelatinosa. Extra- and intracellular recordings from neurons of the substantia gelatinosa are shown in panels A–D, along with the locations of the recording sites, and in D the receptive field. The action potentials in panel A and the EPSP in panel C were evoked by stimulation of the tract of Lissauer. The activity in panel B was spontaneous. The responses in panel D include inhibition due to brushing and squeezing the skin, but excitation due to noxious heating. (From Cervero *et al.*, 1977b.)

Dubuisson *et al.* (1979) and Woolf and Fitzgerald (1983) repeatedly mapped the receptive fields of neurons in the superficial dorsal horn of cats and rats, respectively, and found that the receptive fields were unstable (Fig. 5.14), expanding or contracting over time ("ameboid" receptive fields). This behavior was not observed for receptive fields of neurons in other laminae, and it was eliminated by barbiturate anesthesia. Woolf and Fitzgerald (1983) report that most of the neurons from which they recorded in the outer part of lamina II were of the wide dynamic range type, although some were high-threshold neurons; low- threshold neurons, as well as wide-

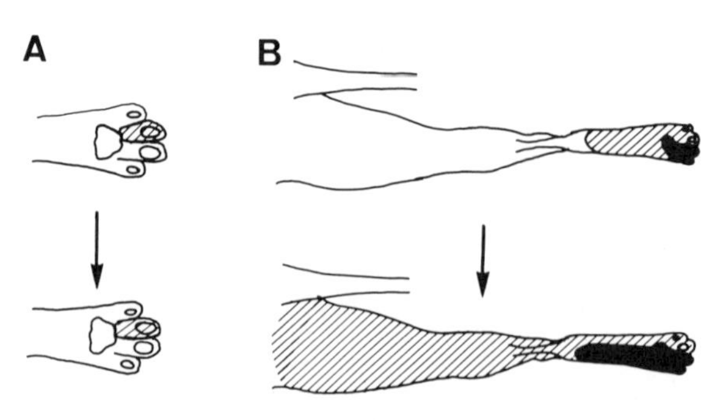

Fig. 5.14. "Ameboid" receptive fields of neurons in the substantia gelatinosa. One cell responded initially to brushing one toe (upper left drawing). Later, the receptive field expanded to include the central pad (lower left). The receptive field of another cell increased from an area on the foot (upper right) to the entire leg (lower right). (From Dubuisson *et al.*, 1979.)

dynamic-range and high-threshold neurons were encountered in the inner part of lamina II. This was true for observations made either extracellularly or intracellularly.

D. D. Price *et al.* (1979) recorded from pairs of neurons in laminae I and II in monkeys. Most of the latter had smaller receptive fields than those of the cells in lamina I, and the initial discharges of these cells had shorter latencies. This observation is consistent with the idea that some cells in lamina II may be excitatory interneurons (Gobel, 1978b; Gobel *et al.*, 1980). However, some neurons in lamina II had only inhibitory receptive fields. Cells in lamina I and the outer part of lamina II were either wide dynamic range or high-threshold neurons; however, only low-threshold cells were found in the inner part of lamina II.

Fitzgerald (1981) recorded from cells in lamina II of cats that were activated by just innocuous stimuli, by both innocuous and noxious stimuli or by just noxious stimuli. Some of the cells had long-lasting discharges, and others showed habituation on repeated stimulation.

Millar and Armstrong-James (1982) examined the responses of substantia gelatinosa neurons that could be excited by iontophoretically applied glutamate. The cells were activated by innocuous mechanical stimuli, but also had a convergent input from higher threshold receptors.

Yoshimura and Jessell (1989a,b) have developed an *in vitro* preparation to investigate synaptic mechanisms in the substantia gelatinosa of rats. Dorsal root stimulation produced early and late monosynaptic EPSPs that could be attributed to activation of Aδ and C fibers. Aδ fibers additionally evoked slow EPSPs, which lasted several minutes. The latter were associated with an increased membrane resistance. A variety of other membrane properties were also described.

Laminae III–VI

The responses of neurons in the neck and base of the dorsal horn of the cat to natural forms of stimulation have been surveyed by Wall and his associates. Cells were typically activated both by rapidly adapting tactile receptors and also by slowly adapting pressure receptors (Wall, 1960). Noxious stimuli added to the response. Cells of this kind were later described as having a "wide dynamic range of response" (Mendell, 1966). Convergence of input from a number of receptor types upon dorsal horn interneurons was confirmed by Wall and Cronly-Dillon (1960).

Wall (1967) later found that the response properties of neurons in different laminae could be distinguished. Cells in lamina IV received a convergent input from mechanoreceptors. They could often be excited by hair movement, touch, and stimulation of tactile domes. Some cells, but not all, could also be excited by pressure and pinch. Thus, lamina IV was found to contain narrow- and wide-dynamic-range cells. There was no input from deep receptors. Neurons in lamina V responded to touch and pressure. Again, no input was noticed from deep receptors. Spinal cord block lowered the threshold for cutaneous stimulation and expanded the receptive field. Neurons in lamina VI responded to both cutaneous stimulation and joint movement. Spinal cord block increased the excitability to cutaneous stimuli and reduced it to proprioceptive stimuli; thus, activity descending from the brain could act as a switch to shift the responsiveness of cells in lamina VI between cutaneous and proprioceptive inputs.

Many investigators have reported that neurons can be found in laminae III and IV (and also deeper layers) that are activated just by weak mechanical stimuli applied to the skin (Kolmodin and Skoglund, 1960; Armett *et al.*, 1962; Fetz, 1968; Wagman and Price, 1969; Gregor and Zimmermann, 1972; Heavner and DeJong, 1973; Tapper *et al.*, 1973; D. D. Price and Browe, 1973; D. D. Price and Mayer, 1974; Handwerker *et al.*, 1975; Besson *et al.*, 1972; Fields *et al.*, 1977a; Menétrey *et al.*, 1977; Egger *et al.*, 1986; Sorkin *et al.*, 1986; Cervero *et al.*, 1988). A convergent input from several types of receptor is typical, although neither invariable nor random (P. B. Brown, 1969; Heavner and DeJong, 1973; Tapper *et al.*, 1973).

Wide-dynamic-range cells have also been found in laminae IV–VI by various investigators (Kolmodin and Skoglund, 1960; Fetz, 1968; Pomeranz *et al.*, 1968; Wagman and Price, 1969;

Besson *et al.*, 1972; Gregor and Zimmermann, 1972; Heavner and DeJong, 1973; D.D. Price and Browe, 1973; D. D. Price and Mayer, 1974; Handwerker *et al.*, 1975; Fields *et al.*, 1977a; Menétrey *et al.*, 1977; Sorkin *et al.*, 1986; Cervero *et al.*, 1988).

Several groups have found interneurons in laminae IV–VI, in addition to those in lamina I, that respond just to intense stimuli (Kolmodin and Skoglund, 1960; Gregor and Zimmermann, 1972; Heavner and DeJong, 1973; D. D. Price and Browe, 1973; Menétrey *et al.*, 1977; Sorkin *et al.*, 1986; Cervero *et al.*, 1988; see also Pomeranz, 1973).

Spike-triggered averaging has been used to show that interneurons in laminae V and VI with cutaneous input can evoke EPSPs or IPSPs in motoneurons and hence are premotor interneurons in reflex pathways (Hongo *et al.*, 1989a,b).

The presence of cells in the deepest part of the dorsal horn that receive a proprioceptive input has been confirmed by several groups (Kolmodin, 1957; Pomeranz *et al.*, 1968; Heavner and DeJong, 1973; Fields *et al.*, 1977b; Menétrey *et al.*, 1977; Jankowska *et al.*, 1981). However, such neurons are not confined to lamina VI, but rather extend from laminae IV to VIII. Neurons with input from group II muscle afferent fibers have been reported to lie as far dorsally as lamina IV (Fukushima and Kato, 1975). Neurons that receive input from group I muscle afferent fibers are known to exert synaptic actions directly on motoneurons (Jankowska and Roberts, 1972a,b; Jankowska and Lindström, 1972; Czarkowska *et al.*, 1981; Brink *et al.*, 1983a); it is not clear whether the same neurons also play a sensory role. A number of studies have been done to define the spectrum of inputs to these interneurons (see, e.g., Hultborn *et al.*, 1976a,b; Jankowska *et al.*, 1981; Brink *et al.*, 1983b; Harrison and Jankowska, 1985a,b; Rudomín *et al.*, 1987; 1990).

Cells that respond to visceral input have been found in laminae IV–V, as well as in lamina I and in the ventral horn at thoracic, lumbar, and sacral levels (Pomeranz *et al.*, 1968; Selzer and Spencer, 1969; Fields *et al.*, 1970a,b; Hancock *et al.*, 1973; R. D. Foreman, 1977; Guilbaud *et al.*, 1977b; Gokin *et al.*, 1977; R. D. Foreman and Ohata, 1980; D. D. Price *et al.*, 1981; Cervero, 1982a; 1983a; S. B. McMahon and Morrison, 1982; M. Takahashi and Yokota, 1983; Cervero and Tattersall, 1985; Kanui, 1985; Ammons, 1986; Tattersall *et al.*, 1986a; Knuepfer *et al.*, 1988; Ness and Gebhart, 1987; 1988; 1989; Alarcon and Cervero, 1990). These cells also receive cutaneous input and can be classified as low-threshold, high-threshold or wide-dynamic-range cells on this basis. Some of the responses to visceral stimulation appear to depend upon a supraspinal loop through the brain stem (Cervero and Wolstencroft, 1984; Cervero *et al.*, 1985).

Neurons in laminae IV–VI (as well as in lamina VIII) can often be activated by articular input (Schaible *et al.*, 1986). These cells commonly have a convergent input from cutaneous and muscle receptors (Schaible *et al.*, 1987a).

Lamina X

A population of neurons that respond to noxious cutaneous stimuli has been found in the vicinity of the central canal of rats (Nahin *et al.*, 1983). The cells are of the high-threshold type and are excited by noxious mechanical and thermal stimuli.

Comparable cells have also been found in the cat (Honda and Perl, 1985; Honda, 1985). However, cells near the central canal of cats include low-threshold and wide-dynamic-range cells, as well as high-threshold ones (Honda and Perl, 1985). Many of the cells have a convergent input from visceral afferent fibers, and a few respond only to visceral stimuli (Honda, 1985).

Summary

In summary, the activity of individual dorsal horn interneurons can be recorded intracellularly to demonstrate synaptic activity or extracellularly to monitor action potentials.

Interneurons can be classified according to their response properties. However, no completely satisfactory classification system has been devised. Experimenters interested in pain mechanisms often refer to neurons that are activated just by innocuous mechanical stimuli as low-threshold (class 1) cells, those excited both by innocuous and noxious stimuli as wide-dynamic-range (class 2) cells, and cells that discharge only if noxious stimuli are used as high-threshold (class 3) cells. The projection pattern of different types of primary afferent fibers influences the kinds of responses made by dorsal horn neurons in different laminae. Cutaneous Aδ nociceptors project to laminae I and V. Aδ thermoreceptors also terminate in lamina I. Cutaneous C fibers, whether mechanoreceptive, nociceptive, or thermoreceptive, end chiefly in lamina II, although some C fibers also terminate in lamina I. Cutaneous Aδ mechanoreceptors distribute mostly to lamina III, whereas Aβ fibers from other mechanoreceptors project to laminae III and IV. Proprioceptive afferent fibers from muscle end chiefly in laminae V and VI. Lamina I neurons often respond to volleys in cutaneous Aδ and sometimes also C fibers. Cutaneous Aβ volleys are occasionally effective, and in the thoracic cord visceral afferent fibers often activate lamina I cells. Natural stimuli that excite lamina I neurons may include noxious mechanical or thermal, innocuous thermal, or sometimes both innocuous and noxious mechanical cutaneous stimuli. Lamina II neurons are excited by C fibers, but often also by Aδ and even by Aβ fibers. Effective natural stimuli include noxious mechanical and thermal, or innocuous mechanical cutaneous stimuli. Unusual responses (compared to responses of neurons in other laminae) include marked habituation to repeated stimuli, prolonged discharges following brief stimuli, predominantly inhibitory responses, and variable ("ameboid") receptive fields. Neurons in laminae III–VI are excited by various combinations of cutaneous Aβ, Aδ, and C fibers. Repetitive C-fiber volleys at rates of ca. 1 Hz may evoke incrementing responses ("windup"). Volleys in visceral or fine muscle afferent fibers often excite neurons of lamina V. Large muscle afferent fibers activate neurons in lamina VI (and VII). Some of the neurons in laminae V and VI are premotor interneurons in reflex pathways, since they are directly connected to motoneurons; whether the same interneurons participate in sensory processing is unclear. Effective natural stimuli include innocuous and/or noxious mechanical stimulation of the skin (depending on whether the cell is of the low-threshold, wide-dynamic-range or high-threshold class), visceral stimuli, and joint movement for cells in deep layers. Neurons in lamina X respond to similar cutaneous inputs and may also have visceral inputs.

RECEPTIVE FIELDS OF DORSAL HORN NEURONS

Wall (1960) found that the excitatory receptive fields of cells in the dorsal horn are ovoid and larger than the receptive fields of primary afferent fibers. The central part of a receptive field is the most sensitive. Strong stimuli applied to the skin adjacent to the excitatory field could produce inhibition. Hillman and Wall (1969) investigated further the receptive fields of cells located in the region of lamina V. The cells had larger receptive fields than did cells characteristic of lamina IV. The receptive fields consisted of three concentric zones (Fig. 5.15). In the center of the receptive field, the movement of hair was sufficient to activate the neuron. Pressure or pinch were required when the stimuli were applied more peripherally. In the surrounding area of skin, tactile stimuli produced inhibition. Electrical stimulation of the dorsal columns or the dorsal roots other than those carrying the excitatory afferent fibers also produced inhibition of neurons in lamina V.

Thus, dorsal horn neurons are often characterized by both excitatory and inhibitory receptive fields.

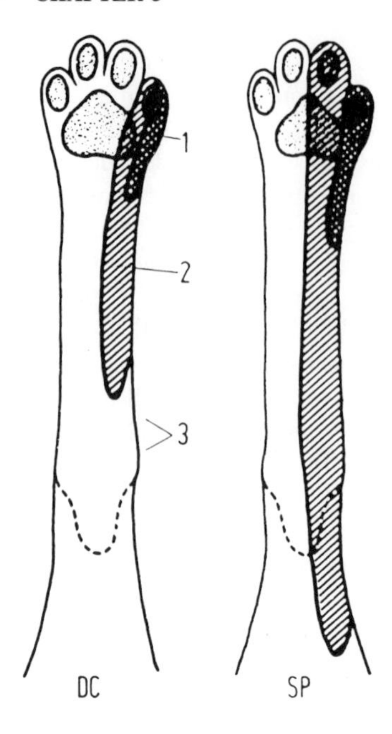

Fig. 5.15. Receptive-field organization for a wide-dynamic-range dorsal horn neuron. In the decerebrate state (DC, left), the receptive field included three zones. The cell responded to innocuous and noxious stimulation in zone 1, but only to noxious stimulation in zone 2. Innocuous stimuli here produced inhibition. In zone 3 (boundary not shown), mechanical stimuli caused inhibition. In the spinalized state (SP, right) produced by cold block of the cord, zones 1 and 2 expanded but zone 3 disappeared. (From Hillman and Wall, 1969.)

Excitatory Receptive Fields

Cascade Theory. It was suggested by Wall (1967) that cells in lamina IV excite those in lamina V and that cells in lamina V excite those in lamina VI. The evidence was that neurons in lamina V had larger receptive fields than did those in lamina IV, and the latency for activation was longer by an average of 1.5 msec. For neurons in lamina VI, the receptive fields on the skin were similar to those of cells in lamina V, but response latencies were longer still. Thus, excitation of neurons in the dorsal horn would be produced by a cascade of excitation. Neurons in lamina VI also receive a direct input from proprioceptors.

The progressive increase in receptive field size that Wall (1967) found for cells at progressively greater depths in laminae IV–VI was not confirmed by P. B. Brown *et al.* (1975; see also P. B. Brown, 1969), nor was the progressively greater latency of response. P. B. Brown *et al.* (1975), therefore, dispute the idea of a cascade of excitation of interneurons in the deeper layers of the dorsal horn by cells located more superficially (however, cf. Egger *et al.*, 1986).

Somatotopic Organization. Wall (1967) described a somatotopic organization of the dorsal horn. The toes were represented medially and the lateral aspect of the foot and more proximal regions were represented laterally. The description of the somatotopic arrangement differed somewhat from that proposed by Bryan *et al* (1973) for spinocervical tract cells. However, P. B. Brown and Fuchs (1975) demonstrated a somatotopic arrangement (Fig. 5.16) that is in reasonable agreement with the ones suggested by Wall (1967) and by Bryan *et al* (1973). Brown and Fuchs found both distoproximal and ventrodorsal gradients in the receptive fields of interneurons located across the mediolateral extent of the dorsal horn. The observation by Devor and Wall (1976) that cells placed laterally in the dorsal horn may have very proximal receptive fields is consistent with the scheme proposed by Brown and Fuchs (1975). Several groups (P. B. Brown and Koerber, 1978; Pubols and Goldberger, 1980; Light and Durkovic, 1984; P. Wilson *et al.*, 1986; Hylden *et al.*, 1987) have confirmed and extended the observations of P. B. Brown and Fuchs (1975). The adult pattern is already present in the neonatal kitten (P. Wilson and Snow, 1988).

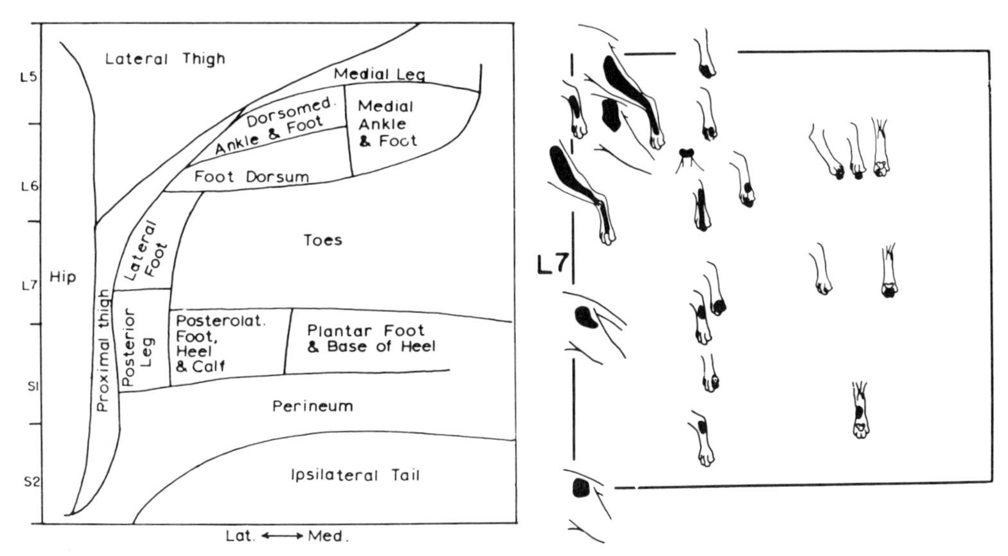

Fig. 5.16. The diagram at the left shows a scheme of the somatotopic organization of the dorsal horn in a horizontal plane through the lumbosacral enlargement of the cat. The abscissa indicates the lateral-to-medial extent of the dorsal horn, while the ordinate shows the segmental level. Receptive fields at segmental level L7 are shown in the summary diagram at the right. (From P. B. Brown and Fuchs, 1975.)

The somatotopic organization of dorsal horn interneurons was proposed by P. B. Brown and his colleagues to be a consequence of the organization of the cutaneous projections to the dorsal horn. Transganglionic labeling of the distribution of the afferent projections of various cutaneous nerves is generally consistent with the somatotopic organization demonstrated by recordings from postsynaptic neurons of the dorsal horn (P. B. Brown and Culberson, 1981; Koerber and Brown, 1980; 1982; Molander and Grant, 1985, 1986; Swett and Woolf, 1985; Woolf and Fitzgerald, 1986). However, the afferent projections appear to extend to some degree into nonsomatotopically appropriate regions (Koerber and Brown, 1982; Meyers and Snow, 1984; 1986; Meyers *et al.*, 1984; C. C. LaMotte *et al.*, 1989; Shortland *et al.*, 1989). Such projections could form the basis of "relatively ineffective synapses" on dorsal horn neurons (Basbaum and Wall, 1976; Devor and Wall, 1981a), "long-ranging" afferent fibers (Mendell *et al.*, 1978; Wall and Werman, 1976) and the "low-probability firing fringe" of the receptive fields of some dorsal horn interneurons (Woolf and King, 1989).

The somatotopic organization of primary afferent projections and of interneurons in the cervical enlargement of the cat is similar to that seen in the lumbosacral enlargement (Koerber, 1980; Nyberg and Blomqvist, 1985; Sorkin *et al.*, 1986). The same statement can be made for the cervical spinal cord of monkeys (Florence *et al.*, 1988; P. B. Brown *et al.*, 1989). There is a simpler somatotopic organization in the thoracic cord (G. Grant and Ygge, 1981; C. L. Smith, 1983; Ygge and Grant, 1983).

Bryan *et al.* (1973, 1974) suggested that the somatotopic organization of the dorsal horn reflects the embryological development of the innervation of the limbs (cf. Sherrington, 1893). The embryonic spinal cord initially has a simple somatotopic arrangement, in which neurons in the lateral part of the dorsal horn receive projections of afferent fibers supplying receptive fields in the dorsal parts of the corresponding dermatomes and neurons in the medial dorsal horn from the ventral parts of the dermatomes; this arrangement is maintained in the adult thoracic cord (G. Grant and Ygge, 1981; C. L. Smith, 1983; Ygge and Grant, 1983; cf. Cervero and Tattersall,

1985). However, in the enlargements, as the limb bud elongates and rotates, the simple dermatomal pattern is distorted. Afferent fibers from more rostral dermatomes continue to project to more rostral segments, and afferent fibers from the dorsal parts of the embryonic dermatomes maintain their projection to the lateral dorsal horn, but the location of the embryonic dorsal surface of the limb shifts. For example, in the hindlimb of the adult cat, the dorsal embryonic surface forms the lateral aspect of the thigh and leg, but the dorsal surface of the foot (Fig. 5.17). Therefore, it would be predicted that afferent fibers from the dorsal skin of the foot would project more laterally and afferent fibers from the plantar surface of the foot more medially in the dorsal horn (cf. Fig. 5.16) (P. B. Brown and Fuchs, 1975; Sorkin *et al.*, 1986).

Recently, it has been found that neurons in the deep layers of the dorsal horn having input from deep tissues are somatotopically organized (Xian-Min and Mense, 1990). Neurons that are excited by inputs from more distal parts of the limb are located medially to neurons with input from more proximal parts. If there is convergent input from the skin, the cutaneous input to these neurons is not somatotopically organized, suggesting that these neurons function to signal deep rather than superficial input.

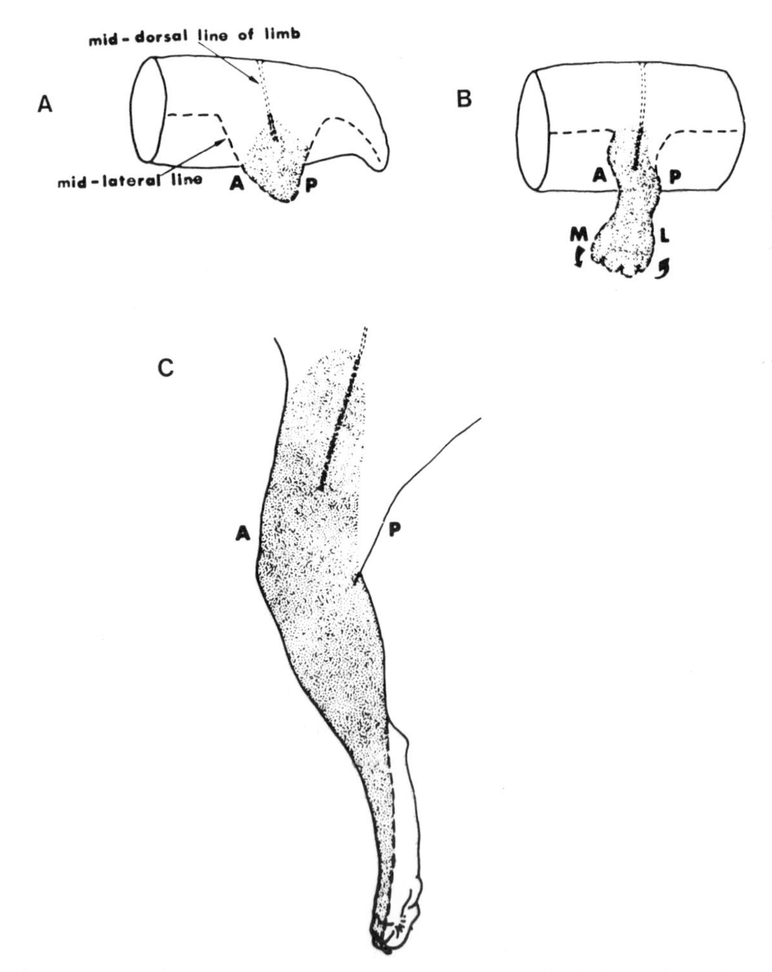

Fig. 5.17. Relationship of the developing limb-bud to the adult dermatomes in the cat hindlimb. The midlateral line is indicated in panels A–B. The dorsal surface of the L6 dermatome is stippled. Panels A–C show the progressive elongation and medial rotation of the distal limb. (From Bryan *et al.*, 1973.)

For further discussion of somatotopy with respect to the ascending tracts, see Chapters 7–9.

Plasticity. The receptive fields of at least some dorsal horn neurons are modifiable. This implies that the sensations evoked by stimulation of these receptive fields can change under certain conditions. Many sensory abnormalities are known to occur in pathologic conditions, including paresthesias, dysesthesias, pain, hyperalgesia, and allodynia. Presumably, many of these examples of sensory pathology are due to the plasticity of dorsal horn circuits.

The "ameboid" receptive fields of many neurons in lamina II have already been described (see above). However, changeable receptive fields have also been found in deeper laminae. Devor and Wall (1981a) mapped the excitatory receptive fields of neurons in laminae IV and V of decerebrate, spinalized cats. The neurons in lamina IV had stable receptive fields, but those in lamina V had receptive fields that varied spontaneously up to 30% in size. These changes made the fields resemble the "ameboid" fields of neurons in the substantia gelatinosa, except that the changes were not as large (Fig. 5.14). Tetanic electrical stimulation often caused a reduction in receptive field size, followed by gradual recovery over several minutes. Blocking or cutting a peripheral nerve caused loss of the entire receptive field, loss of part of the receptive field, or no change. In a few cases, a new field developed. By contrast, stimulation of C fibers can cause an increase in responsiveness of dorsal horn neurons, according to Cervero *et al.* (1984b) and A. J. Cook *et al.* (1987).

Pharmacologic evidence for plasticity of the receptive fields of dorsal horn neurons comes from the experiments of Zieglgänsberger and Herz (1971), who showed that release of excitatory amino acids in the vicinity of dorsal horn neurons causes an increase in the sizes of their receptive fields.

Noxious stimuli have been found to produce long-lasting alterations in the receptive fields of dorsal horn neurons. For example, injury to the skin proximal to the receptive field of a dorsal horn neuron causes the expansion of that receptive field toward the area of damage (S. B. McMahon and Wall, 1983a). Noxious heat causes an enhancement of the responses of spinothalamic neurons to repeated noxious heat stimuli (Kenshalo *et al.*, 1979; Ferrington *et al.*, 1987d) and to innocuous mechanical stimuli (Kenshalo *et al.*, 1982). Simone *et al.* (1989b, 1990) showed that intradermal injections of capsaicin cause an expansion of the receptive fields of dorsal horn neurons, including spinothalamic neurons, to innocuous mechanical stimulation. This change parallels the development of secondary hyperalgesia in human subjects given similar injections (Simone *et al.*, 1989a). A similar change is produced by application of mustard oil to the skin (Woolf and King, 1988), intradermal injection of Freund's adjuvant (Hylden *et al.*, 1989), or noxious stimulation of deep tissue (Hoheisel and Mense, 1989). Experimentally induced acute arthritis also leads to increases in the responsiveness of dorsal horn neurons (Schaible *et al.*, 1987b; Calvino *et al.*, 1987; Neugebauer and Schaible, 1990).

Cervero *et al.* (1988) found that repeated noxious pinching of the skin causes a change in the responses of nociceptive neurons in the dorsal horn of rats. Further exploration of this revealed that wide-dynamic-range neurons showed a dramatic increase in receptive field size, whereas there were either no changes or only small increases in the receptive fields of high-threshold neurons. It was proposed that the responses of wide-dynamic-range neurons are more readily changed than are those of high-threshold neurons. Laird and Cervero (1989) report similar findings for neurons in the dorsal horn of the sacral spinal cord with receptive fields on the tail.

A number of laboratories have investigated the influence of partial deafferentation of the spinal cord on receptive-field properties of dorsal horn neurons. One means of deafferentation has been interruption of one or more dorsal roots, another is by hemisection of the spinal cord, and a third is by sectioning one or more peripheral nerves.

Basbaum and Wall (1976) deafferented the lumbosacral spinal cords of cats by cutting all of the dorsal roots below L3 except S1 (spared-root paradigm of C. N. Liu and Chambers, 1958). At 24 hr after surgery, no cells were found in L4 or L5 that could be excited monosynaptically by cutaneous stimulation. However, by 1 week after surgery, such neurons could be found. After 1 month, the numbers of such cells were stable. The receptive fields were abnormal in the following ways: (1) their locations were on the S1 dermatome or on dermatomes rostral to L4; (2) they were often split; (3) their size was unusually variable; (4) they exhibited less convergence than is normally seen; (5) inhibitory fields were rare; and (6) habituation was common and might be different in one part of the receptive field than in another. It was proposed that the new receptive fields were due either to sprouting of primary afferent fibers or to the unmasking of previously ineffective synapses.

Pubols and Goldberger (1980) also used the spared-root paradigm to examine possible changes in the somatotopic organization of the dorsal horn (see also Goldberger and Murray, 1974, and Murray and Goldberger, 1974). All of the dorsal roots of the lumbosacral enlargement were severed, except L6. Soon after the dorsal rhizotomies, there was a loss of responsiveness of dorsal horn neurons, especially laterally in the dorsal horn, as well as a loss of proximal receptive fields. In chronic preparations, there was recovery of function, with an abnormally high proportion of strictly distal receptive fields. They suggested that this recovery could have been due to presynaptic changes, such as local sprouting or strengthening of synaptic connections, or to postsynaptic changes, such as denervation supersensitivity.

Brinkhus and Zimmermann (1983) also studied the effect of chronic partial deafferentation of dorsal horn neurons in cats by interruption of one or more dorsal roots. Cord dorsum potentials evoked by stimulating peripheral nerves of the hindlimb were unchanged when a single dorsal root was cut, but were smaller after sectioning of L5, L6, and most of L7. The response properties of dorsal horn neurons were little affected by partial deafferentation. The only substantial change was an increase in the background activity of a subpopulation of heat-sensitive neurons. More evidence of enhanced responses was expected, since deafferentation has been proposed as a mechanism of chronic pain (Loeser and Ward, 1967; Lombard *et al.*, 1979; Wall *et al.*, 1979b; Wiesenfeld and Lindblom, 1980; Levitt and Heybach, 1981; Levitt and Levitt, 1981; cf. Wynn Parry, 1980).

Devor and Wall (1981a) compared the somatotopic map in animals with intact nerves (controls) with that in animals with an acute or a chronic transection of the sciatic and saphenous nerves (see also Devor and Wall, 1981b; Ovelmen-Levitt *et al.*, 1984; however, cf. L. M. Pubols and Brenowitz, 1981; L. M. Pubols, 1984). Most of the neurons in the control state had receptive fields on the toes and foot. Those with receptive fields extending proximal to the ankle were in the lateral part of the dorsal horn. After acute transection of the nerves, cells with proximal receptive fields were still found laterally in the dorsal horn (Fig. 5.18, right), but most cells in the medial dorsal horn were unresponsive. After chronic transection of the nerves (6–105 days), many neurons found in the medial part of the dorsal horn had proximal receptive fields (Fig. 5.18, left). Other neurons were unresponsive. It was concluded that neurons with input exclusively from the toes or foot are in some way converted to neurons that respond to stimulation of proximal skin after chronic transection of a peripheral nerve. The changes could not be attributed to peripheral sprouting of the severed nerves, nor to the elimination of inhibition. Reasons were given to doubt the development of long-distance central sprouting (cf. C. N. Liu and Chambers, 1958; Goldberger and Murray, 1974; Devor and Claman, 1980; however, see below for evidence of sprouting). Instead, it was hypothesized that under normal conditions the neurons in the medial dorsal horn have latent proximal receptive fields (Wall and Werman, 1976), but that these are not expressed because the synapses are normally ineffective (because of inhibition, low synaptic efficacy, distal dendritic connection, or other reasons; cf. Merrill and Wall, 1972; Wall, 1977; Markus *et al.*, 1984; Hylden *et al.*, 1987; Markus and Pomeranz, 1987). It was proposed that nerve section causes the ineffective synapses to become effective. Mechanisms might include removal of presynaptic inhibition (cf. Horch and Lisney, 1981; Wall and Devor, 1981), increased quantal release, increased density of postsynaptic receptors, and changes in dendritic geometry or membrane conductance. Alternatively, there

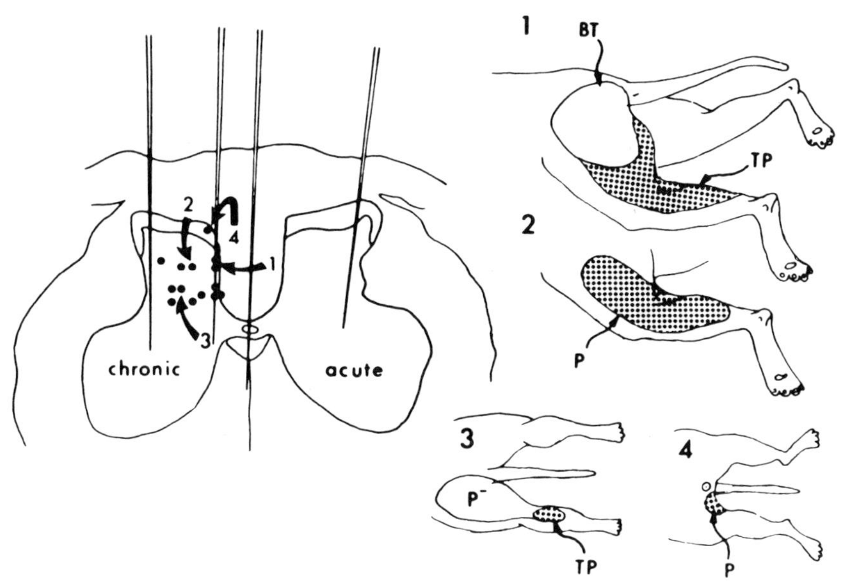

Fig. 5.18. Plasticity of receptive-field organization in the dorsal horn of the cat after chronic section of the sciatic and saphenous nerves. The nerves were cut 61 days earlier on the left and acutely on the right. Four electrode tips were left in place at the end of the experiment. The electrode at the right shows the track in which proximal receptive fields were first found in a mapping experiment that began in the medial dorsal horn. Proximal fields were found on the left at positions 1–4, as illustrated at the right by drawings, and also at the other positions shown by dots. Abbreviations: BT, brush and touch; TP, touch and pressure; P, firm pressure; P^-, inhibition by firm pressure. (From Devor and Wall, 1981; reproduced with permission.)

might be short-range sprouting, for example with displacement of ineffective synapses to more effective sites.

Woolf and Wall (1982) observed that chronic transection of the sciatic nerve in rats reduces the ability of volleys in A fibers to inhibit the responses of dorsal horn neurons to volleys in either A or C fibers. It was proposed that this reduction in inhibition (see below; see also Wall and Devor, 1981; Wall, 1982) may play a role in the unmasking of ineffective synapses and also in the development of chronic pain states.

Lisney (1983) cut the sciatic and saphenous nerves, but then allowed them to regenerate. He found the same receptive field changes that were described by Devor and Wall (1981a) at 1 month. By 9 months neurons in the medial dorsal horn had receptive fields on the toes or foot and not on the proximal limb. However, the somatotopic map had not been restored.

L. M. Pubols *et al.* (1986) presented evidence for the presence of relatively ineffective synapses in the cat dorsal horn by using electrical stimulation of the sural nerve to activate the cells. However, it was not possible to determine whether the ineffective synapses were monosynaptic or polysynaptic.

Using intracellular recordings from dorsal horn interneurons in rats, Woolf and King (1989) were able to demonstrate that the receptive fields of some neurons were surrounded by a "low-probability firing fringe," stimulation of which evoked EPSPs but few if any discharges. Such synaptic potentials could be considered to result from the activity of "relatively ineffective synapses."

A number of morphologic studies have demonstrated evidence for sprouting of primary afferent fibers following dorsal rhizotomies or peripheral nerve injury (C. N. Liu and Chambers,

1958; Hulsebosch and Coggeshall, 1981; Goldberger and Murray, 1982; Molander *et al.*, 1988; C. C. LaMotte *et al.*, 1989; Fitzgerald *et al.*, 1990; Polistina *et al.*, 1990; however, cf. Rodin *et al.*, 1983; Seltzer and Devor, 1984; L. M. Pubols and Bowen, 1988). In addition, both morphologic and pharmacologic studies have provided evidence of transsynaptic changes in dorsal horn interneurons (Gobel and Binck, 1977; Wright and Roberts, 1978; P. B. Brown *et al.*, 1979; Macon, 1979; Nakata *et al.*, 1979; Gobel, 1984; Sugimoto and Gobel, 1984; Westrum *et al.*, 1984; Kapadia and LaMotte, 1987; Iadarola *et al.*, 1988; Ruda *et al.*, 1988; Kajander *et al.*, 1990; see the review by Aldskogius *et al.*, 1985). Thus, both sprouting and transsynaptic changes in dorsal horn neurons continue to offer explanations for the changes in physiologic responses of dorsal horn neurons and presumably of pathologic sensations following such injuries.

Inhibitory Receptive Fields

As mentioned earlier, it is well known that dorsal horn neurons have inhibitory, as well as excitatory, receptive fields (Wall, 1960; Hillman and Wall, 1969). These can be demonstrated by electrical stimulation of peripheral nerves or of the dorsal column, by electroacupuncture (Hongo *et al.*, 1966; Hillman and Wall, 1969; Handwerker *et al.*, 1975; Pomeranz and Cheng, 1979; Fitzgerald, 1983b; Woolf and Wall, 1982; Tomlinson *et al.*, 1983; Woolf, 1983; R. Morris, 1987; Paik *et al.*, 1988; Tsuruoaka *et al.*, 1990), or by natural stimulation of the skin or other organs (Besson *et al.*, 1974; Cadden *et al.*, 1983; Fitzgerald, 1983b; Tomlinson *et al.*, 1983; Kanui, 1987). It is reasonable to assume that such inhibitory receptive fields may play a role in maneuvers that produce analgesia, such as counterirritation, acupuncture, transcutaneous electrical nerve stimulation, vibratory stimulation, and dorsal column stimulation (Gammon and Starr, 1941; Wall and Sweet, 1967; Shealy *et al.*, 1970; Nashold and Friedman, 1972; Kaada, 1974; D. M. Long and Hagfors, 1975; Melzack, 1975; Lundeberg *et al.*, 1984).

Inhibition of dorsal horn neurons appears to result from activation of at least two different inhibitory mechanisms. One system depends upon input from large myelinated afferent fibers and the other from fine myelinated and unmyelinated afferent fibers. The properties of the inhibitions produced by these systems are quite different. The inhibition produced by large afferent fibers tends to be of the "surround" type, and it can be produced by brushing the hair or using other weak mechanical stimuli. It depends upon segmental mechanisms and is the type of inhibition that is a major element of the gate theory of pain (see next section). The inhibition evoked by fine afferent fibers can be produced by stimulation anywhere on the body surface and results from activation of nociceptors. This system involves both a segmental and a suprasegmental mechanism. The latter is termed "diffuse noxious inhibitory controls" or DNIC (see below).

Gate Theory of Pain. In 1965, Melzack and Wall published a model for the dorsal horn circuitry responsible for pain transmission. They called the model the "gate control system." The gate theory has stimulated a great deal of research and has had a major impact on clinical thinking. The theory has been criticized by Nathan (1976), and a rebuttal of some of the criticisms has been published (Wall, 1978). Additional criticisms have been advanced by Cervero and Iggo (1980).

In the gate theory, Melzack and Wall (1965) proposed that activity in large afferent fibers inhibits synaptic transmission in a system otherwise activated by small afferent fibers carrying the signal for pain. A similar proposal, based on clinical findings, was made by Noordenbos (1959). Melzack and Wall drew a specific circuit for the organization of a spinal cord gating mechanism (Fig. 5.19). They proposed that inhibitory neurons in the substantia gelatinosa act as a gating mechanism to control afferent input before it affects nociceptive transmission (T) cells. The output of the T cells results in activation of the neural mechanisms that cause pain responses and perception.

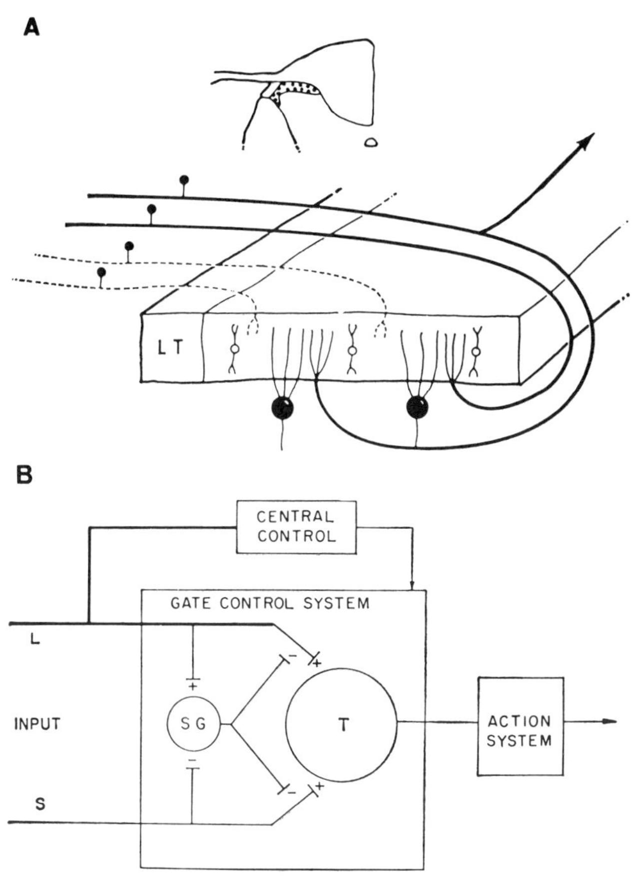

Fig. 5.19. (A) Representation of the manner in which small (dashed lines) and large (heavy lines) afferent fibers terminate in the superficial dorsal horn. Some of the large afferents also give off collaterals that ascend in the dorsal column (arrow). The substantia gelatinosa contains small intrinsic neurons (somata shown by open circles), and the dendrites of neurons whose cell bodies are in deeper laminae. Abbreviation: LT, tract of Lissauer. (B) Circuit proposed by Melzack and Wall (1965) for a gating system that might determine the output of the "action system." Abbreviations: L, large; S, small; SG, substantia gelatinosa; T, transmission cell. (From Melzack and Wall, 1965; copyright 1965 by the AAAS.)

Evidence for the involvement of the substantia gelatinosa in the control of afferent input came from the experiments of Wall (1962) and Mendell and Wall (1964) on dorsal root potentials. Recordings from wide-dynamic-range dorsal horn neurons (candidate T cells) showed that large afferent fibers produce a short-term excitation, whereas the addition of small-fiber activity results in greatly increased activity (Wall, 1958, 1959; Mendell and Wall, 1965; Mendell, 1966). Negative and positive feedback loops were proposed to account for the different responses (Melzack and Wall, 1965). The feedback was thought to occur at a presynaptic level, although postsynaptic actions were not excluded.

For the gate to operate properly, there would have to be (1) ongoing activity in small myelinated and unmyelinated fibers, (2) stimulus-evoked activity, and (3) a relative balance of activity in large and small afferent fibers. The tonic activity in small fibers would keep the gate partly open, whereas an input over large fibers would close the gate, limiting the discharge of the T cells. An increased stimulus intensity would shift the balance of large- and small-fiber inputs, thereby increasing the output through the T cells. A prolonged stimulus would result in adaptation of the large afferent fibers, and the small fibers would get the upper hand, opening the gate further. The gate could be returned toward a closed position by addition of large-fiber input.

An important aspect of the gate theory is the role of descending control systems, which can alter the gate. The activity in descending pathways must be appropriate to the situation, and so a central control trigger was proposed. The dorsal column pathway and the spinocervicothalamic pathway were considered as likely candidates to provide the discriminative information needed for a central decision to alter the sensitivity of the gating mechanism.

Criticisms of the gate theory center around findings in three areas since the theory was proposed: (1) the properties of cutaneous receptors, (2) the dorsal root potentials produced by large and small afferent fibers, and (3) the response characteristics of dorsal horn neurons.

Although the proportion of peripheral-nerve afferent fibers that could be called nociceptors was thought to be small in 1965, further investigations have shown that there are in fact large numbers of such fibers, in both the small myelinated and unmyelinated size ranges (see Chapter 2). However, in the absence of damage, these afferent fibers are silent, in contradiction to one of the requirements of the gate theory.

The circuit proposed to explain the negative and positive feedback in the operation of the gate control mechanism is shown in Fig. 5.19. The neurons of the substantia gelatinosa were thought to be excited by large afferent fibers, but they then produced presynaptic inhibition of the transmission from the large afferent fibers and so limited the excitation of the T cells. The PAD produced in large cutaneous afferent fibers by the activity in other large afferent fibers is in keeping with this model (Koketsu, 1956; J. C. Eccles and Krnjević, 1959; J. C. Eccles *et al.*, 1963a; Hodge, 1972; Whitehorn and Burgess, 1973). However, small afferent fibers were suggested by Melzack and Wall (1965) to inhibit the tonic activity of gelatinosa neurons and thus to cause presynaptic disinhibition. The dorsal root potential associated with such an event would be positive, instead of the usual negative wave. Positive dorsal root potentials have been described by a number of investigators (Lloyd, 1952; Wall, 1964; Dawson *et al.*, 1970; Mendell, 1970, 1972, 1973; R. E. Burke *et al.*, 1971; Hodge, 1972). However, in contrast to the predictions of the gate theory, the activation of unmyelinated afferent fibers is associated with the generation of negative dorsal root potentials (Zimmermann, 1968; Franz and Iggo, 1968; Jänig and Zimmermann, 1971; Gregor and Zimmermann, 1973). Significantly, the activation of C fibers connected with thermal nociceptors produces a negative dorsal root potential (Fig. 5.20) (R. E. Burke *et al.*, 1971), even though positive dorsal root potentials could be evoked in the same animal, ruling out the condition of the animal as a factor. Since many small myelinated and unmyelinated fibers are connected with sensitive mechanoreceptors in cats, the finding that negative dorsal root potentials result when a natural noxious stimulus is used appears to be more meaningful than the finding of positive dorsal root potentials following electrical stimulation of small fibers.

Further evidence concerning the presynaptic modulation of afferent fibers comes from the investigation of Whitehorn and Burgess (1973), who showed that the excitability of the terminals of large afferent fibers was increased whether stimulation was at an innocuous or noxious level of intensity. Primary afferent hyperpolarization was not seen except as an off response. Noxious mechanical stimuli produced primary afferent depolarization of small myelinated nociceptive afferent fibers, although noxious heat had little effect.

The other major finding since the introduction of the gate theory has been the demonstration of a population of dorsal horn neurons that respond specifically to noxious stimuli (Christensen and Perl, 1970; Kumazawa *et al.*, 1975). Thus, specific nociceptors occur in both the peripheral nerves and in the spinal cord.

Nathan (1976) cautions that the gate theory is based not upon the results of noxious stimuli but upon the potentials that are seen when nerves are stimulated electrically. Furthermore, the

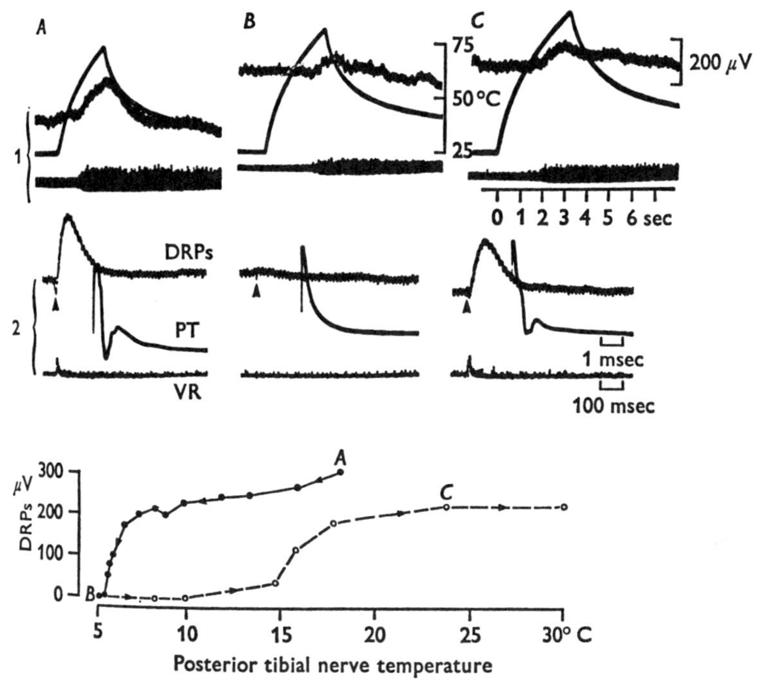

Fig. 5.20. The top traces in panels A–C are dorsal root potentials (negativity upwards). The middle traces in 1 are temperature measurements from the skin, while those in 2 are neurograms of volleys evoked by electrical stimulation of the plantar nerve before (A), during (B), and after (C) cold block of the nerve. The lower traces show activity in the ventral root. The graph below shows the size of the plantar nerve A fiber volley before (A), during (B), and after (C) application of the cold block. Heat pulses applied to the footpad by a radiant heat source evoked negative dorsal root potentials. At least part of this effect was mediated by C fibers, since the blockade of conduction in A fibers did not eliminate the DRP produced by the heat pulses. (From R. E. Burke *et al.*, 1971.)

results of stimulation of primary afferent fibers by microneurography fails to support the gate theory (see Chapter 10). Evidence that is consistent with the gate theory has been reviewed by Wall (1978). The fact that many cells in the dorsal horn receive convergent input from mechanoreceptors and nociceptors has already been discussed in the section on Responses to Natural Stimulation. The therapeutic value of stimulating large afferent fibers for the relief of pain has received considerable attention in recent years (Wall and Sweet, 1967; Shealy *et al.*, 1970; Nashold and Friedman, 1972). Wall (1978) admits that the weight of the evidence obtained since 1965 requires a modification of some of the details of the gate theory, but he believes that it still provides a useful model. Perhaps the most important point is that there is a mechanism in the dorsal horn that allows activity in large myelinated afferent fibers to produce an inhibition of the responses of dorsal horn neurons to noxious stimuli.

Diffuse Noxious Inhibitory Controls. Le Bars *et al.* (1979a,b) observed that wide-dynamic-range neurons in rats anesthetized with a mixture of halothane, nitrous oxide, and oxygen are subject to a powerful and prolonged inhibition when noxious stimuli are applied to any part of the body and face outside of the excitatory receptive field. This form of inhibition was attributed to DNIC. Low-threshold and high-threshold neurons, as well as proprioceptive neurons, are not subject to this form of inhibition. The amount of inhibition is graded according to stimulus intensity (Le Bars *et al.*, 1981a). DNIC are dependent upon activation of a supraspinal loop and a descending pathway, since interruption of the spinal cord at a more rostral level abolishes DNIC (Le Bars *et al.*, 1979b; see also Morton *et al.*, 1987). However, a weaker, short-term segmental inhibition can still be observed after spinal transection (Cadden *et al.*, 1983). The ascending pathway that triggers the DNIC mechanism is in the ventrolateral quadrant, contralateral to the noxious input (Villanueva *et al.*, 1986b). The descending pathway is in the dorsal lateral funiculus (Villanueva *et al.*, 1986a).

Le Bars and Chitour (1983) discuss the possible role of DNIC in nociceptive signaling. They point out that wide-dynamic-range neurons are likely to signal pain, yet are responsive to innocuous stimuli (as well as to noxious ones). One possibility is that an innocuous stimulus applied to the body surface would activate only a few WDR cells, whereas a noxious stimulus would activate many because of the structure of their receptive fields (Fig. 5.21 panel I). Pain might result when the population response exceeded some threshold centrally. However, another possibility is that noxious stimuli activate one set of WDR cells by activating their excitatory receptive fields while simultaneously inhibiting all other WDR cells through the DNIC mechanism. If the central nervous system can recognize the contrast between the activated and the silenced populations of WDR cells, this contrast could serve as a nociceptive signal (Fig. 5.21 panel II; see also J. D. Talbot *et al.*, 1989).

A number of studies have examined the mechanisms of DNIC. Chemical transmitters that may be involved are mentioned below in the section on pharmacology of dorsal horn neurons. A postsynaptic inhibitory action is suggested by the observation that the responses of wide-dynamic-range dorsal horn neurons to pulses of glutamate are reduced by DNIC, whereas those of low-threshold, high-threshold, and proprioceptive neurons are not (Villanueva *et al.*, 1984).

Evidence has been obtained by Willer *et al.* (1989) that there is a DNIC-like mechanism in humans (see also Willer *et al.*, 1984; Roby-Brami *et al.*, 1987; J. D. Talbot *et al.*, 1987; 1989).

Summary

In summary, dorsal horn interneurons often have both excitatory and inhibitory receptive fields. It has been suggested that the excitatory receptive fields of neurons located more dorsally in the dorsal horn are smaller than those of more ventrally placed cells and that the latencies of the initial response are progressively greater for deeper-lying neurons. This led to the idea of an excitatory cascade in which cells in a given lamina excite those in the adjacent deeper lamina. However, this idea is disputed. Dorsal horn neurons have a somatotopic organization. Rostral dermatomes project to rostral segments and caudal dermatomes to caudal segments. The dorsal part of a dermatome projects to the lateral dorsal horn and the ventral part to the medial dorsal horn. For the enlargements, input from the distal extremity predominates. The ventral (glabrous) surface of the toes is represented most medially, the dorsal (hairy) surface more laterally, and the dorsal aspect of the proximal limb most laterally. The responses of dorsal horn neurons are plastic; that is, they can often be shown to change. Alterations can result from stimulation or from lesions such as dorsal rhizotomy or peripheral nerve section. Some of the changes in responsiveness have been attributed to the unmasking of previously "ineffective" synapses. Mechanisms for such changes are still poorly understood. Alternatively or additionally, sprouting of primary afferent fibers may play a role in the plasticity of dorsal horn neurons. Inhibitory receptive fields are of two general types. Weak mechanical stimuli may produce a "surround" type of inhibition through a segmental mechanism; this inhibition can also be evoked by stimulation of large myelinated fibers. The gate theory of pain describes a possible circuit for this inhibitory mechanism that involves the substantia gelatinosa. However, details of this circuit have been shown to be incorrect. Another inhibitory mechanism is activated by noxious stimuli applied anywhere on the body surface or by volleys in fine myelinated and unmyelinated afferent fibers. This inhibition is powerful and long-lasting and involves both segmental and supraspinal pathways. The supraspinal component is called "diffuse noxious inhibitory controls" or (DNIC).

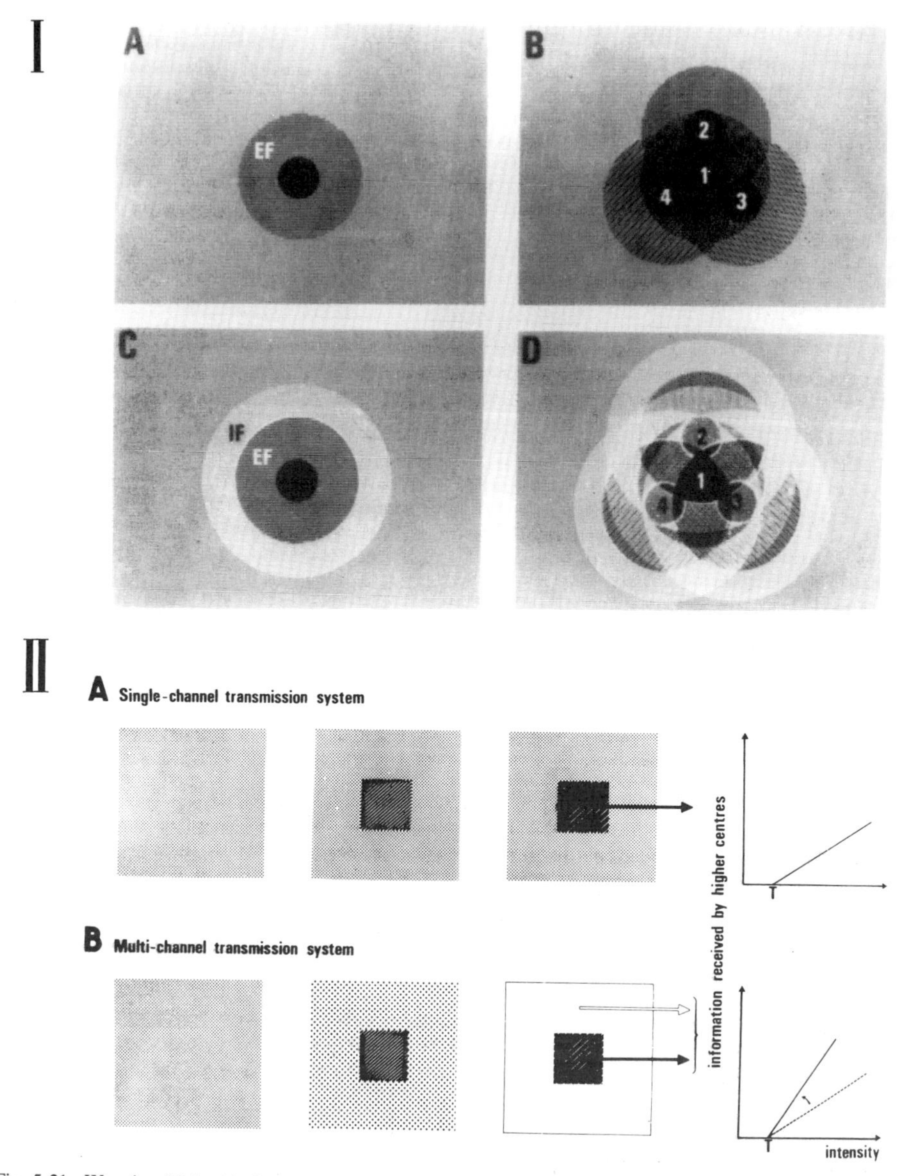

Fig. 5.21. Ways in which wide dynamic range neurons might signal nociceptive information. I. The receptive field of a WDR neuron is shown diagrammatically in IA; the cell responds to innocuous and noxious stimuli applied to the center, but only to noxious stimuli in the periphery. In IB, overlapping receptive fields of several WDR neurons are shown. A noxious stimulus is signaled by all, but an innocuous one by only a few. IC adds an inhibitory field, which in ID is shown to limit the discharges of many WDR cells. II. The activity of a pool of nociceptive neurons activated by different intensities of noxious stimuli is shown in IIA. The dark region indicates the particular set of neurons whose receptive fields are stimulated. In IIB the DNIC mechanism is activated by the pool of neurons whose receptive fields are excited. DNIC suppresses the activity of WDR neurons whose receptive fields are not stimulated, resulting in a greater contrast in the signal reaching the brain. (From Le Bars and Chitour, 1983.)

DESCENDING CONTROL OF DORSAL HORN INTERNEURONS

An important element of the gate control mechanism (see above) is its regulation by pathways descending from the brain. Furthermore, a crucial feature of DNIC is a supraspinal loop that results in descending inhibitory activity. It is a general principle that the brain regulates its sensory input. This is true of all of the somatosensory pathways that have been investigated, although perhaps the greatest current interest is in the inhibitory regulation of nociceptive inputs, since this may lead to a reduction in pain. The history of the concept of centrifugal control of sensation is reviewed by Willis (1982).

One of the main difficulties in the study of the actions of descending pathways on spinal cord interneurons is the problem of distinguishing between interneurons that serve primarily a sensory function and those that are engaged chiefly in motor activity. This problem can be circumvented by investigating the descending control of ascending somatosensory tract cells (see Chapters 7–9). Most investigators, however, have been content to record from unidentified neurons of the dorsal horn with the view that these probably influence sensory behavior by serving in either excitatory or inhibitory pathways affecting the ascending sensory pathways.

Another major issue in experiments on the effects of descending control systems is whether the axons of neurons in the brain exert their effects directly upon spinal neurons or indirectly through a polysynaptic linkage. Even when the particular descending control system being studied is known to project to the vicinity of the recording electrode, it is still possible that the pathway excites or inhibits the dorsal horn interneuron being examined polysynaptically. Rarely are serious attempts made to determine whether the action is direct or indirect. The conduction velocity of the descending pathway is often estimated based on an assumed monosynaptic linkage, but such an exercise can be misleading.

Excitatory Actions

An important pathway originating in the brain that excites many dorsal horn interneurons is the pyramidal tract. It is well known that some corticospinal axons terminate in the dorsal horn (Coulter and Jones, 1977; Flindt-Egebak, 1977; see reviews by Armand, 1982; Kuypers, 1981). In the monkey, corticospinal terminals can be found in laminae III–VI of the dorsal horn; projections to discrete regions of the dorsal horn originate from cortical cytoarchitectonic regions 4, 3a, 3b, 1, 2, and 5 (Coulter and Jones, 1977). Anterograde tracing experiments have also demonstrated a direct projection from the somatosensory cortex to laminae I and II in cats and monkeys (Cheema *et al.*, 1984b). In addition to these direct projections, the cerebral cortex can affect spinal neurons by way of activity relayed in the brain stem (Carpenter *et al.*, 1963; Hongo and Jankowska, 1967).

Wall (1967) and Fetz (1968) observed that stimulation of the pyramidal tract can inhibit or excite dorsal horn interneurons. Wall (1967) was unable to find any effect of stimulation of the pyramidal tract while recording from interneurons in the area of lamina IV, whereas comparable stimuli produced excitation and inhibition of most cells in laminae V and VI. Inhibition was more prominent in lamina V and excitation in lamina VI. Fetz (1968), on the other hand, observed that two-thirds of the cells in lamina IV were inhibited by pyramidal tract volleys and other cells showed an excitation followed by inhibition. Fetz is in agreement with Wall that inhibition is more prominent dorsally and excitation ventrally in the dorsal horn.

In addition to the excitation and inhibition of interneurons, corticospinal volleys have been shown to produce PAD of cutaneous afferent fibers and also group Ib and II muscle afferent fibers (but not group Ia muscle afferent fibers) (Carpenter *et al.*, 1963; P. Andersen *et al.*, 1964e). The implication is that the corticospinal tract controls spinal cord activity in part through presynaptic inhibition.

Dubuisson and Wall (1980) observed that many interneurons in laminae I–III were excited when the dorsal lateral funiculus was stimulated in decerebrate cats. A higher proportion of high-threshold and wide-dynamic-range neurons were excited than of low-threshold neurons. The latency of the excitatory action could be short enough to result from a pathway conducting at as high a velocity as 56 m/sec. Although the pathway could not be identified, Dubuisson and Wall doubted that it was the pyramidal tract, since at that time it was not known that there were corticospinal projections to the superficial dorsal horn. Instead, it was speculated that it might originate from the raphe nuclei of the brain stem. However, in light of the new evidence for a direct corticospinal projection to laminae I and II (Cheema *et al.*, 1984b), the findings of Dubuisson and Wall should be reevaluated.

Several other pathways originating from subcortical nuclei have been reported to excite dorsal horn interneurons, including the vestibulospinal and reticulospinal tracts (Carpenter *et al.*, 1966; Erulkar *et al.*, 1966; W. A. Cook *et al.*, 1969a,b). Of particular interest in the context of pain are excitatory effects of reticulospinal projections on spinal cord nociceptive neurons (Cervero and Wolstencroft, 1984; cf. Haber *et al.*, 1978; 1980; Giesler *et al.*, 1981b; Tattersall *et al.*, 1986b) (see Chapter 9).

S. B. McMahon and Wall (1988) have found that stimulation of the dorsal lateral funiculus in rats causes excitation of lamina I neurons that project in the contralateral dorsal lateral funiculus (generally following a brief inhibition). Repetitive stimulation could produce discharges that lasted up to 5 min, a facilitation of responses to peripheral stimuli with a similar prolonged time course, an increase in the size of the receptive field, and a lowering of the threshold. Other cells were instead inhibited or unaffected. The excitation appeared to be mediated chiefly by axons descending in the dorsal lateral funiculus, although there was some contribution by antidromically activated collaterals of the ascending-tract cells. By contrast, cells in the deep layers of the dorsal horn were inhibited following stimulation of the dorsal lateral funiculus. It was suggested that neurons in lamina I might activate brain stem mechanisms that inhibit neurons in the deep layers of the dorsal horn.

Inhibitory Actions

Tonic Descending Inhibition. Activity descending from the brain stem has been shown to inhibit nociceptive neurons of the dorsal horn tonically (reviewed by Willis, 1988). One experimental approach that has been used by many investigators to demonstrate tonic descending inhibition is reversible cold block (Wall, 1967; A. G. Brown, 1971). The activity of a neuron at a location caudal to the area to be blocked, generally in the lumbosacral spinal cord, is examined before, during, and after the block. If the responses of the neuron increase as a result of the block, this is taken as evidence that tonic descending inhibition had interfered with the responses until the inhibition was prevented by the block (Fig. 5.22).

Recordings have been made from nociceptive neurons affected by tonic descending inhibition in laminae I (Cervero *et al.*, 1976; Necker and Hellon, 1978; however, cf. S. B. McMahon and Wall, 1988) and IV-VI (Wall, 1967; Hillman and Wall, 1969; Besson *et al.*, 1975; Handwerker *et al.*, 1975; Duggan *et al.*, 1981a; Dickhaus *et al.*, 1985). Neurons in lamina II do not appear to be subject to tonic descending inhibition (Cervero *et al.*, 1979c).

Lesions have been made in a variety of brain stem structures in an attempt to localize the source of tonic descending inhibition of nociceptive dorsal horn neurons. Complete lesions of the pontine and medullary raphe nuclei do not alter the inhibition, nor do lesions of the periaqueductal gray or nucleus gigantocellularis (Hall *et al.*, 1981, 1982). However, bilateral lesions of the ventrolateral medulla in the region of the lateral reticular nucleus do eliminate the inhibition (Hall *et al.*, 1982; Morton *et al.*, 1983). It appears, however, that these successful lesions may have interrupted the descending projections of the tonic descending inhibitory neurons, since

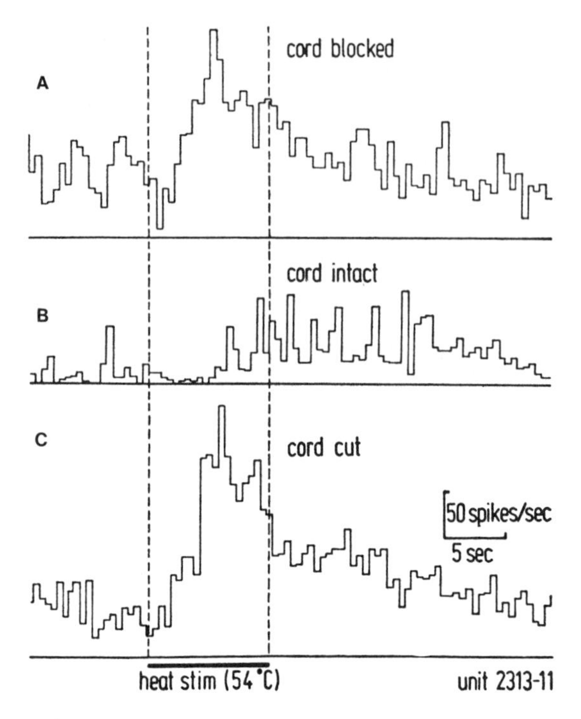

Fig. 5.22. Background discharge and responses of a WDR dorsal horn neuron to noxious heat. (A) The transmission of signals descending from the brain was blocked by cold. (B) The cord was intact, and there was no cold block. (C) The cord was transected. (From Handwerker *et al.*, 1975.)

lesions placed more rostrally in the region of the nucleus paragigantocellularis lateralis (just ventral to the facial nucleus) are also effective (Foong and Duggan, 1986). Furthermore, microinjections of a γ-aminobutyric acid (GABA) agonist in this second location reduced the tonic inhibition, whereas similar microinjections near the lateral reticular nucleus were ineffective. The drug would presumably have acted only in regions containing neuronal cell bodies and would not have affected axons of passage.

Stimulation-Evoked Inhibition. There has been considerable interest in the possible role of descending pathways in producing analgesia by inhibiting nociceptive transmission in the spinal cord (see reviews by D. J. Mayer and Price, 1976; Basbaum and Fields, 1978; Fields and Basbaum, 1978; Willis, 1982, 1988; Besson and Chaouch, 1987). Recordings from nociceptive interneurons in the dorsal horn have been used to demonstrate the inhibitory effects of stimulating at many sites in the brain, including the periaqueductal gray (Liebeskind *et al.*, 1973; Oliveras *et al.*, 1974; Bennett and Mayer, 1979; Carstens *et al.*, 1979, 1980a,b, 1981a,b; Duggan and Griersmith, 1979; Hayes *et al.*, 1979; Gebhart *et al.*, 1983a,b; B. G. Gray and Dostrovsky, 1983a; Light *et al.*, 1986; Sandkuehler and Zimmermann, 1987; Sandkuehler *et al.*, 1988), nucleus raphe magnus (Engberg *et al.*, 1968b; Fields *et al.*, 1977a; Guilbaud *et al.*, 1977a; Willis *et al.*, 1977; Belcher *et al.*, 1978; Duggan and Griersmith, 1979; Rivot *et al.*, 1980; Griffith and Gatipon, 1981; Gebhart *et al.*, 1983a; B. G. Gray and Dostrovsky, 1983a; Edeson and Ryall, 1983; Mokha *et al.*, 1985, 1986; Light *et al.*, 1986; Tattersall *et al.*, 1986a; Pretel *et al.*, 1988), nucleus reticularis gigantocellularis (B. G. Gray and Dostrovsky, 1983a; Tattersall *et al.*, 1986a; Pretel *et al.*, 1988), the region of the lateral reticular nucleus (Morton *et al.*, 1983), nucleus paragigantocellularis lateralis (B. G. Gray and Dostrovsky, 1985), the parabrachial region, locus coeruleus and subcoeruleus (Hodge *et al.*, 1981, 1983, 1986; Mokha *et al.*, 1985, 1986; Mokha and Iggo, 1987; S. L. Jones and Gebhart, 1986, 1987), anterior pretectal nucleus (Rees and Roberts, 1987), hypothalamus (Carstens, 1982), and other areas of the forebrain (Carstens *et al.*, 1982). In some studies, it has been shown that stimulation of many of the same areas of the brain evoke IPSPs (Giesler *et al.*, 1981a; Light *et al.*, 1986; Mokha and Iggo, 1987) and/or PAD (presumably causing presynaptic inhibition in the spinal cord) (Carpenter *et al.*, 1966; Engberg *et al.*, 1968a; Hentall and Fields, 1979; R. F. Martin *et al.*, 1979; Mokha and Iggo, 1987). Not only interneurons, but also nociceptive ascending-tract cells are inhibited by these descending systems (see Chapters 7–9).

Descending Pathways

Many of the inhibitory effects resulting from stimulation in the brain stem are likely to be mediated by raphe and reticulospinal projections. There is a spinal projection from the raphe nuclei of the lower brain stem (Kuypers and Maisky, 1975; Basbaum *et al.*, 1978, 1986a; R. F. Martin *et al.*, 1978; Basbaum and Fields, 1979; Kneisley *et al.*, 1978; Tohyama *et al.*, 1979a,b; G. F. Martin *et al.*, 1982; Zemlan *et al.*, 1984; Carlton *et al.*, 1985a). The nucleus raphe magnus projects to the dorsal horn through the dorsal lateral funiculus (Kuypers and Maisky, 1977; Basbaum *et al.*, 1978; Leichnetz *et al.*, 1978; R. F. Martin *et al.*, 1978; Tohyama *et al.*, 1979b). Projections from individual raphe-spinal axons have been traced into the dorsal horn (Light, 1985; Light and Kavookjian, 1985). There are numerous serotonin-containing neurons in the raphe magnus nuclei (see the section on pharmacology of dorsal horn neurons below). Many of these project to the spinal cord (Bowker *et al.*, 1981). However, there are also nonserotonergic raphe-spinal neurons, and many raphe-spinal axons contain peptides in addition to serotonin (Bowker *et al.*, 1983; Cassini *et al.*, 1989).

In addition to raphe-spinal axons, there are descending projections to the dorsal horn that originate in other brain stem nuclei, including the hypothalamus, the region of the Edinger-Westphal nucleus, the red nucleus, the locus coeruleus and nucleus subcoeruleus, Koelliker-Fuse and parabrachial nuclei, the nucleus gigantocellularis, the nucleus paragigantocellularis lateralis, the lateral reticular nucleus, and the solitary nucleus (Kuypers and Maisky, 1975, 1977; Hancock, 1976; Hancock and Fougerousse, 1976; Basbaum *et al.*, 1978; Kneisley *et al.*, 1978; Tohyama *et al.*, 1979a; Westlund and Coulter, 1980; Zemlan *et al.*, 1984; Carlton *et al.*, 1985a; Tan and Holstege, 1986; Cechetto and Saper, 1988; R. H. Liu *et al.*, 1989). Many of the projections from the locus coeruleus, subcoeruleus, and Koelliker-Fuse nucleus are noradrenergic (Westlund and Coulter, 1980; R. T. Stevens *et al.*, 1982). Some of the neurons contain peptides, such as enkephalin and substance P (Cassini *et al.*, 1989). Many of the neurons in the rostral ventrolateral medulla that project to the spinal cord contain GABA (Reichling and Basbaum, 1990).

The inhibitory effects of brain stem stimulation are often mediated polysynaptically through poorly understood circuits. Since few neurons in the periaqueductal gray project directly to the spinal cord, it has been proposed that the inhibitory effects of stimulation of the periaqueductal gray are relayed through a synaptic linkage in the nucleus raphe magnus and adjacent reticular formation (Basbaum and Fields, 1978). However, the descending effects of periaqueductal gray stimulation are also mediated in part by pathways through more lateral parts of the lower brainstem, as well as through the medial medulla (Gebhart *et al.*, 1983b; Morton *et al.*, 1984).

Summary

The activity of dorsal horn interneurons is controlled by pathways descending from the brain. These controls can be either excitatory or inhibitory. The corticospinal tract can have either action, but excitation is more common for interneurons in the deeper layers of the dorsal horn. A pathway descending in the dorsal lateral funiculus can excite nociceptive neurons in the superficial layers of the dorsal horn. There are other excitatory pathways as well. Many dorsal horn neurons are under tonic inhibitory control. This can be revealed experimentally through use of a reversible cold block of the descending pathways. Neurons responsible for the tonic descending inhibition appear to be localized to the nucleus reticularis paragigantocellularis lateralis, just ventral to the facial nucleus. Inhibition of dorsal horn neurons, especially of their responses to noxious stimuli, can be produced by stimulation in many locations of the brain, including the periaqueductal gray, nucleus raphe magnus, nucleus reticularis gigantocellularis, the region of the lateral reticular nucleus, nucleus reticularis paragigantocellularis lateralis, the region of parabrachial, locus coeruleus and subcoeruleus nuclei, anterior pretectal nucleus, hypothalamus, and others. Such inhibition is usually thought to relate to analgesia. The pathways responsible for the inhibition are likely to include raphe-spinal and reticulospinal tracts, but many other pathways could be involved as well. Some of the descending systems use serotonin or norepinephrine as transmitters, but other substances, including peptides, are also present in many of the descending axons. It should be kept in mind that the effects of descending pathways on a given type of neuron may be direct or indirect, involving excitatory or inhibitory interneurons at the spinal cord level.

PHARMACOLOGY OF DORSAL HORN INTERNEURONS

A number of pharmacologic agents have been found to affect the activity of dorsal horn interneurons. Some of these, including the excitatory amino acids and several peptides, have been implicated in synaptic transmission from primary afferent fibers. Excitatory amino acids and peptides have also been proposed as excitatory neurotransmitters or modulators that are used by dorsal horn interneurons or by the axons of pathways descending from the brain. Inhibitory transmitters, including amino acids such as GABA and glycine, as well as peptides such as enkephalin, may be released secondarily to the activation of spinal cord interneurons by primary afferent inputs or indirectly by the activation of pathways descending from the brain. Other transmitters, such as serotonin and several catecholamines, are directly released by descending control systems.

A large number of studies have (1) demonstrated the presence of various neuroactive chemicals in synaptic terminals in the spinal cord, (2) shown that these substances are released following activation of neural pathways, (3) demonstrated pharmacologic actions of these substances on spinal neurons, and (4) provided evidence that blocking agents prevent such actions. These studies go far towards meeting the criteria required for proving that these substances are neurotransmitters or neuromodulators (Werman, 1966). However, the evidence may not be complete for a given substance or for a particular synaptic connection. This field is a highly active one.

Excitatory Amino Acids

Excitatory Action of Glutamate and Aspartate. There is a strong possibility that glutamate (and/or asparate) is the excitatory transmitter used by many primary afferent fibers (Curtis *et al.*, 1960; Curtis and Watkins, 1960; Graham *et al.*, 1967; Duggan and Johnston, 1970; J. L. Johnson and Aprison, 1970; J. L. Johnson, 1972; P. J. Roberts *et al.*, 1973; Zieglgänsberger and Puil, 1973; Duggan, 1974; Biscoe *et al.*, 1976; J. Davies and Watkins, 1983; Peet *et al.*, 1983; S. P. Schneider and Perl, 1985, 1988; Jessell *et al.*, 1986; Sillar and Roberts, 1988; see reviews by Puil, 1981; Watkins and Evans, 1981; Salt and Hill, 1983; M. L. Mayer and Westbrook, 1987).

The excitatory action of glutamic and aspartic acid (also homocysteic acid) on spinal cord interneurons, as well as on other neurons of the central nervous system, was first demonstrated by iontophoretic release of these amino acids into the vicinity of the neurons (Curtis *et al.*, 1959a, 1960; Curtis and Watkins, 1960). Since then, many investigators have shown that spinal cord interneurons are generally, but not invariably, responsive to glutamate. Schneider and Perl (1985, 1988) surveyed interneurons in the dorsal horn in a slice preparation from hamster spinal cord. In their sample, only about 30% of the units were excited by iontophoretically applied or bath-applied glutamate or aspartate. Neurons that responded to excitatory amino acids were concentrated in laminae I and II (see also Yoshimura and Jessell, 1990), whereas insensitive units were distributed in laminae I–V. These observations suggest that other excitatory substances besides the excitatory amino acids may play an important role in transmission within the spinal cord. However, a larger proportion of the neurons found by King *et al.* (1988) in the deep layers of the dorsal horn in an *in vitro* preparation could be activated by excitatory amino acids.

Presence of Excitatory Amino Acids in the Spinal Cord. Glutamate is known to be present in peripheral nerves, dorsal root ganglia, dorsal roots, and the dorsal horn (Graham *et al.*, 1967; Duggan and Johnston, 1970). However, dorsal rhizotomy does not change the concentration of glutamate in the dorsal horn (P. J. Roberts and Keen, 1974). Since glutamate is a major constituent of neuronal tissue because of its role in metabolism, the presence of glutamate is not in itself indicative of a role in neurotransmission. Glutamate has been shown to be released from dorsal root ganglia (P. J. Roberts, 1974b). However, again there is some difficulty in interpreting the release of glutamate as necessarily related to synaptic transmission. Recently, it has become feasible to identify glutamate by immunocytochemical means, and this substance can be shown to occur in certain synaptic terminals but not others (De Biasi and Rustioni, 1988). Assuming that the concentration of metabolic glutamate is too low to be recognized by the immunocytochemical technique, this suggests that terminals identified by their content of glutamate may in fact use this amino acid in neurotransmission.

Excitatory Amino Acid Receptors. Glutamate receptors have been subdivided into several classes based on their responses to different glutamate agonists, including kainic acid, quisqualic acid, and *n*-methyl-D-aspartic acid (NMDA), and antagonism by several substances that are more or less selective for one or more glutamate receptor types (McLennan and Lodge, 1979; McLennan and Liu, 1982; McLennan, 1983; Watkins and Evans, 1981; Peet *et al.*, 1983; Childs *et al.*, 1988). Kainate and quisqualate receptors have been reported to be more responsive to glutamate than to aspartate, whereas NMDA receptors are thought to be more responsive to aspartate (Peet *et al.*, 1983). Receptors of all three types are concentrated in lamina II of the human dorsal horn (Jansen *et al.*, 1990).

Kynurenic acid has been found to antagonize glutamate receptors on spinal neurons in a non-selective fashion (Jahr and Jessell, 1985). Recently, the quinoxalinediones, DNQX and CNQX, have been proposed as selective non-NMDA receptor blockers (Drejer and Honoré, 1988; Honoré *et al.*, 1988). NMDA receptors are selectively antagonized by 2-amino-5-phosphovalerate (APV or AP5; Fig. 5.23), 2-amino-7-phosphonoheptanoate (AP7 or 2-APH), 3-(2-carboxypiperazin-4-yl)propyl-1-phosphonic acid (CPP), 5-methyl-10,11-dihydro-5H-dibenzo[a,d]cyclohepten-5,10-imine maleate (MK-801), and Mg^{++} (Ault *et al.*, 1980; J. D. Davies *et al.*, 1981; J. Davies and Watkins, 1979, 1982, 1983; McLennan and Liu, 1982; Peet *et al.*, 1983; Childs *et al.*, 1988).

Magnuson *et al.* (1988) tested the excitatory actions of a number of analogs of glutamate and also their susceptibility to antagonism by APV and kynurenate. The results are consistent with the notion "that compounds reacting with the NMDA receptor do so in an extended configuration whereas the QUIS receptor has a more folded template."

Ketamine has been proposed as an NMDA antagonist (Anis *et al.*, 1983). However, Headley *et al.* (1987) applied ketamine iontophoretically to dorsal horn interneurons and found that the currents needed to block nociceptive responses were so high that the ketamine antagonized not only NMDA but also quisqualate. A similar problem was seen with a nonselective antagonist, which blocked the responses to applied amino acids better than synaptic responses. Ketamine also failed to block synaptic responses substantially when administered systemically at moderate doses.

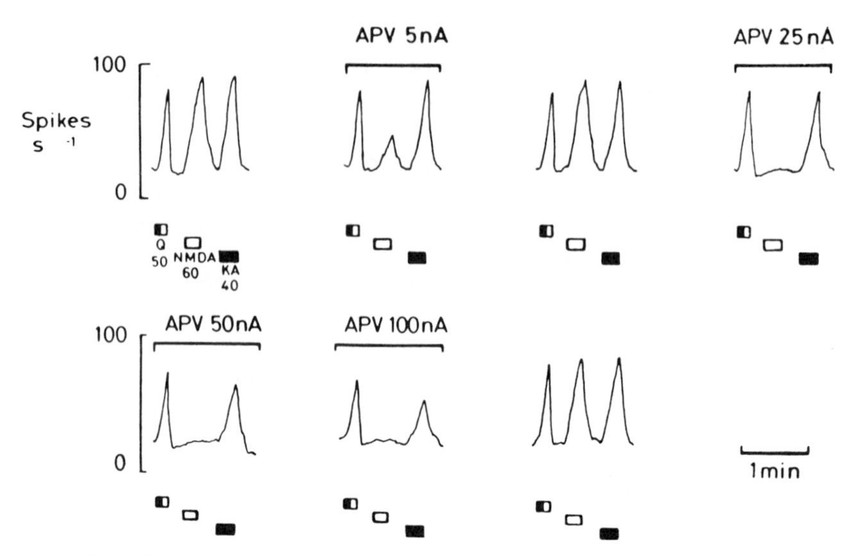

Fig. 5.23. Effects of excitatory amino acids on a spinal cord interneuron and the selective blocking action of the responses to NMDA by APV. The peaks of activity in ratemeter records are the responses, in order, to quisqualate (Q), NMDA, and kainate (KA). APV was administered in increasing doses, as indicated. APV produced a selective decrease in the response to NMDA at the lower currents. High currents affected the responses to all three excitatory amino acids. (From J. Davies and Watkins, 1983.)

Monosynaptic versus Polysynaptic Transmission. The actions of the excitatory amino acids on spinal cord interneurons differ for effects mediated through kainate and quisqualate receptors and those mediated through NMDA receptors. When the effects of antagonists are observed on monosynaptic versus polysynaptic transmission to interneurons in the dorsal horn, it is found that specific NMDA antagonists, such as APV, block polysynaptic transmission, whereas less specific glutamate receptor antagonists, such as γ-D-glutamylglycine (γ-DGG), decrease both monosynaptic and polysynaptic transmission (J. Davies and Watkins, 1983; see Peet *et al.*, 1983). These findings are consistent with the proposal that glutamate released from primary afferent fibers produces monosynaptic excitation of spinal neurons by an action on kainate/quisqualate receptors and that aspartate released from interneurons produces polysynaptic excitation through an action on NMDA receptors (cf. R. A. Davidoff *et al.*, 1967; Duggan and Johnston, 1970; Curtis and Johnston, 1974).

Schouenborg and Sjölund (1986) tested the effects of a relatively nonselective excitatory amino acid antagonist (γ-DGG) on the initial part of the C-fiber evoked field potential recorded in lamina I, as well as on the responses of lamina II interneurons to the same C-fiber volley. Since the early C-fiber field potential is likely to be monosynaptic (see above), the depression of this potential and of the associated interneuronal activity by γ-DGG is consistent with the proposal of J. Davies and Watkins (1983) that monosynaptic transmission involves glutamate released from primary afferent fibers.

S. N. Davies and Lodge (1987) found that ketamine (a presumed NMDA antagonist) prevents wind-up in wide-dynamic-range dorsal horn interneurons but does not reduce the initial response to a series of volleys in C (plus A) fibers. By contrast, kynurenate (a nonselective excitatory amino acid antagonist) reduces both the initial response and wind-up. The conclusions were that NMDA receptors do not contribute to the initial response to afferent volleys, but that they do contribute to wind-up.

King *et al.* (1988) found that the polysynaptic responses to dorsal root stimulation recorded intracellularly from neurons in laminae III–VI in a rat spinal cord slice preparation were blocked by APV.

S. P. Schneider and Perl (1988) also examined the effect of excitatory amino acids and their antagonists on synaptic transmission in a spinal cord slice preparation, but in this case the slice was from hamster spinal cord and was in the horizontal plane so as to include the superficial dorsal horn. The connection from a dorsal root remained intact, allowing activation of large and small primary afferent fibers. Relatively nonselective antagonists of excitatory amino acid receptors, such as kynurenic acid, blocked the short latency responses of dorsal horn neurons to dorsal root volleys, as well as to bath application of glutamate. Antagonists of NMDA receptors, such as APV, were less effective.

R. Morris (1989) tested the effects of excitatory amino acid antagonists on the responses of dorsal horn neurons to peripheral nerve stimulation in an isolated, hemisected rat spinal cord preparation. The initial evoked spike was blocked by the nonselective antagonist kynurenic acid and also by a quisqualate receptor antagonist, CNQX (6-cyano-2,3-dihydroxy-7-nitroquinoxaline). CNQX had little effect on the later spikes of the response. Conversely, an NMDA antagonist, CPP, did not block the initial spike; instead, it reduced the number of spikes that occurred later in the response.

Gerber and Randić (1989a) observed the effects of excitatory amino acid antagonists on the EPSPs evoked in neurons of the deep layers of the dorsal horn in slice preparations of the rat spinal cord by dorsal root stimulation. Kynurenate and CNQX reduced monosynaptic EPSPs, whereas APV had only a small effect on these. Polysynaptic EPSPs were increased when Mg^{++} was omitted from the bathing medium, and they were reduced by APV, as well as by kynurenate and CNQX. It was suggested that polysynaptic EPSPs have both non-NMDA and NMDA components; however, polysynaptic pathways begin with monosynaptic transmission, and so these results do not rule out the initiation of polysynaptic EPSPs by non-NMDA transmission at the first synapse of primary afferent fibers and NMDA transmission at a later synapse in the pathway.

Dickenson and Sullivan (1990) found that the nonselective excitatory amino acid antagonist DGG blocked the responses of most dorsal horn neurons to volleys in both Aβ and C fibers; it also blocked wind-up. At the time of peak inhibition, the cells were insensitive to natural stimulation of their receptive fields. The NMDA antagonist APV had little effect on neurons in lamina I, but reduced the responses of neurons in the deep layers of the dorsal horn to Aβ- and C-fiber volleys and increased the responses of cells in the substantia gelatinosa to C-fiber volleys. Wind-up was blocked by APV in cells of lamina I and the deep dorsal horn, but enhanced in the substantia gelatinosa. These findings are consistent with the idea that non-NMDA receptors are involved in monosynaptic transmission from primary afferent fibers and that NMDA receptors are responsible for polysynaptic effects, including wind-up. The converse effects of the antagonists on cells in an area thought to be the substantia gelatinosa call to mind the "inverse" receptive field types of many of the cells in this region (see above).

Membrane Effects of Excitatory Amino Acids. A number of investigations of the effect of glutamate on spinal cord neurons using intracellular recording have shown that the excitatory action of this amino acid can produce depolarizations that are accompanied by increases, decreases, or biphasic changes in membrane conductance (MacDonald *et al.*, 1982; S. P. Schneider and Perl, 1985, 1988; King *et al.*, 1988). These paradoxical responses can be explained by the complex kinetics of NMDA receptors.

Westbrook and Mayer (1984) examined spinal cord neurons in cultures by using voltage clamp. Bath-applied excitatory amino acids produced depolarizations of the neurons. However, responses to glutamate differed from those to aspartate and NMDA. Glutamate responses did not change in amplitude as the membrane potential was reduced from −80 to −40 mV; it then decreased at still more depolarized levels. NMDA and aspartate responses increased as the membrane potential was changed from −80 to −40 mV, reaching a peak at −25 to −35 mV. Responses to kainate and quisqualate were maximal at the most negative membrane potentials and decreased with depolarization. It was hypothesized that the glutamate response curve reflected activation of both NMDA and non-NMDA receptors. APV made the glutamate current-voltage curve more linear, so that it resembled the curves for kainate and quisqualate. The apparent absence of a conductance increase when glutamate is applied is due to the negative-slope conductance of the NMDA receptor. It was easy to demonstrate that glutamate produces a substantial increase in conductance in the presence of APV.

The membrane effects of excitatory amino acids have been studied further by the patch-clamp technique using neurons of the chick spinal cord in culture (Vlachová *et al.*, 1987). A step application of NMDA produces an initial peak current that reaches its maximum in less than 0.5 sec, followed by a decline to a plateau (Fig. 5.24A). When the NMDA application is terminated, there is an after-current that reaches a maximum in 2 sec and then decays over 6–7 sec. Similar

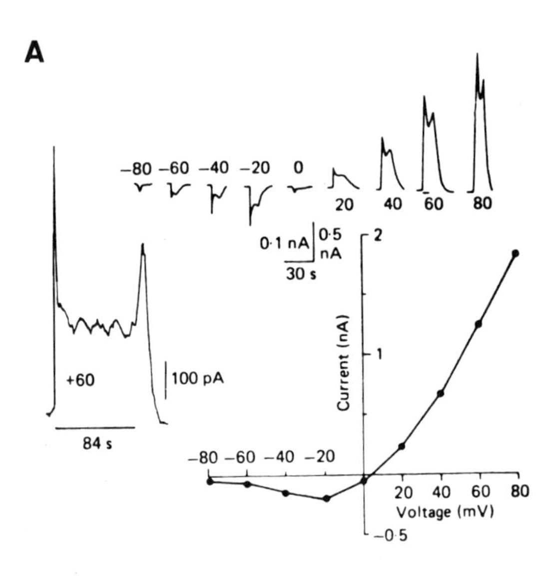

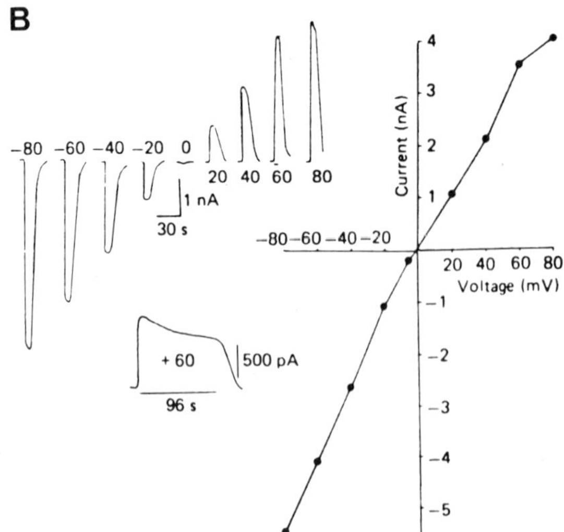

Fig. 5.24. Membrane currents produced by application of excitatory amino acids to cultured spinal neurons in the chick. The currents were recorded during voltage clamp. (A) Currents produced by rapid application of NMDA at various holding potentials for one neuron and at 60 mV for 84 sec for another neuron. The graph shows the amplitude of the initial peak response plotted against voltage. (B) Membrane currents resulting from application of kainate at various holding potentials for 1 cell and at 60 mV for 96 sec for another neuron. The graph is comparable to that in panel A. (From Vlachová *et al.*, 1987.)

responses are produced by aspartate, glutamate, and quisqualate. This response pattern is independent of synaptic events, since it is insensitive to tetrodotoxin and can be recorded in cultured neurons that do not receive any synaptic connections and in the presence of solutions containing 5 mM Co^{++}. The decay in the response could be explained by occlusion of channels by the amino acid or by desensitization.

Responses to kainate differ in that they are monophasic and decay slowly; there is no initial peak current nor any after-current (Fig. 5.24B). Apparently, kainate responses do not desensitize.

NMDA and kainate responses also differ in voltage dependence. Currents produced by NMDA in the presence of Mg^{++} are voltage dependent, rectifying at positive membrane potentials, and have a negative slope conductance at holding potentials of −20 to −80 mV (Fig. 5.24A). The chord conductance at +80 mV is 20 nS. By contrast, the current-voltage relationship for the response to kainate is linear (Fig. 5.24B), with a conductance of 60 nS. The reversal potential for all responses to the excitatory amino acids is near zero, indicating that the responses of all the excitatory amino acids depend on a similar ionic mechanism. When quisqualate and kainate or NMDA and kainate are applied simultaneously, there is no summation of responses. In fact, in the presence of quisqualate, there is a reduction in the response to kainate. These observations suggest that the channels activated by the excitatory amino acids are not independent. Single-channel recordings with glutamate in the electrode reveal channels with a mean conductance of 52 pS.

Involvement of Second-Messenger Systems. Activation of NMDA receptors results in an influx of Ca^{++} into the cytoplasm of neurons (MacDermott *et al.*, 1986; Womack *et al.*, 1988). Kainic acid is less effective in causing Ca^{++} influx. At least part of the Ca^{++} appears to enter through the channels associated with the NMDA receptor.

Woolf (1987b) found that intrathecal administration of glutamate or aspartate increases the active form of glycogen phosphorylase in the dorsal horn. The similar effect of stimulating nociceptive C fibers by applying mustard oil to the skin is prevented by APV. Substance P (SP) has no effect, suggesting that NMDA receptors are responsible for the change in enzymatic activity. As mentioned earlier in the section on Changes in Metabolic Activity, glycogen phosphorylase activity is enhanced by second messengers, and so the effects observed by Woolf (1987b) are likely to result from the activation of second-messenger systems through excitation of NMDA receptors.

Involvement in Generation of PAD. Evans and Long (1989) examined the effect of excitatory amino acid antagonists on the generation of dorsal root potentials following stimulation of an adjacent dorsal root in an *in vitro* preparation of rat spinal cord. NMDA antagonists (AP5, CPP) had little effect, but non-NMDA antagonists (CNQX) caused a reduction in the dorsal root potential. Presumably, the reduction is due to an action at the initial synapse in the pathway responsible for evoking the dorsal root potential. Any other excitatory synapses in the pathway appear not to use NMDA receptors.

Involvement in Excitation by Descending Pathways. In addition to its role in mediating excitatory transmission from primary afferent fibers and interneurons, glutamate may be used as a neurotransmitter in some descending pathways. For example, Rizzoli (1968) found that the concentration of glutamate in the dorsal lateral funiculus of the spinal cord decreases by 3 weeks following a more rostral spinal cord transection, suggesting that axons descending in that sector of the white matter contain a large quantity of glutamate.

ATP

Another candidate excitatory transmitter in the dorsal horn is ATP (see the review by Sawynok and Sweeney, 1989). ATP has been found to depolarize some dorsal horn neurons in an *in vitro* preparation (Jahr and Jessell, 1983). Fyffe and Perl (1984) have shown that ATP excites neurons in laminae II and III that receive mechanoreceptive or mechanoreceptive and nociceptive inputs (low-threshold and wide-dynamic-range cells), but not cells receiving just nociceptive inputs (high-threshold neurons). It was proposed that ATP might mediate the central actions of mechanoreceptive afferent fibers. Salter and Henry (1985) confirmed that most low-threshold cells, but only a small fraction of wide-dynamic-range cells, are excited by ATP, supporting the suggestion that ATP mediates the input from mechanoreceptors.

Excitatory Neuropeptides

Other compounds that are likely to serve as transmitters or modulators in the central nervous system are the neuropeptides (see reviews by Otsuka and Takahashi, 1977; Emson, 1979). Unlike many neurotransmitters, the peptides must be synthesized in the cell bodies of neurons and transported to the axonal terminals for storage and later release. Furthermore, there is no reuptake mechanism, and so replenishment after release depends upon synthesis and transport. On the other hand, only small quantities of peptides are needed since they are quite potent.

The terminals of small-diameter primary afferent fibers in the dorsal horn may contain one or more peptides, such as SP, calcitonin gene-related peptide (CGRP), somatostatin, or vasoactive intestinal polypeptide (VIP), in addition to an excitatory amino acid neurotransmitter (see Chapter 4). In addition, many spinal cord interneurons contain peptides (S. P. Hunt *et al.*, 1981a; Schroder, 1983; Fuji *et al.*, 1985; Shimosegawa *et al.*, 1986; Leah and Menétrey, 1989), as do the terminals of some of the pathways descending from the brain (O. Johansson *et al.*, 1981). Peptides are likely to play a modulatory role by enhancing or inhibiting synaptic transmission through any of several mechanisms (Emson, 1979).

Substance P. SP is thought to be a neurotransmitter or modulator in nociceptive afferent fibers (see review by Nicoll *et al.*, 1980). SP has been demonstrated chemically to occur in dorsal roots and in the spinal cord (T. Takahashi and Otsuka, 1975). It is concentrated in the dorsal horn, especially the superficial layers. Dorsal rhizotomy causes a marked decrease in the levels of SP in the dorsal horn. Furthermore, SP accumulates distal to a ligature of a dorsal root, indicating that it is transported to the spinal cord from the DRG.

Immunohistochemical studies have demonstrated SP-like immunoreactivity in many small DRG cells, as well as in synaptic terminals in the dorsal horn (see, e.g., Hökfelt *et al.*, 1975b,c, 1976; Chan-Palay and Palay, 1977a,b; Pickel *et al.*, 1977; Barber *et al.*, 1979; see Chapter 4). It is not clear what types of primary afferent fibers contain SP (Leah *et al.*, 1985a,b). SP appears to be stored in dense-core synaptic vesicles (Cuello *et al.*, 1977; Pickel *et al.*, 1977; Plenderleith *et al.*, 1990), although immunoreactivity is also found in association with other structures in the synaptic terminals (DeLanerolle and LaMotte, 1983). The SP-containing endings are especially concentrated in laminae I and II (DeLanerolle and LaMotte, 1983). SP is also found in intrinsic neurons of the spinal cord dorsal horn (Tessler *et al.*, 1980, 1981). In addition, SP is contained in the terminals of some of the projections to the spinal cord from the brainstem (O. Johansson *et al.*, 1981). The SP in descending projections may be colocalized with serotonin (Chan-Palay *et al.*, 1978; Hökfelt *et al.*, 1978; O. Johansson *et al.*, 1981).

One problem in immunohistochemical studies designed to map the distribution of SP-containing neurons or processes is that antibodies to SP may cross-react with a related tachykinin, such as one of the neurokinins (S. Kimura *et al.*, 1983; Kanazawa *et al.*, 1984). An alternative approach is to use *in situ* hybridization to localize the mRNA for SP or other tachykinins. In rats, mRNA for SP is contained in DRG cells and in many neurons of laminae I and II, and the lateral spinal nucleus, as well as in some neurons of laminae V, VII, and X (Warden and Young, 1988). mRNA for neurokinin B is chiefly in neurons of lamina III, and it is also present in some neurons of laminae I, II, and X, but not in DRG cells. The presence of mRNA for SP and the lack of mRNA for neurokinin B in the DRG is consistent with the observation that the concentration of SP, but not of neurokinin B, in the spinal cord is reduced by dorsal rhizotomy (Ogawa *et al.*, 1985). The concentration of mRNA for SP is very high in many DRG cells in rats and rabbits when compared with central neurons, despite the low levels of SP as demonstrated by immunocytochemistry; this may reflect a high turnover of SP in DRG cells (Boehmer *et al.*, 1989).

SP receptors have also been found in the spinal cord (Ninkovic *et al.*, 1985). Recently, the distributions of three different tachykinin receptors have been mapped in the rat spinal cord (Yashpal *et al.*, 1990). NK-1 receptors (which respond preferentially to SP) are concentrated in laminae I, II, and X; NK-2 receptors (responsive to neurokinin A) are found mainly along the dorsal and ventromedial borders of the dorsal horn; and NK-3 receptors (having greatest affinity to neurokinin B and eledoisin) are in the most superficial part of the dorsal horn.

SP has been shown to be released from the spinal cord following stimulation of unmyelinated afferent fibers; following noxious mechanical, thermal, or chemical stimulation of the skin; or following administration of capsaicin (Fig. 5.25C) (Angelucci, 1956; Otsuka and Konishi, 1976; Gamse *et al.*, 1979; Bergstrom *et al.*, 1983; Brodin *et al.*, 1987; Go and Yaksh, 1987; Duggan and Hendry, 1986; Duggan *et al.*, 1987, 1988a). SP can also be released by stimulation in the brain stem (Takano *et al.*, 1984). Since SP and serotonin are colocalized in the terminals of many raphe-spinal axons, in fact in the same dense-core vesicles (Pelletier *et al.*, 1977), it is instructive that SP can be released independently of serotonin by intrathecal administration of

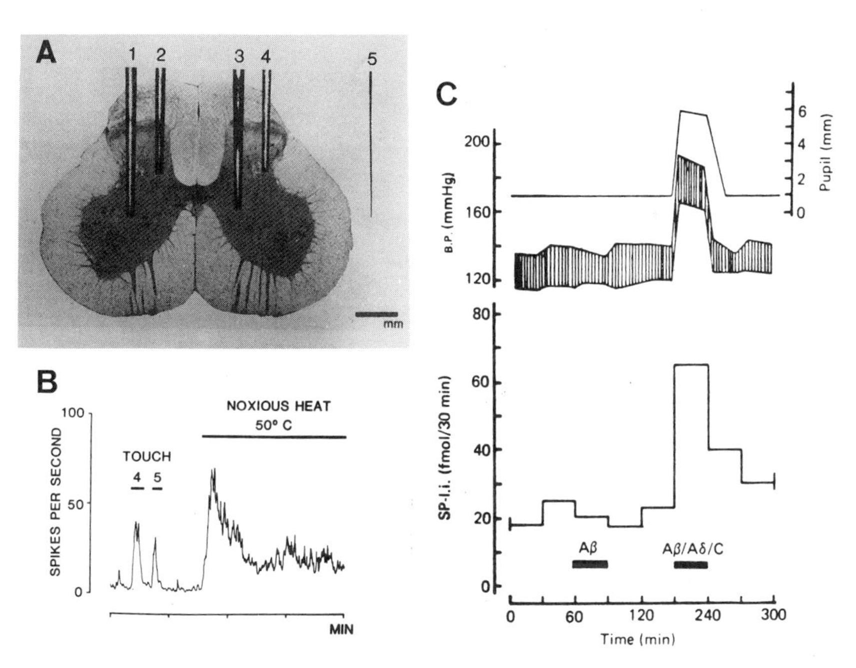

Fig. 5.25. Release of SP from the spinal cord following noxious stimulation. (A) Several antibody microprobes in relation to a transverse section of the cat spinal cord. Microprobe 1 shows that no SP was released when the paw was immersed in water at 37°C, whereas microprobe 2 shows release of SP in laminae I and V when the paw was put in water at 50°C. Microprobe 3 shows the effect of 52°C water and microprobe 4 shows the result of holding the limb in air. The actual size of the microprobes is shown by 5. (B) Responses of neurons in the dorsal horn recorded through microprobe 2 when the skin was touched on digits 4 and 5 and then when the foot was immersed in hot water. (A and B from Duggan *et al.*, 1987.) (C) The lower graph shows that SP (measured by radioimmunoassay) was released from the spinal cord of a cat when A and C fibers were stimulated, but not when Aβ fibers alone were stimulated. The volleys in A and C fibers also caused an elevation in the blood pressure (hatched) and in pupillary diameter (upper graph). (From Go and Yaksh, 1987.)

capsaicin or of *p*-chloroamphetamine (Bergstrom *et al.*, 1983). The implication is that there may be independence of release under normal conditions.

Duggan and his colleagues have introduced a novel technique, the antibody microprobe, which allows the determination of the sites of SP release with considerable spatial resolution (Duggan and Hendry, 1986; Duggan *et al.*, 1987, 1988a,b). The antibody microprobe is a glass micropipette coated with antibody molecules. When the microprobe is introduced into the spinal cord, SP released in the vicinity of the microprobe binds to SP antibodies on the microprobe. The microprobe is withdrawn from the spinal cord and reacted with radiolabeled SP. The distribution of the unlabeled zone to which neurally released SP had bound is determined by autoradiography and related to the layers of the dorsal horn. A major finding is that noxious stimuli cause the release of SP in the substantia gelatinosa (Fig. 5.25A). SP is also released in the dorsal horn in response to movements of the knee joint after the development of an acute arthritis (Schaible *et al.*, 1990).

The antibody microprobe technique has also shown that noxious stimuli cause the release of neurokinin A in the spinal cord (Duggan *et al.*, 1990; Hope *et al.*, 1990). However, there are major differences in the release of SP and of neurokinin A. There is a substantial basal level of neurokinin A, but not of SP, in the absence of stimulation. SP is released focally in the superficial dorsal horn, whereas neurokinin A is released diffusely in the cord gray matter. SP remains in the extracellular space for only a brief period, whereas neurokinin A persists for at least 30 min after stimulation.

When SP is iontophoresed into the vicinity of nociceptive dorsal horn neurons, these cells show a prolonged excitation, often lasting several minutes (Henry *et al.*, 1975; Henry, 1976; Randić and Miletić, 1977; Sastry, 1979b; Zieglgänberger and Tulloch, 1979a; J. Davies and Dray, 1980; Miletić and Randić, 1982). Sometimes, high doses of SP can depress the responses of dorsal horn neurons to glutamate (Henry *et al.*, 1975; J. Davies and Dray, 1980). SP antagonists have been reported to block the action of SP, at least in an *in vitro* system (Urban and Randić, 1984).

Murase *et al.* (1982) examined the responses of dorsal horn neurons to SP using intracellular recordings in a spinal cord slice preparation. SP caused a depolarization of the neurons. The depolarization had a slow time course and it was often associated with an increase in membrane resistance. A similar slow depolarization could be evoked in about half of the dorsal horn neurons examined by repetitive stimulation of the dorsal root (Urban and Randić, 1984). The depolarization produced by dorsal root stimulation could be prevented by adding SP or an SP antagonist to the bathing solution, suggesting that it is due to the action of SP (Urban and Randić, 1984). Similarly, antibodies to SP block the depolarizations produced either by SP or by dorsal root stimulation (Randić *et al.*, 1986).

The mechanism of action of SP is being investigated. One proposal is that SP inhibits the M-current in dorsal horn neurons (Hösli *et al.*, 1981; Nowak and MacDonald, 1982), as it does in sympathetic ganglion cells (Adams *et al.*, 1983). The M-current is a voltage-sensitive K^+ current, and so inhibition of this current would cause a depolarization accompanied by an increased membrane resistance. Another proposal is that SP blocks inward rectification (Stanfield *et al.*, 1985). Alternatively, SP may cause a depolarization through an increase in Na^+ or Ca^{++} conductance (Murase and Randić, 1984). Voltage clamp experiments on dorsal horn neurons in a slice preparation suggest that the application of SP augments an inward Ca^{++} current and decreases the M-current (Murase *et al.*, 1986). However, Womack *et al.* (1988) found that application of SP to isolated dorsal horn neurons *in vitro* causes an increase in cytoplasmic Ca^{++} concentration. Since this effect was independent of extracellular Ca^{++}, the change must have been due to mobilization of Ca^{++} from intracellular stores.

Murase *et al.* (1989a) confirmed that SP enhances a voltage-dependent inward Ca^{++} current. Other conductances were also affected, including an increase in the voltage-gated K^+ conductance and the Ca^{++}-activated K^+ conductance, as well as a calcium-sensitive, voltage-independent inward current carried by Na^+ and K^+ ions. There was no change in the inward (anomalous) rectifier current. Neurokinin A had similar complex actions. These multiple effects of SP can account for both the depolarization and variable alterations in membrane resistance that are produced when SP is applied to dorsal horn neurons.

According to Murase *et al.* (1989a), the conductance changes evoked by SP are likely to be mediated by the phosphoinositide second-messenger system (cf. Mantyh *et al.*, 1984). Two second messengers that would be produced following binding of SP to its receptor are inositol 1,4,5-triphosphate and diacylglycerol. Inositol 1,4,5,-triphosphate causes release of Ca^{++} from intracellular stores, and diacylglycerol activates protein kinase C, which in turn phosphorylates ion channels, including voltage-dependent Ca^{++} channels (Nowycky *et al.*, 1985).

SP has also been reported to exert presynaptic actions (Randić *et al.*, 1982). The excitability of primary afferent axons of C and Aβ fiber caliber within the spinal cord was tested by using electrical pulses to activate the fibers antidromically. Changes in excitability were often found for both Aβ and C fibers, but the changes could be reductions or increases in excitability or biphasic changes. There was no change in SP release when the spinal cord was blocked more rostrally by cold, indicating that tonic descending inhibition is not mediated by presynaptic inhibition of SP release (Duggan *et al.*, 1988b).

Kellstein *et al.* (1990) have found that superfusion of the spinal cord with SP has little effect on the background activity or responses to A fiber volleys of dorsal horn neurons. However, the responses to C fiber volleys are enhanced. The changes in the responses to C fiber volleys can be prevented with a SP antagonist.

Using microiontophoretic application of several amino acid agonists, Aanonsen *et al.* (1990) observed that NMDA, QUIS, and AMPA preferentially activate nociceptive dorsal horn neurons, whereas kainate preferentially excites low threshold cells.

Calcitonin Gene-Related Peptide. Another peptide that is found in many small DRG cells, as well as in some larger ones, is CGRP (Rosenfeld *et al.*, 1983; Skofitsch and Jacobowitz, 1985a; Yokokawa *et al.*, 1986; Ju *et al.*, 1987b; Molander *et al.*, 1987). Frequently, CGRP is colocalized with SP in the same DRG cells (Wiesenfeld-Hallin *et al.*, 1984; Gibbins *et al.*, 1985; 1987b; Skofitsch and Jacobowitz, 1985c). In fact, CGRP and SP can be found in the same dense-core vesicles (Gulbenkian *et al.*, 1986; Merighi *et al.*, 1988; Plenderleith *et al.*, 1990), and as many as four different peptides can be found in the same dorsal root ganglion cell (Leah *et al.*, 1985b; Cameron *et al.*, 1988).

CGRP is found in Aδ and C fibers in the dorsal roots and in the tract of Lissauer (Gibson *et al.*, 1984b; Carlton *et al.*, 1987b; Harmann *et al.*, 1988b; D. L. McNeill *et al.*, 1988a,b), as well as in the terminals of primary afferent fibers. The greatest concentration of CGRP-containing terminals is in laminae I–II and V, although there are a few CGRP-containing fibers in laminae III, IV, VI, VII, and X (Carlton *et al.*, 1987b; D. L. McNeill *et al.*, 1988a; Harmann *et al.*, 1988b; Traub *et al.*, 1989b). Unlike SP, CGRP does not appear to occur in interneurons of the dorsal horn (Gibson *et al.*, 1984b; Kawai *et al.*, 1985; Skofitsch and Jacobowitz, 1985b; Traub *et al.*, 1989a,b; however, see Conrath *et al.*, 1989) or in axons descending from the brain. mRNA for CGRP is found in DRG cells and motoneurons, but not in interneurons of the dorsal horn (Gibson *et al.*, 1988). CGRP is reduced dramatically by dorsal rhizotomy, especially if multiple, bilateral rhizotomies are performed (Gibson *et al.*, 1984b; K. Chung *et al.*, 1988; Traub *et al.*, 1989a,b). Thus, CGRP is a marker for primary afferent terminals in the dorsal horn (Carlton *et al.*, 1988). However, it should be noted that CGRP is also contained in the cell bodies of motoneurons and preganglionic autonomic neurons (Gibson *et al.*, 1984b; Senba and Tohyama, 1988).

CGRP-binding sites are concentrated in laminae I–VI of the dorsal horn and also around the central canal (Skofitsch and Jacobowitz, 1985a).

It has been determined that there are two forms of CGRP in DRG cells: α-CGRP and β-CGRP (Amara *et al.*, 1985; Steenbergh *et al.*, 1985). Small and medium-sized DRG cells may contain the mRNA for either or both types of CGRP, whereas large DRG cells contain mRNA chiefly of α-CGRP (Noguchi *et al.*, 1990a,b). SP and both forms of CGRP may coexist in small DRG cells.

The release of CGRP in the spinal cord has been demonstrated (Saria *et al.*, 1986; Diez Guerra *et al.*, 1988). Using the antibody microprobe technique, Morton and Hutchison (1989) showed that noxious mechanical or thermal stimulation of the skin caused the release of CGRP in the substantia gelatinosa. CGRP was also released when fine afferent fibers in the tibial nerve were stimulated electrically, but not when only large myelinated fibers were stimulated. Concurrent release of SP and CGRP following noxious thermal stimulation could also be shown.

The functions of CGRP may include enhancement of the action of SP. For example, CGRP has been found to inhibit the enzymatic degradation of SP (Greves *et al.*, 1985), and it also potentiates the release of SP from terminals (Oku *et al.*, 1987), perhaps by increasing Ca^{++} influx into synaptic terminals (Oku *et al.*, 1988). Ryu *et al.*, (1988b) have shown that CGRP increases both the amplitude and the duration of Ca^{++} spikes in DRG cells. They propose that enhanced Ca^{++} currents in primary afferent fibers help explain the enhancement of synaptic transmission produced in the dorsal horn by CGRP.

In addition to presynaptic modulatory effects, CGRP has a postsynaptic action. Miletić and Tan (1988) have applied CGRP iontophoretically onto dorsal horn neurons and have found that this peptide activates low-threshold and wide-dynamic-range neurons, but not high-threshold neurons. The time course of the excitation is slow, with an onset that takes 30 sec to 3 min and a duration of up to 10 min.

Ryu *et al.* (1988a) applied CGRP to slice preparations of the dorsal horn and DRG of rat spinal cord while recording intracellularly from dorsal horn neurons. The peptide often caused a slow, dose-dependent depolarization (sometimes preceded by a transient hyperpolarization) that was not prevented when voltage-gated Na^+ and K^+ currents were blocked by tetrodotoxin and tetraethylammonium, respectively. Membrane conductance was reduced during the depolarization, and excitability was enhanced. The action of SP was stronger than that of CGRP when the two were compared on the same neurons.

Combined Action of Excitatory Amino Acids and Peptides. At some synapses on dorsal horn interneurons, stimulation of primary afferent fibers causes a fast excitation followed by a slow excitation with two components (Gerber and Randić, 1989b). The initial excitation is due to monosynaptic transmission, often as a result of an action of an excitatory amino acid on non-NMDA receptors (see above). The first part of the slow excitation lasts several seconds and appears to result from activation of NMDA receptors, perhaps in addition to non-NMDA receptors, by excitatory amino acids released from interneurons (Gerber and Randić, 1989b). The later part of the slow excitation, lasting several minutes, seems to be peptidergic (Randić *et al.*, 1986; Urban and Randić, 1984).

The responses of dorsal horn neurons *in vitro* to NMDA are enhanced by CGRP and also by glycine (Murase *et al.*, 1989b).

Application of SP or CGRP to an *in vitro* spinal cord preparation causes an increased release of glutamate and aspartate (Kangrga *et al.*, 1990).

Co-application of excitatory amino acids and SP to dorsal horn neurons, including spinothalamic tract cells, can produce a long-lasting enhancement of the responsiveness of these cells to later applications of an excitatory amino acid or to peripheral input (Dougherty and Willis, 1990; Randić *et al.*, 1990).

Other Excitatory Peptides. Besides SP and CGRP, several other peptides have been reported to have an excitatory action on dorsal horn neurons. These include VIP, neurotensin, cholecystokinin, and thyrotropin-releasing hormone (TRH). Much less is known about the role of these substances in the dorsal horn than about SP and CGRP.

VIP is found in the terminals of small-diameter primary afferent fibers in the dorsal horn, especially of the sacral spinal cord (Gibson *et al.*, 1981; Basbaum and Glazer, 1983; Honda *et al.*, 1983). Many VIP-containing afferent fibers are likely to supply pelvic visceral organs (C.

Morgan *et al.*, 1981; Nadelhaft *et al.*, 1983). However, VIP-containing afferent fibers are also found at other spinal cord levels (C. C. LaMotte and DeLanerolle, 1986). Terminals are especially concentrated in laminae I, II, and X. VIP-containing neurons have been observed in lamina X and in the intermediolateral cell column (Fuji *et al.*, 1985; C. C. LaMotte and DeLanerolle, 1986).

Iontophoretic application of VIP causes excitation of nociceptive and non-nociceptive dorsal horn neurons (Jeftinija *et al.*, 1982; Salt and Hill, 1983).

Neurotensin has not been found in primary afferent neurons (see Chapters 3 and 4). However, it is found in interneurons located in the inner part of the substantia gelatinosa (Uhl *et al.*, 1979b; Seybold and Elde, 1982; Yaksh *et al.*, 1982b) and in terminals in the superficial layers of the superficial dorsal horn (Seybold and Elde, 1980). Neurotensin-binding sites are present in the dorsal horn (Ninkovic *et al.*, 1981), and neurotensin is released from the spinal cord following stimulation of the sciatic nerve (Yaksh *et al.*, 1982b). Iontophoretic application of neurotensin produces an excitation of nociceptive and nonnociceptive neurons in laminae I–III, but not in laminae IV-VII (Miletić and Randić, 1979; Stanzione and Zieglgänsberger, 1983). The excitatory actions of neurotensin and glutamate are additive.

Cholecystokinin-like immunoreactivity has been reported in a population of DRG cells and in neurons of several laminae of the dorsal horn (Schroder, 1983; Fuji *et al.*, 1985). It has been questioned whether cholecystokinin-immunoreactivity actually represents cross-reactivity with CGRP (Ju *et al.*, 1986). However, the precursor peptide for cholecystokinin (pro-cholecystokinin) is present in the spinal cord (Beinfeld, 1985). Jeftinija *et al.* (1981a) found that many of the neurons tested in laminae I–VII could be excited by iontophoretic application of cholecystokinin in both the intact rat spinal cord and in an *in vitro* preparation. All classes of neurons were affected. Since cholecystokinin was still effective in the absence of Ca^{++} in the bathing solution, it was concluded that cholecystokinin has a direct postsynaptic excitatory action. However, cholecystokinin may have an indirect inhibitory action by causing the release of GABA, perhaps through an excitatory action on GABA-ergic inhibitory interneurons (Rodriguez *et al.*, 1987).

TRH is found not only in the ventral horn but also in neuronal cell bodies and terminals in the dorsal horn (Coffield *et al.*, 1986; Harkness and Brownfield, 1986; Ulfhake *et al.*, 1987), and there are TRH-binding sites in the dorsal horn (Manaker *et al.*, 1985a,b; Mantyh and Hunt, 1985a). Iontophoretically applied TRH facilitates the responses of wide-dynamic-range and high-threshold neurons in laminae II–V to glutamate (Jackson and White, 1988). Often, facilitation is preceded by inhibition of wide-dynamic-range neurons.

Inhibitory Amino Acids

GABA and Glycine. Both GABA and glycine are candidate inhibitory neurotransmitters in the spinal cord. These amino acids and their synthetic enzymes are distributed widely in the spinal cord (Aprison and Werman, 1965; Curtis and Watkins, 1965; Davidoff *et al.*, 1967; Graham *et al.*, 1967; Johnston, 1968; Aprison *et al.*, 1969; Graham and Aprison, 1969; Johnston and Vitali, 1969; McLaughlin *et al.*, 1975). GABA is concentrated more dorsally, whereas glycine is concentrated more ventrally in the spinal cord gray matter (Graham *et al.*, 1967). However, both are found in the dorsal horn. GABA has been shown to occur in a large fraction (24–33%) of the interneurons in laminae I, II, and III of the dorsal horn, as well as in axons of the dorsal and lateral funiculus and tract of Lissauer (S. P. Hunt *et al.*, 1981a; Barber *et al.*, 1982; Magoul *et al.*, 1987; A. J. Todd and McKenzie, 1989; Carlton and Hayes, 1990). A combined technique using the Golgi stain in combination with immunohistochemistry has enabled A. J. Todd and McKenzie (1989) to show that many central (islet) cells of the substantia gelatinosa contain GABA, whereas some central cells and also limiting (stalked) cells do not. A detailed analysis of synaptic endings immunocytochemically stained for GABA has demonstrated numerous GABAergic presynaptic dendrites in the superficial laminae of the primate dorsal horn

(Carlton and Hayes, 1990). The GABAergic dendrites received synaptic connections from the central terminals of glomeruli and often participated in reciprocal synapses.

In addition to its presence in interneurons, GABA is contained in neurons of the rostral ventral lateral medulla that project to the spinal cord (Reichling and Basbaum, 1990).

Glycine-containing neurons and synaptic terminals are also found in the superficial laminae of the dorsal horn (A. J. Todd, 1990).

It has been found that neurons of laminae I–III can take up tritiated GABA (Ribeiro-DaSilva and Coimbra, 1980), and tritiated glycine is taken up by spinal cord slices (Neal, 1971). These substances are released in a calcium-dependent fashion following stimulation (Aprison, 1970; Hopkin and Neal, 1970; Hammerstad *et al.*, 1971; G. G. S. Collins, 1974). Iontophoresis of GABA and glycine results in the inhibition of the activity of dorsal horn interneurons (Curtis *et al.*, 1959b, 1967a,b, 1968, 1977; Bruggencate and Engberg, 1968; Werman *et al.*, 1968), including those in the substantia gelatinosa (Zieglgänsberger and Sutor, 1983). The inhibitory action of GABA is blocked by the GABA antagonists, picrotoxin and bicuculline (Curtis *et al.*, 1969; 1971a; Engberg and Thaller, 1970; Straughan *et al.*, 1971; Benoist *et al.*, 1972; Game and Lodge, 1975), whereas that of glycine is prevented by the glycine receptor antagonist, strychnine (Curtis, 1962; Curtis *et al.*, 1969, 1971b; R. A. Davidoff *et al.*, 1969; Game and Lodge, 1975). GABA and glycine are both candidate transmitters to account for IPSPs in dorsal horn interneurons that result from stimulation of primary afferent fibers (Graham *et al.*, 1967; Bruggencate and Engberg, 1968; Curtis *et al.*, 1968, 1971a,b; R. A. Davidoff *et al.*, 1969; Game and Lodge, 1975; Rudomín *et al.*, 1990). These amino acids would be released by inhibitory interneurons to act at postsynaptic sites to generate the IPSPs. The ionic mechanism responsible for the IPSPs is an increased chloride conductance (J. L. Barker and Nicoll, 1973).

There is considerable evidence that the synaptic transmitter involved in many forms of presynaptic inhibition is GABA: (1) PAD is at least partly blocked by the GABA antagonists, picrotoxin and bicuculline (J. C. Eccles *et al.*, 1963b; R. F. Schmidt, 1963; Besson *et al.*, 1971; Curtis *et al.*, 1971a, 1977; Levy *et al.*, 1971; R. A. Davidoff, 1972; Benoist *et al.*, 1972, 1974; Levy and Anderson, 1972; DeGroat *et al.*, 1972; J. L. Barker and Nicoll, 1973; Levy, 1974, 1975; McLaughlin *et al.*, 1975; Gmelin and Cerletti, 1976; Repkin *et al.*, 1976; Barber *et al.*, 1978; Gallagher *et al.*, 1978; Hackman *et al.*, 1982; Mokha *et al.*, 1983; Rudomín *et al.*, 1990; see reviews by R. F. Schmidt, 1971; Nicoll and Alger, 1979); (2) PAD is also decreased by depletion of GABA by semicarbazide (Bell and Anderson, 1972; Repkin *et al.*, 1976); (3) PAD is enhanced when GABA-transaminase is inhibited (R. A. Davidoff *et al.*, 1973); and (4) group Ia afferent fibers are depolarized by iontophoretically applied GABA (Curtis and Lodge, 1982; Curtis *et al.*, 1982; however, cf. Curtis and Ryall, 1966).

The depolarization of primary afferent fibers by GABA appears to involve an ionic mechanism different from that which produces IPSPs. Evidence for involvement of Na^+ ions in the generation of PAD in the isolated frog spinal cord has been reported, as well as evidence against a prominent role of Cl^- ions (J. L. Barker and Nicoll, 1973). Curtis and Gynther (1987) found that a number of divalent cations can reduce the PAD produced by GABA without affecting postsynaptic inhibition by GABA. This suggests the possibility that GABA depolarizes primary afferent terminals through a mechanism involving the influx of calcium ions (cf. Robertson and Taylor, 1986).

Potassium Release. An alternative mechanism for PAD is the release of K^+ into the extracellular space (Vyklický *et al.*, 1972, 1975, 1976; Krnjević and Morris, 1972, 1974b; Kríz *et al.*, 1974, 1975; Bruggencate *et al.*, 1974; Somjen and Lothman, 1974; Lothman and Somjen, 1975; Syková *et al.*, 1976; Syková and Vyklický, 1977, 1978; Deschenes *et al.*, 1976; Deschenes and Feltz, 1976; Nicoll, 1979; R. A. Davidoff *et al.*, 1980, 1988; Evans, 1980; Syková *et al.*, 1980; Czéh *et al.*, 1981; Bagust *et al.*, 1982; Urban *et al.*, 1985; Svoboda *et al.*, 1988). The main region undergoing large changes in K^+ concentration is the intermediate region, rather than the superficial dorsal horn or ventral horn.

An increase in extracellular K^+ concentration does not account for the specific organization of PAD (Rudomín *et al.*, 1981), nor for the susceptibility of PAD to GABA antagonists such as picrotoxin and bicuculline (see above). Thus, it appears that there are two independent processes for generating PAD: an early, synaptically generated component that can be decreased by GABA antagonists, and a late component that is secondary to K^+ release and is unaffected by GABA antagonists. K^+ release would also have important postsynaptic effects (Czéh *et al.*, 1981).

The uptake of the excess K^+ seems to be through active transport and may produce primary afferent hyperpolarization by means of electrogenic pumping of Na^+ (R. A. Davidoff and Hackman, 1980; Czéh *et al.*, 1981).

The mechanism of K^+ release in the spinal cord following repetitive stimulation of primary afferent fibers has been investigated by several groups. Activation of motoneurons does not appear to contribute, since antidromic activation of motoneurons fails to cause a detectable change in K^+ levels (Nicoll, 1979). Using the isolated frog spinal cord, R. A. Davidoff *et al.* (1988) were able to show that K^+ release was reduced by 85% when the Mg^{++} concentration in the bathing solution was increased to 20 mM, a level that would prevent synaptic transmission from primary afferent fibers. This led them to conclude that 15% of the K^+ release is from activity in primary afferent fibers and 85% from activity of interneurons. K^+ release could also be produced by application of excitatory amino acids and, to a lesser extent, by tachykinin peptides. Kynurenate prevented most of the K^+ release caused by excitatory amino acids, whereas APV was less effective, and an SP antagonist had no effect. The actions of excitatory amino acids and tachykinins in causing K^+ release were prevented by tetrodotoxin, and so it was concluded that the K^+ release by these agents depended upon the excitation of interneurons.

Acute nociceptive stimuli cause a transient increase in extracellular K^+ concentration lasting 5–30 sec, but chemical or thermal injury causes a prolonged increase, beginning 5–15 min after injury and lasting more than 2 hr (Svoboda *et al.*, 1988). This change in extracellular K^+ concentration could play a role in sensory changes that follow injury.

Monoamines

Several monoamines are thought to be neurotransmitters in pathways descending to the spinal cord from the brain. These include serotonin, norepinephrine and acetylcholine.

Serotonin. Serotonin is present in the axons and terminals of raphe-spinal neurons in the dorsal horn, especially in laminae I and II, but also in laminae III–V; serotonin-containing endings are abundant in the intermediolateral cell column and the ventral horn (Dahlström and Fuxe, 1965; Segu and Calas, 1978; G. F. Martin *et al.*, 1982; Ruda *et al.*, 1982; Bowker *et al.*, 1982; C. C. LaMotte and DeLanerolle, 1983b; Miletić *et al.*, 1984). The morphology of serotonin-containing synapses has been described previously (Ruda *et al.*, 1982; Light *et al.*, 1983; C. C. LaMotte and De Lanerolle, 1983b). Serotonin-containing synapses have been found on enkephalin-containing interneurons in the dorsal horn (Glazer and Basbaum, 1984).

The origin of the serotonergic projection to the dorsal horn is chiefly the nucleus raphe magnus (Dahlström and Fuxe, 1964; Oliveras *et al.*, 1977; Basbaum *et al.*, 1978; Bourgoin *et al.*, 1980; Goode *et al.*, 1980; G. F. Martin *et al.*, 1982; Skagerberg and Bjorklund, 1985), although several other raphe nuclei, including raphe pallidus, obscurus, and pontis, also project to the spinal cord (Bowker *et al.*, 1981, 1987; Skagerberg and Bjorklund, 1985; see review by Willis, 1984; cf. C. C. LaMotte *et al.*, 1982). However, these nuclei tend to project to more ventral targets, including the intermediolateral cell column and ventral horn. Raphe-spinal axons from the nucleus raphe magnus are concentrated in the dorsal lateral funiculus (Kuypers and Maisky, 1977; Leichnetz *et al.*, 1978; R. F. Martin *et al.*, 1978; Basbaum and Fields, 1979; Tohyama *et al.*, 1979b; G. F. Martin *et al.*, 1982; Light and Kavookjian, 1985; Basbaum *et al.*, 1986a). Almost all of the serotonergic raphe-spinal axons are unmyelinated, both in rats and cats (Dahlström and Fuxe, 1965; Basbaum *et al.*, 1988).

Serotonin and any of several peptides, including SP, CGRP enkephalin, somatostatin, and TRH, may be colocalized in the same raphe neurons and their terminals (Hökfelt *et al.*, 1978; Chan-Palay *et al.*, 1978; Bowker *et al.*, 1981; 1983; O. Johansson *et al.*, 1981; Leger *et al.*, 1986; Menétrey and Basbaum, 1987; Wessendorf and Elde, 1987; Tashiro and Ruda, 1988; Chiba and Masuko, 1989; Millhorn *et al.*, 1989; Arvidsson *et al.*, 1990). Serotonin may also be colocalized with GABA (Millhorn *et al.*, 1987a).

The proportion of axons containing both serotonin and SP is much lower in the superficial dorsal horn than in the intermediolateral cell column, ventral horn or lamina X (Wessendorf and Elde, 1987; Tashiro and Ruda, 1988). Only 1–3% of serotonergic terminals in the superficial laminae of the dorsal horn have colocalized SP, whereas 99% of those in the ventral horn show such colocalization (Wessendorf and Elde, 1987). These differences support the proposal of Skagerberg and Bjorklund (1985) that the B3 cell group (which is in the nucleus raphe magnus) projects to the superficial dorsal horn, whereas the B1 and B2 groups (in the nuclei raphe pallidus and obscurus) project to the ventral horn. Wessendorf and Elde (1987) suggest that the colocalized serotonin and SP play a role in the excitation of motoneurons.

Stimulation of the dorsal lateral funiculus, the sciatic nerve, a branch of the trigeminal nerve, or the medulla causes release of serotonin from the spinal cord (Tyce and Yaksh, 1981; Hammond *et al.*, 1985). Cold block of the spinal cord prevents the effect of sciatic nerve stimulation, indicating that a supraspinal loop is involved. This observation is consistent with evidence that serotonin plays a role in the mechanism of DNIC (Dickenson *et al.*, 1981; Chitour *et al.*, 1982; Kraus *et al.*, 1982; see above).

More localized release of serotonin from the dorsal horn can be demonstrated by using microdialysis (Sorkin *et al.*, 1988a). With this approach, release of serotonin is found following stimulation in the nucleus raphe magnus or the periaqueductal gray (Abhold and Bowker, 1990; Sorkin *et al.*, 1990). Serotonin turnover is increased in the spinal cord following stimulation in the nucleus raphe magnus (Bourgoin *et al.*, 1980; Rivot *et al.*, 1982). This is prevented by pretreatment with *p*-chlorophenylalanine (PCPA) (Rivot *et al.*, 1982).

Iontophoretic release of serotonin in the vicinity of dorsal horn neurons, especially nociceptive ones, generally causes inhibition (Engberg and Ryall, 1966; Weight and Salmoiraghi, 1966; Randić and Yu, 1976; Belcher *et al.*, 1978; Headley *et al.*, 1978; Griersmith and Duggan, 1980; Jeftinija *et al.*, 1986; El-Yassir *et al.*, 1988), although excitatory effects have also been reported (Weight and Salmoiraghi, 1966; Belcher *et al.*, 1978; A. J. Todd and Millar, 1983; Jeftinija *et al.*, 1986). A. J. Todd and Millar (1983) observed long-lasting (up to 10 min) excitation of many neurons in laminae I–III by iontophoretically applied serotonin; they suggested that these neurons are inhibitory interneurons.

Inhibition of dorsal horn nociceptive neurons by stimulation in the nucleus raphe magnus is reduced following pretreatment with PCPA, suggesting that part of the inhibition is due to serotonin release (Rivot *et al.*, 1980). Experiments showing that serotonin antagonists or PCPA pretreatment reduce the inhibition of dorsal horn neurons by stimulation in various sites in the brain indicate that there is a serotonergic component to inhibition from the periaqueductal gray (Carstens *et al.*, 1981b) and the medial diencephalon and basal forebrain (Carstens *et al.*, 1983).

Most studies have been concerned with postsynaptic actions of serotonin. However, serotonin may also have a presynaptic action (Carstens *et al.*, 1981c; Proudfit and Anderson, 1974; Proudfit *et al.*, 1980; cf. Hamon *et al.*, 1989), although no PAD was found when serotonin was applied to an *in vivo* preparation that did show PAD in response to GABA (J. E. Davies and Roberts, 1981).

Experiments with selective agonists and antagonists of serotonin receptors indicate that several different receptors are involved in the actions of serotonin in the spinal cord (El-Yassir *et al.*, 1988; El-Yassir and Fleetwood-Walker, 1990). Evidence was obtained that different serotonin receptors mediate a selective inhibition of nociceptive responses (5-HT1B receptor), a nonselective inhibition (5-HT1A receptor), or a nonselective excitation (receptor undetermined).

5-HT2 receptors appeared not to be involved. 5-HT3 receptors have been demonstrated on capsaicin-sensitive primary afferent fibers (Hamon *et al.*, 1989).

Depletion of serotonin by treatment of cats with PCPA or increasing the extracellular levels of serotonin by administration of the serotonin uptake inhibitor fluoxetine failed to affect tonic descending inhibition, and so it was proposed that this form of inhibition does not depend upon a serotonergic mechanism (Soja and Sinclair, 1980). However, systemic administration of serotonin receptor antagonists causes an enhancement of the responses of many dorsal horn interneurons in rats to volleys in C fibers, suggesting that serotonin may indeed play a role in tonic descending inhibition (Rivot *et al.*, 1987).

Serotonin can also have indirect effects. For example, it may interfere with the responses of dorsal horn interneurons to SP (J. E. Davies and Roberts, 1981). It was suggested that this effect is mediated by an action of serotonin on SP receptors or on the receptor-excitation coupling mechanism. Conversely, SP can affect serotonin autoreceptors (R. Mitchell and Fleetwood-Walker, 1981). Serotonin has also been shown to cause the release of adenosine from dorsal horn synaptosomes (Sweeney *et al.*, 1988).

Depletion of serotonin by pretreatment with PCPA enhances the effectiveness of SP in exciting dorsal horn neurons and increases the proportion of neurons that are excited by iontophoretically released serotonin (Jeftinija *et al.*, 1986). The mechanism of these actions is unclear.

Norepinephrine. Synaptic terminals containing norepinephrine are found in the dorsal horn in laminae I, II, and IV–VI, as well as in the intermediate region, lamina X, and the ventral horn (Dahlström and Fuxe, 1965; Westlund and Coulter, 1980; Westlund *et al.*, 1983; Higihira *et al.*, 1990). The noradrenergic projection originates from neurons in several nuclear groups of the pons, including the locus coeruleus, subcoeruleus, parabrachial and Koelliker-Fuse nuclei (Nygren and Olson, 1977; Ader *et al.*, 1979; Westlund and Coulter, 1980; Westlund *et al.*, 1981, 1983; R. T. Stevens *et al.*, 1982). Norepinephrine-containing axons descend in the lateral funiculus (Dahlström and Fuxe, 1965; Westlund and Coulter, 1980; R. T. Stevens *et al.*, 1982). Those from the locus coeruleus in rats descend chiefly within laminae I and II and end in the dorsal horn and intermediate region, whereas those from the A5 and A7 noradrenergic groups descend in the dorsal lateral funiculus and more ventral white matter and supply autonomic preganglionic neurons and motoneurons (Fritschy and Grzanna, 1990). None of the noradrenergic neurons of the medulla project to the spinal cord.

Stimulation of the dorsal lateral funiculus, the sciatic nerve, a branch of the trigeminal nerve, or the medulla causes the release of norepinephrine from the spinal cord (Tyce and Yaksh, 1981; Hammond *et al.*, 1985). As with serotonin release (see above), cold block of the spinal cord prevents norepinephrine release following sciatic nerve stimulation.

Electrical stimulation in the medullary reticular formation causes release of norepinephrine in the spinal cord, as detected by *in vivo* dialysis; stimulation in the nucleus raphe magnus also causes release of norepinephrine, as well as serotonin (Abhold and Bowker, 1990). Stimulation of comparable sites is well known to cause inhibition of nociceptive dorsal horn neurons. However, it is unclear whether the inhibition is caused by release of either norepinephrine or serotonin. Stimulation in the region of the lateral pons containing a large population of noradrenergic neurons can produce inhibition of dorsal horn neurons that does not depend upon release of catecholamines in the spinal cord (Hodge *et al.*, 1983; Zhao and Duggan, 1988).

Iontophoretic application of norepinephrine near dorsal horn interneurons generally causes inhibition of the background activity of these cells, as well as of their responses to excitatory amino acids (Biscoe *et al.*, 1966; Engberg and Ryall, 1966; Belcher *et al.*, 1978; Headley *et al.*, 1978; Fleetwood-Walker *et al.*, 1985; Howe and Zieglgänsberger, 1987), although excitatory effects have been reported (Weight and Salmoiraghi, 1966; Howe and Zieglgänsberger, 1987). The inhibition can be selective for nociceptive inputs (Belcher *et al.*, 1978). Howe and Zieglgänsberger (1987) found that norepinephrine tends to inhibit low-threshold and wide-dynamic-range neurons in more superficial parts of the dorsal horn, but to produce a long-lasting excitation of proprioceptive neurons at the base of the dorsal horn.

Some neurons in laminae I and II are excited by norepinephrine (A. J. Todd and Millar, 1983). Millar and Williams (1989) confirm that the excited units tend to have small action potentials and are low-threshold cells; high-threshold cells in lamina I and wide-dynamic-range neurons in deeper laminae are inhibited by norepinephrine (Fig. 5.26). Similar excitatory or inhibitory effects are produced in the same units by stimulation in the periaqueductal gray. It is suggested that the low-threshold neurons are inhibitory interneurons that synapse on the HT and WDR neurons.

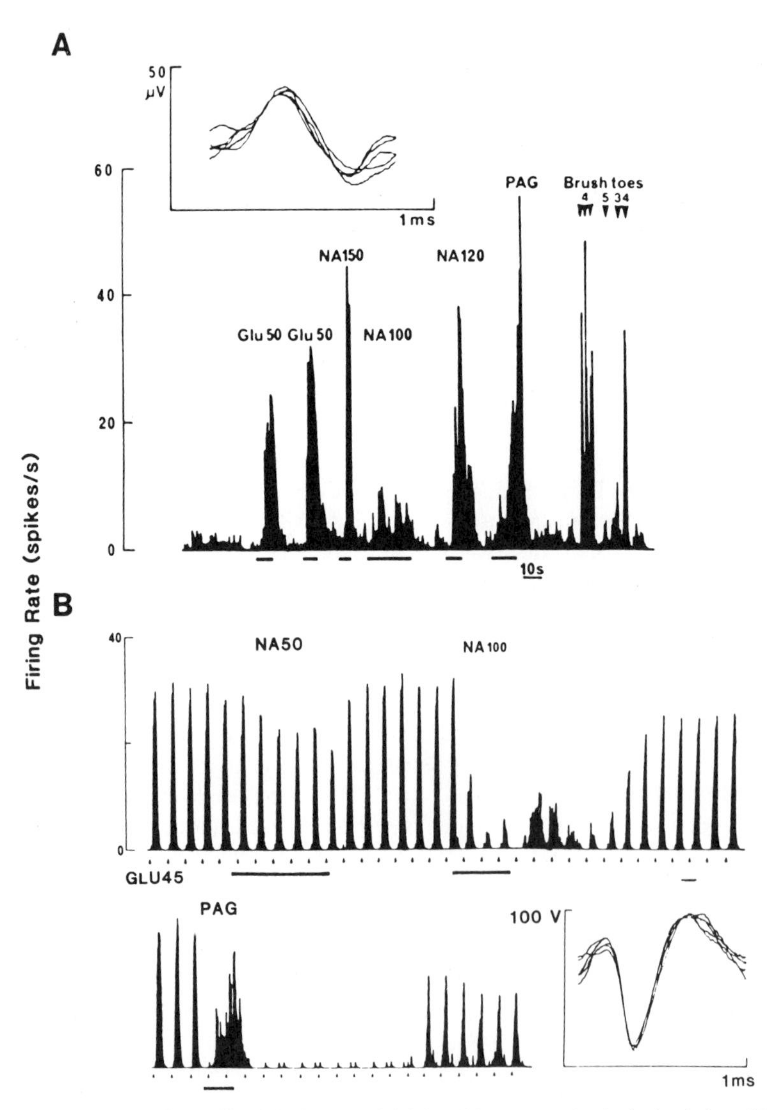

Fig. 5.26. (A) Responses of a small unit in the superficial dorsal horn to mechanical stimulation of its receptive field (brush applied to toe 4) and to iontophoretic application of glutamate (Glu) and norepinephrine (NA). The excitation produced by stimulation in the periaqueductal gray (PAG) is also shown. (B) Inhibitory effects of NA application and PAG stimulation on a large unit. The action potentials of the units are shown in the insets. (From Millar and Williams, 1989; reproduced with permission.)

North and Yoshimura (1984) examined the action of norepinephrine on substantia gelatinosa neurons, from which they recorded intracellularly in a slice preparation. Most of the neurons were hyperpolarized by norepinephrine, apparently because of an increase in K^+ conductance. The hyperpolarization could be prevented by antagonists of α_2 receptors.

There is also evidence for a presynaptic role of norepinephrine (Jeftinija *et al.*, 1981b; Carstens *et al.*, 1987). However, it has not been determined whether the presynaptic action of norepinephrine is a direct or an indirect one. Systemically administered clonidine causes an increased excitability of the presynaptic terminals of C fibers, but not A fibers; this action was blocked by antagonists of α_2 adrenoreceptors (Calvillo and Ghignone, 1986). Norepinephrine prevents the release of SP from the spinal cord (Kuraishi *et al.*, 1985b; Pang and Vasko, 1986). Furthermore, clonidine prevents the release of CGRP from an *in vitro* preparation of the spinal cord (Solomon *et al.*, 1989).

Other Monoamines. There is evidence for both dopaminergic and adrenergic projections to the spinal cord (Blessing and Chalmers, 1979; Skagerberg and Linvall, 1985; Carlton *et al.*, 1987a). The densest projections are to the intermediolateral cell columns, but there are also projections to the dorsal horn, especially its superficial laminae (Skagerberg *et al.*, 1982). The dopaminergic projection originates from the A11 cell group in the diencephalon (Bjorklund and Skagerberg, 1979; Skagerberg *et al.*, 1982; Skagerberg and Lindvall, 1985). The adrenergic projection arises from neurons in the medulla (Ross *et al.*, 1981; Carlton *et al.*, 1987a, 1989a). Both dopamine and epinephrine have been reported to inhibit dorsal horn neurons (Biscoe *et al.*, 1966). Acetylcholine may be a transmitter in some raphe-spinal neurons, since acetylcholinesterase has been demonstrated in some of these neurons (Bowker *et al.*, 1983). Choline acetyltransferase is abundant in the dorsal horn, especially in the superficial laminae (Aquilonius *et al.*, 1981; H. Kimura *et al.*, 1981); the enzyme is mostly in nerve terminals but may also be in interneurons (Aquilonius *et al.*, 1981; Kayaalp and Neff, 1980). In an early study, iontophoretically released acetylcholine was found to have no effect on interneurons other than Renshaw cells (Curtis *et al.*, 1961). However, acetylcholine and its agonists were later found to depress the responses of interneurons to excitatory amino acids (Curtis *et al.*, 1966) or to cause either excitation or depression (Weight and Salmoiraghi, 1966). More recently, it has been shown that cholinergic agents can depolarize many dorsal horn neurons and hyperpolarize others in a slice preparation (Urban *et al.*, 1989). These effects are mediated by both nicotinic and muscarinic receptors. Muscarinic antagonists interfere with the inhibition of dorsal horn neurons produced by stimulation of the medial medulla (Zhuo and Gebhart, 1990).

Purines

It has been proposed that purines, such as adenosine, serve as inhibitory neurotransmitters in the dorsal horn (Salter and Henry, 1985, 1987; see the review by Sawynok and Sweeney, 1989). There are two types of adenosine receptors (A1 and A2); both are present in the spinal cord, although A1 receptors appear to be more abundant (Choca *et al.*, 1987; cf. K. S. Lee and Reddington, 1986). Adenosine receptors are coupled to adenylate cyclase (Choca *et al.*, 1987) and G-proteins (Fredholm and Dunwiddie, 1988). Both postsynaptic and presynaptic effects of adenosine have been reported, and this purine may play a role in the inhibitory actions of morphine, serotonin, and norepinephrine (Sawynok and Sweeney, 1989).

Inhibitory Peptides

Opioid Peptides. The role of opiates and endogenous opioids in the central nervous system has been reviewed (see, e.g., Frederickson and Geary, 1982; Duggan and North, 1984; W. R. Martin, 1984; Yaksh and Noueihed, 1985; Olson *et al.*, 1989).

Several opioid peptides are known to be present in interneurons and synaptic terminals of the dorsal horn; these include leu- and met-enkephalin and dynorphin (Hökfelt *et al.*, 1977a,b; S. P. Hunt *et al.*, 1980; Aronin *et al.*, 1981; Glazer and Basbaum, 1981, 1982; C. C. LaMotte and DeLanerolle, 1983a; Charnay *et al.*, 1984; Schoenen *et al.*, 1985a). Many enkephalin-containing neurons are found in laminae I and II (Glazer and Basbaum, 1981; Bennett *et al.*, 1982). Dynorphin-containing interneurons are found in laminae I, II, and V (Ruda *et al.*, 1988; Cho and Basbaum, 1988; Carlton and Hayes, 1989). CGRP-containing terminals, presumably from primary afferent fibers, make synaptic contact with dynorphin-containing cells in lamina II (Carlton and Hayes, 1989). Some enkephalin-containing terminals belong to axons descending from the ventral medulla; the same endings may also contain somatostatin (Millhorn *et al.*, 1987b).

Met-enkephalin has been shown to be released from the spinal cord in response to stimulation of fine afferent fibers and to perfusion with K^+ or SP (Yaksh and Elde, 1980, 1981; Cesselin *et al.*, 1985, 1989). Le Bars *et al.* (1987a) found that met-enkephalin is not released from a particular region of the spinal cord by stimulation of a related area of the body, but instead is released when noxious stimuli are applied to distant areas of the body. They suggest that endogenous opiates may be involved in DNIC, a proposal that is supported by the observations that DNIC is naloxone reversible (Le Bars *et al.*, 1981b; see section on serotonin, which also plays a role in DNIC), and that the release of met-enkephalin in the lumbar cord is blocked by lesions of the dorsal lateral funiculi (Le Bars *et al.*, 1987b; see also Woolf, 1984). However, noxious heat has the opposite effect, causing release of met-enkephalin from the segmentally related region of spinal cord and not a remote region (Cesselin *et al.*, 1989). This is unexpected, since stimulation with noxious heat evokes DNIC, just as does noxious mechanical stimulation. Interestingly, morphine can reduce the amount of met-enkephalin that is released from the spinal cord (Jhamandas *et al.*, 1984). This effect is prevented by naloxone.

Using antibody microprobes, Hutchison *et al.* (1990) showed that stimulation of C fibers, but not A fibers, in a peripheral nerve causes the release of dynorphin A in lamina I of the dorsal horn. Dynorphin release could be prevented by transection of the spinal cord, suggesting that the release was under supraspinal control.

Several types of opiate receptors, including μ, δ and κ receptors, have been found in the spinal cord by specific binding (Atweh and Kuhar, 1977; Czlonkowski *et al.*, 1983; Gouarderes and Cros, 1984; Gouarderes *et al.*, 1985, 1986; Zarr *et al.*, 1986; Faull and Villiger, 1987; Hunter *et al.*, 1989; Sales *et al.*, 1989). Opiate receptors are especially concentrated in the superficial dorsal horn.

Opiate receptor binding in the superficial dorsal horn decreases significantly following dorsal rhizotomy (C. C. LaMotte *et al.*, 1976). This may indicate that there are opiate receptors on primary afferent fibers, although loss of opiate receptor binding secondary to postsynaptic changes cannot be ruled out. Such presynaptic receptors could account for the ability of opiates to prevent SP release from afferent terminals (Jessell and Iversen, 1977; Yaksh *et al.*, 1980; Hirota *et al.*, 1985). These receptors appear to be of the μ and δ, but not κ, subtypes (Fields *et al.*, 1980; Hirota *et al.*, 1985).

Iontophoretic release of enkephalin (or the stable analog, met-enkephalinamide) can produce a selective inhibition of nociceptive dorsal horn interneurons (Duggan *et al.*, 1976, 1977, 1981b; Randić and Miletić, 1978; Zieglgänsberger and Tulloch, 1979b; Sastry and Goh, 1983). This inhibition can be blocked by the opiate receptor antagonist naloxone.

Duggan *et al.* (1977) were able to inhibit the responses of dorsal horn neurons by iontophoretic application of met-enkephalinamide near the cell bodies in laminae IV and V, but the effect was relatively nonspecific, affecting responses to both noxious and innocuous stimuli, like the action of glycine or GABA. However, application of met-enkephalinamide in the substantia gelatinosa produced a more selective inhibition of the responses to noxious stimuli (Fig. 5.27). It was suggested that enkephalin has a postsynaptic inhibitory effect when applied to the cell bodies of dorsal horn neurons. The more selective action of enkephalin released in the substantia gelatinosa could be explained by a presynaptic action. If the opioid action is due to a

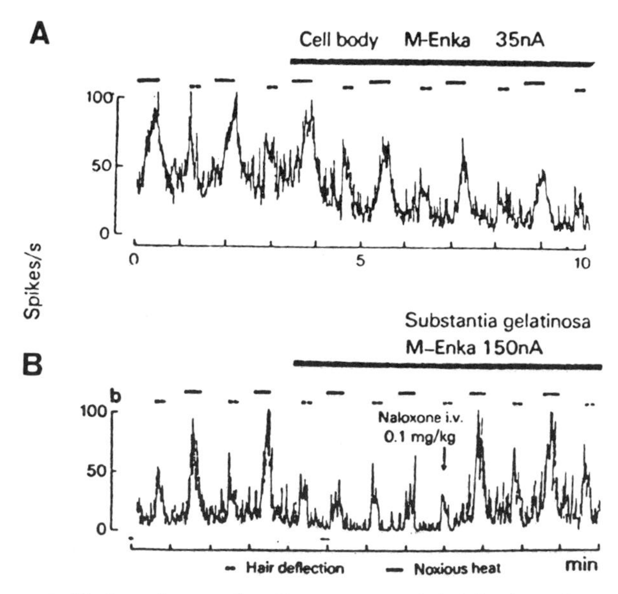

Fig. 5.27. Responses of wide-dynamic-range dorsal horn neurons to hair deflection and to noxious heat are shown in the ratemeter records. (A) Met-enkephalinamine was applied iontophoretically to the soma of a neuron. The responses to both innocuous and noxious stimulation were reduced. (B) Met-enkephalinamide was released in the substantia gelatinosa, while recordings were made from the neuron by an electrode in lamina IV at a location 220 μm from the drug barrel. The responses to noxious stimuli were reduced much more than those to innocuous stimuli, and naloxone antagonized this effect. (From Duggan *et al.*, 1977.)

decrease in the release of SP from fine afferent terminals, it would be anticipated that iontophoresis of SP at the same site as release of enkephalin would prevent the inhibitory action of the latter; however, simultaneous release of SP does not change the inhibitory action of enkephalin (Duggan *et al.*, 1979). On the other hand, the release of SP from the spinal cord by capsaicin is prevented by opioids acting at either μ or δ (but not κ) receptors (Aimone and Yaksh, 1989).

The responses of the dorsal horn neurons to excitatory amino acids are also blocked by opioids, suggesting that the inhibition produced by opioids is at least partly postsynaptic. J. Davies and Dray (1978) found that enkephalin (and morphine) had little or no effect when applied iontophoretically onto dorsal horn neurons, although morphine released in the substantia gelatinosa caused a decrease in nociceptive responses in deeper lying neurons. Sastry and Goh (1983) found that interneurons of the substantia gelatinosa are excited by enkephalin (and also by morphine), whereas deeper-lying nociceptive cells are inhibited; they suggest that the lamina II cells are inhibitory interneurons that mediate part of the depressant action on the deeper-lying dorsal horn neurons.

Zieglgänsberger and Tulloch (1979b) found that iontophoretically applied enkephalin reduced the background activity of dorsal horn neurons and also their responses to excitatory amino acids, presumably indicating a postsynaptic inhibition. The responses to noxious and innocuous stimuli were also nonselectively inhibited. Further evidence for a postsynaptic action of enkephalin has been obtained by use of spinal cord slice preparations (Murase *et al.*, 1982; Yoshimura and North, 1983; Jeftinija *et al.*, 1987). Many dorsal horn neurons are hyperpolarized by enkephalin, and their membrane resistance is reduced.

Dickenson *et al.* (1986) provide evidence that the inhibitory actions of enkephalin on dorsal horn neurons are mediated by δ opiate receptors. Jeftinija (1988), on the other hand, offers evidence to support a role for μ receptors.

Evidence for a presynaptic action of enkephalin has been reported (Sastry, 1978, 1979a; see also Jeftinija, 1988). Sastry suggests that enkephalin potentiates presynaptic inhibition. Mudge *et al.* (1979) find that enkephalin prevents the release of SP from DRG cells in culture; this effect is accompanied by a reduction in the duration of the action potentials of these neurons. Naloxone reduces the ability of volleys in A fibers to increase the excitability of C-fiber terminals (Woolf and Fitzgerald, 1982). β-Endorphin and met-enkephalinamide have been found to depress dorsal root potentials in an *in vitro* preparation of rat spinal cord (Suzue and Jessell, 1980).

Dynorphin applied to the surface of the spinal cord can produce either excitatory or inhibitory actions on the responses of nociceptive dorsal horn neurons to C-fiber volleys (Knox and Dickenson, 1987). The κ-receptor agonist U50488H has a similar effect. The actions are insensitive to naloxone. Knox and Dickenson suggest that the excitatory effects of dynorphin may be mediated through SP release.

Other Peptides. Somatostatin is found in primary afferent fibers and in interneurons within the dorsal horn (Hökfelt *et al.*, 1976; Seybold and Elde, 1980; S. P. Hunt *et al.*, 1981a), as well as in terminals of axons descending from the medulla (Millhorn *et al.*, 1987b). Noxious thermal stimuli, but not noxious mechanical stimuli, applied to the skin cause release of somatostatin in the dorsal horn (Kuraishi *et al.*, 1985a; Morton *et al.*, 1988, 1989). However, SP is released by both noxious mechanical and thermal stimuli (see previous section), suggesting that somatostatin is released from heat nociceptors, whereas SP is released from polymodal nociceptors. There may also be release of somatostatin from interneurons (Morton *et al.*, 1989). As for SP, somatostatin release is mainly in the region of the substantia gelatinosa, as shown by the antibody microprobe technique (Morton *et al.*, 1989). Somatostatin has been reported to cause a selective inhibition of nociceptive dorsal horn interneurons following its iontophoretic application (Randić and Miletić, 1978). It is not evident whether release of this substance from primary afferent fibers causes inhibition; equally likely is that the inhibitory effects are related to effects at synapses formed by interneurons containing this peptide.

Bombesin and the related peptides neuromedin B and neuromedin C, also have inhibitory actions. Bombesin-like immunoreactivity has been found in small DRG cells and in the superficial layers of the dorsal horn, and the immunoreactivity in the dorsal horn decreases following dorsal rhizotomy (Fuxe *et al.*, 1983; Massari *et al.*, 1985; Panula *et al.*, 1983). Iontophoretic release of these peptides near interneurons of the dorsal horn inhibits the responses of these cells, especially responses to noxious stimuli (De Koninck and Henry, 1989).

Summary

A number of candidate neurotransmitters and neuromodulators have been identified that may play important roles in synaptic transmission in pathways involving dorsal horn interneurons. Excitatory substances include the excitatory amino acids (glutamate, aspartate, and their agonists) and several peptides, including SP and CGRP. There are several receptors for excitatory amino acids, including ones particularly sensitive to kainate, quisqualate, and NMDA. Differences in the action of excitatory amino acids can be explained on the basis of differences in effectiveness in activating non-NMDA and NMDA receptors. Such an analysis is facilitated by the availability of selective antagonists, especially for the NMDA receptor. It has been suggested that the excitatory action of many primary afferent fibers is mediated by release of an excitatory amino acid, such as glutamate, with an action on non-NMDA receptors and that the excitatory action of many interneurons is produced by release of aspartate and an action on NMDA receptors. NMDA has a special action because of a voltage, Mg^{++}- and glycine-sensitive receptor that probably engages a second messenger system. SP and CGRP are found in primary afferent endings of fine fibers, including C fibers. SP is also present in interneurons and in the axons of some bulbospinal neurons. These peptides may be colocalized with an excitatory amino acid. SP is released by noxious stimuli, and so it may be involved in pain mechanisms. The membrane effects of SP include a slow depolarization due to a complex action on several conductance mechanisms and activation of second messengers. Excitation of dorsal horn neurons by primary afferent fibers may involve early and late components as a result of successive actions by excitatory amino acids and peptides. CGRP appears to enhance the action of SP by inhibiting its breakdown and by promoting its release. It can also produce a slow depolarization of dorsal horn neurons. Other excitatory peptides include neurotensin, cholecystokinin, and thyrotropin-releasing hormone. Inhibitory substances in the dorsal horn include GABA, glycine, monoamines, and several peptides, including the endogenous opioid enkephalin. These substances are found in small interneurons, including many cells in the superficial layers of the dorsal horn. GABA and glycine both produce postsynaptic inhibition through an increase in membrane conductance to Cl^- ions. GABA also produces PAD, apparently by enhancing cation conductance. PAD can also be produced by the release of K^+ ions in the intermediate region. Serotonin and several catecholamines are in the terminals of brain stem neurons that project to the spinal cord. Serotonin may be colocalized with any of several peptides, especially in the ventral horn. Serotonin can be released by stimulation of a peripheral nerve. This effect can be prevented by cold block of spinal cord pathways, indicating that a supraspinal loop is involved. This observation is consistent with a role of serotonin in DNIC. Serotonin can also be released in the spinal cord by stimulation in the nucleus raphe magnus. Serotonin applied iontophoretically causes inhibition of nociceptive dorsal horn cells in the deeper laminae, but it has been reported that serotonin excites many neurons in laminae I–III. These may be inhibitory interneurons. Several serotonin receptors have been implicated in the spinal cord actions of this amine, including 5-HT1B, 5-HT1A and 5-HT3 receptors. Norepinephrine is also present in the dorsal horn in terminals of descending projection neurons. It is released by peripheral-nerve stimulation, and this is prevented by cold block of the cord. Iontophoretically applied norepinephrine is generally inhibitory, although it excites proprioceptive neurons and also some cells of laminae I and II. The latter may be inhibitory interneurons. Norepinephrine causes a hyperpolarization of dorsal horn neurons *in vitro* by increasing K^+ conductance. This action is mediated by α_2 receptors. Other monoamines, including dopamine, epinephrine, and acetylcholine, can also inhibit dorsal horn neurons. Opioid peptides in the dorsal horn include enkephalin and dynorphin. Enkephalin is released when noxious stimuli are applied to the body. This release may play a role in the DNIC mechanism, although some contradictory findings have been reported. Opiate receptors in the spinal cord include μ, δ, and κ receptors. Some of the μ and δ receptors appear to be on primary afferent terminals and may account for the ability of opiates to prevent SP release. Iontophoretically applied enkephalin generally inhibits dorsal horn neurons. The inhibition tends to be nonselective when the enkephalin is released near cell bodies, perhaps because of a postsynaptic inhibitory action. However, it is more selective in decreasing nociceptive responses when released in the substantia gelatinosa. This may relate to a presynaptic action or be due to an action on interneurons. It has been reported that interneurons in the substantia gelatinosa may be excited by enkephalin. Enkephalin causes a membrane hyperpolarization of dorsal horn neurons *in vitro* by a reduction in membrane resistance. Dynorphin applied to the surface of the spinal cord may either excite or inhibit dorsal horn neurons. Another peptide with an inhibitory action on dorsal horn neurons is somatostatin. It is not clear whether release of somatostatin from primary afferent fibers produces inhibition by a postsynaptic effect. If so, this would be a unique action.

DORSAL HORN CELLS IN UNANESTHETIZED, BEHAVING ANIMALS

The activity of interneurons is influenced by the presence or absence of anesthetic drugs and also by several pathways descending from the brain. Before it is possible to relate the activity of particular classes of interneurons to behavioral events, it is necessary to evaluate the discharge patterns of such neurons in unanesthetized animals. However, the usual experimental preparations available that avoid anesthesia (decerebrate or spinalized animals) have the drawback that there is an abnormal set in the activity of the descending control systems. For example, decerebrate animals have an increased activity in the brain stem pathways that produce a tonic inhibition of flexion reflex pathways (see, e.g., R. M. Eccles and Lundberg, 1959b), whereas spinalized animals completely lack such descending control.

An approach to the solution of this problem is to record from unanesthetized, behaving animals. Such an approach has been very successful in studies of the activity of neurons in the brain. Technical difficulties have slowed the application of this approach to the spinal cord. However, several groups have now described methods that have been successfully used to record from spinal cord neurons in awake, behaving rats, cats, and monkeys (Wall *et al.*, 1967; Courtney and Fetz, 1973; Bromberg and Fetz, 1977; J. G. Collins, 1985, 1987; Sorkin *et al.*, 1988b). The initial studies using this approach suggest that the activity of spinal cord interneurons in the awake animal is highly regulated by inhibitory pathways (Wall *et al.*, 1967; Courtney and Fetz, 1973; J. G. Collins, 1983, 1987; J. G. Collins and Ren, 1987; Sorkin *et al.*, 1988b), and this modulation can be abolished by anesthesia (J. G. Collins and Ren, 1987; J. G. Collins *et al.*, 1990).

CONCLUSIONS

1. Interneurons of the dorsal horn are involved in complex interactions that affect the signalling of ascending somatosensory tract cells (as well as of motor activity).

2. The activity of populations of dorsal horn interneurons is reflected in field potentials that can be recorded from the surface of the spinal cord or within the cord. After an initial afferent volley evoked by an electrical stimulus applied to a peripheral nerve, there may be several slow potentials due to excitatory activity, followed by one associated with PAD. When recording from the surface of the spinal cord, the excitatory field potentials are negative and the one associated with PAD is positive. The details of the field potentials vary with the nerve stimulated and the stimulus strength.

3. Activity in the dorsal horn can also be demonstrated anatomically by using the 2-deoxyglucose technique, histochemical staining for glycogen phosphorylase activity, or immunocytochemical staining for the presence of c-*fos* protein. The 2-deoxyglucose response may reflect increased glucose metabolism in synaptic terminals. Increases in glycogen phosphorylase activity probably reflect activation of second messenger systems. c-*fos* protein is expressed following induction of its gene as a result of activity in some, but not all, types of neurons.

4. Interneuronal activity can be recorded either intracellularly or extracellularly. Intracellular recording allows an appraisal of synaptic events and changes in membrane properties, whereas extracellular recording permits the monitoring of activity for a prolonged period while a variety of experimental procedures are done.

5. A number of schemes for classifying dorsal horn neurons have been proposed. These have been based on a variety of approaches, such as whether particular inputs are mono- or polysynaptic, the location of the neurons, the responses to innocuous and noxious stimuli, and the particular classes of sensory receptors that activate the cells. A commonly used classification is one which designates neurons as low-threshold (class 1) neurons if they respond just to innocuous mechanical stimuli, wide-dynamic-range (class 2) neurons if they respond to both innocuous and noxious stimuli, and high-threshold (class 3) neurons if they respond just to noxious stimuli. Improvements on this scheme may result from statistical analyses of the responses to defined stimuli.

6. The activity of neurons in the different layers of the dorsal horn reflects the input to these layers from particular classes of primary afferent fibers. Neurons in lamina I respond chiefly to inputs conveyed by Aδ and C fibers, although some neurons respond also to Aβ fibers. According to some investigators, cells in lamina II, the substantia gelatinosa, are activated chiefly by C fibers. However, other investigators find that there are also inputs to this lamina from Aδ and even Aβ fibers, as well as from C fibers. Some of these actions must be polysynaptic. Laminae IV–VI contain cells receiving a variety of combinations of inputs.

7. The functions of neurons in different laminae are better reflected in their responses to natural forms of stimulation than to electrically evoked afferent volleys. Neurons in lamina I may respond chiefly or only to noxious or thermal stimuli, although some wide-dynamic-range cells are also found in this layer. Inputs are from skin, viscera, or joints. The substantia gelatinosa contains neurons with unusual responses, including habituation, prolonged discharges following brief stimuli, strong inhibitory inputs, and variable or "ameboid" receptive fields. Cells in laminae III–VI include low-threshold, wide-dynamic-range, and high-threshold neurons, based on cutaneous inputs. Proprioceptive stimulation can activate neurons in lamina VI (as well as some cells in adjacent laminae). Visceral and joint afferents are often effective in exciting neurons of laminae IV–VI. Nociceptive and visceral responses have been reported for cells of lamina X.

8. Dorsal horn neurons typically have both excitatory and inhibitory receptive fields. Wide-dynamic-range neurons often have a central low-threshold excitatory receptive field, surrounded by a high-threshold excitatory field and an inhibitory field.

9. It has been proposed that neurons in lamina IV excite those in lamina V, and that cells in lamina V in turn excite neurons in lamina VI. This "cascade theory" has been disputed.

10. Neurons in the upper layers of the dorsal horn are somatotopically organized. In the lumbar enlargement, the toes are represented medially and the lateral foot and more proximal regions laterally. A similar arrangement is found in the cervical enlargement. This pattern can be related to the embryologic development of the limbs and their dermatomal innervation.

11. Receptive fields are not necessarily hard-wired. Often, receptive fields can be shown to change, depending on experimental circumstances. Changes in receptive field size or location following transection of peripheral nerves have been attributed to the strengthening of previously ineffective synapses. Possible mechanisms of changes in receptive fields are under investigation.

12. Inhibitory receptive fields are of particular interest in the context of pain mechanisms, since inhibition of nociceptive transmission may lead to analgesia. Inhibitory systems include one that is activated by innocuous stimuli, has a surround organization, and is mediated segmentally and another that is activated by noxious stimuli applied to most of the body surface and depends upon a supraspinal loop.

13. The gate theory of pain was proposed because of the need to account for many features of pain that were difficult to predict from knowledge current at the time. The circuitry suggested involved the substantia gelatinosa as a source of presynaptic inhibition of input to nociceptive transmission cells in the dorsal horn. The inhibition would be evoked by increases in activity in large afferent fibers, but decreased by inputs over small fibers. There are a number of difficulties with the details of the mechanism as proposed, but clearly the activation of large, mechanoreceptive fibers can lead to inhibition of nociceptive transmission. An example is the use of dorsal column stimulation for the relief of pain.

14. Another powerful inhibitory system is activated by noxious mechanical or thermal stimulation of essentially the entire body and face. The only neurons that are inhibited are wide-dynamic-range neurons. The main component of the inhibitory mechanism involves the activation of neurons in the brain stem, which in turn project to the spinal cord. There is also a less effective segmental component. The functional role of this inhibitory mechanism (DNIC) is unclear. One possibility is that it serves as a contrast-enhancing device.

15. A number of pathways descending from the brain to the spinal cord influence the activity of dorsal horn interneurons. Both excitatory and inhibitory effects have been described. Some excitatory actions of the pyramidal tract have been noted. Other excitatory effects are mediated by reticulospinal and other pathways. Greater attention, however, has been paid to descending inhibitory pathways. There is a tonic descending inhibitory control that particularly affects nociceptive transmission in the spinal cord. The source of this is now thought to be in a reticular formation region ventral to the facial nucleus. In addition, inhibition can be evoked in a phasic manner following stimulation in many sites in the brain stem, including the periaqueductal gray, nucleus raphe magnus, nucleus gigantocellularis, and several other nuclei of the reticular formation, the locus coeruleus and adjacent nuclei in the parabrachial region, anterior pretectal nucleus, hypothalamus, and other sites in the forebrain. The inhibitory actions are mediated in part through raphe-spinal and reticulospinal pathways and in part through other projections.

16. Many neurotransmitters and neuromodulators affect the activity of dorsal horn neurons. Excitatory substances include the excitatory amino acids, ATP, and several peptides. Inhibitory substances include GABA, glycine, serotonin, norepinephrine, other monoamines, adenosine and several peptides, including the endogenous opioids.

17. Glutamate and aspartate are likely to serve as excitatory neurotransmitters in many primary afferent fibers, interneurons, and descending projections. These amino acids are found in synaptic endings, and they excite specific receptors in the dorsal horn. The excitatory amino acid receptors are classified by their responses to several agonists, including kainate, quisqualate, and NMDA, as well as by the ability of several more or less specific antagonists to block these responses.

18. It has been proposed that monosynaptic transmission in the dorsal horn involves non-NMDA receptors, activated perhaps by glutamate released from primary afferent terminals, and that polysynaptic transmission depends upon NMDA receptors, perhaps excited by aspartate released from interneurons.

19. The dose-response curve for NMDA receptors includes a negative-slope conductance due to the activation of a voltage-dependent channel. This negative-slope conductance can make measurements of membrane conductance changes in response to excitatory amino acids difficult to interpret. NMDA receptors can also activate second-messenger systems.

20. SP is a peptide contained in many primary afferent neurons, interneurons of the dorsal horn, and some descending projection fibers. SP is released in the dorsal horn, particularly in the substantia gelatinosa, following noxious mechanical, thermal, or chemical stimulation of the skin. Application of SP to nociceptive dorsal horn interneurons results in a prolonged excitation.

21. In vitro studies of the action of SP indicate that this peptide produces a prolonged depolarization due to changes in a number of different conductance mechanisms. It also activates second messengers.

22. CGRP is another excitatory peptide. In the dorsal horn, CGRP seems to be found only in the terminals of primary afferent fibers. CGRP is released by noxious mechanical or thermal stimulation of the skin. It enhances the action of SP and probably also has its own postsynaptic action.

23. The combined action of excitatory amino acids and SP is likely to account for several phases of excitation of dorsal horn neurons. The earliest phase may be due to activation of non-NMDA receptors, a later phase to the additional activation of NMDA receptors, and a very late phase, lasting several minutes, to activation of peptide receptors.

24. Other excitatory peptides include VIP, neurotensin, cholecystokinin, and TRH.

25. Inhibitory amino acids include GABA and glycine. Both substances inhibit dorsal horn interneurons, and GABA produces primary afferent depolarization and presynaptic inhibition. GABA is antagonized by picrotoxin and by bicuculline, whereas glycine is antagonized by strychnine.

26. An alternative mechanism for producing PAD is release of K^+ ions into the extracellular space. An increased extracellular K^+ concentration can depolarize not only primary afferent fibers but also dorsal horn interneurons, and so the effects of K^+ may be quite important, especially since damaging stimuli produce increases in K^+ concentration that last for hours.

27. Serotonin is one of several amines that cause inhibition of dorsal horn interneurons. This substance is contained in the terminals of raphe-spinal neurons. It is sometimes colocalized with a peptide, although this is more common for raphe-spinal axons that project to the ventral horn than to the dorsal horn. Serotonin can be released by stimulation of a peripheral nerve by a mechanism that may be identical to that producing DNIC. It can also be released by stimulation in the raphe nuclei. Iontophoretic application of serotonin generally causes inhibition of dorsal horn interneurons, although excitation of neurons in laminae I–III has been reported. The excited neurons in the superficial dorsal horn may be inhibitory in function.

28. Norepinephrine is also present in the dorsal horn. It is in the terminals of axons descending chiefly from the locus coeruleus. Norepinephrine is released following strong peripheral nerve stimulation. Iontophoretic application of norepinephrine has effects similar to those of serotonin. Most nociceptive neurons are inhibited, but some in laminae I and II are excited. The inhibition is probably due to a hyperpolarization caused by an increased K^+ conductance via α_2 receptors.

29. Other monoamines with inhibitory actions in the dorsal horn include dopamine, epinephrine, and acetylcholine.

30. Opioid peptides, including enkephalin and dynorphin, are present in the dorsal horn. Although these substances may be in descending projections and perhaps dynorphin in a few primary afferent fibers, most of the opioid content is probably in interneurons and their projections. Met-enkephalin is released in the spinal cord in response to noxious stimulation or application of SP to the cord. Enkephalin, like monoamines, may play a role in DNIC. Some of the opiate receptors in the spinal cord may be on primary afferent terminals, and so opioid effects may be mediated by presynaptic, as well as postsynaptic, actions. Iontophoretic application of enkephalin produces an inhibition of nociceptive dorsal horn neurons. This inhibition is more selective if the enkephalin is released in the substantia gelatinosa rather than near the cell bodies of neurons in the neck of the dorsal horn. The nonselective inhibitory action of release of enkephalin near the cell bodies may result from a postsynaptic conductance change. Some interneurons in the substantia gelatinosa are excited by enkephalin. If these are inhibitory interneurons, the selective effects of enkephalin released in the substantia gelatinosa may be due to an action of these interneurons on presynaptic terminals or on excitatory interneurons.

31. Dynorphin applied topically to the spinal cord can produce excitation or inhibition of dorsal horn neurons.

32. Other inhibitory peptides include somatostatin and bombesin-related peptides.

33. Techniques for monitoring the activity of dorsal horn neurons in awake, behaving animals are now available. The initial studies with this approach indicate that dorsal horn interneurons are subject to powerful descending controls in the absence of anesthesia.

6 Ascending Sensory Pathways in the Cord White Matter

INFORMATION TRANSMITTED BY THE ASCENDING PATHWAYS

Clues about the functions of sensory pathways come both from human and from animal studies. The evidence from human subjects is of particular importance, since a subjective report of alterations in sensory experience can be obtained and sometimes correlated with objective tests of changes in reactions to particular sensory stimuli. However, investigations on humans are limited by the degree to which interventions are permissible. In clinical studies, most disease processes and many surgical interventions are insufficiently localized to allow the assignment of a particular deficit to the interruption of a particular neural pathway in an unambiguous fashion, even when postmortem examination is possible. Lesion studies in animals have the advantage of a greater precision in the performance of surgical damage to specific neural structures and in the postmortem verification of the area damaged, but the role of an ascending pathway in sensory experience can be evaluated only indirectly. Two ways in which the effects of lesions in animals can be investigated are studies of behavioral changes that are produced by lesions and recordings of alterations in neuronal activity.

The following discussion will consider evidence for the functions of sensory pathways that ascend in the posterior column, posterior lateral funiculus, and anterior quadrant of the spinal cord (posterior and anterior become dorsal and ventral in animals).

POSTERIOR COLUMN

The traditional view of the functions of the human posterior column is that ascending fibers in this structure are responsible for transmitting the information required for discriminative touch, vibratory sensibility, and position sense. Evidence supporting this has been reviewed by Nathan *et al.* (1986). They point out that many clinical investigations have shown that tactile sensibility depends upon two different pathways in the human spinal cord, an uncrossed one in the posterior column and a crossed one in the anterolateral quadrant. However, the tactile pathway in the posterior column is especially important for stimulus localization, two-point discrimination, stereognosis, and graphesthesia (cf. Bender *et al.*, 1982).

Lesions of the Posterior Column

Positive Evidence of Sensory Deficits. Nathan *et al.* (1986) refer to a number of case studies to document the functions of the posterior column. A key case was one reported by Gans (1916). The posterior columns, but not other sensory pathways, were completely interrupted at an upper cervical level by a syphilitic gumma, as proven at autopsy (Fig. 6.1). Pain and thermal sensations were intact. The patient had paresthesias of the upper extremities. Tactile stimuli applied to the hands were sometimes felt, but often they were not, especially when applied in certain areas such as the palm. A glass lens felt smooth and hard, and a piece of rubber soft. However, the patient thought that a handkerchief placed in his hand was a piece of wood, and he could not distinguish between scissors and a coin. Two-point discrimination was disturbed, and he had no position sense in his hands, fingers, or toes, although position sense was retained in the

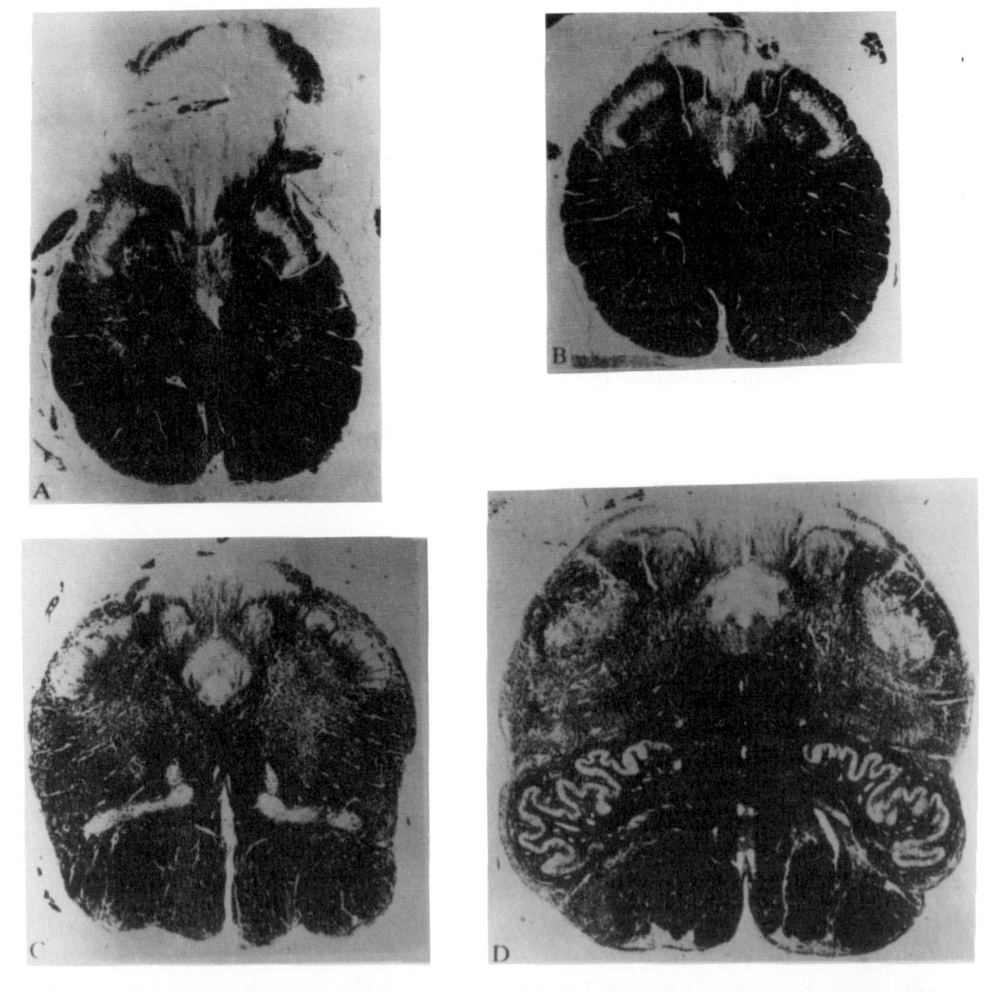

Fig. 6.1. Histologic sections of the spinal cord in a case in which the posterior columns were transected by a syphilitic gumma. The gumma is shown in panel A as it impinges on the posterior columns. The other sections trace the degenerated posterior columns through the lower medulla. (From Gans, 1916, as quoted by Nathan *et al.*, 1986; reproduced with permission.)

large joints. The sensory deficits severely affected the everyday life of the patient, causing him to give up his occupation. Another of Gans' patients with a lesion presumed to be in the posterior columns at the junction of the spinal cord and medulla had lost two-point discrimination, position sense in the upper extremities, and vibratory sensation in the hands.

Nathan *et al.* (1986) also refer to the experiments of Foerster (1936), who severed the posterior columns of patients surgically. Touch and pressure sensations persisted, but became abnormal. Patients could not necessarily distinguish between being touched by cotton or by a finger. Intensity of pressure could not be distinguished well. Spatial perception was particularly disturbed, a change not seen after anterolateral cordotomy. When a line was drawn on the skin, the patient could not tell its direction or when it was parallel to or at right angles to the long axis of the limb. Similarly, the patient could not recognize numbers or letters written on the skin. Position sense and two-point discrimination were disturbed. Other disturbances in spatial and temporal resolution attributed to the posterior columns by Nathan *et al.* (1986) include changes in threshold for sensation with repeated tactile stimulation, persistence of sensation after stimulation was stopped, and tactile and postural hallucinations.

The posterior columns also seem to have some role in pain. Transection of the posterior columns produces an enhancement in the sensations of pain, tickle, warmth, and cold, suggesting that activity in the posterior columns in humans tends to interfere tonically with input to the higher centers via the anterolateral quadrant (Nathan *et al.*, 1986). The use of posterior column stimulation to relieve pain will be mentioned below in the section on Effects of Stimulation of Posterior Columns. At the stimulus intensity used, the patient feels paresthesias, but not pain. The ability of posterior column stimulation to relieve pain is presumably related to the enhancement of pain following posterior column transection.

By contrast, Kroll (1930), in examining Foerster's cases, found that there was an increased threshold of pain spots, in addition to hyperpathia, in the lower extremities after the fasciculus gracilis was sectioned. Furthermore, several neurosurgeons have noticed that stimulation of the posterior column may produce severe pain (Foerster and Gagel, 1932; Sweet *et al.*, 1950). However, it is unclear if the pain is due to volleys ascending in the posterior column or to antidromically conducted volleys in posterior column afferents that activate other sensory pathways, such as the spinothalamic tract (R. D. Foreman *et al.*, 1976). The fact that no analgesia is produced by transection of the posterior column, whereas profound analgesia results from anterolateral cordotomy, argues against an important role for the posterior columns in pain transmission, but a minor role cannot be ruled out.

Dissociated Sensory Loss. Some clinical and experimental evidence that contradicts the traditional view of posterior column function has been reviewed by Wall (1970). For example, a variety of clinical cases have shown a distinct loss of vibratory sense with preservation of position sense or *vice versa* in patients with lesions of the posterior column (Davison and Wechsler, 1936; Weinstein and Bender, 1940; J. C. Fox and Klemperer, 1942; Netsky, 1953; see also Calne and Pallis, 1966). This can be explained if the axons carrying the information needed for vibratory sensation and position sense are in somewhat separate locations in the posterior column.

Difficulties in Interpretation. Wall (1970) also cautions that disease states, such as the Brown-Séquard syndrome from cord hemisection, tabes dorsalis, Friedreich's ataxia, and subacute combined degeneration, may not be directly relevant to a demonstration of posterior column function, since the posterior columns are not interrupted in isolation. Wall (1970) points out, for instance, that in the Brown-Séquard syndrome, not only is the posterior column interrupted, but also the posterior lateral funiculus, which contains an important pathway for somatic sensation from the ipsilateral body, the spinocervical tract, is interrupted. Tabes dorsalis has been used as a model of posterior column disease, since the loss of many of the axons in the posterior columns in neuropathologic material is striking. However, the disease process affects primary afferent neurons that provide input to all of the ascending sensory pathways of the spinal cord, and the loss of afferent fibers is not restricted to the large ones Calne and Pallis, 1966). Similarly, Friedreich's ataxia and subacute combined degeneration affect other pathways in addition to the posterior columns (Calne and Pallis, 1966).

Interruption of Posterior Columns. Deliberate lesions of the human posterior columns have been made surgically (Browder and Gallagher, 1948; Rabiner and Browder, 1948; A. W. Cook and Browder, 1965) in an attempt to relieve phantom limb pain. Of the five patients with thoracic posterior cordotomies, two had a transient and minimal reduction in tactile sensibility and one had a temporary slight decrease in appreciation of passive movement. The three patients with cervical dorsal cordotomies also showed little deficit. One had a short-lived zone of hypersensitivity of the arm and upper thorax; another had a slight transient decrease in the sense of passive movement; and the third had a permanent reduction in two-point discrimination and a temporary reduction in stereognosis and position sense. There was no loss in vibratory sensibility.

Before reaching the conclusion that the human dorsal column makes little contribution to sensory experience on the basis of these observations, several points should be made. The first is that there was no postmortem confirmation of the extent of the lesions. If even a small number of fibers remained intact, there may well have been little clinical deficit. In cats, more than 86% of the cross-sectional area of the dorsal column must be sectioned before motor and sensory deficits can be demonstrated (Dobry and Casey, 1972a). In monkeys, only a small part of the dorsal column can prevent the appearance of a tactile deficit despite the interruption of the rest of the dorsal column plus the contralateral ventral quadrant (Vierck, 1974). Even if the lesions in the human subjects were complete, testing could not be done in a normal way, because the patients all had a complete or partial amputation of a limb, causing the phantom-limb pain for which the surgery was done. Nathan *et al.* (1986) point out that the main sensory deficits seen when the posterior columns are interrupted occur distally, and typical deficits were seen in the one patient who had only a partial amputation.

Residual Function after Interruption of Anterior Cord Pathways

The sensory information that can be transmitted through the posterior part of the spinal cord despite interruption of the anterior two-thirds of the spinal cord was described by Nathan *et al.* (1986). One case involved a patient who had an occlusion of the anterior spinal artery. As expected (see the section below on the anterolateral quadrant), there was a loss of pain and of temperature sensation caudal to the T12 segment. Vibration was sometimes, but not always, sensed. Touch, two-point discrimination, graphesthesia, and position sense were all normal. At autopsy, the anterior two-thirds of the spinal cord was found to be infarcted. There was also some loss of fibers in the posterior lateral funiculus on one side and in the posterior columns. Similar findings were obtained in seven patients with very extensive bilateral anterolateral cordotomies. Position sense, tactile threshold, graphesthesia, and vibratory sense were all normal in these patients.

Effects of Stimulation of Posterior Columns

Additional evidence about the function of the posterior columns in humans can be obtained from stimulation of these columns. A technique was developed by Shealy for relieving pain by stimulating the posterior column electrically through chronically implanted electrodes (Shealy *et al.*, 1967, 1970). During posterior column stimulation at a midthoracic level by using high frequencies of pulses, patients reported a buzzing or tingling sensation that radiated down the body to the feet (Shealy *et al.*, 1970; Nashold *et al.*, 1972). With low frequencies, the sensation was described as beating or thumping. High stimulation intensities produced an unpleasant sensation that patients wished to avoid but that was not described as painful (Nashold *et al.*,

1972). Posterior column stimulation did not consistently interfere grossly with tactile, position, or vibratory sensibility or with pain elicited by pinprick on clinical testing (Shealy *et al.*, 1967, 1970; Nashold and Friedman, 1972), but a blunting or even suppression of awareness of touch, mild pinprick, and passive movement of the toes could be demonstrated in many cases (Nashold *et al.*, 1972).

Since the rate of stimulation generally chosen by patients employing posterior column stimulation for pain relief is 100–200 Hz (Shealy *et al.*, 1970), a case can be made that the buzzing sensation felt by such patients is due to the activation of primary afferent axons within the dorsal column that are connected peripherally with Pacinian corpuscles, which are responsive to high rates of vibration (C. C. Hunt, 1961; McIntyre, 1962; McIntyre *et al.*, 1967; Silfvenius, 1970; see also Calne and Pallis, 1966) (see Chapter 2). However, caution is needed in the interpretation of sensations produced by electrical stimulation of large numbers of sensory axons, since the pattern of evoked neural discharge need not produce a normal sensation. Furthermore, it is unlikely that the sensory experience produced by posterior column stimulation results just from the ascending traffic of impulses within the dorsal column itself. The same fibers give off collaterals at the segmental level that excite neurons of other ascending tracts, and so posterior column stimulation would cause antidromically conducted volleys that would activate at least some neurons in other major sensory pathways (R. D. Foreman *et al.*, 1976).

POSTERIOR LATERAL FUNICULUS

Lesions of Posterior Lateral Funiculus

Pathways in the posterior lateral funiculus do not appear to have a major sensory role in humans. Rabiner and Browder (1948) report that posterior lateral cordotomies in their patients failed to affect touch or deep sensibility. Similarly, Nathan *et al.* (1986) compared the effects of cordotomies on each side of the cord in cases in which one cordotomy extended into the posterior quadrant or was entirely in that region. In a case in which there was a lesion restricted to one posterior lateral funiculus, touch, position sense, and vibratory sense were normal, and there were no hallucinations. In several cases in which a cordotomy extended into the posterior lateral funiculus on one side, no sensory changes were seen that could be attributed to this component of the lesion. This argues against the notion that proprioception in the lower extremities of humans depends entirely upon axons ascending in the posterior lateral funiculus (cf. Chapter 8). Nathan *et al.* (1986) suggest that proprioception in humans may depend upon fibers ascending in both the posterior column and the posterior lateral funiculus. Similarly, it is possible that other mechanoreceptive functions are shared by the posterior column and posterior lateral funiculus in humans.

There is some evidence for a role of fibers in the posterior lateral funiculus in pain. Stimulation of the posterior lateral funiculus causes patients to report pain referred to the ipsilateral side (White *et al.*, 1950). Moossy *et al.* (1967) reported in an abstract that in some cases in which a percutaneous cordotomy successfully relieved pain, the lesion was found at autopsy to extend into the posterior lateral funiculus. If confirmed, this observation would be consistent with a role for axons in the posterior lateral funiculus in pain (see Chapter 9); however, it could not be ruled out that the analgesia depended on the interruption of fibers in the anterior part of the cord. Moffie (1975) describes a case in which a cordotomy at a high cervical level and confined to the posterior lateral funiculus produced analgesia and loss of temperature sensation; in another case in which a cordotomy involving the posterior lateral funiculus produced analgesia, the lesion also extended well into the anterior quadrant. On the other hand, Nathan (1990) describes a case in which bilateral lesions of the posterior lateral funiculi failed to affect deep pain or warm sensations. One of the lesions extended into the anterior lateral funiculus and was associated with a contralateral reduction in cold sensation and pinprick.

ANTEROLATERAL QUADRANT

Lesions of the Anterolateral Quadrant

Interruption due to Disease. Pain and temperature sensations depend primarily on fibers ascending in the anterolateral white matter of the spinal cord. This was dramatically demonstrated by Spiller (1905) in a patient who had lost pain and temperature but not tactile sensation over the lower part of the body. At autopsy, tuberculomas were found to have disrupted the anterolateral quadrants bilaterally. This finding led to the first deliberate anterolateral cordotomy for pain relief (Spiller and Martin, 1912). Others have observed that unilateral lesions of the anterolateral quadrant produced by neurologic disorders cause a contralateral analgesia and thermoanesthesia (Brown-Séquard, 1860; Gowers, 1878; Petren, 1902; Head and Thompson, 1906).

Surgical Interruption. Since the report by Spiller and Martin (1912), there have been numerous clinical experiences with anterolateral cordotomy, which, when done unilaterally, produces a loss of pain and temperature sensations on the contralateral side of the body (see, e.g., Frazier, 1920; Horrax, 1929; F. C. Grant, 1930, 1932; Foerster and Gagel, 1932; Kahn, 1933; Hyndman and Van Epps, 1939; Walker, 1940; Hyndman and Wolkin, 1943; Falconer and Lindsay, 1946; Kuru, 1949; White *et al.*, 1950, 1956; Nathan, 1963; Nathan and Smith, 1979; cf. Nathan *et al.*, 1986). More recently, the technique of percutaneous cordotomy has often been used, rather than the open procedure, with a resultant lowering of the operative mortality and postoperative morbidity (Mullan *et al.*, 1963; Mullan, 1966; Rosomoff *et al.*, 1965, 1966). However, there is considerable risk (10% mortality) with bilateral percutaneous cordotomies (Ischia *et al.*, 1984).

Some neurosurgeons have interrupted the spinothalamic tract in the brain stem to obtain higher segmental levels of analgesia than can be obtained by cordotomy (H. G. Schwartz and O'Leary, 1941; White, 1941; Walker, 1942a,b), but the success rate is decreased for higher levels of surgically placed lesions (White and Sweet, 1955).

Location of Pathways for Pain and Temperature. Cordotomies have provided information about the distribution of axons required for pain and temperature sensations within the spinal cord, the effects of interruption of these fibers, and the sensory experiences produced by stimulation in the anterolateral quadrant. Early results with anterolateral cordotomy suggested that axons responsible for pain ascend in a compact bundle in the lateral funiculus (Fig. 6.2A and B). However, it is now recognized that adequate pain relief requires an incision to within a few millimeters of the anterior median fissure (Fig. 6.2D–F; Kahn and Peet, 1948; White *et al.*, 1956; White and Sweet, 1969).

Kuru (1949) described two small bundles of axons which he named the dorsolateral and the ventromedial spinothalamic tracts (Fig. 6.3). Lesions of the dorsolateral spinothalamic tract were correlated with pain relief and chromatolysis in neurons in the marginal layer on the contralateral side of the spinal cord. Lesions of the ventromedial spinothalamic tract were, instead, related to tactile loss and chromatolysis of neurons in the contralateral nucleus proprius. Kuru's results must be reevaluated in the light of the progression illustrated in Fig. 6.2. Evidently, lesions required to produce maximal analgesia would have to interrupt both of the bundles described by Kuru.

Somatotopic Organization. The axons that must be severed to produce pain relief appear to be distributed somatotopically, with those conveying information from sacral and lumbar levels lying dorsolaterally and those from the thoracic and cervical levels ventromedially (Fig. 6.2D–F and 6.4) (Hyndman and Van Epps, 1939; Walker, 1940; White and Sweet, 1969). However, some mixing of the fibers from all levels is thought to occur (Nathan, 1963; White and Sweet, 1969). Axons carrying information leading to pain sensation have been generally thought

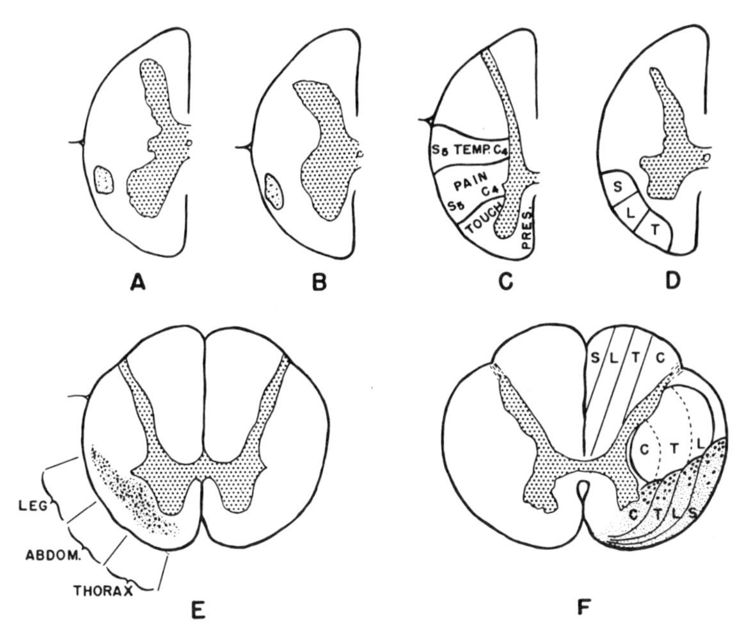

Fig. 6.2. Location and somatotopic organization of the human spinothalamic tract within the spinal cord according to a variety of authors (see text). Abbreviations: S, sacral; L, lumbar; T, thoracic; C, cervical. (From White, copyright © 1954, American Medical Association.)

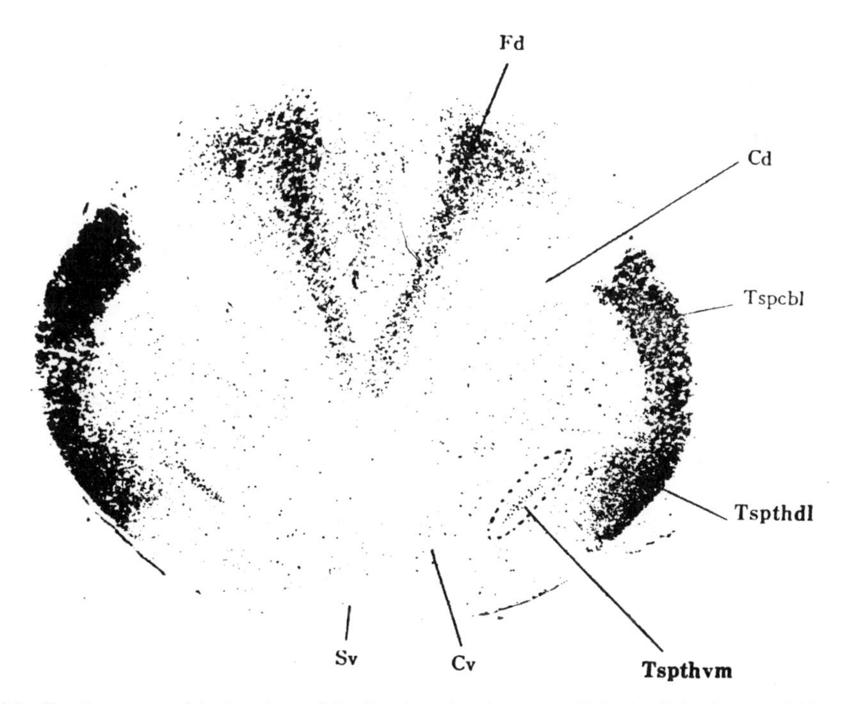

Fig. 6.3. Kuru's concept of the locations of the dorsolateral and ventromedial spinothalamic tracts (abbreviated Tspthdl and Tspthvm). (From Kuru, 1949.)

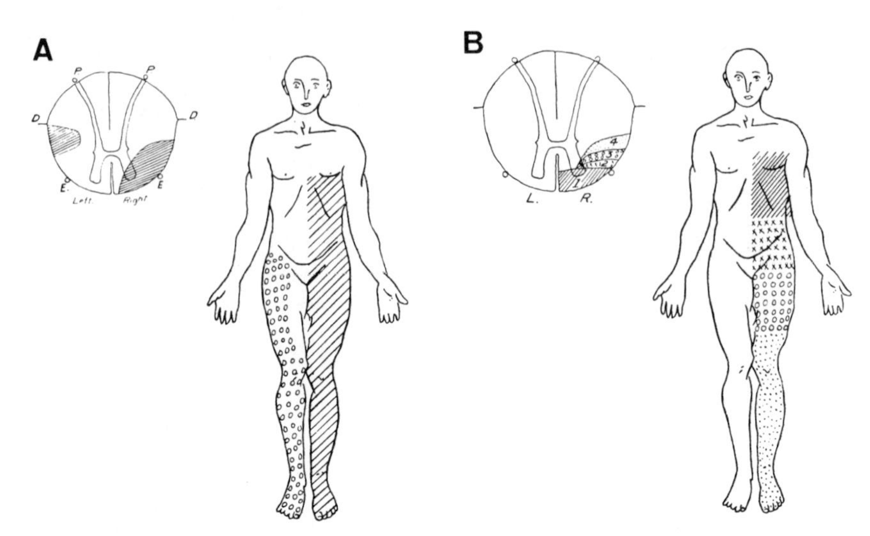

Fig. 6.4. Results of partial cordotomies. (A) The cross-hatched area on the left side of the body is devoid of pain and temperature sense as a result of a lesion of the right side of the cord, whereas the circles indicate a partial loss of these sensations on the right lower extremity following a more restricted lesion on the left side of the cord. (B) There was complete loss of pain and temperature sensations over an area below the left nipple. The different symbols show the progressive increments in sensory loss as the cordotomy was extended. (From Hyndman and Van Epps, 1939; copyright © 1939 American Medical Association.)

to ascend several segments before crossing the midline and moving laterally in the anterolateral quadrant (White *et al.*, 1956), although Foerster and Gagel (1932) suggested that the decussation occurred within one segment. The discrepancy between the level of sensory loss and that of the cordotomy can be accounted for, in part, by the ascent of information by way of the dorsolateral fasciculus of Lissauer for a few segments (Hyndman, 1942).

Sensory Deficits Produced by Anterolateral Cordotomy. When an anterolateral cordotomy is performed properly, the patient loses all forms of pain and temperature sensations completely on the contralateral side of the body below the segmental level of the lesion. In a few rare cases, the sensory loss has been reported to be ipsilateral to the cordotomy (French and Peyton, 1948; Sweet *et al.*, 1950; Voris, 1951, 1957). The level of the sensory loss depends upon the level of the cordotomy, as well as the location of the cordotomy within the anterolateral quadrant. A complete anterolateral cordotomy at a high cervical level may produce a sensory loss as high as the uppermost cervical dermatomes, but in some cases only to the level of the upper thorax. A complete upper thoracic anterolateral cordotomy results in a sensory loss to a level between the nipple and the umbilicus (White and Sweet, 1969). The sensory loss persists for 1 year or more in more than half the patients, but in many cases the sensory deficit becomes smaller or pain recurs within several months. The reason for recurrence of pain is a matter of speculation. Among the suggestions are regeneration of the sensory pathways (unlikely) and the development of other sensory channels for pain information (White and Sweet, 1969). In some cases, the pain has a dysesthetic quality and is a form of central neurogenic or "deafferentation" pain (Pagni, 1989; Tasker and Dostrovsky, 1989).

Following anterolateral cordotomy, visceral pain is lost, as well as somatic pain, and pain from pressure on bone is reduced, as well as cutaneous pain (White and Sweet, 1969). It is possible to elicit pain in patients with successful unilateral cordotomies by such strong stimuli as electrical shocks applied to the skin, testicular compression, and distension of the renal pelvis (Hyndman and Wolkin, 1943; White *et al.*, 1950). However, bilateral anterolateral cordotomies completely eliminate the pain of testicular compression and of renal pelvis distension (Hyndman and Wolkin, 1943). Itch is also lost following anterolateral cordotomy (Hyndman and Wolkin, 1943; White *et al.*, 1950), suggesting that the pathways responsible for itch are in the vicinity of those conveying pain. Tickle may (Foerster and Gagel, 1932; H. G. Schwartz and O'Leary, 1941) or may not (Hyndman and Wolkin, 1943; White *et al.*, 1950) be lost.

In addition to the loss of pain sensation, patients who have had an anterolateral cordotomy lose thermal sensation (Frazier, 1920; Foerster and Gagel, 1932; Hyndman and Wolkin, 1943; Kuru, 1949). Thermoanesthesia is generally more complete than is analgesia, and thermal sense is less likely to recur than pain (White and Sweet, 1969). Shivering is abolished as well (Hyndman and Wolkin, 1943). There is some evidence for a different distribution within the anterolateral quadrant of axons conveying pain and temperature information. For example, some lesions produced surgically or by disease have resulted in a differential loss of pain or of thermal sensation (Head and Thompson, 1906; G. Wilson and Fay, 1929; Stookey, 1929; Foerster and Gagel, 1932; Sherman and Arieff, 1948; Nathan, 1990). Most of the results can be explained if the axons related to temperature sensation are just posterior to those conveying pain (Fig. 6.2C) (cf. Foerster and Gagel, 1932); however, Stookey (1929) suggested the opposite arrangement. The regions of analgesia and thermoanesthesia following cordotomy are generally superimposable (Kuru, 1949; White and Sweet, 1955). The fact that there can be an independent loss of pain and thermal sense or, within thermal sense, of warm sensation without a change in cold sensation, or *vice versa*, indicates that different axons convey pain, warm, and cold sensations (see Nathan, 1990).

Bilateral anterolateral cordotomy may abolish erection, ejaculation, and orgasm in the male, orgasm in the female, as well as libidinous sensations in both (Foerster and Gagel, 1932; Hyndman and Wolkin, 1943); such changes are not necessarily found after a unilateral procedure (White *et al.*, 1950).

Touch is not dramatically affected by anterolateral cordotomy, although careful testing shows a slight increase in the tactile threshold (Foerster and Gagel, 1932; Kuru, 1949; White *et al.*, 1950). Vibratory sense, position sense, tactile localization, stereognosis, and graphesthesia are little if at all affected (Hyndman and Van Epps, 1939; Hyndman and Wolkin, 1943; White *et al.*, 1950; White and Sweet, 1969; Nathan *et al.*, 1986). Two-point discrimination is impaired when the compass points are sharp, but there is only a slight deficit when the points are blunt (White *et al.*, 1950). Changes in tactile sensation are not increased when the cordotomy is extended across most of the anterior funiculus (White *et al.*, 1956), arguing against the presence of a ventral spinothalamic tract with a tactile function (cf. Fig. 6.2C).

Interruption of Entire Cord except One Anterolateral Quadrant

A unique case of spinal cord injury with sensory deficits was reported by Noordenbos and Wall (1976). The patient had been stabbed in the back with a knife, with the result that all of the cord was transected except for the left anterolateral quadrant. The lesion was verified surgically (Fig. 6.5). Following stabilization, the patient was able to cooperate in a careful sensory examination. The patient could detect touch and pressure stimuli applied bilaterally below the level of the lesion, although the threshold was generally elevated compared with normal. Localization was rather good. Position sense was accurate to within 5° for the left knee and ankle, although it was lost for the left toes and for the whole right lower extremity. The fact that proprioception was partially spared may relate to the fact that a portion of the dorsal spinocerebellar tract may have been intact (see Chapter 8). Vibratory sense was absent. Temperature sensation was lost on the left side of the body, but intact on the right. Pinprick applied to the right side had a pricking quality and a tendency to radiate. Pinprick on the left side could not be identified as such, but was unpleasant.

A second case of a patient who had an extensive lesion of the posterior part of the spinal

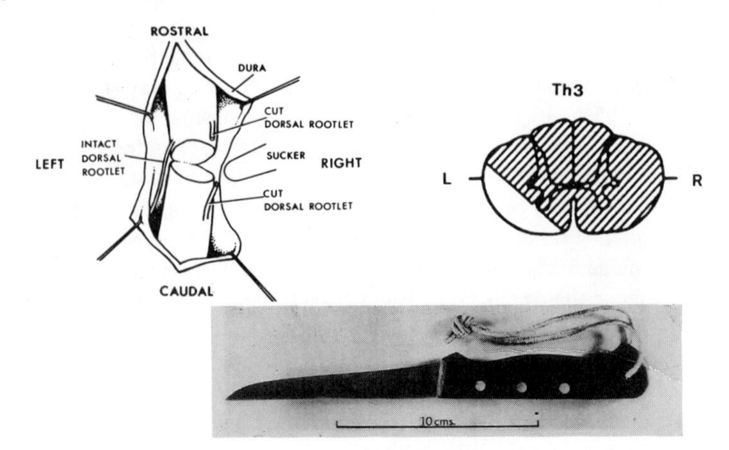

Fig. 6.5. The sketch shows a spinal cord that had been nearly transected by a knife wound. The extent of the lesion was demonstrated by surgical exposure. The drawing of the spinal cord in cross-section indicates the extent of the lesion. The weapon is shown below. (From Noordenbos and Wall, 1976.)

cord was described by Wall and Noordenbos (1977). The posterior columns were completely interrupted, and this was verified surgically. The anterior and lateral funiculi were presumed to be spared on one side and partially spared on the other. Position sense was present but impaired. Two-point discrimination was absent on the side with the intact anterolateral quadrant and diminished on the other side. Tactile sensation was good, but graphesthesia was absent.

Central Cord Lesions

Since the axons of neurons responsible for pain sensation decussate at the segmental level, it is not surprising that lesions in the central part of the spinal cord, such as those often produced in syringomyelia (cf. Netsky, 1953), cause a bilateral loss of pain and temperature sensations over the corresponding region of the body. Syringomyelia may also produce central neurogenic or deafferentation pain in the area that is analgesic to evoked pain (Pagni, 1989).

Commissural myelotomy has been used by neurosurgeons to treat intractable pain originating from both sides of the body (see, e.g., A. W. Cook and Kawakami, 1977; Gildenberg and Hirshberg, 1984; see the review by A. W. Cook *et al.*, 1984). Surgical intervention causes major damage to the posterior columns. Thus, the sensory deficits resulting from commissural myelotomy represent a combination of posterior column and anterolateral quadrant deficits. Sensory losses include a disturbance in position sense, as well as loss of pain and temperature. In regions in which there is correspondence between the posterior column and anterolateral quadrant loss, touch may be completely absent. Curiously, clinical pain is sometimes relieved without a demonstrable loss of cutaneous pain sense. Even when there is analgesia on testing, the region of clinical pain loss may extend much more rostrally and caudally than expected from the distribution of analgesia. This observation is reminiscent of that of Hitchcock (1970, 1974), who found that a commissural myelotomy at an upper cervical level could produce pain loss as far caudally as below the knee. One possible explanation is interruption of a descending pathway that influences the pain transmission system (A. W. Cook *et al.*, 1984).

Stimulation of Anterolateral Quadrant Axons

Stimulation of axons within the anterolateral white matter of the human spinal cord may result in thermal or pain sensations (Foerster and Gagel, 1932; Sweet *et al.*, 1950; D. J. Mayer *et al.*, 1975). The sensations are generally referred to the contralateral side of the body, although on occasion the referral is to the same side of the body or bilaterally.

The stimulus factors required to produce pain by stimulation within the anterolateral quadrant have been determined by D. J. Mayer *et al.* (1975) during percutaneous cordotomies in conscious subjects. The threshold stimulus strengths were generally below 300 μ, provided that the stimulus frequency was high enough. No pain was produced by stimulus rates of less than 5 Hz, although some patients reported tingling or warmth. When the intensity was raised during stimulation at 5 Hz, only some patients reported pain. High-frequency stimuli (50–500 Hz) below pain threshold produced a tingling paresthesia. By using a double-pulse technique, it was possible to estimate the refractory period of the fibers carrying pain information. Usually, a value between 1 and 1.5 msec was obtained. On the basis of a comparison of the human data with results obtained by using monkeys to determine the refractory periods of the axons of different classes of spinothalamic tract neurons, it was suggested that stimulation of the axons of "wide-dynamic-range" spinothalamic neurons can evoke pain (D. D. Price and Mayer, 1975). This assumes that the observations in the monkey spinal cord are immediately applicable to humans.

Summary

In summary, clinical evidence indicates that the posterior column is responsible for discriminative touch, vibratory sensibility, and position sense. The discriminative functions include two-point discrimination, stereognosis, and graphesthesia. Loss of posterior column function may result not only in loss of these sensations, but may also result in a poor ability to detect repeated stimuli, a reduced capability for recognizing gradations of pressure stimulation, tactile and positional hallucinations, and increases in pain, thermal, and other sensations mediated by the anterolateral quadrant. Stimulation of the posterior column produces a "buzzing sensation," but sensory data would reach the brain in such cases over several different sensory pathways. Clinical evidence does not clearly demonstrate any special mechanosensory functions for the posterior lateral funiculus in man. It is likely that position sense depends upon both the posterior columns and the posterior lateral funiculus, and it is possible that axons in this part of the cord make a minor contribution to pain sensation, at least in some individuals. The functions of the anterolateral quadrant are shown by anterolateral cordotomy to include contralateral pain and temperature sensations, as well as itch and libidinous sensation. The recurrence of pain in some cases suggests that there are also pathways with the potential to mediate pain sensation that ascend in other sectors of the spinal cord. Cordotomies produce a slight reduction in tactile function. The reports of a patient who had interruption of the sensory pathways of the posterior spinal cord, but not of an anterolateral quadrant, shows that the anterolateral quadrant is sufficient for pain and temperature sensations and that it contributes to touch and pressure as well.

ANIMAL STUDIES: ALTERATIONS IN BEHAVIORAL MEASURES

Animal experiments involving lesions interrupting one or more of the ascending sensory tracts of the spinal cord provide useful information that can be correlated with clinical findings. Such experiments have the advantage that the lesions can be made in discrete regions of the spinal cord in otherwise healthy subjects, the effects of additional lesions can be determined, and postmortem evaluation of the extent of the lesions can readily be done. However, changes in the sensory experience of animals must be inferred from behavioral tests, and the results of such tests can be difficult to interpret.

Experiments designed to determine the functions of the sensory pathways of the spinal cord have included studies in which transections of any of the following have been made: dorsal column; all of the cord except the dorsal column; the dorsal lateral funiculus; the dorsal column plus the dorsal lateral funiculus; and the ventrolateral quadrant.

Dorsal Column

Mechanoreceptive Functions. *Negative Findings.* A number of investigations of the effects of interruption of the dorsal column in animals have found either no change, a minimal change, or only a transient change in the behavioral responses to the mechanical stimuli that were examined. Behavioral measures in such studies have included, for cats, tactile placing (A. Lundberg and Norrsell, 1960), tactile localization (Diamond *et al.*, 1964), and roughness discrimination (Kitai and Weinberg, 1968); for dogs, conditioned reflexes to tactile stimuli (Norrsell, 1966b); and for monkeys, weight discrimination (DeVito *et al.*, 1964), proprioceptive and tactile placing (Christiansen, 1966), two-point discrimination (Levitt and Schwartzman, 1966), limb position (Vierck, 1966; Schwartzman and Bogdonoff, 1969), vibration (Schwartzman and Bogdonoff, 1968, 1969), and tactile discrimination (A. S. Schwartz *et al.*, 1972).

Interruption of Cord except Dorsal Columns. One explanation for these negative or nearly negative findings is that other sensory pathways outside the dorsal column can transmit the information required for the behavioral responses that were tested. An alternative approach is to see what functional capabilities remain after interruption of all of the cord except the dorsal columns. Wall (1970) tried this in rats. The animals became paraplegic. Intense stimulation of the hindlimbs failed to produce orientation, vocalization, or changes in respiration in awake animals; similar stimuli did not awaken sleeping animals. Wall suggested on the basis of these experiments that the dorsal columns are more important for exploratory behavior than for tactile sensation; an absence of responses to stimuli that are received passively is in keeping with such a role.

However, different conclusions were reached when a similar experiment was done in cats (Myers *et al.*, 1974; Frommer *et al.*, 1977). Sensation was tested by operant conditioning in the study by Myers *et al.* (1974) and found to return to normal soon after surgery. Frommer *et al.* (1977) trained cats to perform a tactile discrimination task. Following transection of all of the spinal cord except the dorsal columns, the cats were still able to discriminate between tactile stimuli. However, like Wall's rats, the cats that had cord lesions sparing only the dorsal columns did not show orienting responses to somatic stimuli applied below the lesion. It is likely that orientation and arousal reactions require input to the reticular formation and that such input is lacking in animals whose ventral white matter is interrupted.

Behavioral Deficits after Dorsal Column Lesions. Experiments in which more rigorous tests of sensory performance were used have demonstrated alterations in function following lesions of the dorsal columns.

Dobry and Casey (1972a) found a deficit in roughness discrimination in cats, provided that more than 86% of the cervical dorsal columns was interrupted. On the basis of these experiments, it seems likely that an insufficient interruption of the dorsal columns contributed to some of the negative results in previous investigations.

Vierck (1973) found a substantial impairment in the ability of monkeys to distinguish between different-sized plastic disks palpated with the foot after a lesion of the fasciculus gracilis (Fig. 6.6A). Recovery occurred but took months. A dorsal column lesion alone did not impair the performance of monkeys in a movement-detection test (Vierck, 1974). However, the dorsal column lesion eliminated the ability of the monkeys to distinguish the direction of a stimulus moved across the skin (Fig. 6.6B). Since the deficit was in discrimination of a passively detected stimulus, the role of the dorsal column is evidently not limited to exploratory activity.

Azulay and Schwartz (1975) required their monkey subjects to discriminate among plastic disks of various geometric patterns. The animals had to palpate disks actively to distinguish

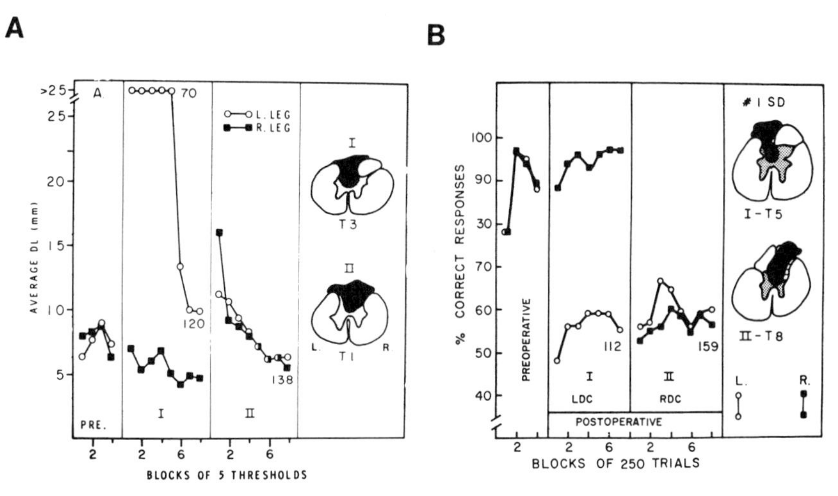

Fig. 6.6. Deficits in mechanoreception in monkeys produced by lesions of the dorsal columns. (A) The animal was trained to distinguish between different-sized plastic cylinders with its foot. Prior to surgery, correct responses were made 90% of the time when the stimuli differed by 6 mm or more. When the dorsal column was sectioned on the left side (I), discrimination was markedly impaired on that side, but partially recovered within 120 days. A second lesion that interrupted both dorsal columns (II) impaired discrimination, especially on the right side, but there was nearly complete recovery by 138 days. (B) A monkey was trained to recognize the direction of a stroking stimulus applied to the medial part of each calf. After a lesion of the left dorsal column (I), the ability to perform the task was lost on the left, and after a second lesion that interrupted the right dorsal column (II), the animal could not perform the task on the right. (From Vierck, 1973, 1974.)

between different ones. Lesions of the dorsal columns made it very difficult or impossible to make the discrimination. It was concluded that the dorsal columns play a unique role in the discrimination of tactile stimuli requiring sequential or spatiotemporal analysis.

Nociception. Melzack *et al.* (1958) did not notice any obvious changes in the responses of cats to noxious thermal stimuli or pinprick after lesions of the dorsal columns. However, Kennard (1954) observed a transient loss of responses to noxious stimuli applied to the hindlimbs in cats following bilateral interruption of the dorsal quadrants of the spinal cord. In two animals, the intent was to interrupt just the dorsal columns, but the lesions actually extended into the dorsal lateral funiculi, and so the sensory changes cannot be attributed solely to the loss of transmission through the dorsal columns. Casey and Morrow (1988) reported results consistent with those of Kennard; lesions of the dorsal columns and dorsal parts of the dorsal lateral funiculi in cats reduced responses to noxious thermal stimuli applied to the hindlimb. Vierck *et al.* (1971) found that a lesion of the dorsal column in monkeys reduced their reactivity to painful stimuli.

Motor Behavior. A number of investigations in which the dorsal columns were cut have shown striking changes in motor behavior on testing. Ferraro and Barrera (1934) reported that monkeys with dorsal column lesions had deficits in grasping movements and in placing and hopping reflexes. If the lesion of the dorsal column was at a cervical level, the deficits were more severe for the upper than for the lower extremities. Pain and temperature appeared to be intact. The deficit in grasping and in contact placing was confirmed by DeVito *et al.* (1964).

Gilman and Denny-Brown (1966) found similar deficits in their lesioned animals and concluded that dorsal column lesions produce a severe disorder in exploratory movements. A lack of attention was also found to be characteristic of monkeys with transected dorsal columns (Schwartzman and Bogdonoff, 1968, 1969). Diamond *et al.* (1964) also found a severe motor disorder in their cats.

Studies by Melzack's group have also emphasized the part played by the dorsal columns in motor activity. Melzack and Bridges (1971) found that dorsal column lesions interfered with the ability of cats to perform coordinated walking and turning on a narrow beam or to jump to a moving target. Interestingly, performance is less affected when the lesion is placed in the dorsal column nuclei rather than in the dorsal columns (cf. Heckmann and Bourassa, 1981). Melzack and Southmayd (1974) also found deficits in "anticipatory" motor behavior. This was tested by having the animals use a conveyor belt. Sometimes a barrier was placed partway along the belt. After a dorsal column lesion, a cat that would normally get on and off the belt and jump the barrier without difficulty would now experience problems in stepping onto the belt and would often be carried by the belt into the barrier before trying to jump over it. These experiments suggested that the information carried by the dorsal columns is needed for selecting the appropriate motor programs. Visual information alone did not seem to be adequate for such "anticipatory" motor behavior.

Experiments consistent with the work of Melzack and his associates have been done by Dubrovsky *et al.* (1971) and Dubrovsky and Garcia-Rill (1973). They showed a considerable deficit in the performance of serial-order acts by cats with dorsal column lesions. The cats were trained to jump vertically to release a piece of liver attached to a rotating wheel placed above them. Performance was judged in terms of efficiency (proportion of successful attempts), accuracy, tracking of the falling liver, and searching when the cat failed to release the liver. All of these measures showed deficits after a dorsal column lesion, although the efficiency rating could be improved by overtraining before surgery. It was suggested that the deficits were due to the interruption of proprioceptive input from the forelimbs, with a resultant incoordination.

A proprioceptive deficit was also found by Reynolds *et al.* (1972) in dogs following a dorsal column lesion. The dogs were trained to maintain a symmetric stance, but their ability to do this was impaired after the lesion. However, the dogs responded in the same way to perturbations whether or not they had a dorsal column lesion.

Alstermark *et al.* (1986) found that transection of the dorsal column at C2 in cats resulted in a severe deficit in the use of the upper extremity to reach and to retrieve food from a tube.

Vierck (1982) also observed an impairment in motor performance in monkeys following a dorsal column lesion. However, a dorsal rhizotomy extending the length of the cervical enlargement produced a more severe deficit.

Heckmann and Bourassa (1981) found that the full motor impairment seen in cats after lesions of the dorsal column is not seen with a lesion confined to the main parts of the dorsal column nuclei, but is seen when the lesion is enlarged to include the lateral cuneate nuclei. Purely sensory deficits were observed when the lesions were confined to the main dorsal column nuclei or the medial lemniscus.

Dorsal Lateral Funiculus

There is less evidence concerning the behavioral changes that result from lesions of the dorsal lateral funiculus.

Mechanoreception. The tactile placing reaction was abolished by a lesion of the dorsal lateral funiculus, but not by lesions in either the dorsal column or the ventrolateral quadrant (A. Lundberg and Norrsell, 1960). However, the pathway involved proved not to be the spinocervical-thalamic path (see Chapter 8), since there was no deficit in tactile placing when the cervicothalamic tract was interrupted at C1 (Norrsell and Voorhoeve, 1962).

Norrsell (1966b) found that a lesion of the dorsal lateral funiculus in dogs caused only a transient impairment in tactile conditioned reflexes. However, a severe impairment resulted from a combined lesion of the dorsal lateral funiculus and dorsal column (Fig. 6.7). A lesion of just the dorsal column had no effect.

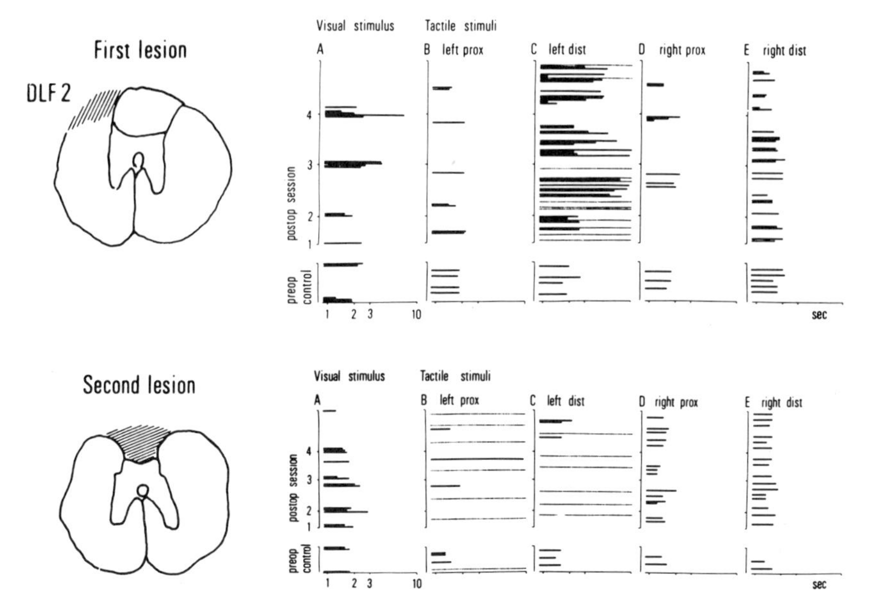

Fig. 6.7. Responses of a dog to conditioned stimuli before and after lesions of the dorsal lateral funiculus and of the dorsal columns. The lesions are shown in the drawings at the left. The responses are indicated by horizontal bars, with successive trials starting at the bottom of each column. The lines are interrupted at a point corresponding to the latency of the response, unless there was no response within 10 sec, in which case the horizontal line is a dotted one. There was no change in the responses to control stimuli (visual; tactile stimuli applied to the right lower extremity, either proximally or distally). In the first part of the experiment (top), a lesion of the dorsal lateral funiculus produced an impairment in the response to stimulation of the distal part of the left hind limb, but there was little change in the response to more proximal stimuli. However, when the dorsal columns were also sectioned (bottom), there was a severe deficit in conditioned responses to tactile stimulation of either the distal or proximal left hind limb. (From Norrsell, 1966b.)

Two-point discrimination was impaired in monkeys following a combined lesion of the dorsal column and the dorsal lateral funiculus, whereas there was no change with just a dorsal column lesion or with combined lesions of the dorsal column and ventrolateral quadrant (Levitt and Schwartzman, 1966).

Kennard (1954) found that bilateral lesions of the dorsal quadrants of the spinal cord at a lower thoracic level in cats produced a marked loss of proprioception in the lower extremities. Changes in proprioceptive and tactile placing in monkeys were more lasting following a lesion of the dorsal column and dorsal lateral funiculus than with a lesion just of the dorsal column (Christiansen, 1966).

Tactile roughness discrimination in cats was severely impaired following a lesion of the cervicothalamic tract or a combined lesion of this pathway and the dorsal column, but was only minimally affected by just a dorsal column lesion (Kitai and Weinberg, 1968).

Cats trained in a conditioned avoidance paradigm to respond to activation of a single tactile dome showed an unaltered response when the dorsal column was sectioned, but they no longer responded to such a stimulus after a combined lesion of the dorsal column and dorsal lateral funiculus (Tapper, 1970).

Vierck (1973) examined the ability of monkeys to discriminate between different-sized plastic cylinders. After a combined lesion of the dorsal column and dorsal lateral funiculus, the threshold for discrimination remained elevated, whereas a lesion of just the dorsal column produced a change of threshold with recovery to preoperative levels after several months. Vierck (1974) also observed that a dorsal quadrant lesion produced a more severe impairment of movement detection in monkeys than did a dorsal column lesion. However, the monkeys recovered from this deficit.

Nociception. Kennard (1954) found that cats having bilateral lesions of the dorsal quadrants of the spinal cord showed hypalgesia, suggesting that some nociceptive information is transmitted in the dorsal part of the spinal cord (see section on behavioral deficits after dorsal column lesions; however, cf. Ranson and Hess, 1915). Similar results were obtained by Casey and Morrow (1988). It is not clear whether the dorsal columns or dorsal lateral funiculi are more important for nociceptive transmission in cats.

Vierck *et al.* (1971) found an enhanced reactivity to painful stimuli in monkeys following a lesion of the dorsal lateral funiculus. This may have been due to interruption of tonic descending inhibitory pathways.

Casey and Morrow (1988) observed that bilateral lesions of the dorsal lateral funiculi in cats caused an increase in movements and in interruption of feeding in response to noxious thermal stimuli applied to the hindlimb. They concluded that these nociceptive responses are under tonic inhibitory control by axons descending in the dorsal lateral funiculi.

Ventral Quadrant

Nociception. Transection of the ventral quadrant of the spinal cord in the dog produced an increased threshold for nociception (Cadwalader and Sweet, 1912). This observation was used as experimental support for the introduction of cordotomy for the treatment of pain in humans (Spiller and Martin, 1912).

It is interesting that reactions to painful stimuli applied to one side of the body in cats are not prevented by hemisecting the contralateral cord (Ranson and Hess, 1915; Kennard, 1954) and that the dorsal lateral funiculus appears to be more important for the transmission of nociceptive information than is the ventral quadrant in this animal (Kennard, 1954; however, cf. Ranson and Hess, 1915). One difference between the cat and the dog may be a more prominent spinothalamic tract in the latter (Hagg and Ha, 1970). However, nociceptive information does seem to be transmitted in the ventral quadrant of spinal cord of the cat. Kennard (1954) found that the analgesia observed immediately after bilateral lesions of the dorsal quadrants diminished to hypalgesia with time, indicating that nociceptive information must be able to reach the brain through the ventral white matter in cats. Neither Ranson and Hess (1915) nor Melzack et al. (1958) observed alterations in responses to noxious stimuli following lesions of the dorsal half of the cord in cats. Casey and Morrow (1988) found that a bilateral lesion of the ventral half of the cat spinal cord reduced but did not eliminate movements and interruption of feeding produced by noxious thermal stimulation of the hindlimb.

Several investigators have found that ventrolateral cordotomy produced analgesia on the side contralateral to the lesion in monkeys (Yoss, 1953; Kennard, 1954; Poirier and Bertrand, 1955; Christiansen, 1966; Vierck *et al.*, 1971; Vierck and Luck, 1979). Enduring analgesia required lesions that involved both the ventral lateral funiculus and the ventral funiculus (Vierck and Luck, 1979). The most effective lesion was a ventral hemisection, which produced a reduction in escape responses that lasted at least 300 days. Secondary lesions of the dorsal columns, the tract of Lissauer or the dorsal lateral funiculus did not alter the level of analgesia. Vierck and Luck (1979) concluded that the dorsal pathways are not crucial for pain perception in monkeys and that for spinal cord lesions to produce a long-lasting reduction in pain sensitivity in monkeys, they must be bilateral and must include the ventral lateral and ventral funiculi.

The pathway used in rats for nociceptive transmission was investigated by Peschanski *et al.*

(1986). They transected different parts of the spinal cord at a cervical level, using hindpaw withdrawal and vocalization as tests of nociceptive transmission. No lesion modified the threshold for hindpaw withdrawal, but the threshold for vocalization increased significantly when the contralateral ventrolateral quadrant of the spinal cord was transected.

Thermoreception. Norrsell (1979, 1983, 1989a,b) has examined the effects of lesions of the spinal cord on temperature discrimination in trained cats. A lesion of the middle part of the lateral funiculus at C5 impairs the ability of the cat to discriminate temperatures with the contralateral paws. No effect was seen when the dorsal column and ventral quadrant were interrupted either individually, in combination, or bilaterally. The thermoreceptive pathway in cats appears to ascend in the ventral part of the dorsal lateral funiculus, suggesting the possible identification of the pathway with the dorsal spinothalamic tract originating from lamina I in cats (M. W. Jones *et al.*, 1985, 1987; cf. Craig and Kniffki, 1985).

Mechanoreception. The ventral quadrant of the spinal cord also seems to contain fibers that contribute to weight discrimination (DeVito *et al.*, 1964) and position sense (Vierck, 1966) in monkeys.

Summary

In summary, behavioral evidence from animal experiments suggests that the dorsal columns are especially important in spatiotemporal analysis of mechanoreceptive stimuli. Major deficits in roughness discrimination in cats and discrimination of stimulus size or geometric pattern and recognition of stimulus direction in monkeys are produced by lesions of the dorsal funiculus. The dorsal columns also appear to play some role in nociception. Dorsal column lesions produce striking changes in motor behavior in cats and monkeys. However, lesions of the main dorsal column nuclei in cats produce little motor deficit, whereas lesions that include the lateral cuneate nuclei produce a severe motor deficit, suggesting that the motor deficits produced by dorsal column lesions are mediated largely because of deafferentation of the pathway through the lateral cuneate nuclei. The dorsal lateral funiculi are involved in parallel with the dorsal funiculi in a number of sensory functions. One of the technical difficulties in lesion studies is that most of the dorsal column must be interrupted to produce a deficit. However, a danger for interpretation of the results of a dorsal column lesion is that a large lesion may also damage the dorsal lateral funiculus. Therefore, the conclusions drawn by many investigators must be evaluated with caution. The dorsal lateral funiculus of the cat seems to be involved, along with the dorsal column, in mechanoreception. Fibers in the dorsal lateral funiculus are required for tactile placing (but the spinocervical-thalamic path is not required). Conditioned responses to tactile stimuli in dogs depend more upon the dorsal lateral funiculus than on the dorsal column, but this function is shared by fibers in both areas. Two-point discrimination, proprioceptive and tactile placing, and spatial discrimination in the monkey depend upon the dorsal lateral funiculus, perhaps in addition to the dorsal column. Tactile roughness discrimination and responses to type I slowly adapting receptors in cats appear to result primarily from transmission through the dorsal lateral funiculus. Tactile size discrimination in monkeys is more impaired after a combined lesion of the dorsal lateral and dorsal funiculi than after lesion of the dorsal funiculus alone. The dorsal lateral funiculus may participate in nociceptive transmission in cats, but it is also involved in antinociception, since nociceptive responses in cats and monkeys are enhanced followed lesions of the dorsal lateral funiculi. Pathways ascending in the ventral quadrant of the spinal cord are important in nociception in rats and monkeys, and these are also functional in dogs and cats. In addition, the middle part of the lateral funiculus is important in thermoreception in cats, and the ventral quadrants make a contribution to weight discrimination and position sense in monkeys.

ANIMAL STUDIES: ALTERATIONS IN NEURAL ACTIVITY

The effects of lesions that interrupt one or more of the sensory pathways of the spinal cord have also been studied by recordings of activity in the thalamus or in the cerebral cortex.

Evoked Potentials

Cortical evoked potentials have been used by a number of investigators to monitor the effects of lesions that interrupt the dorsal column, the dorsal lateral funiculus, or the ventral quadrant. Although some early reports suggested that sectioning the dorsal columns eliminates the potentials evoked in the sensorimotor cortex by peripheral nerve stimulation (Ruch *et al.*, 1952; Bohm, 1953), most later studies indicate that small changes at most are seen following transection of just the dorsal column.

The spinal pathways carrying information from cutaneous receptors, as revealed by studies in which evoked potentials were altered by selective lesions of the spinal cord, will be discussed first. Pathways responsible for information from receptors in other organs will then be considered.

In cats, the shortest-latency cortical potentials evoked by stimulation of cutaneous afferent fibers depend partly on the dorsal column and partly on the dorsal lateral funiculus (Gardner and Haddad, 1953; Morin, 1955; Catalano and Lamarche, 1957; Mark and Steiner, 1958; Norrsell and Voorhoeve, 1962; Andersson, 1962; Norrsell and Wolpow, 1966; Oscarsson and Rosen, 1966; cf. Landgren *et al.*, 1965). The same is true for dogs (Gambarian, 1960; Norrsell, 1966a). Interestingly, when the dorsal lateral funiculus is cut, the potentials evoked by stimulating hindlimb cutaneous nerves increase in latency, whereas no change in latency occurs following a lesion of the dorsal column (Fig. 6.8) (Mark and Steiner, 1958; Norrsell and Voorhoeve, 1962; Andersson, 1962; Norrsell and Wolpow, 1966). This indicates that the overall conduction velocity of the pathway in the dorsal lateral funiculus is higher than that of the dorsal column pathway,

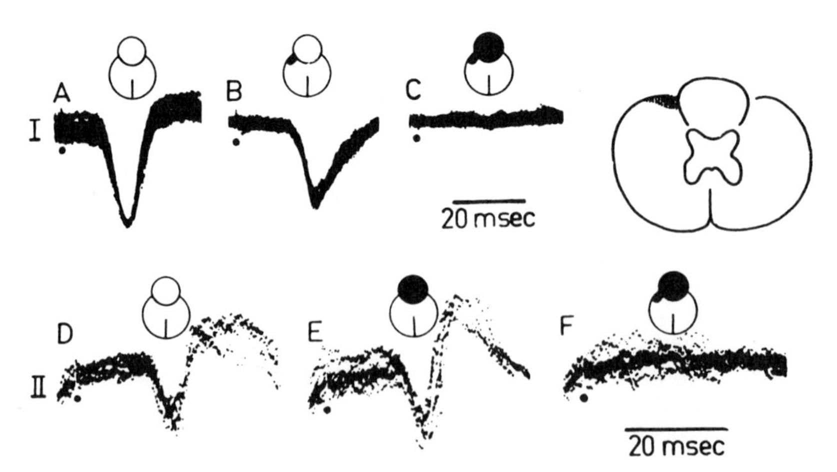

Fig. 6.8. Evoked potentials recorded in the right SI region following movement of hair on the left hindlimb. For the animal in panels A–C, the first lesion interrupted the dorsal lateral funiculus and the second the dorsal columns. In the other animal, panels D–F, the initial lesion was of the dorsal column and was followed by one of the dorsal lateral funiculus. The drawing at the upper right shows the extent of the smallest lesion needed to interrupt the tactile tract in the dorsal lateral funiculus. Note the latency difference for the evoked potentials in panels B and E. (From Norrsell and Voorhoeve, 1962.)

despite the presence of an extra synapse (see Chapter 8). However, the latencies of the potentials evoked via the two pathways are the same when forelimb nerves are stimulated (Andersson, 1962) or the dorsal column path is faster (Oscarsson and Rosen, 1966). The short-latency potentials evoked via the dorsal lateral funiculus appear to depend on the spinocervical-thalamic path, since the potentials disappear following a combined lesion of the dorsal column and of the cervicothalamic tract (Fig. 6.9) (Morin, 1955; Norrsell and Voorhoeve, 1962).

Evidence disputing the concept that cortical evoked potentials due to cutaneous nerve stimulation depend upon both the dorsal column and the dorsal lateral funiculus was presented by Whitehorn *et al.* (1969) and Ennever and Towe (1974). These investigators found that a complete transection of the dorsal column essentially eliminated the short-latency potentials evoked by cutaneous stimulation. Direct stimulation of the dorsal column produced an evoked potential like that which results from cutaneous stimulation, but direct stimulation of the dorsal lateral funiculus had a minimal effect, provided that stimulus spread to the dorsal column was minimized by inserting an insulating sheet between the dorsal column and the dorsal lateral funiculus. Previous studies were considered to have erred by not producing complete lesions of the dorsal column. However, Andersson and Leissner (1975) were able to evoke a cortical potential by stimulating the dorsal lateral funiculus, even though stimulus spread was minimized by an insulating sheet. They suggested that the failure of Towe's group to produce a cortical evoked potential via the dorsal lateral funiculus was due to damage of the dorsal lateral funiculus or of the lateral cervical nucleus.

There is evidence that cortical potentials evoked by particular receptor types may utilize one or the other of these pathways. For example, McIntyre (1962) found that the potential evoked in the secondary somatosensory cortex (SII) by stimulating large fibers of the interosseous nerve of the hindlimb disappears when the dorsal column is cut (see also Norrsell and Wolpow, 1966). He gave evidence that this potential was due to excitation of afferents from Pacinian corpuscles. This supposition was supported by a later study in which it was demonstrated that individually dissected Pacinian corpuscles could produce detectable cortical evoked potentials when mechanically stimulated (Fig. 6.10) (McIntyre *et al.*, 1967). These findings have been confirmed and extended to the forelimb (Silfvenius, 1970). High-threshold fibers of the interosseous nerves still produce evoked potentials after dorsal column lesions, but not after dorsal quadrant lesions (McIntyre, 1962; Norrsell and Wolpow, 1966; Silfvenius, 1970).

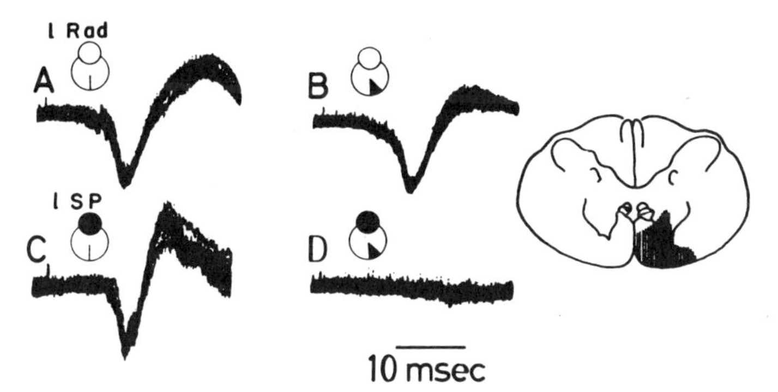

Fig. 6.9. Evoked potentials recorded from the right SI cortex in response to stimulation of the left radial nerve (A and B) and of the left superficial peroneal nerve (C and D). Lesions interrupted the dorsal columns at a midthoracic level and then the right ventral funiculus at C1 (shown in the drawing at the right). The forelimb evoked potential was changed only slightly by the lesion at C1. The hindlimb evoked potential was not greatly affected by the dorsal column lesion, but it was abolished by the addition of the lesion at C1, which interrupted the cervicothalamic pathway. (From Norrsell and Voorhoeve, 1962.)

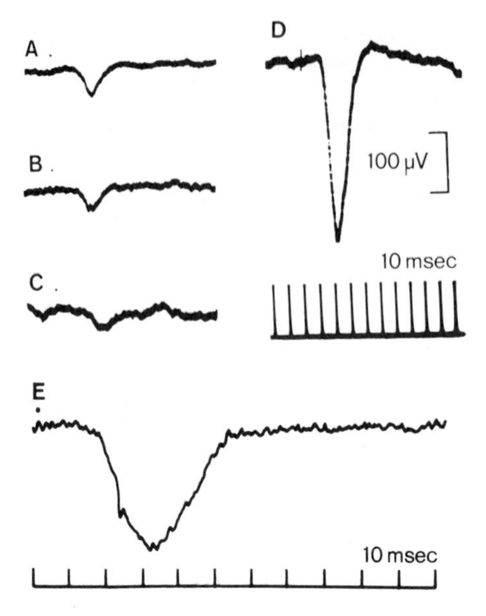

Fig. 6.10. (A–C) Potentials evoked in the contralateral SII region of the cerebral cortex by single impulses in isolated Pacinian corpuscles. (D) Evoked potential that resulted from stimulation of the interosseous nerve. (E) Signal-averaged record of the evoked potential produced by activating a Pacinian corpuscle 20 times. (From McIntyre *et al.*, 1967.)

Lesions interrupting a part of the dorsal column were found by Mann *et al.* (1972) not to alter the cortical evoked potential produced by activating a single type I slowly adapting receptor, whereas a discrete lesion in the dorsal lateral funiculus eliminated the evoked potential.

Although cutaneous short-latency cortical evoked potentials are abolished by lesions interrupting both the dorsal column and the dorsolateral funiculus in cats, it is still possible to observe small, late evoked potentials in cats as a result of conduction of information in more ventrally located pathways (Norrsell, 1966a; Norrsell and Wolpow, 1966; Oscarsson and Rosen, 1966).

The results of similar studies with monkeys reveal that short-latency cortical evoked potentials in primates depend not only on the dorsal column and dorsal lateral funiculus but also on the ventral quadrant (Gardner and Morin, 1953; 1957; Eidelberg and Woodbury, 1972; Andersson *et al.*, 1972). The pathways in the dorsal part of the cord are ipsilateral to the stimulus, whereas that in the ventral cord is contralateral (Fig. 6.11). By contrast with the cat, the major early evoked potentials transmitted outside of the dorsal column in monkeys are due to fibers in the contralateral ventral quadrant, presumably in the spinothalamic tract (see Chapter 9) (Eidelberg and Woodbury, 1972; Andersson *et al.*, 1972).

Potentials evoked in the cortex by stimulation of muscle nerves of the hindlimb were found to depend on the lateral funiculus rather than the dorsal column in the cat (Gardner and Noer, 1952; Gardner and Haddad, 1953). However, this appears to be true just of the lowest-threshold muscle afferent fibers, since high-threshold muscle afferent fibers can evoke cortical potentials by way of either the dorsal column or the dorsal lateral funiculus (Norrsell and Wolpow, 1966). The observation that high-threshold muscle afferent fibers can utilize the dorsal column pathway seems firm, since a cortical evoked potential was elicited even in an animal that had all of the cord sectioned except the dorsal column (Norrsell and Wolpow, 1966). Since no afferent fibers of this kind have been found to project directly up the dorsal column, it seems likely that postsynaptic dorsal column neurons are responsible for transmission of this input. However, the possibility of a direct projection has to be reconsidered in light of recent evidence of unmyelinated fibers that reach the medulla by way of the dorsal funiculus (see Chapter 7). The absence of cortical evoked potentials due to activity transmitted through the dorsal columns by low-threshold muscle afferent fibers is consistent with evidence cited later that proprioceptive

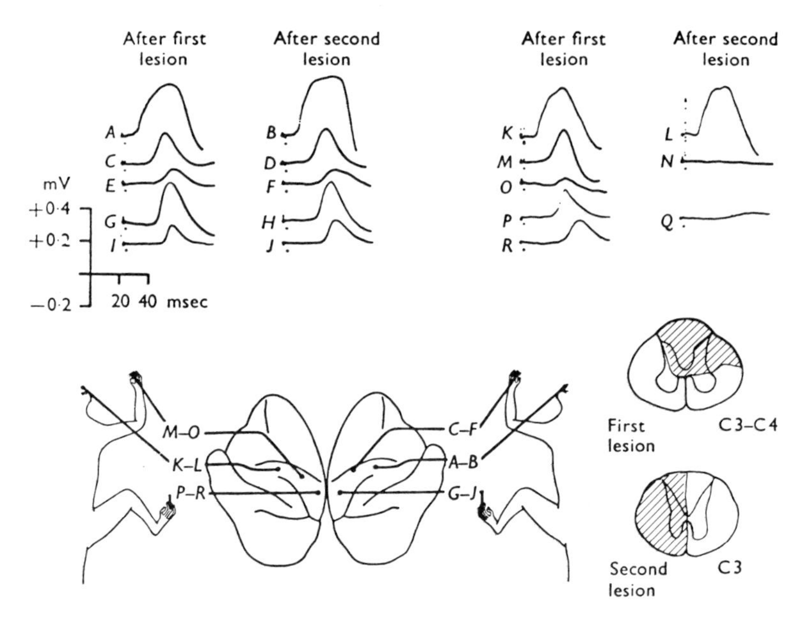

Fig. 6.11. Averaged evoked potentials in a monkey after the lesions shown at the bottom right. The records above at the left were recorded from the right cerebral hemisphere and those at the right from the left cerebral hemisphere. In each case, stimuli were applied contralaterally at the sites indicated in the drawing. There was no change in the responses to stimulation of the face (A, B, K, L) or of the left forelimb (C–F) or hindlimb (G–J). However, the potentials evoked by stimulation of the right forelimb and hindlimb (M–Q) were abolished. The pathways available to the left limbs were in the contralateral ventral quadrant. (From Andersson *et al.*, 1972).

information from the hindlimb in cats is transmitted via the spinomedullothalamic pathway, which ascends in the dorsal lateral funiculus and synapses in nucleus z (see Chapter 8).

By contrast to the findings concerning low-threshold muscle afferent fibers from the hindlimb, it has been shown that group I muscle afferent fibers of the cat forelimb produce evoked potentials in SI and SII via the dorsal column (Oscarsson and Rosen, 1963, 1966). The group I fibers include group Ia fibers from muscle spindles, but group Ib fibers from Golgi tendon organs were not excluded.

Large visceral afferent fibers in the splanchnic nerve of cats produce cortical evoked potentials that are often abolished by lesions of the dorsal column (Amassian, 1951; Gardner *et al.*, 1955; see also Aidar *et al.*, 1952). However, there seems to be a pathway for such activity in the lateral funiculus as well (Gardner *et al.*, 1955). No early evoked potentials can be recorded in monkey cortex following splanchnic nerve stimulation, presumably because of the small number of large myelinated fibers in this nerve in monkeys (Gardner *et al.*, 1955). Small myelinated splanchnic nerve afferent fibers evoke potentials by way of the ventrolateral quadrants as well as the dorsal column (Amassian, 1951).

In experiments utilizing a different approach, Curry and Gordon (1972) tried to determine which ascending pathways were responsible for evoking potentials in the somatosensory thalamus of cats. They stimulated the dorsal column, the dorsal lateral funiculus, or the ventral quadrant after making suitable lesions to prevent the spread of activity through other pathways. The contralateral dorsal column produced evoked potentials both in the ventral posterior lateral nucleus and in the medial part of the posterior complex. The contralateral dorsal lateral funiculus produced smaller evoked potentials in the same nuclei. The ipsilateral ventral quadrant produced small evoked potentials in the medial part of the posterior complex. It was concluded that the main input to the nucleus of the medial posterior complex is from the dorsal funiculus.

Unit Activity

There have been several studies concerning the effects of interruption of spinal cord sensory pathways upon the responses of neurons in the thalamus or cortex in cats and monkeys.

Following transection of the dorsal columns, it is still possible to find units in the thalamus or sensory cortex that can be activated by innocuous or strong stimulation of the skin (Gaze and Gordon, 1955; Andersson, 1962; Levitt and Levitt, 1968; Dobry and Casey, 1972b; Millar, 1973b; Dreyer *et al.*, 1974). Some units could be excited by subcutaneous receptors (Andersson, 1962; Millar, 1973b). However, the proportion of cells in the sensory cortex responding to weak mechanical stimuli decreases and the responses change in pattern (Levitt and Levitt, 1968; Dobry and Casey, 1972b; Dreyer *et al.*, 1974; but, cf. Eidelberg *et al.*, 1975).

When all of the cord is sectioned except the dorsal columns, neurons can be found in the thalamus and cortex that respond to hair movement, vibration, tap, or stimulation of subcutaneous receptors. The cells may have small contralateral receptive fields and demonstrate surround inhibition, or they may have large receptive fields and be activated by both weak and strong stimuli (Gaze and Gordon, 1955; Andersson, 1962; Levitt and Levitt, 1968; Curry, 1972). Altered receptive field locations are seen after chronic lesions of all of the cord but the dorsal columns (Frommer *et al.*, 1977).

Following a selective lesion of the dorsal lateral funiculus, units can be found in the thalamus that respond to hair movement, tap, or pressure (Curry, 1972), and neurons in much of the sensory cortex of the monkey behave normally (Dreyer *et al.*, 1974). When combined lesions are used to permit conduction only in one dorsal lateral funiculus, units can be found that have small or large contralateral receptive fields and that can be activated either just by weak mechanical stimuli, by weak and strong cutaneous stimuli, or by stimulation of subcutaneous receptors (Andersson, 1962). No surround inhibition is seen. There is an increase in the proportion of cortical units that respond to tap in such preparations (Levitt and Levitt, 1968).

Combined lesions of the dorsal columns and one or both dorsal lateral funiculi produce more drastic changes in unit activity (Andersson, 1962; Levitt and Levitt, 1968). Nevertheless, units can still be found in the thalamus and cortex that are activated by somatic stimuli, especially in monkeys (Gaze and Gordon, 1955; Whitlock and Perl, 1959, 1961; Perl and Whitlock, 1961; Curry, 1972; Dreyer *et al.*, 1974; Andersson *et al.*, 1975; Eidelberg *et al.*, 1975). The responses seen in monkeys are often just to strong stimuli, but neurons activated by weak stimuli, such as hair movement, touch, tap, or stimulation of subcutaneous receptors, can also be found (Fig. 6.12). Some of these units have restricted contralateral receptive fields (Perl and Whitlock, 1961; Andersson *et al.*, 1975). In addition, proprioceptive responses can be found that are mediated through a ventral pathway (Perl and Whitlock, 1961; Eidelberg *et al.*, 1975).

Lesions of the spinal cord have been made in monkeys during recordings from nociceptive neurons of the ventral posterior lateral nucleus (Kenshalo *et al.*, 1980; J. M. Chung *et al.*, 1986a). Usually, the nociceptive response was eliminated when the ventral quadrant of the spinal cord was sectioned ipsilateral to the thalamic unit and contralateral to the receptive field (Fig. 6.13). However, in some cases the nociceptive response was lost when a lesion was placed in the dorsal half of the cord (J. M. Chung *et al.*, 1986a).

Cold block of axons in either the ventrolateral or the dorsolateral funiculus in cats can prevent the responses of neurons in the ventrobasal complex to noxious stimuli (Martin *et al.*, 1990). This observation is consistent with evidence that nociception in cats depends on pathways located in both the dorsal and the ventral pat of the cord.

Summary

In summary, most investigators agree that the short-latency cortical evoked potentials produced in cats by stimulation of cutaneous afferent fibers can be transmitted by activity in both

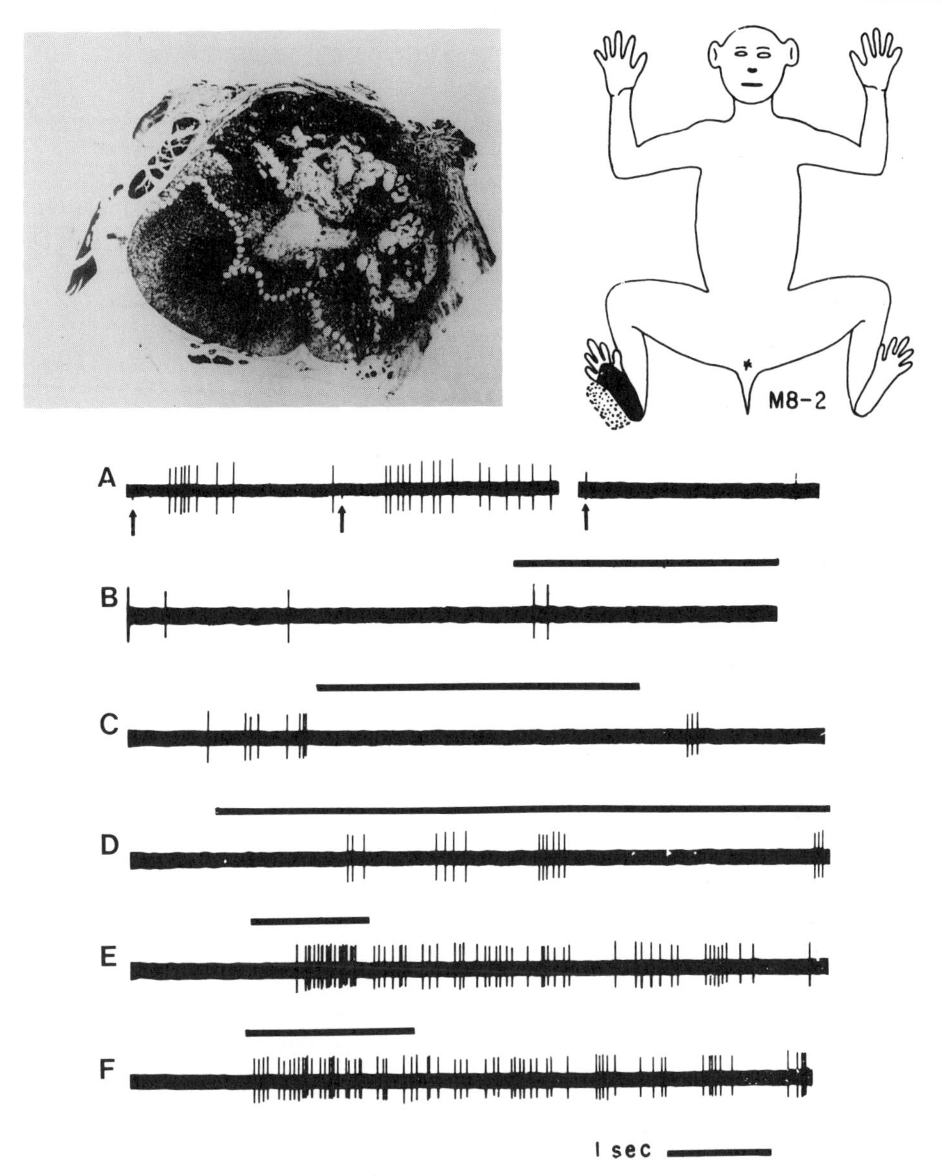

Fig. 6.12. At the top left is a photograph of a lesion of all of the spinal cord except one ventral quadrant at a high cervical level. The drawing at the top right shows the receptive field of a thalamic neuron in a monkey with such a lesion. The records in A–F show the responses of the neuron to electrical stimulation of the right and left plantar nerves (arrows, A), tactile stimulation of the right sole (B and C), and pinch of the right sole (E and F). The duration of mechanical stimuli is shown by bars above traces in B–F. (From Whitlock and Perl, 1961; Perl and Whitlock, 1961.)

the dorsal column and the dorsal lateral funiculus. A similar early evoked potential can also result from activation of a crossed ventral quadrant pathway in monkeys. Some kinds of afferent fibers project preferentially in a particular pathway. Pacinian corpuscles project in the dorsal column, and information from type I slowly adapting afferent fibers reaches the brain largely through the dorsal lateral funiculus. Group I muscle afferent fibers from the hindlimb relay to a pathway in

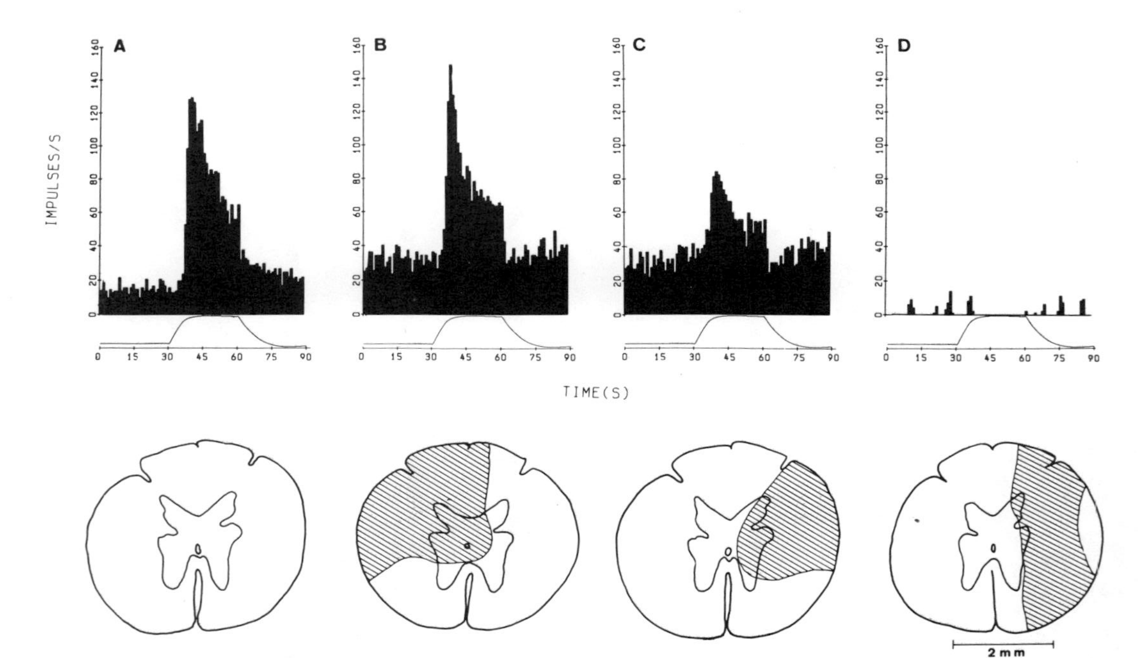

Fig. 6.13. Effect of lesions of the spinal cord on the responses of a nociceptive neuron in the ventral posterior lateral nucleus of the thalamus in a monkey. Panel A shows the excitation of the cell by noxious heat. The response was not affected by a lesion of the dorsal quadrant of the spinal cord contralateral to the neuron (ipsilateral to the excitatory receptive field), as shown in panel B. However, a lesion of the lateral funiculus, especially of the ventral quadrant, reduced or abolished the response, as shown in panels C and D. (From Kenshalo *et al.*, 1980.)

the dorsal lateral funiculus, while those from the forelimb utilize the dorsal column. Large visceral afferent fibers of the splanchnic nerve of cats project through the dorsal column. Transection of the dorsal column appears to produce some changes in the proportion of units in the sensory cortex that respond to somatic stimuli, but many normally behaving units remain. Responses with just the dorsal columns intact are generally like those seen normally, including the presence of surround inhibition. A lesion of the dorsal lateral funiculus does not produce an obvious change. Interruption of all of the cord except the dorsal lateral funiculus alters the responses of cortical neurons. One result is the absence of surround inhibition. Despite lesions that transect both the dorsal columns and the dorsal lateral funiculi, units can still be found in the thalamus and cortex of monkeys that have small, contralateral receptive fields and that are activated by weak mechanical stimuli. Proprioceptive units with input through ventral pathways can also be found. Nociceptive input to units in the ventral posterior lateral nucleus of monkeys is generally conveyed by fibers in the ventral quadrant ipsilateral to the thalamic unit, but sometimes a pathway located in the dorsal part of the spinal cord is responsible.

CONCLUSIONS

1. The traditional concept that the posterior columns in humans are responsible for discriminative touch, vibratory sensibility, and position sense is supported by evidence from patients who have had selective lesions of the posterior column. Sensory functions that are particularly affected by posterior column lesions include two-point discrimination, stereognosis, and graphesthesia. Other results of such lesions are poor ability to detect repeated stimuli, reduced capacity to recognize gradations of pressure stimuli, tactile and positional hallucinations, and increases in pain and temperature sensations.

2. Some disease states that affect the posterior columns are less helpful in distinguishing the functions of this pathway because other sensory paths are also affected. Conditions of this kind include the Brown-Séquard syndrome, tabes dorsalis, Friedreich's ataxia, and subacute combined degeneration. Cases in which there is a differential loss of position sense but not vibratory sense, or *vice versa*, indicate that the fibers responsible for mediating these sensory experiences travel separately.

3. Surgical lesions of the posterior columns in humans suffering from phantom-limb pain produce only minimal sensory deficits. However, the lesions may have been incomplete. Furthermore, it would have been difficult to demonstrate the sensory deficits usually seen with interruption of the posterior columns, since these affect primarily the distal limb, which was absent in these amputees.

4. Patients who have had an interruption of the ventral part of the spinal cord because of occlusion of the anterior spinal artery or large, bilateral cordotomies retain normal tactile thresholds, graphesthesia, vibratory sense, and position sense.

5. Stimulation of the posterior column in humans produces a buzzing sensation. This may be due to ascending volleys in dorsal column fibers, but a contribution may also be made by descending dorsal column volleys that excite other somatosensory pathways.

6. Little is known about possible sensory pathways in the posterior lateral funiculus of humans. Lesions in this area do not seem to impair touch, position sense, or vibratory sense. However, it is possible that some of these functions are shared with the posterior columns. A role in pain cannot be ruled out, since there are occasional successful cordotomies in which the lesion was in the posterior lateral funiculus. Stimulation in this region of the cord evokes pain, but this is referred ipsilaterally, whereas the analgesia produced by cordotomy is contralateral.

7. The anterolateral quadrant of the human spinal cord contains pathways that are crucial for pain and temperature sensations. The results of cordotomies indicate that these pathways

cross near the segmental level of their input and that the pathways have a somatotopic arrangement (lower segmental levels represented dorsolaterally and upper levels ventromedially). Itch and libidinous sensation are also conveyed by a similar route. Tactile sensation is slightly reduced by cordotomy, but tactile localization, stereognosis, graphesthesia, vibratory sense, and position sense are not altered.

8. Recovery of pain sensation after cordotomy sometimes seems to be due to the realization of the potential for pathways in other parts of the cord than the contralateral anterolateral quadrant to transmit pain information. In other cases, the pain is of a new type due to the development of a central neurogenic or "deafferentation" pain syndrome.

9. The case of a human patient with a transection of all of the spinal cord except one anterolateral quadrant shows that this part of the cord carries sufficient information to permit some degree of touch and pressure sensations bilaterally, normal pain and temperature sensations from the opposite side of the body, and position sense at some joints ipsilaterally.

10. Lesions of the central part of the spinal cord, as in syringomyelia, cause a bilateral loss of pain and temperature sensations over dermatomes corresponding to segments in which decussating axons were interrupted. Syringomyelia may also produce central neurogenic or "deafferentation" pain. Commissural myelotomies may have a similar effect, but in addition may damage the posterior columns and so may also produce tactile deficits. There are reports that commissural myelotomies of limited extent can produce widespread relief of clinical pain, although there is no explanation for this.

11. Stimulation of the human anterolateral quadrant results in pain sensation, provided that the stimuli are of sufficient intensity and frequency. The axons responsible for evoking this pain have refractory periods of 1 to 1.5 msec, indicating that they are large myelinated axons. Weaker stimuli may produce thermal sensations.

12. Interruption of the dorsal columns in animals has been reported to produce surprisingly little alteration in a number of behavioral tests of sensation, including tactile localization, tactile placing, tactile conditioned reflexes, two-point discrimination, vibratory sensibility, roughness discrimination, proprioceptive placing, weight discrimination, and limb position. In some studies, negative results may have reflected incomplete interruption of the dorsal column, rather than the lack of an important role of the dorsal column pathway in a particular function. In other cases, another pathway may have supported the function, either completely or in parallel with the dorsal column. An additional sensory pathway in cats is probably located in the dorsal lateral funiculus. In monkeys, it is more likely to be in the contralateral ventral quadrant.

13. Cutaneous stimulation below a lesion interrupting all of the cord except the dorsal columns results in transmission of sensory information but fails to evoke orienting responses or arousal.

14. Dorsal column lesions in cats or monkeys have been found to impair both sensory and motor functions when the lesions are sufficiently large and when appropriate tests are used. Sensory deficits have been found to include decreases in position sense, tactile placing, roughness discrimination, recognition of stimulus movement direction, and tactile discrimination requiring spatiotemporal analysis. Motor deficits include impairment of grasping movements, hopping reflexes, exploratory movements, coordinated walking and turning, "anticipatory" motor behavior, serial-order acts, and maintenance of a symmetric stance. There is typically also a loss of attention in monkeys. Motor deficits in cats are much less apparent if lesions are made in the main part of the dorsal column nuclei, rather than in the dorsal columns, although sensory deficits are present. The full motor deficits appear when the dorsal column nuclear lesion includes the lateral cuneate nuclei. This suggests that sensory functions depend upon the main dorsal column nuclei and motor functions upon the pathway through the lateral cuneate nucleus.

15. The dorsal column pathway may have some nociceptive function. A lesion of the dorsal

column in monkeys reduces reactivity to painful stimuli, and a lesion of the dorsal columns and dorsal parts of the lateral funiculi has a similar effect in cats.

16. The dorsal lateral funiculus contains pathways that transmit information similar to that conveyed in the dorsal column or in the ventral quadrant. In cats and to some extent in monkeys, tactile deficits are generally made much worse when combined lesions are made in the dorsal column and dorsal lateral funiculus than when a lesion is placed in either alone, indicating that the dorsal column and the dorsal lateral funiculus both contribute to tactile conditioned reflexes, two-point discrimination, tactile and proprioceptive placing, roughness discrimination, and size discrimination. The dorsal lateral funiculus may also be important in nociception, since a lesion here in cats may produce hypalgesia. On the other hand, a bilateral lesion of the dorsal lateral funiculi in cats or monkeys can increase reactivity to noxious stimuli, perhaps because of interruption of descending inhibitory pathways.

17. The ventral quadrant contains at least part of the nociceptive pathway in cats and dogs and is the main route by which nociceptive responses are mediated in rats and monkeys. It also contributes to weight discrimination and position sense in monkeys. The thermoreceptive pathway in cats is in the middle part of the lateral funiculus; it actually seems to be in the ventral part of the dorsal lateral funiculus, rather than in the ventral quadrant.

18. Short-latency cortical evoked potentials due to stimulation of cutaneous nerves depend upon both the dorsal column and the dorsal lateral funiculus in cats and on these plus the contralateral ventral quadrant in monkeys. Potentials evoked in cats by Pacinian corpuscles depend on just the dorsal column, whereas those produced by type I slowly adapting receptors depend only on the dorsal lateral funiculus.

19. Short-latency cortical evoked potentials due to stimulation of group I muscle afferent fibers in cats depend upon the dorsal lateral funiculus for hindlimb nerves and upon the dorsal columns for forelimb nerves. High-threshold muscle afferents of the hindlimb can, however, evoke a cortical potential via the dorsal column or the dorsal lateral funiculus.

20. Short-latency cortical evoked potentials due to stimulation of large visceral afferent fibers of the greater splanchnic nerve in cats depend upon the dorsal and lateral columns. Those due to small visceral afferent fibers depend upon the ventral quadrant, as well as upon the dorsal columns.

21. Short-latency thalamic evoked potentials due to stimulation of the dorsal columns or the dorsal lateral funiculus in cats can be recorded in the ventral posterior lateral nucleus and the medial part of the posterior complex, whereas stimulation of the ventral quadrant evokes only a small potential in the medial part of the posterior complex.

22. The activity of single units with small, contralateral receptive fields to tactile stimulation can be recorded in the thalamus or sensory cortex after lesions of the dorsal columns, the dorsal lateral funiculus, or both. However, surround inhibition is absent when the dorsal columns are cut. Interruption of the dorsal half of the cord drastically reduces the number of responsive units in cats. However, activity mediated by way of the ventral part of the spinal cord is still present in monkeys. The responsive neurons in monkeys include many that can be activated by weak mechanical stimulation of the skin or by proprioceptors.

23. Nociceptive neurons in the ventral posterior lateral nucleus in monkeys usually lose their responses to noxious stimuli if pathways in the ipsilateral ventral quadrant of the spinal cord are interrupted; however, some nociceptive responses are blocked by lesions of the dorsal part of the contralateral cord. Nociceptive responses of ventrobasal thalamic neurons in cats can be prevented by cold block of the ipsilateral dorsolateral or ventrolateral funiculus.

7 Sensory Pathways in the Dorsal Funiculus

DORSAL COLUMN–MEDIAL LEMNISCUS PATHWAY

The best-studied sensory tract originating in the spinal cord is the dorsal column–medial lemniscus pathway (Fig. 7.1). The initial part of this pathway consists of branches of primary afferent fibers that ascend to the medulla oblongata in the dorsal funiculus. The dorsal funiculus is subdivided into two components known as the fasciculus gracilis and the fasciculus cuneatus (Latin for thin and wedge-shaped bundles, respectively). The fasciculus gracilis contains the ascending branches of primary afferent fibers from levels caudal to the midthoracic region, whereas the fasciculus cuneatus contains the branches of afferent fibers from midthoracic to upper cervical levels.

The fasciculi gracilis and cuneatus terminate in nuclei of the same names in the caudal medulla, the nucleus gracilis and the nucleus cuneatus. Collectively, these nuclei (along with the lateral cuneate nucleus) are called the dorsal column nuclei. The nuclei gracilis and cuneatus project, among other places, to the contralateral thalamus by way of the medial lemniscus. For this reason, this sensory pathway may be referred to as the dorsal column–medial lemniscus pathway. The dorsal column nuclei also receive synaptic input from neurons whose cell bodies are located in the spinal cord gray matter (postsynaptic dorsal column pathway).

Although the dorsal column nuclei are sometimes called "relay" nuclei, this term should not be taken to imply a simple organization or function (Gordon, 1973).

TAXONOMIC DISTRIBUTION

A dorsal column–medial lemniscus pathway has been found in amphibia (Joseph and Whitlock, 1968a; Ebbesson, 1969; Hayle, 1973; Corvaja *et al.*, 1978; Silvey *et al.*, 1974; Neary and Wilcznski, 1977; Antal *et al.*, 1980), reptiles (Kruger and Witkovsky, 1961; Goldby and Robinson, 1962; Ebbesson, 1967, 1969; Joseph and Whitlock, 1968b; V. L. Jacobs and Sis, 1980; Kusuma and ten Donkelaar, 1980; Ebbesson and Goodman, 1981; Kuenzle and Woodson, 1983; Pritz, 1983; Pritz and Stritzel, 1986), and birds (Sinn, 1913; Craigie, 1928; Wild, 1985, 1989), but it is most prominent in mammals. Its absence in fish (Zeehandelaar, 1921; Hayle, 1973; however, cf. Finger, 1978; Ebbesson and Hodde, 1981) suggests that it may have evolved initially with the emergence of vertebrates from an aquatic environment, where much of the sensory stimulation of the body surface is processed by the lateral line system. Its prominence in mammals suggests further evolution along with the sensory apparatus of the hairy skin. Representatives of all orders of mammals that have been studied have a substantial dorsal column–medial lemniscus pathway;

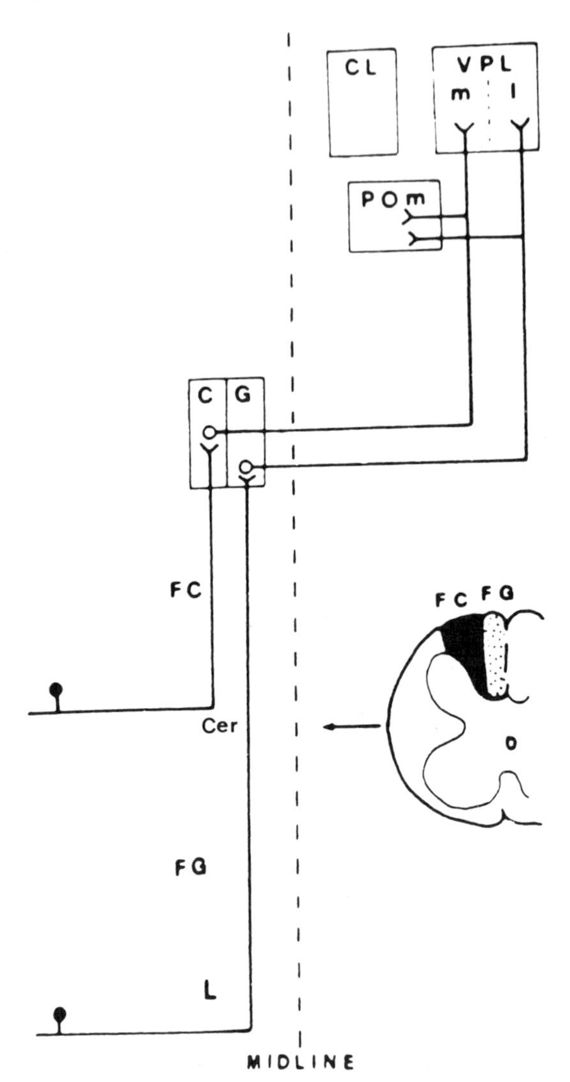

Fig. 7.1. The dorsal column pathway. Axons of primary afferent neurons are shown entering the cord through cervical (Cer) and lumbar (L) dorsal roots and turning rostrally to ascend in the fasciculus cuneatus (FC) or fasciculus gracilis (FG) to end in the nucleus cuneatus (C) or nucleus gracilis (G). The second-order neurons in the dorsal column nuclei then project contralaterally through the medial lemniscus to end in the medial part of the posterior complex (PO_m) and in the ventral posterior lateral (VPL) nucleus. The nucleus cuneatus projects to the medial (m) and the nucleus gracilis to the lateral portions (l) of the VPL nucleus. The drawing of a cross-section of the spinal cord indicates the locations of the FC and FG at a cervical level.

these include the opossum (Clezy *et al.*, 1961; Rockel *et al.*, 1972; Hamilton and Johnson, 1973; Walsh and Ebner, 1973; T. S. Gray *et al.*, 1981; Culberson, 1987), hedgehog (Jane and Schroeder, 1971), raccoon (J. I. Johnson *et al.*, 1968; Rasmusson, 1988, 1989), cat (Imai and Kusama, 1969), sheep (Woudenberg, 1970), aquatic mammals and edentates (Kappers *et al.*, 1936), squirrel (Albright *et al.*, 1983; Ostapoff *et al.*, 1983), rat (Valverde, 1966; Lund and Webster, 1967a), tree shrew (Schroeder and Jane, 1971), lesser bushbaby (Albright and Haines, 1978), monkey (W. E. L. Clark, 1936; Walker and Weaver, 1942; Chang and Ruch, 1947a; Bowsher, 1958), chimpanzee (Kappers *et al.*, 1936), and human (Ferraro and Barrera, 1935a,b).

The dorsal column nuclei include the nucleus gracilis and nucleus cuneatus in many species. However, in forms that have small hindlimbs, such as cetaceans, the nucleus gracilis is poorly developed (Kappers *et al.*, 1936). Many species with a prominent tail have an additional midline dorsal column nucleus called the nucleus of Bischoff. Examples of animals with a nucleus of Bischoff include the alligator, some birds, kangaroo, rat, shrew, great anteater, and several species of monkeys (Kappers *et al.*, 1936). However, other animals with prominent tails, including *Ateles* and *Cebus* monkeys, lack a nucleus of Bischoff.

DEVELOPMENT

The ontogenesis of the dorsal column pathway in human embryos has been examined by A. Hughes (1976). The fasciculus cuneatus appears prior to the fasciculus gracilis. The fasciculus cuneatus is first recognizable at 8 weeks, which is the time when the dorsal root ganglia of the cervical and thoracic segments differentiate. The tract can be seen to terminate on a cell mass at the level of the obex by 9 weeks. This cell mass later forms the dorsal column nuclei.

The nucleus gracilis does not develop in the opossum when the corresponding hindlimb is removed neonatally (J. I. Johnson *et al.*, 1972).

DORSAL FUNICULUS

Composition

The dorsal funiculus contains branches of primary afferent fibers that ascend from the dorsal root entry level all the way to the medulla. As afferent fibers approach the spinal cord through a dorsal root, they course medially into the dorsal funiculus, where they bifurcate into branches that turn rostrally and caudally (Fig. 7.2; Ramón y Cajal, 1909; Sprague and Ha, 1964; Sterling and Kuypers, 1967; Carpenter *et al.*, 1968; Shriver *et al.*, 1968; Imai and Kusama, 1969; Réthelyi and Szentágothai, 1973).

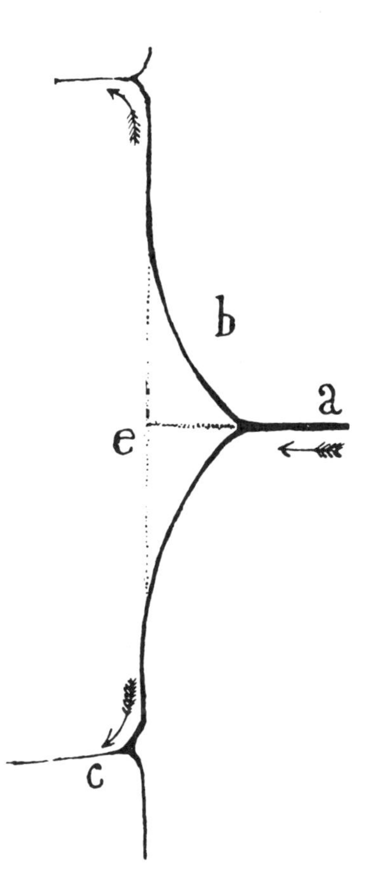

Fig. 7.2. Trajectory of a dorsal root afferent fiber as it enters the spinal cord. The view is from the dorsal aspect showing the fiber as if it were in a horizontal section. The central process of a DRG cell is shown entering the cord (a). As it reaches the dorsal funiculus, it bifurcates. There is an ascending branch (b) and a descending branch (c). Collaterals are given off by both branches to the segmental gray matter. (From Ramón y Cajal, 1909.)

The fasciculus gracilis of monkeys contains the ascending branches of primary afferent fibers of dorsal roots T8 and caudally (Walker and Weaver, 1942; Carpenter *et al.*, 1968), whereas the fasciculus cuneatus contains afferent fibers from dorsal roots T7 and rostrally (Shriver *et al.*, 1968). A few fibers from segments as low as L1 in monkeys (Carpenter *et al.*, 1968) and L5 or even lower in cats (Rustioni and Macchi, 1968) project to the nucleus cuneatus. Where there is a nucleus of Bischoff, it may receive input from afferent fibers supplying the tail (Kappers *et al.*, 1936). In rats, the pudendal nerve projects bilaterally to the nucleus gracilis (Ueyama *et al.*, 1985).

Only a fraction (22% in cats, according to Glees and Soler [1951]) of the primary afferent fibers entering the fasciculus gracilis actually reach the medulla. The others end in the gray matter of the spinal cord. At segmental levels, dorsal root fibers terminate in various parts of the gray matter, including the dorsal horn, intermediate region, and ventral horn (Fig. 7.3) (Sprague and Ha, 1964; Sterling and Kuypers, 1967; Shriver *et al.*, 1968; Carpenter *et al.*, 1968; Réthelyi and Szentágothai, 1973). Some ascending fibers of the fasciculus gracilis end in Clarke's column (C. N. Liu, 1956; Carpenter *et al.*, 1968; Réthelyi and Szentágothai, 1973). Comparable fibers in the fasciculus cuneatus end in the lateral cuneate nucleus (Ranson *et al.*, 1932; C. N. Liu, 1956; Shriver *et al.*, 1968). The dorsal funiculus also includes the descending branches of primary afferent fibers (Fig. 7.2) (Sterling and Kuypers, 1967; Shriver *et al.*, 1968; Carpenter *et al.*, 1968; Imai and Kusama, 1969; Réthelyi and Szentágothai, 1973).

In addition to primary afferent fibers, the dorsal funiculus contains axons that belong to propriospinal neurons (see the review by Nathan and Smith, 1959), the axons that form the postsynaptic dorsal column pathway (Uddenberg, 1968b; Rustioni, 1973, 1974; Angaut-Petit, 1975a,b), and several pathways that descend from the brain (Erulkar *et al.*, 1966; Kerr, 1968;

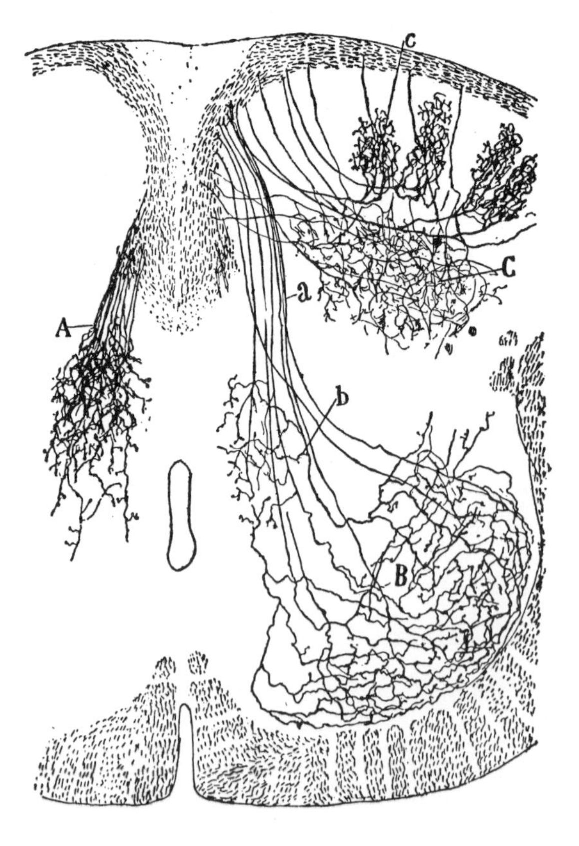

Fig. 7.3. Course of primary afferent fiber collaterals from the dorsal columns into the spinal cord gray matter. The primary afferent fibers were demonstrated by the Golgi technique in the newborn rat. A, projections into the intermediate region; B, projections to the motor nucleus; C, projections to the dorsal horn. a represents a bundle of reflex afferents coursing through the medial dorsal horn; b are collaterals of these in the intermediate region; c are collaterals that ascend toward the substantia gelatinosa. (From Ramón y Cajal, 1909.)

Hancock, 1982), including, in some animals such as rats, the corticospinal tract (Ranson, 1913a; Valverde, 1966). Fibers have also been found descending in the dorsal and dorsal lateral funiculi to the spinal cord from cells in the dorsal column nuclei (Dart, 1971; Burton and Loewy, 1977; Bromberg *et al.*, 1981; Enevoldson and Gordon, 1984).

Systems of Primary Afferent Fibers

The distance over which a given ascending primary afferent fiber travels in the dorsal funiculus depends upon its peripheral receptor type (Fig. 7.4). Three systems of primary afferent

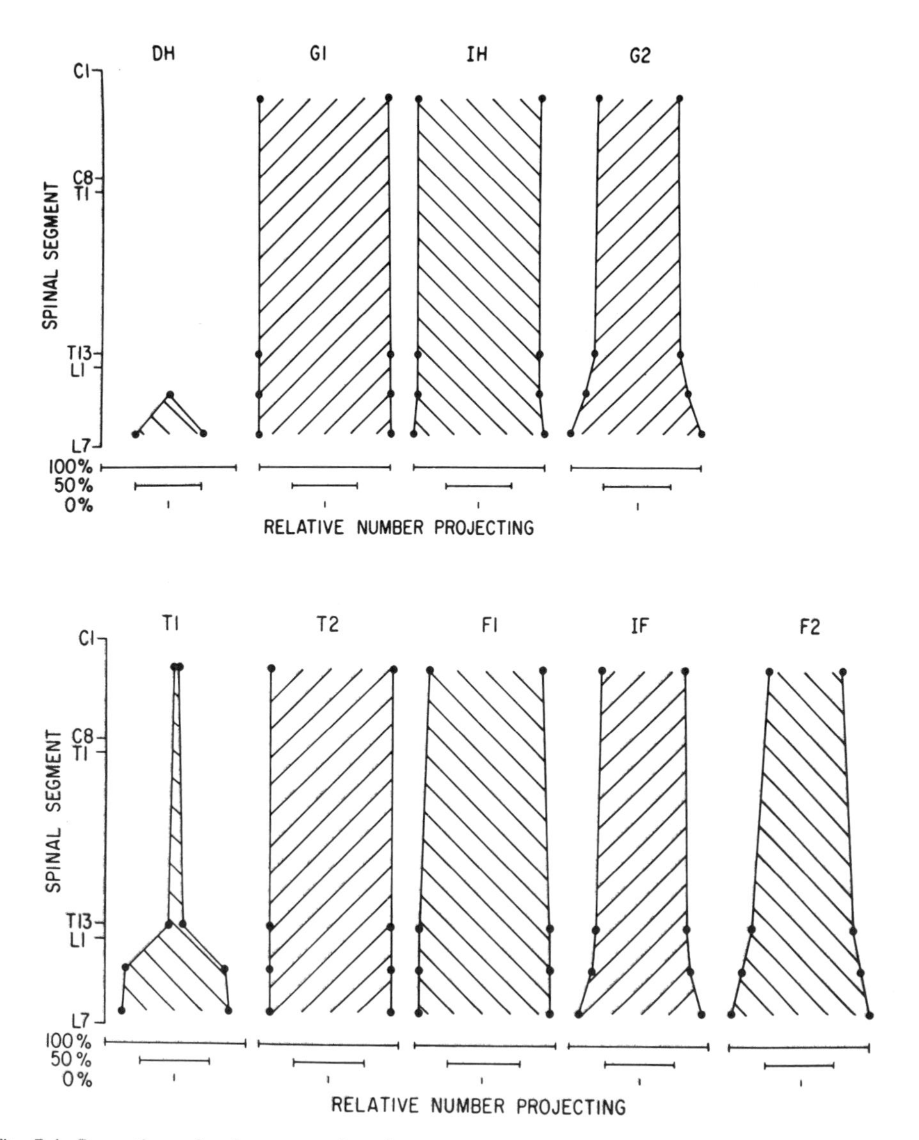

Fig. 7.4. Proportions of various types of mechanoreceptors that project to different levels of the spinal cor through the dorsal column. Abbreviations: DH, D hair follicle receptors; G_1, IH, and G_2, G_1, intermediate, and G_2 hair follicle receptors; T1 and T2, SA I and SA II slowly adapting receptors; F1, IF, and F2 = F1, intermediate and F2 field receptors. (From Horch *et al.*, 1976.)

fibers that ascend from the hindlimb have been described: a short system, an intermediate system, and a long system (Horch *et al.*, 1976).

The short system terminates within one or two segments. Although most fine afferent fibers enter the dorsolateral fasciculus of Lissauer or the dorsal part of the lateral funiculus (see Chapter 4) (Coggeshall *et al.*, 1981; K. Chung and Coggeshall, 1983; K. Chung *et al.*, 1985), some join the dorsal funiculus (Horch *et al.*, 1976; Langford and Coggeshall, 1981b; K. Chung *et al.*, 1985, 1987; K. Chung and Coggeshall, 1985; Briner *et al.*, 1988). Thus, the short system includes small myelinated and unmyelinated fibers, but only a few large myelinated fibers (Burgess and Horch, 1978).

The intermediate system of primary afferent fibers in the dorsal funiculus projects rostrally 4 to 12 segments (Horch *et al.*, 1976). This system includes large myelinated fibers that innervate type I slowly adapting receptors, several types of rapidly adapting receptors (G_2 hair follicle receptors, intermediate and F2 field receptors), and rapidly conducting nociceptors. The intermediate system also includes muscle stretch receptors and slowly adapting joint receptors that project to Clarke's column (Lloyd and McIntyre, 1950; F. J. Clark, 1972; F. J. Clark *et al.*, 1973; Kuno *et al.*, 1973). Few small myelinated fibers belong to the intermediate system (Burgess and Horch, 1978). Recently, it has been shown that a given unmyelinated visceral afferent fiber can project to more than five segments (Sugiura *et al.*, 1989), and so these fibers would belong to the intermediate system.

The long system of primary afferent fibers of the dorsal funiculus projects all the way to the medulla; these fibers synapse in the dorsal column nuclei (Horch *et al.*, 1976). Most of the axons of the long system are large myelinated fibers. Their receptor types include type II slowly adapting mechanoreceptors and several kinds of rapidly adapting receptors (G_1, intermediate, and G_2 hair follicle receptors; F1, intermediate, and F2 field receptors). Only a few small myelinated fibers belong to the long system (Burgess and Horch, 1978). Recently, it has been shown that the long system also includes unmyelinated primary afferent fibers (D. L. McNeill *et al.*, 1988b; Tamatani *et al.*, 1989; Patterson *et al.*, 1989, 1990). The unmyelinated axons and their synapses in the dorsal column nuclei appear to contain peptides, including substance P and calcitonin gene-related peptide (see below). The receptor types that are supplied by these fibers have not been determined, but it is possible that some are from nociceptors.

It is important to note that the conduction velocities of axons in the intermediate and long projection systems of the dorsal funiculus slow near the point of entry of the axons into the spinal cord from the dorsal roots, suggesting that the fibers become narrower as collaterals are given off at this level (Horch *et al.*, 1976). Furthermore, the long-system axons slow again near the level of the cervical enlargement, suggesting that collaterals are given off here as well. Thus, the intermediate system of afferent fibers from the hindlimb can presumably affect the activity of neurons at the level of the spinal cord corresponding to the dorsal root entry level of the afferents, as well as the activity of neurons in more rostral segments, such as those in Clarke's column. The long system from the hindlimb can affect segmental neurons and neurons in the cervical enlargement, as well as neurons in the dorsal column nuclei (Horch *et al.*, 1976).

The slowed conduction velocity of the long-fiber system within the cord is consistent with the observation that most of the axons in the fasciculus gracilis at the third cervical level are of Aδ size (Hildebrand and Skoglund, 1971; Hwang *et al.*, 1975), although peripherally most of the same axons would be of Aβ size.

Sensory Representation

The fasciculus gracilis contains a sensory representation of the lower part of the trunk and the lower extremity, whereas the fasciculus cuneatus contains a representation of the upper part of the trunk and the upper extremity. A finer-grained topographic organization can be demonstrated within each of these tracts.

The organization of the fibers within the caudal parts of the dorsal column tracts is dermatomal (see Werner and Whitsel, 1967), whereas that in the rostralmost parts (at least of the fasciculus gracilis, but probably also of the fasciculus cuneatus) is somatotopic. The distinction between a dermatomal and a somatotopic organization is illustrated in Fig. 7.5 (see B. H. Pubols *et al.*, 1965). As shown, the primary afferent fibers from a given area of skin, such as from one of the digits of a forepaw, may enter the spinal cord over any of several dorsal roots, and a given dorsal root may innervate a considerable expanse of skin, including that over several digits. For example, Fig. 7.5 shows afferent fibers from digit 2 in the raccoon entering the spinal cord over C7 and C8; conversely, C7 supplies parts of digits 2, 3, and 4. All of the skin innervated by a given dorsal root belongs to the dermatome of that segment. Thus, recordings from the C7 dorsal root in the raccoon will sample the activity of axons with receptors on digits 2, 3, or 4. Such a sample reveals the dermatomal organization of the dorsal root. A similar organization will be revealed if the activity in the axons of the caudal part of the fasciculus gracilis is sampled. Neighboring axons may innervate digit 2, 3, or 4. However, as the axons ascend toward the medulla, their relative positions within the fasciculus gracilis shift, so that in the rostral part of the tract the axons sampled with a recording electrode will all have receptive fields on adjacent parts of the skin. For instance, all axons within a short distance of each other may supply digit 2. If the electrode is moved slightly, the representation of a new set of axons may shift over the body

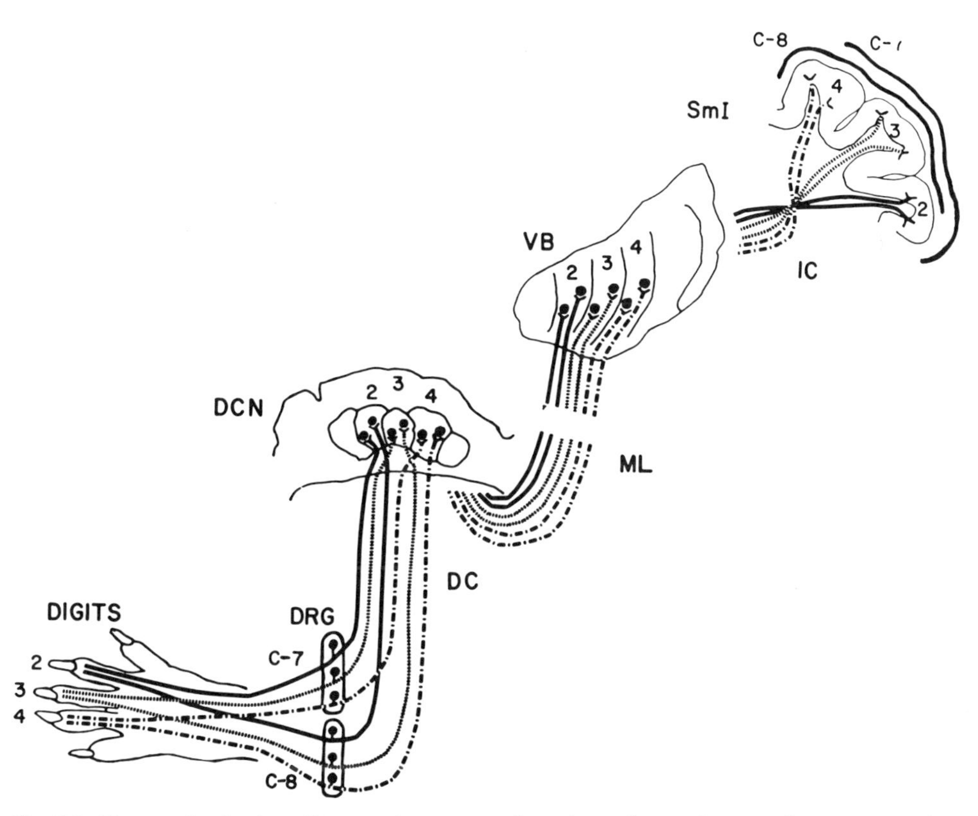

Fig. 7.5. Diagram showing how fiber resorting accounts for a change from a dermatomal to a somatotopic organization in the dorsal column pathway. Abbreviations: DRG, dorsal root ganglion; DC, dorsal column; DCN, dorsal column nuclei; ML, medial lemniscus; VB, ventrobasal complex of thalamus; IC, internal capsule; SmI, sensorimotor area I. (From B. H. Pubols *et al.*, 1965.)

surface to an adjacent area, for example to digit 3. However, no axons supplying digit 4 will be found between axons innervating digits 2 and 3. Thus, the organization of the rostral fasciculus gracilis is somatotopic, rather than dermatomal.

Fiber resorting in the dorsal column pathway in the squirrel monkey was investigated by Whitsel *et al.* (1970). At the lumbar level, the fasciculus gracilis has a dermatomal organization, whereas in the cervical cord it has a somatotopic one (Fig. 7.6). A similar study of the fasciculus cuneatus of the raccoon (L. M. Pubols and Pubols, 1973) showed an incomplete resorting of fibers between the lower and upper cervical levels. Presumably, the final resorting in the fasciculus cuneatus is done at the level of the nucleus cuneatus.

Because of this fiber resorting, it becomes understandable that the pattern of degeneration within the dorsal funiculus following dorsal rhizotomy is complex. The pattern of degeneration has been described for both the fasciculus gracilis and the fasciculus cuneatus (Walker and Weaver, 1942; Carpenter *et al.*, 1968; Shriver *et al.*, 1968; Whitsel *et al.*, 1970). The degeneration is seen caudally as a sequence of bands of fibers for successive dorsal roots, but with some overlap of the fibers from adjacent roots. However, at more rostral levels, the amount of overlap increases (Fig. 7.6) (Walker and Weaver, 1942; Whitsel *et al.*, 1970). The degenerating fibers from a given root may actually split into two different bands as the axons ascend in the fasciculus cuneatus (Albright *et al.*, 1983; Culberson and Albright, 1984).

Functional Types of Primary Afferent Fibers

The functional categories of primary afferent axons ascending in the dorsal funiculus have been examined by single-unit recordings in experiments involving natural stimulation of receptors.

S. Yamamoto and Sugihara (1956) investigated the response characteristics of axons in both

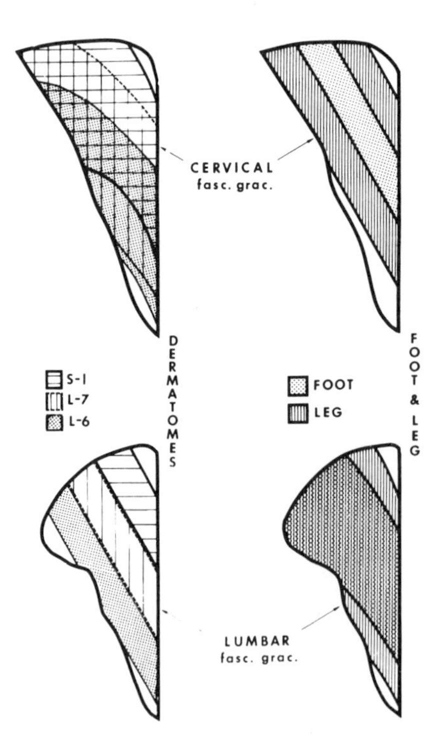

Fig. 7.6. Fiber sorting in the fasciculus gracilis of the monkey. The distribution of primary afferent fibers supplying the L6, L7, and S1 dermatomes as demonstrated by degeneration following dorsal rhizotomy is shown on an outline of the fasciculus gracilis at the lumbar level (below) and the cervical level (above) at the left. Note that there is little overlap at the lumbar level but considerable overlap at the cervical level. By contrast, the representation of the leg and foot, as determined by microelectrode recordings (right), shows considerable overlap at the lumbar level, but good resolution at the cervical level. (From Whitsel *et al.*, 1970.)

components of the dorsal funiculus at an upper cervical level in cats. Rapidly adapting tactile responses are the most common responses found. These are produced by hair movement or by contact with a pad. Other units show slowly adapting responses to indentation, elevation, or pinching of the skin. Units are also observed that discharge when joints are moved or when pressure is applied to joint capsules, tendons, or muscle. Some units supplying the chest wall discharge with respiration. (Similar units were found in the human dorsal funiculus by Puletti and Blomqvist [1967].) Finally, slowly adapting discharges are seen in response to bladder distention. There is a topographic organization: units from the bladder are located along the paramedian, superficial part of the fasciculus gracilis; the hindlimb and lower body are represented in the main part of the fasciculus gracilis; and the forelimb and upper body are supplied by axons in the fasciculus cuneatus. The observation that there are visceral afferent fibers in the dorsal funiculus is in agreement with the findings of others who used electrical stimulation (Amassian, 1951; Aidar *et al.*, 1952). The fasciculus gracilis contains more tactile than proprioceptive fibers, whereas the opposite is true of the fasciculus cuneatus.

Fasciculus Gracilis. The fiber composition of the fasciculus gracilis of the squirrel monkey was studied by Werner and Whitsel (1967) and by Whitsel *et al.* (1969, 1970). Werner and Whitsel (1967) found that at the lumbar level, axons of the fasciculus gracilis carrying information from cutaneous receptors are intermingled with those carrying information from deep receptors. The proportions of axons from cutaneous and deep receptors vary according to the locations of the receptors in the hindlimb (Werner and Whitsel, 1967; Whitsel *et al.*, 1969). There are more fibers representing skin than deep tissues of the distal extremity, whereas there are more deep than cutaneous receptors from the proximal extremity. This presumably reflects the density of innervation of the skin and deep structures of the distal *vs.* the proximal limb. The organization of the fasciculus gracilis at an upper cervical level is rearranged compared with that at the lumbar level (see the previous description of dermatomal *vs.* somatotopic organization), and the proportion of afferents from the skin is much greater than that from deep structures (Whitsel *et al.*, 1969, 1970). The Whitsel group failed to find any axons that respond to joint bending or that are slowly adapting in the cervical fasciculus gracilis of the squirrel monkey (Whitsel *et al.*, 1969). Presumably, the decrease in the number of axons from deep structures reflects the termination of many intermediate system axons, including muscle stretch receptors, in Clarke's column.

Several studies provide detailed information about the types of receptors of the hindlimb whose axons project through the fasciculus gracilis as far as the upper cervical spinal cord in the cat. A. G. Brown (1968) restricted his survey to axons connected with cutaneous receptors, although he often found units that could be activated by joint movement. The types of axons that he found are compared with the composition of the spinocervical tract and the postsynaptic dorsal column pathway in Fig. 7.7. The most common units supply hair follicle receptors. Some respond like Pacinian corpuscle afferents. A number of other kinds of rapidly adapting receptors are also represented. In addition, axons that behave like type I and type II slowly adapting receptors are present. The most striking negative finding is the absence of afferent fibers with peripheral conduction velocities in the Aδ range: down hair follicle afferents and nociceptors. However, this can be explained on the basis of either a lack of such fibers in the dorsal funiculus or of the difficulty in recording the activity of such fibers.

Petit and Burgess (1968) recorded from primary afferent fibers in peripheral nerves that could be activated antidromically by stimulation of the dorsal funiculus at a high cervical level. They obtained results similar to those of A. G. Brown (1968). The most common receptor types in the fasciculus gracilis are G_1 and G_2 hair follicle afferents. There are also numerous field receptor afferent fibers (cf. Burgess *et al.*, 1968). In addition, they encountered type II slowly adapting units and Pacinian corpuscle-like units. Horch *et al.* (1976) later suggested that the latter are misidentified hair units. The dorsal column sample did not include any nociceptors, but Petit

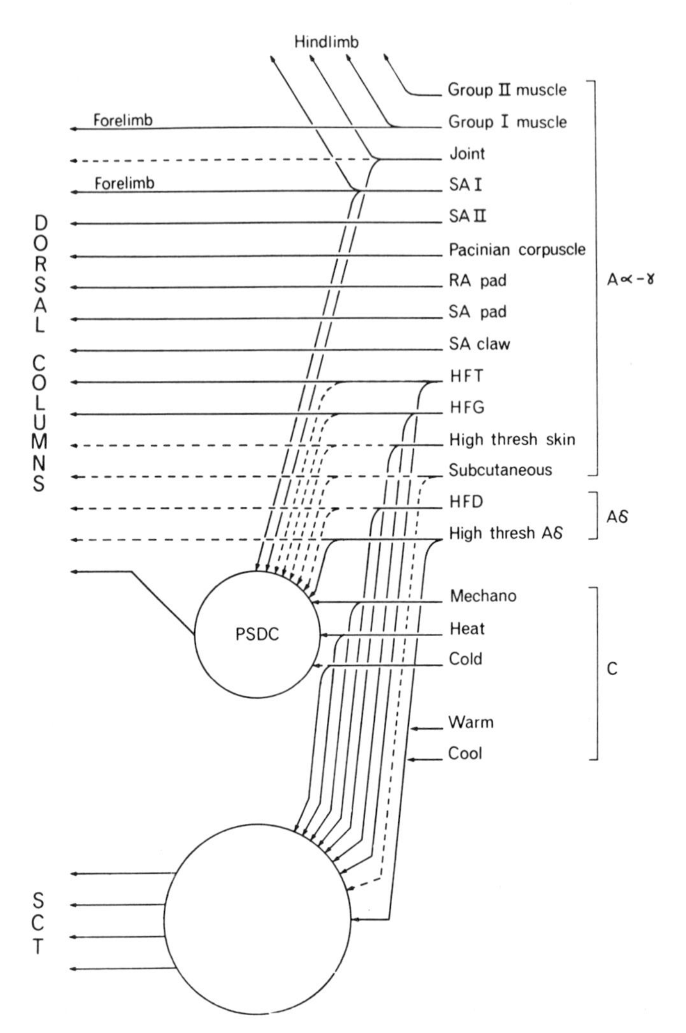

Fig. 7.7. Receptor types represented in the dorsal column and spinocervical tract. The postsynaptic dorsal column (PSDC) pathway is shown, as well as the primary afferent projection. HFT, HFG, and HFD are abbreviations for tylotrich, guard, and down hair follicle afferent fibers, respectively. (From A. G. Brown, 1973.)

and Burgess (1968) did find a few D hair follicle units in the plantar nerve that projected to the upper cervical cord. Unlike Brown, they did not record the activity of any type I slowly adapting units. Since touch domes can activate postsynaptic dorsal column neurons (see below), it has been suggested that Brown recorded type I-like activity from second-order axons rather than from primary afferent fibers (Angaut-Petit, 1975b).

Fasciculus Cuneatus. Uddenberg (1968a) did not subdivide the receptor types represented in the fasciculus cuneatus as finely as did A. G. Brown (1968) and Petit and Burgess (1968) for the fasciculus gracilis. He found hair follicle units, touch units, and Pacinian corpuscle-like units. He also observed large numbers of muscle afferent fibers, most of which could be classified as group I on the basis of their low thresholds to electrical stimulation. The hair follicle units are located superficially in the dorsal funiculus, whereas the muscle units are intermediate in depth and touch units are deep (and therefore hard to excite by a surface electrode).

The fasciculus cuneatus at an upper cervical level contains large numbers of axons from deep structures in the forelimb, as well as from the skin (Whitsel *et al.*, 1969). These are presumably destined both for the main cuneate and for the lateral cuneate nucleus. (See Chapter 8 for a discussion of the pathway through nucleus z by which proprioceptive information from the hindlimb reaches the thalamus.)

L. M. Pubols and Pubols (1973) investigated the afferent fiber types in the fasciculus cuneatus of the raccoon. They limited their sample to afferent fibers supplying the glabrous skin, although they did observe activity in axons that are activated by stimulation of hairy skin and by joint movements. The classes of receptors that are represented in the fasciculus cuneatus include Pacinian corpuscles, rapidly adapting receptors, and two kinds of slowly adapting receptors. There are proportionately many more rapidly adapting than slowly adapting receptors in the dorsal column as compared with peripheral nerve or dorsal root. The relationship between the two types of slowly adapting receptors found and the type I and II receptors of other species is unclear, but one of the receptor types in the raccoon did fulfill most of the criteria generally used to identify type I receptors.

Bromberg and Whitehorn (1974) examined the receptor types that project in the fasciculus cuneatus of the cat. Although their sample was small, they did find that three of the four type I units that they identified in the superficial radial nerve projected to the caudal medulla.

Evidently, the fasciculus cuneatus differs from the fasciculus gracilis in containing not only large numbers of group I muscle afferents but also type I slowly adapting cutaneous mechanoreceptors. Some of the slowly adapting afferents, at least from muscle, undoubtedly end in the lateral cuneate nucleus, but many must terminate in the cuneate nucleus, judging from the properties of some cuneate neurons (see below). It would be interesting to know whether slowly adapting joint afferents from the forelimb project up the fasciculus cuneatus, since only rapidly adapting joint afferents from the hindlimb project in the fasciculus gracilis (Burgess and Clark, 1969a).

Summary

In summary, the dorsal columns contain a long system of primary afferent fibers in the dorsal funiculus that continue rostrally to the medulla, where they synapse in the dorsal column nuclei. Short and intermediate primary afferent fiber systems serve other purposes, such as providing information to neurons at segmental levels and to such pathways as the dorsal spinocerebellar tract. The dorsal columns are subdivided into the fasciculus gracilis, which terminates in the nucleus gracilis, and the fasciculus cuneatus, which ends in the nucleus cuneatus. Fibers of the fasciculus cuneatus also synapse in the lateral cuneate nucleus, which relays information chiefly to the cerebellum. The dorsal columns have a topographic organization. The fasciculus gracilis contains a sensory representation of the hindlimb and caudal trunk, whereas the fasciculus cuneatus represents the upper trunk and forelimb. Axons of the dorsal funiculus have a dermatomal organization caudally, but through fiber resorting they acquire a somatotopic organization by the time they terminate in the dorsal column nuclei. The functional types of axons contained in the fasciculus gracilis at the upper cervical level differ from those of the fasciculus cuneatus in that there is a much greater proportion of axons in the fasciculus gracilis from cutaneous than from deep receptors. The absence of proprioceptive afferent fibers in the fasciculus gracilis at this level is presumably because these end in Clarke's column (proprioceptive signals from the hindlimb reach the thalamus via a relay in nucleus z [see Chapter 8]). By contrast, the fasciculus cuneatus contains a high proportion of proprioceptive afferents, some of which end in the cuneate and some in the lateral cuneate nucleus. Not all cutaneous receptors are represented in the population of primary afferent axons of the dorsal funiculus that project directly to the brainstem. The largest proportion of cutaneous afferent fibers are from rapidly adapting receptors, especially hair follicle receptors. A few type II slowly adapting receptors project to the medulla in the fasciculus gracilis, and both type I and II receptors are found in the fasciculus cuneatus. Receptors with fine afferent fibers (D hair follicle afferents, C mechanoreceptors, thermoreceptors, and nociceptors) are thought to be poorly represented, if at all, in the dorsal funiculus. However, recent evidence indicates that some unmyelinated primary afferent fibers reach the dorsal column nuclei by way of the dorsal funiculus. Their receptor types are unknown. The dorsal funiculus also contains some visceral afferent fibers and proprioceptive afferent fibers with a respiratory rhythm.

POSTSYNAPTIC DORSAL COLUMN PATHWAY

In addition to the branches of primary afferent fibers, the dorsal funiculus contains the ascending axons of tract cells of the dorsal horn (Uddenberg, 1968a,b; Petit, 1972; L. M. Pubols and Pubols, 1973; Rustioni, 1973, 1974; Angaut-Petit, 1975a,b; Rustioni *et al.*, 1979).

Cells of Origin of Postsynaptic Dorsal Column Pathway

Rat. The distribution of the cells of origin of the postsynaptic dorsal column pathway in rats has been mapped by labeling them retrogradely with horseradish peroxidase or wheat germ agglutinin-conjugated horseradish peroxidase injected into the dorsal column nuclei (Giesler *et al.*, 1984d,e; de Pommery *et al.*, 1984). The cells are in the nucleus proprius just below the substantia gelatinosa. The projection is somatotopic, since the nucleus gracilis receives an input from cells in the lumbar enlargement and the nucleus cuneatus from the cervical enlargement. Giesler *et al.* (1984) used selective lesions to show that the postsynaptic dorsal column neurons in the lumbar enlargement of the rat all project through the dorsal funiculus, whereas about 90% of the cells in the cervical enlargement project through the dorsal funiculus and 10% through the dorsal lateral funiculus (see below).

Cat. The cells of origin of the postsynaptic dorsal column pathway in cats have been labeled retrogradely by injections of tracer into the dorsal column nuclei (Fig. 7.8) (Rustioni and Kaufman, 1977). The cells are located in lamina IV and the medial parts of laminae V and VI in the lumbosacral enlargement and chiefly in lamina IV in the cervical enlargement. Lesions show that the projection is through the dorsal quadrant of the spinal cord.

A. G. Brown and Fyffe (1981) injected postsynaptic dorsal column cells intracellularly with horseradish peroxidase. The cells were identified by antidromic activation of their axons in the dorsal funiculus. The injected cells were located in laminae III, IV, and V (Fig. 7.9).

Bennett *et al.* (1983) confirmed that postsynaptic dorsal column neurons in the cat are concentrated in a band in lamina IV and that many of these cells are also found in laminae V and VI in the medial part of the dorsal horn. However, they also found a population of the cells in lamina VII.

Enevoldson and Gordon (1989a) agreed that postsynaptic dorsal column neurons in the cat are concentrated in lamina IV. They also found some of these neurons in laminae III, V, and VI.

Monkey. Postsynaptic dorsal column neurons in monkeys are located chiefly in laminae IV–VI (Rustioni *et al.*, 1979). Interestingly, there is a distinct tendency for the cells to be concentrated in the medial dorsal horn in the cervical enlargement but in the lateral dorsal horn in the lumbar enlargement. There are some in the lateral part of the ventral horn. According to Bennett *et al.*, (1983), they are found slightly more superficially than in cats and many of the cells are in lamina III. Some are in lamina VII, and some in laminae I and II.

Number of Postsynaptic Dorsal Column Neurons

Rat. Giesler *et al.* (1984) estimated that there are at least 750–1000 postsynaptic dorsal column neurons in the cervical enlargement of the rat and about 500–700 in the lumbar

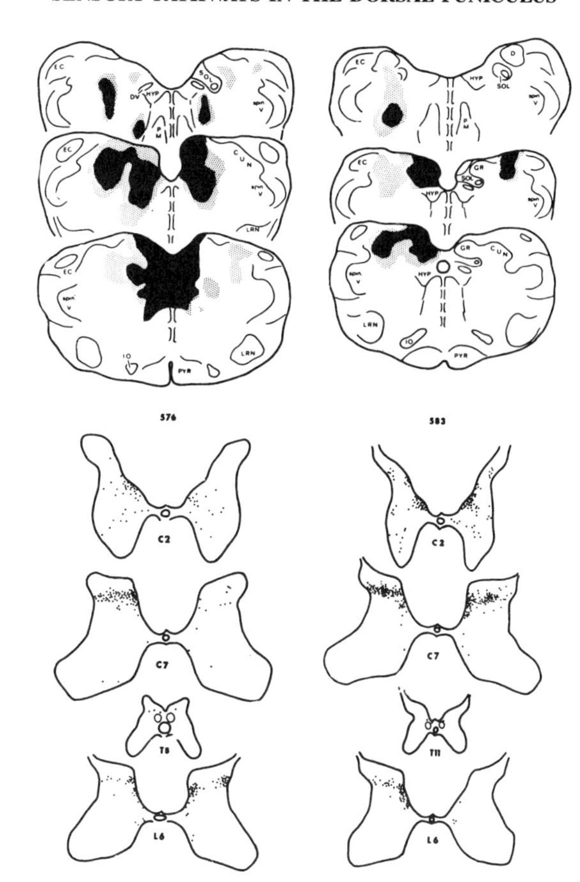

Fig. 7.8. Locations of the cells of origin of the postsynaptic dorsal column pathway. The injection sites in the medulla for horseradish peroxidase labeling are shown above for two experiments. The locations of the labeled neurons at several levels of the spinal cord are plotted below. (From Rustioni and Kaufman, 1977b.)

enlargement (densities of 91–122 and 57–86/mm, respectively). The number of postsynaptic dorsal column neurons in the cervical enlargement is about 38% of the total number of neurons at spinal levels (including dorsal root ganglion [DRG] cells) that project to the nucleus cuneatus; the comparable value for the lumbar cord and nucleus gracilis is 29%.

Cat and Monkey. Bennett *et al.* (1983) estimate that there are about 1000 postsynaptic dorsal column neurons in the lumbar enlargement of cats and monkeys. The density of these cells is 10–50/mm. These values are likely to underestimate the total numbers of these cells, since the labeling was produced by horseradish peroxidase-containing gel implants in the dorsal funiculus. Presumably not all of the axons of the cells took up the label, and axons in the dorsal lateral funiculus were unaffected. A possible confounding issue was that it would have been impossible to distinguish long propriospinal from postsynaptic dorsal column neurons in these experiments.

Enevoldson and Gordon (1989a) also evaluated the number of postsynaptic dorsal column neurons in the cat. They estimated that there are 800–1200 in the lumbosacral enlargement and 1700–2000 in the cervical enlargement. Since they also used horseradish peroxidase-containing gel implants in the dorsal funiculus, these are likely to be underestimates. Clearly, the postsynaptic dorsal column projection is as large as or larger than the spinocervical tract in cats (cf. Chapter 8).

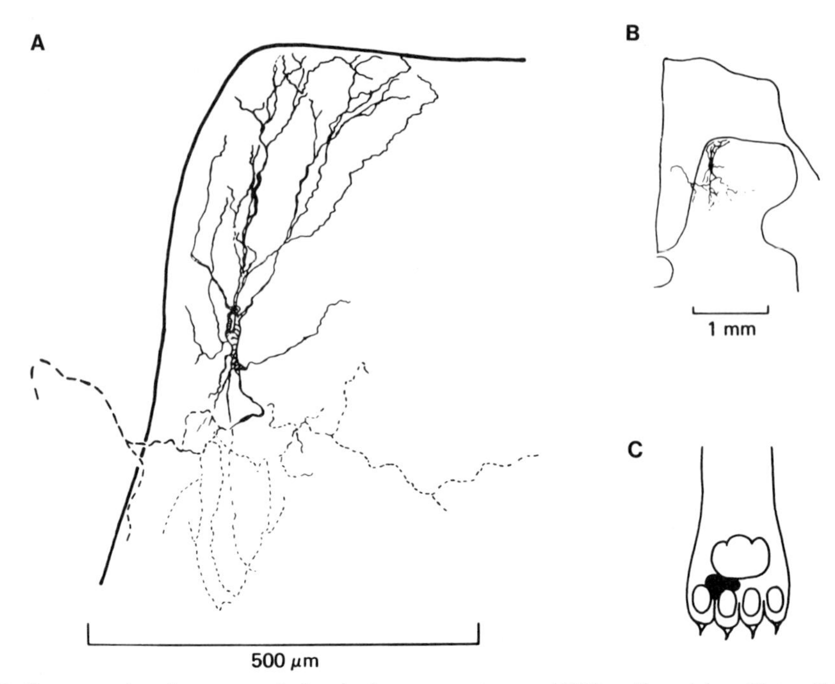

Fig. 7.9. Reconstruction of a postsynaptic dorsal column neuron in a cat. (A) The cell was injected intracellularly with horseradish peroxidase. The soma was in lamina III, and the dendrites extended into lamina I. The axon gave off four collaterals (dotted lines) and entered the dorsal funiculus. (B) Lower-power view. (C) Receptive field. (From A. G. Brown and Fyffe, 1981.)

Morphology of Individual Postsynaptic Dorsal Column Neurons

The morphology of a sample of postsynaptic dorsal column neurons injected intracellularly with horseradish peroxidase in cats was examined by A. G. Brown and Fyffe (1981). The cell bodies range in diameter from 12 to 60 μm. The form of the dendritic trees varies with the position of the soma in the dorsal horn. Cells with somas in laminae III and superficial IV have dendritic trees that are dorsally directed, with a limited transverse or rostrocaudal spread. The dendrites extend into lamina II and even I (Fig. 7.9), unlike spinocervical tract cells (see Chapter 8). Cells with somas in lamina V and the deep part of IV have extensive dendritic fields in the transverse plane, but with little rostrocaudal spread. Cells of the third type have somas in lamina IV and dorsally directed dendrites that sometimes enter lamina II. These cells have extensive dendritic trees in the transverse plane and a substantial longitudinal spread.

The dendritic trees of postsynaptic dorsal column neurons in the nucleus proprius of cats and monkeys are generally elongated dorsoventrally, but narrow mediolaterally and rostrocaudally (Bennett *et al.*, 1983), although some have dendrites with a longitudinal span of 300–450 μm, resembling spinocervical tract cells (see Chapter 8). The mean soma diameter is 27.2 μm in cats and 17.7 μm in monkeys. Postsynaptic dorsal column cells injected intracellularly with horseradish peroxidase were found by Bennett *et al.* (1984) to differ from those labeled by A. G. Brown and Fyffe (1981). The dendritic trees are oriented primarily rostrocaudally, rather than dorsoventrally, and the dendrites rarely extend as far dorsally as lamina II. It is unclear why there is such a discrepancy in the findings of the two groups.

The ultrastructure of synaptic endings on labeled postsynaptic dorsal column neurons has been reported by Bannatyne *et al.* (1987). The synaptic arrangements include contacts by small boutons containing spherical or elongated vesicles, sometimes by the central elements of glomeruli, and axoaxonal complexes. These arrangements differ from those seen on spinocervical tract neurons (see Chapter 8).

Enevoldson and Gordon (1989a) were able to distinguish 3 types of postsynaptic dorsal column neurons on the basis of the location of the soma and the configuration of the dendritic tree. Type C neurons are the most numerous. The somas are in lamina IV, and they have extensive rostrocaudally oriented dendritic trees with a limited mediolateral spread. Type B neurons are large cells in medial lamina V with long, straight dendrites oriented in the transverse plane. Type A cells are in medial laminae III and IV, with restricted rostrocaudal and mediolateral dendritic trees.

Axons of Postsynaptic Dorsal Column Neurons

A. G. Brown and Fyffe (1981) were able to reconstruct the initial courses of the axons of postsynaptic dorsal column neurons injected intracellularly with horseradish peroxidase in cats. Most (85%) give off collaterals, and the collaterals usually arborize ventral to the soma (Fig. 7.9). Some axons go fairly directly into the dorsal funiculus, but a few enter the dorsal lateral funiculus first and then recross the gray matter as they ascend and join the dorsal funiculus. Bennett *et al.* (1984) also observed collaterals of the axons of postsynaptic dorsal column neurons that ramify locally.

The axons of postsynaptic dorsal column neurons labeled retrogradely in cats are observed to follow a tortuous course ventrally to the cell body and to give off collaterals (Enevoldson and Gordon, 1989a). Confirming the observations by A. G. Brown and Fyffe (1981), some of the axons enter the dorsal lateral funiculus before joining the dorsal funiculus.

The conduction velocities of the axons of postsynaptic dorsal column neurons in rats range from 11–44 m/sec (mean 19.4 m/sec; Giesler and Cliffer, 1985) and in cats from 16–80 m/sec (means, 38-55 m/sec) (Uddenberg, 1968b; Angaut-Petit, 1975a; A. G. Brown and Fyffe, 1981; A. G. Brown *et al.*, 1983a; Jankowska *et al.*, 1979; Lu *et al.*, 1983), indicating that the axons are small to medium-sized myelinated fibers.

Background Activity of Postsynaptic Dorsal Column Neurons

A. G. Brown *et al.* (1983a) found that the axons of most postsynaptic dorsal column neurons in their sample from the cat L5 segment had a background discharge consisting of short high-frequency bursts separated by longer intervals. A similar observation was made by Noble and Riddell (1989).

Responses to Electrical Stimulation

Uddenberg (1968b) observed that most postsynaptic dorsal column neurons in the cervical spinal cord of the cat respond to stimulation of more than one peripheral nerve. Petit (1972) found that postsynaptic dorsal column neurons can be activated by stimulation of group III muscle afferent fibers.

Jankowska *et al.* (1979) recorded intracellularly from postsynaptic dorsal column neurons in the cat. Stimulation of the lateral funiculus evokes excitatory postsynaptic potentials (EPSPs) in most of the cells. These were attributed to axons ending or originating in the upper thoracic to upper cervical cord, the upper cervical cord, or fibers of the corticospinal tract. The EPSPs from stimulating fibers of the upper cervical segments could have been due to activation of the axons of spinocervical tract cells. Electrical stimulation of peripheral nerves demonstrated the following connections: monosynaptic and disynaptic EPSPs from low-threshold cutaneous afferents; disynaptic inhibitory postsynaptic potentials (IPSPs) from low-threshold cutaneous afferents; monosynaptic and disynaptic EPSPs from group I muscle afferents; disynaptic IPSPs from group I muscle afferents; and EPSPs from group II and from high-threshold muscle, skin and joint afferents.

A. G. Brown *et al.* (1983a) produced inhibition of the responses of postsynaptic dorsal column neurons to peripheral nerve volleys by stimulation of another peripheral nerve. The time course of inhibition resembles that of primary afferent depolarization, with a peak at 20–40 msec and a duration of 100–200 msec. Sometimes the inhibition has an even slower time course.

Responses to Natural Stimulation

Rat. Giesler and Cliffer (1985) examined the response properties of postsynaptic dorsal column neurons in unanesthetized, decerebrate, partially spinalized rats and found that 64% of the cells responded exclusively to innocuous stimuli. The remainder responded to similar stimuli but best to strong mechanical stimuli. Both types of neurons had restricted receptive fields. Repeated noxious heating of the skin caused no response in 93% of the neurons examined, suggesting that these cells do not receive an input from C polymodal nociceptors, heat nociceptors, mechanical-heat nociceptors or mechanical nociceptors, that can be sensitized by noxious heat. The results were taken as evidence that this pathway is not important for nociception in the rat.

Cat. Uddenberg (1968b) recorded from the axons of postsynaptic dorsal column neurons in the cervical spinal cord and found convergent inputs from several different receptor types. All of the units could be activated by hair movement, and these responses were rapidly adapting. Pressure on the skin also evoked slowly adapting responses, probably by activation of touch receptors. The discharges were increased by pinching the skin or by intense stimulation of deep receptors of the forelimb.

Angaut-Petit (1975a,b) identified postsynaptic dorsal column units in the lumbosacral spinal cord by their discharge properties and also by antidromic activation from the cervical dorsal funiculus. The recordings were from axons located deep to the primary afferent fibers from the skin but more superficial than the proprioceptive afferents. Some neurons (16%) were activated in a rapidly adapting fashion just by innocuous stimuli, such as hair movement or tapping. Most (77%) received a convergent input from sensitive mechanoreceptors and nociceptors. A few of the cells (6%) were excited just by noxious mechanical (but not thermal) stimuli. The convergent or wide-dynamic-range neurons were activated in a rapidly adapting fashion by hair movement and in a slowly adapting fashion by maintained tactile stimulation. The latter response could often be produced by stimulation of tactile domes. An additional response was obtained by pinching a fold of skin with forceps or by heating the skin to 45°C or more. A firing rate as high as 350 Hz could result from a noxious heat stimulus. The receptive fields of postsynaptic dorsal column neurons were fairly small when on the distal limb, but larger when proximal. No inhibitory receptive fields were observed.

A. G. Brown and Fyffe (1981) mapped the receptive fields of postsynaptic dorsal column neurons in the cat and found that they vary from 0.2 cm^2 on the toes to 15–20 cm^2 more proximally on the hindlimb. A subliminal receptive field could often be demonstrated when intracellular recordings were made. Hair movement activated some of the cells, with no further effects when pressure was applied to the skin or the skin was pinched. Most cells were activated by hair movement, pressure, and pinch. Some responded to displacement of tactile domes. Two-thirds of the cells had receptive fields that included glabrous skin. These showed a rapidly adapting discharge to displacement of the glabrous skin and were partially entrained by a 500-Hz tuning fork, suggesting that Pacinian corpuscles provide an input to some of these neurons. Inhibitory receptive fields were found, in some cases extending over a large area of skin.

Brown *et al.* (1983a) recorded in the dorsal funiculus at L5 from the axons of postsynaptic dorsal column neurons in cats. Most (90%) of the neurons had convergent inputs, from either hairy and glabrous skin (67%) or low- and high-threshold mechanoreceptors (60%), or both (Fig. 7.10). Only one cell responded just to noxious stimuli, and only three just to innocuous stimuli. Two of the latter had an input just from Pacinian corpuscles; the other was excited by hair movement. Many cells had discontinuous receptive fields, and in some the fields increased in size during the period of observation following stimulation of the skin, of peripheral nerves, or of the dorsal funiculus at an upper cervical level (Fig. 7.10). Almost half of the cells had inhibitory receptive fields. These could be small and within or near the excitatory field or large and usually proximal to the excitatory field.

About half of the postsynaptic dorsal column neurons examined by Bennett *et al.* (1984) responded to noxious stimuli, and all responded to innocuous mechanical stimuli.

Kamogawa and Bennett (1986) tested the effects of repeated noxious heat stimulation on the responses of postsynaptic dorsal column neurons in cats. Unlike cells of this pathway in rats (Giesler and Cliffer, 1985), sensitization of the skin in cats led to responses to noxious heat pulses. It was concluded that many of these neurons in cats receive an input from Aδ nociceptors.

Noble and Riddell (1988) recorded from the axons of postsynaptic dorsal column neurons in cats. They observed both excitatory and inhibitory receptive fields. Most (80%) of the units were excited by innocuous and noxious stimuli, usually including noxious heat. Many of the units had excitatory fields on the hairy and the glabrous skin, unlike spinocervical tract cells in cats, which rarely have receptive fields on glabrous skin.

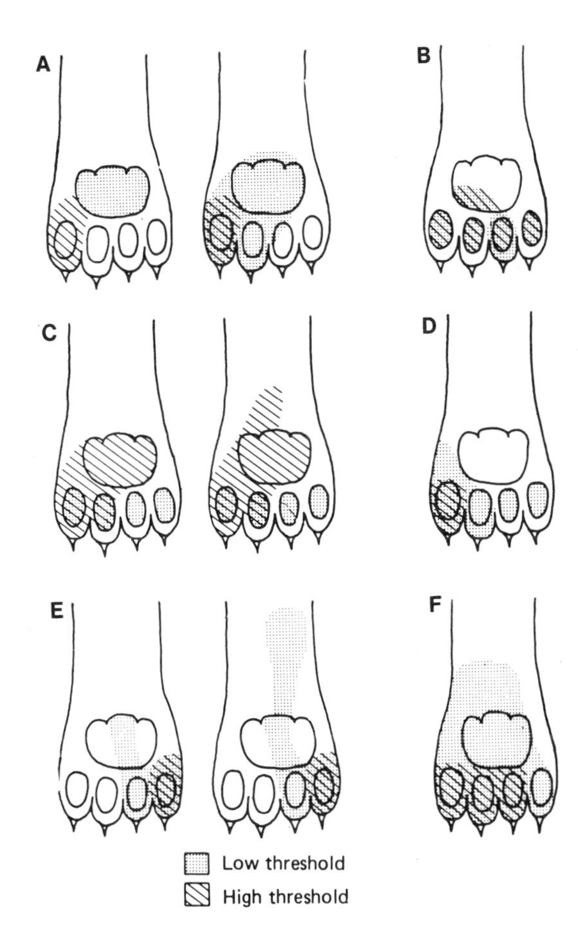

Fig. 7.10. Excitatory receptive fields of postsynaptic dorsal column neurons in the cat. The fields are distinguished for low- versus high-threshold mechanoreceptors. Both hairy and glabrous skin could be included. Some of the fields were discontinuous, as shown in panels B, C, and D. In panels A, C, and E, the fields expanded from those seen at the left to those at the right following experimental manipulations. (From A. G. Brown *et al.*, 1983a.)

Projection Target of Postsynaptic Dorsal Column Pathway

The termination of the postsynaptic dorsal column pathway has been investigated by using cats that first had chronic dorsal rhizotomies over the length of either the lumbosacral or the cervical enlargement (Rustioni, 1973, 1974). Acute transection of the dorsal funiculus then produces an isolated ascending degeneration of the postsynaptic dorsal column pathway. The ascending degeneration from the lumbosacral enlargement can be followed into the nucleus gracilis, the rostral part of the nucleus cuneatus, and nucleus z (Rustioni, 1973). Within the nucleus gracilis, the termination zone differs from that of primary afferent fibers but is similar to the area of termination of the projection to the nucleus gracilis from the cerebral cortex (Kuypers and Tuerck, 1964). Similarly, the postsynaptic dorsal column projection from the cervical cord is largely to the rostral part and base of the cuneate nucleus, avoiding the cluster region (Fig. 7.11) (Rustioni, 1974).

The projection of the postsynaptic dorsal column pathway in rats has now been examined by anterograde tracing of the axons of these cells labeled by injection of *Phaseolus vulgaris* leucoagglutinin (PHA-L) into the dorsal horn of the spinal cord (Cliffer and Giesler, 1989). Neurons in the dorsal column nuclei that project to the thalamus were also labeled retrogradely by injections of Fluoro-Gold into the thalamus. The projection from the cervical enlargement is to the cuneate nucleus, from the thoracic cord to the medial cuneate and lateral gracile nuclei, from the lumbar enlargement to the gracile nucleus, and from the sacral cord to the medial gracile nucleus. Postsynaptic dorsal column axon terminals are dense in areas of the dorsal column nuclei that contain a high concentration of thalamic relay neurons and terminations of primary afferent fibers, suggesting an organization different from that in cats.

In monkeys, the postsynaptic dorsal column pathway from the lumbar level ends in the nucleus gracilis throughout its length (Rustioni *et al.*, 1979). Some of the fibers ascend into the restiform body and then turn caudally to descend into the nucleus gracilis. The projection from the cervical cord is to the pars triangularis of the cuneate nucleus, as well as to the lateral cuneate nucleus.

The projection of individual axons of postsynaptic dorsal column neurons to the cuneate nucleus in cats has been traced with PHA-L or by intra-axonal injection of HRP (Pierce *et al.*, 1990). Endings were concentrated around the rim and in the ventral region of the nucleus.

Pathway in Dorsal Lateral Funiculus to Dorsal Column Nuclei

Another pathway that ascends to the dorsal column nuclei from the spinal cord travels in the dorsal lateral funiculus (Tomasulo and Emmers, 1972; Hazlett *et al.*, 1972; Dart and Gordon, 1973; Rustioni, 1973; 1974; Rustioni and Molenaar, 1975; Nijensohn and Kerr, 1975; Rustioni *et al.*, 1979; Gordon and Grant, 1982; Enevoldson and Gordon, 1989b). Terminals of this projection have been found in the same regions in which the postsynaptic dorsal column pathway ends in cats (cf. Rustioni, 1973, 1974). However, in monkeys the pathway in the dorsal lateral funiculus includes fibers destined for the lateral cuneate nucleus (Rustioni *et al.*, 1979).

Enevoldson and Gordon (1989b) labeled only a small number of these cells when they implanted horseradish peroxidase-containing gel in the dorsal column nuclei of cats after sectioning the dorsal funiculus. It was not clear whether the reason for the small number of cells is insufficient uptake of horseradish peroxidase or the presence of only a small population of neurons projecting to the dorsal column nuclei through the dorsal lateral funiculus.

There is evidence that some neurons can be activated antidromically from both the dorsal funiculus and the dorsal lateral funiculus (Lu *et al.*, 1985). It is possible that the axons of some postsynaptic dorsal column neurons have branches that travel in both the dorsal and the lateral funiculi. It is less likely that the same neurons belong to both the spinocervical and the

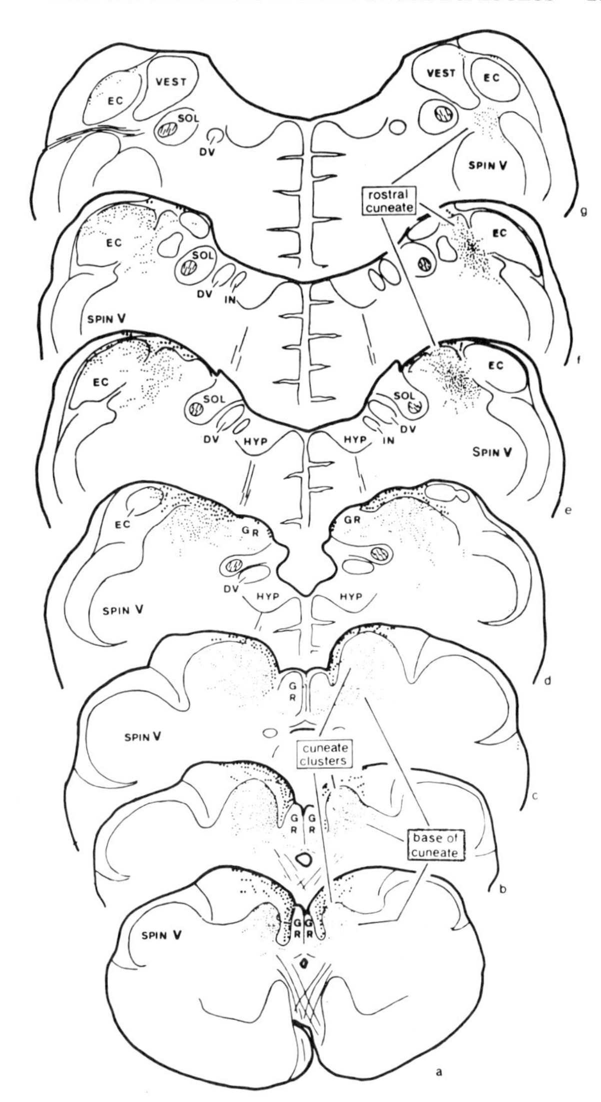

Fig. 7.11. Projections of the postsynaptic dorsal column pathway to the cuneate nucleus in the cat are shown on the right. The projections on the left were a combination of direct dorsal root and postsynaptic dorsal column projections. The degeneration was produced by a lesion of the dorsal funiculus after a chronic dorsal rhizotomy of the cervical enlargement on the right side. Anterograde degeneration was traced by using a silver method. (From Rustioni, 1974.)

postsynaptic dorsal column pathways, since there are a number of differences in the morphology and functional properties of cells in these two pathways (A. G. Brown and Fyffe, 1981; Brown *et al.*, 1983a; Enevoldson and Gordon, 1989b). However, it cannot be ruled out that some of the axons in the dorsal lateral funiculus give off collaterals to the lateral cervical nucleus en route to the dorsal column nuclei (cf. Enevoldon and Gordon, 1989b).

The pathway to the dorsal column nuclei that is contained in the dorsal lateral funiculus can activate neurons that relay to the thalamus, as well as neurons that are not activated antidromically from the medial lemniscus (Dart and Gordon, 1973). Some of these "interneurons" may project to the cerebellum (cf. H. Johansson and Silfvenius, 1977c). Neurons of the dorsal column nuclei respond to natural stimuli even after the dorsal funiculus has been sectioned (Dart and Gordon, 1973; Dostrovsky and Millar, 1977). This observation indicates that it cannot be assumed that transection of the dorsal funiculus will prevent the participation of the dorsal column nuclei in somesthesis. Furthermore, behavioral responses in an animal with a transected dorsal funiculus that disappear after a lesion of the dorsal lateral funiculus cannot automatically be ascribed to the spinocervical tract (see Chapter 6).

The response properties of neurons that project to the dorsal column nuclei by way of the dorsal lateral funiculus need to be evaluated.

Descending Control of Postsynaptic Dorsal Column Neurons

Noble and Riddell (1989) investigated the influence of descending pathways on the responses of postsynaptic dorsal column neurons in cats. Cold block was used to interfere with the tonic activity of descending pathways. Receptive fields to innocuous tactile stimuli are little affected by cold block, whereas the responses of many of the neurons to noxious mechanical and thermal stimuli are enhanced during cold block. Inhibition from peripheral stimulation is unaffected by cold block, indicating that the inhibition depends upon segmental mechanisms. Background activity often also increases during cold block.

It is possible that there is an inhibitory projection by serotonergic neurons in the brain stem onto postsynaptic dorsal column neurons, since these cells have been shown to receive synaptic contacts that can be stained immunocytochemically for serotonin (Nishikawa *et al.*, 1983).

Summary

In summary, the dorsal column contains not only the ascending branches of primary afferent fibers, but also a projection from second order neurons of the dorsal horn. This postsynaptic dorsal column pathway originates from neurons in the nucleus proprius (laminae III or IV), as well as laminae V and VI. This is a major sensory pathway made up of thousands of neurons. The dendrites of some of the neurons extend dorsally into lamina II and even lamina I. The axons often give off local collaterals. The funicular branches are fine or medium-sized myelinated fibers. Postsynaptic dorsal column neurons generate monosynaptic and polysynaptic EPSPs, as well as disynaptic IPSPs, following peripheral nerve stimulation. They may also be subject to presynaptic inhibition. Postsynaptic dorsal column neurons respond to innocuous or to both innocuous and noxious stimulation of the skin. However, in rats nearly all fail to respond to noxious heat, even after sensitization of the skin. In cats, some postsynaptic dorsal column neurons can be activated both by hair movement and also by pinching the skin and by noxious heat. Furthermore, sensitization of the skin can lead to responses to noxious heat in cells that were initially unresponsive. Some postsynaptic dorsal column neurons in cats respond specifically to noxious mechanical stimuli (but not to heat). Evidently, the postsynaptic dorsal column pathway in cats differs from that in rats and is likely to contribute to nociception. Responses to stimulation of SA I receptors have been reported in several investigations of postsynaptic dorsal column neurons in cats. Many of these neurons have receptive fields on the glabrous skin. Responses to activation of Pacinian corpuscles have also been identified. Thus, postsynaptic dorsal column neurons in cats show a number of differences in their responses to those of spinocervical tract cells, which are not activated by SA I receptors or Pacinian corpuscles and which usually lack receptive fields on the glabrous skin (see Chapter 8). The postsynaptic dorsal column pathway projects in a somatotopic fashion to the dorsal column nuclei. Postsynaptic dorsal column neurons in the lumbosacral enlargement terminate in the nucleus gracilis and those in the cervical enlargement in the nucleus cuneatus. There is also a postsynaptic projection to the dorsal column nuclei in the dorsal lateral funiculus. It is unclear whether collaterals from this pathway terminate in the lateral cervical nucleus. The response properties of these neurons have not been investigated. However, since this projection would be able to provide somatosensory information to the dorsal column nuclei even after complete transection of the dorsal column, this pathway is potentially of considerable importance.

DORSAL COLUMN NUCLEI

Termination of Primary Afferent Fibers

The axons of the long–system of the dorsal funiculus from the lower trunk and hindlimb ascend in the fasciculus gracilis and terminate in the nucleus gracilis (or nucleus of Goll), and comparable primary afferent fibers from the upper thoracic and cervical levels ascend in the fasciculus cuneatus and synapse in the nucleus cuneatus (or nucleus of Burdach) and the lateral (external, accessory) cuneate nucleus (or nucleus of Clarke-Monakow or of Monakow) (Fig. 7.12). These projections have been demonstrated by anterograde degeneration techniques and more recently by transganglionic anterograde labeling (Ferraro and Barrera, 1935a,b; Walker and Weaver, 1942; Chang and Ruch, 1947a; Glees and Soler, 1951; Hand, 1966; Carpenter *et al.*, 1968; Rustioni and Macchi, 1968; Shriver *et al.*, 1968; Keller and Hand, 1970; Blomqvist and Westman, 1975; H. Johansson and Silfvenius, 1977c; G. Grant *et al.*, 1979; Beck, 1981; Nyberg and Blomqvist, 1982; Abrahams and Swett, 1986; Nyberg, 1988; Culberson and Brushart, 1989; Rasmusson, 1989) or labeling from the dorsal root ganglion (Cliffer and Giesler, 1989). In some animals, there is also a midline nucleus of Bischoff.

The sensory system includes just the nuclei gracilis and cuneatus; the lateral cuneate nucleus is usually considered a part of the motor system, since it projects to the cerebellum (Ferraro and Barrera, 1935a,b). However, there is now evidence for a thalamic projection from the lateral cuneate nucleus in rats and monkeys, in contrast to cats, which lack such a projection (Albe-Fessard *et al.*, 1975; Boivie *et al.*, 1975; Fukushima and Kerr, 1979; Boivie and Boman,

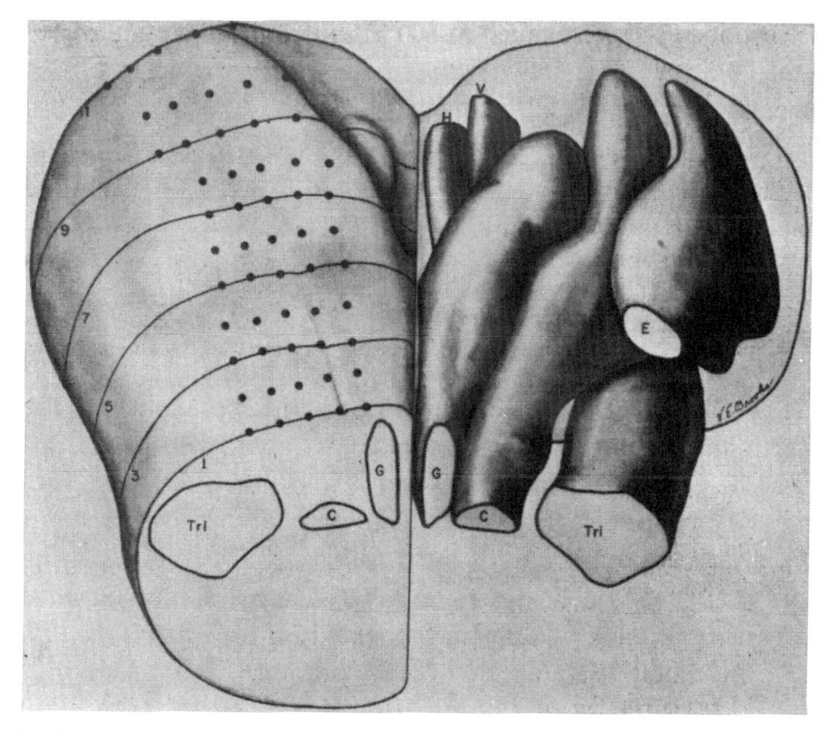

Fig. 7.12. Three-dimensional representation of the dorsal column and adjacent nuclei in the medulla oblongata of the monkey. Abbreviations: C, cuneate nucleus; G, gracile nucleus; E, external cuneate nucleus; Tri, spinal nucleus of the trigeminal nerve; H, hypoglossal nucleus; V, dorsal motor nucleus of the vagus. (From Biedenbach, 1972.)

1981; Massopust *et al.*, 1985; Mantle-St. John and Tracey, 1987; Cliffer and Giesler, 1989). Presumably, neurons of the lateral cuneate nucleus that have a thalamic projection belong to the somatosensory system. However, these neurons have not as yet been investigated electrophysiologically.

The pattern of termination of fibers of the fasciculi gracilis and cuneatus within the dorsal column nuclei has been described by many investigators (Ferraro and Barrera, 1935b; Walker and Weaver, 1942; Glees and Soler, 1951; Hand, 1966; Shriver *et al.*, 1968; Carpenter *et al.*, 1968; Rustioni and Macchi, 1968; Keller and Hand, 1970; Basbaum and Hand, 1973). Although it was once claimed that lumbosacral fibers end in the caudal part of the nucleus gracilis and that thoracic fibers end in the rostral part (Ferraro and Barrera, 1935b; Walker and Weaver, 1942), workers using modern staining techniques for tracing degenerating axons to their terminals, which was not possible with the Marchi technique, disagree with this (e.g., Hand, 1966). Instead, it appears that projections from any given dorsal root end at all rostrocaudal levels of the dorsal column nuclei. However, the overlap of dermatomal projections is greater in the rostral than in the more caudal levels of the dorsal column nuclei (Hand, 1966; Rustioni and Macchi, 1968).

Somatotopic Organization

The sensory representation in the dorsal column nuclei is arranged in the transverse plane, with the caudal parts of the body represented medially in the nucleus gracilis and the rostral part laterally in the nucleus cuneatus; the trunk representation is in a region between the two nuclei. The distal extremities are represented dorsally and the proximal body ventrally. This somatotopic organization has been described for a number of species, including rats, cats, and monkeys, in experiments using electrophysiological recordings to map receptive fields or transganglionic anterograde transport of horseradish peroxidase or wheat germ agglutinin-conjugated horseradish peroxidase to map the projections of different peripheral nerves (Fig. 7.13) (Kruger *et al.*,

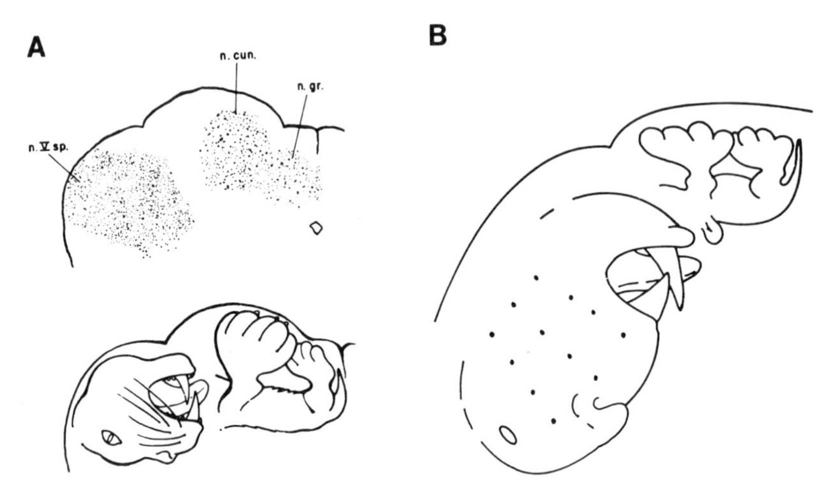

Fig. 7.13. (A) (top) The dorsal column nuclei of the cat (n. cun., nucleus cuneatus; n. gr., nucleus gracilis) and the spinal nucleus of the trigeminal, pars caudalis (n. V sp.). (bottom) A "feliculus" showing the somatotopic organization of the somatosensory nuclei of the caudal medulla. (From Kruger *et al.*, 1961.) (B) Similar somatotopic arrangement for the rat, although in this animal the representation of the face is proportionately larger than in cats (From Nord, 1967; reproduced with permission.)

1961; Nord, 1967; Millar and Basbaum, 1975; Beck, 1981; Nyberg and Blomqvist, 1982; Hummelsheim *et al.*, 1985; Nyberg, 1988; Culberson and Brushart, 1989; Florence *et al.*, 1989). Perhaps the most detailed map is that for the raccoon (Fig. 7.14) (J. I. Johnson *et al.*, 1968; cf. Rasmusson, 1989).

Essentially the same somatotopic map can be demonstrated at various rostrocaudal levels of the dorsal column nuclei. This is consistent with the manner in which axons in the dorsal funiculus terminate in the dorsal column nuclei (Ramón y Cajal, 1909; Glees and Soler, 1951; Valverde, 1966; Gulley, 1973). As the dorsal funiculus approaches the dorsal column nuclei, it divides into several bundles that turn to project ventrally into the dorsal column nuclei (Fig. 7.15). A given axon may give off several collaterals as it ascends along the dorsal aspect of the nucleus in which it terminates (Figs. 7.5 and 7.16) (Gulley, 1973; Fyffe *et al.*, 1986a). An axon that terminates in a dorsal column nucleus will tend to remain in a parasagittal plane at the level

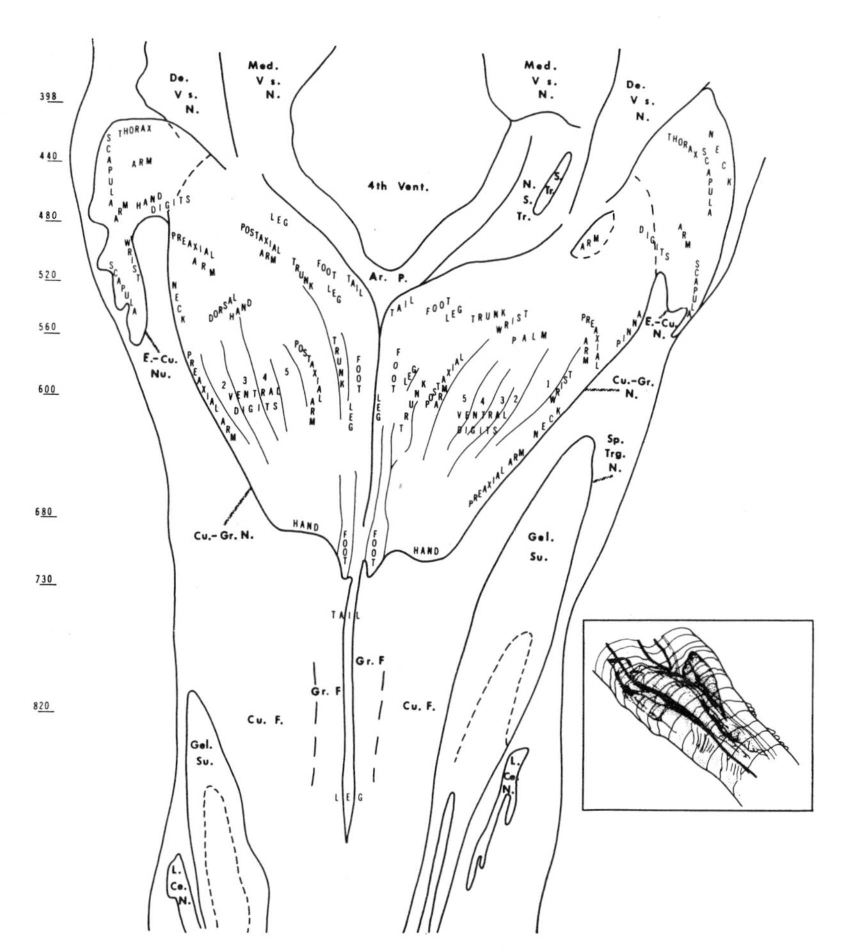

Fig. 7.14. (A) Somatotopic organization of the dorsal column nuclei of the raccoon. The drawing was made from a horizontal section through the widest extent of the dorsal column nuclei. Abbreviations: Cu.F., Gr. F., Cu.-Gr. N., cuneate and gracile fasciculi and nuclei; E.-Cu. Nu, external cuneate nucleus; Gel. Su., substantia gelatinosa; L. Ce. N., lateral cervical nucleus; Sp. Trg. N., spinal nucleus of the trigeminal; Ar. P., area postrema; N. and Tr. S., nucleus and tractus solitarius; De. and Med. Vs. N., descending and medial vestibular nuclei. (From Johnson, 1968; reproduced with permission.)

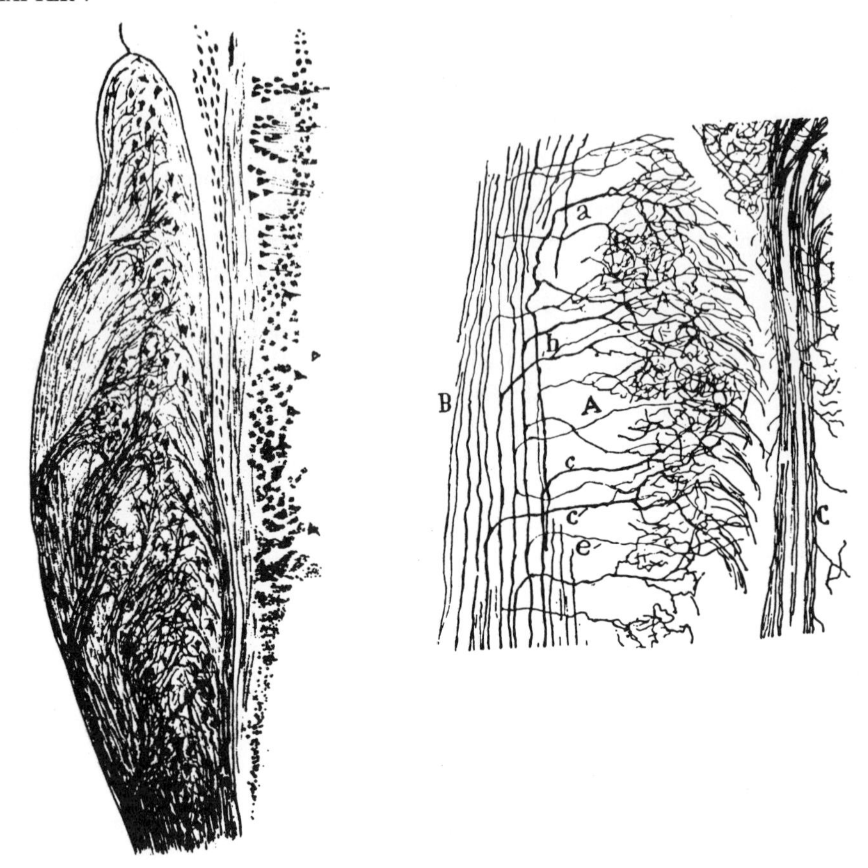

Fig. 7.15. (Left) A sagittal section through the nucleus gracilis and the termination of the fasciculus gracilis. (From Glees and Soler, 1951.) (Right) Sagittal section through the cuneate fasciculus (B) and nucleus (A) showing collaterals of the afferent fibers entering the nucleus. (From Ramón y Cajal, 1909.)

of the caudal and middle parts of the dorsal column nuclei, and so the same somatotopic information is distributed longitudinally over a considerable distance. In the rostralmost part of the dorsal column nuclei, the precision of the projection is lost, since here individual afferent fibers may bend transversely in either direction (Gulley, 1973). Correspondingly, receptive fields tend to be somewhat larger for more rostrally located neurons in the cuneate nucleus (Rowinski *et al.*, 1981). A few axons even cross the midline to terminate in the contralateral dorsal column nuclei (Hand, 1966; Rustioni and Macchi, 1968). However, the bulk of the afferent projection to the dorsal column nuclei is suited to transmit somatotopic information ipsilaterally in an orderly fashion.

Cytoarchitecture

Ramón y Cajal (1909) recognized three types of neurons in the dorsal column nuclei (Fig. 7.17): (1) neurons with short, bushy dendrites located in cell "islands" or "nests"; (2) neurons with long dendrites located around the margins of the nuclei; and (3) stellate, fusiform, or triangular neurons with radiating dendrites placed between the cell nests or ventrally in the nuclei.

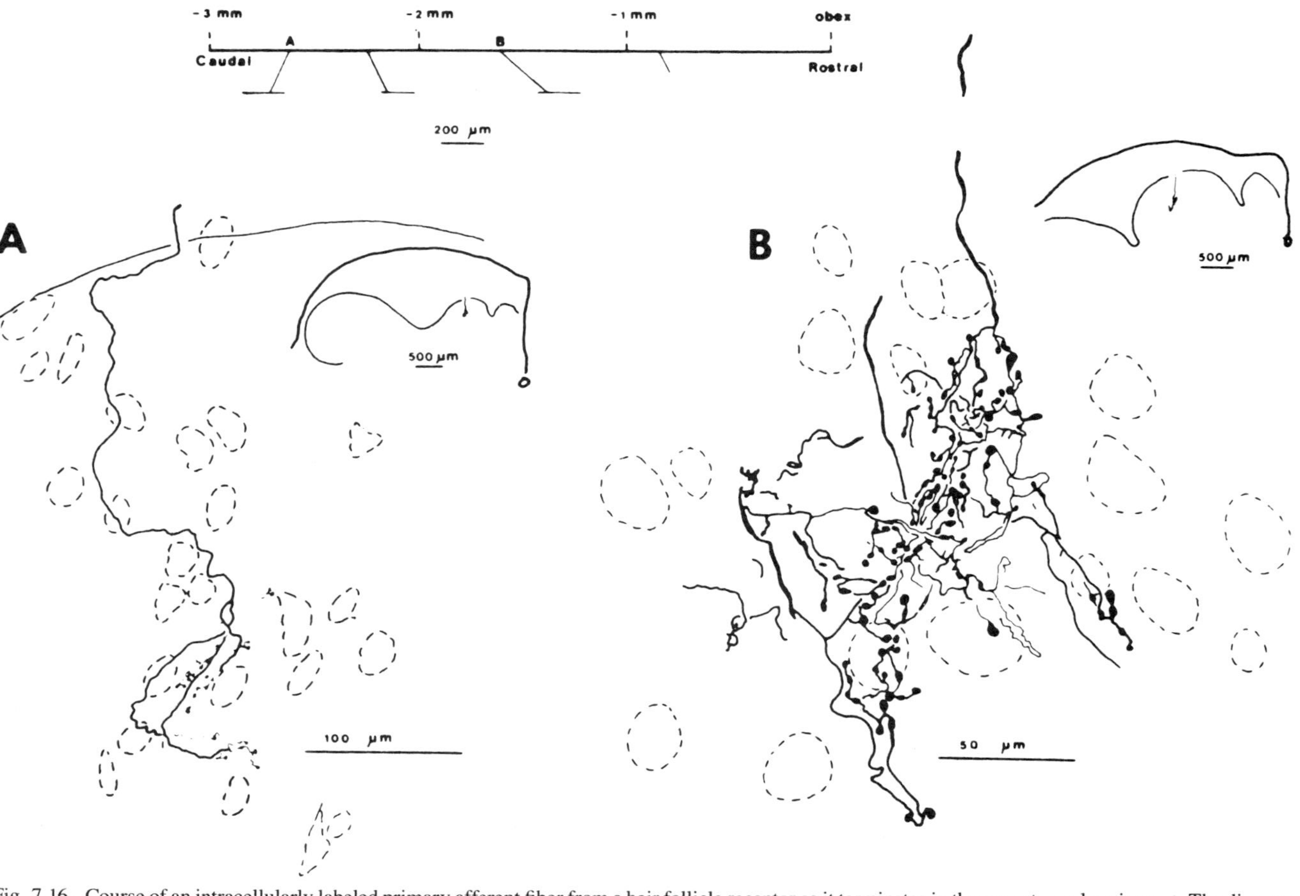

Fig. 7.16. Course of an intracellularly labeled primary afferent fiber from a hair follicle receptor as it terminates in the cuneate nucleus in a cat. The diagram at the top shows the longitudinal extent of the fiber and the locations of four collaterals. Panels A and B are reconstructions of two of the collaterals plotted on a transverse section through the cuneate nucleus. The dashed lines indicate cell bodies of cuneate neurons. (From Fyffe *et al.*, 1986a.)

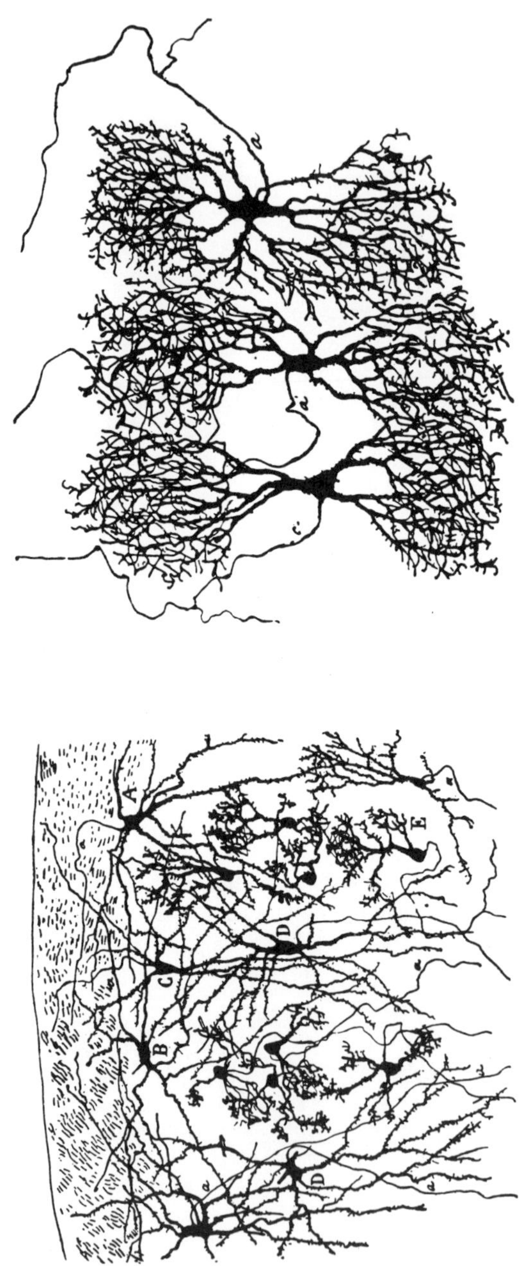

Fig. 7.17. (Left) Drawing of a transverse section through the cuneate nucleus of a kitten (Golgi stain). A–C are marginal neurons (note the axons projecting into the dorsal column). The neurons labeled D are large multipolar cells ("cellules de cloisons"). The smaller neurons that are arranged in clusters, E, are the round stellate cells of the cell nests. The last two cell types project into the medial lemniscus. (Right) View of a cell nest from human fetal material is shown at the right. (From Ramón y Cajal, 1909.)

Recent studies using the horseradish peroxidase retrograde labeling technique have shown that the highest density of cells in the cat that project to the thalamus are in the cell nests, but neurons in other parts of the dorsal column nuclei are also labeled following injections of horseradish peroxidase into the thalamus (Fig. 7.18) (Berkley, 1975; Blomqvist and Westman, 1975; Cheek *et al.*, 1975; Blomqvist, 1980; Ellis and Rustioni, 1981). Some of the large neurons project caudally into the spinal cord (Burton and Loewy, 1977), and others project to the cerebellum or to the inferior olivary nucleus (Cheek *et al.*, 1975; Berkley, 1975).

The dorsal column nuclei have several cytoarchitectural zones (Ramón y Cajal, 1909; Taber, 1961; Kuypers and Turck, 1964; Hand, 1966; Keller and Hand, 1970; Biedenbach, 1972; Basbaum and Hand, 1973). These are less well defined in rats than in other animals, such as cats and monkeys.

Rat. Basbaum and Hand (1973) subdivided the rat cuneate nucleus into two cytoarchitectural areas, a caudal and a rostral zone. The caudal zone consists of a single population of "round" cells arranged in "slabs" or "bricks" oriented dorsoventrally and from side to side. The slablike arrangement can be seen in horizontal or sagittal, but not in transverse, sections. The rostral zone is composed of "round", "spindle-shaped," and "multipolar" cells arranged in a dispersed fashion. When individual dorsal roots are sectioned, degenerating terminal fields are found to occupy vertically oriented bands in the caudal part of the nucleus at levels corresponding to the cellular slabs. The cranialmost roots are distributed ventrolaterally and the caudalmost ones dorsomedially. In the rostral zone, the degeneration is diffuse rather than focal and the topographic arrangement, while present, shows more overlap than in the caudal zone.

The Golgi picture of the rat nucleus gracilis is complementary to this description of the rat cuneate nucleus. The dendrites of the neurons in the caudal part of the nucleus gracilis form vertically oriented "dendritic columns" (Gulley, 1973). The collaterals of afferent fibers of the dorsal funiculus project vertically into the nucleus and appear to contact the dendritic columns. A single afferent fiber may give off collaterals to a series of 4–6 columns arranged in a caudorostral sequence. The dendrites of the neurons in the rostral part of the nucleus, on the other hand, are oriented longitudinally. Dorsal funiculus afferent fibers may terminate in a horizontal arborization along such dendrites, but often the fibers bend abruptly in a transverse plane and end on dendrites of an adjacent "somatotopic strip." Further details about the patterns of termination of Golgi-stained dorsal column axons in rats are given by Odutola (1977b).

Cat. Experimenters using cats subdivide the dorsal column nuclei into caudal, middle, and rostral zones (Hand, 1966). In cats, the middle zone of the dorsal column nuclei contains the cell cluster region. Dorsal and ventral to the cell clusters are regions that more closely resemble the caudal and rostral zones (Kuypers and Tuerck, 1964). When dorsal roots are cut, degenerating fibers can be traced to all three zones of the dorsal column nuclei of the cat, although the density of degeneration is greatest in the cell cluster region (Hand, 1966).

However, dorsal rhizotomy interrupts both cutaneous and muscle afferent fibers. These can be distinguished by using the transganglionic anterograde tracing technique. When cutaneous afferent fibers of the superficial radial nerve are labeled, they are found to project most densely to the cell cluster region; they also end in the rostral and caudal zones, but not in the ventral part of the middle zone (Fig. 7.19) (Nyberg and Blomqvist, 1982). By contrast, muscle afferent fibers of the deep radial nerve terminate in the ventral part of the middle zone, but not in the cell cluster area. Muscle afferent fibers also project to the lateral cuneate nucleus. The pattern of labeling from other forelimb nerves is consistent with that of these branches of the radial nerve.

Individual axons of hair follicle afferent fibers, type I slowly adapting afferent fibers, and group Ia muscle spindle afferent fibers have been injected intra-axonally with horseradish peroxidase by Fyffe *et al.* (1986a) and traced to their terminals within the cuneate nucleus of the cat (Fig. 7.16). The cutaneous afferent fibers give off collaterals that end within cell clusters at several rostrocaudal levels of the nucleus. The muscle afferent fibers end chiefly in regions ventral or dorsal to the cluster zone. A similar study of the projections of individual G hair

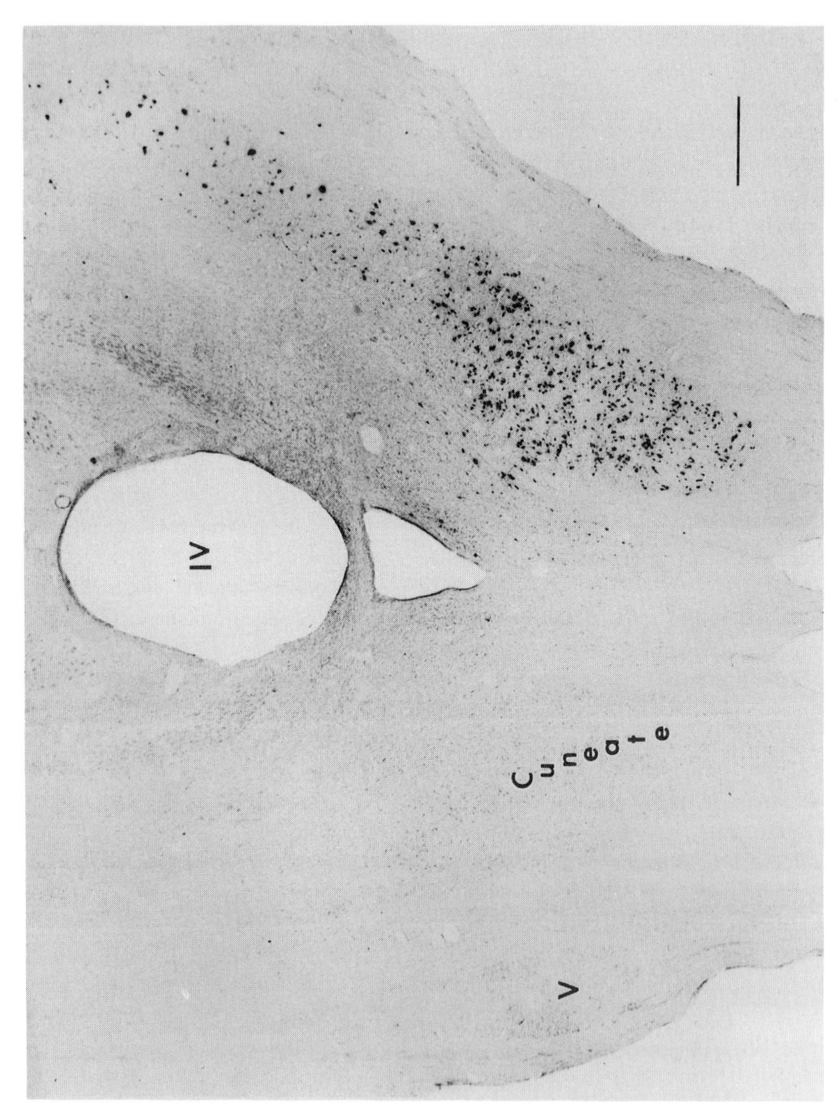

Fig. 7.18. Neurons of the dorsal column nuclei that project to the thalamus of cat. Horseradish peroxidase was injected into the ventrobasal complex on the side contralateral to the labeled neurons. Note the higher density of labeled cells in the cell cluster region than in the rostral part of the nuclei. (From Ellis and Rustioni, 1981.)

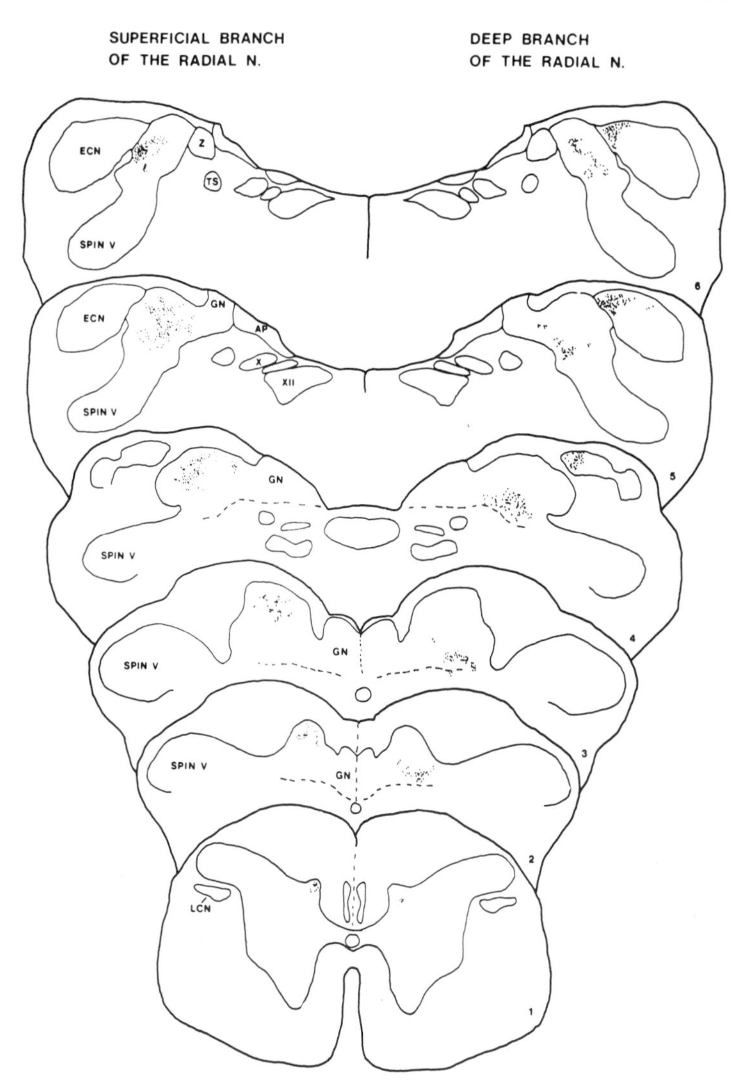

Fig. 7.19. Projections of the superficial radial nerve (a purely cutaneous nerve) and of the deep radial nerve (a purely muscular nerve) to the dorsal column nuclei of the cat. The terminals of these nerves were labeled by the transganglionic horseradish peroxidase method. Abbreviations: GN, gracile nucleus; ECN, external cuneate nucleus; SPIN V, spinal nucleus of V. (From Nyberg and Blomqvist, 1982.)

follicle afferents and group Ia afferents to the cuneate nucleus in cats has recently been reported (Weinberg *et al.*, 1990).

Monkey. In monkeys, the middle zone of the cuneate nucleus is subdivided into a pars rotundus and a pars triangularis (Ferraro and Barrera, 1935b; Boivie, 1978; Rustioni *et al.*, 1979). Cutaneous afferent fibers end in the pars rotundus, whereas muscle afferent fibers project to the pars triangularis, as well as to the deep part of the caudal zone and to the lateral cuneate nucleus (Hummelsheim *et al.*, 1985; Culberson and Brushard, 1989). Thus, the pars rotundus of the monkey corresponds to the cell cluster region in the cat. The nucleus gracilis includes a medial and a lateral zone in the caudal two-thirds of the nucleus; these merge rostrally in a region that may be equivalent to nucleus z in cats (Rustioni *et al.*, 1979). The medial zone of the gracile nucleus appears to be a region associated with input from the skin of the distal limb, comparable to the pars rotundus of the cuneate nucleus.

Biedenbach (1972) has distinguished five cell types in the cuneate nucleus of the monkey. The most common cells are the "round" cells, synonymous with the "cluster" or "nest" cells found by others in cats (Fig. 7.17) (Ramón y Cajal, 1909; Kuypers and Tuerck, 1964; Hand, 1966; Keller and Hand, 1970). The dendrites of these cells are short and bushy (Kuypers and Tuerck, 1964), giving the cells an idiodendritic pattern (Ramón-Moliner and Nauta, 1966) typical of sensory relay neurons. The "round" cells are concentrated in the middle zone of the monkey cuneate nucleus and are grouped in clusters or nests (Biedenbach, 1972). They constitute a large fraction of the cuneate neurons that project to the thalamus. "Large" cells make up a minor fraction of the population of cuneate neurons. Their somata range in size up to 60 μm, and they are relatively abundant in the rostral zone of the nucleus. "Small" cells, whose somas typically have a diameter of about 10 μm, are scattered throughout the three zones of the nucleus, although there are proportionately more of these neurons caudally. "Spindle" and "polygonal" cells are the other two categories of cuneate neurons. Some of the "polygonal" cells have somas with diameters as large as 50 μm. "Polygonal" cells are the most common type in the caudal zone. "Spindle" cells are also relatively abundant caudally. The dendrites of "spindle" and "polygonal" cells are long and branch sparsely, and hence they can be described as isodendritic (Ramón-Moliner and Nauta, 1966), like reticular neurons (cf. Fig. 7.17).

Thalamic Relay Neurons and Interneurons

Blomqvist and Westman (1975) labeled thalamic projection neurons in the nucleus gracilis of cats retrogradely with horseradish peroxidase and then examined the anterograde degeneration produced by dorsal rhizotomies. Degenerating terminals were found near both labeled and unlabeled neurons.

Using the Golgi technique, Blomqvist and Westman (1976) were able to demonstrate the presence of neurons in the nucleus gracilis that have axons which end close to the cell body. They suggested that these are interneurons. In addition, they observed collaterals from the axons of projection neurons that terminate within the nucleus.

Ellis and Rustioni (1981) also examined neurons of the dorsal column nuclei that were labeled retrogradely in cats. All but 8% of the neurons in the cluster region could be labeled from the thalamus. They suggested that some of the unlabeled neurons might be interneurons. However, they did not rule out the possibility that some of these neurons project out of the dorsal column nuclei elsewhere than to the thalamus.

Fyffe *et al.* (1986b) injected cuneothalamic neurons in the cat intracellularly with horseradish peroxidase. The cells were in the cluster region (Fig. 7.20). All of the cells examined responded to hair movement, and half also responded to other forms of stimuli. The dendrites of a given neuron extended beyond a typical cell cluster, reaching as far as 500 μm from the soma. Half of the injected cells had axon collaterals, most of which ended in a region ventral to the soma.

Number of Neurons in the Dorsal Column Nuclei

Biedenbach (1972) estimated that the total number of neurons in the monkey cuneate nucleus is 48,000.

Ultrastructure of the Dorsal Column Nuclei

The synaptic endings in the dorsal column nuclei of cats are often very large, up to 7 μm in diameter, and have a surface area that can exceed 20 μm^2 (Fig. 7.21) (Rozsos, 1958; Walberg,

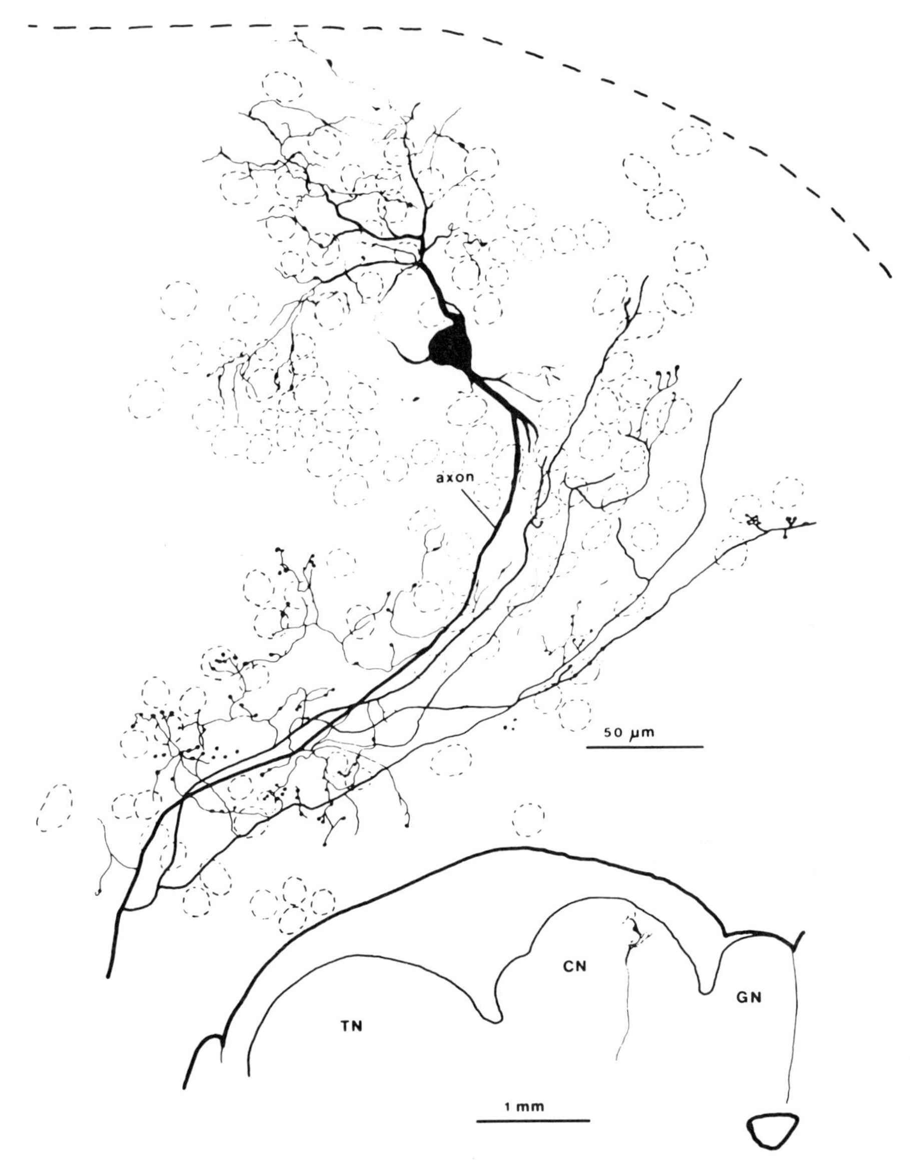

Fig. 7.20. Cuneothalamic relay neuron injected intracellularly with horseradish peroxidase. The dendritic tree extends into the vicinity of several groups of cuneate neurons of the cell cluster region. The axon collaterals terminate ventrally among several other groups of neurons. (From Fyffe *et al.*, 1986b.)

1965; 1966; Ellis and Rustioni, 1981). Many contain numerous round vesicles, but others have flattened vesicles, and some contain dense core vesicles (Walberg, 1965, 1966; Ellis and Rustioni, 1981; Fyffe *et al.*, 1986a). The endings appear at the light-microscopic level to be on both dendrites and cell bodies in the cell cluster region, but mostly on dendrites in the rostral zone (Hand, 1966). However, axodendritic synapses are much more common than axosomatic ones in the cluster zone when examined at the electron-microscopic level (Walberg, 1965; Rustioni and Sotelo, 1974; Fyffe *et al.*, 1986a). Dendrodendritic synapses have also been found in the cuneate nucleus of monkeys (Wen *et al.*, 1977).

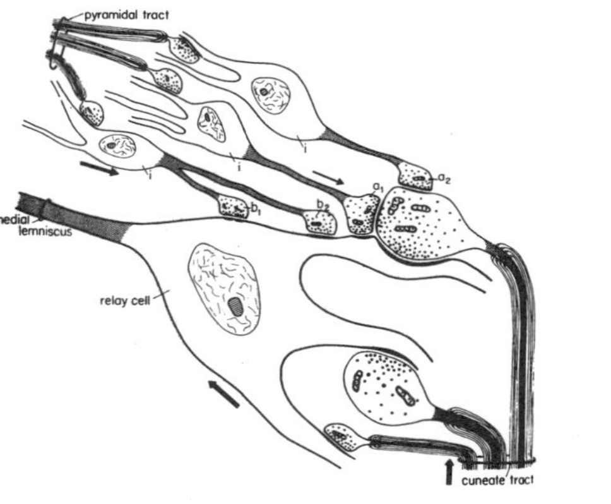

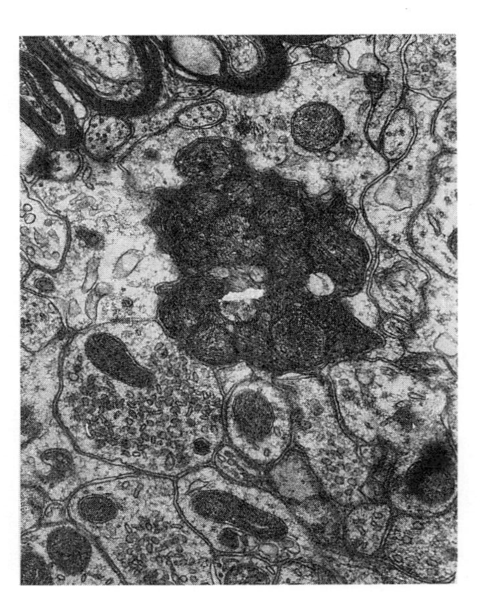

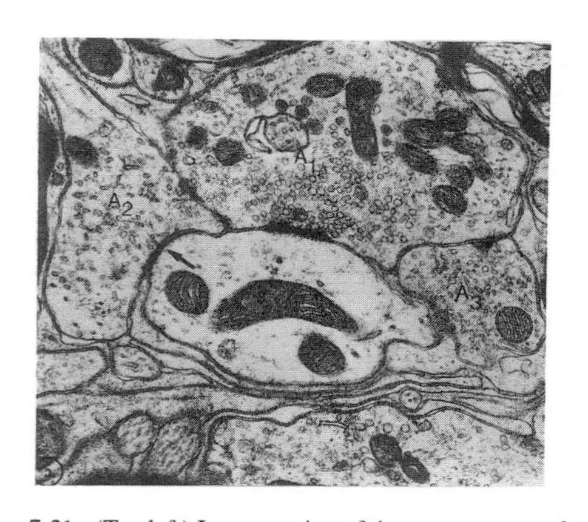

Fig. 7.21. (Top left) Interpretation of the arrangement of synaptic contacts in the cuneate nucleus for afferents in the fasciculus cuneatus and also for fibers descending in the pyramidal tract. The cuneate terminals make axodendritic contacts with cuneothalamic relay cells. The pyramidal tract fibers terminate on interneurons, which in turn form axoaxonic synapses with the afferent endings and also axodendritic endings on the cuneothalamic relay cells. (From Walberg, 1965.) (Right) Electron micrograph showing a degenerating bouton 48 hr after dorsal rhizotomy. (Bottom left) Electron micrograph showing the type of bouton, A1, which originates from a dorsal column afferent and synapses with a dendrite. A1 is also postsynaptic to A2 and A3 in an axoaxonic complex. A2 also synapses with the dendrite. Note that A1 contains spherical synaptic vesicles, while A2 and A3 contain flattened vesicles. (From Rustioni and Sotelo, 1974.)

Dorsal rhizotomies result in the degeneration of large terminals up to 3.1 μm in diameter, forming axodendritic synapses (Fig. 7.21). These degenerating terminals account for about 18% of the synaptic population (Walberg, 1966). Ellis and Rustioni (1981) had similar results, although they estimated that 25% of the synapses are from primary afferent fibers. Presumably, the synapses that do not degenerate belong to interneurons, the collaterals of the axons of projection cells, and corticofugal and other fibers projecting into the dorsal column nuclei (Walberg, 1966; Blomqvist and Westman, 1976; Rustioni and Sotelo, 1974).

Axoaxonic synapses are found upon the large terminals made by the dorsal funiculus afferent fibers (Fig. 7.21) (Walberg, 1965, 1966; Rustioni and Sotelo, 1974; Ellis and Rustioni, 1981; Fyffe *et al.*, 1986a). The presynaptic elements of these axoaxonic complexes contain flattened vesicles. They also form synapses with the same dendrites that are contacted by the primary afferent endings (Fig. 7.21) (Rustioni and Sotelo, 1974; Ellis and Rustioni, 1981). The terminals of group Ia fibers from muscle spindles in the cuneate nucleus form axosomatic synapses, but appear not to receive axoaxonic synapses (Fyffe *et al.*, 1986a). Axoaxonal synapses are also seen in the rat cuneate nucleus (C. K. Tan and Lieberman, 1974).

Blomqvist (1981) performed a quantitative analysis of the synaptic contacts on the somas of gracile neurons, including those labeled retrogradely from the thalamus. Some (8.5%) of the boutons contained large, round vesicles, but most had small, pleomorphic vesicles. There were more endings with large, round vesicles on large than on small gracile neurons. If many of these endings are from primary afferent fibers, this suggests a greater synaptic input from primary afferent fibers to large than to small gracile neurons.

Projection Targets of the Dorsal Column Nuclei

Projections to the Thalamus. The dorsal column nuclei project to the contralateral thalamus by way of the internal arcuate fibers and the medial lemniscus in a variety of species (Fig. 7.22) (Mott, 1895; Ranson and Ingram, 1932; W. E. L. Clark, 1936; Rasmussen and Peyton, 1948; Matzke, 1951; Bowsher, 1958; 1961; Clezy *et al.*, 1961; Hand and Liu, 1966; Lund and Webster, 1967a; Ralston, 1969; Boivie, 1971b, 1978; Jane and Schroeder, 1971; Schroeder and Jane, 1971; Hazlett *et al.*, 1972; Rockel *et al.*, 1972; E. G. Jones and Burton, 1974; Blomqvist and Westman, 1975; Groenewegen *et al.*, 1975; Berkley, 1975, 1980; Cheek *et al.*, 1975; Hand and van Winkle, 1977.

Rat. The projections of the dorsal column nuclei to the contralateral thalamus in the rat end in the ventral posterior lateral (VPL) nucleus (Lund and Webster, 1967a; Massopust *et al.*, 1985). The distribution of the terminals is somatotopic, with the projection from the nucleus gracilis terminating more laterally than that from the nucleus cuneatus (Lund and Webster, 1967a). There are also nonsomatotopically organized projections to the magnocellular region of the medial geniculate complex and to the zona incerta.

Injections of horseradish peroxidase into the contralateral side of the thalamus label neurons throughout the length of the dorsal column nuclei, but with the greatest concentration of cells near the obex in the "slab region," which may correspond to the cluster region in cats (C. K. Tan and Lieberman, 1978; Massopust *et al.*, 1985; Mantle-St. John and Tracey, 1987).

Cat. The thalamic nuclei to which the dorsal column nuclei project in cats include the contralateral VPL nucleus and medial part of the posterior complex (PO_m) (Boivie, 1971b; Berkley, 1975, 1980; Hand and van Winkle, 1977). Another diencephalic projection is to the zona incerta. The projection to the VPL nucleus is somatotopic: the nucleus gracilis projects to the lateral segment (VPL_l) and the nucleus cuneatus to the medial segment (VPL_m) (Boivie, 1971b; Groenewegen *et al.*, 1975). The projections show a complete inversion in space (Fig. 7.23), since the dorsal part of the cuneate nucleus (the more lateral of the dorsal column nuclei) projects to the ventral part of VPL_m, the ventral cuneate to the dorsal VPL_m, the lateral part of the cuneate

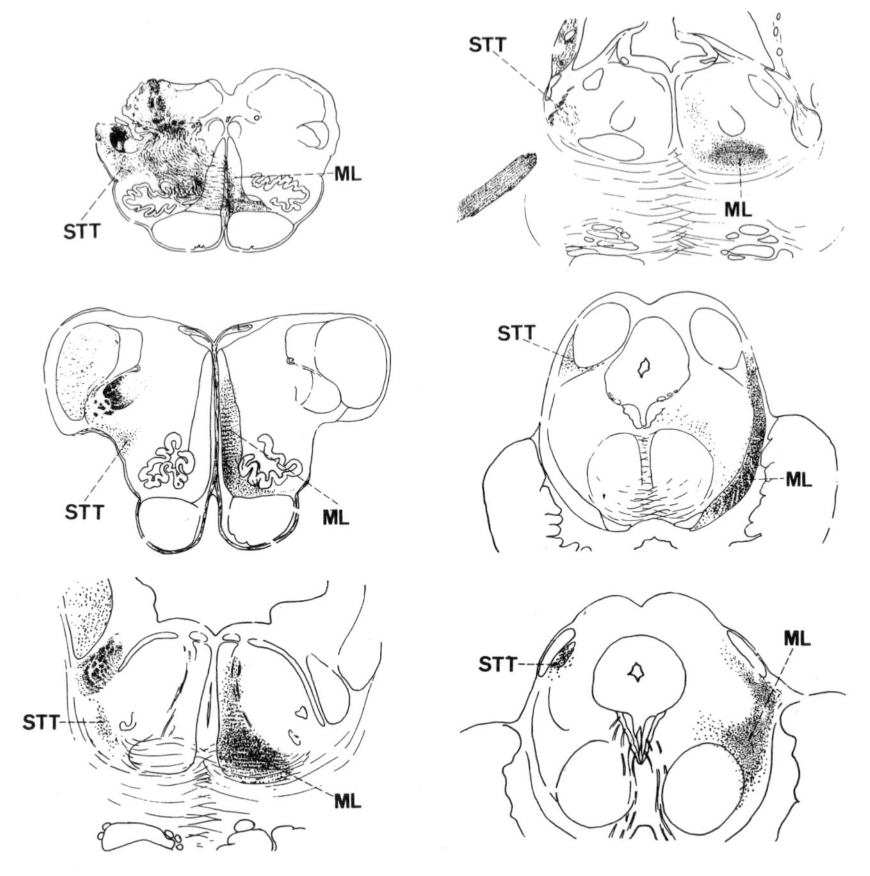

Fig. 7.22. Degeneration ascending through a human brain stem following lesions that damaged the spinothalamic tract and the dorsal column nuclei on the left. A Marchi stain demonstrates the course of the left spinothalamic tract and the right medial lemniscus. (From Rasmussen and Peyton, 1948.)

nucleus to the medial VPL_m and the medial part of the cuneate nucleus to the lateral VPL_m (Hand and van Winkle, 1977). Similarly, the projection from the nucleus gracilis to VPL_l is inverted: i.e., the lateral part of the nucleus gracilis projects to medial VPL_l and the dorsal gracilis to ventral VPL_l (Berkley and Hand, 1978).

An intriguing suggestion has been made by Hand and van Winkle (1977). This is that neurons in a particular part of the cell cluster region provide input to a comparable group of cells in the VPL nucleus, which then connect to a functional cell column in the cerebral cortex (Fig. 7.23).

The ultrastructure of the synaptic terminals of dorsal column neurons in the VPL nucleus has been described following lesions placed in the dorsal column nuclei (Boivie and Westman, 1968; Ralston, 1969). The terminals are large and make asymmetric synapses, mostly on dendrites.

As mentioned previously, cells that label retrogradely following injection of horseradish peroxidase into the contralateral thalamus are distributed throughout the dorsal column nuclei, but labeled cells are especially concentrated in the cell cluster region (Fig. 7.18) (Berkley, 1975; Cheek *et al.*, 1975; Ellis and Rustioni, 1981; Bull and Berkley, 1984; cf. Blomqvist, 1980).

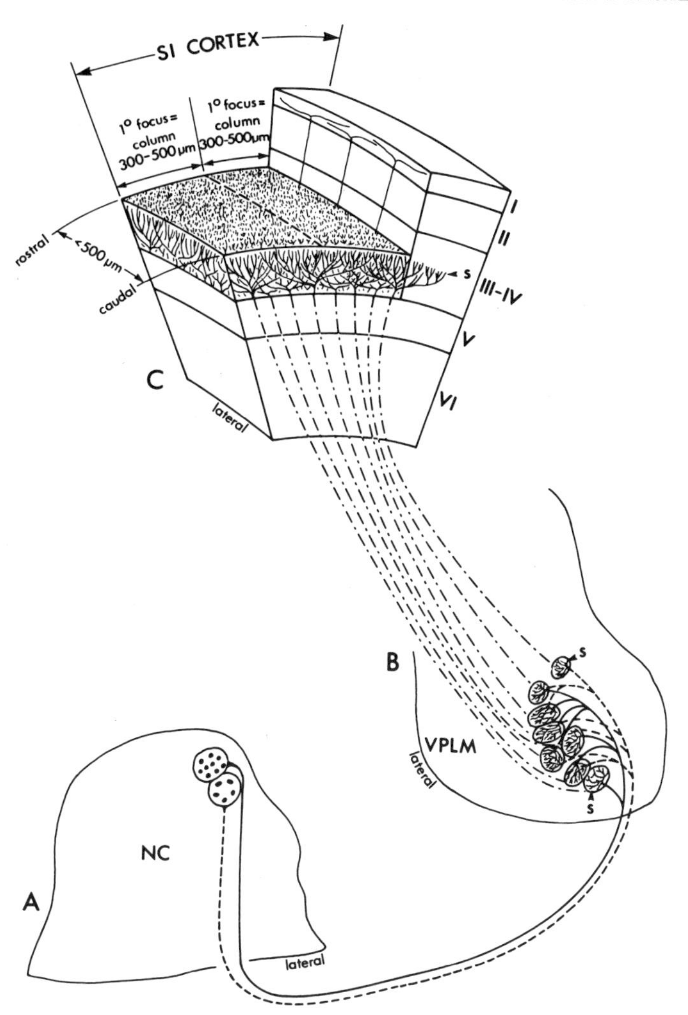

Fig. 7.23. Diagram showing the inversion of topographic relationships between the cuneate nucleus and VPL. Neurons in the dorsolateral part of the cuneate nucleus connect with neurons in the ventromedial VPL. In addition, the possible relationships between clusters of cuneate and VPL neurons and cortical columns is shown. (From Hand and van Winkle, 1977; reproduced with permission.)

Monkey. The dorsal column nuclei in monkeys project to the contralateral VPL (VPL_c) and PO_m nuclei, as well as to the zona incerta (Boivie, 1978; Berkley, 1980). The projection to the VPL nucleus is somatotopic, but not that to the other nuclei (Boivie, 1978). The nucleus gracilis projects to the lateral one-third of the VPL nucleus, whereas the nucleus cuneatus projects to the medial two-thirds (Fig. 7.24).

Neurons are labeled throughout the rostrocaudal length and also across all of the transverse extent of the dorsal column nuclei following injection of horseradish peroxidase into the contralateral thalamus (Albe-Fessard *et al.*, 1975; Rustioni *et al.*, 1979).

Other Projections of the Dorsal Column Nuclei. The dorsal column nuclei project to many parts of the central nervous system other than the diencephalon (Fig. 7.25) (Berkley *et al.*, 1986). There are projections to the dorsal and medial accessory olivary nuclei (Hand and Liu,

A

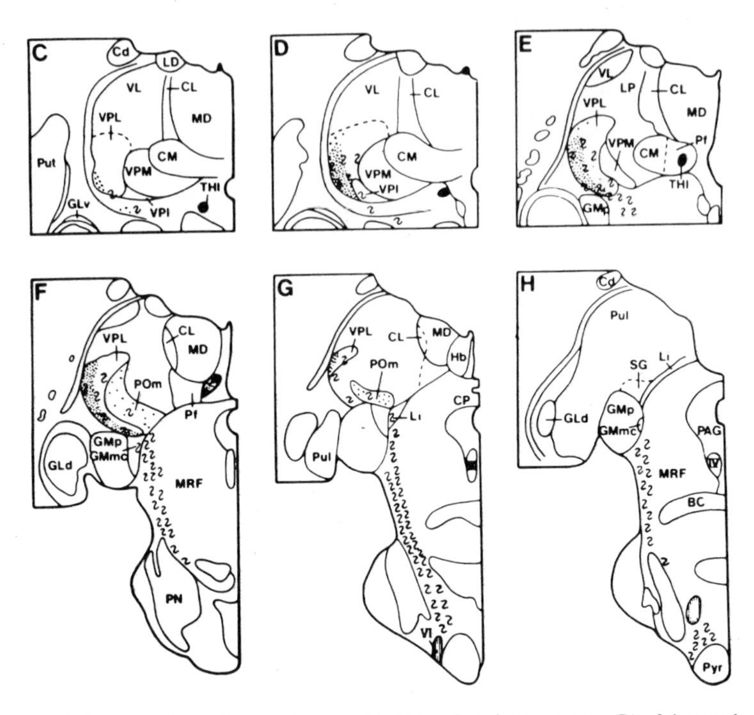

Fig. 7.24. Diencephalic projections of the nucleus gracilis (A) and nucleus cuneatus (B) of the monkey. Lesions were made in the dorsal column nuclei and the anterograde degeneration traced by a silver technique. (From Boivie, 1978; reproduced with permission.)

1966; Morest, 1967; Ebbesson, 1968; Jane and Schroeder, 1971; Schroeder and Jane, 1971; Hazlett *et al.*, 1972; Walsh and Ebner, 1973; Berkley, 1975; Boesten and Voogd, 1975; Groenewegen *et al.*, 1975; Hand and van Winkle, 1977), pontine nuclei (Kosinski *et al.*, 1986), red nucleus (Hand and Liu, 1966; Hand and van Winkle, 1972; Berkley and Hand, 1978), and cerebellum (Gordon and Seed, 1961; Gordon and Horrobin, 1967; Cheek *et al.*, 1975; Rinvik and Walberg, 1975; H. Johansson and Silfvenius, 1977c; Hayes and Rustioni, 1979; Massopust *et al.*, 1985). Dorsal column nuclear projections also reach a number of midbrain structures, including the intercollicular nucleus, superior colliculus and pretectum (Kuypers and Tuerck, 1964; Hand and Liu, 1966; Lund and Webster, 1967a; Jane and Schroeder, 1971; Schroeder and Jane, 1971; Hazlett *et al.*, 1972; Walsh and Ebner, 1973; Hand and van Winkle, 1977; Berkley and Hand, 1978; Blomqvist *et al.*, 1978; Berkley *et al.*, 1980; Bull and Berkley, 1984; Massopust *et al.*, 1985). Descending projections distribute to the spinal cord (Dart, 1971; Kuypers and Maisky, 1975; Burton and Loewy, 1977; Enevoldson and Gordon, 1984).

The different projections of the dorsal column nuclei can be associated in part with different cytoarchitectural regions of the dorsal column nuclei (Fig. 7.25). In cats, the thalamic projection originates from cells distributed over much of the longitudinal extent of the dorsal column nuclei, including the rostral zone, where they are much less concentrated (Berkley, 1975; Blomqvist and Westman, 1975; Cheek *et al.*, 1975). The largest concentration of cells projecting to the thalamus is in the cell nests (Berkley, 1975; Blomqvist and Westman, 1975; Cheek *et al.*, 1975; Blomqvist *et al.*, 1978). Cells that do not project to the VPL nucleus include neurons in the ventral parts of

B

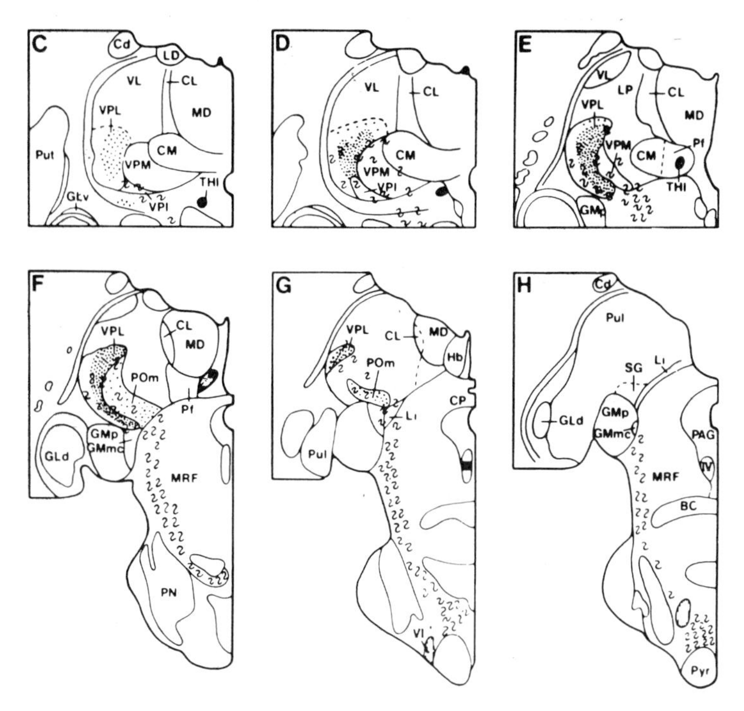

Fig. 7.24. (*Continued*)

the caudal and middle zones, the dorsal part of the middle zone, and much of the rostral zone (Berkley *et al.*, 1986). Neurons that project to the border region between the ventral posterior lateral and ventral lateral nuclei include cells in the ventral parts of the caudal and middle zones. The zona incerta receives input from the rostral zone (Berkley *et al.*, 1986). In the cuneate nucleus, the cells that project to the cerebellum are located only in the rostral zone (Cheek *et al.*, 1975; Rinvik and Walberg, 1975). Midbrain projections are from the rostral zone and the ventral part of the middle and caudal zones (Berkley *et al.*, 1986). The spinal projection originates mainly from cells of the ventral part of the middle zone and cells located between the gracile and cuneate nuclei (Burton and Loewy, 1977; Enevoldson and Gordon, 1984; Berkley *et al.*, 1986). Much of the projection to the inferior olivary nucleus comes from small cells scattered longitudinally along the length of the dorsal column nuclei (Berkley, 1975).

In addition to the complicated cytoarchitecture and multiple projections, several other features of the dorsal column nuclei add to the complexity of the circuitry. It has been suggested that some of the neurons project to more than one locus (Gordon and Seed, 1961; Berkley, 1975; Cheek *et al.*, 1975). Second, the axons of many projection neurons give off recurrent collaterals that end within the dorsal column nuclei and presumably affect the activity of adjacent neurons. Third, there are interneurons within the dorsal column nuclei (Gulley, 1973; Blomqvist and Westman, 1976), although, according to Valverde (1966), these are few in rodents.

The issue of multiple projections of a given neuron of the dorsal column nuclei has been addressed by use of the double-labeling technique (Berkley *et al.*, 1980; Bull and Berkley, 1984;

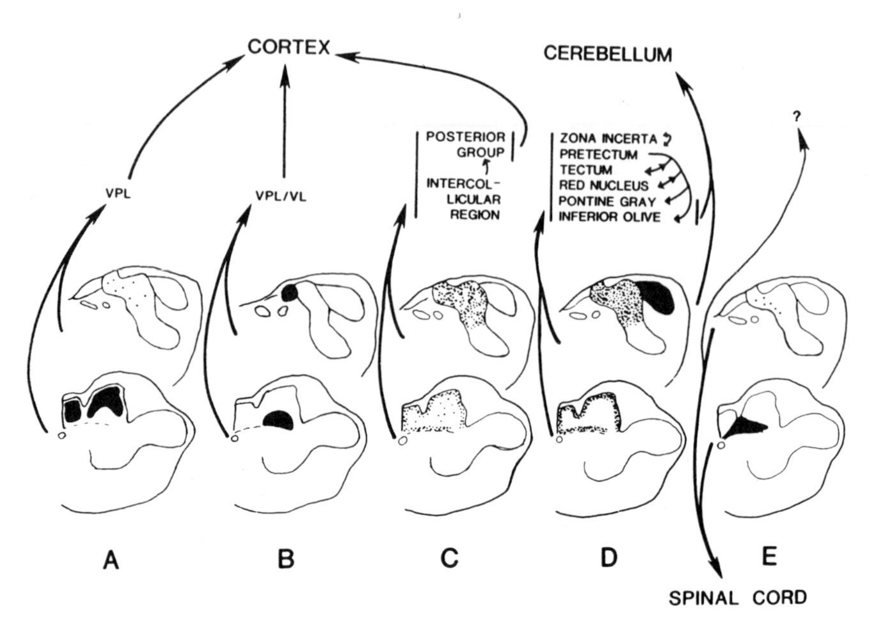

Fig. 7.25. Three projection systems of the dorsal column nuclei of the cat. The projections shown in panels A–C reach the cerebral cortex. (A) Connection between the cell cluster region and the contralateral VPL nucleus. This seems to have a largely tactile function. (B) Projections from the ventral part of the cuneate nucleus and nucleus z to the border zone between VP and VL; this pathway is proprioceptive. (C) Projections of the cells in the rostral zone and in areas surrounding the cell clusters to thalamic and midbrain nuclei that may play a role in orienting behavior and nociception. (D) Cerebellar system involving neurons in the rostral zone and surrounding the cluster region and connecting with precerebellar nuclei. (E) A spinal system that was shown to originate from neurons located ventrally between the gracile and cuneate nuclei. (From Berkley *et al.*, 1986.)

Mantle-St. John and Tracey, 1987). The evidence from this approach is that hardly any dorsal column neurons project to multiple targets. For example, in cats, injection of different tracers into the thalamus and midbrain tectum or pretectum labels almost entirely separate populations of dorsal column neurons (Berkley *et al.*, 1980; Bull and Berkley, 1984; Berkley *et al.*, 1986). Similarly, few double-labeled cells are found in the rat dorsal column nuclei after injections of different tracers into the thalamus and cerebellum (Mantle-St. John and Tracey, 1987), and no double-labeled cells are seen after injections into the ventral posterior lateral nucleus and spinal cord (Berkley *et al.*, 1986).

Summary

In summary, the long system of primary afferent axons in the fasciculi gracilis and cuneatus terminate in an orderly fashion in the nuclei gracilis and cuneatus. A given primary afferent fiber distributes collaterals within a parasagittal plane along the length of a dorsal column nucleus. This accounts for the similarity of the somatotopic map at different rostrocaudal levels of the dorsal column nuclei. However, the pattern of termination of the afferent fibers is less precise at rostral levels of the nuclei. The dorsal column nuclei have a somatotopic organization. The sensory representation of the lower extremity and caudal parts of the body is in the nucleus gracilis and of the upper extremity and rostral parts in the nucleus cuneatus. The distal extremities are represented dorsally and the proximal extremities and trunk ventrally. The cytoarchitecture of the dorsal column nuclei in several species reflects highly ordered caudal and middle zones and a less well ordered rostral zone. The terminals of the afferent fibers of the dorsal funiculus are large, mainly axodendritic synapses that contain spherical vesicles. Axoaxonic synapses containing flattened vesicles terminate on the primary afferent endings and on dendrites. The output from the dorsal column nuclei includes a highly ordered projection to the contralateral ventral posterior lateral thalamic nucleus. There are also projections to the medial part of the posterior complex and the zona incerta in the diencephalon, various mesencephalic nuclei, parts of the inferior olivary nuclei, the cerebellum, and the spinal cord. There is evidence that different projections originate in part from different cytoarchitectural regions of the dorsal column nuclei, although the organization is complex. A given dorsal column projection neuron connects either with the contralateral thalamus or with one of the other targets in the brain stem, but rarely both. Besides projection neurons, the dorsal column nuclei contain interneurons that make local connections within the nuclei.

RESPONSE PROPERTIES OF NEURONS IN THE DORSAL COLUMN NUCLEI

Background Activity

Amassian and Giblin (1974) found that cuneate neurons often have a background discharge consisting of brief, high-frequency bursts and irregular individual discharges. They were able to demonstrate that single, slowly adapting afferent fibers can determine the periodicity of the discharges of tactile and proprioceptive neurons of the cuneate nucleus. The cells were shown to project to the thalamus by antidromic activation from the medial lemniscus. The discharge pattern of a given neuron reflects both the period of discharge of a single afferent fiber and the mixture of inputs from several afferent fibers. A differently coded output would be expected for an input from a population of slowly adapting afferent fibers excited, for example, by a rough surface as compared with a smooth surface. Such differences in coding might contribute to the discriminative ability of a population of dorsal column neurons known not to be subject to afferent inhibition (Perl *et al.*, 1962; Gordon and Jukes, 1964a).

Surmeier and Towe (1987a,b) examined the background activity of proprioceptive neurons in the cuneate nucleus of the cat. The main input to these neurons appeared to be from muscle spindles. Based on the pattern of periodicity of the spike trains, several different classes of neurons could be recognized. Analysis of the spike trains indicated that most of the cells have at least two and as many as six major periodic inputs, suggesting strong synaptic connections from two to six individual muscle spindle afferent fibers. Many of the cells had a convergent input from the skin. The discharge properties of the cells appeared to depend not only upon the synaptic discharges of certain afferent fibers but also on intrinsic membrane properties.

B. H. Pubols *et al.* (1989) analyzed the background discharges of neurons in the cuneate nucleus of the raccoon. Forty-one percent of cells with cutaneous receptive fields projecting to the thalamus have background activity. These cells tend to fire in bursts of two to five spikes. Twenty-two percent of cuneate neurons projecting to the cerebellum have background discharges that are often irregular. It was suggested that neither the background activity nor the tendency of cuneate neurons to fire in bursts is imposed by extrinsic inputs to the dorsal column nuclei.

Responses to Electrical Stimulation

Population Response. Electrical stimulation of peripheral nerves or of the dorsal columns results in a sequence of field potentials that can be recorded from the dorsal column nuclei (Fig. 7.26). For example, in recordings from the cuneate nucleus following stimulation of cutaneous nerves of the forelimb, there is an initial spikelike potential, and then slow negative (N) and positive (P) waves (Therman, 1941; P. Andersen *et al.*, 1964a). The spike potential is attributable

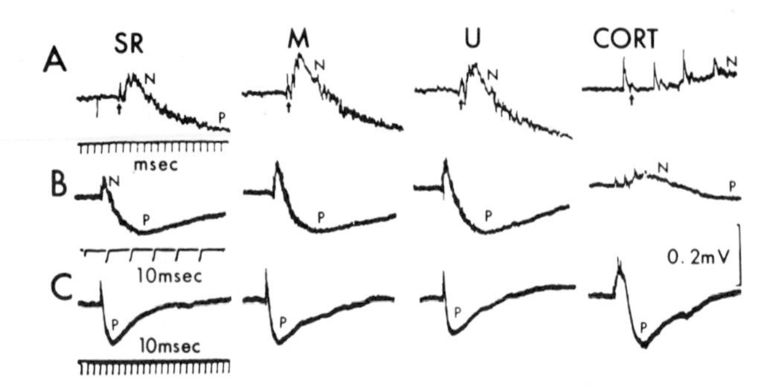

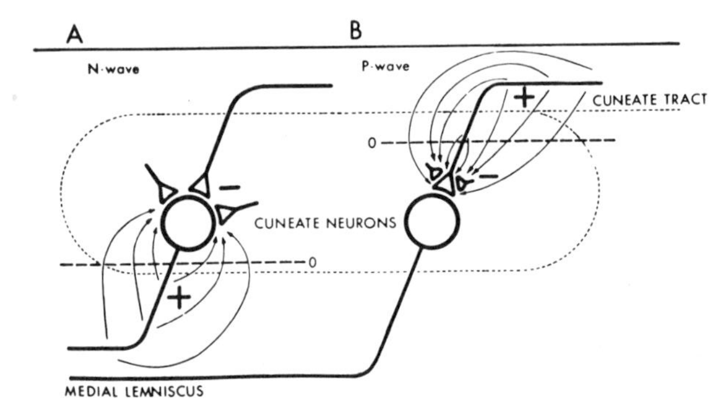

Fig. 7.26. Field potentials recorded from the surface of the cuneate nucleus of the cat. Structures stimulated included the superficial radial (SR), median (M) and ulnar (U) nerves, and the sensorimotor cortex (CORT). The negative (N) and positive (P) field potentials are shown on progressively slower time bases in panels A–C. The diagram at the bottom shows the pattern of current distribution that might account for the field potentials. (From P. Andersen *et al.*, 1964b.)

to the arrival of the afferent volley in the fastest fibers of the fasciculus cuneatus, whereas the N wave is the sum of an afferent volley in slower afferent fibers and the extracellular excitatory postsynaptic currents and action potentials that are triggered in cuneate neurons by the excitatory postsynaptic potentials (P. Andersen *et al.*, 1964a; 1972a; Andres-Trelles *et al.*, 1976). The P wave is considered to reflect the depolarization of the terminals of primary afferent fibers ending in the cuneate nucleus (primary afferent depolarization [PAD]) resulting from activation of an interneuronal pathway that forms axoaxonic synapses on the primary afferent terminals (P. Andersen *et al.*, 1964a,b; Walberg, 1965; Rustioni and Sotelo, 1974). PAD would presumably cause presynaptic inhibition of transmission through the dorsal column nuclei (P. Andersen *et al.*, 1964d; J. C. Eccles, 1964; R. F. Schmidt, 1971).

Krnjević and Morris (1976) examined the input-output relationship of the cuneate nucleus by stimulating the dorsal funiculus and recording the antidromically conducted volley in primary afferent fibers from a peripheral nerve (input) and the relayed compound action potential in the medial lemniscus (output). The input-output curve so derived was a power function with a very steep rise, indicating a high safety factor for transmission.

Rigamonti and Hancock (1974) recorded a negative field potential centered in the ventrolateral gracile nucleus and the cell bridge between the gracile and cuneate nuclei when they stimulated Aβ fibers in the greater splanchnic nerve of the cat. This region corresponds to the trunk representation.

Single-Unit Activity. Many studies of single-unit activity in the dorsal column nuclei have failed to distinguish between neurons that transmit information to the thalamus and those that project elsewhere or that serve as interneurons. Thalamic relay neurons can be identified by antidromic activation following stimulation in the contralateral medial lemniscus (Fig. 7.27) (Amassian and DeVito, 1957; Gordon and Seed, 1961; Gordon and Jukes, 1964a; P. Andersen *et al.*, 1964c; Gordon and Horrobin, 1967; Rosén, 1969). The collision test (Paintal, 1959; Bishop *et*

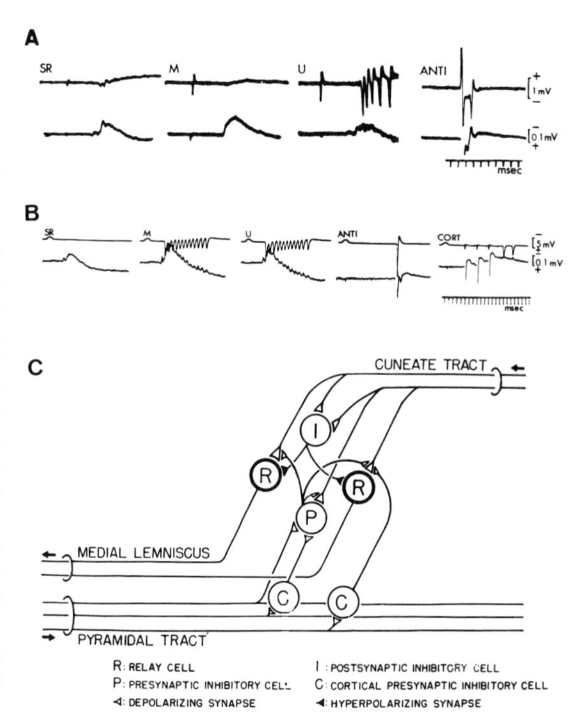

Fig. 7.27. (A) Records made from a cuneothalamic relay cell (upper traces) and the dorsal surface of the cuneate nucleus (lower traces). The neuron was activated orthodromically by stimulation of the ulnar nerve (U), but not of the superficial radial (SR) or median (M) nerves. Identification was provided by antidromic activation (ANTI) following stimulation of the contralateral medial lemniscus. (B) Activity of an interneuron that was excited by stimulation of either the median or ulnar nerves or of the sensorimotor cortex (CORT), but that was not antidromically activated from the medial lemniscus. (C) Possible circuit that accounts for convergence of excitatory input from peripheral nerves and the sensorimotor cortex onto cuneate interneurons, which then produce presynaptic inhibition of excitatory pathways to cuneothalamic relay neurons. A postsynaptic inhibitory pathway through a different interneuron is also shown. (From P. Andersen *et al.*, 1964d.)

al., 1962; Darian-Smith *et al.*, 1963; see Fuller and Schlag, 1976) has been used in some of the studies to help distinguish between orthodromic and antidromic activation (Gordon and Jukes, 1964a; Rosén, 1969). Interneurons will obviously not be activated antidromically from any of the projection targets of the dorsal column nuclei. However, it would be difficult to stimulate in all of these targets, and so firm identification of a given neuron as an interneuron is problematical.

The excitatory effects of electrically evoked volleys on single units in the cuneate nucleus of the cat were examined by Amassian and DeVito (1957). In addition to monosynaptic excitation, repetitive discharges were observed that were attributed to interconnections within the nucleus through axon collaterals. However, P. Andersen *et al.* (1964d) found that the EPSPs produced by the primary afferent volley were large enough to account for the repetitive discharges. In fact, Amassian and DeVito (1957) observed that some cuneate neurons were discharged by a small afferent volley representing as little as 0.3% of the dorsal column fibers (fewer than eight fibers). P. Andersen *et al.* (1964d) agreed that the intracellularly recorded EPSPs produced by single afferent fibers were quite large and that summation of only a few would be needed to cause a discharge. Amassian and Giblin (1974) showed that cuneate neurons could discharge virtually every time a single afferent fiber discharged, suggesting complete synaptic security for the linkage between a single afferent fiber and a single relay neuron. Similarly, Ferrington *et al.* (1986a, 1987a) have found that individual Pacinian corpuscle afferent fibers can cause one-to-one discharges in dorsal column neurons (Fig. 7.28). Such powerful excitatory effects may correlate with the large sizes of the synapses made by dorsal column afferent fibers upon neurons in the dorsal column nuclei (Fig. 7.21) (Rozsos, 1958; Walberg, 1965, 1966).

However, the tendency for dorsal column neurons to discharge repetitively is not due entirely to their synaptic input from primary afferent fibers, since the cells also tend to discharge in doublets or in short bursts when activated by the iontophoretic application of glutamate (Galindo *et al.*, 1968). The repetitive discharge is not the result of an excitatory feedback through recurrent collaterals of the axons of relay cells, since antidromic activation by stimulation in the

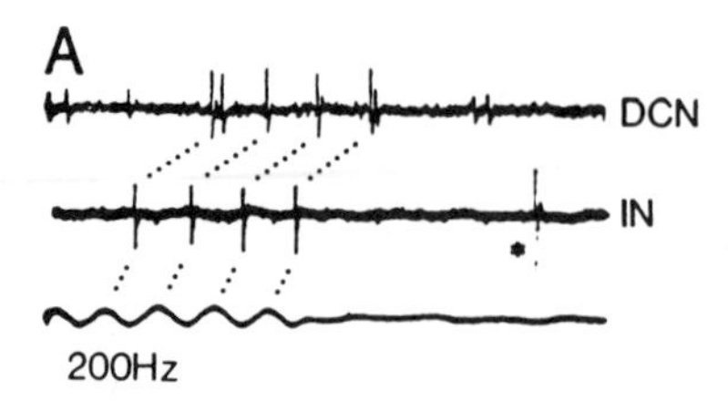

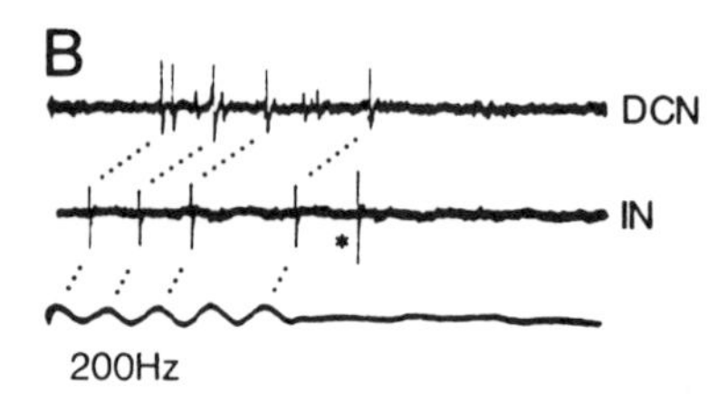

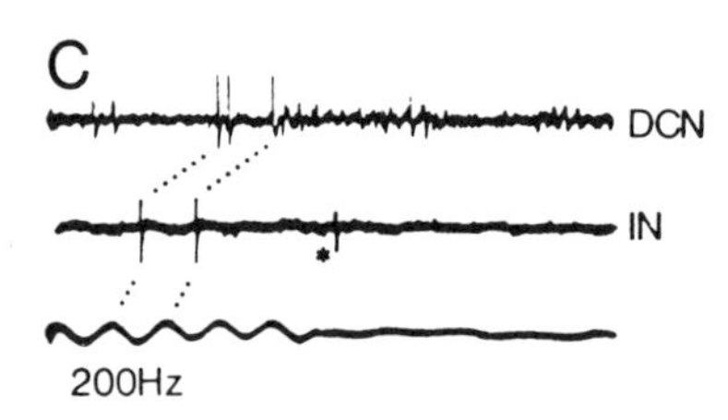

Fig. 7.28. Excitation of neurons in the cat gracile nucleus by the discharges of individual Pacinian corpuscle afferent fibers. The upper trace of each set shows the discharges of a dorsal column neuron. The middle trace was recorded from the interosseous nerve and shows the discharges of a Pacinian corpuscle afferent fiber (asterisks indicate spikes from other afferent fibers). The lower trace is a monitor of the vibratory stimulus. (From Ferrington *et al.*, 1986a.)

medial lemniscus does not result in a repetitive discharge (Gordon and Seed, 1961). It should be noted, however, that Blum *et al.* (1975) and Rowinski *et al.* (1985) have reported that some relay cells and interneurons of the cuneate nucleus can be activated synaptically by stimulation in the medial lemniscus. The tendency toward a repetitive discharge appears to be an intrinsic property of cells in the dorsal column nuclei (Galindo *et al.*, 1968), possibly as a result of the retriggering of the initial segment by a dendritic spike (cf. Calvin and Loeser, 1975).

Rosén (1967, 1969) investigated the input to the cuneate nucleus of the cat from group I muscle afferent fibers. The responses of both cuneothalamic neurons and unidentified cuneate neurons were recorded. Cells activated by group I muscle afferent fibers tended to be deeper than most of the cells excited by cutaneous input. Usually, only a single muscle nerve activated a given neuron, although sometimes there was a convergent input from several muscle nerves. There could also be a convergent input from group II muscle afferent fibers and from cutaneous fibers.

Single-unit recordings from the gracile nucleus in cats showed a convergent input from visceral afferent fibers of the greater splanchnic nerve and cutaneous afferent fibers supplying the trunk (Rigamonti and Hancock, 1978). Many of the cells could be activated antidromically from the contralateral medial lemniscus.

Inhibition. Inhibition of background or evoked activity in the dorsal column nuclei has been reported by many groups (Therman, 1941; Amassian, 1952; Gordon and Paine, 1960; Gordon and Jukes, 1964a; P. Andersen *et al.*, 1964d, 1970, 1972b; Jabbur and Banna, 1968; 1970; Rosén, 1969; Biedenbach *et al.*, 1971; Davidson and Smith, 1972; Silvey *et al.*, 1974; Bromberg *et al.*, 1975; Aoki, 1981). Part of the inhibition in the cat can be ascribed to a postsynaptic inhibitory action, since inhibitory postsynaptic potentials (IPSPs) have been recorded intracellularly from dorsal column neurons, and the excitability of dorsal column neurons to direct electrical stimulation decreases during inhibition (P. Andersen *et al.*, 1964d, 1970; Schwartzkroin *et al.*, 1974). The durations of most of the IPSPs are only about 15–20 msec, and yet the time course of inhibition of dorsal column neurons often exceeds 100–200 msec. This more prolonged inhibition can be attributed in part to occasional long-lasting IPSPs, but is also due in part to presynaptic inhibition (P. Andersen *et al.*, 1964d, 1970, 1972b; Jabbur and Banna, 1968, 1970; Bromberg *et al.*, 1975). In the frog, there is no evidence for postsynaptic inhibition, but there is for presynaptic inhibition (Silvey *et al.*, 1974).

A number of lines of evidence suggest the importance of presynaptic inhibition in the dorsal column nuclei. The responses of dorsal column neurons to electrical evoked afferent volleys are often complicated by the occurrence of a "dorsal column reflex" (Hursh, 1940; Andersen *et al.*, 1964b). The "dorsal column reflex" is a discharge in the afferent fibers of the dorsal funiculus and is equivalent to the "dorsal root reflex" (Toennies, 1938; J. C. Eccles *et al.*, 1961; J. C. Eccles, 1964), which is due to PAD. The dorsal column reflex can be generated either at the segmental level or at the level of the dorsal column nuclei (P. Andersen *et al.*, 1964b). Excitability testing shows that the afferent terminals within the dorsal column nuclei are depolarized as a result of activity evoked by other afferent fibers ascending in the dorsal funiculus (Wall, 1958; P. Andersen *et al.*, 1964b,d, 1970, 1972b; Jabbur and Banna, 1968;1970; Silvey *et al.*, 1974; Bromberg *et al.*, 1975). The depolarization results in a negative field potential near the terminals of the dorsal column afferent fibers but a positive field potential at the dorsal surface of the medulla (Therman, 1941; Andersen *et al.*, 1964a,b; Silvey *et al.*, 1974).

Putnam and Whitehorn (1973) investigated the organization of the PAD system in the gracile nucleus to see whether PAD in rapidly adapting receptor terminals is generated primarily by inputs in rapidly adapting afferent fibers and PAD in slowly adapting afferent terminals by slowly adapting afferent fibers, as in the spinal cord (Jänig *et al.*, 1968b). There was no evidence for this arrangement. Instead, PAD is produced in the afferent endings of G_1 and G_2 hair follicle receptors, field receptors, and type II slowly adapting receptors by the same form of natural stimulation: hair movement by puffs of air but not steady indentation of the skin. They concluded that the receptors responsible for the PAD were G_2 and–or G_1 hair follicle afferents. However, in view of the evidence of Bystrzycka *et al.* (1977) that second-order neurons of the cuneate nucleus are inhibited by Pacinian corpuscles, which are so sensitive that they can be activated by hair movement, the receptors responsible for PAD in primary afferent fibers to the dorsal column nuclei must be re-evaluated.

An effort was made by P. Andersen *et al.*, (1964c) to identify the interneurons that might be responsible for producing inhibition in the cuneate nucleus. Since stimulation of the cerebral cortex also results in PAD (P. Andersen *et al.*, 1964a), one criterion for identification of the interneurons is a convergent excitation from both peripheral-nerve and corticofugal volleys (Fig. 7.27). P. Andersen *et al.* (1972b) used the technique of cooling the cuneate nucleus to alter the inhibitory activity as a way to assist in the identification of the interneurons. Cooling enhances and prolongs PAD in the cuneate nucleus (P. Andersen *et al.*, 1972b). However, postsynaptic inhibition is not changed. When the discharges of interneurons are recorded, cooling depresses the activity in half the cases, but causes a prolongation of the discharges in the other half. The same interneurons display a poor ability to follow repetitive stimulation, with a curtailed discharge at stimulus rates of 10/sec or greater, a stimulus rate that severely reduces PAD. It seems likely that these interneurons are involved in generating PAD.

Other investigations in which an attempt has been made to identify the interneurons responsible for PAD in the cuneate nucleus are those of Bromberg *et al.* (1975) and Blum *et al.* (1975). They observed that PAD can occur with either a fast or a slow time course. Taking into consideration the time course of the PAD from the cerebral cortex and the pattern of convergence, they concluded that separate populations of interneurons are likely to be responsible for the different PADs.

Recurrent inhibition of cuneothalamic relay neurons has been found following stimulation of the medial lemniscus (Davidson and Smith, 1972). The inhibition is probably in part presynaptic, since PAD is produced concurrently. Interneurons of the cuneate nucleus may be activated by recurrent volleys as well as by input following stimulation of ipsilateral and contralateral peripheral nerves. It is possible that these interneurons form a final common pathway for PAD in the cuneate nucleus.

Field potentials and single-unit activity have been recorded from a slice preparation of the rat nucleus gracilis (Newberry and Simmonds, 1984b,c). Stimulation of the dorsal funiculus evokes a brief latency spike, followed by an N wave. The N wave is decreased by superfusion with the calcium antagonist cadmium, showing its dependence on synaptic transmission. A P wave is not observed. The N wave has a bicuculline sensitive component, but this may reflect postsynaptic rather than presynaptic inhibition. No action potentials could be recorded from primary afferent fibers. Postsynaptic units were subdivided into several classes, depending upon their responses to increasing intensities of stimulation.

Responses to Natural Stimuli

Receptive Fields. According to Gordon and Paine (1960), most of the cells in the nucleus gracilis in cats respond to mechanical stimulation of hair or of the skin, whereas only 10% respond to stimulation of deep tissues, including joints. Receptive-field sizes vary from less than 0.5 cm^2 to the entire hindlimb and lower half of the ipsilateral trunk. Cells with large receptive fields tend to be concentrated in the rostral part of the nucleus, whereas cells with small receptive fields are in the middle zone and those with intermediate-sized fields are in the caudal zone. Cells in the middle zone are most likely to show surround inhibition. Only a small fraction of the cells in the rostral zone are activated antidromically from the contralateral medial lemniscus, whereas 80–90% of those in the middle zone are. These observations fit anatomic findings that the largest concentration of thalamic projection neurons is in the middle zone, whereas cells in the rostral zone tend to project elsewhere, such as to the cerebellum (see above).

The receptive fields of units in the nucleus gracilis of rats were investigated by McComas (1963), who applied brief mechanical pulses to the pads of the hind foot. Cells within the rostral, middle and caudal parts of the nucleus were compared with respect to receptive-field size, threshold, latency, and ability to respond to repeated stimulation at a high frequency. The rostral cells have high thresholds, large receptive fields, long latencies, and poor ability to follow high frequency stimulation. The opposite is true for cells in the middle and caudal regions of the nucleus. Furthermore, cells in the middle part of the nucleus are more likely to have inhibitory receptive fields than are cells in the rostral part.

Classification of Dorsal Column Neurons. Kruger *et al.* (1961) subdivided the units within the dorsal column nuclei of cats according to the forms of natural stimuli that activate them. Kruger *et al.* (1961) could activate most of the neurons by hair movement or by tactile stimulation of the skin. Pressure units may be activated in at least some cases by subcutaneous receptors. Joint units respond to small changes in joint angle but not to cutaneous stimulation or manipulation of muscle. Hair-activated units are usually rapidly adapting, whereas touch, pressure, and joint units can be rapidly or slowly adapting. Different classes of cells are intermingled. Receptive-field size varies with the position of the field on the limb: the smallest fields are found distally, and the largest are on the trunk. Kruger *et al.* (1961) found no relationship between receptive-field size and the rostrocaudal location of the cell (cf. Gordon and Paine, 1960). No attempt was made to identify the units by antidromic activation, and no distinction was made between the properties of neurons in the gracile versus the cuneate nucleus.

Perl *et al.* (1962) tried to relate the discharge properties of neurons of the cat nucleus gracilis to the responsible receptors. They subdivided their units into three groups. (1) The first class is activated by hair movement. Care had to be taken to distinguish these units from units activated by vibration. The hair units are rapidly adapting and discharge when hair is bent and again when it is allowed to straighten. Receptive-field size varies from 10 mm^2 to more than 2000 mm^2. Receptive fields are smallest distally and larger more proximally. Surround inhibition can be demonstrated. (2) The second class is activated by tactile stimulation of the skin and also by cooling. The responses to touch are slowly adapting. (Note that the cooling responses presumably reflect the activation of slowly adapting mechanoreceptors by such stimuli, rather than activation of cold receptors; see Chapter 2). Receptive-field sizes are larger than for hair-activated units, ranging from 228 to 28,900 mm^2. Surround inhibition cannot be demonstrated. (3) The third class is activated by vibratory stimulation (cf. Amassian and DeVito, 1957), following 100–300-Hz stimuli. These units sometimes discharge with each cardiac cycle. The receptors appear to be in the abdominal cavity and are presumably Pacinian corpuscles. In addition, a few neurons can be excited by joint bending or application of pressure to deep tissues of the foot.

The most elaborate classification of the response properties of neurons in the dorsal column nuclei is that of Gordon and Jukes (1964a) for cells in the cat nucleus gracilis that receive input from cutaneous receptors. Eight classes were found: (1) cells activated by hair movement, usually with a rapidly adapting response; the receptive-field size depends upon position within the rostrocaudal extent of the nucleus and location on the limb, with the smallest fields for cells in the middle zone of the nucleus and on the distal extremity; (2) cells unaffected by hair movement but giving either rapidly or slowly adapting responses to weak tactile stimulation of the skin; they are located in the middle zone and have small receptive fields; (3) cells responsive both to hair movement and to tactile stimulation; most are rapidly adapting, but some are slowly adapting; most are in the rostral zone, but some in the middle zone; receptive-field sizes vary from 1 to 50 cm^2; (4) cells giving slowly adapting responses to claw movement or to stimulation of a small area at the base of a claw; almost all are in the middle zone; (5) cells that respond to light touch or pressure with a slowly adapting discharge; located throughout the nucleus, but those in the middle zone tend to be deep; receptive fields are often large, but are smaller for cells in middle than in the rostral zone; (6) cells with a rapidly adapting response to hair movement and a slowly adapting response to touch and pressure; they are found throughout the nucleus and have large receptive fields for pressure; (7) cells with rapidly or slowly adapting responses to subcutaneous stimuli; they are located in the middle zone, but have large receptive fields; (8) cells sensitive to vibration caused by applying a 100-Hz tuning fork to skin or bone; they are found in the middle zone and probably activated by Pacinian corpuscles, including some in the abdominal cavity; a subpopulation is also activated by hair movement or touch. Inhibition is often demonstrated following stimulation of hair, pads, claws, or subcutaneous tissue. Touch-pressure units are not inhibited. Antidromic activation from the contralateral medial lemniscus is possible for at least some units of each class. Most of these antidromically activated units are in the middle zone. However, only 31% of the touch-pressure units in the middle zone are activated antidromically from the medial lemniscus.

Gordon and Jukes (1964a) conclude that the nucleus gracilis of the cat has a dual organization. Hair-sensitive, pad-sensitive, pad- and hair-sensitive, claw-sensitive, and subcutaneous units have small receptive fields, are located in the middle zone of the nucleus, are subject to surround inhibition, are somatotopically organized, and project into the contralateral medial lemniscus. Other hair-sensitive, pad- and hair-sensitive, and touch-pressure units have large receptive fields, are located in the rostral zone or deep in the middle zone, are not subject to surround inhibition, are not somatotopically organized, and often do not project into the contralateral medial lemniscus.

The response properties of axons of the cat medial lemniscus have been described by A. G. Brown *et al.* (1974). These authors believed that most of their units are the axons of neurons in the dorsal column nuclei projecting to the thalamus. However, it is possible that some belong to the cervicothalamic or the spinothalamic tract. Ten classes of units can be recognized, in addition to an unidentified class: (1) units with rapidly adapting responses to movement of tylotrich hairs, small receptive fields, and surround inhibition, (2) units with rapidly adapting responses to tactile stimulation of one or more footpads and often inhibition; (3) units responding to claw movement or pressure on the skin adjacent to a claw and inhibitory fields; (4) units excited by touch corpuscles, with inhibitory fields; (5) units having a slowly adapting response to stimulation of a single spot (presumably over an SA II receptor) and no inhibitory fields; (6) units with rapidly adapting responses to all types of hairs and with inhibitory fields; (7) units excited by hair movement and pad stimulation, rapidly adapting or both rapidly and slowly adapting responses, and usually with inhibitory fields; (8) units with rapidly adapting responses to hair movement and slowly adapting responses to skin displacement, either low or high thresholds for the skin displacement, and usually inhibitory fields; (9) units excited in a rapidly adapting fashion by hair movements, probably owing to guard hairs; (10) units with slowly adapting responses to joint movement or stimulation of deep receptors.

A different approach to the classification of neurons within the nucleus cuneatus was attempted by Blum *et al.* (1975). In addition to identifying the cells by their reactions to natural stimuli applied to the skin and to stimulation of the medial lemniscus, these workers divided the cells in two other ways. Using both natural and electrical stimulation, they could distinguish between two subsets of neurons. One subset was characterized by an ability to follow rapidly repeated stimuli ("strong synaptic connection"), while the other had a poor ability to do so ("weak connection"). Another subdivision was according to whether a given cell was activated from just the ipsilateral forelimb or whether it received a convergent input from more than one limb. Cuneothalamic relay cells might have strong or weak synaptic drive. Interneurons might have strong or weak input, and they might be nonconvergent or convergent. Excitatory inputs from other regions of the brain, such as the cerebral cortex, reticular formation, or cerebellum, activate the convergent interneurons (either directly or through the reticular formation).

Further Evidence by Using Mechanical Stimulation of Skin. The time required for the arrival of activity from different kinds of afferent fibers in the skin to the level of the gracile nucleus was studied by Whitehorn *et al.* (1972). The ranges of conduction times are characteristic for different classes of receptor and fall within these ranges despite differing conduction distances.

Bystrzycka *et al.* (1977) reexamined the excitatory and inhibitory responses of neurons of the nucleus cuneatus in unanesthetized, decerebrate cats. They used carefully controlled natural stimuli to activate particular types of afferents selectively. The cells were subdivided into slowly and rapidly adapting classes by their responses to step indentations of the skin. The rapidly adapting cells were further subdivided into those that were most sensitive to vibratory stimuli in the frequency range of 20–50 Hz and those that were fired best by vibrations of 200–300 Hz. The threshold for the latter were often of the order of 1–2 μm of displacement, and the receptive fields were very large. It was suggested that the cells responsive to 300-Hz stimulation have a specific input from Pacinian corpuscles. Hair movement also activated these cells, but this kind of stimulus will excite Pacinian corpuscles as well as hair follicle receptors. All three kinds of neurons could be inhibited by Pacinian corpuscle input and probably not by other inputs. Some previous workers who did not find inhibition of slowly adapting neurons in the dorsal column nuclei used barbiturate anesthesia, which may have interfered with the inhibition. The findings of Bystrzcka *et al.* (1977) are consistent with evidence for a strong projection of Pacinian corpuscle afferents through the dorsal column-medial lemniscus system to the cerebral cortex (see previous section) (McIntyre, 1962; McIntyre *et al.*, 1967; Silfvenius, 1970).

Golovchinsky (1980) examined the responses of neurons in the cuneate nucleus of the cat, by using controlled mechanical stimuli. He was able to recognize inputs to different cells from guard hairs, field receptors, slowly adapting mechanoreceptors, and Pacinian corpuscles. No units were found that could clearly be activated by down hair afferent fibers or high-threshold mechanoreceptors. Only a few neurons (2%) had properties that were distinctly different from those of primary afferent fibers. Inhibition was rarely observed.

Dykes *et al.* (1982) explored the rostrocaudal length of the dorsal column nuclei in cats and found evidence to support the hypothesis that afferent fibers of different modalities project to different rostrocaudal zones. Afferent fibers from Pacinian corpuscles were found to activate a subset of neurons in the caudal zone, whereas rapidly and slowly adapting cutaneous mechanoreceptors activated different populations of neurons in the middle zone.

Rowinski *et al.* (1985) investigated the response properties of cuneothalamic neurons in the raccoon. Three subsets of neurons were recognized: rapidly adapting (RA, 56%), slowly adapting (SA, 24%) and Pacinian (PC, 20%) units, depending on their response properties. Most of the cells had receptive fields on the glabrous skin, in contrast to cells of the dorsal column nuclei that projected to the cerebellum, which were generally responsive to muscle stretch (Haring *et al.*, 1984).

The ability of neurons in the dorsal column nuclei of the rat to respond to stimuli moving across the skin was studied by Castiglioni and Kruger (1985). The stimulus was an air jet that was moved across the receptive field and back and then again at right angles to the first trajectory. The responses reflected the position of the stimulus within the receptive field, as well as the intensity and velocity of the stimulus. There was no evidence for feature extraction, such as recognition of stimulus direction.

Pertovaara *et al.* (1986) studied the responses to foot pad stimulation of RA, SA, and PC classes of neurons in the cuneate nucleus of the cat. Many of the neurons had stimulus response functions that resembled those of primary afferent fibers. Noxious stimuli failed to influence the six neurons tested.

Ferrington *et al.* (1986a, 1987a,b,c) investigated the capacity of Pacinian corpuscles to activate neurons in the nucleus gracilis and cuneatus of the cat. In some experiments, the activity of individual Pacinian corpuscle afferents was monitored in a recording from the interosseous nerve in the hindlimb (Fig. 7.28). A given gracile neuron could be activated by any of several different Pacinian corpuscle afferents, and a given Pacinian corpuscle afferent could excite several different gracile neurons. This divergence and the ability of a single afferent spike to trigger a repetitive discharge in the gracile neuron were considered to be mechanisms for amplification of signals originating from Pacinian corpuscles. Analysis of phase locking indicated that only one of a few Pacinian corpuscles had a predominant control over a given gracile neuron.

Inhibitory Receptive Fields. Primary afferent depolarization is produced in the cuneate nucleus by blowing on or brushing hairs (P. Andersen *et al.*, 1970). Comparable stimuli can also evoke IPSPs which sometimes last as long as 160 msec.

Jänig *et al.* (1977) recorded from cuneothalamic relay neurons in the cat. They classified the cells according to whether they were excited by bending hairs (H neurons), by vibration (P neurons), or by constant-force stimuli (SA neurons). The H and SA cells have small receptive fields and the P neurons large ones. Air pulses evoke inhibition which lasts for about 80–100 msec in H and SA neurons and more than 250 msec in P neurons. The inhibitory receptive field overlaps the excitatory field, but spreads more widely. It was concluded that the effect of the inhibitory field would be to enhance contrast.

Aoki (1981) finds inhibition of only 18% of cuneate neurons. However, the proportion of neurons receiving inhibition is higher (44%) for the subpopulation of cuneothalamic neurons. Dynamic stimuli (vibration, hair movement) cause inhibition both of rapidly and slowly adapting units, whereas steady pressure causes inhibition just of slowly adapting neurons. The inhibitory receptive fields are eccentric rather than of the surround type. It is suggested that this arrangement may provide information useful for detecting the direction of a moving stimulus.

Activation of Proprioceptors. S. Yamamoto and Miyajima (1961) studied proprioceptive units within the dorsal column nuclei. There are proportionately more of these in the nucleus cuneatus than in the nucleus gracilis. Some of the proprioceptive neurons discharge with a respiratory rhythm.

The projection of forelimb group I muscle afferent fibers to the nucleus cuneatus has been examined by using natural stimuli (Rosén and Sjölund, 1973a,b). Cuneothalamic relay neurons are activated by group Ia muscle spindle afferent fibers, but not by group Ib Golgi tendon organ afferent fibers, as shown by a low threshold to passive stretch of muscle, by a pause in discharge during twitch contractions, by excitation during longitudinal vibration of the muscle (oscillations with a 50-μm amplitude), and by activation following the administration of succinylcholine. Interneurons in the cuneate nucleus can be activated by Golgi tendon organ afferent fibers, but most are excited by Ia afferent fibers. Cuneothalamic neurons and interneurons are usually excited by input from just a single muscle, although some are excited by afferent fibers from several muscles, almost always synergistic ones. Few cells are inhibited, even by muscles antagonistic to those producing excitation.

W. J. Williams *et al.* (1973) studied the responses of neurons in the nucleus gracilis to bending the knee joint in cats in which the hindlimb was denervated except for the medial and posterior articular nerves. The responses are relatively rapidly adapting and signal either velocity or acceleration of changes in joint angle. The lack of slowly adapting neurons signaling position is consistent with evidence that slowly adapting afferent fibers from the knee joint do not project all the way up the fasciculus gracilis (Burgess and Clark, 1969a).

Millar (1979a) activated neurons in the cat cuneate nucleus by stimulating the elbow joint nerve. Cells excited by this stimulus are found in the dorsolateral part of the cuneate nucleus (as well as in the lateral cuneate nucleus). Thus, the cuneate cells that respond to joint input are distinct from those in the base of the nucleus that respond to muscle afferent input. Some cuneate cells respond in either a slowly or a rapidly adapting fashion to elbow movement, and many receive a convergent input from sensitive cutaneous mechanoreceptors in the region of the cubital fossa (see also Millar, 1979b).

Tracey (1980) has also examined the responses of cuneate neurons to joint input in cats. The cuneate fasciculus contains both branches of primary afferent fibers of the wrist joint nerve and the axons of neurons of the postsynaptic dorsal column pathway that respond to stimulation of the wrist joint nerve. Units responsive to stimulation of the wrist joint nerve can be recorded in and ventral to the cuneate nucleus. Phasic responses are seen in many of the cells when the wrist is moved; however, a few cuneate neurons show tonic responses. Convergent input from cutaneous and–or muscle afferent fibers is common. Only a few of the cuneate neurons activated by wrist joint afferent fibers can be shown to project to the thalamus.

Nociception. Angaut-Petit (1975b) found cells in the deep middle zone of the nucleus gracilis that responded like postsynaptic dorsal column neurons. No attempt was made to determine whether they project to the thalamus. The cells received a convergent input from sensitive mechanoreceptors (with a rapidly adapting response to hair movement and a slowly adapting one to maintained touch) and from mechanical and thermal nociceptors. Similar units can be found after all of the spinal cord except the dorsal funiculus is transected, suggesting that the responses are due, at least in part, to the postsynaptic dorsal column pathway (although it is now known that there are fine primary afferent fibers in the dorsal funiculus that could conceivably account for the nociceptive responses [see above]).

Ferrington *et al.* (1988) recorded from thalamic projection neurons in the nucleus gracilis of monkeys. In addition to units that responded with a rapidly adapting discharge to either low- or high-frequency stimulation and units responding with a slowly adapting discharge (Fig. 7.29), a few gracile neurons were of the wide-dynamic-range type. That is, they responded to tactile input but better to noxious stimuli. Effective noxious stimuli included both intense mechanical and thermal stimuli. Some low-threshold neurons, activated best by hair movement, showed a convergent input from nociceptors (Fig. 7.30).

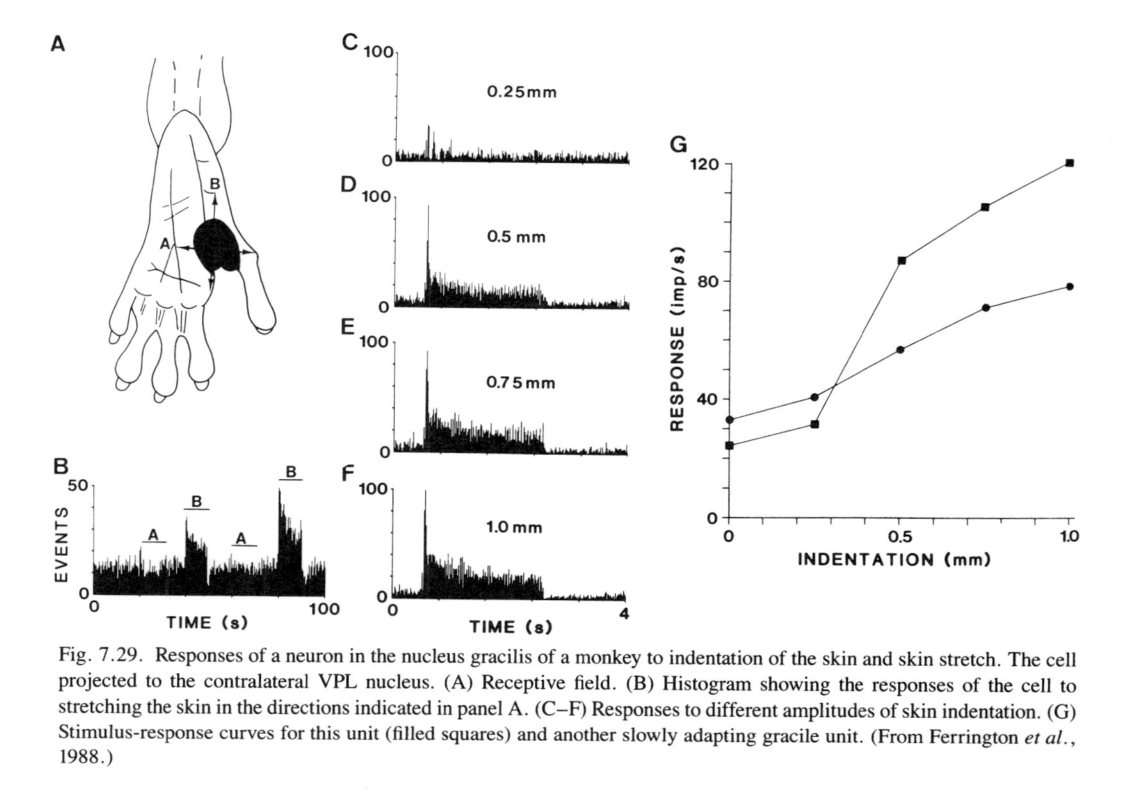

Fig. 7.29. Responses of a neuron in the nucleus gracilis of a monkey to indentation of the skin and skin stretch. The cell projected to the contralateral VPL nucleus. (A) Receptive field. (B) Histogram showing the responses of the cell to stretching the skin in the directions indicated in panel A. (C–F) Responses to different amplitudes of skin indentation. (G) Stimulus-response curves for this unit (filled squares) and another slowly adapting gracile unit. (From Ferrington *et al.*, 1988.)

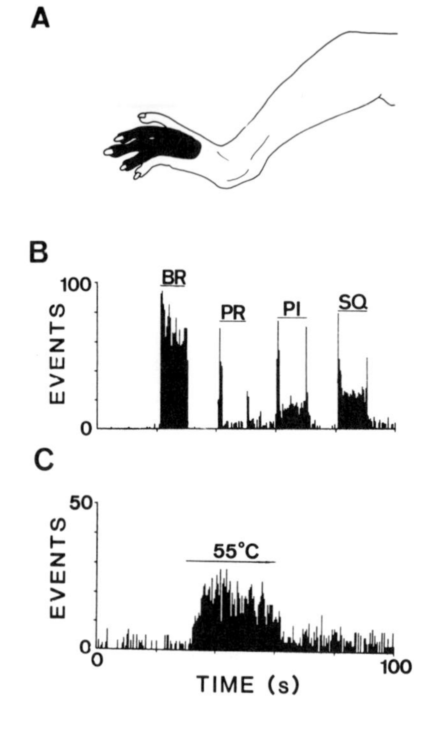

Fig. 7.30. Responses of a gracile neuron in a monkey to mechanical and noxious thermal stimulation. The cell was activated antidromically from the contralateral VPL nucleus. (A) Receptive field. (B) Responses to brushing (BR), pressure (PR), pinch (PI), and squeeze (SQ) are shown in B. The PR, PI, and SQ stimuli represent graded compressive stimuli ranging from near the pain threshold in humans to severely painful. Note the graded maintained responses, in addition to the rapidly adapting component of the responses. (C) Activation by noxious heat. (From Ferrington *et al.*, 1988.)

Development

Ferrington and Rowe (1982) and Connor *et al.* (1984) examined the tactile coding properties of neurons in the cuneate nucleus of kittens. In neonatal kittens, the neurons could be subdivided into a slowly adapting class and two rapidly adapting classes, indicating that the highly specific pattern of projection seen in the adult is already established at birth (Ferrington and Rowe, 1982). The functional capabilities of the activity of the neurons in the newborn kitten were limited when compared with the adult, and maturation of the adult properties required 2–3 months (Connor *et al.*, 1984).

Plastic Changes in Receptive Fields

The effects of partial or complete deafferentation of the lumbosacral enlargement by dorsal rhizotomy on the receptive-field properties of neurons in the nucleus gracilis have been investigated (Dostrovsky *et al.*, 1976; Millar *et al.*, 1976; S. B. McMahon and Wall, 1983b). When all but one of the dorsal roots are cut, a large number of cells unresponsive to stimulation of the skin are found, as expected. However, in addition, a number of cells with receptive fields on the trunk or with split receptive fields (e.g., one on the limb and one on the trunk) appear. The findings are similar whether the roots are cut acutely or chronically, but are exaggerated in chronically rhizotomized animals (Millar *et al.*, 1976). When all of the dorsal roots of the lumbosacral enlargement are cut, there are again many more cells than normal that are responsive to receptive fields on the trunk, especially the abdomen. Another change is a greatly expanded inhibitory field. Similar changes occur whether the deafferentation is acute or chronic. A cold block of the spinal cord results in an abdominal receptive field for cells normally responsive to receptive fields just on the limb. The best explanation for these findings is that dorsal column neurons receive ineffective synapses from afferents that are disinhibited when the normal input is removed.

Sectioning a peripheral nerve can also produce changes in the receptive fields of neurons in the dorsal column nuclei (Kalaska and Pomeranz, 1982), although this is disputed (S. B. McMahon and Wall, 1983b).

Summary

In summary, the response properties of most dorsal column neurons reflect two distinctive patterns of organization. There are cells that appear to transmit information from one or a few specific receptor types, and there are cells that receive a convergent input from a variety of receptor types. Among the responses of the more specific neurons, the activity of the following kinds of receptors can be recognized: tylotrich hair, rapidly adapting pad receptors, type I and II slowly adapting receptors (including claw receptors), and Pacinian corpuscles. In addition, in the nucleus cuneatus there are cells that respond specifically to input from group Ia muscle spindle afferent fibers. Other neurons receive a convergent input from hair follicle, tactile, and pressure receptors and nociceptors. There are also "interneurons" within the dorsal column nuclei which may receive either rather specific input or a convergent input. The convergent input can come not only from different kinds of primary afferent fibers but also from different sources in the brain and spinal cord. One difficulty in the experiments on neurons of the dorsal column nuclei is that many cells considered "interneurons" may actually be misidentified relay neurons whose projection has not been identified.

IMMUNOCYTOCHEMISTRY OF THE DORSAL COLUMN NUCLEI

The chemical neuroanatomy of the dorsal column nuclei has been reviewed recently by Rustioni and Weinberg (1989).

Westman *et al.* (1984) have found synaptic terminals within the dorsal column nuclei of the cat that contain immunoreactivity for glutamic acid decarboxylase (GAD). These are generally presynaptic to large terminals that presumably belong to primary afferent fibers, although some are on dendrites. The heaviest labeling is in the cell cluster region. GAD-positive terminals are also found in the rostral and the ventral part of the middle zones. These are mostly on dendrites.

Rustioni *et al.* (1984) and Barbaresi *et al.* (1986) examined the dorsal column nuclei in cats and rats treated with colchicine for cells that contain GAD immunoreactivity. GAD-positive cells are found at all rostrocaudal levels of the nuclei. GAD-positive cells are smaller than GAD-negative ones. These neurons are presumably interneurons. The GAD-positive synaptic endings in the dorsal column nuclei must arise from these cells, since the neurons that project into the dorsal column nuclei from the brain stem do not contain GAD (Weinberg and Rustioni, 1985).

Roettger *et al.* (1989) confirmed the presence of GABA-ergic neurons within the rat cuneate nucleus, observing γ-aminobutyric acid (GABA) immunoreactivity in small neurons distributed throughout much of the nucleus.

Substance P (SP)-immunoreactive synaptic endings have been observed in the dorsal column nuclei (Ljungdahl *et al.*, 1978; F. L. Douglas *et al.*, 1982; Westman *et al.*, 1984; Tamatani *et al.*, 1989; Conti *et al.*, 1990). Westman *et al.* (1984) described these as lying mainly in the ventral part of the middle zone, in the rostral zone, and between cell clusters. The SP endings are much larger than the GAD-positive ones. They generally synapse on dendrites, but are often postsynaptic to small boutons. It has been suggested that the SP endings might be primary afferent terminals, perhaps of muscle afferents. However, Conti *et al.* (1990) proposed that the SP projection to the dorsal column nuclei originates in part from neurons of the dorsal horn.

Another excitatory peptide that is present in the dorsal column nuclei is calcitonin gene-related peptide (CGRP) (Kawai *et al.*, 1985; Kruger *et al.* 1988b; Tamatani *et al.*, 1989; Fabri and

Conti, 1990). Most of the unmyelinated axons in the dorsal column that contain CGRP disappear following dorsal rhizotomies, indicating that these axons belong to primary afferent neurons (Patterson *et al.*, 1990). Furthermore, CGRP-containing dorsal root ganglion (DRG) cells can be labeled retrogradely from the dorsal column nuclei (Fabri and Conti, 1990).

Peptides other than SP and CGRP that have been found in the dorsal column nuclei include enkephalin (Ibuki *et al.*, 1989), vasoactive intestinal polypeptide (VIP) (Abrams *et al.*, 1985), neurotensin (Kessler *et al.*, 1987), corticotropin releasing factor (Sakanaka *et al.*, 1987; Foote and Cha, 1988), neuropeptide Y (Halliday *et al.*, 1988) and cholecystokinin (C. A. Hunt *et al.*, 1987).

Steinbusch (1981) and J. C. Pearson and Goldfinger (1987) have described the distribution of serotonin-immunoreactive terminals in the dorsal column nuclei of rats and cats. The labeled endings are present in all parts of the nuclei. Willcockson *et al.* (1987) observed serotonin-immunoreactive terminals in apposition to neurons of the dorsal column nuclei that project to the thalamus of the rat. They provided evidence from double-labeling experiments that the serotonergic projections originate from neurons in several of the raphe nuclei.

PHARMACOLOGY OF DORSAL COLUMN NEURONS

A number of studies have been done to determine the responsiveness of neurons in the dorsal column nuclei to a variety of putative neurotransmitters or the effects of blocking agents. Others have observed the release of active substances from the dorsal column nuclei. Some of this work has been reviewed by Banna and Jabbur (1989).

Systemic Drug Administration

In several studies, drugs have been administered systemically and the effects on synaptic transmission in the dorsal column nuclei observed. E. S. Boyd *et al.* (1966) found that picrotoxin and pentylenetetrazol reduce the amount of inhibition of the medial lemniscus evoked potential in conditioning testing experiments. They attributed this to blocking presynaptic inhibition. Strychnine (plus mephenesin) had a comparable action that was attributed to blocking postsynaptic inhibition.

Banna and Jabbur (1971) administered semicarbazide to cats to deplete GABA stores. After about 1 hr, they observed a depression in the P wave recorded from the surface of the cuneate nucleus, and the wave was nearly eliminated in 3 hr. This finding is consistent with a role of GABA in the generation of PAD. Semicarbazide also reduced the excitability increase that normally results from conditioning stimulation of the forepaw. Similar effects were observed with intravenous picrotoxin (Banna and Jabbur, 1969) and bicuculline (Banna *et al.*, 1972). A similar effect in depleting GABA and reducing the P wave resulted from administration of another inhibitor of GAD 3-mercaptopropionate (F. Roberts *et al.*, 1978).

Topical Application of Drugs

Davidson and Southwick (1971) applied a number of substances to the surface of the cuneate nucleus. They observed nonspecific effects of glycine and glutamate, but found that GABA has a specific action in depolarizing primary afferent terminals.

Simmonds (1978; 1980; 1982) and Newberry and Simmonds (1984a) used a slice preparation of the rat dorsal column nuclei to study the pharmacology of PAD. A GABA-mediated depolarization of afferent terminals can be observed in this preparation, and this is antagonized in different ways by bicuculline and by picrotoxin.

Iontophoretic Application

Another approach to the pharmacology of the dorsal column nuclei has been the use of microiontophoresis to determine how the cells respond to putative neurotransmitters, their agonists or antagonists.

Steiner and Meyer (1966) found that iontophoretically applied glutamate activates 50% of the cells identified by peripheral stimulation. Some of the unresponsive units may have been afferent fibers. Acetylcholine inhibits a few units and excites others. Dopamine inhibits most of the cells tested.

Galindo *et al.* (1967) found that all cuneate neurons tested in cats are strongly excited by iontophoretically applied glutamate. Hair-activated cells are more responsive than other classes. ATP is also excitatory in most cases. Usually ATP is less effective than glutamate, but sometimes it is more powerful. There is no apparent effect when SP is applied. Acetylcholine, norepinephrine, serotonin, epinephrine, dopamine, and histamine were all tried, but none had a strong or consistent excitatory action. Inhibition could be produced by GABA, glycine, acetylcholine, norepinephrine, histamine, and sometimes serotonin. In monkeys, all dorsal column neurons tested are excited by glutamate and depressed by GABA. Catecholamines and serotonin have a weakly depressant action.

Galindo *et al.* (1968) were also able to excite neurons of the cuneate nucleus in cats by iontophoretic application of glutamate and to inhibit the neurons with GABA. They found that the neurons could be excited by iontophoretically applied gallamine triethiodide (Flaxedil) and that systemic administration of this paralytic agent can have an excitatory action on somatosensory transmission in the dorsal column nuclei.

Kelly and Renaud (1973a,b) examined the inhibitory action of iontophoretically applied GABA and glycine on cuneothalamic neurons in cats. Iontophoretically applied bicuculline antagonizes the depressant action of GABA (and GABA agonists) but not that of glycine (Kelly and Renaud, 1973a; however, cf. R. G. Hill *et al.*, 1976). Picrotoxin has a similar effect. Conversely, strychnine interferes with the inhibitory action of glycine but not that of GABA or an agonist (Kelly and Renaud, 1973b). Inhibition from a variety of sources is resistant to iontophoretically or systemically administered strychnine but is easily blocked by iontophoretic bicuculline or picrotoxin (Kelly and Renaud, 1973c). This suggests that much of the inhibition within the cuneate nucleus is mediated by GABA. On the basis of the sensitivity of inhibition in the cuneate to intravenous bicuculline, it was proposed that there are at least two types of GABA receptors responsible for the inhibition in the cuneate nucleus.

J. Davies and Watkins (1973a,b) found that 1-hydroxy-3-aminopyrrolid-2-one (HA-966) antagonizes the actions of the excitatory amino acids glutamate and aspartate, as well as excitatory synaptic transmission within the cuneate nucleus. Since the action of this substance is not altered by bicuculline or strychnine, and since it does not alter the amplitude of the action potential, it is argued that the compound might be an antagonist of excitatory amino acid receptors.

Contrary to the findings of Galindo *et al.* (1967), Krnjević and Morris (1974a) and S. Fox *et al.* (1978) find that application of SP onto neurons of the dorsal column nuclei results in a slow excitation.

Release Experiments

The release of substances from the dorsal column nuclei has been studied by a few groups.

P. J. Roberts (1974a) found that stimulation of the dorsal funiculus in a superfused preparation of the dorsal column nuclei in rats results in the release of glutamate and GABA into the perfusate. By contrast, stimulation of the medial lemniscus causes release of GABA and glycine. Superfusion with a solution containing a high potassium ion concentration causes release of all of these amino acids. A lesion of the dorsal column reduces the uptake of glutamate and aspartate into the dorsal column nuclei (F. Roberts and Hill, 1978; N. Kojima and Kanazawa, 1987).

Goldfinger *et al.* (1984) and Goldfinger (1985) measured the release of substances from the dorsal column nuclei by using a push-pull perfusion system. A number of peaks are detected by high-performance liquid chromatography. Those tentatively identified on the basis of their elution times in one study include homovanillic acid and 5-hydroxyindoleacetic acid (Goldfinger *et al.*, 1984), and those identified in the other study include glutamate, serine, glutamine, and glycine (Goldfinger, 1985).

Summary

In summary, the synaptic endings of axons in the dorsal columns may use an excitatory amino acid, such as glutamate or aspartate, as an excitatory neurotransmitter. Another candidate neurotransmitter is ATP, which has a powerful excitatory action on some neurons in the dorsal column nuclei. SP is also present in synaptic endings in areas outside of the cluster zone. Inhibitory substances in synapses in the dorsal column nuclei include GABA and serotonin. GABA-ergic synapses are likely to be involved in both pre- and postsynaptic inhibition. Other agents that produce inhibitory effects when applied iontophoretically include glycine, acetylcholine, norepinephrine, and histamine. Stimulation of the dorsal columns evokes a release of glutamate and GABA from the dorsal column nuclei, whereas stimulation of the medial lemniscus results in the release of GABA and glycine.

DESCENDING CONTROL OF DORSAL COLUMN NUCLEI

Hagbarth and Kerr (1954) showed that stimulation within the reticular formation inhibits activity ascending in the lateral and ventral funiculi of the spinal cord and also through the dorsal column nuclei. Similar effects are produced by stimulation of the SI and SII areas of the sensorimotor cortex, the anterior part of the cingulate gyrus, and the anterior cerebellar vermis. The depression of transmission through the dorsal column nuclei has also been shown by the reduction in evoked potentials following stimulation of the reticular formation, cerebral cortex, or pyramidal tract (Hernández-Péon *et al.*, 1956; Guzman-Flores *et al.*, 1962; Magni *et al.*, 1959; Satterfield, 1962).

The functional significance of the various control pathways to the dorsal column nuclei is conjectural. The inhibitory actions common to many of the inputs are thought to explain the changes in transmission through the dorsal column nuclei that occur (1) as a result of polysensory stimulation (lights, clicks) (Atweh *et al.*, 1974), (2) during sleep (Carli *et al.*, 1967a,b,c; Favale *et al.*, 1965; Gherlarducci *et al.*, 1970), (3) during voluntary movement (Coquery *et al.*, 1971; O'Keefe and Gaffan, 1971; Ghez and Lenzi, 1971; Ghez and Pisa, 1972; Coulter, 1974; Dyhre-Poulsen, 1975; Chapman *et al.*, 1988), and (4) during seizures experimentally induced in the sensory cortex (Schwartzkroin *et al.*, 1974). The feedback during movement could sum with peripheral feedback (Evarts, 1971), or it could substitute for peripheral feedback in its absence (E. Taub and Berman, 1968). The inhibitory feedback might serve as a kind of corollary discharge by counteracting that part of the sensory input predicted from a voluntary movement, yet allowing information not due simply to the movement to be transmitted (von Holst, 1954; Teuber, 1960). Alternatively, the inhibition might interfere with input that would disturb preprogrammed ballistic movements (Dyhre-Poulsen, 1975). More studies are needed to determine the role of centrifugal control of the dorsal column nuclei during exploratory movements, preferably in primates.

Descending Projections to Dorsal Column Nuclei

Anatomic studies have shown that the dorsal column nuclei receive projections from a number of structures in the brain stem, but the greatest emphasis has been on the cortical input. The projection pathway from the cerebral cortex to the dorsal column nuclei is well established anatomically (see, e.g., W. W. Chambers and Liu, 1957; Walberg, 1957; Kuypers, 1958; Kuypers and Tuerck, 1964; Levitt *et al.*, 1964; Valverde, 1966; Shriver and Noback, 1967; McComas and Wilson, 1968; Gordon and Miller, 1969; G. F. Martin *et al.*, 1971; Weisberg and Rustioni, 1976; 1977; Cheema *et al.*, 1983, 1984a). In cats, the projection originates from the sensory cortex, partly from cells in cytoarchitectonic area 3a and partly from cells in areas 4, 3b, 1, and 2 and from SII (Gordon and Miller, 1969; Weisberg and Rustioni, 1976). In monkeys, it also originates from the supplemental sensory and motor areas (Weisberg and Rustioni, 1977). The projection to the nucleus gracilis comes from the medial part of the sensorimotor cortex and that to the nucleus cuneatus from the lateral part, suggesting a somatotopic arrangement comparable to that of the ascending pathways (Kuypers, 1958; Levitt *et al.*, 1964; however, cf. Walberg, 1957). The region of termination of the cortical projection is largely in the rostral and ventral portions of the dorsal column nuclei, rather than in the cell cluster zone (Kuypers and Tuerck, 1964). Presumably, the cortical inhibitory effects are brought about by way of interneurons located within the dorsal column nuclei or just below these nuclei.

Anterograde and retrograde tracing of the cortical projection to the monkey cuneate nucleus indicates that the precentral gyrus projects chiefly to the brain stem tegmentum just ventral to the cuneate nucleus, whereas the postcentral gyrus projects into the cuneate nucleus proper (Fig. 7.31) (Cheema *et al.*, 1985).

A double-labeling study shows that many of the axons of cortical neurons that project to the cuneate nucleus also project to the spinal cord (Bentivoglio and Rustioni, 1986). These represent as many as 60% of corticocuneate projection cells, and the cells are found chiefly in areas 1 and 2.

Other inhibitory input to the dorsal column nuclei is from the reticular formation. A projection from the nucleus gigantocellularis has been described (Sotgiu and Margnelli, 1976; Sotgiu and Marini, 1977; Odutola, 1977a). However, recent retrograde labeling studies have shown that projections to the dorsal column nuclei from the reticular formation are relatively sparse (Willcockson *et al.*, 1987; Weinberg and Rustioni, 1989), and anterograde labeling studies confirm this (Weinberg and Rustioni, 1989). The nucleus paragigantocellularis lateralis seems to have a more substantial input to the dorsal column nuclei than does the nucleus gigantocellularis (Weinberg and Rustioni, 1989).

The dorsal column nuclei also receive inputs from the red nucleus (Edwards, 1972; Weinberg and Rustioni, 1989); the trigeminal, vestibular, and cochlear nuclei; neurons in the upper cervical spinal cord (Weinberg and Rustioni, 1989); and the nuclei raphe obscurus and magnus (Willcockson *et al.* 1987; Weinberg and Rustioni, 1989). However, there is also a substantial intrinsic projection between different parts of the dorsal column nuclei. This projection is presumably made by interneurons. It has been suggested that interneurons at the base of the dorsal column nuclei may mediate many of the inhibitory actions within the dorsal column nuclei (Weinberg and Rustioni, 1989).

Functional Studies

The global effects of cortical stimulation and the possible functional roles of corticofugal control have already been discussed (see above). Recordings from neurons within the dorsal column nuclei show that cortical stimulation results in the excitation of some neurons and inhibition of others (Towe and Jabbur, 1961; Jabbur and Towe, 1961; Levitt *et al.*, 1964; Felix and Wiesendanger, 1970). The excitation is chiefly of "interneurons," whereas the inhibition is most

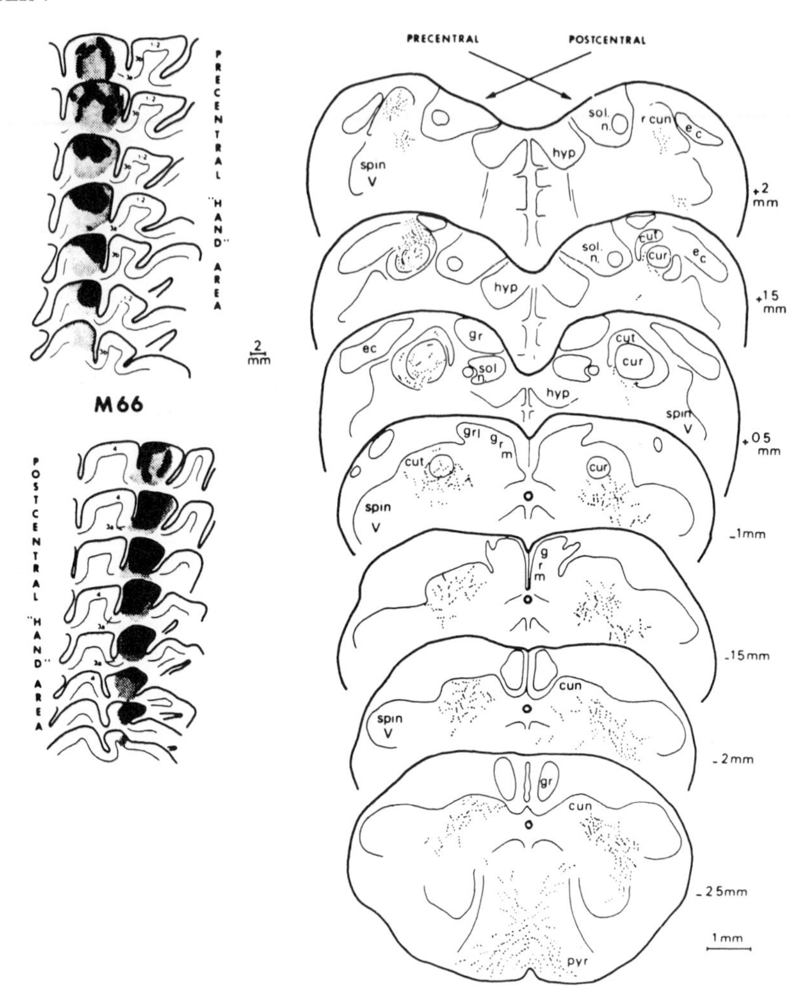

Fig. 7.31. Projections from the pre- and postcentral gyri to the vicinity of the cuneate nucleus in the monkey. Horseradish peroxidase was injected into the arm representation in the precentral gyrus on one side and into the postcentral gyrus on the other. The injection sites are shown to the left and the terminal zones at the right. (From Cheema *et al.*, 1985; reproduced with permission.)

prominently seen in cuneothalamic relay neurons (Gordon and Jukes, 1964b; P. Andersen *et al.*, 1964a). However, both excitation and inhibition were commonly seen in individual neurons by Winter (1965). The inhibition is presumably in part presynaptic, since cortical volleys produce PAD in the dorsal column nuclei (P. Andersen *et al.*, 1964a,b). The excitatory action is mediated by way of the pyramids, whereas the inhibition can still be obtained after the pyramids are sectioned at the level of the trapezoid body (Jabbur and Towe, 1961; Levitt *et al.*, 1964). The inhibition of the dorsal column cells that relay to the thalamus would presumably be mediated in part by interneurons that are directly excited by pyramidal tract fibers and in part by neurons in the reticular formation that are activated by the pyramidal fibers.

Reticular formation stimulation results in PAD in the dorsal column nuclei (Cesa-Bianchi *et al.*, 1968). Inhibition of dorsal column thalamic relay neurons can be demonstrated at the single unit level, whereas "interneurons" of the cuneate nucleus are excited by comparable stimuli

(Cesa-Bianchi and Sotgiu, 1969). Thus, the inhibitory effects of the reticular formation may be mediated by interneurons located in the dorsal column nuclei.

Stimulation of cerebellar nuclei and of the nonspecific nuclei in the thalamus results in PAD of afferents to the cuneate nucleus (Sotgiu and Cesa-Bianchi, 1970), inhibition of many cuneothalamic relay cells, and excitation often followed by inhibition of many cuneate "interneurons" (Sotgiu and Cesa-Bianchi, 1972). Cuneate neurons, especially cuneothalamic relay neurons, are also inhibited by stimulation of the caudate nucleus (Jabbur *et al.*, 1977). Presumably, many of these inhibitory actions are mediated by way of inhibitory interneurons in the dorsal column nuclei.

A projection has been described anatomically from the red nucleus to the rostral cuneate nucleus and also to the gracile nucleus (Edwards, 1972; Weinberg and Rustioni, 1989), and B. G. Gray and Dostrovsky (1983b) have described effects of red nucleus stimulation on neurons of the cuneate nucleus.

There is also a minor projection to the dorsal column nuclei from the raphe nuclei (J. C. Pearson and Goldfinger, 1987; Willcockson *et al.*, 1987; Weinberg and Rustioni, 1989). This may produce a relatively weak inhibition of transmission through the dorsal column nuclei (Gerhart *et al.*, 1981a; Dostrovsky, 1980; however, cf. Jundi *et al.*, 1982).

Summary

In summary, transmission of sensory information through the dorsal column–medial lemniscus pathway is under the control of pathways from other parts of the brain, including the cerebral cortex, cerebellum, and reticular formation. Pathways from the cerebral cortex originate in areas located both in the primary sensory and motor cortex, as well as in the supplementary sensory and motor cortex. An important component is from area 3a. The cortical projection is somatotopic, but ends chiefly in the rostral and ventral parts of the dorsal column nuclei, rather than in the cell cluster zone. Areas of the brain stem that project to the dorsal column nuclei include the nucleus gigantocellularis, nucleus paragigantocellularis lateralis, nucleus raphe obscurus and magnus, and red nucleus. Cortical stimulation excites some neurons of the dorsal column nuclei and inhibits others. It has been proposed that the excitation is of interneurons that then inhibit neurons projecting to the thalamus. However, this has been disputed. A similar argument applies to the reticular formation projection to the dorsal column nuclei.

CONCLUSIONS

1. Branches of primary afferent fibers ascending in the dorsal funiculus form the initial part of the dorsal column–medial lemniscus pathway. These synapse on neurons in the dorsal column nuclei, many of which project through the medial lemniscus to the contralateral thalamus. Other sensory projections originating from neurons of the spinal cord dorsal horn reach the dorsal column nuclei by way of the dorsal funiculus (postsynaptic dorsal column pathway) or the dorsal lateral funiculus.
2. A dorsal column–medial lemniscus pathway seems to be present in amphibia, reptiles, and birds, but is most prominent in mammals.
3. The dorsal funiculus contains not only the ascending branches of primary afferent fibers and the axons of postsynaptic dorsal column neurons, but also descending branches of primary afferent fibers, propriospinal axons, and descending pathways from the brain (including the corticospinal tract in certain animals, such as rodents).
4. The primary afferent fibers in the dorsal funiculus can be subdivided into three systems. The short system ends within a few segments and includes mostly fine afferent fibers. The

intermediate system projects 4–12 segments; it includes large proprioceptive afferents that terminate in Clarke's column, and its axons give off segmental collaterals. There may be some fine afferents in the intermediate system. The long system projects all the way to the medulla; however, collaterals are given off segmentally and into the gray matter of the spinal cord at a distance from the dorsal root of entry. Most of the axons in the long system are large, myelinated fibers in the periphery, but there is now evidence that some unmyelinated fibers belong to this system as well.

5. The dorsal funiculus is subdivided into the fasciculus gracilis, which contains axons from sensory receptors of the lower extremity and trunk, and the fasciculus cuneatus, which innervates the upper extremity and trunk.

6. At caudal levels of the dorsal funiculus, the sensory axons are organized in a dermatomal fashion, whereas at rostral levels the axons resort and assume a somatotopic organization.

7. The functional categories of primary afferent axons in the dorsal funiculus differ when the fasciculus gracilis is compared with the fasciculus cuneatus. At the cervical level of the fasciculus gracilis, there are few or no axons that respond to joint bending and few are from slowly adapting receptors. By contrast, the fasciculus cuneatus contains many proprioceptive afferent fibers. The difference is explainable on the basis that the proprioceptive pathway from the hindlimb relays in Clarke's column and then in nucleus z, rather than in the nucleus gracilis (see Chapter 8).

8. The cutaneous afferent fibers in the fasciculus gracilis include axons that supply a variety of rapidly adapting receptors (G_1 and G_2 hair follicle receptors, field receptors, and Pacinian corpuscles), as well as some SA II receptors. Those in the fasciculus cuneatus include similar afferent fibers. In addition, the fasciculus cuneatus contains SA I receptors, as well as group I muscle afferent fibers.

9. In addition to the primary afferent fiber projection, the dorsal columns contain the axons of cells located in the dorsal horn of the spinal cord. These axons form the postsynaptic dorsal column pathway. The cells of origin of the postsynaptic dorsal column pathway are concentrated in laminae III–V. A few cells are located more ventrally. In cats, these cells are more ventral in the dorsal horn than spinocervical tract cells.

10. It is estimated that there are at least 750–1000 postsynaptic dorsal column cells in the cervical enlargement and 500–700 in the lumbar enlargement of rats. The numbers in cats are about 1700–2000 in the cervical enlargement and 1000 in the lumbar enlargement. Monkeys have about the same number in the lumbar enlargement as cats.

11. Several types of postsynaptic dorsal column cells can be recognized on the basis of dendritic morphology. Some of these cells have dendrites that extend dorsally into lamina II or even lamina I. The dendritic trees are distinct from those of spinocervical tract cells.

12. The axons of postsynaptic dorsal column cells give off collaterals in the vicinity of the soma. Most enter the dorsal funiculus directly, but some pass first into the dorsal lateral funiculus. The conduction velocities of the axons indicate that they are small to medium-sized myelinated fibers.

13. Postsynaptic dorsal column neurons often have convergent inputs from several peripheral nerves and from low-threshold cutaneous and muscle afferents, as well as from high-threshold muscle, skin, and joint afferents.

14. Most postsynaptic dorsal column cells in rats are activated exclusively by innocuous stimuli. Some respond best to strong mechanical stimuli, but very few can be activated by noxious heat, even after sensitization of the skin. Presumably these cells are not involved in nociception in rats.

15. Postsynaptic dorsal column neurons in cats may be of the low-threshold, the wide-dynamic-range, or the high-threshold class. Wide-dynamic-range cells can be excited by innocuous mechanical stimuli, but they respond best to noxious stimuli, including noxious

mechanical or thermal stimuli. Innocuous stimuli that are effective may include stimulation of tactile domes. Some cells respond to stimuli that selectively activate Pacinian corpuscles. The receptive fields may be on glabrous or hairy skin. Receptive-field size can increase reversibly after experimental manipulations. These functional properties distinguish postsynaptic dorsal column neurons from spinocervical tract cells (see Chapter 8).

16. Neurons of the postsynaptic dorsal column pathway are somatotopically organized. Cells in the lumbosacral enlargement project to the nucleus gracilis and cells in the cervical enlargement to the nucleus cuneatus. In cats, the projection is concentrated in the rostral and ventral parts of the dorsal column nuclei, but in rats the projection is more widespread.

17. Some neurons that are presumably equivalent to postsynaptic dorsal column neurons project to the dorsal column nuclei by way of the dorsal lateral funiculus. It is not yet clear how many cells there are of this type or what their functional properties are. It is possible that the axons of some of these neurons give off collaterals to the lateral cervical nucleus, but the neurons should be considered distinct from spinocervical tract cells.

18. The background activity and the responses of postsynaptic dorsal column cells to noxious stimuli are increased by cold block of the spinal cord, indicating that these neurons are subject to descending inhibitory control systems.

19. The axons of the fasciculus gracilis synapse largely in the nucleus gracilis and those of the fasciculus cuneatus in the nucleus cuneatus. In some animals, there is also a nucleus of Bischoff, which receives input from the tail. The lateral cuneate nucleus projects largely to the cerebellum, although some of its neurons in rats and monkeys do project to the thalamus. The response properties of these cells are not known.

20. The axons of the dorsal funiculus that end in the dorsal column nuclei project to all rostrocaudal levels of the nuclei. A given axon gives off a series of collaterals that penetrate ventrally to their termination zone within a dorsal column nucleus. For this reason, the somatotopic representation tends to be repeated over a considerable rostrocaudal distance. The somatotopic representation at a given level is as follows: the distal parts of the extremities are represented dorsally and the trunk ventrally. Caudal segments are medial and rostral ones lateral.

21. The dorsal column nuclei are not homogeneous. The dorsal column nuclei of cats can be subdivided into caudal, middle, and rostral zones. Within a special area called the "cluster region" in the middle zone, there are many neurons with short, bushy dendrites. Dorsal and ventral to the cluster region and in the caudal and rostral zones, the neurons tend to have long dendrites. In rats, the dorsal column nuclei are divided into caudal and rostral zones. The neurons of the caudal zone of the rat are organized in "slabs" (equivalent to the cell clusters of other animals). In monkeys, the middle zone is divided into a pars rotundus and a pars triangularis. The pars rotundus appears to be equivalent to the cluster region of cats.

22. Primary afferent fibers from the dorsal funiculus terminate on cells both in the cluster region and also in the other parts of the dorsal column nuclei. However, transganglionic labeling studies show that cutaneous afferent fibers terminate densely in the cluster region and in the rostral and caudal zones, but not in the ventral part of the middle zone. By contrast, muscle afferent fibers end in the ventral part of the middle zone, but not in the cluster region. (Muscle afferent fibers also project to the lateral cuneate nucleus.)

23. Most of the neurons in the cluster region project to the contralateral thalamus. Some neurons in the caudal and rostral zones also project to the thalamus. Other neurons within the dorsal column nuclei have axons that terminate near their own cell bodies. These are presumably interneurons. The axons of thalamic relay neurons give off collaterals within the dorsal column nuclei.

24. The cuneate nucleus of the monkey contains about 48,000 neurons.

25. The synapses in the dorsal column nuclei include large endings containing round vesicles and making chiefly axodendritic connections, smaller endings with flattened vesicles

that make axoaxonic and axodendritic connections, and some endings with dense-core vesicles. Dorsal rhizotomies cause degeneration of about 18–25% of the large endings with round vesicles, indicating that these belong to primary afferent fibers.

26. The projection to the contralateral thalamus is by way of internal arcuate fibers and the medial lemniscus. The diencephalic projection targets in rats, cats and monkeys are the VPL nucleus, the posterior complex, and the zona incerta. The projection to the VPL nucleus is somatotopically organized, whereas the other projections are not.

27. The dorsal column nuclei also project to the inferior olive, red nucleus, cerebellum, intercolliculus nucleus, superior colliculus, pretectum, and spinal cord. These projections arise from different neurons than those that connect with the thalamus.

28. Neurons of the dorsal column nuclei often exhibit background activity. The discharges sometimes exhibit periodicity reflecting input from a few individual primary afferent fibers. A single afferent fiber can excite a cell on a one-for-one basis. However, the discharge properties of the neurons may also reflect intrinsic membrane properties.

29. Electrical stimulation of peripheral nerves evokes a sequence of field potentials in the dorsal column nuclei. Following a spikelike potential, there is a negative and then a positive field potential. The negative wave is presumed to reflect excitatory events in the dorsal column nuclei and the positive wave PAD. PAD is associated with presynaptic inhibition.

30. Dorsal column neurons often tend to discharge in bursts. These bursts appear to be due to membrane properties of the neurons, since bursting is seen not only spontaneously but also after iontophoretic application of glutamate. Recurrent excitation is ruled out by the observation that the cells generally do not burst when activated antidromically.

31. Many neurons in the cuneate nucleus can be activated by volleys in muscle afferent fibers.

32. Some neurons in the gracile nucleus and its gray matter bridge with the cuneate nucleus are excited by stimulation of the greater splanchnic nerve. The same neurons also have cutaneous receptive fields.

33. Many neurons in the dorsal column nuclei can be inhibited following peripheral nerve stimulation. The mechanisms of inhibition include both presynaptic and postsynaptic inhibition. Generally, presynaptic inhibition lasts 100–200 msec and postsynaptic inhibition 15–20 msec, but sometimes long-lasting inhibitory postsynaptic potentials can be recorded from dorsal column neurons. The PAD underlying presynaptic inhibition is presumed to be responsible for the occurrence of the dorsal column reflex. Interneurons that may be responsible for presynaptic inhibition within the dorsal column nuclei have been identified.

34. Neurons in the nucleus gracilis of the cat respond chiefly to cutaneous stimulation; only 10% respond to stimulation of deep tissues. The receptive fields are small for neurons in the middle zone, large for cells of the rostral zone, and intermediate for neurons in the caudal zone. Most cells in the cluster region can be activated antidromically from the contralateral medial lemniscus.

35. A number of types of neurons can be distinguished in the nucleus gracilis. Some have response properties that resemble those of sensory receptors. These include neurons that respond to hair movement, tactile contact with the skin, or vibration. The adaptation rates are high for hair- and vibration-activated cells and low for the tactile units. Some neurons have convergent inputs from rapidly and slowly adapting mechanoreceptors and a few even from nociceptors. Inhibition, often of the surround type, is seen most commonly for cells in the cluster zone that are excited by hair movement, touch or both. Some slowly adapting units do not have inhibitory fields. A few gracile neurons respond to input from deep receptors, including rapidly adapting joint afferent fibers.

36. Pacinian corpuscles provide a powerful excitatory input to a population of dorsal column neurons. They are also an important source of inhibitory input to many dorsal column neurons.

37. Inhibitory receptive fields of dorsal column neurons may overlap the excitatory fields. However, they extend beyond the excitatory fields, thus assuming a "surround" form. Inhibitory fields are thought to enhance contrast.

38. Individual dorsal column neurons do not appear to be capable of feature extraction, such as recognition of stimulus directionality.

39. Neurons in the nucleus cuneatus may have cutaneous input, like most cells in the nucleus gracilis, or they may respond to input from muscle receptors. Many cells are excited by group Ia fibers from muscle spindles, and some by group Ib fibers from Golgi tendon organs. Usually, the input is from only a single muscle, although in some cases there is convergence from more than one muscle. These proprioceptive neurons tend not to have inhibitory fields.

40. Wide-dynamic-range neurons have been found in the cat nucleus gracilis. Comparable cells projecting to the contralateral thalamus have also been found in the primate nucleus gracilis. These neurons may contribute to nociception.

41. The major types of neurons (those responding to rapidly adapting or Pacinian corpuscle afferents or to slowly adapting afferents) are found in the neonatal kitten. However, the full signaling capabilities of these neurons are not seen until the kittens have matured for 2–3 months.

42. Unusual receptive fields can be observed in neurons of the nucleus gracilis following dorsal rhizotomies in the lumbosacral enlargement. These may be explained on the basis of disinhibition of otherwise ineffective synapses.

43. Immunohistochemical studies have demonstrated the presence of immunoreactivity for GAD, the enzyme responsible for synthesis of GABA, in many synaptic endings in the dorsal column nuclei. These form axoaxonic complexes or are on dendrites. GAD-positive synapses are most abundant in the cell cluster region, but they are also found in other zones. These appear to arise from GAD-positive neurons distributed throughout the dorsal column nuclei, since extrinsic projections do not come from GAD-positive cells. The GAD-positive cells of the dorsal column nuclei are presumably inhibitory interneurons, and they presumably mediate both pre- and postsynaptic inhibition within these nuclei. The same neurons can also be stained immunohistochemically with antibodies against GABA.

44. The dorsal column nuclei also contain synaptic terminals that stain for SP. These endings are mainly in the ventral part of the middle zone, the rostral zone, and between cell clusters. These synapses may provide a slow excitatory input to dorsal column neurons outside of the cluster region.

45. Fast excitation in the dorsal column nuclei may be mediated by glutamate and/or aspartate. Iontophoretic application of glutamate (or aspartate) causes excitation of dorsal column neurons. This excitation is antagonized by HA-966, an excitatory amino acid antagonist. Stimulation of the dorsal funiculus, but not of the medial lemniscus, causes release of glutamate from the dorsal column nuclei.

46. Other excitatory agents affecting neurons of the dorsal column nuclei include SP which causes a slow excitation, ATP, and gallamine triethiodide (Flaxedil).

47. Pre- and postsynaptic inhibition in the dorsal column nuclei appears to be due largely to the release of GABA. This is shown by antagonism of both forms of inhibition by using picrotoxin or bicuculline, decrease in the inhibition following GABA depletion by inhibitors of GAD, depolarization of primary afferent terminals by application of GABA, and inhibition of the neurons in the dorsal column nuclei by iontophoretic application of GABA. Furthermore, GABA is released following stimulation of the dorsal funiculus or of the medial lemniscus.

48. Other inhibitory agents that affect the neurons of the dorsal column nuclei include glycine, acetylcholine, norepinephrine, histamine, and serotonin.

49. The dorsal column nuclei are under strong descending control. The best-studied control system originates from the cerebral cortex. Stimulation of the cerebral cortex produces inhibition of dorsal column neurons relaying information to the thalamus. Cortical stimulation

also excites many neurons in the dorsal column nuclei. It has been proposed that the excited neurons are inhibitory interneurons. However, a given neuron in the dorsal column nuclei may be excited and also inhibited by cortical stimulation.

50. The cortical input to the dorsal column nuclei originates from the sensorimotor region. The motor cortex projects largely to the tegmentum just ventral to the dorsal column nuclei. The sensory cortex projects into the dorsal column nuclei. The particular areas of sensory cortex providing this projection vary with species. In cats, the main input to the dorsal column nuclei is from area 3a, whereas in monkeys it is from areas 1 and 2.

51. The dorsal column nuclei also receive inputs from the reticular formation (nucleus gigantocellularis and paragigantocellularis lateralis), the raphe nuclei, the red nucleus, and the vestibular and cochlear nuclei. There is a substantial internal projection of interneuronal connections within the dorsal column nuclei (presumably formed by the inhibitory interneurons). These connections may be responsible for the inhibitory effects on dorsal column neurons of stimulation in such diverse sites as the reticular formation, the cerebellum, nonspecific nuclei of the thalamus, caudate nucleus, and raphe nuclei, as well as recurrent inhibition from stimulation in the contralateral medial lemniscus.

8 Sensory Pathways in the Dorsal Lateral Funiculus

Two major sensory pathways that ascend entirely in the dorsal lateral funiculus are the spinocervical tract and the spinomedullary tract that ends in nucleus z. In addition, the dorsal lateral funiculus contains parts of the postsynaptic pathway to the dorsal column nuclei (see Chapter 7), of the spinothalamic tract, and of the spinomesencephalic tract (see Chapter 9).

SPINOCERVICOTHALAMIC PATHWAY

The spinocervicothalamic pathway was first recognized by Morin (1955). The cells of origin of the spinocervical tract (SCT) are in the gray matter of the spinal cord dorsal horn (Fig. 8.1). SCT cells receive peripheral input, and their axons project in the dorsal part of the ipsilateral lateral funiculus to synapse on cells of the lateral cervical nucleus (LCN) (Brodal and Rexed, 1953; Nijensohn and Kerr, 1975). The LCN is a special group of neurons (Fig. 8.2) within the white matter just ventrolateral to the dorsal horn in the uppermost cervical level (Cl–C3). The axons of most LCN neurons decussate and then ascend into the brain stem to join the medial lemniscus. The projection is in part to the ventral posterior lateral (VPL) nucleus and the medial part of the posterior complex of the thalamus and in part to the midbrain.

Originally, it was proposed that the LCN projects to the cerebellum, since LCN neurons showed retrograde chromatolysis following large lesions of the cerebellum (Rexed, 1951; Rexed and Brodal, 1951). However, later experiments failed to confirm that there is chromatolysis in the LCN after cerebellar lesions (Morin and Catalano, 1955; G. Grant *et al.*, 1968). Furthermore, if the dorsal spinocerebellar tract is first interrupted chronically, there is no anterograde degeneration into the cerebellum after lesions of the LCN. The material on which the original opinion was based has been reexamined, and the most likely source of error was interference to the blood supply of the LCN by the large cerebellar lesions (G. Grant *et al.*, 1968).

The spinocervicothalamic path forms a rapidly conducting system for the transmission of information from the skin to the cerebral cortex (Catalano and Lamarche, 1957). The latency of the evoked potential recorded from the cerebral cortex and transmitted by way of the spinocervicothalamic pathway is shorter than that transmitted by the dorsal column–medial lemniscus pathway in cats (see Chapter 6; Mark and Steiner, 1958; Norrsell and Voorhoeve, 1962; Landgren *et al.*, 1965). However, transmission in the dorsal column–medial lemniscus pathway appears to be faster in monkeys (Downie *et al.*, 1988).

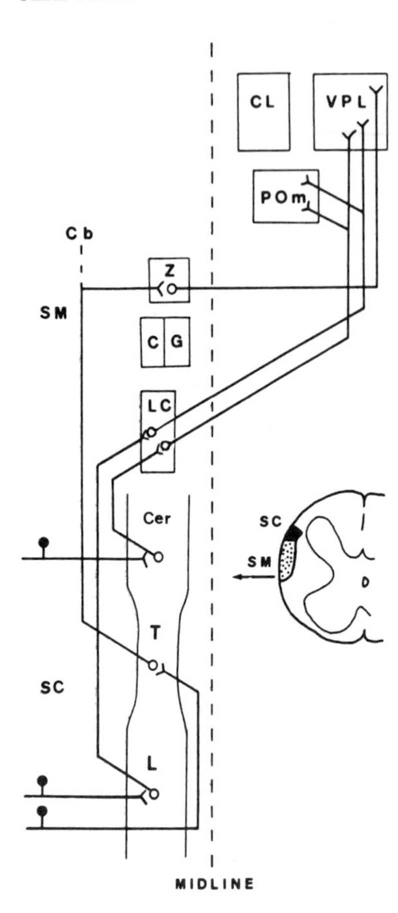

Fig. 8.1. Sensory pathways of the DLF. The organization of the spinocervical and spinomedullothalamic pathways is shown. Primary afferent fibers end upon cells of the dorsal horn in the cervical (Cer) and lumbar (L) enlargements. The second-order neurons project through the spinocervical tract (SC) to the lateral cervical (LC) nucleus. The axons of third-order cells decussate and project through the medial lemniscus to the PO_m and the VPL nucleus of the thalamus. Other primary afferent fibers ascend in the dorsal column to the thoracic cord (T), where they synapse upon cells in Clarke's column, which in turn project in the spinomedullary tract (SM) to nucleus z. Some of these axons also project to the cerebellum (Cb). Nucleus z projects contralaterally to the VPL nucleus and adjacent parts of the ventral lateral nucleus. The transverse section of the spinal cord indicates the locations of the SC and SM pathways.

TAXONOMIC DISTRIBUTION

The most distinctive neuroanatomical component of the spinocervicothalamic pathway is the LCN (Fig. 8.2). This nucleus is prominent in the upper cervical spinal cord in a variety of mammals: cat, dog, sheep, seal, whale, raccoon, tree shrew, lemur, slow loris, galago, and several species of monkeys (Ramon y Cajal, 1909; Rexed, 1951, 1954; Brodal and Rexed, 1953; Gardner and Morin, 1957; Ha and Morin, 1964; Ha *et al.*, 1965; Kitai *et al.*, 1965; Mizuno *et al.*, 1967; Kircher and Ha, 1968; Shriver *et al.*, 1968; Ha, 1971). However, its relative size varies. For instance, the lateral cervical nucleus is larger in carnivores (cat, dog, raccoon) than in primates (Ha *et al.*, 1965; Kitai *et al.*, 1965; Mizuno *et al.*, 1967).

A column of cells in the white matter just ventral to the dorsal horn and extending the entire length of the spinal cord has been found in several species, including the rat, guinea pig, rabbit, ferret, and hedgehog. It was suggested that this column of cells is equivalent to the lateral cervical nucleus (Gwyn and Waldron, 1968, 1969; Waldron, 1969), but no evidence was advanced that this cell column has a thalamic projection. Others have shown that rats have a distinct lateral cervical nucleus in the upper cervical segments (Lund and Webster, 1967a; Giesler

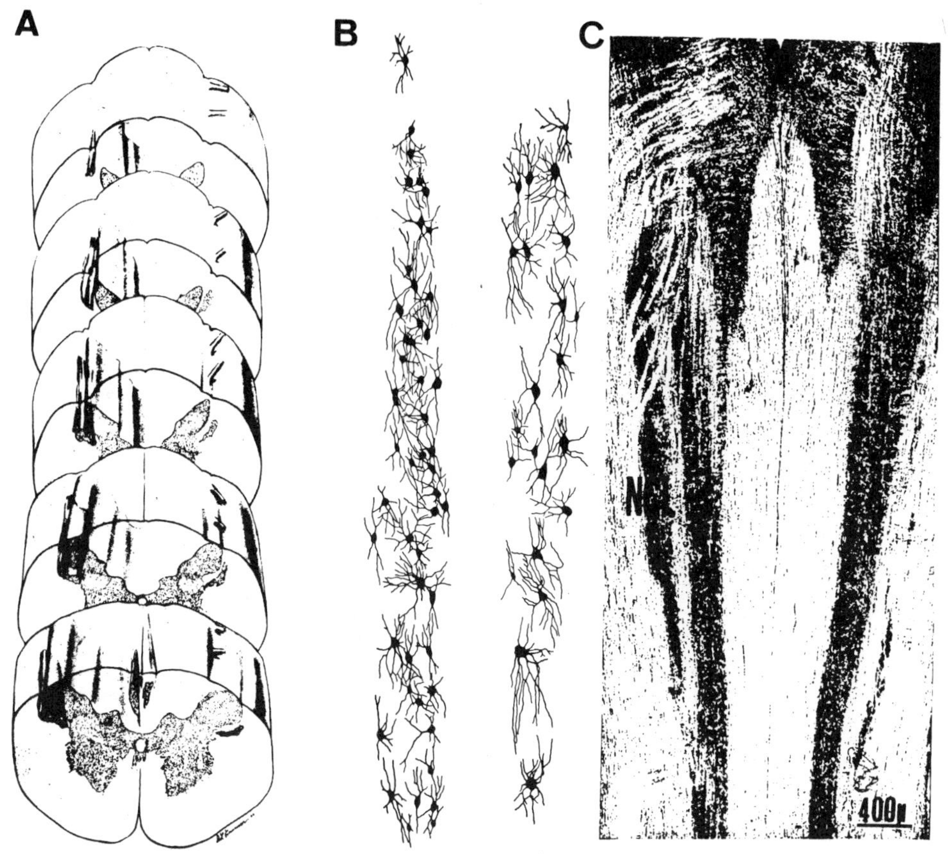

Fig. 8.2. (A) Three-dimensional reconstruction of the lateral cervical nucleus in the C1 and C2 spinal cord segments of a cat spinal cord. (From Oswaldo-Cruz and Kidd, 1964.) (B) The lateral cervical nucleus of a cat is shown bilaterally in a horizontal section of material stained by the Golgi technique. (C) A similar demonstration of the lateral cervical nucleus (LCN) in a horizontal section at an upper cervical level of a cat spinal cord stained by the Nissl technique. The lateral cervical nucleus is seen well just on the left side because of the angle of the section. (From Ha and Liu, 1966; reproduced with permission.)

et al., 1979a, 1988; M. L. Baker and Giesler, 1984; Giesler and Elde, 1985), as do mice, guinea pigs, and rabbits (Giesler *et al.*, 1987). The remainder of the column of cells, extending the length of the dorsal lateral funiculus (DLF) in some species, is the lateral spinal nucleus. This nucleus has been found to project to the midbrain (see Chapter 9) and hypothalamus (Burstein *et al.*, 1990).

The presence of an LCN in humans is controversial. None was observed by several investigators (Rexed, 1951; Brodal and Rexed, 1953; Gardner and Morin, 1957; Seki, 1962). However, others have found an LCN in at least some human spinal cords (Ha and Morin, 1964; Mizuno *et al.*, 1967; Kircher and Ha, 1968; Truex *et al.*, 1965). Possibly the nucleus is not distinctly separate from the dorsal horn in some specimens. An abundance of degenerating axons ascending in the spinal cord and terminating in the human LCN in a victim of a diving accident was observed by Kircher and Ha (1968).

ANATOMY OF THE SCT

Cells of Origin of the SCT

The distribution of the cells of origin of the SCT in rats, cats, and dogs has been mapped following retrograde labeling with horseradish peroxidase (HRP) injected into the LCN.

M. L. Baker and Giesler (1984) made small injections of HRP restricted to the LCN of rats. Labeled SCT cells are found in the ipsilateral nucleus proprius at all levels of the spinal cord. A few SCT cells have also been found in the substantia gelatinosa (Giesler *et al.*, 1978; M. L. Baker and Giesler, 1984). The cells are relatively small, with mean maximum soma diameters of 16.7–18.9 μm. Selective lesions demonstrate that the axons ascend in the dorsal lateral funiculus.

Craig (1976, 1978) has mapped the distribution of SCT cells in cats and dogs following injection of HRP into the LCN at C1 or C2 (Fig. 8.3). Labeled cells are found at all levels of the ipsilateral spinal cord, predominantly in lamina IV, although some labeled cells are in medial lamina V and a few in laminae I, VI, and VII at cervical levels. There are also a few labeled cells on the contralateral side. Cells outside of the ipsilateral laminae IV and V are considered by Craig to be propriospinal neurons projecting to the dorsal horn at C1–C2. There are no labeled neurons in Clarke's column, the central cervical nucleus, or the parts of the ventral horn expected to contain ventral spinocerebellar tract cells. This argues against the view that the axons of the SCT are collaterals of the spinocerebellar tract (cf. Ha and Liu, 1966).

A. G. Brown *et al.* (1980a) have also mapped the distribution of SCT cells in the lumbosacral enlargement of the cat after injection of HRP into the LCN. The HRP spread beyond the LCN, and so presumably neurons other than SCT cells were also labeled. Most of the labeled cells were in laminae III–V in the ipsilateral dorsal horn, but some were in the ipsilateral marginal zone and there were a few in laminae I, III–V, and VII–VIII contralaterally. Brown *et al.* compare the locations of HRP labeled neurons on one side of the cord with the locations of electrophysiologically identified SCT cells on the other. The locations correspond well, with 25% of the cells in lamina III, 60% in IV, 10% in V and 5% in other layers. Brown *et al.* suggest that the labeled cells in the marginal zone probably do not belong to the SCT.

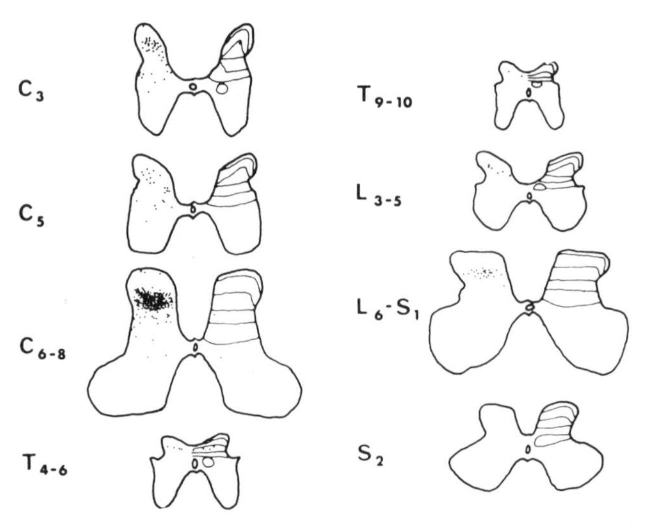

Fig. 8.3. Distribution of SCT cells retrogradely labeled with HRP that was injected into the lateral cervical nucleus of a cat. (From Craig, 1978; reproduced with permission.)

Morphology of Individual SCT Cells

Most of the SCT cells retrogradely labeled in lamina IV in the study by Craig (1978) have large, round somata with diameters of 30–40 μm. The dendrites tend to project dorsally. There are also smaller cells, with diameters of 10–20 μm.

A. G. Brown *et al.* (1976, 1977b, 1980a,b), Jankowska *et al.* (1976), Sedivec *et al.* (1986), and Enevoldson and Gordon (1989b) have studied the morphology of SCT cells, either following intracellular injections with procion yellow or HRP or retrograde labeling by HRP gels implanted in the DLF. The dendrites of the labeled cells project chiefly dorsally and are oriented longitudinally (Fig. 8.4). The rostrocaudal spread of the dendrites ranges from 550 to 2000 μm, whereas the transverse spread is from 200 to 500 μm (A. G. Brown *et al.*, 1977b). The dendrites rarely extend dorsally into lamina II; instead, they run rostrocaudally along the border between laminae II and III (Fig. 8.4). Many SCT cells have dendritic spines.

Sedivec *et al.* (1986) have compared the dendritic trees of SCT cells in intact cats with those in animals with extensive dorsal rhizotomies, with or without a spared root. Partial denervation results in an expansion of the surface area and volume of the dendrites due to an increase in dendritic diameter, as well as an increase in the maximum branch order of the dendritic trees of some of the neurons. Thus, deafferentation may produce some growth and reorganization of the dendrites of SCT cells, rather than atrophy.

Ultrastructure of SCT Cells

Maxwell *et al.* (1982b, 1984b) have injected SCT cells intracellularly with HRP and then have examined them in the electron microscope. The SCT cells were found to have dendritic spines, in confirmation of light-microscopic observations. Two main types of synapses are seen on the cells (Fig. 8.5A and B). The most common type is an asymmetric synapse that contains round vesicles and a few dense-core vesicles. The other type, about half as frequent, contains somewhat flattened vesicles and is also asymmetric. Occasional boutons with highly flattened vesicles are seen. Flattened vesicle endings are more common on proximal than on distal dendrites. No axoaxonal synapses or glomeruli are found in apposition to SCT cells. It is suggested by Maxwell *et al.* (1982b) that the synapses containing round vesicles originate from hair follicle afferent fibers and that those with flattened vesicles are inhibitory.

To test for the origin of the different types of synapses, Maxwell *et al.* (1984b) performed extensive dorsal rhizotomies (L3-S2) and then examined SCT cells during and after degeneration of the primary afferent fibers. Terminals with round vesicles undergo filamentous degeneration. Postrhizotomy, many areas of dendrites are completely free of synapses, especially the distal dendrites. The round vesicle terminals that degenerate are presumably from hair follicle receptors with monosynaptic connections to the SCT cells. Terminals that survive dorsal rhizotomy include some with round vesicles and others with flattened vesicles. It is presumed that most of these are from interneurons (Fig. 8.5C), although descending pathways could be another source.

Number of SCT Cells

M. L. Baker and Giesler (1984) have found about 140 SCT cells in alternate sections through the cervical enlargement and 60 in the lumbar enlargement of the rat. The density of SCT cells is about 10–13 cells/mm of cord length in the lumbar enlargement.

A. G. Brown *et al.* (1980a) estimate that there are about 550–800 SCT cells on one side of the cat lumbosacral enlargement, with a density of 20–40 cells/mm.

On the basis of fiber counts, van Beusekom (1955) estimates that the SCT of the cat includes

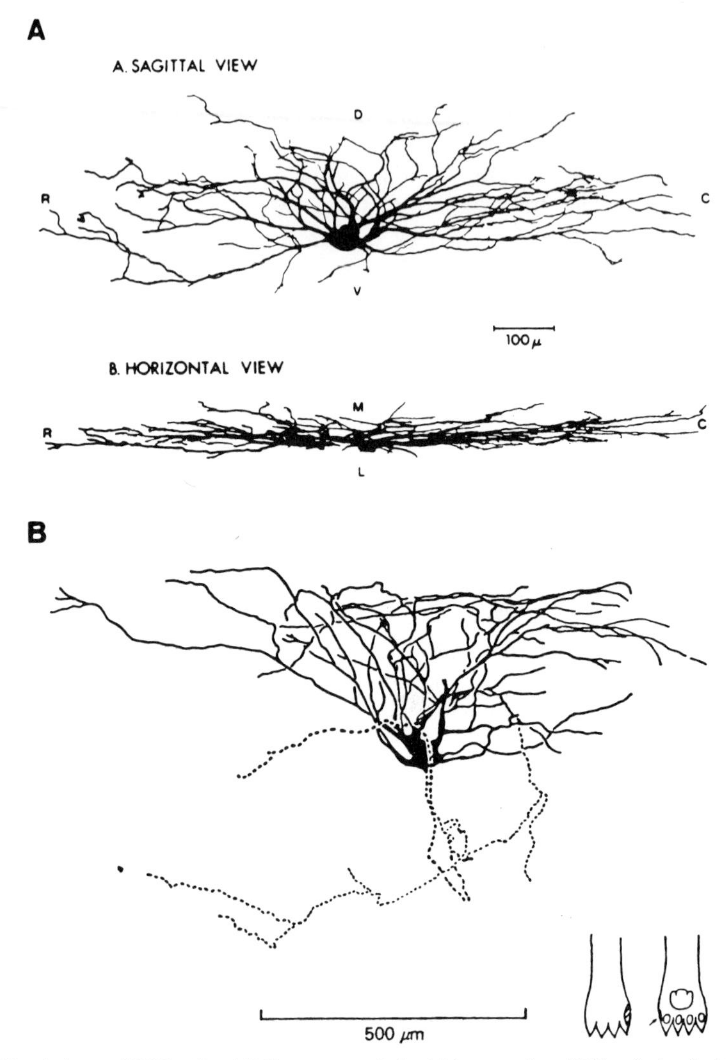

Fig. 8.4. Morphology of SCT cells. (A) Reconstructed dendritic tree of an SCT cell that had been injected intracellularly with horseradish peroxidase. The cell is viewed in both the sagittal and the horizontal planes. (From Sedivec *et al.*, 1986.) (B) Another SCT cell reconstructed in the sagittal plane and showing the axon and its collaterals (dashed lines). This neuron responded to hair movement and skin pressure in the receptive field shown below. (From A. G. Brown *et al.*, 1977b.)

2000–3000 axons on each side of the cord. In recordings from SCT axons in the cervical cord, J. P. Heath (1978) found that half have receptive fields on the forelimb, 34% on the hindlimb and 16% on the tail and trunk. Using a count of 750 SCT cells in the lumbosacral enlargement and the proportions of SCT cells at different levels determined by Heath, A. G. Brown *et al.* (1980a) estimate that the cat has a total of 2200 SCT cells on each side of the cord.

Enevoldson and Gordon (1989b) have labeled SCT cells and other tract cells retrogradely by implanting HRP in the DLF at C4 or C5 after interruption of the dorsal funiculus. The SCT cells are identified as labeled neurons in laminae III and IV with dorsally directed dendrites having a rostrocaudal orientation and with limitation of dendritic spread to the border of laminae II and III. The numbers of these cells at different levels are as follows: 700 in the lumbosacral

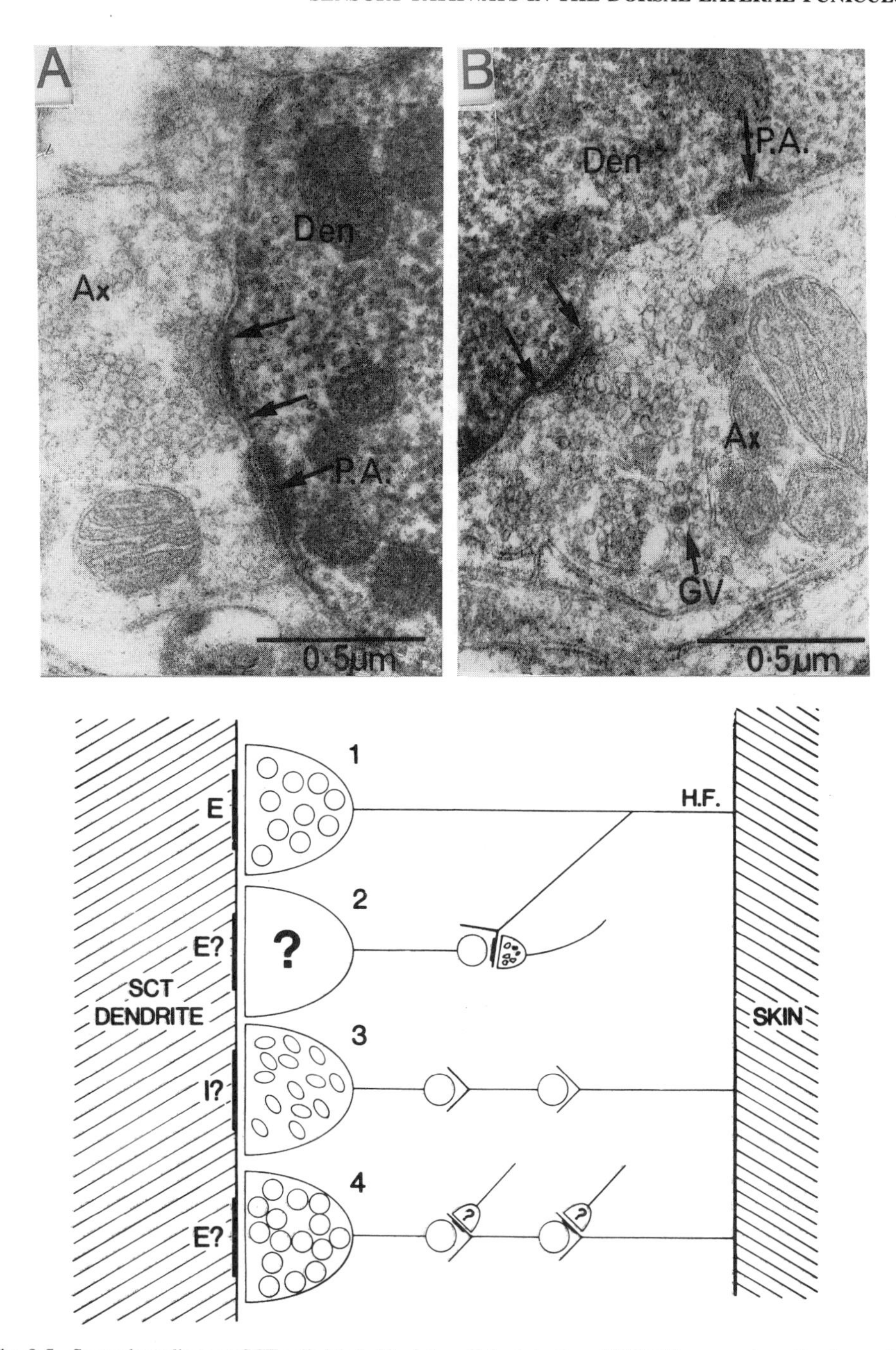

Fig. 8.5. Synaptic endings on SCT cells labeled by intracellular injection of HRP. The synaptic ending in panel A contains round vesicles and forms an asymmetric junction with an SCT dendrite. That in panel B contains flattened vesicles and also makes an asymmetric contact with an SCT dendrite. The diagram in panel C shows (1) a round-vesicle-containing ending of a hair follicle (H.F.) afferent fiber on an SCT cell dendrite; (2) a polysynaptic pathway from hair follicle afferent fibers that can be inhibited presynaptically; (3) an interneuronal connection with flattened vesicles, presumably forming an inhibitory input; and (4) an interneuronal connection with round vesicles, presumably forming an excitatory input that can be modulated presynaptically. (From Maxwell *et al.*, 1984; reprinted with permission.)

enlargement, 450 in the upper lumbar and thoracic cord, and 1100 in the cervical enlargement. In addition to these, the axons of other neurons retrogradely labeled by HRP implanted in the dorsal column nuclei are seen to give off collaterals to the LCN. The cell bodies of these neurons are either in the middle part or at the lateral edge of lamina IV. The ones in lamina IV have dendritic trees that are like those of SCT cells, and those found laterally have a more prominent ventral dendritic arborization than do SCT cells.

Axons of SCT Cells

The initial parts of the axons of SCT cells can be labeled with HRP (A. G. Brown *et al.*, 1977b; Rastad *et al.*, 1977; Enevoldson and Gordon, 1989b). The axons of SCT cells originate from the soma or a proximal dendrite and may start out in any direction. In some cases, the axon ascends for a short distance in the dorsal funiculus before entering the DLF (Enevoldson and Gordon, 1989b). The axon reaches the ipsilateral DLF within about 250 μm rostrocaudally from the cell body and turns rostrally in the superficial part of the medial DLF (A. G. Brown *et al.*, 1977b). Most SCT axons give off one or more local collaterals near the cell body before entering the white matter (see Fig. 8.4B), as well as a few that reenter the gray matter as the axon ascends in the DLF. The terminals of the collaterals are generally between the level of the cell body and 500 μm ventrally. Evidently, SCT cells not only transmit sensory information to the brain but also influence the activity of other dorsal horn neurons.

Rastad (1981a) and Maxwell and Koerber (1986) have examined the contacts made by the collaterals of SCT cells injected intracellularly with HRP on the somata and proximal dendrites of postsynaptic dorsal column neurons labeled retrogradely with HRP. The synaptic boutons contain round or somewhat flattened vesicles, and the contacts are of the asymmetric type. Some are components of axoaxonal complexes. Most of the terminals are axodendritic, but some are axosomatic (Rastad, 1981b). On the basis of the physiologic evidence of Jankowska *et al.* (1979), which indicates disynaptic excitation of postsynaptic dorsal column neurons by way of a circuit containing SCT cells, it is suggested that these synapses are excitatory.

The calculated conduction velocities of the axons of SCT cells vary widely over the myelinated range: for the cat, 7–103 m/sec (A. G. Brown and Franz, 1969; Bryan *et al.*, 1973; Cervero *et al.*, 1977a); for the monkey, 7–71 m/sec (Bryan *et al.*, 1974; Downie *et al.*, 1988).

Projection Targets of the SCT

Axons of SCT cells terminate in the LCN, entering it from all sides, but chiefly from the dorsolateral and lateral aspects (Ramón y Cajal, 1909; Ha and Liu, 1966; Westman, 1968a). Some of the fibers entering the nucleus can be shown to be collaterals of larger fibers that continue to ascend in the lateral funiculus (Ramón y Cajal, 1909; Ha and Liu, 1966; Enevoldson and Gordon, 1989b). It is not yet clear whether collaterals or direct projections of axons in the DLF also provide input to the large group of neurons within the gray matter of the upper cervical segments (especially the ipsilateral ventral horn) that form a substantial component of the spinothalamic, spinoreticular and spinomesencephalic tracts (see Chapter 9) (cf. Nijensohn and Kerr, 1975). Furthermore, the projection targets of the DLF axons that continue to the brain stem after giving off collaterals to the LCN are incompletely described. At least some of these axons appear to synapse in the dorsal column nuclei (Enevoldson and Gordon, 1989b). The SCT projection of cats is somatotopically organized. Svensson *et al.* (1985a) have injected wheat germ agglutinin-conjugated HRP into different segments of the spinal cord and have mapped the distribution of anterogradely labeled terminal fields in the LCN. Injections into the cervical cord result in dense labeling in the rostral two-thirds and the medial region of the ipsilateral LCN. Lumbar injections label the dorsocaudal and lateral portions of the LCN. Injections into other segments produce labeling in intermediate areas. There is also sparse contralateral labeling.

Craig *et al.* (1987) are not convinced that there is a crossed projection to the LCN. In contrast to the organization described by Svensson *et al.* (1985a), they did not find consistent differences in the longitudinal extent of labeling from different spinal levels. However, cervical injections of HRP result in a concentration of label in the ventromedial LCN, and lumbar injections produce labeling in the dorsolateral LCN. There is a projection to the medial LCN from all spinal cord levels.

Giesler *et al.* (1988) were unable to detect evidence for a somatotopic organization of the SCT to the LCN in rats.

Lu *et al.* (1985) have been able to activate neurons in the dorsal horn antidromically both from the dorsal column and from the DLF. They suggest that some SCT cells may also project to the dorsal column nuclei. However, it is also possible that the axonal branch in the dorsal lateral funiculus projects elsewhere than to the LCN.

FUNCTIONAL PROPERTIES OF SCT NEURONS

Background Activity

All SCT cells in spinalized cats and most of those observed in decerebrate cats have background activity (A. G. Brown and Franz, 1969; Cervero *et al.*, 1977a). This consists of bursts of two to five spikes separated by silent intervals. The mean frequency of the background discharge depends on the preparation (greater in spinalized than in decerebrate or anesthetized animals) and on the cell type. In spinalized animals, the mean frequency varies from 5 Hz for hair-activated units to 11 Hz for units excited by pressure and pinch (A. G. Brown and Franz, 1969).

Responses to Electrical Stimulation

A. Taub and Bishop (1965) have identified axons of SCT cells in the DLF by antidromic activation from the upper cervical spinal cord. The axons respond to stimulation of the DLF at C4. Either there is no response to stimulation at a level rostral to C1 or the conduction velocity slows considerably, suggesting collateralization at the level of the LCN. The SCT cells so identified are activated monosynaptically by stimulation of the sural nerve. Effective volleys are in both Aβ and Aδ fibers.

Mendell (1966) recorded from axons of the DLF at L4 or L5 that he assumes belong to the SCT. The neurons could be activated monosynaptically by volleys in Aβ fibers. They were also excited by C-fiber volleys and probably also by Aδ fibers. When C-fiber volleys were repeated at certain rates, the responses increased, a phenomenon called "windup." Convergence of A- and C-fiber inputs onto SCT cells has been confirmed by Fetz (1968), A. G. Brown *et al.* (1973a,b, 1975) and Cervero *et al.* (1977a).

Intracellular recordings have been made by Hongo *et al.* (1968) from tract cells assumed to belong to the SCT. The cells were activated antidromically following stimulation of the DLF at a lower thoracic level. Stimulation of cutaneous nerves evoked monosynaptic excitatory postsynaptic potentials (EPSPs) and disynaptic inhibitory postsynaptic potentials (IPSPs). Stimulation of high-threshold muscle and cutaneous afferent fibers evoked polysynaptic EPSPs and/or IPSPs.

A. G. Brown *et al.* (1975) have used an anodal blocking technique to demonstrate the effects of C-fiber volleys on SCT cells in the absence of A-fiber input. They found that 29% of SCT cells were excited just by A fibers, while 71% were excited by both A and C fibers.

Hamann *et al.* (1978) have tested the effects of volleys in muscle and cutaneous nerves on SCT cells. They found that if cutaneous activation is produced only by hair follicle afferent fibers, the same cell can often also be excited by group III (and sometimes group II) muscle afferent fibers. Group III and IV muscle afferent fibers are effective in activating SCT cells that respond to both hair movement and skin pressure. The cutaneous receptive fields of cells excited by muscle afferent volleys generally overlay the muscle. Muscle afferent volleys can also produce inhibition.

Harrison and Jankowska (1984) have recorded intracellularly from SCT cells to examine the synaptic inputs from muscle, joint, and interosseus nerves, in addition to cutaneous nerves. Group II muscle afferent fibers evoke a disynaptic EPSP. Some muscle afferent volleys produce IPSPs. Afferent fibers of the interosseus nerve and of the posterior nerve to the knee joint can produce either excitation or inhibition. The shortest-latency actions from any of the noncutaneous nerves are disynaptic. This observation is in agreement with the suggestion by Maxwell *et al.* (1984b) that degenerating dorsal root afferent fibers seen in contact with SCT cells by electron microscopy are likely to supply the skin.

Responses to Natural Stimuli

Characteristics of Cells Likely to Be SCT Cells. Wall (1960) has described the receptive-field properties of dorsal horn neurons that have axons in the dorsal lateral funiculus. The cells are most responsive to stimulation in the centers of their receptive fields, with a gradient in sensitivity out to the periphery of the field. Although weak stimuli fail to inhibit the cells, intense stimuli can do so. The responses to tactile stimuli are rapidly adapting, but there are also slowly adapting responses to pressure and to noxious stimulation.

A. Lundberg and Oscarsson (1961) recorded from axons, many of which presumably belonged to the SCT, identifying them by direct activation following stimulation of the lateral funiculus below Clarke's column, but not antidromically from the anterior lobe of the cerebellum. One population of axons discharged in response to hair movement, but there was no response to pressure or pinch. The receptive fields were small distally but large proximally. No inhibition resulted from touch, pressure, or pinch. Other axons in the DLF could be activated by touch and also by pressure and pinch or just by pressure and pinch. A third pathway was activated after a long latency by ipsilateral and contralateral flexion reflex afferents.

Responses have also been recorded from axons of the DLF just rostral to the lumbar enlargement by A. Taub (1964). Most of the receptive fields were on the toes (60%), whereas only 25% were on the ankle and foot and 15% on more proximal structures or the tail. The cells were excited by hair movement, but not by tactile stimulation of the glabrous skin. Inhibition could be produced by strong ipsilateral or contralateral stimulation of the hindlimbs, forelimbs, face, or tail. Innocuous stimulation outside the excitatory receptive field could also produce inhibition.

Hongo *et al.* (1968) agree with Taub that tactile stimulation can inhibit presumed SCT cells. They demonstrated IPSPs in intracellular recordings when hair was moved in a region adjacent to the excitatory receptive field. Sometimes tactile stimulation of a pad or pressure on deep tissue produced IPSPs, as well. The inhibitory receptive fields were eccentric to the excitatory field, rather than being of the surround type.

Hongo and Koike (1975) recorded intracellularly from cells activated antidromically from the upper lumbar level that they presume were SCT cells. Large, all-or-nothing EPSPs were evoked by weak cutaneous stimuli. The EPSPs included both monosynaptic and polysynaptic components. The monosynaptic EPSPs were quite large (1–5 mV, ranging up to 18 mV), and the polysynaptic EPSPs were even larger (up to 27 mV). The receptors responsible for evoking the EPSPs were judged to be hair follicle receptors with rapidly conducting afferent fibers.

Response Classes of SCT Cells. A. G. Brown and Franz (1969) have studied the effects of specific receptors on identified SCT cells. They recorded from axons of the DLF identified by activation from the C3 level but not the C1 level, or, if activated from C1, the calculated conduction velocity had to slow by 50% or more, consistent with collateralization at the level of the LCN. In cats having a spinal transection (thereby eliminating tonic descending inhibition), three classes of neurons could be recognized: (1) units excited just by hair movement; (2) those excited by hair movement and by pressure; (3) axons excited by pressure and pinch applied to the skin or by activation of receptors in subcutaneous tissue. The first class of units (30% of 69 units) are rapidly adapting and have small receptive fields on the distal limb (Fig. 8.6A). All three kinds of hair follicle afferents are effective in exciting these cells. Some units can be inhibited by squeezing the skin. The second class of cell (48%) has a rapidly adapting response to hair movement and a slowly adapting response to pressure (Fig. 8.6B). Again, all three types of hair follicle afferents are effective. There is no evidence that the slowly adapting responses depend upon slowly adapting mechanoreceptors, and so presumably they are due to nociceptors. Most units can be inhibited, often by squeezing the contralateral ankle or by hair movement, pressure or squeezing the skin. The third class of unit (21%) responds to pressure or pinch with a slowly adapting discharge. There is an afterdischarge lasting as long as 30 sec. Inhibitory receptive fields are like those of the second class of unit. If the terminology introduced by Mendell (1966) is applied, these three classes of SCT cells would be termed low-threshold, wide-dynamic-range, and high-threshold cells. Evidently, in spinalized cats, 70% of SCT cells receive excitatory nociceptive input.

Five classes of SCT cells are found in anesthetized or decerebrate cats: (1) units excited by guard hairs; (2) those excited by tylotrich hairs; (3) cells excited by all three types of hair (guard, tylotrich, and down); (4) units excited by pressure or pinch applied to the skin or activation of subcutaneous receptors; and (5) units not affected by peripheral stimulation (A. G. Brown and Franz, 1969). Guard hair units can sometimes also be activated by pressure. Inhibition of the first two classes of cells is generally not observed, but class 3 units can be inhibited. Noxious heat and cold often excite SCT cells.

An analysis of the distribution of interspike intervals of SCT cells activated by maintained mechanical and thermal stimuli has been made by A. G. Brown and Franz (1970). The discharge

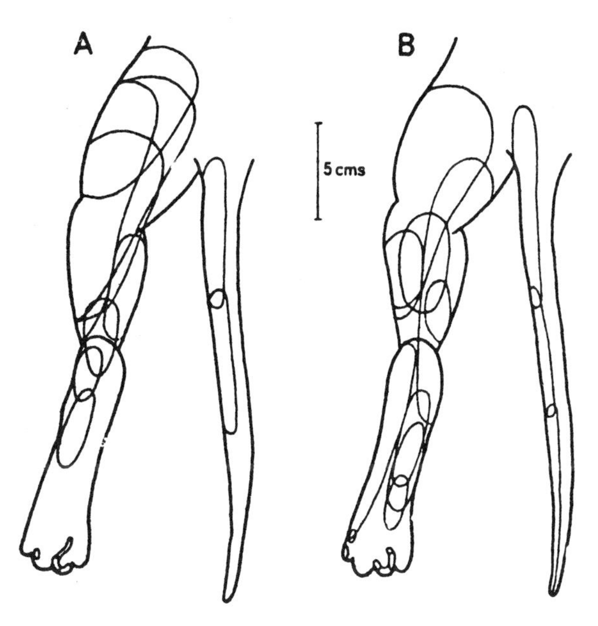

Fig. 8.6. Excitatory receptive fields of SCT cells in spinalized cats. The fields in panel A are those of SCT cells excited by hair movement, and those in panel B of cells excited by hair movement and by skin pressure. (From A. G. Brown and Franz, 1969.)

patterns of different cells of the same class produced by the same stimulus can vary markedly; furthermore, a given neuron can show quite distinct patterns of discharge in response to different stimuli without changing the mean discharge rate.

The receptive fields of SCT cells in cats and in monkeys have been examined by Bryan *et al.* (1973, 1974). The cells were identified by antidromic activation from the DLF at C3 and in some instances (but not all) by failure of antidromic activation from C1 or rostrally. Most of the cells responded to hair movement or to tactile stimulation of the skin. Some also responded to pressure, and a few responded just to pressure or pinch. The receptive fields in monkeys were often located on glabrous skin (Bryan *et al.*, 1974), although comparable receptive fields were not found by A. G. Brown and Franz (1969) or Bryan *et al.* (1973) on the glabrous skin of cats.

Cervero *et al.* (1977a) have reinvestigated the convergence of tactile and nociceptive inputs onto SCT cells in the cat. Almost all of these cells were excited by innocuous stimuli. However, most were also excited (61%) or excited and inhibited (19%) by noxious stimuli, and rarely they were excited exclusively by noxious stimuli (see Fig. 8.11).

Kunze *et al.* (1987) have been able to demonstrate that some SCT cells located in the medial part of the dorsal horn in cats have receptive fields on the glabrous skin of the footpads.

Downie *et al.* (1988) examined the receptive field properties of a few SCT cells in the monkey spinal cord. The cells were identified by antidromic activation from C3 but not C1 or above. Small receptive fields were seen on both hairy and glabrous skin (cf. Bryan *et al.*, 1974). The cells could be classified as low-threshold, wide-dynamic-range, or high-threshold cells. Entrainment to vibratory stimuli at frequencies of 5–30 Hz could be demonstrated. Most of the cells responded to noxious heat stimuli, indicating that they received an input from nociceptors.

Input from Muscle. Evidence for a convergence of inputs from the skin and from high threshold muscle afferents onto SCT cells has been obtained by Kniffki *et al.* (1977). The stimuli used have included intra-arterial injections of algesic chemicals such as potassium ions, bradykinin, and serotonin into the circulation of the triceps surae muscles. A higher percentage of neurons responded to muscle input in spinalized than in decerebrate, nonspinalized animals (29 versus 4%, respectively).

Input from Viscera. Cervero and Iggo (1978) have reported that activation of afferent fibers from the bladder fails to influence the activity of SCT cells.

Receptive-Field Organization. A. G. Brown *et al.* (1980a,b) have proposed that SCT neurons with similar receptive fields form longitudinally oriented columns in laminae III and IV. SCT neurons that are adjacent to each other in the rostrocaudal direction should commonly have overlapping dendritic fields, whereas SCT cells that are adjacent in the medial-lateral direction are less likely to have interdigitating dendrites. Primary afferent fibers from hair follicle receptors would form sagittally oriented sheets of terminals in lamina III that should project onto the dendrites of SCT cells in the rostrocaudal columns. SCT cells with overlapping dendritic fields should share inputs from primary afferent fibers, as shown by overlapping receptive fields, whereas those with nonoverlapping dendritic fields should not.

This hypothesis has been directly tested and verified by A. G. Brown *et al.* (1980b) by recording from adjacent pairs of SCT cells, mapping of receptive fields, and subsequent intracellular injections of horseradish peroxidase and reconstruction of the dendritic fields of the cells (Fig. 8.7). On the basis of the density of SCT cells and the geometry of their dendritic trees, it is estimated that hair movement on the dorsal aspect of one or two toes in the cat can activate as much as 20% of the population of SCT cells on one side of the lumbar enlargement.

A. G. Brown and Noble (1982) have demonstrated anatomic connections between intracellularly labeled hair follicle afferent fibers and spinocervical tract cells. When both units share the same receptive field, there are contacts between the afferent fiber and the SCT cell. However, when the receptive fields are separate, no contacts are found.

The responsiveness of individual SCT cells to movements of hair in different parts of the

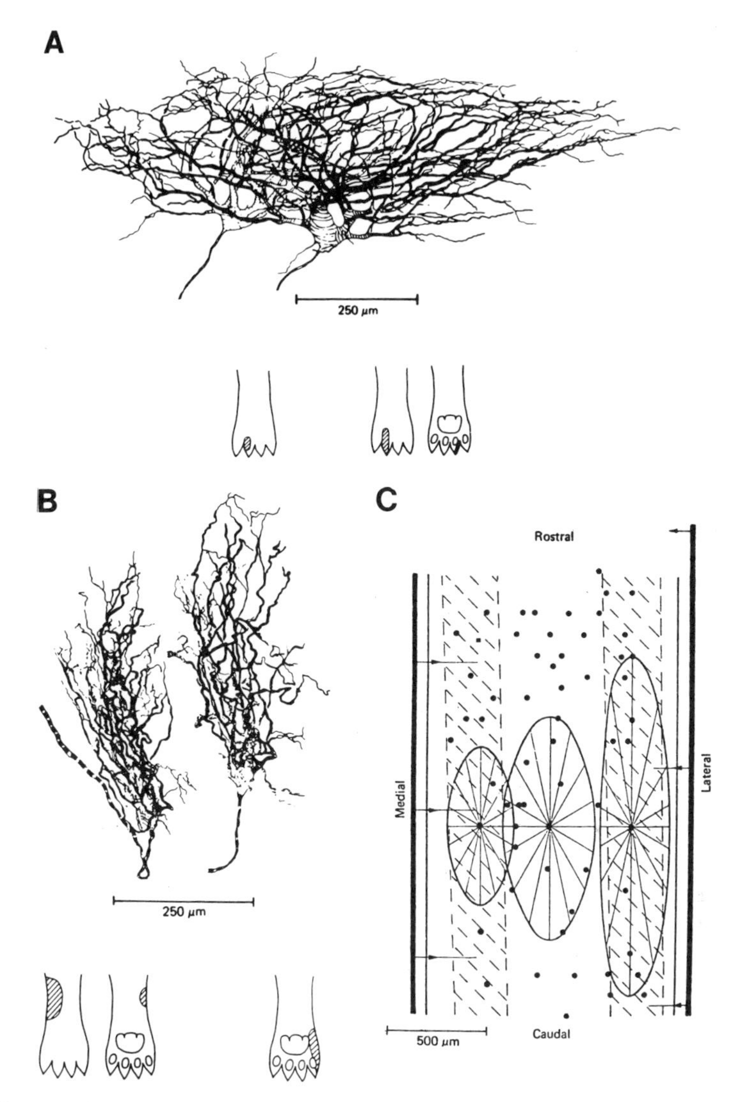

Fig. 8.7. Correspondence between overlap in dendrites and in receptive fields of SCT cells. (A) Overlapping dendritic fields of two SCT cells injected intracellularly with HRP. The overlapping receptive fields are shown below. (B) Nonoverlapping dendrites and receptive fields. (C) Schematic diagram showing the expected dendritic fields of three SCT cells as viewed from a dorsal perspective in the L7 segment. Dots represent the locations of 60 SCT cells retrogradely filled with HRP. The territory expected to be occupied by two hair follicle afferent fibers is shown by the hatched areas. (From A. G. Brown *et al.*, 1980b.)

receptive field has been tested, using air puffs, by A. G. Brown *et al.* (1986a). There is a gradient of sensitivity, with a peak response to stimuli applied near the center of the receptive field and smaller responses to stimuli applied peripherally in the receptive field (Fig. 8.8). Paired stimuli fail to evoke any larger response than the larger of the responses to the individual stimuli. It is concluded that for a population of SCT cells, application of a stimulus at a point on the skin should produce the greatest excitation of cells whose receptive fields are centered on the stimulus and less for cells whose receptive fields are offset from the stimulus. Various explanations for the

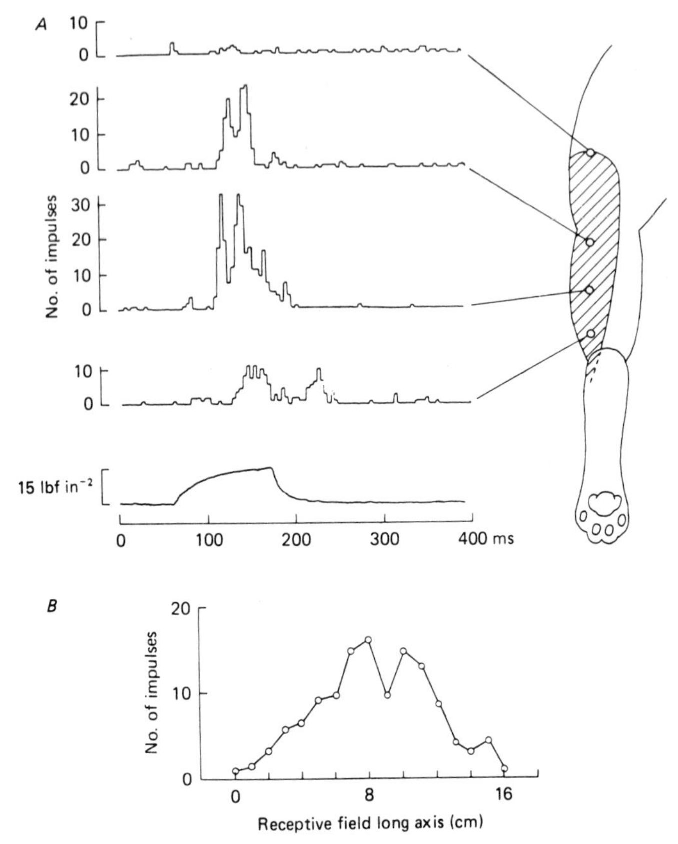

Fig. 8.8. Excitability gradient in the receptive field of an SCT cell. (A) Histograms showing the responses of an SCT cell to standardized puffs of air applied to different parts of the receptive field, as indicated at the right. (B) Graph showing the average number of discharges evoked by air puff stimuli applied at different sites along the long axis of the field. (From A. G. Brown *et al.*, 1986a.)

lack of summation of the responses to paired stimuli have been considered, and it is argued that the most likely is inhibition.

In a series of papers on the synaptic interactions of hair follicle afferent fibers and individual SCT cells, A. G. Brown *et al.* (1987a,b,c) continue the analysis of the receptive-field organization of these neurons. In the first two papers, they observe that a single impulse in a single group II hair follicle afferent in the middle part of the receptive field will cause one or more impulses in an SCT cell in half of the trials. The latencies of the impulses usually indicate a polysynaptic connection, although some discharges can be attributed to monosynaptic linkages. Following the excitation, there is a long period of reduced responsiveness, lasting some 1500 msec. The last paper of the series is an intracellular analysis. The receptive fields of SCT cells are complex (cf. Hongo *et al.*, 1968). They include a firing zone, in which hair movements elicit EPSPs that cause the cell to discharge, as well as IPSPs (Fig. 8.9). Adjacent to the firing zone is an inhibitory zone from which EPSPs but not action potentials can be elicited, as well as IPSPs. Whereas hair movements result in EPSPs (often causing action potentials) and IPSPs, intracellular stimulation of the dorsal root ganglion (DRG) cells of single hair follicle afferents results in mono- and polysynaptic EPSPs, but not IPSPs. It is concluded that the IPSPs probably result from movement of down hairs. With paired stimuli, the second response is much smaller than the first.

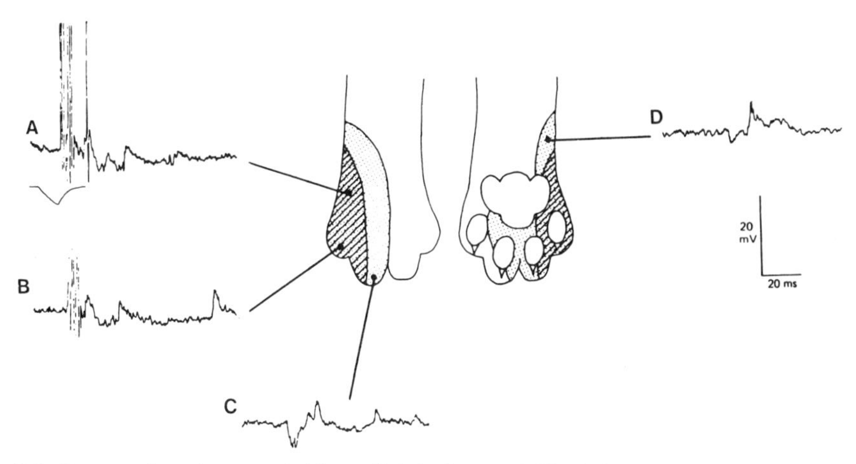

Fig. 8.9. Postsynaptic actions recorded intracellularly from an SCT cell in response to jets of air applied at different sites in the receptive field. The impulse firing zone is indicated by the hatched area and the inhibitory zone by the stippled area on the drawing of the foot. EPSPs and IPSPs were elicited from both zones, but the EPSPs reached threshold only in the impulse firing zone. (From A. G. Brown *et al.*, 1987c.)

Activation of SCT cells does not result in recurrent excitation or inhibition of other SCT cells (indicating that the axon collaterals of SCT cells affect neurons other than SCT cells).

Recently, Short *et al.* (1990) have examined the inhibition of SCT cells produced by air-jet stimuli applied either within the excitatory receptive field ("in-field" inhibition) or outside it ("out-of-field" inhibition). In-field inhibition was strongest when the conditioning and testing stimuli were near each other, and it lasted for as long as 1 sec. Out-of-field inhibition had two phases, one lasting approximately the duration of the stimulus and the other occurring after a delay.

Somatotopic Organization. Wall (1960) has recorded the activity of interneurons in the lumbosacral enlargement of the cat spinal cord, some of which may belong to the SCT. The locations of the cells have a somatotopic relationship to their receptive fields. The distal leg is represented medially in the dorsal horn, and the proximal leg is represented laterally. The anterior leg is represented by cells at the L4 segmental level, whereas the posterior leg and perineum are represented by cells at S2.

Bryan *et al.* (1973, 1974) have also found a somatotopic relationship between the locations of the SCT cells in the dorsal horn of cats and monkeys and the position of the receptive fields on the limb. The rostrocaudal location of the cells is dermatomal. Laterally placed cells have receptive fields on the dorsal surface of the limb, whereas medially placed cells have fields on the ventral surface.

A. G. Brown *et al.* (1980a) have described the somatotopic organization of SCT cells in detail (Fig. 8.10). There are continuous receptive-field representations in longitudinal columns in the dorsal horn. However, small transverse shifts in recording sites across the dorsal horn are associated with discontinuities in receptive-field representations. In L7, the representation of the toes occupies the medial three-quarters of the dorsal horn. The tips of the toes and their ventral surfaces are represented medially, and the dorsal surface of the toes and foot are represented in the middle of the dorsal horn. The proximal foot, leg, and thigh are represented in the lateral one-quarter of the dorsal horn. About 60% of the SCT cells in L6–S1 have receptive fields on the toes (cf. A. Taub, 1964).

Plasticity. The plasticity of the receptive fields of SCT cells following interruption of primary afferent fiber inputs has been examined by A. G. Brown *et al.* (1983b; 1984). Following dorsal rhizotomy, no rearrangements are observed in the somatotopic map of SCT cells in the

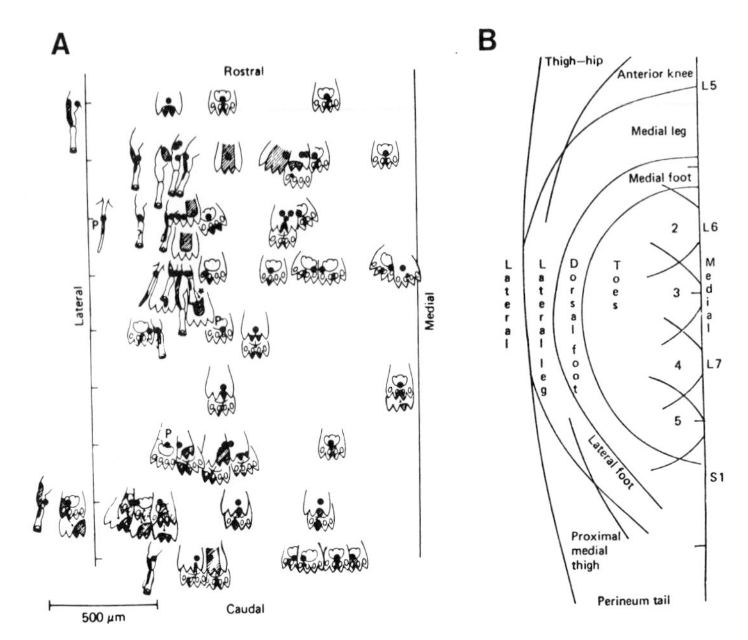

Fig. 8.10. Somatotopic organization of SCT cells. (A) Receptive fields of 60 SCT cells in the L7 segment of the cat. The locations of the cells are indicated as if viewed from a dorsal perspective. The arrangement of the cells in longitudinal columns is evident. P indicates cells that did not respond to hair movement. (From A. G. Brown *et al.*, 1980a.) (B) Schematic representation of the somatotopic organization of SCT cells throughout the lumbosacral enlargement, again from a dorsal view. (From A. G. Brown *et al.*, 1984, as modified from A. G. Brown *et al.*, 1980a.)

lumbosacral enlargement (A. G. Brown *et al.*, 1983b). However, this may be due to the tight coupling of afferent input to SCT cells. In another study, the sciatic and saphenous nerves were sectioned (A. G. Brown *et al.*, 1984) and the receptive fields of SCT neurons examined. The locations of a number of the SCT neurons were determined by intracellular injection of HRP. SCT cells in the medial half of the dorsal horn deafferented by the peripheral nerve lesions would be likely candidates for developing inappropriate receptive fields. However, such fields are not observed, even 55 days following nerve section. Nor are inappropriate fields found for interneurons, in contrast to the reports of Devor and Wall (1978, 1981a,b) and Lisney (1983). Possible explanations for the discrepancy include differences in anesthetics, microelectrodes, or history of stimulation of the skin.

By contrast, Sedivec *et al.* (1983) have observed a reorganization of the receptive fields of SCT cells following dorsal rhizotomies at the lumbosacral level. Following transection of the dorsal roots, except L7, of the lumbosacral enlargement, there is an immediate reduction in the responsiveness of SCT cells. However, after 6 weeks the cells are as responsive as in the normal animal, and a higher proportion of the SCT cells than normal can be activated by noxious stimuli.

Summary

In summary, most SCT cells in the lumbosacral enlargement in cats respond vigorously to hair movement on the dorsal aspect of the toes or foot. The input is in part monosynaptic and in part polysynaptic. A single stimulus to a toe can excite up to 20% of the total population of SCT cells on one side of the lumbosacral enlargement. The cutaneous receptive fields of neurons of the cat SCT tend to be small, and they are somatotopically organized. In addition, there are adjacent inhibitory receptive fields that potentially help enhance contrast. SCT cells also have a potential input from muscle receptors, but this input is under tonic inhibitory control when the neuraxis is intact. There is no evidence for visceral input to SCT cells. These observations suggest that the SCT is a pathway capable of discriminating and reacting strongly to small movements of hair on the dorsal aspect of the paw in the cat and thus is likely to be an important tactile pathway. In addition, many SCT cells are capable of responding to noxious stimuli. If these responses can be distinguished by the brain from responses to tactile stimuli, the SCT could contribute to nociception. There seem to be important species differences in the organization and perhaps utilization of the SCT. For example, the spinocervicothalamic path is a major somatosensory pathway in carnivores, but may be less significant in other forms such as rats, monkeys, and humans. In monkeys, the receptive fields of many SCT cells extend onto the glabrous skin, whereas only a few SCT cells in the medial dorsal horn in cats have receptive fields on the glabrous skin of the footpads. This may reflect a different use of the pathway in primates as compared with cats.

PHARMACOLOGY OF SCT CELLS

Zieglgänsberger and Herz (1971) have examined the effects of iontophoretically applied excitatory and inhibitory amino acids on presumed SCT cells identified by antidromic activation from C3. Cells activated by both hair movement and pressure are excited and their receptive fields expand during iontophoretic application of glutamate. On the other hand, glycine and γ-aminobutyric acid (GABA) depress the responses of these cells and reduce the sizes of their receptive fields. Cells that respond just to hair movement are excited by glutamate in a dose-related fashion at low iontophoretic currents, but are inhibited when the current is high. It is hypothesized that the inhibition results from the activation of inhibitory interneurons in the vicinity of the SCT cells by the glutamate. These neurons are also inhibited by glycine.

Fleetwood-Walker *et al.* (1985) have found that iontophoretically released norepinephrine results in a selective inhibition of the responses of SCT cells to noxious stimuli. A similar action is produced by α_2 agonists, such as clonidine, and the effect is blocked by α_2 antagonists. Fleetwood-Walker *et al.* (1988b) have also shown that iontophoretically applied dopamine or a D_2 dopamine agonist produces a selective inhibition of the nociceptive responses of SCT cells that can be blocked by a D_2 dopamine antagonist.

In addition to amino acids and amines, the activity of SCT cells can be influenced by opioids. Fleetwood-Walker *et al.* (1988a) have shown that the κ opioids, dynorphin and U50488H, produce a selective inhibition of nociceptive responses when these substances are released iontophoretically near the cell bodies of SCT neurons, whereas agonists of μ and δ opioid receptors fail to have such an action. However, μ agonists, but not κ or δ agonists, released in the substantia gelatinosa also produce a selective antinociceptive action.

DESCENDING CONTROL OF SCT CELLS

A. Lundberg and Oscarsson (1961) have found that a pathway descending in the DLF can inhibit the responses of the axons of presumed SCT cells to stimulation of high-threshold muscle afferent fibers but not of cutaneous afferent fibers. Stimulation of the sensorimotor cortex has only a weak action (A. Lundberg *et al.*, 1963).

A. Taub (1964) has reported that activity in axons presumed to belong to the SCT can be inhibited by stimulation in the deep cerebellar nuclei or the brainstem tegmentum.

Fetz (1968) has examined the effects of pyramidal tract stimulation on dorsal horn interneurons, including cells projecting in the DLF. Most of the cells are inhibited by pyramidal tract volleys. Some cells are weakly excited and then inhibited, and a few are unaffected. Fetz suggests that the inhibition originates from the postcruciate cortex and the excitation from the precruciate cortex.

The finding by A. G. Brown (1970, 1971) that the response properties of SCT neurons are altered depending on activity conveyed by descending pathways has already been mentioned. A. G. Brown *et al.* (1973b,c) have further shown that stimulation of axons in the DLF, ventral funiculus, and dorsal funiculus can inhibit SCT cells. The inhibition following brief trains of stimuli peaks at 20–40 msec and lasts 150–250 msec. It is proposed that the inhibition from the DLF and ventral funiculus is produced by descending pathways, whereas that from the dorsal funiculus is secondary to the activation of spinal cord inhibitory circuits by the antidromic volleys in dorsal column afferent fibers. The time course of the inhibition and the associated cord dorsum P waves suggests that the inhibition is presynaptic. A. G. Brown and Martin (1973) provide evidence that part of the inhibition produced by dorsal funiculus stimulation involves the cerebellum and brain stem.

Cervero *et al.* (1977a) have used cold block to demonstrate that many SCT cells, especially those with excitatory input from nociceptors, are under tonic descending inhibitory control (Fig. 8.11).

Stimulation of the SI and SII regions of the cat cerebral cortex has been found by A. G. Brown and Short (1974) to inhibit SCT cells, as well as to evoke cord dorsum P waves. There is also weak excitation of some neurons from stimulation of areas 5 and 7. The inhibition is abolished when the DLF is sectioned bilaterally and reduced by barbiturate anesthesia. The effect of barbiturate anesthesia would presumably account for the negative results of A. Lundberg *et al.* (1963) in otherwise similar experiments.

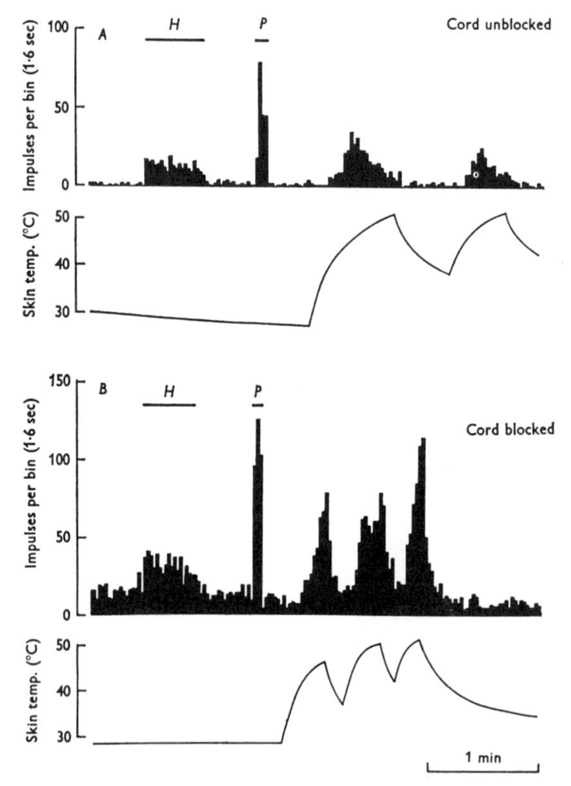

Fig. 8.11. Responses of an SCT cell before and after blocking tonic descending inhibitory pathways with cold. The cells were excited by hair movement (H), pinprick (P), and noxious heat (note the temperature monitor in the lower trace). The background activity and the evoked responses were enhanced following cold block (B) as compared with the controls (A). (From Cervero *et al.*, 1977a.)

A. G. Brown *et al.* (1977a) have used intracortical microstimulation to investigate the cytoarchitectonic areas responsible for inhibiting SCT cells in the cat. Inhibitory effects are evoked by stimulation in areas 4, 3a, 3b, 1, and 5.

Using cold block to interfere with tonic descending inhibitory control systems, Hong *et al.* (1979) have shown that the background activity and the responses of SCT cells to high-threshold muscle afferent fibers become enhanced. For example, in the decerebrate state, only a little more than one-third of SCT cells are excited by intra-arterial injection of algesic substances (bradykinin, serotonin, KCl), whereas during cold block of the descending inhibitory pathways, most SCT cells respond to such stimuli. Furthermore, the responses of cells excited in the decerebrate state are increased during spinalization.

In intracellular recordings from SCT cells, Harrison and Jankowska (1984) have been able to observe up to seven different monosynaptic unitary EPSPs in a given cell following single-shock stimulation of the pyramid, and they suggest that each unitary EPSP is attributable to the action of a different axon in the corticospinal tract. Repetitive stimulation often produces polysynaptic IPSPs or both IPSPs and EPSPs. There does not appear to be any direct input to SCT cells from the red nucleus.

B. G. Gray and Dostrovsky (1983a) have investigated the inhibitory effects of stimulation in the periaqueductal gray, nucleus cuneiformis, nucleus raphe magnus, and medullary reticular formation on dorsal horn neurons, including SCT cells. Most dorsal horn neurons are inhibited from all of the sites, whether they are classified as low-threshold, wide-dynamic-range, or high-threshold cells.

Kajander *et al.* (1984) have observed that stimulation in the periaqueductal gray or the nucleus raphe magnus causes an inhibition of the responses of SCT cells to innocuous stimuli.

Fleetwood-Walker *et al.* (1988b) have been able to inhibit the responses of SCT cells by stimulating in the A11 dopaminergic cell group. The inhibition is blocked by a D_2 dopamine receptor antagonist.

ANATOMY OF THE LATERAL CERVICAL NUCLEUS

Topography of the LCN

In cats, the LCN extends from the junction of the spinal cord and medulla to the middle of C3 (Fig. 8.2) (Rexed and Brodal, 1951; Brodal and Rexed, 1953). A detailed description of the nucleus is given by Craig and Burton (1979). According to them, the nucleus is first detectable rostrally as a small group of cells just ventral to the subnucleus caudalis of the spinal nucleus of the trigeminal nerve. More caudally, it becomes a comma-shaped structure ventral to the superficial dorsal horn at C1. The nucleus has its largest cross-sectional area about 5 mm caudal to the obex, extending nearly to the pia at this level. From mid-C1 to mid-C2, the nucleus is close to the lateral margin of laminae IV–VI. The LCN contains large (70%) and small (30%) cells. Although the small cells are distributed throughout the nucleus, there is a concentration of them (20% of the LCN population) in the medial part of the nucleus.

Flink and Westman (1986) did not include the rostral population of cells described by Craig and Burton (1979) as part of the LCN. They stressed that this group was clearly distinguishable from the LCN by strands of white matter (cf. Hockfield and Gobel, 1978).

The postnatal development of the LCN in the cat has been described by Broman *et al.* (1987).

Morphology of Individual LCN Neurons

Brodal and Rexed (1953) state that the somas of cat LCN neurons in frozen transverse sections stained by the Glees method are generally round, oval, or pear shaped, with diameters of

24–30 μm by 40–54 μm. A few smaller and larger cells are also seen. The long axis of the cells is oriented rostrocaudally (cf. Fig. 8.2B).

The synapses made by the axons of SCT cells on LCN neurons have been described by Ha and Liu (1963), who used material stained by the Golgi technique.

Westman (1968a) has described the neurons of the LCN as visualized by the Golgi technique. Most of the cells are medium sized to large, although a few are small. Typical cells have four to nine main dendrites. The dendrites branch and are relatively symmetric, although there is a tendency for them to have a longitudinal orientation. There are numerous dendritic spines. The axon usually originates from the soma, but it is sometimes from a dendrite. The initial course of the axon may be in any direction, but eventually it turns toward the gray matter of the spinal cord. Axon collaterals branch and end within the LCN. Neurons in the rostromedial LCN may have atypical appearances. They have long, slender dendrites without many branches.

Ultrastructure of LCN Neurons

The LCN has been described electron microscopically by Westman (1968b). Most LCN neurons are about 30 μm in diameter in the transverse plane and 30–50 μm in diameter in the longitudinal plane. However, some have a diameter of less than 20 μm. The organelles are appropriate to neurons in general. Somatic spines are often found. Synaptic endings can have round or ovoid vesicles and sometimes one or two dense-core vesicles. Dendritic spines occur at intervals of about 30 μm. Much of the dendritic surface is covered with synaptic boutons that are usually similar to those on the soma. However, a special type of synapse, an elongated giant bouton, with dimensions of 8 by 1 μm (exceptionally as large as 10 by 3 μm), can be found on dendrites of LCN neurons. The synapses on dendritic shafts can be asymmetric or symmetric, but those on the spines are asymmetric.

Flink and Westman (1986) have made additional electron-microscopic observations of LCN neurons labeled retrogradely from either the thalamus or the midbrain. They note little difference in the ultrastructure of these two populations of neurons. Small, unlabeled neurons differ in several ways, including having a lower density of synaptic terminals and no somatic spines.

After interruption of the SCT, degenerating synaptic boutons are seen in the LCN (Westman, 1969). The dense type of degeneration predominates, although some synapses undergoing filamentous degeneration are also observed. However, most synaptic endings are normal; only 15% of the synapses are found in the process of degenerating. Degenerating synapses are found on somatic spines and on dendritic shafts and spines of the large cells of the LCN, but none are seen on the small cells. The climbing giant boutons totally disappear after the SCT is interrupted.

Following lesions of the midbrain in kittens, there is a loss of 89% of the neurons in the contralateral LCN (G. Grant and Westman, 1969). No descending degenerating axons projecting to the LCN are observed. At the ultrastructural level, degenerating dendrites can be seen as dark profiles with synaptic contacts on their surface.

Westman (1971) estimated the proportion of the surface membrane of LCN neurons that is covered by synaptic boutons. About 42% of the somatic and 48% of the dendritic membrane of projection cells, but only about 10% of the somatic membrane of interneurons, were covered by synapses.

A stereologic comparison of the volumes of constituents of the LCN in normal cats and after interruption of the SCT was made by Griph and Westman (1977). The volume of synaptic boutons was decreased by 66% in the operated animals; this was attributed to a reduced number of synapses, since the size did not appear to change (at least in cross-sections). This suggested that the previous estimate of a loss of 15% of LCN synapses after interruption of the SCT was a gross underestimate. The dendritic volume also decreased by 64%, indicating a substantial transsynaptic degeneration.

The terminals in the LCN of SCT neurons can be labeled anterogradely by injecting wheat germ agglutinin-HRP into the spinal cord (Svensson *et al.*, 1987). Labeled boutons average 2.2 μm in their longest diameter, which is larger than that of the unlabeled boutons. Most of the labeled endings contain round or oval synaptic vesicles, but a few have flattened vesicles. The vesicle content of the labeled boutons is less than that of the unlabeled population. The labeled boutons represent 14% of the bouton volume of the area of the LCN. Most postsynaptic contacts are on dendrites, although some (16%) are on somas. There are numerous somatic spines on LCN neurons. It is estimated that each SCT cell in the lumbar cord gives rise to an average of 4400 synaptic boutons in the LCN.

Number of LCN Neurons

Counts have been made of the number of LCN neurons in several species. In rats, the LCN nucleus contains about 300–500 neurons (M. L. Baker and Giesler, 1984; see Giesler and Elde, 1985).

Craig and Burton (1979) find a total of 4777, 5402 and 7465 LCN neurons on one side in representative cats. Flink and Westman (1986) have counted the cells in the LCN in semithin sections and find a larger population than that reported previously (Blomqvist *et al.*, 1978; Craig and Burton, 1979) based on the use of frozen sections. The average number of cells in the counts of Flink and Westman (1986) is 8300.

Kitai *et al.* (1965) estimate that there are about 6000 neurons in the dog LCN, and Ha *et al.* (1965) count about 6000 cells on one side of the LCN in the raccoon. They consider that the LCN in dogs and raccoons is substantially larger than in cats. Recounts made by using semithin sections would be desirable.

Immunocytochemical Studies of the LCN

Giesler and Elde (1985) have contrasted the immunocytochemical staining properties of the LCN and the lateral spinal nucleus in rats. They have been unable to demonstrate staining in the LCN for many substances for which there is immunoreactivity in the lateral spinal nucleus, including dynorphin, met-enkephalin, substance P (SP), somatostatin, and a peptide that may prove to be neuropeptide Y. In cats, immunoreactivity for SP is found only in the most medial part of the LCN.

Blomqvist *et al.* (1985b) have examined the LCN in cats for the presence of a reaction to antibodies to glutamic acid decarboxylase (GAD) and to SP. GAD-positive synaptic endings are present throughout the nucleus, whereas SP-containing boutons are restricted to the ventromedial LCN.

Maxwell *et al.* (1989) have confirmed the presence of GAD-positive synaptic endings in the cat LCN. They also find GABA-like immunoreactivity in synaptic boutons in this nucleus. No axoaxonal complexes are seen in apposition to LCN neurons, just axodendritic and axosomatic ones.

Broman and Blomqvist (1989a,b) find that SP-immunoreactive fibers and terminals are distributed evenly throughout the LCN in monkeys and rats but are concentrated in the medial part of the nucleus in cats.

Axons of LCN Cells

The axons of LCN neurons in cats, dogs, and monkeys have been tracked by the anterograde degeneration technique (G. Grant *et al.*, 1968; Hagg and Ha, 1970; Ha, 1971), as well as by retrograde labeling with HRP in cats (Craig and Burton, 1979). Some of the axons give off collaterals within the LCN. The axons cross the gray matter in a ventromedial direction through the intermediate zone, decussate in the ventral white commissure, ascend in the ventral funiculus, pass just lateral to the inferior olivary nucleus, and join the medial lemniscus. Some of the axons give off collaterals within the LCN.

Horrobin (1966) has been able to activate the axons of LCN neurons from the contralateral medial lemniscus at a midbrain level, but not from the anterior cerebellum. Craig and Tapper (1978) have also activated LCN neurons antidromically in the cat from the contralateral medial lemniscus, and Kajander and Giesler (1987a) did this from the contralateral ventral funiculus. Downie *et al.* (1988) activated LCN neurons antidromically in the monkey from the contralateral ventral posterior lateral thalamic nucleus.

The conduction velocities of the axons of LCN neurons average 27.8 m/sec (range, 6.4–87.5 m/sec) (Craig and Tapper, 1978) or 37 m/sec (Kajander and Giesler, 1987a) in cats and 17 m/sec in monkeys (Downie *et al.*, 1988).

Projection Targets of LCN Neurons

The main projections of the LCN are to the contralateral thalamus (Fig. 8.12) and also to the midbrain. In addition, there is a small descending projection to the spinal cord (Svensson *et al.*, 1985b). The cervicothalamic and cervicomesencephalic pathways have been studied experimentally in rats (Giesler *et al.*, 1988), cats (Morin and Catalano, 1955; Landgren *et al.*, 1965; Boivie, 1970; Craig and Burton, 1979; Berkley, 1980; Flink *et al.*, 1983; Flink and Westman, 1986;

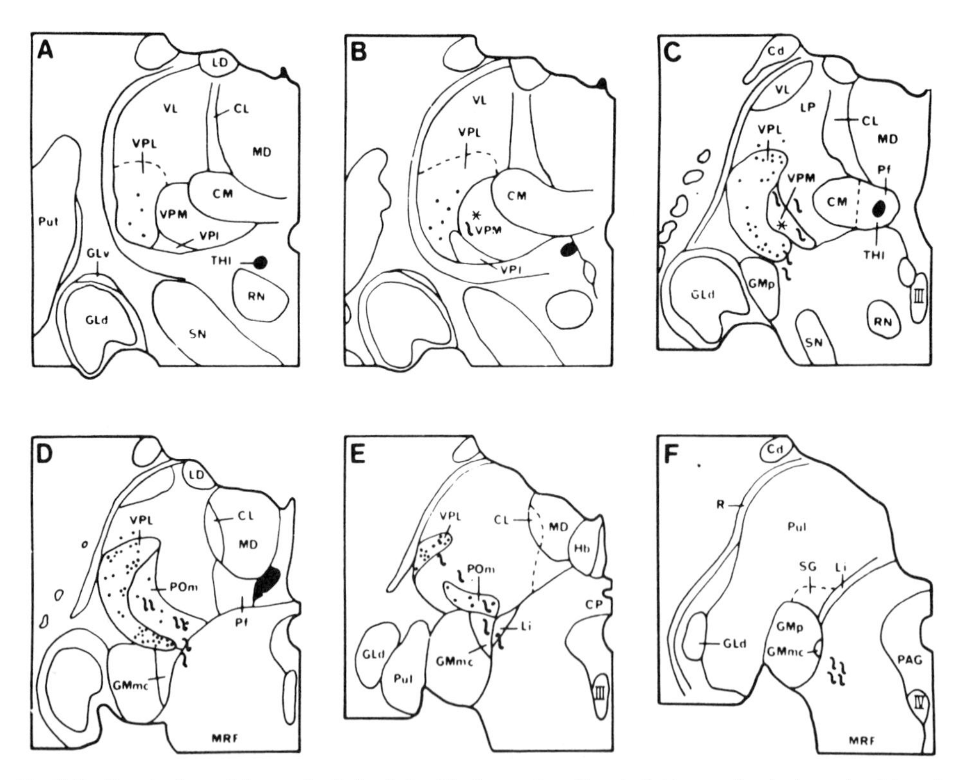

Fig. 8.12. Terminations of the cervicothalamic tract in the monkey. Terminal degeneration is shown by dots and degenerating fibers of passage by wavy lines. The lesion was in the contralateral LCN. (From Boivie, 1980.)

Metherate *et al.*, 1986), dogs (Hagg and Ha, 1970), and monkeys (Ha, 1971; Berkley, 1980; Boivie, 1980).

Anterograde tracing studies show that the thalamic nuclei that receive cervicothalamic terminals include the ventral posterior lateral (VPL) nucleus and the medial part of the posterior complex (PO_m) in rats, cats and monkeys (Fig. 8.12) (Berkley, 1980; Boivie, 1970, 1980; Giesler *et al.*, 1988). There may also be projections to other nuclei, but if they exist, they are sparse (Berkley, 1980; Boivie, 1980). There is extensive overlap in the VPL termination zone of the projection from the LCN with that from the dorsal column nuclei both in cats and monkeys (Berkley, 1980). However, the terminals of axons from the dorsal column nuclei are distributed throughout the VPL nucleus, whereas those from the LCN are mainly in restricted regions (Blomqvist *et al.*, 1985a). The morphology of both sets of terminals is similar.

The distribution of cells in the LCN with different projection targets has been investigated by retrograde tracing methods. Large injections of HRP into the thalamus of cats label more than 90% of LCN neurons (Craig and Burton, 1979). Although most of the labeled cells are contralateral to the injection site, a few (1–4%) are ipsilateral. Of the unlabeled cells, most are small and located in the medial part of the nucleus. However, small cells in other parts of the LCN are also labeled. Flink and Westman (1986) report similar figures: an average of 94% of LCN neurons project to the thalamus and 5% of LCN cells project to the ipsilateral thalamus. Unlabeled cells are in the ventromedial part of the LCN.

Small injections of HRP into different parts of the thalamus show that the LCN has a topographic organization in cats (Craig and Burton, 1979). The dorsolateral part of the LCN projects to the lateral part of the ventral posterior lateral nucleus (VPL_l), whereas the ventromedial part of the LCN projects to the medial part (VPL_m). A few cells in the medial LCN appear to project to the ventral posterior medial (VPM) nucleus. Cells in the LCN are also labeled following injections into the region of the PO_m nucleus; the source of the projection to PO_m is uncertain, but may have been the medial part of the LCN.

The cervicomesencephalic tract in the rat and cat terminates in the contralateral intercollicular nucleus and the posterior part of the superior colliculus (Flink *et al.*, 1983; Giesler *et al.*, 1988; see also Edwards *et al.*, 1979; Menétrey *et al.*, 1982). There is also a weak projection to the periaqueductal gray in the rat and cat and to the nucleus of the brachium of the inferior colliculus. The monkey shows a similar pattern to that in the cat, but with the addition of minor projections to the parabrachial nucleus, the posterior pretectal nucleus, and the nucleus of Darkschewitz (Wiberg *et al.*, 1987).

LCN neurons can also be labeled retrogradely from the midbrain. About 90% of the midbrain-projecting LCN neurons in the cat (5000 cells) are contralateral and 10% (500 cells) ipsilateral to the injection site (Flink *et al.*, 1983). Flink and Westman (1986) have observed that the largest proportion of LCN neurons projecting to the midbrain is 49%. Only 3% of LCN cells project to the ipsilateral midbrain. LCN neurons can also be labeled retrogradely from the midbrain in monkeys (Wiberg *et al.*, 1987). In the best example in a monkey, 2000 labeled LCN cells are found contralaterally and 500 ipsilaterally.

G. Grant and Westman (1969) have found that 89% of the cells of the LCN are lost following a lesion of the contralateral midbrain in kittens. They suggest that some of the residual neurons could be interneurons.

Berkley *et al.* (1980) used retrograde labeling with different tracers to examine the possibility that some LCN neurons project both to the thalamus and the midbrain. Neurons labeled from the thalamus or the midbrain are intermingled in the LCN. A total of 69% of the cells are labeled only from the thalamus, 18% only from the midbrain, and 12% from both sites when HRP and tritiated apo-HRP are used. However, a larger proportion of doubly labeled neurons is observed when fluorescent tracers are used: 39% and 48% in two cases.

Injections of HRP into the spinal cord label a few LCN neurons retrogradely, in both cats (Svensson *et al.* 1985b) and monkeys (Burton and Loewy, 1976). Most of the cells in cats are in the rostroventral part of the LCN ipsilateral to the injection site; more project to the cervical than to the lumbar enlargement (Svensson *et al.*, 1985b). On the basis of counts of the number of cells labeled with different-sized injections, it seems likely that a given cell terminates at several levels of the spinal cord. Lesions of the spinal cord provide evidence that the axons descend in the dorsomedial part of the DLF. The population of descending projection cells has been estimated to be about 500, which represents about 10% of the population of LCN neurons in cats (Svensson *et al.*, 1985b).

Functional Properties of LCN Neurons

Responses to Electrical Stimulation. In response to stimulation of a cutaneous nerve (but not a muscle nerve), a rapidly conducted nerve volley followed by a negative field potential in the LCN was observed (Rexed and Ström, 1952). The responsible fibers are the lowest-threshold group in the nerve. The effects are elicited only from the ipsilateral side.

Inhibitory events within the LCN have been examined by Fedina *et al.* (1968), using activity evoked by stimulating the DLF. Medial lemniscus volleys are inhibited following stimulation of the skin, of cutaneous nerves or of high-threshold muscle afferent fibers. However, these experiments are not conclusive, since some of the medial lemniscus evoked potential may have resulted from activity in neurons of the dorsal column nuclei, rather than of the LCN (cf. Nijensohn and Kerr, 1975; Enevoldson and Gordon, 1989b). However, intracellular recordings from LCN neurons have demonstrated IPSPs, as well as EPSPs, following peripheral nerve stimulation. Cutaneous and high-threshold muscle afferent fibers are responsible for IPSPs. There is also evidence for presynaptic inhibition of input from the SCT, since EPSP depression can be observed, as well as enhanced excitability of SCT axon terminals. These actions occur in decerebrate animals and thus do not require connections with the rostral part of the brain. However, it should be noted that axoaxonic synapses have not been found in the LCN (see above), and so the morphologic basis for presynaptic inhibition in this nucleus is unclear.

Rigamonti and DeMichelle (1977) have been able to excite neurons of the LCN in cats by electrical stimulation of the greater splanchnic nerve. The effective volleys are in the fine myelinated fibers, but not the Aβ fibers.

Responses to Natural Stimulation. *Rat.* Giesler *et al.* (1979b) have examined the response properties of LCN cells in urethane-anesthetized rats. All of the neurons can be activated by hair movement. The receptive fields are often large, and in 40% of cases they include a contralateral component. Responses are observed to noxious mechanical stimuli in 27% of the cells, most of which are also activated by noxious heat. Some LCN neurons can be excited by intraperitoneal injection of hypertonic saline, suggesting that they have a visceral receptive field (cf. Rigamonti and DeMichelle, 1977).

Cat. Morin *et al.* (1963) have recorded the activity of single LCN neurons of the cat in response to tactile stimulation of the skin, to pressure, or to joint movements. Most of the cells are excited by touch, pressure, or both, but some units respond to joint movement. Most of the tactile and pressure units have receptive fields on the limbs. Although some of the receptive fields are small, most are large or very large. The tactile responses are rapidly adapting. The pressure units may have been nociceptive. The responses to joint movement reflect both the degree and rate of the movement. A similar organization is found in dogs (Kitai *et al.*, 1965). The LCN of raccoons differs in that no units responding to joint movements are found (Ha *et al.*, 1965).

Oswaldo-Cruz and Kidd (1964) found that 73% of the LCN neurons in their sample could be excited by hair movement. Other neurons could be activated only by cutaneous or deep stimulation. Surround inhibition was not observed, but a few neurons could be inhibited by stimulation of distant parts of the body. Most cells had a background discharge of 10–15 Hz.

Horrobin (1966) recorded from neurons in the cat LCN activated antidromically from the medial lemniscus in the contralateral midbrain. About 90% of the cells were activated by tactile stimuli, although noxious stimuli over the entire ipsilateral surface of the body and head was the effective stimulus in some cases. Inhibition could sometimes be produced by noxious stimulation on the contralateral side or over most of the body surface. Some neurons were activated by both touch and pressure. More receptive fields were found on the rostral than on the caudal part of the body. The LCN appeared to be somatotopically organized, with the hindlimb representation medial to the forelimb representation (see below).

Craig and Tapper (1978) also recorded from neurons of the LCN in the cat. Most of the cells (535 of 556) could be activated antidromically from the contralateral medial lemniscus. Receptive-field sizes varied from a single digit to half of the body surface (Fig. 8.13). Half of the cells had receptive fields on the forelimb. A few cells had receptive fields on the face. The LCN had a topographic organization (Fig. 8.14), with the forelimb representation medial and the hindlimb representation lateral and superficial in the nucleus (note that this is opposite to the organization described by Horrobin (1966) but in agreement with anatomical findings). A special group of neurons with deep, visceral, or nociceptive response properties was found in the most ventromedial part of the LCN. Most LCN neurons, however, responded to hair movements, in particular to movements of guard hairs. Usually, there was convergence of inputs from several different types of mechanoreceptors (especially guard hair, field receptor, and down hair receptors). A rare unit had an input from SA I or SA II mechanoreceptors. Input from muscle and viscera was uncommon, but did occur. A few nociceptive units were also found. Most of these were in the ventromedial LCN.

Metherate *et al.* (1986) recorded from LCN neurons in barbiturate-anesthetized cats. Most of the cells responded to input from guard hairs and none to activation of afferent fibers supplying Pacinian corpuscles.

Kajander and Giesler (1987a) have found that LCN neurons in decerebrated, spinalized cats are much more likely to have nociceptive responses (Fig. 8.15) than are neurons in anesthetized cats with an intact neuraxis. In the former preparation, 41% of LCN neurons are categorized as low-threshold neurons, 49% as wide-dynamic-range neurons, and 9% as high-threshold neurons. Low-threshold neurons do not respond to noxious heat stimuli, whereas wide-dynamic-range and high-threshold neurons often respond in a graded fashion to graded heat pulses. Receptive fields vary in size; more than half are on the rostral part of the body; all but one are exclusively ipsilateral; and no receptive fields are on glabrous skin. The topographic organization corresponds to that found by Craig and Tapper (1978), rather than that found by

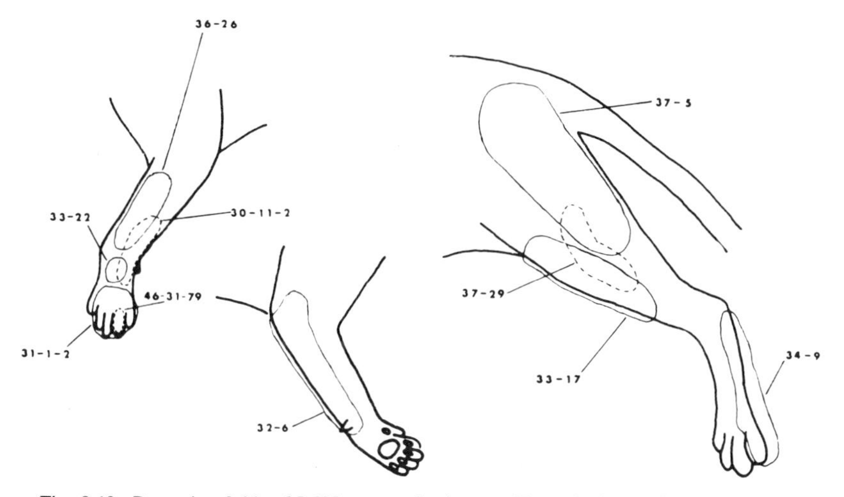

Fig. 8.13. Receptive fields of LCN neurons in the cat. (From Craig and Tapper, 1978.)

Fig. 8.14. Locations of LCN neurons and positions of their receptive fields. The drawing shows a transverse section through the C2 segment, and the LCN is outlined with a dotted line. The symbols indicate the recording sites for neurons with receptive fields on the forelimb (X), trunk (○), or hindlimb (●). (From Craig and Tapper, 1978.)

Horrobin (1966). When barbiturates are given, nociceptive responses are reduced, whereas tactile responses remain. In decerebrated cats that are not spinalized, most LCN neurons are classified as low threshold, but an occasional cell is of the wide-dynamic-range type. The dominance of tactile inputs suggests that there is a descending inhibition of nociceptive input to LCN neurons in decerebrate cats. Most LCN neurons found in animals anesthetized with urethane are also of the low-threshold type.

Kajander and Giesler (1987b) find that nociceptive LCN neurons in decerebrate, spinalized cats show enhanced responses to noxious heat pulses following repeated noxious heat stimulation, suggesting an input from Aδ nociceptors.

Monkey. Downie *et al.* (1988) recorded from LCN neurons in anesthetized monkeys with an intact neuraxis. The cells selected were activated antidromically from the contralateral ventral posterior lateral thalamic nucleus. More of the receptive fields were on the hindlimb than on the forelimb. Most were cutaneous and on the hairy skin, but occasional cells had deep receptive fields and several had at least part of the receptive field on glabrous skin. The receptive fields tended to be restricted in size, but some were large. The cells were classified as follows: low threshold, 45%; wide dynamic range, 47.5%; high threshold, 7.5%. LCN neurons tended to be entrained to vibratory stimuli at frequencies of 10–30 Hz. Most LCN neurons were excited by noxious heat stimuli. Sensitization of the skin by noxious heating resulted in an enhanced response to repeated noxious heat stimuli in the lower part of the stimulus range, but a depressed response to stronger noxious heat pulses. Responses to weak mechanical stimuli, but not to strong ones, were also increased.

Descending Control of LCN Neurons

Craig (1978) has demonstrated the existence of descending connections from the dorsal column nuclei to the LCN. The function of this pathway is as yet unclear.

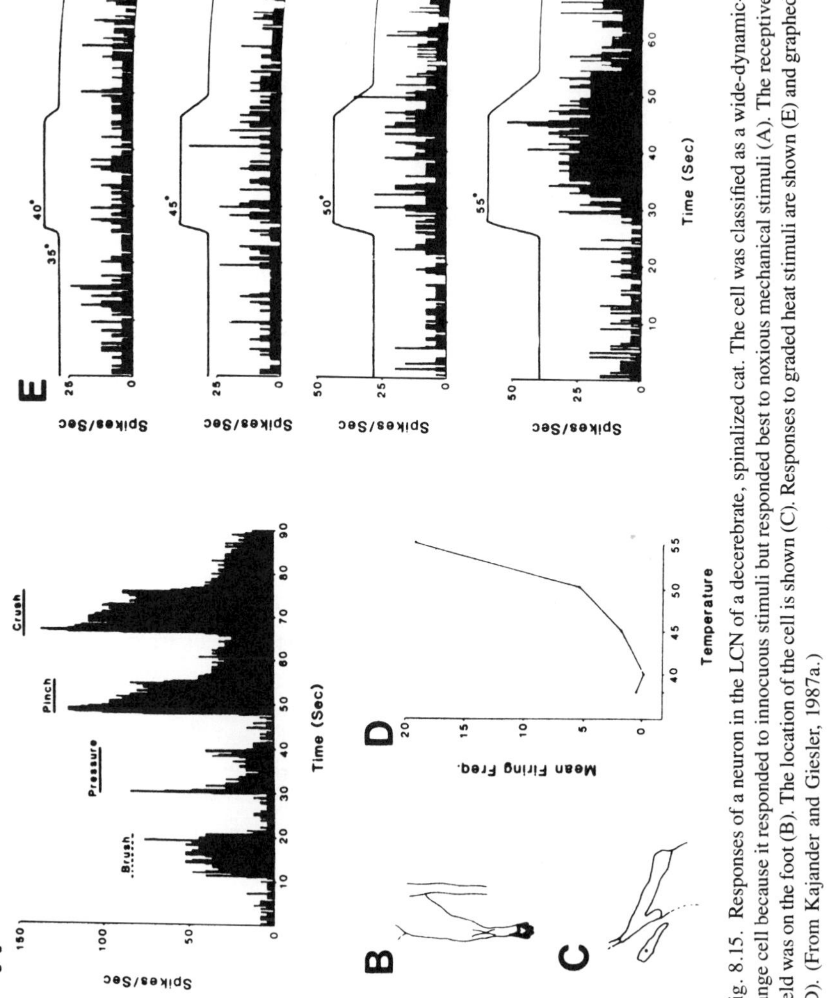

Fig. 8.15. Responses of a neuron in the LCN of a decerebrate, spinalized cat. The cell was classified as a wide-dynamic-range cell because it responded to innocuous stimuli but responded best to noxious mechanical stimuli (A). The receptive field was on the foot (B). The location of the cell is shown (C). Responses to graded heat stimuli are shown (E) and graphed (D). (From Kajander and Giesler, 1987a.)

Peto (1980) has examined the effects of cortical stimulation on LCN neurons. Stimulation in the face representation of area 3a causes excitation of about 25% and inhibition of all LCN neurons. The functional meaning of the interaction between the face representation and nonsomatotopically related parts of the LCN is unclear, but it is suggested that the LCN may be involved in tasks that require face-limb coordination.

Dostrovsky (1984) has found that the responses of LCN neurons to peripheral inputs are more readily inhibited by stimulation in the periaqueductal gray, nucleus raphe magnus, or medullary reticular formation than are the responses of the same cells to volleys in the dorsal lateral funiculus. He concludes that much (or possibly all) of the inhibition takes place at the level of the spinal cord. However, stimulation of the pyramidal tract inhibits responses to stimulation of the dorsal lateral funiculus, and so pyramidal tract volleys may also act at the level of the LCN.

Summary

In summary, the LCN relays signals from the SCT to the contralateral thalamus and midbrain. The vast majority of LCN cells respond to tactile stimuli. However, many also respond to noxious stimuli. Nociceptive responses are most commonly seen in the cat LCN when descending control systems are interrupted and when anesthesia is not present. However, nociceptive responses are readily found in the LCN of anesthetized rats and monkeys. The receptive fields of LCN neurons tend to be larger than those of neurons of the dorsal column nuclei, and even for LT cells, the usual case is of convergent inputs from several different kinds of sensory receptors.

SPINOMEDULLOTHALAMIC PATHWAY THROUGH NUCLEUS Z

Group I afferent fibers from hindlimb muscles provide information to the cerebral cortex by way of the spinomedullothalamic pathway (Fig. 8.1), which relays in nucleus z (Landgren and Silfvenius, 1969, 1971). The group I afferent fibers synapse on neurons in Clarke's column and also in the gray matter caudal to Clarke's column. Axons from these cells ascend in the ipsilateral dorsal lateral funiculus in company with or as part of the dorsal spinocerebellar tract (DSCT). The pathway synapses in a nucleus that was once considered a part of the vestibular complex called nucleus z. The cells of nucleus z project contralaterally by way of the medial lemniscus to the thalamus, terminating in parts of the ventral posterior lateral nucleus and its boundary with the ventral lateral nucleus. The pathway is specific to the hindlimb; group I afferent fibers from the forelimb project through the dorsal funiculus to the cuneate and lateral cuneate nuclei (see Chapter 7).

TAXONOMIC DISTRIBUTION

Most of the work on the spinomedullothalamic pathway through nucleus z has been done with cats. However, a nucleus comparable to the cat nucleus z has been found in rats (Low *et al.*, 1986), monkeys (Nijensohn and Kerr, 1975) and humans (Sadjadpour and Brodal, 1968).

MORPHOLOGY OF PATHWAY THROUGH NUCLEUS Z

A pathway from the spinal cord to nucleus z was first recognized by Pompeiano and Brodal (1957). Previous authors had considered these fibers to be spinovestibular projections, since they accompany ascending axons to the vestibular complex proper. Although nucleus z was listed among the nuclei of the vestibular complex by Brodal and Pompeiano (1957), this was because of topographic rather than functional relationships. Since nucleus z receives no vestibular afferent fibers (Walberg *et al.*, 1958) and projects to the thalamus (G. Grant *et al.*, 1973), it is appropriate to consider it a part of the somatosensory system.

The cells of origin of the spinomedullothalamic pathway may be largely in Clarke's column, since the pathway seems to enter the dorsal lateral funiculus at L3 or above (Landgren and Silfvenius, 1971). However, Pompeiano and Brodal (1957) found ample ascending degeneration in nucleus z following midlumbar lesions, suggesting that part of the pathway may originate below Clarke's column. Cells of Clarke's column may contribute projections both to the cerebellum and to nucleus z (Aoyama *et al.*, 1973; H. Johansson and Silfvenius, 1977a).

The pathway ascending to nucleus z is located in the dorsal part of the lateral funiculus (Pompeiano and Brodal, 1957; Landgren and Silfvenius, 1971; Rustioni, 1973; Nijensohn and Kerr, 1975). There is no change in responses to muscle afferent volleys at the level of the nucleus z or the cortex when the dorsal funiculus is interrupted, but such responses disappear after a lesion of the dorsal lateral funiculus (Landgren and Silfvenius, 1969, 1971). The axons of neurons in nucleus z project to the contralateral thalamus by way of the medial lemniscus, and they terminate in the rostral part of the ventral posterior lateral nucleus (lateral segment) and in the ventrolateral part of the ventral lateral nucleus (Fig. 8.16) (G. Grant *et al.*, 1973). Projections from this part of the thalamus ascend to the cerebral cortex, accounting for the group I activity from hindlimb muscle nerves that can be recorded from the sensorimotor region (Landgren *et al.*, 1967; Landgren and Silfvenius, 1971). Presumably, the spinomedullothalamic pathway is

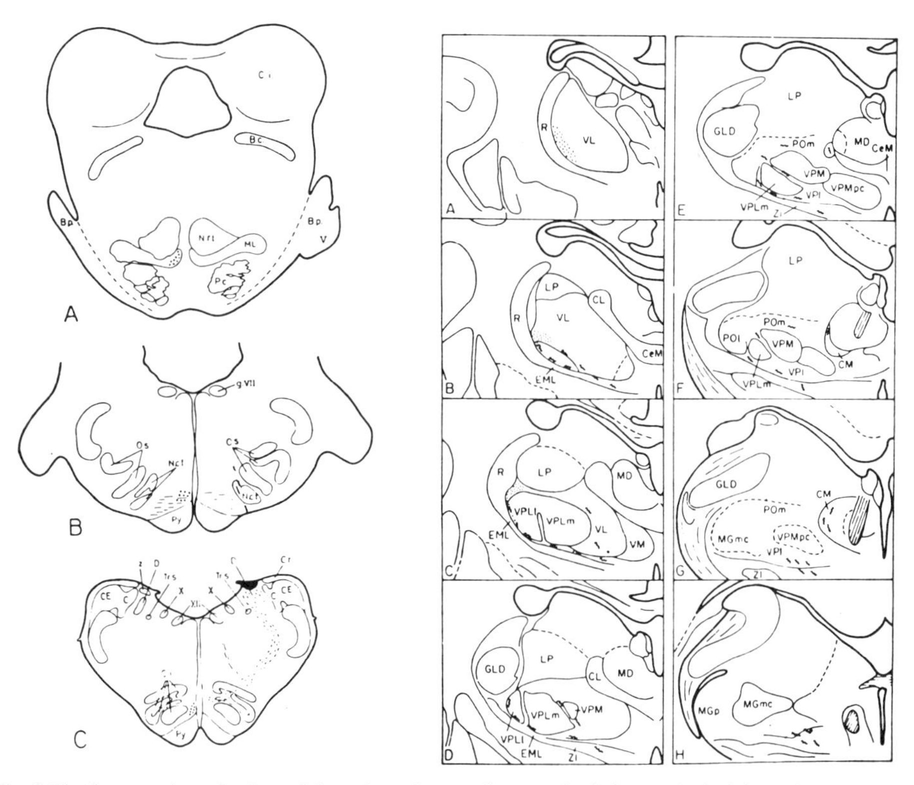

Fig. 8.16. Course and terminations of the pathway from nucleus z to the thalamus. At the left are shown the lesion (black) in the medulla and the initial course of the pathway from nucleus z at medullary, pontine, and midbrain levels. At the right are shown degenerating fibers of passage (wavy lines) and terminals (dots) in the contralateral thalamus. The experiment was performed on a cat. (From Grant *et al.*, 1973.)

also responsible for conveying hindlimb group I afferent input to the sensorimotor cortex in primates (see, e.g., Hore *et al.*, 1976). Low *et al.* (1986) find that neurons projecting to nucleus z in rats are located in Clarke's column and in lamina X of the spinal cord. They estimate that as many as 92% of the spinal afferents to nucleus z are collaterals of dorsal spinocerebellar tract neurons and that about 3% of dorsal spinocerebellar tract cells project both to the cerebellum and to nucleus z.

RESPONSE PROPERTIES OF NEURONS THAT PROJECT TO NUCLEUS Z

Recordings were made by Magherini *et al.* (1974, 1975) from the axons of some ascending-tract cells at the level of nucleus z. The axons had a conduction velocity of about 100 m/sec. They responded monosynaptically to sinusoidal stretch of hindlimb muscle at amplitudes of stretch that selectively excite group Ia muscle spindle afferent fibers (Fig. 8.17A). The responses of some units increased as the stimulus recruited secondary endings, indicating a convergence of group Ia and group II afferent fibers on some tract cells.

RESPONSES OF NEURONS IN NUCLEUS Z

Landgren and Silfvenius (1971) found that volleys in group I afferent fibers of hindlimb muscles evoked by electrical stimulation activated cells in nucleus z. Group II afferent fibers contributed an additional excitation of some cells. There was also a convergent input in some cases from cutaneous afferent fibers. The muscle input was powerful, since many cells could follow repetitive stimulation at rates of 150–200 Hz. Most cells were activated by stimulation of just one muscle nerve, some by just cutaneous input, and some both by muscle and cutaneous volleys. Some of the neurons were antidromically activated following stimulation in the contralateral medial lemniscus.

Magherini *et al.* (1974, 1975) used sinusoidal stretch of hindlimb muscles to test the effects

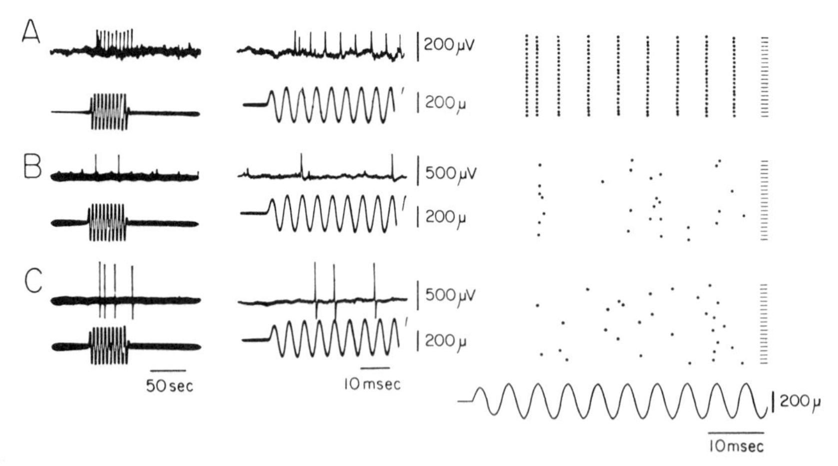

Fig. 8.17. Responses of units in the nucleus z of a cat to muscle vibration. Panels A–C show the activity of three different units. The action potentials (top traces) and a monitor of the vibratory stimulus (bottom traces) are shown on a slow (left) and a fast (right) sweep. The raster plot at the right illustrates the responses of the same units to several trials of stimulation. The unit at the top (A) followed the stimulus more closely than did the others, and the latency of the first spike was earlier; therefore, this unit was thought to be an axon of the spinomedullary tract and thus an afferent to nucleus z. The other units were considered intrinsic to nucleus z. (From Magherini *et al.*, 1975.)

of group Ia and group II muscle afferent fibers upon cells in nucleus z (Fig. 8.17B and C). Such stimuli excited nucleus z cells or produced excitation followed by inhibition. The latencies of discharges indicated that at least in some instances there was a disynaptic connection between primary afferent fibers and cells of nucleus z. Alterations in the amplitudes of the stretch stimuli showed that most units were excited by group Ia afferent fibers, although some were also activated by group II fibers (in a few cases by group II alone). Dynamic but not static stretch also excited the neurons. Only two of the twelve units tested were antidromically activated from the contralateral medial lemniscus.

The question of whether the input to the neurons of nucleus z is at least in part by way of collaterals of the dorsal spinocerebellar tract was investigated by H. Johansson and Silfvenius (1977a). They stimulated within the cerebellum and looked for responses of neurons in nucleus z, especially in cells proven to project to the thalamus by antidromic activation. Most of the nucleus z cells were excited by cerebellar stimulation, and a collision test indicated that the excitation in at least half the cases was transmitted by the DSCT. H. Johansson and Silfvenius (1977b) also studied cells in nucleus z that were not antidromically activated from stimulation of the medial lemniscus. The responses were often like those of relay cells, but many of these neurons were excited by cutaneous afferent fibers or by low-threshold joint afferent fibers. It is likely that many of these cells actually project to the thalamus but that technical factors prevented the demonstration of antidromic activation. The ones excited by joint afferent fibers may account for the information that reaches the cerebral cortex from the joints via the dorsal lateral funiculus (F. J. Clark *et al.*, 1973). Cells of the DSCT have been described that are activated by joint afferent fibers (Lindström and Takata, 1972; Kuno *et al.*, 1973).

DESCENDING CONTROL OF PATHWAY TO NUCLEUS Z

Virtually nothing is known about a possible descending control of the spinomedullo-thalamic tract. H. Johannson and Silfvenius (1977a) observed a synaptic excitation of some nucleus z units following stimulation in the contralateral diencephalon. Otherwise, this issue does not seem to have been investigated.

CONCLUSIONS

1. Sensory pathways ascending in the dorsal lateral funiculus include the SCT and the spinomedullary tract that relays in nucleus z. The DLF also contains axons belonging to pathways that are predominantly in other funiculi (postsynaptic pathway to the dorsal column nuclei, spinothalamic tract, spinomesencephalic tract).
2. The SCT originates from cells of the spinal cord dorsal horn and terminates in the LCN, which is a nucleus located just ventrolateral to the dorsal horn in segments C1–C3. Most of the neurons in the LCN project to the contralateral thalamus, but many project to the midbrain, and a few to the spinal cord.
3. The spinocervicothalamic pathway forms a rapidly conducting system for transmission of information from the skin to the contralateral cerebral cortex and midbrain. In cats, this pathway is slightly faster than the dorsal column–medial lemniscus pathway, although the opposite is the case in monkeys.
4. An LCN is found in all orders of mammals that have been investigated. It is especially well developed in carnivores.
5. The cells of origin of the SCT are mostly in laminae III and IV at all levels of the cord. A few SCT cells are also found in other laminae.
6. The dendritic trees of SCT cells have a characteristic morphology. The dendrites extend

predominantly in a dorsal direction, and they have a pronounced longitudinal orientation. The dorsally directed dendrites extend only to the border between laminae II and III, where they turn in a rostrocaudal direction. The maximum dimensions of the dendritic trees are a rostrocaudal length of 2 mm and a transverse spread of 0.5 mm. The dendrites have numerous spines.

7. The synaptic endings on SCT cells are of two main types: endings with round vesicles and endings with flattened vesicles. Axoaxonal synapses are not seen in apposition to SCT cells. Many of the round-vesicle endings disappear after dorsal rhizotomy. These are thought to be primary afferent terminals, probably from hair follicle receptors.

8. The population of SCT cells on one side of the rat spinal cord is probably several hundred, and that in cats is about 2000. The cat lumbar enlargement probably has about 700–750 SCT cells on each side.

9. The initial course of the axons of SCT cells in cats is unpredictable, but within a distance of a few hundred micrometers the axons enter the dorsal lateral funiculus. Several local collaterals synapse in the gray matter ventral to the cell bodies of SCT cells. These collaterals may form excitatory contacts with postsynaptic dorsal column cells.

10. The axons of SCT cells have conduction velocities throughout the range of both large and small myelinated fibers.

11. The projection of the SCT to the LCN is somatotopically organized in cats. The hindlimb is represented dorsolaterally and the forelimb ventromedially. There may also be some somatotopic relationship in the rostrocaudal dimension of the LCN, but this is controversial. Somatotopy is not apparent in rats.

12. SCT cells exhibit background activity consisting of brief bursts separated by longer silent intervals. The frequency of the background activity depends on the nature of the animal preparation and on the cell type.

13. SCT cells can be excited by volleys in Aβ, Aδ, and C fibers of cutaneous nerves. Some cells can also be excited by high-threshold muscle and joint afferent fibers. Intracellular recordings reveal monosynaptic and polysynaptic EPSPs in response to hair movement or stimulation of cutaneous nerves. IPSPs can also be evoked.

14. SCT cells can be excited by natural forms of cutaneous stimulation. The responses in cats can be classified on the basis of the receptor types that activate the SCT cells. However, the types of SCT cells observed depend upon the presence or absence of tonic descending inhibition. In spinalized animals, three types of SCT cells are observed: (1) cells activated just by hair movement; (2) cells excited by hair movement and by pressure or pinch applied to the skin; and (3) cells responsive just to pressure and pinch. Given that the input from pressure and pinch is mediated by nociceptors, 71% of SCT cells in spinalized animals should be considered nociceptive.

15. Stimulation of high-threshold muscle afferent fibers by algesic chemicals injected intra-arterially excites 29% of SCT cells when the conduction in descending pathways of the spinal cord is blocked with cold. With conduction intact, the proportion of SCT cells activated by muscle afferent fibers is only 4%.

16. SCT cells with overlapping dendritic trees have overlapping receptive fields; conversely, if the dendritic trees do not overlap, neither do the receptive fields. Correspondingly, morphologic contacts are made by hair follicle afferent fibers with SCT cells that have overlapping receptive fields but not with SCT cells that have nonoverlapping fields.

17. It is estimated that movement of hair on one or two toes on the cat foot should activate as much as 20% of the total population of SCT cells on one side of the lumbar enlargement.

18. The receptive fields of SCT cells can be subdivided into a firing zone and an adjacent inhibitory zone. However, IPSPs can be evoked from either area. Monosynaptic and polysynaptic EPSPs appear to be evoked by single hair follicle afferents supplied by Aβ fibers, whereas IPSPs are probably attributable to down hair afferents.

19. Whereas the receptive fields of most SCT cells in cats are restricted to the hairy skin, this is not true in monkeys. Receptive fields on both hairy and glabrous skin are observed in recordings from many primate SCT cells.

20. There is a somatotopic organization of the population of SCT cells. This is similar to the somatotopic organization of dorsal horn interneurons in general. About 60% of the SCT cells in the L6–S1 segments have receptive fields on the toes.

21. The somatotopic map of the population of SCT cells in the lumbar enlargement does not change following dorsal rhizotomy or peripheral nerve section. Many neurons become deafferented, but inappropriate synaptic connections do not appear up to 55 days following the lesions. However, the responsiveness of SCT neurons recovers after dorsal rhizotomies, and more of the cells can be activated by noxious stimuli.

22. Glutamate applied iontophoretically excites SCT cells and expands their receptive fields. Glycine and GABA inhibit SCT cells. Inhibition of hair-activated SCT cells by glutamate applied by high iontophoretic currents is probably mediated by inhibitory interneurons.

23. Norepinephrine and dopamine produce a selective inhibition of the responses of SCT cells to noxious stimuli. A similar effect is produced by κ opioid agonists released iontophoretically near the cell bodies of SCT cells and by μ agonists released in the substantia gelatinosa.

24. SCT cells are under tonic descending inhibitory control. Their activity can be inhibited phasically by stimulation in the DLF, ventral funiculus, dorsal funiculus, periaqueductal gray, nucleus cuneiformis, nucleus raphe magnus, medullary reticular formation, deep cerebellar nuclei, pyramid, and the sensorimotor and parietal association cortex. Excitation sometimes results from stimulation of the cerebral cortex or pyramid.

25. Both large and small neurons are found in the LCN. The dendrites tend to be oriented rostrocaudally. LCN neurons have numerous dendritic spines and also somatic spines.

26. The synaptic endings on LCN neurons contain round or ovoid synaptic vesicles. Elongated giant boutons are found on dendrites; these disappear after lesions of the dorsal lateral funiculus. Only about 15% of the synapses on LCN neurons degenerate following DLF lesions. A similar proportion of terminations (14% of the bouton volume) in the LCN can be labeled anterogradely with wheat germ agglutinin-HRP applied in the dorsal lateral funiculus. However, stereological measurements indicate that there is actually a loss of two-thirds of the synaptic volume in the LCN after interruption of the SCT. It is estimated that the average SCT neuron forms 4400 synaptic boutons in the LCN.

27. There are as many as 500 LCN neurons on one side in rats, 8300 in cats, 6000 in dogs, and 6000 in raccoons.

28. Immunoreactivity for a number of peptides is lacking in the rat LCN. SP immunoreactivity is found in the most medial part of the cat LCN but is distributed evenly throughout the LCN in monkeys. GAD and GABA-like immunoreactivity are present throughout the cat LCN.

29. The axons of LCN neurons decussate in the ventral white commissure, turn rostrally in the ventral funiculus and join the medial lemniscus. Some collaterals are given off within the LCN. Conduction velocities of LCN axons indicate that they are medium sized myelinated fibers.

30. The LCN projects to the contralateral thalamus. Specific thalamic targets include the VPL nucleus and the medial part of the posterior complex. The projection to the VPL nucleus overlaps that of the dorsal column nuclei.

31. More than 90% of LCN neurons in cats can be retrogradely labeled from the contralateral thalamus. A few (about 5%) LCN neurons ipsilateral to a thalamic injection are also labeled. The thalamic projections are somatotopically organized: the dorsolateral LCN projects to the lateral segment of the VPL nucleus, and the ventromedial LCN projects to the medial segment of the VPL nucleus.

32. The LCN also projects to the contralateral midbrain. The main target nuclei include the

intercollicular nucleus and the posterior part of the superior colliculus. There are also minor projections to the periaqueductal gray, the nucleus of the brachium of the inferior colliculus, and in some species the parabrachial nucleus, the posterior pretectal nucleus, and the nucleus of Darkschewitz.

33. About half of the neurons in the LCN can be retrogradely labeled from the contralateral midbrain (and only 3% from the ipsilateral midbrain). Many LCN neurons project to both the thalamus and the midbrain.

34. Some LCN neurons (about 10%) project to the spinal cord.

35. LCN neurons can be activated by volleys in cutaneous nerve fibers and (in one report) visceral afferent fibers of the greater splanchnic nerve.

36. LCN neurons generally respond to hair movement. They may also respond to skin pressure, noxious heat, and in some cases to joint movement, muscle stimulation, or visceral afferent stimulation. Nociceptive responses are much more prominent in the absence of tonic descending inhibition. In cats, they are reduced considerably by barbiturate anesthesia; however, nociceptive responses are prominent in rats and monkeys even in the presence of anesthesia.

37. The excitatory receptive fields of LCN neurons may be small but are often large. There may also be inhibitory receptive fields. The receptive fields are somatotopically organized in cats: cells in the medial part of the nucleus respond to stimulation of the forelimb and cells in the lateral part of the nucleus to stimulation of the hindlimb.

38. The LCN is probably under descending controls, at least from the dorsal column nuclei and the cerebral cortex. However, inhibition of the responses of LCN neurons by stimulation in the medial brain stem results in large part from inhibition of SCT cells.

39. Group I muscle afferent volleys reach the contralateral cerebral cortex by way of a collateral pathway of the dorsal spinocerebellar tract. The spinomedullary path relays in nucleus z, whose neurons then project to the contralateral thalamus through the medial lemniscus.

40. A spinomedullothalamic pathway through nucleus z is found in rats, cats, monkeys, and probably humans.

41. The thalamic targets of nucleus z are parts of the ventral posterior lateral and ventral lateral nuclei.

42. Neurons in nucleus z respond to sinusoidal stretch of hindlimb muscles. The muscle afferent fibers involved appear to include both group Ia and group II fibers from muscle spindles. There are also inputs from cutaneous and joint afferent fibers to nucleus z neurons.

9 Sensory Pathways in the Ventral Quadrant

This chapter will be concerned with several pathways in the white matter of the ventral quadrant of the spinal cord: the spinothalamic, spinoreticular, and spinomesencephalic tracts. Components of the spinothalamic and spinomesencephalic tracts that ascend in the dorsal lateral funiculus will also be considered.

SPINOTHALAMIC TRACT

The spinothalamic tract (STT) arises largely from neurons in the dorsal horn of the spinal cord, although some STT cells are in the intermediate region and ventral horn. The axon of a given STT neuron decussates at a level near the cell body and ascends in the lateral funiculus to the thalamus (Fig. 9.1). Associated with the STT are other pathways that may synapse on neurons which in turn project to the thalamus. Collectively, these ventral quadrant pathways form a system of direct and indirect projections to the thalamus. The indirect pathways include the spinoreticular and spinomesencephalic tracts (Fig. 9.1).

TAXONOMIC DISTRIBUTION

Although several neurologists of the last century, including Brown-Séquard (1860) and Gowers (1878), reported clinical evidence that pain and temperature sensations depend on activation of tracts ascending in the anterolateral white matter of the spinal cord, Edinger (1889, 1890) is generally regarded as the first to suggest that there is a direct projection from the spinal cord to the thalamus (Keele, 1957) on the basis of studies of poikilothermic vertebrates and cats.

Although several investigators were unable to confirm the presence of an STT in the cat by using the Marchi technique (Patrick, 1896; Chang and Ruch, 1947b; Morin *et al.*, 1951), it is now known that there is at least a moderate-sized STT in this animal, since projections can be traced from the spinal cord to the thalamus by using anterograde degeneration or transport methods (M. Probst, 1902; Getz, 1952; Anderson and Berry, 1959; Mehler, 1966; Boivie, 1971a; E. G. Jones and Burton, 1974; Berkley, 1980; Craig and Burton, 1981, 1985; Mantyh, 1983b). Other mammalian species that have been shown to have an STT include the marsupial phalanger (Rockel *et al.*, 1972), opossum (Mehler, 1969; Hazlett *et al.*, 1972), hedgehog (Jane and Schroeder, 1971), pig (Breazile and Kitchell, 1968), sheep (Rao *et al.*, 1969), raccoon (Craig and Burton, 1985), dog (M. Probst, 1902; Hagg and Ha, 1970), rabbit (Kohnstamm, 1900; Wallen-

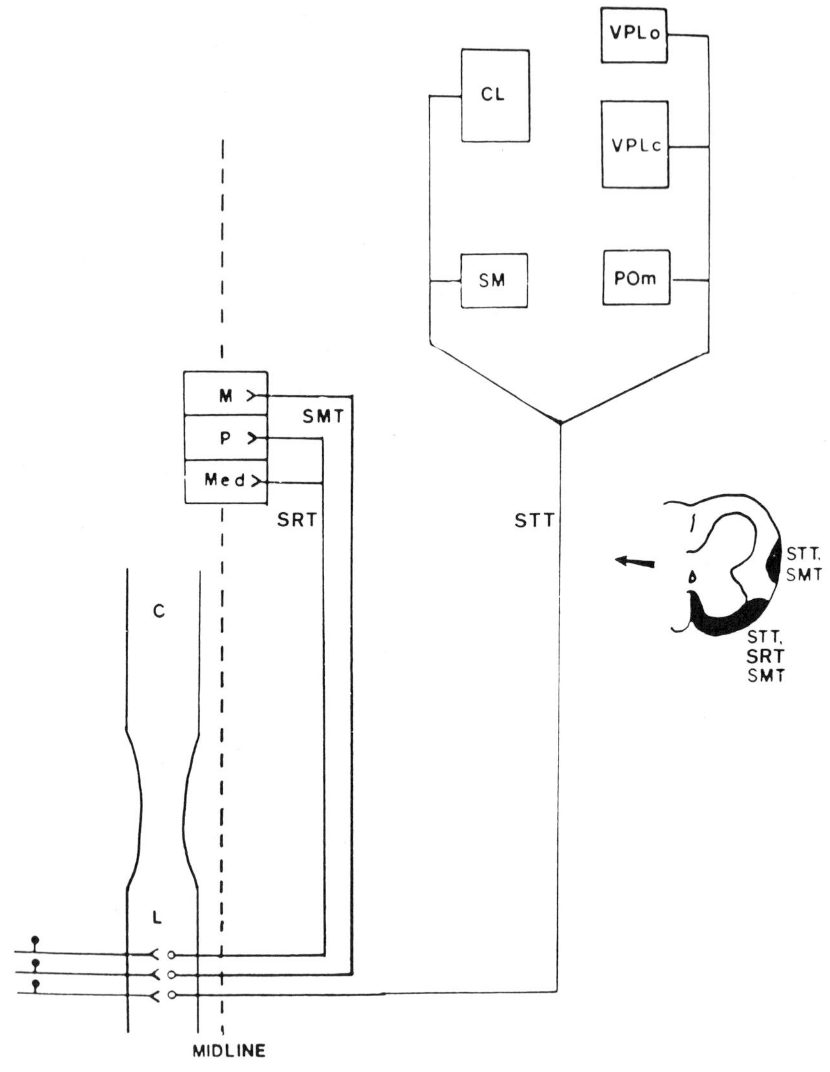

Fig. 9.1. Ascending pathways in the ventral quadrant. These include the STT, SMT and SRT. The drawing shows these for the lumbar enlargement (L) only. The cell bodies receive primary afferent input and project their axons across the midline and rostrally to the brain. The SRT ends in the reticular formation of the medulla (Med) and pons (P), whereas the SMT ends in the midbrain (M). The STT projects to several thalamic nuclei, including the VPL_c and VPL_o, the PO_m, the CL, and the nucleus submedius (SM). The transverse section of the spinal cord at the right shows the locations of the tracts. Note that a component of the STT and SMT may be present in the DLF (at least in some species), whereas the SRT is confined to the ventral quadrant.

berg, 1900; Gerebtzoff, 1939), rat (Lund and Webster, 1967b; Mehler, 1969; Zemlan *et al.*, 1978; Peschanski *et al.*, 1983; Ma *et al.*, 1986), and a variety of primates (Mott, 1895; W. E. L. Clark, 1936; Walker, 1938; Weaver and Walker, 1941; Chang and Ruch, 1947b; Morin *et al.*, 1951; Poirier and Bertrand, 1955; Bowsher, 1961; Mehler, 1966; 1969; Schroeder and Jane, 1971; Kerr and Lippman, 1974; Kerr, 1975c; Mehler, 1966; Boivie, 1979; Berkley, 1980; Pearson and Haines, 1980; Mantyh, 1983a), including humans (Quensel, 1898; Thiele and Horsley, 1901;

Collier and Buzzard, 1903; Goldstein, 1910; Foerster and Gagel, 1932; Walker, 1940; Gardner and Cuneo, 1945; Kuru, 1949; Glees and Bailey, 1951; Bowsher, 1957; Poirier and Bertrand, 1955; Mehler, 1962, 1974).

As pointed out by Boivie (1971a), some of the degeneration studies failed to take into account damage to the cervicothalamic tract, but in many investigations the spinal cord lesions producing anterograde degeneration were made below the level of the lateral cervical nucleus. Thus, an STT has been found in all orders of mammals that have been examined, although its size and complexity are greatest in primates (Mehler, 1969).

Since the studies of Edinger (1889, 1890), several investigations have addressed the issue of a direct STT in nonmammalian vertebrates. No STT was found in cyclostomes (Northcutt and Ebbeson, 1980; Ronan and Northcutt, 1981) or representative species of elasmobranch and teleost fish (Hayle, 1973). However, Ebbesson and Hodde (1981) found a spinothalamic projection in nurse sharks. Burr (1928) claimed to have traced an STT through the brain stem of a teleost, the deep-sea sunfish; however, confirmation of such a pathway by using experimental techniques is needed before it can be accepted that a true STT occurs in any teleost fish. Herrick (1939) found an STT in Golgi-stained preparations of tiger salamander larvae in Golgi-stained preparations. However, this was not confirmed in an experimental study of axolotls (Nieuwenhuys and Cornelisz, 1971), and no direct spinothalamic path has been found in frogs (Ebbesson, 1969; Hayle, 1973; Neary and Wilczynski, 1977, 1979). Goldby and Robinson (1962) failed to find an STT in one species of lizard. However, direct spinal projections to the thalamus have now been found in several reptiles, including caiman, alligator, turtle, lizard, and snake (Ebbesson, 1967, 1969; Riss *et al.*, 1972; Pritz and Northcutt, 1980; Ebbesson and Goodman, 1981; Hoogland, 1981; Kuenzle and Woodson, 1982). There is also an STT in birds, as shown by work on the pigeon (Karten, 1963; Karten and Revzin, 1966; A. Schneider and Necker, 1989).

A discussion of phylogenetic theories concerning the evolution of a direct STT in more advanced vertebrate forms is presented by Kevetter and Willis (1984).

CELLS OF ORIGIN

Retrograde Labeling

The locations of the cells of origin of the STT have been mapped in rats, cats, and monkeys by using retrograde tracing methods. In some experiments, the tracer substance has been injected into a large volume of tissue including both lateral and medial nuclei of the thalamus. In other experiments, the objective was to determine whether different populations of STT cells project to different parts of the thalamus by targeting the injections of retrograde marker to limited regions of the thalamus. In yet other studies, collateral projections have been demonstrated by double retrograde labeling.

Rat. The locations of STT cells that label when horseradish peroxidase (HRP) is injected into much of the medial-lateral extent of the rat thalamus are shown in Fig. 9.2 (Giesler *et al.*, 1979a; see also Giesler *et al.*, 1981c; Kevetter and Willis, 1982; 1983; Granum, 1986; Kemplay and Webster, 1986; Harmann *et al.*, 1988a; Huang, 1989b; Lima and Coimbra, 1989; Burstein *et al.*, 1990).

The largest concentration of labeled cells is in the upper cervical segments (Fig. 9.2, C1 and C2). In addition to neurons of the nucleus gracilis and the lateral cervical nucleus, the labeled cells include numerous STT neurons both contralateral and ipsilateral to the side of the thalamic injection. According to Granum (1986) and Kemplay and Webster (1986), STT cells in the upper cervical segments account for half or more of the population in the entire spinal cord of the rat. However, these authors appear to have included the neurons of the lateral cervical nucleus (see

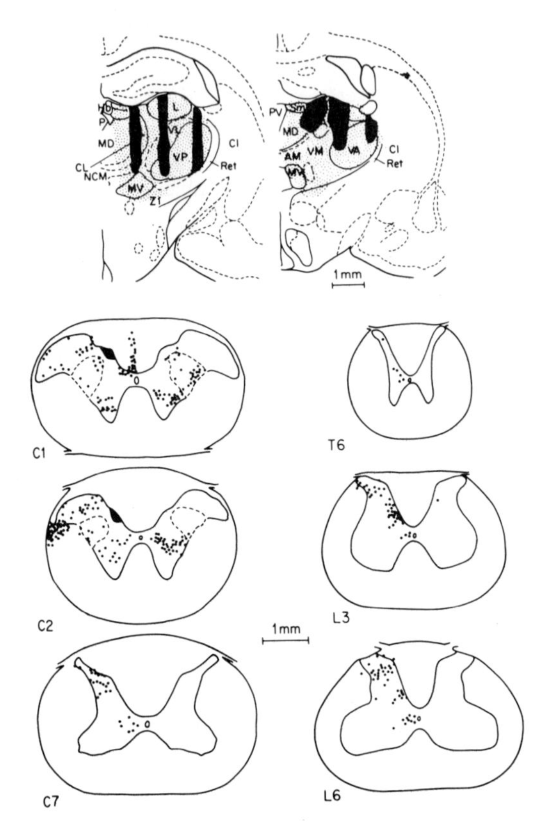

Fig. 9.2. Locations of the cells of origin of the STT in a rat. The site of injection of HRP is shown in the drawings of sections through the thalamus at the top. Labeled cells within 400-μm blocks at several segmental levels are shown. (From Giesler *et al.*, 1979; reproduced with permission.)

Chapter 8) in their counts. A particularly dense concentration of labeled cells is found in C1 and C2 along the medial edge of the dorsal horn in an area equivalent to Cajal's internal basilar nucleus.

In segments below the upper cervical cord, STT cells are found chiefly in the cervical and lumbosacral enlargements, with the greatest concentrations in the marginal zone, nucleus proprius, and the medial intermediate gray (Fig. 9.2, C7–L6). Almost all of the labeled cells at these levels are contralateral to the thalamic injection site (Giesler *et al.*, 1979a, 1981c; Kevetter and Willis, 1983; Granum, 1986; Kempley and Webster, 1986; Burstein *et al.*, 1990).

An injection of HRP restricted to the lateral part of the thalamus labels more STT neurons than does an injection into the medial thalamus (Giesler *et al.*, 1979a). The populations labeled from the lateral thalamus at levels below the upper cervical segments include cells in the marginal zone, nucleus proprius, and the medial part of the base of the dorsal horn. STT neurons projecting to the medial thalamus are in the medial base of the dorsal horn and the intermediate gray. The axons of STT neurons projecting to the lateral thalamus ascend through the lateral funiculus, whereas the axons projecting to the medial thalamus ascend in the ventral funiculus (Giesler *et al.*, 1981c).

The adult pattern of distribution of STT cells is already present in 14–18-day-old rats (Huang, 1989b).

Cat. The distribution of STT neurons in the cat spinal cord labeled by a large injection of HRP into the lateral and medial thalamus is shown in Fig. 9.3 (Carstens and Trevino, 1978a). As in the rat, large numbers of labeled neurons are found bilaterally in the upper cervical segments.

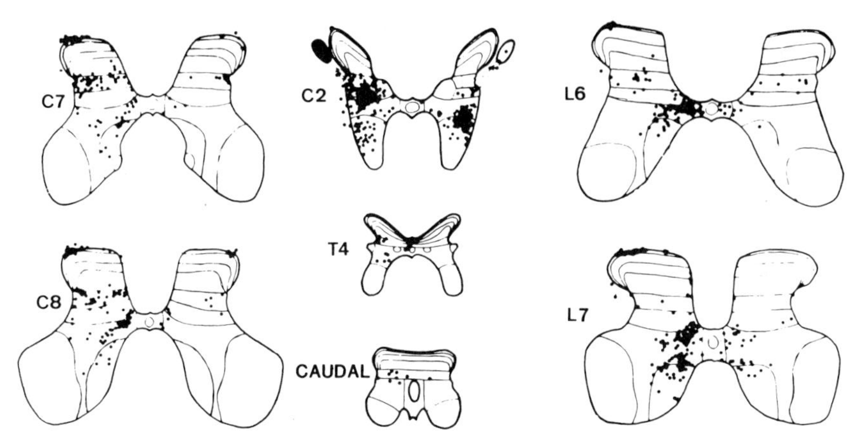

Fig. 9.3. Locations of the cells of origin of the STT in a cat. HRP was injected unilaterally into the caudal thalamus. The dots represent cells observed in a given segment. Most of the labeled cells are contralateral to the injection. (From Carstens and Trevino, 1978; reproduced with permission.)

Apart from the lateral cervical nucleus, labeled cells are concentrated in laminae V–VII contralateral and in laminae VII and VIII ipsilateral to the thalamic injection (Carstens and Trevino, 1978b; see also Comans and Snow, 1981a).

In segments below the upper cervical level, labeled neurons are concentrated in three zones: lamina I, laminae IV–VI, and laminae VII–VIII (Trevino and Carstens, 1975; Carstens and Trevino, 1978a; Craig *et al.*, 1989a). Unlike the arrangement in the rat, the STT neurons in the deeper layers of the dorsal horn are not concentrated medially, but rather are distributed across the width of the dorsal horn. The proportion of labeled cells in laminae IV–VI versus VII–VIII depends upon the segmental level. In the cervical enlargement, there are approximately the same number of STT neurons in each of these regions, whereas in the lumbar enlargement there are more STT neurons in laminae VII–VIII than in laminae IV–VI (Carstens and Trevino, 1978a; see also M. W. Jones *et al.*, 1987; Craig *et al.*, 1989a). Most of the labeled cells are contralateral to the injection site. Craig *et al.* (1989a) find that more than half of the STT cells in the cervical enlargement of the cat are in lamina I, whereas more than half of those in the lumbosacral enlargement are in laminae VII–VIII.

Small injections of marker have been used to distinguish separate populations of STT neurons according to their thalamic targets. When HRP is injected into the border region between the ventrobasal complex and the ventral lateral nucleus of the thalamus, most of the labeled neurons are in lamina I, with a smaller number in medial lamina VII and other locations (Carstens and Trevino, 1978a). An injection into the region of the posterior complex labels some neurons in laminae IV–VI and a few in lamina I (Carstens and Trevino, 1978a). No neurons are labeled in the cat following an injection into the ventral posterior lateral (VPL) nucleus (Carstens and Trevino, 1978a). When HRP is injected into the intralaminar complex, most of the labeled cells are in laminae VII and VIII (Carstens and Trevino, 1978a; Comans and Snow, 1981b). On the other hand, an injection into a region that includes the nucleus submedius labels neurons chiefly in lamina I (Craig and Burton, 1981).

A surprising finding by M. W. Jones *et al.* (1985, 1987) is that one-quarter of the STT neurons in the cat appear to project to the thalamus through the dorsal lateral funiculus (DLF) (see later discussion). According to these authors, most of the cells that do this are in lamina I (Fig. 9.4). The part of the tract that ascends in the ventrolateral quadrant originates almost exclusively from neurons in laminae IV–VI and VII–VIII.

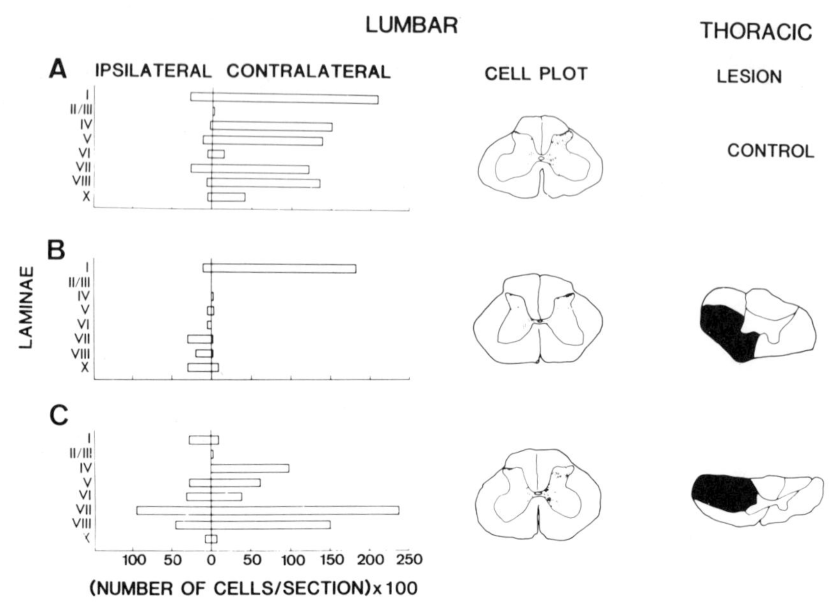

Fig. 9.4. (A) Bar graph and a drawing of a transverse section of the spinal cord indicating the distribution of HRP-labeled neurons in the lumbar spinal cord after an injection into the thalamus. Most of the labeled cells are contralateral to the injection, and they are concentrated in laminae I, IV–VI and VII–VIII. (B) Pattern of labeling after the ventral quadrant was lesioned at a thoracic level. (C) Pattern of labeling after a lesion of the DLF. (From Jones *et al.*, 1987; reproduced with permission.)

Monkey. Fig. 9.5 shows the distribution of the cells of origin of the STT of a monkey following a large injection of wheat germ agglutinin-conjugated horseradish peroxidase (WGA-HRP) into one side of the thalamus (Apkarian and Hodge, 1989a). A similar pattern has been reported previously (Trevino and Carstens, 1975; Trevino, 1976; Willis *et al.*, 1979; Hayes and Rustioni, 1980; cf. Albe-Fessard *et al.*, 1975).

Of the total population of labeled primate STT cells, 35% are in the upper cervical segments (Apkarian and Hodge, 1989a). Many of these are ipsilateral to the thalamic injection site. The distribution of the labeled cells observed in the upper cervical segments of monkeys resembles that in cats. Contralateral to the injection, STT cells are found in laminae IV–VIII, with the greatest concentrations in laminae VI and VII. Ipsilaterally, most of the labeled cells are in lamina VIII.

Below the upper cervical segments, 90% or more of the labeled STT neurons are contralateral to the thalamic injection site (Willis *et al.*, 1979; Hayes and Rustioni, 1980; Apkarian and Hodge, 1989a). In the sacral spinal cord, the proportion of ipsilateral STT neurons is higher (23%) than in the enlargements; many of the ipsilaterally projecting neurons are in Stilling's nucleus (Willis *et al.*, 1979).

In segments below the upper cervical spinal cord, 18% of the entire population of STT cells are in the lower cervical spinal cord, 19% in the thoracic cord, 19% in the lumbar cord, and 8% in the sacral and coccygeal cord (Apkarian and Hodge, 1989a). STT cells in the enlargements are concentrated in laminae I and the lateral part of lamina V, although some labeled cells are also found in laminae VII and VIII (Fig. 9.5) (Trevino and Cartens, 1975; Trevino, 1976; Willis *et al.*, 1979; Hayes and Rustioni, 1980; Apkarian and Hodge, 1989a). A few STT neurons can be found in laminae II and III (Willis *et al.*, 1978; 1979; Hayes and Rustioni, 1980; Apkarian and Hodge,

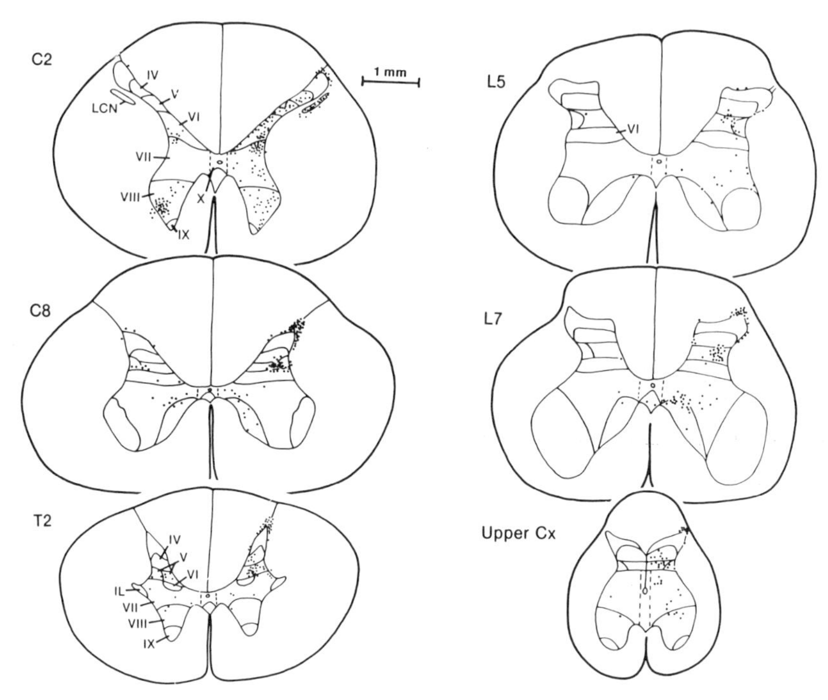

Fig. 9.5. Distribution of STT cells labeled retrogradely after an injection of WGA-HRP into one side of the thalamus of a monkey. Most of the labeled cells are contralateral to the injection. Representative levels are shown. (From Apkarian and Hodge, 1989a; reproduced with permission.)

1989a). There is a group of STT cells in the lateral parts of laminae VII and IX in segments L3–L6; some of these appear in the location usually attributed to spinal border cells (Fig. 9.5) (Willis *et al.*, 1979; Hayes and Rustioni, 1980; Apkarian and Hodge, 1989a; cf. Cooper and Sherrington, 1940).

When injections of HRP are made into restricted regions of the monkey thalamus, more neurons are labeled in laminae VII and VIII from the region of the medial thalamus than from the lateral thalamus (Willis *et al.*, 1979). Otherwise, the populations of neurons labeled from these regions of the thalamus overlap substantially, including many cells in lamina I. Albe-Fessard *et al.* (1975) report retrograde labeling of neurons in lamina I following injection of HRP into the intralaminar complex and in both lamina I and laminae IV–V after HRP injection into the VPL nucleus. Furthermore, lamina I neurons that project to the intralaminar complex have been demonstrated electrophysiologically (Giesler *et al.*, 1981b; Ammons *et al.*, 1985a). A projection of lamina I to the intralaminar complex represents a species difference with respect to cats (cf. Carstens and Trevino, 1978a; Comans and Snow, 1981b).

As in the cat, part of the STT of monkeys appears to ascend in the DLF (the dorsolateral STT) (Apkarian and Hodge, 1989b,d; see section on axons of STT cells). The cells in the lumbar enlargement that contribute to this projection make up about one-quarter of the total population, and they are located chiefly in lamina I (although in squirrel monkeys, many STT cells in laminae IV-VI are also said to project in the dorsolateral STT). A larger proportion of lumbar lamina I STT cells project through the DLF in squirrel monkeys (99%) than in macaques (78%). In an anterograde-labeling study, it has been found that the dorsolateral and ventral components of the STT terminate in the same thalamic nuclei (Apkarian and Hodge, 1989c). This includes the central lateral nucleus, again indicating a projection of neurons of lamina I to the intralaminar complex of the monkey.

Double-Labeling Studies

In several studies, STT axons have been shown in double retrograde labeling experiments to collateralize, terminating in more than one nucleus, either in the thalamus or in the brain stem. For example, in rats, about 15–20% of STT cells project to both the lateral and the medial thalamus (Kevetter and Willis, 1983). Most of these neurons are located bilaterally in the upper cervical segments, but some are in lamina V on the side contralateral to the thalamic injection site. A few STT neurons can be labeled from both sides of the thalamus (Kevetter and Willis, 1983).

Some STT cells in the rat and monkey can also be retrogradely labeled from the periaqueductal gray (R. F. Liu, 1986; Harmann *et al.*, 1988a; Zhang *et al.*, 1990) or the rhombencephalic reticular formation (Kevetter and Willis, 1982; 1983). Hayes and Rustioni (1980) find that a few spinothalamic tract cells in monkeys can also be labeled from the dorsal column nuclei, indicating that these cells project in both the spinothalamic and the postsynaptic pathway to the dorsal column nuclei.

On the basis of a double-labeling study with fluorescent tracers, Craig *et al.* (1989a) conclude that 62% of the STT neurons in cats project only to the medial thalamus, 25% only to the lateral thalamus, and 13% to both. This general pattern is true of the subpopulations of STT neurons in laminae I and VII–VIII, but most of the neurons of laminae V–VI project to the lateral thalamus. Within the population of neurons in lamina I, those in the lateral part of the lamina project only to the medial thalamus, whereas those in the medial part project chiefly to the lateral thalamus. In a parallel study, Craig *et al.* (1989b) observed that about twice as many STT cells are labeled in lamina I when fluorescent tracers are used rather than HRP or WGA-HRP.

R. T. Stevens *et al.* (1989), also using fluorescent retrograde labels, have examined the origins of STT cells projecting to the lateral thalamus, intralaminar complex, and medial thalamus (including the nucleus submedius) in cats and evaluated the double labeling from more than one target. Of the total population of labeled STT cells contralateral to the injection site, 33% are in laminae I–III, 13% in laminae IV–VI, and 54% in laminae VII–X. Lesions of the ventral quadrant or of the DLF at a thoracic level allowed them to confirm that most STT cells projecting in the dorsolateral STT are in laminae I–III and that most of those projecting in the ventral STT are in deeper laminae. About 40% of the labeled STT cells in the lumbar enlargement project to the lateral thalamus, 48% to the medial thalamus, and 12% to both. Of STT cells in lamina I, 51% project laterally, 36% medially, and 13% both laterally and medially. Of STT cells in laminae IV-VI, 57% project laterally, 40% medially, and 3% both laterally and medially. Of STT cells in laminae VII-X, 36% project laterally, 52% medially, and 11% both laterally and medially. Only a few STT cells in lamina I are labeled after injections into the intralaminar complex, in contrast to STT cells in laminae VII–X, which make up a large proportion of STT cells that are labeled from the intralaminar region.

Antidromic Mapping

The distribution of STT cells has also been demonstrated by using antidromic activation of their axons in the thalamus of rats (Dilly *et al.*, 1968; Giesler *et al.*, 1976), cats (Dilly *et al.*, 1968; Trevino *et al.*, 1972; Albe-Fessard *et al.*, 1974a) and monkeys (Trevino *et al.*, 1973; Albe-Fessard *et al.*, 1974a,b; Willis *et al.*, 1974; Giesler *et al.*, 1981b; Ammons *et al.*, 1985a). In general, the locations of antidromically activated spinothalamic cells agree well with maps produced by using retrograde tracers, although the antidromic technique undoubtedly underestimates the number of small neurons, particularly in laminae I–III.

In rats, STT cells have been activated antidromically from the medial lemniscus, posterior thalamus, VPL nucleus, and central lateral (CL) nucleus (Giesler *et al.*, 1976). In cats, STT neurons have been activated antidromically from the "shell region" around the VPL nucleus and from the intralaminar complex (Trevino *et al.*, 1972; Holloway *et al.*, 1978; Meyers and Snow, 1982b; Blair *et al.*, 1984; Craig and Kniffki, 1985; Ferrington *et al.*, 1986b). STT cells in monkeys have been activated antidromically from the medial lemniscus, posterior thalamus, the caudal (VPL_c) and rostral (VPL_0) parts of the VPL nucleus, the rostral part of the ventral lateral (VL_0) nucleus, the CL nucleus, and the parafascicular nucleus (Trevino *et al.*, 1973; Albe-Fessard *et al.*, 1974a,b; A. E. Applebaum *et al.*, 1975, 1979; Giesler *et al.*, 1981b; Ammons *et al.*, 1985a). In several studies, STT neurons have been activated antidromically from more than one site, indicating that the same neuron projected to multiple targets. Combinations of antidromic activation sites in the monkey have included VPL_c and CL (Fig. 9.6) (A. E. Applebaum *et al.*, 1979; Giesler *et al.*, 1981b; Ammons *et al.*, 1985a), VPL_c bilaterally (K. D. Gerhart, R. P. Yezierski and W. D. Willis, unpublished data), VPL_c and periaqueductal gray (Fig. 9.6) (D. D. Price *et al.*, 1978; Yezierski *et al.*, 1987); VPL_c and rhombencephalic reticular formation (Giesler *et al.*, 1981b; Blair *et al.*, 1984).

The areas of termination of individual STT axons have been mapped by antidromic microstimulation in monkeys (A. E. Applebaum *et al.*, 1979). Lamina I STT axons have been followed to their terminations in the VPL_c nucleus (Fig. 9.7), as have branches of the axon of an STT neuron in laminae IV–VI to VPL_c and CL (Fig. 9.6).

Number of STT Cells

Rat. In two studies in which the total number of STT cells labeled after injection of HRP into one side of the thalamus have been counted, it has been estimated that there are about 1000

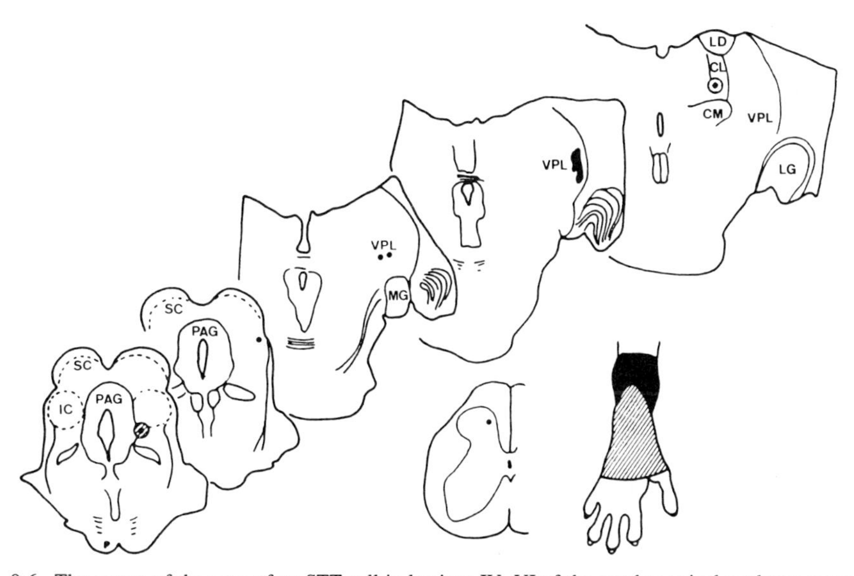

Fig. 9.6. The course of the axon of an STT cell in laminae IV–VI of the monkey spinal cord was mapped by microstimulation. The locations of the soma and of the receptive field are shown in the drawings at the bottom right. The dots on the transverse sections of the midbrain and diencephalon show the sites at which antidromic activation was produced with the minimum current. The shaded circles indicated the estimated stimulus spread. The irregular black area in the VPL nucleus was the site at which antidromic activation occurred using an electrode whose position was fixed. The other antidromic activation sites were determined with a roving electrode. Note that this neuron probably projected to the periaqueductal gray, VPL nucleus and CL nucleus. (From the data of A. E. Applebaum *et al.*, 1979.)

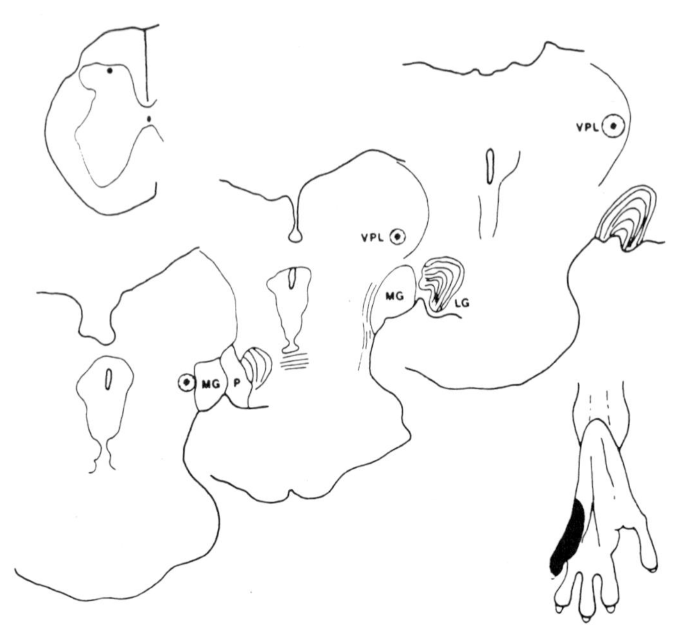

Fig. 9.7. The course of the axon of a lamina I STT cell in a monkey was mapped by microstimulation. The location of the cell body is shown at the top left and of the receptive field at the bottom right. The dots drawn on transverse sections of the rostral midbrain and diencephalon show the position of the stimulating electrode used for antidromic activation when the minimum stimulus currents were used to excite the axon. The circles indicate the approximate amount of stimulus spread. (From the data of A. E. Applebaum *et al.*, 1979.)

STT cells in the rat spinal cord (Granum, 1986; Kemplay and Webster, 1986). However, in recent studies, the number of STT cells in rats has been reexamined by using more sensitive retrograde markers. Lima and Coimbra (1989) count about 6000 STT cells in the spinal cord enlargements labeled with cholera toxin subunit B. Using fluorogold, Burstein *et al.* (1990) estimate that the total number of STT cells that can be labeled retrogradely from one side of the thalamus exceeds 9500.

Cat. Craig *et al.* (1989a) estimate that about 5000 STT cells can be labeled retrogradely from one side of the cat thalamus.

Monkey. On the basis of counts of the labeled cells in their best case, Apkarian and Hodge (1989a) estimate that more than 18,000 STT cells project to one side of the thalamus in macaque monkeys.

Morphology of Individual STT Cells

The configurations of the somas and proximal dendrites of STT cells retrogradely labeled with HRP have been described (Willis *et al.*, 1979; Apkarian and Hodge, 1989a). In transverse sections from the lumbosacral cord, the smallest cells, as judged by measurements of the area of the cell bodies, are in lamina I and the largest in laminae V and VII–IX (Willis *et al.*, 1979). Measurements of soma diameter show that the smallest STT cells throughout the spinal cord are in lamina I; the largest in the cervical enlargement are in lamina VIII and in the lumbar cord in laminae IV–VI (Apkarian and Hodge, 1989a).

STT cells in lamina I are sometimes large and fusiform in transverse sections; these are often referred to as Waldeyer cells (Fig. 9.8A) (cf. Waldeyer, 1888). Similar neurons are

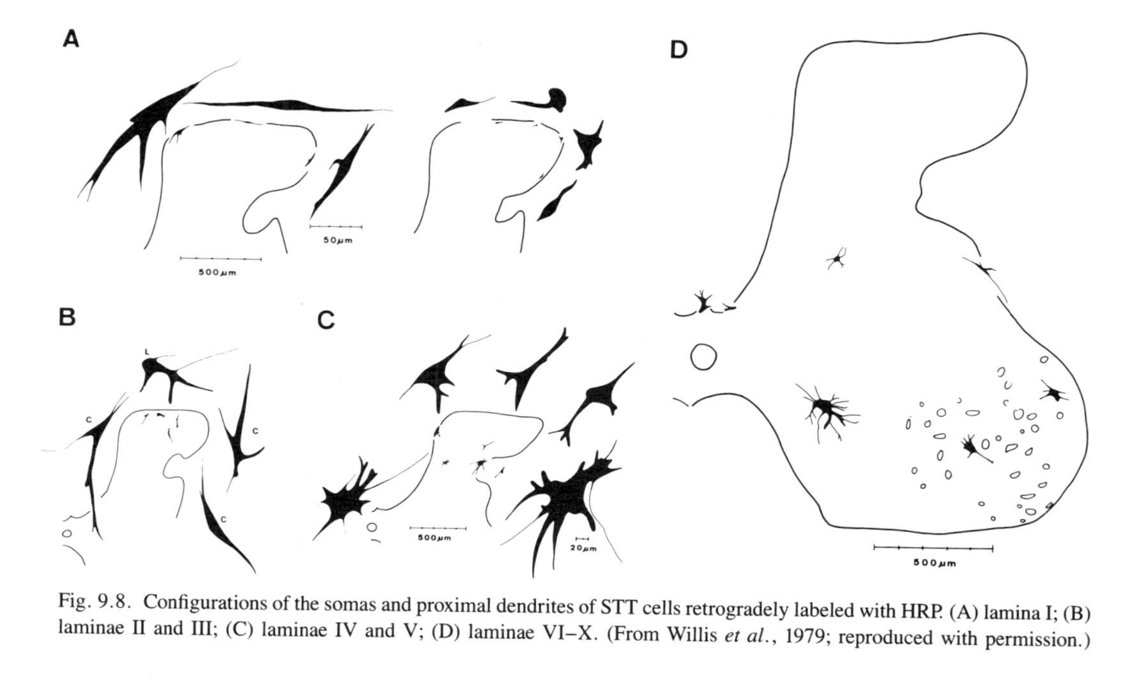

Fig. 9.8. Configurations of the somas and proximal dendrites of STT cells retrogradely labeled with HRP. (A) lamina I; (B) laminae II and III; (C) laminae IV and V; (D) laminae VI–X. (From Willis *et al.*, 1979; reproduced with permission.)

observed along the lateral boundary of lamina V (Willis *et al.*, 1979). Other STT cells in lamina I are fusiform, pyriform, or triangular in transverse sections (Fig. 9.8A); these may in fact be large fusiform cells when viewed in longitudinal sections. STT cells in laminae II and III have the appearance of limitrophe (stalked) and central (islet) cells (Fig. 9.8B) (Willis *et al.*, 1979; cf. Ramon y Cajal, 1909; Gobel, 1978b). Labeled cells in lamina IV are polygonal or elongate; many in lamina V are polygonal, although some are flattened (Fig. 9.8C). Those in lateral laminae VII and IX resemble motoneurons and may be regarded as part of the population of "spinal border cells" (Fig. 9.8D) (Cooper and Sherrington, 1940). STT cells in Stilling's nucleus resemble the neurons of Clarke's column (Willis *et al.*, 1979; cf. Snyder *et al.*, 1978). Very large, polygonal STT cells are sometimes found in the medial ventral horn (Fig. 9.8D).

Intracellular injections of HRP have been made into a number of STT cells (Surmeier *et al.*, 1988; Meyers and Snow, 1982a; D. Zhang, C. Owens and W. D. Willis, unpublished data). It is clear from reconstructions of injected cells that retrograde labeling reveals only a small part of the dendritic trees of STT neurons. The somas of STT cells labeled by intracellular injection have been in laminae IV–VI in the monkey spinal cord (Fig. 9.9) (Surmeier *et al.*, 1988; Zhang *et al.*, unpublished) and laminae VII and VIII in the cat (Fig. 9.10) (Meyers and Snow, 1982a).

Intracellularly labeled STT cells typically have long dendrites that spread across several adjacent laminae. The dendrites of STT cells in laminae IV–VI of the monkey dorsal horn may

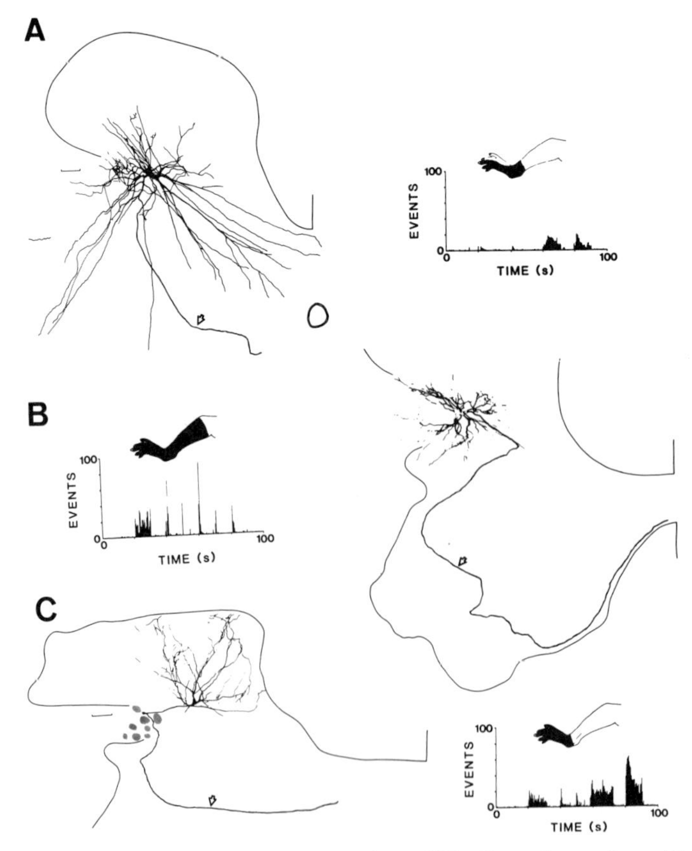

Fig. 9.9. Reconstructions of three intracellularly labeled primate STT cells are shown, along with their responses to graded mechanical stimulation and their receptive fields. Cell types: A, high threshold; B, low threshold; C, wide dynamic range. (From Surmeier *et al.*, 1988.)

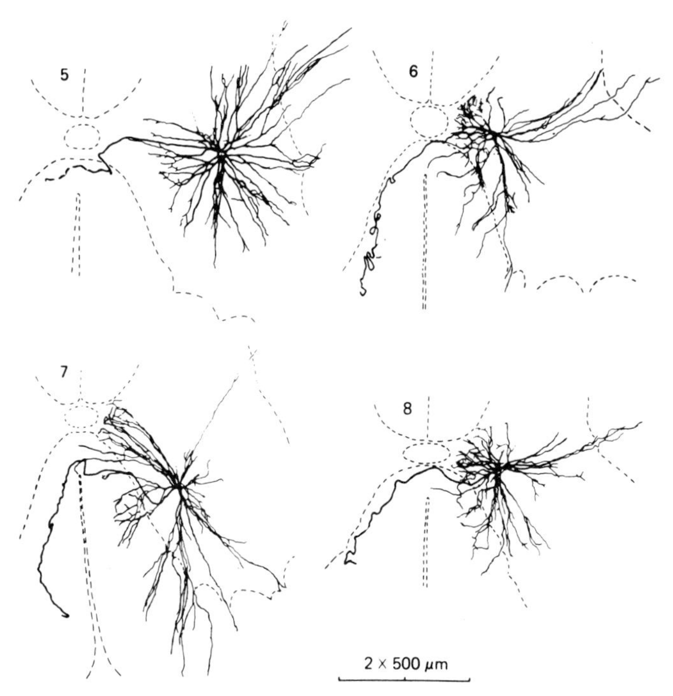

Fig. 9.10. Reconstructions of four cat STT cells in the ventral horn. The cells were injected intracellularly with HRP. (From Meyers and Snow, 1982a.)

extend dorsally as far as lamina I, laterally into the lateral funiculus, and/or ventromedially to near the border between laminae VII and X (Fig. 9.9). The dendrites of STT cells in laminae VII and VIII in cats often spread laterally into the lateral funiculus or medially into lamina VII or even X (Fig. 9.10).

Ultrastructure of STT Neurons

Electron micrographs have been made of STT neurons labeled either retrogradely or by intracellular injection of HRP. The only unusual feature of the cytology of these cells is the presence of somatic spines (Ruda *et al.*, 1984; Carlton *et al.*, 1989b). Somatic spines have also been seen on cells in the lateral cervical nucleus (see Chapter 8). The terminals contacting STT neurons of lamina V have been subdivided into six types (Carlton *et al.*, 1989b), depending upon their content of round clear vesicles, flattened clear vesicles, and different numbers of large or small dense-core vesicles per profile (cf. Snow and Meyers, 1981).

Immunocytochemical Studies of STT Neurons

The cell bodies of some STT neurons in rats contain neuropeptides. Nahin (1988) observed that a subpopulation of STT cells located chiefly in the lateral spinal nucleus of the rat and projecting to the lateral and medial thalamus contains vasoactive intestinal polypeptide (VIP). Leah *et al.* (1988) have also observed STT cells in the lateral spinal nucleus that contain bombesin. Other STT neurons in laminae VI, VII, and X and projecting to the medial thalamus stain for enkephalin or for dynorphin (Coffield and Mileti, 1987; Nahin, 1988). STT cells in lamina X react with antibodies to cholecystokinin, bombesin, and/or galanin (Ju *et al.*, 1987a; Leah *et al.*, 1988).

Synaptic endings on STT neurons have been shown to contain several different neuroactive substances, including enkephalin (Ruda *et al.*, 1984), substance P (SP) (Carlton *et al.*, 1985b), calcitonin gene-related peptide (CGRP) (Carlton *et al.*, 1990), serotonin (C. C. LaMotte *et al.*, 1988), and dopamine-ß-hydroxylase (Westlund *et al.*, 1990).

Axons of STT Cells

As deduced from their conduction velocities, the axons of STT cells have a wide range of diameters, including large or small myelinated axons and unmyelinated axons. The conduction velocities of myelinated STT axons of cells in different laminae of the primate spinal cord are shown in Fig. 9.11A (Trevino *et al.*, 1973). Very long conduction delays have been demonstrated for many STT cells in lamina I of the cat spinal cord, indicating that these neurons may have very small myelinated or unmyelinated axons (Fig. 9.11B) (Craig and Kniffki, 1985).

The small axons of lamina I STT cells have not been satisfactorily traced. However, the large axons from STT neurons in deeper laminae have been visualized, either in retrogradely labeled material or after intracellular injections of label. Typically, the axons cross in the ventral white commissure within a short distance of the cell body (Fig. 9.12; see also Fig. 9.9). No STT axons have been seen crossing in the dorsal commissure (Willis *et al.*, 1979; Apkarian and

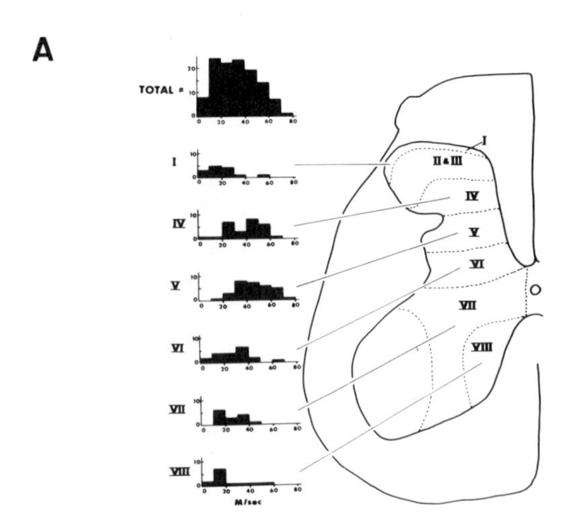

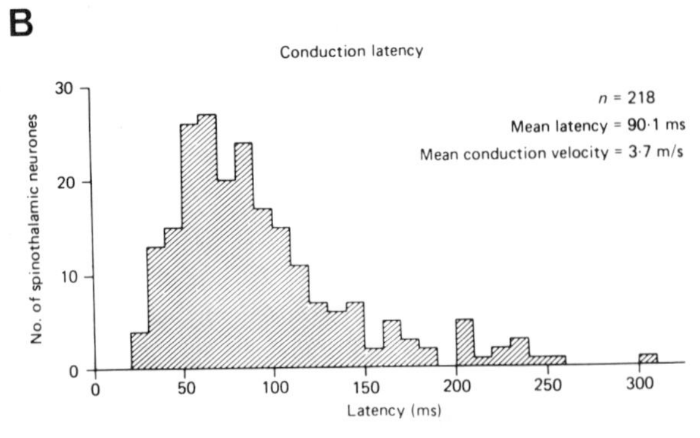

Fig. 9.11. Conduction velocities of STT axons. (A) Conduction velocities of STT axons in monkeys. The histogram at the top is for the entire population sampled, whereas those below are for neurons whose cell bodies were in the laminae indicated. (From Trevino *et al.*, 1973.) (B) Histogram showing the antidromic latencies of the axons of STT cells in lamina I of cats. The mean latency of 90.1 msec corresponds to a conduction velocity of 3.7 m/sec. Note that many of the axons conducted at velocities appropriate for unmyelinated fibers (latencies above 133 msec). The highest conduction velocity was about 17 m/sec. (From Craig and Kniffki, 1985.)

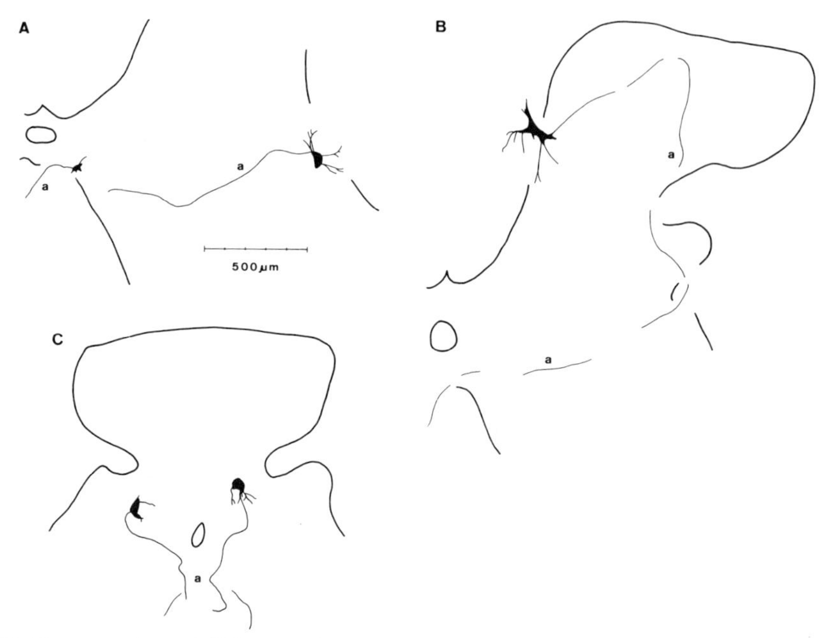

Fig. 9.12. The axons of several STT neurons (a) are shown in reconstructions from retrogradely labeled material. Note that the axons decussate in the ventral white commissure near the level of the cell body. In some cases, the route followed by the axon is rather indirect (e.g., that in panel B). (From Willis *et al.*, 1979; reproduced with permission.)

Hodge, 1989a). Sometimes an axon will travel dorsally and then loop back ventromedially to reach the ventral commissure (Fig. 9.12B) (Willis *et al.*, 1979; Apkarian and Hodge, 1989a). In experiments in which primate and cat STT cells have been injected intracellularly, branching of the axon is observed in only a few cases and there do not appear to be any local collaterals (Meyers and Snow, 1982a; Surmeier *et al.*, 1988; D. Zhang, Owens and Willis, unpublished).

After decussating, axons from many STT neurons turn rostrally in the ventral funiculus. As the axons ascend the spinal cord, they migrate laterally into the lateral funiculus. Evidence for this is that STT axons from the primate lumbosacral cord have been identified electrophysiologically in the ventral part of the lateral funiculus at an upper lumbar level (Fig. 9.13) (Applebaum *et al.*, 1975). Indirect evidence for a dorsolateral shift in the position of STT axons as they ascend the cord comes from the results of cordotomies in humans (Weaver and Walker, 1941) (see Chapter 6).

Numerous retrogradely labeled STT axons have been observed in the ventral lateral and ventral funiculi in rats (Giesler *et al.*, 1981c) and monkeys (Willis *et al.*, 1979; Apkarian and Hodge, 1989b). No retrogradely labeled axons have been observed in the dorsal lateral funiculus of the monkey, suggesting that any axons belonging to the dorsolateral STT must be quite small (Apkarian and Hodge, 1989b). However, retrogradely labeled STT axons have been observed in the dorsolateral, as well as the ventrolateral, white matter in cats (M. W. Jones *et al.*, 1985).

Recently, Craig (1989) has done an anterograde tracing study, labeling the axons of lamina I neurons by injections of *Phaseolus* leucoagglutinin into lamina I in cats and monkeys. It can be presumed that many of the lamina I neurons that were labeled projected to the thalamus. The distribution of the axons of lamina I cells in each of several experiments is shown in Fig. 9.14. Some of the axons are located in the DLF and others are in the ventral lateral funiculus. However,

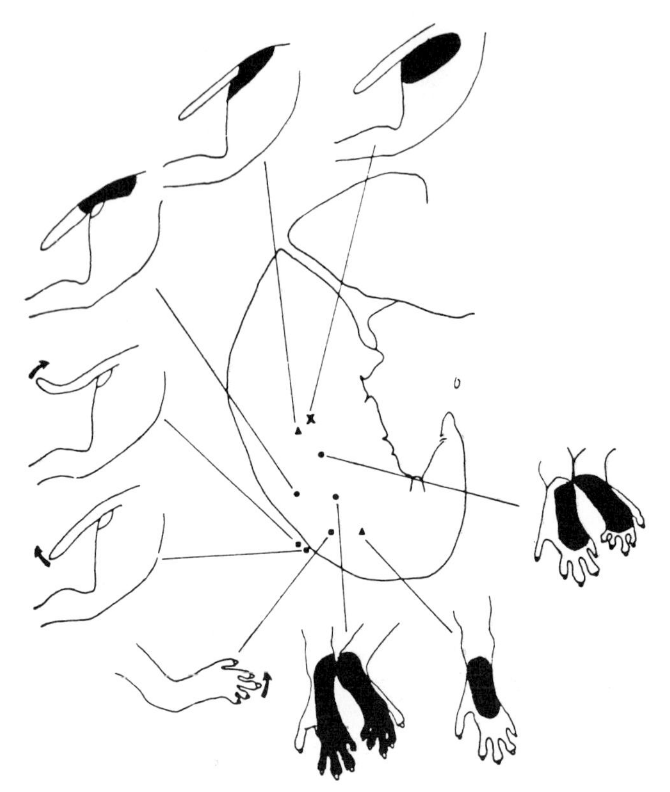

Fig. 9.13. Somatotopic organization of the primate STT at an upper lumbar level. Receptive fields on the skin are indicated by the blackened areas; receptive fields in deep tissues demonstrated by joint movements are indicated by the arrows. (From A. E. Applebaum *et al.*, 1975.)

there is a distinct tendency for the axons to ascend within the middle part of the lateral funiculus. It is important to determine whether this tendency applies to unmyelinated as well as to myelinated axons of neurons in lamina I.

The STT in the lateral funiculus has a somatotopic organization (Fig. 9.13). Clinical evidence shows that axons from the caudalmost cord assume the most dorsolateral position at a high cervical level, with axons from progressively more rostral levels ascending in progressively more ventromedial positions in the ventral lateral funiculus (Horrax, 1929; Foerster and Gagel, 1932; Hyndman and Van Epps, 1939; Walker, 1940; Weaver and Walker, 1941) (see Chapter 6). It is unclear whether the ventral STT tract described by Kerr (1975c) in experiments in which he made lesions in the ventral funiculus at a cervical level is a separate projection or whether these axons should be regarded as a component of the ventrolateral STT.

The course of the STT through the brain stem has been described in a number of anterograde degeneration studies (Walker, 1940; Morin *et al.*, 1951; Mehler *et al.*, 1960; Mehler, 1962, 1969; Kerr, 1975c). It passes just dorsolateral to the inferior olivary nucleus in the medulla and then ascends dorsolateral to the medial lemniscus through higher levels of the brain stem to the thalamus. The somatotopic arrangement is thought to be preserved in the brain stem (Weaver and Walker, 1941; Walker, 1942a; Morin *et al.*, 1951). Collateral projections to brain stem nuclei have already been mentioned.

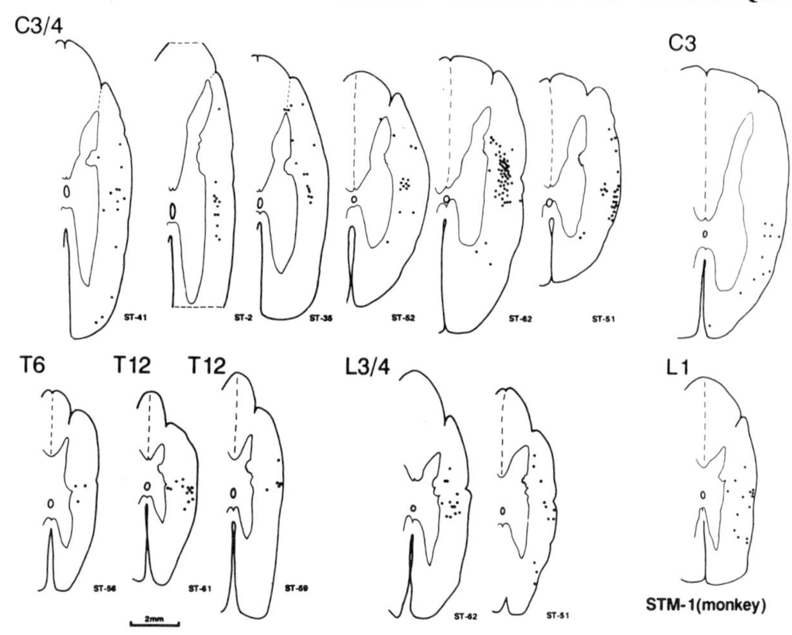

Fig. 9.14. Locations of the axons of lamina I projection neurons in cats and monkeys. The axons were labeled anterogradely by injections of *Phaseolus* leucoagglutinin into lamina I on the contralateral side. Note that the axons tend to be concentrated in the middle part of the lateral funiculus. (From Craig, unpublished.)

Thalamic Projection Targets of STT Cells

Rat. The STT of rats terminates chiefly in the VPL nucleus, the CL nucleus of the intralaminar complex (as well as adjacent parts of the medial dorsal and parafascicular nuclei), and the posterior complex (Lund and Webster, 1967b; Mehler, 1969; Zemlan *et al.*, 1978; Peschanski *et al.*, 1983). A double-label anterograde tracing study has shown that the STT ends in the same region of the VPL nucleus as the medial lemniscus, although the STT projection is smaller than that of the medial lemniscus (Ma *et al.*, 1986).

The terminals of STT axons in the VPL nucleus are of the same ultrastructural type as those of medial lemniscus and trigeminothalamic axons: large endings containing round vesicles that make asymmetric synaptic contacts, usually with the dendrites of the thalamic neurons (Ralston, 1983; Peschanski *et al.*, 1985).

Cat. Unlike the case in rats and primates, there is only a small spinal projection to the VPL nucleus in cats (Boivie, 1971a; E. G. Jones and Burton, 1974; Berkley, 1980; Mantyh, 1983b; Craig and Burton, 1981, 1985). STT terminals in the lateral thalamus of cats are found chiefly in a "shell region" surrounding the VPL nucleus. The nuclei included in the shell region are the ventral posterior inferior nucleus, parts of the posterior medial nucleus, the ventral margin of the ventrobasal complex, and the ventral lateral and lateral posterior nuclei near their boundaries with the VPL nucleus. Other major projections are to the posterior complex, the CL nucleus (and adjacent parts of the medial dorsal nucleus), and the nucleus submedius. Minor projections have been traced to the zona incerta, central medial nucleus, parafascicular nucleus, and other parts of the thalamus. Interestingly, spinal afferents to the CL nucleus overlap cells that project to area 4 of the cerebral cortex but not cells that project to area 5 (Molinari *et al.*, 1986).

Monkey. Like rats, but unlike cats, a major thalamic target of the STT in monkeys is the VPL nucleus (Fig. 9.15) (Mehler *et al.*, 1960; Bowsher, 1961; Mehler, 1969; Kerr and Lippman,

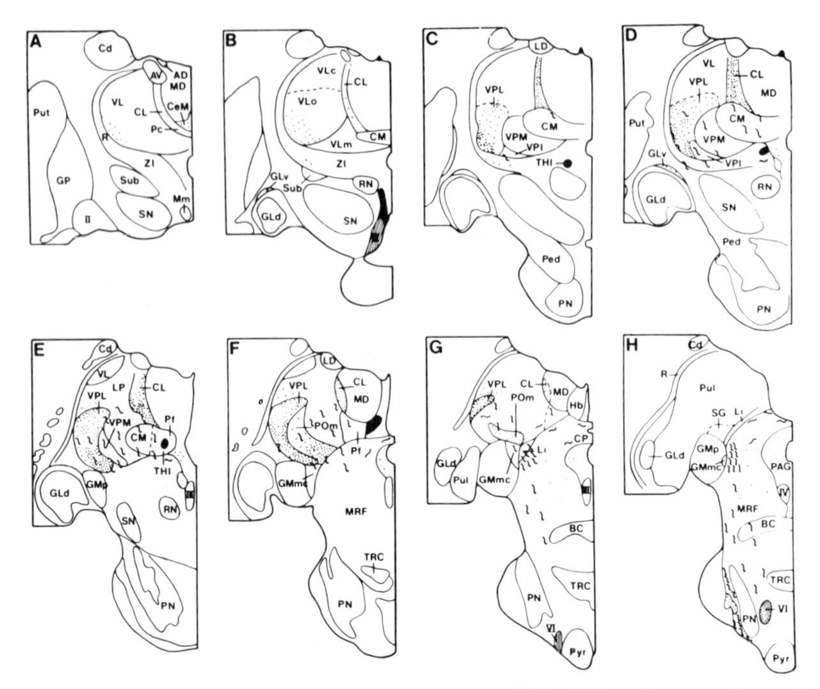

Fig. 9.15. The thalamic terminations of the STT in a macaque monkey. The STT was interrupted at a midcervical level by a lesion, and the anterograde degeneration was traced using a silver impregnation method. Transverse sections are arranged from rostral (A) to caudal (H). (From Boivie, 1979; reproduced with permission.)

1974; Kerr, 1975c; Boivie, 1979; Berkley, 1980; Mantyh, 1983a; Apkarian and Hodge, 1989c). According to the nomenclature introduced by Olszewski (1952), the somatosensory part of the monkey VPL nucleus is its caudal portion, VPL_c.

The termination zone of the STT in the VPL_c nucleus is somatotopically organized. STT axons originating from the lumbosacral spinal cord synapse in the lateral part of the nucleus, whereas those from the cervical enlargement synapse in the medial part. The area of the VPL_c nucleus in which the STT ends overlaps the area which receives the projection from the dorsal column nuclei (Berkley, 1980), although the STT projection is less dense. Actually, the STT ends in a number of focal zones that in transverse sections appear as "bursts" or "clusters" of terminals scattered in "archipelagolike" fashion across the VPL_c nucleus, following Mehler's description (Mehler *et al.*, 1960; Mehler, 1969; cf. Bowsher, 1961; Kerr and Lippman, 1974; Kerr, 1975c; Boivie, 1979; Mantyh, 1983a). Boivie (1979) points out that these "clusters" correspond to areas that extend for some distance rostrocaudally. Furthermore, they have a consistent pattern from animal to animal. The largest cluster zone is in the ventromedial part of the VPL_c nucleus; others are in the dorsolateral part, along the lateral and medial borders, and at the border between the forelimb and hindlimb representations of the VPL_c nucleus. Thus, the STT terminates in variably sized "rodlike" areas in the VPL_c nucleus (Boivie, 1979; Mantyh, 1983a). This suggests the possibility that each thalamic rod that receives STT input projects to a particular cortical zone (cf. E. G. Jones, 1985).

Other major target nuclei of the STT in the monkey (Fig. 9.15) include the posterior complex, the oral part of the VPL nucleus (VPL_o), and the CL nucleus (and neighboring parts of the medial dorsal nucleus). Minor projections are to the central medial nucleus, the paracentral nucleus, the parafascicular nucleus, the ventral lateral nucleus, the zona incerta, the nucleus submedius, and others.

A few STT axons in primates decussate in the posterior commissure (Mehler *et al.*, 1960; Bowsher, 1961; Mehler, 1969), as shown by anterograde degeneration after a lesion of the lateral funiculus. These axons appear to project chiefly to the intralaminar complex contralateral to the lesioned tract, although some may also end in the VPL nucleus.

The synaptic endings of STT axons in the primate VPL nucleus are of the large, round vesicle type, as in rats (Ralston, 1983, 1984). However, they have a triadic relationship with thalamic relay neurons and the presynaptic dendrites of interneurons. The structural arrangement suggests that activity in STT axons would result in excitation of a relay neuron and also of a presynaptic dendrite, which would in turn inhibit the relay neuron.

Human. The thalamic nuclei in which the human STT terminates have been described by Mehler (1962, 1974). The major projections are to the VPL nucleus, the CL nucleus (and adjacent parts of the medial dorsal nucleus), and the posterior complex. Mehler suggests that the area of PO receiving STT terminations in humans should actually be regarded as a caudal extension of the VPL_c nucleus (Mehler, 1974). The STT endings in the VPL nucleus are in "bursts," similar to those in monkeys.

Summary

In summary, the projections of three of the subdivisions of the STT originating in the lumbosacral enlargement are shown in Fig. 9.16, on the basis of evidence from rats, cats, monkeys, and humans. One component of the STT (Fig. 9.16A) originates largely from lamina I (although a few STT neurons in laminae II and III also contribute [Willis *et al.*, 1978, 1979; Apkarian and Hodge, 1989b]). According to Apkarian and Hodge (1989b), many of these cells project to the thalamus through the contralateral DLF in the dorsolateral STT. It should be emphasized that a dorsolateral STT has not been identified in humans, and Craig's evidence suggests that most of the axons of lamina I STT cells in cats and monkeys ascend in the middle

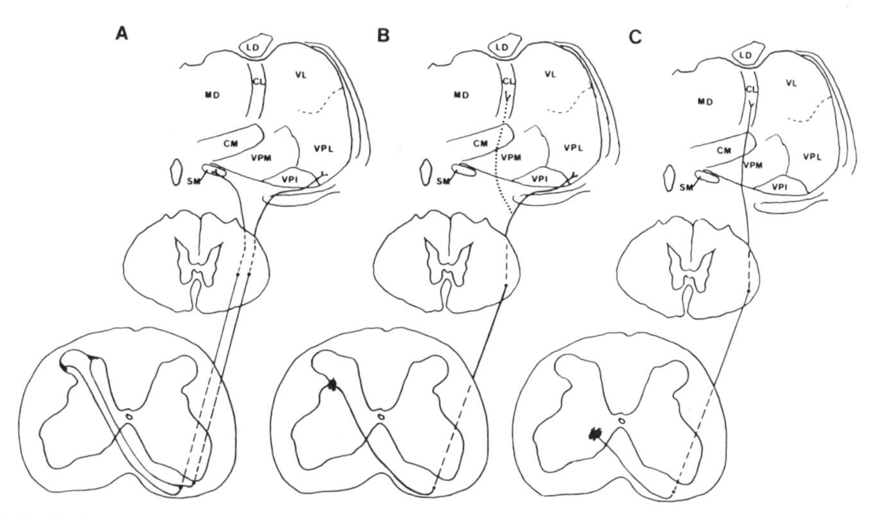

Fig. 9.16. Projections of different components of the STT. (A) Drawing in A shows the projections of STT cells in lamina I to the VPL nucleus and the nucleus submedius (SM). (B) Projection of STT cells in laminae IV–VI to the VPL nucleus; sometimes there is also a collateral to the central lateral (CL) nucleus (dotted line). (C) Projection of STT cells in deeper laminae to the CL nucleus.

part of the lateral funiculus (Craig, 1989) (Fig. 9.14). The target nuclei in the thalamus include the VPL_c nucleus and the nucleus submedius. In monkeys, lamina I also projects to the intralaminar complex (Albe-Fessard *et al.*, 1975; Giesler *et al.*, 1981b; Ammons *et al.*, 1985a; Apkarian and Hodge, 1989c). In cats, cells in the medial part of lamina I project to the lateral thalamus, whereas cells in lateral lamina I project to the medial thalamus, presumably to the nucleus submedius (Craig *et al.*, 1989a). A second component of the STT (Fig. 9.16B) originates from cells in laminae IV–VI and projects to the thalamus through the contralateral ventral quadrant. The axons end in the VPL_c nucleus and some also end in the CL nucleus (cf. Kevetter and Willis, 1983; Giesler *et al.*, 1981b; Ammons *et al.*, 1985a). A third component of the STT (Fig. 9.16C) arises from neurons in laminae VII–X and projects to the CL nucleus through the contralateral ventral quadrant (Willis *et al.*, 1979; Giesler *et al.*, 1981b). The STT also projects to the posterior complex. Other subdivisions of the STT include a substantial population of neurons in the upper cervical segments that project to both the contralateral and the ipsilateral thalamus. The thalamic targets of these cells have not yet been defined, although Apkarian and Hodge (1989a) suggest that most project to areas away from the lateral and ventral border of the lateral thalamus. Special groups of STT cells have been observed in the lateral part of lamina VII at upper lumbar levels (Willis *et al.*, 1979; Hayes and Rustioni, 1980; Apkarian and Hodge, 1989a) and in Stilling's nucleus (Willis *et al.*, 1979). Neurons in these areas are known to project to the cerebellum (Petras, 1977; Snyder *et al.*, 1978; Xu and Grant, 1988), suggesting the possibility of parallel or even collateral pathways to both thalamus and cerebellum. It will be of interest to know whether axons of these cells terminate in the VPL_o nucleus.

FUNCTIONAL PROPERTIES OF STT NEURONS

Background Activity

Most STT cells discharge in the absence of deliberate stimulation. However, the firing rates differ for different populations of STT cells and for different animals. For example, only 10% of lamina I STT cells in cats have ongoing activity when first isolated (Craig and Kniffki, 1985). The firing rates of those with background activity average 3 Hz. On the other hand, 86% of lamina I STT cells in monkeys demonstrate background activity, with a mean rate of 2.3 Hz (Ferrington *et al.*, 1987d). For STT cells in laminae IV and V of cats, 85% of the cells have background activity, with a mean rate of 1.4 Hz (Ferrington *et al.*, 1986b), whereas comparable STT cells in monkeys have been reported to discharge at an average rate of 13.5 Hz (Willis *et al.*, 1975) or 15 Hz (Giesler *et al.*, 1981b). STT cells projecting to the medial but not the lateral thalamus in monkeys have lower background firing rates, averaging 4 Hz (Giesler *et al.*, 1981c), whereas STT cells with deep receptive fields discharge at high rates that depend upon the static position of the limb (Willis *et al.*, 1975). Most (84%) STT cells in laminae VII and VIII of cats discharge spontaneously at rates of 1–67 Hz (Meyers and Snow, 1982b). Snow and Meyers (1983) propose that the background discharges may depend upon excitation by synaptic connections on STT cell dendrites in the lateral and/or ventral funiculi. Other factors that may influence the level of background activity would be tonic descending inhibition, anesthetic level, and damage due to surgery or to noxious stimulation.

Surmeier *et al.*, (1989) have analyzed the patterns of background discharges of a sample of primate STT cells of laminae IV–VI. One pattern is a regular discharge lacking in short intervals; a second pattern is dominated by short discharge intervals; the third has both short and long intervals between discharges. They suggest that the background discharges of STT cells in laminae IV–VI are determined in large part by the intrinsic membrane properties of these neurons.

Electrophysiological Properties

Only a limited number of intracellular recordings from STT cells *in vivo* have been published (Trevino *et al.*, 1972, 1973; R. D. Foreman *et al.*, 1975, 1976; Giesler *et al.*, 1981a; Milne *et al.*, 1982; Zhang *et al.*, 1988). These recordings indicate that STT cells generate action potentials with an inflection on the rising phase of the spike (A-B or IS-SD inflection: cf. Brock *et al.*, 1953) and an afterpotential sequence that includes a brief followed by a prolonged afterhyperpolarization (Trevino *et al.*, 1972, 1973; Giesler *et al.*, 1981a). Background and evoked EPSPs and IPSPs have been observed (Trevino *et al.*, 1973; R. D. Foreman *et al.*, 1975, 1976; Giesler *et al.*, 1981a; Milne *et al.*, 1982; Zhang *et al.*, 1988). The hyperpolarizing potentials evoked by stimulation in the nucleus raphe magnus were shown to be IPSPs by the fact that they can be reversed when the membrane potential is artificially hyperpolarized, made larger when the membrane is depolarized, and reversed when chloride ions are allowed to leak from the recording electrode (Giesler *et al.*, 1981a; cf. Coombs *et al.*, 1955).

A more quantitative approach to the analysis of the membrane properties of STT cells has been taken by Huang (1987), who marked the neurons retrogradely with rhodamine-labeled latex beads injected into the thalamus in very young rats and then isolated the labeled cells for study *in vitro*. The resting membrane potentials of the isolated STT cells range from −65 to −75 mV. Action potentials are triggered by depolarizations beyond -50 mV. Repetitive firing resulted when long depolarizing pulses were used (Fig. 9.17). Action potentials have amplitudes of 70–90 mV. The spike is followed by an afterhyperpolarization. The spike depends in part on an inward sodium current, since voltage clamping demonstrates a fast, inactivating tetrodotoxin-sensitive inward current which is reduced when the extracellular Na^+ concentration is lowered. This current has a reversal potential of about +37 mV. There is also inward current attributable to calcium ions. Both voltage-dependent and calcium-dependent K^+ currents are also found. The calcium current has been analyzed further and shown to consist of three components: a transient calcium current and two types of slowly inactivating calcium currents (Huang, 1989a).

Responses to Electrical Stimulation

The responses of primate STT neurons to volleys evoked by electrical stimulation of peripheral-nerve fibers have been studied by using either extracellular or intracellular recordings (R. D. Foreman *et al.*, 1975; Beall *et al.*, 1977; J. M. Chung *et al.*, 1979; Zhang *et al.*, 1988). Many of the neurons discharge in separate bursts in response to volleys in Aβ, Aδ and C fibers. For example, in Fig. 9.18A, there are two burst discharges in association with volleys in Aβ and Aδ fibers. When the Aβ volley is blocked by anodal current, only the late burst is seen (Fig. 9.18B). Some STT neurons respond only to the Aδ volley and not the Aβ volley, such as the lamina I neuron whose activity is illustrated in Fig. 9.18C.

When a stimulus evokes a C-fiber volley, as well as an A-fiber volley, there is generally a late response at the appropriate latency for an action of C fibers (Fig. 9.19). Anodal blockade of the A-fiber volley leaves just the response to the C-fiber volley (Fig. 9.20). C-fiber responses are observed in most STT cells, although the sizes of the responses vary considerably (J. M. Chung *et al.*, 1979).

The effects of barbiturates on the responses of neurons presumed to be STT cells have been tested in decerebrated monkeys (Hori *et al.*, 1984). The cells could be activated antidromically by stimulation of the contralateral lateral funiculus at an upper cervical level. Small doses of pentobarbital do not interfere with the activation of the cells by volleys in C fibers; in fact, the responses are enhanced. However, large doses reduce the responses. Intracellular recordings show that EPSPs underlie the responses of STT neurons to volleys in A and C fibers (Fig. 9.21)

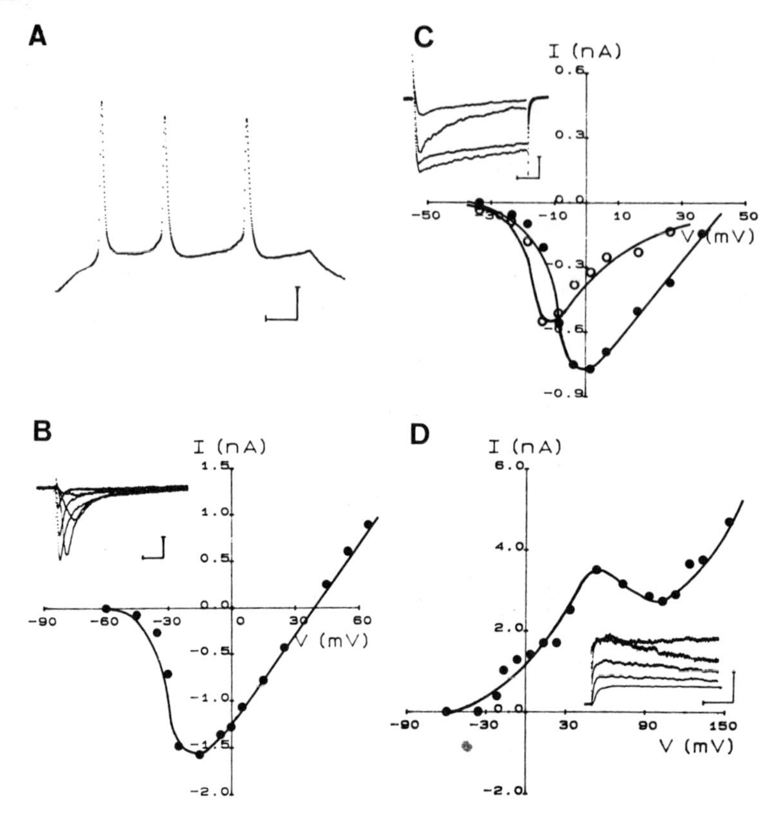

Fig. 9.17. Electrophysiological properties of isolated thalamic projection neurons. (A) Several action potentials triggered by a 5 msec depolarizing pulse. (B) Current-voltage curve for the peak INa^+; the inset shows INa^+ recordings (in a solution with low Na^+, low temperature, Cs, and tetraethylammonium). (C) Voltage-dependent current carried by barium ions (measured in tetrodotoxin, $BaCl_2$, and tetraethylammonium in the external solution and CsF and EGTA in the internal solution). (D) Current-voltage curve for the Ca^{++}-dependent K^+ current (measured in tetrodotoxin solution). (From Huang, 1987.)

(Zhang *et al.*, 1988). The EPSPs may be very long lasting, especially when repetitive stimulation is used.

As discussed in Chapter 2, the particular kinds of afferent nerve fibers that are responsible for the central actions of peripheral-nerve volleys cannot be determined in experiments in which electrical evoked volleys are used, and so it is generally difficult to relate the results of such experiments to specific sensory functions. For example, Aδ fibers include afferent fibers supplying down hair receptors, Aδ nociceptors, and cold receptors. However, in monkeys few C mechanoreceptors have been found in peripheral nerves of the hindlimb, and so volleys in afferent C fibers are likely to be conducted largely in C polymodal nociceptors (plus afferent fibers supplying warm receptors).

In addition to their responses to stimulation of peripheral nerves containing cutaneous afferent fibers, STT cells can also be excited by volleys in muscle and visceral nerves. There have been no reports of the effects of volleys in joint afferent fibers on STT cells. The muscle afferent fibers that are most effective belong to groups II, III, and IV, although some STT cells can be excited by group I afferent fibers (R. D. Foreman *et al.*, 1979a). Both monosynaptic and polysynaptic group I actions are seen. The group I fibers responsible for exciting STT cells are

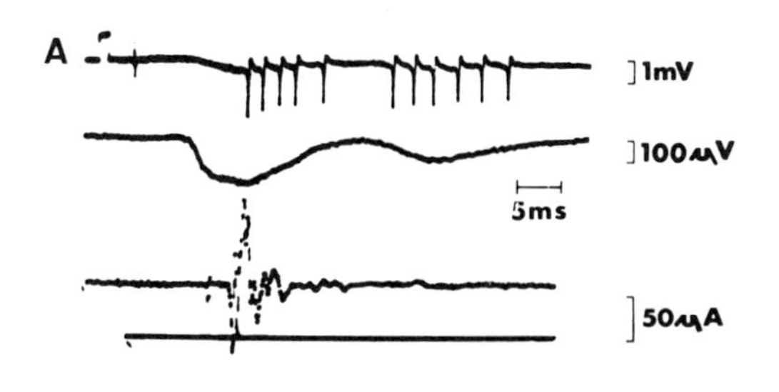

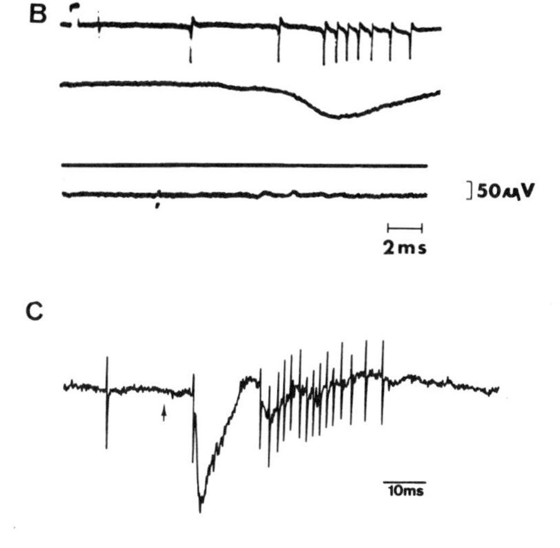

Fig. 9.18. Burst discharges evoked in an STT cell by volleys in Aβ and Aδ afferent nerve fibers in the sural nerve of a monkey. (A) The spikes are shown in the top trace (the pulse at the start of the trace is a calibration pulse). The second trace shows the cord dorsum potentials. The third trace is a neurogram of the volleys. The fourth trace is a current monitor for anodal blockade; no current was applied at this time. (B) The anodal current was on, and the Aβ volley was blocked, resulting in the loss of the early burst discharge and most of the early negative cord dorsum potential. (From Beall *et al.*, 1977.) (C) Discharges of a lamina I STT cell activated by Aδ fibers of the sural nerve. The time of the stimulus is indicated by the arrow (From Zhang *et al.*, in press.)

likely to supply Golgi tendon organs, since intra-arterial injections of succinylcholine that excite group Ia fibers from muscle spindles have little or no effect on the discharges of STT cells (R. D. Foreman *et al.*, 1979b). Both Aδ- and C-fiber volleys in visceral nerves (greater splanchnic nerve, ansa subclavia, sympathetic chain, renal nerve) activate many of the STT cells located at levels of the spinal cord receiving afferent input from these nerves (Hancock *et al.*, 1975; Blair *et al.*, 1981; R. D. Foreman *et al.*, 1981; Foreman *et al.*, 1984; Rucker and Holloway, 1982; Rucker *et al.*, 1984; Ammons, 1987; R. D. Foreman, 1989).

Responses to Natural Stimulation

Mechanical Stimulation of Skin. Most STT cells that have been studied have cutaneous receptive fields and respond to mechanical stimulation of the skin. There have been relatively few reports of the response properties of STT cells in rats (Giesler *et al.*, 1976) and cats (Dilly *et al.*, 1968; Trevino *et al.*, 1972; Hancock *et al.*, 1975; McCreery and Bloedel, 1975, 1976; McCreery *et al.*, 1979a; R. E. Fox *et al.*, 1980; Craig and Kniffki, 1985; Ferrington *et al.*, 1987d). However, STT cells in rats and cats usually respond to noxious mechanical stimulation and often to innocuous mechanical stimulation as well, although some low-threshold neurons have also been found. One study has described the response properties of STT cells in the ipsilateral ventral horn at an upper cervical level in the cat (Carstens and Trevino, 1978b). Although a large proportion of these neurons could not be activated under the conditions of the experiment, many had very large, bilateral receptive fields of the wide-dynamic-range type.

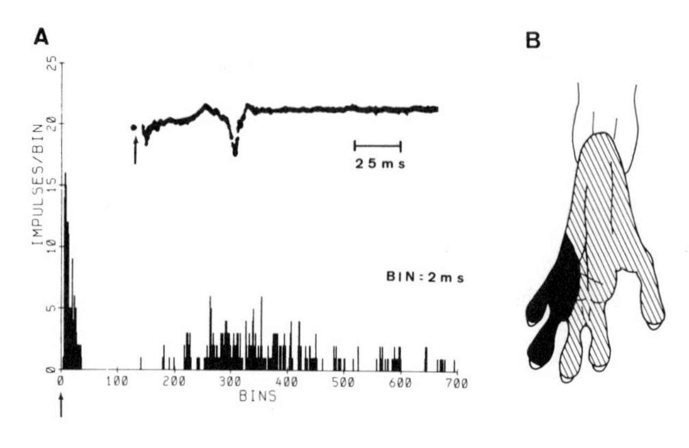

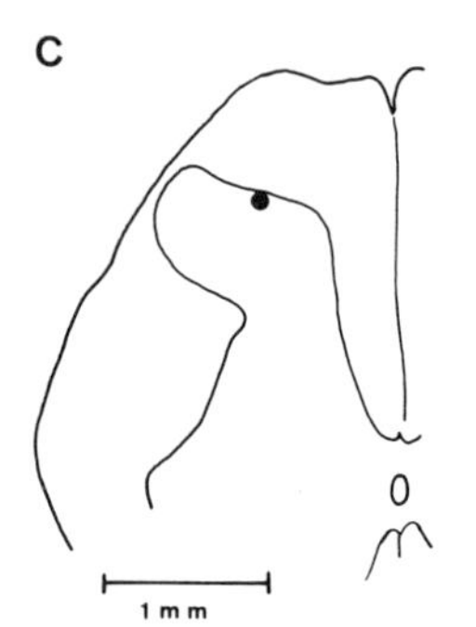

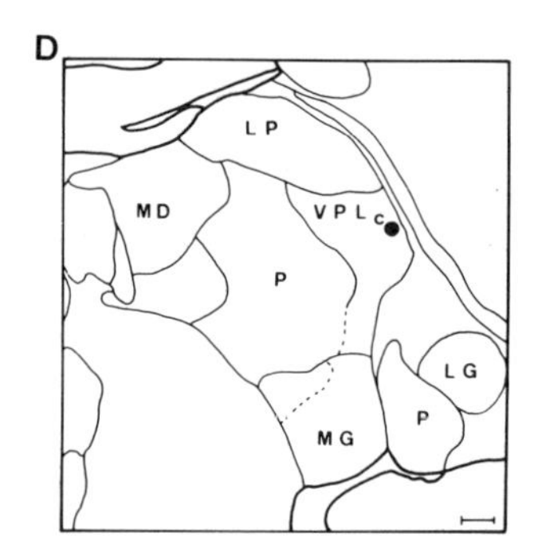

Fig. 9.19. Responses of an STT neuron to volleys in A and C fibers in the sural nerve. (A) Histogram showing an early and a late discharge in response to A- and C-fiber volleys. The inset is a recording of the compound action potential from the sural nerve showing the C-fiber volley (the A-fiber volley is partially obscured by the shock artifact). (B) Receptive field. (C) Location of the cell in lamina I. (D) Site from which antidromic activation was evoked in the VPL_c nucleus is indicated in D. (From J. M. Chung *et al.*, 1979.)

Others had receptive fields of restricted size. The function of the STT cells in the upper cervical segments is unclear. Since more work has been done on STT cells in monkeys, the following will emphasize results from primate STT cells.

The sizes of the receptive fields of primate STT neurons vary considerably. However, there are correlations between receptive-field size and other properties of the cells. For example, STT neurons in lamina I have smaller receptive fields than do STT cells in deeper laminae (Fig. 9.22) (Applebaum *et al.*, 1975; Ferrington *et al.*, 1987d; Willis, 1989). STT neurons in laminae IV–VI have receptive fields restricted to the ipsilateral limb, but the sizes range from relatively small to most of the surface of a limb (Fig. 9.22) (Giesler *et al.*, 1981b). STT neurons that project to the CL nucleus of the thalamus but not to the VPL_c nucleus often have very large, bilateral receptive fields (Fig. 9.23A) (Giesler *et al.*, 1981b; Ammons *et al.*, 1985a). If the spinal cord is transected at an upper cervical level, the receptive fields of such neurons are reduced to include just a single limb, suggesting that much of the receptive field depends upon a spino-bulbo-spinal pathway (Fig. 9.23B,C) (Giesler *et al.*, 1981b).

In monkeys, receptive fields are commonly observed on the glabrous skin or both glabrous and hairy skin, as well as on the hairy skin (Willis *et al.*, 1974). By contrast, the receptive fields of STT cells in laminae IV–V of the cervical enlargement of the cat spinal cord appear to be limited to the hairy skin (Ferrington *et al.*, 1986b). This may reflect selective sampling, or it may be a species difference.

There is a somatotopic relationship between the distribution of the receptive fields of STT cells and their locations in the superficial layers of the dorsal horn of the monkey (Fig. 9.24) (Willis *et al.*, 1974). STT neurons in the lateral part of laminae I–IV have receptive fields on the extensor surface of the hindlimb, whereas cells in the medial part of the dorsal horn have fields on

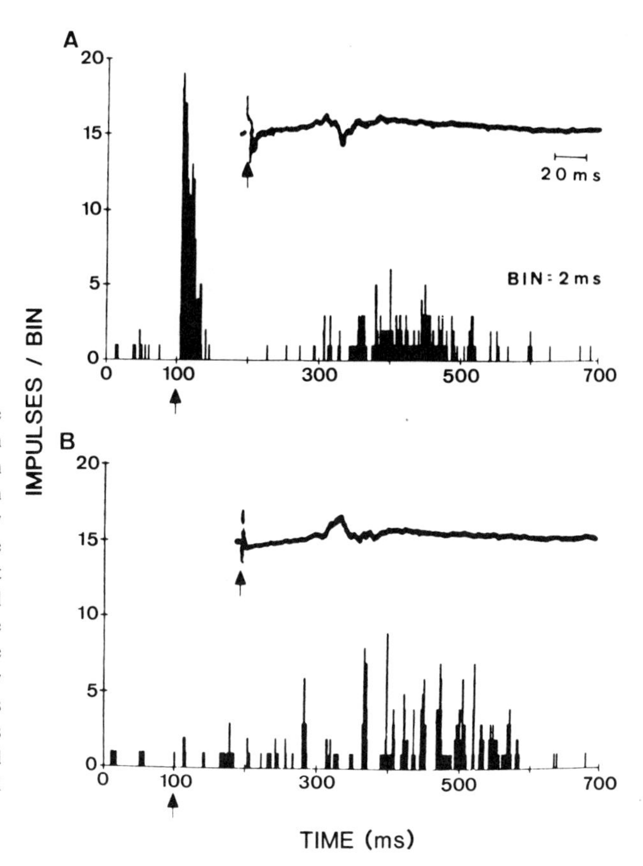

Fig. 9.20. Histograms showing the responses of an STT cell to volleys in the A and C fibers of the sural nerve in a monkey. The times of stimulation are indicated by the arrows. (A) Early and late discharges produced by the A- and C-fiber volleys. The inset shows the compound action potential recorded from the sural nerve. The C-fiber volley is evident, although the A-fiber volley is partially obscured by the shock artifact. (B) Discharges produced when the A-fiber volley has been blocked by anodal current and the C-fiber volley remains and still elicits the late discharge. (From J. M. Chung *et al.*, 1979.)

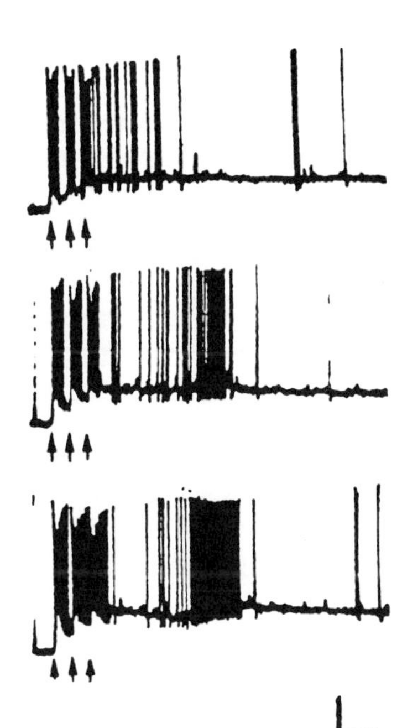

Fig. 9.21. Excitatory postsynaptic potentials and action potentials recorded intracellularly in an STT neuron in response to A- and C-fiber volleys in the sural nerve of a monkey. The nerve was stimulated with a brief train of three shocks at a strength sufficient to activate the A fibers (upper trace), some C fibers (middle trace), and most C fibers (lower trace). (From Zhang *et al.*, in press.)

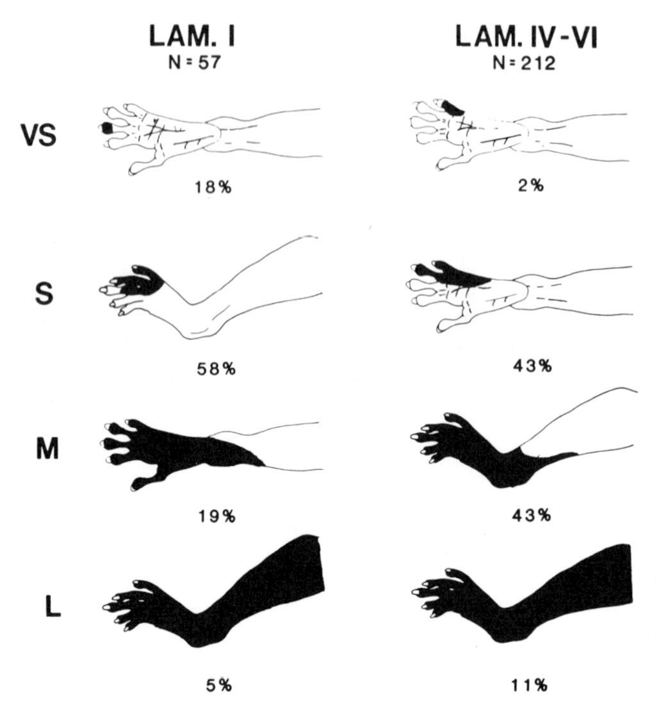

Fig. 9.22. Sizes of the receptive fields of STT cells in lamina I or in laminae IV–VI of the monkey spinal cord. Very small (VS) receptive fields had an area equivalent to or less than a digit; small (S) receptive fields occupied an area corresponding to that of the surface of the foot or less. Medium (M) receptive fields were larger than the foot but smaller than the foot plus leg. Large (L) receptive fields had an area greater than the foot and leg. (From Willis, 1989; reproduced with permission.)

the flexor surface. A similar somatotopic organization has been reported for neurons of the dorsal horn in general (see Chapter 5). This relationship is not seen for STT cells in lamina V and VI, where most STT cell bodies are concentrated laterally no matter where the receptive field is located.

When the receptive fields of STT neurons are stimulated mechanically, a few of the cells can be activated best by innocuous stimuli, such as movement of hair or gentle brushing of the glabrous skin; others are activated only by noxious mechanical stimuli; still others are excited by innocuous stimuli but best by noxious ones (Willis *et al.*, 1974, 1975; D. D. Price *et al.*, 1978; J. M. Chung *et al.*, 1979, 1986b; Blair *et al.*, 1981; Surmeier *et al.*, 1988). Using the terminology introduced by Mendell (1966), these patterns of response have led to the classification of many STT neurons as low-threshold, wide-dynamic-range, or high- threshold cells (J. M. Chung *et al.*, 1979). An additional category used by Foreman's group is high-threshold-inhibitory for cells that are inhibited by innocuous stimulation and excited only by noxious intensities of stimulation (Blair *et al.*, 1981). The latter category of cell appears to be more common in the thoracic and sacral spinal cord than in the lumbar enlargement.

Another approach to the classification of STT neurons on the basis of their responses to mechanical stimuli depends upon a multivariate statistical analysis called k means cluster analysis (J. M. Chung *et al.*, 1986b; Surmeier *et al.*, 1988; C. Owens, D. Zhang and W. D. Willis, unpublished data). The responses chosen in these studies have been elicited by a standard series of stimuli, each lasting 10 sec, using brushing (an innocuous stimulus), pressure (due to application to a fold of skin of an arterial clip that produces a sense of firm pressure or minimal pain in humans), pinch (application of an arterial clip with a strong grip to a fold of skin, a painful stimulus in humans), and squeeze (with serrated forceps, an overtly damaging stimulus). Depending on how the responses are compared, STT cells can be subdivided into several classes. For example, if each response is taken as a percentage of the sum of the four responses to the

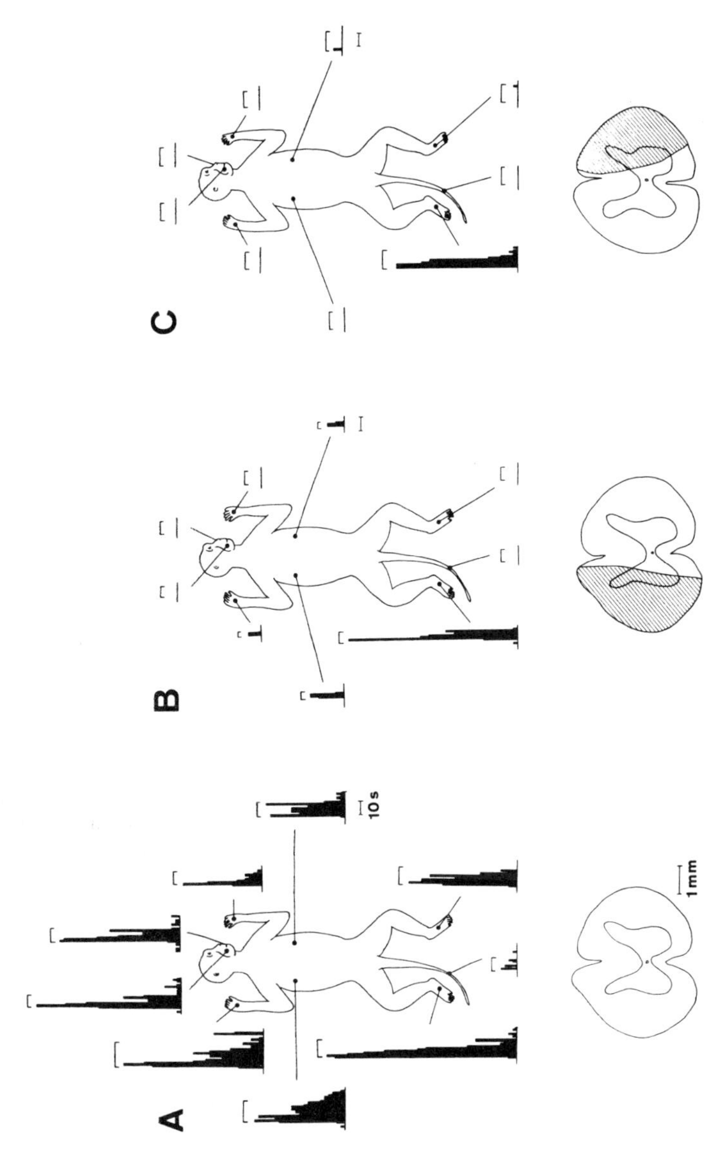

Fig. 9.23. Complex receptive field of an STT neuron that projected to the CL nucleus in the intralaminar complex of the medial thalamus. The neuron was excited when the skin was squeezed anywhere on the surface of the body or face (A). When the spinal cord was transected in 2 stages (except for the ventral and parts of the dorsal funiculi), the receptive field was diminished to include just the hindlimb ipsilateral to the recording site (B and C). (From Giesler *et al.*, 1981b.)

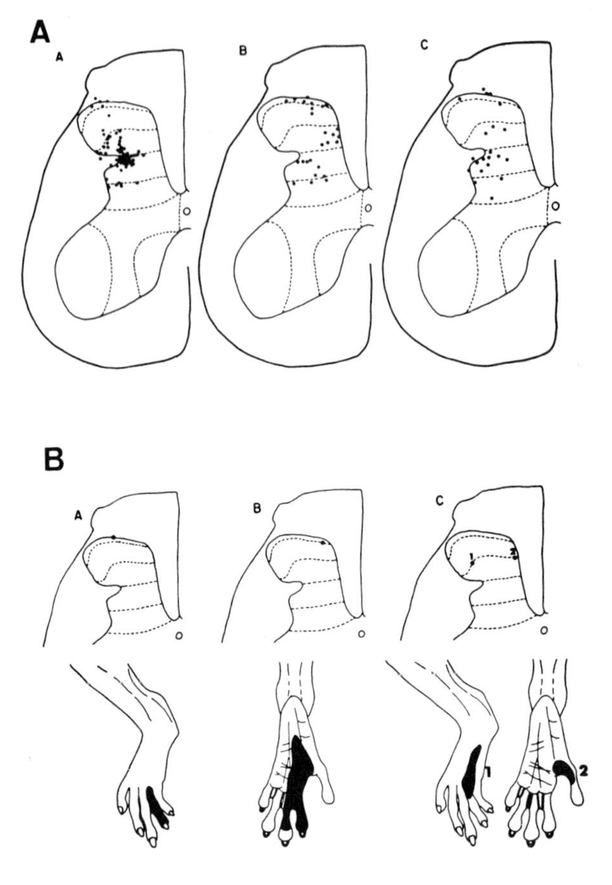

Fig. 9.24. Somatotopic organization of STT cells. (A) Recording sites of STT cells in the superficial dorsal horn and their receptive fields on the distal hindlimb. (B) Distribution of recording sites of STT cells with receptive fields on the extensor and/or flexor surface of the hindlimb. (From Willis *et al.*, 1974.)

standard stimuli, a large population of STT cells can be subdivided into three classes (C. Owens, D. Zhang and W. D. Willis, unpublished data). The average responses of each of these classes to the four stimuli are shown in Fig. 9.25, along with sample peristimulus time histograms of the responses of representative neurons. One class responds best to the brush stimulus; this group represents 15% of the sample. The other two groups respond only weakly to the brush and pressure stimuli. They differ in that class 2 cells (59% of the sample) discharge maximally to the pinch stimulus, whereas class 3 cells (26% of the sample) discharge maximally to the squeeze stimulus. Some class 3 cells are inhibited by the brush stimulus; these correspond to the high-threshold-inhibitory category used by Foreman's group.

Based on the assumption that class 1 cells have a tactile function and class 2 and 3 cells a nociceptive function, it appears that 15% of STT cells projecting to the VPL_c nucleus from the monkey dorsal horn are tactile and 85% are nociceptive (Owens *et al.*, unpublished). However, almost all of the tactile cells are located in laminae IV–VI, and so essentially all lamina I STT cells in monkeys that have been examined so far are nociceptive. This statement is based on responses to mechanical stimuli. However, there is also a population of lamina I STT cells that are thermoreceptive (see below). These have been characterized best in cats (Craig and Kniffki, 1985).

Thermal Stimulation of Skin. STT neurons often respond to noxious thermal stimulation of the skin (Willis *et al.*, 1974; Price *et al.*, 1978; Kenshalo *et al.*, 1979; Surmeier *et al.*, 1986a,b; Ferrington *et al.*, 1987d; Simone *et al.*, 1991). For example, Fig. 9.26A shows a graph of the

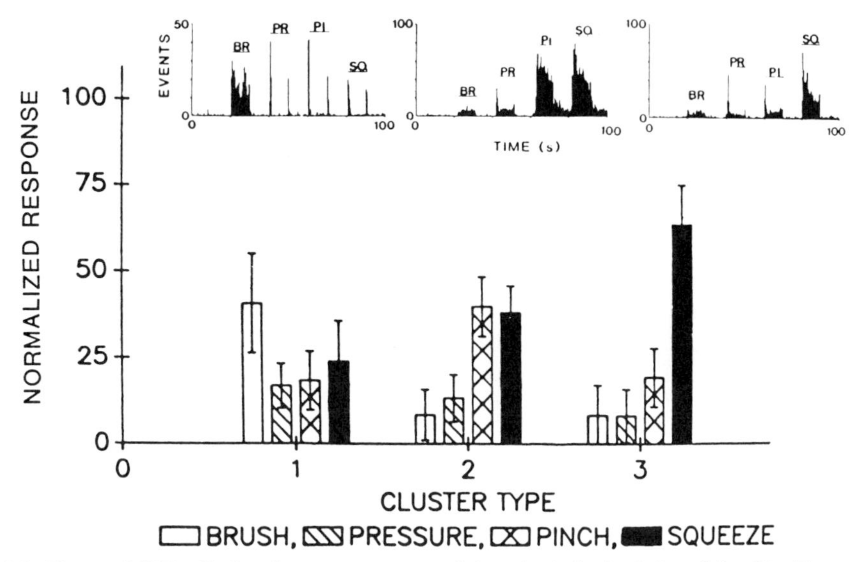

Fig. 9.25. Classes of STT cells, based on responses to graded mechanical stimulation of the skin. The mean responses to brush, pressure, pinch, and squeeze stimuli for a population of more than 300 STT cells could be subdivided into three different classes by cluster analysis. The response profiles for types 1, 2, and 3 cells are shown in the bar graph. Variability is indicated by the standard deviation lines. Examples of the responses of individual STT neurons belonging to the three classes are shown at the top. (From Owens *et al.*, unpublished.)

responses of several individual STT neurons in lamina I of the primate spinal cord to 30 sec thermal stimuli ranging from 5 to 55°C, and Fig. 9.26B shows the mean responses for 12 wide-dynamic-range and 9 high-threshold lamina I STT cells (Ferrington *et al.*, 1987d). The responses of primate STT cells to noxious heat stimuli have been classified by using cluster analysis (Surmeier *et al.*, 1986a,b). Some STT cells have steep input-output curves for their responses to noxious heat pulses; these cells correspond to a subpopulation of STT neurons that also have steep input-output curves for noxious mechanical stimulation and are likely to be classified as wide-dynamic-range cells (Surmeier *et al.*, 1986b). Other STT cells do not respond vigorously to noxious heat stimuli until the temperature exceeds 49–50°C. These cells tend to be of the high-threshold type (Surmeier *et al.*, 1986b). Low-threshold STT cells tend not to respond to noxious heat stimuli.

It was proposed that the STT neurons with steep input-output curves provide a warning that a potentially damaging stimulus is occurring, whereas STT neurons that respond vigorously only when very intense stimuli are applied signal the presence of damage (Surmeier *et al.*, 1986b). This proposal is consistent with the results of studies of the responses of medullary dorsal horn neurons, including trigeminothalamic neurons, in awake, behaving monkeys (Maixner *et al.*, 1986, 1989). Wide-dynamic-range neurons in the caudal nucleus of the spinal trigeminal complex have been shown to have a capability for discriminating between noxious heat pulses that is equivalent to the behavioral discriminative capacity of the animal, whereas high-threshold neurons do not.

A subpopulation of STT neurons in lamina I of the cat spinal cord responds to innocuous warming or cooling (Craig and Kniffki, 1985; cf. Christensen and Perl, 1970). Some of these neurons have a convergent input from mechanical nociceptors. Most thermoreceptive STT neurons can be activated antidromically from an electrode placed in the medial thalamus, but not from one in the lateral thalamus, indicating that these neurons project medially rather than

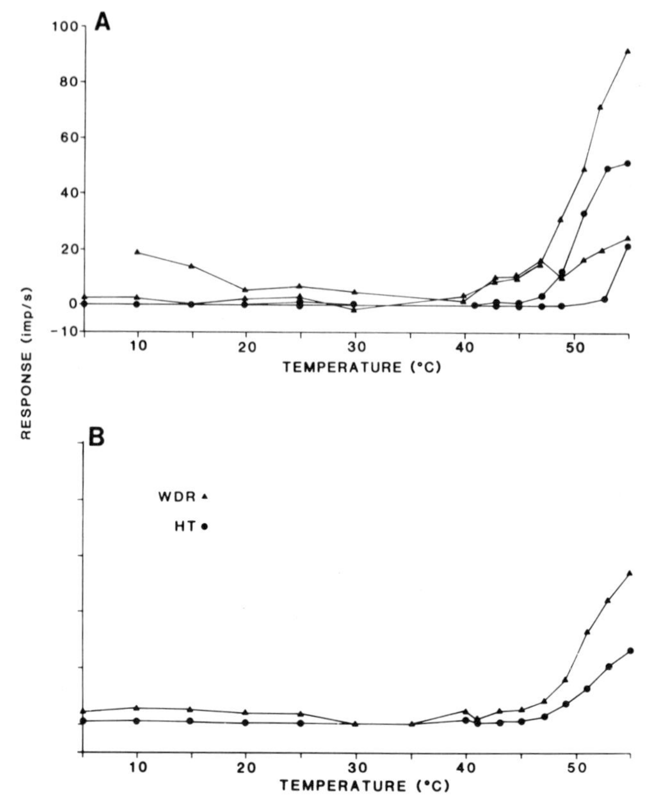

Fig. 9.26. (A) Responses of several individual wide-dynamic-range (WDR) (▲) and high-threshold (●) STT cells in lamina I of the monkey spinal cord for thermal stimuli ranging from 5 to 55°C. (B) Averaged responses of 12 wide-dynamic-range and 9 high-threshold cells. (From Ferrington *et al.*, 1987d.)

laterally. The main target of lamina I STT neurons in the medial thalamus of cats is the nucleus submedius (see above), suggesting that this nucleus could play a role in thermoception. It is conceivable that some of the thermoreceptive lamina I neurons project to the hypothalamus as part of the spinohypothalamic tract (Burstein *et al.*, 1987). Such a projection could provide an input to the thermoregulatory system (Hensel, 1973b). Norrsell (1979, 1983, 1989a,b) (see Chapter 6) has provided behavioral evidence that thermosensitivity in cats depends upon activity in axons that ascend in the middle part of the lateral funiculus, the location of many of the axons of lamina I STT cells according to Craig (1989).

Chemical Stimulation of Skin. STT neurons can be activated by injection of small amounts of capsaicin into the skin (Simone *et al.*, 1991). Fig. 9.27A shows the excitation of an STT cell following an intradermal injection of capsaicin, and Fig. 9.27B shows the time course of the mean human pain rating following a comparable injection (Simone *et al.*, 1989a; R. H. LaMotte *et al.*, 1991).

Injection of bradykinin into the femoral artery in monkeys activates a large proportion of the population of wide-dynamic-range STT cells (Levante *et al.*, 1975). It is unclear what receptors are activated by intra-arterially injected bradykinin, but it is known that such injections in humans are painful.

Stimulation of Muscle Afferent Fibers. Many primate STT cells have receptive fields not only in the skin but also in the underlying musculature. The responses of such neurons have been

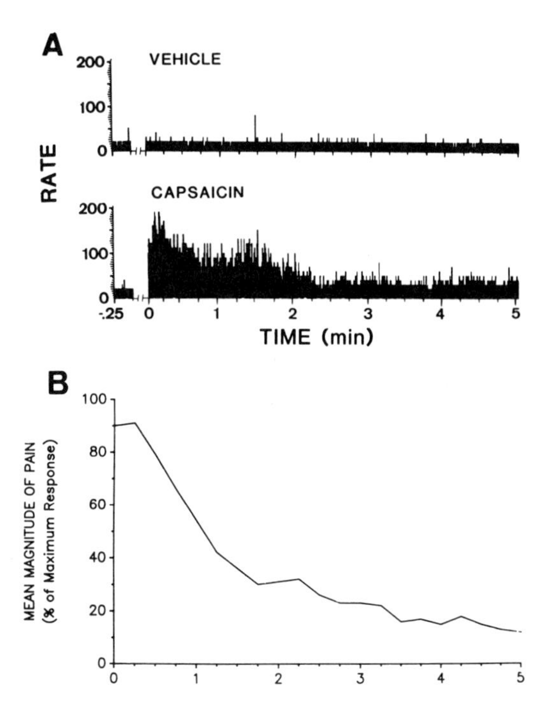

Fig. 9.27. (A) Histograms showing the responses of a primate STT cell to an intradermal injection of vehicle (above) and of capsaicin (below). (B) Graph showing the time course of pain in human subjects given a similar intradermal injection of capsaicin. (From Simone *et al.*, 1990.)

studied in animals in which all of the nerves of the distal hindlimb have been severed except for those innervating the triceps surae muscles (R. D. Foreman *et al.*, 1977, 1979a,b). Direct mechanical stimulation of the triceps surae muscles and also intra-arterial injection of algesic chemicals into the circulation of the muscle (potassium ions, bradykinin, serotonin) are all effective in activating STT neurons. Injections of hypertonic saline into muscle or into the Achilles tendon also produce excitation of STT neurons. The duration of such excitation is comparable to that reported for pain produced in human subjects by similar injections (Lewis, 1942).

Recordings from STT neurons in Stilling's nucleus suggest that these neurons have a proprioceptive function (Milne *et al.*, 1982). STT cells in Stilling's nucleus are unusual in lacking a cutaneous receptive field. They respond to bending the tail at particular joints, and their discharges appear to encode tail position. The responses are reminiscent of those of muscle spindles, and these neurons can be activated when the muscles between bony segments of the tail are probed. The neurons are monosynaptically activated by large afferent fibers in the appropriate dorsal root on one side and inhibited by comparable stimulation of the contralateral dorsal root. Proprioceptive responses have also been recorded from STT neurons in the lumbar enlargement (Willis *et al.*, 1974) and from STT axons located at the periphery of the ventrolateral funiculus at an upper lumbar level (A. E. Applebaum *et al.*, 1975).

Stimulation of Visceral Afferent Fibers. STT cells have been shown to respond to a variety of visceral stimuli. A series of studies has been performed on the responses of STT neurons in the upper thoracic cord of the cat and monkey to stimulation of cardiopulmonary visceral afferent fibers (Ammons *et al.*, 1984a,b,c, 1985a,b; Blair *et al.*, 1981, 1982, 1984; R. D. Foreman and Weber, 1980; reviewed by R. D. Foreman, 1989). Such neurons can often be excited by occlusion of a coronary artery. There may be an immediate response to the occlusion, suggesting an input from mechanically responsive afferent fibers supplying the arterial wall, or there may be a delayed response corresponding to the development of ischemia in cardiac muscle; injection of bradykinin into the coronary circulation (Fig. 9.28C) also causes the excitation of STT neurons (Blair *et al.*, 1984).

Upper thoracic STT neurons that project to the VPL nucleus and that respond to cardiopulmonary afferent fibers have cutaneous receptive fields in an area that resembles the distribution of referred pain in humans experiencing angina pectoris (Fig. 9.28A and B). This observation supports the convergence-projection theory of pain referral that was proposed by Ruch (1946). There is also convergence of somatic and cardiopulmonary input onto upper thoracic STT cells that project to the intralaminar complex (Ammons *et al.*, 1985a). The somatic receptive fields of these neurons are large and often bilateral.

Many STT neurons in the thoracic cord can also be excited by stimulation of the greater splanchnic nerve (Hancock *et al.*, 1975; R. D. Foreman *et al.*, 1981) or distention of the gall bladder (Ammons *et al.*, 1984a), kidney (Ammons, 1987) or ureter (Ammons, 1989). Evidently, there are visceral afferent inputs to thoracic STT neurons from viscera in different parts of the body.

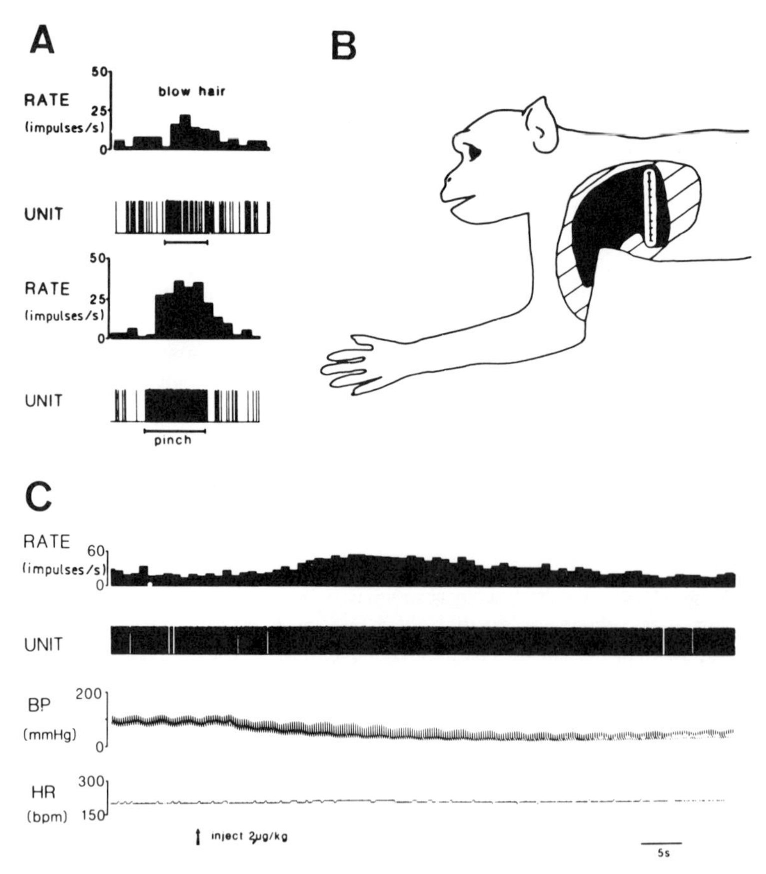

Fig. 9.28. Responses of primate STT cells in the upper thoracic spinal cord to somatic and visceral afferent fibers. Panel A shows the responses to hair movement (above) and pinching of the skin and muscle (below) in the somatic receptive field shown in panel B. (From Ammons *et al.*, 1984c). Panel C shows the response of an STT cell to injection of bradykinin into the left atrium. A similar bradykinin injection into the aorta had no effect on the activity of the cell. (From Blair *et al.*, 1982; reproduced with permission.)

STT neurons in the upper lumbar and sacral segments have also been tested for visceral inputs (Milne *et al.*, 1981). Distention of the urinary bladder often excites STT neurons at the thoracolumbar level and in the sacral spinal cord, although inhibitory effects are also observed. Noxious stimulation of the testicle causes an excitation of STT neurons at the thoracolumbar junction but not in the sacral cord. The same neurons have cutaneous receptive fields in areas similar to the distribution of referred pain in humans experiencing bladder or testicular pain (Head, 1893).

Plastic Changes

There have not yet been any studies to determine whether the receptive fields of STT cells change following deafferentation. However, several investigations have demonstrated alterations in the responsiveness of STT cells after damage to the skin. When the skin is subjected to a series of noxious heat stimuli, leading to a mild burn, the responses of primate STT cells to further noxious heat stimuli are enhanced, at least for lower intensities of noxious heat, and the threshold for a response is diminished (Kenshalo *et al.*, 1979; Ferrington *et al.*, 1987b). These changes presumably parallel the development of primary hyperalgesia in humans (Lewis, 1942; Hardy *et al.*, 1952) and are likely to be due at least in part to sensitization of nociceptive afferent fibers (see Chapter 2).

A mild burn also enhances the responses of primate STT cells to innocuous mechanical stimuli (Kenshalo *et al.*, 1982; Ferrington *et al.*, 1987d). This observation suggests that changes in the responses of primate STT cells may serve as an animal model for the condition in humans called allodynia in which previously nonpainful stimuli become painful following damage. In the study by Kenshalo *et al.* (1982), the responses to innocuous mechanical stimuli were enhanced for stimuli applied not only to the burned area but also to undamaged skin (Fig. 9.29). This latter change resembles the development of secondary hyperalgesia in humans (Lewis, 1942; Hardy *et al.*, 1952). Secondary hyperalgesia occurs in a zone surrounding a damaged area and is characterized by tenderness to previously innocuous stimuli. Following damage to a small area of skin, secondary hyperalgesia develops over a progressively greater area during a period of about 15 minutes and lasts for hours. Lewis (1942) proposed that secondary hyperalgesia is due to sensitization of a system of "nocifensor" nerves within the skin through axon reflexes. However, Hardy *et al.* (1952) concluded that secondary hyperalgesia is due to a central process leading to enhanced excitability of STT cells.

Primary and secondary hyperalgesia have been investigated further by use of intradermal injections of capsaicin in human subjects and in monkeys (Simone *et al.*, 1987, 1989a, 1991; R. H. LaMotte *et al.*, 1991; Baumann *et al.*, 1991; cf. Simone *et al.*, 1989b). Intradermal injections of capsaicin produce pain that lasts for 10–15 min (Fig. 9.27B), followed by the development of primary hyperalgesia at the injection site and secondary hyperalgesia in the surrounding skin. Sensory changes include a lowering of the threshold for heat pain in the area of the injection but not in the surrounding skin, but tenderness to previously nonpainful mechanical stimuli in the area of secondary hyperalgesia. The responses of STT cells in monkeys have been examined before and after intradermal injections of capsaicin (or control injections of vehicle) to determine whether these cells show altered responses that might account for primary and secondary hyperalgesia (Simone *et al.*, 1990). Injection of capsaicin, but not vehicle, results in a decrease in the threshold for a response to noxious heat at the injection site but little change in the surrounding skin (Fig. 9.30). Furthermore, the responses of the surrounding skin to mechanical stimuli are enhanced (Fig. 9.31). Evidently, the changes in the responses of STT cells mimic those that must occur in humans to produce primary and secondary hyperalgesia.

A further question is whether a central change in the excitability of STT cells helps account for secondary hyperalgesia. This has been tested by recording the responses of STT cells to

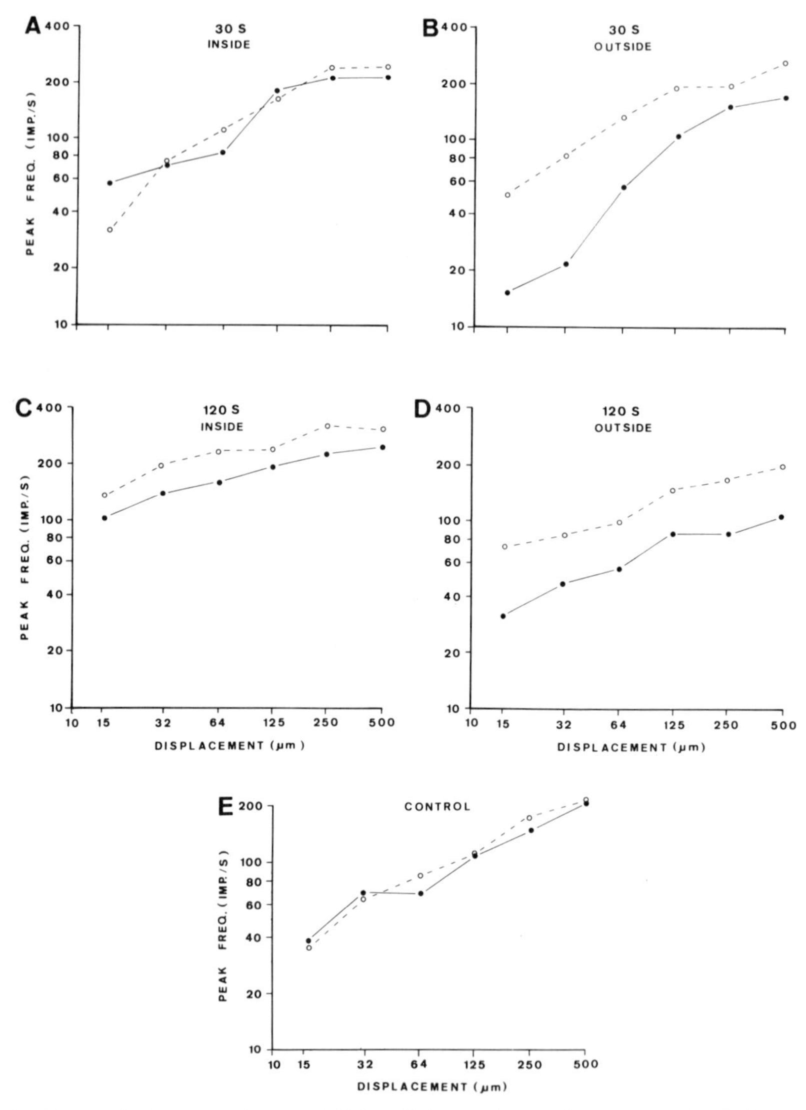

Fig. 9.29. Changes in the responsiveness of STT cells in monkeys to innocuous mechanical stimuli (skin indentations of up to 500 μm) following a mild burn. Symbols: ●, averaged responses before heating; ○, averaged responses after heating. Mechanical stimuli were applied either within the area exposed to noxious heat (A, C) or 1 cm away from the heated area (B, D). For eight cells, the skin was exposed to a graded series of 30 sec heat pulses with intensities up to 50°C (A, B). The other nine cells were tested before and after 120-sec heat pulses of the same intensities. Panel E is a control in which four cells were tested before and after the thermal stimulator was placed on the skin; no heat pulses were given. (From Kenshalo *et al.*, 1982.)

volleys in the largest myelinated fibers in dorsal rootlets (Simone *et al.*, 1991). The dorsal rootlets are disconnected from the periphery, and so changes in the skin cannot affect the activity of the axons in the dorsal root filament. Injection of capsaicin into the skin causes increases in the responses of STT cells to volleys in axons of the dorsal root filament, indicating that there is change in the excitability of the STT cells in response to the capsaicin injection. Presumably, this change is due to the central action of the C fibers stimulated by the capsaicin. It is possible that

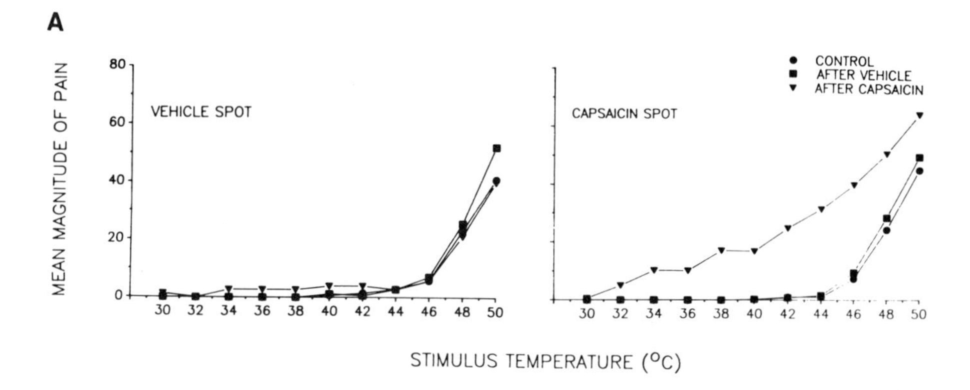

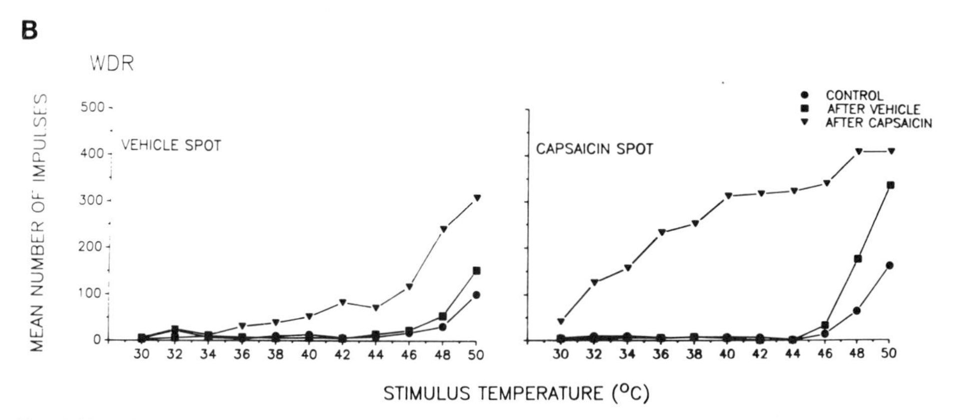

Fig. 9.30. Primary hyperalgesia following intradermal injection of capsaicin. (A) Averaged psychophysical curves relating the intensity of 5 sec heat pulses to estimates of the magnitude of pain in human subjects. The threshold for heat pain is lowered and the intensity increased for a given suprathreshold stimulus in the area of the capsaicin injection, but not in the area of vehicle injection. (B) Averaged stimulus-response functions for primate STT cells to graded 5 sec heat pulses. The responses are changed in parallel to those seen in human hyperalgesia (although the vehicle injection appears to have a slight effect in the monkeys). (From Simone *et al.*, 1991.)

capsaicin produces a sensitization of primary afferent fibers, leading to a continuous discharge in these fibers that might account for the enhanced excitability of STT cells. However, Baumann *et al.* (1991) have been unable to demonstrate long-enduring discharges in primary afferent fibers following capsaicin injections into monkey skin.

Recently, it has been found that iontophoretic release of both an excitatory amino acid (*N*-methyl-D-aspartic acid [NMDA] or quisqualate, depending upon the cell) and substance P (SP) near STT cells can lead to a prolonged increase in their excitability (Dougherty and Willis, 1991). This is demonstrated by increases in the responsiveness of the STT cells both to later applications of excitatory amino acids and to mechanical stimulation of the skin. The change often develops over a period of about 15 min and may last hours (Fig. 9.32). This enhanced responsiveness of the STT cells presumably reflects the activation of a second-messenger system in the STT cells. Similar changes in interneurons or in STT cells may occur in response to the C-fiber discharges that are triggered by intradermal capsaicin injections, since the C fibers would be expected to release excitatory amino acids and peptides in the dorsal horn (see Chapter 5). Such an enhancement in the activity of primate STT cells may provide an explanation for the development of secondary hyperalgesia.

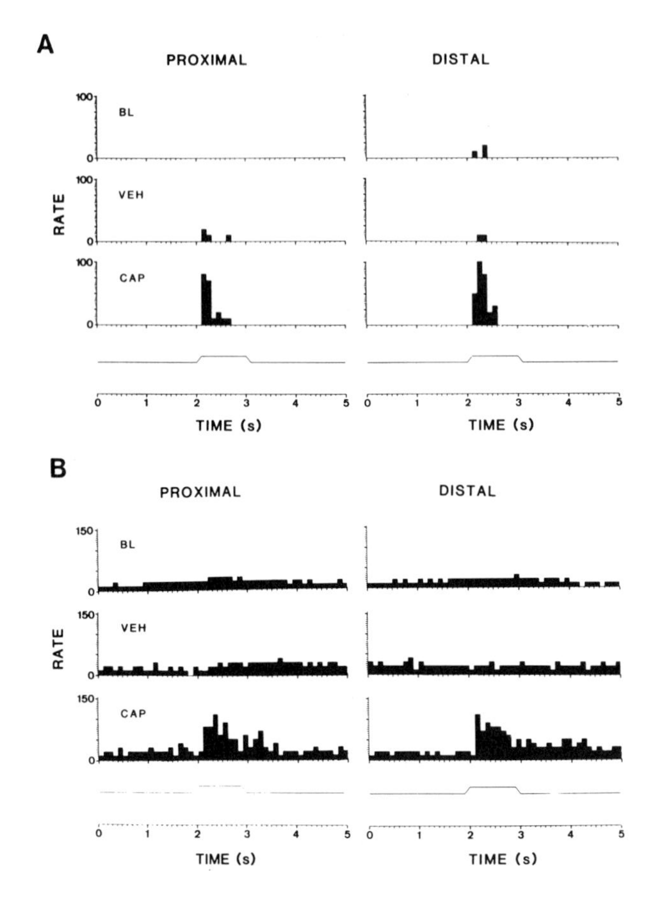

Fig. 9.31. Neural model of allodynia. Panels A and B show the responses of two different primate STT cells before and after an injection of vehicle and after an injection of capsaicin into the skin. The responses were evoked by weak tactile stimuli applied by stroking the skin proximal and distal to the injection site with a spring-loaded cotton-tipped applicator. (From Simone *et al.*, 1991.)

Inhibitory Receptive Fields of STT Cells

STT cells have inhibitory receptive fields, as well as excitatory ones. The inhibition of some high-threshold STT cells by innocuous mechanical stimulation of the skin has already been mentioned (Blair *et al.*, 1981; see also Willis *et al.*, 1974; McCreery *et al.*, 1979a; Milne *et al.*, 1981). Inhibition of STT cells by mechanoreceptive visceral afferent fibers has also been reported (Milne *et al.*, 1981).

The most powerful inhibition of STT cells from stimulation of the skin results from noxious stimuli, either mechanical or thermal (Fig. 9.33) (Gerhart *et al.*, 1981b). The inhibitory receptive fields often include all of the surface of the body and face, apart from the excitatory receptive field (Gerhart *et al.*, 1981b; cf. McCreery and Bloedel, 1976; McCreery *et al.*, 1979a; Brennan *et al.*, 1989). Wide-dynamic-range STT cells have more effective inhibitory receptive fields than do high-threshold STT cells. This pattern is reminiscent of that of the diffuse noxious inhibitory control (DNIC) described for dorsal horn interneurons in the rat by Le Bars *et al.* (1979a,b) (see Chapter 5). However, there are several important differences. The inhibition produced by DNIC is very long-lasting, in contrast to the inhibition of primate STT cells. Furthermore, the inhibition due to DNIC disappears when the spinal cord is transected, whereas the inhibition of primate STT cells was only partly reduced by interruption of the spinal cord (cf. Cadden *et al.*, 1983). However, the inhibition of STT cells by noxious stimuli may correspond to DNIC attenuated by the anesthetic regimen used in these experiments.

Inhibition of STT cells has also been investigated by using electrical stimulation of

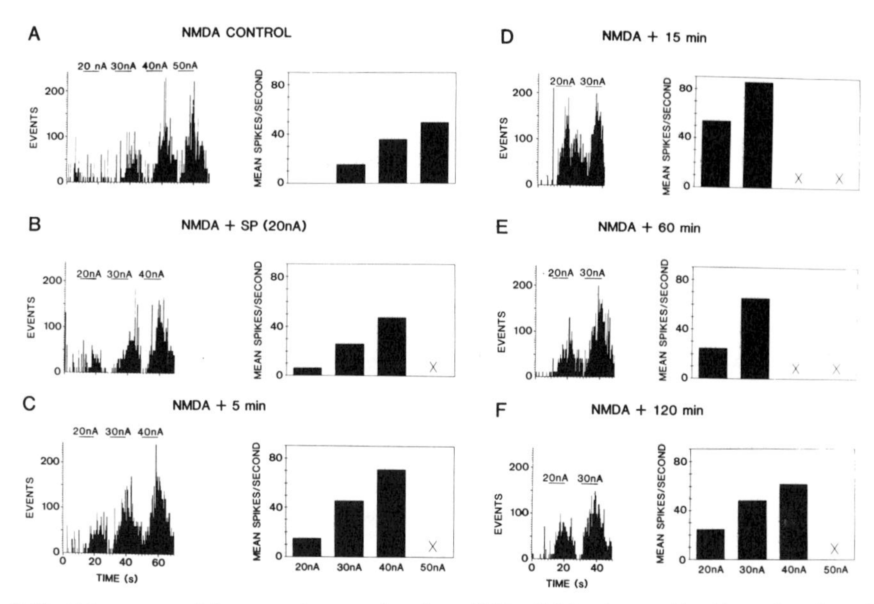

Fig. 9.32. Enhancement of the responsiveness of a primate STT cell following combined iontophoretic application of NMDA and SP. The initial responses of the cell to graded iontophoretic currents are shown in panel A. The responses are increased during the concurrent application of SP (B). Following cessation of the SP application, the responses to NMDA continued to increase at 5 (C) and 15 (D) min, and then declined over a period of more than 2 hr (E and F). High currents were not used when the responses were greatly enhanced to avoid possible cytotoxicity (X's in histograms). (From the results of Dougherty and Willis, unpublished.)

peripheral nerves or the dorsal funiculus (R. D. Foreman *et al.*, 1976; J. M. Chung *et al.*, 1984a,b). Stimulation of a peripheral nerve or of the dorsal funiculus with a single shock results in inhibition or excitation followed by inhibition of STT cells (R. D. Foreman *et al.*, 1976). The inhibition lasts about 150 msec. The inhibition from dorsal funiculus stimulation results from segmental interactions, since the inhibition is blocked by a lesion of the dorsal column between the stimulating electrode and the recorded STT cell. It has been proposed that the inhibition is due to a combination of inhibitory postsynaptic potentials, primary afferent depolarization, and possibly collision with afferent input.

A prolonged inhibition (lasting several minutes) can be produced by repetitive stimulation of peripheral nerves (J. M. Chung *et al.*, 1984a,b). The best inhibition is produced when Aδ fibers are stimulated; Aβ fibers cause only a weak inhibition. Activation of C as well as Aδ fibers enhances the inhibition; it is not known what the effect of volleys restricted to C fibers would be. A similar inhibition of STT cells can be produced when a transcutaneous electrical nerve stimulator is used (K. H. Lee *et al.*, 1985).

Inhibition of STT cells can be produced by stimulation of visceral afferent fibers, as well as by stimulation of the skin or of peripheral nerves (Foreman *et al.*, 1977, 1981; Milne *et al.*, 1981; Brennan *et al.*, 1989). The probability that visceral stimulation will produce inhibition increases as the distance between the visceral inflow and the segmental level of the STT cells examined increases (Brennan *et al.*, 1989).

Summary

In summary, STT cells in lamina I often have small receptive fields. This suggests that these cells provide information useful for stimulus localization. Such information would presumably

A

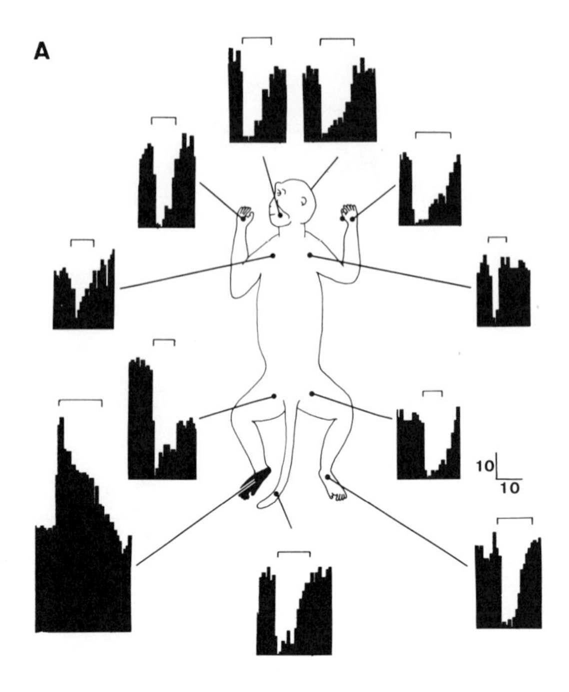

B

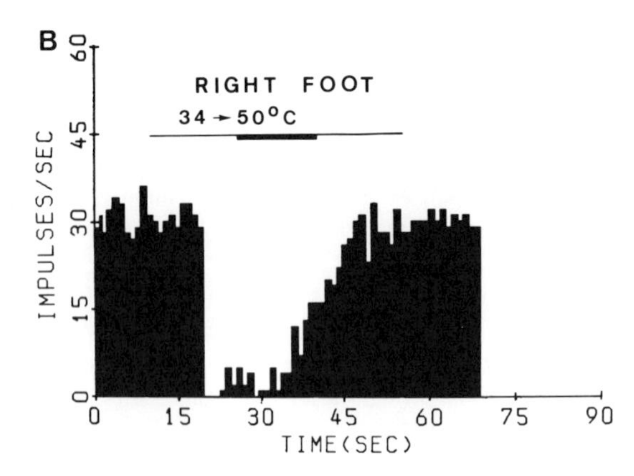

Fig. 9.33. Inhibitory receptive field of a wide-dynamic-range STT cell in the left side of the lumbar enlargement of the spinal cord of a monkey. (A) Excitatory receptive field on the left hindlimb. Inhibition was produced when a noxious mechanical stimulus was applied anywhere on the body surface outside of the excitatory receptive field. (B) Inhibition by noxious heat stimulation of the right foot. (From Gerhart *et al.*, 1981b.)

be processed in the VPL nucleus, since neurons in the nucleus submedius, the other major projection target of lamina I STT cells, have receptive fields that extend over the entire body (Dostrovsky and Guilbaud, 1988). Many STT cells of lamina I are of the high-threshold type, but others are wide-dynamic-range cells. High-threshold STT neurons may signal damage, and wide-dynamic-range STT cells may provide a warning of impending harm. The latter cell type is more likely to be involved in making sensory discriminations for stimuli in the noxious range. A subpopulation of lamina I STT cells is thermoreceptive. Although some thermoreceptive STT cells project to the lateral thalamus, most project medially. The destination of the latter is unknown. STT cells in laminae IV–VI have been investigated extensively. These neurons have receptive fields that are restricted to a portion of a limb contralateral to their thalamic projection. The receptive fields vary in size from small to quite large. Most of these STT cells are of the wide-dynamic-range class, but many are high-threshold cells and a few are low-threshold cells.

Responses similar to those of STT cells of laminae IV–VI have been recorded from the VPL_c nucleus in monkeys (Kenshalo *et al.*, 1980; J. M. Chung *et al.*, 1986b), and the responses are often eliminated by interruption of the ventral part of the lateral funiculus at a thoracic level, suggesting that this projection contributes to the response properties of some thalamic neurons. One difference between STT cells and VPL_c neurons is that the receptive-field sizes are smaller for the latter (J. M. Chung *et al.*, 1986b). Perhaps inhibitory processes shape the receptive fields of the thalamic neurons. Some STT cells in laminae IV–VI send collateral projections to the CL nucleus. A population of STT cells in the deeper layers of the gray matter of the monkey spinal cord project just to the CL nucleus. Many of these neurons have receptive fields that occupy the entire surface of the body and face. Most of these STT cells are of the high-threshold type. The information that these neurons signal is unlikely to be useful for sensory discrimination. It may instead be utilized to trigger motivational-affective responses. Recordings have not yet been made from STT cells in the lateral part of lamina IX. However, it is possible that the proprioceptive STT axons found at the margin of the ventrolateral funiculus originate from these cells. STT cells in Stilling's nucleus have receptive fields in muscle, but not skin, and appear to be proprioceptive, encoding the position of the tail.

PHARMACOLOGY OF STT CELLS

The responses of primate STT cells to a variety of compounds have been tested by microiontophoresis (Jordan *et al.*, 1978, 1979; Willcockson *et al.*, 1984a,b, 1986; Dougherty and Willis, 1991). Testing is generally done by using the responses of the neurons to glutamate, since alterations in glutamate excitation can be considered to reflect postsynaptic changes. In addition to glutamate and other excitatory amino acids, such as NMDA and quisqualic acid (see Chapter 5), the most consistently excitatory agent is SP. Inhibitory compounds include γ-aminobutyric acid (GABA), glycine, serotonin, catecholamines (norepinephrine, dopamine, epinephrine), leu- and met-enkephalin, and metenkephalinamide (a stable analog of enkephalin), cholecysto-kinin, phencyclidine, and SKF 10047. An example of an inhibitory action of iontophoretically released metenkephalinamide on an STT cell is shown in Fig. 9.34. The same neuron is also inhibited by phencyclidine. It is possible that phencyclidine has this effect by blocking NMDA receptors. Morphine and dynorphin are often inhibitory, but sometimes these opiates produce excitatory effects, and they often affect the spiking mechanism in STT cells.

Whereas serotonin is consistently inhibitory when applied to nociceptive STT cells, it has an excitatory action on a subpopulation of STT cells that seems to respond preferentially to proprioceptive input (Jordan *et al.*, 1979).

DESCENDING CONTROL OF STT NEURONS

The descending control of responses of STT cells to noxious stimuli is of considerable interest, since many studies have demonstrated that pain can be suppressed by pathways descending from the brain to the spinal cord (see reviews by D. J. Mayer and Price, 1976; Basbaum and Fields, 1978, 1984; Willis, 1982, 1988; Besson and Chaouch, 1987).

STT cells can be inhibited by stimulation in a number of regions of the brain, including the nucleus raphe magnus (Beall *et al.*, 1976; Willis *et al.*, 1977; McCreery *et al.*, 1979b; Gerhart *et al.*, 1981a; Giesler *et al.*, 1981a; Ammons *et al.*, 1984b), the medullary reticular formation (McCreery and Bloedel, 1975; McCreery *et al.*, 1979b; Haber *et al.*, 1978, 1980; Gerhart *et al.*, 1981a), the periaqueductal gray and adjacent midbrain reticular formation (Hayes *et al.*, 1979; Gerhart *et al.*, 1984; Carstens, 1988; Zhang *et al.*, 1988), the subcoeruleus-parabrachial region

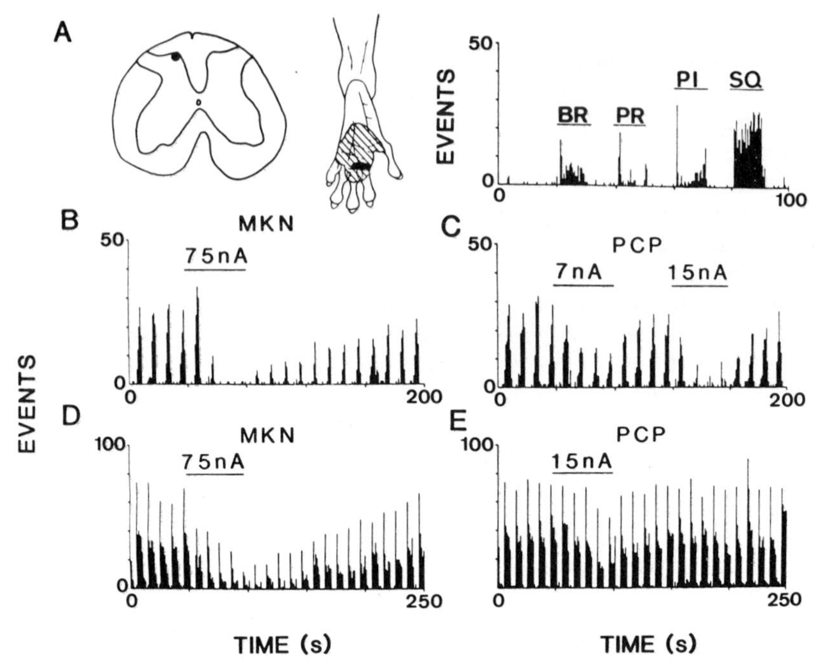

Fig. 9.34. Inhibitory actions of metenkephalinamide (MKN) and phencyclidine (PCP) on the responses of a primate STT cell. The location of the recording site in lamina I, the receptive field, and the responses to graded mechanical stimuli are shown in panel A. The mechanical stimuli included BR (brushing the skin), PR (pressure by application of an arterial clip with a weak grip to the skin), PI (pinch using an arterial clip with a strong grip), and SQ (squeezing the skin with forceps). In panels B and C, the STT cell was activated repeatedly by iontophoretic application of pulses of glutamate. Metenkephalinamide and phencyclidine were applied during the times indicated by the horizontal bars. Both substances reduced the responses to glutamate. In panels D and E, the skin was squeezed intermittently. Iontophoretic release of metenkephalinamide and phencyclidine inhibited the responses to squeezing. (From Willcockson *et al.*, 1986.)

(Brennan *et al.*, 1987; Girardot *et al.*, 1987), the periventricular gray (Ammons *et al.*, 1986), the VPL_c nucleus (Gerhart *et al.*, 1981c, 1983), and the Rolandic cerebral cortex (Coulter *et al.*, 1974; Yezierski *et al.*, 1983). Stimulation of afferent fibers of the trigeminal nerve can also inhibit STT cells, presumably by activation of pathways descending from the brain stem (McCreery and Bloedel, 1976; McCreery *et al.*, 1979a).

Excitation of STT cells has also been observed following stimulation in the brain. Some sites within the medullary reticular formation are found to excite primate STT cells (Haber *et al.*, 1978, 1980), and strong stimulation in the medullary raphe and reticular formation excites and then inhibits cat STT cells (McCreery *et al.*, 1979b). It is unclear what structures are responsible for the excitation. The descending pathway producing this excitation appears to travel in the ventral quadrant of the spinal cord, since lesions of the DLF do not prevent the excitation (Haber *et al.*, 1980).

Excitation of STT cells in monkeys has also resulted from stimulation of the cerebral cortex in the vicinity of the central sulcus (Coulter *et al.*, 1974). A mapping study, using intracortical microstimulation, has shown that the cytoarchitectural area responsible for the excitation is area 4, the primary motor cortex (Yezierski *et al.*, 1983). Excitation from this area is often followed by inhibition. The excitation (and at least part of the inhibition) can be eliminated by a lesion of the DLF at an upper cervical level, suggesting involvement of the corticospinal tract.

The descending pathways that are likely to be responsible for the inhibitory effects of brain stem stimulation in the monkey have been demonstrated by using retrograde labeling with HRP (see, e.g., Carlton *et al.*, 1985a). It has been suggested that the periaqueductal gray exerts its effect on the spinal cord by way of a relay in the nucleus raphe magnus and adjacent reticular formation (Basbaum and Fields, 1978, 1984). There is a substantial projection from the periaqueductal gray (PAG) to the medial medulla in the monkey (see, e.g., J. M. Chung *et al.*, 1983a), but the details of the mechanisms of inhibition, including the microcircuitry within the spinal cord, are still unclear. For example, whereas stimulation in the nucleus raphe magnus produces IPSPs in many STT neurons (Giesler *et al.*, 1981a), stimulation in the PAG often does not evoke an IPSP. Instead, PAG stimulation can produce a hyperpolarization that is not associated with an increased membrane conductance and that is increased rather than reversed when the membrane is artificially hyperpolarized (Zhang *et al.*, 1988). In experiments in which systemic injections of serotonin antagonists are made, the inhibition from stimulation in the nucleus raphe magnus is only slightly affected, whereas the inhibition from PAG stimulation is nearly completely blocked (Yezierski *et al.*, 1982). The role of presynaptic inhibition also must be clarified, since stimulation in the nucleus raphe magnus or the medullary reticular formation results in primary afferent depolarization of at least some nociceptive afferent fibers in both cats and monkeys (R. F. Martin *et al.*, 1979).

SPINORETICULAR TRACT

Projections of spinal cord neurons to the brain stem reticular formation are collectively called the spinoreticular tract (SRT). Most of the axons of the SRT ascend in the ventrolateral white matter. At least two major components of the SRT should be distinguished: a projection to the lateral reticular nucleus, which is a precerebellar nucleus (see Oscarsson, 1973), and one to the medial pontomedullary reticular formation. Since the cerebellum plays no essential role in sensation, cerebellar connections will not be reviewed. The medial part of the SRT ends on neurons that either project caudally to the spinal cord, and thus form part of the descending sensorimotor control systems, or rostrally to higher levels of the nervous system, including the midbrain and diencephalon. A spino-reticulo-thalamo-cortical pathway is often proposed as an important afferent pathway to the forebrain, especially with reference to pain mechanisms. Recently, another SRT projection has been demonstrated to relay in the dorsal reticular nucleus of the medulla.

TAXONOMIC DISTRIBUTION

Spinoreticular projections have been found in all vertebrate classes and in all mammalian orders that have been examined, including elasmobranch and teleost fish (Ebbesson and Hodde, 1981; Hayle, 1973), amphibia and reptiles (Goldby and Robinson, 1962; Ebbesson, 1967, 1969; Ebbesson and Goodman, 1981), birds (Karten, 1963; Schneider and Necker, 1989), opossum (Mehler, 1969; Hazlett *et al.*, 1972), pig (Breazile and Kitchell, 1968), sheep (Rao *et al.*, 1969), cat (Morin *et al.*, 1951; G. F. Rossi and Brodal, 1957; Bowsher and Westman, 1970), rat (Lund and Webster, 1967b; Mehler, 1969; Zemlan *et al.*, 1978), monkey (Mehler *et al.*, 1960; Kerr and Lippman, 1974; Kerr, 1975c), and humans (Bowsher, 1957; Mehler, 1974). Mehler (1966) emphasized the similarity in the projections of spinoreticular fibers in the various mammalian species. The most significant change in the more "advanced" species is the greater number of fibers that ascend past the pons to end in the midbrain and diencephalon.

ANATOMY OF THE SRT

Cells of Origin

The cells of origin of the spinoreticular tract have been mapped with the retrograde tracer technique in several species.

Rat. Large injections of HRP into the medial reticular formation, including the nucleus reticularis gigantocellularis, nucleus pontis caudalis, nucleus paragigantocellularis, or central nucleus of the medulla in the rat, label cells at all levels of the spinal cord (Andrezik *et al.*, 1981; Kevetter and Willis, 1982, 1983; Chaouch *et al.*, 1983; Peschanski and Besson, 1984). Most of the cells (68% overall and 85% of SRT cells caudal to C3, according to Chaouch *et al.*, 1983; 74% of SRT cells in the lumbar enlargement, according to Kevetter and Willis, 1983) are contralateral to the injection.

Most SRT cells are located in laminae V, VII, and VIII. There is almost no label in the superficial dorsal horn when tracer is injected into the medial reticular formation (Kevetter and Willis, 1982, 1983; Chaouch *et al.*, 1983). Although one study with HRP has reported that few cells are labeled in lamina X (Chaouch *et al.*, 1983), other investigations using HRP or a fluorescent tracer demonstrate SRT cells in this lamina (Kevetter and Willis, 1982, 1983; Peschanski and Besson, 1984; Nahin *et al.*, 1986; Nahin and Micevych, 1986).

The uppermost cervical segments are exceptional in having more labeled cells, including a large ipsilateral group in the ventral horn. According to Kevetter and Willis (1983), there are twice as many labeled cells per section in the upper cervical segments than in either enlargement. More restricted injections result in a similar distribution of cells, but fewer cells are labeled. Cells projecting to both the reticular formation and thalamus can be labeled with two different retrograde tracers (Kevetter and Willis, 1982, 1983). Approximately 10% of the labeled cells project to both sites.

Peschanski and Besson (1984) have injected WGA-HRP into the nucleus gigantocellularis in rats to label SRT cells retrogradely, as well as to trace the projection from this nucleus to the thalamus anterogradely. The SRT cells are mostly in laminae VII, VIII, and X, although there are some in the neck of the dorsal horn. The projections to the thalamus are chiefly to the intralaminar nuclei, although some other terminations are also noted. These findings are consistent with the notion of a spino-reticulo-thalamic pathway suited to play a role in the motor responses to noxious stimuli.

A separate SRT has now been demonstrated that terminates in the dorsal reticular nucleus of the medulla (S. B. McMahon and Wall, 1985a; Lima, 1990; Lima and Coimbra, 1990). This nucleus contains neurons that respond to noxious stimulation of large areas of the body and to volleys in Aδ and C fibers (Villanueva *et al.*, 1988; cf. Villanueva *et al.*, 1990). The cells of origin of this projection are chiefly in laminae I and X, although some are also in laminae II–IV and in deeper laminae. The projection from the dorsal horn is mostly ipsilateral, but that from laminae VII and X is bilateral. On the basis of the reduction in retrograde labeling following lesions of the spinal cord white matter, Lima (1990) suggests that the ascending pathway from the dorsal horn is mainly in the dorsal columns. However, the pathway from deeper laminae appears to ascend in the ventral quadrant. The latter pathway appears to be more important for activation of nociceptive neurons in the dorsal reticular nucleus than does the dorsal pathway (Bing *et al.*, 1990).

Cat. Abols and Basbaum (1981) find a similar distribution of labeled cells to that described above for rats after injecting HRP into the medial reticular formation of the rostral medulla in cats.

Monkey. Large injections of HRP into the medial reticular formation of monkeys label cells at all levels of the spinal cord, although most of the labeled cells are in the uppermost cervical segments (Kevetter *et al.*, 1982). Most of the cells are in lamina VII, but some are also in lamina VIII and in the dorsal horn, especially the lateral part of lamina V (Fig. 9.35). Few labeled cells are found in the superficial dorsal horn. The laterality of the projection has been examined after small, unilateral injections of HRP. Although the projection from the cervical and lumbar enlargements is chiefly crossed, more ipsilateral cells are labeled in the cervical than in the lumbar enlargement. This may in part account for the observation of Kerr and Lippman (1974) that a commissural myelotomy in the cervical spinal cord produces little anterograde degeneration in the reticular formation of monkeys.

Number of SRT Cells

In both rats and monkeys, only a modest number of SRT cells are labeled by large injections of HRP into the medial rhombencephalic reticular formation (Kevetter and Willis, 1982, 1983; Kevetter *et al.*, 1982; Chaouch *et al.*, 1983). For instance, Kevetter and Willis (1982) counted 1481 SRT cells in alternate sections through the upper cervical segments, the cervical and lumbar enlargements, and parts of the thoracic and upper lumbar cord. Recognizing that a more thorough count at all levels of the spinal cord and larger injections of HRP would undoubtedly label more SRT cells, it is nevertheless surprising that the total population of SRT cells is of a size comparable to that of the STT (cf. Willis *et al.*, 1979). The cell counts are in marked contrast to the difference in volume of the anterograde degeneration in these pathways after interruption of the ventrolateral quadrant of the spinal cord (Mehler, 1969; Bowsher, 1976). Perhaps the massive anterograde degeneration seen when the SRT is interrupted (Fig. 9.36) reflects the amount of collateralization of this pathway rather than the number of cells giving rise to the SRT (Kevetter *et al.*, 1982).

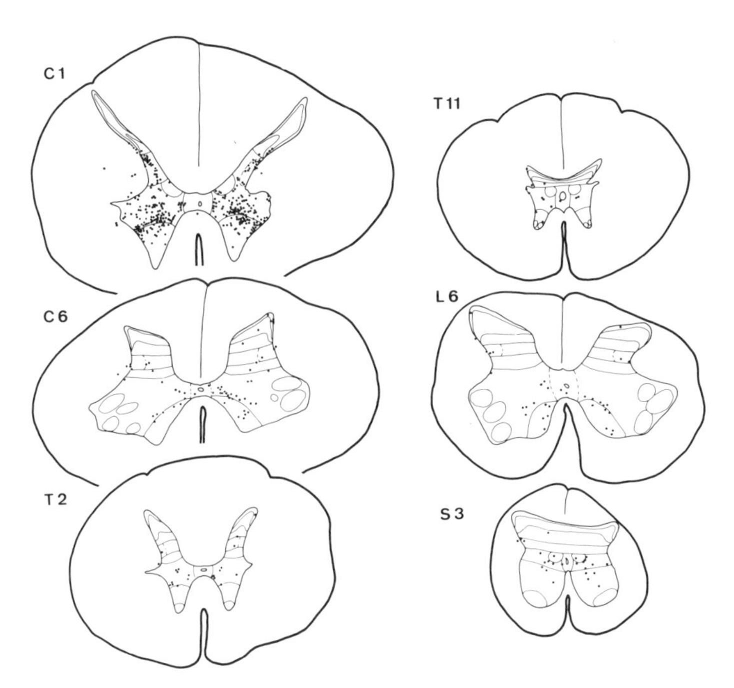

Fig. 9.35. Distribution of spinoreticular tract cells labeled retrogradely by a large, bilateral injection of HRP into the rhombencephalic reticular formation of a monkey. (From Kevetter *et al.*, 1982; reproduced with permission.)

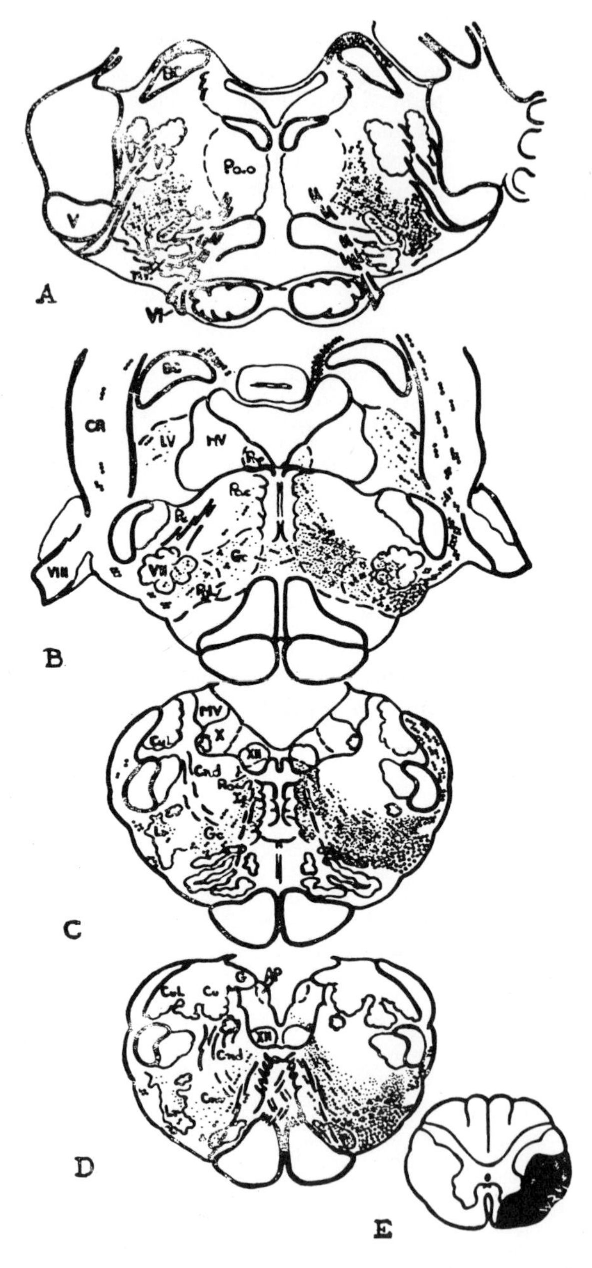

Fig. 9.36. Distribution of degenerating fibers in the rhombencephalon of a monkey following an anterolateral cordotomy at an upper cervical level. (From Mehler *et al.*, 1960; reproduced with permission.)

Antidromic Mapping

The cells of origin of the SRT have also been mapped in physiologic experiments on animals of several species by using antidromic activation (Levante and Albe-Fessard, 1972; Albe-Fessard *et al.*, 1974a; Fields *et al.*, 1975, 1977b; Maunz *et al.*, 1978; Menétrey *et al.*, 1980; Haber *et al.*, 1982; Thies and Foreman, 1983; Blair *et al.*, 1984; Cervero and Wolstencroft, 1984; S. B.

McMahon and Wall, 1985a; Thies, 1985; Ammons, 1987). In general, the locations of the antidromically activated SRT neurons coincide with those predicted from the anatomic studies. Some SRT neurons in cats can also be activated antidromically from the midbrain (Fields *et al.*, 1977b) or thalamus (Ammons, 1987; Blair *et al.*, 1984) and in monkeys from the thalamus (Giesler *et al.*, 1981b; Haber *et al.*, 1982).

Morphology of Individual SRT Cells

In monkeys, most SRT cells observed in retrogradely labeled preparations are multipolar, although some are fusiform and a few round (Kevetter *et al.*, 1982). The dendrites can be very long and straight, extending for as much as 900 μm in a single section. Dendrites of SRT cells in the medial ventral horn can cross the midline or can extend laterally and enter the lateral funiculus. The mean area of the somas of a population of SRT cells in lamina VII of the lumbar enlargement is significantly smaller (329 ± 180 μm^2) than that of a population of STT cells in the same lamina (544 ± 324 μm^2; $n = 25$).

The SRT neurons of lamina I that project to the dorsal reticular nucleus of the rat medulla are generally of the multipolar type, although some are of the pyramidal or flattened types (Lima and Coimbra, 1990).

Immunocytochemical Studies of SRT Cells

Some SRT neurons in the vicinity of the central canal (laminae VII and X), identified by retrograde labeling with the fluorescent marker True Blue, have been shown to stain immunocytochemically with an antibody to enkephalin (Nahin and Micevych, 1986).

Axons of SRT Neurons

The axons of SRT neurons ascend in the white matter of the ventral part of the spinal cord (G. F. Rossi and Brodal, 1957; Anderson and Berry, 1959; Mehler *et al.*, 1960; Kerr, 1975c; Nahin *et al.*, 1986). Numerous labeled axons are seen in the lateral funiculus ipsilateral to an injection of HRP into the reticular formation, but few contralaterally (Kevetter and Willis, 1983). Since the cell bodies are largely contralateral to their reticular formation targets, the axons must decussate near the level of the cell bodies (cf. Bowsher, 1961). The axons of almost all of the SRT cells that have been studied are myelinated, having conduction velocities of 2–93 m/sec (Fields *et al.*, 1977b), 16–96 m/sec (Maunz *et al.*, 1978), and 17–63 m/sec (Thies, 1985) in cats and 9–54 m/sec in monkeys (Haber *et al.*, 1982).

Projection Targets of SRT

Following interruption of the ventrolateral quadrant of the spinal cord, degenerating axons of the SRT can be followed into the brain stem. As the SRT enters the medulla, the axons form a prominent laterally placed bundle. Degenerating fibers leave this bundle, coursing medially into the reticular formation. Degenerating terminals are found in the following reticular nuclei (following the terminology of Meessen and Olszewski, 1949; Olszewski and Baxter, 1954; Olszewski, 1954; Brodal, 1957; G. F. Rossi and Zanchetti, 1957; Taber, 1961): nucleus medullae oblongatae centralis, lateral reticular nucleus, nucleus reticularis gigantocellularis, nucleus reticularis pontis caudalis and oralis, nucleus paragigantocellularis dorsalis and lateralis, and nucleus subcoeruleus (Fig. 9.30) (Anderson and Berry, 1959; Mehler *et al.*, 1960; Westman and Bowsher, 1971; Bowsher and Westman, 1970; Jane and Schroeder, 1971; Schroeder and Jane, 1971; Kerr, 1975c). There is no obvious somatotopic organization.

FUNCTIONAL PROPERTIES OF SRT NEURONS

Background Activity

Many SRT cells in anesthetized cats have very low rates of background discharge. Thies and Foreman (1983) report that about half of their sample have background activity, whereas the others are either silent or fire at rates below 1 Hz. Thies (1985) reports that 77% of SRT neurons in cats have background discharges of 0.2–10 Hz; the others are silent.

Responses to Electrical Stimulation

Maunz *et al.* (1978) find that SRT neurons in the cat lumbosacral enlargement can be excited following electrical stimulation of a variety of peripheral nerves of both hindlimbs. Input is from both cutaneous and high-threshold muscle afferents. The ipsilateral input is more effective than the contralateral input. Inhibitory actions are also observed. Thies (1985) demonstrates activation of SRT neurons by stimulation of muscle and skin nerves. However, effective muscle afferents include group II fibers.

Thies and Foreman (1983) have tested the effects of visceral afferents on SRT neurons of the upper thoracic cord in the cat. Stimulation of Aδ afferents and sometimes C fibers in sympathetic nerves causes excitation (cf. Thies, 1985; Blair *et al.*, 1984). The excitatory input can be unilateral or bilateral. On the other hand, stimulation of the vagus nerve, in either the neck or the right cardiac branches, often produces inhibition of SRT neurons. A few SRT neurons are instead excited by vagal stimulation.

Electrical stimulation of the renal nerve excites most SRT neurons in segments T12–L2 of the cat spinal cord, although a few of the cells are inhibited (Ammons, 1987). Both Aδ and C fibers are effective.

Responses to Natural Stimuli

Mechanical Stimulation of Somatic Structures. Fields *et al.* (1975) have found that SRT neurons in the cat lumbosacral spinal cord have various combinations of ipsilateral or contralateral excitatory or inhibitory receptive fields on the hindlimbs, although most of the cells have excitatory inputs from the ipsilateral hindlimb (see also Albe-Fessard *et al.*, 1974a). The inputs can come from the skin or from deep tissue. Bending the joints is often effective. Innocuous stimulation of the skin can excite or inhibit a particular cell, but often noxious stimuli are excitatory.

Fields *et al.* (1977b) have subdivided their sample into (1) SR neurons, which have superficial, restricted receptive fields; (2) DR neurons, which have deep, restricted receptive fields; and 3) CE neurons, which have complex, extensive receptive fields.

SR neurons have cutaneous excitatory receptive fields, but there can be a surrounding or overlapping inhibitory field as well. Innocuous stimuli are sometimes maximally effective (low-threshold neurons), but in most cases a maximal discharge requires noxious stimuli (that is, these cells are often of the wide-dynamic-range or of the high-threshold class, depending on whether there is a convergent input from mechanoreceptors). SR neurons are located in the dorsal horn.

DR neurons are excited by stimulation of deep structures, such as a joint capsule, periosteum, muscle, or tendon. It is possible for some of these cells to have a convergent input from the skin. Maximal discharge is attained for some neurons with innocuous deep stimuli, but in many cases noxious stimuli are required. These cells are widely distributed.

CE neurons have very large receptive fields, often with both excitatory and inhibitory components. Some respond to joint rotation. The cells are located deep within the spinal cord gray matter.

Results consistent with these have been obtained by Maunz *et al.*, (1978), Thies and Foreman (1983), Cervero and Wolstencroft (1984), Thies (1985), and Ammons (1987) in cats and by Haber *et al.* (1982) in monkeys. However, Haber *et al.* (1982) find that nearly half of the SRT neurons sampled are unresponsive to any natural stimulus tried.

Stimulation of Visceral Structures. Blair *et al.* (1984) report that occlusion of a coronary artery or injection of bradykinin into the coronary circulation often excites SRT neurons in the upper thoracic spinal cord of cats.

DESCENDING CONTROL OF SRT NEURONS

Both inhibitory and excitatory actions on SRT neurons have been found following stimulation within the brain stem.

Fields *et al.* (1975) report that in the cat stimulation in the nucleus reticularis gigantocellularis can powerfully inhibit SRT neurons projecting to the same nucleus (see also Fields *et al.*, 1977b). The inhibition of SRT neurons by vagal stimulation (Thies and Foreman, 1983) is presumably mediated by the activation of descending pathways from the brain stem by vagal afferents.

Haber *et al.* (1982) find that reticular formation stimulation can either excite or inhibit SRT neurons in monkeys and that excitatory effects are more common than inhibitory ones. The excitation of STT neurons projecting to the medial thalamus by reticular formation stimulation has already been discussed (Giesler *et al.*, 1981b), who concluded that part of the complex excitatory receptive fields of these neurons is mediated by way of a supraspinal loop through the reticular formation. Presumably, a similar contribution is made by a supraspinal loop to the complex receptive fields of many SRT neurons. There is evidence that supraspinal loops provide a means for visceral input to activate many spinal neurons by way of descending excitatory pathways (Cervero, 1983b; Cervero *et al.*, 1985). An alternative or additional role for such a positive feedback loop has been proposed by Cervero and Wolstencroft (1984), who find that many neurons in laminae VII and VIII of the lumbar spinal cord of cats, including SRT cells, are excited following stimulation in the reticular formation. These authors suggest that an excitatory supraspinal loop underlies the prolonged activation characteristic of the descending inhibitory control system (DNIC).

Chandler *et al.* (1989) have recently shown that electrical stimulation in the PAG and midbrain reticular formation causes an inhibition of the responses of SRT neurons in the thoracic spinal cord to input from cardiopulmonary afferent fibers. However, chemical stimulation by injection of glutamate is effective only when the injection is into the PAG but not when it is into the midbrain reticular formation. This suggests that inhibition from stimulation in the PAG is due, at least in part, to activation of neurons in that nucleus, but that inhibition from stimulation in the midbrain reticular formation might be due to activation of fibers of passage.

SPINOMESENCEPHALIC TRACT

The spinomesencephalic tract (SMT) is actually a collection of pathways from the spinal cord to several different midbrain target zones. The term "spinotectal tract" is best applied to the part of the SMT that synapses in the superior colliculus. The "spinoannular tract" projects to the periaqueductal gray (PAG).

TAXONOMIC DISTRIBUTION

Spinomesencephalic projections are found in the same vertebrate forms mentioned previously as having spinoreticular projections (see the section on the SRT); (Antonetty and Webster, 1975; Robards *et al.*, 1976; Pritz and Strizel, 1989).

ANATOMY OF THE SMT

Cells of Origin

Rat. The distribution of spinal neurons that project to the midbrain tegmentum in the rat has been mapped by Menétrey *et al.* (1982), following injections of HRP. Labeled cells are concentrated in the marginal zone (lamina I), the lateral spinal nucleus, the lateral reticular part of the neck of the dorsal horn (lamina V), and the dorsal gray commissure (lamina X) at all levels of the spinal cord. Most of the labeling is contralateral. Additional cell groups are labeled in the upper cervical segments, including the lateral cervical nucleus and parts of the ventral horn bilaterally.

R. P. Liu (1983) has injected HRP or a fluorescent marker into different parts of the PAG of the rat and determined the locations of spinoannular neurons. Only a few cells are labeled from the dorsal, medial, or dorsolateral PAG. However, a large number of cells can be found after HRP injections into the ventrolateral PAG. Most are contralateral. They are located in laminae I, III–V, VII, and X and the lateral spinal nucleus at all levels of the cord and in the lateral cervical nucleus.

Swett *et al.* (1985) have made injections of WGA-HRP into the midbrain tegmentum and mapped the distribution of labeled lamina I cells (cells were also noted in other laminae). Most labeled lamina I cells (85%) were contralateral to the injection site. The midbrain sites yielding the largest numbers of retrogradely labeled lamina I cells were the PAG and nucleus cuneiformis. However, none of the injections were in the region of the parabrachial nucleus.

Pechura and Liu (1986) found collateral projections of neurons in the rat spinal cord to both the periaqueductal gray and the medullary reticular formation by using a double-labeling technique.

Harmann *et al.* (1988a) have performed a double labeling study in which one fluorescent tracer was injected into the ventrobasal complex and another into the PAG, bilaterally. Most of the cells labeled from the PAG are in laminae I and V, as well as the lateral spinal nucleus. Double-labeled cells are in these same regions. Only 7.5% of the cells that project to the PAG are simultaneously labeled from the thalamus and 1.7% of those projecting to the thalamus from the PAG. However, this is certainly an underestimate, since the injections were small.

Lima and Coimbra (1989) have been able to label a large population of lamina I neurons retrogradely from the caudal (but not the rostral) PAG and cuneiform nucleus and also from the parabrachial nuclei (cf. Cechetto *et al.*, 1985). In addition, there are many labeled cells in laminae V, VII, and X and in the lateral spinal nucleus. They confirm that most of the SMT cells are contralateral to the injection site.

Yezierski *et al.* (1991) have examined the pattern of labeling of SMT neurons following injections of retrograde label into both the contralateral and ipsilateral midbrain. SMT cells projecting either contralaterally or ipsilaterally are seen in close proximity, and some neurons project bilaterally. About 74% of the cells project contralaterally and 26% ipsilaterally. There is also a population of cells, located chiefly in lamina X, that have descending spinal as well as mesencephalic projections. SMT cells are located in the marginal zone, nucleus proprius and neck of the dorsal horn, the ventral horn, the central gray, and the lateral spinal nucleus.

Cat. The cells of origin of the SMT have been mapped by the retrograde transport method by Wiberg and Blomqvist (1984). The injection sites include the intercollicular nucleus, PAG, posterior superior colliculus, and in some cases the parabrachial nucleus. The greatest number of labeled cells are in the upper cervical segments; the labeled cell groups include the lateral cervical nucleus and cells in laminae VII and VIII. The ventral horn cells are ipsilateral. In the enlargements, most of the cells are in laminae I, IV, and V on the side contralateral to the injection site. Smaller injections have also been made into the superior colliculus, posterior pretectal nucleus, and nucleus of Darkschewitsch. Labeled cells following injections into the superior colliculus are distributed like those seen after intercollicular injections, although fewer cells are labeled. There is sparse labeling with a similar distribution after injection of the posterior pretectal nucleus; most of the cells observed are in the upper cervical cord in contralateral lamina VI and the ipsilateral ventral horn. Injection of the nucleus of Darkschewitsch results in labeling in the upper cervical cord, mostly in deep laminae, including the ipsilateral ventral horn, and chiefly in lamina I in the enlargements.

Panneton and Burton (1985) have examined the neurons that project to the parabrachial region (including the parabrachial and other adjacent nuclei) in the cat by using WGA-HRP or HRP. The labeled cells are mostly in lamina I, but some are in laminae V–VIII, bilaterally.

Monkey. Trevino (1976) and Willis *et al.* (1979) have labeled SMT neurons in a single monkey by injecting HRP into the intercollicular region of the midbrain. In both studies, large numbers of SMT cells are observed in the enlargements in lamina I, with fewer cells in laminae IV–VIII. At upper cervical levels, Trevino (1976) finds labeled cells bilaterally in the ventral horns, as well as mostly contralaterally in the dorsal horn and lateral cervical nucleus. Wiberg *et al.* (1987) have obtained similar results (Fig. 9.37). Mantyh (1982) finds that injections of HRP into the medial PAG labels spinal cord cells mainly in deeper laminae (V–X), whereas injections into the lateral PAG labels cells both in these same laminae and in lamina I. Most of the cells are contralateral.

Zhang *et al.* (1990) have injected different retrograde tracers into the ventral posterior lateral nucleus of the thalamus and the PAG in the monkey. Double-labeled cells are found mostly in contralateral laminae I, V, VII, and X. Double-labeled neurons represent 14.7% of the SMT neurons and 6% of the STT cells.

Antidromic Mapping

Menétrey *et al.* (1980) have recorded from what they term "spinoreticular neurons" in rats. However, many of the cells are identified by antidromic activation from the midbrain and so can also be called SMT cells. The locations of the cells are in the marginal zone, the neck of the dorsal horn, and the lateral spinal nucleus. Most of the cells (61%) are contralateral to the stimulus site. However, SMT neurons in the lateral spinal nucleus often have bilateral projections.

Yezierski and Schwartz (1986) have activated SMT cells antidromically in cats. SMT cells projecting to the nucleus of Darkschewitsch and adjacent reticular formation are located contralaterally in laminae V and VII–VIII. Cells that can be backfired from the rostral PAG and deep superior colliculus are in contralateral laminae V–VII. Cells that can be activated antidromically from the PAG, reticular formation, and deep layers of the superior colliculus at the intercollicular level are in contralateral laminae I–VIII. Those projecting to the caudal PAG and adjacent reticular formation are in the contralateral laminae I and VI–VIII. Some SMT neurons can be activated antidromically from more than one site in the midbrain, suggesting collateralization to several targets.

Hylden *et al.* (1985, 1986a) have recorded in cats from lamina I neurons activated antidromically from the parabrachial region, including the nucleus cuneiformis and lateral PAG. These cells can project to either side or bilaterally. They also find that a given SMT cell can sometimes be activated antidromically from more than one site.

D. D. Price *et al.* (1978) and Yezierski *et al.* (1987) demonstrate that some spinal neurons in

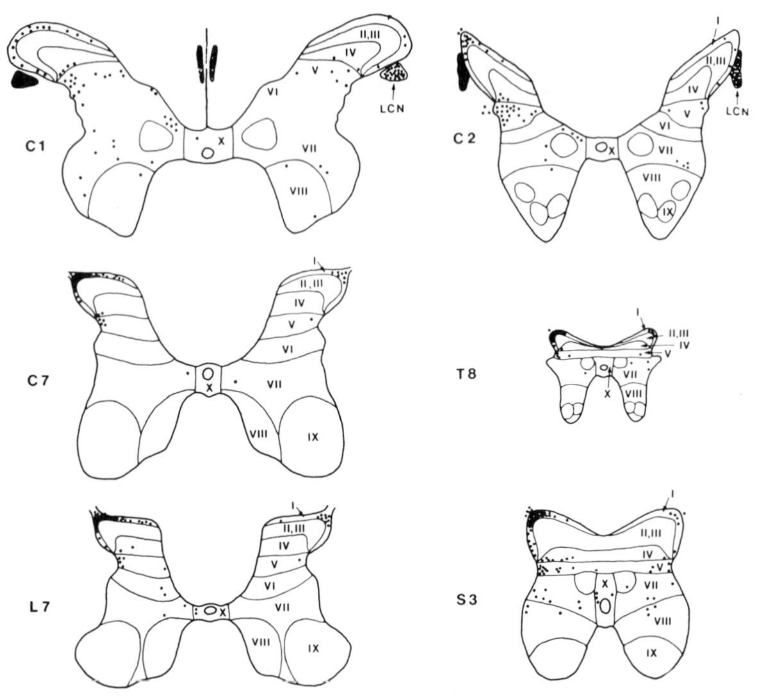

Fig. 9.37. Distribution of SMT cells labeled retrogradely following injection of WGA-HRP into the intercollicular region and superior colliculus in a monkey. Blackened areas indicate a large number of labeled cells in a small area. (From Wiberg *et al.*, 1987; reproduced with permission.)

monkeys project both to the PAG and to the VPL nucleus of the thalamus. Similarly, Hylden *et al.* (1986a) find that some SMT neurons in cats also project to the thalamus.

Number of SMT Cells

Wiberg *et al.* (1987) estimate that the total number of SMT cells retrogradely labeled in their study in the monkey is about 10,000 (excluding the LCN). About 75% are contralateral to the injection site.

Morphology of Individual SMT Cells

The morphology of retrogradely labeled cell bodies and proximal dendrites of SMT cells have been described by several investigators. Menétrey *et al.* (1980) report that SMT cells in lamina I of the rat are generally small or medium sized and round, oval, or fusiform in appearance in transverse sections. Rarely, large transversely oriented fusiform neurons are labeled (Waldeyer cells). SMT cells of the nucleus of the dorsal lateral funiculus are medium sized and oval or fusiform, whereas those in lamina V are large and multipolar. Labeled cells in lamina X have various sizes and are fusiform or oval.

A more detailed description of the SMT cells of lamina I in the rat spinal cord has been given by Lima and Coimbra (1989). They have used cholera toxin subunit B to provide greater filling of the dendritic trees of the retrogradely labeled cells, which they examined in several planes of section. There are two distinct types of lamina I SMT cells: spindle-shaped neurons (type A or fusiform cells), oriented rostrocaudally and concentrated in the lateral third of lamina I

(Fig. 9.38A), and triangular-shaped cells (type B or pyramidal cells), also oriented rostrocaudally and present throughout the mediolateral extent of lamina I (Fig. 9.38B). It is suggested that the cells with dendrites oriented in a medial to lateral direction seen in transverse sections and termed Waldeyer cells belong to the group of pyramidal cells. The fusiform cells project largely to the contralateral parabrachial region and the pyramidal cells to the contralateral periaqueductal gray.

Hylden *et al.* (1986a) have injected lamina I SMT neurons in cats intracellularly with HRP (Fig. 9.38A). The somata have mean diameters of 13–34 μm. The dendrites of reconstructed cells are oriented rostrocaudally and extend for 850–1900 μm. Most of the cells have dendritic spines. The axons are myelinated, and some give off collaterals. These cells belong to the type A class of Lima and Coimbra (1989).

Immunocytochemical Studies of SMT Cells

Standaert *et al.* (1986) demonstrated that most of the lamina I SMT neurons projecting to the parabrachial region contain immunoreactivity for dynorphin or enkephalin.

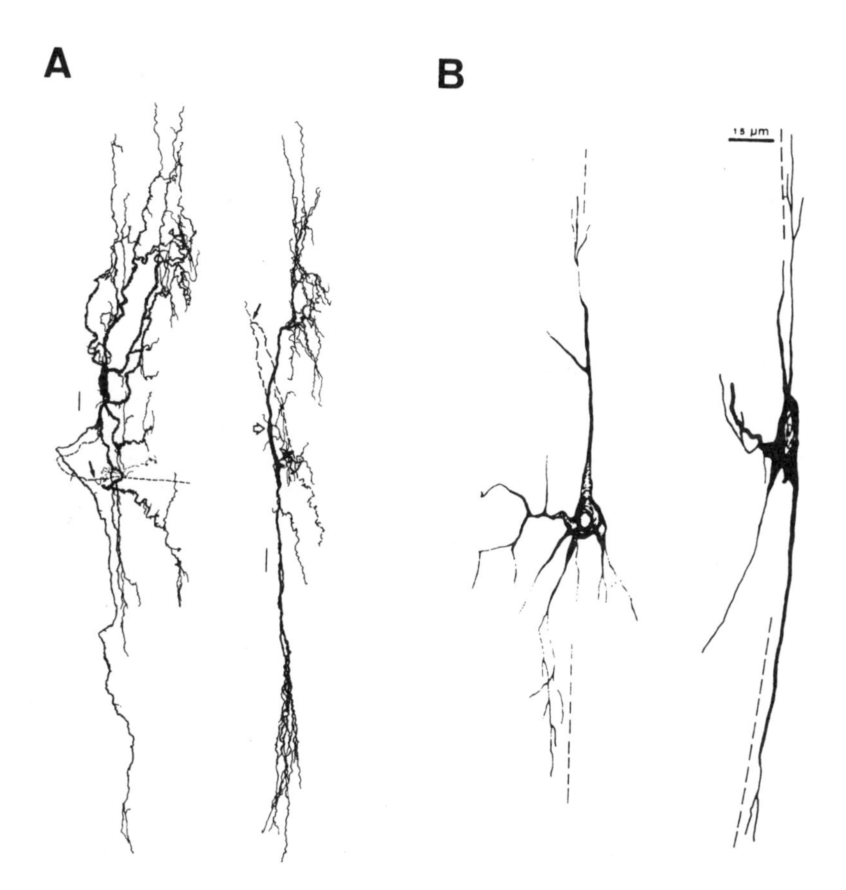

Fig. 9.38. Lamina I SMT cells in horizontal sections. (A) Drawings of two high-threshold lamina I SMT cells in a cat. The cells were identified by antidromic activation from the parabrachial region and injected intracellularly with HRP. (From Hylden *et al.*, 1986; reproduced with permission 1986.) (B) Drawings of two lamina I SMT cells retrogradely labeled in rats by injection of cholera toxin subunit B injected into a region that included the PAG. (From Lima and Coimbra, 1989; reproduced with permission.)

Axons of SMT Cells

The axons of many SMT cells ascend in the white matter of the ventral half of the spinal cord, in company with the STT and SRT (Mehler *et al.*, 1960; Kerr, 1975c). However, the part of the SMT originating from neurons of lamina I appears to ascend in the dorsal part of the lateral funiculus (Zemlan *et al.*, 1978; S. B. McMahon and Wall, 1983c, 1985a).

The axons of the SMT vary in size. Their conduction velocities have a wide range of values. The neurons studied by Menétrey *et al.* (1980) in the rat dorsal horn, which include SMT cells, have axons with conduction velocities of 3.6–40 m/sec; however, the cells of the lateral spinal nucleus have more slowly conducting axons, including unmyelinated ones (0.6–20 m/sec). The conduction velocities of the SMT axons studied by Yezierski and Schwartz (1986) in the cat range from 7.8 to 102.8 m/sec. However, cells in the superficial dorsal horn have slower axons (mean, 14.1 m/sec) by comparison with cells in laminae III–VI (mean, 43.6 m/sec) or VII–VIII (mean, 56.3 m/sec). The mean conduction velocity of the axons of a population of 24 SMT cells recorded by Yezierski *et al.* (1987) in the monkey is 47.8 m/sec.

The SMT cells in lamina I that project to the parabrachial region have axons with relatively low conduction velocities (1–18 m/sec), according to Hylden *et al.* (1985).

Projections of SMT

Spinomesencephalic axons ascend in the white matter of the ventral half of the spinal cord, in company with the spinothalamic and spinoreticular tracts (Mehler *et al.*, 1960; Kerr, 1975c). In addition, SMT cells in lamina I and the lateral spinal nucleus project through the dorsal lateral funiculus (Zemlan *et al.*, 1978; S. B. McMahon and Wall, 1983c, 1985a; Swett *et al.*, 1985; cf. M. L. Baker and Giesler, 1984).

The midbrain nuclei in which many of the axons of the SMT terminate in rats, cats, and monkeys can be subdivided into three groups by rostrocaudal level: (1) nuclei receiving SMT projections at the pontomesencephalic junction include the nucleus cuneiformis, the parabrachial nucleus, and the periaqueductal gray; (2) target nuclei in the intercollicular region include the nucleus cuneiformis, periaqueductal gray, intercollicular nucleus, and deep layers of the superior colliculus; (3) projections in the rostral midbrain are to the periaqueductal gray, the nucleus of Darkschewitsch, the anterior and posterior pretectal nuclei, the red nucleus, the Edinger-Westphal nucleus, and the interstitial nucleus of Cajal (Yezierski, 1988). The heaviest projections are contralateral ones. These findings are consistent with the observations made by different groups in studies on rats (Lund and Webster, 1967b; Antonetty and Webster, 1975; Mehler, 1969; Zemlan *et al.*, 1978), cats (Björkeland and Boivie, 1984; Wiberg and Blomqvist, 1984), and monkeys (Kerr, 1975c; Mehler *et al.*, 1960; Mehler, 1969; Wiberg *et al.*, 1987).

The SMT has at least a roughly somatotopic organization in that the projections from the lumbosacral enlargement to the caudal and middle zones of the midbrain tend to end more caudally than do the projections from the cervical enlargement; trigeminal projections are more rostral still (Fig. 9.39) (Wiberg *et al.*, 1987).

FUNCTIONAL PROPERTIES OF SMT CELLS

Responses to Natural Stimuli: Mechanical Stimulation of Somatic Structures

The rat SMT (and SRT) neurons described by Menétrey *et al.* (1980) include cells that respond only to innocuous mechanical stimulation (low-threshold cells). These neurons are in the nucleus proprius. Other cells are activated by innocuous mechanical stimuli but maximally

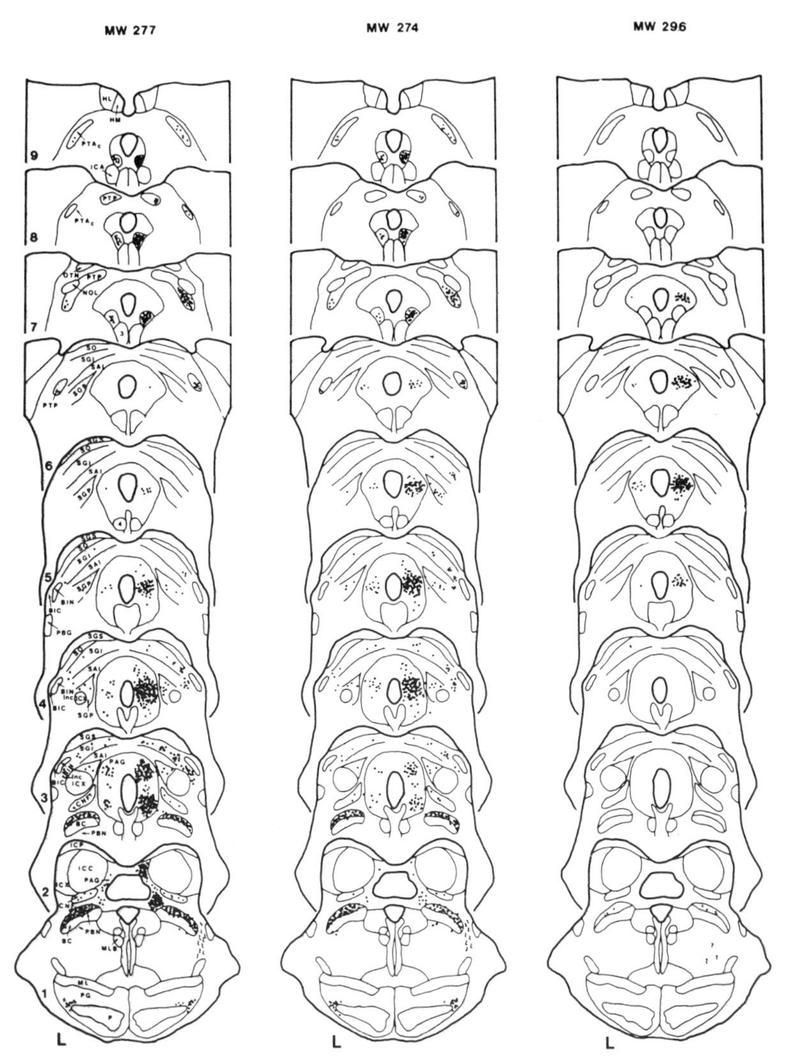

Fig. 9.39. Distribution of anterogradely transported WGA-HRP in the midbrain after injection into the lumbar enlargement (left column), cervical enlargement (middle column), and nucleus caudalis of the spinal trigeminal complex (right column). (From Wiberg *et al.*, 1987; reproduced with permission.)

by noxious mechanical stimuli (wide-dynamic-range cells). Most such neurons also respond to noxious heat. These cells are either in the nucleus proprius or in two cases the marginal zone. Another group of cells are excited just by noxious mechanical stimuli and not by innocuous mechanical stimuli (high-threshold cells). Some are also excited by noxious heat. These cells are in the nucleus proprius or marginal zone. Most of the neurons in the lateral spinal nucleus have poorly defined receptive fields, some apparently including deep tissues.

The SMT cells examined in the cat by Yezierski and Schwartz (1986) can generally be classified as wide-dynamic-range or high-threshold neurons. Many wide-dynamic-range cells have a convergent input from muscle or joints. Some wide-dynamic-range cells have restricted excitatory receptive fields, often with an additional extensive inhibitory receptive field; others have complex excitatory and inhibitory receptive fields, with excitatory inputs from several regions of the body. Some of these neurons show prolonged afterdischarges following noxious stimulation of several parts of the body or face (Fig. 9.40). High-threshold SMT cells include those with a restricted excitatory receptive field (and often an extensive inhibitory field), those with an inhibitory receptive field on the ipsilateral hindlimb (inverse high-threshold cells), and often an additional extensive inhibitory field, and high-threshold cells with complex excitatory and inhibitory fields. A few SMT cells respond only to tapping, presumably by activation of deep receptors, and a substantial number of cells are unresponsive to any of the stimuli tried.

Similar results have been obtained in the monkey by Yezierski *et al.* (1987). SMT cells that project to the thalamus as well as to the midbrain tend to have restricted excitatory receptive fields, whereas those projecting only to the midbrain have complex receptive fields.

Most of the lamina I SMT cells that project to the parabrachial region in the cat (including the nucleus cuneiformis and lateral PAG) are classified as high-threshold neurons; only a few are wide-dynamic-range cells (Hylden *et al.*, 1985, 1986a). The receptive fields are restricted in size. Recordings have been made from high-threshold neurons in the parabrachial region that may receive their input over the SMT (Bernard and Besson, 1990). Many of these parabrachial neurons can be activated antidromically from the central nucleus of the amygdala, suggesting that a spinopontoamygdaloid pathway exists that might mediate some of the motivational-affective aspects of the pain response.

DESCENDING CONTROL OF SMT NEURONS

The SRT neurons (including SMT cells) in the rat dorsal horn that have been studied by Menétrey *et al.* (1980) are powerfully inhibited following electrical stimulation in the nucleus raphe magnus or adjacent reticular formation. Similarly, the SMT cells investigated by Yezierski and Schwartz (1986) in the cat are inhibited following electrical stimulation in the PAG or the midbrain reticular formation.

Yezierski (1990) has recently described the effects on SMT cells of electrical stimulation in the midbrain and also at several locations in the medulla, including the nucleus raphe magnus, nucleus reticularis gigantocellularis, and nucleus reticularis magnocellularis. Excitation, inhibition, or excitation followed by inhibition is observed. The inhibition is mediated chiefly by projections descending in the DLF and the excitation by pathways in the ventral quadrant.

CONCLUSIONS

1. The sensory pathways that ascend in the white matter of the ventral quadrant of the spinal cord include the STT, the SRT, and the SMT. Some STT and SMT axons also ascend in the dorsal lateral funiculus.

2. The STT is present in elasmobranch fish, amphibia, reptiles, birds, and mammals. It is most highly developed in primates, including humans.

3. The STT originates from cell groups distributed along the entire length of the spinal cord. There is a particularly large concentration of STT cells in the uppermost cervical segments (apart from the lateral cervical nucleus). These cells include a large ipsilateral group in the ventral horn. There is a similar concentration of SRT and SMT cells in the upper cervical cord, suggesting the possibility that the same neurons project to a variety of brain structures.

4. Below the upper cervical segments, more than 90% of STT cells are contralateral to their thalamic targets.

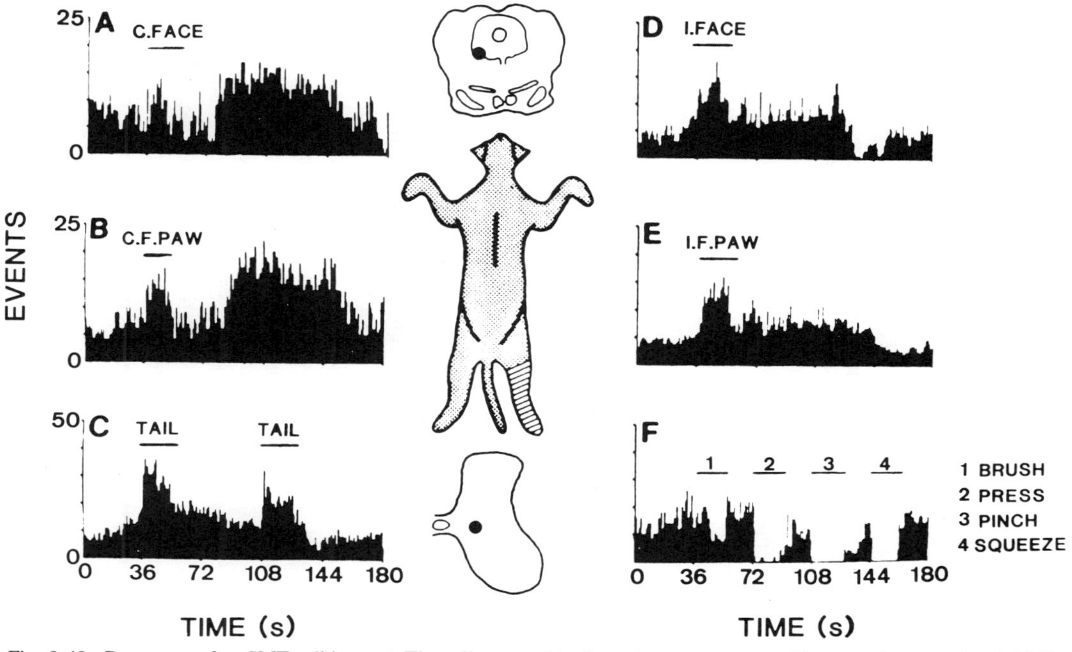

Fig. 9.40. Responses of an SMT cell in a cat. The cell was a wide-dynamic-range neuron with a complex receptive field. It was activated antidromically from the site indicated in the PAG, and was located in lamina VII. The receptive field is shown in the drawing: the hatched area is an inhibitory receptive field on the ipsilateral hindlimb; the stippled area is a large excitatory field. Responses to noxious squeezing of various parts of the body are shown in panels A–E. Note the prolonged afterdischarges in many cases. Inhibitory responses to graded intensities of mechanical stimuli on the ipsilateral hindlimb are shown in panel F. (From Yezierski and Schwartz, 1986.)

5. There is a controversy about whether STT cells of lamina I project largely through the DLF or the middle of the lateral funiculus to the VPL nucleus. There may be species differences in this regard. Lamina I STT cells also project to the CL nucleus (at least in monkeys) and the nucleus submedius. The latter projection in cats is from cells in the lateral part of lamina I.

6. Estimates of the total number of STT cells projecting to one side of the thalamus are 9500 in rats, 5000 in cats, and 18,000 in monkeys.

7. STT neurons include a variety of morphologic types, including small, medium, and large sizes and fusiform and polygonal shapes. They are contacted by several different types of synapses, distinguished by their content of round, flat, and dense-core synaptic vesicles. Some chemical substances contained in the synaptic endings on STT cells include SP, enkephalin, CGRP, serotonin, and dopamine-β-hydroxylase. Small subpopulations of STT cells contain any of a variety of peptides, including VIP, enkephalin, dynorphin, cholecystokinin, and galanin.

8. The axons of STT cells have a wide range of sizes, from unmyelinated (at least for some lamina I STT cells in cats) to large myelinated ones. The axons of STT cells in laminae IV and deeper decussate very near the soma and ascend in the ventral quadrant. The axons of some lamina I STT cells ascend in the DLF, but anterogradely labeled axons of lamina I STT cells tend to concentrate in the middle of the lateral funiculus.

9. The STT projects to several different thalamic nuclei. There is a somatotopic projection to the VPL nucleus in rats and in primates, including humans. However, there is only a sparse projection to the VPL nucleus in cats; instead, the STT ends in a shell region around the VPL nucleus. There are also terminations of the STT in the posterior complex, the intralaminar complex (especially the CL nucleus), the nucleus submedius and several other nuclei of the medial thalamus.

10. Some STT cells project to another target besides the thalamus, such as the rhombencephalic reticular formation, periaqueductal gray, or dorsal column nuclei. STT neurons may also project to more than one thalamic nucleus (e.g., VPL and CL).

11. STT cells show a background discharge. Several patterns of background discharge can be detected in recordings from STT cells of the monkey. The discharge may depend largely on intrinsic membrane properties of STT cells in laminae IV–VI.

12. Intracellular recordings show that the action potentials of STT cells have an inflection on the rising phase and a brief followed by a prolonged afterhyperpolarization. Excitatory and inhibitory postsynaptic potentials can be recorded from these cells. Isolated STT cells investigated by using the patch clamp technique have several voltage-dependent channels associated with action potentials, including a fast inward sodium current, voltage-dependent and calcium-dependent potassium currents, and one fast and two slow calcium currents.

13. STT cells can often be excited by volleys in Aβ, Aδ, and C fibers of cutaneous nerves. However, some are not activated by Aβ volleys. The excitation by Aδ and C fibers is produced by long-lasting excitatory postsynaptic potentials. Volleys in group II, III, and IV fibers of muscle nerves (and sometimes group I fibers) and volleys in Aδ and C fibers of visceral nerves can also excite some STT cells.

14. Most STT cells have cutaneous receptive fields. The receptive fields of primate STT cells in the dorsal horn of the lumbosacral enlargement and projecting to the VPL nucleus are restricted to the hindlimb, and those of STT cells in lamina I are smaller than those of STT cells in laminae IV–VI. However, STT cells in deeper laminae that project to the CL nucleus may have very large receptive fields, occupying much of the surface of the body and face.

15. The cutaneous receptive fields of STT neurons with cell bodies in laminae I–IV are somatotopically organized. Those of STT cells in lamina V and in deeper laminae do not seem to have a somatotopic organization.

16. Several classification schemes have been used to describe the responses of STT cells. Low-threshold cells respond best to tactile stimuli; wide-dynamic-range cells have a tactile input

but respond best to noxious stimuli; high-threshold cells respond only to noxious stimuli. Groups somewhat different in character from these have been distinguished on the basis of a multivariate statistical analysis. Class 1 cells respond best to tactile stimuli; class 2 cells respond weakly to innocuous and noxious stimuli, but equally well to intermediate and intensely noxious stimuli; class 3 cells are similar but respond best to the most intense stimuli.

17. STT cells often respond in a graded fashion to graded noxious thermal stimuli. Cells can be distinguished that have a steep stimulus-response curve and others that have a shallow curve until high intensity stimuli are used. The former may help in sensory discrimination, and the latter may signal damage.

18. Some STT cells in lamina I of the cat dorsal horn respond to input from warm or cold receptors. Most of these thermoreceptive neurons project toward the medial thalamus.

19. STT cells can be activated by chemical stimulation, such as the injection of capsaicin into the skin or bradykinin into the femoral artery.

20. Many STT cells can be activated by mechanical or chemical stimulation of muscle or visceral afferent fibers. Some of these inputs are clearly mediated by nociceptors, but for STT cells in Stilling's nucleus, the most likely receptors to be involved are muscle spindles in the muscles of the tail.

21. Plastic changes in the responses of STT cells can be demonstrated following damage to the skin. These changes parallel the development of primary and secondary hyperalgesia in humans and may serve as an animal model of these sensory phenomena.

22. STT cells commonly have inhibitory receptive fields. Extensive cutaneous inhibitory receptive fields are most easily demonstrated for STT cells of the wide-dynamic-range type. This pattern resembles that of the diffuse noxious inhibitory control system, but unlike the latter at least some of the inhibition of STT cells remains after spinal cord transection.

23. Inhibition of STT cells results from electrical stimulation of the dorsal columns or of peripheral nerves. The most effective peripheral-nerve volleys are those in Aδ fibers, although Aβ fibers can produce weak inhibition of STT cells and C fibers add to the inhibition caused by Aδ fibers. Visceral afferent volleys from segmentally distant regions of the body are also very effective in inhibiting many STT cells.

24. STT cells can be excited by iontophoretic application of glutamate and other excitatory amino acids, and such responses are facilitated by SP. STT cells are inhibited by GABA, glycine, serotonin, several catecholamines, enkephalin, cholecystokinin, phencyclidine, and SKF 10047. Mixed effects are produced by morphine and dynorphin.

25. STT cells are subject to powerful descending controls. They can be inhibited following electrical stimulation in the nucleus raphe magnus, medullary reticular formation, periaqueductal gray and midbrain reticular formation, parabrachial region, periventricular gray, VPL nucleus, and the primary (SI) sensory and posterior parietal cortex. STT cells are sometimes excited following stimulation in the medial brain stem, especially in the medullary reticular formation, and often following stimulation of the motor cortex or pyramid.

26. The part of the SRT that synapses in the medial rhombencephalic reticular formation is likely to be involved in sensory processing, unlike the lateral reticular nucleus, which is a precerebellar nucleus. There is also an SRT projection to the dorsal reticular nucleus, apparently providing nociceptive input to that nucleus.

27. The SRT is prominent in all classes of vertebrates examined.

28. The SRT originates from cells distributed along the length of the spinal cord. However, there is a prominent group of SRT neurons in the upper cervical segments, including a number in the ipsilateral ventral horn. A similar distribution is seen for the STT and SMT.

29. Below the upper cervical segments, most SRT cells are contralateral to their reticular formation targets (especially in the lumbar enlargement), and the majority are located in laminae V, VII, VIII, and X. It is presumably significant that few neurons in lamina I project in the part of

the SRT that distributes to the medial reticular formation. However, neurons in the ipsilateral lamina I do project to the dorsal reticular nucleus by way of a dorsally located pathway; SRT cells in deeper laminae of the dorsal horn, the ventral horn, and lamina X also project to the dorsal reticular nucleus. The dominant input is mediated by axons ascending in the ventral quadrant.

30. Most SRT cells are multipolar neurons. Some in laminae VII and X contain enkephalin.

31. The axons of SRT cells decussate near the somata. Most SRT axons are myelinated.

32. In mammals, the SRT projects to the central nucleus of the medulla, the lateral reticular nucleus, the nucleus reticularis gigantocellularis, nuclei pontis caudalis and oralis, nucleus paragigantocellularis, and nucleus subcoeruleus. There is no obvious somatotopic organization. Some SRT cells also project to the thalamus.

33. A large fraction of SRT cells exhibit background activity, but some are silent or discharge at a low rate.

34. There is a convergent input to these cells from axons in a variety of peripheral nerves, including a broad spectrum of sizes of cutaneous afferent fibers and high-threshold muscle afferent fibers. Aδ and sometimes C fibers in visceral nerves are effective in activating SRT neurons.

35. SRT neurons in the dorsal horn have restricted excitatory receptive fields. They may be low-threshold, wide-dynamic-range, or high-threshold cells. SRT neurons responsive to deep inputs are widely distributed in the gray matter. These neurons may respond to innocuous or noxious stimuli applied to deep receptors. Other SRT neurons have complex, extensive receptive fields. These cells are in deep laminae. Some SRT cells do not respond to any peripheral stimuli.

36. SRT neurons are under the influence of descending modulatory systems. Some SRT neurons are inhibited by stimulation in the reticular formation (or by stimulation of the vagus nerve). Others are excited by reticular formation stimulation. Such neurons may in turn excite other spinal cord neurons and thus account for part of the excitatory receptive fields of neurons such as the STT cells that project to the CL nucleus, SRT neurons with complex receptive fields, and many neurons receiving visceral afferent input. Alternatively or additionally, excitation of SRT neurons may reflect activity in spino-bulbo-spinal loops that produce a prolonged action in inhibitory circuits, such as the diffuse noxious inhibitory control system.

37. The SMT is present in all classes of vertebrates that have been examined.

38. The SMT originates from cells that are distributed along the whole length of the spinal cord. There is a prominent group of SMT cells in the uppermost cervical segments resembling similar groups of STT and SRT cells. These include many SMT cells that are ipsilateral to their midbrain projections.

39. Below the upper cervical cord, most SMT cells are located contralaterally to their projections in laminae I, V, VII, X and the nucleus of the DLF (lateral spinal nucleus) in rats; in laminae I, IV, and V in cats; and in laminae I and IV–VIII in monkeys. Lamina I SMT cells project to the parabrachial nuclei bilaterally and to the periaqueductal gray contralaterally, as well as to the nucleus of Darkschewitz.

40. The number of SMT cells that can be labeled from one side of the midbrain of monkeys is about 10,000 (75% contralaterally).

41. SMT cells can be small, medium, or large in size and fusiform, oval, round, or multipolar in shape. Lamina I SMT cells can be subdivided into a class of fusiform cells in the lateral one-third of the lamina, projecting to the parabrachial region, and a class of pyramidal cells scattered across the lamina, projecting to the periaqueductal gray.

42. Lamina I SMT cells projecting to the parabrachial region may contain dynorphin or enkephalin.

43. The axons of SMT cells may ascend in the ventrolateral white matter or, for lamina I cells, in the DLF. The axons have a wide range of sizes. Those from lamina I cells may be unmyelinated or myelinated; deeper cells give rise to myelinated axons.

44. The SMT terminates in the following nuclei of the midbrain: nucleus cuneiformis, parabrachial nucleus, periaqueductal gray, intercolliculus nucleus, deep layers of the superior colliculus, nucleus of Darkschewitz, anterior and posterior pretectal nuclei, red nucleus, Edinger-Westphal nucleus, and interstitial nucleus of Cajal. There is a rough somatotopic organization of the SMT: the projection from the cervical enlargement tends to end somewhat more rostrally than that from the lumbosacral enlargement. The parabrachial projection appears to be in part to neurons that in turn project to the central nucleus of the amygdala.

45. Some SMT cells also project to the thalamus.

46. SMT cells can be of the low-threshold, wide-dynamic-range, or high-threshold classes. Some have restricted receptive fields, whereas others have complex fields that include large areas of the surface of the body and face. The latter sometimes have long-lasting afterdischarges following noxious stimulation. Some SMT cells are unresponsive to peripheral stimulation. SMT cells that also project to the thalamus are likely to have restricted receptive fields, whereas those that project only to the midbrain tend to have complex receptive fields.

47. SMT cells can be inhibited or excited by stimulation in the nucleus raphe magnus, medullary reticular formation, PAG or midbrain reticular formation. Frequently, the inhibition is preceded by excitation.

10 The Sensory Channels

As discussed in Chapter 1, the mechanism for transmission of information concerning a specific modality or submodality of sensation can be called a sensory channel. A sensory channel would include a set of sensory receptors, spinal cord processing circuits, one or more spinal cord sensory pathways, and the parts of the brain (including the thalamus and cerebral cortex) that then use the information to produce perception. This chapter will consider what is known or can be deduced about the initial parts of the sensory channels for sensations arising from the body. Emphasis will be on evidence derived from experiments on monkeys and on human subjects.

MECHANORECEPTION

Two sensory channels concerned with mechanoreception give rise to the sensations of touch–pressure and flutter–vibration. It has been estimated that there are about 17,000 mechanoreceptors with myelinated afferent fibers in the glabrous skin of the human hand that are available for touch–pressure and flutter–vibration sensations (R. S. Johansson and Vallbo, 1979a, 1983). About half of these (44%) are slowly adapting, and about half (56%) are rapidly adapting. The slowly adapting (SA) receptors presumably correspond to the SA I and SA II mechanoreceptors (Merkel cell and Ruffini endings, respectively) and the fast-adapting (FA) ones to FA I (Meissner corpuscle) and FA II (Pacinian and paciniform corpuscle) endings (R. S. Johansson and Vallbo, 1979a; Schady *et al.*, 1983). All of these morphological receptor types are present in the glabrous skin of the human hand (M. R. Miller *et al.*, 1958). The hairy skin of humans would contain FA I (field receptors), FA II, SA I, and SA II receptors, in addition to hair follicle receptors (Vallbo *et al.*, 1979).

C mechanoreceptors do not contribute to mechanoreception in the distal extremities of primates, including humans, since these receptors do not appear to occur in this region (see Chapter 2). However, they have now been identified in the supraorbital nerve and so may contribute to sensation in the face (Nordin, 1990).

TOUCH–PRESSURE

The sensations of touch and pressure reflect a continuum of stimulus intensity, and so they can be considered as a single sensation of touch–pressure. The sensation meant here is one that has a low threshold and that continues as long as the stimulus is maintained (up to 1 or 2 min; Horch *et al.*, 1975). The tactile experiences associated with mechanical transients will be discussed in the section on flutter–vibration.

Receptors

SA I Receptors. The sensory receptors thought to signal information about the intensity and duration of prolonged mechanical contact of a stimulus with the skin are the SA I mechanoreceptors. SA II mechanoreceptors may also play a role, but this is less clear (see below). SA I receptors in humans are not always associated with recognizable tactile domes (K.R. Smith, 1970), and so they are sometimes difficult to visualize on the surface of the human skin; however, they do involve Merkel cell complexes (Pinkus, 1902; K. R. Smith, 1970).

Properties. Recordings have been made from the primary afferent fibers supplying SA I receptors in human skin by dissecting small filaments of the superficial radial nerve in volunteers (Hensel and Boman, 1960). Spotlike receptive fields are found, and the thresholds for activating the receptors are between 0.15 and 0.6 g, consistent with the thresholds for touch sensation in the same subjects (Fig. 10.1).

Other groups have recorded the activity of SA I afferent fibers in humans by using microneurography (see, e.g., Hagbarth *et al.*, 1970; Knibestöl and Vallbo, 1970; Gybels and Van Hees, 1971; Knibestöl, 1975; R. S. Johansson, 1976, 1978; Vallbo, 1981; Järvilehto *et al.*, 1976, 1981; see review by Vallbo *et al.*, 1979). The axons of SA I receptors in human glabrous skin have conduction velocities of 58.7 ± 2.3 m/sec (Knibestöl, 1975). The receptive fields average 44.7 mm^2 (range, 2–451 mm^2), and tend to be smaller on the distal phalanx than on the palm (Fig. 10.2A). Amplitude thresholds average 0.51 mm (range, 0.15–1.35 mm). There is usually only a single spotlike field, but occasionally there can be several low-threshold areas. Detailed maps of the receptive fields of SA I receptors show that these may have multiple low-threshold zones that presumably correspond to the locations of several Merkel cell complexes (R. S. Johansson, 1978).

Skin indentation evokes a dynamic followed by a static discharge (Knibestöl, 1975). Other features of SA I receptors in human skin include absence of spontaneous activity, a poor responsiveness to stretching the skin, high dynamic sensitivity, and a high variability of

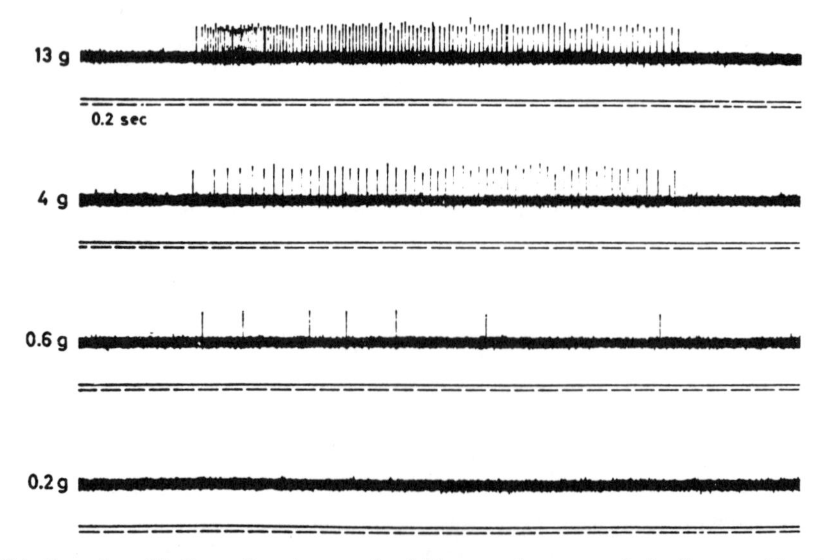

Fig. 10.1. Recording of discharges from the axon of an SA I receptor in an unanesthetized human subject. The unit was in the distal cut stump of a filament of the superficial radial nerve. The stimulus was indentation of the skin with a plastic rod with a diameter of 1 mm. (From Hensel and Boman, 1960.)

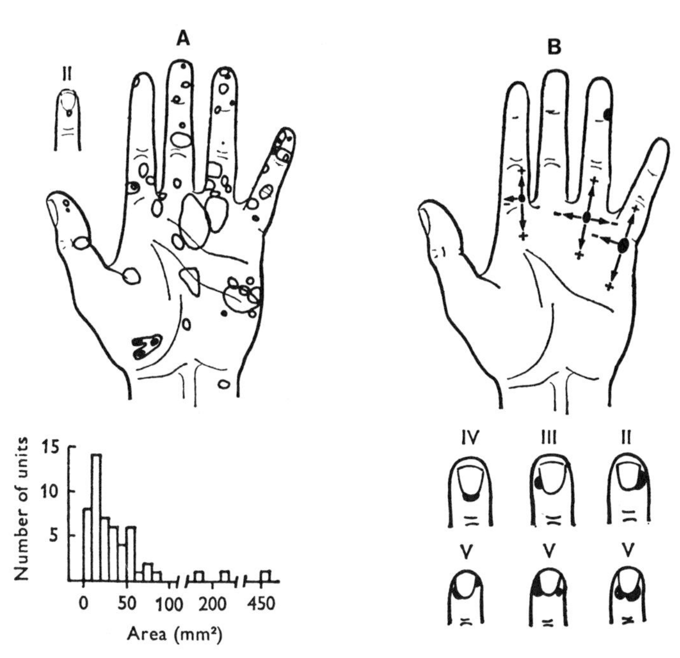

Fig. 10.2. Receptive fields of slowly adapting mechanoreceptors in human skin. The receptive fields in panel A are for SA I receptors, and the histogram shows the distribution of receptive field sizes for them. The receptive fields in panel B are for SA II endings. The arrows indicate responses to stretching (+, excitation; −, reduced firing). (From Knibestöl, 1975.)

interspike intervals during a maintained stimulus (Knibestöl and Vallbo, 1970; Knibestöl, 1975; Vallbo, 1981; Järvilehto *et al.*, 1981). The characteristics of SA I receptors in humans are thus very similar to those described in animals (see Chapter 2).

Sensation Evoked by SA I Receptors. Harrington and Merzenich (1970) have observed that the sensation of pressure is lost when the skin is anesthetized, suggesting that subcutaneous receptors are not involved in the sensation being studied. Rapidly adapting receptors cannot account for a maintained sensation of pressure, and in their experience SA I receptors do not produce any sensation in human subjects (see also Järvilehto *et al.*, 1976). Harrington and Merzenich conclude that SA II receptors must be responsible for pressure sensation (the sensation referred to here as touch–pressure).

However, in experiments in which the axons of identified receptor types in the glabrous skin of the human hand have been stimulated through a microneurography electrode (Torebjörk and Ochoa, 1980; Vallbo, 1981; see the review by Torebjörk *et al.*, 1987), it has been found that stimulation of SA I receptors often evokes a sensation of touch or pressure, whereas stimulation of individual SA II receptors usually does not produce any sensation (Vallbo, 1981; Ochoa and Torebjörk, 1983; Schady *et al.*, 1983; Vallbo *et al.*, 1984; Macefield *et al.*, 1990). Sensation evoked by activity in an SA I unit is continuous, not intermittent as when FA I or FA II units are stimulated (Ochoa and Torebjörk, 1983; see the section on flutter–vibration). Threshold for sensation is reached when the stimulus frequency is at least 3–10 Hz, and the intensity of the pressure sensation increases with the stimulus frequency (Ochoa and Torebjörk, 1983). Therefore, SA I receptors can be presumed to contribute to touch–pressure, and they may well be the only receptors to do so (however, see section on SA II receptors). Furthermore, the perceived intensity of pressure is a function of the discharge rate of the afferent fibers.

Intensity Coding. SA endings in primates have been found by Werner and Mountcastle

(1965) and Mountcastle *et al.* (1966) to respond to skin indentations of less than 10 μm. The stimulus-response functions of SA endings in hairy skin can be fitted by a power function with an exponent of less than unity. Werner and Mountcastle (1965) have plotted the mean stimulus–response curve for 10 SA fibers and find that the exponent for the best fitting power function is 0.52 (Fig. 10.3A). A calculation based on information theory suggests that the information transmitted by such fibers would allow the discrimination of about six or seven steps of stimulus intensity, an estimate that is consistent with values reported in human psychophysical experi-

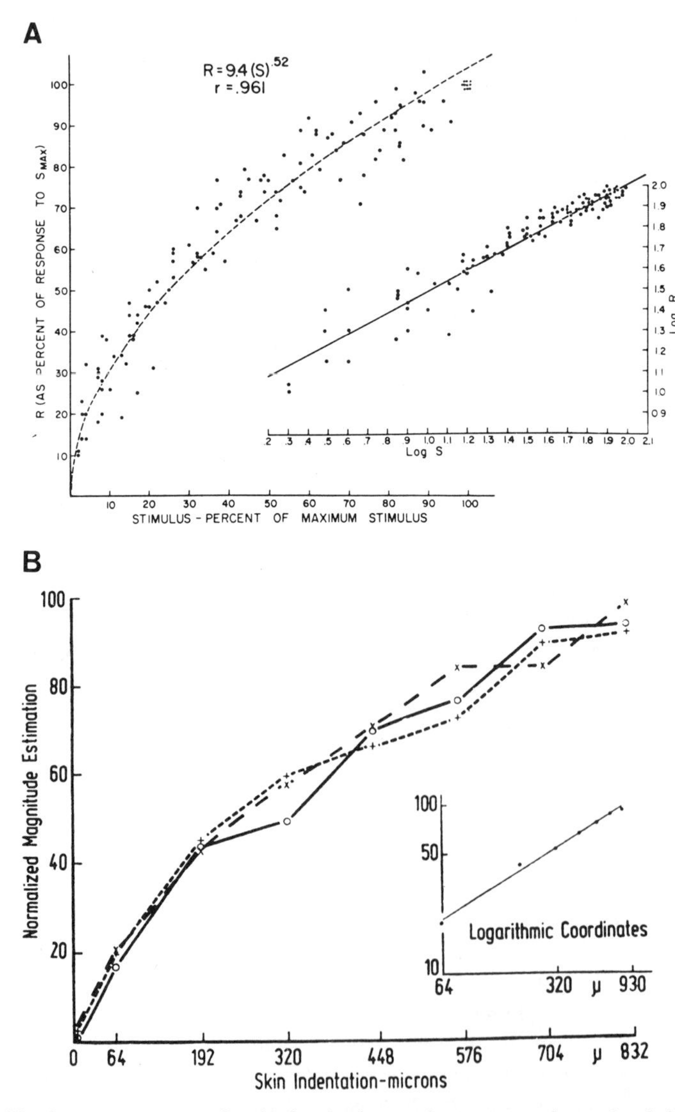

Fig. 10.3. (A) Stimulus-response curves for slowly adapting mechanoreceptors in monkey hairy skin. (From Werner and Mountcastle, 1965.) (B) Relationship between estimates of stimulus intensity and the amount of indentation of hairy skin in human subjects. Three different rates of skin indentation are shown to give similar results. The pooled results are plotted on logarithmic coordinates in the inset (slope = 0.59). (From Harrington and Merzenich, 1970.)

ments (Fig. 10.3B) (G. A. Miller, 1956; F. N. Jones, 1960). Werner and Mountcastle (1965) conclude that the overall neural transform between the level of the first order tactile neurons and the level of perception must be linear to give such good agreement between the stimulus-response properties of the receptors and sensory experience. Harrington and Merzenich (1970) agree with this conclusion.

Mountcastle *et al.* (1966) found that the stimulus-response functions of SA receptors of the glabrous skin in the monkey hand are linear (the exponents of power functions fitted to the data are approximately 1.0). The information transmitted by these fibers would permit the recognition of six to eight steps of stimulus intensity.

Harrington and Merzenich (1970) agree that a power function provides a good fit for the stimulus-response relationships of both SA I and SA II receptors in the hairy skin of monkeys (the ranges of values for the exponent vary from 0.35 to 0.77 for SA I receptors and from 0.39 to 0.75 for SA II receptors). When human subjects are asked to estimate the magnitudes of comparable stimuli, the human psychophysical curves are described by power functions with exponents of 0.4 and 0.9 for the hairy and glabrous skin, respectively (cf. Mountcastle *et al.*, 1966; Werner and Mountcastle, 1965).

Kruger and Kenton (1973) criticize some of the propositions put forward by Mountcastle and his associates. Kruger and Kenton find linear relationships for most of the slowly adapting receptors from hairy skin in cats. They contest the notion that the central nervous system necessarily operates linearly upon the data from slowly adapting cutaneous mechanoreceptors, and they give an example of an SA I fiber that is capable of signaling 17 or 18 levels of stimulus intensity (Fig. 10.4). Furthermore, they estimate that the channel capacity for SA I and SA II fibers can range as high as 42 discrete steps if increments of a single impulse are recognized by the central nervous system. Evidently, the information transmitted by SA receptors in monkeys is more than sufficient to account for the human sensation.

Several groups have described the stimulus-response curves of SA receptors in human skin and compared these with psychophysical magnitude estimation functions in the same or other human subjects. This approach has the advantage of avoiding the need for a cross-species comparison.

Gybels and Van Hees (1971) find that power functions provide the best fit for both the stimulus-response functions of SA receptors in human skin and magnitude estimation functions in the same subjects. The neural responses are linearly related to subjective estimates of the magnitude of sensation.

Järvilehto *et al.* (1976) describe the stimulus-response curves of SA I receptors in the hairy skin. The curves can be fitted with power functions having exponents of about 0.5 (cf. Werner and Mountcastle, 1965; Harrington and Merzenich, 1970). The exponents are similar to those determined for the amplitudes of somatosensory evoked responses and for subjective estimates of stimulus intensity in human subjects tested by using touch stimuli applied to the glabrous skin (Franzén and Offenloch, 1969; see also Harrington and Merzenich, 1970).

Knibestöl (1975) finds that the stimulus-response curves of SA I receptors in the glabrous skin of the human hand are fitted better with log tanh functions than with power functions. However, Knibestöl and Vallbo (1980) have reexamined the issue of intensity coding by SA I receptors in the human glabrous skin. They compare the stimulus-response functions of SA I receptors with plots of psychophysical magnitude estimates in the same subjects. The psychophysical functions are approximately linear (power function with an exponent near 1.0), as has been reported by Mountcastle *et al.* (1966) and by Harrington and Merzenich (1970). However, unlike the case in the monkey, the stimulus-response curves of human SA I receptors in glabrous skin can be described best by power functions that have on average an exponent of about 0.7. This difference in the power functions for psychophysical magnitude estimation and receptor stimulus-response curves implies that the central nervous system does not operate linearly on the

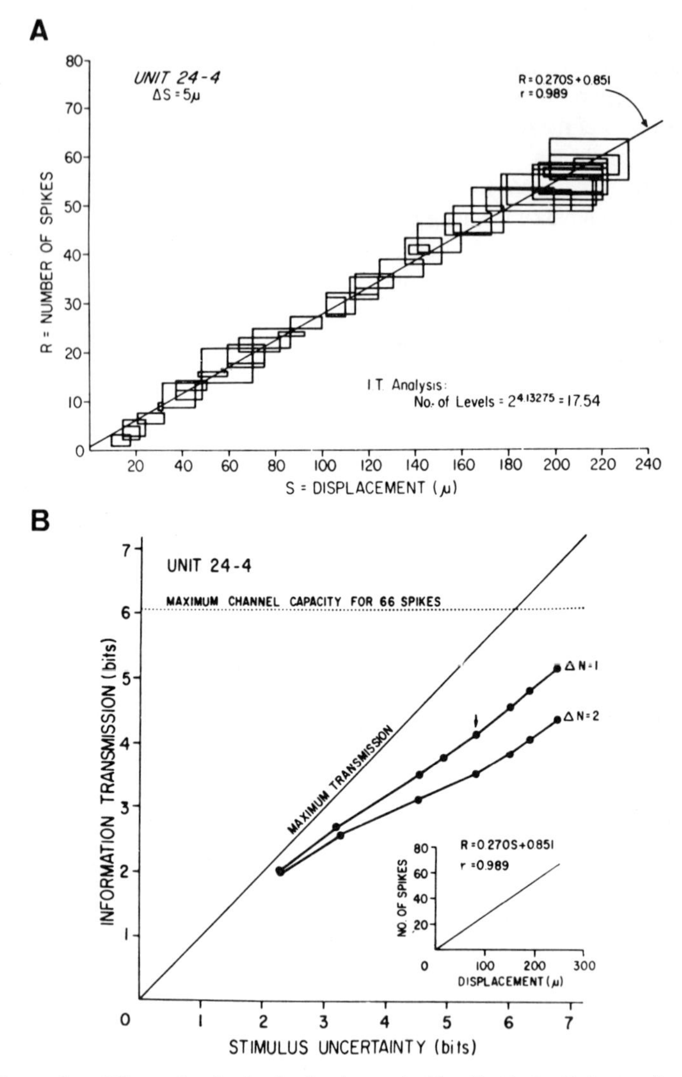

Fig. 10.4. (A) The number of discrete levels of activation that can be identified in the discharge of an SA I receptor, using a graphical solution. (B) Information theory analysis confirming that more than 17–18 discrete levels can be recognized. (From Kruger and Kenton, 1973.)

outputs of individual receptors. One speculation to account for the nonlinear transform is that increasing intensities of stimulation will recruit additional receptors.

Threshold for Contact Recognition. In hairy skin, a good correspondence is found between subjective thresholds for tactile stimuli and the thresholds for activating several classes of sensory receptors, including SA I endings (Hämäläinen and Järvilehto, 1981; Järvilehto *et al.*, 1981). Thus, SA I receptors in humans have the capability of providing information about the threshold for recognizing contact, at least in hairy skin. By contrast, contact recognition in glabrous skin appears to depend on rapidly adapting receptors (Fig. 10.5) (see section on flutter–vibration).

Two-point Discrimination. The receptive fields of SA I receptors are smallest on the

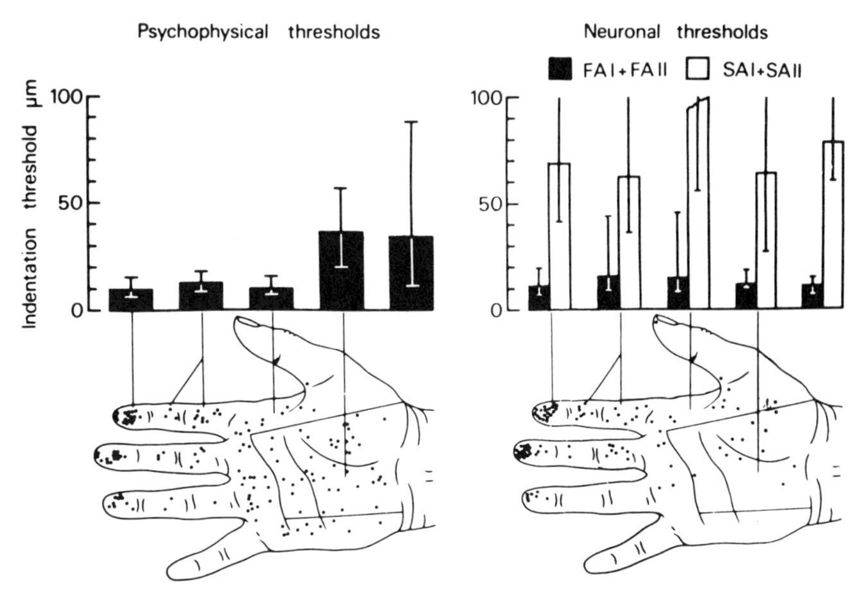

Fig. 10.5. (Left) Graph showing the gradient of psychophysical detection thresholds to triangular indentation of the glabrous skin of the hand. Areas tested are indicated on the drawing. (Right) Graph showing the thresholds of FA I, FA II, SA I, and SA II receptors on the glabrous skin. Since the FA units had similar thresholds, the results are pooled; similarly, the response thresholds of the SA units are pooled. Note that the FA receptors can account for the psychophysical thresholds, but the SA receptors cannot. (From R. S. Johansson and Vallbo, 1983.)

fingertips and larger more proximally (Fig. 10.2A). Projected receptive fields of afferent fibers evoking touch-pressure sensation when stimulated through a microneurography electrode usually coincide with the receptive fields of SA I units (Fig. 10.6) (Ochoa and Torebjörk, 1983; Schady and Torebjörk, 1983; Schady *et al.*, 1983). The gradient of receptive-field size parallels the gradient of the human capacity for tactile discrimination. Furthermore, the density gradient of SA I receptors (as well as of FA I receptors) in the glabrous skin of the human hand corresponds to the gradient of threshold for two-point discrimination (Fig. 10.7) (R. S. Johansson and Vallbo, 1979a, 1983). An analysis of the overlap of receptive fields of SA I and FA I receptors in the human hand indicates that these two classes of receptors can provide the information required for two-point discrimination (R. S. Johansson and Vallbo, 1980; see also Schady *et al.*, 1983). By contrast, the SA II and FA II receptors cannot.

Dynamic Sensitivity. SA I receptors have both a dynamic and a static response (see Chapter 2). In humans, SA I receptors can be entrained by vibratory stimuli, but they do not follow frequencies as high as 100 Hz (Järvilehto *et al.*, 1976). SA I units are most sensitive to frequencies of 2–32 Hz and could contribute to the detection of stimuli having frequency components below 4 Hz (R. S. Johansson *et al.*, 1982a). Their sensitivity to low frequency-vibratory stimuli is greatly enhanced when the edge of the probe of a stimulator is placed over the receptive field (Fig. 10.8) (R. S. Johansson *et al.*, 1982b; R. S. Johansson and Vallbo, 1983; cf. Phillips and Johnson, 1981a). Presumably, this enables SA I receptors to detect tactile contours during exploratory movements of the hand.

The responsiveness of SA I receptors to low-frequency vibration could imply a role for these receptors in flutter sensation. However, stimulation of single afferent fibers from SA I receptors through a microneurographic electrode has usually been found to evoke a sensation of weak, sustained pressure or touch on the skin surface, although sometimes no sensation is detected

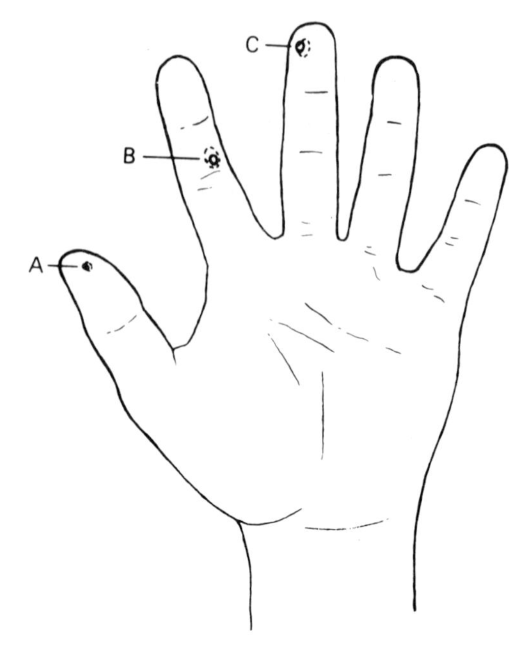

Fig. 10.6. Correspondence between the receptive fields of mechanoreceptors in the glabrous skin of the hand and the projected fields for sensation evoked by stimulation of the same afferent fiber through a microneurography electrode. The units whose fields are shown in B and C were SA I receptors, and that in A was an FA I unit. Stimulation of the SA I units produced a sensation of pressure and of the FA I unit intermittent tapping. Receptive fields are shown by circles and projected fields by dashed lines. (From Ochoa and Torebjörk, 1983.)

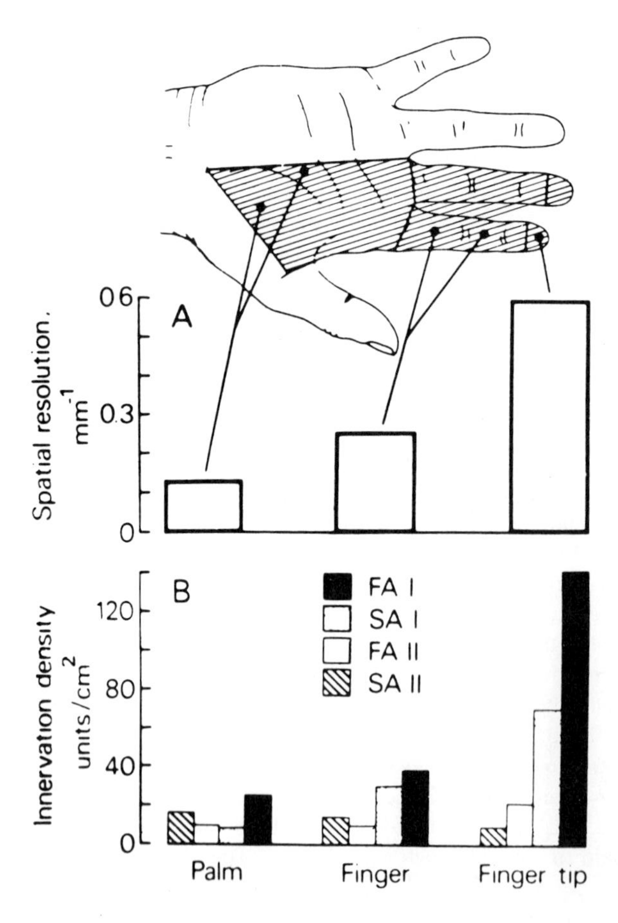

Fig. 10.7. Spatial resolution in a two-point discrimination test compared with innervation densities of four types of mechanoreceptor in the glabrous skin of the human hand. (A) Plot of the reciprocals of the thresholds for two-point discrimination on three parts of the hand. The threshold is lowest on the fingertip. (B) Plot of the densities of FA I, SA I, FA II, and SA II units in the same regions. (From R. S. Johansson and Vallbo, 1983.)

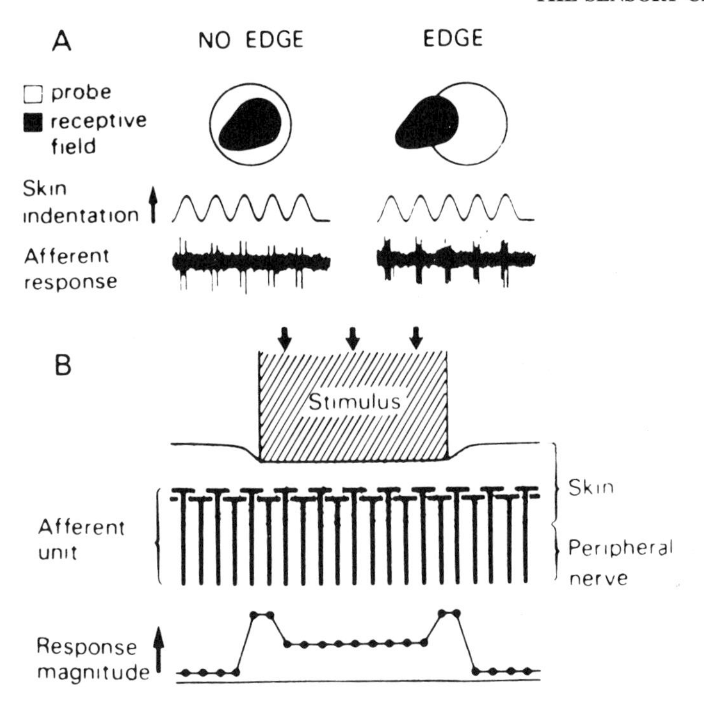

Fig. 10.8. Effects of an edge on the sensitivity of an SA I receptor in human skin. (A) Responses of an SA I unit to vibratory stimuli when the stimulator probe either covers the receptive field or crosses it (no edge versus edge conditions). (B) Plot of the responses of afferent units located outside of, at the edge of, or under a stimulus probe. (From R. S. Johansson and Vallbo, 1983.)

(Vallbo, 1981; Ochoa and Torebjörk, 1983; Schady *et al.*, 1983). No sensation of vibration or of movement is perceived even though the stimuli are repeated at 50 or 100 Hz (Vallbo, 1981).

Edge Detection. A detailed investigation of tactile spatial discrimination has been done by K. O. Johnson and J. R. Phillips (K. O. Johnson and Phillips, 1981; Phillips and Johnson, 1981a,b). Stationary stimuli are used to avoid possible confusion of spatial and temporal coding. In addition to two-point discrimination, these investigators have performed psychophysical tests of the ability of human subjects to discriminate gaps, gratings, and letters with the glabrous skin of the index finger (Fig. 10.9). The goal is to determine to what degree the results of a particular test depend on a spatial code rather than on an intensity code. The two-point discrimination task is a modification of the traditional one; there is no gap between the two stimulus probes, each of which has a diameter of 0.5 mm. Subjects presumably judge whether one or two stimuli are applied on the basis of the area of contact, and so the discrimination may depend just on an intensity code, such as the number of afferent fibers activated or the total number of action potentials elicited. In the gap test, the dimensions of the objects discriminated are the same, but the subject is asked to recognize the presence of a gap. Different-sized gaps were used to measure discrimination of gap size. The information used for this discrimination could have included responses enhanced by the presence of edges, as well as spatial information. In the grating task, the size of the object, the contact area, and edge content are constant and the subject is asked about the direction along which the grating is aligned. Although intensive information may result from the orientation of the edges, this is regarded as a test of spatial discrimination; however, see below for evidence from animal experiments. The final task, recognition of raised letters, is also considered to provide a test of spatial discrimination.

Phillips and Johnson (1981a) have recorded from mechanoreceptors innervating the

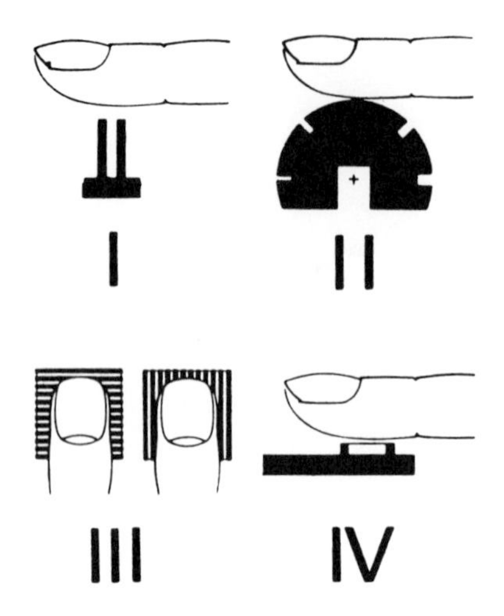

Fig. 10.9. Diagrams showing stimuli used in psychophysical tests of tactile spatial resolution. (I) Two-point discrimination; (II) gap detection; (III) grating resolution; (IV) letter recognition. (From K. O. Johnson and Phillips, 1981.)

glabrous skin (mostly digital) of monkeys and have used stimuli similar to those that were used in the human psychophysical study to evoke responses. The responses of SA receptors are found to be greatly enhanced when the receptive field contacts an edge and suppressed when it is placed across a gap (Fig. 10.10). For grids, the enhancement of the responses is greater when the bars are aligned with skin ridges than when they are at right angles. No evidence has been found for edge enhancement of FA I receptors. The results of the experiments with grids indicate that the psychophysical results obtained by using grid orientation could have reflected intensive cues rather than spatial ones, and so the only reliable test of spatial resolution among the four tests used is letter recognition. Phillips and Johnson (1981b) have developed a model of the skin that predicts the distribution of stresses and strains within the skin in relation to mechanoreceptor terminals. Satisfactory predictions of the behavior of SA receptors are obtained by using the profiles of maximum compressive strain and predictions for FA I receptors by using the maximum horizontal tensile strain.

Shape Discrimination. R. H. LaMotte and Srinivasan (1987) have studied the responses of SA receptors to passive movement of sinusoidally shaped plates. The responses of the SA receptors vary with the amount and rate of change of skin curvature, as well as the amount and velocity of indentation. It is concluded that the discharges of SA receptors contain enough information to allow the determination of the shape of an object's surface.

Precision Grip. Recordings have been made from the four types of mechanoreceptors in the human glabrous skin during a precision grip task (Westling and Johansson, 1984), which consists of lifting an object from a support, holding it up for a short period, and then replacing it (Westling and Johansson, 1987; R. S. Johansson and Westling, 1987). Input from cutaneous mechanoreceptors is necessary for maintaining an appropriate grip force to provide enough friction to prevent slippage without undue fatigue or the danger of breaking a fragile object (R. S. Johansson and Westling, 1984). Most SA I units discharge during the lift and continue to discharge during the loading and holding phases. They are silent during unloading, but some discharge during release. Presumably, the SA I afferent fibers contribute to the information needed by the central nervous system about contact with the lifted object (Westling and Johansson, 1987). When the object slips slightly, SA I units, as well as FA I and FA II units, but not SA II units, respond. Presumably, input from these receptors contributes to reflex adjustments preventing further slippage of the object being gripped (R. S. Johansson and Westling, 1987).

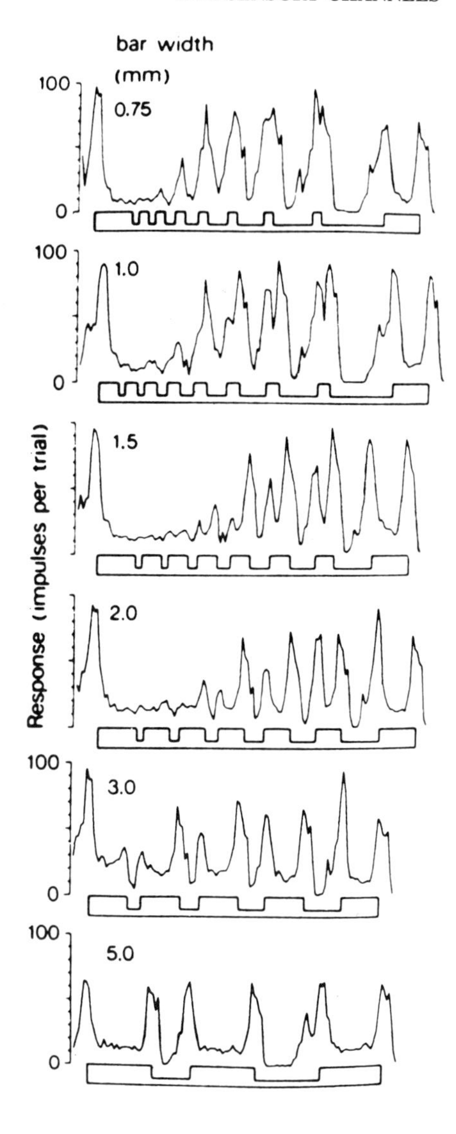

Fig. 10.10. Effects of bar width on the responses of an SA receptor in the glabrous skin of a monkey's hand. The bars were oriented in parallel to the ridges of the skin, and the skin was indented for 1 sec by 1 mm at each position of the bars, which were shifted in steps of 200 μm. (From Phillips and Johnson, 1981a.)

SA II Receptors. SA II receptors may play a rather different role than SA I receptors. They could be involved in touch–pressure, as suggested by Harrington and Merzenich (1970), but another hypothesis is that they contribute to position sense (Knibestöl and Vallbo, 1970; Knibestöl, 1975).

Properties. Recordings have been made from SA II receptors by microneurography (Vallbo and Hagbarth, 1968; Hagbarth *et al.*, 1970; Knibestöl and Vallbo, 1970; Vallbo, 1981). The characteristics of SA II receptors in humans are very similar to those in animals (see Chapter 2). These units may be silent or may have a background discharge. They respond not only to indentation of the skin but also to skin stretch (Fig. 10.2B). Some have receptive fields near joints or the fingernails (Fig. 10.2B, lower drawings) (R. S. Johansson and Vallbo, 1979a) and may respond to bending the neighboring joint (Knibestöl, 1975). They have a low dynamic sensitivity, and their discharges are regular. The stimulus-response curves resemble those of SA I units (Harrington and Merzenich, 1970; Knibestöl, 1975). SA II units in human glabrous skin respond to low-frequency vibration in a fashion that is comparable to the response of SA I units (R. S. Johansson *et al.*, 1982a). There appears to be only a single low-threshold region for each SA II axon, suggesting that an axon supplies only one receptor organ (R. S. Johansson, 1978).

Sensation Evoked by SA II Receptors. Stimulation of the axons of individual SA II receptors through microneurography electrodes usually does not produce a sensation (Ochoa and Torebjörk, 1983; Schady and Torebjörk, 1983; Schady *et al.*, 1983). Evidently, if SA II receptors contribute to conscious experience, spatial summation is needed. Recently, stimulation of two different SA II afferent fibers with receptive fields near nailbeds resulted in a sensation of joint movement, and stimulation of another SA II afferent with a receptive field in the webbing between the thumb and index finger caused a sensation of sustained pressure (Macefield *et al.*, 1990). These observations imply a role of SA II receptors in both proprioception and touch–pressure.

Two-Point Discrimination. There is not much of a density gradient for SA II receptors located on the distal versus the proximal hand (Fig. 10.7) (R. S. Johansson and Vallbo, 1979a, 1983). Furthermore, the receptive fields are very large. An analysis of the number and spatial distribution of the receptive fields of SA II units indicates that they would not be able to provide the information required for two-point discrimination (R. S. Johansson and Vallbo, 1980).

Precision Grip. SA II units discharge during a precision grip task, but they do not respond during the initial grip or during release (Westling and Johansson, 1987). The responses are consistent with some role in the reflex control of the grip. However, SA II units do not respond to slips during the grip task (R. S. Johansson and Westling, 1987).

Spinal Pathways. *Projections of SA I and SA II Primary Afferent Fibers.* Somc information about the spinal cord processing of information from SA I and SA II receptors is available from animal experiments. In cats, the primary afferent axons of SA I and SA II receptors project into the nucleus proprius (Fig. 10.11) (A. G. Brown, 1977; A.G. Brown *et al.*, 1978). SA I endings are distributed in a series of elliptical zones arranged rostrocaudally in a sagittal plane through laminae III, IV, and dorsal V (Fig. 10.12).

Spinal Cord Processing. Interneuronal responses to input from SA I receptors in cats have been investigated in Tapper's laboratory. Some interneurons discharge in response to the activity of a single SA I afferent fiber (Tapper and Mann, 1968; Tapper and Wiesenfeld, 1980; Tapper *et al.*, 1983). In fact, a single SA I fiber can produce excitation, inhibition, or combinations of these in dorsal horn interneurons (P. B. Brown *et al.*, 1973; cf. Craig and Tapper, 1985). Although some interneurons are affected just by SA I fibers, many receive a convergent input from hair follicle receptors (Tapper *et al.*, 1973).

Ascending Pathways. Transmission of touch–pressure information seems to involve pathways in the dorsal half of the spinal cord white matter (Fig. 10.13). Both the dorsal columns and the dorsal lateral funiculus (DLF) transmit information from SA receptors. The DLF is needed for the transmission of information from individual SA I receptors in the hindlimb required for a behavioral task in cats (Tapper, 1970; Mann *et al.*, 1972). Two-point discrimination in monkeys is unimpaired by a dorsal column lesion, but is lost following a combined lesion of the dorsal column and the DLF (Levitt and Schwartzman, 1966).

The fasciculus gracilis in the cat cervical spinal cord does not seem to contain collaterals of primary afferent fibers from SA I receptors (Petit and Burgess, 1968). However, the fasciculus cuneatus in cats and raccoons does (Uddenberg, 1968a; Pubols and Pubols, 1973; Bromberg and Whitehorn, 1974). Thus, the rostral fasciculus gracilis in carnivores differs from the fasciculus cuneatus in lacking slowly adapting primary afferent fibers from touch corpuscles, just as it lacks large primary afferent fibers from muscle stretch receptors (see Chapter 7 and the section on proprioception below). On the other hand, the fasciculus gracilis in cats does include the axons of neurons belonging to the postsynaptic dorsal column pathway that are activated by SA I receptors (Angaut-Petit, 1975a,b; A. G. Brown and Fyffe, 1981).

The pathway in the DLF that transmits information from SA I afferent fibers is likely to be

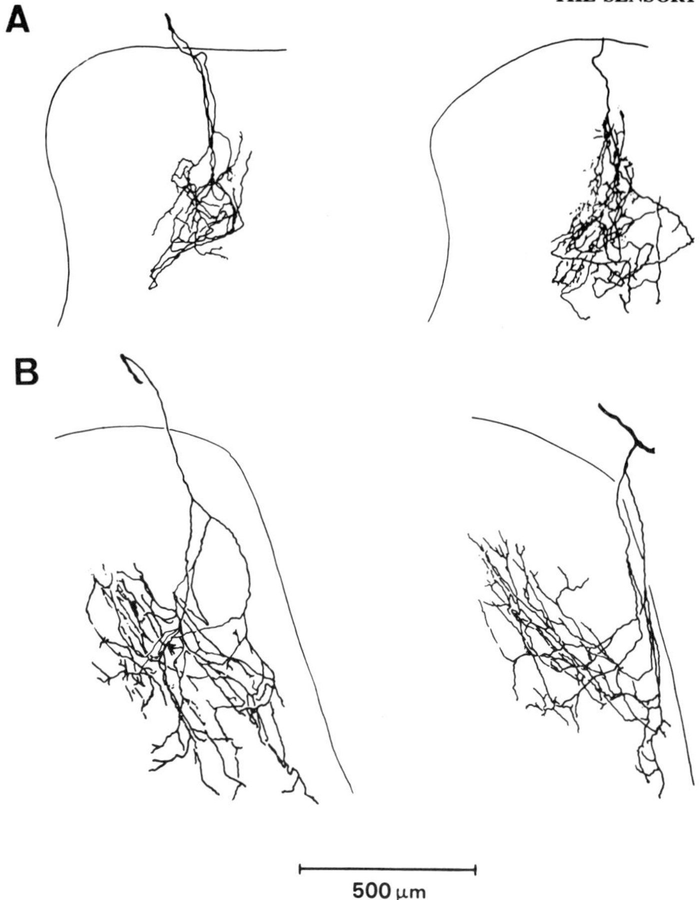

Fig. 10.11. The terminal distributions of two collaterals of an SA I afferent fiber (A) and of an SA II afferent fiber (B). (From Brown, 1977; reproduced with permission.)

the one that ascends to the dorsal column nuclei (see Chapter 7). However, this has not yet been firmly established (cf. Mann *et al.*, 1971). The spinocervical tract is not a candidate, since spinocervical tract cells in the cat do not respond to input from slowly adapting cutaneous mechanoreceptors (A. G. Brown and Franz, 1969).

It is unlikely that touch–pressure information can be transmitted through ventral pathways in cats. The spinothalamic tract in monkeys has been reported to include some neurons that show slowly adapting responses to cutaneous stimulation (Willis *et al.*, 1974). However, input to such cells from SA I or SA II afferent fibers has never been satisfactorily demonstrated, and slowly adapting responses in spinothalamic neurons are more likely to be mediated by nociceptors (Willis *et al.*, 1975).

The fact that information from SA I and SA II receptors reaches the dorsal column nuclei and activates neurons there has been documented for cats and monkeys, and at least some of these neurons of the dorsal column nuclei project to the contralateral ventrobasal complex (see Chapter 7).

Clinical Correlations. Patients with lesions of the posterior columns whose cases are reviewed by Nathan *et al.* (1986) had disturbances of two-point discrimination and difficulty in judging the intensity of pressure stimuli. They had severe deficits in stereognosis and in graphesthesia. These deficits may have been due in part to interruption of the touch-pressure

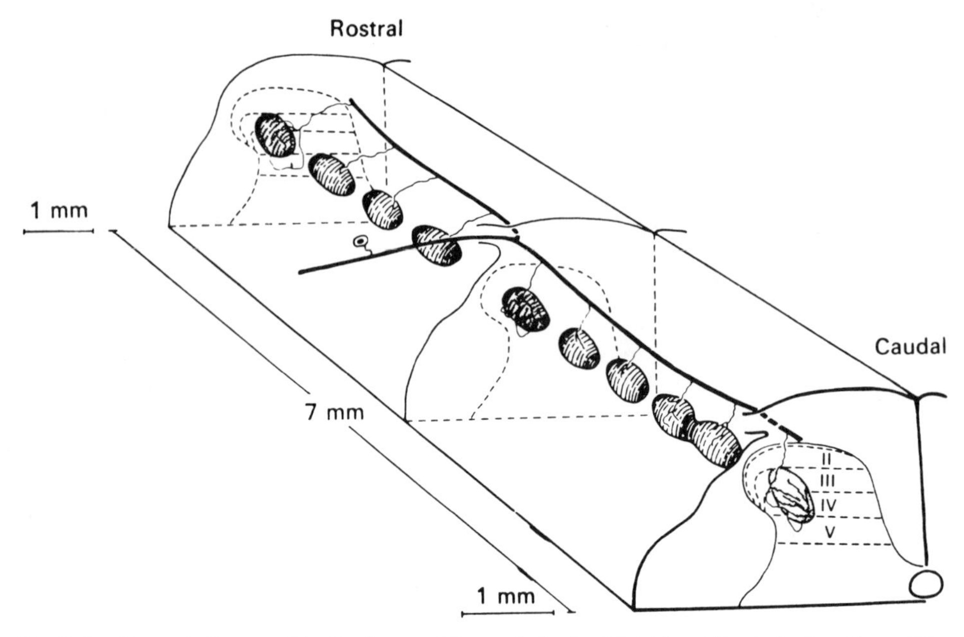

Fig. 10.12. Schematic diagram of the distribution of the terminals of an SA I afferent fiber in the dorsal horn of the cat spinal cord. (From A. G. Brown *et al.*, 1978.)

channel. The patient of Noordenbos and Wall (1976) with a transection of all of the spinal cord except for one anterolateral quadrant could still sense mechanical contact bilaterally. However, this probably depended upon rapidly adapting mechanoreceptors, since the stimuli were delivered by von Frey hairs.

Summary

In summary, the sensation of touch–pressure seems to depend chiefly upon input to the central nervous system from SA I receptors. It cannot be ruled out that SA II receptors also contribute, but there is insufficient evidence to affirm this; alternatively, SA II receptors may be proprioceptors. The intensity of touch–pressure sensation is a function of the firing rate of SA I afferent fibers. In the glabrous skin of monkeys, the stimulus-response relationship is linear; in hairy skin, the stimulus-response relationship may be described by a power function or a log tanh function. However, in the glabrous skin of the human hand, the relationship is a power function

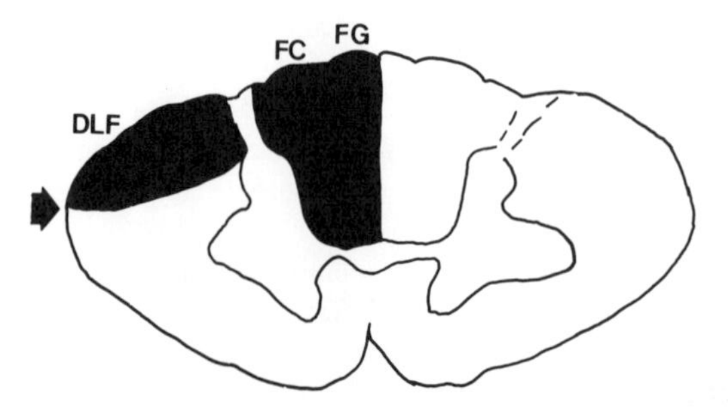

Fig. 10.13. The blackened areas show the locations of the pathways that appear to mediate touch–pressure. The arrow at the left indicates the side of the input. The pathways include the fasciculus gracilis and cuneatus (FG, FC) and the dorsal lateral funiculus (DLF).

with an exponent of about 0.7. Psychophysical magnitude estimate curves are linear for the glabrous skin, suggesting that there is a central transformation of the input from individual SA I receptors. The dynamic component of the responses of SA I receptors corresponds to an enhanced responsiveness when the edge of a stimulus crosses the spot-like receptive field. This property enables SA I receptors to play a major role in two-point discrimination and in the analysis of spatially patterned stimuli, such as gaps, gratings and raised letters. The high density of these receptors, as well as of FA I receptors, on distal glabrous skin appears to be related to the gradient of thresholds for two-point discrimination in the hand. Threshold for recognition of contact with the hairy skin may be signalled in part by SA I receptors, but it is likely that FA I receptors are responsible for this capability in glabrous skin. In cats, SA I receptors in the hindlimb do not appear to project directly to the dorsal column nuclei, but rather relay in the dorsal horn. However, there is probably a direct projection from SA I receptors of the upper extremity to the dorsal column nuclei. Information from SA I receptors of the hindlimb (at least in cats) reaches the dorsal column nuclei through the postsynaptic dorsal column pathway and probably also through a second-order pathway in the DLF. The spinocervical tract (at least in cats) does not seem to convey inputs from slowly adapting cutaneous mechanoreceptors. In primates, information from slowly adapting cutaneous mechanoreceptors is also transmitted through both the dorsal columns and the DLF. However, in humans the dorsal columns appear to be the most important route for touch–pressure information. There is no compelling evidence that this information is conveyed by pathways in the anterolateral quadrants.

FLUTTER–VIBRATION

Another channel involved in mechanoreception is that responsible for the sensations of flutter and vibration. When an oscillating stimulus is applied to the human skin, which of these two different sensations are felt depends upon the frequency of the oscillation. In the range of 5–40 Hz, the sensation is described as "flutter," whereas oscillations at frequencies above 60 Hz produce a sensation of "vibration" (W. H. Talbot *et al.*, 1968). Although these sensory experiences can be separated, they merge psychologically into a frequency continuum and so can be referred to as flutter–vibration. The threshold for evoking these sensations is low, and the sensation adapts rapidly if the stimulus is maintained.

Receptors

FA I and FA II Receptors. It is clear from human psychophysical studies that two different receptor populations are responsible for the sensations of flutter and of vibration. The sense of flutter is localized accurately to an area of skin, and the threshold is elevated by an order of magnitude when the skin is anesthetized. Vibration is poorly localized to deep tissues and remains intact after the skin is anesthetized (W. H. Talbot *et al.*, 1968). Therefore, the receptors that are responsible for flutter found within the skin, whereas those responsible for vibration are in tissue deep to the skin. For both sensations, the thresholds are sufficiently low to make it evident that sensitive mechanoreceptors must be involved.

Monkeys can be trained to recognize stimuli comparable to those that give rise to flutter–vibration sensation in humans, and the psychophysical curves relating detection threshold to frequency are identical in monkeys and humans (Mountcastle *et al.*, 1972). The ability of monkeys and humans to discriminate between sinusoidal stimuli of different amplitudes is also similar (R. H. LaMotte and Mountcastle, 1975). Detection appears to be related to the presence of a stimulus that exceeds threshold for the afferent fibers, and recognition of stimulus frequency depends on stimuli that can entrain the discharges of the afferent fibers. There is an "atonal" interval between the detection and entrainment thresholds.

The most obvious types of receptors that are candidates for the sense of flutter are the rapidly adapting sensitive mechanoreceptors, i.e., FA I receptors (Meissner's corpuscles in glabrous skin and hair follicle and field receptors in hairy skin). Slowly adapting mechanoreceptors have also been considered, since these have discharge patterns with a velocity-sensitive component; however, stimulation of the axons of SA I afferent fibers evokes a sensation of touch–pressure and not flutter–vibration (see above). The obvious candidate receptor for vibratory sensation is the FA II receptor (Pacinian corpuscle).

Lindblom and Lund (1966) have confirmed that FA I receptors are located within the skin and that FA II receptors are located in subcutaneous tissue of monkeys. The evidence that FA I receptors in glabrous skin correspond to Meissner's corpuscles and FA II receptors to Pacinian corpuscles is summarized by Lindblom and Lund (1966) and Knibestöl (1973). Bolton *et al.* (1966) have made counts of the numbers of Meissner's corpuscles in human glabrous skin.

Properties. A number of investigators have recorded from the primary afferent fibers supplying rapidly adapting receptors in humans (see, e.g., Hensel and Boman, 1960; Hagbarth *et al.* 1970; Knibestöl and Vallbo, 1970; Knibestöl, 1973). Knibestöl (1973) has studied the responses of FA I receptors in the glabrous skin of the human hand and finds that these receptors have small receptive fields with relatively distinct borders (Fig. 10.14A). The axons of FA I

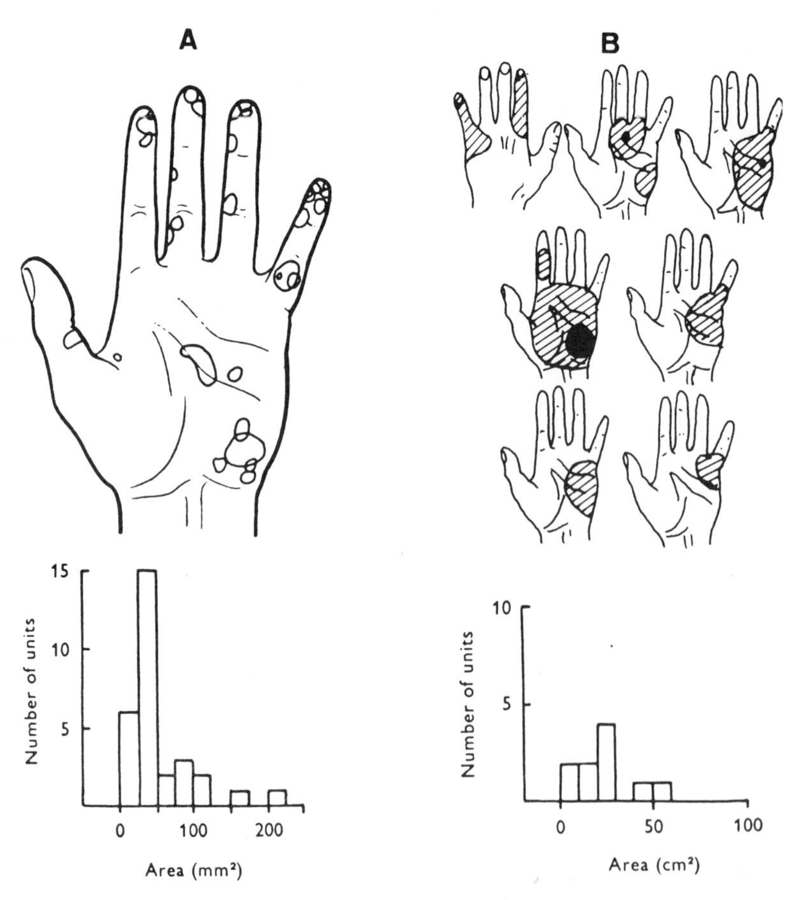

Fig. 10.14. The receptive fields in panel A are for FA I receptors (Meissner's corpuscles) in the glabrous skin of the human hand, whereas those in panel B are for FA II receptors (Pacinian corpuscles). Note the changes in scale for the abscissas in the graphs. (From Knibestöl, 1973.)

receptors have an average conduction velocity of 55.3 ± 3.4 m/sec. The receptive fields average 54.9 ± 8.6 mm^2. The fields are smaller on the distal phalanx than on the palm. Amplitude thresholds are lower than 0.5 mm and sometimes lower than 0.1 mm. Velocity thresholds are 0.4–39.3 mm/sec. Knibestöl (1973) has also described the responses of FA II receptors in the glabrous skin of the human hand. The conduction velocities of the axons from FA II receptors average 46.9 ± 3.5 m/sec. The receptive fields are large, ranging from 4 to 44 cm^2, and have indistinct borders (Fig. 10.14B). This is an order of magnitude larger than the receptive fields of FA I receptors. Amplitude thresholds are commonly 50–250 μm, which is lower than for most FA I receptors, but velocity thresholds resemble those of FA I receptors. The responses of FA II receptors to step indentations consist of both on and off discharges (Fig. 10.15). By contrast, only half of the FA I receptors show off responses. Stimulus-response curves show a clear relationship between the velocity of the indentation and the frequency of discharge for only some of the units. The best fit for some units is a log tanh function and for others a log function. As mentioned earlier, Pacinian corpuscles are better suited to signal stimulus acceleration than stimulus velocity (see Chapter 2).

Careful mapping of the receptive fields of FA I receptors often shows that a given field may have several low-threshold zones. These probably correspond to sets of Meissner's corpuscles that are supplied by the same axon (R. S. Johansson, 1978). The receptive fields of FA II units have just a single low-threshold zone, corresponding to the innervation of single Pacinian corpuscles by individual axons.

Rapidly adapting receptors in the hairy skin of the human hand have also been studied (Järvilehto *et al.*, 1976). Conduction velocities of the axons of a small sample of fibers average 28.6 m/sec. The receptive fields in hairy skin are larger than in glabrous skin, in the order of several square centimeters. The best stimulus is stroking, and one fiber was found that responded preferentially to stroking in one direction. Thresholds range from 0.3 to 1.7 g. These are generally higher than the thresholds for SA receptors in the human hairy skin. Movement of hair evoked a burst discharge and caused a weak tactile sensation (see also Järvilehto *et al.*, 1981).

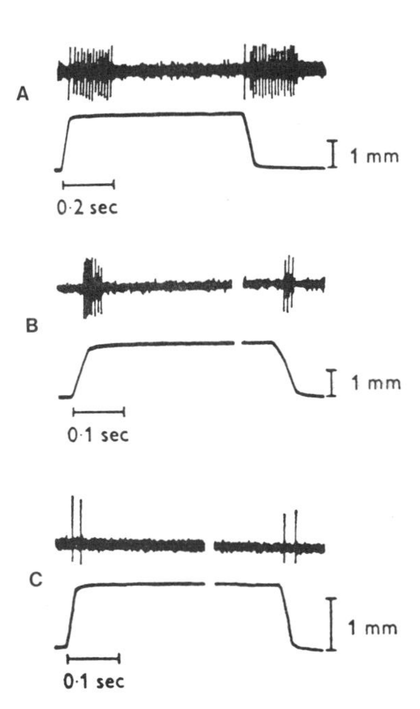

Fig. 10.15. Responses of FA II receptors in the human glabrous skin to the onset and also the termination of step indentations of the skin. (From Knibestöl, 1973.)

Stimulus-response functions have been plotted for several fibers and were found to be linear or log functions. Vibratory stimuli entrained some of the receptors. One followed 40–400 Hz stimuli and evidently supplied a Pacinian corpuscle. Another responded best to low frequencies of vibration and so was presumably an FA I receptor.

R. S. Johansson *et al.* (1982a) have been able to record from two classes of rapidly adapting receptors in the glabrous skin of the human hand which correspond to FA I and FA II receptors. The FA I receptors respond best to sinusoidal stimuli in the range of 8–64 Hz, but some are activated even at 400 Hz. FA II receptors respond best at frequencies of 128 to 400 Hz, but strong stimuli can activate them even at low frequencies. Both kinds of unit tend to discharge both on indentation and retraction, but FA II units often respond better on retraction of the stimulus probe.

Sensations Evoked by FA I and FA II Receptors. According to Vallbo (1981), stimulation of the axons of individual rapidly adapting receptors of the glabrous skin in humans through a microneurography electrode results in a sensation variously described as touch, pressure, vibration, or tickle. However, in later studies (Torebjörk and Ochoa, 1980; Schady *et al.*, 1983; Vallbo *et al.*, 1984) the subjects have reported sensations of tapping, flutter, or vibration, depending upon the stimulus frequency. The projected receptive fields of the sensation correspond in location to the receptive fields of rapidly adapting receptors (Fig. 10.6A) (Ochoa and Torebjörk, 1983; Schady *et al.*, 1983). However, the projected fields are generally somewhat smaller than the receptive fields, perhaps because of central processing (Schady and Torebjörk, 1983; Vallbo *et al.*, 1984).

Ochoa and Torebjörk (1983; see also Macefield *et al.*, 1990) find that stimulation of units identified as FA I receptors produces a sensation of intermittent tapping when the rate of stimulation is low (1–10 Hz) and a sensation of flutter–vibration when the rate is high (above 100 Hz). The intensity of sensation does not change as the rate is increased until a frequency of 100 Hz is reached; above this level, the sensation no longer changes in frequency but instead increases in intensity. Single-shock stimulation of FA I units supplying the fingertip can evoke a sensation of a tap, but no sensation is produced when FA I units supplying the palm are stimulated until the rate of stimulation exceeds 5 Hz. Stimulation of FA II units usually results in a sensation of vibration, provided that the rate exceeds 10–80 Hz. Lower rates fail to evoke a sensation. As the frequency increases, so does the perceived frequency up to 200–300 Hz. In a few cases, stimulation of FA II units causes a tickling sensation rather than vibration. Schady and Torebjörk (1983) find that stimulation of the axons of rapidly adapting receptors innervating the hairy skin often produces a buzzing sensation.

Frequency Coding. W. H. Talbot *et al.* (1968) compare the curve relating threshold for flutter–vibration sensation to stimulus frequency in the glabrous skin of the human hand with the tuning curves of "quickly adapting" and Pacinian corpuscle afferent fibers of the monkey hand (Fig. 10.16). The tuning curves consist of plots of thresholds for one-to-one entrainment of a receptor at different frequencies of sinusoidal stimulation. The quickly adapting (Meissner's) receptors have best frequencies of about 30 Hz (20–40 Hz), whereas the Pacinian corpuscles have best frequencies of about 250 Hz. The tuning curves for the quickly adapting receptors overlie the low-frequency end of the curve based on human psychophysical data (Fig. 10.16A), whereas the tuning curves for the Pacinian corpuscles overlie the high-frequency end (Fig. 10.16B). This finding suggests that flutter in the glabrous skin results from input to the central nervous system from Meissner's corpuscles and vibration from Pacinian corpuscles. There is also a sense of "roughness" at amplitudes of vibration below those that produce one-for-one entrainment of the receptors but that do produce phase locking of responses to the stimuli. Frequency or pitch discrimination is poor in human skin, presumably because individual members of the receptor population have similar tuning curves. The significant signal is location, rather than frequency.

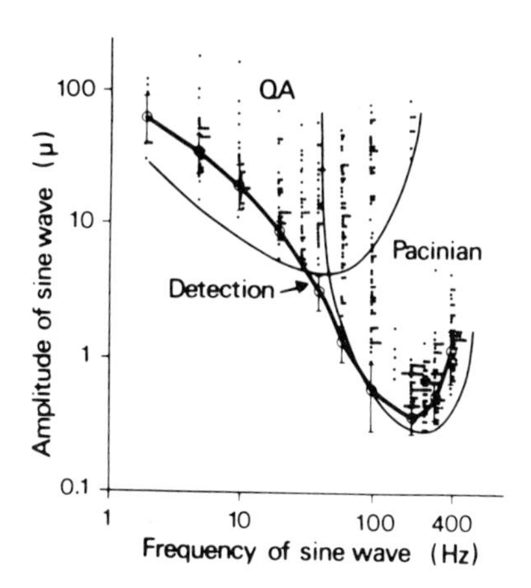

Fig. 10.16. The heavy line shows the frequency–threshold function for the detection of sinusoidal stimuli applied to the glabrous skin of monkeys. The dots show the thresholds of FA I (QA) and FA II (Pacinian) afferent fibers innervating the same region of skin. The thin lines show the boundaries for the FA I and FA II responses. (From Mountcastle *et al.*, 1972.)

Merzenich and Harrington (1969) have extended the findings of W. H. Talbot *et al.* (1968) to the hairy skin. Thresholds for vibration applied to the distal ventral forearm are higher than on glabrous skin, and the minimum threshold is at 200 Hz, which is a lower best frequency than that for glabrous skin. Exploration of the hairy skin reveals low-threshold spots over hair follicles; furthermore, oscillatory stimuli at 100 Hz applied to touch corpuscles do not produce a sensation of vibration. The responses of both rapidly and slowly adapting afferent fibers of the hairy skin have been investigated. G_2 hair follicle afferent fibers can be excited by the vibratory stimuli, whereas G_1 receptors cannot. D hair follicle receptors are quite sensitive to vibratory stimuli over a wide range of frequencies. Pacinian corpuscles behave like those under glabrous skin. Tuning curves show the following best frequencies: G_2 hair follicle receptors, 10–40 Hz; Pacinian corpuscles, 100–400 Hz; D hair follicle receptors, 5–100 Hz. It is proposed that G_2 hair follicle receptors largely account for flutter and Pacinian corpuscles for vibration in hairy skin. The D hair follicle receptors have a much lower threshold than is seen for flutter sensation, and so they presumably do not contribute to this sensation. Similarly, it has been found that the discharges of slowly adapting receptors are modulated by the sinusoidal stimuli used at intensities below the threshold for flutter sensation. Therefore, these receptors can be discounted as candidates to explain this sensation. Furthermore, slowly adapting receptors in humans do not follow high-frequency stimuli that elicit a sensation of vibration (Järvilehto *et al.*, 1976).

Mountcastle *et al.* (1972) compare the thresholds for recognition of various frequencies of sinusoidally varying stimuli in monkeys and humans, and they compare the psychophysical thresholds with the thresholds for entraining FA I and FA II units recorded in the monkeys. Detection thresholds are similar for monkeys and humans over the range of 2–400 Hz. The best frequencies are 200–250 Hz. Two thresholds are recognized for activity in afferent fibers: one at which sinusoidal stimuli produce any activity and one at which the responses are entrained (tuning threshold). The tuning thresholds of FA I units account for the sensations evoked by stimuli of 2–40 Hz and the FA II units for those produced at 60–400 Hz. The activity produced by stimuli below the tuning threshold evokes a sensation of mechanical contact but not a sensation of flutter–vibration (cf. R. H. LaMotte and Mountcastle, 1975).

Lofvenberg and Johansson (1984) describe a third sensation at frequencies of stimulation below 5 Hz. This is an awareness of slow oscillations of the skin, which they attribute to the activity of SA I receptors (see the section above on dynamic sensitivity of SA I receptors). They agree that flutter sensations at stimulus rates of 5–60 Hz and vibration sensations at rates above 60 Hz are attributable to FA I and FA II receptors, respectively.

Intensity Coding. A major question left unanswered by W. H. Talbot *et al*. (1968) is how stimulus intensity is encoded in the flutter–vibration channel. Intensity of sensation has been found to vary linearly with the amplitude of the stimulus in human psychophysical experiments, yet the discharges of individual receptors increase along a discontinuous curve. One possibility is that the number of receptors that participate in the response grows in proportion to stimulus intensity.

Knibestöl (1973) has observed that FA I receptors respond to ramp stimuli with a series of discharges that can be constant for all velocities tried or that, in some cases, can actually decrease with higher velocities (Fig. 10.17). The stimulus-response relationships of these receptors are best described by a log tanh function, rather than by a power function. As mentioned above, the Mountcastle group finds a discontinuous relationship between the amplitude of indentation and the number of spikes generated per cycle of sinusoidal stimuli (W. H. Talbot *et al*., 1968; Merzenich and Harrington, 1969; K. O. Johnson, 1974). However, this is not inconsistent with the results of Knibestöl (1973), who uses ramp stimuli. Knibestöl (1973) suggests that the central nervous system would have to determine the discharge rate of the first few spikes evoked by a ramp stimulus in order to determine stimulus velocity.

The question of how the receptors for flutter–vibration encode stimulus intensity is

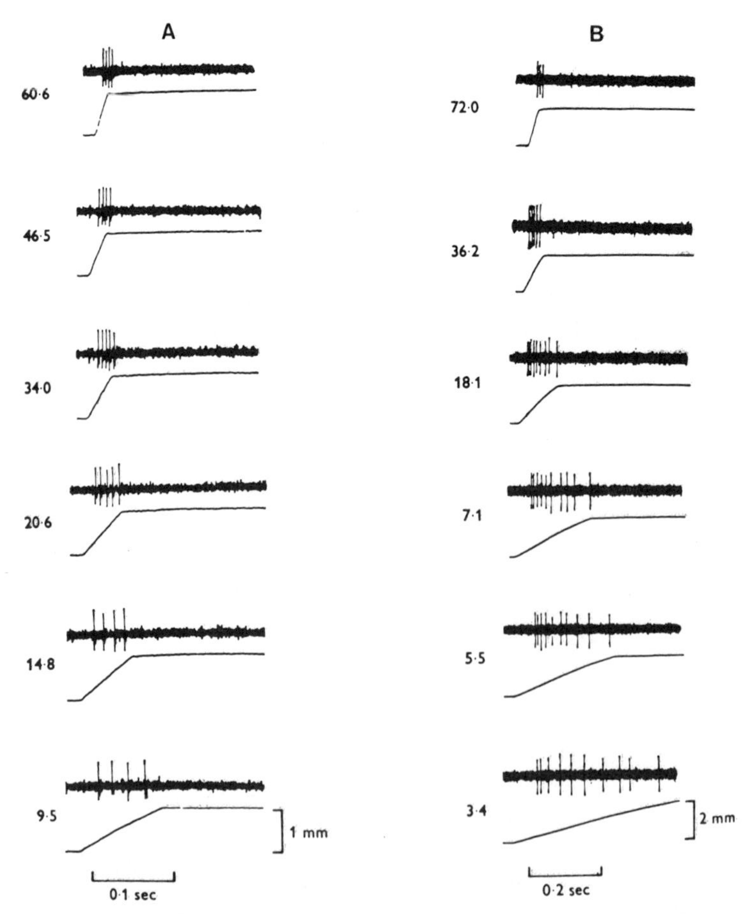

Fig. 10.17. Discharges of two different FA I receptors in the glabrous skin of the human hand in response to graded ramp displacements (velocities given in millimeters per second). Note the constant number of spikes for the unit in panel A and the increased number for lower velocity stimuli in panel B. (From Knibestöl, 1973.)

addressed by K. O. Johnson (1974). The study is limited to the population of rapidly adapting receptors in the glabrous skin. The responses of single units are defined, both with 40-Hz stimuli of variable amplitude applied to the point of maximum sensitivity and with the same stimuli at various distances from this point. By using a mathematical analysis, it is possible to predict the behavior of the population of receptors to stimuli of various amplitude (Fig. 10.18). Several ways in which the stimulus intensity can be encoded by the receptor population are considered: (1) total number of active fibers; (2) total activity of the receptor population; and (3) total activity of the receptors under or near the stimulus probe. All of these measures are linearly related to stimulus intensity and provide a good fit to the psychophysical data. A further outcome of this study is the finding that the size and position of the stimulus are well represented by the activity of the receptor population (Fig. 10.18).

Threshold for Contact Recognition. The thresholds at which monkeys and humans recognize sinusoidal stimuli have been examined by Mountcastle *et al.* (1972) and compared with the thresholds of FA I and FA II units recorded from the monkeys. The lowest thresholds are for stimuli with frequencies of 200–250 Hz. At such frequencies, stimuli with amplitudes of about 1 μm can be recognized. FA II units (Pacinian corpuscles) are responsible for conveying the information.

Lindblom (1974) has measured the threshold amplitudes for tactile sensation evoked by brief displacements of the finger pads of human subjects. The threshold is about 5 μm for pulses having a rate of rise of 0.3 mm/sec or more and is attributed to the activation of Pacinian corpuscles (FA II receptors), since these have the requisite sensitivity and FA I receptors have thresholds of more than 50 μm for comparable stimuli applied to human glabrous skin (Knibestöl, 1973).

R. S. Johansson *et al.* (1980) have tested the thresholds of FA I, FA II, SA I, and SA II receptors in the glabrous skin of the human hand with von Frey hairs. The most sensitive units are the FA I and FA II receptors (median thresholds, 0.58 and 0.54 mN, respectively). SA I and SA II receptors have higher thresholds (median thresholds, 1.3 and 7.5 mN, respectively). Force thresholds are correlated with amplitude thresholds. Most FA I and FA II receptors have thresholds that are low enough to account for the psychophysical detection threshold (cf. Lindblom, 1974).

R. S. Johansson and Vallbo (1979a,b) have done psychophysical tests of the threshold for recognition of a triangular indentation of the glabrous skin by human subjects and have compared these with the thresholds for activating FA I, FA II, SA I, and SA II units recorded microneurographically. The thresholds of the FA I and FA II units are lower than those of the SA I and SA II units and are equal to or lower than the psychophysical thresholds (Fig. 10.5). It is concluded that activation of a single FA I receptor would be sufficient to account for the threshold

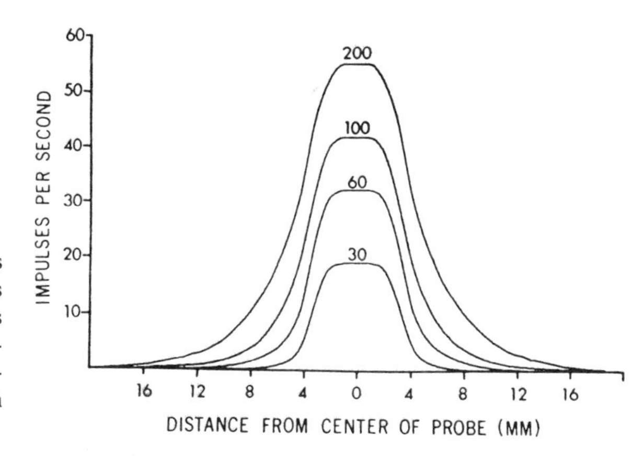

Fig. 10.18. Average firing rate at various distances from the center of the stimulus probe for different stimuli (displacements given in micrometers). The curves indicate the form of the expect spatial recruitment at sites away from the probe. (From K. O. Johnson, 1974.)

for contact recognition in glabrous skin. Higher thresholds are found for contact recognition in some areas of the hand, and yet the receptor thresholds are the same as in the zones of greatest sensitivity. This suggests that individual receptors in these areas may be unable to evoke a sensation.

Hämäläinen and Järvilehto (1981) find that the thresholds for contact and for "touch" sensation in response to brief tactile pulses in human subjects are higher on the hairy than on the glabrous skin. Järvilehto *et al.* (1981) point out that there are few FA I receptors with spotlike receptive fields in the hairy skin of humans. Hair follicle units are sparsely distributed on the back of the hand, and so recognition of threshold contact on the hairy skin of the hand depends on SA receptors, except when hair follicle receptors are activated (see the section on touch-pressure above).

R. H. LaMotte and Whitehouse (1986) have compared the ability of human subjects to recognize the presence of a raised dot on a smooth surface (delivered at a preselected velocity) with the responses of SA, FA I, and FA II receptors to this stimulus in monkeys. The dot height threshold for humans is 1–3 μm for a stroke velocity of 10 mm/sec. The threshold is 2–4 μm for monkey FA I receptors, more than 21 μm for FA II receptors, and more than 8 μm for SA receptors. It is concluded that the human capacity for recognizing a small raised dot depends on the activation of FA I receptors. The intensity of the sensation depends on the number of discharges evoked in individual FA I receptors and on the number of these receptors that are excited.

Similarly, slippage of a glass plate in contact with the fingertip can be detected if FA I receptors are activated by surface protrusions (Srinivasan *et al.*, 1990). Skin stretch is detected with or without slip by SA receptors. On the other hand, recognition of a textured surface composed of dots of a size below the human detection threshold is due to activation of FA II receptors apparently by vibrations set up in the skin.

Two-Point Discrimination. R. S. Johansson and Vallbo (1979a, 1983) report that the density of FA I receptors in the human glabrous skin corresponds to the gradient of sensitivity for two-point discrimination, as does that of SA I receptors (Fig. 10.7) (cf. Bolton *et al.*, 1966). An analysis of receptive field overlap is consistent with the hypothesis that both FA I and SA I receptors contribute to two-point discrimination (R. S. Johansson and Vallbo, 1980).

Edge Detection. Phillips and Johnson (1981a) find that FA I receptors in the glabrous skin of monkeys do not respond well to static contact with edges or with raised letters, in contrast to SA receptors. They suggest that SA receptors are more suited for tactile spatial discrimination than are FA I receptors.

However, R. S. Johansson *et al.* (1982b) have observed that FA I units in the human glabrous skin, like SA I units, are more responsive to sinusoidal displacements when the edge of a stimulus probe is over the receptive field than when the receptive field is covered by the stimulator. This is consistent with the idea that FA I receptors play an important role in tactile spatiotemporal discrimination.

Precision Grip. FA I receptors are very active during the initiation of precision grip. These units tend to stop firing while the grip is maintained, but discharge again during release (Westling and Johansson, 1987). These receptors also respond during slips (R. S. Johansson and Westling, 1987).

Motion. The contribution of various mechanoreceptors to the perception of motion of the skin across a textured surface has been investigated using a vibrotactile stimulator called an OPTACON to simulate such motion (Gardner and Palmer, 1989a,b, 1990; Palmer and Gardner, 1990). This stimulator has been used as a sensory substitute for the visually or hearing impaired. It produces spatial and temporal patterns of mechanical stimuli by movement of a grid of 144 miniature probes under computer control. With the appropriate patterns, the human subject senses motion of the stimuli across the skin. Recordings from mechanoreceptors supplying the primate hand show that FA I and FA II receptors, but not SA receptors, respond to OPTACON stimuli. Spatial resolution is much better for FA I than for FA II receptors. It is suggested that the poorer performance of blind people using the OPTACON than with braille may reflect the inability of the OPTACON to activate SA receptors and the fact that FA II receptors, which have poor spatial resolution, are readily activated.

Other Receptors. As mentioned in connection with touch–pressure, C mechanoreceptors cannot be considered important for flutter–vibration sensations arising from the human hand, since no recordings have yet demonstrated such receptors in this region, either in monkeys or in humans (see Chapter 2). However, C mechanoreceptors have been found in more proximal hairy skin in monkeys (Kumazawa and Perl, 1977), and so these receptors could contribute to sensations arising from proximal parts of the body in primates. Nordin (1990) suggests that C mechanoreceptors supplying the face may contribute to tickle.

Muscle receptors are also unlikely to be involved in flutter–vibration. Although muscle spindle primary endings are very sensitive to small-amplitude stretches of the muscle containing them (Bianconi and Van der Meulen, 1963; M. C. Brown *et al.*, 1967), it is doubtful that small-amplitude vibrations of the skin would activate them under normal circumstances. The controversy surrounding the sensory contribution of muscle spindle afferent fibers will be reviewed below in the section on position sense. The only joint receptors that are likely to play a role in flutter–vibration are paciniform corpuscles. These undoubtedly contribute to the sensation of vibration in the same fashion as do subcutaneous Pacinian corpuscles.

Spinal Pathways. The pathways carrying information from FA I receptors in glabrous skin and from hair follicles and field receptors in hairy skin will be considered first, since these receptors appear to be responsible for flutter sensation. Then the pathways for FA II receptors will be described. These receptors are presumed to mediate vibration sensation.

Projections of FA I and FA II Primary Afferent Fibers. Many hair follicle afferent fibers project directly to the dorsal column nuclei by means of collaterals ascending in the dorsal columns. In addition, hair follicle afferent fibers synapse in the dorsal horn in "flame-shaped arbors," recurrent branches of afferent fibers that first descend into deeper layers of the dorsal horn and then turn dorsally to end in laminae III and IV (Fig. 10.19) (A. G. Brown *et al.*, 1977c). The terminals form a continuous rostrocaudal column (Fig. 10.20). A few axons of FA I receptors in the glabrous skin of cats have also been injected with horseradish peroxidase (HRP) (A. G. Brown *et al.*, 1977c). Labeled terminals were found in laminae III and IV. The primary afferent fibers of Pacinian corpuscles terminate in the medial part of the dorsal horn (Fig. 10.21A and B) (A. G. Brown *et al.*, 1980c). The endings form a series of elliptical terminal zones arranged in a rostrocaudal plane in laminae III–VI (Fig. 10.21C).

Ascending Pathways. Hair follicle afferent fibers have access to the sensory processing mechanisms of the brain through several parallel ascending pathways (Fig. 10.22A). In addition to a direct projection through the fasciculi gracilis and cuneatus to the dorsal column nuclei, hair follicle afferent fibers activate neurons that project in the spinocervical tract, the postsynaptic dorsal column path, and the spinothalamic tract (see Chapters 7–9). Many neurons in the dorsal column and lateral cervical nuclei that project to the contralateral ventrobasal thalamus respond to input from hair follicle afferent fibers, and so information from these neurons, as well as from the spinothalamic tract, reaches higher sensory centers.

FA I receptors in the glabrous skin and field receptors in the hairy skin also have access to the dorsal column nuclei through direct projections in the dorsal columns, as well as by way of several ascending tracts. In cats, few neurons of the spinocervical tract have receptive fields on the glabrous skin (A. G. Brown and Franz, 1969; A. G. Brown, 1981). However, comparable neurons in the monkey do (Bryan *et al.*, 1974). Some neurons of the postsynaptic dorsal column pathway in cats have receptive fields on the glabrous skin (A. G. Brown and Fyffe, 1981). Spinothalamic tract cells in the monkey can often be excited in a rapidly adapting fashion by low-intensity mechanical stimulation of the glabrous skin (Willis *et al.*, 1974).

Animal experiments indicate that the dorsal columns are needed for sequential and spatiotemporal analysis of mechanical stimuli. For example, after a dorsal column lesion,

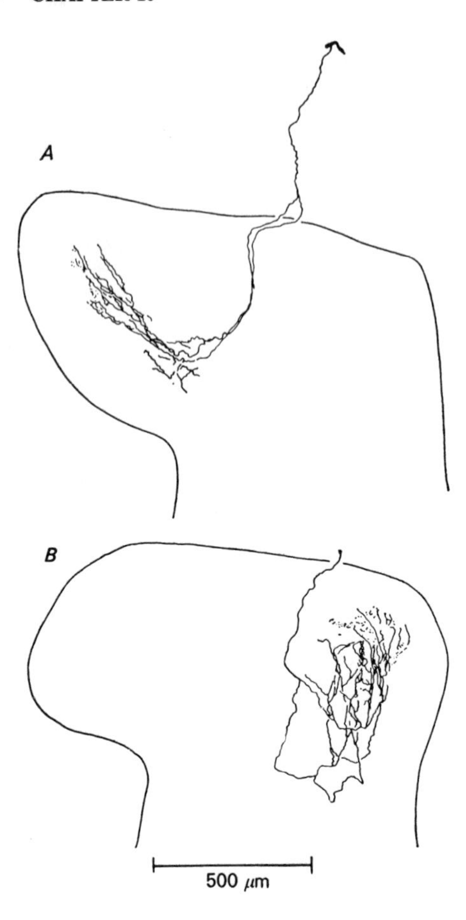

Fig. 10.19. Terminations of two hair follicle afferent fibers in the dorsal horn of cats. The fibers descend into the neck of the dorsal horn and then recurve, forming flame-shaped arbors. (From A. G. Brown *et al.*, 1977c.)

monkeys can still detect a moving stimulus, but they cannot distinguish the direction of stimulus movement (Vierck, 1974). For many behavioral tests, especially in cats, information about mechanical stimuli is conveyed in both the dorsal columns and the DLF. In some cases, the functional pathway in the DLF is the spinocervical tract, but in other cases it is an alternative pathway, possibly the projection to the dorsal column nuclei through the DLF (see Chapter 6).

Vibratory sensation depends upon input from Pacinian corpuscles to the central nervous system. Pacinian corpuscles send collaterals up the dorsal columns to the dorsal column nuclei, and they also activate ascending-tract cells. Neurons of the postsynaptic dorsal column pathway in the cat have been found to respond to a 500-Hz tuning fork, suggesting an input from Pacinian corpuscles (A. G. Brown and Fyffe, 1981). Some axons of postsynaptic dorsal column cells respond exclusively to Pacinian corpuscle input (A. G. Brown *et al.*, 1983a). However, there is no evidence for the activation of either spinocervical tract cells (Chapter 8) or spinothalamic tract cells (Chapter 9) by Pacinian corpuscles.

Clinical Correlations. Clinical observations indicate that recognition of contact with the skin is mediated not only by the posterior columns but also by axons of the anterolateral quadrant (Fig. 10.22A) (see Chapter 6). The patient of Noordenbos and Wall (1976) who had all of the spinal cord interrupted except for one anterolateral quadrant could detect mechanical stimuli applied to either side below the level of the lesion. The ability to localize the stimulus was relatively good. Presumably, the information for contact recognition involves rapidly adapting receptors, since testing was done with von Frey hairs. The patients with posterior column lesions reviewed by Nathan *et al.* (1986) could still feel tactile stimuli, but they had defective two-point

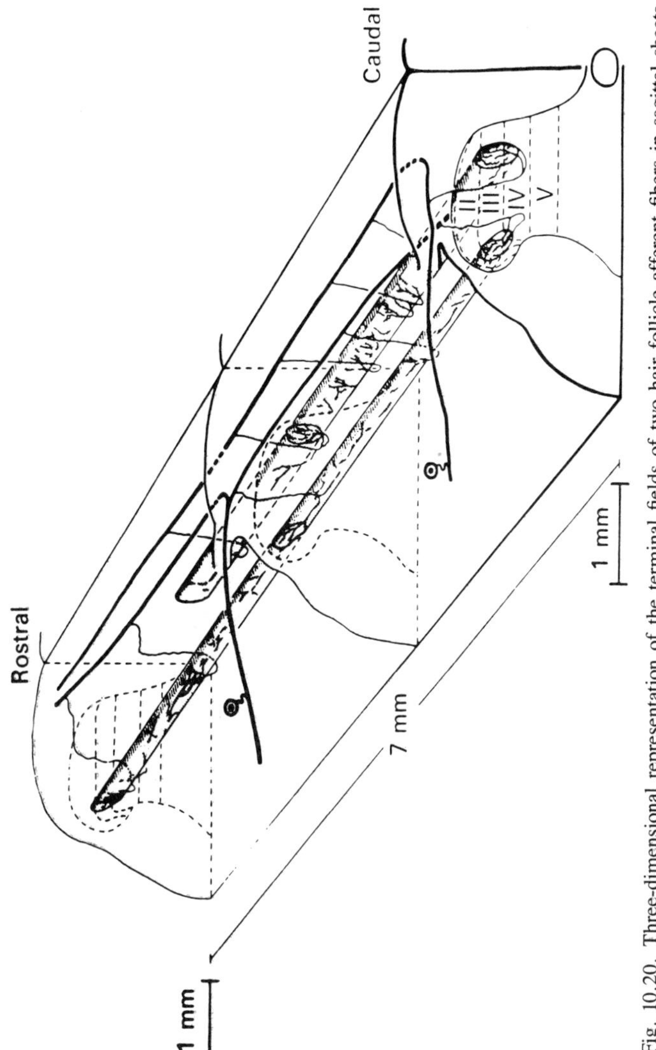

Fig. 10.20. Three-dimensional representation of the terminal fields of two hair follicle afferent fibers in sagittal sheets centered in lamina III of the cat dorsal horn. One ending is shown in the lateral and one in the medial dorsal horn. (From A. G. Brown *et al.*, 1977c.)

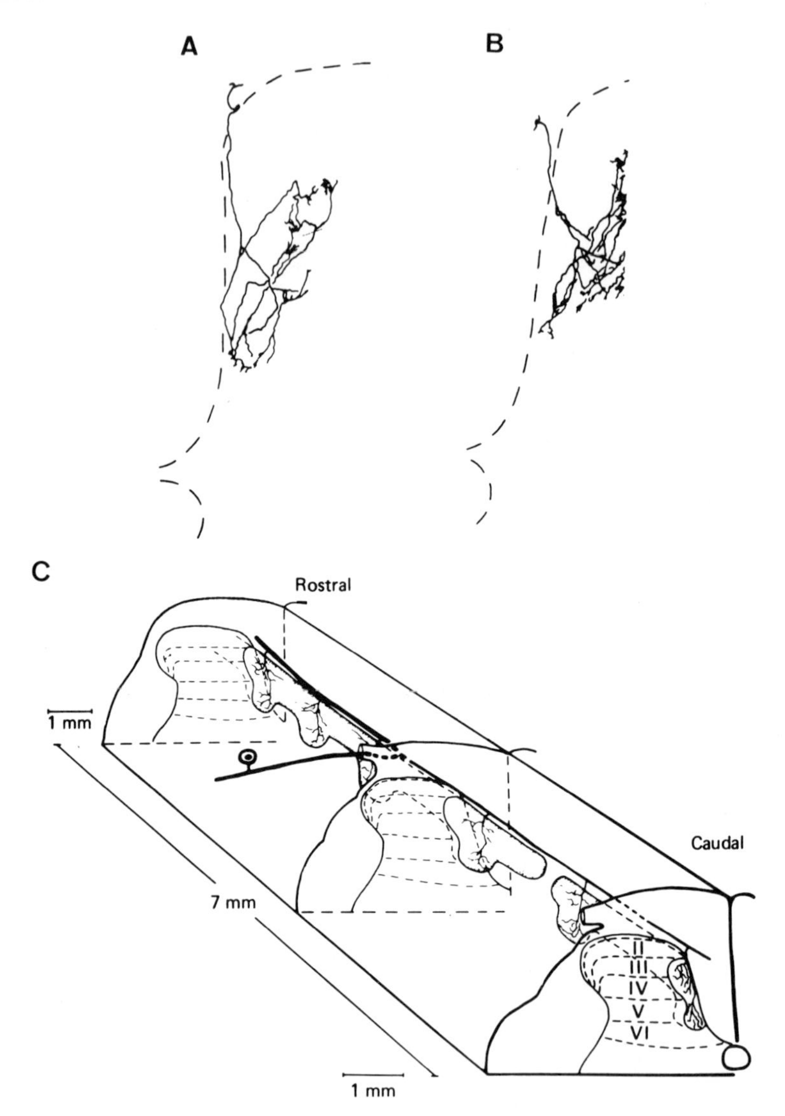

Fig. 10.21. (A and B) Reconstruction of the terminations of two of a series of collaterals given off by a Pacinian corpuscle afferent fiber that also ascended in the dorsal column of a cat. (C) Three-dimensional view of the termination zone of Pacinian corpuscle afferent fibers in a series of rostrocaudally arranged zones in laminae III–VI. (From A. G. Brown *et al.*, 1980c.)

discrimination, astereognosis, and loss of graphesthesia. Thus, information needed for complex spatiotemporal analysis of tactile stimuli depends in humans upon the posterior columns; the spinothalamic tract is sufficient for contact recognition.

Clinical evidence indicates that vibratory sensation in humans depends upon transmission of information in the posterior columns (Fig. 10.22B). The patient of Noordenbos and Wall (1976) lacked vibratory sense, as did the patients with posterior column lesions reviewed by Nathan *et al.* (1986). Conversely, the patients with lesions of the anterolateral quadrants described by Nathan *et al.* (1986) retained vibratory sensation. The patients who retained vibratory sensation after posterior cordotomies for phantom pain by A. W. Cook and Browder

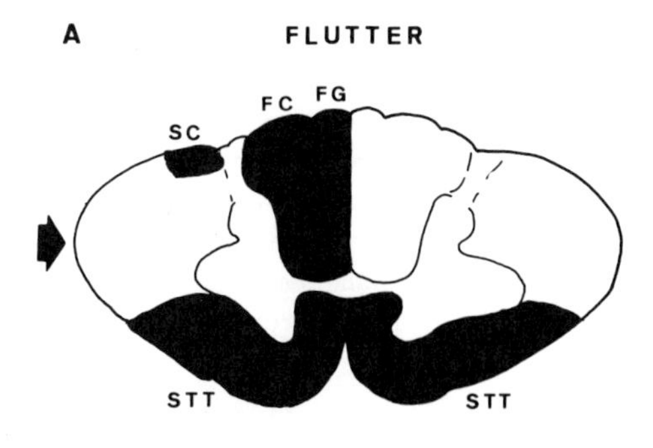

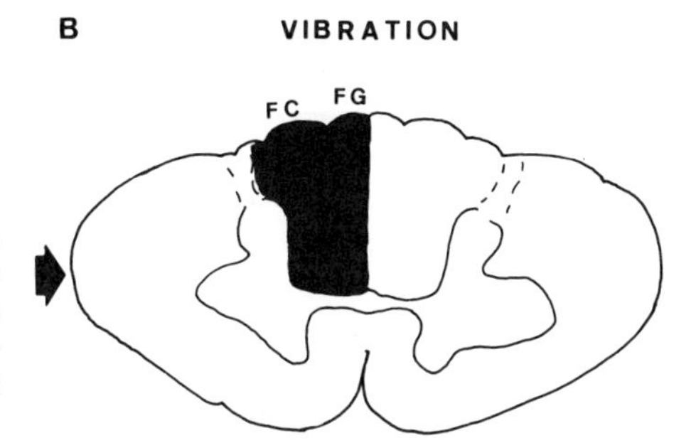

Fig. 10.22. Pathways for flutter (A) and for vibration sense (B). The pathways for flutter include the fasciculus cuneatus and gracilis (FC, FG), the spinocervical tract (SC), and the spinothalamic tract (STT). The pathway for vibration is principally the dorsal column pathway (FC and FG). The arrows indicate the side of the input.

(1965) and animals that retained vibratory sensibility in the face of dorsal column lesions (Schwartzman and Bogdonoff, 1968, 1969) may have had incomplete lesions.

Summary

In summary, the sensation of flutter depends upon several sets of receptors, including FA I receptors (Meissner's corpuscles) in the glabrous skin and hair follicle and field receptors in the hairy skin. Vibratory sensation can be attributed to FA II receptors (Pacinian corpuscles). Tuning curves indicate that FA I receptors respond best at about 30 Hz and FA II units at 250 Hz, accounting for the lower and upper ends of the psychophysical tuning curve for flutter–vibration. Coding for stimulus location depends upon the small receptive fields of FA I receptors. Coding for stimulus intensity is largely attributable to a progressive increase in the activity of a population of rapidly adapting receptors, rather than upon information conveyed by individual members of the population. Direct evidence of this is that stimulation of individual FA I and FA II units at progressively increasing rates causes an increase in the frequency of the perceived sensation; however, for FA I units, stimulation at rates above 100 Hz does cause an increase in the intensity of the sensation. The thresholds of FA I and FA II receptors in the glabrous skin are low enough to account for the threshold for psychophysical detection of contact; in hairy skin, SA I units probably contribute as well. FA I receptors are not as sensitive as SA I receptors in detecting edges of statically applied stimuli, but their responses to edges are enhanced when stimuli are moving; therefore, FA I receptors are likely to play an important role in tactile spatiotemporal discrimination. Flutter sensation probably results from information transmitted by several different routes, including direct projections in the dorsal columns, as well as by activation of neurons belonging to the postsynaptic dorsal column pathway, the spinocervical tract, and the spinothalamic tract. Vibratory sensation appears to depend upon direct projections in the dorsal columns and upon activation of neurons of the postsynaptic dorsal column pathway.

PROPRIOCEPTION

Proprioception depends on information arising from an appendage. Two different sensations can be regarded as submodalities of proprioception: position sense and kinesthesia (Goodwin *et al.*, 1972b; McCloskey, 1973; Horch *et al.*, 1975). Position sense is knowledge of the location of an appendage in space; kinesthesia is an awareness of the movement of an appendage. A further distinction can be made between sensations of active and of passive movements and a sense of the force applied during a voluntary contraction (Roland and Ladegaard-Pedersen, 1977).

Receptors

There has been a longstanding controversy over which receptors are responsible for position sense and kinesthesia. The history of this controversy has been amply reviewed (Goodwin *et al.*, 1972b; P. B. C. Matthews, 1977, 1981, 1982; McCloskey, 1981; Burgess *et al.*, 1982; Proske *et al.*, 1988). Although Sherrington (1900) attributed position sense to both joint and muscle receptors, later investigators have argued against an involvement of muscle spindles, since information from these has been thought not to project to the cerebral cortex and their operation is complex and perhaps not suited to provide unambiguous signals about joint position. This left joint receptors, cutaneous receptors, and the "sense of effort" (corollary discharge) (cf. Merton, 1964; McCloskey, 1981) as candidate mechanisms to explain position sense. However, muscle spindles now appear to be the main receptors for position sense. Information from muscle spindles does reach the cerebral cortex in humans (see Chapter 7) (Starr *et al.*, 1981; Gandevia *et al.*, 1984; Gandevia, 1985), and the complexity of operation of muscle spindles does not exclude their participation in position sense.

A difficulty with the suggestion that muscle receptors are important to position sense was the negative result of a study by Gelfan and Carter (1967) (see also Moberg, 1983). They examined patients with tendons exposed at the wrist. When the tendons were pulled so that distal structures in the hand moved, the subjects perceived the movements in a normal fashion. However, when the tendons were pulled in the opposite direction, stretching the muscles, the subjects did not report any sensations referable to the muscles, although they did notice sensations, including pain, arising in the skin. There are a number of serious objections to these experiments, including the fact that the subjects were not asked about sensations of joint position when the muscles of the forearm were stretched (Goodwin *et al.*, 1972b). When this experiment was repeated under more favorable conditions (Matthews and Simmonds, 1974; McCloskey *et al.*, 1983), a sensation of joint movement was reported when a tendon was pulled to stretch the muscles without allowing movement of the joint.

For a time, joint receptors seemed to be the best candidates for proprioceptive signals (Skoglund, 1973). Slowly adapting discharges can be recorded from joint nerves at all joint angles, although the greatest activity occurs when a joint is held at the extremes of flexion or of extension (see Chapter 2). The slowly adapting joint receptors were identified with Ruffini endings. However, it has since been found that Ruffini endings (at least in the knee joint of cats) discharge only when tension is produced in the joint capsule. Therefore, these endings appear to signal joint torque rather than joint position, and it has been proposed that they contribute to a sensation of joint pressure rather than position (Burgess and Clark, 1969b; F. J. Clark and Burgess, 1975; F. J. Clark, 1975; Grigg, 1975, 1976; Grigg and Greenspan, 1977; see also Eklund and Skoglund, 1960, and the review by Burgess *et al.*, 1982). Their role might be to act as "limit detectors" (A. Rossi and Grigg, 1982). The few slowly adapting receptors with afferent fibers in the posterior articular nerve of the cat knee joint that do discharge when the joint is at an intermediate position are actually muscle spindle afferent fibers originating in the popliteus muscle (F. J. Clark and Burgess, 1975; McIntyre *et al.*, 1978; see also Grigg and Greenspan, 1977; Rossi and Grigg, 1982).

The most likely cutaneous receptors to contribute to position sense are the SA II afferent fibers (see the section on touch–pressure). These respond in a slowly adapting fashion to stretching of the skin, and so SA II receptors in the neighborhood of joints are in a favorable location to signal changes in joint position. Stimulation of the afferent fibers of individual SA II receptors does not usually produce any sensation (Ochoa and Torebjörk, 1983; Schady and Toreböրk, 1983; Schady *et al.*, 1983); however, it is reasonable to suppose that with sufficient spatial summation, a sensation could result from activation of these receptors. A few SA II receptors near nailbeds have recently been shown to evoke a sensation of movement of finger joints (Macefield *et al.*, 1990).

It does not appear that a "corollary discharge" is sufficient for position sense, since a blockade of peripheral input causes a loss of position sense but would not prevent the generation of a corollary discharge (McCloskey and Torda, 1975; McCloskey, 1978, 1981; P. B. C. Matthews, 1981; Ferrell and Milne, 1989). However, if position sense depends in part upon input from muscle spindles, there is an important role for corollary discharges in helping to make the distinction between passive and active movements. For example, during a voluntary motor command in human subjects, there is an increased activity in group Ia afferent fibers, presumably because of an increase in fusimotor activity (Hagbarth and Vallbo, 1969; see below). The sensory consequences of this must in some way be compensated for, presumably by a central computation involving both sensory input from muscle spindle afferent fibers and corollary discharges associated with the motor command (P. B. C. Matthews, 1981).

Properties of Human Proprioceptors. Recordings have been made from the afferent fibers supplying primary endings, secondary endings, and Golgi tendon organs in several different nerves of human subjects (Hagbarth and Vallbo, 1969; Hagbarth *et al.*, 1975; see the review by Vallbo *et al.*, 1979). The axons of primary muscle spindle endings are recognized by their high dynamic sensitivity to passive muscle stretch, irregular static discharges (when present), silence during muscle shortening, silence during the rising phase and discharge during the falling phase of a twitch contraction, and discharge during relaxation from an isometric voluntary contraction. Secondary muscle spindle endings have less dynamic sensitivity than do primary endings; they have a static discharge with regular intervals between spikes; and they cease discharging during a twitch. Golgi tendon organs are silent at intermediate muscle lengths; they have low dynamic sensitivity to stretch; and they discharge in proportion to the force of contraction and during the rising phase of a twitch. These properties parallel the behavior of similar endings in animals (see Chapter 2).

Muscle Spindle Afferent Activity. Recordings from the axons of muscle spindle receptors when the muscle is completely relaxed generally show no activity, indicating no background fusimotor drive (Hagbarth and Vallbo, 1968; Hagbarth *et al.*, 1970; D. Burke *et al.*, 1976a; D. Burke and Eklund, 1977). The sensitivity of human primary endings to changing joint positions is much lower than in the cat (Vallbo, 1974a,b), and blocking the muscle nerve with local anesthetic does not change the responses to stretch or vibration (D. Burke *et al.*, 1976a,b). Thus, there appears to be no fusimotor tone to relaxed muscles in healthy human subjects. However, muscle spindle primary endings in humans have a high dynamic sensitivity, for example to vibration, even when the muscle nerve is anesthetized (D. Burke *et al.*, 1976a,b).

During isometric contractions, human muscle spindle afferent fibers usually show an increased discharge after sufficient force is generated (D. Burke *et al.*, 1978). This behavior is presumably due to activation of the fusimotor system. Evidence for this is that the increase is prevented by partial anesthetic block of the muscle nerve (thought to block γ motor axons but not α motor axons) (D. Burke *et al.*, 1976b; Hagbarth *et al.*, 1970, 1975). It is of considerable interest that the accelerated activity of the muscle spindle afferent fibers begins after the start of the muscle contraction, indicating that the fusimotor system is activated concurrently with α motor neurons and not before, as would be expected if the α motor neurons were activated through the γ loop (Vallbo, 1971; Hagbarth *et al.*, 1975; D. Burke *et al.*, 1978). When a pressure block is used that interrupts conduction in α motor axons, as shown by a reduction in the strength of voluntary contractions by more than 90%, it is still possible to activate muscle spindle afferent fibers (D. Burke *et al.*, 1979). This observation is interpreted to indicate that the muscle spindle excitation is mediated by γ motorneurons, rather than by collaterals of α motor axons (β innervation).

Recordings from primary and secondary muscle spindle afferent fibers from finger extensor muscles in human subjects during an isotonic holding task do not show activity related to joint position (Vallbo *et al.*, 1981); instead, under the conditions of the experiment, the afferent fibers give a negative position response (Hulliger *et al.*, 1982). This is attributed to fusimotor activity, since the same afferent fibers produce a position-related response during passive movements. This does not rule out a role for muscle receptors in position sense, since activity in muscle afferent fibers in the antagonistic muscles might provide key information, information from Golgi tendon organs is available, and the central nervous system may use computations involving corollary discharges to determine joint position.

Some muscle spindle afferent fibers in human muscle show in-parallel behavior during a twitch contraction of the whole muscle but in-series responses during submaximal contractions (D. Burke *et al.*, 1987). This is attributed to mechanical coupling between some muscle spindles and extrafusal muscle fibers. Such responses may contribute to the differences in the behavior of human and cat muscle spindles.

Joint Afferent Fibers. D. Burke *et al.* (1988) have recorded from afferent fibers supplying human finger joints. Some of the afferent fibers discharge at intermediate joint positions, but most discharge only at the ends of the normal range of joint movement. It is their opinion that human finger joint afferent fibers have only a limited capacity to provide proprioceptive information, since most of the joint afferent fibers do not distinguish the direction of joint movement. However, a population response might permit extraction of kinesthetic information. Similar conclusions were reached by Edin (1990).

Cutaneous Afferent Fibers. Recordings have been made from FA I, FA II, SA I, and SA II afferent fibers supplying the glabrous skin of the human hand during voluntary finger movements (Hulliger *et al.*, 1979). Receptors of all of these types are activated, although FA II and SA II units are more responsive than the other types of receptor. Static responses are seen chiefly in SA II units. Since many units would be activated by movements, it is likely that they provide important information about the movements. However, it is considered unlikely that the activity of these receptors contributes to position sense, except perhaps the SII units.

Sensation Evoked by Muscle and Joint Receptors. *Muscle Spindle Afferent Fibers.* Gandevia (1985) finds that electrical stimulation of large muscle afferent fibers, presumably supplying primary endings of muscle spindles, in the ulnar nerves of human subjects evokes illusions of movements of the fingers. This observation is regarded as direct evidence for a role of muscle spindle afferent discharges in kinesthesia. However, stimulation of axons from individual muscle spindles fails to evoke any sensation (Macefield *et al.*, 1990), suggesting that spatial summation is required.

Joint and Cutaneous Afferent Fibers. Repetitive stimulation of digital nerves at the level of the middle phalanx in human subjects elicits a sensation of oscillatory movements or torsion of the distal interphalangeal joint (Gandevia, 1985). This supports the suggestion that joint or cutaneous afferent fibers contribute to kinesthesia in distal joints. Furthermore, stimulation of the axons of individual joint afferents can elicit a sensation of joint movement (Macefield *et al.*, 1990).

Proximal Joints. The receptors that signal joint position may differ for different joints. Evidence in favor of an involvement of muscle spindles in position sense in proximal joints comes from experiments on human subjects in which illusions of movement are produced by vibrating the tendons of various muscles (Goodwin *et al.*, 1972b; Eklund, 1972; Craske, 1977). The illusions can be quantified by asking the blindfolded subjects to track the vibrated arm with the control arm (Fig. 10.23) (Goodwin *et al.*, 1972b). Such stimulation is thought by Goodwin *et al.* (1972b) to be relatively specific for activation of muscle spindle afferents, since illusions do

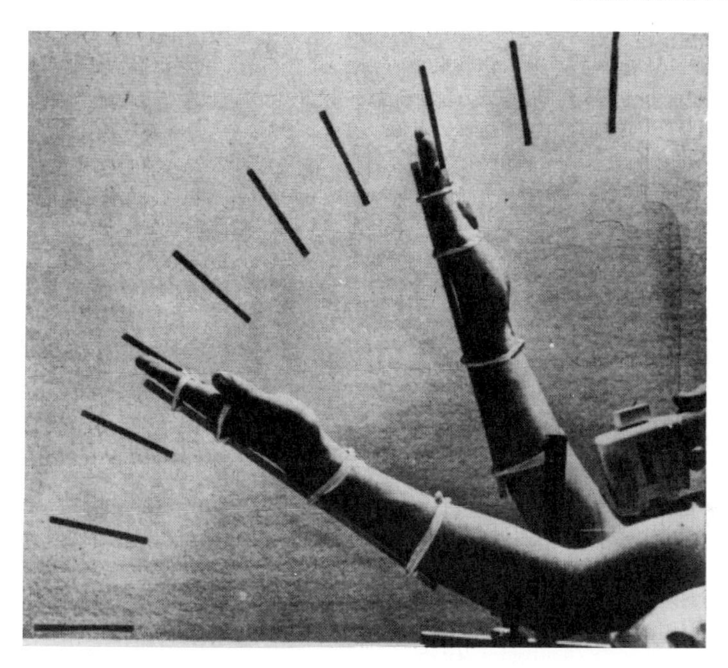

Fig. 10.23. The misalignment in the two arms reflects the extent of the illusion in position sense produced in a subject by vibration of the biceps tendon. The scale markings are 10°. The subject was instructed to align the arms in the absence of visual cues. (From Goodwin *et al.*, 1972; reproduced with permission.)

not result when the vibrator is applied to the skin over joints or bone. The sensory experience is largely one of continued movement in the direction that would stretch the muscle being vibrated. The illusions do not depend upon the presence of a muscle contraction, and so it can be concluded that activation of muscle spindle afferent fibers produces the illusions. It is suspected that the primary endings are largely responsible for the illusions of movement. There is also a false sense of position that can be due to activation of secondary endings. Since no proprioceptive illusions occur during voluntary contractions, the muscle spindle input that occurs during such contractions because of fusimotor activity must not be perceived. Presumably, perception is prevented by integration of input due to fusimotor activity with corollary discharges.

McCloskey (1973) has extended these experiments to an analysis of illusions both of movement and of static position of the elbow joint when the biceps brachii tendon is vibrated. Experimental variables include the frequency and amplitude of vibration, the amount of loading of the muscle, and fatigue. Vibration at 100 Hz makes the subject feel that the joint is being extended whether the muscle is loaded or not, although loading reduces the error in matching the vibrated limb with the control limb. Lower-frequency vibration produces an illusion that the arm is positioned as if the muscle has been stretched. If the vibratory rate is low enough, there is no illusion of movement, but rather just one of position. In the case of such illusions of position, loading increases the error. Since illusions of movement and of position can be produced independently and are affected in opposite directions by loading, there must be separate mechanisms for these sensations. It is speculated that the muscle spindle primary endings contribute to kinesthesia and that other receptors (perhaps including both primary and secondary endings) evoke static position sense.

Evidence that joint afferent fibers are not required for position sense at proximal joints has been obtained by Grigg *et al.* (1973), who find that position sense is almost normal in the hips of patients after total hip joint replacement. The thresholds for detection of abduction are sometimes elevated, and the precision of estimates of small changes in position is reduced by the surgery, but the patients can judge large passive movements normally and can position the hip perfectly well. The subjective feeling of abduction is similar to that on the normal side, although the intensity of the sensation is lower. It is concluded that hip joint position sense can be mediated by extracapsular receptors. Others have reported similar results for joint removal (Cross and McCloskey, 1973; Barrack *et al.*, 1983a; Karanjia and Ferguson, 1983) and for joint anesthesia (F. J. Clark *et al.*, 1979, 1985; Barrack *et al.*, 1983b). However, it is difficult to be sure that all joint afferent fibers are blocked by anesthetics (F. J. Clark *et al.*, 1979; Ferrell, 1980), and absence of deficits after removal of a joint does not rule out participation of joint afferent fibers in proprioception.

Slowly adapting cutaneous afferent fibers are not suited for signaling position sense in the knee joint, since the sensation produced by maintained indentation of the skin fades in 1–2 min, whereas static position sense in the human knee is accurate to within 2–3° even when the position is changed so slowly that there is no sense of an alteration in skin position (Horch *et al.*, 1975). Furthermore, anesthesia of the skin around the knee joint, without or with concurrent anesthesia of joint afferents, does not alter static position sense at that joint (F. J. Clark *et al.*, 1979).

Distal Joints. The joints of the digits may differ from more proximal joints. For example, static position sense in the toes and fingers is lost within a few minutes, and so it is possible that cutaneous receptors are involved in these distal joints (see above). In recordings from human subjects, SA II afferent fibers have been described that discharge at rates proportionate to the amount of finger flexion (Knibestöl, 1975).

Anesthesia of the skin and joints of the digits produces a severe deficit in position sense, at least for slow movements (Browne *et al.*, 1954; Provins, 1958; Goodwin *et al.*, 1972a,b). However, when faster movements are tried, subjects can recognize joint position, especially if the forearm muscles are tensed (Goodwin *et al.*, 1972a,b). These experiments suggest that muscle receptors can contribute to position sense in distal joints, although joint or cutaneous receptors are also likely to contribute.

The reduction in position sense in the fingers by local anesthesia could be due to the loss of input from receptors that contribute specifically to position sense, or it may reflect the loss of a more general facilitation of central pathways activated by muscle afferent fibers. Cross and McCloskey (1973) have tried to approach this question by examining patients who have had complete removal and replacement of digital joints (metacarpophalangeal or metatarsophalangeal joints). In all the cases, position sense is normal despite complete removal of the joint capsule. This is true soon enough after surgery that reinnervation by joint afferent fibers could not have occurred. In one patient, the intrinsic muscles of the foot had been disconnected from a toe. Side-to-side movement of this toe could not be perceived (presumably because of disconnection of the intrinsic muscles), although flexion and extension movements could. Evidently, the perceptual cues from the long extensor and flexor muscles are more significant for position sense than is joint input.

Others have confirmed that after anesthesia of the hand or of a finger, there can still be an accurate sense of the position of a finger in some subjects (J. K. Collins, 1976; Gandevia and McCloskey, 1976; Roland and Ladegaard-Pedersen, 1977; Rymer and D'Almeida, 1980), although the subjective clarity of the sensation is impaired when the muscles are relaxed. Voluntary contractions of the muscles moving the joint improve position sense. The sensory change may be due to the removal of a facilitation provided by input from cutaneous and joint afferent fibers (P. B. C. Matthews, 1981).

Roland and Ladegaard-Pedersen (1977) had subjects judge the strength of a spring squeezed between the thumb and finger of an anesthetized hand. This could be done with nearly normal accuracy, indicating that the subjects received information about muscle force as well as length. The discrimination of spring strength is not affected by anesthesia of the skin or joints. It is proposed that the tension is sensed by Golgi tendon organs.

Rymer and D'Almeida (1980) have evaluated the effects of variations in muscle force on the perception of the position of the proximal interphalangeal joint of the index finger of human subjects. When the loaded finger is able to move freely, position is perceived accurately. However, when the subject is asked to generate isometric forces across a deflected joint, position sense is grossly inaccurate. The errors are in the direction of the force (opposite the direction of illusions produced by vibration). Anesthesia of the finger does not affect the results. These results are interpreted to indicate that force signals derived from Golgi tendon organs contribute to joint position sense, in addition to length signals from muscle spindles. The force feedback may help the nervous system compensate for fusimotor bias, although errors result when conflicting information is provided by muscle spindle and Golgi tendon organ afferent fibers, as when the subject generates isometric force.

F. J. Clark *et al.* (1986) have been able to demonstrate static position sense in the metacarpophalangeal joint but not in the proximal interphalangeal joint of the human index finger. Kinesthesia at the proximal interphalangeal joint involves cutaneous receptors, since anesthesia of the skin blunts the perception of joint movements.

Ferrell and Smith (1987) have repeated the experiment of Rymer and D'Almeida (1980) and find a striking deficit in position sense after anesthesia of the finger. The difference in the results may have been the use of a wider range of joint angles in the study by Ferrell and Smith (1987).

Ferrell *et al.* (1987) have studied position sensation in the distal interphalangeal joint of the middle finger when the hand is placed in a position that functionally disconnects the muscles controlling the joint. Even though the muscles are functionally disconnected, joint position is accurately sensed (see also Gandevia and McCloskey, 1976). Injection of local anesthetic into the joint produces an impairment of position sense. It is concluded that both joint and muscle receptors contribute to proprioception. A possible additional contribution from cutaneous receptors is not ruled out. A similar conclusion has been reached by Ferrell and Smith (1988) in a study in which the position of an index finger is matched by the position of the contralateral index finger. These observations are consistent with the conclusions of Gandevia *et al.* (1983) that muscle, skin, and joint afferent fibers are all required for normal proprioception at distal joints.

Ferrell and Smith (1989) have extended these observations by showing that the errors made after anesthesia of a digit are worse at extremes of the normal range of movement of the joint than at intermediate positions. They propose that cutaneous afferent fibers contribute to position sense at intermediate joint positions and that joint afferent fibers contribute to position sense at the extremes. It is concluded that although muscle receptors contribute to proprioception at distal joints, input from joint and cutaneous receptors is also used. Additional evidence for this viewpoint is provided by Ferrell and Milne (1989).

Spinal Pathways. *Projection of Proprioceptive Afferent Fibers.* Muscle afferent fibers project to laminae I, V, and VI of the dorsal horn, as well as to the ventral horn, Clarke's column, and the external cuneate nucleus (Craig and Mense, 1983; Nyberg and Blomqvist, 1984; Hongo *et al.*, 1987). The afferent fibers that project to laminae I and V include muscle nociceptive afferent fibers, as shown by injection of these with HRP (Mense and Prabhakar, 1986). Of the proprioceptive muscle afferent fiber types (Fig. 10.24), group Ia afferent fibers have been shown to terminate in laminae VI, VII, and IX of the spinal cord (A. G. Brown and Fyffe, 1978; A. G. Brown, 1981), group II afferent terminals in laminae IV–VII and IX (Fyffe, 1979; A. G. Brown, 1981), and group Ib afferent fibers from Golgi tendon organs in laminae VI and VII (A. G. Brown and Fyffe, 1979; A. G. Brown, 1981). Many neurons in the deep layers of the dorsal horn and in the intermediate region are proprioceptive (see Chapter 5). Some may contribute to sensory experience, while others are involved in reflex activity.

Ascending Pathways. In the cat, slowly adapting joint and muscle receptors do not ascend to the medulla in the fasciculus gracilis (Lloyd and McIntyre, 1950; Burgess and Clark, 1969a). The pathway used for proprioceptive information from the hindlimbs in the cat is the spinomedullothalamic pathway that relays in nucleus z (Fig. 10.25) (see Chapter 8). This pathway is

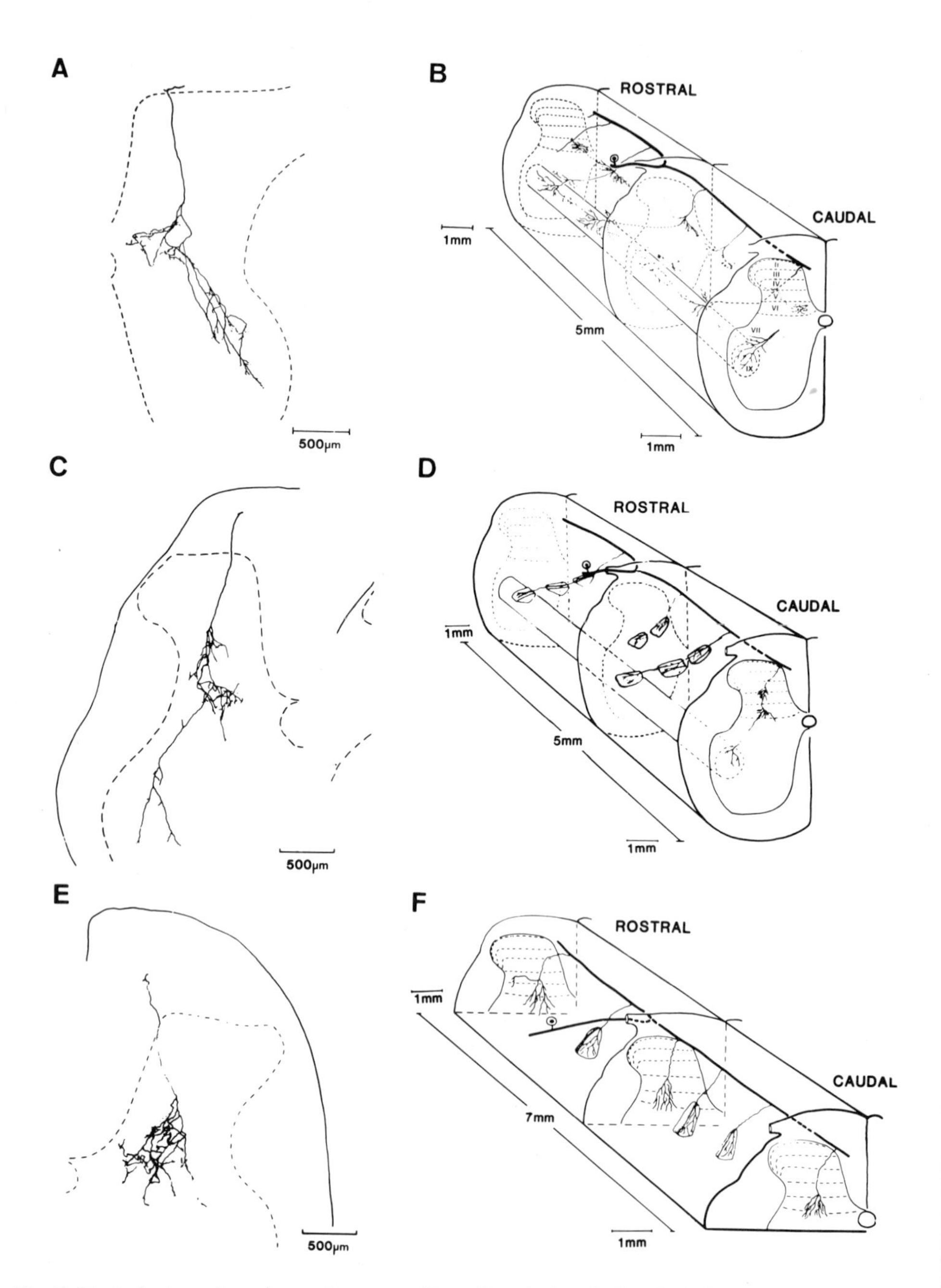

Fig. 10.24. Projections of muscle stretch receptor afferent fibers in the spinal cord. (A) Course of a group Ia fiber of the lateral gastrocnemius-soleus (LGS) muscle in a cat, as reconstructed after injection of the axon with HRP. (B) Representation of the distribution of group Ia fiber collaterals (triceps surae muscles) in the lumbosacral enlargement. (A and B from A. G. Brown and Fyffe, 1978.) (C) Group II axon of the LGS muscle. (D) Schematic diagram of the distribution of a group II axon in the lumbar enlargement. (E) Group Ib afferent fiber. (F) Distribution of a group Ib fiber in the lumbar enlargement. (C–F from A. G. Brown, 1981.)

formed at least in part by collaterals of the dorsal spinocerebellar tract. There is a nucleus z in humans (Sadjapour and Brodal, 1968), suggesting that a similar pathway exists in humans. On the other hand, the pathway for proprioceptive information from the forelimb in cats involves the fasciculus cuneatus and the main cuneate nucleus (Fig. 10.25) (see Chapter 7).

It is possible that the spinothalamic tract makes a contribution to position sense, although there is no deficit in proprioception in humans following anterolateral cordotomy (White and Sweet, 1969). In monkeys, some spinothalamic tract cells respond in a slowly adapting fashion to changes in joint position, and the axons of these cells ascend in a peripheral layer along the ventrolateral aspect of the ventral quadrant contralateral to the receptive fields (Willis *et al.*, 1974; A. E. Applebaum *et al.*, 1975). A special group of proprioceptive spinothalamic neurons in monkeys has been found in Stilling's nucleus, which is located in the sacral spinal cord (Milne *et al.*, 1982). These cells appear to encode the position of the tail in space. They lack cutaneous receptive fields, but receive a reciprocal innervation from muscle receptors on opposite sides of the tail.

Clinical Correlations. The traditional view is that the main pathway for position sense is the dorsal column–medial lemniscus system (Head and Thompson, 1906). However, diseases that affect dorsal column function may produce a dissociation between vibratory and position sense. For instance, lesions at thoracic or lumbar levels can produce a reduction in vibratory sensation without a deficit in proprioception, whereas cervical lesions may do the opposite (Weinstein and Bender, 1940).

Proprioception in humans is lost following lesions of pathways ascending in the dorsal part of the spinal cord (Nathan *et al.*, 1986), but not following anterolateral cordotomy (White and Sweet, 1969). For the lower extremity, ascending fibers in both the posterior column and the posterior lateral column may be involved. The striking loss of proprioception in tabes dorsalis should be attributed to the effects of the disease on dorsal root ganglion (DRG) cells, rather than on the demyelinated posterior columns (see Chapter 6). In the patient of Noordenbos and Wall (1976), position sense remained intact in the knee and ankle joints ipsilaterally to the intact anterolateral quadrant. However, position sense was lost in the toes. It is possible that enough of the spinomedullary (dorsal spinocerebellar) tract was left intact to allow signals from proprioceptors in some of muscles of the lower extremity to provide signals to nucleus z. The loss of proprioception in the toes may have resulted from the interruption of signals from joint and cutaneous afferent fibers transmitted in part through the posterior column, from partial interruption of the spinomedullary tract, or both.

Summary

In summary, proprioception includes both position sense and kinesthesia. These sensations depend upon input from muscle stretch receptors (primary and secondary endings of muscle spindles and Golgi tendon organs), especially for proximal joints, but some information is also

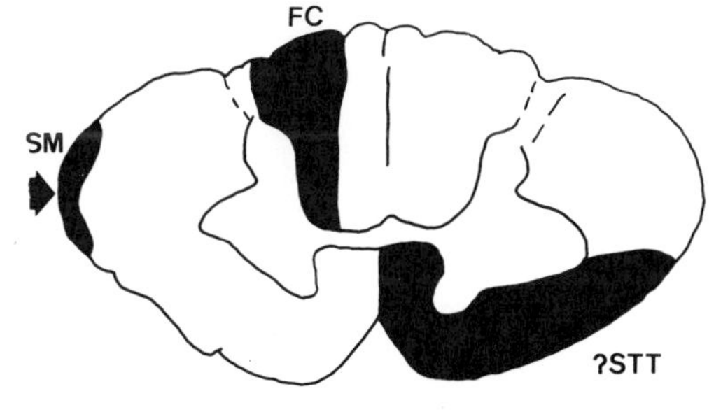

Fig. 10.25. Pathways for position sense include the fasciculus cuneatus (FC), but not the fasciculus gracilis, and the spinomedullothalamic pathway through nucleus z (SM). The spinothalamic tract (STT) in monkeys also carries proprioceptive information.

provided by joint and perhaps cutaneous afferent fibers, especially for distal joints. The signals from muscle spindles that are produced by fusimotor activity appear to be compensated for in some way by integrative mechanisms involving "corollary discharges" that accompany the motor commands. Input from muscle proprioceptors is processed in laminae IV–VII and IX. In addition, large muscle afferent fibers at cervical levels project directly to the nucleus cuneatus through the dorsal funiculus. At lumbar levels, proprioceptive information is relayed through Clarke's column and reaches the contralateral thalamus after a synapse in nucleus z. The spinothalamic tract may transmit some proprioceptive information, but this is not crucial since anterolateral cordotomies do not interfere with proprioception.

PAIN

Pain will be treated here as a specific sensation with its own sensory channel (Perl, 1971; Zimmermann, 1976; D. D. Price and Dubner, 1977; Willis, 1985; Besson and Chaouch, 1987), albeit a complex one (Melzack and Wall, 1965; Melzack, 1973).

Several types of pain can be recognized. Lewis (1942) described two basic kinds of pain: superficial and deep. Superficial pain results from intense stimulation of the skin and can be well localized. Deep (or aching) pain arises from skeletal muscles, tendons, periosteum, and joints and is poorly localized. Visceral pain shares many of the attributes of deep pain, including a tendency to be referred to superficial structures and to induce powerful autonomic responses. Superficial pain can be further subdivided into pricking pain and burning pain. Synonyms are first and second pain, since a noxious stimulus applied to some parts of the body may evoke the two kinds of superficial pain in temporal succession (early evidence is reviewed by Lewis, 1942). It has been proposed that first and second pain are produced by activation of Aδ and C fibers, respectively, and that the temporal lag between the pains is due to different peripheral conduction velocities of the nociceptive afferent fibers involved (Lewis and Pochin, 1938a; Sinclair and Stokes, 1964; D. D. Price *et al.*, 1977; Campbell and LaMotte, 1983).

In addition to the sensory experience called pain, strong stimuli produce other behavioral events, including flexor withdrawal reflexes (Sherrington, 1906), autonomic responses (R. F. Schmidt and Weller, 1970), arousal, aversive behavior, endocrine changes, and other motivational–affective responses (Magoun, 1963; Melzack and Casey, 1968). These more global consequences of painful stimuli will not be discussed in detail here, but they are very important aspects of the pain reaction and in many respects are more important in human disease than is the sensation of pain.

Stimuli that evoke the pain reaction are said to be noxious (Sherrington, 1906), meaning that they threaten or actually produce damage. This is an important consideration in experiments designed to investigate the neural basis of pain. Animal subjects cannot report pain, nor can neurons. It is reasonable to expect that neural responses will occur in the pain channel when strong but not overtly damaging stimuli are applied, since such responses can serve as a warning to the individual that harm is imminent. However, maximum activity in the pain system is likely to result only when frank damage is produced. Observations consistent with these suppositions have been made in studies of peripheral nociceptors (see, e.g., Burgess and Perl, 1967).

Another major consideration in experiments on nociceptive neurons is that some neurons may play a role in one aspect or another of the total pain reaction, but not all such neurons need contribute to all aspects of the reaction. Thus, it is possible to separate the sensory mechanism from other aspects of the reaction to painful stimuli, such as motivational–affective mechanisms. For example, patients who have had a frontal lobotomy may feel pain but are not concerned by it (W. Freeman and Watts, 1950), and the flexion reflex is vigorous below the level of a spinal transection, which eliminates pain sensation (Sherrington, 1906).

In addition to pain that is produced by activation of nociceptors and the rest of the pain channel, pain has been experienced in the absence of obvious activity in nociceptors in some patients. This is common following damage to peripheral nerves or to certain parts of the central nervous system (Leijon *et al.*, 1989; Boivie *et al.*, 1989; see reviews by Cassinari and Pagni, 1969; Fields, 1987; Pagni, 1989; Tasker and Dostrovsky, 1989). Such neurogenic pain is poorly understood and will not be considered here. It presumably results from plastic changes in the sensory channels for pain.

Receptors

Nociceptors. Nociceptors, whether in the skin, muscle, joints, or viscera, seem to terminate in unencapsulated endings and to be supplied with either finely myelinated (Aδ) or unmyelinated (C) afferent fibers (although a few nociceptors supplying the skin have conduction velocities as high as 40–50 m/sec; Burgess and Perl, 1967; Georgopoulos, 1976; cf. Willer and Albe-Fessard, 1983).

There are no obvious structural differences between the endings of various kinds of nociceptors, yet several have quite distinctive response properties. For example, cutaneous mechanical nociceptors respond just to intense mechanical stimuli, whereas polymodal nociceptors can be activated by several different forms of noxious stimuli, including mechanical, thermal or chemical stimuli (see Chapter 2).

Recordings from human cutaneous or mixed nerves have shown that several of the types of nociceptors that have been identified in animal experiments are also present in humans, including cutaneous Aδ mechanical nociceptors, Aδ mechanoheat nociceptors, and C polymodal nociceptors (Fig. 10.26) (see Chapter ,2) (cf. Van Hees and Gybels, 1972, 1981; Torebjörk, 1974; Torebjörk and Hallin, 1974; Gybels *et al.*, 1979; Hallin *et al.*, 1981; Adriaensen *et al.*, 1983; Bromm *et al.*, 1984; Ochoa and Torebjörk, 1989; Torebjörk and Ochoa, 1990).

Sensations Evoked by Stimulation of Nociceptors. A single impulse or even a low-frequency discharge in a single nociceptive afferent fiber appears to be insufficient to evoke a sensation of pain (Van Hees and Gybels, 1972, 1981; Torebjörk and Hallin, 1973). However, activation of a population of Aδ or C fibers in a human cutaneous or mixed nerve produces considerable pain, even if the activity of Aα,β fibers is blocked (Fig. 10.27) (Heinbecker *et al.*, 1933; D. Clark *et al.*, 1935; Lewis and Pochin, 1938b; Pattle and Weddell, 1948; W. F. Collins *et al.*, 1960; Torebjörk and Hallin, 1973; MacKenzie *et al.*, 1975; Hallin and Torebjörk, 1976). Conversely, pain is abolished when Aδ and C fiber activity is blocked, e.g., by local anesthetic (Fig. 10.28) (Torebjörk and Hallin, 1976; MacKenzie *et al.*, 1975; Hallin and Torebjörk, 1976).

Nociceptors are also found in muscle, joint, and at least some visceral nerves (see Chapter 2). The properties of most of these have not been studied in detail in humans as yet. However, stimulation of muscle nerves produces a deep, cramping pain (Torebjörk and Ochoa, 1980; Torebjörk *et al.*, 1984a).

Stimulation of Aδ nociceptors in human nerves by using microneurography electrodes is thought to produce pricking pain and stimulation of C polymodal nociceptors generally produces dull or burning pain, depending on whether the units innervate glabrous or hairy skin (Konietzny *et al.*, 1981; Ochoa and Torebjörk, 1989; Torebjörk and Ochoa, 1990). The sensation remains the same after conduction in A fibers is blocked. The projected fields increase in size with stimulus intensity, suggesting that the sensations do not depend upon a unitary input from a single afferent fiber but rather upon activation of small groups of axons. During repetitive stimulation, no sensation is evoked until the stimulus frequency exceeds 3 Hz. Above this threshold, the intensity of the sensation is proportional to stimulus frequency. The sensation during repetitive stimulation is sustained rather than intermittent. Stimulation of some C polymodal nociceptors gives rise to a sensation of itch, rather than of pain (Torebjörk and Ochoa, 1981, 1983).

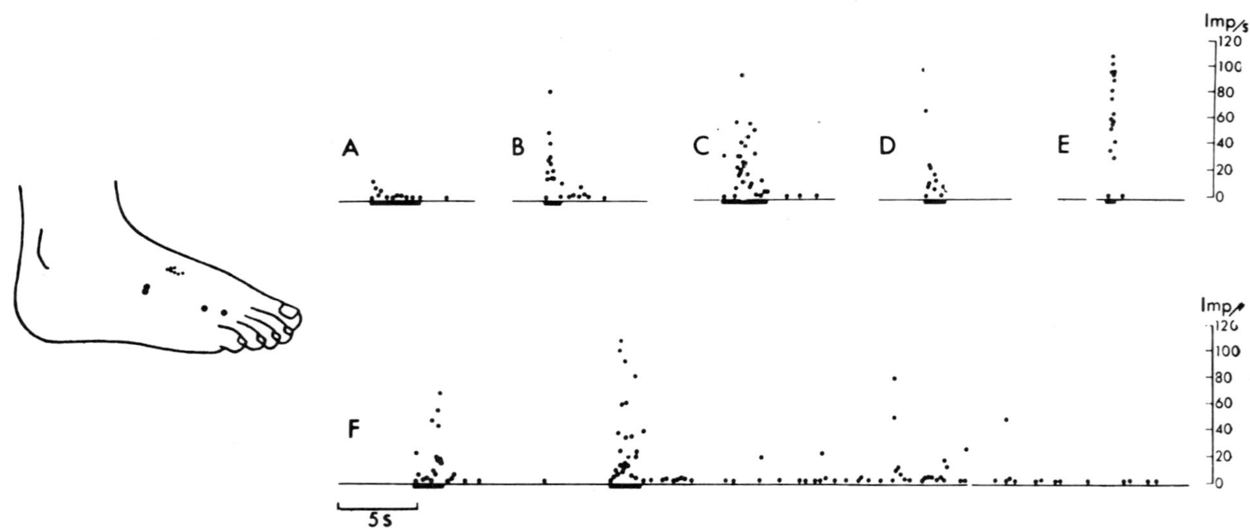

Fig. 10.26. The drawing shows the receptive fields of several C polymodal nociceptors in the human foot. The circles indicate the locations of four receptive fields in a schematic way, while the dots show seven foci of sensitivity for a single receptor that had the largest receptive field of the units studied. The responses in the graph were from a single unit and resulted from the following stimuli: A, pointed stimulus probe; B, firm stroke; C, squeezing the skin with forceps; D, needle prick; E, touching the skin with a glowing match; F, puncturing the skin with a hypodermic needle at the left and then injecting 0.02 ml of 5% KCl. The heavy bars under the records indicate the times when the subject reported a sensation, generally pain. (From Torebjörk, 1974.)

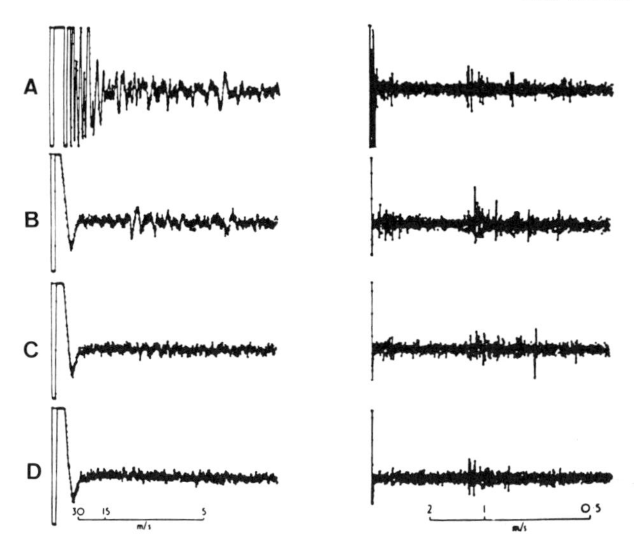

Fig. 10.27. Changes in the afferent volley and in evoked sensation during a progressive nerve block by using pressure. The records in the left column show the compound action potential recorded from myelinated axons of the superficial radial nerve in a human subject by using a microneurography electrode. The column at the right shows the activity in unmyelinated axons (near the middle of the traces), as well as in myelinated fibers (at the far left of the traces). (A) The activity was not yet blocked and perception was normal except for some dysesthesia. (B) The Aα,β volley was blocked, and the Aδ volley was reduced. Light touch was absent, and pinprick and cold were impaired. (C) Most or all of the myelinated fibers were blocked, and pinprick and cold were absent. However, warm and pain sensations were felt. (D) The unmyelinated fiber volley was reduced, and warm and pain sensations were impaired. (From Mackenzie *et al.*, 1975.)

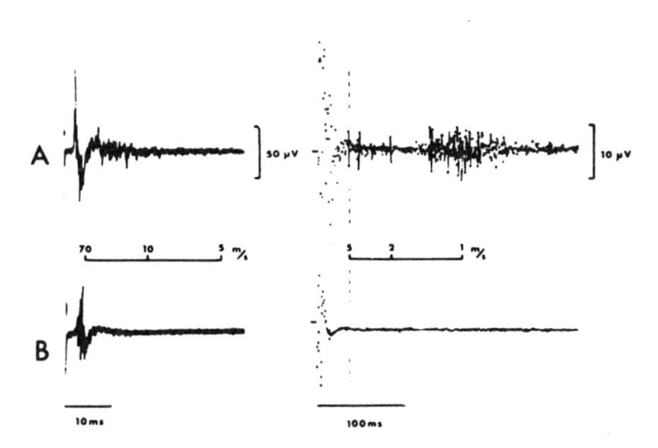

Fig. 10.28. Activity recorded from the human radial nerve in response to electrical stimulation of the dorsum of the hand. (A) The volley is seen to include Aδ and C fibers (note the two different sweeps and the bars indicating conduction velocity). The stimulus was painful. (B) The Aδ and C-fiber activity was abolished by local anesthetic, but some Aα,β activity remained. The subject no longer felt pain, but could still feel touch. (From Hallin and Torebjörk, 1976; reproduced with permission.)

The projected receptive fields of C polymodal receptors stimulated through a microneurography electrode average 92 mm^2 (Jorum *et al.*, 1989). The receptive fields mapped for single C polymodal receptors average 57 mm^2. With suprathreshold stimuli, the projected fields can increase 10-fold, suggesting that additional nociceptive afferents are activated and that these have adjacent receptive fields. Only occasionally do distant receptive fields appear as the stimulus strength is increased. These observations suggest that C polymodal nociceptors from a given area of the skin travel together for long distances in peripheral nerves. The mapped and projected receptive fields often overlap, but on average the centers are separated by 24.8 mm. When A fibers are blocked ischemically and the subject asked to localize a painful heat stimulus, the subject is accurate to within a mean distance of 17.1 mm (compared with an accuracy of tactile localization of 11.8 mm under normal conditions).

Properties. Konietzny *et al.* (1981) have recorded the activity of a small sample (three units) of Aδ mechanical nociceptors in human subjects by microneurography. The receptive field of one of the units was 7 by 5 mm in size, with 12 spotlike areas of greatest sensitivity. Two units responded to squeezing the skin with forceps and pinprick, but not to weak mechanical stimuli, cooling, or heating the skin to 50°C.

Adriaensen *et al.* (1980, 1983) have also recorded from a small sample of Aδ nociceptors in humans. Many of the units are activated by noxious chemical or thermal stimuli, as well as by noxious mechanical ones, suggesting that there is a substantial population of nociceptors with Aδ fibers in human skin that respond to more than one modality of noxious stimulation. The responses to noxious heat can occur without prior heating. Furthermore, the discharge rate during noxious heating can be substantially higher than that shown by C polymodal nociceptors. Presumably, the Aδ nociceptors in human skin that respond to noxious heat can account for the pricking pain that heat pulses provoke (Lewis, 1942). Other Aδ nociceptors in human skin resemble Aδ mechanical nociceptors in animals in that they respond to noxious mechanical but not to noxious chemical or thermal stimuli.

Bromm *et al.* (1984) have recorded the activity of a few Aδ nociceptors by microneurography. One of these could be activated by brief heat pulses produced by a CO_2 laser. It is classified as a mechanothermal unit.

C polymodal nociceptors in the skin of the human hand and foot respond to noxious mechanical, thermal, and chemical stimuli (Fig. 10.23) (Van Hees and Gybels, 1972, 1981; Torebjörk, 1974; Torebjörk and Hallin, 1974; Adriaensen *et al.*, 1980; Hallin *et al.*, 1981; Konietzny *et al.*, 1981; Bromm *et al.*, 1984; see also the review by Vallbo *et al.*, 1979). The receptive fields can be larger than those of C polymodal nociceptors in animals, varying in overall dimensions from 1 by 1 mm to 6 by 17 mm. Unlike the fields of comparable receptors in animals, the receptive fields of human C polymodal nociceptors often include two to seven spotlike areas. Thresholds for von Frey hairs range from 0.7 to 13.1 g. The units do not show spontaneous activity. Weak mechanical stimuli that activate mechanoreceptors, such as air puffs, hair bending, stroking with cotton, or skin stretch are ineffective. Moderate-intensity, nonpainful mechanical stimuli may cause discharges at low rates, but noxious stimuli produce the highest discharge rates seen in these units (50–60 impulses/sec). The responses are slowly adapting, and afterdischarges are sometimes observed following strong stimulation. Repeated stimulation causes receptor fatigue. Cooling is rarely effective in activating these units (although a few discharge when skin temperature is lowered below 23°C). Warming to 40–45°C can excite some units. Effective chemical stimuli include intradermal injection of KCl, application of itch powder (cowhage) or nettles to the skin, or coating the skin with paint remover (methylene chloride dissolved in methanol). Injection of histamine or papain can also activate C polymodal nociceptors, but the mechanical stimulus provided by the needle makes the interpretation of the responses difficult.

Although low rates of discharge in C polymodal nociceptors can occur without pain, higher levels of activity in these receptors are found to correlate well with pain in human subjects when the afferent fibers are activated by noxious heat or chemical stimuli, but not when they are activated by mechanical stimuli (Van Hees and Gybels, 1981). It is suggested that the concomitant activation of Aβ fibers by mechanical stimuli results in inhibitory interactions that prevent pain (see Chapter 5). Noxious heat and chemical stimuli would activate the nociceptors without exciting Aβ fibers, and so pain can result from such stimuli without interference from the inhibitory system.

Intensity Coding. Psychophysical tests of the relationship between the intensity of a painful stimulus, such as noxious heat pulses, and the estimate of pain magnitude can result in nonlinear stimulus-response functions that can often be fit with power functions having exponents of more than 1.0 (Fig. 10.29A) (Melzack *et al.*, 1963; Adair *et al.*, 1968; S. S. Stevens, 1970; R. H.

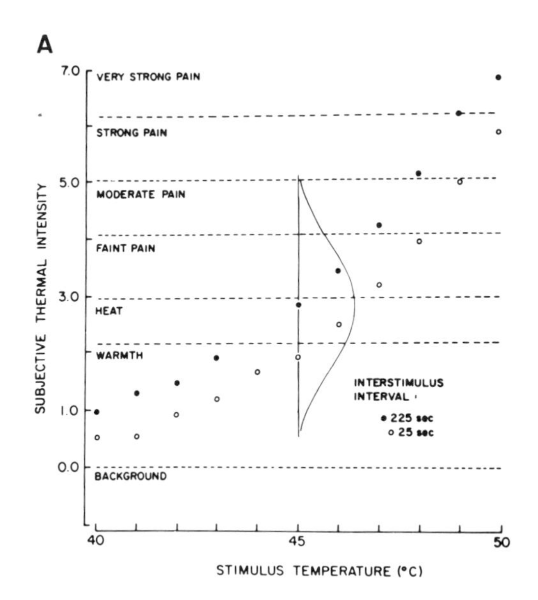

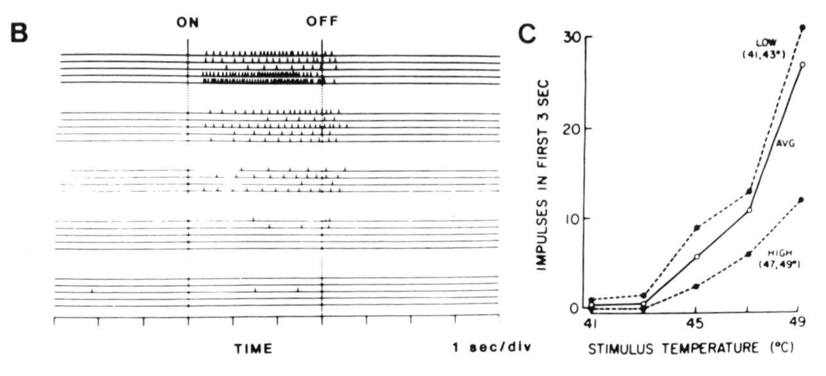

Fig. 10.29. (A) Graph showing the intensity of thermal sensations produced by different thermal stimuli as judged by a human subject. Symbols: •, responses to stimuli delivered at a long interval; ○, responses to stimuli delivered at a short interval. (B) Activity of a C polymodal nociceptor recorded in a monkey in response to graded noxious heat stimuli. (C) Stimulus-response functions from the same data. The dotted curves show the effects of plotting the responses when the preceding stimulus was either low (41, 43°C) or high (47, 49°C). (From R. H. LaMotte and Campbell, 1978.)

LaMotte and Campbell, 1978). It seems probable that an accelerating response helps account for the overwhelming nature of pain reactions.

Graded stimulation of C polymodal nociceptors by using noxious heat pulses produces a linear stimulus-response relationship in cats (Beck *et al.*, 1974), but a power function with an exponent greater than 1 in monkeys (Fig. 10.29B and C) (Croze *et al.*, 1976; R. H. LaMotte and Campbell, 1978). However, the relationship for individual C polymodal nociceptors may be linear in humans (R. H. LaMotte *et al.*, 1982). Presumably, an accelerating stimulus-response curve in human subjects reflects the addition of more nociceptive afferent fibers to the population as the intensity of the noxious stimulus increases. Since Aδ mechanoheat receptors have a higher threshold for noxious heat pulses than do C polymodal nociceptors (Fig. 10.30) (R. H. LaMotte *et al.*, 1982), C polymodal nociceptors must be responsible for the discrimination of heat pain from threshold (43–45°C) to levels that may damage the skin (above 50°C).

Direct comparisons of psychophysical magnitude estimation curves and the stimulus-response curves of individual C polymodal nociceptors in the same human subjects have been made (Van Hees, 1976a,b; Gybels *et al.*, 1979). When discharge rates in C polymodal nociceptors are correlated with sensation, it is found that rates below 0.3 impulses/sec are not associated with pain, whereas rates of 0.4 impulses/sec usually are, and 1.5 impulses/sec invariably are (Van Hees, 1976a,b). In the study by Gybels *et al.* (1979), nociceptive afferent units have thermal thresholds of about 45°C, and can also be activated by squeezing of the skin or by pinprick. The discharge rates of the units correlate well with pain ratings, but poorly with ratings of warmth. The subjects are able to predict stimulus intensity better than individual afferent fibers can. Presumably, magnitude ratings depend upon the integrated input from a number of nociceptors. The results of other studies by the same group suggest that cutaneous pain in humans depends upon spatial and perhaps temporal summation of the activity of nociceptors (Adriaensen *et al.*, 1980).

With repeated noxious heat stimulation in human subjects, pain sensation decreases. The responses of C polymodal nociceptors to the same stimuli are also decreased, perhaps accounting for the reduction in sensation (Adriaensen *et al.*, 1984a). However, during prolonged mechanical compression of the skin, pain tends to increase, despite adaptation of the discharges of C polymodal nociceptors (Adriaensen *et al.*, 1984b). Several explanations are proposed: (1) the amount of pain depends not just upon the activity of nociceptors but also on the ratio between

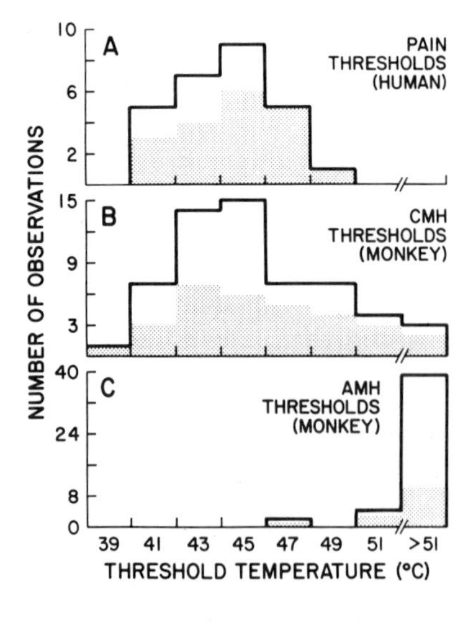

Fig. 10.30. Relationship between the responses of nociceptors to heat stimuli and pain sensation. (A) Threshold for heat pain in human subjects. (B and C) Distribution of thresholds for discharges in C polymodal and Aδ nociceptors in monkeys to heat stimuli. (From R. H. LaMotte *et al.*, 1982.)

the activity of C polymodal nociceptors and Aβ mechanoreceptive afferent fibers, assuming that the latter inhibit nociceptive transmission; (2) the enhanced pain is due to recruitment of more C polymodal nociceptors or to activity of Aδ nociceptors; or (3) summation of the input centrally results in enhancement of pain despite the adaptation of the discharges of nociceptors during maintained noxious stimulation. Several of these explanations support the idea that there may not be a direct correspondence between input in nociceptors and the intensity of pain sensation.

Bromm *et al.* (1984) have been able to avoid receptor adaptation in their studies by activating C polymodal nociceptors with brief laser pulses. They find that no pain results when the nociceptors discharge only a few times, even at instantaneous frequencies of 75 impulses/sec. Temporal summation is required for pain to result. Stimulus intensity corresponds more closely to the total number of discharges than to interspike intervals.

Hyperalgesia. A stimulus that damages the skin may result in hyperalgesia, a condition in which pain can be produced by weaker stimuli than in undamaged skin and pain evoked by suprathreshold stimuli is more intense (Lewis, 1942; Hardy *et al.*, 1952) (see Chapter 9). Two different forms of hyperalgesia may be produced. Primary hyperalgesia occurs in the immediate area of damage, and secondary hyperalgesia develops for a time in a surrounding area. Repeated application of a noxious stimulus may sensitize nociceptors (Bessou and Perl, 1969; Beck *et al.*, 1974; Croze *et al.*, 1976; Fitzgerald and Lynn, 1977) or suppress their responses, depending upon the intensity of the stimulus. Sensitization of nociceptors may account for primary hyperalgesia (Fig. 10.31) (Meyer and Campbell, 1981; R. H. LaMotte *et al.*, 1982, 1983). There is a controversy about the mechanism of secondary hyperalgesia. Lewis (1942) thought that this

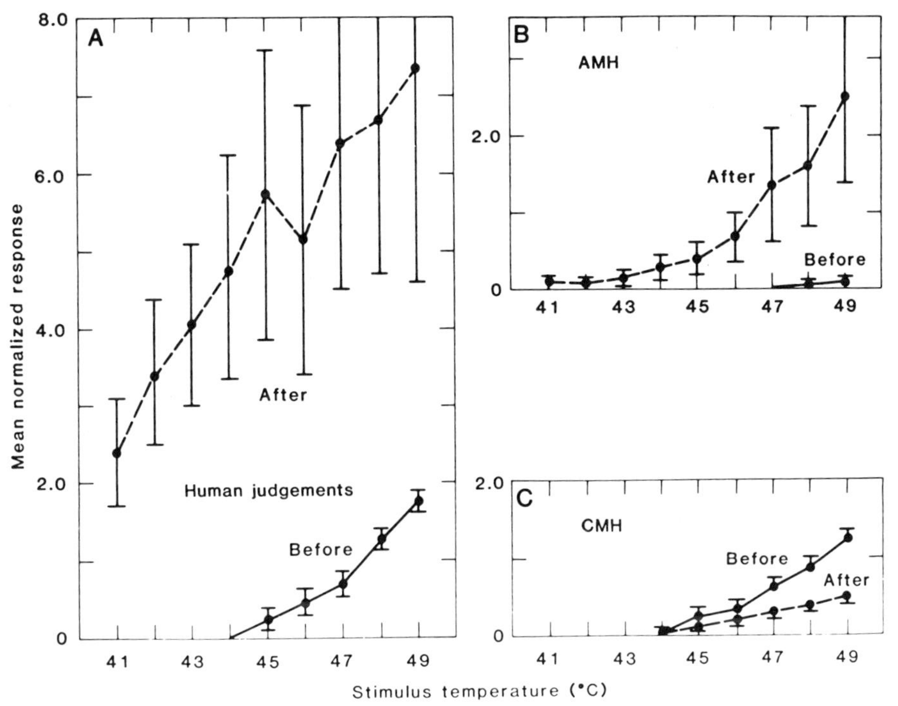

Fig. 10.31. Responses to noxious heat stimuli before and after thermal injury of the skin. (A) Human judgments of pain intensity. (B) Mean responses of a population of Aδ nociceptors. In C are the mean responses of a population of C polymodal nociceptors in monkey skin. (From Meyer and Campbell, 1981; copyright 1981 by the AAAS.)

phenomenon was due to a peripheral mechanism, whereas Hardy *et al.* (1952) proposed a central mechanism. Recent evidence suggests that the secondary hyperalgesia depends at least in part on a central mechanism (R. H. LaMotte *et al.*, 1991; Baumann *et al.*, 1991; Simone *et al.*, 1991).

Chemical Mediation. It is likely that sensitization of nociceptors occurs as part of the process of inflammation. In fact, it has been proposed that the adequate stimulus for nociceptors is a chemical one (Lim, 1970). Damage releases such algesic chemicals as K^+, polypeptides such as bradykinin, monoamines such as serotonin and histamine, and prostaglandins. Administration of bradykinin causes heat nociceptors to respond more vigorously to noxious heat stimulation (Fig. 10.32) (Zimmermann, 1976). Similarly, prostaglandins sensitize nociceptors (Schaible and Schmidt, 1988; Cohen and Perl, 1990). Another substance that causes sensitization of nociceptors is leukotriene B_4 (H. A. Martin, 1990). Inflammation induced by injection of carrageenan results in the enhancement of the responses of polymodal nociceptors to bradykinin (Kirchhoff *et al.*, 1990).

The fact that some nociceptors fail to respond to chemical stimuli might be a consequence of a special morphologic arrangement that might prevent chemical stimuli from having an effect. Alternatively, the membranes of different forms of nociceptors may be specialized in different ways for chemical, mechanical, or thermal stimuli. Evidence for membrane specialization comes from the observation that chemically sensitive nociceptors have different receptor sites for particular chemical agents, such as serotonin and bradykinin (Beck and Handwerker, 1974; Mense and Schmidt, 1974; Hiss and Mense, 1976; Fock and Mense, 1976). Although slowly adapting mechanoreceptors may also respond to algesic chemicals (Fjällbrant and Iggo, 1961; Beck and Handwerker, 1974), it seems reasonable to suppose that the release of algesic agents in an area of damage contributes to the sensitization process (Beck and Handwerker, 1974; Zimmermann, 1976).

Pain Referral. A clinically important aspect of pain is the tendency of patients to refer visceral or deep pain to superficial structures (Head, 1893; Lewis, 1942). A number of theories have been proposed to account for pain referral (Head, 1893; MacKenzie, 1893; Ruch, 1946; Sinclair *et al.*, 1948; see Cervero and Tattersall, 1986). Current evidence appears to support the convergence-projection theory of Ruch (1946) as the most likely mechanism (see Chapter 9). This theory assumes that somatic and visceral nociceptive afferent fibers converge on a common pool of neurons in the spinal cord. These neurons transmit information concerning noxious stimuli to the brain. Visceral pain occurs only rarely, whereas somatic pain is common. In the convergence–projection theory, it is assumed that information transmitted by the neurons receiving convergent inputs from somatic and visceral nociceptors is associated through learning with somatic stimuli, and so when visceral damage does occur, the information arising from visceral nociceptors is misinterpreted to originate from somatic structures. There is now ample

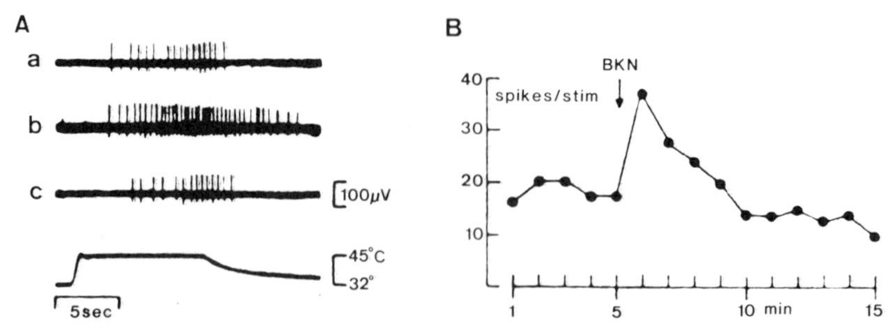

Fig. 10.32. The discharges of a C polymodal nociceptor are shown in A before (a), 1 min after (b), and 6 min after (c) an intra-arterial injection of bradykinin (10 μg). The stimulus was a heat pulse of 45°C. The graph in B shows the increase in discharge as a result of the bradykinin injection. (From Beck and Handwerker, 1974.)

evidence that many nociceptive neurons of the spinal cord, including spinothalamic tract cells, are activated by both somatic and visceral afferent fibers (see Chapter 9). Conversely, few spinal cord neurons respond exclusively to visceral stimulation (Cervero and Tattersall, 1986). In addition to pain referral, viscerosomatic convergence can help explain the superficial tenderness that may develop secondary to visceral disease. Perhaps a change in central excitability occurs, similar to that found in secondary hyperalgesia following damage to the skin.

Inhibitory Interactions. Inhibitory interactions at the spinal cord level are undoubtedly of importance in pain reactions. It is well known that strong stimulation of the skin (sometimes called "counterirritation") helps alleviate pain. Phenomena of this kind form part of the basis for the gate theory of pain (Melzack and Wall, 1965) (see Chapter 5) and led to the reintroduction (Kane and Taub, 1975) of electrical stimulation of peripheral nerves, either directly or transcutaneously, or of the dorsal columns for the treatment of pain (Wall and Sweet, 1967; Shealy *et al.*, 1967, 1970; Nashold and Friedman, 1972; Nielson *et al.*, 1975; Sweet and Wepsic, 1968; D. M. Long, 1973; D. M. Long and Hagfors, 1975; Loeser *et al.*, 1975; Sternbach *et al.*, 1976).

It is still not clear how pain relief is achieved by stimulation of large afferent fibers. The details of the gate theory have been questioned, and it is possible that inhibitory interactions occur in part at other levels of the nervous system than the spinal cord. Furthermore, noxious stimuli or volleys in fine afferent fibers are more effective in inhibiting nociceptive transmission than are volleys in large afferent fibers or innocuous stimuli (J. M. Chung *et al.*, 1984a; Le Bars *et al.*, 1979a,b). The functional meaning of the system of "diffuse noxious inhibitory controls" (DNIC) is still under study, but it appears to involve the activation of pathways descending from the brain stem.

Spinal Pathways. *Projection of Nociceptive Afferent Fibers.* The primary afferent fibers from nociceptors end in the dorsal horn of the spinal cord. Cutaneous Aδ nociceptive fibers end in laminae I and V (Fig. 10.33) (Light and Perl, 1979b), whereas C polymodal nociceptors terminate in laminae I and II (Fig. 10.34) (Sugiura *et al.*, 1986). Many of the neurons in laminae I, II, and V are nociceptive (see Chapter 5), presumably in part owing to direct input from these primary afferent nociceptive fibers. Nociceptive interneurons must play an important role in nociceptive responses, as well as in inhibitory interactions.

Ascending Pathways. Neurons in several different ascending pathways have been found to respond to noxious stimuli (Fig. 10.35) (see Chapters 7–9). Most neurons of the spinothalamic tract in monkeys are nociceptive (Willis *et al.*, 1974; D. D. Price *et al.*, 1978; J. M. Chung *et al.*, 1979; Kenshalo *et al.*, 1979; Surmeier *et al.*, 1988). Some of these cells have a convergent input from rapidly adapting mechanoreceptors (wide-dynamic-range neurons), whereas others respond just to input from nociceptors (nociceptive-specific or high-threshold neurons).

There is evidence that wide-dynamic-range neurons are particularly important for discrimination of different intensities of painful stimulation (Dubner *et al.*, 1989; Kenshalo *et al.*, 1989; Maixner *et al.*, 1986, 1989). Furthermore, there is evidence from human studies that neurons with large myelinated axons in the anterolateral quadrant, presumably wide-dynamic-range spinothalamic neurons, account for the pain that can be evoked by electrical stimulation through a cordotomy needle during percutaneous cordotomy (D. J. Mayer *et al.*, 1975). The activity of wide-dynamic-range neurons shows parallels with second-pain sensation (D. D. Price, 1972; D. D. Price *et al.*, 1977). For example, such neurons show wind-up (Mendell, 1966; Wagman and Price, 1969), which could account for the increment in second-pain sensation that is produced by repetitive stimulation of C fibers. Stimulation of the axons of presumed wide-dynamic-range spinothalamic tract neurons in the human results in sensations of burning, aching, cramping, but sometimes sharp pain (D. J. Mayer *et al.*, 1975).

Spinoreticular and spinomesencephalic tract neurons are also commonly nociceptive, as are spinohypothalamic tract neurons (see Chapter 9). However, these pathways are less likely to be involved in the sensory component of pain than in the motivational–affective component

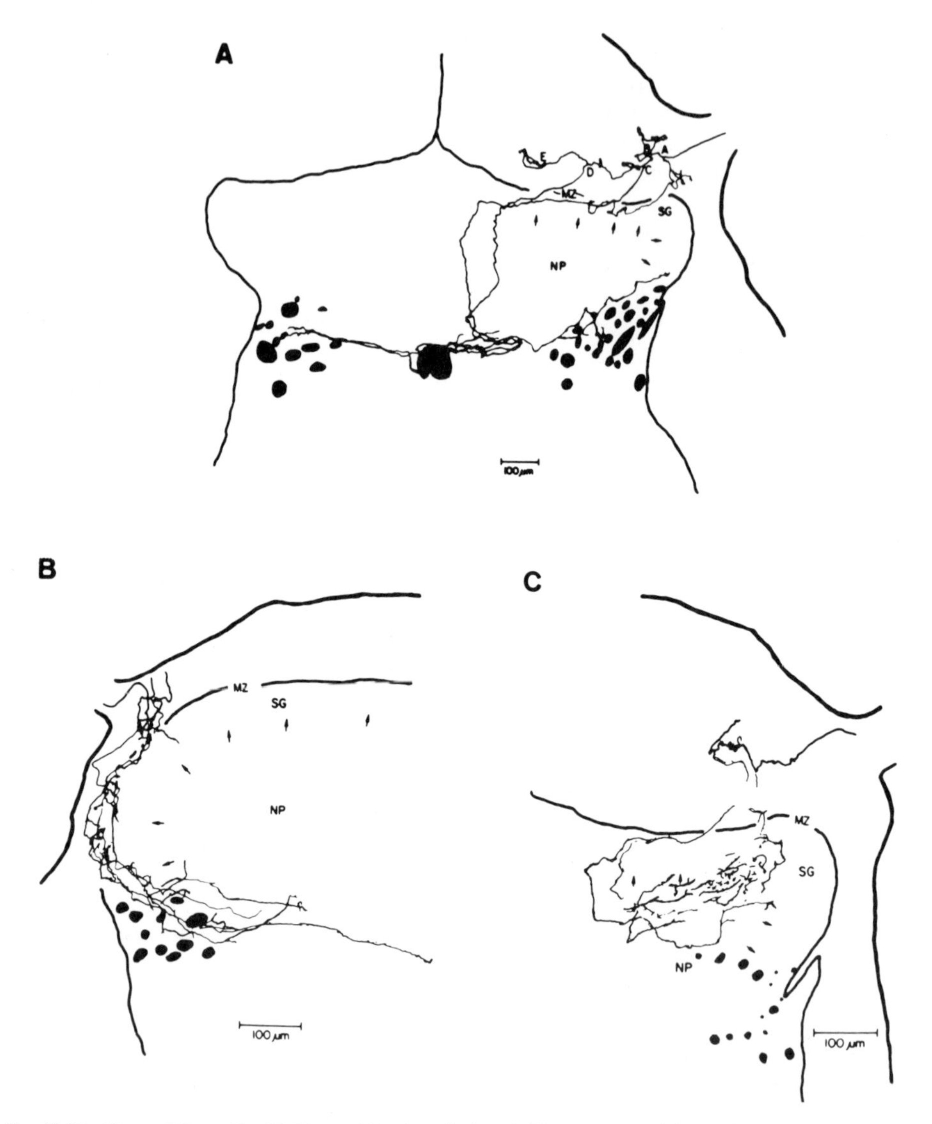

Fig. 10.33. Course followed by Aδ fibers within the spinal cord. The axons were injected intracellularly with HRP after functional identification. The axons in panels A and B were Aδ mechanical nociceptors, and that in panel C was a D hair follicle receptor. Panel A was from a monkey, and panels B and C were from cats. (From Light and Perl, 1979; reproduced with permission.)

(Willis, 1985). There are also neurons in the spinocervical tract and the postsynaptic dorsal column pathway that are nociceptive, and there may even be direct projections of nociceptors through the dorsal columns to the dorsal column nuclei (see Chapters 7 and 8).

Spinothalamic neurons in monkeys provide a model for pain transmission in humans. Many stimuli that cause pain in humans activate spinothalamic neurons strongly. For instance, spinothalamic tract cells in monkeys are excited by noxious mechanical, thermal, and chemical stimulation of the skin; noxious mechanical and chemical stimulation of muscle; and noxious mechanical and chemical stimulation of viscera (reviewed by Willis, 1985). The spinothalamic

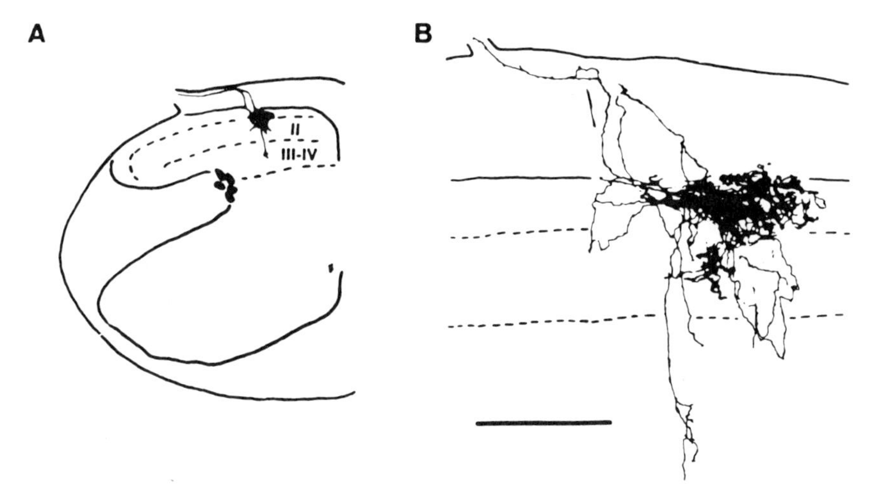

Fig. 10.34. Projection of a C polymodal nociceptor to the dorsal horn in a guinea pig. (A) Terminal arborization of one of the branches of the fiber in laminae I and II. (B) Enlargement of panel A. (From Sugiura *et al.*, 1986; copyright 1986.)

projection to the ventral posterior lateral nucleus is likely to play a role in pain sensation (Kenshalo *et al.*, 1980; J. M. Chung *et al.*, 1986a; Kenshalo and Willis, 1990), whereas the projections to the intralaminar complex and nucleus submedius may be involved in arousal or other motivational–affective behavior (Giesler *et al.*, 1981b; Dostrovsky and Guilbaud, 1988).

In monkeys, ventrolateral cordotomies produce a decreased responsiveness to painful stimuli that lasts for a variable period, depending upon the size and location of the lesion (Vierck and Luck, 1979). The most enduring reduction in nociceptive responsiveness occurs when the entire ventral half of the spinal cord is interrupted. This suggests that one route for pain information outside of the anterolateral quadrant contralateral to the noxious input may be in the ipsilateral anterolateral quadrant. This route may involve the ipsilateral component of the

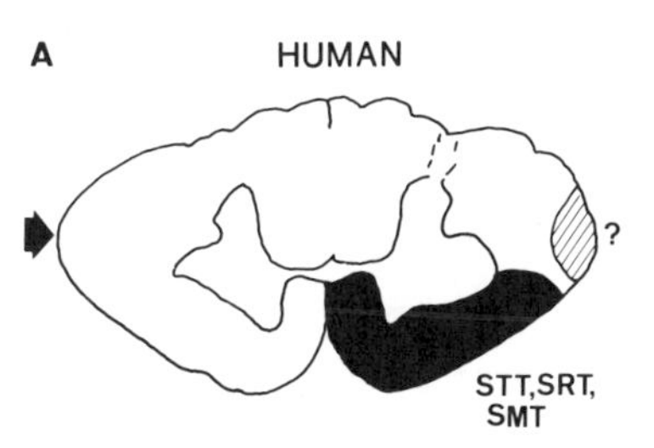

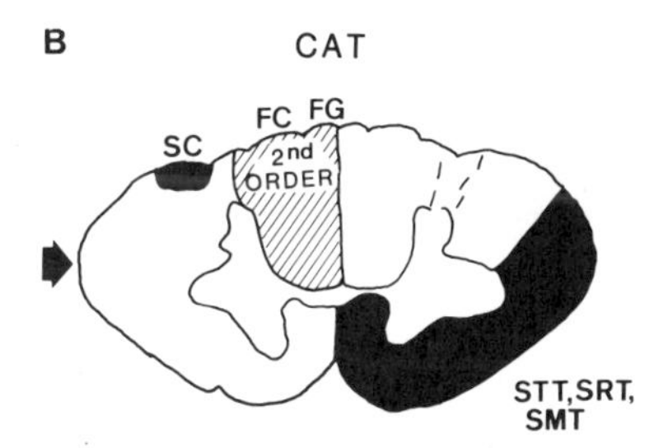

Fig. 10.35. Nociceptive pathways. The major nociceptive somatosensory pathways in humans are shown in panel A to lie in the anterolateral quadrant contralateral to the input (arrow) and include the spinothalamic tract (STT), the spinoreticular tract (SRT), and the spinomesencephalic tract (SMT). The possibility of a dorsolateral STT in humans is indicated by the hatched area. Other pathways, which assume more significance in animals, are shown in panel B to include the spinocervical tract (SC) and the postsynaptic dorsal column pathway (second order) in the fasciculus cuneatus and gracilis (FC, FG) ipsilateral to the input.

spinothalamic tract (see Chapter 9). Other pathways that include nociceptive neurons and that ascend in the dorsal part of the cord could also be involved in the pain that returns following an initially successful anterolateral cordotomy. These have been studied most thoroughly in cats and include the spinocervical tract, the postsynaptic dorsal column pathway, and the dorsal spino-thalamic tract (Fig. 10.32B). The small number of nociceptive neurons that have been found in the dorsal column nuclei in monkeys could have been activated by way of the postsynaptic dorsal column pathway or possibly even by direct projections of collaterals of nociceptors up the dorsal column (see Chapter 7). The recently described dorsal spinothalamic tract is a crossed pathway in the DLF of monkeys that may be important in transmitting information about pain (Apkarian and Hodge, 1989b,d). However, according to Craig (1989, most of the axons of lamina I cells actually ascend in the middle part of the lateral funiculus, rather than in the DLF.

Clinical Correlations. Clinical studies indicate that the chief pathway carrying nociceptive information to the human brain is crossed and ascends in the anterolateral quadrant (Fig. 10.32A) (see Chapter 6). For example, anterolateral cordotomies can produce long-lasting analgesia on the contralateral side of the body below the lesion (see the review by White and Sweet, 1969). This observation indicates that the anterolateral quadrant of the spinal cord is normally necessary for human pain sensation. The area of pain relief corresponds somatotopically to the sector of spinal cord interrupted, and all forms of pain sensation (as well as itch) are abolished if the cordotomy is done bilaterally. The patient of Noordenbos and Wall (1976) who had a transection of all of the spinal cord except one anterolateral quadrant could still identify pinpricks applied to the lower extremity contralateral to the intact anterolateral quadrant, indicating that the pathways in this part of the cord are sufficient for pain sensation.

There are presumably alternative pathways that can assume a role in the mediation of pain sensation, since in about half of the cases an anterolateral cordotomy provides relief of pain for a matter of months to a year or so, even though the surgical lesion interrupts most of the anterolateral quadrant (White and Sweet, 1969). If there is a component of the spinothalamic tract in the posterior lateral funiculus in humans, this could contribute to the return of pain; other possibilities include the ipsilateral component of the spinothalamic tract, the spinocervical tract, and the postsynaptic posterior column pathway.

Summary

In summary, there are several types of pain, including superficial pricking and burning pain, deep pain, and visceral pain. Besides the sensation of pain, the pain reaction includes somatic and autonomic reflexes, endocrine changes, and motivational–affective responses, including aversive behavior and arousal. Painful stimuli are noxious; i.e., they threaten or produce damage. Neurons that are involved in the pain reaction may contribute to one or more of the aspects of the total reaction. Pain secondary to damage to the nervous system is important but poorly understood. Nociceptors are involved in signaling painful stimuli. There are several kinds of cutaneous nociceptors that can be activated by one or more kinds of noxious stimuli: mechanical, thermal, or chemical. Other nociceptors are found in muscle, joints, and viscera. Dull or burning pain is produced by stimulation of small groups of C polymodal nociceptors through a microneurography electrode. The intensity of the pain is proportionate to the number of nerve impulses in nociceptors, at least for brief stimuli. However, central mechanisms may be involved in the pain produced by prolonged stimuli, since the pain may increase despite a reduction in afferent discharges. Single stimuli or low frequencies of stimulation do not produce pain, suggesting that central summation is required for pain sensation. Some C polymodal nociceptors evoke itch rather than pain. The psychophysical response to noxious heat pulses applied to the skin is a power function with an exponent of 1.0 or more; a similar curve relates stimulus strength to the responses of C polymodal nociceptors. Sensitization of nociceptors may occur following damage and may be responsible for primary hyperalgesia. Secondary hyperalgesia may involve a central mechanism. The adequate stimulus for nociceptors may be a chemical event, although explanations are required for differences between different kinds of nociceptors. The primary afferent fibers of nociceptors terminate in laminae I, II, and V of the spinal cord dorsal horn. Nociceptive neurons include wide-dynamic-range and high-threshold cells. There is evidence that discrimination of painful stimuli is likely to depend on signals carried by wide-dynamic-range neurons. The main pathway for pain sensation in humans (and monkeys) crosses at the segmental level and ascends in the anterolateral quadrant. The spinothalamic tract is a good candidate for the sensory pathway for pain, although the spinoreticular and spinomesencephalic tracts presumably also contribute to the motivational–affective aspects of the pain reaction. Recurrence of pain after an initially successful anterolateral cordotomy is presumably due to nociceptive transmission in pathways such as the ipsilateral spinothalamic tract and the spinocervical tract and pathways in the posterior columns. If there is spinothalamic tract in the posterior lateral funiculus of humans, this could also contribute. Referral of visceral pain is best accounted for by the convergence–projection theory, which is based on the convergence of viscerosomatic information onto nociceptive transmission cells, such as those of the spinothalamic tract. Inhibitory interactions are an important aspect of pain mechanisms and offer an avenue for therapy.

TEMPERATURE

There are two distinct thermal sensations: cold and warm (Hensel, 1973a, 1974). When the skin is kept at 32–34°C, no thermal sensation is noted (Hensel, 1973a). However, when the temperature of the skin is altered in either direction from this indifferent level, a sense of warming or of cooling results. After the skin temperature has been changed to a new level between 28 and 37.5°C, thermal sensation adapts within about 25 min (Kenshalo and Scott, 1966).

The threshold for thermal sensation depends upon a number of factors, including the rate of temperature change, the amount of change, the surface area affected, and the original or "adapting" temperature (Hensel, 1950; cf. Molinari *et al.*, 1977). Two steady-state thermal stimuli that differ by less than 0.5°C can be distinguished when the stimuli are applied simultaneously (Erickson and Poulos, 1973). Actually, discrimination is 0.5°C when the steady-state temperatures are near midrange (27–33°C), but decreases to 0.3°C at higher and lower steady-state temperatures. Erickson and Poulos (1973) suggest that this difference in sensitivity results from the nature of the population response, since cold receptors show little change in discharge at midrange, but sizable changes at higher and lower temperatures.

At the extremes of the temperature range (above 45°C and below 18°C) the subject feels heat pain (Hardy *et al.*, 1951; Neisser, 1959; R. H. LaMotte and Campbell, 1978) or cold pain (Wolf and Hardy, 1941). Thermal pain is evoked by a different set of receptors from those responsible for thermal sensation (see the section on pain).

In addition to thermal sensation, changes in surface temperature result in thermoregulatory responses in homoiothermic animals (Benzinger, 1969; Nadel and Horvath, 1969; Hensel, 1973b; Dykes, 1975). Despite their importance, such thermoregulatory responses will not be considered here. Nor will the activity of reptilian infrared detectors be discussed (see Hartline, 1974).

Receptors

Thermal sensation depends upon two different kinds of receptors: cold receptors and warm receptors. Cold and warm sensations can be evoked by focal stimulation of cold and warm spots (see Fig. 1.1) (Blix, 1884; Donaldson, 1885; von Frey, 1906; Rein, 1925; Dallenbach, 1927; Strughold and Porz, 1931), as well as by changing the temperature of broad areas of the skin. Recordings have been made in humans from afferent fibers innervating cold or warm receptors (Hensel and Bowman, 1960; Knibestöl and Vallbo, 1970; Torebjörk, 1974; Konietzny and Hensel, 1975, 1977, 1979, 1983; Järvilehto and Hämäläinen, 1979; Hallin *et al.*, 1981; Konietzny, 1984). The response properties of human thermoreceptors resemble those of cold and warm receptors in animals (see Chapter 2).

Cold receptors in humans are thought to be innervated by Aδ fibers and warm receptors by C fibers (Fruhstorfer *et al.*, 1974; MacKenzie *et al.*, 1975; Konietzny and Hensel, 1975; Fruhstorfer, 1976; Norrsell and Ullman, 1978; Hallin *et al.*, 1981). However, some cold receptors in monkeys are supplied by C fibers (Duclaux and Kenshalo, 1973), and the conduction velocities of fibers supplying cold receptors in human subjects were in the C-fiber range (Konietzny, 1984). Evidently, cold receptors in primates, including humans, are supplied by Aδ and C fibers.

Thermoreceptors are insensitive to mechanical stimuli, responding little or not at all even to intense mechanical stimuli (see Chapter 2). However, chemical agents can evoke cold or warm sensations by activating thermoreceptors. For example, menthol produces a cold sensation whether applied topically or injected intravenously (Hensel, 1973a) and has been shown to excite cold receptors in animals (Hensel and Zotterman, 1951c; Dodt *et al.*, 1953).

Several observations argue against the possibility that the sense of warmth is due to a reduction in the activity of cold receptors by warming. One line of evidence is that cold receptors stop discharging when the skin temperature approaches 40°C, yet the sense of warmth increases in the range between 40 and 45°C (see Konietzny and Hensel, 1975). Furthermore, warm sensation, but not cold sensation, can be eliminated by a level of local anesthesia that blocks C fibers (Fruhstorfer *et al.*, 1974). A complementary experiment is ischemic nerve block, which reaches a stage that eliminates all sensations except warmth and pain because of blocked conduction in A fibers (D. Clark *et al.*, 1935; MacKenzie *et al.*, 1975). Conversely, very little information about cooling can be signaled by a reduction in the discharges of warm receptors, since these cease discharging with cooling. There is a significant change in the discharge rate of warm receptors only when the adapting temperature is 35°C or higher, and then only for cooling pulses of up to 2°C (K. O. Johnson *et al.*, 1973). Finally, spinal cord disease can produce a dissociated loss of either cold or warm sensation (see Chapter 6) (Head and Thompson, 1906). Thus, the evidence clearly indicates independence of the channels for warm and cold sensations.

In addition to actions on specific thermoreceptors, temperature changes affect mechanoreceptors (Werner and Mountcastle, 1965; Iggo, 1969; Duclaux and Kenshalo, 1972; Booth and Hahn, 1974; Burton *et al.*, 1972; H. A. Martin and Manning, 1969, 1972; Poulos and Lende, 1970b; Hahn, 1971; see Konietzny, 1984). For instance, cutaneous SA I and SA II receptors can be excited by cooling. These thermally induced changes may account for Weber's illusion that objects feel heavier if cold (Weber, 1846; Hensel and Zotterman, 1951b; Witt and Hensel, 1959). However, the responses of slowly adapting mechanoreceptors cannot account for thermal sensation. For example, K. O. Johnson *et al.* (1973) observed that the stimulus-response curves of slowly adapting mechanoreceptors have a form similar to that of cold receptors, but that the mechanoreceptors are 20 times less sensitive than are the thermoreceptors. Using a mathematical analysis, they have determined how the combined activity of a population of mechanoreceptors might code for thermal stimulus intensity (see below for the results of a similar analysis applied to thermoreceptors). It is calculated that it would take more than 4000 slowly adapting receptors to account for the human discriminative capacity, and yet there are probably fewer than 100 slowly adapting receptors in the area stimulated. Furthermore, the peak of thermal responsiveness and the changes in responses produced by alterations in the adapting temperature in the responses of SA I receptors, at least, do not match psychophysical responses in cold sensation (Kenshalo, 1970; Duclaux and Kenshalo, 1972). In humans, thermal stimuli are generally unable to activate SA I or SA II receptors, although such stimuli could modify responses to mechanical stimuli; SA I receptors are affected only by large temperature changes of 10–20°C; and there is no report of a thermal sensation when SA I receptors are activated (Konietzny, 1984). Pressure causes no change in thermal sensation; that is, there is no reverse Weber effect (Zimmermann and Stevens, 1982). Finally, touch can still be felt after local anesthesia has blocked cold and warm sensations, indicating that SA receptors are not sufficient to evoke thermal sensation (Fruhstorfer *et al.*, 1974).

Cold Receptors. The receptive fields of cold receptors are small spots about 1 mm in diameter, and in some species a given thermoreceptive afferent fiber supplies just one spot (Chapter 2). In monkeys, a thermoreceptive afferent fiber may innervate two to eight spots, and the largest dimension of a spot may vary from 1 to 5 mm (Duclaux and Kenshalo, 1973, 1980; Darian-Smith *et al.*, 1973; Dubner *et al.*, 1975; Kenshalo and Duclaux, 1977; R. R. Long, 1977). However, in human subjects the receptive fields of the cold receptors so far investigated have consisted of a single spot (Järvilehto and Hämäläinen, 1979). The receptive fields of the specific thermoreceptors correlate well with the cold and warm spots of human psychophysical experiments.

Anatomic findings in the cat indicate that the cold receptors have a distinctive morphology (Hensel *et al.*, 1974). Terminals that can be identified as cold receptors are in the basal layer of the epidermis, an observation that is consistent with the prediction that cold receptors are located about 0.18 mm from the surface of the skin, based on studies investigating latencies of responses of thermoreceptors to the application of known thermal gradients to the tongue of the cat (Hensel *et al.*, 1951). Similar studies in which sensory responses in human subjects were used as the endpoint suggest that cold receptors are on average 0.15 mm from the surface of the skin (Bazett *et al.*, 1930).

Intensity Coding. The afferent fibers of cold receptors have a background discharge at neutral skin temperatures. In animals, when the temperature is lowered, the discharges of cold receptors increase in frequency (Fig. 10.36A and B) until a temperature of about 25°C is reached, below which the discharge rate decreases. When the temperature is elevated above the neutral level, cold receptors show a decreased discharge (see Chapter 2). Thus, the curve relating static discharge rate to temperature for cold receptors is not monotonic but bell shaped (see Fig. 2.12). The curve for the static discharges of cold receptors is very broad, with discharges in some cold receptors occurring at temperatures as low as 5°C and as high as 43°C. However, most cold receptors cease discharging at 10°C and at 40°C. The maximum discharge rate for a given receptor occurs at a temperature somewhere between 18 and 34°C (Iggo, 1969; see Chapter 2 and reviews by Hensel, 1973a, 1974). A similar behavior is seen in cold fibers of humans (Konietzny and Hensel, 1983).

Some cold fibers also discharge more vigorously when the temperature is elevated to 45–50°C. This response is called the "paradoxical response" of cold receptors (Dodt and Zotterman, 1952b; Kenshalo and Duclaux, 1977), and it presumably accounts for the "paradoxical cold" sensation which is reported by human subjects when the skin is heated above 45°C (Hensel, 1973a). The tendency for a given cold receptor to show a paradoxical response appears to be related to the core body temperature, an effect apparently mediated by changes in vasomotor tone (Long, 1977). A paradoxical response has been recorded from at least one cold receptor in a human subject (Konietzny, 1984).

The bell-shaped stimulus-response curve for cold fibers makes it difficult to understand how a human subject can discriminate the intensity of thermal stimuli (cf. Järvilehto, 1973). For example, the same mean frequency of discharge will be recorded from the afferent fiber of a given cold receptor when the temperature is either above or below that which elicits the maximum discharge (neutral zone). One theory to account for discrimination of temperatures above and below the neutral zone is based on the observation that when the temperature is

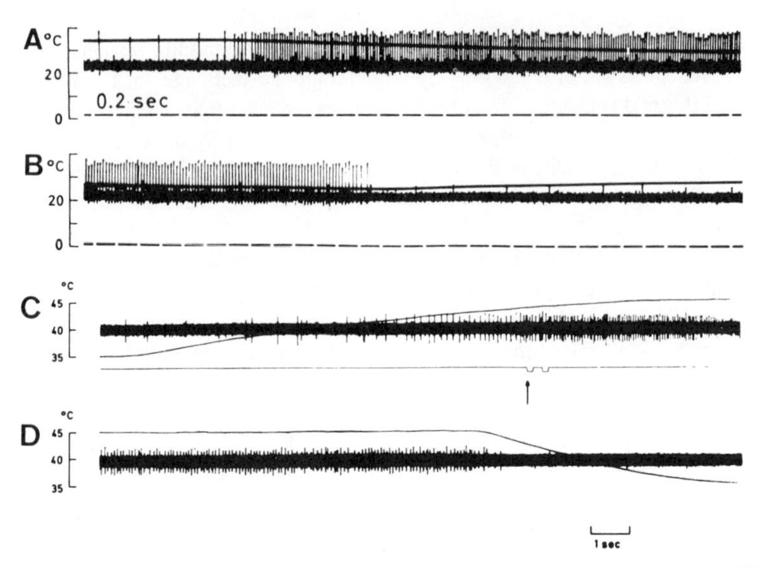

Fig. 10.36. (A and B) Response of a cold receptor in human skin during cooling from 34 to 26°C and then rewarming. (From Hensel and Boman, 1960.) (C and D) Activity of a warm receptor from human skin during warming from 35 to 45°C and then recooling. The arrow in panel C is the time at which the subject reported a warm sensation. (From Konietzny and Hensel, 1975.)

lowered sufficiently, the afferent fibers from cold receptors tend to discharge in bursts (see Fig. 2.11). (Iggo, 1969; Dubner *et al.*, 1975; Kenshalo and Duclaux, 1977; Long, 1977) (see Chapter 2). It has been proposed that the bursts code for levels of temperature below that required to evoke the maximum discharge rate in cold receptors (Iggo, 1969; Poulos, 1971; Dykes, 1975). Although the average firing rate may be the same for two different temperatures above and below the level that evokes the peak firing rate, the number of spikes within a burst increases linearly as the temperature is lowered below that evoking the maximum firing rate. However, there are several objections to this idea. One is that few of the afferent fibers from human cold receptors show this bursting behavior (Konietzny and Hensel, 1983; Konietzny, 1984). Instead, most show a regular discharge. Furthermore, in animal experiments, recordings of the thermal responses of central neurons generally do not reflect the bursting activity of the afferent fibers (Poulos, 1975; Iggo and Ramsey, 1976; Dostrovsky and Hellon, 1978; Kanui, 1988).

Other explanations of intensity coding are (1) that the intensity depends upon the total amount of neural activity that is evoked by a thermal stimulus (Erickson, 1973) and (2) that the recruitment of other types of cold receptors (Duclaux *et al.*, 1980; R. H. LaMotte and Thalhammer, 1982) with monotonic stimulus-response curves at low temperatures (10–30°C) accounts for the ability to discriminate between these temperatures and temperatures above neutral.

Detection Threshold. Cold receptors are sensitive not only to static changes in temperature, but also to the rate of thermal change. Cold receptors have dynamic responses resulting in overshoots and undershoots in their discharge rate when rapid thermal stimuli are applied (see Fig. 2.11). The sensation of cooling correlates well with the dynamic responses of cold receptors. For example, the threshold for detection of cooling transients is about 0.02–0.05°C in humans (Hardy and Oppel, 1938; Kenshalo *et al.*, 1968), well below the threshold for discrimination of static temperature levels. The thresholds of thermoreceptors in primates are capable of accounting for psychophysical thresholds in this range (Darian-Smith *et al.*, 1973; K. O. Johnson *et al.*, 1973). Furthermore, at a given adapting temperature, the response of a cold receptor is linearly related to the amplitude of a step or a ramp change in temperature (Darian-Smith *et al.*, 1973; Kenshalo and Duclaux, 1977).

Activation of a single cold receptor may not evoke a cold sensation unless the discharge rate reaches a high level. This presumably reflects the need for summation to reach the "central threshold" for cold sensation (cf. Hensel, 1952). Järvilehto (1973) estimates that a cold sensation evoked from a cold spot can be correlated with a discharge frequency of 80 impulses/sec in a single cold fiber.

Thermal Discrimination. The information conveyed by a single cold receptor to the central nervous system is insufficient to account for the human capacity to discriminate temperature (K. O. Johnson *et al.*, 1973), which must therefore depend upon the analysis of the activity of a population of receptors. K.O. Johnson *et al.* (1973) calculate that the information from at least 16 cold fibers is required, assuming that all of the fibers operate independently. They estimate that the stimulus that they use in their experiments activates some 50–70 cold fibers, which is a large enough number to account for the human discriminative capacity, even if a part of the population of cold fibers does not discharge independently.

In human psychophysical tests, decisions can be made about which of two cold stimuli is colder within 2 sec (K. O. Johnson *et al.*, 1973). Thus, the decision is made on the basis of the initial neural activity generated at the receptor level. The dynamic response is presumably the effective component. When ramp stimuli are used, a change in the slope of the ramp does not alter the estimate of intensity, but rather the time at which the maximum intensity is experienced (Kenshalo and Duclaux, 1977). Thus, the neural system for judging thermal stimulus intensity acts like an integrator, summing the neural activity during the dynamic response.

Reaction Time. The reaction time for the development of a sensation of coolness is distinctly slower than that for a sensation of warmth; this can be accounted for on the basis of the response latencies of cold and warm fibers (Kenshalo and Duclaux, 1977; Duclaux and Kenshalo, 1980).

Warm Receptors. Warm receptors in human subjects have spot- like receptive fields (Konietzny and Hensel, 1979; Hallin *et al.*, 1981; Konietzny, 1984) (cf. Chapter 2). Their structure is unknown. Warm receptors have been estimated to lie about 0.6 mm below the surface of human skin (Bazett *et al.*, 1930).

Intensity Coding. When the skin temperature is elevated from a neutral level of 32°C to 43°C, warm receptors discharge more vigorously (Fig. 10.36C and D), and the intensity of the sensation of warmth increases in parallel (Konietzny and Hensel, 1977; Hallin *et al.*, 1981; Konietzny, 1984). As in monkeys (Iggo, 1969; Duclaux and Kenshalo, 1980), human warm receptors can be subdivided into two populations, a low-threshold group that has static discharges at 32°C and a high-threshold group that begins to discharge only when the skin temperature is raised to 35-38°C. The low-threshold warm receptors have bell-shaped stimulus-response curves with maxima of 38–43°C. The high-threshold warm fibers have maximum firing rates at or above 45°C (Konietzny and Hensel, 1979; Konietzny, 1984). As the temperature increases into the noxious range, the discharges of the high-threshold warm receptors decrease (cf. Iggo, 1969; Duclaux and Kenshalo, 1980; R. H. LaMotte and Campbell, 1978; Konietzny and Hensel, 1979; Hallin *et al.*, 1981) (see Chapter 2). Thus, as for cold receptors, the stimulus-response curves of warm receptors are not monotonic; however, activity in two populations of warm receptors may permit discrimination of temperatures between the neutral level and the noxious level of intensity.

K. O. Johnson *et al.* (1979) relate the stimulus-responses curves of warm fibers innervating the palms of monkeys to human psychophysical response functions. A satisfactory match between the curves is obtained provided that the responses of a population of warm fibers are combined. It is concluded that activity in a population of warm fibers is just sufficient to encode the intensity of warm stimuli.

It is unlikely that the high-threshold warm receptors contribute to pain sensation, since these receptors are active at temperatures well below the pain threshold (Hensel and Iggo, 1971) and stop discharging as the stimulus intensity is increased. However, high-threshold warm receptors may evoke the sensation of hot (Konietzny and Hensel, 1979; Konietzny, 1984). Heat pain in the range of 43-45°C can be attributed to C polymodal nociceptors (see the section on pain).

Cutaneous warm receptors do not show a paradoxical response to extreme cooling (Hensel, 1973a).

Detection Threshold. Like cold receptors, warm receptors have dynamic responses to changes in skin temperature. The dynamic responses appear to account for the ability of human subjects to detect very small thermal stimuli. However, when a small area of the skin is stimulated, input from a few thermoreceptors may be insufficient to elicit a thermal sensation; this may reflect a "central threshold" for thermal sensation that requires summation of input from thermoreceptors (Hensel, 1952; Konietzny and Hensel, 1979).

K. O. Johnson *et al.* (1979) provide evidence in support of these proposals. A statistical model is used to compare the capacity of human subjects to discriminate different intensities of warm stimuli with the information conveyed by warm fibers. It is concluded that individual warm fibers do not provide sufficient information to make the discriminations; instead, the information carried by a population of some 50 warm fibers would be required.

Spinal Pathways. *Projections of Thermoreceptors.* Cutaneous Aδ and C fibers terminate in laminae I, II, and V. Therefore, it is likely that the afferent fibers from cold and warm receptors also end in one or more of these layers, especially the superficial layers.

Thermoreceptive Processing. Recordings have been made from neurons in laminae I–III (and also from some deeper cells) in rats, cats, and monkeys that respond as if they receive input from cold or warm receptors (Christensen and Perl, 1970; Courtney *et al.*, 1972; Hellon and Misra, 1973a; Kumazawa *et al.*, 1975; Kumazawa and Perl, 1976; Iggo and Ramsey, 1976; Kanui, 1988; cf. Poulos, 1975; Dostrovsky and Hellon, 1978). The second-order neurons receiving input from cold receptors rarely have the bursting firing pattern so characteristic of many cold receptors at certain temperatures, perhaps because the second-order cells receive a convergent input from many receptors that are presumably firing out of phase (Poulos, 1975; Iggo and Ramsey, 1976; Dostrovsky and Hellon, 1978; Kanui, 1988). Cold pulses may produce dynamic and then static responses in the second-order cells; these responses are graded with stimulus intensity (Iggo and Ramsey, 1976). However, some units have only static or only dynamic responses (Hellon and Misra, 1973a).

Ascending Pathways. In some studies, a high proportion of these central thermoreceptive neurons of the dorsal horn are found to project contralaterally at least to the level of the cervical spinal cord (Kumazawa *et al.*, 1975; Iggo and Ramsey, 1976). Cells in lamina I identified as spinothalamic neurons have now been observed that respond to small temperature changes in rats, cats, and monkeys (Craig and Kniffki, 1985; Ferrington *et al.*, 1987d; Kanui, 1988). Many of these spinothalamic neurons also receive a convergent input from receptors other than specific thermoreceptors. However, some spinothalamic cells in cats are exclusively thermoreceptive (Craig and Kniffki, 1985).

The pathway conveying information about thermal stimulation of the skin in humans is generally thought to accompany that mediating pain, since anterolateral cordotomies usually produce contralateral deficits in temperature and pain sensations with the same segmental distribution (White and Sweet, 1969). However, lesions of the spinal cord sometimes produce a differential loss of thermal or pain sensation (Head and Thompson, 1906). This has been attributed to a partial separation of the temperature and pain pathways, with the thermal projection lying more posterior than the pathway mediating pain sensation (Foerster and Gagel, 1932). Norrsell (1979, 1983) has demonstrated that the ascending thermosensory pathway is located in the middle part of the lateral funiculus. This may relate to the finding that the axons of neurons of lamina I ascend more dorsally than do those of spinothalamic neurons of deeper laminae, both in cats and in monkeys (M. W. Jones *et al.*, 1985, 1987; Apkarian and Hodge, 1989a,b; Craig, 1989). It will be important to find whether the axons of lamina I spinothalamic tract cells in humans also ascend more posteriorly in the lateral funiculus than do the axons of spinothalamic neurons of deeper laminae (Fig. 10.37).

One destination of the thermal pathways in the brain is the thalamus. Units have been found in the rat ventrobasal complex that respond to the warming of scrotal skin (Hellon and Misra, 1973b; Hellon and Mitchell, 1975), and there have been other reports of thalamic units in the cat that respond to thermal stimulation of the skin of the extremities or of the tongue (see, e.g., Landgren, 1960; Poulos and Benjamin, 1968; Burton *et al.*, 1970; H. A. Martin and Manning, 1971). It is not known whether the pathway to the thalamus in these cases is direct. The thermoreceptive spinothalamic cells found in cat generally project to the medial rather than to the lateral thalamus (Craig and Kniffki, 1985).

Clinical Correlations. In humans, thermal sensation is lost on the contralateral side below the lesion following anterolateral cordotomy (see Chapter 6). A single anterolateral quadrant is sufficient to permit thermal sensation on the contralateral side (Noordenbos and Wall, 1976). Furthermore, thermal sensations may be evoked by stimulation in the anterolateral quadrant of the cord (Sweet *et al.*, 1950). Thus, thermal sensation in humans is mediated by a crossed pathway in the anterolateral quadrant, presumably the spinothalamic tract (Fig. 10.37).

As mentioned above, there is some evidence that the axons carrying temperature information are at least partially segregated from those carrying pain information (see also Chapter 6).

The tract of Lissauer appears to distribute thermal information over several segments along the ipsilateral side of the cord, as it does pain information, since Lissauer tractotomy can alter the level of thermoanesthesia when combined with contralateral cordotomy (Hyndman, 1942).

Summary

In summary, there are two thermal sensations, cold and warm. These are signaled by two kinds of receptors, cold and warm receptors. Cold receptors in humans appear to be supplied by both Aδ and C fibers and warm receptors by C fibers. One thermoreceptive afferent fiber supplies a single spotlike receptive field in humans. Each of these appears to correspond to the cold and warm spots that can be mapped in psychophysical studies. The morphology of cold receptors, at least in the cat, is distinctive; the structure of warm receptors is unknown. Thermoreceptors do not respond to mechanical stimuli, but may be activated by certain chemical agents. Slowly adapting mechanoreceptors, while affected by thermal changes, cannot account for thermal sensation. Thermoreceptors may have a static discharge at neutral skin temperatures. Lowering the skin temperature increases the discharges of cold receptors and slows the discharges of warm

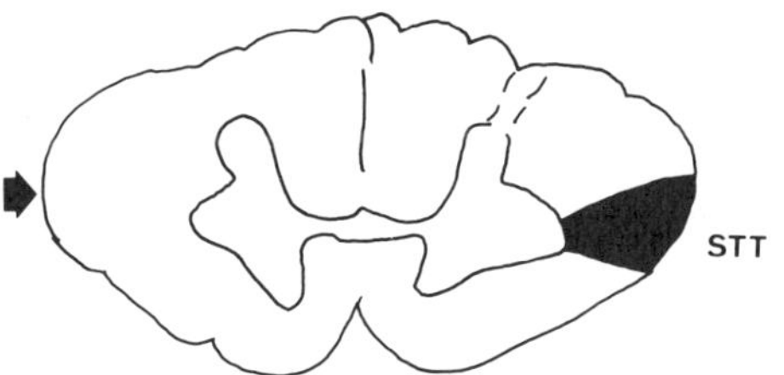

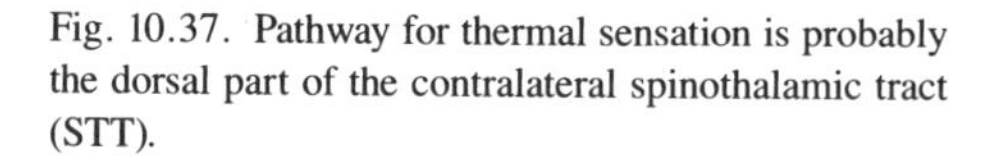

Fig. 10.37. Pathway for thermal sensation is probably the dorsal part of the contralateral spinothalamic tract (STT).

receptors; raising the skin temperature does the opposite, except at high skin temperatures, which may evoke a paradoxical discharge in cold receptors. The bursting discharges of cold receptors may serve as a code for the lower range of temperature; alternatively, the intensity of cold sensation may depend upon the total activity in the population of cold receptors or on recruitment of a high-threshold population of cold receptors as the skin temperature is lowered. Dynamic responses help account for the low thresholds of humans for recognizing thermal transients. The dynamic response is probably responsible for the speed at which thermal discriminations are made. The sensory pathway carrying thermal information to the brain in humans is located in the anterolateral quadrant and is crossed. Presumably, it is the spinothalamic tract. Spinal cord lesions can interrupt thermal sensation independently of pain sensation, suggesting that the pathways for thermal and pain senses are at least partially separate.

VISCERAL SENSE

The most prominent visceral sense is pain. This has already been discussed. The other visceral sensations mediated by afferent fibers entering the spinal cord include visceral fullness and satiation (for a review, see Leek, 1972). There is apparently no thermal sensation in most viscera, although there may be abdominal thermoreceptors that contribute to thermoregulation (Riedel, 1976).

Receptors

Pacinian corpuscles are abundant in the abdominal cavity, at least in cats, and some of these discharge in phase with cardiovascular pulsations; however, their potential contribution to circulatory regulation has not been completely explored (Gammon and Bronk, 1935; Gernandt and Zotterman, 1946; Leitner and Perl, 1964).

The receptors found in the mesentery and along the serosal surface of many visceral organs are activated by movement and by distention of these organs (Paintal, 1957; Todd, 1964; Bessou and Perl, 1966; Ranieri *et al.*, 1973; Morrison, 1973, 1977; Floyd and Morrison, 1974; Floyd *et al.*, 1976). These do not seem to discharge in any precise relationship to changes in intraluminal pressure or volume (Morrison, 1977), but they could contribute to a sense of fullness or, alternatively, to the pain of excessive distention.

Other receptors found in the smooth muscle of such organs as the gastrointestinal tract and the bladder can be activated either by distention or by contraction (Iggo, 1955; Winter, 1971; see Leek, 1972). Such "in-series" receptors may trigger reflex emptying, but it is reasonable to suppose that they also contribute to sensations associated with emptying or to a sense of fullness; in pathologic states, they may also produce pain (Leek, 1972). There are also receptors that supply the mucosal linings of visceral organs. Some of these may be nociceptors.

Spinal Pathways. *Projections of Visceral Afferent Fibers.* Visceral afferent fibers have been traced by transganglionic labeling with HRP into laminae I, II, V–VII, and X of the sacral spinal cord (Fig. 10.38) (De Groat *et al.*, 1978, 1981; C. Morgan *et al.*, 1981; Nadelhaft *et al.*, 1983; Nadelhaft and Booth, 1984; Kawatani *et al.*, 1990). In the thoracic cord, the most dense projections of visceral afferent fibers of the greater splanchnic nerve are to laminae I and V in the thoracic spinal cord, although there are also some terminals in the outer part of lamina II (Cervero and Connell, 1984a,b; Neuhuber *et al.*, 1986). Some of the terminals are contralateral. Intracellular injection of *Phaseolus vulgaris* lectin into DRG cells of visceral C fibers has allowed individual axons to be traced into laminae I, II, IV–V, and X (Sugiura *et al.*, 1989).

Ascending Pathways. Some large afferent fibers, including those supplying the abdominal Pacinian corpuscles, ascend the spinal cord in the dorsal columns and terminate in the dorsal

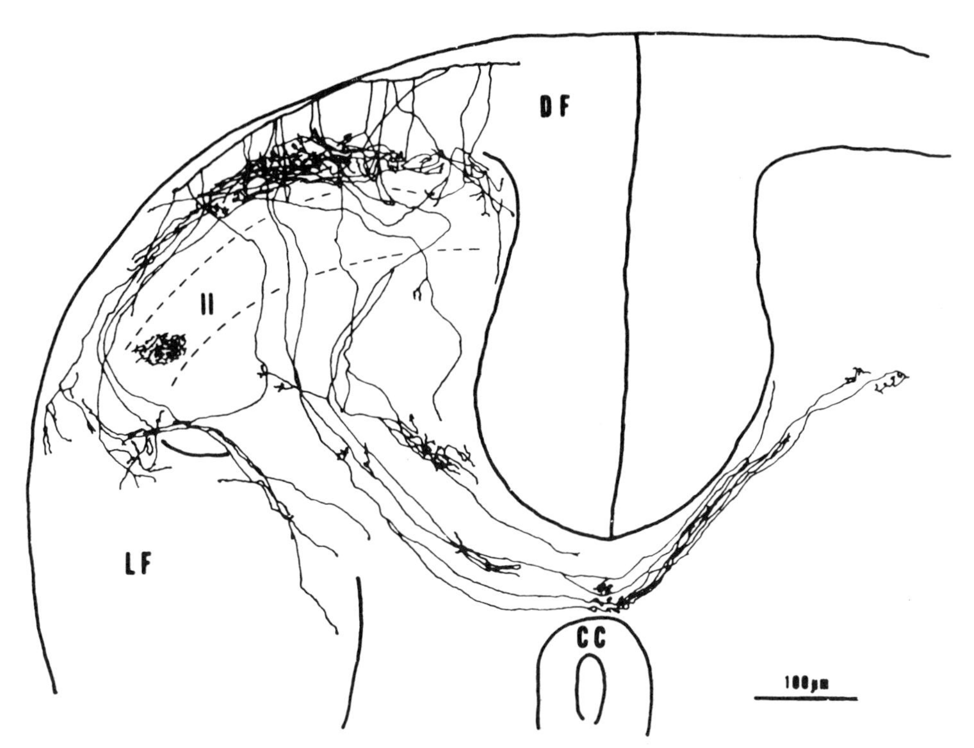

Fig. 10.38. Projection of a visceral afferent fiber of C calibre to the spinal cord gray matter. The DRG cell was injected intracellularly with *Phaseolus* lectin. The projection was reconstructed in a transverse plane, but the fiber distributed longitudinally over six segments (T10–L2). (From Sugiura *et al.*, 1989.)

column nuclei (Fig. 10.39) (Amassian, 1951; Aidar *et al.*, 1952; Perl *et al.*, 1962). Visceral afferent fibers of smaller caliber activate neurons of the spinothalamic and spinoreticular tracts (see Chapter 9). However, it is not known what visceral sensations are mediated by the dorsal columns or by pathways in the lateral and ventral funiculi.

Summary

In summary, the most important visceral sensation from a clinical viewpoint is pain. However, other more vague sensations, including the sense of fullness, also arise from viscera. These are not yet fully understood, although recordings from a variety of visceral afferent fibers provide evidence for at least part of the sensory mechanism. Visceral information is processed by neurons in many laminae of the spinal cord gray matter, including I, II, V–VII, and X. Ascending visceral pathways include the dorsal columns and the spinothalamic and spinoreticular tracts.

CENTRIFUGAL CONTROL OF SOMATOVISCERAL SENSATION

The modulation of sensory input by efferent pathways from the brain has received a great deal of attention in recent years. A considerable amount of work has been done on the descending control of nociception (see reviews by Willis, 1982, 1988). However, a number of investigations

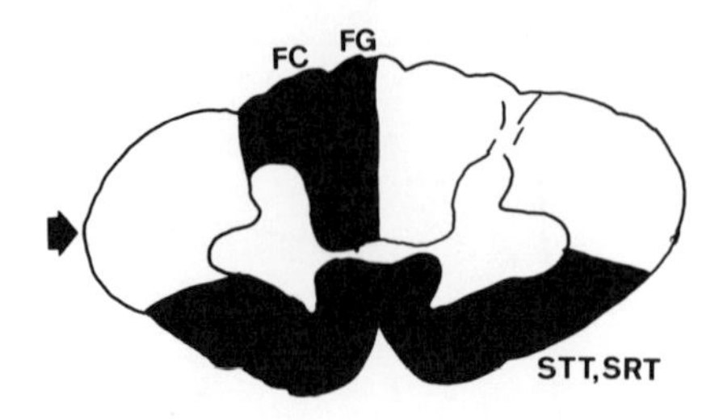

Fig. 10.39. Pathways carrying information from visceral receptors include the fasciculi cuneatus and gracilis (FC, FG) ipsilaterally to the stimulus and anterior quadrant pathways such as the spinothalamic (STT) and spinoreticular (SRT) tracts bilaterally.

have also been done on other aspects of the descending control of somatosensory transmission (Dawson, 1958; Hagbarth, 1960; Towe, 1973).

Some of the ways in which the descending control of a sensory pathway might be expressed are summarized by Wall and Dubner (1972): (1) a change in gain in the afferent pathway; (2) a change in the degree of selectivity among several kinds of input; (3) a change in receptive field size of central neurons; (4) an alteration in inhibitory surround; (5) the appearance of habituation; and (6) switching of sensory inputs or of the distribution of sensory information to higher centers. Another possibility is that the descending control might serve as a corollary discharge by inhibiting the part of the sensory input that would be predicted to result from a voluntary movement (von Holst, 1954; Teuber, 1960).

One source of centrifugal control of the somatovisceral sensory pathways is the sensorimotor cortex. Descending commands from the cortex may be transmitted directly to the spinal cord through the corticospinal tract, or they may involve a relay in the brain stem. Cortical control has been demonstrated for interneurons of the dorsal horn (see Chapter 5), the dorsal column nuclei (see Chapter 7), the spinocervical tract (see Chapter 8), the spinothalamic tract, and probably the spinoreticular tract (see Chapter 9).

Another source of descending control is the reticular formation of the brain stem. Stimulation within the reticular formation or the cerebellum results in primary afferent depolarization in the lumbar cord (Carpenter *et al.*, 1966; Cangiano *et al.*, 1969). Comparable stimulation affects the discharges of neurons in the dorsal column nuclei either directly or through the generation of primary afferent depolarization (see Chapter 7). Also modulated are the spinocervical tract (see Chapter 8), and the spinothalamic, spinoreticular, and spinomesencephalic tracts (see Chapter 9). It seems likely that the reticular formation serves to integrate information from many sources, which thus have access to the sensory pathways by way of the reticular formation. For example, at least a part of the corticofugal action on the dorsal column nuclei is probably mediated by way of the reticular formation, as is the effect on the same nuclei of stimulation in such diverse structures as the deep cerebellar nuclei, the nonspecific nuclei of the thalamus, and the caudate nucleus (see Chapter 7).

A particularly important descending control system originates near the midline of the caudal brain stem. This is the "dorsal reticulospinal system" of Lundberg's group (see reviews by Willis, 1982, 1988). This system affects not only spinal cord circuits but also several ascending tracts, including the spinocervical tract (see Chapter 8) and the spinothalamic, spinoreticular, and spinomesencephalic tracts (see Chapter 9). The pathway involves, at least in part, a raphe-spinal projection originating in the nucleus raphe magnus.

Interest in the dorsal reticulospinal system has been heightened recently because of the growing evidence that the analgesia produced by stimulation in the region of the periaqueductal and periventricular gray may depend upon information transmitted to the spinal cord, at least in part, by way of the nucleus raphe magnus (see reviews by Basbaum and Fields, 1978; Besson and Chaouch, 1987).

In addition to the possibility that electrical stimulation of certain brain regions will prove useful for therapeutic intervention in cases of chronic pain, a part of the "intrinsic analgesia system" can be activated by opiate receptors (see the review by Besson and Chaouch, 1987). In addition to an action in the brain stem, the opiates can also affect neurons in the spinal cord directly. This area of research has already proved to be of considerable interest and significance. It is reasonable to expect that a number of other descending pathways will play an important role in the modulation of sensory input from the body and viscera (see Chapters 5 and 7–9).

CONCLUSIONS

1. The neural mechanisms underlying somatovisceral sensation can be discussed with reference to sensory channels, which are the sets of primary afferent fibers, spinal cord processing circuits, ascending pathways, and brain circuits responsible for particular modalities of sensation.
2. Mechanical senses include touch–pressure and flutter–vibration. Touch–pressure is a sensation with a low threshold and persists during maintained stimulation. Flutter–vibration is a continuum of sensation reflecting oscillations of the stimulus. Flutter refers to the sensation accompanying the lower frequencies and vibration the higher frequencies of such an oscillating stimulus. Flutter–vibration sensation has a low threshold, and the sensation adapts rapidly if the stimulus does not oscillate but is maintained.
3. SA I receptors are largely responsible for touch–pressure sensation, although it is possible that SA II receptors also contribute. SA I receptors in humans usually have a single spotlike receptive field (although several such fields can occur). The afferent fibers respond with a dynamic followed by a static response when the skin over the receptor is indented, but they do not respond to skin stretch. When stimulated through a microneurography electrode at rates above 3–10 Hz, SA I axons often evoke a sensation of touch or pressure. The sensation is continuous, not intermittent. Perceived intensity increases with stimulus frequency.
4. Stimulus-response curves of SA I receptors in human glabrous skin are power functions with exponents of above 0.7, and human psychophysical magnitude estimation curves for glabrous skin are power functions with exponents near 1.0. The psychophysical function may result from recruitment of receptors with higher intensities of stimulation.
5. SA I receptors may be partly responsible for the threshold for contact recognition in the hairy skin, but not in glabrous skin, where contact recognition has a lower threshold than do SA I receptors. FA I receptors account for contact threshold on glabrous skin and probably share this function with SA I receptors in hairy skin.
6. SA I and FA I receptors have density gradients in the human hand that parallel the gradient of two-point discrimination. Presumably these two receptor groups account for two-point discrimination.
7. The dynamic responses of SA I receptors appear to contribute to the sensitivity of these receptors during exploratory movements of the hand. For example, contact of an SA I receptor with an edge results in a greatly enhanced response. This would enable SA I receptors to help signal such stimuli as gaps and grids, as well as recognition of raised letters.
8. SA I receptors are activated during a precision grip task. Their activity would presumably contribute both to the sensory experience and to reflex adjustments.
9. SA II receptors in humans may have a background discharge. Their receptive fields are large, since they respond to indentation and to skin stretch. When stimulated through a microneurography electrode, no sensation is elicited. This may reflect a need for summation centrally to evoke a sensation. In a few instances, sensations of movement of a finger joint or of cutaneous pressure have been elicited.

10. SA II receptors do not have a low enough threshold to account for the threshold for contact recognition, and their density gradient suggests that they do not contribute to two-point discrimination. They may play a role in precision grip.

11. The primary afferent fibers from SA I and SA II receptors synapse in laminae III–V. Some interneurons can be excited by single impulses in SA I receptors. Ascending pathways carrying information from SA I receptors include the dorsal column for cervical levels and the postsynaptic dorsal column pathway and a pathway in the DLF (perhaps that relaying in the nucleus gracilis) for the lumbar levels.

12. The receptors responsible for flutter are in the skin and those for vibration are deep to the skin, as shown by the loss of the former but not the latter when the skin is anesthetized.

13. The receptors for flutter appear to be the FA I receptors, which include Meissner's corpuscles in glabrous skin and hair follicle and field receptors in hairy skin.

14. FA I receptors in glabrous skin (Meissner's corpuscles) have small receptive fields that appear to overlie small groups of receptor organs. These receptors respond best to sinusoidal stimuli at rates of 8–64 Hz. When stimulated through a microneurography electrode, the axons of individual FA I receptors can produce sensations of tap, flutter, or vibration, depending upon the frequency of stimulation. The intensity of the sensation does not increase with frequency until rates above 100 Hz are reached.

15. FA II receptors in glabrous skin (Pacinian corpuscles) have large receptive fields. A single axon supplies an individual receptor organ. These receptors respond best at rates of 128–400 Hz. When they are stimulated through a microneurography electrode at rates exceeding 10–80 Hz, a sensation of vibration is usually evoked (sometimes there is a sensation of tickle). Perceived frequency increases with stimulus frequency.

16. Tuning curves constructed from the thresholds for flutter–vibration sensation in human subjects can be accounted for on the basis of the combination of tuning curves for the thresholds for one-for-one entrainment of FA I and FA II receptors, the former for lower and the latter for higher frequencies. Detection thresholds are lower than entrainment thresholds, and stimuli just above detection threshold evoke a sensation of mechanical contact, rather than of flutter–vibration.

17. The intensity of flutter–vibration can be accounted for by the population responses of FA I and FA II receptors.

18. The threshold for contact recognition in glabrous skin is best accounted for by the activity of FA I and FA II receptors.

19. The responses of FA I receptors are enhanced by edges when the stimuli are moving. This suggests a role for FA I receptors in tactile spatiotemporal discrimination.

20. FA I receptors are involved in precision grip.

21. C mechanoreceptors do not contribute to flutter sensation in the human hand, since this receptor type does not appear to be represented in the nerves supplying the hand. However, C mechanoreceptors innervate the face and may contribute to tickle.

22. Information from FA I receptors is processed in laminae III and IV and information from FA II receptors in laminae III–VI. Ascending pathways carrying information from FA I receptors include direct projections of collaterals in the dorsal column, the postsynaptic dorsal column path, the spinocervical tract, and the spinothalamic tract. FA II receptors also have collaterals that ascend the dorsal column, and these receptors activate some postsynaptic dorsal column neurons. However, they do not excite spinocervical or spinothalamic neurons.

23. Proprioception includes both position sense and kinesthesia. The receptors include muscle stretch receptors and, at least for distal joints, joint afferents, and perhaps some cutaneous receptors (possibly SA II receptors).

24. To provide information about joint position during voluntary movements, the strong responses of muscle spindle afferent fibers to fusimotor drive must be compensated for by integrative processing involving corollary discharges.

25. Stimulation of digital nerves in the human gives rise to a sensation of oscillatory movements of finger joints. Stimulation of individual afferent fibers supplying finger joints can produce a sensation of joint movement.

26. Muscle stretch receptors terminate in deep layers of the dorsal horn and in the ventral horn, as well as in Clarke's column and the external cuneate nucleus. Ascending pathways conveying proprioceptive information to the thalamus include the dorsal column for cervical levels and the spinomedullothalamic tract through nucleus z for lumbar levels. Some proprioceptive information is also carried in the spinothalamic tract, at least in monkeys, but this route is not critical in humans since anterolateral cordotomies do not alter proprioception.

27. Pain includes several submodalities: superficial pain (pricking pain and burning pain), deep (aching) pain, and visceral pain. Accompanying pain sensation are a number of motivational–affective responses that may have at least partially separate neural mechanisms.

28. Nociceptors are generally supplied by fine afferent fibers of Aδ or C calibre. Cutaneous nociceptors in humans include Aδ mechanical nociceptors, Aδ mechanoheat nociceptors, and C polymodal nociceptors. The Aδ nociceptors have receptive fields that include several spotlike areas. Some are polymodal. C polymodal nociceptors in humans, unlike in animals, also often have several spotlike components of their receptive fields. Cooling rarely activates these units, but heating does. Itch-provoking stimuli have been shown to activate some C polymodal nociceptors. Single impulses in nociceptors are not associated with pain. However, repetitive stimulation of C polymodal nociceptors through a microneurography electrode (probably small groups of afferent fibers are activated, rather than single axons) evokes a dull or burning pain sensation. The sensation is unchanged when A-fiber conduction is blocked. Stimulation of some C polymodal nociceptors results in an itch sensation. Stimulation of the axons of Aδ nociceptors causes pricking pain.

29. Psychophysical magnitude estimation curves obtained by using noxious heat pulses have been fitted with power functions with exponents of 1.0 or more. C polymodal nociceptors have similar stimulus-response curves. However, during prolonged mechanical stimulation of the skin, pain increases and yet the activity of C polymodal nociceptors adapts. Several explanations include the use of a ratio code for activity in nociceptors and Aβ fibers, recruitment of additional nociceptors, or central changes.

30. Damage to the skin results in primary hyperalgesia in the area of damage and secondary hyperalgesia in the surrounding area. Hyperalgesia refers to a lower threshold for pain and an enhanced magnitude of pain evoked by suprathreshold stimuli. Primary hyperalgesia may be caused by sensitization of nociceptors. Secondary hyperalgesia appears to depend in part on a central mechanism.

31. Sensitization of nociceptors appears to result from the action of chemicals released by damage. Algesic agents include potassium ions, bradykinin, serotonin, histamine, and prostaglandins.

32. Pain arising from visceral or deep structures may be referred to superficial structures. This may be explained by the convergence of visceral and somatic inputs onto common neurons in the central nervous system, such as spinothalamic neurons. Since visceral pain is much less common during growth and development, an individual may learn to associate activity in neurons receiving viscerosomatic convergence with somatic stimulation. Thus, when visceral damage occurs, the accompanying sensation is referred to a somatic structure.

33. Inhibitory interactions are important in pain mechanisms and account for the usefulness of a number of therapeutic procedures.

34. Nociceptive information is initially processed in laminae I, II, and V of the spinal cord. Ascending nociceptive pathways include the spinothalamic tract, but also the spinoreticular and spinomesencephalic tracts. Some nociceptive information may also be transmitted by way of the spinocervical and postsynaptic dorsal column pathways.

35. Thermal sensations include both cold and warm sensations. Thresholds for these

depend upon several factors, including the rate of temperature change, the amount of change, and the adapting temperature. If skin temperature changes excessively, cold or heat pain results.

36. Cold and warm sensations can be elicited independently by stimulation of cold or warm spots. In humans, cold receptors are supplied by Aδ or C fibers and warm receptors by C fibers. They are insensitive to mechanical stimuli, but cold receptors can be activated by certain chemicals, such as menthol. The activity of some mechanoreceptors can be altered by thermal stimuli, but these receptors do not appear to contribute to thermal sensation.

37. Cold receptors in humans have spotlike receptive fields. The receptor is located about 0.18 mm from the surface of the skin. In animals, cold receptors have a distinctive morphology.

38. The mechanism by which cold receptors code stimulus intensity is unclear. The intensity of perceived cold increases as the temperature is lowered. However, most cold receptors have a bell-shaped stimulus-response function, with the maximum discharge rate occurring at neutral temperatures that cause no sensation or at which sensation adapts. As the temperature is lowered or raised from the neutral zone, the discharges of cold receptors decrease. Since many cold receptors discharge in bursts at temperatures below neutral and the number of discharges in the bursts is a linear function of the temperature, it has been proposed that the bursting activity codes for stimulus intensities below the neutral level. However, central temperature-sensitive neurons do not show this bursting activity, casting doubt on this coding mechanism. An alternative hypothesis is that intensity coding depends upon a population response. Still another hypothesis is that a high–threshold cold receptor type that has recently been described is recruited as the temperature is lowered. Intensity coding would then depend upon a population response that includes the activity of both low- and high-threshold cold receptors.

39. The detection threshold for cold appears to depend upon the dynamic response of cold receptors.

40. Thermal discrimination depends upon the population response. Activity in at least 16 cold receptors is required to account for the human capacity to discriminate temperature.

41. Reaction time for a response to a thermal stimulus can be accounted for on the basis of the response latencies of cold and warm fibers.

42. Warm receptors in humans have spotlike receptive fields. Their structure is unknown, but they lie about 0.6 mm below the surface of the skin.

43. Intensity coding for warmth appears to depend upon a population response in low- and high-threshold warm receptors.

44. Detection threshold for warmth is signalled by the dynamic responses of warm receptors.

45. The afferent fibers from cold and warm receptors project to the superficial layers of the dorsal horn. The ascending pathway responsible for thermal sensation is the spinothalamic tract. The thermoreceptive axons may lie more posteriorly in the lateral funiculus than those conveying pain sensation.

46. Visceral sensations include pain and also the sense of fullness. A variety of mechanoreceptors have been described, some of which may explain nonpainful visceral sensations. Ascending visceral pathways include the dorsal column and the spinothalamic and spinoreticular tracts.

47. Descending control of sensory transmission is an important mechanism common to all of the somatovisceral pathways. Sources of descending control include the cerebral cortex, brain stem reticular formation, raphe nuclei, and periaqueductal and periventricular gray, among other structures. Of particular clinical interest is the involvement of these descending control systems in limiting activity in nociceptive pathways. Analgesia systems involve both opioid and nonopioid neurotransmitter systems.

References

Aanonsen, L. M., Lei, S. and Wilcox, G. L., 1990, Excitatory amino acid receptors and nociceptive neurotransmission in rat spinal cord. *Pain* **41**, 309–321.

Abdel-Maguid, T. E. and Bowsher, D., 1984, Interneurons and proprioneurons in the adult human spinal grey matter and in the general somatic and visceral afferent cranial nerve nuclei. *J. Anat.* **139**, 9–19.

Abhold, R. H. and Bowker, R. M., 1990, Descending modulation of dorsal horn biogenic amines as determined by in vivo dialysis. *Neurosci. Lett.* **108**, 231–236.

Abols, I. A. and Basbaum, A. I., 1981, Afferent connections of the rostral medulla of the cat: a neural-substrate for midbrain-medullary interactions in the modulation of pain. *J. Comp. Neurol.* **201**, 285–297.

Abrahams, V. C. and Swett, J. E., 1986, The pattern of spinal and medullary projections from a cutaneous nerve and a muscle nerve of a forelimb of the cat: a study using the transganglionic transport of HRP. *J. Comp. Neurol.* **246**, 70–84.

Abrahams, V. C., Lynn, B. and Richmond, F. J. R., 1984a, Organization and sensory properties of small myelinated fibres in the dorsal cervical rami of the cat. *J. Physiol.* **347**, 177–187.

Abrahams, V. C., Richmond, F. J. and Keane, J., 1984b, Projections from C2 and C3 nerves supplying muscles and skin of the cat neck: a study using transganglionic transport of horseradish peroxidase. *J. Comp. Neurol.* **230**, 142–154.

Abram, S. E. and Kostreva, D. R., 1986, Spinal cord metabolic response to noxious radiant heat stimulation of the cat hind footpad. *Brain Res.* **386**, 143–147.

Abrams, G. M., Nilaver, G. and Zimmermann, E. A., 1985, VIP-containing neurons. In: *Handbook of Chemical Neuroanatomy*, Vol. 4, *GABA and Neuropeptides in the CNS*, Part I, pp. 335–354. A. Bjorklund and T. Hokfelt (eds.). Elsevier, Amsterdam.

Adair, E. R., Stevens, J. C. and Marks, L. E., 1968, Thermally induced pain: the dol scale and the psychophysical power law. *Amer. J. Psychol.* **81**, 147–164.

Adams, P. R., Brown, D. A. and Jones, S. W., 1983, Substance P inhibits the M-current in bullfrog sympathetic neurones. *Brit. J. Pharmacol.* **79**, 330–333.

Ader, J. P., Postema, F. and Korf, J., 1979, Contribution of the locus coeruleus to the adrenergic innervation of the rat spinal cord: a biochemical study. *J. Neural Trans.* **44**, 159–173.

Adriaensen, H., Gybels, J., Handwerker, H. O. and Van Hees, J., 1980, Latencies of chemically evoked discharges in human cutaneous nociceptors and of the concurrent subjective sensations. *Neurosci. Lett.* **20**, 55–59.

Adriaensen, H., Gybels, J., Handwerker, H. O. and Van Hees, J., 1983, Response properties of thin myelinated (A-δ) fibers in human skin nerves. *J. Neurophysiol.* **49**, 111–122.

Adriaensen, H., Gybels, J., Handwerker, H. O. and Van Hees, J., 1984a, Suppression of C-fibre discharges upon repeated heat stimulation may explain characteristics of concomitant pain sensations. *Brain Res.* **302**, 203–211.

Adriaensen, H., Gybels, J., Handwerker, H. O. and Van Hees, J., 1984b, Nociceptor discharges and sensations due to prolonged noxious mechanical stimulation–a paradox. *Human Neurobiol.* **3**, 53–58.

Adrian, E. D., 1928, *The Basis of Sensation. The Action of the Sense Organs*. Reprinted by Hafner Publishing Co., New York, 1964.

Adrian, E. D., 1946, *The Physical Background of Perception*. Oxford Univ. Press, Oxford.

Adrian, E.D. and Zotterman, Y., 1926a, The impulses produced by sensory nerve-endings. Part 2. The response of a single end-organ. *J. Physiol.* **61**, 151–171.

Adrian, E. D. and Zotterman, Y., 1926b, The impulses produced by sensory nerve endings. Part 3. Impulses set up by touch and pressure. *J. Physiol.* **61**, 465–483.

Aidar, O., Geohegan, W. A. and Ungewitter, L. H., 1952, Splanchnic afferent pathways in the central nervous system. *J. Neurophysiol.* **15**, 131–138.

Aimone, L. D. and Yaksh, T. L., 1989, Opioid modulation of capsaicin-evoked release of substance P from at spinal cord in vivo. *Peptides* **10**, 1127–1131.

Ainsworth, A., Hall, P., Wall, P. D., Allt, G., MacKenzie, M., Gibson, S. and Polak, J. M., 1981, Effects of capsaicin applied locally to adult peripheral nerve. II. Anatomy and enzyme and peptide chemistry of peripheral nerve and spinal cord. *Pain* **11**, 379–388.

Aitken, S. C. and Lal, S., 1982a, The spatial distribution and functional properties of type I mechanoreceptor units of the sural nerve of the rabbit. *Brain Res. Rev.* **4**, 45–56.

Aitken, S. C. and Lal, S., 1982b, The functional properties and innervation density of type II mechanoreceptor units of the sural nerve of the rabbit. *Brain Res. Rev.* **4**, 57–64.

Alarcon, G. and Cervero, F., 1990, The effects of electrical stimulation of A and C visceral afferent fibres on the excitability of viscerosomatic neurones in the thoracic spinal cord of the cat. *Brain Res.* **509**, 24–30.

Albe-Fessard, D., Levante, A. and Lamour, Y., 1974a, Origin of spinothalamic and spinoreticular pathways in cats and monkeys. *Adv. Neurol.* **4**, 157–166.

Albe-Fessard, D., Levante, A. and Lamour, Y., 1974b, Origin of spino-thalamic tract in monkeys. *Brain Res.* **65**, 503–509.

Albe-Fessard, D., Boivie, J., Grant, G. and Levante, A., 1975, Labelling of cells in the medulla oblongata and the spinal cord of the monkey after injections of horseradish peroxidase in the thalamus. *Neurosci. Lett.* **1**, 75–80.

Albright, B. C. and Haines, D. E., 1978, Dorsal column nuclei in a prosimian primate (Galago senegalensis): II. Cuneate and lateral cuneate nuclei: morphology and primary afferent fibers from cervical and upper thoracic spinal segments. *Brain Behav. Evol.* **15**, 165–184.

Albright, B. C., Johnson, J. I. and Ostapoff, E. M., 1983, The projection of cervical primary fibers to the DCN of the squirrel, *Sciurus niger*: fiber sorting in the dorsal columns. *Brain Behav. Evol.* **22**, 118–131.

Aldskogius, H. and Rising, M., 1981, Effect of sciatic neurectomy on neuronal number and size distribution in the L7 ganglion of kittens. *Exp. Neurol.* **74**, 597–604.

Aldskogius, H., Arvidsson, J. and Grant, G., 1985, The reaction of primary sensory neurons to peripheral nerve injury with particular emphasis on transganglionic changes. *Brain Res. Rev.* **10**, 27–46.

Allen, J. M., Gibson, S. J., Adrian, T. E., Polak, J. M. and Bloom, S. R., 1984, Neuropeptide Y in human spinal cord. *Brain Res.* **308**, 145–148.

Alles, A. and Dom, R. N., 1985, Peripheral sensory nerve fibers that dichotomize to supply the brachium and the pericardium in the rat: a possible morphological explanation for referred cardiac pain? *Brain Res.* **342**, 382–385.

Aloisi, A. M., Carli, G. and Rossi, A., 1988, Response of hip joint afferent fibers to pressure and vibration in the cat. *Neurosci. Lett.* **90**, 130–134.

Alstermark, B., Gorska, T., Johannisson, T. and Lundberg, A., 1986, Effects of dorsal column transection in the upper cervical segments on visually guided forelimb movements. *Neurosci. Res.* **3**, 462–466.

Altman, J. and Bayer, S. A., 1984, The development of the rat spinal cord. *Adv. Embryol. Cell Biol.* **85**, 1–166.

Alvarez, F. J., Cervantes, C., Blasco, I., Villalba, R., Martinez-Murillo, R., Polak, J. M. and Rodrigo, J., 1988, Presence of calcitonin gene-related peptide (CGRP) and substance P (SP) immunoreactivity in intraepidermal free nerve endings of cat skin. *Brain Res.* **442**, 391–395.

Amara, S. G., Arriza, J. L., Leff, S. E., Swanson, L. W., Evans, R. M. and Rosenfeld, M. G., 1985, Expression in brain of a messenger RNA encoding a novel neuropeptide homologous to calcitonin gene-related peptide. *Science* **229**, 1094–1097.

Amassian, V. E., 1951, Fiber groups and spinal pathways of cortically represented visceral afferents. *J. Neurophysiol.* **14**, 445–460.

Amassian, V. E., 1952, Interaction in the somatovisceral projection system. *Res. Publ. Assoc. Res. Nerv. Ment. Dis.* **30**, 371–402.

Amassian, V. E. and DeVito, J. L., 1957, La transmission dans le noyau de Burdach (nucleus cuneatus). Etude analytique par unites isolees d'un relais somatosensoriel primaire. *Colloq. Int. Cent. Nat. Rech. Sci.* **67**, 353–393.

Amassian, V. E. and Giblin, D., 1974, Periodic components in steady-state activity of cuneate neurones and their possible role in sensory coding. *J. Physiol.* **246**, 353–385.

Ambrose, W. W. and McNeill, M. E., 1978, Graphic representation of the distribution of acetylcholinesterase in cat dorsal root ganglion neurons. *Histochem. J.* **10**, 711–720.

Ammann, B., Gottschall, J. and Zenker, W., 1983, Afferent projections from the rat longus capitis muscle studied by transganglionic transport of HRP. *Anat. Embryol.* **166**, 275–289.

Ammons, W. S., 1986, Renal afferent input to thoracolumbar spinal neurons of the cat. *Amer. J. Physiol.* **250**, R435–R443.

Ammons, W. S., 1987, Characteristics of spinoreticular and spinothalamic neurons with renal input. *J. Neurophysiol.* **50**, 480–495.

Ammons, W. S., 1989, Primate spinothalamic cell responses to ureteral occlusion. *Brain Res.* **496**, 124–130.

Ammons, W. S., Blair, R. W. and Foreman, R. D., 1984a, Responses of primate T1-T5 spinothalamic neurons to gallbladder distension. *Am. J. Physiol.* **247** (Regulatory Integrative Comp. Physiol. 16), R995–R1002.

Ammons, W. S., Blair, R. W. and Foreman, R. D., 1984b, Raphe magnus inhibition of primate T1-T4 spinothalamic cells with cardiopulmonary visceral input. *Pain* **20**, 247–260.

Ammons, W. S., Blair, R. W. and Foreman, R. D., 1984c, Greater splanchnic excitation of primate T_1-T_5 spinothalamic neurons. *J. Neurophysiol.* **51**, 592–603.

Ammons, W. S., Girardot, M. N. and Foreman, R. D., 1985a, T2–T5 spinothalamic neurons projecting to medial thalamus with viscerosomatic input. *J. Neurophysiol.* **54**, 73–89.

Ammons, W. S., Girardot, M. N. and Foreman, R. D., 1985b, Effects of intracardiac bradykinin on T2–T5 medial spinothalamic cells. *Am. J. Physiol.* **249** (Regulatory Integrative Comp. Physiol. 17), R147–R152.

Ammons, W. S., Girardot, M. N. and Foreman, R. D., 1986, Periventricular gray inhibition of thoracic spinothalamic cells projecting to medial and lateral thalamus. *J. Neurophysiol.* **55**, 1091–1103.

Anand, P., Gibson, S. J., McGregor, G. P., Blank, M. A., Ghatei, M. A., Bacarese-Hamilton, A. J., Polak, J. M. and Bloom, S. R., 1983, A VIP-containing system concentrated in the lumbosacral region of human spinal cord. *Nature* **305**, 143–145.

Andersen, H. T., Koerner, L., Landgren, S. and Silfvenius, H., 1967, Fibre components and cortical projections of the elbow joint nerve in the cat. *Acta Physiol. Scand.* **69**, 373–382.

Andersen, P., Eccles, J. C., Schmidt, R. F. and Yokota, T., 1964a, Slow potential waves produced in the cuneate nucleus by cutaneous volleys and by cortical stimulation. *J. Neurophysiol.* **27**, 78–91.

Andersen, P., Eccles, J. C., Schmidt, R. F. and Yokota, T., 1964b, Depolarization of presynaptic fibers in the cuneate nucleus. *J. Neurophysiol.* **27**, 92–106.

Andersen, P., Eccles, J. C., Schmidt, R. F. and Yokota, T., 1964c, Identification of relay cells and interneurons in the cuneate nucleus. *J. Neurophysiol.* **27**, 1080–1095.

Andersen, P., Eccles, J. C., Oshima, T. and Schmidt, R. F., 1964d, Mechanisms of synaptic transmission in the cuneate nucleus. *J. Neurophysiol.* **27**, 1096–1116.

Andersen, P., Eccles, J. C. and Sears, T. A., 1964e, Cortically evoked depolarization of primary afferent fibers in the spinal cord. *J. Neurophysiol.* **27**, 63–77.

Andersen, P., Etholm, B. and Gordon, G., 1970, Presynaptic and post-synaptic inhibition elicited in the cat's dorsal column nuclei by mechanical stimulation of skin. *J. Physiol.* **210**, 433–455.

Andersen, P., Gjerstad, L. and Pasztor, E., 1972a, Effect of cooling on synaptic transmission through the cuneate nucleus. *Acta Physiol. Scand.* **84**, 433–447.

Andersen, P., Gjerstad, L. and Pasztor, E., 1972b, Effects of cooling on inhibitory processes in the cuneate nucleus. *Acta Physiol. Scand.* **84**, 448–461.

Anderson, F. D. and Berry, C. M., 1959, Degeneration studies of long ascending fiber systems in the cat brain stem. *J. Comp. Neurol.* **111**, 195–229.

Andersson, S. A., 1962, Projection of different spinal pathways to the second somatic sensory area in cat. *Acta Physiol. Scand.* **56** (Suppl. 194), 1–74.

Andersson, S. A. and Leissner, P. E., 1975, Does the spinocervical pathway exist? *Brain Res.* **98**, 359–363.

Andersson, S. A., Norrsell, K. and Norrsell, U., 1972, Spinal pathways projecting to the cerebral first somatosensory area in the monkey. *J. Physiol.* **225**, 589–597.

Andersson, S. A., Finger, S. and Norrsell, U., 1975, Cerebral units activated by tactile stimuli via a ventral spinal pathway in monkeys. *Acta Physiol. Scand.* **93**, 119–128.

Andres, K. H., 1961, Untersuchungen uber den Feinbau von Spinalganglien. *Zeit. f. Zellforsch.* **55**, 1–48.

Andres, K. H., 1966, Ueber die Feinstruktur der Rezeptoren an Sinushaaren. *Z. Zellforsch. Mikrosk. Anat.* **75**, 339–365.

Andres, K. H. and Düring, M. von, 1973, Morphology of cutaneous receptors. In: *Handbook of Sensory Physiology*, Vol. II, *Somatosensory System*, pp. 3–28. A. Iggo (ed.). Springer, New York.

Andres-Trelles, F., Cowan, C. M. and Simmonds, M. A., 1976, The negative potential wave evoked in cuneate nucleus by stimulation of afferent pathways: its origins and susceptibility to inhibition. *J. Physiol.* **258**, 173–186.

Andrew, B. L., 1954, The sensory innervation of the medial ligament of the knee joint. *J. Physiol.* **123**, 241–250.

Andrew, B. L. and Dodt, E., 1953, The deployment of sensory endings at the knee joint of the cat. *Acta Physiol. Scand.* **28**, 287–296.

Andrezik, J. A., Chan-Palay, V. and Palay, S. L., 1981, The nucleus paragigantocellularis lateralis in the rat. Demonstration of afferents by retrograde transport of horseradish peroxidase. *Anat. Embryol.* **161**, 373–390.

Angaut-Petit, D., 1975a, The dorsal column system: I. Existence of long ascending postsynaptic fibres in the cat's fasciculus gracilis. *Exp. Brain Res.* **22**, 457–470.

Angaut-Petit, D., 1975b, The dorsal column system: II. Functional properties and bulbar relay of the postsynaptic fibres of the cat's fasciculus gracilis. *Exp. Brain Res.* **22**, 471–493.

Angelucci, L., 1956, Experiments with perfused frog's spinal cord. *Brit. J. Pharmacol.* **11**, 161–170.

Anis, N. A., Berry, S. C., Burton, N. R. and Lodge, D., 1983, The dissociative anaesthetics, ketamine and phencyclidine, selectively reduce excitation of central mammalian neurones by N-methyl-asparate. *Br. J. Pharmacol.* **79**, 565–575.

Antal, M., Tornai, I. and Székely, G., 1980, Longitudinal extent of dorsal root fibers in the spinal cord and brain stem of the frog. *Neuroscience* **5**, 1311–1322.

Antonetty, C. M. and Webster, K. E., 1975, The organization of the spinotectal projection. An experimental study in the rat. *J. Comp. Neurol.* **163**, 449–466.

Aoki, M., 1981, Afferent inhibition on various types of cat's cuneate neurons induced by dynamic and steady tactile stimuli. *Brain Res.* **221**, 257–269.

Aoyama, M., Hongo, T. and Kudo, N., 1973, An uncrossed ascending tract originating from below Clarke's column and conveying group I impulses from the hindlimb muscles in the cat. *Brain Res.* **62**, 237–241.

Apkarian, A. V. and Hodge, C. J., 1989a, The primate spinothalamic pathways: I. A quantitative study of the cells of origin of the spinothalamic pathway. *J. Comp. Neurol.* **288**, 447–473.

Apkarian, A. V. and Hodge, C. J., 1989b, The primate spinothalamic pathways: II. The cells of origin of the dorsolateral and ventral spinothalamic pathways. *J. Comp. Neurol.* **288**, 474–492.

Apkarian, A. V. and Hodge, C. J., 1989c, The primate spinothalamic pathways: III. Thalamic terminations of the dorsolateral and ventral spinothalamic pathways. *J. Comp. Neurol.* **288**, 493–511.

Apkarian, A. V. and Hodge, C. J., 1989d, A dorsolateral spinothalamic tract in macaque monkey. *Pain* **37**, 323–333.

Appelberg, B., Bessou, P. and Laporte, Y., 1966, Action of static and dynamic fusimotor fibres on secondary endings of cat's spindles. *J. Physiol.* **185**, 160–171.

Applebaum, A. E., Beall, J. E., Foreman, R. D. and Willis, W. D., 1975, Organization and receptive fields of primate spinothalamic tract neurons. *J. Neurophysiol.* **38**, 572–586.

Applebaum, A. E., Leonard, R. B., Kenshalo, D. R., Jr., Martin, R. F. and Willis, W. D., 1979, Nuclei in which functionally identified spinothalamic tract neurons terminate. *J. Comp. Neurol.* **188**, 575–586.

Applebaum, M. L., Clifton, G. L., Coggeshall, R. E., Coulter, J. D., Vance, W. H. and Willis, W. D., 1976, Unmyelinated fibres in the sacral 3 and caudal 1 ventral roots of the cat. *J. Physiol.* **256**, 557–572.

Aprison, M. H., 1970, Evidence of the release of [^{14}C]glycine from hemisectioned toad spinal cord with dorsal root stimulation. *Pharmacologist* **12**, 222P.

Aprison, M. H. and Werman, R., 1965, The distribution of glycine in cat spinal cord and roots. *Life Sci.* **4**, 2075–2083.

Aprison, M. H., Shank, R. P. and Davidoff, R. A., 1969, A comparison of the concentration of glycine, a transmitter suspect, in different areas of the brain and spinal cord in seven different vertebrates. *Comp. Biochem. Physiol.* **28**, 1345–1355.

Aquilonius, S. M., Eckernas, S. A. and Gillberg, P. G., 1981, Topographical localization of choline acetyltransferase within the human spinal cord and a comparison with some other species. *Brain Res.* **211**, 329–340.

Arbuthnott, E. R., Boyd, I. A. and Kalu, K. U., 1975, Ultrastructure and conduction velocity of small, myelinated peripheral nerve fibers. In: *The Somatosensory System*, pp. 168–175. H. H. Kornhuber (ed.). Thieme, Stuttgart.

Arimatsu, Y., Seto, A. and Amano, T., 1981, An atlas of α-bungarotoxin binding sites and structures containing acetylcholinesterase in the mouse central nervous system. *J. Comp. Neurol.* **198**, 603–631.

Arluison, M., Conrath-Verrier, M., Tauc, M., Mailly, P., De la Manche, I. S., Cesselin, F., Bourgoin, S. and Hamon, M., 1983a, Different localizations of met-enkephalin-like immunoreactivity in rat forebrain and spinal cord using hydrogen peroxide and triton X-100. Light microscopic study. *Brain Res. Bull.* **11**, 555–571.

Arluison, M., Conrath-Verrier, M., Tauc, M., Mailly, P., De la Manche, I. S., Dietl, M., Cesselin, F., Bourgoin, S. and Hamon, M.,, 1983b, Met-enkephalin-like immunoreactivity in rat forebrain and spinal cord using hydrogen peroxide and triton X-100. Ultrastructural study. *Brain Res. Bull.* **11**, 573–586.

Armand, J., 1982, The origin, course and terminations of corticospinal fibers in various mammals. *Prog. Brain Res.* **57**, 329–360.

Armett, C. J., Gray, J. A. B. and Palmer, J. F., 1961, A group of neurones in the dorsal horn associated with cutaneous mechanoreceptors. *J. Physiol.* **156**, 611–622.

Armett, C. J., Gray, J. A. B., Hunsperger, R. W. and Lal, S., 1962, The transmission of information in primary receptor neurones and second-order neurones of a phasic system. *J. Physiol.* **164**, 395–421.

Aronin, N., Difiglia, M., Liotta, A. S. and Martin, J. B., 1981, Ultrastructural localization and biochemical features of immunoreactive leu-enkephalin in monkey dorsal horn. *J. Neurosci.* **1**, 561–577.

Arvidsson, V., Schalling, M., Cullheim, S., Ulfhake, B., Terenius, L., Verhofstad, A. and Hokfelt, T., 1990, Evidence for coexistence between calcitonin gene-related peptide and serotonin in the bulbospinal pathway in the monkey. *Brain Res.* **532**, 47–57.

Atkinson, M. E. and Shehab, S. A. S., 1986, Peripheral axotomy of the rat mandibular trigeminal nerve leads to an increase in VIP and decrease of other primary afferent neuropeptides in the spinal trigeminal nucleus. *Reg. Peptides* **16**, 69–82.

Atweh, S. F. and Kuhar, M. J., 1977, Autoradiographic localization of opiate receptors in rat brain. I. Spinal cord and lower medulla. *Brain Res.* **124**, 53–67.

Atweh, S. F., Banna, N. R., Jabbur, S. J. and To'mey, G. F., 1974, Polysensory interactions in the cuneate nucleus. *J. Physiol.* **238**, 343–355.

Ault, B., Evans, R. H., Francis, A. A., Oakes, D. J. and Watkins, J. C., 1980, Selective depression of excitatory amino acid induced depolarizations by magnesium ions in isolated spinal cord preparations. *J. Physiol.* **307**, 413–428.

Austin, G. M. and McCouch, G. P., 1955, Presynaptic component of intermediary cord potential. *J. Neurophysiol.* **18**, 441–451.

Azerad, J., Hunt, C. C., Laporte, Y., Pollin, B. and Thiesson, D., 1986, Afferent fibers in cat ventral roots: electrophysiological and histological evidence. *J. Physiol.* (Lond.) **379**, 229–243.

Azulay, A. and Schwartz, A. S., 1975, The role of the dorsal funiculus of the primate in tactile discrimination. *Exp. Neurol.* **46**, 315–332.
Back, S. A. and Gorenstein, C., 1989, Fluorescent histochemical localization of neutral endopeptidase-24.11 (enkephalinase) in the rat spinal cord. *J. Comp. Neurol.* **280**, 436–450.
Bagust, J., Forsythe, I. D., Kerkut, G. A. and Loots, J. M., 1982, Synaptic and non-synaptic components of the dorsal horn potential in isolated hamster spinal cord. *Brain Res.* **233**, 186–194.
Bahr, R., Blumberg, H. and Jänig, W., 1981, Do dichotomizing afferent fibers exist which supply visceral organs as well as somatic structures? *Neurosci. Lett.* **24**, 25–28.
Baik-Han, E. J., Kim, K. J. and Chung, J. M., 1989, Electrophysiological evidence for the presence of looping myelinated afferent fibers in the rat ventral root. *Neurosci. Lett.* **104**, 65–70.
Baker, D. G., Coleridge, H. M., Coleridge, J. C. G. and Nerdrum, T., 1980, Search for a cardiac nociceptor: stimulation by bradykinin of sympathetic nerve endings in the heart of the cat. *J. Physiol.* 306, 519–536.
Baker, M. L. and Giesler, G. J., 1984, Anatomical studies of the spinocervical tract of the rat. *Somatosensory Res.* **2**, 1–18.
Bakker, D. A., Richmond, F. J. R. and Abrahams, V. C., 1984, Central projections from cat suboccipital muscles: a study using transganglionic transport of horseradish peroxidase. *J. Comp. Neurol.* **228**, 409–421.
Banna, N. R. and Jabbur, S. J., 1969, Pharmacological studies on inhibition in the cuneate nucleus of the cat. *Int. J. Neuropharmacol.* **8**, 299–307.
Banna, N. R. and Jabbur, S. J., 1971, The effects of depleting GABA on cuneate presynaptic inhibition. *Brain Res.* **33**, 530–532.
Banna, N. R. and Jabbur, S. J., 1989, Neurochemical transmission in the dorsal column nuclei. *Somatosensory & Motor Res.* **6**, 237–251.
Banna, N. R., Naccache, A. and Jabbur, S. J., 1972, Picrotoxin-like action of bicuculline. *Europ. J. Pharmacol.* **17**, 301–302.
Bannatyne, B. A., Maxwell, D. J., Fyffe, R. E. W. and Brown, A. G., 1984, Fine structure of primary afferent axon terminals of slowly adapting cutaneous receptors in the cat. *Quart. J. Exp. Physiol.* **69**, 547–557.
Bannatyne, B. A., Maxwell, D. J. and Brown, A. G., 1987, Fine structure of synapses associated with characterized postsynaptic dorsal column neurons in the cat. *Neuroscience* **23**, 597–612.
Barbaresi, P., Rustioni, A. and Cuénod, 1985, Retrograde labeling of dorsal root ganglion neurons after injection of tritiated amino acids in the spinal cord of rats and cats. *Somatosensory Res.* **3**, 57–74.
Barbaresi, P., Spreafico, R., Frassoni, C. and Rustioni, A., 1986, GABAergic neurons are present in the dorsal column nuclei but not in the ventroposterior complex of rats. *Brain Res.* **282**, 305–326.
Barber, R. P., Vaughn, J. E., Saito, K., McLaughlin, B. J. and Roberts, E., 1978, GABAergic terminals are presynaptic to primary afferent terminals in the substantia gelatinosa of the rat spinal cord. *Brain Res.* **141**, 35–55.
Barber, R. P., Vaughn, J. E., Slemmon, J. R., Salvaterra, P. M., Roberts, E. and Leeman, S. E., 1979, The origin, distribution and synaptic relationships of substance P axons in rat spinal cord. *J. Comp. Neurol.* **184**, 331–352.
Barber, R. P., Vaughn, J. E. and Roberts, E., 1982, The cytoarchitecture of GABAergic neurons in rat spinal cord. *Brain Res.* **238**, 305–328.
Barber, R. P., Phelps, P. E., Houser, C. R., Crawford, G. D., Salvaterra, P. M. and Vaughn, J. E., 1984, The morphology and distribution of neurons containing choline acetyltransferase in the adult rat spinal cord: an immunocytochemical study. *J. Comp. Neurol.* **229**, 329–346.
Barker, D., 1962, The structure and distribution of muscle receptors. In: *Symposium on Muscle Receptors*, pp. 227–240. D. Barker (ed.). Hong Kong Univ. Press, Hong Kong.
Barker, D., Hunt, C. C. and McIntyre, A. K., 1974, *Handbook of Sensory Physiology*. III/2 *Muscle Receptors*. Springer, Berlin.
Barker, J. L. and Nicoll, R. A., 1973, The pharmacology and ionic dependency of amino acid responses in the frog spinal cord. *J. Physiol.* **228**, 259–277.
Barrack, R. L., Skinner, H. B., Cook, S. D. and Haddad, R. J., 1983a, Effect of articular disease and total knee arthroplasty on knee joint-position sense. *J. Neurophysiol.* **50**, 684–687.
Barrack, R. L., Skinner, H. B., Brunet, M. E. and Haddad, R. J., 1983b, Functional performance of the knee after intrarticular anesthesia. *Amer. J. Sports Med.* **11**, 258–261.
Barron, D. H. and Matthews, B. H. C., 1938, The interpretation of potential changes in the spinal cord. *J. Physiol.* **92**, 276–321.
Basbaum, A. I., 1985, Functional analysis of the cytochemistry of the spinal dorsal horn. *Adv. Pain Res. Ther.* **9**, 149–175.
Basbaum, A. I., 1988, Distribution of glycine receptor immunoreactivity in the spinal cord of the rat: cytochemical evidence for a differential glycinergic control of lamina I and V nociceptive neurons. *J. Comp. Neurol.* **278**, 330–336.
Basbaum, A. I. and Fields, H. L., 1978, Endogenous pain control mechanisms: review and hypothesis. *Ann. Neurol.* **4**, 451–462.
Basbaum, A. I. and Fields, H. L., 1979, The origin of descending pathways in the dorsolateral funiculus of the spinal cord of the cat and rat: further studies on the anatomy of pain modulation. *J. Comp. Neurol.* **187**, 513–532.

Basbaum, A. I. and Fields, H. L., 1984, Endogenous pain control systems: brainstem spinal pathways and endorphin circuitry. *Ann. Rev. Neurosci.* **7**, 309–338.

Basbaum, A. I. and Glazer, E. J., 1983, Immunoreactive vasoactive intestinal polypeptide is concentrated in the sacral spinal cord: A possible marker for pelvic visceral afferent fibers. *Somatosensory Res.* **1**, 69–82.

Basbaum, A. I. and Hand, P. J., 1973, Projections of cervicothoracic dorsal roots to the cuneate nucleus of the rat, with observations on cellular "bricks". *J. Comp. Neurol.* **148**, 347–360.

Basbaum, A. I. and Wall, P. D., 1976, Chronic changes in the response of cells in the adult cat dorsal horn following partial deafferentation: The appearance of responding cells in a previously non-responsive region. *Brain Res.* **116**, 181–204.

Basbaum, A. I., Clanton, C. H. and Fields, H.L ., 1978, Three bulbospinal pathways from the rostral medulla of the cat: an autoradiographic study of pain modulating systems. *J. Comp. Neurol.* **178**, 209–224.

Basbaum, A. I., Ralston, D. D. and Ralston, H. J., 1986a, Bulbospinal projections in the primate: a light and electron microscopic study of a pain modulating system. *J. Comp. Neurol.* **250**, 311–323.

Basbaum, A. I., Cruz, L. and Weber, E., 1986b, Immunoreactive dynorphin B in sacral primary afferent fibers of the cat. *J. Neurosci.* **6**, 127–133.

Basbaum, A. I., Glazer, E. J. and Oertel, W., 1986c, Immunoreactive glutamic acid decarboxylase in the trigeminal nucleus caudalis of the cat: a light- and electron-microscopic analysis. *Somatosensory Res.* **4**, 77–94.

Basbaum, A. I., Zahs, K., Lord, B. and Lakos, S., 1988, The fiber caliber of 5-HT immunoreactive axons in the dorsolateral funiculus of the spinal cord of the rat and cat. *Somatosensory Res.* **5**, 177–185.

Battaglia, G. and Rustioni, A., 1988, Coexistence of glutamate and substance P in dorsal root ganglion neurons of the rat and monkey. *J. Comp. Neurol.* **277**, 302–312.

Battaglia, G., Rustioni, A., Altschuler, R. A. and Petrusz, P., 1987, Glutamic acid coexists with substance P in some primary sensory neurons. In: *Fine Afferent Nerve Fibers and Pain*, pp. 77–84. R. F. Schmidt, H. G. Schaible, and C. Vahle-Hinz (eds.). VCH Verlagsgesellschaft mbH, Weinheim, Federal Republic of Germany.

Baumann, T. K., Simone, D. A., Shain, C. N. and LaMotte, R. H., 1991, Neurogenic hyperalgesia: The search for the primary cutaneous afferent fibers that contribute to capsaicin-induced pain and hyperalgesia. *J. Neurophysiol.*, in press.

Baxendale, R. H. and Ferrell, W. R., 1983, Discharge characteristics of the elbow joint nerve of the cat. *Brain Res.* **261**, 195–203.

Bazett, H. C., McGlone, B. and Brocklehurst, R. J., 1930, The temperatures in the tissues which accompany temperature sensations. *J. Physiol.* **69**, 88–112.

Beal, J. A., 1979a, Serial reconstruction of Ramon y Cajal's large primary afferent complexes in laminae II and III of the adult monkey spinal cord: a Golgi study. *Brain Res.* **166**, 161–165.

Beal, J. A., 1979b, The ventral dendritic arbor of marginal (Lamina I) neurons in the adult primate spinal cord. *Neurosci. Lett.* **14**, 201–206.

Beal, J. A., 1983, Identification of presumptive long axon neurons in the substantia gelatinosa of rat lumbosacral spinal cord: a Golgi study. *Neurosci. Lett.* **41**, 9–14.

Beal, J. A. and Bicknell, H. R., 1981, Primary afferent distribution pattern in the marginal zone (lamina I) of adult monkey and cat lumbosacral spinal cord. *J. Comp. Neurol.* **202**, 255–263.

Beal, J. A. and Cooper, M. H., 1978, The neurons in the gelatinosal complex (laminae II and III) of the monkey (*Macaca mulatta*): a Golgi study. *J. Comp. Neurol.* **179**, 89–122.

Beal, J. A. and Fox, C. A., 1976, Afferent fibers in the substantia gelatinosa of the adult monkey (*Macaca mulatta*): a Golgi study. *J. Comp. Neurol.* **168**, 113–144.

Beal, J. A., Penny, J. E. and Bicknell, H. R., 1981, Structural diversity of marginal (lamina I) neurons in the adult monkey (*Macaca mulatta*) lumbosacral spinal cord: a Golgi study. *J. Comp. Neurol.* **202**, 237–254.

Beal, J. A., Russell, C. T. and Knight, D. S., 1988, Morphological and development characterization of local-circuit neurons in lamina III of rat spinal cord. *Neurosci. Lett.* **86**, 1–5.

Beall, J. E., Martin, R. F., Applebaum, A. E. and Willis, W. D., 1976, Inhibition of primate spinothalamic tract neurons by stimulation in the region of the nucleus raphe magnus. *Brain Res.* **114**, 328–333.

Beall, J. E., Applebaum, A. E., Foreman, R. D. and Willis, W. D., 1977, Spinal cord potentials evoked by cutaneous afferents in the monkey. *J. Neurophysiol.* **40**, 199–211.

Beattie, M. S., Bresnahan, J. C., Mawe, G. M. and Finn, S., 1987, Distribution and ultrastructure of ventral root afferents to lamina I of the cat sacral spinal cord. *Neurosci. Lett.* **76**, 1–6.

Beck, C. H. M., 1981, Mapping of forelimb afferents to the cuneate nuclei of the rat. *Brain Res. Bull.* **6**, 503–516.

Beck, P. W. and Handwerker, H. O., 1974, Bradykinin and serotonin effects on various types of cutaneous nerve fibres. *Pfluegers Arch.* **347**, 209–222.

Beck, P. W., Handwerker, H. O. and Zimmermann, M., 1974, Nervous outflow from the cat's foot during noxious radiant heat stimulation. *Brain Res.* **67**, 373–386.

Beinfeld, M. C., 1985, Cholecystokinin (CCK) gene-related peptides; distribution and characterization of immunoreactive pro-CCK and an amino-terminal pro-CCK fragment in rat brain. *Brain Res.* **344**, 351–355.

Belcher, G., Ryall, R. W. and Schaffner, R., 1978, The differential effects of 5-hydroxytryptamine, noradrenaline and raphe stimulation on nociceptive and non-nociceptive dorsal horn interneurones in the cat. *Brain Res.* **151**, 307–321.

Bell, C., 1811, *Idea of a New Anatomy of the Brain*. Strahan and Preston, London. Reprinted in Cranefield, P.F. (ed.), 1974.
Bell, J. A. and Anderson, E. G., 1972, Semicarbazide induced depletion of γ-aminobutyric acid and blockade of presynaptic inhibition. *Brain Res.* **43**, 161–169.
Belmonte, C. and Gallego, R., 1983, Membrane properties of cat sensory neurones with chemoreceptor and baroreceptor endings. *J. Physiol.* **342**, 603–614.
Bender, M. B., Stacy, C. and Cohen, J., 1982, Agraphesthesia: a disorder of directional cutaneous kinesthesia or a disorientation in cutaneous space. *J. Neurol. Sci.* **53**, 531–555.
Bennett, G. J. and Mayer, D. J., 1979, Inhibition of spinal cord interneurons by narcotic microinjection and focal electrical stimulation in the periaqueductal central gray matter. *Brain Res.* **172**, 243–257.
Bennett, G. J., Hayashi, H., Abdelmoumene, M. and Dubner, R., 1979, Physiological properties of stalked cells of the substantia gelatinosa intracellularly stained with horseradish peroxidase. *Brain Res.* **164**, 285–289.
Bennett, G. J., Abdelmoumene, M., Hayashi, H. and Dubner, R., 1980, Physiology and morphology of substantia gelatinosa neurons intracellularly stained with horseradish peroxidase. *J. Comp. Neurol.* **194**, 809–827.
Bennett, G. J., Abdelmoumene, M., Hayashi, H., Hoffert, M. J. and Dubner, R., 1981, Spinal cord layer I neurons with axon collaterals that generate local arbors. *Brain Res.* **209**, 421–426.
Bennett, G. J., Ruda, M. A., Gobel, S. and Dubner, R., 1982, Enkephalin immunoreactive stalked cells and lamina IIb islet cells in cat substantia gelatinosa. *Brain Res.* **240**, 162–166.
Bennett, G. J., Seltzer, Z., Lu, G. W., Nishikawa, N. and Dubner, R., 1983, The cells of origin of the dorsal column postsynaptic projection in the lumbosacral enlargements of cats and monkeys. *Somatosensory Res.* **1**, 131–149.
Bennett, G. J., Nishikawa, N., Lu, G. W., Hoffert, M. J. and Dubner, R., 1984, The morphology of dorsal column postsynaptic (DCPS) spino-medullary neurons in the cat. *J. Comp. Neurol.* **224**, 568–578.
Benoist, J. M., Besson, J. M., Conseiller, C. and Le Bars, D., 1972, Action of bicuculline on presynaptic inhibition of various origins in the cat's spinal cord. *Brain Res.* **43**, 672–676.
Benoist, J. M., Besson, J. M. and Boissier, J. R., 1974, Modifications of presynaptic inhibition of various origins by local application of convulsant drugs on cat's spinal cord. *Brain Res.* **71**, 172–177.
Bentivoglio, M. and Rustioni, A., 1986, Corticospinal neurons with branching axons to the dorsal column nuclei in the monkey. *J. Comp. Neurol.* **253**, 260–276.
Benzinger, T. H., 1969, Heat regulation: homeostasis of central temperature in man. *Physiol. Rev.* **49**, 671–759.
Bergstrom, L., Hammond, D. L., Go, V. L. W. and Yaksh, T. L., 1983, Concurrent measurement of substance P and serotonin in spinal superfusates: failure of capsaicin and p-chloroamphetamine to co-release. *Brain Res.* **270**, 181–184.
Berkley, K. J., 1975, Different targets of different neurons in nucleus gracilis of the cat. *J. Comp. Neurol.* **163**, 285–304.
Berkley, K. J., 1980, Spatial relationships between the terminations of somatic sensory and motor pathways in the rostral brainstem of cats and monkeys. I. Ascending somatic sensory inputs to lateral diencephalon. *J. Comp. Neurol.* **193**, 283–317.
Berkley, K. J. and Hand, P. J., 1978, Efferent projections of the gracile nucleus in the cat. *Brain Res.* **153**, 263–283.
Berkley, K. J., Blomqvist, A., Pelt, A. and Flink, R., 1980, Differences in the colateralization of neuronal projections from the dorsal column nuclei and lateral cervical nucleus to the thalamus and tectum in the cat: an anatomical study using two different double-labeling techniques. *Brain Res.* **202**, 273–290.
Berkley, K. J., Budell, R. J., Blomqvist, A. and Bull, M., 1986, Output systems of the dorsal column nuclei in the cat. *Brain Res. Rev.* **11**, 199–225.
Berkley, K. J., Hotta, H., Robbins, A. and Sato, Y., 1990, Functional properties of afferent fibers supplying reproductive and other pelvic organs in pelvic nerve of female rat. *J. Neurophysiol.* **63**, 256–272.
Bernard, C.; 1858, *Lecons sur la physiologie et la pathologie du systeme nerveux.*, pp. 20-112. J.B. Bailliere et Fils, Paris.
Bernard, J. F. and Besson, J. M., 1990, The spino(trigemino)-pontoamygdaloid pathway: electrophysiological evidence for an involvement in pain processes. *J. Neurophysiol.* **63**, 473–490.
Bernhard, C. G., 1953, The spinal cord potentials in leads from the cord dorsum in relation to peripheral source of afferent stimulation. *Acta Physiol. Scand.* **29** (Suppl. 106), 1–29.
Bernhard, C. G. and Widén, L., 1953, On the origin of the negative and positive spinal cord potentials evoked by stimulation of low threshold cutaneous fibres. *Acta Physiol. Scand.* **29** (Suppl. 106), 42–54.
Bernhard, C. G., Lindblom, U. F. and Ottosson, J. O., 1953, The longitudinal distribution of the negative cord dorsum potential following stimulation of low threshold cutaneous fibres. *Acta Physiol. Scand.* 29 (Suppl. 106), 170–179.
Besson, J. M. and Chaouch, A., 1987, Peripheral and spinal mechanisms of nociception. *Physiol. Rev.* **67**, 67–186.
Besson, J. M., Rivot, J. P. and Aleonard, P., 1971, Action of picrotoxin on presynaptic inhibition of various origins in the cat's spinal cord. *Brain Res.* **26**, 212–216.
Besson, J. M., Conseiller, C., Hamann, K. F. and Maillard, M. C., 1972, Modifications of dorsal horn cell activities in the spinal cord, after intra-arterial injection of bradykinin. *J. Physiol.* **221**, 189–205.
Besson, J. M., Catchlove, R. F. H., Feltz, P. and Le Bars, D., 1974, Further evidence for postsynaptic inhibitions on lamina 5 dorsal horn interneurons. *Brain Res.* **66**, 531–536.

Besson, J. M., Guilbaud, G. and Le Bars, D., 1975, Descending inhibitory influences exerted by the brain stem upon the activities of dorsal horn lamina V cells induced by intra-arterial injection of bradykinin into the limbs. *J. Physiol.* **248**, 725–739.

Bessou, P. and Laporte, Y., 1961, Etude des recepteurs musculaires innerves par les fibres afferentes du groupe III (fibres myelinisees fines), chez le chat. *Arch. Ital. Biol.* **99**, 293–321.

Bessou, P. and Perl, E. R., 1966, A movement receptor of the small intestine. *J. Physiol.* **182**, 404–426.

Bessou, P. and Perl, E. R., 1969, Response of cutaneous sensory units with unmyelinated fibers to noxious stimuli. *J. Neurophysiol.* **32**, 1025–1043.

Bessou, P., Laporte, Y. and Pages, B., 1968, Frequencygrams of spindle primary endings elicited by stimulation of static and dynamic fusimotor fibres. *J. Physiol.* **196**, 47–63.

Bessou, P., Burgess, P. R., Perl, E. R. and Taylor, C. B., 1971, Dynamic properties of mechanoreceptors with unmyelinated (C) fibers. *J. Neurophysiol.* **34**, 116–131.

Bianconi, R. and van der Meulen, J. P., 1963, The responses to vibration of the end organs of mammalian muscle spindles. *J. Neurophysiol.* **26**, 177–190.

Bicknell, H. R. and Beal, J. A., 1981, Star shaped neurons in the substantia gelatinosa of the adult cat spinal cord. *Neurosci. Lett.* **22**, 37–41.

Bicknell, H. R. and Beal, J. A., 1984, Axonal and dendritic development of substantia gelatinosa neurons in the lumbosacral spinal cord of the rat. *J. Comp. Neurol.* **226**, 508–522.

Biedenbach, M. A., 1972, Cell density and regional distribution of cell types in the cuneate nucleus of the rhesus monkey. *Brain Res.* **45**, 1–14.

Biedenbach, M. A., Jabbur, S. J. and Towe, A. L., 1971, Afferent inhibition in the cuneate nucleus of the rhesus monkey. *Brain Res.* **27**, 179–183.

Bing, Z., Villanueva, L. and Le Bars, D., 1990, Ascending pathways in the spinal cord involved in the activation of subnucleus reticularis dorsalis neurons in the medulla of the rat. *J. Neurophysiol.* **63**, 424–438.

Biscoe, T. J., Curtis, D. R. and Ryall, R. W., 1966, An investigation of catecholamine receptors of spinal interneurones. *Int. J. Neuropharmacol.* **5**, 429–434.

Biscoe, T. J., Headley, P. M., Lodge, D., Martin, M. R. and Watkins, J. C., 1976, The sensitivity of rat spinal interneurones and Renshaw cells to L-glutamate and L-aspartate. *Exp. Brain Res.* **26**, 547–551.

Bishop, P. O., Burke, W. and Davis, R., 1962, Single-unit recording from antidromically activated optic radiation neurones. *J. Physiol.* **162**, 432–450.

Björkeland, M. and Boivie, J., 1984, The termination of spinomesencephalic fibers in cat. An experimental anatomical study. *Anat. Embryol.* **170**, 265–277.

Björklund, A. and Hökfelt, T., 1983, *Handbook of Chemical Neuroanatomy*. Elsevier, New York.

Bjorklund, A. and Skagerberg, G., 1979, Evidence for a major spinal cord projection from the diencephalic A11 dopamine cell group in the rat using transmitter-specific fluorescent retrograde tracing. *Brain Res.* **177**, 170–175.

Blair, R. W., Weber, R. N. and Foreman, R. D., 1981, Characteristics of primate spinothalamic tract neurons receiving viscerosomatic convergent inputs in the T3-T5 segments. *J. Neurophysiol.* **46**, 797–811.

Blair, R. W., Weber, R. N. and Foreman, R. D., 1982, Responses of thoracic spinothalamic neurons to intracardiac injection of bradykinin in the monkey. *Circ. Res.* **51**, 83–94.

Blair, R. W., Ammons, W. S. and Foreman, R. D., 1984, Responses of thoracic spinothalamic and spinoreticular cells to coronary artery occlusion. *J. Neurophysiol.* **51**, 636–648.

Blessing, W. W. and Chalmers, J. P., 1979, Direct projection of catecholamine (presumably dopamine)-containing neurons from hypothalamus to spinal cord. *Neurosci. Lett.* **11**, 35–40.

Blix, M., 1884, Experimentelle Beitraege zur Losung der Frage ueber die specifische Energie der Hautnerven. *Z. Biol.* **20**, 141–156.

Blomqvist, A., 1980, Gracilo-diencephalic relay cells: a quantitative study in the cat using retrograde transport of horseradish peroxidase. *J. Comp. Neurol.* **193**, 1097–1125.

Blomqvist, A., 1981, Morphometric synaptology of gracilo-diencephalic relay cells: an electron microscopic study in the cat using retrograde transport of horseradish peroxidase. *J. Neurocytol.* **10**, 709–724.

Blomqvist, A. and Westman, J., 1975, Combined HRP and Fink-Heimer staining applied on the gracile nucleus in the cat. *Brain Res.* **99**, 339–342.

Blomqvist, A. and Westman, J., 1976, Interneurons and initial axon collaterals in the feline gracile nucleus demonstrated with the rapid Golgi technique. *Brain Res.* **111**, 407–410.

Blomqvist, A., Flink, R., Bowsher, D., Griph, S. and Westman, J., 1978, Tectal and thalamic projections of dorsal column and lateral cervical nuclei: a quantitative study in the cat. *Brain Res.* **141**, 335–341.

Blomqvist, A., Flink, R., Westman, J. and Wiberg, M., 1985a, Synaptic terminals in the ventroposterolateral nucleus of the thalamus from neurons in the dorsal column and lateral cervical nuclei: an electron microscopic study in the cat. *J. Neurocytol.* **14**, 869–886.

Blomqvist, A., Westman, J., Koehler, C. and Wu, J.Y., 1985b, Immunocytochemical localization of glutamic acid decarboxylase and substance P in the lateral cervical nucleus: a light and electron microscopic study in the cat. *Neurosci. Lett.* **56**, 229–233.

Blum, P., Bromberg, M. B. and Whitehorn, D., 1975, Population analysis of single units in the cuneate nucleus of the cat. *Exp. Neurol.* **48**, 57–78.

Blumenkopf, B., 1988, Neurochemistry of the dorsal horn. In: *Applied Neurophysiology*, vol. 51, pp. 89-103. N.C. Durham (ed.). Karger, A.G., Basel.

Boehmer, C. G., Norman, J., Catton, M., Fine, L. G. and Mantyh, P. W., 1989, High levels of mRNA coding for substance P, somatostatin and alpha-tubulin are expressed by rat and rabbit dorsal root ganglia neurons. *Peptides* **10**, 1179–1194.

Boesten, A. J. P. and Voogd, J., 1975, Projections of the dorsal column nuclei and the spinal cord on the inferior olive in the cat. *J. Comp. Neurol.* **161**, 215–238.

Bohm, E., 1953, An electro-physiological study of the ascending spinal anterolateral fibre system connected to coarse cutaneous afferents. *Acta Physiol. Scand.* **29** (Suppl. 106), 106–137.

Boivie, J., 1970, The termination of the cervicothalamic tract in the cat. An experimental study with silver impregnation methods. *Brain Res.* **19**, 333–360.

Boivie, J., 1971a, The termination of the spinothalamic tract in the cat. An experimental study with silver degeneration methods. *Exp. Brain Res.* **12**, 331–353.

Boivie, J., 1971b, The termination in the thalamus and the zona incerta of fibers from the dorsal column nuclei (DCN) in the cat. An experimental study with silver impregnation methods. *Brain Res.* **28**, 459–490.

Boivie, J., 1978, Anatomical observations on the dorsal column nuclei, their thalamic projection and the cytoarchitecture of some somatosensory thalamic nuclei in the monkey. *J. Comp. Neurol.* **178**, 17–48.

Boivie, J., 1979, An anatomical reinvestigation of the termination of the spinothalamic tract in the monkey. *J. Comp. Neurol.* **186**, 343–370.

Boivie, J., 1980, Thalamic projections from lateral cervical nucleus in monkey. A degeneration study. *Brain Res.* **198**, 13–26.

Boivie, J. and Boman, K., 1981, Termination of a separate (proprioceptive?) cuneothalamic tract from external cuneate nucleus in monkey. *Brain Res.* **224**, 235–246.

Boivie, J. and Perl, E. R., 1975, Neural substrates of somatic sensation. In: *MTP International Review of Science*, Physiology Series One, Vol. 3, *Neurophysiology*, pp. 303-411. C. C. Hunt (ed.). University Park Press, Baltimore.

Boivie, J. and Westman, J., 1968, Electron microscopy of medial lemniscal terminal degeneration in the ventral posterolateral thalamic nucleus of the cat. *Experientia* **24**, 159–160.

Boivie, J., Grant, G., Albe-Fessard, D. and Levante, A., 1975, Evidence for a projection to the thalamus from the external cuneate nucleus in the monkey. *Neurosci. Lett.* **1**, 308.

Boivie, J., Leijon, G. and Johansson, I., 1989, Central post-stroke pain—a study of the mechanisms through analyses of the sensory abnormalities. *Pain* **37**, 173–185.

Bolton, C. F., Winkelmann, R. K. and Dyck, P. J., 1966, A quantitative study of Meissner's corpuscles in man. *Neurology* **16**, 1–9.

Booth, C. S. and Hahn, J. F., 1974, Thermal and mechanical stimulation of type II receptors and field receptors in cat. *Exp. Neurol.* **44**, 49–59.

Borges, L. F. and Iversen, S. D., 1986, Topography of choline acetyltransferase immunoreactive neurons and fibers in the rat spinal cord. *Brain Res.* **362**, 140–148.

Borges, L. F. and Moskowitz, M. A., 1983, Do intracranial and extracranial trigeminal afferents represent divergent axon collaterals? *Neurosci. Lett.* **35**, 265–270.

Boring, E. G., 1916, Cutaneous sensation after nerve-division. *Quart. J. Exp. Physiol.* **10**, 1–95.

Boring, E. G., 1942, *Sensation and Perception in the History of Experimental Psychology*. Appleton-Century-Crofts, New York.

Bostock, H., 1981, U-turns in rat ventral roots. *J. Physiol.* (Lond.) **312**, 49P–50P.

Bourgoin, S., Oliveras, J. L., Bruxelle, J., Hamon, M. and Besson, J. M., 1980, Electrical stimulation of the nucleus raphe magnus in the rat. Effects on 5-HT metabolism in the spinal cord. *Brain Res.* **194**, 377–389.

Bouthenet, M. L., Ruat, M., Sales, N., Garbarg, M. and Schwartz, J. C., 1988, A detailed mapping of histamine H_1-receptors in guinea-pig central nervous system established by autoradiography with [^{125}I]iodobolpyramine. *Neuroscience* **26**, 553–600.

Boutry, J. M. and Novikoff, A. B., 1975, Cytochemical studies on Golgi apparatus, GERL, and lysosomes in neurons of dorsal root ganglia in mice. *Proc. Natl. Acad. Sci.* **72**, 508–512.

Bowery, N. G., Hudson, A. L. and Price, G. W., 1987, GABAa and GABAb receptor site distribution in the rat central nervous system. *Neuroscience* **20**, 365–383.

Bowker, R. M., 1986, Intrinsic 5HT-immunoreactive neurons in the spinal cord of the fetal non-human primate. *Dev. Brain Res.* **28**, 137–143.

Bowker, R. M., Westlund, K. N. and Coulter, J. D., 1981, Origins of serotonergic projections to the spinal cord in rat: an immunocytochemical-retrograde transport study. *Brain Res.* **226**, 187–199.

Bowker, R. M., Westlund, K. N., Sullivan, M. C. and Coulter, J. D., 1982, Organization of descending serotonergic projections to the spinal cord. In: *Descending Pathways to the Spinal Cord*. H. G. J. M. Kuypers and G. F. Martin (eds.). *Progress in Brain Res.* **57**, 239–265.

Bowker, R. M., Westlund, K. N., Sullivan, M. C., Wilber, J. F. and Coulter, J. D., 1983, Descending serotonergic, peptidergic and cholinergic pathways from the raphe nuclei: a multiple transmitter complex. *Brain Res.* **288**, 33–48.

Bowker, R. M., Reddy, V. K., Fung, S. J., Chan, J. Y. H. and Barnes, C. D., 1987, Serotonergic and non-

serotonergic raphe neurons projecting to the feline lumbar and cervical spinal cord: a quantitative horseradish peroxidase-immunocytochemical study. *Neurosci. Lett.* **75**, 31–37.
Bowsher, D., 1957, Termination of the central pain pathway in man: the conscious appreciation of pain. *Brain* **80**, 606–622.
Bowsher, D., 1958, Projection of the gracile and cuneate nuclei in *Macaca mulatta*: an experimental degeneration study. *J. Comp. Neurol.* **110**, 135–155.
Bowsher, D., 1961, The termination of secondary somatosensory neurons within the thalamus of *Macaca mulatta*: an experimental degeneration study. *J. Comp. Neurol.* **117**, 213–227.
Bowsher, D., 1976, Role of the reticular formation in responses to noxious stimulation. *Pain* **2**, 361–378.
Bowsher, D. and Abdel-Maguid, T. E., 1984, Superficial dorsal horn of the adult human spinal cord. *Neurosurgery* **15**, 893–899.
Bowsher, D. and Westman, J., 1970, The gigantocellular reticular region and its spinal afferents: a light and electron microscopic study in the cat. *J. Anat.* **106**, 23–36.
Boyd, E. S., Meritt, D. A. and Gardner, L. C., 1966, The effect of convulsant drugs on transmission through the cuneate nucleus. *J. Pharm. Exp. Ther.* **154**, 398–409.
Boyd, I. A., 1954, The histological structure of the receptors in the knee-joint of the cat correlated with their physiological response. *J. Physiol.* **124**, 476–488.
Boyd, I. A. and Roberts, T. D. M., 1953, Proprioceptive discharges from stretch-receptors in the knee-joint of the cat. *J. Physiol.* **122**, 38–58.
Braas, K. M., Newby, A. C., Wilson, V. S. and Snyder, S. H., 1986, Adenosine-containing neurons in the brain localized by immunocytochemistry. *J. Neurosci.* **6**, 1952–1961.
Bradley, K. and Eccles, J. C., 1953, Analysis of the fast afferent impulses from thigh muscles. *J. Physiol.* **122**, 462–473.
Brain, S. D., Williams, T. J., Tippins, J. R., Morris, H. R. and McIntyre, I., 1985, Calcitonin gene-related peptide is a potent vasodilator. *Nature* **313**, 54–56.
Breazile, J. E. and Kitchell, R. L., 1968, Ventrolateral spinal cord afferents to the brain stem in the domestic pig. *J. Comp. Neurol.* **133**, 363–372.
Brennan, T. J., Oh, U. T., Girardot, M. N., Ammons, W. S. and Foreman, R. D., 1987, Inhibition of cardiopulmonary input to thoracic spinothalamic tract cells by stimulation of the subcoeruleus-parabrachial region in the primate. *J. Auton. Nerv. Syst.* **18**, 61–72.
Brennan, T. J., Oh, U. T., Hobbs, S. F., Garrison, D. W. and Foreman, R. D., 1989, Urinary bladder and hindlimb afferent input inhibits activity of primate T2-T5 spinothalamic tract neurons. *J. Neurophysiol.* **61**, 573–588.
Bresnahan, J. C., Ho, R. H. and Beattie, M. S., 1984, A comparison of the ultrastructure of substance P and enkephalin-immunoreactive elements in the nucleus of the dorsal lateral funiculus and laminae I and II of the rat spinal cord. *J. Comp. Neurol.* **229**, 497–511.
Briner, R. P., Carlton, S. M., Coggeshall, R. E. and Chung, K., 1988, Evidence for unmyelinated sensory fibres in the posterior columns in man. *Brain* **111**, 999–1007.
Brink, E., Harrison, P. J., Jankowska, E., McCrea, D. and Skoog, B., 1983a, Post-synaptic potentials in a population of motoneurones following activity in single interneurones in the cat. *J. Physiol.* 343, 341–359.
Brink, E., Jankowska, E., McCrea, D. and Skoog, B., 1983b, Inhibitory interactions between interneurones in reflex pathways from group Ia afferents in the cat. *J. Physiol.* **343**, 361–379.
Brink, E., Jankowska, E. and Skoog, B., 1984, Convergence onto interneurones subserving primary afferent depolarization of group I afferents. *J. Neurophysiol.* **51**, 432–449.
Brinkhus, H. B. and Zimmermann, M., 1983, Characteristics of spinal dorsal horn neurons after partial chronic deafferentation by dorsal root transection. *Pain* **15**, 221–236.
Brock, L. G., Coombs, J. S. and Eccles, J. C., 1953, Intracellular recording from antidromically activated motoneurones. *J. Physiol.* **122**, 429–461.
Brodal, A., 1957, *The Reticular Formation of the Brain Stem. Anatomical Aspects and Functional Correlations.* Oliver and Boyd, Edinburgh.
Brodal, A. and Pompeiano, O., 1957, The vestibular nuclei in the cat. *J. Anat.* **91**, 438–454.
Brodal, A. and Rexed, B., 1953, Spinal afferents to the lateral cervical nucleus in the cat. *J. Comp. Neurol.* **98**, 179–211.
Brodin, E., Linderoth, B., Gazelius, B. and Ungerstedt, U., 1987, In vivo release of substance P in cat dorsal horn studied with microdialysis. *Neurosci. Lett.* **76**, 357–362.
Broman, J. and Blomqvist, A., 1989a, GABA-immunoreactive neurons and terminals in the lateral cervical nucleus of the cynomologus monkey. *J. Comp. Neurol.* **283**, 415–424.
Broman, J. and Blomqvist, A., 1989b, Substance P-like immunoreactivity in the lateral cervical nucleus of the owl monkey (*Aotus trivirgatus*): a comparison with the cat and rat. *J. Comp. Neurol.* **289**, 111–117.
Broman, J., Flink, R. and Westman, J., 1987, Postnatal development of the feline lateral cervical nucleus: I. A quantitative light and electron microscopic study. *J. Comp. Neurol.* **260**, 539–551.
Bromberg, M. B. and Fetz, E. E., 1977, Responses of single units in cervical spinal cord of alert monkeys. *Exp. Neurol.* **55**, 469–482.
Bromberg, M. B. and Whitehorn, D., 1974, Myelinated fiber types in the superficial radial nerve of the cat and their central projections. *Brain Res.* **78**, 157–163.

Bromberg, M. B., Blum, P. and Whitehorn, D., 1975, Quantitative characteristics of inhibition in the cuneate nucleus of the cat. *Exp. Neurol.* **48**, 37–56.

Bromberg, M. B., Burnham, J. A. and Towe, A. L., 1981, Doubly projecting neurons of the dorsal column nuclei. *Neurosci. Lett.* **25**, 215–220.

Bromm, B., Jahnke, M. T. and Treede, R. D., 1984, Responses of human cutaneous afferents to CO_2 laser stimuli causing pain. *Exp. Brain Res.* **55**, 158–166.

Browder, J. and Gallagher, J. P., 1948, Dorsal cordotomy for painful phantom limb. *Ann. Surg.* **128**, 456–469.

Brown, A. G., 1968, Cutaneous afferent fibre collaterals in the dorsal columns of the cat. *Exp. Brain Res.* **5**, 293–305.

Brown, A. G., 1970, Descending control of the spinocervical tract in decerebrate cats. *Brain Res.* **17**, 152–155.

Brown, A. G., 1971, Effects of descending impulses on transmission through the spinocervical tract. *J. Physiol.* **219**, 103–125.

Brown, A. G., 1973, Ascending and long spinal pathways: dorsal columns, spinocervical tract and spinothalamic tract, pp. 315-338. In: *Handbook of Sensory Physiology*, vol. II, *Somatosensory System*, pp. 315–338. A. Iggo (ed.). Springer, New York.

Brown, A. G., 1977, Cutaneous axons and sensory neurones in the spinal cord. *Brit. Med. Bull.* **33**, 109–112.

Brown, A. G., 1981, *Organization in the Spinal Cord: The Anatomy and Physiology of Identified Neurones.* Springer-Verlag, Berlin.

Brown, A. G. and Franz, D. N., 1969, Responses of spinocervical tract neurones to natural stimulation of identified cutaneous receptors. *Exp. Brain Res.* **7**, 231–249.

Brown, A. G. and Franz, D. N., 1970, Patterns of response in spinocervical tract neurones to different stimuli of long duration. *Brain Res.* **17**, 156–160.

Brown, A. G. and Fyffe, R. E. W., 1978, The morphology of group Ia afferent fibre collaterals in the spinal cord of the cat. *J. Physiol.* **274**, 111–127.

Brown, A. G. and Fyffe, R. E. W., 1979, The morphology of group Ib afferent fibre collaterals in the spinal cord of the cat. *J. Physiol.* **296**, 215–228.

Brown, A. G. and Fyffe, R. E. W., 1981, Form and function of dorsal horn neurones with axons ascending the dorsal columns in cat. *J. Physiol.* **321**, 31–47.

Brown, A. G. and Hayden, R. E., 1971, The distribution of cutaneous receptors in the rabbit's hind limb and differential electrical stimulation of their axons. *J. Physiol.* **213**, 495–506.

Brown, A. G. and Iggo, A., 1967, A quantitative study of the cutaneous receptors and afferent fibres in the cat and rabbit. *J. Physiol.* **193**, 707–733.

Brown, A. G. and Martin, H. F., 1973, Activation of descending control of the spinocervical tract by impulses ascending the dorsal columns and relaying through the dorsal column nuclei. *J. Physiol.* **235**, 535–550.

Brown, A. G. and Noble, R., 1982, Connexions between hair follicle afferent fibres and spinocervical tract neurones in the cat: the synthesis of receptive fields. *J. Physiol.* **232**, 77–91.

Brown, A. G. and Short, A. D., 1974, Effects from the somatic sensory cortex on transmission through the spinocervical tract. *Brain Res.* **74**, 338–341.

Brown, A. G., Iggo, A. and Miller, S., 1967, Myelinated afferent nerve fibers from the skin of the rabbit ear. *Exp. Neurol.* **18**, 338–349.

Brown, A. G., Hamann, W. C. and Martin, H. F., 1973a, Interactions of cutaneous myelinated (A) and non-myelinated (C) fibres on transmission through the spinocervical tract. *Brain Res.* **53**, 222–226.

Brown, A. G., Hamann, W. C. and Martin, H. F., 1973b, Descending influences on spinocervical tract cell discharges evoked by non-myelinated cutaneous afferent nerve fibres. *Brain Res.* **53**, 218–221.

Brown, A. G., Kirk, E. J. and Martin, H. F., 1973c, Descending and segmental inhibition of transmission through the spinocervical tract. *J. Physiol.* **230**, 689–705.

Brown, A. G., Gordon, G. and Kay, R. H., 1974, A study of single axons in the cat's medial lemniscus. *J. Physiol.* **236**, 225–246.

Brown, A. G., Hamann, W. C. and Martin, H. F., 1975, Effects of activity in non-myelinated afferent fibres on the spinocervical tract. *Brain Res.* **98**, 243–259.

Brown, A. G., House, C. R., Rose, P. K. and Snow, P. J., 1976, The morphology of spinocervical tract neurones in the cat. *J. Physiol.* **260**, 719–738.

Brown, A. G., Coulter, J. D., Rose, P. K., Short, A. D. and Snow, P. J., 1977a, Inhibition of spinocervical tract discharges from localized areas of the sensorimotor cortex in the cat. *J. Physiol.* **264**, 1–16.

Brown, A. G., Rose, P. K. and Snow, P. J., 1977b, The morphology of spinocervical tract neurones revealed by intracellular injection of horseradish peroxidase. *J. Physiol.* **270**, 747–764.

Brown, A. G., Rose, P. K. and Snow, P. J., 1977c, The morphology of hair follicle afferent fibre collaterals in the spinal cord of the cat. *J. Physiol.* **272**, 779–797.

Brown, A. G., Rose, P. K. and Snow, P. J., 1978, Morphology and organization of axon collaterals from afferent fibres of slowly adapting type I units in cat spinal cord. *J. Physiol.* **277**, 15–27.

Brown, A. G., Fyffe, R. E. W., Noble, R., Rose, P. K. and Snow, P. J., 1980a, The density, distribution and topographical organization of spinocervical tract neurones in the cat. *J. Physiol.* **300**, 409–428.

Brown, A. G., Rose, P. K. and Snow, P. J., 1980b, Dendritic trees and cutaneous receptive fields of adjacent spinocervical tract neurones in the cat. *J. Physiol.* **300**, 429–440.

Brown, A. G., Fyffe, R. E. W. and Noble, R., 1980c, Projections from Pacinian corpuscles and rapidly adapting mechanoreceptors of glabrous skin to the cat's spinal cord. *J. Physiol.* **307**, 385–400.

Brown, A. G., Fyffe, R. E. W., Rose, P. K. and Snow, P. J., 1981, Spinal cord collaterals from axons of type II slowly adapting units in the cat. *J. Physiol.* **316**, 469–480.

Brown, A. G., Brown, P. B., Fyffe, R. E. W. and Pubols, L. M., 1983a, Receptive field organization and response properties of spinal neurones with axons ascending the dorsal columns in the cat. *J. Physiol.* **337**, 575–588.

Brown, A. G., Brown, P. B., Fyffe, R. E. W. and Pubols, L. M., 1983b, Effects of dorsal root section on spinocervical tract neurones in the cat. *J. Physiol.* **337**, 589–608.

Brown, A. G., Fyffe, R. E. W., Noble, R. and Rowe, M. J., 1984, Effects of hind limb nerve section on lumbosacral dorsal horn neurones in the cat. *J. Physiol.* **354**, 375–394.

Brown, A. G., Noble, R. and Rowe, M. J., 1986a, Receptive field profiles and integrative properties of spinocervical tract cells in the cat. *J. Physiol.* **374**, 335–348.

Brown, A. G., Noble, R. and Riddell, J. S., 1986b, Relations between spinocervical and postsynaptic dorsal column neurones in the cat. *J. Physiol.* **381**, 333–349.

Brown, A. G., Koerber, H. R. and Noble, R., 1987a, Excitatory actions of single impulses in single hair follicle afferent fibres on spinocervical tract neurones in the cat. *J. Physiol.* **382**, 291–312.

Brown, A. G., Koerber, H. R. and Noble, R., 1987b, Actions of trains and pairs of impulses from single primary afferent fibres on single spinocervical tract cells in cat. *J. Physiol.* **382**, 313–329.

Brown, A. G., Koerber, H. R. and Noble, R., 1987c, An intracellular study of spinocervical tract cell responses to natural stimuli and single hair afferent fibres in cats. *J. Physiol.* **382**, 331–354.

Brown, A. G., Fyffe, R. E. W., Noble, R. and Rowe, M., 1984,. Effects of hindlimb nerve section on lumbosacral dorsal horn neurones in the cat. *J. Physiol.* **354**, 375–394.

Brown, A. M., 1967, Excitation of afferent cardiac sympathetic fibres during myocardial ischaemia. *J. Physiol.* **190**, 35–53.

Brown, M. C., Engberg, I. and Matthews, P. B. C., 1967, The relative sensitivity to vibration of muscle receptors of the cat. *J. Physiol.* **192**, 773–800.

Brown, P. B., 1969, Response of cat dorsal horn cells to variations of intensity, location, and area of cutaneous stimuli. *Exp. Neurol.* **23**, 249–265.

Brown, P. B. and Culberson, J. L., 1981, Somatotopic organization of hindlimb cutaneous dorsal root projections to cat dorsal horn. *J. Neurophysiol.* **45**, 137–143.

Brown, P. B. and Fuchs, J. L., 1975, Somatotopic representation of hindlimb skin in cat dorsal horn. *J. Neurophysiol.* **38**, 1–9.

Brown, P. B. and Koerber, H. R., 1978, Cat hindlimb tactile dermatomes determined with single unit recordings. *J. Neurophysiol.* **41**, 260–267.

Brown, P. B., Moraff, H. and Tapper, D. N., 1973, Functional organization of the cat's dorsal horn: spontaneous activity and central cell response to single impulses in single type I fibers. *J. Neurophysiol.* **36**, 827–839.

Brown, P. B., Fuchs, J. L. and Tapper, D. N., 1975, Parametric studies of dorsal horn neurons responding to tactile stimulation. *J. Neurophysiol.* **38**, 19–25.

Brown, P. B., Busch, G. R. and Whittington, J., 1979, Anatomical changes in cat dorsal horn cells after transection of a single dorsal root. *Exp. Neurol.* **64**, 453–468.

Brown, P. B., Brushart, T. M. and Ritz, L. A., 1989, Somatotopy of digital nerve projections to the dorsal horn in the monkey. *Somatosensory and Motor Res.* **6**, 309–317.

Browne, K., Lee, J. and Ring, P. A., 1954, The sensation of passive movement at the metatarso-phalangeal joint of the great toe in man. *J. Physiol.* **126**, 448–458.

Brown-Séquard, C. E., 1860, *Course of Lectures on the Physiology and Pathology of the Central Nervous System.* Lippincott, Philadelphia.

Bruggencate, G. ten and Engberg, I., 1968, Analysis of glycine actions on spinal interneurones by intracellular recording. *Brain Res.* **11**, 446–450.

Bruggencate, G. ten, Lux, H. D. and Liebl, L., 1974, Possible relationships between extracellular potassium activity and presynaptic inhibition in the spinal cord of the cat. *Pfluegers Arch.* **349**, 301–317.

Brushart, T. M. and Mesulam, M.-M., 1980, Transganglionic demonstration of central sensory projections from skin and muscle with HRP-lectin conjugates. *Neurosci. Lett.* **17**, 1–6.

Bryan, R. N., Trevino, D. L., Coulter, J. D. and Willis, W. D., 1973, Location and somatotopic organization of the cells of origin of the spino-cervical tract. *Exp. Brain Res.* **17**, 177–189.

Bryan, R. N., Coulter, J. D. and Willis, W. D., 1974, Cells of origin of the spinocervical tract in the monkey. *Exp. Neurol.* **42**, 574–586.

Buck, S. H., Walsh, J. H., Yamamura, H. I. and Burks, T. F., 1982, Neuropeptides in sensory neurons. *Life Sci.* **30**, 1857–1866.

Bull, M. S. and Berkley, K. J., 1984, Differences in the neurons that project from the dorsal column nuclei to the diencephalon, pretectum, and tectum in the cat. *Somatosensory Res.* **1**, 281–300.

Bullitt, E., 1989, Induction of c-fos-like protein within the lumbar spinal cord and thalamus of the rat following peripheral stimulation. *Brain Res.* **493**, 391–397.

Burgess, P. R. and Clark, F. J., 1969a, Dorsal column projection of fibres from the cat knee joint. *J. Physiol.* **203**, 301–315.

Burgess, P. R. and Clark, F. J., 1969b, Characteristics of knee joint receptors in the cat. *J. Physiol.* **203**, 317–335.
Burgess, P. R. and Horch, K. W., 1978, The distinction between the short and intermediate ascending pathways in the fasciculus gracilis of the cat. *Brain Res.* **151**, 579–580.
Burgess, P. R. and Perl, E. R., 1967, Myelinated afferent fibres responding specifically to noxious stimulation of the skin. *J. Physiol.* **190**, 541–562.
Burgess, P. R. and Perl, E. R., 1973, Cutaneous mechanoreceptors and nociceptors. In: *Handbook of Sensory Physiology*, Vol. II, *Somatosensory System*, pp. 29–78. A. Iggo (ed.). Springer, New York.
Burgess, P. R., Petit, D. and Warren, R. M., 1968, Receptor types in cat hairy skin supplied by myelinated fibers. *J. Neurophysiol.* **31**, 833–848.
Burgess, P. R., Howe, J. F., Lessler, M. J. and Whitehorn, D., 1974, Cutaneous receptors supplied by myelinated fibers in the cat. II. Number of mechanoreceptors excited by a local stimulus. *J. Neurophysiol.* **37**, 1373–1386.
Burgess, P. R., Wei, J. Y., Clark, F. J. and Simon, J., 1982, Signalling of kinesthetic information by peripheral sensory receptors. *Ann. Rev. Neurosci.* **5**, 171–187.
Burke, D. and Eklund, G., 1977, Muscle spindle activity in man during standing. *Acta Physiol. Scand.* **100**, 187–199.
Burke, D., Hagbarth, K. E., Lofstedt, L. and Wallin, B. G., 1976a, The responses of human muscle spindle endings to vibration of non-contracting muscles. *J. Physiol.* **261**, 673–694.
Burke, D., Hagbarth, K. E., Lofstedt, L. and Wallin, B. G., 1976b, The responses of human muscle spindle endings to vibration during isometric contraction. *J. Physiol.* **261**, 695–711.
Burke, D., Hagbarth, K. E. and Skuse, N. F., 1978, Recruitment order of human spindle endings in isometric voluntary contractions. *J. Physiol.* **285**, 101–112.
Burke, D., Hagbarth, K. E. and Skuse, N. F., 1979, Voluntary activation of spindle endings in human muscles temporarily paralyzed by nerve pressure. *J. Physiol.* **287**, 329–336.
Burke, D., Aniss, A. M. and Gandevia, S. C., 1987, In-parallel and in-series behavior of human muscle spindle endings. *J. Neurophysiol.* **58**, 417–426.
Burke, D., Gandevia, S. C. and Macefield, G., 1988, Responses to passive movement of receptors in joint, skin and muscle of the human hand. *J. Physiol.* **402**, 347–361.
Burke, R. E., Rudomin, P., Vyklický, L. and Zajac, F. E., 1971, Primary afferent depolarization and flexion reflexes produced by radiant heat stimulation of the skin. *J. Physiol.* **213**, 185–214.
Burnweit, C. and Forssman, W. G., 1979, Somatostatinergic nerves in the cervical spinal cord of the monkey. *Cell Tiss. Res.* **200**, 83–90.
Burr, H. S., 1928, The central nervous system of *Orthagoriscus mola*. *J. Comp. Neurol.* **45**, 33–128.
Burstein, R., Cliffer, K. D. and Giesler, G. J., 1987, Direct somatosensory projections from the spinal cord to the hypothalamus and telencephalon. *J. Neurosci.* **7**, 4159–4164.
Burstein, R., Dado, R. J. and Giesler, G. J., 1990, The cells of origin of the spinothalamic tract of the rat: a quantitative reexamination. *Brain Res.* **511**, 329–337.
Burt, A. M., 1971, The histochemical localization of choline acetyltransferase. In: *Progress in Brain Research*, pp. 327–335. O. Eränkö (ed.). Elsevier, Amsterdam.
Burton, H. and Loewy, A. D., 1976, Descending projections from the marginal cell layer and other regions of the monkey spinal cord. *Brain Res.* **116**, 485–491.
Burton, H. and Loewy, A. D., 1977, Projections to the spinal cord from medullary somatosensory relay nuclei. *J. Comp. Neurol.* **173**, 773–792.
Burton, H., Forbes, D. J. and Benjamin, R. M., 1970, Thalamic neurons responsive to temperature changes of glabrous hand and foot skin in squirrel monkey. *Brain Res.* **24**, 179–190.
Burton, H., Terashima, S. I. and Clark, J., 1972, Response properties of slowly adapting mechanoreceptors to temperature stimulation in cats. *Brain Res.* **45**, 401–416.
Burton, W. and McFarlane, J. J., The organization of the seventh lumbar spinal ganglion of the cat. *J. Comp. Neurol.* **149**, 215–232, 1973.
Bystrzycka, E., Nail, B. S. and Rowe, M., 1977, Inhibition of cuneate neurones: its afferent source and influence on dynamically sensitive 'tactile' neurones. *J. Physiol.* **268**, 251–270.
Cadden, S. W., Villanueva, L., Chitour, D. and LeBars, D., 1983, Depression of activities of dorsal horn convergent neurones by propriospinal mechanisms triggered by noxious inputs; comparison with diffuse noxious inhibitory controls (DNIC). *Brain Res.* **275**, 1–11.
Cadwalader, W. B. and Sweet, J. E., 1912, Experimental work on the function of the anterolateral column of the spinal cord. *J.A.M.A.* **56**, 1490–1493.
Calas, A., Dupery, J. J., Gannerari, H., Gonella, J., Mourre, C., Condamin, M., Pellissier, J. F. and van den Bosch, P., 1981, Radioautographic investigation of serotonin cells; serotonin—current aspects of neurochemistry and function. In: *Advances in Experimental Medicine and Biology*, pp. 51–66. B. Haber, S. Gabay, N. R. Essidorides and S. G. A. Alivisatos (eds.). Plenum Press, New York.
Calne, D. B. and Pallis, C. A., 1966, Vibratory sense: a critical review. *Brain* **89**, 723–746.
Calvillo, O. and Ghignone, M., 1986, Presynaptic effect of clonidine on unmyelinated afferent fibers in the spinal cord of the cat. *Neurosci. Lett.* **64**, 335–339.
Calvillo, O., Madrid, J. and Rudomin, P., 1982, Presynaptic depolarization of unmyelinated primary afferent fibers in the spinal cord of the cat. *Neuroscience* **7**, 1389–1400.

Calvin, W. H. and Loeser, J. D., 1975, Doublet and burst firing patterns within the dorsal column nuclei of cat and man. *Exp. Neurol.* **48**, 406–426.

Calvino, B., Villanueva, L. and Le Bars, D., 1987, Dorsal horn (convergent) neurones in the intact anaesthetized arthritic rat. I. Segmental excitatory influences. *Pain* **28**, 81–98.

Cameron, A. A., Leah, J. D. and Snow, P. J., 1986, The electrophysiological and morphological characteristics of feline dorsal root ganglion cells. *Brain Res.* **362**, 1–6.

Cameron, A. A., Leah, J. D. and Snow, P. J., 1988, The coexistence of neuropeptides in feline sensory neurons. *Neuroscience* **27**, 969–979.

Campbell, J. N. and LaMotte, R. H., 1983, Latency to detection of first pain. *Brain Res.* **266**, 203–208.

Campistron, G., Buijs, R. M. and Geffard, M., 1986, Glycine neurons in the brain and spinal cord. Antibody production and immunocytochemical localization. *Brain Res.* **376**, 400–405.

Cangiano, A., Cook, W. A. and Pompeiano, O., 1969, Primary afferent depolarization in the lumbar cord evoked from the fastigial nucleus. *Arch. Ital. Biol.* **107**, 321–340.

Cangro, C. B., Sweetnam, P. M., Wrathall, J. R., Haser, W. B., Curthoys, N. P. and Neale, J. H., 1985, Localization of elevated glutaminase immunoreactivity in small DRG neurons. *Brain Res.* **336**, 158–161.

Cangro, C. B., Namboodiri, M. A. A., Sklar, L. A., Corigliano-Murphy, A. and Neale, J. H., 1987. Immunohistochemistry and biosynthesis of n-acetylaspartylglutamate in spinal sensory ganglia. *J. Neurochem.* **49**, 1579–1588.

Carli, G., Diete-Spiff, K. and Pompeiano, O., 1967a, Transmission of sensory information through the lemniscal pathway during sleep. *Arch. Ital. Biol.* **105**, 31–51.

Carli, G., Diete-Spiff, K. and Pompeiano, P., 1967b, Presynaptic and postsynaptic inhibition of transmission of somatic afferent volleys through the cuneate nucleus during sleep. *Arch. Ital. Biol.* **105**, 52–82.

Carli, G., Diete-Spiff, K. and Pompeiano, O., 1967c, Vestibular influences during sleep. V. Vestibular control on somatic afferent transmission in the cuneate nucleus during desynchronized sleep. *Arch. Ital. Biol.* **105**, 83–103.

Carli, G., Farabollini, F. and Fontani, G., 1975, Static characteristics of slowly adapting hip joint receptors in the cat. *Exp. Brain Res.* Suppl. 23, 36.

Carli, G., Farabollini, F., Fontani, G. and Meucci, M., 1979, Slowly adapting receptors in cat hip joint. *J. Neurophysiol.* **42**, 767–779.

Carli, G., Dontani, G. and Meucci, M., 1981, Static characteristics of muscle afferents from gluteus medius muscle: comparison with joint afferents of hip in cats. *J. Neurophysiol.* **45**, 1085–1095.

Carlton, S. M. and Hayes, E. S., 1989, Dynorphin A (1-8) immunoreactive cell bodies, dendrites and terminals are postsynaptic to calcitonin gene-related peptide primary afferent terminals in the monkey dorsal horn. *Brain Res.* **504**, 124–128.

Carlton, S. M. and Hayes, E. S., 1990, Light microscopic and ultrstructural analysis of GABA-immunoreactive profiles in the monkey spinal cord. *J. Comp. Neurol.* **300**, 162–182.

Carlton, S. M, Chung, J. M., Leonard, R. B. and Willis, W. D., 1985a, Funicular trajectories of brainstem neurons projecting to the lumbar spinal cord in the monkey (*Macaca fascicularis*): a retrograde labeling study. *J. Comp. Neurol.* **241**, 382–404.

Carlton, S. M., LaMotte, C. C., Honda, C. N., Surmeier, D. J., DeLanerolle, N. C. and Willis, W. D., 1985b, Ultrastructural analysis of substance P and other synaptic profiles innervating an identified primate spinothalamic tract neuron. *Neurosci. Abstr.* **11**, 578.

Carlton, S. M., Honda, C. N., Denoroy, L. and Willis, W. D., 1987a, Descending phenylethanolamine-N-methyltransferase projections to the monkey spinal cord: an immunohistochemical double labeling study. *Neurosci. Lett.* **76**, 133–139.

Carlton, S. M., McNeill, D. L., Chung, K. and Coggeshall, R. E., 1987b, A light and electron microscopic level analysis of calcitonin gene-related peptide (CGRP) in the spinal cord of the primate: An immunohistochemical study. *Neurosci. Lett.* **82**, 145–150.

Carlton, S. M., McNeill, D. L., Chung, K. and Coggeshall, R. E., 1988, Organization of calcitonin gene-related peptide-immunoreactive terminals in the primate dorsal horn. *J. Comp. Neurol.* **276**, 527–536.

Carlton, S. M., Honda, C. N. and Denoroy, L., 1989a, Distribution of phenylethanolamine-N-methyl transferase (PNMT) cell bodies, axons and terminals in the monkey brainstem: An immunohistochemical mapping study. *J. Comp. Neurol.* **287**, 273–285.

Carlton, S. M., LaMotte, C. C., Honda, C. N., Surmeier, D. J., DeLanerolle, N. and Willis, W. D., 1989b, Ultrastructural analysis of axosomatic contacts on functionally identified primate spinothalamic tract neurons. *J. Comp. Neurol.*, in press.

Carlton, S. M., Westlund, K. N., Zhang, D., Sorkin, L. S. and Willis, W. D., 1990, Calcitonin gene-related peptide containing primary afferent fibers synapse on primate spinothalamic tract cells. *Neurosci. Lett.* **109**, 76–81.

Carpenter, C. S., 1965, A histochemical study of oxidative enzymes in the nervous system of vitamin E-deficient rats. *Neurology* **15**, 328–332.

Carpenter, D., Lundberg, A. and Norrsell, U., 1963, Primary afferent depolarization evoked from the sensorimotor cortex. *Acta Physiol. Scand.* **59**, 126–142.

Carpenter, D., Engberg, I. and Lundberg, A., 1966, Primary afferent depolarization evoked from the brain stem and cerebellum. *Arch. Ital. Biol.* **104**, 73–85.

Carpenter, M. B., Stein, B. M. and Shriver, J. E., 1968, Central projections of spinal dorsal roots in the monkey. II. Lower thoracic, lumbosacral and coccygeal dorsal roots. *Amer. J. Anat.* **123**, 75–118.

Carr, P. A., Yamamoto, T., Karmy, G., Baimbridge, K. G. and Nagy, J. I., 1989, Parvalbumin is highly colocalized with calbindin D28k and rarely with calcitonin gene-related peptide in dorsal root ganglia neurons of rat. *Brain Res.* **497**, 163–170.

Carstens, E., 1982, Inhibition of spinal dorsal horn neuronal responses to noxious skin heating by medial hypothalamic stimulation in the cat. *J. Neurophysiol.* **48**, 808–822.

Carstens, E., 1988, Inhibition of rat spinothalamic tract neuronal responses to noxious skin heating by stimulation in midbrain periaqueductal gray or lateral reticular formation. *Pain* **33**, 215–224.

Carstens, E. and Trevino, D. L., 1978a, Laminar origins of spinothalamic projections in the cat as determined by the retrograde transport of horseradish peroxidase. *J. Comp. Neurol.* **182**, 151–166.

Carstens, E. and Trevino, D. L., 1978b, Anatomical and physiological properties of ipsilaterally projecting spinothalamic neurons in the second cervical segment of the cat's spinal cord. *J. Comp. Neurol.* **182**, 167–184.

Carstens, E., Yokota, T. and Zimmermann, M., 1979, Inhibition of spinal neuronal responses to noxious skin heating by stimulation of mesencephalic periaqueductal gray in the cat. *J. Neurophysiol.* **42**, 558–568.

Carstens, E., Klumpp, D. and Zimmermann, M., 1980a, Time course and effective sites for inhibition from midbrain periaqueductal gray of spinal dorsal horn neuronal responses to cutaneous stimuli in the cat. *Exp. Brain Res.* **38**, 425–430.

Carstens, E., Klumpp, D. and Zimmermann, M., 1980b, Differential inhibitory effects of medial and lateral midbrain stimulation on spinal neuronal discharges to noxious skin heating in the cat. *J. Neurophysiol.* **43**, 332–342.

Carstens, E., Bihl, H., Irvine, D. R. F. and Zimmermann, M., 1981a, Descending inhibition from medial and lateral midbrain of spinal dorsal horn neuronal responses to noxious and nonnoxious cutaneous stimuli in the cat. *J. Neurophysiol.* **45**, 1029–1042.

Carstens, E., Fraunhoffer, M. and Zimmermann, M., 1981b, Serotonergic mediation of descending inhibition from midbrain periaqueductal gray, but not reticular formation, of spinal nociceptive transmission in the cat. *Pain* **10**, 149–167.

Carstens, E., Klumpp, D., Randić, M. and Zimmermann, M., 1981c, Effect of iontophoretically applied 5-hydroxytryptamine on the excitability of single primary afferent C- and A-fibers in the cat spinal cord. *Brain Res.* **220**, 151–158.

Carstens, E., MacKinnon, J. D. and Guinan, M. J., 1982, Inhibition of spinal dorsal horn neuronal responses to noxious skin heating by medial preoptic and septal stimulation in the cat. *J. Neurophysiol.* **48**, 981–991.

Carstens, E., MacKinnon, J. D. and Guinan, M. J., 1983, Serotonin involvement in descending inhibition of spinal nociceptive transmission produced by stimulation of medial diencephalon and basal forebrain. *J. Neurosci.* **3**, 2112–2120.

Carstens, E., Gilly, H., Schreiber, H. and Zimmermann, M., 1987, Effects of midbrain stimulation and iontophoretic application of serotonin, noradrenaline, morphine and GABA on electrical thresholds of afferent C- and A-fibre terminals in cat spinal cord. *Neuroscience* **21**, 395–406.

Casey, K. L. and Morrow, T. J., 1988, Supraspinal nocifensive responses of cats: spinal cord pathways, monoamines, and modulation. *J. Comp. Neurol.* **270**, 591–605.

Cassinari, V. and Pagni, C. A., 1969, *Central Pain. A Neurosurgical Survey*. Harvard Univ. Press, Cambridge.

Cassini, P., Ho, R. H. and Martin, G. F., 1989, The brainstem origin of enkephalin- and substance-P-like immunoreactive axons in the spinal cord of the North American opossum. *Brain Behav. Evol.* **34**, 212–222.

Castiglioni, A. J. and Kruger, L., 1985, Excitation of dorsal column nucleus neurons by air-jet moving across the skin. *Brain Res.* **346**, 348–352.

Catalano, J. V. and Lamarche, G., 1957, Central pathway for cutaneous impulses in the cat. *Amer. J. Physiol.* **189**, 141–144.

Cauna, N., 1956, Nerve supply and nerve endings in Meissner's corpuscles. *Amer. J. Anat.* **99**, 315–350.

Cauna, N. and Mannan, G., 1958, The structure of human digital Pacinian corpuscles (corpuscula lamellosa) and its functional significance. *J. Anat.* **92**, 1–24.

Cauna, N. and Naik, N. T., 1963, The distribution of cholinesterase in the sensory ganglia of some mammals. *J. Histochem. Cytochem.* **11**, 129–138.

Cauna, N. and Ross, L. L., 1960, The fine structure of Meissner's touch corpuscles of human fingers. *J. Biophys. Biochem. Cytol.* **8**, 467–482.

Cechetto, D. F. and Saper, C. B., 1988, Neurochemical organization of the hypothalamic projection to the spinal cord in the rat. *J. Comp. Neurol.* **272**, 579–604.

Cechetto, D. F., Standaert, D. G. and Saper, C. B., 1985, Spinal and trigeminal dorsal horn projections to the parabrachial nucleus in the rat. *J. Comp. Neurol.* **240**, 153–160.

Cervero, F., 1982a, Noxious intensities of visceral stimulation are required to activate viscero-somatic multireceptive neurons in the thoracic spinal cord of the cat. *Brain Res.* **240**, 350–352.

Cervero, F., 1982b, Afferent activity evoked by natural stimulation of the biliary system in the ferret. *Pain* **13**, 137–151.

Cervero, F., 1983a, Somatic and visceral inputs to the thoracic spinal cord of the cat: effects of noxious stimulation of the biliary system. *J. Physiol.* **337**, 51–67.

Cervero, F., 1983b, Supraspinal connections of neurones in the thoracic spinal cord of the cat: ascending projections and effects of descending impulses. *Brain Res.* **275**, 251–261.

Cervero, F. and Connell, L. A., 1984a, Fine afferent fibers from viscera do not terminate in the substantia gelatinosa of the thoracic spinal cord. *Brain Res.* **294**, 370–374.

Cervero, F. and Connell, L. A., 1984b, Distribution of somatic and visceral primary afferent fibres within the thoracic spinal cord of the cat. *J. Comp. Neurol.* **230**, 88–98.

Cervero, F. and Iggo, A., 1978, Natural stimulation of urinary bladder afferents does not affect transmission through lumbosacral spinocervical tract neurons in the cat. *Brain Res.* **156**, 375–379.

Cervero, F. and Iggo, A., 1980, The substantia gelatinosa of the spinal cord. A critical review. *Brain* **103**, 717–772.

Cervero, F. and Sharkey, K. A., 1988, An electrophysiological and anatomical study of intestinal afferent fibres in the rat. *J. Physiol.* **401**, 381–397.

Cervero, F. and Tattersall, J. E. H., 1985, Cutaneous receptive fields of somatic and viscerosomatic neurones in the thoracic spinal cord of the cat. *J. Comp. Neurol.* **237**, 325–332.

Cervero, F. and Tattersall, J. E. H., 1986, Somatic and visceral sensory integration in the thoracic spinal cord. In: *Visceral Sensation.* F. Cervero and J. F. B. Morrison (eds.). *Prog. Brain Res.* **67**, 189–205. Elsevier, Amsterdam.

Cervero, F. and Tattersall, J. E. H., 1987, Somatic and visceral inputs to the thoracic spinal cord of the cat: marginal zone (lamina I) of the dorsal horn. *J. Physiol.* **383**, 383–395.

Cervero, F. and Wolstencroft, J. H., 1984, A positive feedback loop between spinal cord nociceptive pathways and antinociceptive areas of the cat's brain stem. *Pain* **20**, 125–138.

Cervero, F., Iggo, A. and Ogawa, H., 1976, Nociceptor-driven dorsal horn neurones in the lumbar spinal cord of the cat. *Pain* **2**, 5–14.

Cervero, F., Iggo, A. and Molony, V., 1977a, Responses of spinocervical tract neurones to noxious stimulation of the skin. *J. Physiol.* **267**, 537–558.

Cervero, F., Molony, V. and Iggo, A., 1977b, Extracellular and intracellular recordings from neurones in the substantia gelatinosa Rolandi. *Brain Res.* **136**, 565–569.

Cervero, F., Iggo, A. and Molony, V., 1979a, An electrophysiological study of neurones in substantia gelatinosa Rolandi of the cat's spinal cord. *Quart. J. Exp. Physiol.* **64**, 297–314.

Cervero, F., Iggo, A. and Molony, V., 1979b, Ascending projections of nociceptor-driven lamina I neurones in the cat. *Exp. Brain Res.* **35**, 135–149.

Cervero, F., Molony, V. and Iggo, A., 1979c, Supraspinal linkage of substantia gelatinosa neurones: effects of descending impulses. *Brain Res.* **175**, 351–355.

Cervero, F., Connell, L. A. and Lawson, S. N., 1984a, Somatic and visceral primary afferents in the lower thoracic dorsal root ganglia of the cat. *J. Comp. Neurol.* **228**, 422–431.

Cervero, F., Schouenborg, J., Sjölund, B. H. and Waddell, P. J., 1984b, Cutaneous inputs to dorsal horn neurones in adult rats treated at birth with capsaicin. *Brain Res.* **301**, 47–57.

Cervero, F., Lumb, B. M. and Tattersall, J. E. H., 1985, Supraspinal loops that mediate visceral inputs to thoracic spinal cord neurones in the cat: involvement of descending pathways from raphe and reticular formation. *Neurosci. Lett.* **56**, 189–194.

Cervero, F., Handwerker, H. O. and Laird, J. M. A., 1988, Prolonged noxious mechanical stimulation of the rat's tail: responses and encoding properties of dorsal horn neurones. *J. Physiol.* **404**, 419–436.

Cesa-Bianchi, M. G. and Sotgiu, M. L., 1969, Control by brain stem reticular formation of sensory transmission in Burdach nucleus. Analysis of single units. *Brain Res.* **13**, 129–139.

Cesa-Bianchi, M. G., Mancia, M. and Sotgiu, M. L., 1968, Depolarization of afferent fibers to the Goll and Burdach nuclei induced by stimulation of the brain-stem. *Exp. Brain Res.* **5**, 1–15.

Cesselin, F., LeBars, D., Bourgoin, S., Artaud, F., Gozlan, H., Clot, A. M., Besson, J. M. and Hamon, M., 1985, Spontaneous and evoked release of methionine-enkephalin-like material from the rat spinal cord in vivo. *Brain Res.* **339**, 305–313.

Cesselin, F., Bourgoin, S., Clot, A. M, Hamon, M. and LeBars, D., 1989, Segmental release of met-enkephalin-like material from the spinal cord of rats, elicited by noxious thermal stimuli. *Brain Res.* **484**, 71–77.

Chambers, M. R., Andres, K. H., Düring, M. von and Iggo, A., 1972, The structure and function of the slowly adapting type II mechanoreceptor in hairy skin. *Quart. J. Exp. Physiol.* **57**, 417–445.

Chambers, W. W. and Liu, C. N., 1957, Cortico-spinal tract of the cat. An attempt to correlate the pattern of degeneration with deficits in reflex activity following neocortical lesions. *J. Comp. Neurol.* **108**, 23–56.

Chandler, M. J., Garrison, D. W., Brennan, T. J. and Foreman, R. D., 1989, Effects of chemical and electrical stimulation of the midbrain on feline T2-T6 spinoreticular and spinal cell activity evoked by cardiopulmonary afferent input. *Brain Res.* **496**, 148–164.

Chang, H. T. and Ruch, T. C., 1947a, Organization of the dorsal columns and their nuclei in the spider monkey. *J. Anat.* **81**, 140–149.

Chang, H. T. and Ruch, T. C., 1947b, Topographical distribution of spinothalamic fibres in the thalamus of the spider monkey. *J. Anat.* **81**, 150–164.

Chan-Palay, V. and Palay, S. L., 1977a, Ultrastructural identification of substance P cells and their processes in rat sensory ganglia and their terminals in the spinal cord by immunocytochemistry. *Proc. Natl. Acad. Sci. USA* **74**, 4050–4054.

Chan-Palay, V. and Palay, S. L., 1977b, Immunocytochemical identification of substance P cells and their processes in rat sensory ganglia and their terminals in the spinal cord: light microscopic studies. *Proc. Natl. Acad. Sci. USA* **74**, 3597–3601.

Chan-Palay, V., Jonsson, G. and Palay, S. L., 1978, Serotonin and substance P coexist in neurons of the rat's central nervous system. *Proc. Natl. Acad. Sci. USA* **75**, 1582–1586.

Chaouch, A., Menetrey, D., Binder, D. and Besson, J. M., 1983, Neurons at the origin of the medial component of the bulbopontine spinoreticular tract in the rat: an anatomical study using horseradish peroxidase retrograde transport. *J. Comp. Neurol.* **214**, 309–320.

Chapman, C. E., Jiang, W. and Lamarre, Y., 1988, Modulation of lemniscal input during conditioned arm movements in the monkey. *Exp. Brain Res.* **72**, 316–334.

Charlton, C. G. and Helke, C., 1985a, Autoradiographic localization and characterization of spinal cord substance P binding sites: high densities in sensory, autonomic, phrenic, and Onuf's motor nuclei. *J. Neurosci.* **5**, 1653–1661.

Charlton, C. G. and Helke, C. J., 1985b, Characterization and segmental distribution of [125]Bolton-Hunter labeled substance P binding sites in rat spinal cord. *J. Neurosci.* **5**, 1293–1299.

Charnay, Y., Paulin, C., Chayvialle, J.-A. and Dubois, P. M., 1983, Distribution of substance P-like immunoreactivity in the spinal cord and dorsal root ganglia of the human foetus and infant. *Neuroscience* **10**, 41–55.

Charnay, Y., Paulin, C., Dray, F. and Dubois, P. M., 1984, Distribution of enkephalin in human fetus and infant spinal cord: An immunofluorescence study. *J. Comp. Neurol.* **223**, 415–423.

Charnay, Y., Chayviale, J., Said, S. I. and Dubois, P., 1985, Localization of vasoactive intestinal peptide immunoreactivity in human foetus and newborn infant spinal cord. *Neuroscience* **14**, 195–205.

Charnay, Y., Chayvialle, J. A., Pradayrol, L., Bouvier, R., Paulin, C. and Dubois, P. M., 1987, Ontogeny of somatostatin-like immunoreactivity in the human fetus and infant spinal cord. *Dev. Brain Res.* **36**, 63–73.

Cheek, M. D., Rustioni, A. and Trevino, D. L., 1975, Dorsal column nuclei projections to the cerebellar cortex in cats as revealed by the use of the retrograde transport of horseradish peroxidase. *J. Comp. Neurol.* **164**, 31–46.

Cheema, S., Whitsel, B. L. and Rustioni, A., 1983, The corticocuneate pathway in the cat: relation among terminal distribution patterns, cytoarchitecture, and single functional properties. *Somatosensory Res.* **1**, 169–205.

Cheema, S., Fyffe, R. and Rustioni, A., 1984a, Arborizations of single corticofugal axons in the feline cuneate nucleus stained by iontophoretic injection of horseradish peroxidase. *Brain Res.* **290**, 158–164.

Cheema, S., Rustioni, A. and Whitsel, B. L., 1984b, Light and electron microscopic evidence for a direct corticospinal projection to superficial laminae of the dorsal horn in cats and monkeys. *J. Comp. Neurol.* **225**, 276–290.

Cheema, S., Rustioni, A. and Whitsel, B. L., 1985, Sensorimotor cortical projections to the primate cuneate nucleus. *J. Comp. Neurol.* **240**, 196–211.

Chiba, T. and Masuko, S., 1989, Coexistence of varying combinations of neuropeptides with 5-hydroxytryptamine in neurons of the raphe pallidus et obscurus projecting to the spinal cord. *Neurosci. Res.* **7**, 13–23.

Chiba, T., Masuko, S. and Kawano, H., 1986, Correlation of mitochondrial swelling after capsaicin treatment and substance P and somatostatin immunoreactivity in small neurons of dorsal root ganglion in the rat. *Neurosci. Lett.* **64**, 311–316.

Childs, A. M., Evans, R.H. and Watkins, J. C., 1988, The pharmacological selectivity of three NMDA antagonists. *Eur. J. Pharmacol.* **145**, 81–86.

Chitour, D., Dickenson, A. H. and LeBars, D., 1982, Pharmacological evidence for the involvement of serotonergic mechanisms in diffuse noxious inhibitory controls (DNIC). *Brain Res.* **236**, 329–337.

Ch'ng, J. L. C., Christofides, N. D., Anand, P., Gibson, S. J., Allen, Y. S., Su, H. C., Tatemoto, K., Morrison, J. F. B., Polak, J. M. and Bloom, S. R., 1985, Distribution of galanin immunoreactivity in the central nervous system and the responses of galanin-containing neuronal pathways to injury. *Neuroscience* **16**, 343–354.

Cho, H. J. and Basbaum, A. I., 1988, Increased staining of immunoreactive dynorphin cell bodies in the deafferented spinal cord of the rat. *Neurosci. Lett.* **84**, 125–130.

Cho, H. J. and Basbaum, A. I., 1989, Ultrastructural analysis of dynorphin B-immunoreactive cells and terminals in the superficial dorsal horn of the deafferented spinal cord of the rat. *J. Comp. Neurol.* **281**, 193–205.

Choca, J. I., Proudfit, H. K. and Green, R. D., 1987, Identification of A1 and A2 receptors in the rat spinal cord. *J. Pharmacol. Exp. Ther.* **242**, 905–910.

Choca, J. I., Green, R. D. and Proudfit, H. K., 1988, Adenosine A1 and A2 receptors of the substantia gelatinosa and located predominantly on intrinsic neurons: an autoradiograph study. *J. Pharmacol. Exp. Ther.* **247**, 757–764.

Chou, D. K. H., Dodd, J., Jessell, T. M., Costello, C. E. and Jungalwala, F. B., 1988, Identification of α-galactose (α-fucose)-asialo-G_{ml} glycolipid expressed by subsets of rat dorsal ganglion neurons. *J. Biol. Chem.* **264**, 3409–3415.

Christensen, B. N. and Perl, E. R., 1970, Spinal neurons specifically excited by noxious or thermal stimuli: marginal zone of the dorsal horn. *J. Neurophysiol.* **33**, 293–307.

Christiansen, J., 1966, Neurological observations of macaques with spinal cord lesions. *Anat. Rec.* **154**, 330.

Chronwall, B. M., DiMaggio, D. A., Massari, V. J., Pickel, V. M., Ruggiero, D. A. and O'Donohue, T. L., 1985, The anatomy of neuropeptide-Y-containing neurons in rat brain. *Neuroscience* **15**, 1159–1181.

Chung, J. M., Kenshalo, D. R., Jr., Gerhart, K. D. and Willis, W. D., 1979, Excitation of primate spinothalamic neurons by cutaneous C-fiber volleys. *J. Neurophysiol.* **42**, 1354–1369.

Chung, J. M., Kevetter, G. A., Yezierski, R. P., Haber, L. H., Martin, R. F. and Willis, W. D., 1983a, Midbrain nuclei projecting to the medial medulla oblongata in the monkey. *J. Comp. Neurol.* **214**, 93–102.

Chung, J. M., Lee, K. H., Endo, K. and Coggeshall, R. E., 1983b, Activation of central neurons by ventral root afferents. *Science* **222**, 934–935.

Chung, J. M., Fang, Z. R., Hori, Y., Lee, K. H. and Willis, W. D., 1984a, Prolonged inhibition of primate spinothalamic tract cells by peripheral nerve stimulation. *Pain* **19**, 259–275.

Chung, J. M., Lee, K. H., Hori, Y., Endo, K. and Willis, W. D., 1984b, Factors influencing peripheral nerve stimulation produced inhibition of primate spinothalamic tract cells. *Pain* **19**, 277–293.

Chung, J. M., Lee, K. H., Kim, J. and Coggeshall, R. E., 1985, Activation of dorsal horn cells by ventral root stimulation in the cat. *J. Neurophysiol.* **54**, 261–272.

Chung, J. M., Lee, K. H., Surmeier, D. J., Sorkin, L. S., Kim, J. and Willis, W. D., 1986a, Response characteristics of neurons in the ventral posterior lateral nucleus of the monkey thalamus. *J. Neurophysiol.* **56**, 370–390.

Chung, J. M., Surmeier, D. J., Lee, K. H., Sorkin, L. S., Honda, C. N., Tsong, Y. and Willis, W. D., 1986b, Classification of primate spinothalamic and somatosensory thalamic neurons based on cluster analysis. *J. Neurophysiol.* **56**, 308–327.

Chung, J. M., Kim, J. and Shin, H. K., 1986c, Blood pressure response evoked by ventral root afferent fibres in the cat. *J. Physiol.* **370**, 255–265.

Chung, K. and Coggeshall, R. E., Primary afferent axons in the tract of Lissauer in the cat. *J. Comp. Neurol.* **186**, 451–463, 1979.

Chung, K. and Coggeshall, R. E., 1982, Quantitation of propriospinal fibers in the tract of Lissauer of the rat. *J. Comp. Neurol.* **211**, 418–426.

Chung, K. and Coggeshall, R. E., 1983, Numbers of axons in lateral and ventral funiculi of rat sacral spinal cord. *J. Comp. Neurol.* **214**, 72–78.

Chung, K. and Coggeshall, R. E., 1984, The ratio of dorsal root ganglion cells to dorsal root axons in sacral segments of the cat. *J. Comp. Neurol.* **225**, 24–30.

Chung, K. and Coggeshall, R. E., 1985, Unmyelinated primary afferent fibers in dorsal funiculi of cat sacral spinal cord. *J. Comp. Neurol.* **238**, 365–369.

Chung, K., Langford, L. A., Applebaum, A. E. and Coggeshall, R. E., 1979, Primary afferent fibers in the tract of Lissauer in the rat. *J. Comp. Neurol.* **184**, 587–598.

Chung, K., Kevetter, G. A., Willis, W. D. and Coggeshall, R. E., 1984, An estimate of the ratio of propriospinal to long tract neurons in the sacral spinal cord of the rat. *Neurosci. Lett.* **44**, 173–177.

Chung, K., Sharma, J. and Coggeshall, R. E., 1985, Numbers of myelinated and unmyelinated axons in the dorsal, lateral, and ventral funiculi of the white matter of the S2 segment of cat spinal cord. *J. Comp. Neurol.* **234**, 117–121.

Chung, K., Langford, L. A. and Coggeshall, R. E., 1987, Primary afferent and propriospinal fibers in the rat dorsal and dorsolateral funiculi. *J. Comp. Neurol.* **263**, 68–75.

Chung, K., Lee, W. T. and Carlton, S. M., 1988, The effects of dorsal rhizotomy and spinal cord isolation on calcitonin gene-related peptide containing terminals in the rat lumbar dorsal horn. *Neurosci. Lett.* **90**, 27–32.

Chung, K., Briner, R. P., Carlton, S. M. and Westlund, K. N., 1989a, Immunohistochemical localization of seven different peptides in the human spinal cord. *J. Comp. Neurol.* **280**, 158–170.

Chung, K., McNeill, D. L., Hulsebosch, C. E. and Coggeshall, R. E., 1989b, Changes in dorsal horn synaptic disc numbers following unilateral dorsal rhizotomy. *J. Comp. Neurol.* **283**, 568–577.

Clark, D., Hughes, J. and Gasser, H. S., 1935, Afferent function in the group of nerve fibers of slowest conduction velocity. *Amer. J. Physiol.* **114**, 69–76.

Clark, F. J., 1972, Central projection of sensory fibers from the cat knee joint. *J. Neurobiol.* **3**, 101–110.

Clark, F. J., 1975, Information signaled by sensory fibers in medial articular nerve. *J. Neurophysiol.* **38**, 1464–1472.

Clark, F. J. and Burgess, P. R., 1975, Slowly adapting receptors in cat knee joint: can they signal joint angle? *J. Neurophysiol.* **38**, 1448–1463.

Clark, F. J., Landgren, S. and Silfvenius, H., 1973, Projections to the cat's cerebral cortex from low threshold joint afferents. *Acta Physiol. Scand.* **89**, 504–521.

Clark, F. J., Horch, K. W., Bach, S. M. and Larsen, G. F., 1979, Contributions of cutaneous and joint receptors to static knee-position sense in man. *J. Neurophysiol.* **42**, 877–888.

Clark, F. J., Burgess, R. C., Chapin, J. W. and Lipscomb, W. T., 1985, Role of intramuscular receptors in the awareness of limb position. *J. Neurophysiol.* **54**, 1529–1540.

Clark, F. J., Burgess, R. C. and Chapin, J. W., 1986, Proprioception with the proximal interphalangeal joint of the index finger. *Brain* **109**, 1195–1208.

Clark, S. L., 1926, Nissl granules of primary afferent neurones. *J. Comp. Neurol.* **41**, 423–451.

Clark, S. L., 1931, Innervation of the pia mater of the spinal cord and medulla. *J. Comp. Neurol.* **53**, 129–145.

Clark, W. E. L., 1936, The termination of ascending tracts in the thalamus of the macaque monkey. *J. Anat.* **71**, 7–40.

Clarke, J. L., 1859, Further researches on the grey substance of the spinal cord. *Phil. Trans. Roy. Soc. Lond.* **149**, 437–467.

Clezy, J. K. A., Dennis, B. J. and Kerr, D. I. B., 1961, A degeneration study of the somaesthetic afferent systems in the marsupial phalanger Trichosurus vulpecula. *Austral. J. Exp. Biol.* **39**, 19–28.

Cliffer, K. D. and Giesler, G. J., 1989, Postsynaptic dorsal column pathway of the rat. III. Distribution of ascending afferent fibers. *J. Neurosci.* **9**, 3146–3168.

Clifton, G. L., Coggeshall, R. E., Vance, W. H. and Willis, W. D., 1976, Receptive fields of unmyelinated ventral root afferent fibres in the cat. *J. Physiol.* **256**, 573–600.

Coffield, J. A. and Miletić, V., 1987, Immunoreactive enkephalin is contained within some trigeminal and spinal neurons projecting to the rat medial thalamus. *Brain Res.* **425**, 380–383.

Coffield, J. A., Miletić, V., Zimmermann, E., Hoffert, M. J. and Brooks, B. R., 1986, Demonstration of thyrotropin-releasing hormone immunoreactivity in neurons of the mouse spinal dorsal horn. *J. Neurosci.* **6**, 1194–1197.

Coggeshall, R. E., 1980, The law of separation of function of the spinal roots. *Physiol. Rev.* **60**, 716–755.

Coggeshall, R. E. and Ito, H., 1977, Sensory fibres in ventral roots L7 and S1 in the cat. *J. Physiol.* **267**, 215–235.

Coggeshall, R. E., Coulter, J. D. and Willis, W. D., 1974, Unmyelinated axons in the ventral roots of the cat lumbosacral enlargement. *J. Comp. Neurol.* **153**, 39–58.

Coggeshall, R. E., Chung, K., Chung, J. M. and Langford, L. A., 1981, Primary afferent axons in the tract of Lissauer in the monkey. *J. Comp. Neurol.* **196**, 431–442.

Coggeshall, R. E., Hong, K. A. P., Langford, L. A., Schaible, H. G. and Schmidt, R. F., 1983, Discharge characteristics of fine medial articular afferents at rest and during passive movements of inflamed knee joints. *Brain Res.* **272**, 185–188.

Cohen, L. A., 1955, Activity of knee joint proprioceptors recorded from the posterior articular nerve. *Yale J. Biol. Med.* **28**, 225–232.

Cohen, R. H. and Perl, E. R., 1990, Contributions of arachidonic acid derivatives and substance P to the sensitization of cutaneous nociceptors. *J. Neurophysiol.* **64**, 457–464.

Coimbra, A., Magalhaes, M. M. and Sodre-Borges, B. P., 1970, Ultrastructural localization of acid phosphatase in synaptic terminals of the rat substantia gelatinosa Rolandi. *Brain Res.* **22**, 142–146.

Coimbra, A., Sodre-Borges, B. P. and Magalhaes, M. M., 1974, The substantia gelatinosa Rolandi of the rat. Fine structure, cytochemistry (acid phosphatase) and changes after dorsal root section. *J. Neurocytol.* **3**, 199–217.

Coimbra, A., Ribeiro-da-Silva, A. and Pignatelli, D., 1984, Effects of dorsal rhizotomy on the several types of primary afferent terminals in laminae I-III of the rat spinal cord. *Anat. Embryol.* **170**, 279–287.

Coimbra, A., Ribeiro-da-Silva, A. and Pignatelli, D., 1986, Rexed's laminae and the acid phosphatase (FRAP)-band in the superficial dorsal horn of the neonatal rat spinal cord. *Neurosci. Lett.* **71**, 131–136.

Collier, J. and Buzzard, E. F., 1903, The degenerations resulting from lesions of posterior nerve roots and away from transverse lesions of the spinal cord in man. A study of twenty cases. *Brain* **26**, 559–591.

Collins, G. G. S., 1974, The spontaneous and electrically evoked release of ^{3}H-γ-aminobutyric acid (^{3}H-GABA) from spinal cord. *Brit. J. Pharmacol.* **47**, 641P.

Collins, J. G., 1983, Neuronal activity recorded from the spinal cord dorsal horn of physiologically intact, awake, drug-free, restrained cats: a preliminary report. *Brain Res.* **322**, 301–304.

Collins, J. G., 1985, A technique for chronic extracellular recording of neuronal activity in the dorsal horn of the lumbar spinal cord in drug-free, physiologically intact cats. *J. Neurosci. Meth.* **12**, 277–287.

Collins, J. G., 1987, A descriptive study of spinal dorsal horn neurons in the physiologically intact, awake, drug-free cat. *Brain Res.* **416**, 34–42.

Collins, J. G. and Ren, K., 1987, WDR response profiles of spinal dorsal horn neurons may be unmasked by barbiturate anesthesia. *Pain* **28**, 369–378.

Collins, J. G., Ren, K., Saito, Y., Iwasaki, H. and Tang, J., 1990, Plasticity of some spinal dorsal horn neurons as revealed by pentobarbital-induced disinhibition. *Brain Res.* **525**, 189–197.

Collins, J. K., 1976, Proprioceptive space perception after anaesthetization of the mid-interphalangeal joint of the finger. *Perception and Psychophysics* **20**, 45–48.

Collins, W. F., Nulsen, F. E. and Randt, C. T., 1960, Relation of peripheral nerve fiber size and sensation in man. *Arch. Neurol.* **3**, 381–385.

Colmant, H. J., 1959, Aktivitatsschwankunger der sauren Phosphatase im Ruckenmark und den Spinalganglien der Ratte nach Durchschneidung des Nervus ischiadicus. *Arch. f. Psy. u. Zeitsch. f. d. ges. Neurologie* **199**, 60–71.

Comans, P. E. and Snow, P. J., 1981a, Rostrocaudal and laminar distribution of spinothalamic neurons in the high cervical spinal cord of the cat. *Brain Res.* **223**, 123–127.

Comans, P. E. and Snow, P. J., 1981b, Ascending projections to nucleus parafascicularis of the cat. *Brain Res.* **230**, 337–341.

Connor, K. M., Ferrington, D. G. and Rowe, M. J., 1984, Tactile sensory coding during development: signalling capacities of neurons in kitten dorsal column nuclei. *J. Neurophysiol.* **52**, 86–98.

Conradi, S., 1969, On motoneuron synaptology in adult cats. *Acta Physiol. Scand.* Suppl. 332, 1–115.

Conrath, M., Couraud, J. Y. and Pradelles, P., 1988, Anti-idiotypic antibodies as a tool for cytochemical identification of substance P receptors in the central nervous system. *J. Histochem. Cytochem.* **36**, 1397–1401.

Conrath, M., Taquet, H., Pohl, M. and Carayon, A., 1989, Immunocytochemical evidence for calcitonin gene-related peptide-like neurons in the dorsal horn and lateral spinal nucleus of the rat cervical spinal cord. *J. Chem. Neuroanat.* **2**, 335–347.

Conrath-Verrier, M., Dietl, M., Arluison, M., Cesselin, F., Bourgoin, S. and Hamon, M., 1983, Localization of met-enkephalin-like immunoreactivity within pain-related nuclei of cervical spinal cord, brainstem and midbrain in the cat. *Brain Res. Bull.* **11**, 587–604.

Conrath-Verrier, M., Dietl, M. and Tramu, G., 1984, Cholecystokinin-like immunoreactivity in the dorsal horn of the spinal cord of the rat: a light and electron microscopic study. *Neuroscience* **13**, 871–885.

Conti, F., DeBiasi, S., Giuffrida, R. and Rustioni, A., 1990, Substance P-containing projections in the dorsal columns of rats and cats. *Neuroscience* **34**, 607–621.

Cook, A. J., Woolf, C. J., Wall, P. D. and McMahon, S. B., 1987, Dynamic receptive field plasticity in rat spinal cord dorsal horn following C-primary afferent input. *Nature* **325**, 151–153.

Cook, A. W. and Browder, E. J., 1965, Function of posterior columns in man. *Arch. Neurol. Psychiat.* **12**, 72–79.

Cook, A. W. and Kawakami, Y., 1977, Commissural myelotomy. *J. Neurosurg.* **47**, 1–6.

Cook, A. W., Nathan, P. W. and Smith, M. C., 1984, Sensory consequences of commissural myelotomy. A challenge to traditional anatomical concepts. *Brain* **107**, 547–568.

Cook, W. A., Cangiano, A. and Pompeiano, O., 1969a, Dorsal root potentials in the lumbar cord evoked from the vestibular system. *Arch. Ital. Biol.* **107**, 275–295.

Cook, W. A., Cangiano, A. and Pompeiano, O., 1969b, Vestibular control of transmission in primary afferents to the lumbar spinal cord. *Arch. Ital. Biol.* **107**, 296–320.

Coombs, J. S., Eccles, J. C. and Fatt, P., 1955, The specific ionic conductances and the ionic movements across the motoneuronal membrane that produce the inhibitory post-synaptic potential. *J. Physiol.* **130**, 326–373.

Coombs, J. S., Curtis, D. R. and Landgren, S., 1956, Spinal cord potentials generated by impulses in muscle and cutaneous afferent fibres. *J. Neurophysiol.* **19**, 452–467.

Cooper, S. and Sherrington, C. S., 1940, Gower's tract and spinal border cells. *Brain* **63**, 123–134.

Coquery, J. M., Malcuit, G. and Coulmance, M., 1971, Alterations de la perception d'un stimulus somesthesique durant un mouvement volontaire. *Comptes Rendus* **165**, 1946–1951.

Corvaja, N., Pellegrini, M. and Buisseret-Delmas, C., 1978, Ultrastructure of supraspinal dorsal root projections in the toads. I. The obex region. *Brain Res.* **142**, 413–424.

Coulter, J. D., 1974, Sensory transmission through the lemniscal pathway during voluntary movement in the cat. *J. Neurophysiol.* **37**, 831–845.

Coulter, J. D. and Jones, E. G., 1977, Differential distribution of corticospinal projections from individual cytoarchitectonic fields in the monkey. *Brain Res.* **129**, 335–340.

Coulter, J. D., Maunz, R. A. and Willis, W. D., 1974, Effects of stimulation of sensorimotor cortex on primate spinothalamic neurons. *Brain Res.* **65**, 351–356.

Coupland, R. E. and Holmes, R. L., 1957, The use of cholinesterase techniques for the demonstration of peripheral nerve structures. *J. Microscop. Soc.* **98**, 327–330.

Courtney, K. R. and Fetz, E. E., 1973, Unit responses recorded from cervical spinal cord of awake monkey. *Brain Res.* **53**, 445–450.

Courtney, K., Brengelman, K. and Sundsten, J. W., 1972, Evidence for spinal cord unit activity responsive to peripheral warming in the primate. *Brain Res.* **43**, 657–661.

Craig, A. D., 1976, Spinocervical tract cells in cat and dog, labeled by the retrograde transport of horseradish peroxidase. *Neurosci. Lett.* **3**, 173–177.

Craig, A. D., 1978, Spinal and medullary input to the lateral cervical nucleus. *J. Comp. Neurol.* **181**, 729–744.

Craig, A. D., 1989, Ascending lamina I axons in the cat are concentrated in the middle of the lateral funiculus. *Neurosci. Abstr.* **15**, 1190.

Craig, A. D. and Burton, H., 1979, The lateral cervical nucleus in the cat: anatomic organization of cervicothalamic neurons. *J. Comp. Neurol.* **185**, 329–346.

Craig, A. D. and Burton, H., 1981, Spinal and medullary lamina I projection to nucleus submedius in medial thalamus: a possible pain center. *J. Neurophysiol.* **45**, 443–466.

Craig, A. D. and Burton, H., 1985, The distribution and topographical organization in the thalamus of anterogradely-transported horseradish peroxidase after spinal injections in cat and raccoon. *Exp. Brain Res.* **58**, 227–254.

Craig, A. D. and Kniffki, K. D., 1985, spinothalamic lumbosacral lamina I cells responsive to skin and muscle stimulation in the cat. *J. Physiol.* **365**, 197–221.

Craig, A. D. and Mense, S., 1983, The distribution of afferent fibers from the gastrocnemius-soleus muscle in the dorsal horn of the cat, as revealed by the transport of horseradish peroxidase. *Neurosci. Lett.* **41**, 233–238.

Craig, A. D. and Tapper, D. N., 1978, Lateral cervical nucleus in the cat: functional organization and characteristics. *J. Neurophysiol.* **41**, 1511–1534.

Craig, A. D. and Tapper, D. N., 1985, A dorsal spinal neural network in cat. III. Dynamic nonlinear analysis of responses to random stimulation of single type I cutaneous fibers. *J. Neurophysiol.* **53**, 995–1015.

Craig, A. D., Sailer, S. and Kniffki, K. D., 1987, Organization of anterogradely labeled spinocervical tract terminations in the lateral cervical nucleus of the cat. *J. Comp. Neurol.* **263**, 214–222.

Craig, A. D., Heppelmann, B. and Schaible, H. G., 1988, Projection of the medial and posterior articular nerves of the cat's knee to the spinal cord. *J. Comp. Neurol.* **276**, 279–288.

Craig, A. D., Linington, A. J. and Kniffki, K. D., 1989a, Cells of origin of spinothalamic projections to medial and/or lateral thalamus in the cat. *J. Comp. Neurol.* **289**, 568–585.
Craig, A. D., Linington, A. J. and Kniffki, K. D., 1989b, Significant differences in the retrograde labeling of spinothalamic tract cells by horseradish peroxidase and the fluorescent tracers fast blue and diamidino yellow. *Exp. Brain Res.* **74**, 431–436.
Craigie, E. H., 1928, Observations on the brain of the humming bird (*Chrysolampis mosquitus* Linn. and *Chlorostilbon caribaeus* Lawr.). *J. Comp. Neurol.* **45**, 377–481.
Cranefield, P. F. (ed.), 1974, *The Way in and the Way Out*. Futura Publishing Co., Mount Kisco, New York.
Craske, B., 1977, Perception of impossible limb positions induced by tendon vibration. *Science* **196**, 71–73.
Crockett, D. P., Smith, W. K., Proshansky, E., Kauer, J. S., Stewart, W. B., Woodward, D. J., Schlusselberg, D. S. and Egger, M. D., 1989, Computer-assisted three-dimensional reconstructions of [14C]-2-deoxy-D-glucose metabolism in cat lumbosacral spinal cord following cutaneous stimulation of the hindfoot. *J. Comp. Neurol.* **288**, 326–338.
Cross, M. J. and McCloskey, D. I., 1973, Position sense following surgical removal of joints in man. *Brain Res.* **55**, 443–445.
Crowe, R. and Burnstock, G., 1988, An increase of vasoactive intestinal polypeptide-, but not neuropeptide Y-, substance P- or catecholamine-containing nerves in the iris of the streptozotocin-induced diabetic rat. *Exp. Eye Res.* **47**, 751–759.
Croze, S., Duclaux, R. and Kenshalo, D. R., 1976, The thermal sensitivity of the polymodal nociceptors in the monkey. *J. Physiol.* **263**, 539–562.
Cruz, F., Lima, D. and Coimbra, A., 1987, Several morphological types of terminal arborizations of primary afferents in laminae I-II of the rat spinal cord, as shown after HRP labeling and Golgi impregnation. *J. Comp. Neurol.* **261**, 221–236.
Cruz, L. and Basbaum, A. I., 1985, Multiple opioid peptides and the modulation of pain: immunohistochemical analysis of dynorphin and enkephalin in the trigeminal nucleus caudalis and spinal cord of the cat. *J. Comp. Neurol.* **240**, 331–348.
Csillik, B. and Knyihár-Csillik, E., 1986, *The Protean Gate*. Akademai Kaido, Budapest.
Csillik, B. and Savay, G., 1954, Contributions to the histochemistry of cholinesterase activity in the nervous system. *Acta Morphol. Acad. Sci. Hung.* **4**, 103–109.
Csillik, B., Knyihár-Csillik, E. and Bezzegh, A., 1986, Comparative electron histochemistry of thiamine monophosphatase and substance P in the upper dorsal horn. *Acta Histochem.* **80**, 125–134.
Cuello, A. C. and Kanazawa, I., 1978, The distribution of substance P immunoreactive fibers in the rat central nervous system. *J. Comp. Neurol.* **178**, 129–156.
Cuello, A. C., Jessell, T. M., Kanazawa, I. and Iversen, L. L., 1977, Substance P: localization in synaptic vesicles in rat central nervous system. *J. Neurochem.* **29**, 747–751.
Cuello, A. C., Del Fiacco, M. and Paxinos, G., 1978, The central and peripheral ends of the substance P-containing sensory neurones in the rat trigeminal system. *Brain Res.* **152**, 499–509.
Culberson, J. L., 1987, Projection of cervical dorsal root fibers to the medulla oblongata in the brush-tailed possum (*Trichosurus vulpecula*). *Amer. J. Anat.* **179**, 232–242.
Culberson, J. L. and Albright, B. C., 1984, Morphologic evidence for fiber sorting in the fasciculus cuneatus. *Exp. Neurol.* **85**, 358–370.
Culberson, J. L. and Brushart, T. M., 1989, Somatotopy of digital nerve projections to the cuneate nucleus in the monkey. *Somatosensory Motor Res.* **6**, 319–330.
Curry, M. J., 1972, The exteroceptive properties of neurones in the somatic part of the posterior group (PO). *Brain Res.* **44**, 439–462.
Curry, M. J. and Gordon, G., 1972, The spinal input to the posterior group in the cat. An electrophysiological investigation. *Brain Res.* **44**, 417–437.
Curtis, D. R., 1962, The depression of spinal inhibition by electrophoretically administered strychnine. *Int. J. Neuropharmacol.* **1**, 239–250.
Curtis, D. R. and Gynther, B. D., 1987, Divalent cations reduce depolarization of primary afferent terminations by GABA. *Brain Res.* **422**, 192–195.
Curtis, D. R. and Johnston, G. A. R., 1974, Amino acid transmitters in the mammalian central nervous system. *Ergebn. Physiol.* **69**, 97–188.
Curtis, D. R. and Lodge, D., 1982, The depolarization of feline ventral horn group Ia spinal afferent terminations by GABA. *Exp. Brain Res.* **46**, 215–233.
Curtis, D. R. and Ryall, R. W., 1966, Pharmacological studies upon spinal presynaptic fibres. *Exp. Brain Res.* **1**, 195–204.
Curtis, D. R. and Watkins, J. C., 1960, The excitation and depression of spinal neurones by structurally related amino acids. *J. Neurochem.* **6**, 117–141.
Curtis, D. R. and Watkins, J. C., 1965, The pharmacology of amino acids related to γ-aminobutyric acid. *Pharm. Rev.* **17**, 347–391.
Curtis, D. R., Phillis, J. W. and Watkins, J. C., 1959a, Chemical excitation of spinal neurones. *Nature* **183**, 611–612.
Curtis, D. R., Phillis, J. W. and Watkins, J. C., 1959b, The depression of spinal neurones by γ-amino-n-butyric acid and β-alanine. *J. Physiol.* **146**, 185–203.

Curtis, D. R., Phillis, J. W. and Watkins, J. C., 1960, The chemical excitation of spinal neurones by certain acidic amino acids. *J. Physiol.* **150**, 656–682.

Curtis, D. R., Phillis, J. W. and Watkins, J. C., 1961, Cholinergic and non-cholinergic transmission in the mammalian spinal cord. *J. Physiol.* **158**, 296–323.

Curtis, D. R., Ryall, R. W. and Watkins, J. C., 1966, The action of cholinomimetics on spinal interneurones. *Exp. Brain Res.* **2**, 97–106.

Curtis, D. R., Hösli, L. and Johnston, G. A. R., 1967a, Inhibition of spinal neurones by glycine. *Nature* **215**, 1502–1503.

Curtis, D. R., Hösli, L., Johnston, G. A. R. and Johnston, I. H., 1967b, Glycine and spinal inhibition. *Brain Res.* **5**, 112–114.

Curtis, D. R., Hösli, L. and Johnston, G. A. R., 1968, A pharmacological study of the depression of spinal neurones by glycine and related amino acids. *Exp. Brain Res.* **6**, 1–18.

Curtis, D. R., Duggan, A. W. and Johnston, G. A. R., 1969, Glycine, strychnine, picrotoxin and spinal inhibition. *Brain Res.* **14**, 759–762.

Curtis, D. R., Duggan, A. W., Felix, D. and Johnston, G. A. R., 1971a, Bicuculline, an antagonist of GABA and synaptic inhibition in the spinal cord of the cat. *Brain Res.* **32**, 69–96.

Curtis, D. R., Duggan, A. W. and Johnston, G. A. R., 1971b, The specificity of strychnine as a glycine antagonist in the mammalian spinal cord. *Exp. Brain Res.* **12**, 547–565.

Curtis, D. R., Lodge, D. and Brand, S. J., 1977, GABA and spinal afferent terminal excitability in the cat. *Brain Res.* **130**, 360–363.

Curtis, D. R., Lodge, D., Bornstein, J. C., Peet, M. J. and Leah, J. D., 1982, The dual effects of GABA and related amino acids on the electrical threshold of ventral horn group Ia afferent terminations in the cat. *Exp. Brain Res.* **48**, 387–400.

Czarkowska, J., Jankowska, E. and Sybirska, E., 1981, Common interneurones in reflex pathways from group Ia and Ib afferents of knee flexors and extensors in the cat. *J. Physiol.* **310**, 367–380.

Czéh, G., Kříž, N. and Syková, E., 1981, Extracellular potassium accumulation in the frog spinal cord induced by stimulation of the skin and ventrolateral columns. *J. Physiol.* **320**, 57–72.

Czlonkowski, A., Costa, T., Przewlocki, R., Pasi, A. and Herz, A., 1983, Opiate receptor binding sites in human spinal cord. *Brain Res.* **267**, 392–396.

Dahlström, A. and Fuxe, K., 1964, Evidence for the existence of monoamine-containing neurons in the central nervous system. *Acta Physiol. Scand.* **62**, Suppl. 232, 1–55.

Dahlström, A. and Fuxe, K., 1965, Evidence for the existence of monoamine neurons in the central nervous system. II. Experimentally induced changes in the intraneuronal amine levels of bulbospinal neuron systems. *Acta Physiol. Scand.* **64**, Suppl. 247, 1–36.

Dallenbach, K. M., 1927, The temperature spots and end-organs. *Amer. J. Psychol.* **39**, 402–427.

Dalsgaard, C. J., Hökfelt, T., Johansson, O. and Elde, R., 1981, Somatostatin immunoreactive cell bodies in the dorsal horn and the parasympathetic intermediolateral nucleus of the rat spinal cord *Neurosci. Lett.* **27**, 335–339.

Dalsgaard, C. J., Risling, M. and Cuello, C., 1982a, Immunohistochemical localization of substance P in the lumbosacral spinal pia mater and ventral roots of the cat. *Brain Res.* **246**, 168–171.

Dalsgaard, C. J., Vincent, S. R., Hökfelt, T., Lundburg, J. M., Dahlström, A., Schultzberg, M., Dockray, G. J. and Cuello, A. C., 1982b, Coexistence of cholecystokinin and substance P-like peptides in neurons of the dorsal root ganglia of the rat. *Neurosci. Lett.* **33**, 159–163.

Dalsgaard, C. J., Jonsson, C. E., Hökfelt, T. and Cuello, A. C., 1983, Localization of substance P-immunoreactive nerve fibres in the human digital skin. *Experientia* **39**, 1018–1020.

Dalsgaard, C., Ygge, J., Vincent, S., Ohrling, M., Dockray, G. and Elde, R., 1984, Peripheral projections and neuropeptide coexistence in a subpopulation of fluoride-resistant acid phosphatase reactive spinal primary sensory neurons. *Neurosci. Lett.* **51**, 139–144.

Dalsgaard, C. J., Haegerstrand, A., Theodorsson-Norheim, E., Brodin, E. and Hökfelt, T., 1985, Neurokinin A-like immunoreactivity in rat primary sensory neurons; coexistence with substance P. *Histochemistry* **83**, 37–39.

Darian-Smith, I., Phillips, G. and Ryan, R. D., 1963, Functional organization in the trigeminal main sensory and rostral spinal nuclei of the cat. *J. Physiol.* **168**, 129–146.

Darian-Smith, I., Johnson, K. O. and Dykes, R., 1973, "Cold" fiber population innervating palmar and digital skin of the monkey: response to cooling pulses. *J. Neurophysiol.* **36**, 325–346.

Dart, A. M., 1971, Cells of the dorsal column nuclei projecting down into the spinal cord. *J. Physiol.* **219**, 29–30P.

Dart, A. M. and Gordon, G., 1973, Some properties of spinal connections of the cat's dorsal column nuclei which do not involve the dorsal columns. *Brain Res.* **58**, 61–68.

Daval, G., Verge, D., Basbaum, A. I., Bourgoin, S. and Hamon, M., 1987, Autoradiographic evidence of serotonin 1 binding sites on primary afferent fibers in the dorsal horn of the rat spinal cord. *Neurosci. Lett.* **83**, 71–76.

Davidoff, M. S., Galabov, P. G. and Kaufmann, P., 1986, Localization of substance P-like immunoreactive fibers in the thoracic spinal cord of guinea pig. *Cell Tiss. Res.* **246**, 653–665.

Davidoff, R. A., 1972, The effects of bicuculline on the isolated spinal cord of the frog. *Exp. Neurol.* **35**, 179–193.

Davidoff, R. A. and Hackman, J. C., 1980, Hyperpolarization of frog primary afferent fibres caused by activation of a sodium pump. *J. Physiol.* **302**, 297–309.

Davidoff, R. A., Graham, L. T., Shank, R. P., Werman, R. and Aprison, M. H., 1967, Changes in amino acid concentrations associated with loss of spinal interneurones. *J. Neurochem.* **14**, 1025–1031.

Davidoff, R. A., Aprison, M. H. and Werman, R., 1969, The effects of strychnine on the inhibition of interneurons by glycine and γ -aminobutyric acid. *Internat. J. Neuropharmacol.* **8**, 191–194.

Davidoff, R. A., Grayson, V. and Adair, R., 1973, GABA-transaminase inhibitors and presynaptic inhibition in the amphibian spinal cord. *Amer. J. Physiol.* **224**, 1230–1234.

Davidoff, R. A., Hackman, J. C. and Osorio, I., 1980, Amino acid antagonists do not block the depolarizing effects of potassium ions on frog primary afferents. *Neuroscience* **5**, 117–126.

Davidoff, R. A., Hackman, J. C., Holohean, A. M., Vega, J. L. and Zhang, D. X., 1988, Primary afferent activity, putative excitatory transmitters and extracellular potassium levels in frog spinal cord. *J. Physiol.* **397**, 291–306.

Davidson, N. and Smith, C. A., 1972, A recurrent collateral pathway for presynaptic inhibition in the rat cuneate nucleus. *Brain Res.* **44**, 63–71.

Davidson, N. and Southwick, C. A. P., 1971, Amino acids and presynaptic inhibition in the rat cuneate nucleus. *J. Physiol.* **219**, 689–708.

Davies, J., and Dray, A., 1978, Pharmacological and electrophysiological studies of morphine and enkephalin on rat supraspinal neurones and cat spinal neurones. *Br. J. Pharmacol.* **63**, 87–96.

Davies, J. and Dray, A., 1980, Depression and facilitation of synaptic responses in cat dorsal horn by substance P administered into substantia gelatinosa. *Life Sci.* **27**, 2037–2042.

Davies, J. and Watkins, J. C., 1973a, Antagonism of synaptic and amino acid induced excitation in the cuneate nucleus of the cat by HA-966. *Neuropharmacology* **12**, 637–640.

Davies, J. and Watkins, J. C., 1973b, Microelectrophoretic studies on the depressant action of HA-966 on chemically and synaptically excited neurones in the cat cerebral cortex and cuneate nucleus. *Brain Res.* **59**, 311–322.

Davies, J. and Watkins, J. C., 1979, Selective antagonism of amino acid-induced and synaptic excitation in the cat spinal cord. *J. Physiol.* **297**, 621–635.

Davies, J. and Watkins, J. C., 1982, Actions of D and L forms of 2-amino-5-phosphonovalerate and 2-amino-4-phosphonobutyrate in the cat spinal cord. *Brain Res.* **235**, 378–386.

Davies, J. and Watkins, J. C., 1983, Role of excitatory amino acid receptors in mono- and polysynaptic excitation in the cat spinal cord. *Exp. Brain Res.* **49**, 280–290.

Davies, J. D., Francis, A. A., Jones, A. W. and Watkins, J. C., 1981, 2-amino-5-phosphonovalerate as a potent and selective antagonist of amino acid-induced and synaptic excitation. *Neurosci. Lett.* **21**, 77–82.

Davies, J. E. and Roberts, M. H. T., 1981, 5-hydroxytryptamine reduces substance P responses on dorsal horn interneurones: a possible interaction of neurotransmitters. *Brain Res.* **217**, 399–404.

Davies, S. N. and Lodge, D., 1987, Evidence for involvement of N-methylaspartate receptors in "wind-up" of class 2 neurones in the dorsal horn of the rat. *Brain Res.* **424**, 402–406.

Davison, C. and Wechsler, I. S., 1936, Amyotrophic lateral sclerosis with involvement of posterior column and sensory disturbances. *Arch. Neurol. Psychiat.* **35**, 229–239.

Dawson, G. D., 1958, The central control of sensory inflow. *Proc. Roy. Soc. Med.* **51**, 531–535.

Dawson, G. D., Merrill, E. G. and Wall, P. D., 1970, Dorsal root potentials produced by stimulation of fine afferents. *Science* **167**, 1385–1387.

De Biasi, S. and Rustioni, A., 1988, Glutamate and substance P coexist in primary afferent terminals in the superficial laminae of spinal cord. *Proc. Natl. Acad. Sci. USA* **85**, 7820–7824.

DeGroat, W. C., 1986, Spinal cord projections and neuropeptides in visceral afferent neurons. *Prog. in Brain Res.* **67**, 165–187.

DeGroat, W. C., 1987, Neuropeptides in pelvic afferent pathways. *Experientia* **43**, 801–813.

DeGroat, W. C., Lalley, P. M. and Saum, W. R., 1972, Depolarization of dorsal root ganglia in the cat by GABA and related amino acids: antagonism by picrotoxin and bicuculline. *Brain Res.* **44**, 273–277.

DeGroat, W. C., Nadelhaft, I., Morgan, C. and Schauble, T., 1978, Horseradish peroxidase tracing of visceral efferent and primary afferent pathways in the sacral spinal cord of the cat using benzidine processing. *Neurosci. Lett.* **10**, 103–108.

DeGroat, W. C., Nadelhaft, I., Milne, R. J., Booth, A. M., Morgan, C. and Thor, K., 1981, Organization of the sacral parasympathetic reflex pathways to the urinary bladder and large intestine. *J. Auton. Nerv. System* **3**, 135–160.

DeGroat, W. C., Kawatani, M., Hisamitsu, T., Lowe, I., Morgan, C., Roppolo, J., Booth, A. M. and Nadelhaft, I., 1983, The role of neuropeptides in the sacral autonomic reflex pathways of the cat. *J. Auton. Nerv. System* **7**, 339–350.

De Koninck, Y. and Henry, J. L., 1989, Bombesin, neuromedin B and neuromedin C selectively depress superficial dorsal horn neurones in the cat spinal cord. *Brain Res.* **498**, 105–117.

DeLanerolle, N. C. and LaMotte, C. C., 1982, The human spinal cord: Substance P and methionine-enkephalin immunoreactivity. *J. Neurosci.* **2**, 1369–1386.

DeLanerolle, N. C. and LaMotte, C. C., 1983, Ultrastructure of chemically defined neuron systems in the dorsal horn of the monkey. I. Substance P immunoreactivity. *Brain Res.* **274**, 31–49.

Del Fiacco, M. and Cuello, A. C., 1980, Substance P- and enkephalin-containing neurones in the rat trigeminal system. *Neuroscience* **5**, 803–815.

Denny-Brown, D., Kirk, E. J. and Yanagisawa, N., 1973, The tract of Lissauer in relation to sensory transmission in the dorsal horn of spinal cord in the macaque monkey. *J. Comp. Neurol.* **151**, 175–200.

de Pommery, J., Roudier, F. and Menétrey, D., 1984, Postsynaptic fibers reaching the dorsal column nuclei in the rat. *Neurosci. Lett.* **50**, 319–323.

De Quidt, M. E. and Emson, P. C., 1986, Distribution of neuropeptide Y-like immunoreactivity in the rat central nervous system. II. Immunohistochemical analysis. *Neuroscience* **18**, 545–618.

Deschenes, M. and Feltz, P., 1976, GABA-induced rise of extracellular potassium in rat dorsal root ganglia: an electrophysiological study in vivo. *Brain Res.* **118**, 494–499.

Deschenes, M., Feltz, P. and Lamour, Y., 1976, A model for an estimate in vivo of the ionic basis of presynaptic inhibition: an intracellular analysis of the GABA-induced depolarization of rat dorsal root ganglia. *Brain Res.* **118**, 486–493.

DeVito, J. L., Ruch, T. C. and Patton, H. D., 1964, Analysis of residual weight discriminatory ability and evoked cortical potentials following section of dorsal columns in monkeys. *Indian J. Physiol. Pharmacol.* **8**, 117–126.

Devor, M. and Claman, D., 1980, Mapping and plasticity of acid phosphatase afferents in the rat dorsal horn. *Brain Res.* **190**, 17–28.

Devor, M. and Wall, P. D., 1976, Dorsal horn cells with proximal cutaneous receptive fields. *Brain Res.* **118**, 325–328.

Devor, M. and Wall, P. D., 1978, Reorganization of spinal cord sensory map after peripheral nerve injury. *Nature* **267**, 75–76.

Devor, M. and Wall, P. D., 1981a, Effect of peripheral nerve injury on receptive fields of cells in the cat spinal cord. *J. Comp. Neurol.* **199**, 277–291.

Devor, M. and Wall, P. D., 1981b, Plasticity in the spinal cord sensory map following peripheral nerve injury in rats. *J. Neurosci.* **1**, 679–684.

Devor, M., Wall, P. D. and McMahon, S. B., 1984, Dichotomizing somatic nerve fibers exist in rats but they are rare. *Neurosci. Lett.* **49**, 187–192.

Diamond, I. T., Randall, W. and Springer, L., 1964, Tactual localization in cats deprived of cortical areas SI and SII and the dorsal columns. *Psychon. Sci.* **1**, 261–262.

Dickenson, A. H. and Sullivan, A., 1990, Differential effects of excitatory amino acid antagonists on dorsal horn nociceptive neurones in the rat. *Brain Res.* **506**, 31–39.

Dickenson, A. H., Rivot, J. P., Chaouch, A., Besson, J. M. and LeBars, D., 1981, Diffuse noxious inhibitory controls (DNIC) in the rat with or without pCPA pretreatment. *Brain Res.* **216**, 313–321.

Dickenson, A. H., Sullivan, A., Feeney, C., Fournie-Zaluski, M. C. and Roques, B. P., 1986, Evidence that endogenous enkephalins produce delta-opiate receptor mediated neuronal inhibitions in rat dorsal horn. *Neurosci. Lett.* **72**, 179–182.

Dickhaus, H., Pauser, G. and Zimmermann, M., 1985, Tonic descending inhibition affects intensity coding of nociceptive responses of spinal dorsal horn neurones in the cat. *Pain* **23**, 145–158.

Diculescu, I., Onicescu, D. and Casian-Joandrea, C., 1968, Recherches histochimiques sur l'avitaminose B1 experimentale (*). *Ann. Histochem.* **13**, 83–88.

Diez Guerra, F. J., Zaidi, M., Bevis, P., MacIntyre, I. and Emson, P. C., 1988, Evidence for release of calcitonin gene-related peptide and neurokinin A from sensory nerve endings in vivo. *Neuroscience* **25**, 839–846.

DiFiglia, M., Aronin, N. and Leeman, S. E., 1982a, Immunocytochemical study of neurotensin localization in the monkey spinal cord. *Ann. N.Y. Acad. Sci.* **400**, 405–408.

DiFiglia, M., Aronin, N. and Leeman, S. E., 1982b, Light microscopic and ultrastructural localization of immunoreactive substance P in the dorsal horn of monkey spinal cord. *Neuroscience* **7**, 1127–1139.

DiFiglia, M., Aronin, N. and Leeman, S. E., 1984, Ultrastructural localization of immunoreactive neurotensin in the monkey superficial dorsal horn. *J. Comp. Neurol.* **225**, 1–12.

Dilly, P. N., Wall, P. D. and Webster, K. E., 1968, Cells of origin of the spinothalamic tract in the cat and rat. *Exp. Neurol.* **21**, 550–562.

Dimsdale, J. A. and Kemp, J. M., 1966, Afferent fibres in ventral nerve roots in the rat. *J. Physiol.* (Lond.) **187**, 25–26P.

DiTirro, F. J., Ho, R. H. and Martin, G. F., 1981, Immunohistochemical localization of substance-P, somatostatin, and methionine-enkephalin in the spinal cord and dorsal root ganglia of the North American opossum, *Didelphis virginia*. *J. Comp. Neurol.* **198**, 351–363.

DiTirro, F. J., Martin, G. F. and Ho, R. H., 1983, A developmental study of substance-P, somatostatin, enkephalin, and serotonin immunoreactive elements in the spinal cord of the North American opossum. *J. Comp. Neurol.* **213**, 241–261.

Dixon, A. D., 1968, Cholinesterase distribution in trigeminal ganglionic neurons. *J. Cell Biol.* **39**, 166A.

Dobry, P. J. K. and Casey, K. L., 1972a, Roughness discrimination in cats with dorsal column lesions. *Brain Res.* **44**, 385–397.

Dobry, P. J. K. and Casey, K. L., 1972b, Coronal somatosensory unit responses in cats with dorsal column lesions. *Brain Res.* **44**, 399–416.

Dockray, G. J. and Sharkey, K. A., 1986, Neurochemistry of visceral afferent neurones. *Prog. in Brain Res.* **67**, 133–148.

Dodd, J. and Jessell, T. M., 1985, Lactoseries carbohydrates specify subsets of dorsal root ganglion neurons projecting to the superficial dorsal horn of rat spinal cord. *J. Neurosci.* **5**, 3278–3294.

Dodd, J. and Jessell, T. M., 1986, Cell surface glycoconjugates and carbohydrate-binding proteins: possible recognition signals in sensory neurone development. *J. Exp. Biol.* **124**, 225–238.

Dodd, J., Jahr, C. E., Hamilton, P. N., Heath, M. J. S., Matthew, W. D. and Jessell, T. M., 1983, Cytochemical and physiological properties of sensory and dorsal horn neurons that transmit cutaneous sensation. *Cold Spring Harbor Symposium on Quantitative Biology* **48**: 685–695.

Dodd, J., Solter, D. and Jessell, T. M., 1984, Monoclonal antibodies against carbohydrate differentiation antigens identify subsets of primary sensory neurones. *Nature* **311**, 469–472.

Dodt, E., 1952, The behaviour of thermoreceptors at low and high temperatures with special reference to Ebbecke's temperature phenomena. *Acta Physiol. Scand.* **27**, 295–314.

Dodt, E. and Zotterman, Y., 1952a, Mode of action of warm receptors. *Acta Physiol. Scand.* **26**, 345–357.

Dodt, E. and Zotterman, Y., 1952b, The discharge of specific cold fibres at high temperatures. *Acta Physiol. Scand.* **26**, 358–365.

Dodt, E., Skouby, A. P. and Zotterman, Y., 1953, The effects of cholinergic substances on the discharges from thermal receptors. *Acta Physiol. Scand.* **28**, 101–114.

Dogiel, A. S., 1896, Der Bau der Spinalganglien bei den Saeugetieren. *Anat. Anz.* **12**, 140–152.

Dogiel, A. S., 1908, *Der Bau der spinalganglien des menschen und der Saugetiere*. Gustav Fisher, Jena, 151 pp.

Donaldson, H. H., 1885, On the temperature sense. *Mind* **10**, 399–416.

Dostrovsky, J. O., 1980, Raphe and periaqueductal gray induced suppression of non-nociceptive neuronal responses in the dorsal column nuclei and trigeminal sub-nucleus caudalis. *Brain Res.* **200**, 184–189.

Dostrovsky, J. O., 1984, Brainstem influences on transmission of somatosensory information in the spinocervicothalamic pathway. *Brain Res.* **292**, 229–238.

Dostrovsky, J. O. and Guilbaud, G., 1988, Noxious stimuli excite neurons in nucleus submedius of the normal and arthritic rat. *Brain Res.* **460**, 269–280.

Dostrovsky, J. O. and Hellon, R. F., 1978, The representation of facial temperature in the caudal trigeminal nucleus of the cat. *J. Physiol.* **277**, 29–47.

Dostrovsky, J. O. and Millar, J., 1977, Receptive fields of gracile neurons after transection of the dorsal columns. *Exp. Neurol.* **56**, 610–621.

Dostrovsky, J. O., Millar, J. and Wall, P. D., 1976, The immediate shift of afferent drive of dorsal column nucleus cells following deafferentation: a comparison of acute and chronic deafferentation in gracile nucleus and spinal cord. *Exp. Neurol.* **52**, 480–495.

Dougherty, P. M. and Willis, W. D., 1991, Enhancement of spinothalamic neuron responses to chemical and mechanical stimuli following combined microiontophoretic application of N-methyl-D-aspartic acid and substance P. *Pain*, in press.

Douglas, F. L., Palkovits, M. and Brownstein, M. J., 1982, Regional distribution of substance P-like immunoreactivity in the lower brainstem of the rat. *Brain Res.* **245**, 376–378.

Douglas, W. W. and Ritchie, J. M., 1957, Non-medullated fibres in the saphenous nerve which signal touch. *J. Physiol.* **139**, 385–399.

Downie, J. W., Ferrington, D. G., Sorkin, L. S. and Willis, W. D., 1988, The primate spinocervicothalamic pathway: responses of cells of the lateral cervical nucleus and spinocervical tract to innocuous and noxious stimuli. *J. Neurophysiol.* **59**, 861–885.

Drejer, J. and Honoré, T., 1988, New quinoxalinediones show potent antagonism of quisqualate responses in cultured mouse cortical neurons. *Neurosci. Lett.* **87**, 104–108.

Drew, J. P., Westrum, L. E. and Ho, R. H., 1986, Mapping of the normal distribution of substance P-like immunoreactivity in the spinal trigeminal nucleus of the cat. *Exp. Neurol.* **93**, 168–179.

Dreyer, D. A., Schneider, R. J., Metz, C. B. and Whitsel, B. L., 1974, Differential contributions of spinal pathways to body representation in postcentral gyrus of *Macaca mulatta*. *J. Neurophysiol.* **37**, 119–145.

Droz, B. and Kazimierczak, J., 1987, Carbonic anhydrase in primary sensory neurons of dorsal root ganglia. *J. Biochem. Physiol.* **88B**, 713–717.

Dubner, R., Sumino, R. and Wood, W. I., 1975, A peripheral "cold" fiber population responsive to innocuous and noxious thermal stimuli applied to monkey's face. *J. Neurophysiol.* **38**, 1373–1389.

Dubner, R., Kenshalo, D. R., Jr., Maixner, W., Bushnell, M. C. and Oliveras, J. L., 1989, The correlation of monkey medullary dorsal horn neuronal activity and the perceived intensity of noxious heat stimuli. *J. Neurophysiol.* **62**, 450–457.

Dubrovsky, B. and Garcia-Rill, E., 1973, Role of dorsal columns in sequential motor acts requiring precise forelimb projection. *Exp. Brain Res.* **18**, 165–177.

Dubrovsky, B., Davelaar, E. and Garcia-Rill, E., 1971, The role of dorsal columns in serial order acts. *Exp. Neurol.* **33**, 93–102.

Dubuisson, D. and Wall, W. D., 1980, Descending influences on receptive fields and activity of single units recorded in laminae 1, 2 and 3 of cat spinal cord. *Brain Res.* **199**, 283–298.

Dubuisson, D., Fitzgerald, M. and Wall, P. D., 1979, Ameboid receptive fields of cells in laminae 1, 2 and 3. *Brain Res.* **177**, 376–378.

Duce, I. R. and Keen, P., 1977, An ultrastructural classification of the neuronal cell bodies of rat dorsal root ganglion using zinc iodine—osmium impregnation. *Cell Tiss. Res.* **185**, 263–277.

Duce, I. R. and Keen, P., 1983, Selective uptake of [^{3}H] glutamine and [^{3}H] glutamate into neurons and satellite cells of dorsal root ganglia in vitro. *Neuroscience* **8**, 861–866.

Duclaux, R. and Kenshalo, D. R., 1972, The temperature sensitivity of the type I slowly adapting mechanoreceptors in cats and monkeys. *J. Physiol.* **224**, 647–664.

Duclaux, R. and Kenshalo, D. R., 1973, Cutaneous receptive fields of primate cold fibers. *Brain Res.* **55**, 437–442.

Duclaux, R. and Kenshalo, D. R., 1980, Response characteristics of cutaneous warm receptors in the monkey. *J. Neurophysiol.* **43**, 1–15.

Duclaux, R., Schaefer, K. and Hensel, H., 1980, Response of cold receptors to low skin temperatures in nose of the cat. *J. Neurophysiol.* **43**, 1571–1577.

Duggan, A. W., 1974, The differential sensitivity to L-glutamate and L-aspartate of spinal interneurones and Renshaw cells. *Exp. Brain Res.* **19**, 522–528.

Duggan, A. W. and Griersmith, B. T., 1979, Inhibition of the spinal transmission of nociceptive information by supraspinal stimulation in the cat. *Pain* **6**, 149–161.

Duggan, A. W. and Hendry, I. A., 1986, Laminar localization of the sites of release of immunoreactive substance P in the dorsal horn with antibody-coated microelectrodes. *Neurosci. Lett.* **68**, 134–140.

Duggan, A. W. and Johnston, G. A. R., 1970, Glutamate and related amino acids in cat spinal roots, dorsal root ganglia, and peripheral nerves. *J. Neurochem.* **17**, 1205–1208.

Duggan, A. W. and North, R. A., 1984, Electrophysiology of opioids. *Pharmacol. Rev.* **35**, 219–281.

Duggan, A. W., Hall, J. G. and Headley, P. M., 1976, Morphine, enkephalin and the substantia gelatinosa. *Nature* **264**, 456–458.

Duggan, A. W., Hall, J. G. and Headley, P. M., 1977, Enkephalins and dorsal horn neurones of the cat: effects on responses to noxious and innocuous skin stimuli. *Brit. J. Pharmacol.* **61**, 399–408.

Duggan, A. W., Griersmith, B. T., Headley, P. M. and Hall, J. G., 1979, Lack of effect by substance P at sites in the substantia gelatinosa where met-enkephalin reduces the transmission of nociceptive impulses. *Neurosci. Lett.* **12**, 313–317.

Duggan, A. W., Griersmith, B. T. and Johnson, S. M., 1981a, Supraspinal inhibition of the excitation of dorsal horn neurones by impulses in unmyelinated primary afferents: lack of effect by strychnine and bicuculline. *Brain Res.* **210**, 231–241.

Duggan, A. W., Johnson, S. M. and Morton, C. R., 1981b, Differing distributions of receptors for morphine and met5-enkephalinamide in the dorsal horn of the cat. *Brain Res.* **229**, 379–387.

Duggan, A. W., Morton, C. R., Zhao, Z. Q. and Hendry, I. A., 1987, Noxious heating of the skin releases immunoreactive substance P in the substantia gelatinosa of the cat: a study with antibody microprobes. *Brain Res.* **403**, 345–349.

Duggan, A. W., Hendry, I. A., Morton, C. R., Hutchison, W. D. and Zhao, Z. Q., 1988a, Cutaneous stimuli releasing immunoreactive substance P in the dorsal horn of the cat. *Brain Res.* **451**, 261–273.

Duggan, A. W., Morton, C. R., Hutchison, W. D. and Hendry, I. A., 1988b, Absence of tonic supraspinal control of substance P release in the substantia gelatinosa of the anaesthetized cat. *Exp. Brain Res.* **71**, 597–602.

Duggan, A. W., Hope, P. J., Jarrott, B., Schaible, H. G. and Fleetwood-Walker, S. M., 1990, Release, spread and persistence of immunoreactive neurokinin A in the dorsal horn of the cat following noxious cutaneous stimulation. Studies with antibody microprobes. *Neuroscience* **35**, 195–202.

Duncan, D. and Keyser, L. L., 1936, Some determinations of the ratio of nerve fibers to nerve cells in the thoracic dorsal roots and ganglia of the cat. *J. Comp. Neurol.* **64**, 303–311.

Duncan, D. and Morales, R., 1973, Location of large cored synaptic vesicles in the dorsal grey matter of the cat and dog spinal cord. *Amer. J. Anat.* **136**, 123–127.

Duncan, D. and Morales, R., 1978, Relative numbers of several types of synaptic connections in the substantia gelatinosa of the cat spinal cord. *J. Comp. Neurol.* **182**, 601–610.

Dyck, P. J., Jedrzejowska, H., Karnes, J., Kawamura, Y., O'Brien, P. C., Offord, K., Ohnishi, A., Ott, M., Pollock, M. and Stevens, J. C., 1979, Reconstruction of motor, sensory, and autonomic neurons based on morphometric study of sampled levels. *Muscle and Nerve* **2**, 399–405.

Dyhre-Poulsen, P., 1975, Increased vibration threshold before movements in human subjects. *Exp. Neurol.* **47**, 516–522.

Dykes, R. W., 1975, Coding of steady and transient temperatures by cutaneous "cold" fibers serving the hand of monkeys. *Brain Res.* **98**, 485–500.

Dykes, R. W., Rasmusson, D. D., Sretavan, D. and Rehman, N. B., 1982, Submodality segregation and receptive-field sequences in cuneate, gracile, and external cuneate nuclei of the cat. *J. Neurophysiol.* **47**, 389–416.

Earle, K. M., 1952, The tract of Lissauer and its possible relation to the pain pathway. *J. Comp. Neurol.* **96**, 93–109.

Ebbesson, S. O. E., 1967, Ascending axon degeneration following hemisection of the spinal cord in the Tegu lizard (*Tupinambis nigropuynctatus*). *Brain Res.* **5**, 178–206.

Ebbesson, S. O. E., 1968, A connection between the dorsal column nuclei and the dorsal accessory olive. *Brain Res.* **8**, 393–397.

Ebbesson, S. O. E., 1969, Brainstem afferents from the spinal cord in a sample of reptilian and amphibian species. *Ann. N.Y. Acad. Sci.* **167**, 80–101.

Ebbesson, S. O. E. and Goodman, D. C., 1981, Organization of ascending spinal projections in Caiman crocodilus. *Cell Tiss. Res.* **215**, 383–395.

Ebbesson, S. O. E. and Hodde, K. C., 1981, Ascending spinal systems in the nurse shark, *Ginglymostoma cirratum*. *Cell Tiss. Res.* **216**, 313–331.

Eccles, J. C., 1964, *The Physiology of Synapses*. Springer, New York.

Eccles, J. C. and Krnjević, K., 1959, Potential changes recorded inside primary afferent fibres within the spinal cord. *J. Physiol.* **149**, 250–273.

Eccles, J. C., Fatt, P., Landgren, S. and Winsbury, G. J., 1954, Spinal cord potentials generated by volleys in the large muscle afferents. *J. Physiol.* **125**, 590–606.

Eccles, J. C., Fatt, P. and Landgren, S., 1956, Central pathway for direct inhibitory action of impulses in largest afferent nerve fibres to muscle. *J. Neurophysiol.* **19**, 75–98.

Eccles, J. C., Eccles, R. M. and Lundberg, A., 1957, Synaptic actions on motoneurones in relation to the two components of the group I muscle afferent volley. *J. Physiol.* **136**, 527–546.

Eccles, J. C., Eccles, R. M. and Lundberg, A., 1960, Types of neurone in and around the intermediate nucleus of the lumbosacral cord. *J. Physiol.* **154**, 89–114.

Eccles, J. C., Kozak, W. and Magni, F., 1961, Dorsal root reflexes of muscle group I afferent fibres. *J. Physiol.* **159**, 128–146.

Eccles, J. C., Kostyuk, P. G. and Schmidt, R. F., 1962a, Central pathways responsible for depolarization of primary afferent fibres. *J. Physiol.* **161**, 237–257.

Eccles, J. C., Magni, F. and Willis, W. D., 1962b, Depolarization of central terminals of group I afferent fibres from muscle. *J. Physiol.* **160**, 62–93.

Eccles, J. C., Schmidt, R. F. and Willis, W. D., 1963a, Depolarization of the central terminals of cutaneous afferent fibers. *J. Neurophysiol.* 26, 646–661.

Eccles, J. C., Schmidt, R. F. and Willis, W. D., 1963b, Pharmacological studies on presynaptic inhibition. *J. Physiol.* **168**, 500–530.

Eccles, R. M., 1965, Interneurones activated by higher threshold group I muscle afferents. In: *Studies in Physiology*, pp. 59–64. D. R. Curtis and A.K. McIntyre (eds.). Springer, New York.

Eccles, R. M. and Lundberg, A., 1959a, Synaptic actions in motoneurones by afferents which may evoke the flexion reflex. *Arch. Ital. Biol.* **97**, 199–221.

Eccles, R. M. and Lundberg, A., 1959b, Supraspinal control of interneurones mediating spinal reflexes. *J. Physiol.* **147**, 565–584.

Eckenstein, F., 1988, Transient expression of NFG-receptor-like immunoreactivity in postnatal rat brain and spinal cord. *Brain Res.* **446**, 149–154.

Edeson, R. O. and Ryall, R. W., 1983, Systematic mapping of descending inhibitory control by the medulla of nociceptive spinal neurones in cats. *Brain Res.* **271**, 251–262.

Edin, B. B., 1990, Finger joint movement sensitivity of non-cutaneous mechanoreceptor afferents in the human radial nerve. *Exp. Brain Res.* **82**, 417–422.

Edinger, L., 1889, Vergleichend-entwicklungsgeschichtliche und anatomische Studien im Bereiche des Central-nervensystems. 2. Ueber die Fortsetzung der hinteren Rueckenmarkswurzeln zum Gehirn. *Anat. Anz.* **4**, 121–128.

Edinger, L., 1890, Einiges vom Verlauf der Gefuehlsbahnen im centralen Nervensysteme. *Deut. Med. Woch.* **16**, 421–426.

Edwards, S. B., 1972, The ascending and descending projections of the red nucleus in the cat: an experimental study using an autoradiographic tracing method. *Brain Res.* **48**, 45–63.

Edwards, S. B., Ginsburgh, C. L., Henkel, C. K. and Stein, B. E., 1979, Sources of subcortical projections to the superior colliculus in the cat. *J. Comp. Neurol.* **184**, 309–330.

Egger, M. D., Freeman, N. C. G., Jacquin, M., Proshansky, E. and Semba, K., 1986, Dorsal horn cells in the cat responding to stimulation of the plantar cushion. *Brain Res.* **383**, 68–82.

Eidelberg, E. and Woodbury, C. M., 1972, Apparent redundancy in the somatosensory system in monkeys. *Exp. Neurol.* **37**, 573–581.

Eidelberg, E., Kreinick, C. J. and Langescheid, C., 1975, On the possible functional role of afferent pathways in skin sensation. *Exp. Neurol.* **47**, 419–432.

Eklund, G., 1972, Position sense and state of contraction; the effects of vibration. *J. Neurol. Neurosurg. Psychiat.* **35**, 606–611.

Eklund, G. and Skoglund, S., 1960, On the specificity of the Ruffini like joint receptors. *Acta Physiol. Scand.* **49**, 184–191.

Eldred, E., Yellin, H., DeSantis, M. and Smith, C. M., 1977, Supplement to bibliography on muscle receptors: their morphology, pathology, physiology, and pharmacology. *Exp. Neurol.* **55** (No. 3, part 2), 1–118.

Ellis, L. C. and Rustioni, A., 1981, A correlative HRP, Golgi and EM study of the intrinsic organization of the feline dorsal column nuclei. *J. Comp. Neurol.* **197**, 341–367.

El-Yassir, N. and Fleetwood-Walker, S. M., 1990, A 5-HT1-type receptor mediates the antinociceptive effect of nucleus raphe magnus stimulation in the rat. *Brain Res.* **523**, 92–99.

El-Yassir, N., Fleetwood-Walker, S. M. and Mitchell, R., 1988, Heterogeneous effects of serotonin in the dorsal horn of rat: the involvement of 5-HT1 receptor subtypes. *Brain Res.* **456**, 147–158.

Emery, D. G., Ito, H. and Coggeshall, R. E., 1977, Unmyelinated axons in thoracic ventral roots of the cat. *J. Comp. Neurol.* **172**, 37–48.

Emson, P. C., 1979, Peptides as neurotransmitter candidates in the mammalian CNS. *Prog. Neurobiol.* **13**, 61–116.

Emson, P. C., Goedert, M., Williams, B., Ninkovic, M. and Hunt, S. P., 1982, Neurotensin: regional distribution, characterization, and inactivation. *Ann. N.Y. Acad. Sci.* **400**, 198–215.

Endo, K., Kang, Y., Kayano, F., Kojima, H. and Hori, Y., 1985, Synaptic actions of the ventral root afferents on cat hindlimb motoneurons. *Neurosci. Lett.* **58**, 201–205.

Enevoldson, T. P. and Gordon, G., 1984, Spinally projecting neurones in the dorsal column nuclei: distribution, dendritic trees and axonal projections. *Exp. Brain Res.* **54**, 538–550.

Enevoldson, T. P. and Gordon, G., 1989a, Postsynaptic dorsal column neurons in the cat: a study with retrograde transport of horseradish peroxidase. *Exp. Brain Res.* **75**, 611–620.

Enevoldson, T. P. and Gordon, G., 1989b, Spinocervical neurons and dorsal horn neurons projecting to the dorsal column nuclei through the dorsolateral fascicle: a retrograde HRP study in the cat. *Exp. Brain Res.* **75**, 621–630.

Engberg, I. and Ryall, R. W., 1966, The inhibitory action of noradrenaline and other monoamines on spinal neurones. *J. Physiol.* **185**, 298–322.

Engberg, I. and Thaller, A., 1970, On the interaction of picrotoxin with GABA and glycine in the spinal cord. *Brain Res.* **19**, 151–154.

Engberg, I., Lundberg, A. and Ryall, R. W., 1968a, Reticulospinal inhibition of transmission in reflex pathways. *J. Physiol.* **194**, 201–223.

Engberg, I., Lundberg, A. and Ryall, R. W., 1968b, Reticulospinal inhibition of interneurones. *J. Physiol.* **194**, 225–236.

Ennever, J. A. and Towe, A. L., 1974, Response of somatosensory cerebral neurons to stimulation of dorsal and dorsolateral spinal funiculi. *Exp. Neurol.* **43**, 124–142.

Erickson, R. P., 1973, On the intensive aspect of the temperature sense. *Brain Res.* **61**, 113–118.

Erickson, R. P. and Poulos, D. A., 1973, On the qualitative aspect of the temperature sense. *Brain Res.* **61**, 107–112.

Erlanger, J. and Gasser, H. S., 1937, *Electrical Signs of Nervous Activity*. Univ. Pennsylvania Press, Philadelphia.

Erulkar, S. D., Sprague, J. M., Whitsel, B. L., Dogan, S. and Jannetta, P. J., 1966, Organization of the vestibular projection to the spinal cord of the cat. *J. Neurophysiol.* **29**, 626–664.

Evans, R. H., 1980, Evidence supporting the indirect depolarization of primary afferent terminals in the frog by excitatory amino acids. *J. Physiol.* **298**, 25–35.

Evans, R. H. and Long, S. K., 1989, Primary afferent depolarization in the rat spinal cord is mediated by pathways utilizing NMDA and non-NMDA receptors. *Neurosci. Lett.* **100**, 231–236.

Evarts, E. V., 1971, Feedback and corollary discharge: a merging of the concepts. *Neurosci. Res. Prog. Bull.* **9**, 86–112.

Fabri, M. and Conti, F., 1990, Calcitonin gene-related peptide-positive neurons and fibers in the cat dorsal column nuclei. *Neuroscience* **35**, 167–174.

Falconer, M. A. and Lindsay, J. S. B., 1946, Painful phantom limb treated by high cervical chordotomy. Report of two cases. *Brit. J. Surg.* **33**, 19–306.

Faull, R. L. M. and Villiger, J. W., 1986, Benzodiazepine receptors in the human spinal cord: a detailed anatomical and pharmacology study. *Neuroscience* **17**, 791–802.

Faull, R. L. M. and Villiger, J. W., 1987, Opiate receptors in the human spinal cord: a detailed anatomical study comparing the autoradiographic localization of [^{3}H]diprenorphine binding sites with laminar pattern of substance P, myelin and Nissl staining. *Neuroscience* **20**, 395–407.

Faull, R. L. M., Villiger, J. W. and Dragunow, M., 1989, Neurotensin receptors in the human spinal cord: a quantitative autoradiographic study. *Neuroscience* **29**, 603–613.

Favale, E., Loeb, C., Manfredi, M. and Sacco, G., 1965, Somatic afferent transmission and cortical responsiveness during natural sleep and arousal in the cat. *EEG Clin. Neurophysiol.* **18**, 354–368.

Fedina, L., Gordon, G. and Lundberg, A., 1968, The source and mechanisms of inhibition in the lateral cervical nucleus of the cat. *Brain Res.* **11**, 694–696.

Felix, D. and Wiesendanger, M., 1970, Cortically induced inhibition in the dorsal column nuclei of monkeys. *Pflugers Arch.* **320**, 285–288.

Fernandez de Molina, A. and Gray, J. A. B., 1957, Activity in the dorsal spinal grey matter after stimulation of cutaneous nerves. *J. Physiol.* **137**, 126–140.

Ferrarese, C., Iadarola, M. J., Yang, H.-Y. and Costa, E., 1986, Peripheral and central origin of the Phe-Met-Arg-Phe-amide immunoreactivity in rat spinal cord. *Reg. Peptides* **13**, 245–252.

Ferraro, A. and Barrera, S. E., 1934, Effects of experimental lesions of the posterior columns in Macacus rhesus monkeys. *Brain* **57**, 307–332.

Ferraro, A. and Barrera, S. E., 1935a, The nuclei of the posterior funiculi in Macacus rhesus. An anatomic and experimental investigation. *Arch. Neurol. Psychiat.* **33**, 262–275.

Ferraro, A. and Barrera, S. E., 1935b, Posterior column fibers and their termination in Macacus rhesus. *J. Comp. Neurol.* **62**, 507–530.

Ferrell, W. R., 1980, The adequacy of stretch receptors in the cat knee joint for signalling joint angle throughout a full range of movement. *J. Physiol.* **299**, 85–99.
Ferrell, W. R. and Milne, S. E., 1989, Factors affecting the accuracy of position matching at the proximal interphalangeal joint in human subjects. *J. Physiol.* **411**, 575–583.
Ferrell, W. R. and Smith, A., 1987, The effect of digital nerve block on position sense at the proximal interphalangeal joint of the human index finger. *Brain Res.* **425**, 369–371.
Ferrell, W. R. and Smith, A., 1988, Position sense at the proximal interphalangeal joint of the human index finger. *J. Physiol.* **399**, 49–61.
Ferrell, W. R. and Smith, A., 1989, The effect of loading on position sense at the proximal interphalangeal joint of the human index finger. *J. Physiol.* **418**, 145–161.
Ferrell, W. R., Nade, S. and Newbold, P. J., 1986, The interrelation of neural discharge, intra-articular pressure, and joint angle in the knee of the dog. *J. Physiol.* **373**, 353–365.
Ferrell, W. R., Gandevia, S. C. and McCloskey, D. I., 1987, The role of joint receptors in human kinaesthesia when intramuscular receptors cannot contribute. *J. Physiol.* **386**, 63–67.
Ferrington, D. G. and Rowe, M. J., 1982, Specificity of connections and tactile coding capacities in cuneate nucleus of the neonatal kitten. *J. Neurophysiol.* **47**, 622–640.
Ferrington, D. G., Rowe, M. J. and Tarvin, R. P. C., 1986a, High gain transmission of single impulses through dorsal column nuclei of the cat. *Neurosci. Lett.* **65**, 277–282.
Ferrington, D. G., Sorkin, L. S. and Willis, W. D., 1986b, Responses of spinothalamic tract cells in the cat cervical spinal cord to innocuous and graded noxious stimuli. *Somatosensory Res.* **3**, 339–358.
Ferrington, D. G., Rowe, M. J. and Tarvin, R. P. C., 1987a, Actions of single sensory fibres on cat dorsal column nuclei neurones: vibratory signalling in a one-to-one linkage. *J. Physiol.* **386**, 293–309.
Ferrington, D. G., Rowe, M. J. and Tarvin, R. P. C., 1987b, Integrative processing of vibratory information in cat dorsal column nuclei neurones driven by identified sensory fibres. *J. Physiol.* **386**, 311–331.
Ferrington, D. G., Horniblow, S. and Rowe, M. J., 1987c, Temporal patterning in the responses of gracile and cuneate neurones in the cat to cutaneous vibration. *J. Physiol.* **386**, 277–291.
Ferrington, D. G., Sorkin, L. S. and Willis, W. D., 1987d, Responses of spinothalamic tract cells in the superficial dorsal horn of the primate lumbar spinal cord. *J. Physiol.* **388**, 681–703.
Ferrington, D. G., Downie, J. W. and Willis, W. D., 1988, Primate nucleus gracilis neurons: responses to innocuous and noxious stimuli. *J. Neurophysiol.* **59**, 886–907.
Fetz, E. E., 1968, Pyramidal tract effects on interneurons in the cat lumbar dorsal horn. *J. Neurophysiol.* **31**, 69–80.
Fields, H. L., 1987, *Pain.* McGraw-Hill, New York.
Fields, H. L. and Basbaum, A. I., 1978, Brainstem control of spinal pain-transmission neurons. *Ann. Rev. Physiol.* **40**, 217–248.
Fields, H. L., Meyer, G. A. and Partridge, L. D., 1970a, Convergence of visceral and somatic input onto spinal neurons. *Exp. Neurol.* **26**, 36–52.
Fields, H. L., Partridge, L. D. and Winter, D. L., 1970b, Somatic and visceral receptive field properties of fibers in ventral quadrant white matter of the cat spinal cord. *J. Neurophysiol.* **33**, 827–837.
Fields, H. L., Wagner, G. M. and Anderson, S. D., 1975, Some properties of spinal neurons projecting to the medial brain-stem reticular formation. *Exp. Neurol.* **47**, 118–134.
Fields, H. L., Basbaum, A. I., Clanton, C. H. and Anderson, S. D., 1977a, Nucleus raphe magnus inhibition of spinal cord dorsal horn neurons. *Brain Res.* **126**, 441–453.
Fields, H. L., Clanton, C. H. and Anderson, S. D., 1977b, Somatosensory properties of spinoreticular neurons in the cat. *Brain Res.* **120**, 49–66.
Fields, H. L., Emson, P. C., Leigh, B. K., Gilbert, R. F. T. and Iversen, L. L., 1980, Multiple opiate receptor sites on primary afferent fibres. *Nature* **284**, 351–353.
Finger, T. E., 1978, Cerebellar afferents in teleost cat fish (Ictaluridae). *J. Comp. Neurol.* **181**, 173–182.
Finley, J. C. W., Maderdrut, J. L., Roger, L. J. and Petrusz, P., 1981a, The immunocytochemical localization of somatostatin-containing neurons in the rat central nervous system. *Neuroscience* **6**, 2173–2192.
Finley, J. C. W., Maderdrut, J. L. and Petrusz, P., 1981b, The immunocytochemical localization of enkephalin in the central nervous system of the rat. *J. Comp. Neurol.* **198**, 541–565.
Fitzgerald, M., 1981, A study of the cutaneous afferent input to substantia gelatinosa. *Neuroscience* **6**, 2229–2237.
Fitzgerald, M., 1983a, Capsaicin and sensory neurones—a review. *Pain* **15**, 109–130.
Fitzgerald, M., 1983b, Influences of contralateral nerve and skin stimulation on neurones in the substantia gelatinosa of the rat spinal cord. *Neurosci. Lett.* **36**, 139–143.
Fitzgerald, M., 1989, The course and termination of primary afferent fibers. In: *Textbook of Pain*, pp. 46–62. P. D. Wall and R. Melzack, (eds.). Churchill Livingstone, New York.
Fitzgerald, M. and Lynn, B., 1977, The sensitization of high threshold mechanoreceptors with myelinated axons by repeated heating. *J. Physiol.* **265**, 549–563.
Fitzgerald, M. and Wall, P. D., 1980, The laminar organization of dorsal horn cells responding to peripheral C fibre stimulation. *Exp. Brain Res.* **41**, 36–44.
Fitzgerald, M. and Woolf, C. J., 1981, Effects of cutaneous nerve and intraspinal conditioning on C-fibre afferent terminal excitability in decerebrate spinal rats. *J. Physiol.* **318**, 25–39.

Fitzgerald, M., Woolf, C. J. and Shortland, P., 1990, Collateral sprouting of the central terminals of cutaneous primary afferent neurons in the rat spinal cord: pattern, morphology, and influence of targets. *J. Comp. Neurol.* **300**, 370–385.

Fjällbrant, N. and Iggo, A., 1961, The effect of histamine, 5-hydroxytryptamine and acetylcholine on cutaneous afferent fibres. *J. Physiol.* **156**, 578–590.

Fleetwood-Walker, S. M., Mitchell, R., Hope, P. J., Molony, V. and Iggo, A., 1985, An alpha2 receptor mediates the selective inhibition by noradrenaline of nociceptive responses of identified dorsal horn neurones. *Brain Res.* **334**, 243–254.

Fleetwood-Walker, S. M., Hope, P. J. and Mitchell, R., 1988a, Antinociceptive actions of descending dopaminergic tracts on cat and rat dorsal horn somatosensory neurones. *J. Physiol.* **399**, 335–348.

Fleetwood-Walker, S. M., Hope, P. J., Mitchell, R., El-Yassir, N. and Molony, V., 1988b, The influence of opioid receptor subtypes on the processing of nociceptive inputs in the dorsal horn of the cat. *Brain Res.* **451**, 213–226.

Fleischer, E., Handwerker, H. O. and Joukhadar, S., 1983, Unmyelinated nociceptive units in two skin areas of the rat. *Brain Res.* **267**, 81–92.

Flindt-Egebak, P., 1977, Autoradiographical demonstration of the projections from the limb areas of the feline sensorimotor cortex to the spinal cord. *Brain Res.* **136**, 153–156.

Flink, R. and Westman, J., 1986, Different neuron populations in the feline lateral cervical nucleus: a light and electron microscopic study with the retrograde axonal transport technique. *J. Comp. Neurol.* **250**, 265–281.

Flink, R., Wiberg, M. and Blomqvist, A., 1983, The termination in the mesencephalon of fibres from the lateral cervical nucleus. An anatomical study in the cat. *Brain Res.* **259**, 11–20.

Florence, S. L., Wall, J. T. and Kaas, J. H., 1988, The somatotopic pattern of afferent projections from the digits to the spinal cord and cuneate nucleus in macaque monkeys. *Brain Res.* **452**, 388–392.

Florence, S. L., Wall, J. T. and Kaas, J. H., 1989, Somatotopic organization of inputs from the hand to the spinal gray and cuneate nucleus of monkeys with observations on the cuneate nucleus of humans. *J. Comp. Neurol.* **286**, 48–70.

Floyd, K. and Morrison, J. F. B., 1974, Splanchnic mechanoreceptors in the dog. *Quart. J. Exp. Physiol.* **59**, 361–366.

Floyd, K., Hick, V. E. and Morrison, J. F. B., 1976, Mechanosensitive afferent units in the hypogastric nerve of the cat. *J. Physiol.* **259**, 457–471.

Floyd, K., Hick, V. E., Koley, J. and Morrison, J. F. B., 1977, The effects of bradykinin on afferent units in intra-abdominal sympathetic nerve trunks. *Quart. J. Exp. Physiol.* **62**, 19–25.

Fock, S. and Mense, S., 1976, Excitatory effects of 5-hydroxytryptamine, histamine and potassium ions on muscular group IV afferent units: a comparison with bradykinin. *Brain Res.* **105**, 459–469.

Foerster, O., 1936, Symptomatologie der Erkrankungen des Rueckenmarks und seiner Wurzeln. In: *Handbuch der Neurologie*, Vol. 5, pp. 1–403. O. Bumke and O. Foerster (eds.). Berlin, Springer.

Foerster, O. and Gagel, O., 1932, Die Vorderseitenstrangdurchschneidung beim Menschen. Eine klinisch-pathophysiologisch-anatomische Studie. *Z. Ges. Neurol. Psychiat.* **138**, 1–92.

Fontaine-Perus, J., Chanconie, M. and LeDouarin, N. M., 1985, Embryonic origin of substance P containing neurons in cranial and spinal sensory ganglia of the avian embryo. *Devel. Biol.* **107**, 227–238.

Foong, F. W. and Duggan, A. W., 1986, Brainstem areas tonically inhibiting dorsal horn neurons: studies with microinjection of the GABA analogue piperidine-4-sulphonic acid. *Pain* **27**, 361–372.

Foote, S. L. and Cha, C. I., 1988, Distribution of corticotropin-releasing factor-lime immunoreactivity in brainstem of two monkey species (*Saimiri sciureus* and *Macaca fascicularis*): an immunocytochemical analysis. *J. Comp. Neurol.* **276**, 239–264.

Foreman, J. C., Jordan, C. C., Oehme, P. and Renner, H., 1983, Structure-activity relationships for substance P-related peptides that cause wheal and flare reactions in human skin. *J. Physiol.* **335**, 449–465.

Foreman, R. D., 1977, Viscerosomatic convergence onto spinal neurons responding to afferent fibers located in the inferior cardiac nerve. *Brain Res.* **137**, 164–168.

Foreman, R. D., 1989, Organization of the spinothalamic tract as a relay for cardiopulmonary sympathetic afferent fiber activity. In: *Progress in Sensory Physiology* 9, pp. 1–51. D. Ottoson (ed.). Springer-Verlag, New York.

Foreman, R. D. and Ohata, C. A., 1980, Effects of coronary artery occlusion on thoracic spinal neurons receiving viscerosomatic inputs. Amer. *J. Physiol.* **238**, H667–H674.

Foreman, R. D. and Weber, R. N., 1980, Responses from neurons of the primate spinothalamic tract to electrical stimulation of afferents from the cardiopulmonary region and somatic structures. *Brain Res.* **186**, 463–468.

Foreman, R. D., Applebaum, A. E., Beall, J. E., Trevino, D. L. and Willis, W. D., 1975, Responses of primate spinothalamic tract neurons to electrical stimulation of hindlimb peripheral nerve. *J. Neurophysiol.* **38**, 132–145.

Foreman, R. D., Beall, J. E., Applebaum, A. E., Coulter, J. D. and Willis, W. D., 1976, Effects of dorsal column stimulation on primate spinothalamic tract neurons. *J. Neurophysiol.* **39**, 534–546.

Foreman, R. D., Schmidt, R. F. and Willis, W. D., 1977, Convergence of muscle and cutaneous input onto primate spinothalamic tract neurons. *Brain Res.* **124**, 555–560.

Foreman, R. D., Kenshalo, D. R., Jr., Schmidt, R. F. and Willis, W. D., 1979a, Field potentials and excitation of primate spinothalamic neurons in response to volleys in muscle afferents. *J. Physiol.* **286**, 197–213.

Foreman, R. D., Schmidt, R. F. and Willis, W. D., 1979b, Effects of mechanical and chemical stimulation of fine muscle afferents upon primate spinothalamic tract cells. *J. Physiol.* **286**, 215–231.

Foreman, R. D., Hancock, M. B. and Willis, W. D., 1981, Responses of spinothalamic tract cells in the thoracic spinal cord of the monkey to cutaneous and visceral inputs. *Pain* **11**, 149–162.

Foreman, R. D., Blair, R. W. and Weber, R. N., 1984, Viscerosomatic convergence onto T2-T4 spinoreticular, spinoreticular-spinothalamic, and spinothalamic tract neurons in the cat. *Exp. Neurol.* **85**, 597–619.

Forssman, W. G., 1978, A new somatostatinergic system in the mammalian spinal cord. *Neurosci. Lett.* **10**, 293–297.

Forssmann, W. G., Burnweit, C., Shehab, T. and Triepel, J., 1979, Somatostatin-immunoreactive nerve cell bodies and fibers in the medulla oblongata et spinalis. *Histochem. Cytochem.* **27**, 1391–1393.

Fox, J. C. and Klemperer, W. W., 1942, Vibratory sensibility: a quantitative study of its threshold in nervous disorders. *Arch. Neurol. Psychiat.* **48**, 622–645.

Fox, R. E., Holloway, J. A., Iggo, A. and Mokha, S. S., 1980, Spinothalamic neurones in the cat: some electrophysiological observations. *Brain Res.* **182**, 186–190.

Fox, S., Krnjević, K., Morris, M. E., Puil, E. and Werman, R., 1978, Action of baclofen on mammalian synaptic transmission. *Neuroscience* **3**, 495–515.

Franco-Cereceda, A., Henke, H., Lundberg, J. M., Peterman, J. B., Hökfelt, T. and Fisher, J. A., 1986, Calcitonin gene-related peptide (CGRP) in capsaicin-sensitive substance P-immunoreactive sensory neurons in animals and man: distribution and release by capsaicin. *Peptides* **8**, 399–410.

Frank, K. and Fuortes, M. G. F., 1956, Unitary activity of spinal interneurones of cats. *J. Physiol.* **131**, 425–435.

Frankenhaeuser, B., 1949, Impulses from a cutaneous receptor with slow adaptation and low mechanical threshold. *Acta Physiol. Scand.* **18**, 68–74.

Franz, M. and Iggo, A., 1968, Dorsal root potentials and ventral root reflexes evoked by nonmyelinated fibers. *Science* **162**, 1140–1142.

Franz, M. and Mense, S., 1975, Muscle receptors with group IV afferent fibres responding to afferent fibres responding to application of bradykinin. *Brain Res.* **92**, 369–383.

Franzén, O. and Offenloch, K., 1969, Evoked response correlates of psychophysical magnitude estimates for tactile stimulation in man. *Exp. Brain Res.* **8**, 1–18.

Frazier, C. H., 1920, Section of the anterolateral columns of the spinal cord for the relief of pain. A report of six cases. *Arch. Neurol. Psychiat.* **4**, 137–147.

Frederickson, R. C. A. and Geary, L. E., 1982, Endogenous opioid peptides: review of physiological, pharmacological and clinical aspects. *Prog. Neurobiol.* **19**, 19–69.

Fredholm, B. B. and Dunwiddie, T. V., 1988, How does adenosine inhibit transmitter release? *Trends Pharmacol. Sci.* **9**, 130–134.

Freeman, A. W. and Johnson, K. O., 1982a, Cutaneous mechanoreceptors in macaque monkey: temporal discharge patterns evoked by vibration, and a receptor model. *J. Physiol.* **323**, 21–41.

Freeman, A. W. and Johnson, K. O., 1982b, A model accounting for effects of vibratory amplitude on responses of cutaneous mechanoreceptors in macaque monkey. *J. Physiol.* **323**, 43–64.

Freeman, M. A. R. and Wyke, B., 1967, The innervation of the knee joint. An anatomical and histological study in the cat. *J. Anat.* **101**, 502–532.

Freeman, W. and Watts, J. W., 1950, *Psychosurgery in the Treatment of Mental Disorders and Intractable Pain*. Thomas, Springfield.

French, L. A. and Peyton, W. T., 1948, Ipsilateral sensory loss following cordotomy. *J. Neurosurg.* **5**, 403–404.

Fritschy, J. M. and Grzanna, R., 1990, Demonstration of two separate descending noradrenergic pathways to the rat spinal cord: evidence for an intragriseal trajectory of locus coeruleus axons in the superficial layers of the dorsal horn. *J. Comp. Neurol.* **291**, 553–582.

Frommer, G. P., Trefz, B. R. and Casey, K. L., 1977, Somatosensory function and cortical unit activity in cats with only dorsal column fibers. *Exp. Brain Res.* **27**, 113–129.

Frostholm, A. and Rotter, A., 1985, Glycine receptor distribution in mouse CNS: autoradiographic localization of [^{3}H]strychnine binding sites. *Brain Res. Bull.* **15**, 473–486.

Fruhstorfer, H., 1976, Conduction in the afferent thermal pathway of man. In: *Sensory Functions of the Skin in Primates, with Special Reference to Man*, pp. 355–366. Y. Zotterman (ed.). Pergamon Press, Oxford.

Fruhstorfer, H., Zenz, M., Nolte, H. and Hensel, H., 1974, Dissociated loss of cold and warm sensibility during regional anaesthesia. *Pfluegers Arch.* **349**, 73–82.

Frykholm, R., 1951, Cervical root compression resulting from disc degeneration and root-sleeve fibrosis. A clinical investigation. *Acta Chir. Scand.* (Suppl.), 160.

Frykholm, R., Hyde, J., Norlen, G. and Skoglund, C. R., 1953, On pain sensations produced by stimulation of ventral roots in man. *Acta Physiol. Scand.* (Suppl. 106) **29**, 455–469.

Fu, T. G., Santini, M. and Schomburg, E. D., 1974, Characteristics and distribution of spinal focal synaptic potentials generated by group II muscle afferents. *Acta Physiol. Scand.* **91**, 298–313.

Fuji, K., Senba, E., Ueda, Y. and Tohyama, M., 1983, Vasoactive intestinal polypeptide (VIP)-containing neurons in the spinal cord of the rat and their projections. *Neurosci. Lett.* **37**, 51–55.

Fuji, K., Senba, E., Fujii, S., Nomura, I., Wu, J. Y., Ueda, Y. and Tohyama, M., 1985, Distribution, ontogeny and projections of cholecystokinin-8, vasoactive intestinal polypeptide and γ-aminobutyrate-containing neuron systems in the rat spinal cord: an immunohistochemical analysis. *Neuroscience* **14**, 881–894.

Fukushima, K. and Kato, M., 1975, Spinal interneurons responding to group II muscle afferent fibers in the cat. *Brain Res.* **90**, 307–312.

Fukushima, T. and Kerr, F. W. L., 1979, Organization of trigeminothalamic tracts and other thalamic afferent systems of the brainstem in the rat: presence of gelatinosa neurons with thalamic connections. *J. Comp. Neurol.* **183**, 169–184.

Fuller, J. H. and Schlag, J. D., 1976, Determination of antidromic excitation by the collision test: problems of interpretation. *Brain Res.* **112**, 283–298.

Fuxe, K., Agnati, L. F., McDonald, T., Locatelli, V., Hökfelt, T., Dalsgaard, C. J., Battistini, N., Yanaihara, N., Mutt, V. and Cuello, A. C., 1983, Immunohistochemical indications of gastrin releasing peptide bombesin-like immunoreactivity in the nervous system of the rat. Codistribution with substance P-like immunoreactive terminal systems and coexistence with substance P-like immunoreactivity in dorsal root ganglion cell bodies. *Neurosci. Lett.* **37**, 17–22.

Fuxe, K., Härfstrand, A., Agnati, L. F., Yu, Z. Y., Cintra, A., Wikström, A. C., Okret, S., Cantoni, E. and Gustafsson, J. A., 1985, Immunocytochemical studies on the localization of glucocorticoid receptor immunoreactive nerve cells in the lower brain stem and spinal cord of the male rat using a monoclonal antibody against rat liver glucocorticoid receptor. *Neurosci. Lett.* **60**, 1–6.

Fyffe, R. E. W., 1979, The morphology of Group II muscle afferent fibre collaterals. *J. Physiol.* **296**, 39–40.

Fyffe, R. E. W., 1984, Afferent fibers. In: *Handbook of the Spinal Cord*, pp. 79-136. R. A. Davidoff (ed.). Dekker, New York, Basle.

Fyffe, R. E. W. and Perl, E. R., 1984, Is ATP a central synaptic mediator for certain primary afferent fibers from mammalian skin? *Proc. Natl. Acad. Sci.* **81**, 6890–6893.

Fyffe, R. E. W., Cheema, S. S. and Rustioni, A., 1986a, Intracellular staining study of the feline cuneate nucleus. I. Terminal patterns of primary afferent fibers. *J. Neurophysiol.* **56**, 1268–1283.

Fyffe, R. E. W., Cheema, S. S., Light, A. R. and Rustioni, A., 1986b, Intracellular staining study of the feline cuneate nucleus. II. Thalamic projecting neurons. *J. Neurophysiol.* **56**, 1284–1296.

Galindo, A., Krnjević, K. and Schwartz, S., 1967, Microiontophoretic studies on neurones in the cuneate nucleus. *J. Physiol.* **192**, 359–377.

Galindo, A., Krnjević, K. and Schwartz, S., 1968, Patterns of firing in cuneate neurones and some effects of flaxedil. *Exp. Brain Res.* **5**, 87–101.

Gallagher, J. P., Higashi, H. and Nishi, S., 1978, Characterization and ionic basis of GABA-induced depolarizations recorded in vitro from cat primary afferent neurones. *J. Physiol.* **275**, 263–282.

Gallego, R. and Eyzaguirre, C., 1978, Membrane and action potential characteristics of A and C nodose ganglion cells studied in whole ganglia and in tissue slices. *J. Neurophysiol.* **41**, 1217–1232.

Gambarian, L. S., 1960, Spinal paths for the cortical projection of proprioceptive signals. *Fiziol. Zh. SSSR* **46**, 1098–1104. Eng. Trans.: *Sechenov Physiol. J. USSR* **46**, 1283–1290, 1960.

Game, C. J. A. and Lodge, D., 1975, The pharmacology of the inhibition of dorsal horn neurones by impulses in myelinated cutaneous afferents in the cat. *Exp. Brain Res.* **23**, 75–84.

Gammon, G. D. and Bronk, D. M., 1935, The discharge of impulses from Pacinian corpuscles in the mesentery and its relation to vascular changes. *Amer. J. Physiol.* **114**, 77–84.

Gammon, G. D. and Starr, I., 1941, Studies on the relief of pain by counterirritation. *J. Clin. Invest.* **20**, 13–20.

Gamse, R. and Saria, A., 1985, Potentiation of tachykinin-induced plasma protein extravasation by calcitonin gene-related peptide. *Eur. J. Pharmacol.* **114**, 61–66.

Gamse, R., Molnar, A. and Lembeck, F., 1979, Substance P release from spinal cord slices by capsaicin. *Life Sci.* **25**, 629–636.

Gandevia, S. C., 1985, Illusory movements produced by electrical stimulation of low-threshold muscle afferents from the hand. *Brain* **108**, 965–981.

Gandevia, S. C. and McCloskey, D. I., 1976, Joint sense, muscle sense, and their combination as position sense, measured at the distal interphalangeal joint of the middle finger. *J. Physiol.* **260**, 387–407.

Gandevia, S. C., Hall, L. A., McCloskey, D. I. and Potter, E. K., 1983, Proprioceptive sensation at the terminal joint of the middle finger. *J. Physiol.* **335**, 507–517.

Gandevia, S. C., Burke, D. and McKeon, B., 1984, The projection of muscle afferents from the hand to cerebral cortex in man. *Brain* **107**, 1–13.

Gans, A., 1916, Ueber Tastblindheit und ueber Stoerungen der raeumlichen Wahrnehmungen der Sensibilitaet. *Z. f. d. Gesamte Neurol. Psychiat.* **31**, 303–428.

Gardner, E., 1944, The distribution and termination of nerves in the knee joint of the cat. *J. Comp. Neurol.* **80**, 11–32.

Gardner, E., 1948, Conduction rates and dorsal root inflow of sensory fibers from the knee joint of the cat. *Amer. J. Physiol.* **152**, 436–445.

Gardner, E. and Cuneo, H. M., 1945, Lateral spinothalamic tract and associated tracts in man. *Arch. Neurol. Psychiat.* **53**, 423–430.

Gardner, E. and Haddad, B., 1953, Pathways to the cerebral cortex for afferent fibers from the hindleg of the cat. *Amer. J. Physiol.* **172**, 475–482.

Gardner, E. and Morin, F., 1953, Spinal pathways for projection of cutaneous and muscular afferents to the sensory and motor cortex of the monkey (*Macaca mulatta*). *Amer. J. Physiol.* **174**, 149–154.

Gardner, E. and Morin, F., 1957, Projection of fast afferents to the cerebral cortex of monkey. *Amer. J. Physiol.* **189**, 152–158.

Gardner, E. and Noer, R., 1952, Projection of afferent fibers from muscles and joints to the cerebral cortex of the cat. *Amer. J. Physiol.* **168**, 437–441.

Gardner, E., Thomas, L. M. and Morin, F., 1955, Cortical projections of fast visceral afferents in the cat and monkey. *Amer. J. Physiol.* **183**, 438–444.

Gardner, E. P. and Palmer, C. I., 1989a, Simulation of motion on the skin. I. Receptive fields and temporal frequency coding by cutaneous mechanoreceptors of OPTACON pulses delivered to the hand. *J. Neurophysiol.* **62**, 1410–1436.

Gardner, E. P. and Palmer, C. I., 1989b, Simulation of motion on the skin. II. Cutaneous mechanoreceptor coding of the width and texture of bar patterns displaced across the OPTACON. *J. Neurophysiol.* **62**, 1437–1460.

Gardner, E. P. and Palmer, C. I., 1990, Simulation of motion on the skin. III. Mechanisms used by rapidly adapting cutaneous mechanoreceptors in the primate hand for spatiotemporal resolution and two-point discrimination. *J. Neurophysiol.* **63**, 841–859.

Gasser, H. S., 1950, Unmedullated fibers originating in dorsal root ganglia. *J. Gen. Physiol.* **3**, 651–690.

Gasser, H. S. and Graham, H. T., 1933, Potentials produced in the spinal cord by stimulation of dorsal roots. *Amer. J. Physiol.* **103**, 303–320.

Gaze, R. M. and Gordon, G., 1955, Some observations on the central pathway for cutaneous impulses in the cat. *Quart. J. Exp. Physiol.* **40**, 187–194.

Gazelius, B., Edwall, B., Olgart, L., Lundberg, J. M., Hökfelt, T. and Fischer, J. A., 1987, Vasodilatory effects and coexistence of calcitonin gene-related peptide (CGRP) and substance P in sensory nerves of cat dental pulp. *Acta Physiol. Scand.* **130**, 33–40.

Gebhart, G. F., Sandkühler, J., Thalhammer, J. G. and Zimmermann, M., 1983a, Quantitative comparison of inhibition in spinal cord of nociceptive information by stimulation in periaqueductal gray or nucleus raphe magnus of the cat. *J. Neurophysiol.* **50**, 1433–1445.

Gebhart, G. F., Sandkühler, J., Thalhammer, J. G. and Zimmermann, M., 1983b, Inhibition of spinal nociceptive information by stimulation in the midbrain of the cat is blocked by lidocaine microinjected in nucleus raphe magnus and the medullary reticular formation. *J. Neurophysiol.* **50**, 1446–1457.

Geiger, J. D. and Nagy, J. I., 1985, Localization of [^{3}H]nitrobenzylthionosine binding sites in rat spinal cord and primary afferent neurons. *Brain Res.* **347**, 321–327.

Geldard, F. A., 1953, *The Human Senses*. John Wiley & Sons, Inc., New York.

Gelfan, S. and Carter, S., 1967, Muscle sense in man. *Exp. Neurol.* **18**, 469–473.

Georgopoulos, A. P., 1976, Functional properties of primary afferent units probably related to pain mechanisms in primate glabrous skin. *J. Neurophysiol.* **39**, 71–83.

Georgopoulos, A. P., 1977, Stimulus-response relations in high-threshold mechanothermal fibers innervating primate glabrous skin. *Brain Res.* **128**, 547–552.

Gerber, G. and Randić, M., 1989a, Excitatory amino acid-mediated components of synaptically evoked input from dorsal roots to deep dorsal horn neurons in the rat spinal cord slice. *Neurosci. Lett.* **106**, 211–219.

Gerber, G. and Randić, M., 1989b, Participation of excitatory amino acid receptors in the slow excitatory synaptic transmission in the rat spinal dorsal horn in vitro. *Neurosci. Lett.* **106**, 220–228.

Gerebetzoff, M., 1939, Les voies centrales de la sensibilite et du gout et leurs terminaisons thalamiques. *Cellule* **48**, 91–146.

Gerebetzoff, M. A., 1959, Cholinesterases—modern trends in physical science. In: *International Monographs on Pure and Applied Biological Science,* pp. 1–105. Pergamon Press, New York.

Gerebtzoff, M. A., 1966, Detection histochimique d'isoenzymes de la lactate deshydrogenase dans le nerf et le ganglion spinal. *Comp. Rend. d'Sci. Soc. Biol.* **160**, 1323–1325.

Gerebtzoff, M. A. and Maeda, T., 1968, Caracteres et localisation histochimique d'un isoenzyme fluororesistant de la phosphatase acide dans la moelle epiniere du rat. *Comp. Rend. Soc. Biol.* **162**, 2032–2035.

Gerhart, K. D., Wilcox, T. K., Chung, J. M. and Willis, W. D., 1981a, Inhibition of nociceptive and nonnociceptive responses of primate spinothalamic cells by stimulation in medial brain stem. *J. Neurophysiol.* **45**, 121–136.

Gerhart, K. D., Yezierski, R. P., Giesler, G. J. and Willis, W. D., 1981b, Inhibitory receptive fields of primate spinothalamic tract cells. *J. Neurophysiol.* **46**, 1309–1325.

Gerhart, K. D., Yezierski, R. P., Wilcox, T. K., Grossman, A. E. and Willis, W. D., 1981c, Inhibition of primate spinothalamic tract neurons by stimulation in ipsilateral or contralateral ventral posterior lateral (VPL^c) thalamic nucleus. *Brain Res.* **229**, 514–519.

Gerhart, K. D., Yezierski, R. P., Fang, Z. R. and Willis, W. D., 1983, Inhibition of primate spinothalamic tract neurons by stimulation in ventral posterior lateral (VPL_c) thalamic nucleus: possible mechanisms. *J. Neurophysiol.* **49**, 406–423.

Gerhart, K. D., Yezierski, R. P., Wilcox, T. K. and Willis, W. D., 1984, Inhibition of primate spinothalamic tract neurons by stimulation in periaqueductal gray or midbrain reticular formation. *J. Neurophysiol.* **51**, 450–466.

Gernandt, B. and Zotterman, Y., 1946, Intestinal pain: an electrophysiological investigation on mesenteric nerves. *Acta Physiol. Scand.* **12**, 56–72.

Getz, B., 1952, The termination of spinothalamic fibres in the cat as studied by the method of terminal degeneration. *Acta Anat.* **16**, 271–290.

Gherlarducci, B., Pisa, M. and Pompeiano, O., 1970, Transformation of somatic afferent volleys across the prethalamic and thalamic components of the lemniscal system during the rapid eye movements of sleep. *EEG Clin. Neurophysiol.* **29**, 348–357.

Ghez, C. and Lenzi, G.L., 1971, Modulation of sensory transmission in cat lemniscal system during voluntary movement. *Pfluegers Arch.* **323**, 273–278.

Ghez, C. and Pisa, M., 1972, Inhibition of afferent transmission in cuneate nucleus during voluntary movement in the cat. *Brain Res.* **40**, 145–151.

Giacobini, E., 1956, Histochemical demonstration of ACHE activity in isolated nerve cells. *Acta Physiol. Scand.* **36**, 276–289.

Giacobini, E., 1958, The intracellular distribution of cholinesterase in the nerve cell studied by means of a quantitative micromethod. *Acta Physiol. Scand.* **42**, 49–50.

Giacobini, E., 1959a, The distribution and localization of cholinesterases in nerve cells. *Acta Physiol. Scand.* (Suppl. 156), **45**, 1–47.

Giacobini, E., 1959b, Quantitative determination of cholinesterase in individual spinal ganglion cells. *Acta Physiol. Scand.* (Suppl. 145) **45**, 238–254.

Giacobini, G. L., 1960, The intracellular localization of cholinesterase. *J. Histochem. Cytochem.* **8**, 419–424.

Giacobini, E. and Kerpel-Fronius, S., 1970, Histochemical and biochemical correlations of monoamine oxidase activity in autonomic and sensory ganglia of the cat. *Acta Physiol. Scand.* **78**, 522–528.

Gibbins, I. and Morris, J., 1987, Co-existence of neuropeptides in sympathetic, cranial autonomic and sensory neurons innervating the iris of the guinea-pig. *J. Autonom. Nerv. System* **21**, 67–82.

Gibbins, I. L., Furness, J. B., Costa, M., MacIntyre, I., Hillyard, C. J. and Girgis, S., 1985, Co-localization of calcitonin gene-related peptide-like immunoreactivity with substance P in cutaneous, vascular and visceral sensory neurons of guinea pigs. *Neurosci. Lett.* **57**, 125–130.

Gibbins, I. L., Wattchow, D. and Coventry, B., 1987a, Immunohistochemically identified populations of calcitonin gene-related peptide-immunoreactive (CGRP-IR) axons in human skin. *Brain Res.* **414**, 143–148.

Gibbins, I. L., Furness, J. B. and Costa, M., 1987b, Pathway-specific patterns of the co-existence of substance P, calcitonin gene-related peptide, cholecystokinin and dynorphin in neurons of the dorsal root ganglia of the guinea-pig. *Cell Tiss. Res.* **248**, 417–437.

Gibson, S. J., Polak, J. M., Bloom, S. R. and Wall, P. D., 1981, The distribution of nine peptides in rat spinal cord with special emphasis on the substantia gelatinosa and on the area around the central canal (lamina X). *J. Comp. Neurol.* **201**, 65–79.

Gibson, S. J., Polak, J. M., Anand, P., Blank, M. A., Morrison, J. F. B., Kelly, J. S. and Bloom, S. R., 1984a, The distribution and origin of VIP in the spinal cord of six mammalian species. *Peptides* **5**, 201–207.

Gibson, S. J., Polak, J. M., Bloom, S. R., Sabate, I. M., Mulderry, P. M., Ghatei, M. A., McGregor, G. P., Morrison, J. F., Kelly, J. S., Evans, R. M. and Rosenfeld, M. G., 1984b, Calcitonin gene-related peptide immunoreactivity in the spinal cord of man and eight other species. *J. Neurosci.* **4**, 3101–3111.

Gibson, S. J., Polak, J. M., Allen, J. M., Adrian, T. E., Kelly, J. S. and Bloom, S. R., 1984c, The distribution and origin of a novel brain peptide, neuropeptide Y, in the spinal cord of several mammals. *J. Comp. Neurol.* **227**, 78–91.

Gibson, S. J., Polak, J. M., Giaid, A. Hamid, Q. A., Kar, S., Jones, P. M., Denny, P., Legon, S., Amara, S. G., Craig, R. K., Bloom, S. R., Penketh, R. J. A., Rodek, C., Ibrahim, N. B. N. and Dawson, A., 1988, Calcitonin gene-related peptide messenger RNA is expressed in sensory neurones of the dorsal root ganglia and also in spinal motoneurones in man and rat. *Neurosci. Lett.* **91**, 283–288.

Giesler, G. J. and Cliffer, K. D., 1985, Postsynaptic dorsal column pathway of the rat. II. Evidence against an important role in nociception. *Brain Res.* **326**, 347–356.

Giesler, G. J. and Elde, R. P., 1985, Immunocytochemical studies of the peptidergic content of fibers and terminals within the lateral spinal and lateral cervical nuclei. *J. Neurosci.* **5**, 1833–1841.

Giesler, G. J., Menétrey, D., Guilbaud, G. and Besson, J. M., 1976, Lumbar cord neurons at the origin of the spinothalamic tract in the rat. *Brain Res.* **118**, 320–324.

Giesler, G. J., Cannon, J. T., Urca, G. and Liebeskind, J. C., 1978, Long ascending projections from substantia gelatinosa Rolandi and the subjacent dorsal horn in the rat. *Science* **202**, 984–986.

Giesler, G. J., Menétrey, D. and Basbaum, A. I., 1979a, Differential origins of spinothalamic tract projections to medial and lateral thalamus in the rat. *J. Comp. Neurol.* **184**, 107–126.

Giesler, G. J., Urca, G., Cannon, J. T. and Liebeskind, J. C., 1979b, Response properties of neurons of the lateral cervical nucleus in the rat. *J. Comp. Neurol.* **186**, 65–78.

Giesler, G. J., Gerhart, K. D., Yezierski, R. P., Wilcox, T. K. and Willis, W. D., 1981a, Postsynaptic inhibition of primate spinothalamic neurons by stimulation in nucleus raphe magnus. *Brain Res.* **204**, 184–188.

Giesler, G. J., Yezierski, R. P., Gerhart, K. D. and Willis, W. D., 1981b, Spinothalamic tract neurons that project to medial and/or lateral thalamic nuclei: evidence for a physiologically novel population of spinal cord neurons. *J. Neurophysiol.* **46**, 1285–1308.

Giesler, G. J., Spiel, H. R. and Willis, W. D., 1981c, Organization of spinothalamic tract axons within the rat spinal cord. *J. Comp. Neurol.* **195**, 243–252.

Giesler, G. J., Nahin, R. L. and Madsen, A. M., 1984, Postsynaptic dorsal column pathway of the rat. I. Anatomical studies. *J. Neurophysiol.* **51**, 260–275.

Giesler, G. J., Miller, L. R., Madsen, A. M. and Katter, J. T., 1987, Evidence for the existence of a lateral cervical nucleus in mice, guinea pigs, and rabbits. *J. Comp. Neurol.* **263**, 106–112.

Giesler, G. J., Björkeland, M., Xu, Q. and Grant, G., 1988, Organization of the spinocervicothalamic pathway in the rat. *J. Comp. Neurol.* **268**, 223–233.

Gildenberg, P. L. and Hirshberg, R. M., 1984, Limited myelotomy for the treatment of intractable cancer pain. *J. Neurol. Neurosurg. Psychiat.* **47**, 94–96.

Gillberg, P. G. and Aquilonius, S. M., 1985, Cholinergic, opioid and glycine receptor binding sites localized in human spinal cord by in vitro autoradiography. *Acta Neurol. Scand.* **72**, 299–306.

Gillberg, P. G. and Wiksten, B., 1986, Effects of spinal cord lesions and rhizotomies on cholinergic and opiate receptor binding sites in rat spinal cord. *Acta Physiol. Scand.* **126**, 575–582.

Gillberg, P. G., Nordberg, A. and Aquilonius, S. M., 1984, Muscarinic binding sites in small homogenates and in autoradiographic sections from rat and human spinal cord. *Brain Res.* **300**, 327–333.

Gillberg, P. G., d'Argy, R. and Aquilonius, S. M., 1988, Autoradiographic distribution of [^{3}H]acetylcholine binding sites in the cervical spinal cord of man and some other species. *Neurosci. Lett.* **90**, 197–202.

Gilman, S. and Denny-Brown, D., 1966, Disorders of movement and behaviour following dorsal column lesions. *Brain* **89**, 397–418.

Girardot, M. N., Brennan, T. J., Ammons, W. S. and Foreman, R. D., 1987, Effects of stimulating the subcoeruleus-parabrachial region on the non-noxious and noxious responses of T2-T4 spinothalamic tract neurons in the primate. *Brain Res.* **409**, 19–30.

Glazer, E. J. and Basbaum, A. I., 1981, Immunohistochemical localization of leucine-enkephalin in the spinal cord of the cat: enkephalin-containing marginal neurons and pain modulation. *J. Comp. Neurol.* **196**, 377–389.

Glazer, E. J. and Basbaum, A. I., 1982, Opioid neurons and pain modulation: an ultrastructural analysis of enkephalin in cat superficial dorsal horn. *Neuroscience* **10**, 357–376.

Glazer, E. J. and Basbaum, A. I., 1984, Axons which take up [^{3}H]serotonin are presynaptic to enkephalin immunoreactive neurons in cat dorsal horn. *Brain Res.* **298**, 386–391.

Glees, P. and Bailey, R. A., 1951, Schichtung und Fasergroesse des Tractus spino-thalamicus des Menschen. *Msch. Psychiat. Neurol.* **122**, 129–141.

Glees, P. and Soler, J., 1951, Fibre content of the posterior column and synaptic connections of nucleus gracilis. *Z. Zellforsch.* **36**, 381–400.

Gmelin, G. and Cerletti, A., 1976, Electrophoretic studies on presynaptic inhibition in the mammalian spinal cord. *Experientia* **32**, 756.

Go, V. L. W. and Yaksh, T. L., 1987, Release of substance P from the cat spinal cord. *J. Physiol.* **391**, 141–167.

Gobel, S., 1974, Synaptic organization of the substantia gelatinosa glomeruli in the spinal trigeminal nucleus of the adult cat. *J. Neurocytol.* **3**, 219–243.

Gobel, S., 1975, Golgi studies of the substantia gelatinosa neurons in the spinal trigeminal nucleus of the adult cat. *Brain Res.* **83**, 333–338.

Gobel, S., 1976, Dendroaxonic synapses in the substantia gelatinosa of the spinal trigeminal nucleus of the cat. *J. Comp. Neurol.* **167**, 165–176.

Gobel, S., 1978a, Golgi studies of the neurons in layer I of the dorsal horn of the medulla (trigeminal nucleus caudalis). *J. Comp. Neurol.* **180**, 375–394.

Gobel, S., 1978b, Golgi studies of the neurons in layer II of the dorsal horn of the medulla (trigeminal nucleus caudalis). *J. Comp. Neurol.* **180**, 395–414.

Gobel, S., 1979, Neural circuitry in the substantia gelatinosa of Rolando: anatomical insights. *Adv. Pain Res. Ther.* **3**, 175–195.

Gobel, S., 1984, An electron microscopic analysis of the transynaptic effects of peripheral nerve injury subsequent to tooth pulp extirpations on neurons in laminae I and II of the medullary dorsal horn. *J. Neurosci.* **4**, 2281–2290.

Gobel, S. and Binck, J. M., 1977, Degenerative changes in primary trigeminal axons and in neurons in nucleus caudalis following tooth pulp extirpations in the cat. *Brain Res.* **132**, 347–354.

Gobel, S. and Falls, W. M., 1979, Anatomical observations of horseradish peroxidase-filled terminal primary axonal arborizations in layer II of the substantia gelatinosa of Rolando. *Brain Res.* **175**, 335–340.

Gobel, S., Falls, W. M. and Hockfield, S., 1977, The division of the dorsal and ventral horns of the mammalian caudal medulla into eight layers using anatomical criteria. In: *Pain in the Trigeminal Region*, pp. 443-453. D. J. Anderson and B. Matthews (eds.). Elsevier, Amsterdam.

Gobel, S., Falls, W. M., Bennett, G. J., Hayashi, H. and Humphrey, E., 1980, An EM analysis of the synaptic connections of horseradish peroxidase-filled stalked cells and islet cells in the substantia gelatinosa of adult cat spinal cord. *J. Comp. Neurol.* **194**, 781–807.

Gobel, S., Falls, W. M. and Humphrey, E., 1981, Morphology and synaptic connections of ultrafine primary axons in lamina I of the spinal dorsal horn: candidates for the terminal axonal arbors of primary neurons with unmyelinated (C) axons. *J. Neurosci.* **1**, 1163–1179.

Gokin, A. P., Kostyuk, P. G. and Preobrazhensky, N. N., 1977, Neuronal mechanisms of interactions of high-threshold visceral and somatic afferent influences in spinal cord and medulla. *J. Physiol.* Paris **73**, 319–333.

Goldberger, M. and Murray, M., 1974, Restitution of function and collateral sprouting in cat spinal cord: The deafferented animal. *J. Comp. Neurol.* **158**, 37–54.
Goldberger, M. and Murray, M., 1982, Lack of sprouting and its presence after lesions of the cat spinal cord. *Brain Res.* **241**, 227–239.
Goldby, F. and Robinson, L. R., 1962, The central connexions of dorsal spinal nerve roots and the ascending tracts in the spinal cord of *Lacerta viridis*. *J. Anat.* **96**, 153–170.
Goldfinger, M. D., 1985, III. Detection by HPLC-EC of primary amines recovered in aqueous push-pull perfusates from cat cuneate nucleus. *Life Sci.* **37**, 1765–1774.
Goldfinger, M. D., Simpson, C. W. and Resch, G. E., 1984, Recovery by push-pull perfusion of neurochemicals released within the cuneate nucleus of the cat by somatosensory stimulation. *Pharm. Biochem. Behav.* **21**, 117–123.
Goldscheider, A., 1884, The specific energy of the sensory nerves of the skin. Transl. from German by Biederman-Thorson in Handwerker, H. O. and Brune, K. (eds.), 1987, pp. 47–67, *Classical German Contributions to Pain Research*, Vth World Congress on Pain, Hamburg. Original German: 3. Die Specifische Energie der Gefuehlsnerven der Haut. Monatschefte fuer practische Dermatologie. III Band. Nr. 9 und 10.
Goldstein, K., 1910, Ueber die aufsteigende Degeneration nach Querschnittsunterbrechung des Rueckenmarks (Tractus spino-cerebellis posterior, Tractus spino-olivaris, Tractus spino-thalamicus. *Neurol. Centr.* **29**, 898–911.
Golovchinsky, V., 1980, Patterns of responses of neurons in cuneate nucleus to controlled mechanical stimulation of cutaneous velocity receptors in the cat. *J. Neurophysiol.* **43**, 1673–1699.
Gonzalez-Lima, F., 1986, Activation of substantia gelatinosa by midbrain reticular stimulation demonstrated with 2-deoxyglucose in the rat spinal cord. *Neurosci. Lett.* **65**, 326–330.
Goode, G. E., Humbertson, A. O. and Martin, G. F., 1980, Projections from the brain stem reticular formation to laminae I and II of the spinal cord. Studies using light and electron microscopic techniques in the North American opossum. *Brain Res.* **189**, 327–342.
Goodman, R. R. and Snyder, S. H., 1982, Autoradiographic localization of adenosine receptors in rat brain using [^{3}H]cyclohexyladenosine. *J. Neurosci.* **2**, 1230–1241.
Goodwin, G. M., McCloskey, D. I. and Matthews, P. B. C., 1972a, The persistence of appreciable kinesthesia after paralysing joint afferents but preserving muscle afferents. *Brain Res.* **37**, 326–329.
Goodwin, G. M., McCloskey, D. I. and Matthews, P. B. C., 1972b, The contribution of muscle afferents to kinaesthesia shown by vibration induced illusions of movement and by the effects of paralysing joint afferents. *Brain* **95**, 705–748.
Gordon, G., 1973, The concept of relay nuclei. In: *Handbook of Sensory Physiology*, Vol. II, *Somatosensory System*, pp. 137–150. A. Iggo (ed.). Springer-Verlag, New York.
Gordon, G. and Grant, G., 1982, Dorsolateral spinal afferents to some medullary sensory nuclei: an anatomical study in the cat. *Exp. Brain Res.* **46**, 12–23.
Gordon, G. and Horrobin, D., 1967, Antidromic and synaptic responses in the cat's gracile nucleus to cerebellar stimulation. *Brain Res.* **5**, 419–421.
Gordon, G. and Jukes, M. G. M., 1964a, Dual organization of the exteroceptive components of the cat's gracile nucleus. *J. Physiol.* **173**, 263–290.
Gordon, G. and Jukes, M. G. M., 1964b, Descending influences on the exteroceptive organization of the cat's gracile nucleus. *J. Physiol.* **173**, 291–319.
Gordon, G. and Miller, R., 1969, Identification of cortical cells projecting to the dorsal column nuclei of the cat. *Quart. J. Exp. Physiol.* **54**, 85–98.
Gordon, G. and Paine, C. H., 1960, Functional organization in nucleus gracilis of the cat. *J. Physiol.* **153**, 331–349.
Gordon, G. and Seed, W. A., 1961, An investigation of nucleus gracilis of the cat by antidromic stimulation. *J. Physiol.* **155**, 589–601.
Gottschaldt, K. M., Iggo, A. and Young, D. W., 1973, Functional characteristics of mechanoreceptors in sinus hair follicles of the cat. *J. Physiol.* **235**, 287–315.
Gouarderes, C. and Cros, J., 1984, Opiate binding sites in different levels of rat spinal cord. *Neuropeptides* **5**, 113–116.
Gouarderes, C., Cros, J. and Quirion, R., 1985, Autoradiographic localization of mu, delta and kappa opioid receptor binding sites in rat and guinea-pig spinal cord. *Neuropeptides* **6**, 331–342.
Gouarderes, C., Kopp, N., Cros, J. and Quirion, R., 1986, Kappa opioid receptors in human lumbo-sacral spinal cord. *Brain Res. Bull.* **16**, 355–361.
Gowers, W. R., 1878, A case of unilateral gunshot injury to the spinal cord. *Trans. Clin. London* **11**, 24–32.
Graham, L. T. and Aprison, M. H., 1969, Distribution of some enzymes associated with the metabolism of glutamate, aspartate, γ-aminobutyric acid and glutamine in cat spinal cord. *J. Neurochem.* **16**, 559–566.
Graham, L. T., Shank, R. P., Werman, R. and Aprison, M. H., 1967, Distribution of some synaptic transmitter suspects in cat spinal cord: glutamic acid, aspartic acid, γ-aminobutyric acid, glycine, and glutamine. *J. Neurochem.* **14**, 465–472.
Grant, F. C., 1930, Value of cordotomy for the relief of pain. *Ann. Surg.* **92**, 998–1006.
Grant, F. C., 1932, Results with cordotomy for relief of intractable pain due to carcinoma of the pelvic organs. *Amer. J. Obst. Gynec.* **24**, 620–625.

Grant, G. and Westman, J., 1969, The lateral cervical nucleus in the cat. IV. A light and electron microscopical study after midbrain lesions with demonstration of indirect Wallerian degeneration at the ultra-structural level. *Exp. Brain Res.* **7**, 51–67.
Grant, G. and Ygge, J., 1981, Somatotopic organization of the thoracic spinal nerve in the dorsal horn demonstrated with transganglionic degeneration. *J. Comp. Neurol.* **202**, 357–364.
Grant, G., Boivie, J. and Brodal, A., 1968, The question of a cerebellar projection from the lateral cervical nucleus re-examined. *Brain Res.* **9**, 95–102.
Grant, G., Boivie, J. and Silfvenius, H., 1973, Course and termination of fibres from the nucleus z of the medulla oblongata. An experimental light microscopical study in the cat. *Brain Res.* **55**, 55–70.
Grant, G., Arvidsson, J., Robertson, B. and Ygge, J., 1979, Transganglionic transport of horseradish peroxidase in primary sensory neurons. *Neurosci. Lett.* **12**, 23–28.
Granum, S. L., 1986, The spinothalamic system of the rat. I. Locations of cells of origin. *J. Comp. Neurol.* **247**, 159–180.
Gray, B. G. and Dostrovsky, J. O., 1983a, Descending inhibitory influences from periaqueductal gray, nucleus raphe magnus, and adjacent reticular formation. I. Effects on lumbar spinal cord nociceptive and nonnociceptive neurons. *J. Neurophysiol.* **49**, 932–947.
Gray, B. G. and Dostrovsky, J. O., 1983b, Modulation of the sensory responses of cat trigeminal and cuneate neurons by electrical stimulation of the red nucleus. *Neurosci. Abstr.* **9**, 247.
Gray, B. G. and Dostrovsky, J. O., 1985, Inhibition of feline spinal cord dorsal horn neurons following electrical stimulation of nucleus paragigantocellularis lateralis. A comparison with nucleus raphe magnus. *Brain Res.* **348**, 261–273.
Gray, E. G., 1963, Electron microscopy of presynaptic organelles of the spinal cord. *J. Anat.* **97**, 101–106.
Gray, T. S., Hazlett, J. C. and Martin, G. F., 1981, Organization of projection from gracile, medial cuneate and lateral cuneate nuclei in the North American opossum. Horseradish peroxidase study of the cells projecting to the cerebellum, thalamus and spinal cord. *Brain Behav. Evol.* **18**, 140–156.
Green, T. and Dockray, G., 1987, Calcitonin gene-related peptide and substance P in afferents to the upper gastrointestinal tract in the rat. *Neurosci. Lett.* **76**, 151–156.
Greenamyre, T., Young, A. B. and Penney, J. B., 1984, Quantitative autoradiographic distribution of L-[^{3}H]glutamate-binding sites in rat central nervous system. *J. Neurol. Sci.* **4**, 2133–2144.
Gregor, M. and Zimmermann, M., 1972, Characteristics of spinal neurones responding to cutaneous myelinated and unmyelinated fibres. *J. Physiol.* **221**, 555–576.
Gregor, M. and Zimmermann, M., 1973, Dorsal root potentials produced by afferent volleys in cutaneous group III fibres. *J. Physiol.* **232**, 413–425.
Greves, P. L., Nyberg, F., Terenius, L. and Hökfelt, T., 1985, Calcitonin gene-related peptide is a potent inhibitor of substance P degradation. *Eur. J. Pharmacol.* **115**, 309–311.
Griersmith, B. T. and Duggan, A. W., 1980, Prolonged depression of spinal transmission of nociceptive information by 5-HT administered in the substantia gelatinosa: antagonism by methysergide. *Brain Res.* **187**, 231–236.
Griffith, J. L. and Gatipon, G. B., 1981, A comparative study of selective stimulation of raphe nuclei in the cat in inhibiting dorsal horn neuron responses to noxious stimulation. *Brain Res.* **229**, 520–524.
Grigg, P., 1975, Mechanical factors influencing response of joint afferent neurons from cat knee. *J. Neurophysiol.* **38**, 1473–1484.
Grigg, P., 1976, Response of joint afferent neurons in cat medial articular nerve to active and passive movements of the knee. *Brain Res.* **118**, 482–485.
Grigg, P. and Greenspan, B. J., 1977, Response of primate joint afferent neurons to mechanical stimulation of knee joint. *J. Neurophysiol.* **40**, 1–8.
Grigg, P. and Hoffman, A. H., 1982, Properties of Ruffini afferents revealed by stress analysis of isolated sections of cat knee capsule. *J. Neurophysiol.* **47**, 41–54.
Grigg, P. and Hoffman, A. H., 1984, Ruffini mechanoreceptors in isolated joint capsule: responses correlated with strain energy density. *Somatosensory Res.* **2**, 149–162.
Grigg, P., Finerman, G. A. and Riley, L. H., 1973, Joint-position sense after total hip replacement. *J. Bone Joint Surg.* **55A**, 1016–1025.
Grigg, P., Hoffman, A. H. and Fogarty, K. E., 1982, Properties of Golgi-Mazzoni afferents in cat knee joint capsule, as revealed by mechanical studies of isolated joint capsule. *J. Neurophysiol.* **47**, 31–40.
Grigg, P., Schaible, H. G. and Schmidt, R. F., 1986, Mechanical sensitivity of group III and IV afferents from posterior articular nerve in normal and inflamed cat knee. *J. Neurophysiol.* **55**, 635–643.
Griph, S. and Westman, J., 1977, Volume composition of the lateral cervical nucleus in the cat. I. A stereological and electron microscopical study of normal and deafferentated animals. *J. Neurocytol.* **6**, 723–743.
Groenewegen, H. J., Boesten, A. J. P. and Voogd, J., 1975, The dorsal column nuclear projections to the nucleus ventralis posterior lateralis thalami and the inferior olive in the cat. An autoradiographic study. *J. Comp. Neurol.* **162**, 505–518.
Guilbaud, G., Oliveras, J. L., Giesler, G. and Besson, J. M., 1977a, Effects induced by stimulation of the centralis inferior nucleus of the raphe on dorsal horn interneurons in cat's spinal cord. *Brain Res.* **126**, 355–360.
Guilbaud, G., Benelli, G. and Besson, J. M., 1977b, Responses of thoracic dorsal horn interneurons to cutaneous

stimulation and to the administration of algogenic substances into the mesenteric artery in the spinal cat. *Brain Res.* **124**, 437–448.

Guilford, J. P. and Lovewell, E. M., 1936, The touch spots and the intensity of the stimulus. *J. Gen. Psychol.* **15**, 149–159.

Gulbenkian, S., Merighi, A., Wharton, J., Varndell, I. M. and Polak, J. M., 1986, Ultrastructural evidence for the coexistence of calcitonin gene-related peptide and substance P in secretory vesicles of peripheral nerves in the guinea pig. *J. Neurocytol.* **15**, 535–542.

Gulley, R. L., 1973, Golgi studies of the nucleus gracilis in the rat. *Anat. Rec.* **177**, 325–342.

Gundlach, A. L., Largent, B. L. and Snyder, S. H., 1986, Autoradiographic localization of sigma receptor binding sites in guinea pig and rat central nervous system with (+)^{3}H-3-(3-hydroxyphenyl)-N-(1-porpyl)piperdine. *J. Neurosci.* **6**, 1757–1770.

Guzman-Flores, C., Buendia, N., Anderson, C. and Lindsley, D. B., 1962, Cortical and reticular influences upon evoked responses in dorsal column nuclei. *Exp. Neurol.* **5**, 37–46.

Gwyn, D. G. and Flumerfelt, B. A., 1971, Acetylcholinesterase in non-cholinergic neurons—a histochemical study of dorsal root ganglion cells in the rat. *Brain Res.* **34**, 193–198.

Gwyn, D. G. and Waldron, H. A., 1968, A nucleus in the dorsolateral funiculus of the spinal cord of the rat. *Brain Res.* **10**, 342–351.

Gwyn, D. G. and Waldron, H. A., 1969, Observations on the morphology of a nucleus in the dorsolateral funiculus of the spinal cord of the guinea-pig, rabbit, ferret and cat. *J. Comp. Neurol.* **136**, 233–236.

Gybels, J. and Van Hees, J., 1971, Unit activity from mechanoreceptors in human peripheral nerve during intensity discrimination of touch. *Excerpta Med. Int. Congress Series No.* 253, 198-206.

Gybels, J., Handwerker, H. O. and Van Hees, J., 1979, A comparison between the discharges of human nociceptive nerve fibres and the subject's ratings of his sensations. *J. Physiol.* **292**, 193–206.

Ha, H., 1971, Cervicothalamic tract in the Rhesus monkey. *Exp. Neurol.* **33**, 205–212.

Ha, H. and Liu, C. N., 1963, Synaptology of spinal afferents in the lateral cervical nucleus of the cat. *Exp. Neurol.* **8**, 318–327.

Ha, H. and Liu, C. N., 1966, Organization of the spino-cervico-thalamic system. *J. Comp. Neurol.* **127**, 445–470.

Ha, H. and Morin, F., 1964, Comparative anatomical observations of the cervical nucleus, N. cervicalis lateralis, of some primates. *Anat. Rec.* **148**, 374–375.

Ha, H., Kitai, S. T. and Morin, F., 1965, The lateral cervical nucleus of the raccoon. *Exp. Neurol.* **11**, 441–450.

Haapanen, L., Kolmodin, G. M. and Skoglund, C. R., 1958, Membrane and action potentials of spinal interneurons in the cat. *Acta Physiol. Scand.* **43**, 315–348.

Haber, L. H., Martin, R. F., Chatt, A. B. and Willis, W. D., 1978, Effects of stimulation in nucleus reticularis gigantocellularis on the activity of spinothalamic tract neurons in the monkey. *Brain Res.* **153**, 163–168.

Haber, L. H., Martin, R. F., Chung, J. M. and Willis, W. D., 1980, Inhibition and excitation of primate spinothalamic tract neurons by stimulation in region of nucleus reticularis gigantocellularis. *J. Neurophysiol.* **43**, 1578–1593.

Haber, L. H., Moore, B. D. and Willis, W. D., 1982, Electrophysiological response properties of spinoreticular neurons in the monkey. *J. Comp. Neurol.* **207**, 75–84.

Häbler, H. J., Jänig, W., Koltzenburg, M. and McMahon, S. B., 1990, A quantitative study of the central projection patterns of unmyelinated ventral root afferents in the cat. *J. Physiol.* **422**, 265–287.

Hackman, J. C., Auslander, D., Grayson, V. and Davidoff, R. A., 1982, GABA 'desensitization' of frog primary afferent fibers. *Brain Res.* **253**, 143–152.

Hadjiconstantinou, M., Panula, P., Lackovic, Z. and Neff, N. H., 1984, Spinal cord serotonin: a biochemical and immunohistochemical study following transection. *Brain Res.* **322**, 245–254.

Hagbarth, K. E., 1960, Centrifugal mechanisms of sensory control. *Ergenb. Biol.* **22**, 47–66.

Hagbarth, K. E. and Kerr, D. I. B., 1954, Central influences on spinal afferent conduction. *J. Neurophysiol.* **17**, 295–307.

Hagbarth, K. E. and Vallbo, A. B., 1967, Mechanoreceptor activity recorded percutaneously with semi-microelectrodes in human peripheral nerves. *Acta Physiol. Scand.* **69**, 121–122.

Hagbarth, K. E. and Vallbo, A. B., 1968, Discharge characteristics of human muscle afferents during muscle stretch and contraction. *Exp. Neurol.* **22**, 674–694.

Hagbarth, K. E. and Vallbo, A. B., 1969, Single unit recordings from muscle nerves in human subjects. *Acta Physiol. Scand.* **76**, 321–334.

Hagbarth, K. E., Hongell, A., Hallin, R. G. and Torebjörk, H. E., 1970, Afferent impulses in median nerve fascicles evoked by tactile stimuli of the human hand. *Brain Res.* **24**, 423–442.

Hagbarth, K. E., Wallin, G. and Lofstedt, L., 1975, Muscle spindle activity in man during voluntary fast alternating movements. *J. Neurol. Neurosurg. Psychiat.* **38**, 625–635.

Hagg, S. and Ha, H., 1970, Cervicothalamic tract in the dog. *J. Comp. Neurol.* **139**, 357–374.

Hagihara, S., Senba, E., Yoshida, S., Tohyama, M. and Yoshiya, I., 1990, Fine structure of noradrenergic terminals and their synapses in the rat spinal dorsal horn: an immunohistochemical study. *Brain Res.* **526**, 73–80.

Hahn, J. F., 1971, Thermal-mechanical stimulus interactions in low-threshold C-fiber mechanoreceptors of cat. *Exp. Neurol.* **33**, 607–617.

Hahn, J. F. and Wall, J. T., 1975, Frequency responses of down- and guard-hair receptors in the cat. *Amer. J. Physiol.* **229**, 23–27.

Halata, Z., 1975, The mechanoreceptors of the mammalian skin. Ultrastructure and morphological classification. *Adv. Anat. Embryol. Cell Biol.* **50**, 4–77.

Halata, Z., 1977, The ultrastructure of sensory nerve endings in the articular capsule of the knee joint of the domestic cat (Ruffini corpuscles and Pacinian corpuscles). *J. Anat.* **124**, 717–729.

Halata, Z., Badalamente, M. A., Dee, R. and Propper, M., 1984, Ultrastructure of sensory nerve endings in monkey (*Macaca fascicularis*) knee joint capsule. *J. Orthopedic Res.* **2**, 169–176.

Halata, Z., Rettig, T. and Schulze, W., 1985, The ultrastructure of sensory nerve endings in the human knee joint capsule. *Anat. Embryol.* **172**, 265–275.

Hall, J. G., Duggan, A. W., Johnson, S. M. and Morton, C. R., 1981, Medullary raphe lesions do not reduce descending inhibition of dorsal horn neurones of the cat. *Neurosci. Lett.* **25**, 25–29.

Hall, J. G., Duggan, A. W., Morton, C. R. and Johnson, S. M., 1982, The location of brainstem neurones tonically inhibiting dorsal horn neurones of the cat. *Brain Res.* **244**, 215–222.

Halliday, G. M., Li, Y. W., Oliver, J. R., Joh, T. H., Cotton, R. G. H., Howe, P. R. C., Geffen, L. B. and Blessing, W. W., 1988, The distribution of neuropeptide Y-like immunoreactive neurons in the human medulla oblongata. *Neuroscience* **26**, 179–191.

Hallin, R. G. and Torebjörk, H. E., 1976, Studies on cutaneous A and C fibre afferents, skin nerve blocks and perception. In: *Sensory Functions of the Skin in Primates, with Special Reference to Man*, pp. 137–148. Y. Zotterman (ed.). Pergamon Press, New York.

Hallin, R. G., Torebjörk, H. E. and Wiesenfeld, Z., 1981, Nociceptors and warm receptors innervated by C fibres in human skin. *J. Neurol. Neurosurg. Psychiat.* **44**, 313–319.

Hämäläinen, H. and Järvilehto, T., 1981, Peripheral neural basis of tactile sensations in man: I. Effect of frequency and probe area on sensations elicited by single mechanical pulses on hairy and glabrous skin of the hand. *Brain Res.* **219**, 1–12.

Hamann, W. C., Hong, S. K., Kniffki, K. D. and Schmidt, R. F., 1978, Projections of primary afferent fibres from muscle to neurones of the spinocervical tract of the cat. *J. Physiol.* **283**, 369–378.

Hamano, K., Mannen, H. and Ishizuka, N., 1978, Reconstruction of trajectory of primary afferent collaterals in the dorsal horn of the cat spinal cord, using Golgi-stained serial sections. *J. Comp. Neurol.* **181**, 1–16.

Hamburger, V., Brunso-Bechtold, J. K. and Yip, I. W., 1981, Neuronal death in the spinal ganglia of the chick embryo and its reduction by nerve growth factor. *J. Neurosci.* **1**, 60–71.

Hamilton, T. C. and Johnson, J. I., 1973, Somatotopic organization related to nuclear morphology in the cuneate-gracile complex of opossums *Didelphis marsupialis virginiana*. *Brain Res.* **51**, 125–140.

Hammerstad, J. P., Murray, J. E. and Cutler, R. W. P., 1971, Efflux of amino acid neurotransmitters from rat spinal cord slices. II. Factors influencing the electrically induced efflux of [14C]glycine and 3H-GABA. *Brain Res.* **35**, 357–367.

Hammond, D. L., Tyce, G. M. and Yaksh, T. L., 1985, Efflux of 5-hydroxytryptamine and noradrenaline into spinal cord superfusates during stimulation of the rat medulla. *J. Physiol.* **359**, 151–162.

Hamon, M., Gallissot, M. C., Menard, F., Gozlan, F., Bourgoin, S. and Verge, D., 1989, 5-HT3 receptor binding sites are on capsaicin-sensitive fibres in the rat spinal cord. *Eur. J. Pharmacol.* **164**, 315–322.

Hancock, M. B., 1976, Cells of origin of hypothalamo-spinal projections in the rat. *Neurosci. Lett.* **3**, 179–184.

Hancock, M. B., 1982, A serotonin-immunoreactive fiber system in the dorsal columns of the spinal cord. *Neurosci. Lett.* **31**, 247–252.

Hancock, M. B. and Fougerousse, C. L., 1976, Spinal projections from the nucleus locus coeruleus and nucleus subcoeruleus in the cat and monkey as demonstrated by the retrograde transport of horseradish peroxidase. *Brain Res. Bull.* **1**, 229–234.

Hancock, M. B., Rigamonti, D. D. and Bryan, R. N., 1973, Convergence in the lumbar spinal cord of pathways activated by splanchnic nerve and hind limb cutaneous nerve stimulation. *Exp. Neurol.* **38**, 337–348.

Hancock, M. B., Foreman, R. D. and Willis, W. D., 1975, Convergence of visceral and cutaneous input onto spinothalamic tract cells in the thoracic spinal cord of the cat. *Exp. Neurol.* **47**, 240–248.

Hand, P. J., 1966, Lumbosacral dorsal root terminations in the nucleus gracilis of the cat. Some observations on terminal degeneration in other medullary sensory nuclei. *J. Comp. Neurol.* **126**, 137–156.

Hand, P. and Liu, C. N., 1966, Efferent projections of the nucleus gracilis. *Anat. Rec.* **154**, 353–354.

Hand, P. J. and van Winkle, T., 1977, The efferent connections of the feline nucleus cuneatus. *J. Comp. Neurol.* **171**, 83–110.

Handwerker, H. O. and Neher, K. D., 1976, Characteristics of C-fibre receptors in the cat's foot responding to stepwise increase of skin temperature to noxious levels. *Pfluegers Arch.* **365**, 221–229.

Handwerker, H. O., Iggo, A. and Zimmermann, M., 1975, Segmental and supraspinal actions on dorsal horn neurons responding to noxious and non-noxious skin stimuli. *Pain* **1**, 147–165.

Handwerker, H. O., Anton, F. and Reeh, P. W., 1987, Discharge pattern of afferent cutaneous nerve fibers from the rat's tail during prolonged noxious mechanical stimulation. *Exp. Brain Res.* **65**, 493–504.

Hanker, J. S. and Peach, R., 1976, Histochemical and ultrastructural studies of primary sensory neurons in mice with dystonia musculorum. I. Acetylcholinesterase and lysosomal hydrolases. *Neuropath. Appl. Neurobiol.* **2**, 79–97.

Hanker, J. S., Kusyk, C. J., Bloom, F. E. and Pearse, A. G. E., 1973, The demonstration of dehydrogenases and monoamine oxidase by the formation of osmium blacks at the sites of hatchett's brown. *Histochemie* **33**, 205–230.

Hanko, J., Hardebo, J. E., Kahrstrom, J., Owman, C. and Sundler, F., 1986, Existence and cerebrovascular nerve and trigeminal ganglion cells. *Acta Physiol. Scand.* **552**, 29–33.

Hard, W. L. and Peterson, A. C., 1950, The distribution of choline esterase in nerve tissue of the dog. *Anat. Rec.* **108**, 57–65.

Hardy, J. D. and Oppel, T. W., 1938, Studies in temperature sensation. IV. The stimulation of cold sensation by radiation. *J. Clin. Invest.* **17**, 771–778.

Hardy, J. D., Goodell, H. and Wolff, H. G., 1951, The influence of skin temperature upon the pain threshold as evoked by thermal radiation. *Science* **114**, 149–150.

Hardy, J. D., Wolff, H. G. and Goodell, H., 1952, *Pain Sensations and Reactions*. Williams & Wilkins, New York. Reprinted by Hafner, New York, 1967.

Haring, J. H., Rowinski, M. J. and Pubols, B. H., 1984, Electrophysiology of raccoon cuneocerebellar neurons. *Somatosensory Res.* **1**, 247–264.

Harkness, D. H. and Brownfield, M. S., 1986, A thyrotropin-releasing hormone-containing system in the rat dorsal horn separate from serotonin. *Brain Res.* **384**, 323–333.

Harmann, P. A., Carlton, S. M. and Willis, W. D., 1988a, Collaterals of spinothalamic tract cells to the periaqueductal gray: a fluorescent double-labeling study in the rat. *Brain Res.* **441**, 87–97.

Harmann, P. A., Chung, K., Briner, R. P., Westlund, K. N. and Carlton, S. M., 1988b, Calcitonin gene-related peptide (CGRP) in the human spinal cord: A light and electron microscopic analysis. *J. Comp. Neurol.* **269**, 371–380.

Harper, A. A. and Lawson, S. N., 1985a, Conduction velocity is related to morphological cell type in rat dorsal root ganglion neurones. *J. Physiol.* **359**, 31–46.

Harper, A. A. and Lawson, S. N., 1985b, Electrical properties of rat dorsal root ganglion neurones with different peripheral nerve conduction velocities. *J. Physiol.* **359**, 47–63.

Harrington, T. and Merzenich, M. M., 1970, Neural coding in the sense of touch: human sensations of skin indentation compared with the responses of slowly adapting mechanoreceptive afferents innervating the hairy skin of monkeys. *Exp. Brain Res.* **10**, 251–264.

Harrison, P. J. and Jankowska, E., 1984, An intracellular study of descending and non-cutaneous afferent input to spinocervical tract neurones in the cat. *J. Physiol.* **356**, 245–261.

Harrison, P. J. and Jankowska, E., 1985a, Sources of input to interneurones mediating group I non-reciprocal inhibition of motoneurones in the cat. *J. Physiol.* **361**, 379–401.

Harrison, P. J. and Jankowska, E., 1985b, Organization of input to the interneurones mediating group I non-reciprocal inhibition of motoneurones in the cat. *J. Physiol.* **361**, 403–418.

Hartline, P. H., 1974, Thermoreception in snakes. In: *Handbook of Sensory Physiology*, Vol. III/3, *Electroreceptors and Other Specialized Receptors in Lower Vertebrates*, pp. 297-312. A. Fessard (ed.). Springer, New York.

Hayes, N. L. and Rustioni, A., 1979, Dual projections of single neurons are visualized simultaneously: use of enzymatically inactive [^{3}H]HRP. *Brain Res.* **165**, 321–326.

Hayes, N. L. and Rustioni, A., 1980, Spinothalamic and spinomedullary neurons in macaques: a single and double retrograde tracer study. *Neuroscience* **5**, 861–874.

Hayes, R. L., Price, D. D., Ruda, M. A. and Dubner, R., 1979, Suppression of nociceptive responses in the primate by electrical stimulation of the brain or morphine administration: behavioral and electrophysiological comparisons. *Brain Res.* **167**, 417–421.

Hayle, T. H., 1973, A comparative study of the spinal projections to the brain (except cerebellum) in three classes of poikilothermic vertebrates. *J. Comp. Neurol.* **149**, 463–476.

Hazlett, J. C., Dom, R. and Martin, G. F., 1972, Spino-bulbar, spino-thalamic and medial lemniscal connections in the American opossum, *Didelphis marsupialis virginiana*. *J. Comp. Neurol.* **146**, 95–118.

He, X., Schepelmann, K., Schaible, H. G. and Schmidt, R. F., 1990, Capsaicin inhibits responses of fine afferents from the knee joint of the cat to mechanical and chemical stimuli. *Brain Res.* **530**, 147–150.

Head, H., 1893, On disturbances of sensation with especial reference to the pain of visceral disease. *Brain* **16**, 1-132.

Head, H., 1920, *Studies in Neurology*, Vol. I, pp. 55-65 and 225-329. Oxford Univ. Press, London.

Head, H. and Thompson, T., 1906, The grouping of afferent impulses within the spinal cord. *Brain* **29**, 537–741.

Headley, P. M., Duggan, A. W. and Griersmith, B. T., 1978, Selective reduction by noradrenaline and 5-hydroxytryptamine of nociceptive responses of cat dorsal horn neurones. *Brain Res.* **145**, 185–189.

Headley, P. M., Parsons, C. G. and West, D. C., 1987, The role of N-methylaspartate receptors in mediating responses of rat and cat spinal neurones to defined sensory stimuli. *J. Physiol.* **385**, 169–188.

Heath, D. D., Coggeshall, R. E. and Hulsebosch, C. E., 1986, Axon and neuron numbers after forelimb amputation in neonatal rats. *Exp. Neurol.* **92**, 220–233.

Heath, J. P., 1978, *The Cutaneous Input to the Spinocervical Tract and the Corticofugal Modulation of Transmission from the Forelimb Component*. Ph.D. Thesis, University of Edinburgh. Quoted in Brown et al., 1980a.

Heavner, J. E. and DeJong, R. H., 1973, Spinal cord neuron response to natural stimuli. A microelectrode study. *Exp. Neurol.* **39**, 293–306.

Heckmann, T. and Bourassa, C. M., 1981, Lesions of dorsal column nuclei or medial lemniscus of the cat: effect on motor performance. *Brain Res.* **224**, 405–411.

Heii, A. and Emson, P. C., 1986, Regional distribution of neuropeptide K and other tachykinins (Neurokinin A, Neurokinin B and Substance P) in rat central nervous system. *Brain Res.* **399**, 240–249.

Heimer, L. and Wall, P. D., 1968, The dorsal root distribution to the substantia gelatinosa of the rat with a note on the distribution in the cat. *Exp. Brain Res.* **6**, 89–99.

Heinbecker, P., Bishop, G. H. and O'Leary, J. L., 1933, Pain and touch fibers in peripheral nerves. *Arch. Neurol. Psychiat.* **29**, 771–789.

Heinbecker, P., Bishop, G. H. and O'Leary, J. L., 1934, Analysis of sensation in terms of the nerve impulse. *Arch. Neurol. Psychiat.* **31**, 34–53.

Helke, C. J., Charlton, C. G. and Wiley, R. G., 1986, Studies on the cellular localization of spinal cord substance P receptors. *Neuroscience* **19**, 523–533.

Hellon, R. F. and Misra, N. K., 1973a, Neurones in the dorsal horn responding to scrotal skin temperature changes. *J. Physiol.* **232**, 375–388.

Hellon, R. F. and Misra, N. K., 1973b, Neurones in the ventrobasal complex of the rat thalamus responding to scrotal skin temperature changes. *J. Physiol.* **232**, 389–399.

Hellon, R. F. and Mitchell, D., 1975, Convergence in a thermal afferent pathway in the rat. *J. Physiol.* **248**, 359–376.

Hellon, R. F., Hensel, H. and Schaefer, K., 1975, Thermal receptors in the scrotum of the rat. *J. Physiol.* **248**, 349–357.

Henken, D. B., Tessler, A., Chesselet, M.-F., Hudson, A., Baldino, F. and Murray, M., 1988, In situ hybridization of mRNA for β-preprotachynin and proprosomatostatin in adult rat dorsal root ganglia: comparison with immunocytochemical localization. *J. Neurocytol.* **17**, 671–681.

Henry, J. L., 1976, Effects of substance P on functionally identified units in cat spinal cord. *Brain Res.* **114**, 439–451.

Henry, J. L., Krnjevic, K. and Morris, M. E., 1975, Substance P and spinal neurones. *Canad. J. Physiol. Pharmacol.* **53**, 423–432.

Hensel, H., 1950, Temperaturempfindung und intracutane Waermebewegung. *Pfluegers Arch.* **252**, 165–215.

Hensel, H., 1952, Physiologie der thermoreception. *Erbeg. Physiol.* **47**, 166–246.

Hensel, H., 1973a, Cutaneous thermoreceptors. In: *Handbook of Sensory Physiology*, Vol. II. *Somatosensory System*, pp. 79-110. A. Iggo (ed.). Springer, New York.

Hensel, H., 1973b, Neural processes in thermoregulation. *Physiol. Rev.* **53**, 948–1017.

Hensel, H., 1974, Thermoreceptors. *Ann. Rev. Physiol.* **36**, 233–249.

Hensel, H. and Boman, K. K. A., 1960, Afferent impulses in cutaneous sensory nerves in human subjects. *J. Neurophysiol.* **23**, 564–578.

Hensel, H. and Iggo, A., 1971, Analysis of cutaneous warm and cold fibres in primates. *Pfluegers Arch.* **329**, 1–8.

Hensel, H. and Kenshalo, D. R., 1969, Warm receptors in the nasal region of cats. *J. Physiol.* **204**, 99–112.

Hensel, H. and Zotterman, Y., 1951a, The response of the cold receptors to constant cooling. *Acta Physiol. Scand.* **22**, 96–105.

Hensel, H. and Zotterman, Y., 1951b, The persisting cold sensation. *Acta Physiol. Scand.* **22**, 106–113.

Hensel, H. and Zotterman, Y., 1951c, The effect of menthol on the thermoreceptors. *Acta Physiol. Scand.* **24**, 27–34.

Hensel, H., Stroem, L. and Zotterman, Y., 1951, Electrophysiological measurements of depth of thermoreceptors. *J. Neurophysiol.* **14**, 423–429.

Hensel, H., Iggo, A. and Witt, I., 1960, A quantitative study of sensitive cutaneous thermoreceptors with C afferent fibres. *J. Physiol.* **153**, 113–126.

Hensel, H., Andres, K. H. and Düring, M. von, 1974, Structure and function of cold receptors. *Pfluegers Arch.* **352**, 1–10.

Hentall, I., 1977, A novel class of unit in substantia gelatinosa of the spinal cat. *Exp. Neurol.* **57**, 792–806.

Hentall, I. D. and Fields, H. L., 1979, Segmental and descending influences on intraspinal thresholds of single C-fibers. *J. Neurophysiol.* **42**, 1527–1537.

Heppelmann, B., Heuss, C. and Schmidt, R. F., 1988, Fiber size distribution of myelinated and unmyelinated axons in the medial and posterior articular nerves of the cat's knee joint. *Somatosensory Res.* **5**, 273–281.

Heppelmann, B., Messlinger, K., Neiss, W. F. and Schmidt, R. F., 1990, Ultrastructural three-dimensional reconstruction of group III and group IV sensory nerve endings ("free nerve endings") in the knee joint capsule of the cat: evidence for multiple receptive sites. *J. Comp. Neurol.* **292**, 103–116.

Hernández-Peón, R., Scherrer, H. and Velasco, M., 1956, Central influences on afferent conduction in the somatic and visual pathways. *Acta Neurol. Latinoamer.* **2**, 8–22.

Herrick, C. J., 1939, Cerebral fiber tracts of Amblystoma tigrinum in midlarval stages. *J. Comp. Neurol.* **71**, 511–612.

Hertel, H. C., Howaldt, B. and Mense, S., 1976, Responses of group IV and group III muscle afferents to thermal stimuli. *Brain Res.* **113**, 201–205.

Hildebrand, C. and Skoglund, S., 1971, Calibre spectra of some fibre tracts in the feline central nervous system during postnatal development. *Acta Physiol. Scand.*, Suppl. 364, 5–42.

Hill, D. R., Shaw, T. M. and Woodruff, G. N., 1988, Binding sites for ^{25}I-cholecystokinin in primate spinal cord are of the CCK-A subclass. *Neurosci. Lett.* **89**, 133–139.

Hill, R. G., Simmonds, M. A. and Staughan, D. W., 1976, Antagonism of γ-aminobutyric acid and glycine by convulsants in the cuneate nucleus of cat. *Br. J. Pharmacol.* **56**, 9–19.

Hillman, P. and Wall, P. D., 1969, Inhibitory and excitatory factors influencing the receptive fields of lamina 5 spinal cord cells. *Exp. Brain Res.* **9**, 284–306.

Hirota, N., Kuraishi, Y., Hino, Y., Sato, Y., Satoh, M. and Takagi, H., 1985, Met-enkephalin and morphine but not dynorphin inhibit noxious stimulation-induced release of substance P from rabbit dorsal horn in situ. *Neuropharmacology* **24**, 567–570.

Hiss, E. and Mense, S., 1976, Evidence for the existence of different receptor sites for algesic agents at the endings of muscular group IV afferent units. *Pfluegers Arch.* **362**, 141–146.

Hitchcock, E., 1970, Stereotactic cervical myelotomy. *J. Neurol Neurosurg. & Psychiatr.* **33**, 224–230.

Hitchcock, E., 1974, Stereotactic myelotomy. *Proc. Roy. Soc. Med.* **67**, 771–772.

Hiura, A., 1982, Do the dorsal efferent and ventral afferent fibers exist in the L6 spinal nerve roots in the rat? *Fukushima J. Med. Sci.* **28**, 77–81.

Hník, P., Hudlická, O., Kucera, J. and Payne, R., 1969, Activation of muscle afferents by nonproprioceptive stimuli. *Amer. J. Physiol.* **217**, 1451–1458.

Ho, R. H., 1983, Widespread distribution of substance P- and somatostatin-immunoreactive elements in the spinal cord of the neonatal rat. *Cell Tiss. Res.* **232**, 471–486.

Ho, R. H., 1988, Somatostatin immunoreactive structures in the developing rat spinal cord. *Brain Res. Bull.* **21**, 105–116.

Ho, R. H. and Berelowitz, M., 1984, Somatostatin $28_{1\text{-}14}$ immunoreactivity in primary afferent neurons of the rat spinal cord. *Neurosci. Lett.* **46**, 161–166.

Hockfield, S. and Gobel, S., 1978, Neurons in and near nucleus caudalis with long ascending projection axons demonstrated by retrograde labeling with horseradish peroxidase. *Brain Res.* **139**, 333–339.

Hodge, C. J., 1972, Potential changes inside central afferent terminals secondary to stimulation of large- and small-diameter peripheral nerve fibers. *J. Neurophysiol.* **35**, 30–43.

Hodge, C. J., Apkarian, A. V., Stevens, R., Vogelsang, G. and Wisnicki, H. J., 1981, Locus coeruleus modulation of dorsal horn unit responses to cutaneous stimulation. *Brain Res.* **204**, 415–420.

Hodge, C. J., Apkarian, A. V., Stevens, R. T., Vogelsang, G. D., Brown, O. and Frank, J. I., 1983, Dorsolateral pontine inhibition of dorsal horn responses to cutaneous stimulation: lack of dependence on catecholaminergic system in cat. *J. Neurophysiol.* **50**, 1220–1235.

Hodge, C. J., Apkarian, A. V. and Stevens, R. T., 1986, Inhibition of dorsal-horn cell responses by stimulation of the Koelliker-Fuse nucleus. *J. Neurosurg.* **65**, 825–833.

Hoffert, M. J., Miletić, V., Ruda, M. A. and Dubner, R., 1983, Immunocytochemical identification of serotonin axonal contacts on characterized neurons in laminae I and II of the cat dorsal horn. *Brain Res.* **267**, 361–364.

Hoheisel, U. and Mense, S., 1985, Morphological features of dorsal root ganglion cells with groups III and IV afferent fibers. *Neurosci. Lett.*, Suppl. 22, S29.

Hoheisel, U. and Mense, S., 1986, Non-myelinated afferent fibres do not originate exclusively from the smallest dorsal root ganglion cells in the cat. *Neurosci. Lett.* **72**, 153–157.

Hoheisel, U. and Mense, S., 1989, Long-term changes in discharge behaviour of cat dorsal horn neurones following noxious stimulation of deep tissues. *Pain* **36**, 239–247.

Hökfelt, T. and Ljungdahl, Å., 1971, Light and electron microscopic autoradiography on spinal cord slices after incubation with labeled glycine. *Brain Res.* **32**, 189–194.

Hökfelt, T., Elde, R., Johansson, O., Luft, R. and Arimura, A., 1975a, Immunohistochemical evidence for the presence of somatostatin, a powerful inhibitory peptide, in some primary sensory neurons. *Neurosci. Lett.* **1**, 231–235.

Hökfelt, T., Kellerth, J. O., Nilsson, C. and Pernow, B., 1975b, Experimental immunohistochemical studies on the localization and distribution of substance P in cat primary sensory neurons. *Brain Res.* **100**, 235–252.

Hökfelt, T., Kellerth, J. O., Nilsson, G. and Pernow, B., 1975c, Substance P: localization in the central nervous system and in some primary sensory neurons. *Science* **190**, 889–890.

Hökfelt, T., Elde, R., Johansson, O., Luft, R., Nilsson, G. and Arimura, A., 1976, Immunohistochemical evidence for separate populations of somatostatin-containing and substance P-containing primary afferent neurons in the rat. *Neuroscience* **1**, 131–136.

Hökfelt, T., Elde, K., Johansson, O., Terenius, L. and Stein, L., 1977a, The distribution of enkephalin immunoreactive cell bodies in the rat central nervous system. *Neurosci. Lett.* **5**, 25–32.

Hökfelt, T., Ljungdahl, Å., Terenius, L., Elde, R. and Nilsson, G., 1977b, Immunohistochemical analysis of peptide pathways possibly related to pain and analgesia: enkephalin and substance P. *Proc. Natl. Acad. Sci. USA* **74**, 3081–3085.

Hökfelt, T., Ljungdahl, Å., Steinbusch, H., Verhofstad, A., Nilsson, G., Brodin, E., Oernow, B. and Goldstein, M., 1978, Immunohistochemical evidence of substance P-like immunoreactivity in some 5-hydroxytryptamine-containing neurons in the rat central nervous system. *Neuroscience* **3**, 517–538.

Hökfelt, T., Johansson, O., Ljungdahl, Å., Lundberg, J. M. and Schultzberg, M., 1980, Peptidergic neurones. *Nature* **284**, 515–521.

Hökfelt, T., Lundberg, J. M., Terenius, L., Jancso, G. and Kimmel, J., 1981, Avian pancreatic polypeptide (APP) immunoreactive neurons in the spinal cord and spinal trigeminal nucleus. *Peptides* **2**, 81–87.

Hökfelt, T., Wiesenfeld-Hallin, Z., Villar, M. and Melander, T., 1987, Increase of galanin-like immunoreactivity in rat dorsal root ganglion cells after peripheral axotomy. *Neurosci. Lett.* **83**, 217–220.

Holloway, J. A., Fox, R. E. and Iggo, A., 1978, Projections of the spinothalamic tract to the thalamic nuclei of the cat. *Brain Res.* **157**, 336–340.

Holmes, A., 1982, Do the dorsal efferent and ventral afferent fibers exist in the L6 spinal nerve roots in the rat? *Fukushima J. Med. Sci.* **28**, 77–81.

Holtman, J. R., Jr., 1989, Localization of substance P immunoreactivity in phrenic primary afferent neurons. *Peptides* **10**, 53–56.

Holzer-Petsche, U., Rinner, I. and Lembeck, F., 1986, Distribution of choline acetyltransferase activity in rat spinal cord—influence of primary afferents? *J. Neural Trans.* **66**, 85–92.

Homor, G. and Kasa, P., 1978, Acetylcholinesterase resynthesis after DFP poisoning; histochemical and biochemical study. *Acta Histochem.* Bd. 62, 293–301.

Honda, C. N., 1985, Visceral and somatic afferent convergence onto neurons near the central canal in the sacral spinal cord of the cat. *J. Neurophysiol.* **53**, 1059–1078.

Honda, C. N. and Perl, E. R., 1985, Functional and morphological features of neurons in the midline region of the caudal spinal cord of the cat. *Brain Res.* **340**, 285–295.

Honda, C. N., Réthelyi, M. and Petrusz, P., 1983, Preferential immunohistochemical localization of vasoactive intestinal polypeptide (VIP) in the sacral spinal cord of the cat: Light and electron microscopic observations. *J. Neurosci.* **3**, 2183–2196.

Hong, S. K., Kniffki, K. D., Mense, S., Schmidt, R. F. and Wendisch, M., 1979, Descending influences on the responses of spinocervical tract neurones to chemical stimulation of fine muscle afferents. *J. Physiol.* **290**, 129–140.

Hongo, T. and Jankowska, E., 1967, Effects from the sensorimotor cortex on the spinal cord in cats with transected pyramids. *Exp. Brain Res.* **3**, 117–134.

Hongo, T. and Koike, H., 1975, Some aspects of synaptic organizations in the spinocervical tract cell in the cat. In: *The Somatosensory System*, pp. 218–226. H. H. Kornhuber (ed.). Georg Thieme Verlag, Stuttgart.

Hongo, T., Jankowska, E. and Lundberg, A., 1966, Convergence of excitatory and inhibitory action on interneurones in the lumbosacral cord. *Exp. Brain Res.* **1**, 338–358.

Hongo, T., Jankowska, E. and Lundberg, A., 1968, Post-synaptic excitation and inhibition from primary afferents in neurones of the spinocervical tract. *J. Physiol.* **199**, 569–592.

Hongo, T., Kudo, N., Sasaki, S., Yamashita, M., Yoshida, K., Ishizuka, N. and Mannen, H., 1987, Trajectory of group Ia and Ib fibers from the hind-limb muscles at the L3 and L4 segments of the spinal cord of the cat. *J. Comp. Neurol.* **262**, 159–194.

Hongo, T., Kitazawa, S., Ohki, Y., Sasaki, M. and Xi, M. C., 1989a, A physiological and morphological study of premotor interneurones in the cutaneous reflex pathways in cats. *Brain Res.* **505**, 163–166.

Hongo, T., Kitazawa, S., Ohki, Y. and Xi, M. C., 1989b, Functional identification of last-order interneurones of skin reflex pathways in the cat forelimb segments. *Brain Res.* **505**, 167–170.

Honoré, T., Davies, S. N., Drejer, J., Fletcher, E. J., Jacobsen, P., Lodge, D., and Nielsen, F. E., 1988, Quinoxalinediones: potent competitive non-NMDA glutamate receptor antagonists. *Science* **241**, 701–703.

Hooglund, P. V., 1981, Spinothalamic projections in a lizard Varanus exanthematicus: an HRP study. *J. Comp. Neurol.* **198**, 7–12.

Hopkin, J. M. and Neal, M. J., 1970, The release of [^{14}C]glycine from electrically stimulated rat spinal cord slices. *Brit. J. Pharmacol.* **40**, 136–138P.

Hope, P. J., Lang, C. W. and Duggan, A. W., 1990, Persistence of immunoreactive neurokinins in the dorsal horn of barbiturate anaesthetized and spinal cats, following release by tibial nerve stimulation. *Neurosci. Lett.* **118**, 25–28.

Horch, K. W. and Burgess, P. R., 1975, Effect of activation and adaptation on the sensitivity of slowly adapting cutaneous mechanoreceptors. *Brain Res.* **98**, 109–118.

Horch, K. W. and Burgess, P. R., 1976, Responses to threshold and suprathreshold stimuli by slowly adapting cutaneous mechanoreceptors in the cat. *J. Comp. Physiol.* **110**, 307–315.

Horch, K. W. and Lisney, S. J. W., 1981, Changes in primary afferent depolarization of sensory neurones during peripheral nerve regeneration in the cat. *J. Physiol.* **313**, 287–299.

Horch, K. W., Whitehorn, D. and Burgess, P. R., 1974, Impulse generation in type I cutaneous mechanoreceptors. *J. Neurophysiol.* **37**, 267–281.

Horch, K. W., Clark, F. J. and Burgess, P. R., 1975, Awareness of knee joint angle under static conditions. *J. Neurophysiol.* **38**, 1436–1447.

Horch, K. W., Burgess, P. R. and Whitehorn, D., 1976, Ascending collaterals of cutaneous neurons in the fasciculus gracilis of the cat. *Brain Res.* **117**, 1–17.

Horch, K. W., Tuckett, R. P. and Burgess, P. R., 1977, A key to the classification of cutaneous mechanoreceptors. *J. Invest. Dermatol.* **69**, 75–82.

Hore, J., Preston, J. B., Durkovic, R. G. and Cheney, P. D., 1976, Responses of cortical neurons (areas 3a and 4) to ramp stretch of hindlimb muscles in the baboon. *J. Neurophysiol.* **39**, 484–500.

Hori, Y., Lee, K. H., Chung, J. M. and Willis, W. D., 1984, The effects of small doses of barbiturate on the activity of primate nociceptive tract cells. *Brain Res.* **307**, 9–15.

Horrax, G., 1929, Experiences with cordotomy. *Arch. Surg.* **18**, 1140–1164.

Horrobin, D. F., 1966, The lateral cervical nucleus of the cat: an electrophysiological study. *Quart. J. Exp. Physiol.* **51**, 351–371.

Hösli, L., Hösli, E., Zehntner, C. and Landolt, H., 1981, Effects of substance P on neurones and glial cells in cultured rat spinal cord. *Neurosci. Lett.* **24**, 165–168.

Hosobuchi, Y., 1980, The majority of unmyelinated afferent axons in human ventral roots probably conduct pain. *Pain* **8**, 167–180.

Houk, J. and Henneman, E., 1967, Responses of Golgi tendon organs to active contractions of the soleus muscle of the cat. *J. Neurophysiol.* **30**, 466–481.

Houser, C. R., Crawford, G. D., Barber, R. P., Salvaterra, P. M. and Vaughn, J. E., 1983, Organization and morphological characteristics of cholinergic neurons: an immunocytochemical study with a monoclonal antibody to choline acetyltransferase. *Brain Res.* **266**, 97–119.

Howe, J. R. and Zieglgänsberger, W., 1987, Responses of rat dorsal horn neurons to natural stimulation and to iontophoretically applied norepinephrine. *J. Comp. Neurol.* **255**, 1–17.

Howe, J. R., Yaksh, T. L. and Go, V. L. W., 1987, The effect of unilateral dorsal root ganglionectomies or ventral rhizotomies on α2-adrenoceptor binding to, and the substance P, enkephalin, and neurotensin content of, the cat lumbar spinal cord. *Neuroscience* **21**, 385–394.

Howland, B., Lettvin, J. Y., McCulloch, W. S., Pitts, W. and Wall, P. D., 1955, Reflex inhibition by dorsal root interaction. *J. Neurophysiol.* **18**, 1–17.

Huang, L. Y. M., 1987, Electrical properties of acutely isolated, identified rat spinal dorsal horn projection neurons. *Neurosci. Lett.* **82**, 267–272.

Huang, L. Y. M., 1989a, Calcium channels in isolated rat dorsal horn neurons, including labeled spinothalamic and trigeminothalamic cells. *J. Physiol.* **411**, 161–177.

Huang, L. Y. M., 1989b, Origin of thalamically projecting somatosensory relay neurons in the immature rat. *Brain Res.* **495**, 108–114.

Hubbard, S. J., 1958, A study of rapid mechanical events in a mechanoreceptor. *J. Physiol.* **141**, 198–218.

Hughes, A., 1976, The development of the dorsal funiculus in the human spinal cord. *J. Anat.* **122**, 169–175.

Hughes, J. and Gasser, H. S., 1934a, Some properties of the cord potentials evoked by a single afferent volley. *Amer. J. Physiol.* **108**, 295–306.

Hughes, J. and Gasser, H. S., 1934b, The response of the spinal cord to two afferent volleys. *Amer. J. Physiol.* **108**, 307–321.

Hulliger, M., Nordh, E., Thelin, A. E. and Vallbo, A. B., 1979, The responses of afferent fibres from the glabrous skin of the hand during voluntary finger movements in man. *J. Physiol.* **291**, 233–249.

Hulliger, M., Nordy, E. and Vallbo, A. B., 1982, The absence of position response in spindle afferent units from human finger muscles during accurate position holding. *J. Physiol.* **322**, 167–179.

Hulsebosch, C. E. and Coggeshall, R. E., 1981, Sprouting of dorsal root axons. *Brain Res.* **224**, 170–174.

Hultborn, H., Illert, M. and Santini, M., 1976a, Convergence on interneurones mediating the reciprocal Ia inhibition of motoneurones. I. Disynaptic Ia inhibition of Ia inhibitory interneurones. *Acta Physiol. Scand.* **96**, 193–201.

Hultborn, H., Illert, M. and Santini, M., 1976b, Convergence on interneurones mediating the reciprocal Ia inhibition of motoneurones. II. Effects from segmental flexor reflex pathways. *Acta Physiol. Scand.* **96**, 351–367.

Hummelsheim, H., Wiesendanger, R., Wiesendanger, M. and Bianchetti, M., 1985, The projection of low-threshold muscle afferents of the forelimb to the main and external cuneate nuclei of the monkey. *Neuroscience* **16**, 979–987.

Hunt, C. A., Seroogy, K. B., Gall, C. M. and Jones, E. G., 1987, Cholecystokinin innervation of rat thalamus, including fibers to ventroposterolateral nucleus from dorsal column nuclei. *Brain Res.* **426**, 257–269.

Hunt, C. C., 1954, Relation of function to diameter in afferent fibers of muscle nerves. *J. Gen. Physiol.* **38**, 117–131.

Hunt, C. C., 1961, On the nature of vibration receptors in the hind limb of the cat. *J. Physiol.* **155**, 175–186.

Hunt, C. C. and Kuno, M., 1959a, Properties of spinal interneurones. *J. Physiol.* **147**, 346–363.

Hunt, C. C. and Kuno, M., 1959b, Background discharge and evoked responses of spinal interneurones. *J. Physiol.* **147**, 364–384.

Hunt, C. C. and McIntyre, A. K., 1960a, Characteristics of responses from receptors from the flexor longus digitorum muscle and the adjoining interosseous region of the cat. *J. Physiol.* **153**, 74–87.

Hunt, C. C. and McIntyre, A. K., 1960b, Properties of cutaneous touch receptors in cat. *J. Physiol.* **153**, 88–98.

Hunt, C. C. and McIntyre, A. K., 1960c, An analysis of fibre diameter and receptor characteristics of myelinated cutaneous afferent fibres in cat. *J. Physiol.* **153**, 99–112.

Hunt, S. P., 1983, Cytochemistry of the spinal cord. In: *Chemical Neuroanatomy*, pp. 53–84. P. C. Emson (ed.). Raven Press, New York.

Hunt, S. P. and Rossi, J., 1985, Peptide- and non-peptide-containing unmyelinated primary afferents: the parallel processing of nociceptive information. *Phil. Trans. Roy. Soc. Lond. B* **308**, 283–289.

Hunt, S. P., Kelly, J. S. and Emson, P. C., 1980, The electron microscopic localization of methionine-enkephalin within the superficial layers (I and II) of the spinal cord. *Neuroscience* **5**, 1871–1890.

Hunt, S. P., Kelly, J. S., Emson, P. C., Kimmel, J. R., Miller, R. J. and Wu, J. Y., 1981a, An immunohistochemical study of neuronal populations containing neuropeptides or γ-aminobutyrate within the superficial layers of the rat dorsal horn. *Neuroscience* **6**, 1883–1898.

Hunt, S. P., Emson, P. C., Gilbert, R., Goldstein, M. and Kimmell, J. R., 1981b, Presence of avian pancreatic polypeptide-like immunoreactivity in catecholamine and methionine-enkephalin-containing neurones within the central nervous system. *Neurosci. Lett.* **21**, 125–130.

Hunt, S. P., Nagy, J., Ninkovic, M. and Iversen, L. L., 1982, A cytochemical analysis of the interrelationships between the dorsal root and the dorsal horn. In: *Cytochemical Methods in Neuroanatomy*, pp. 165–178. V. Chan-Palay and S.L. Palay (eds.). Alan R. Liss, Inc., New York.

Hunt, S. P., Pini, A. and Evan, G., 1987, Induction of c-fos-like protein in spinal cord neurons following sensory stimulation. *Nature* **328**, 632–634.

Hunter, J. C., Birchmore, B., Woodruff, R. and Hughes, J., 1989, Kappa opioid binding sites in the dog cerebral cortex and spinal cord. *Neuroscience* **31**, 735–743.

Hursh, J. B., 1939, Conduction velocity and diameter of nerve fibers. *Amer. J. Physiol.* **127**, 131–139.

Hursh, J. B., 1940, Relayed impulses in ascending branches of dorsal root fibers. *J. Neurophysiol.* **3**, 166–174.

Hutchison, W. D., Morton, C. R. and Terenius, L., 1990, Dynorphin A: in vivo release in the spinal cord of the cat. *Brain Res.* **532**, 299–306.

Hwang, Y. C., Hinsman, E. J. and Roesel, O. F., 1975, Calibre spectra of fibers in the fasciculus gracilis of the cat cervical spinal cord: a quantitative electron microscopic study. *J. Comp. Neurol.* **162**, 195–204.

Hyden, H., Lovtrup, S. and Pigon, A., 1958, Cytochrome oxidase and succinoxidase in spinal ganglion cells and in glial capsule cells. *J. Neurochem.* **2**, 304–311.

Hylden, J. L. K., Hayashi, H., Bennett, G. J. and Dubner, R., 1985, Spinal lamina I neurons projecting to the parabrachial area of the cat midbrain. *Brain Res.* **336**, 195–198.

Hylden, J. L. K., Hayashi, H., Dubner, R. and Bennett, G. J., 1986a, Physiology and morphology of the lamina I spinomesencephalic projection. *J. Comp. Neurol.* **247**, 505–515.

Hylden, J. L. K., Hayashi, H., Ruda, M. A. and Dubner, R., 1986b, Serotonin innervation of physiologically identified lamina I projection neurons. *Brain Res.* **370**, 401–404.

Hylden, J. L. K., Nahin, R. L. and Dubner, R., 1987, Altered responses of nociceptive cat lamina I spinal dorsal horn neurons after chronic sciatic neuroma formation. *Brain Res.* **411**, 341–350.

Hylden, J. L. K., Nahin, R. L., Traub, R. J. and Dubner, R., 1989, Expansion of receptive fields of spinal lamina I projection neurones in rats with unilateral adjuvant-induced inflammation: the contribution of central dorsal horn mechanisms. *Pain* **37**, 229–243.

Hyndman, O. R., 1942, Lissauer's tract section. A contribution to chordotomy for the relief of pain. *J. Internat. Coll. Surgeons* **5**, 394–400.

Hyndman, O. R. and Van Epps, C., 1939, Possibility of differential section of the spinothalamic tract. A clinical and histological study. *Arch. Surg.* **38**, 1036–1053.

Hyndman, O. R. and Wolkin, J., 1943, Anterior chordotomy. Further observations on physiologic results and optimal manner of performance. *Arch. Neurol. Psychiat.* **50**, 129–148.

Hynes, M. A., Buck, L. B., Gitt, M., Barondes, S., Dodd, J. and Jessell, T. M., 1989, Carbohydrate recognition in neuronal development: Structure and expression of surface oligosaccharides and β-galactoside-binding lectins. In: *Carbohydrate Recognition in Cellular Function*, pp. 189–218. Wiley, Chichester, Ciba Foundation Symposium 145.

Iadarola, M. J., Douglass, J., Civelli, O. and Naranjo, J. R., 1988, Differential activation of spinal cord dynorphin and enkephalin neurons during hyperalgesia: evidence using cDNA hybridization. *Brain Res.* **455**, 205–212.

Ibuki, T., Okamura, H., Miyazaki, M., Yanaihara, N., Zimmerman, E. A. and Ibata, Y., 1989, Comparative distribution of three opioid systems in the lower brainstem of the monkey (Macaca fuscata). *J. Comp. Neurol.* **279**, 445–456.

Iggo, A., 1955, Tension receptors in the stomach and the urinary bladder. *J. Physiol.* **128**, 593–607.

Iggo, A., 1959, Cutaneous heat and cold receptors with slowly conducting (C) afferent fibres. *Quart. J. Exp. Physiol.* **44**, 362–370.

Iggo, A., 1960, Cutaneous mechanoreceptors with afferent C fibres. *J. Physiol.* **152**, 337–353.

Iggo, A., 1961, Non-myelinated afferent fibres from mammalian skeletal muscle. *J. Physiol.* **155**, 52–53P.

Iggo, A., 1963, New specific sensory structures in hairy skin. *Acta Neuroveg.* **24**, 175–180.

Iggo, A., 1969, Cutaneous thermoreceptors in primates and subprimates. *J. Physiol.* **200**, 403–430.

Iggo, A. and Kornhuber, H. H., 1977, A quantitative study of C-mechanoreceptors in hairy skin of the cat. *J. Physiol.* **271**, 549–565.

Iggo, A. and Muir, A. R., 1969, The structure and function of a slowly adapting touch corpuscle in hairy skin. *J. Physiol.* **200**, 763–796.

Iggo, A. and Ogawa, H., 1971, Primate cutaneous thermal nociceptors. *J. Physiol.* **216**, 77–78P.

Iggo, A. and Ogawa, H., 1977, Correlative physiological and morphological studies of rapidly adapting mechanoreceptors in cat's glabrous skin. *J. Physiol.* **266**, 275–296.

Iggo, A. and Ramsey, R. L., 1976, Thermosensory mechanisms in the spinal cord of monkeys. In: *Sensory Functions of the Skin in Primates with Special Reference to Man*, pp. 285–304. Y. Zotterman (ed.). Pergamon, Oxford.

Iggo, A., Molony, V. and Steedman, W. M., 1988, Membrane properties of nociceptive neurones in lamina II of lumbar spinal cord in the cat. *J. Physiol.* **400**, 367–380.

Imai, Y. and Kusama, T., 1969, Distribution of the dorsal root fibers in the cat. An experimental study with the Nauta method. *Brain Res.* **13**, 338–359.

Inagaki, N., Kamisaki, Y., Kiyama, H., Horio, Y., Tohyama, M. and Wada, H., 1987, Immunocytochemical localizations of cytosolic and mitochondrial glutamic oxaloacetic transaminase isoenzymes in rat primary sensory neurons as a marker for the glutamate neuronal system. *Brain Res.* **402**, 197–200.

Inagaki, S., Kito, S., Kubota, Y., Girgis, S., Hillyard, C. J. and MacIntyre, I., 1986, Autoradiographic localization of calcitonin gene-related peptide binding sites in human and rat brain. *Brain Res.* **374**, 287–298.

Ingvar, S., 1927, Zur Morphogeneses der Tabes. *Acta Med. Scand.* **65**, 645–674.

Inyama, C. O., Wharton, J., Su, H. C. and Polak, J. M., 1986, CGRP-immunoreactive nerves in the genitalia of the female rat originate from dorsal root ganglia T11-L3 and L6-Sa: a combined immunocytochemical and retrograde tracing study. *Neurosci. Lett.* **69**, 13–18.

Iriuchijima, J. and Zotterman, Y., 1960, The specificity of afferent cutaneous C fibres in mammals. *Acta Physiol. Scand.* **49**, 267–278.

Ischia, S., Luzzani, A., Ischia, A. and Maffezzoli, G., 1984, Bilateral percutaneous cervical cordotomy: immediate and long-term results in 36 patients with neoplastic disease. *J. Neurol. Neurosurg. Psychiat.* **47**, 141–147.

Ishida-Yamamoto, A., Senba, E. and Tohyama, M., 1988, Calcitonin gene-related peptide- and substance P-immunoreactive nerve fibers in Meissner's corpuscles of rats: an immunohistochemical analysis. *Brain Res.* **453**, 362–366.

Ishida-Yamamoto, A., Senba, E. and Tohyama, M., 1989, Distribution and fine structure of calcitonin gene-related peptide-like immunoreactive nerve fibers in the rat skin. *Brain Res.* **491**, 93–101.

Jabbur, S. J. and Banna, N. R., 1968, Presynaptic inhibition of cuneate transmission by wide-spread cutaneous inputs. *Brain Res.* **10**, 273–276.

Jabbur, S. J. and Banna, N. R., 1970, Widespread cutaneous inhibition in dorsal column nuclei. *J. Neurophysiol.* **33**, 616–624.

Jabbur, S. J. and Towe, A. L., 1961, Cortical excitation of neurons in dorsal column nuclei of cat, including an analysis of pathways. *J. Neurophysiol.* **24**, 499–509.

Jabbur, S. J., Harik, S. I. and Hush, J. A., 1977, Caudate influence on transmission in the cuneate nucleus. *Brain Res.* **120**, 559–563.

Jackson, D. A. and White, S. R., 1988, Thyrotropin releasing hormone (TRH) modifies excitability of spinal cord dorsal horn cells. *Neurosci. Lett.* **92**, 171–176.

Jacobs, J., Carmichael, N. and Cavanagh, J. B., 1975, Ultrastructural changes in the dorsal root and trigeminal ganglia of rats poisoned with methyl mercury. *Neuropath. Appl. Neurobiol.* **1**, 1–19.

Jacobs, V. L. and Sis, R. F., 1980, Ascending projections of dorsal column in a garter snake (*Thamnophis siritalis*): a degeneration study. *Anat. Rec.* **196**, 37–50.

Jahr, C. E. and Jessell, T. M., 1983, ATP excites a subpopulation of rat dorsal horn neurones. *Nature* **304**, 730–733.

Jahr, C. E. and Jessell, T. M., 1985, Synaptic transmission between dorsal root ganglion and dorsal horn neurons in culture: antagonism of EPSPs and glutamate excitation by kynurenate. *J. Neurosci.* **5**, 2281–2289.

Jancsó, G. and Knyihár, E., 1975, Functional linkage between nociception and fluoride-resistant acid phosphatase activity in the Rolando substance. *Neurobiology* **5**, 42–43.

Jancsó, G., Hökfelt, T., Lundberg, J. M., Király, E., Halasz, N., Nilsson, G., Terenius, L., Rehfeld, J., Steinbusch, H., Verhofstad, A., Elder, R., Said, S. and Brown, M., 1981, Immunohistochemical studies on the effect of capsaicin on spinal and medullary peptide and monoamine neurons using antisera to substance P, gastrin/CCK, somatostatin, VIP enkephalin, neurotensin and 5-hydroxytryptamine. *J. Neurocytol.* **10**, 963–980.

Jane, J. A. and Schroeder, D. M., 1971, A comparison of dorsal column nuclei and spinal afferents in the European hedgehog (*Erinaceus europeaus*). *Exp. Neurol.* **30**, 1–17.

Jänig, W., 1971a, The afferent innervation of the central pad of the cat's hind foot. *Brain Res.* **28**, 203–216.

Jänig, W., 1971b, Morphology of rapidly and slowly adapting mechanoreceptors in the hairless skin of the cat's hind foot. *Brain Res.* **28**, 217–231.

Jänig, W. and Zimmermann, M., 1971, Presynaptic depolarization of myelinated afferent fibres evoked by stimulation of cutaneous C fibres. *J. Physiol.* **214**, 29–50.

Jänig, W., Schmidt, R. F. and Zimmermann, M., 1968a, Single unit responses and the total afferent outflow from the cat's foot pad upon mechanical stimulation. *Exp. Brain Res.* **6**, 100–115.

Jänig, W., Schmidt, R. F. and Zimmermann, M., 1968b, Two specific feedback pathways to the central afferent terminals of phasic and tonic mechanoreceptors. *Exp. Brain Res.* **6**, 116–129.

Jänig, W., Schoultz, T. and Spencer, W. A., 1977, Temporal and spatial parameters of excitation and afferent inhibition in cuneothalamic relay neurons. *J. Neurophysiol.* **40**, 822–835.

Jankowska, E. and Lindström, S., 1972, Morphology of interneurones mediating Ia reciprocal inhibition of motoneurones in the spinal cord of the cat. *J. Physiol.* **226**, 805–823.

Jankowska, E. and Roberts, W.J., 1972a, An electrophysiological demonstration of the axonal projections of single spinal interneurones in the cat. *J. Physiol.* **222**, 597–622.

Jankowska, E. and Roberts, W. J., 1972b, Synaptic actions of single interneurones mediating reciprocal Ia inhibition of motoneurones. *J. Physiol.* **222**, 623–642.

Jankowska, E., Rastad, J. and Westman, J., 1976, Intracellular application of horseradish peroxidase and its light and electron microscopical appearance in spinocervical tract cells. *Brain Res.* **105**, 557–562.

Jankowska, E., Rastad, J. and Zarzecki, P., 1979, Segmental and supraspinal input to cells of origin of nonprimary fibres in the feline dorsal columns. *J. Physiol.* **290**, 185–200.

Jankowska, E., Johannisson, T. and Lipski, J., 1981, Common interneurones in reflex pathways from group Ia and Ib afferents of ankle extensors in the cat. *J. Physiol.* **310**, 381–402.

Jansen, K. L. R., Faull, R. L. M., Dragunow, M. and Waldvogel, H., 1990, Autoradiographic localization of NMDA, quisqualate and kainic acid receptors in human spinal cord. *Neurosci. Lett.* **108**, 53–57.

Järvilehto, T., 1973, Neural coding in the temperature sense. *Ann. Acad. Sci. Fenn.* (Ser B) **184**, 1–71.

Järvilehto, T. and Hämäläinen, H., 1979, Touch and thermal sensations: psychophysical observations and unit activity in human skin nerves. In: *Sensory Functions of the Skin of Humans*, pp. 279–295. D. R. Kenshalo (ed.). Plenum Press, New York.

Järvilehto, T., Hämäläinen, H. and Laurinen, P., 1976, Characteristics of single mechanoreceptive fibres innervating hairy skin of the human hand. *Exp. Brain Res.* **25**, 45–61.

Järvilehto, T., Hämäläinen, H. and Soininen, K., 1981, peripheral neural basis of tactile sensations in man: II. Characteristics of human mechanoreceptors in the hairy skin and correlations of their activity with tactile sensations. *Brain Res.* **219**, 13–27.

Jeftinija, S., 1988, Enkephalins modulate excitatory synaptic transmission in the superficial dorsal horn by acting at mu-opioid receptor sites. *Brain Res.* **460**, 260–268.

Jeftinija, S., Miletić, V. and Randić, M., 1981a, Cholecystokinin octapeptide excites dorsal horn neurons both in vivo and in vitro. *Brain Res.* **213**, 231–236.

Jeftinija, S., Semba, K. and Randić, M., 1981b, Norepinephrine reduces excitability of single cutaneous primary afferent C-fibers in the cat spinal cord. *Brain Res.* **219**, 456–463.

Jeftinija, S., Murase, K., Nedeljikov, V. and Randić, M., 1982, Vasoactive intestinal polypeptide excites mammalian dorsal horn neurons both in vivo and in vitro. *Brain Res.* **243**, 158–164.

Jeftinija, S., Raspantini, C., Randić, M., Yaksh, T. L., Go, V. L. W. and Larson, A. A., 1986, Altered responsiveness to substance P and 5-hydroxytryptamine in cat dorsal horn neurons after 5-HT depletion with p-chlorophenylalanine. *Brain Res.* **368**, 107–115.

Jeftinija, S., Urban, L., Kangrga, I., Ryu, P. D. and Randić, M., 1987, Slow excitatory and inhibitory transmission in the rat spinal dorsal horn in vitro and depressant effect of enkephalins. *Neurol. Neurobiol.* **28**, 271–281.

Jenkins, W. L., 1941a, Studies in thermal sensitivity: 15. Effects of stimulus-temperature in seriatim warm-mapping. *J. Exp. Psychol.* **28**, 517–523.

Jenkins, W. L., 1941b, Studies in thermal sensitivity: 16. Further evidence on the effects of stimulus temperature. *J. Exp. Psychol.* **29**, 413–419.

Jennes, L., Stumpf, W. E. and Kalivas, P. W., 1982, Neurotensin: topographical distribution in rat brain by immunohistochemistry. *J. Comp. Neurol.* **210**, 211–224.

Jessell, T. M. and Dodd, J., 1985, Structure and expression of differentiation antigens on functional subclasses of primary sensory neurons. *Phil. Trans. R. Soc. B* **308**, 271–281.

Jessell, T. M. and Dodd, J., 1986, Neurotransmitters and differentiation antigens in subsets of sensory neurons projecting to the spinal dorsal horn. In: *Neuropeptides in Neurologic and Psychiatric Disease*, pp. 111–131. J. B. Martin and J. D. Barchas (eds.). Raven Press, New York.

Jessell, T. M. and Iversen, L. L., 1977, Opiate analgesics inhibit substance P release from rat trigeminal nucleus. *Nature* **68**, 549–551.

Jessell, T. M., Yoshioka, K. and Jahr, C. E., 1986, Amino acid receptor-mediated transmission at primary afferent synapses in rat spinal cord. *J. Exp. Biol.* **124**, 239–258.

Jhamandas, K., Yaksh, T. L. and Go, V. L. W., 1984, Acute and chronic morphine modifies the in vivo release of methionine enkephalin-like immunoreactivity from the cat spinal cord and brain. *Brain Res.* **297**, 91–103.

Johansson, H. and Silfvenius, H., 1977a, Axon-collateral activation by dorsal spinocerebellar tract fibres of group I relay cells of nucleus z in the cat medulla oblongata. *J. Physiol.* **265**, 341–369.

Johansson, H. and Silfvenius, H., 1977b, Input from ipsilateral proprio- and exteroceptive hind limb afferents to nucleus z of the cat medulla oblongata. *J. Physiol.* **265**, 371–393.

Johansson, H. and Silfvenius, H., 1977c, Connexions from large, ipsilateral hind limb muscle and skin afferents to the rostral main cuneate nucleus and to the nucleus X region in the cat. *J. Physiol.* **265**, 395–428.

Johansson, O., 1978, Localization of somatostatin-like immunoreactivity in the Golgi apparatus of central and peripheral neurons. *Histochemistry* **58**, 167–176.

Johansson, O. and Vaalasti, A., 1987, Immunohistochemical evidence for the presence of somatostatin-containing sensory nerve fibres in the human skin. *Neurosci. Lett.* **73**, 225–230.

Johansson, O., Hökfelt, T., Pernow, B., Jeffcoate, S. L., White, N., Steinbusch, H. W. M., Verhofstad, A. A. J., Emson, P. C. and Spindel, E., 1981, Immunohistochemical support for three putative transmitters in one neuron: coexistence of 5-hydroxytryptamine, substance P- and thyrotropin releasing hormone-like immunoreactivity in medullary neurons projecting to the spinal cord. *Neuroscience* **6**, 1857–1881.

Johansson, O., Hökfelt, T. and Elde, R., 1984, Immunohistochemical distribution of somatostatin-like immunoreactivity in the central nervous system of the adult rat. *Neuroscience* **13**, 265–339.

Johansson, R. S., 1976, Receptive field sensitivity profile of mechanoreceptive units innervating the glabrous skin of the human hand. *Brain Res.* **104**, 330–334.

Johansson, R. S., 1978, Tactile sensibility in the human hand: receptive field characteristics of mechanoreceptive units in the glabrous skin area. *J. Physiol.* **281**, 101–123.

Johansson, R. S. and Vallbo, A. B., 1976, Skin mechanoreceptors in the human hand: an inference of some population properties. In: *Sensory Functions of the Skin in Primates*, pp. 171–184. Y. Zotterman (ed.). Pergamon Press, New York.

Johansson, R. S. and Vallbo, A. B., 1979a, Tactile sensibility in the human hand: relative and absolute densities of four types of mechanoreceptive units in glabrous skin. *J. Physiol.* **286**, 283–300.

Johansson, R. S. and Vallbo, A. B., 1979b, Detection of tactile stimuli. Thresholds of afferent units related to psychophysical thresholds in the human hand. *J. Physiol.* **297**, 405–422.

Johansson, R. S. and Vallbo, A. B., 1980, Spatial properties of the population of mechanoreceptive units in the glabrous skin of the human hand. *Brain Res.* **184**, 353–366.

Johansson, R. S. and Vallbo, A. B., 1983, Tactile sensory coding in the glabrous skin of the human hand. *Trends in Neurosci.* **6**, 27–32.

Johansson, R. S. and Westling, G., 1984, Roles of glabrous skin receptors and sensorimotor memory in automatic control of precision grip when lifting rougher or more slippery objects. *Exp. Brain Res.* **56**, 550–564.

Johansson, R. S. and Westling, G., 1987, Signals in tactile afferents from the fingers eliciting adaptive motor responses during precision grip. *Exp. Brain Res.* **66**, 141–154.

Johansson, R. S., Vallbo, A. B. and Westling, G., 1980, Thresholds of mechanosensitive afferents in the human hand as measured with von Frey hairs. *Brain Res.* **184**, 343–351.

Johansson, R. S., Landström, U. and Lundström, R., 1982a, Responses of mechanoreceptive afferent units in the glabrous skin of the human hand to sinusoidal skin displacements. *Brain Res.* **244**, 17–25.

Johansson, R. S., Landström, U. and Lundström, R., 1982b, Sensitivity to edges of mechanoreceptive afferent units innervating the glabrous skin of the human hand. *Brain Res.* **244**, 27–32.

Johnson, J. I., Welker, W. I. and Pubols, B. H., 1968, Somatotopic organization of raccoon dorsal column nuclei. *J. Comp. Neurol.* **132**, 1–44.

Johnson, J. I., Hamilton, T. C., Hsung, J. C. and Ulinski, P. S., 1972, Gracile nucleus absent in adult opossums after leg removal in infancy. *Brain Res.* **38**, 421–424.

Johnson, J. L., 1972, Glutamic acid as a synaptic transmitter in the nervous system. A review. *Brain Res.* **37**, 1–19.

Johnson, J. L. and Aprison, M. H., 1970, The distribution of glutamic acid, a transmitter candidate, and other amino acids in the dorsal sensory neuron of the cat. *Brain Res.* **24**, 285–292.

Johnson, K. O., 1974, Reconstruction of population response to a vibratory stimulus in quickly adapting mechanoreceptive afferent fiber population innervating glabrous skin of the monkey. *J. Neurophysiol.* **37**, 48–72.

Johnson, K. O. and Phillips, J. R., 1981, Tactile spatial resolution. I. Two-point discrimination, gap detection, grating resolution and letter recognition. *J. Neurophysiol.* **46**, 1177–1191.

Johnson, K. O., Darian-Smith, I. and LaMotte, C., 1973, Peripheral neural determinants of temperature discrimination in man: a correlative study of responses to cooling of skin. *J. Neurophysiol.* **36**, 347–370.

Johnson, K. O., Darian-Smith, I., LaMotte, C., Johnson, B. and Oldfield, S., 1979, Coding of incremental changes in skin temperature by a population of warm fibers in the monkey: correlation with intensity discrimination in man. *J. Neurophysiol.* **42**, 1332–1353.

Johnston, G. A. R., 1968, The intraspinal distribution of some depressant amino acids. *J. Neurochem.* **15**, 1013–1017.

Johnston, G. A. R. and Vitali, M. V., 1969, Glycine-producing transaminase activity in extracts of spinal cord. *Brain Res.* **12**, 471–472.

Jonakait, G. M., Markey, K. A., Goldstein, M. and Black, I. B., 1984, Transient expression of selected catecholaminergic traits in cranial sensory and dorsal root ganglia of the embryonic rat. *Develop. Biol.* **101**, 51–60.

Jones, E. G., 1985, *The Thalamus*. Plenum Press, New York.

Jones, E. G. and Burton, H., 1974, Cytoarchitecture and somatic sensory connectivity of thalamic nuclei other than the ventrobasal complex in the cat. *J. Comp. Neurol.* **154**, 395–432.

Jones, F. N., 1960, Some subjective magnitude functions for touch. In: *Symposium on Cutaneous Sensibility*. G. R. Hawkes (ed.). Report No. 424, U.S. Army Med. Res. Lab., Ft. Knox, Kentucky.

Jones, M. W., Hodge, C. J., Apkarian, A. V. and Stevens, R. T., 1985, A dorsolateral spinothalamic pathway in cat. *Brain Res.* **335**, 188–193.

Jones, M. W., Apkarian, A. V., Stevens, R. T. and Hodge, C. J., 1987, The spinothalamic tract: an examination of the cells of origin of the dorsolateral and ventral spinothalamic pathways in cats. *J. Comp. Neurol.* **260**, 349–361.

Jones, S. L. and Gebhart, G. F., 1986, Quantitative characterization of ceruleospinal inhibition of nociceptive transmission in the rat. *J. Neurophysiol.* **56**, 1397–1410.

Jones, S. L. and Gebhart, G. F., 1987, Spinal pathways mediating tonic, coeruleospinal, and raphespinal descending inhibition in the rat. *J. Neurophysiol.* **58**, 1–21.

Jones, S. L. and Light, A. R., 1990, Electrical stimulation in the medullary nucleus raphe magnus inhibits noxious heat-evoked *fos* protein-like immunoreactivity in the rat lumbar spinal cord. *Brain Res.* **530**, 335–338.

Jordan, L. M., Kenshalo, D. R., Jr., Martin, R. F., Haber, L. H. and Willis, W. D., 1978, Depression of primate spinothalamic tract neurons by iontophoretic application of 5-hydroxytryptamine. *Pain* **5**, 135–142.

Jordan, L. M., Kenshalo, D. R., Jr., Martin, R. F., Haber, L. H. and Willis, W. D., 1979, Two populations of spinothalamic tract neurons with opposite responses to 5-hydroxytryptamine. *Brain Res.* **164**, 342–346.

Jorum, E., Lundberg, L. E. R. and Torebjörk, H. E., 1989, Peripheral projections of nociceptive unmyelinated axons in the human peroneal nerve. *J. Physiol.* **416**, 291–301.

Joseph, B. S. and Whitlock, D. G., 1968a, Central projections of selected spinal dorsal roots in anuran amphibians. *Anat. Rec.* **160**, 279–288.

Joseph, B. S. and Whitlock, D. G., 1968b, Central projections of brachial and lumbar dorsal roots in reptiles. *J. Comp. Neurol.* **132**, 469–484.

Józsa, R., Korf, H-W. and Merchenthaler, I., 1987, Growth hormone-release factor (GRF)-like immunoreactivity in sensory ganglia of the rat. *Cell Tiss. Res.* **247**, 441–444.

Ju, G., Hökfelt, T., Fischer, J. A., Frey, P., Rehfeld, J. F. and Dockray, G. J., 1986, Does cholecystokinin-like immunoreactivity in rat primary sensory neurones represent calcitonin gene-related peptide? *Neurosci. Lett.* **305**, 310.

Ju, G., Melander, T., Ceccatelli, S., Hökfelt, T. and Frey, P., 1987a, Immunohistochemical evidence for a spinothalamic pathway co-containing cholecystokinin- and galanin-like immunoreactivities in the rat. *Neuroscience* **20**, 439–456.

Ju, G., Hökfelt, T., Brodin, E., Fahrenkrug, J., Fischer, J. A., Frey, P., Elde, R. P. and Brown, J. C., 1987b, Primary sensory neurons of the rat showing calcitonin gene-related peptide immunoreactivity and their relation to substance P-, somatostatin-, galanin-, vasoactive intestinal polypeptide- and cholecystokinin-immunoreactive ganglion cells. *Cell Tiss. Res.* **247**, 417–431.

Jundi, A., Saadé, G. E., Banna, N. R. and Jabbur, S. J., 1982, Modification of transmission in the cuneate nucleus by raphe and periaqueductal gray stimulation. *Brain Res.* **250**, 349–352.

Kaada, B., 1974, Mechanisms of acupuncture analgesia. *Tidsskr. Norske Laegeforen.* **94**, 422–431.

Kadekaro, M., Crane, A. M. and Sokoloff, L., 1985, Differential effects of electrical stimulation of sciatic nerve on metabolic activity in spinal cord and dorsal root ganglion in the rat. *Proc. Natl. Acad. Sci. USA* **82**, 6010–6013.

Kahn, E. A., 1933, Anterolateral chordotomy for intractable pain. *J.A.M.A.* **100**, 1925–1928.

Kahn, E. A. and Peet, M. M., 1948, The technique of anterolateral cordotomy. *J. Neurosurg.* **5**, 276–283.

Kai-Kai, M. A., 1989, Cytochemistry of the trigeminal and dorsal root ganglia and spinal cord of the rat. *Comp. Biochem. Physiol.* **93A**, 183–193.

Kai-Kai, M. A. and Keen, P., 1985, Localization of 5-hydroxytryptamine to neurons and endoneurial mast cells in rat sensory ganglia. *J. Neurocytol.* **14**, 63–78.

Kai-Kai, M. A., Swann, R. W. and Keen, P., 1985, Localization of chromatographically characterized oxytocin and arginine-vasopressin to sensory neurones in the rat. *Neurosci. Lett.* **55**, 83–88.

Kai-Kai, M. A., Anderton, B. H. and Keen, P., 1986, A quantitative analysis of the interrelationships between subpopulations of rat sensory neurons containing arginine vasopressin or oxytocin and those containing substance P, fluoride-resistant acid phosphatase or neurofilament protein. *Neuroscience* **18**, 475–486.

Kajander, K. C. and Giesler, G. J., 1987a, Responses of neurons in the lateral cervical nucleus of the cat to noxious cutaneous stimulation. *J. Neurophysiol.* **57**, 1686–1704.

Kajander, K. C. and Giesler, G. J., 1987b, Effects of repeated noxious thermal stimuli on the responses of neurons in the lateral cervical nucleus of cats: evidence for an input from A-nociceptors to the spinocervicothalamic pathway. *Brain Res.* **436**, 390–395.

Kajander, K. C., Ebner, T. J. and Bloedel, J. R., 1984, Effects of periaqueductal gray and raphe magnus stimulation on the responses of spinocervical and other ascending projection neurons to non-noxious inputs. *Brain Res.* **291**, 29–37.

Kajander, K. C., Sahara, Y., Iadarola, M. J. and Bennett, G. J., 1990, Dynorphin increases in the dorsal spinal cord in rats with a painful peripheral neuropathy. *Peptides* **11**, 719–728.

Kalaska, J. and Pomeranz, B., 1982, Chronic peripheral nerve injuries alter the somatotopic organization of the cuneate nucleus in kittens. *Brain Res.* **236**, 35–47.

Kalia, M., Mei, S. S. and Kao, F. F., 1981, Central projections from ergoreceptors (C fibers) in muscle involved in cardiopulmonary responses to static exercise. *Circulation* **48**, I–48–I–62.

Kalina, M. and Bubis, J. J., 1968, Histochemical studies on the distribution of acid phosphatases in neurons of sensory ganglia. *Histochemie* **14**, 103–112.

Kalina, M. and Bubis, J. J., 1969, Ultrastructural localization of acetylcholine esterase in neurones of rat trigeminal ganglia. *Experimentia* **25**, 388–389.

Kalina, M. and Wolman, M., 1970, Correlative histochemical and morphological study on the maturation of sensory ganglion cells in the rat. *Histochemie* **22**, 100–108.

Kamogawa, H. and Bennett, G. J., 1986, Dorsal column postsynaptic neurons in the cat are excited by myelinated nociceptors. *Brain Res.* **364**, 386–390.

Kanazawa, I., Ogawa, T., Kimura, S. and Munekata, E., 1984, Regional distribution of substance P, neurokinin α and neurokinin β in rat central nervous system. *Neurosci. Res.* **2**, 111–120.

Kane, K. and Taub, A., 1975, A history of local electrical analgesia. *Pain* **1**, 125–138.

Kangrga, I., Larew, J. S. A. and Randić, M., 1990, The effects of substance P and calcitonin gene-related peptide on the efflux of endogenous glutamate and aspartate from the rat spinal dorsal horn in vitro. *Neurosci. Lett.* **108**, 155–160.

Kanui, T. I., 1985, Responses of spinal cord neurones to noxious and non-noxious stimulation of the skin and testicle of the rat. *Neurosci. Lett.* **58**, 315–319.

Kanui, T. I., 1987, Thermal inhibition of nociceptor-driven spinal cord neurones in the cat: a possible neuronal basis for thermal analgesia. *Brain Res.* **402**, 160–163.

Kanui, T. I., 1988, Receptive field organization and electrophysiological responses of spinal cord thermoreactive neurones in the rat. *Exp. Brain Res.* **71**, 508–514.

Kapadia, S. E. and LaMotte, C. C., 1987, Deafferentation-induced alterations in the rat dorsal horn: I. Comparison of peripheral nerve injury vs. rhizotomy effects on presynaptic, postsynaptic, and glial processes. *J. Comp. Neurol.* **266**, 183–197.

Kappers, C. U. A., Huber, G. C. and Crosby, E. C., 1936, *The Comparative Anatomy of the Nervous System of Vertebrates, Including Man*. Hafner, New York. (Reprinted in 1960).

Kar, S., Gibson, S. J., Scaravilli, F., Jacobs, J. M. and Aber, V. R., 1989, Reduced number of calcitonin gene-related peptide-(CGRP-) and tachykinin-immunoreactive sensory neurones associated with greater enkephalin immunoreactivity in the dorsal horn of a mutant rat with hereditary sensory neuropathy. *Cell Tiss. Res.* **255**, 451–466.

Karanjia, P. N. and Ferguson, J. H., 1983, Passive joint position sense after total hip replacement surgery. *Ann. Neurol.* **13**, 654–657.

Karczmar, A. G., Nishi, S., Minota, S. and Kindel, G., 1980, Electrophysiology, acetylcholine and acetylcholinesterase of immature spinal ganglia of the rabbit -- an experiment study and a review. *Gen. Pharmacol.* **11**, 127–134.

Karten, H. J., 1963, Ascending pathways from the spinal cord in the pigeon (*Columba livia*). Proc. 16th Internat. Congr. Zool., Washington, D.C., **2**, 23.

Karten, H. J. and Revzin, A. M., 1966, The afferent connections of the nucleus rotundus in the pigeon. *Brain Res.* **2**, 368–377.

Katan, S., Gottschall, J. and Neuhuber, W., 1982, Simultaneous visualization of horseradish peroxidase and nuclear yellow in tissue sections for neuronal double labeling. *Neuroscience* **28**, 121–126.

Kato, M. and Hirata, Y., 1968, Sensory neurons in the spinal ventral roots of the cat. *Brain Res.* **7**, 479–482.

Kato, M. and Tanji, J., 1971, Physiological properties of sensory fibers in the spinal ventral roots in the cat. Jpn. *J. Physiol.* **21**, 71–77.

Katoh, S., Hisano, S. and Daikoku, S., 1988a, Ultrastructural localization of immunolabeled substance P and methionine-enkephalin-octapeptide in the surface layer of the dorsal horn of rat spinal cord. *Cell Tiss. Res.* **253**, 55–60.

Katoh, S., Hisano, S., Kawano, H., Kagotani, Y. and Daikoku, S., 1988b, Light- and electron-microscopic evidence of costoring of immunoreactive enkephalins and substance P in dorsal horn neurons of rat. *Cell Tiss. Res.* **253**, 297–303.

Katz, D. M., Markey, K. A., Goldstein, M. and Black, I. B., 1983, Expression of catecholaminergic characteristics by primary sensory neurons in the normal adult rat in vivo. *Proc. Natl. Acad. Sci. USA* **80**, 3526–3530.

Kausz, M. and Réthelyi, M., 1985, Lamellar arrangement of neuronal somata in the dorsal root ganglion of the cat. *Somatosensory Res.* **2**, 193–204.

Kawai, Y., Takami, K., Shiosaka, S., Emson, P. C., Hillyard, C. J., Girgis, S., MacIntyre, I. and Tohyama, M., 1985, Topographic localization of calcitonin gene-related peptide in the rat brain: an immunohistochemical analysis. *Neuroscience* **15**, 747–763.

Kawamura, J. and Dyck, P. J., 1978, Evidence for 3 populations by size in L5 spinal ganglion in man. *J. Neuropath. Exp. Neurol.* **37**, 269–272.

Kawamura, Y., Dyck, P. J., Shimono, M., Okazaki, H., Tateishi, J. and Dio, H., 1981, Morphometric comparison of the vulnerability of peripheral motor and sensory neurons in amyotrophic lateral sclerosis. *J. Neuropath. Exp. Neurol.* **40**, 667–675.

Kawatani, M., Lowe, I. P., Nadelhaft, I., Morgan, C. and DeGroat, W. C., 1983, Vasoactive intestinal polypeptide in visceral afferent pathways to the sacral spinal cord of the cat. *Neurosci. Lett.* **42**, 311–316.

Kawatani, M., Erdman, S. L. and DeGroat, W. C., 1985, Vasoactive intestinal polypeptide and substance P in primary afferent pathways to the sacral spinal cord of the cat. *J. Comp. Neurol.* **241**, 327–347.

Kawatani, M., Nagel, J. and DeGroat, W. C., 1986, Identification of neuropeptides in pelvic and pudendal nerve afferent pathways to the sacral spinal cord of the cat. *J. Comp. Neurol.* **249**, 117–132.

Kawatani, M., Takeshige, C. and DeGroat, W. C., 1990, Central distribution of afferent pathways from the uterus of the cat. *J. Comp. Neurol.* **302**, 294–304.

Kayahara, T., 1986, Synaptic connections between spinal motoneurons and dorsal root ganglion cells in the cat. *Brain Res.* **376**, 299–309.

Kayahara, T., Takimoto, T. and Sakashita, S., 1981, Synaptic junctions in the cat spinal ganglion. *Brain Res.* **216**, 277–290.

Kayahara, T., Sakashita, S. and Takimoto, T., 1984, Evidence for spinal origin of neurons synapsing with dorsal root ganglion cells of the cat. *Brain Res.* **293**, 225–230.

Kazimierczak, J., Sommer, E. W., Philippe, E. and Droz, B., 1984, Carbonic anhydrase and acid phosphatase isoenzymes in dorsal root ganglia of mouse and chicken. I. Methodological approach. *Experientia*, **40**, 628.

Kazimierczak, J., Sommer, E. W., Philippe, E. and Droz, B., 1986, Carbonic anhydrase activity in primary sensory neurons. I. Requirements for the cytochemical localization in the dorsal root ganglion of chicken and mouse by light and electron microscopy. *Cell Tiss. Res.* **245**, 487–495.

Keegan, J. J. and Garrett, F. D., 1948, The segmental distribution of the cutaneous nerves in the limbs of man. *Anat. Rec.* **102**, 409–437.

Keele, K.D., 1957, *Anatomies of Pain*. Charles A. Thomas, Springfield.

Keller, J. H. and Hand, P. J., 1970, Dorsal root projections to nucleus cuneatus of the cat. *Brain Res.* **20**, 1–17.

Kellstein, D. E., Price, D. D., Hayes, R. L. and Mayer, D. J., 1990, Evidence that substance P selectively modulates C-fiber-evoked discharges of dorsal horn nociceptive neurons. *Brain Res.* **526**, 291–298.

Kelly, J. S. and Renaud, L. P., 1973a, On the pharmacology of the γ-aminobutyric acid receptors on the cuneo-thalamic relay cells of the cat. *Br. J. Pharmacol.* **48**, 369–386.

Kelly, J. S. and Renaud, L. P., 1973b, On the pharmacology of the glycine receptors on the cuneo-thalamic relay cells in the cat. *Br. J. Pharmacol.* **48**, 387–395.

Kelly, J. S. and Renaud, L. P., 1973c, On the pharmacology of ascending, descending and recurrent postsynaptic inhibition of the cuneo-thalamic relay cells in the cat. *Br. J. Pharmacol.* **48**, 396–408.

Kelly, J. S., Gottesfeld, Z. and Schon, F., 1973, Reduction in radioactivity from the dorsal lateral region of the deafferented rat spinal cord. *Brain Res.* **62**, 581–586.

Kemplay, S. K. and Webster, K. E., 1986, A qualitative and quantitative analysis of the distributions of cells in the spinal cord and spinomedullary junction projecting to the thalamus of the rat. *Neuroscience* **17**, 769–789.

Kenins, P., 1981, Identification of the unmyelinated sensory nerves which evoke extravasation in response to antidromic stimulation. *Neurosci. Lett.* **25**, 137–141.

Kenins, P., 1988, The functional anatomy of the receptive fields of rabbit C polymodal nociceptors. *J. Neurophysiol.* **59**, 1098–1115.

Kenins, P., Hurley, J. V. and Bell, C., 1984, The role of substance P in the axon reflex in the rat. *Br. J. Dermatol.* **111**, 551–559.

Kennard, M. A., 1954, The course of ascending fibres in the spinal cord essential to the recognition of painful stimuli. *J. Comp. Neurol.* **100**, 511–524.

Kenshalo, D. R., 1970, Psychophysical studies of temperature sensitivity. In: *Contribution to Sensory Physiology*, Vol. 4, pp. 19– 74. W. D. Neff (ed.). Academic Press, New York.

Kenshalo, D. R. and Duclaux, R., 1977, Response characteristics of cutaneous cold receptors in the monkey. *J. Neurophysiol.* **40**, 319–332.

Kenshalo, D. R. and Scott, H. A., 1966, Temporal course of thermal adaptation. *Science* **151**, 1095–1096.

Kenshalo, D. R. and Willis, W. D., 1990, The role of the cerebral cortex in pain sensation. In: *The Cerebral Cortex*. A. Peters and E. G. Jones (eds.). Plenum Press, New York.

Kenshalo, D. R., Holmes, C. E. and Wood, P. B., 1968, Warm and cool thresholds as a function of rate of stimulus temperature change. *Perception and Psychophysics* **3**, 81–84.

Kenshalo, D. R., Jr., Leonard, R. B., Chung, J. M. and Willis, W. D., 1979, Responses of primate spinothalamic neurons to graded and to repeated noxious heat stimuli. *J. Neurophysiol.* **42**, 1370–1389.

Kenshalo, D. R., Giesler, G. J., Leonard, R. B. and Willis, W. D., 1980, Responses of neurons in primate ventral posterior lateral nucleus to noxious stimuli. *J. Neurophysiol.* **43**, 1594–1614.

Kenshalo, D. R., Jr., Leonard, R. B., Chung, J. M. and Willis, W. D., 1982, Facilitation of the responses of primate spinothalamic cells to cold and to tactile stimuli by noxious heating of the skin. *Pain* **12**, 141–152.

Kenshalo, D. R., Jr., Anton, F. and Dubner, R., 1989, The detection and perceived intensity of noxious thermal stimuli in monkey and man. *J. Neurophysiol.* **62**, 429–436.

Kenton, B. and Kruger, L., 1971, Information transmission in slowly adapting mechanoreceptor fibers. *Exp. Neurol.* **31**, 114–139.

Kenton, B., Kruger, L. and Woo, M., 1971, Two classes of slowly adapting mechanoreceptor fibres in reptile cutaneous nerve. *J. Physiol.* **212**, 21–44.

Kerr, F. W. L., 1966, The ultrastructure of the spinal tract of the trigeminal nerve and substantia gelatinosa. *Exp. Neurol.* **16**, 359–376.

Kerr, F. W. L., 1968, The descending pathway to the lateral cuneate nucleus, the nucleus of Clarke and the ventral horn. *Anat. Rec.* **160**, 375.

Kerr, F. W. L., 1970a, The fine structure of the subnucleus caudalis of the trigeminal nerve. *Brain Res.* **23**, 129–145.

Kerr, F. W. L., 1970b, The organization of primary afferents in the subnucleus caudalis of the trigeminal: a light and electron microscopic study of degeneration. *Brain Res.* **23**, 147–165.

Kerr, F. W. L., 1975a, Pain: a central inhibitory balance theory. *Mayo Clinic Proc.* **50**, 685–690.

Kerr, F. W. L., 1975b, Neuroanatomical substrates of nociception in the spinal cord. *Pain* **1**, 325–356.

Kerr, F. W. L., 1975c, The ventral spinothalamic tract and other ascending systems of the ventral funiculus of the spinal cord. *J. Comp. Neurol.* **159**, 335–356.

Kerr, F. W. L. and Lippman, H. H., 1974, The primate spinothalamic tract as demonstrated by anterolateral cordotomy and commissural myelotomy. *Advances in Neurology* **4**, 147–156.

Kessler, J. P., Moyse, E., Kitabgi, P., Vincent, J. P. and Beaudet, A., 1987, Distribution of neurotensin binding sites in the caudal brainstem of the cat: a light microscopic autoradiographic study. *Neuroscience* **23**, 189–198.

Kevetter, G. A. and Willis, W. D., 1982, Spinothalamic cells in the rat lumbar cord with collaterals to the medullary reticular formation. *Brain Res.* **238**, 181–185.

Kevetter, G. A. and Willis, W. D., 1983, Collaterals of spinothalamic cells in the rat. *J. Comp. Neurol.* **215**, 453–464.

Kevetter, G. A. and Willis, W. D., 1984, Colateralization in the spinothalamic tract: new methodology to support or deny phylogenetic theories. *Brain Res. Rev.* **7**, 1–14.

Kevetter, G. A., Haber, L. H., Yezierski, R. P., Chung, J. M., Martin, R. F. and Willis, W. D., 1982, Cells of origin of the spinoreticular tract in the monkey. *J. Comp. Neurol.* **207**, 61–74.

Khachaturian, H., Watson, S. J., Lewis, M. E., Coy, D., Goldstein, A. and Akil, H., 1982, Dynorphin immunocytochemistry in the rat central nervous system. *Peptides* **3**, 941–954.

Khattab, F. I., 1968, A complex synaptic apparatus in spinal cords of cats. *Experientia* **24**, 690–691.

Kim, J., Shin, H. K. and Chung, J. M., 1987, Many ventral root afferent fibers in the cat are third branches of dorsal root ganglion cells. *Brain Res.* **417**, 304–314.

Kimura, H., McGeer, P. L., Peng, J. H. and McGeer, E. G., 1981, The central cholinergic system studied by choline acetyltransferase immunohistochemistry in the cat. *J. Comp. Neurol.* **200**, 151–201.

Kimura, S., Okada, M., Sugita, Y., Kanazawa, I. and Munekata, E., 1983, Novel neuropeptides, neurokinin α and β, isolated from porcine spinal cord. *Proc. Jap. Acad.*, Ser. B **59**, 101–104.

King, A. E., Thompson, S. W. N., Urban, L. and Woolf, C. J., 1988, An intracellular analysis of amino acid induced excitations of deep dorsal horn neurones in the rat spinal cord slice. *Neurosci. Lett.* **89**, 286–292.

Kircher, C. and Ha, H., 1968, The nucleus cervicalis lateralis in primates, including the human. *Anat. Rec.* **160**, 376.

Kirchgessner, A. L., Dodd, J. and Gershon, M. D., 1988, Markers shared between dorsal roots and enteric ganglia. *J. Comp. Neurol.* **276**, 607–621.

Kirchhoff, C., Jung, S., Reeh, P. W. and Handwerker, H. O., 1990, Carrageenan inflammation increases bradykinin sensitivity of rat cutaneous nociceptors. *Neurosci. Lett.* **111**, 206–210.

Kirk, E. J. and Denny-Brown, D., 1970, Functional variation in dermatomes in the macaque monkey following dorsal root lesions. *J. Comp. Neurol.* **139**, 307–320.

Kitai, S. T. and Weinberg, J., 1968, Tactile discrimination study of the dorsal column-medial lemniscal system and spino-cervico-thalamic tract in cat. *Exp. Brain Res.* **6**, 234–246.

Kitai, S. T., Ha, H. and Morin, F., 1965, Lateral cervical nucleus of the dog: anatomical and microelectrode studies. *Amer. J. Physiol.* **209**, 307–311.

Klein, C. M., Westlund, K. N. and Coggeshall, R. E., 1990, Percentages of dorsal root axons immunoreactive for galanin are higher than those immunoreactive for calcitonin gene-related peptide in the rat. *Brain Res.* **519**, 97–101.

Klein, H., 1960, Der Nachweis einer oxydativen Stoffwechseisteigerung in der "primar gereitzen" Spinalganglienzelle am Verhalten der Bernsteinsauredehydrogenase. *Arch. Psych. Zeitsch. Neurol.* **201**, 81–96.

Kneisley, L. W., Biber, M. P. and Lavail, J. H., 1978, A study of the origin of brain stem projections to monkey spinal cord using the retrograde transport method. *Exp. Neurol.* **60**, 367–378.

Knibestöl, M., 1973, Stimulus-response functions of rapidly adapting mechanoreceptors in the human glabrous skin area. *J. Physiol.* **232**, 427–452.

Knibestöl, M., 1975, Stimulus-response functions of slowly adapting mechanoreceptors in the human glabrous skin area. *J. Physiol.* **245**, 63–80.

Knibestöl, M. and Vallbo, A. B., 1970, Single unit analysis of mechanoreceptor activity from the human glabrous skin. *Acta Physiol. Scand.* **80**, 178–195.

Knibestöl, M. and Vallbo, A. B., 1980, Intensity of sensation related to activity of slowly adapting mechanoreceptive units in the human hand. *J. Physiol.* **300**, 251–267.

Kniffki, K. D., Mense, S. and Schmidt, R. F., 1977, The spinocervical tract as a possible pathway for muscular nociception. *J. Physiol. (Paris)* **73**, 359–366.

Kniffki, K. D., Mense, S. and Schmidt, R. F., 1978, Responses of group IV afferent units from skeletal muscle to stretch, contraction and chemical stimulation. *Exp. Brain Res.* **31**, 511–522.

Knox, R. J. and Dickenson, A. H., 1987, Effects of selective and non-selective kappa-opioid receptor agonists on cutaneous C-fibre-evoked responses of rat dorsal horn neurones. *Brain Res.* **415**, 21–29.

Knuepfer, M. M., Akeyson, E. W. and Schramm, L. P., 1988, Spinal projections of renal afferent nerves in the rat. *Brain Res.* **446**, 17–25.

Knyihár, E., 1971, Fluoride resistant acid phosphatase system of nociceptive dorsal root afferents. *Experientia* **27**, 1205–1207.

Knyihár, E. and Csillik, B., 1976, Effect of peripheral axotomy on the fine structure and histochemistry of the Rolando substance: degenerative atrophy of central processes of pseudounipolar cells. *Exp. Brain Res.* **26**, 73–87.

Knyihár, E. and Gerebtzoff, M. A., 1973, Extra-lysosomal localization of acid phosphatase in the spinal cord of the cat. *Exp. Brain Res.* **18**, 383–395.

Knyihár, E., Laszlo, I. and Tornyos, S., 1974, Fine structure and fluoride resistant acid phosphatase activity of electron dense sinusoid terminals in the substantia gelatinosa Rolandi of the rat after dorsal root transection. *Exp. Brain Res.* **19**, 529–544.

Knyihár-Csillik, E. and Csillik, B., 1981, FRAP: Histochemistry of the primary sensory nociceptive neuron. *Prog. Histochem. Cytochem.* **14**, 1–137.

Knyihár-Csillik, E., Csillik, B. and Rakić, P., 1982a, Ultrastructure of normal and degenerating glomerular synaptic terminals of dorsal root axons in the substantia gelatinosa of the rhesus monkey. *J. Comp. Neurol.* **210**, 357–375.

Knyihár-Csillik, E., Csillik, B. and Rakić, P., 1982b, Periterminal synaptology of dorsal root glomerular terminals in the substantia gelatinosa of the spinal cord in the Rhesus monkey. *J. Comp. Neurol.* **210**, 376–399.

Knyihár-Csillik, E., Bezzegh, A., Boti, S. and Csillik, B., 1986, Thiamine monophosphatase: a genuine marker for transganglionic regulation of primary sensory neurons. *J. Histochem. Cytochem.* **34**, 363–371.

Knyihár-Csillik, E., Kreutzberg, G. W. and Csillik, B., 1989, Enzyme translocation in the coourse of regeneration of central primary afferent terminals in the substantia gelatinosa of the adult rodent spinal cord. *J. Neurosci. Res.* **22**, 74–82.

Koelle, G. B., 1951, The elimination of enzymatic diffusion artifacts in the histochemical localization of cholinesterases and a survey of their cellular distributions. *J. Pharm. Exp. Ther.* **103**, 153–171.

Koelle, G. B., 1955, The histochemical identification of acetylcholinesterase in cholinergic, adrenergic and sensory neurons. *J. Histochem. Cytochem.* **103**, 167–184.

Koelle, G. B. and Valk, A. De T., 1954, Physiological implications of the histochemical localization of monoamine oxidase. *J. Physiol.* **126**, 434–447.

Koerber, H. R., 1980, Somatotopic organization of cat brachial spinal cord. *Exp. Neurol.* **69**, 481–492.

Koerber, H. R. and Brown, P. B., 1980, Projections of two hindlimb nerves to cat dorsal horn. *J. Neurophysiol.* **44**, 259–269.

Koerber, H. R. and Brown, P. B., 1982, Somatotopic organization of hindlimb cutaneous nerve projections to cat dorsal horn. *J. Neurophysiol.* **48**, 481–489.

Koerber, H. R. and Mendell, L. M., 1988, Functional specialization of central projections from identified primary afferent fibers. *J. Neurophysiol.* **60**, 1597–1614.

Koerber, H. R., Druzinsky, R. E. and Mendell, L. M., 1988, Properties of somata of spinal dorsal root ganglion cells differ according to peripheral receptor innervated. *J. Neurophysiol.* **60**, 1584–1596.

Kohnstamm, O., 1900, Ueber die Coordinationskerne des Hirnstammes und die absteigenden Spinalbahnen. Nach den Ergebnissen der combinierten Degenerationsmethode. *Mschr. Psychiat. Neurol.* **8**, 261–293.

Kojima, M., Takeuchi, Y., Goto, M. and Sato, Y., 1982, Immunohistochemical study of the distribution of serotonin fibers on the spinal cord of the dog. *Cell Tiss. Res.* **226**, 477–491.

Kojima, M., Takeuchi, Y., Goto, M. and Sato, Y., 1983, Immunohistochemical study on the localization of serotonin fibers and terminals in the spinal cord of the monkey. *Cell Tiss. Res.* **229**, 23–36.

Kojima, N. and Kanazawa, I., 1987, Possible neurotransmitters of the dorsal column afferents: effects of dorsal column transection in the cat. *Neuroscience* **23**, 263–274.

Koketsu, K., 1956, Intracellular potential changes of primary afferent nerve fibers in spinal cords of cats. *J. Neurophysiol.* **19**, 375–392.

Kokko, A., 1965, Histochemical and cytochemical observations on esterases in the spinal ganglion of the rat. *Acta Physiol. Scand.* **66**, 1–76.

Kolmodin, G. M., 1957, Integrative processes in single spinal interneurones with proprioceptive connections. *Acta Physiol. Scand.* **40**, Suppl. 139, 1–89.

Kolmodin, G. M. and Skoglund, C. R., 1958, Slow membrane potential changes accompanying excitation and inhibition in spinal moto- and interneurons in the cat during natural activation. *Acta Physiol. Scand.* **44**, 11–54.

Kolmodin, G. M. and Skoglund, C. R., 1960, Analysis of spinal interneurons activated by tactile and nociceptive stimulation. *Acta Physiol. Scand.* **50**, 337–355.

Kondo, M., Fujiwara, H. and Chikako, T., 1985, Autoradiographic evidence for dopaminergic innervation in guinea pig spinal cord. *Jpn. J. Pharmacol.* **38**, 442–444.

Konietzny, F., 1984, Peripheral neural correlates of temperature sensations in man. *Human Neurobiol.* **3**, 21–32.

Konietzny, F. and Hensel, H., 1975, Warm fiber activity in human skin nerves. *Pfluegers Arch.* **359**, 265–267.

Konietzny, F. and Hensel, H., 1977, The dynamic response of warm units in human skin nerves. *Pfluegers Arch.* **370**, 111–114.

Konietzny, F. and Hensel, H., 1979, The neural basis of the sensory quality of warmth. In: *Sensory Functions of the Skin of Humans*, pp. 241–256. D. R. Kenshalo (ed.). Plenum Press, New York.

Konietzny, F. and Hensel, H., 1983, Static and dynamic properties of cold units in human hairy skin. *J. Term. Biol.* **8**, 11–13.

Konietzny, F., Perl, E. R., Trevino, D., Light, A. and Hensel, H., 1981, Sensory experiences in man evoked by intraneural electrical stimulation of intact cutaneous afferent fibers. *Exp. Brain Res.* **42**, 219–222.

Kopaczyk, F., Zabel, J., Silny, W. and Otulakowski, B., 1974, Some histochemical reactions in the neurocytes of the trigeminal ganglion in rabbits. *Folia Morphol.* **2**, 157–164.

Korhonen, L. K. and Hyyppa, M., 1967, Histochemical localization of carbonic anhydrase activity in the spinal and coeliac ganglia of the rat. *Acta Histochem.* Bd. 26, 75–79.

Kosaka, T., Tauchi, M. and Dahl, J. L., 1988, Cholinergic neurons containing GABA-like and/or glutamic acid decarboxylase-like immunoreactivities in various brain regions of the rat. *Exp. Brain Rev.* **70**, 605–617.

Kosinski, R. J., Neafsey, E. J. and Castro, A. J., 1986, A comparative topographical analysis of dorsal column nuclear and cerebral cortical projections to the basilar pontine gray in rats. *J. Comp. Neurol.* **244**, 163–173.

Kostiuk, P. G., 1960, Electrophysiological characteristics of individual spinal cord neurons. *Sechenov Physiol. J. USSR* **46**, 10–22.

Kraus, E., Besson, J. M. and LeBars, D., 1982, Behavioral model for diffuse noxious inhibitory controls (DNIC): potentiation by 5 hydroxytryptophan. *Brain Res.* **231**, 461–465.

Krause, W., 1859, Ueber Nervenendigungen. *Z. Rat. Med.* **5**, 28–43.

Kříž, N., Syková, E., Ujec, E. and Vyklický, L., 1974, Changes of extracellular potassium concentration induced by neuronal activity in the spinal cord of the cat. *J. Physiol.* **238**, 1–15.

Kříž, N., Syková, E. and Vyklický, L., 1975, Extracellular potassium changes in the spinal cord of the cat and their relation to slow potentials, active transport and impulse transmission. *J. Physiol.* **249**, 167–182.

Krnjević, K. and Morris, M. E., 1972, Extracellular K+ activity and slow potential changes in spinal cord and medulla. *Can. J. Physiol. Pharmacol.* **50**, 1214–1217.

Krnjević, K. and Morris, M. E., 1974a, An excitatory action of substance P on cuneate neurones. *Can. J. Physiol. Pharmacol.* **52**, 736–744.

Krnjević, K. and Morris, M. E., 1974b, Extracellular accumulation of K^+ evoked by activity of primary afferent fibres in the cuneate nucleus and dorsal horn of cats. *Can. J. Physiol. Pharmacol.* **52**, 852–871.

Krnjević, K. and Morris, M. E., 1976, Input-output relation of transmission through cuneate nucleus. *J. Physiol.* **257**, 791–815.

Kroll, F. W., 1930, Schwellenuntersuchungen bei Laesionen der afferenten Leitunsbahnen. *Z. Ges. Neurol. Psychiat.* **128**, 751–776.

Kruger, L., 1987, Morphological correlates of "free" nerve endings—a reappraisal of thin sensory axon classification. In: *Fine Afferent Nerve Fibers and Pain*, pp. 1–13. R. F. Schmidt, H. G. Schaible and C. Vahle-Hinz (eds.). VCH Verlagsgesellschaft mbH, Weinheim, Federal Republic of Germany.

Kruger, L. and Kenton, B., 1973, Quantitative neural and psychophysical data for cutaneous mechanoreceptor function. *Brain Res.* **49**, 1–24.

Kruger, L. and Witkovsky, P., 1961, A functional analysis of neurons in the dorsal column nuclei and spinal nucleus of the trigeminal in the reptile (*Alligator mississippiensis*). *J. Comp. Neurol.* **117**, 97–105.

Kruger, L., Siminoff, R. and Witkovsky, P., 1961, Single neuron analysis of dorsal column nuclei and spinal nucleus of trigeminal in cat. *J. Comp. Neurol.* **24**, 333–349.

Kruger, L., Perl, E. R. and Sedivec, M. J., 1981, Fine structure of myelinated mechanical nociceptor endings in cat hairy skin. *J. Comp. Neurol.* **198**, 137–154.

Kruger, L., Sampogna, S. L., Rodin, B. E., Clague, J., Brecha, N. and Yeh, Y., 1985, Thin-fiber cutaneous innervation and its intraepidermal contribution studied by labeling methods and neurotoxin treatment in rats. *Somatosensory Res.* **2**, 335–356.

Kruger, L., Kumazawa, T., Mizumura, K., Sato, J. and Yeh, Y., 1988a, Observations on electrophysiologically characterized receptive fields of thin testicular afferent axons: a preliminary note on the analysis of fine structural specializations of polymodal receptors. *Somatosensory Res.* **5**, 373–380.

Kruger, L., Sternini, C., Brecha, N. C. and Mantyh, P. W., 1988b, Distribution of calcitonin gene-related peptide immunoreactivity in relation to the rat central somatosensory projection. *J. Comp. Neurol.* **273**, 149–162.

Kruger, L., Silverman, J. D., Mantyh, P. W., Sternini, C. and Brecha, N. C., 1989, Peripheral patterns of calcitonin-gene-related peptide general somatic sensory innervation: cutaneous and deep terminations. *J. Comp. Neurol.* **280**, 291–302.

Krukoff, T. L., 1987, Neuropeptide Y-like immunoreactivity in cat spinal cord with special reference to autonomic areas. *Brain Res.* **415**, 300–308.

Krukoff, T. L., Ciriello, J. and Calaresu, F. R., 1986, Somatostatin-like immunoreactivity in neurons, nerve terminals, and fibers of the cat spinal cord. *J. Comp. Neurol.* **243**, 13–22.

Kuenzle, H. and Woodson, W., 1982, Mesodiencephalic and other target regions of ascending spinal projections in the turtle, *Pseudemys scripta elegans*. *J. Comp. Neurol.* **212**, 349–364.

Kuenzle, H. and Woodson, W., 1983, Primary afferent projections to the spinal cord and the dorsal column nuclear complex in the turtle *Pseudemys*. *Anat. Embryol. (Berl.)* **166**, 229–245.

Kumamoto, T. and Bourne, G. H., 1963a, Experimental studies on the oxidative enzymes and hydrolytic enzymes in spinal neurons. *Acta Anat.* **55**, 255–277.

Kumamoto, T. and Bourne, G. H., 1963b, Histochemical localization of respiratory and other hydrolytic enzymes in neuronal lipopigment (Lipofuscin) in old guinea pigs. *Acta Histochem.* Bd. **16**, 87–100.

Kumazawa, T. and Mizumura, K., 1976, The polymodal C-fiber receptor in the muscle of the dog. *Brain Res.* **101**, 589–593.

Kumazawa, T. and Mizumura, K., 1977a, The polymodal receptors in the testis of dog. *Brain Res.* **136**, 553–558.

Kumazawa, T. and Mizumura, K., 1977b, Thin-fibre receptors responding to mechanical, chemical, and thermal stimulation in the skeletal muscle of the dog. *J. Physiol.* **273**, 179–194.

Kumazawa, T. and Mizumura, K., 1979, Effects of synthetic substance P on unit-discharge of testicular nociceptors of dogs. *Brain Res.* **170**, 553–557.

Kumazawa, T. and Mizumura, K., 1980a, Chemical responses of polymodal receptors of the scrotal contents in dogs. *J. Physiol.* **299**, 219–231.

Kumazawa, T. and Mizumura, K., 1980b, Mechanical and thermal responses of polymodal receptors recorded from the superior spermatic nerve of dogs. *J. Physiol.* **299**, 233–245.

Kumazawa, T. and Mizumura, K., 1983, Temperature dependency of the chemical responses of the polymodal receptor units in vitro. *Brain Res.* **278**, 305–307.

Kumazawa, T. and Perl, E. R., 1976, Differential excitation of dorsal horn marginal and substantia gelatinosa neurons by primary afferent units with fine (A-delta and C) fibers. In: *Sensory Functions of the Skin with Special Reference to Man*, pp. 67–89. Y. Zotterman (ed.). Pergamon, Oxford.

Kumazawa, T. and Perl, E. R., 1977, Primate cutaneous sensory units with unmyelinated (C) afferent fibers. *J. Neurophysiol.* **40**, 1325–1338.

Kumazawa, T. and Perl, E. R., 1978, Excitation of marginal and substantia gelatinosa neurons in the primate spinal cord: indications of their place in dorsal horn functional organization. *J. Comp. Neurol.* **177**, 417–434.

Kumazawa, T., Perl, E. R., Burgess, P. R. and Whitehorn, D., 1975, Ascending projections from marginal zone (lamina I) neurons of the spinal dorsal horn. *J. Comp. Neurol.* **162**, 1–12.

Kumazawa, T., Mizumura, K. and Sato, J., 1987, Response properties of polymodal receptors using in vitro testis superior spermatic nerve preparations of dogs. *J. Neurophysiol.* **57**, 702–711.

Kummer, W. and Heym, C., 1986, Correlation of neuronal size and peptide immunoreactivity in the guinea-pig trigeminal ganglion. *Cell Tiss. Res.* **245**, 657–665.

Kuno, M., Muñoz-Martinez, E. J. and Randic, M., 1973, Sensory inputs to neurones in Clarke's column from muscle, cutaneous and joint receptors. *J. Physiol.* **228**, 327–342.

Kunze, W. A. A., Wilson, P. and Snow, P. J., 1987, Response of lumbar spinocervical tract cells to natural and electrical stimulation of the hindlimb footpads in cats. *Neurosci. Lett.* **75**, 253–258.

Kuo, D. C. and De Groat, W. C., 1985, Primary afferent projections of the major splanchnic nerve to the spinal cord and gracile nucleus of the cat. *J. Comp. Neurol.* **231**, 421–434.

Kuo, D. C., Oravitz, J. J., Eskay, R. and DeGroat, W. C., 1984, Substance P in renal afferent perikarya identified by retrograde transport of fluorescent dye. *Brain Res.* **323**, 168–171.

Kuo, D. C., Kawatani, M. and DeGroat, W. C., 1985, Vasoactive intestinal polypeptide identified in the thoracic dorsal root ganglia of the cat. *Brain Res.* **330**, 178–182.

Kuraishi, Y., Hirota, N., Sato, Y., Hino, Y., Satoh, M. and Takagi, H., 1985a, Evidence that substance P and somatostatin transmit separate information related to pain in the spinal dorsal horn. *Brain Res.* **325**, 294–298.

Kuraishi, Y., Hirota, N., Sato, Y., Kaneko, S., Satoh, M. and Takagi, H., 1985b, Noradrenergic inhibition of the release of substance P from the primary afferents in the rabbit spinal dorsal horn. *Brain Res.* **359**, 177–182.

Kuru, M., 1949, *Sensory Paths in the Spinal Cord and Brain Stem of Man*. Sogensya, Tokyo.

Kusuma, A. and ten Donkelaar, H. J., 1980, Dorsal root projections in various types of reptiles. *Brain Behav. Evol.* **17**, 291–309.

Kusunoki, S., Inoue, K., Iwamori, M., Nagai, Y. and Mannen, T., 1989, Discrimination of human dorsal root ganglion cells by anti-fucosyl GM1 antibody. *Brain Res.* **494**, 391–395.

Kuwayama, Y. and Stone, R. A., 1986a, Cholecystokinin-like immunoreactivity occurs in ocular sensory neurons and partially co-localizes with substance P. *Brain Res.* **381**, 266–274.

Kuwayama, Y. and Stone, R. A., 1986b, Neuropeptide immunoreactivity of pericellular baskets in the guinea pig trigeminal ganglion. *Neurosci. Lett.* **64**, 169–172.

Kuwayama, Y., Terenghi, G., Polak, J. M., Trojanowski, J. Q. and Stone, R. A., 1987, A quantitative correlation of substance P-, calcitonin gene-related peptide- and cholecystokinin-like immunoreactivity with retrogradely labeled trigeminal ganglion cells innervating the eye. *Brain Res.* **405**, 220–226.

Kuypers, H. G. J. M., 1958, An anatomical analysis of cortico-bulbar connexions to the pons and lower brain stem in the cat. *J. Anat.* **92**, 198–218.

Kuypers, H. G. J. M., 1981, Anatomy of the descending pathways. In: *Handbook of Physiology*, Section I, pp. 597–666. V. Brooks (ed.). American Physiological Society, Bethesda.

Kuypers, H. G. J. M. and Maisky, V. A., 1975, Retrograde axonal transport of horseradish peroxidase from spinal cord to brain stem cell groups in the cat. *Neurosci. Lett.* **1**, 9–14.

Kuypers, H. G. J. M. and Maisky, V. A., 1977, Funicular trajectories of descending brain stem pathways in cat. *Brain Res.* **136**, 159–165.

Kuypers, H. G. J. M. and Tuerck, J. D., 1964, The distribution of the cortical fibers within the nuclei cuneatus and gracilis in the cat. *J. Anat.* **98**, 143–162.

Laird, J. M. A. and Cervero, F., 1989, A comparative study of the changes in receptive-field properties of multireceptive and nocireceptive rat dorsal horn neurons following noxious mechanical stimulation. *J. Neurophysiol.* **62**, 854–863.

LaMotte, C. C., 1977, Distribution of the tract of Lissauer and the dorsal root fibers in the primate spinal cord. *J. Comp. Neurol.* **172**, 529–562.

LaMotte, C. C. and DeLanerolle, N. C., 1981, Human spinal neurons: innervation by both substance P and enkephalin. *Neuroscience* **6**, 713–723.

LaMotte, C. C. and DeLanerolle, N. C., 1983a, Ultrastructure of chemically defined neuron systems in the dorsal horn of the monkey. II. Methionine-enkephalin immunoreactivity. *Brain Res.* **274**, 51–63.

LaMotte, C. C. and DeLanerole, N. C., 1983b, Ultrastructure of chemically defined neuron systems in the dorsal horn of the monkey. III. Serotonin immunoreactivity. *Brain Res.* **274**, 65–77.

LaMotte, C. C. and DeLanerolle, N. C., 1986, VIP terminals, axons, and neurons: distribution throughout the length of monkey and cat spinal cord. *J. Comp. Neurol.* **249**, 133–145.

LaMotte, C. C., Pert, C. B. and Snyder, S. H., 1976, Opiate receptor binding in primate spinal cord: distribution and changes after dorsal root section. *Brain Res.* **112**, 407–412.

LaMotte, C. C., Johns, D. R. and DeLanerolle, N. C., 1982, Immunohistochemical evidence of indolamine neurons in monkey spinal cord. *J. Comp. Neurol.* **206**, 359–370.

LaMotte, C. C., Carlton, S. M., Honda, C. N., Surmeier, D. J. and Willis, W. D., 1988, Innervation of identified primate spinothalamic tract neurons: ultrastructure of serotonergic and other synaptic profiles. *Neurosci. Abstr.* **14**, 852.

LaMotte, C. C., Kapadia, S. E. and Kocol, C. M., 1989, Deafferentation-induced expansion of saphenous terminal field labelling in the adult rat dorsal horn following pronase injection of the sciatic nerve. *J. Comp. Neurol.* **288**, 311–325.

LaMotte, R. H. and Campbell, J. N., 1978, Comparison of responses of warm and nociceptive C-fiber afferents in monkey with human judgments of thermal pain. *J. Neurophysiol.* **41**, 509–528.

LaMotte, R. H. and Mountcastle, V. B., 1975, Capacities of humans and monkeys to discriminate between vibratory stimuli of different frequency and amplitude: a correlation between neural events and psychophysical measurements. *J. Neurophysiol.* **38**, 539–559.

LaMotte, R. H. and Srinivasan, M. A., 1987, Tactile discrimination of shape: responses of slowly adapting mechanoreceptive afferents to a step stroked across the monkey fingerpad. *J. Neurosci.* **7**, 1655–1671.

LaMotte, R. H. and Thalhammer, J. G., 1982, Response properties of high-threshold cutaneous cold receptors in the primate. *Brain Res.* **244**, 279–287.

LaMotte, R. H. and Whitehouse, J., 1986, Tactile detection of a dot on a smooth surface: peripheral neural events. *J. Neurophysiol.* **56**, 1109–1128.

LaMotte, R. H., Thalhammer, J. G., Torebjörk, H. E. and Robinson, C. J., 1982, Peripheral neural mechanisms of cutaneous hyperalgesia following mild injury by heat. *J. Neurosci.* **2**, 765–781.

LaMotte, R. H., Thalhammer, J. G. and Robinson, C. J., 1983, Peripheral neural correlates of magnitude of cutaneous pain and hyperalgesia: a comparison of neural events in monkey with sensory judgements in human. *J. Neurophysiol.* **50**, 1–26.

LaMotte, R. H., Shain, C. N., Simone, D. A. and Tsai, E. P., 1991, Neurogenic hyperalgesia: psychophysical studies of underlying mechanisms. *J. Neurophysiol.*, in press.

Landgren, S., 1960, Thalamic neurones responding to cooling of the cat's tongue. *Acta Physiol. Scand.* **48**, 255–267.

Landgren, S. and Silfvenius, H., 1969, Projection to cerebral cortex of group I muscle afferents from the cat's hind limb. *J. Physiol.* **200**, 353–372.

Landgren, S. and Silfvenius, H., 1971, Nucleus z, the medullary relay in the projection path to the cerebral cortex of group I muscle afferents from the cat's hind limb. *J. Physiol.* **218**, 551–571.

Landgren, S., Nordwall, A. and Wengström, C., 1965, The location of the thalamic relay in the spino-cervico-lemniscal path. *Acta Physiol. Scand.* **65**, 164–175.

Landgren, S., Silfvenius, H. and Wolsk, D., 1967, Somato-sensory paths to the second cortical projection area of the group I muscle afferents. *J. Physiol.* **191**, 543–559.

Landon, D. N. (ed.), 1976, *The Peripheral Nerve*. John Wiley and Sons, New York.

Lang, E., Novak, A., Reeh, P. and Handwerker, H. O., 1990, Chemosensitivity of fine afferents from rat skin in vitro. *J. Neurophysiol.* **63**, 887–901.

Langford, L. A., 1983, Unmyelinated axon ratios in cat motor, cutaneous and articular nerves. *Neurosci. Lett.* **40**, 19–22.

Langford, L. A. and Coggeshall, R. E., 1979, Branching of sensory axons in the dorsal root and evidence for the absence of dorsal root efferent fibers. *J. Comp. Neurol.* **184**, 193–204.

Langford, L. A. and Coggeshall, R. E., 1981a, Branching of sensory axons in the peripheral nerve of the rat. *J. Comp. Neurol.* **203**, 745–750.

Langford, L. A. and Coggeshall, R. E., 1981b, Unmyelinated axons in the posterior funiculi. *Science* **211**, 176–177.

Langford, L. A. and Schmidt, R. F., 1983, Afferent and efferent axons in the medial and posterior articular nerves of the cat. *Anat. Rec.* **206**, 71–78.

Larsson, H.-H. and Rehfeld, J. F., 1979, Localization and molecular heterogeneity of cholecystokinin in the central and peripheral nervous system. *Brain Res.* **165**, 201–218.

Laurberg, S. and Sorensen, K. E., 1985, Cervical dorsal root ganglion cells with collaterals to both shoulder skin and the diaphragm. A fluorescent double labelling study in the rat. A model for referred pain? *Brain Res.* **331**, 160–163.

Lawson, S. N., 1979, The postnatal development of large light and small dark neurons in mouse dorsal root ganglion: a statistical analysis of cell numbers and size. *J. Neurocytol.* **8**, 275–294.

Lawson, S. N. and Biscoe, T. J., 1979, Development of mouse dorsal root ganglia: an autoradiographic and quantitative study. *J. Neurocytol.* **8**, 265–274.

Lawson, S. N., Caddy, K. W. T. and Biscoe, T. J., 1974, Development of rat dorsal root ganglion neurones. *Cell Tiss. Res.* **153**, 399–413.

Lawson, S. N., Harper, A. A., Harper, E. I., Garson, J. A. and Anderton, B. H., 1984, A monoclonal antibody against neurofilament protein specifically labels a subpopulation of rat sensory neurones. *J. Comp. Neurol.* **228**, 263–272.

Lawson, S. N., Harper, L. I., Harper, A. A., Garson, J. A., Coakham, H. B. and Randle, B. J., 1985, Monoclonal antibody 2C5: a marker for a subpopulation of small neurons in rat dorsal root ganglia. *Neuroscience* **16**, 365–374.

Leah, J. and Menétrey, D., 1989, Neuropeptides in propriospinal neurones in the rat. *Brain Res.* **495**, 173–177.

Leah, J. D., Cameron, A. A. and Snow, P. J., 1985a, Neuropeptides in physiologically identified mammalian sensory neurones. *Neurosci. Lett.* **56**, 257–264.

Leah, J. D., Cameron, A. A., Kelly, W. L. and Snow, P. J., 1985b, Coexistence of peptide immunoreactivity in sensory neurons of the cat. *Neuroscience* **16**, 683–690.

Leah, J., Menétrey, D. and De Pommery, J., 1988, Neuropeptides in long ascending spinal tract cells in the rat: evidence for parallel processing of ascending information. *Neuroscience* **24**, 195–207.

Le Bars, D. and Chitour, D., 1983, Do convergent neurones in the spinal dorsal horn discriminate nociceptive from non-nociceptive information? *Pain* **17**, 1–19.

Le Bars, D., Dickenson, A. H. and Besson, J. M., 1979a, Diffuse noxious inhibitory controls (DNIC). I. Effects on dorsal horn convergent neurons in the rat. *Pain* **6**, 283–304.

Le Bars, D., Dickenson, A. H. and Besson, J. M., 1979b, Diffuse noxious inhibitory controls (DNIC). II: Lack of effect on non-convergent neurones, supraspinal involvement and theoretical implications. *Pain* **6**, 305–327.

Le Bars, D., Chitour, D. and Clot, A. M., 1981a, The encoding of thermal stimuli by diffuse noxious inhibitory controls (DNIC). *Brain Res.* **230**, 394–399.

Le Bars, D., Chitour, D., Kraus, E., Dickenson, A. H. and Besson, J. M., 1981b, Effect of naloxone upon Diffuse Noxious Inhibitory Controls (DNIC) in the rat. *Brain Res.* **204**, 387–402.

Le Bars, D., Bourgoin, S., Clot, A. M., Hamon, M. and Cesselin, F., 1987a, Noxious mechanical stimuli increase the release of Met-enkephalin-like material heterosegmentally in the rat spinal cord. *Brain Res.* **402**, 188–192.

Le Bars, D., Bourgoin, S., Villanueva, L., Clot, A. M., Hamon, M. and Cesselin, F., 1987b, Involvement of the dorsolateral funiculi in the spinal release of Met-enkephalin-like material triggered by heterosegmental noxious mechanical stimuli. *Brain Res.* **412**, 190–195.

Lechan, R. M., Wu, P. and Jackson, I. M. D., 1987, Immunocytochemical distribution in rat brain of putative peptides derived from thyrotropin-releasing hormone prohormone. *Endocrinology* **121**, 1879–1891.

Lee, K. H., Chung, J. M. and Willis, W. D., 1985, Inhibition of primate spinothalamic tract cells by TENS. *J. Neurosurg.* **62**, 276–287.

Lee, K. H., Chung, K., Chung, J. M. and Coggeshall, R. E., 1986, Correlation of cell body size, axon size, and signal conduction velocity for individually labelled dorsal root ganglion cells in the rat. *J. Comp. Neurol.* **243**, 335–346.

Lee, K. S. and Reddington, M., 1986, Autoradiographic evidence for multiple CNS binding sites for adenosine derivatives. *Neuroscience* **19**, 535–549.

Lee, Y., Kawai, Y., Shiosaka, S., Takami, K., Kiyama, H., Hillyard, C. J., Girgis, S., MacIntyre, I., Emson, P. C. and Tohyama, M., 1985, Coexistence of calcitonin gene-related peptide and substance P-like peptide in single cells of the trigeminal ganglion of the rat: immunohistochemical analysis. *Brain Res.* **330**, 194–196.

Leek, B. F., 1972, Abdominal visceral receptors. In: *Handbook of Sensory Physiology*, Vol. III/1, *Enteroceptors*, pp. 113–160. E. Neil, (ed.). Springer, New York.

Leger, L., Charnay, Y., Dubois, P. M. and Jouvet, M., 1986, Distribution of enkephalin-immunoreactive cell bodies in relation of serotonin-containing neurons in the raphe nuclei of the cat: immunohistochemical evidence for the coexistence of enkephalins and serotonin in certain cells. *Brain Res.* **362**, 63–73.

Lehtosalo, J., 1984, Substance P-like immunoreactive trigeminal ganglion cells supplying the cornea. *Histochemistry* **80**, 273–276.

Leichnetz, G. R., Watkins, L., Griffin, G., Murfin, R. and Mayer, D. J., 1978, The projection from nucleus raphe magnus and other brainstem nuclei to the spinal cord in the rat: a study using the HRP blue-reaction. *Neurosci. Lett.* **8**, 119–124.

Leijon, G., Boivie, J. and Johansson, I., 1989, Central post-stroke pain—neurological symptoms and pain characteristics. *Pain* **36**, 13–25.

Leitner, J. M. and Perl, E. R., 1964, Receptors supplied by spinal nerves which respond to cardiovascular changes and adrenaline. *J. Physiol.* **175**, 254–274.

Lele, P. P. and Weddell, G., 1956, The relationship between neurohistology and corneal sensibility. *Brain* **79**, 119–154.

Lele, P. P. and Weddell, G., 1959, Sensory nerves of the cornea and cutaneous sensibility. *Exp. Neurol.* **1**, 334–359.

Lembeck, F. and Holzer, P., 1979, Substance P as a mediator of antidromic vasodilatation and neurogenic plasma extravasation. *Naunyn Schniedebergs Arch. Pharmacol.* **310**, 175–183.

Lenhossék, M. v., 1895, *Der feinere Bau des Nervensystems im Lichte neuester Forschungen*. Kornfeld, Berlin.

Leob, G. E., 1976, Ventral projections of myelinated dorsal root ganglion cells in the cat. *Brain Res.* **106**, 159–165.

Levante, A. and Albe-Fessard, D., 1972, Localisation dans les couches VII et VIII de Rexed des cellules d'origine d'un faisceau spino-reticulaire croise. *C. R. Acad. Soc. Paris* **274**, 3007–3010.

Levante, A., Lamour, Y., Guilbaud, G. and Besson, J. M., 1975, Spinothalamic cell activity in the monkey during intense nociceptive stimulation: intra-arterial injection of bradykinin into the limbs. *Brain Res.* **88**, 560–564.

Levitt, M. and Heybach, J. P., 1981, The deafferentation syndrome in blind rats: a model of the painful phantom limb. *Pain* **10**, 67–73.

Levitt, M. and Levitt, J., 1968, Sensory hind-limb representation in the SmI cortex of the cat after spinal tractotomies. *Exp. Neurol.* **22**, 276–302.

Levitt, M. and Levitt, J. H., 1981, The deafferentation syndrome in monkeys: dysarthrias of spinal origin. *Pain* **10**, 129–147.

Levitt, M. and Schwartzman, R., 1966, Spinal sensory tracts and two-point tactile sensitivity. *Anat. Rec.* **154**, 377.

Levitt, M., Carreras, M., Liu, C. N. and Chambers, W. W., 1964, Pyramidal and extrapyramidal modulation of somatosensory activity in gracile and cuneate nuclei. *Arch. Ital. Biol.* **102**, 197–229.

Levy, R. A., 1974, GABA: A direct depolarizing action at the mammalian primary afferent terminal. *Brain Res.* **76**, 155–160.

Levy, R. A., 1975, The effect of intravenously administered γ-aminobutyric acid on afferent fiber polarization. *Brain Res.* **92**, 21–34.

Levy, R. A. and Anderson, E. G., 1972, The effect of the GABA antagonists bicuculline and picrotoxin on primary afferent terminal excitability. *Brain Res.* **43**, 171–180.

Levy, R. A., Repkin, A. H. and Anderson, E. G., 1971, The effect of bicuculline on primary afferent terminal excitability. *Brain Res.* **32**, 261–265.

Lewis, T., 1942, *Pain*. MacMillan, New York.

Lewis, T. and Pochin, E. E., 1938a, The double pain response of the human skin to a single stimulus. *Clin. Sci.* **3**, 67–76.

Lewis, T. and Pochin, E. E., 1938b, Effects of asphyxia and pressure on sensory nerves of man. *Clin. Sci.* **3**, 141–155.

Lieberman, A. R., 1976, Sensory ganglia. In: *The Peripheral Nerve*, pp. 188–278. D. N. Landon (ed.). Chapman and Hall, London.

Liebeskind, J. C., Guilbaud, G., Besson, J. M. and Oliveras, J. L., 1973, Analgesia from electrical stimulation of the periaqueductal gray matter in the cat: behavioral observations and inhibitory effects on spinal cord interneurons. *Brain Res.* **50**, 441–446.

Light. A. R., 1985, The spinal terminations of single, physiologically characterized axons originating in the pontomedullary raphe of the cat. *J. Comp. Neurol.* **234**, 536–548.

Light, A. R. and Durkovic, R. G., 1984, Features of laminar and somatotopic organization of lumbar spinal cord units receiving cutaneous inputs from hindlimb receptive fields. *J. Neurophysiol.* **52**, 449–458.

Light, A. R. and Kavookjian, A. M., 1985, The ultrastructure and synaptic connections of the spinal terminations from single, physiologically characterized axons descending in the dorsolateral funiculus from the midline, pontomedullary region. *J. Comp. Neurol.* **234**, 549–560.

Light, A. R. and Kavookjian, A. M., 1988, Morphology and ultrastructure of physiologically identified substantia gelatinosa (lamina II) neurons with axons that terminate in deeper dorsal horn laminae (III-V). *J. Comp. Neurol.* **267**, 172–189.

Light, A. R. and Metz, C. B., 1978, The morphology of the spinal cord efferent and afferent neurons contributing the ventral roots of the cat. *J. Comp. Neurol.* **179**, 501–516.

Light, A. R. and Perl, E. R., 1977, Differential termination of large-diameter and small-diameter primary afferent fibers in the spinal dorsal gray matter as indicated by labeling the horseradish peroxidase. *Neurosci. Lett.* **6**, 59–63.

Light, A. R. and Perl, E. R., 1979a, Reexamination of the dorsal root projection to the spinal dorsal horn including observations on the differential termination of coarse and fine fibers. *J. Comp. Neurol.* **186**, 117–132.

Light, A. R. and Perl, E. R., 1979b, Spinal termination of functionally identified primary afferent neurons with slowly conducting myelinated fibres. *J. Comp. Neurol.* **186**, 133–150.

Light, A. R., Trevino, D. L. and Perl, E. R., 1979, Morphological features of functionally defined neurons in the marginal zone and substantia gelatinosa of the spinal dorsal horn. *J. Comp. Neurol.* **186**, 151–172.

Light, A. R., Kavookjian, A. M. and Petrusz, P., 1983, The ultrastructure and synaptic connections of serotonin-immunoreactive terminals in spinal laminae I and II. *Somatosensory Res.* **1**, 33–50.

Light, A. R., Casale, E. J. and Menétrey, D. M., 1986, The effects of focal stimulation in nucleus raphe magnus and periaqueductal gray on intracellularly recorded neurons in spinal laminae I and II. *J. Neurophysiol.* **56**, 555–571.

Lim, R. K. S., 1970, Pain. *Ann. Rev. Physiol.* **32**, 269–288.

Lim, R. K. S., Liu, C. N., Guzman, F. and Braun, C., 1962, Visceral receptors concerned in visceral pain and the pseudoaffective response to intra-arterial injection of bradykinin and other algesic agents. *J. Comp. Neurol.* **118**, 269–294.

Lima, D., 1990, A spinomedullary projection terminating in the dorsal reticular nucleus of the rat. *Neuroscience* **34**, 577–589.

Lima, D. and Coimbra, A., 1983, The neuronal population of the marginal zone (lamina I) of the rat spinal cord. A study based on reconstructions of serially sectioned cells. *Anat. Embryol.* **167**, 273–288.

Lima, D. and Coimbra, A., 1986, a Golgi study of the neuronal population of the marginal zone (lamina I) of the rat spinal cord. *J. Comp. Neurol.* **244**, 53–71.

Lima, D. and Coimbra, A., 1988, The spinothalamic system of the rat: structural types of retrogradely labeled neurons in the marginal zone (lamina I). *Neuroscience* **27**, 215–230.

Lima, D. and Coimbra, A., 1989, Morphological types of spinomesencephalic neurons in the marginal zone (lamina I) of the rat spinal cord, as shown after retrograde labeling with cholera toxin subunit B. *J. Comp. Neurol.* **279**, 327–339.

Lima, D. and Coimbra, A., 1990, Structural types of marginal (Lamina I) neurons projecting to the dorsal reticular nucleus of the medulla oblongata. *Neuroscience* **34**, 591–606.

Lindblom, U., 1962, The relation between stimulus and discharge in a rapidly adapting touch receptor. *Acta Physiol. Scand.* **56**, 349–361.

Lindblom, U. F., 1965, Properties of touch receptors in distal glabrous skin of the monkey. *J. Neurophysiol.* **28**, 966–985.

Lindblom, U., 1974, Touch perception threshold in human glabrous skin in terms of displacement amplitude on stimulation with single mechanical pulses. *Brain Res.* **82**, 205–210.

Lindblom, U. and Lund, L., 1966, The discharge from vibration sensitive receptors in the monkey foot. *Exp. Neurol.* **15**, 401–417.

Lindblom, U. F. and Ottoson, J. O., 1953a, Localization of the structure generating the negative cord dorsum potential evoked by stimulation of low threshold cutaneous fibres. *Acta Physiol. Scand.* **29** (Suppl. 106), 180–190.

Lindblom, U. F. and Ottoson, J. O., 1953b, Effects of spinal sections on the spinal cord potentials elicited by stimulation of low threshold cutaneous fibres. *Acta Physiol. Scand.* **29** (Suppl. 106), 191–208.

Lindh, B., Dalsgaard, C.-J., Elfvin, L.-G., Hökfelt, T. and Cuello, A. C., 1983, Evidence of substance P immunoreactive neurons in dorsal root ganglia and vagal ganglia projecting to the guinea pig pylorus. *Brain Res.* **269**, 365–369.

Lindström, S. and Takata, M., 1972, Monosynaptic excitation of dorsal spinocerebellar tract neurones from low threshold joint afferents. *Acta Physiol. Scand.* **84**, 430–432.

Lisney, S. J. W., 1983, Changes in somatotopic organization of the cat lumbar spinal cord following peripheral nerve transection and regeneration. *Brain Res.* **259**, 31–39.

Lissauer, H., 1885, Beitrag zur pathologischen Anatomie der Tabes dorsalis und sum Faserverlauf im menschlichen Ruckenmark. *Neurol. Centralblatt.* **4**, 245–246.

Lissauer, H., 1886, Beitrag zum Faserverlauf im Hinterhorn des menschlichen Ruckenmark und zum Verhalten desselben bei Tabes dorsalis. *Arch. Psychiat. Nervenkrankh.* **17**, 377–438.

Liu, C. N., 1956, Afferent nerves to Clarke's and the lateral cuneate nuclei in the cat. *Arch. Neurol. Psychiat.* **75**, 67–77.

Liu, C. N. and Chambers, W. W., 1958, Intraspinal sprouting of dorsal root axons. *Arch. Neurol. Psychiat.* **79**, 46–61.

Liu, R. H., Jing-shi, T. and Zong-lian, H., 1989, Electrophysiological identification of spinally projecting neurons in the lateral reticular nucleus of the rat. *Brain Res.* **481**, 350–355.

Liu, R. P., 1983, Laminar origins of spinal projection to the periaqueductal gray of the rat. *Brain Res.* **264**, 118–122.

Liu, R. P. C., 1986, Spinal neuronal collaterals to the intralaminar thalamic nuclei and periaqueductal gray. *Brain Res.* **365**, 145–150.

Ljungdahl, Å. and Hökfelt, T., 1973, Autoradiographic uptake patterns of [^{3}H]GABA and [^{3}H]glycine in control neurons tissues with special reference to cat spinal cord. *Brain Res.* **62**, 587–595.

Ljungdahl, Å., Hökfelt, T. and Nilsson, G., 1978, Distribution of substance P-like immunoreactivity in the central nervous system of the rat. I. Cell bodies and nerve terminals. *Neuroscience* **3**, 861–943.

Lloyd, D. P. C., 1952, Electrotonus in dorsal nerve roots. *Cold Spring Harbor Symp. Quant. Biol.* **17**, 203–219.

Lloyd, D. P. C. and Chang, H. T., 1948, Afferent fibers in muscle nerves. *J. Neurophysiol.* **11**, 199–208.

Lloyd, D. P. C. and McIntyre, A. K., 1949, On the origins of dorsal root potentials. *J. Gen. Physiol.* **32**, 409–443.

Lloyd, D. P. C. and McIntyre, A. K., 1950, Dorsal column conduction of group I muscle afferent impulses and their relay through Clarke's column. *J. Neurophysiol.* **13**, 39–54.

Loeser, J. D. and Ward, A. A., 1967, Some effects of deafferentation on neurons of the cat spinal cord. *Arch. Neurol.* **17**, 629–636.

Loeser, J. D., Black, R. G. and Christman, A., 1975, Relief of pain by transcutaneous stimulation. *J. Neurosurg.* **42**, 308–314.

Loewenstein, W. R. and Skalak, R., 1966, Mechanical transmission in a Pacinian corpuscle. An analysis and a theory. *J. Physiol.* **182**, 346–378.

Lofvenberg, J. and Johansson, R. S., 1984, Regional differences and interindividual variability in sensitivity to vibration in the glabrous skin of the human hand. *Brain Res.* **301**, 65–72.

Lombard, M. C., Nashold, B. S., Albe-Fessard, D., Salman, N. and Sakr, C., 1979, Deafferentation hypersensitivity in the rat after dorsal rhizotomy: a possible animal model of chronic pain. *Pain* **6**, 163–175.

Long, D. M., 1973, Electrical stimulation for relief of pain from chronic nerve injury. *J. Neurosurg.* **39**, 718–722.

Long, D. M. and Hagfors, N., 1975, Electrical stimulation in the nervous system: the current status of electrical stimulation of the nervous system for relief of pain. *Pain* **1**, 109–123.

Long, R. R., 1977, Sensitivity of cutaneous cold fibers to noxious heat: paradoxical cold discharge. *J. Neurophysiol.* **40**, 489–502.

Longhurst, J. C., Mitchell, J. H. and Moore, M. B., 1980, The spinal cord ventral root: an afferent pathway of the hind-limb pressor reflex in cats. *J. Physiol.* **301**, 467–476.
Loren, I., Aluments, J., Hakanson, R. and Sundler, F., 1979, Distribution of gastrin and CCK-like peptides in rat brain. *Histochemistry* **59**, 249–257.
Lothman, E. W. and Somjen, G. G., 1975, Extracellular potassium activity, intracellular and extracellular potential responses in the spinal cord. *J. Physiol.* **252**, 115–136.
Low, J. S. T., Mantle-St.John, L. A. and Tracey, D. J., 1986, Nucleus z in the rat: spinal afferents from collaterals of dorsal spinocerebellar tract neurons. *J. Comp. Neurol.* **243**, 510–526.
Lu, G. W., Bennett, G. J., Nishikawa, N., Hoffert, M. J. and Dubner, R., 1983, Extra- and intracellular recordings from dorsal column postsynaptic spinomedullary neurons in the cat. *Exp. Neurol.* **82**, 456–477.
Lu, G. W., Bennett, G. J., Nishikawa, N. and Dubner, R., 1985, Spinal neurons with branched axons traveling in both the dorsal and dorsolateral funiculi. *Exp. Neurol.* **87**, 571–577.
Lucas, M. E. and Willis, W. D., 1974, Identification of muscle afferents which activate interneurons in the intermediate nucleus. *J. Neurophysiol.* **37**, 282–293.
Lukas, Z., Cech, S. and Burianek, P., 1970, Cholinesterases and biogenic monoamines in ganglion semilunare (Gasseri). *Histochemie* **22**, 163–168.
Lund, R. D. and Webster, K. E., 1967a, Thalamic afferents from the dorsal column nuclei. An experimental study in the rat. *J. Comp. Neurol.* **130**, 301–312.
Lund, R. D. and Webster, K. E., 1967b, Thalamic afferents from the spinal cord and trigeminal nuclei. An experimental anatomical study in the rat. *J. Comp. Neurol.* **130**, 313–328.
Lundberg, A. and Norrsell, U., 1960, Spinal afferent pathway of the tactile placing reaction. *Experientia* **16**, 123.
Lundberg, A. and Oscarsson, O., 1961, Three ascending spinal pathways in the dorsal part of the lateral funiculus. *Acta Physiol. Scand.* **51**, 1–16.
Lundberg, A., Norrsell, U. and Voorhoeve, P., 1963, Effects from the sensorimotor cortex on ascending spinal pathways. *Acta Physiol. Scand.* **59**, 462–473.
Lundberg, J. M. and Hökfelt, T., 1986, Multiple co-existence of peptides and classical transmitters in peripheral autonomic and sensory neurons -- functional and pharmacological implications. In: *Progress in Brain Research*, pp. 241–263. T. Hökfelt, K. Fuxe and B. Pernow (eds.). Elsevier, Amsterdam.
Lundberg, J. M., Hökfelt, T., Nilsson, G., Terenius, L., Rehfeld, J., Elde, R. and Said, S., 1978, Peptide neurons in the vagus, splanchnic and sciatic nerves. *Acta Physiol. Scand.* **104**, 499–501.
Lundberg, J. M., Hökfelt, T., Anggard, A., Kimmel, J., Goldstein, M. and Markey, K., 1980, Co-existence of an avian pancreatic polypeptide (AAP) immunoreactive substance and catecholamines in some peripheral and central neurons. *Acta Physiol. Scand.* **110**, 107–109.
Lundeberg, T., Nordemar, R. and Ottoson, D., 1984, Pain alleviation by vibratory stimulation. *Pain* **20**, 25–44.
Luo, C. B., Zheng, D. R., Guan, Y. L. and Yew, D. T., 1988, Localization of substance P and enkephalin by immunohistochemistry in the spinal cord of human fetus. *Neuroscience* **27**, 989–993.
Lynch, D. R., Strittmatter, S. M., Venable, J. C. and Snyder, S. H., 1986, Enkephalin convertase: Localization to specific neuronal pathways. *J. Neurosci.* **6**, 1662–1675.
Lynn, B., 1969, The nature and location of certain phasic mechanoreceptors in the cat's foot. *J. Physiol.* **201**, 765–773.
Lynn, B., 1971, The form and distribution of the receptive fields of Pacinian corpuscles found in and around the cat's large foot pad. *J. Physiol.* **217**, 755–771.
Lynn, B. and Carpenter, S. E., 1982, Primary afferent units from the hairy skin of the rat hind limb. *Brain Res.* **238**, 29–43.
Lynn, B. and Hunt, S., 1984, Afferent C-fibres: physiological and biochemical correlations. *TINS* **7**, 186–188.
Ma, W., Peschanski, M. and Besson, J. M., 1986, The overlap of spinothalamic and dorsal column nuclei projections in the ventrobasal complex of the rat thalamus: a double anterograde labeling study using light microscopy analysis. *J. Comp. Neurol.* **245**, 531–540.
MacDermott, A. B., Mayer, M. L., Westbrook, G. L., Smith, S. J. and Barker, J. L., 1986, NMDA-receptor activation increases cytoplasmic calcium concentration in cultured spinal cord neurones. *Nature* **321**, 519–522.
MacDonald, J. F., Porietis, A. V. and Wojtowicz, J. M., 1982, L-aspartic acid induces a region of negative slope conductance in the current-voltage relationship of cultured spinal cord neurons. *Brain Res.* **237**, 248–253.
Macefield, G., Gandevia, S. C. and Burke, D., 1990, Perceptual responses to microstimulation of single afferents innervating joints, muscles and skin of the human hand. *J. Physiol.* **429**, 113–129.
MacKenzie, J., 1893, Some points bearing on the association of sensory disorders and visceral disease. *Brain* **16**, 321–354.
MacKenzie, R. A., Burke, D., Skuse, N. F. and Lethlean, A. K., 1975, Fibre function and perception during cutaneous nerve block. *J. Neurol. Neurosurg. Psychiat.* **38**, 865–873.
Macon, J. B., 1979, Deafferentation hyperactivity in the monkey spinal trigeminal nucleus: neuronal responses to amino acid iontophoresis. *Brain Res.* **161**, 549–554.
Maderut, J. L., Yaksh, T. L., Petruz, P. and Go, V. L. W., 1982, Origin and distribution of cholecystokinin-containing nerve terminals in the lumbar dorsal horn and nucleus caudalis of the cat. *Brain Res.* **243**, 363–368.
Magendie, F., 1822, Experiences sur les fonctions des racines des nerfs rachidiens. *J. Physiol. Exp. Pathol.* **2**, 276–279. Reprinted in Cranefield, P.F. (1974).
Magerl, W., Szolcsanyi, J., Westerman, R. A. and Handwerker, H. O., 1987, Laser doppler measurements of skin

vasodilation elicited by percutaneous electrical stimulation of nociceptors in humans. *Neurosci. Lett.* **82**, 349–354.

Magherini, P. C., Pompeiano, O. and Seguin, J. J., 1974, The response of nucleus z neurons to sinusoidal stretch of hindlimb extensor muscles. *Brain Res.* **73**, 343–349.

Magherini, P. C., Pompeiano, O. and Seguin, J. J., 1975, Responses of nucleus z neurons to vibration of hindlimb extensor muscles in the decerebrate cat. *Arch. Ital. Biol.* **113**, 150–187.

Magni, F., Melzack, R., Moruzzi, G. and Smith, C. J., 1959, Direct pyramidal influences on the dorsal-column nuclei. *Arch. Ital. Biol.* **97**, 357–377.

Magnuson, D. S. K., Curry, K., Peet, M. J. and McLennan, H., 1988, Structural requirements for activation of excitatory amino acid receptors in the rat spinal cord in vitro. *Exp. Brain Res.* **73**, 541–545.

Magnusson, K. R., Larson, A. A., Madl, J. E., Altschuler, R. A. and Beitz, A. J., 1986, Co-localization of fixative-modified glutamate and glutaminase in neurons of the spinal trigeminal nucleus of the rat: an immunohistochemical and immunoradiochemical analysis. *J. Comp. Neurol.* **247**, 477–490.

Magnusson, K. R., Clements, J. R., Larson, A. A., Madl, J. E. and Beitz, A. J., 1987, Localization of glutamate in trigeminothalamic projection neurons: a combined retrograde transport-immunohistochemical study. *Somatosensory Res.* **4**, 177–190.

Magoul, R., Onteniente, B., Geffard, M. and Calas, A., 1987, Anatomical distribution and ultrastructural organization of the GABAergic system in the rat spinal cord. An immunocytochemical study using anti-GABA antibodies. *Neuroscience* **20**, 1001–1009.

Magoun, H. W., 1963, *The Waking Brain*, 2nd ed. Thomas, Springfield.

Mai, J. K., Triepel, J. and Metz, J., 1987, Neurotensin in the human brain. *Neuroscience* **22**, 499–524.

Maixner, W., Dubner, R., Bushnell, M. C., Kenshalo, D. R., Jr. and Oliveras, J. L., 1986, Wide-dynamic-range neurons participate in the encoding process by which monkeys perceive the intensity of noxious heat stimuli. *Brain Res.* **374**, 385–388.

Maixner, W., Bushnell, M. C., Dubner, R., Kenshalo, D. R., Jr. and Oliveras, J. L., 1989, Responses of monkey medullary dorsal horn neurons during the detection of noxious heat stimuli. *J. Neurophysiol.* **62**, 437–449.

Malatova, Z., Longauer, F. and Marsala, J., 1985, Choline acetyltransferase and acetylcholinesterase in canine spinal ganglia: increase of choline acetyltransferase activity following sciatic nerve lesion. *J. Hirnforsch* **26**, 683–688.

Malchiodi, F., Rambourg, A., Clermont, Y. and Caroff, A., 1986, Ultrastructural localization of concanavalin A-binding sites in the Golgi apparatus of various types of neurons in rat dorsal root ganglia: functional implications. *Amer. J. Anat.* **177**, 81–95.

Malinovsky, L., 1966, Variability of sensory nerve endings in foot pads of a domestic cat (Felis ocreata L., F. domestica). *Acta Anat.* **64**, 82–106.

Manaker, S., Winokur, A., Rhodes, C. H. and Rainbow, T. C., 1985a, Autoradiographic localization of thyrotropin-releasing hormone (TRH) receptors in human spinal cord. *Neurology* **35**, 328–332.

Manaker, S., Winokur, A., Rostene, W. H. and Rainbow, T. C., 1985b, Autoradiographic localization of thyrotropin-releasing hormone receptors in the rat central nervous system. *J. Neurosci.* **5**, 167–174.

Mann, M. D., Kasprzak, H. and Tapper, D. N., 1971, Ascending dorsolateral pathways relaying type I afferent activity. *Brain Res.* **27**, 176–178.

Mann, M. D., Kasprzak, H., Hiltz, F. L. and Tapper, D. N., 1972, Activity in single cutaneous afferents: spinal pathways and cortical evoked potentials. *Brain Res.* **39**, 61–70.

Mannen, H., 1975, Reconstruction of axonal trajectory of individual neurons in the spinal cord using Golgi-stained serial sections. *J. Comp. Neurol.* **159**, 357–374.

Mannen, N. and Sugiura, Y., 1976, Construction of neurons of dorsal horn proper using Golgi-stained serial sections. *J. Comp. Neurol.* **168**, 303–312.

Manocha, S. L., 1973, Experimental protein malnutrition in primates histochemical studies on the dorsal root ganglion cells of healthy and malnourished squirrel monkeys, *Saimiri sciureus. Acta Histochem.* **47**, 220–232.

Mantle-St. John, L. A. and Tracey, D. J., 1987, Somatosensory nuclei in the brainstem of the rat: independent projections to the thalamus and cerebellum. *J. Comp. Neurol.* **255**, 259–271.

Mantyh, P. W., 1982, The ascending input to the midbrain periaqueductal gray of the primate. *J. Comp. Neurol.* **211**, 50–64.

Mantyh, P. W., 1983a, The spinothalamic tract in the primate: a re-examination using wheatgerm agglutinin conjugated to horseradish peroxidase. *Neuroscience* **9**, 847–862.

Mantyh, P. W., 1983b, The terminations of the spinothalamic tract in the cat. *Neurosci. Lett.* **38**, 119–124.

Mantyh, P. W. and Hunt, S. P., 1985a, Thyrotropin-releasing hormone (TRH) receptors. *J. Neurosci.* **5**, 551–561.

Mantyh, P. W. and Hunt, S. P., 1985b, The autoradiographic localization of substance P receptors in the rat and bovine spinal cord and the cat spinal trigeminal nucleus pars caudalis and the effects of neonatal capsaicin. *Brain Res.* **332**, 315–324.

Mantyh, P. W., Pinnock, R. D., Downes, C. P., Goedert, M. and Hunt, S. P., 1984, Correlation between inositol phospholipid hydrolysis and substance P receptors in rat CNS. *Nature* **309**, 795–797.

Marchand, R. and Barbeau, H., 1982, Vertically oriented alternating acetylcholinesterase rich and poor territories in laminae VI, VII, VIII of the lumbosacral cord of the rat. *Neuroscience* **7**, 1197–1202.

Marinesco, G., 1909, *La Cellule Nerveuse*. Doin et fils, Paris.

Mark, R. F. and Steiner, J., 1958, Cortical projection of impulses in myelinated cutaneous afferent nerve fibres of the cat. *J. Physiol.* **142**, 544–562.
Markus, H. and Pomeranz, B., 1987, Saphenous has weak ineffective synapses in sciatic territory of rat spinal cord: Electrical stimulation of the saphenous or application of drugs reveal these somatotopically inappropriate synapses. *Brain Res.* **416**, 315–321.
Markus, H., Pomeranz, B. and Krushelnycky, D., 1984, Spread of saphenous projection map in spinal cord and hypersensitivity of the foot after chronic sciatic enervation in adult rat. *Brain Res.* **296**, 27–39.
Marley, P. D., Nagy, J. I., Emson, P. C. and Rehfeld, J. F., 1982, Cholecystokinin in the rat spinal cord: distribution and lack of effect of neonatal capsaicin treatment and rhizotomy. *Brain Res.* **238**, 494–498.
Marti, E., Gibson, S. J., Polak, J. M., Facer, P., Springall, D. R., Van Aswegen, G., Aitchison, M. and Koltzenburg, M., 1987, Ontogeny of peptide- and amine-containing neurones in motor, sensory, and autonomic regions of rat and human spinal cord, dorsal root ganglia, and rat skin. *J. Comp. Neurol.* **266**, 332–359.
Martin, G. F., Megirian, D. and Roebuck, A., 1971, Corticobulbar projections of the marsupial phalanger (*Trichosurus vulpecula*). I. Projections to the pons and medulla oblongata. *J. Comp. Neurol.* **142**, 275–296.
Martin, G. F., Cabana, F. J., Ho, R. H. and Humbertson, A. O., 1982, Raphespinal projections in the North American opossum: Evidence for connectional heterogeneity. *J. Comp. Neurol.* **208**, 67–84.
Martin, H. A., 1990, Leukotriene B_4 induced decrease in mechanical and thermal thresholds of C-fiber mechanonociceptors in rat hairy skin. *Brain Res.* **509**, 273–279.
Martin, H. F. and Manning, J. W., 1969, Rapid thermal cutaneous stimulation: peripheral nerve responses. *Brain Res.* **16**, 524–526.
Martin, H. F. and Manning, J. W., 1971, Thalamic "warming" and "cooling" units responding to cutaneous stimulation. *Brain Res.* **27**, 377–381.
Martin, H. F. and Manning, J. W., 1972, Response of A-delta-fibers of peripheral nerve to warming of cutaneous fields. *Brain Res.* **43**, 653–656.
Martin, R. F., Jordan, L. M. and Willis, W. D., 1978, Differential projections of cat medullary raphe neurons demonstrated by retrograde labelling following spinal cord lesions. *J. Comp. Neurol.* **182**, 77–88.
Martin, R. F., Haber, L. H. and Willis, W. D., 1979, Primary afferent depolarization of identified cutaneous fibers following stimulation in medial brain stem. *J. Neurophysiol.* **42**, 779–790.
Martin, R. J., Apkarian, A. V. and Hodge, C. J., 1990, Ventrolateral and dorsolateral ascending spinal cord pathway influence on thalamic nociception in cat. *J. Neurophysiol.* **64**, 1400–1412.
Martin, W. R., 1984, Pharmacology of opioids. *Pharmacol. Rev.* **35**, 283–323.
Martinez-Rodriguez, R. and Diaz, G., 1987, Immunocytochemical and histoenzymological studies of aspartate aminotransferase (AAT) in the spinal cord of rat. *Cell. Mol. Biol.* **33**, 159–165.
Maruhashi, J., Mizuguchi, K. and Tasaki, I., 1952, Action currents in single afferent nerve fibres elicited by stimulation of the skin of the toad and the cat. *J. Physiol.* **117**, 129–151.
Massari, V. J., Tizabi, Y. T., Park, C. H., Moody, T. W., Helke, C. J. and O'Donohue, T. L., 1983, Distribution and origin of bombesin, substance P and somatostatin in cat spinal cord. *Peptides* **4**, 673–681.
Massari, V. J., Shults, C. W., Park, C. H., Tizabi, Y., Moody, T. W., Cronwall, B. M., Culver, M. and Chase, T. N., 1985, Deafferentation causes a loss of presynaptic bombesin receptors and supersensitivity of substance P receptors in the dorsal horn of the cat spinal cord. *Brain Res.* **343**, 268–274.
Massopust, L. C., Hauge, D. H., Ferneding, J. C., Doubek, W. G. and Taylor, J. J., 1985, Projection systems and terminal localization of dorsal column afferents: an autoradiographic and horseradish peroxidase study in the rat. *J. Comp. Neurol.* **237**, 533–544.
Matsas, R., Kenny, A. J. and Turner, A. J., 1986, An immunohistochemical study of endopeptidase-24.11 ("enkephalinase") in the pig nervous system. *Neuroscience* **18**, 991–1012.
Matsushita, M., 1969, Some aspects of the interneuronal connections in cat's spinal grey matter. *J. Comp. Neurol.* **136**, 57–80.
Matsushita, M., 1970, The axonal pathways of spinal neurons in the cat. *J. Comp. Neurol.* **138**, 391–418.
Matsuura, H., 1967, Histochemical observation of bovine spinal ganglia. *Histochemie* **11**, 152–160.
Matsuura, H., Mori, M. and Kawakatsu, K., 1969, A histochemical and electron-microscopic study of the trigeminal ganglion of the rat. *Arch. Oral Biol.* **14**, 1135–1146.
Matsuura, H., Hirose, I. and Fujita, K., 1970, Electron microscopic localization of alkaline phosphatase in the trigeminal ganglion of the rat. *Histochemie* **23**, 91–97.
Matsuyama, T., Wanaka, A., Yoneda, S., Kimura, K., Kamada, T., Girgis, S., MacIntyre, I., Emson, P. C. and Tohyama, M., 1986, Two distinct calcitonin gene-related-containing peripheral nervous systems: distribution and quantitative differences between the iris and cerebral artery with special reference to substance P. *Brain Res.* **373**, 205–212.
Matthews, M. A., McDonald, G. K. and Hernandez, T. V., 1988, GABA distribution in a pain-modulating zone of trigeminal subnucleus interpolaris. *Somatosensory Res.* **5**, 205–217.
Matthews, P. B. C., 1964, Muscle spindles and their motor control. *Physiol. Rev.* **44**, 219–288.
Matthews, P. B. C., 1972, *Mammalian Muscle Receptors and Their Central Actions*. Williams and Wilkins, Baltimore.
Matthews, P. B. C., 1977, Muscle afferents and kinaesthesia. *Br. Med. Bull.* **33**, 137–142.

Matthews, P. B. C., 1981, Evolving views on the internal operation and functional role of the muscle spindle. *J. Physiol.* **320**, 1–30.
Matthews, P. B. C., 1982, Where does Sherrington's "muscular sense" originate? Muscles, joints, corollary discharges? *Ann. Rev. Neurosci.* **5**, 189–218.
Matthews, P. B. C. and Simmonds, A., 1974, Sensations of finger movement elicited by pulling upon flexor tendons in man. *J. Physiol.* **239**, 27–28P.
Matzke, H. A., 1951, The course of the fibers arising from the nucleus gracilis and cuneatus of the cat. *J. Comp. Neurol.* **94**, 439–452.
Maunz, R. A., Pitts, N. G. and Peterson, B. W., 1978, Cat spinoreticular neurons: locations, responses and changes in responses during repetitive stimulation. *Brain Res.* **148**, 365–379.
Mawe, G. M., Bresnahan, J. C. and Beattie, M. S., 1984, Primary afferent projections from dorsal and ventral roots to autonomic preganglionic neurons in the cat sacral spinal cord: light and electron microscopic observations. *Brain Res.* **290**, 152–157.
Mawe, G. M., Bresnahan, J. C. and Beattie, M. S., 1986, A light and electron microscopic analysis of the sacral parasympathetic nucleus after labelling primary afferent and efferent elements with HRP. *J. Comp. Neurol.* **250**, 33–57.
Maxwell, D. J., 1985, Combined light and electron microscopy of Golgi-labelled neurons in lamina III of feline spinal cord. *J. Anat.* **141**, 155–169.
Maxwell, D. J. and Bannatyne, B. A., 1983, Ultrastructure of muscle spindle afferent terminations in lamina VI of the cat spinal cord. *Brain Res.* **288**, 297–301.
Maxwell, D. J. and Koerber, H. R., 1986, Fine structure of collateral axons originating from feline spinocervical tract neurons. *Brain Res.* **363**, 199–203.
Maxwell, D. J. and Noble, R., 1987, Relationships between hair-follicle afferent terminations and glutamic acid decarboxylase-containing boutons in the cat's spinal cord. *Brain Res.* **408**, 308–312.
Maxwell, D. J. and Réthelyi, M., 1987, Ultrastructure and synaptic connections of cutaneous afferent fibres in the spinal cord. *TINS* **10**, 117–122.
Maxwell, D. J., Bannatyne, B. A., Fyffe, R. E. W. and Brown, A. G., 1982a, Ultrastructure of hair follicle afferent fibre terminations in the spinal cord of the cat. *J. Neurocytol.* **11**, 571–582.
Maxwell, D. J., Fyffe, R. E. W. and Brown, A. G., 1982b, Fine structure of spinocervical tract neurones and the synaptic boutons in contact with them. *Brain Res.* **233**, 394–399.
Maxwell, D. J., Fyffe, R. E. W. and Réthelyi, M., 1983a, Morphological properties of physiologically characterized lamina III neurones in the cat spinal cord. *Neuroscience* **10**, 1–22.
Maxwell, D. J., Leranth, C. and Vertrofstad, A. A., 1983b, Fine structure of serotonin containing axons in the marginal zone of the rat spinal cord. *Brain Res.* **266**, 253–259.
Maxwell, D. J., Bannatyne, B. A., Fyffe, R. E. W. and Brown, A. G., 1984a, Fine structure of primary afferent axon terminal projecting from rapidly adapting mechanoreceptors of the toe and foot pads of the cat. *Quart. J. Exp. Physiol.* **69**, 381–392.
Maxwell, D. J., Fyffe, R. E. W. and Brown, A. G., 1984b, Fine structure of normal and degenerating primary afferent boutons associated with characterized spinocervical tract neurons in the cat. *Neuroscience* **12**, 151–163.
Maxwell, D. J., Christie, W. M. and Somogyi, P., 1989, Synaptic connections of GABA-containing boutons in the lateral cervical nucleus of the cat: an ultrastructural study employing pre- and post-embedding immunocytochemical methods. *Neuroscience* **33**, 169–184.
Mayer, D. J. and Price, D. D., 1976, Central nervous system mechanisms of analgesia. *Pain* **2**, 379–404.
Mayer, D. J., Price, D. D. and Becker, D. P., 1975, Neurophysiological characterization of the anterolateral spinal cord neurons contributing to pain perception in man. *Pain* **1**, 51–58.
Mayer, M. L. and Westbrook, G. L., 1987, The physiology of excitatory amino acids in the vertebrate central nervous system. *Prog. Neurobiol.* **28**, 197–276.
Maynard, C. W., Leonard, R. B., Coulter, J. D. and Coggeshall, R. E., 1977, Central connections of ventral root afferents as demonstrated by the HRP method. *J. Comp. Neurol.* **172**, 601–608.
Mazza, J. P., Hanker, J. A. and Dixon, A. D., 1973, Ultrastructural localization of cholinesterase activity in the trigeminal ganglion of the rat. *J. Anat.* **115**, 65–78.
McCall, W. D., Farias, M. C., Williams, W. J. and BeMent, S. L., 1974, Static and dynamic responses of slowly adapting joint receptors. *Brain Res.* **70**, 221–243.
McCloskey, D. I., 1973, Differences between the senses of movement and position shown by the effects of loading and vibration of muscles in man. *Brain Res.* **63**, 119–131.
McCloskey, D. I., 1978, Kinesthetic sensibility. *Physiol. Rev.* **58**, 763–820.
McCloskey, D. I., 1981, Corollary discharges and motor commands. In: *Handbook of Physiology—The Nervous System III, Motor Control*, pp. 1415-1447. V. B. Brooks (ed.). American Physiological Society, Bethesda.
McCloskey, D. I. and Torda, T. A. G., 1975, Corollary motor discharges and kinaesthesia. *Brain Res.* **100**, 467–470.
McCloskey, D. I., Cross, M. J., Honner, R. and Potte, E. K., 1983, Sensory effects of pulling or vibrating exposed tendons in man. *Brain* **106**, 21–37.
McComas, A. J., 1963, Responses of rat dorsal column system to mechanical stimulation of the hindpaw. *J.Physiol.* **166**, 435–448.

McComas, A. J. and Wilson, P., 1968, An investigation of pyramidal tract cells in the somatosensory cortex of the rat. *J. Physiol.* **194**, 271–288.

McCreery, D. B. and Bloedel, J. R., 1975, Reduction of the response of cat spinothalamic neurons to graded mechanical stimulation of the lower brain stem. *Brain Res.* **97**, 151–156.

McCreery, D. B. and Bloedel, J. R., 1976, Effects of trigeminal stimulation on the excitability of cat spinothalamic neurons. *Brain Res.* **117**, 136–140.

McCreery, D. B., Bloedel, J. R. and Hames, E. G., 1979a, Excitability changes in lumbosacral spinothalamic neurons produced by non-noxious mechanical stimuli and by graded electrical stimuli applied to the face. *Brain Res.* **177**, 253–263.

McCreery, D. B., Bloedel, J. R. and Hames, E. G., 1979b, Effects of stimulating in raphe nuclei and in reticular formation on response of spinothalamic neurons to mechanical stimuli. *J. Neurophysiol.* **42**, 166–182.

McDougal, D. B., Jr., McDougal, S. H. and Johnson, E. M., Jr., 1985, Effect of capsaicin upon fluoride sensitive acid phosphatases in selected ganglia and spinal cord and upon neuronal size and number in dorsal root ganglion. *Brain Res.* **331**, 63–70.

McGregor, G. P., Gibson, S. J., Sabate, I. M., Blank, M. A., Christofides, N. D., Wall, P. D., Polak, J. M. and Bloom, S. R., 1984, Effect of peripheral nerve section and nerve crush on spinal cord neuropeptides in the rat; increased VIP and PHI in the dorsal horn. *Neuroscience* **13**, 207–216.

McIntyre, A. K., 1962, Cortical projection of impulses in the interosseous nerve of the cat's hind limb. *J. Physiol.* **163**, 46–60.

McIntyre, A. K., Holman, M. E. and Veale, J. L., 1967, Cortical responses to impulses from single Pacinian corpuscles in the cat's hind limb. *Exp. Brain Res.* **4**, 243–255.

McIntyre, A. K., Proske, U. and Tracey, D. J., 1978, Afferent fibres from muscle receptors in the posterior nerve of the cat's knee joint. Exp. *Brain Res.* **33**, 415–424.

McLachlan, E. M. and Jänig, W., 1983, The cell bodies of origin of sympathetic and sensory axons in some skin and muscle nerves of the cat hindlimb. *J. Comp. Neurol.* **214**, 115–130.

McLaughlin, B. J., 1972, Dorsal root projections to the motor nuclei in the cat spinal cord. *J. Comp. Neurol.* **144**, 461–474.

McLaughlin, B. J., Barber, R., Saito, K., Roberts, E. and Wu, J. Y., 1975, Immunocytochemical localization of glutamate decarboxylase in rat spinal cord. *J. Comp. Neurol.* **164**, 305–322.

McLennan, H., 1983, Receptors for the excitatory amino acids in the mammalian central nervous system. *Prog. Neurobiol.* **20**, 251–271.

McLennan, H. and Liu, J. R., 1982, The action of six antagonists of the excitatory amino acids on neurones of the rat spinal cord. *Exp. Brain Res.* **45**, 151–156.

McLennan, H. and Lodge, D., 1979, The antagonism of amino acid-induced excitation of spinal neurones in the cat. *Brain Res.* **169**, 83–90.

McMahon, S. B., 1986, The localization of fluoride-resistant acid phosphatase (FRAP) in the pelvic nerves and sacral spinal cord of rats. *Neurosci. Lett.* **64**, 305–310.

McMahon, S. B. and Gibson, S., 1987, Peptide expression is altered when afferent nerves reinnervate inappropriate tissue. *Neurosci. Lett.* **73**, 9–15.

McMahon, S. B. and Moore, C. E. G., 1988, Plasticity of primary afferent acid phosphatase expression following rerouting of afferents from muscle to skin in the adult rat. *J. Comp. Neurol.* **274**, 1–8.

McMahon, S. B. and Morrison, J. F. B., 1982, Two groups of spinal interneurones that respond to stimulation of the abdominal viscera of the cat. *J. Physiol.* **322**, 21–34.

McMahon, S. B. and Wall, P. D., 1983a, Receptive fields of lamina I projection cells to incorporate a nearby region of injury. *Pain* **19**, 235–247.

McMahon, S. B. and Wall, P. D., 1983b, Plasticity in the nucleus gracilis of the rat. *Exp. Neurol.* **80**, 195–207.

McMahon, S. B. and Wall, P. D., 1983c, A system of rat spinal cord lamina I cells projecting through the contralateral dorsolateral funiculus. *J. Comp. Neurol.* **214**, 217–223.

McMahon, S. B. and Wall, P. D., 1985a, Electrophysiological mapping of brainstem projections of spinal cord lamina I cells in the rat. *Brain Res.* **333**, 19–26.

McMahon, S. B. and Wall, P. D., 1985b, The distribution and central termination of single cutaneous and muscle unmyelinated fibres in rat spinal cord. *Brain Res.* **359**, 39–48.

McMahon, S. B. and Wall, P. D., 1987, Physiological evidence for branching of peripheral unmyelinated sensory afferent fibers in the rat. *J. Comp. Neurol.* **261**, 130–136.

McMahon, S. B. and Wall, P. D., 1988, Descending excitation and inhibition of spinal cord lamina I projection neurons. *J. Neurophysiol.* **59**, 1204–1219.

McMahon, S. B., Syková, E., Wall, P. D., Woolf, C. J. and Gibson, S. J., 1984, Neurogenic extravasation and substance P levels are low in muscle as compared to skin in the rat hindlimb. *Neurosci. Lett.* **52**, 235–240.

McNeill, D. L. and Burden, H. W., 1986, Convergence of sensory processes from the heart and left ulnar nerve onto a single afferent perikaryon: a neuroanatomical study in the rat employing fluorescent tracers. *Anat. Rec.* **214**, 441–444.

McNeill, D. L. and Burden, H. W., 1987, Neuropeptides in sensory perikarya projecting to the rat ovary. *Amer. J. Anat.* **179**, 269–276.

McNeill, D. L., Coggeshall, R. E. and Carlton, S. M., 1988a, A light and electron microscopic study of calcitonin gene-related peptide in the spinal cord of the rat. *Exp. Neurol.* **99**, 699–708.

McNeill, D. L., Chung, K., Carlton, S. M. and Coggeshall, R. E., 1988b, Calcitonin gene-related peptide immunostained axons provide evidence for fine primary afferent fibers in dorsal and dorsolateral funiculi of rat spinal cord. *J. Comp. Neurol.* **272**, 303–308.

McNeill, D. L., Chung, K., Hulsebosch, C. E., Bolender, R. P. and Coggeshall, R. E., 1988c, Number of synapses in laminae I-IV of the rat dorsal horn. *J. Comp. Neurol.* **278**, 453–460.

McNeill, M. E. and Norvell, J. E., 1978, Acetylcholinesterase activity of primary sensory neurons and dorsal root fibers in the cat. *Anat. Rec.* **190**, 155–160.

Meessen, H. and Olszewski, J., 1949, *A Cytoarchitectonic Atlas of the Rhombencephalon of the Rabbit.* Karger, New York.

Mehler, W. R., 1962, The anatomy of the so-called "pain tract" in man: an analysis of the course and distribution of the ascending fibers of the fasciculus anterolateralis. In: *Basic Research in Paraplegia*, pp. 26–55. J. D. French and R. W. Porter (eds.). Springfield, Thomas.

Mehler, W. R., 1966, Some observations on secondary ascending afferent systems in the central nervous system. In: *Pain*, pp. 11–32. R. S. Knighton and P. R. Dumke (eds.). Little Brown, Boston.

Mehler, W.R., 1969, Some neurological species differences -- a posteriori. *Ann. N.Y. Acad. Sci.* **167**, 424–468.

Mehler, W. R., 1974, Central pain and the spinothalamic tract. *Adv. Neurol.* **4**, 127–146.

Mehler, W. R., Feferman, M. E. and Nauta, W. J. H., 1960, Ascending axon degeneration following anterolateral cordotomy. An experimental study in the monkey. *Brain* **83**, 718–751.

Meissner, G., 1859, Untersuchungen ueber den Tastsinn. *Z. Rat. Med.* **7**, 92–118.

Melander, T., Hökfelt, T. and Rökaeus, A., 1986, Distribution of galaninlike immunoreactivity in the rat central nervous system. *J. Comp. Neurol.* **248**, 475–517.

Melzack, R., 1973, *The Puzzle of Pain.* Basic Books, New York.

Melzack, R., 1975, Prolonged relief of pain by brief, intense transcutaneous somatic stimulation. *Pain* **1**, 357–373.

Melzack, R. and Bridges, J. A., 1971, Dorsal column contributions to motor behavior. *Exp. Neurol.* **33**, 53–68.

Melzack, R. and Casey, K. L., 1968, Sensory, motivational and central control determinants of pain. In: *The Skin Senses*, pp. 423-443. D. R. Kenshalo (ed.). Thomas, Springfield.

Melzack, R. and Southmayd, S. E., 1974, Dorsal column contributions to anticipatory motor behavior. *Exp. Neurol.* **42**, 274–281.

Melzack, R. and Wall, P. D., 1962, On the nature of cutaneous sensory mechanisms. *Brain* **85**, 331–356.

Melzack, R. and Wall, P. D., 1965, Pain mechanisms: a new theory. *Science* **150**, 971–979.

Melzack, R., Stotler, W. A. and Livingston, W. K., 1958, Effects of discrete brainstem lesions in cats on perception of noxious stimulation. *J. Neurophysiol.* **21**, 353–367.

Melzack, R., Weisz, A. Z. and Sprague, L. T., 1963, Stratagems for controlling pain: contributions of auditory stimulation and suggestion. *Exp. Neurol.* **8**, 239–247.

Mendell, L. M., 1966, Physiological properties of unmyelinated fiber projection to the spinal cord. *Exp. Neurol.* **16**, 316–332.

Mendell, L. M., 1970, Positive dorsal root potentials produced by stimulation of small diameter muscle afferents. *Brain Res.* **18**, 375–379.

Mendell, L., 1972, Properties and distribution of peripherally evoked presynaptic hyperpolarization in cat lumbar spinal cord. *J. Physiol.* **226**, 769–792.

Mendell, L., 1973, Two negative dorsal root potentials evoke a positive dorsal root potential. *Brain Res.* **55**, 198–202.

Mendell, L. M. and Wall, P. D., 1964, Presynaptic hyperpolarization: a role for fine afferent fibres. *J. Physiol.* **172**, 274–294.

Mendell, L. M. and Wall, P. D., 1965, Responses of single dorsal cord cells to peripheral cutaneous unmyelinated fibres. *Nature* **206**, 97–99.

Mendell, L. M., Sassoon, E. M. and Wall, P. D., 1978, Properties of synaptic linkage from long ranging afferents onto dorsal horn neurones in normal and deafferented cats. *J. Physiol.* **285**, 299–310.

Menétrey, D. and Basbaum, A. I., 1987, The distribution of substance P-, enkephalin- and dynorphin-immunoreactive neurons in the medulla of the rat and their contribution to bulbospinal pathways. *Neuroscience* **23**, 173–188.

Menétrey, D., Giesler, G. J. and Besson, J. M., 1977, An analysis of response profiles of spinal cord dorsal horn neurones to nonnoxious and noxious stimuli in the spinal rat. *Exp. Brain Res.* **27**, 15–33.

Menétrey, D., Chaouch, A. and Besson, J. M., 1980, Location and properties of dorsal horn neurons at origin of spinoreticular tract in lumbar enlargement of the rat. *J. Neurophysiol.* **44**, 862–877.

Menétrey, D., Chaouch, A., Binder, D. and Besson, J. M., 1982, The origin of the spinomesencephalic tract in the rat: an anatomical study using the retrograde transport of horseradish peroxidase. *J. Comp. Neurol.* **206**, 193–207.

Menétrey, D., Gannon, J. D., Levine, J. D. and Basbaum, A. I., 1989, Expression of c-fos protein in interneurons and projection neurons of the rat spinal cord in response to noxious somatic, articular, and visceral stimulation. *J. Comp. Neurol.* **285**, 177–195.

Mense, S., 1977, Nervous outflow from skeletal muscle following chemical noxious stimulation. *J. Physiol.* **267**, 75–88.

Mense, S., 1981, Sensitization of group IV muscle receptors to bradykinin by 5-hydroxytryptamine and prostaglandin E2. *Brain Res.* **225**, 95–105.
Mense, S. and Craig, J. R., 1988, Spinal and supraspinal terminations of primary afferent fibers from the gastrocnemius-soleus muscle in the cat. *Neuroscience* **26**, 1023–1055.
Mense, S. and Meyer, H., 1985, Different types of slowly conducting afferent units in cat skeletal muscle and tendon. *J. Physiol.* **363**, 403–417.
Mense, S. and Meyer, H., 1988, Bradykinin-induced modulation of the response behaviour of different types of feline group III and IV muscle receptors. *J. Physiol.* **398**, 49–63.
Mense, S. and Prabhakar, N. R., 1986, Spinal termination of nociceptive afferent fibres from deep tissues in the cat. *Neurosci. Lett.* **66**, 169–174.
Mense, S. and Schmidt, R. F., 1974, Activation of group IV afferent units from muscle by algesic agents. *Brain Res.* **72**, 305–310.
Mense, S. and Stahnke, M., 1983, Responses in muscle afferent fibres of slow conduction velocity to contractions and ischaemia in the cat. *J. Physiol.* **342**, 383–397.
Mense, S., Light, A. R. and Perl, R., 1980, Spinal terminations of subcutaneous high threshold mechanoreceptors. In: *Spinal Cord Sensation*. A. G. Brown and M. Rethelyi (eds.). Scottish Academic Press, Edinburgh.
Merchenthaler, I., 1984, Corticotrophin releasing factor (CRF)-like immunoreactivity in the rat central nervous system. Extrahypothalamic distribution. *Peptides* **5** (Suppl. 1): 53-69.
Merchenthaler, I., Hynes, M. A., Vigh, S., Shally, A. V. and Petrusz, P., 1983, Immunocytochemical localization of corticotropin releasing factor (CRF) in the rat spinal cord. *Brain Res.* **275**, 373–377.
Merighi, A., Polak, J. M., Gibson, S. J., Gulbenkian, S., Valentino, K. L. and Peirone, S. M., 1988, Ultrastructural studies on calcitonin gene-related peptide-, tachykinins- and somatostatin-immunoreactive neurones in rat dorsal root ganglia: evidence for the colocalization of different peptides in single secretory granules. *Cell Tiss. Res.* **254**, 101–109.
Merkel, F., 1875, Tastzellen und Tastkoerperchen bei den Hausthieren und beim Menschen. *Arch. Mikroskop. Anat.* **11**, 636–652.
Merrill, E. G. and Wall, P. D., 1972, Factors forming the edge of a receptive field: the presence of relatively ineffective afferent terminals. *J. Physiol.* **226**, 825–846.
Merton, P. A., 1964, Human position sense and sense of effort. *Symp. Soc. Exp. Biol.* **18**, 387–400.
Merzenich, M. M. and Harrington, T., 1969, The sense of flutter-vibration evoked by stimulation of the hairy skin of primates: comparison of human sensory capacity with the responses of mechanoreceptive afferents innervating the hairy skin of monkeys. *Exp. Brain Res.* **9**, 236–260.
Metherate, R. S., DaCosta, D. C. N., Herron, P. and Dykes, R. W., 1986, A thalamic terminus of the lateral cervical nucleus: the lateral division of the posterior nuclear group. *J. Neurophysiol.* **56**, 1498–1520.
Meyer, R. A. and Campbell, J. N., 1981, Myelinated nociceptive afferents account for the hyperalgesia that follows a burn to the hand. *Science* **213**, 1527–1529.
Meyer, R. A. and Campbell, J. A., 1988, A novel electrophysiological technique for locating cutaneous nociceptive and chemospecific receptors. *Brain Res.* **441**, 81–86.
Meyers, D. E. R. and Snow, P. J., 1982a, The morphology of physiologically identified deep spinothalamic tract cells in the lumbar spinal cord of the cat. *J. Physiol.* **329**, 373–388.
Meyers, D. E. R. and Snow, P. J., 1982b, The responses to somatic stimuli of deep spinothalamic tract cells in the lumbar spinal cord of the cat. *J. Physiol.* **329**, 355–371.
Meyers, D. E. R. and Snow, P. J., 1984, Somatotopically inappropriate projections of single hair follicle afferent fibres to the cat spinal cord. *J. Physiol.* **347**, 59–73.
Meyers, D. E. R. and Snow, P. J., 1986, Distribution of activity in the spinal terminations of single hair follicle afferent fibers to somatotopically identified regions of the cat spinal cord. *J. Neurophysiol.* **56**, 1022–1038.
Meyers, D. E. R., Wilson, P. and Snow, P. J., 1984, Distribution of the central terminals of cutaneous primary afferents innervating a small skin patch: The existence of somatotopically inappropriate projections. *Neurosci. Lett.* **44**, 179–185.
Micevych, P. E., Stoink, A., Yaksh, T. and Go, V. L. W., 1986, Immunochemical studies of substance P and cholecystokinin octapeptide recovery in dorsal horn following unilateral lumbosacral ganglionectomy. *Somatosensory Res.* **33**, 239–260.
Miletić, V. and Randić, M., 1979, Neurotensin excites cat spinal neurones located in laminae I–III. *Brain Res.* **169**, 600–604.
Miletić, V. and Randić, M., 1982, Neonatal rat spinal cord slice preparation: postsynaptic effects of neuropeptides on dorsal horn neurons. *Dev. Brain Res.* **2**, 432–438.
Miletić, V. and Tan, H., 1988, Iontophoretic application of calcitonin gene-related peptide produces a slow and prolonged excitation of neurons in the cat lumbar dorsal horn. *Brain Res.* **446**, 169–172.
Miletić, V., Hoffert, M. J., Ruda, M. A., Dubner, R. and Shigenaga, Y., 1984, Serotoninergic axonal contacts on identified cat spinal dorsal horn neurons and their correlation with nucleus raphe magnus stimulation. *J. Comp. Neurol.* **228**, 129–141.
Millar, J., 1973a, Joint afferent fibres responding to muscle stretch, vibration and contraction. *Brain Res.* **63**, 380–383.
Millar, J., 1973b, The topography and receptive fields of ventroposterolateral thalamic neurons excited by afferents projecting through the dorsolateral funiculus of the spinal cord. *Exp. Neurol.* **41**, 303–313.

Millar, J., 1975, Flexion-extension sensitivity of elbow joint afferents in cat. *Exp. Brain Res.* **24**, 209–214.
Millar, J., 1979a, Loci of joint cells in the cuneate and external cuneate nuclei of the cat. *Brain Res.* **167**, 385–390.
Millar, J., 1979b, Convergence of joint, cutaneous and muscle afferents onto cuneate neurones in the cat. *Brain Res.* **175**, 347–350.
Millar, J. and Armstrong-James, M., 1982, The responses of neurones of the superficial dorsal horn to iontophoretically applied glutamate ion. *Brain Res.* **231**, 267–277.
Millar, J. and Basbaum, A. I., 1975, Topography of the projection of the body surface of the cat to cuneate and gracile nuclei. *Exp. Neurol.* **49**, 281–290.
Millar, J. and Williams, G. V., 1989, Effects of iontophoresis of noradrenaline and stimulation of the periaqueductal gray on single-unit activity in the rat superficial dorsal horn. *J. Comp. Neurol.* **287**, 119–133.
Millar, J., Basbaum, A. I. and Wall, P. D., 1976, Restructuring of the somatotopic map and appearance of abnormal neuronal activity in the gracile nucleus after partial deafferentation. *Exp. Neurol.* **50**, 658–672.
Miller, G. A., 1956, The magical number seven, plus or minus two: some limits on our capacity for processing information. *Psychol. Rev.* **63**, 81–97.
Miller, K. E. and Seybold, V. S., 1987, Comparison of met-enkephalin-, dynorphin A-, and neurotensin-immunoreactive neurons in the cat and rat spinal cords: I. Lumbar cord. *J. Comp. Neurol.* **255**, 293–304.
Miller, K. E. and Seybold, V. S., 1989, Comparison of met-enkephalin, dynorphin A, and neurotensin immunoreactive neurons in the cat and rat spinal cords: II. Segmental differences in the marginal zone. *J. Comp. Neurol.* **279**, 619–628.
Miller, K. E., Clements, J. R., Larson, A. A. and Beitz, A. J., 1988, Organization of glutamate-like immunoreactivity in the rat superficial dorsal horn: light and electron microscopic observations. *Synapse* **2**, 28–36.
Miller, M. R., Ralston, H. J. and Kasahara, M., 1958, The pattern of cutaneous innervation of the human hand. *Amer. J. Anat.* **102**, 183–217.
Millhorn, D. E., Hökfelt, T., Seroogy, K., Oertel, W., Verhofstad, A. A. J. and Wu, J. Y., 1987a, Immunohistochemical evidence for colocalization of γo-aminobutyric acid and serotonin in neurons of the ventral medulla oblongata projecting to the spinal cord. *Brain Res.* **410**, 179–185.
Millhorn, D. E., Seroogy, K., Hökfelt, T., Schmued, L. C., Terenius, L., Buchan, A. and Brown, J. C., 1987b, Neurons of the ventral medulla oblongata that contain both somatostatin and enkephalin immunoreactivities project to nucleus tractus solitarii and spinal cord. *Brain Res.* **424**, 99–108.
Millhorn, D. E., Hökfelt, T., Verhofstad, A. A. J. and Terenius, L., 1989, Individual cells in the raphe nuclei of the medulla oblongata in rat that contain immunoreactivities for both serotonin and enkephalin project to the spinal cord. *Exp. Brain Res.* **75**, 536–542.
Milne, R. J., Foreman, R. D., Giesler, G. J. and Willis, W. D., 1981, Convergence of cutaneous and pelvic visceral nociceptive inputs onto primate spinothalamic neurons. *Pain* **11**, 163–183.
Milne, R. J., Foreman, R. D. and Willis, W. D., 1982, Responses of primate spinothalamic neurons located in the sacral intermediomedial gray (Stilling's nucleus) to proprioceptive input from the tail. *Brain Res.* **234**, 227–236.
Mitchell, D. and Hellon, R. F., 1977, Neuronal and behavioural responses in rats during noxious stimulation of the tail. *Proc. Roy. Soc. Lond. B* **197**, 169–194.
Mitchell, R. and Fleetwood-Walker, S., 1981, Substance P, but not TRH, modulates the 5-HT autoreceptor in ventral spinal cord. *Eur. J. Pharmacol.* **76**, 119–120.
Mizukawa, K., Otsuka, N., McGeer, P. L., Vincent, S. R. and McGeer, E. G., 1988, The ultrastructure of somatostatin-immunoreactive cell bodies, nerve fibers and terminals in the dorsal horn of rat spinal cord. *Arch. Histol. Cytol.* **51**, 443–452.
Mizuno, N., Nakano, K., Imaizumi, M. and Okamoto, M., 1967, The lateral cervical nucleus of the Japanese monkey (*Macaca fuscata*). *J. Comp. Neurol.* **129**, 375–384.
Moberg, E., 1983, The role of cutaneous afferents in position sense, kinaesthesia, and motor function of the hand. *Brain* **106**, 1–19.
Moffie, D., 1975, Spinothalamic fibres, pain conduction and cordotomy. *Clin. Neurol. Neurosurg.* **78**, 261–268.
Mokha, S. S. and Iggo, A., 1987, Mechanisms mediating the brain stem control of somatosensory transmission in the dorsal horn of the cat's spinal cord: an intracellular analysis. *Exp. Brain Res.* **69**, 93–106.
Mokha, S. S., McMillan, J. A. and Iggo, A., 1983, Dorsal root potentials in the cat: effects of bicuculline. *Brain Res.* **259**, 313–318.
Mokha, S. S., McMillan, J. A. and Iggo, A., 1985, Descending control of spinal nociceptive transmission. Actions produced on spinal multireceptive neurones from the nuclei locus coeruleus (LC) and raphe magnus (NRM). *Exp. Brain Res.* **58**, 213–226.
Mokha, S. S., McMillan, J. A. and Iggo, A., 1986, Pathways mediating descending control of spinal nociceptive transmission from the nuclei locus coeruleus (LC) and raphe magnus (NRM) in the cat. *Exp. Brain Res.* **61**, 597–606.
Molander, C. and Grant, G., 1985, Cutaneous projections from the rat hindlimb foot to the substantia gelatinosa of the spinal cord studied by transganglionic transport of WGA-HRP conjugate. *J. Comp. Neurol.* **237**, 476–484.
Molander, C. and Grant, G., 1986, Laminar distribution and somatotopic organization of primary afferent fibers from hindlimb nerves in the dorsal horn. A study by transganglionic transport of horseradish peroxidase in the rat. *Neuroscience* **19**, 297–312.

Molander, C. and Grant, G., 1987, Spinal cord projections from hindlimb muscle nerves in the rat studied by transganglionic transport of horseradish peroxidase, wheat germ agglutinin conjugated horseradish peroxidase, or horseradish peroxidase with dimethylsulfoxide. *J. Comp. Neurol.* **260**, 246–255.

Molander, C., Ygge, J. and Dalsgaard, C. J., 1987, Substance P-, somatostatin- and calcitonin gene-related peptide-like immunoreactivity and fluoride resistant acid phosphatase-activity in relation to retrogradely labeled cutaneous, muscular and visceral primary sensory neurons in the rat. *Neurosci. Lett.* **74**, 37–42.

Molander, C., Kinnman, E. and Aldskogius, H., 1988, Expansion of primary sensory afferent projection following combined sciatic nerve resection and saphenous nerve crush: A horseradish peroxidase study in the adult rat. *J. Comp. Neurol.* **276**, 436–441.

Molinari, H. H., Greenspan, J. D. and Kenshalo, D. R., 1977, The effects of rate of temperature change and adapting temperature on thermal sensitivity. *Sens. Proc.* **1**, 354–362.

Molinari, M., Bentivoglio, M., Minciacchi, D., Granato, A. and Macchi, G., 1986, Spinal afferents and cortical efferents of the anterior intralaminar nuclei: an anterograde-retrograde tracing study. *Neurosci. Lett.* **72**, 258–264.

Molony, V., Steedman, W. M., Cervero, F. and Iggo, A., 1981, Intracellular marking of identified neurons in the superficial dorsal horn of the cat spinal cord. *Quart. J. Exp. Physiol.* **66**, 211–223.

Moossy, J., Sagone, A. and Rosomoff, H. L., 1967, Percutaneous radiofrequency cervical cordotomy: pathologic anatomy. *J. Neuropath. Exp. Neurol.* **26**, 118.

Morest, D. K., 1967, Experimental study of the projections of the nucleus of the tractus solitarius and the area postrema in the cat. *J. Comp. Neurol.* **130**, 277–300.

Morgan, C., Nadelhaft, I. and De Groat, W. C., 1981, The distribution of visceral primary afferents from the pelvic nerve to Lissauer's tract and the spinal gray matter and its relationship to the sacral parasympathetic nucleus. *J. Comp. Neurol.* **201**, 415–440.

Morgan, J. I. and Curran, T., 1986, Role of ion flux in the control of c-fos expression. *Nature* **322**, 552–555.

Morin, F., 1955, A new spinal pathway for cutaneous impulses. *Amer. J. Physiol.* **183**, 245–252.

Morin, F. and Catalano, J. V., 1955, Central connections of a cervical nucleus (nucleus cervicalis lateralis of the cat). *J. Comp. Neurol.* **103**, 17–32.

Morin, F., Schwartz, H. G. and O'Leary, J. L., 1951, Experimental study of the spinothalamic and related tracts. *Acta Psychiat. Neurol.* **26**, 371–396.

Morin, F., Kitai, S. T., Portnoy, H. and Demirjian, C., 1963, Afferent projections to the lateral cervical nucleus: a microelectrode study. *Amer. J. Physiol.* **204**, 667–672.

Morris, B. J. and Herz, A., 1987, Distinct distribution of opioid receptor types in rat lumbar spinal cord. *Arch. Pharmacol.* **336**, 240–243.

Morris, R., 1987, Inhibition of nociceptive responses of laminae V-VII dorsal horn neurones by stimulation of mixed and muscle nerves, in the cat. *Brain Res.* **401**, 365–370.

Morris, R., 1989, Responses of spinal dorsal horn neurones evoked by myelinated primary afferent stimulation are blocked by excitatory amino acid antagonists acting at kainate/quisqualate receptors. *Neurosci. Lett.* **105**, 79–85.

Morrison, J. F. B., 1973, Splanchnic slowly adapting mechanoreceptors with punctate receptive fields in the mesentery and gastrointestinal tract of the cat. *J. Physiol.* **233**, 349–361.

Morrison, J. F. B., 1977, The afferent innervation of the gastrointestinal tract. In: *Nerves and the Gut*, pp. 297–322. F. P. Brooks and P. W. Evers (eds.). C.B. Slack, Thorofare, New Jersey.

Morton, C. R. and Hutchison, W. D., 1989, Release of sensory neuropeptides in the spinal cord: studies with calcitonin gene-related peptide and galanin. *Neuroscience* **31**, 807–815.

Morton, C. R., Johnson, S. M. and Duggan, A. W., 1983, Lateral reticular regions and the descending control of dorsal horn neurones of the cat: selective inhibition by electrical stimulation. *Brain Res.* **275**, 13–21.

Morton, C. R., Duggan, A. W. and Zhao, Z. Q., 1984, The effects of lesions of medullary midline and lateral reticular areas on inhibition in the dorsal horn produced by periaqueductal grey stimulation in the cat. *Brain Res.* **301**, 121–130.

Morton, C. R., Maisch, B. and Zimmermann, M., 1987, Diffuse noxious inhibitory controls of lumbar spinal neurons involve a supraspinal loop in the cat. *Brain Res.* **410**, 347–352.

Morton, C. R., Hutchison, W. D. and Hendry, I. A., 1988, Release of immunoreactive somatostatin in the spinal dorsal horn of the cat. *Neuropeptides* **12**, 189–197.

Morton, C. R., Hutchison, W. D., Hendry, I. A. and Duggan, A. W., 1989, Somatostatin: evidence for a role in thermal nociception. *Brain Res.* **488**, 89–96.

Motavkin, P. A. and Okhotin, V. E., 1985, Histochemistry of choline acetyltransferase in the spinal cord and spinal ganglia of the cat. *Neurosci. Behav. Physiol.* **32**, 307–310.

Mott, F. W., 1895, Experimental enquiry upon the afferent tracts of the central nervous system of the monkey. *Brain* **18**, 1–20.

Mouchet, P., Manier, M., Dietl, M., Feuerstein, C., Berod, A., Arluison, M., Denoroy, L. and Thibault, J., 1986, Immunohistochemical study of catecholaminergic cell bodies in the rat spinal cord. *Brain Res. Bull.* **16**, 341–353.

Mountcastle, V. B., Talbot, W. H. and Kornhuber, H. H., 1966, The neural transformation of mechanical stimuli delivered to the monkey's hand. In: *Touch, Heat and Pain*, pp. 325–345. A. V. S. de Reuck and J. Knight (eds.). Churchill, London.

Mountcastle, V. B., LaMotte, R. H. and Carli, G., 1972, Detection thresholds for stimuli in humans and monkeys: comparison with threshold events in mechanoreceptive afferent nerve fibers innervating the monkey hand. *J. Neurophysiol.* **35**, 122–136.

Mudge, A. W., Leeman, S. E. and Fischbach, G. D., 1979, Enkephalin inhibits release of substance P from sensory neurons in culture and decreases action potential duration. *Proc. Natl. Acad. Sci. USA* **76**, 526–530.

Müller, J., 1840-2, Elements of physiology. Transl. from German, with notes by W. Baly, 2nd ed., Vol. 2, London. Original German: 1838, *Handbuch der Physiologie des Menschen*, 2nd ed., Coblenz.

Mugnaini, E. and Oertel, W. H., 1985, An atlas of the distribution of gabaergic neurons and terminals in the rat CNS as revealed by GAD immunohistochemistry. In: *Handbook of Clinical Neuroanatomy*, pp. 436–595. A. Björklund and T. Hökfelt (eds.). Elsevier, Amsterdam.

Mugnaini, E., Berrebi, A. S., Morgan, J. I. and Curran, T., 1989, Fos-like immunoreactivity induced by seizure in mice is specifically associated with euchromatin in neurons. *Eur. J. Neurosci.* **1**, 46–52.

Mullan, S., 1966, Percutaneous cordotomy for pain. *Surg. Clin. N. Amer.* **46**, 3–12.

Mullan, S., Harper, P. V., Hekmatpanah, J., Torres, H. and Dobbin, G., 1963, Percutaneous interruption of spinal-pain tracts by means of a strontium needle. *J. Neurosurg.* **20**, 931–939.

Munger, B. L., 1965, The intraepidermal innervation of the snout of the opossum. *J. Cell Biol.* **26**, 79–97.

Munger, B. L., 1971, Patterns of organization of peripheral sensory receptors. In: *Handbook of Sensory Physiology*, Vol. 1, pp. 523–556. W. R. Loewenstein (ed.). Springer, Berlin.

Murakami, T., Araki, T., Yamano, M., Wanaka, A., Betz, H. and Tohyama, M., 1988, Localization of the glycine receptors in the rat central nervous system: An immunocytochemical analysis. *Adv. Exp. Med. Biol.* **236**, 71–81.

Murase, K. and Randić, M., 1984, Actions of substance P on rat spinal dorsal horn neurons. *J. Physiol.* **346**, 203–217.

Murase, K., Nedeljkov, V. and Randić, M., 1982, The actions of neuropeptides on dorsal horn neurons in rat spinal cord slice preparation: an intracellular study. *Brain Res.* **234**, 170–176.

Murase, K., Ryu, P. D. and Randić, M., 1986, Substance P augments a persistent slow inward calcium-sensitive current in voltage-clamped spinal dorsal horn neurons of the rat. *Brain Res.* **365**, 369–376.

Murase, K., Ryu, P. D. and Randić, M., 1989a, Tachykinins modulate multiple ionic conductances in voltage-clamped rat spinal dorsal horn neurons. *J. Neurophysiol.* **61**, 854–865.

Murase, K., Ryu, P. D. and Randić, M., 1989b, Excitatory and inhibitory amino acids and peptide-induced responses in acutely isolated rat spinal dorsal horn neurons. *Neurosci. Lett.* **103**, 56–63.

Murray, M. and Goldberger, M. E., 1974, Restitution of function and collateral sprouting in the cat spinal cord: the partially hemisected animal. *J. Comp. Neurol.* **155**, 19–36.

Murray, M. and Goldberger, M., 1986, Replacement of synaptic terminals in lamina II and Clarke's nucleus after unilateral lumbosacral dorsal rhizotomy in adult cats. *J. Neurosci.* **6**, 3205–3217.

Myers, D. A., Hostetter, G., Bourassa, C. M. and Swett, J. E., 1974, Dorsal columns in sensory detection. *Brain Res.* **70**, 350–355.

Nade, S., Newbold, P. J. and Straface, S. F., 1987, The effects of direction and acceleration of movement of the knee joint of the dog on medial articular nerve discharge. *J. Physiol.* **388**, 505–519.

Nadel, E. R. and Horvath, S. M., 1969, Peripheral involvement in thermoregulatory responses to an imposed heat debt in man. *J. Appl. Physiol.* **27**, 484–488.

Nadelhaft, I. and Booth, A. M., 1984, The location and morphology of the preganglionic neurons and the distribution of visceral afferents from the rat pelvic nerve: a horseradish peroxidase study. *J. Comp. Neurol.* **226**, 238–245.

Nadelhaft, I., Roppolo, J., Morgan, C. and De Groat, W. C., 1983, Parasympathetic preganglionic neurons and visceral primary afferents in monkey sacral spinal cord revealed following application of horseradish peroxidase to pelvic nerve. *J. Comp. Neurol.* **216**, 36–52.

Nafe, J. P., 1927, The psychology of felt experience. *Amer. J. Psychol.* **39**, 367–389.

Nafe, J. P., 1929, A quantitative theory of feeling. *J. Gen. Psychol.* **2**, 199-210.

Naftchi, N. E., Abrahams, S. J., St. Paul, H. M., Lowrman, E. W. and Schlosser, W., 1978, Localization and changes of substance P in spinal cord of paraplegic cats. *Brain Res.* **153**, 507–513.

Nagy, J. I. and DaDonna, P. E., 1985, Anatomical and cytochemical relationships of adenosine deaminase-containing primary afferent neurons in the rat. *Neuroscience* **15**, 799–813.

Nagy, J. I. and Hunt, S. P., 1982, Fluoride-resistant acid phosphatase-containing neurones in dorsal root ganglia are separate from those containing substance P or somatostatin. *Neuroscience* **7**, 89–97.

Nagy, J. I., Hunt, S. P., Iverson, L. L. and Emson, P. C., 1981, Biochemical and anatomical observations on the degeneration of peptide containing primary afferent neurons after neonatal capsaicin. *Neuroscience* **6**, 1923–1934.

Nagy, J. I., Buss, M., LaBella, L. A. and DaDonna, P. E., 1984, Immunohistochemical localization of adenosine deaminase in primary afferent neurons of the rat. *Neurosci. Lett.* **48**, 133–138.

Nahin, R. L., 1988, Immunocytochemical identification of long ascending, peptidergic lumbar spinal neurons terminating in either the medial or lateral thalamus in the rat. *Brain Res.* **443**, 345–349.

Nahin, R. L. and Micevych, P. E., 1986, A long ascending pathway of enkephalin-like immunoreactive spinoreticular neurons in the rat. *Neurosci. Lett.* **65**, 271–276.

Nahin, R. L., Madsen, A. M. and Giesler, G. J., 1983, Anatomical and physiological studies of the gray matter surrounding the spinal cord central canal. *J. Comp. Neurol.* **220**, 321–335.
Nahin, R. L., Madsen, A. M. and Giesler, G. J., 1986, Funicular location of the ascending axons of neurons adjacent to the spinal cord central canal in the rat. *Brain Res.* **384**, 367–372.
Nahin, R. L., Hylden, J. L. K., Iadarola, M. J. and Dubner, R., 1989, Peripheral inflammation is associated with increased dynorphin immunoreactivity in both projection and local circuit neurons in the superficial dorsal horn of the rat lumbar spinal cord. *Neurosci. Lett.* **96**, 247–252.
Nakata, Y., Kusaka, Y. and Segawa, T., 1979, Supersensitivity to substance P after dorsal root section. *Life Sci.* **24**, 1651–1654.
Nam, S. C., Kim, K. J., Leem, J. W., Chung, K. and Chung, J. M., 1989, Fiber counts at multiple sites along the rat ventral root after neonatal peripheral neurectomy or dorsal rhizotomy. *J. Comp. Neurol.* **290**, 336–342.
Namba, M., Ghatel, M. A., Gibson, S. J., Polak, J. M. and Bloom, S. R., 1985, Distribution and localization of neuromedin b-like immunoreactivity in pig, cat and rat spinal cord. *Neuroscience* **15**, 1217–1226.
Nandy, K. and Bourne, G. H., 1964, The effects of D-lysergic acid diethylamide tartrate (LSD-25) on the cholinesterases and monoamine oxidase in the spinal cord: a possible factor in the mechanism of hallucination. *J. Neurol. Neurosurg. Psychiat.* **27**, 259–267.
Narotzky, R. A. and Kerr, F. W., 1978, Marginal neurons of the spinal cord: types, afferent synaptology and functional considerations. *Brain Res.* **139**, 1–20.
Nashold, B. S., 1987, Introduction to second international symposium on dorsal root entry zone (DREZ) lesions. *Appl. Neurophysiol.* **51**, 76–77.
Nashold, B. S. and Friedman, H., 1972, Dorsal column stimulation for control of pain. Preliminary report on 30 patients. *J. Neurosurg.* **36**, 590–597.
Nashold, B. S., Somjen, G. and Friedman, H., 1972, Paresthesias and EEG potentials evoked by stimulation of the dorsal funiculi in man. *Exp. Neurol.* **36**, 273–287.
Nathan, P. W., 1963, Results of antero-lateral cordotomy for pain in cancer. *J. Neurol. Neurosurg. Psychiat.* **26**, 353–362.
Nathan, P. W., 1976, The gate-control theory of pain. A critical review. *Brain* **99**, 123–158.
Nathan, P. W., 1990, Comments on 'A dorsolateral spinothalamic tract in macaque monkey' by Apkarian and Hodge. *Pain* **40**, 239–240.
Nathan, P. W. and Smith, M. C., 1959, Fasciculi proprii of the spinal cord in man: review of present knowledge. *Brain* **82**, 610–668.
Nathan, P. W. and Smith, M. C., 1979, Clinico-anatomical correlation in anterolateral cordotomy. *Adv. Pain Res. Ther.* **3**, 921–926.
Nathan, P. W., Smith, M. C. and Cook, A. W., 1986, Sensory effects in man of lesions of the posterior columns and of some other afferent pathways. *Brain* **109**, 1003–1041.
Navaratnam, V. and Lewis, P. R., 1970, Cholinesterase-containing neurones in the spinal cord of the rat. *Brain Res.* **18**, 411–425.
Neal, M. J., 1971, The uptake of [14C]glycine by slices of mammalian spinal cord. *J. Physiol.* **215**, 103–117.
Neary, T. J. and Wilczynski, W., 1977, Ascending thalamic projections from the obex region in ranid frogs. *Brain Res.* **138**, 529–533.
Neary, T. J. and Wilcznski, W., 1979, Anterior and posterior thalamic afferents in the bullfrog, *Rana catesbeiana*. *Neurosci. Abstr.* **5**, 144.
Necker, R. and Hellon, R. F., 1978, Noxious thermal input from the rat tail: modulation by descending inhibitory influences. *Pain* **4**, 231–242.
Neisser, U., 1959, Temperature thresholds for cutaneous pain. *J. Appl. Physiol.* **14**, 368–372.
Ness, T. J. and Gebhart, G. F., 1987, Characterization of neuronal responses to noxious visceral and somatic stimuli in the medial lumbosacral spinal cord of the rat. *J. Neurophysiol.* **57**, 1867–1892.
Ness, T. J. and Gebhart, G. F., 1988, Characterization of neurons responsive to noxious colorectal distension in the T13-L2 spinal cord of the rat. *J. Neurophysiol.* **60**, 1419–1438.
Ness, T. J. and Gebhart, G. F., 1989, Characterization of superficial T13-L2 dorsal horn neurons encoding for colorectal distension in the rat: comparison with neurons in deep laminae. *Brain Res.* **486**, 301–309.
Netsky, M. G., 1953, Syringomyelia. A clinicopathologic study. *Arch. Neurol. Psychiat.* **70**, 741–777.
Neugebauer, V. and Schaible, H. G., 1990, Evidence for a central component in the sensitization of spinal neurons with joint input during development of acute arthritis in cat's knee. *J. Neurophysiol.* **64**, 299–311.
Neuhuber, W. L., Sandoz, P. A. and Fryscak, T., 1986, The central projections of primary afferent neurons of greater splanchnic and intercostal nerves in the rat: a horseradish peroxidase study. *Anat. Embryol. (Berl.)* **174**, 123–144.
Newberry, N. R. and Simmonds, M. A., 1984a, The rat gracile nucleus in vitro: I. Evidence for a GABA-mediated depolarisation of the dorsal column afferents. *Brain Res.* **303**, 41–49.
Newberry, N. R. and Simmonds, M. A., 1984b, The rat gracile nucleus in vitro: II. Field potentials and their conditioned depression. *Brain Res.* **303**, 51–57.
Newberry, N. R. and Simmonds, M. A., 1984c, The rat gracile nucleus in vitro: III. Unitary spike potentials and their conditioned inhibition. *Brain Res.* **303**, 59–65.

Newton, B. W. and Hamill, R. W., 1988, The morphology and distribution of rat serotoninergic intraspinal neurons: an immunohistochemical study. *Brain Res. Bull.* **20**, 349–360.

Nicoll, R. A., 1979, Dorsal root potentials and changes in extracellular potassium in the spinal cord of the frog. *J. Physiol.* **290**, 113–127.

Nicoll, R. A. and Alger, B. E., 1979, Presynaptic inhibition: transmitter and ionic mechanisms. *Int. Rev. Neurobiol.* **21**, 217–258.

Nicoll, R. A., Schenker, C. and Leeman, S. E., 1980, Substance P as a transmitter candidate. *Ann. Rev. Neurosci.* **3**, 227–268.

Nielson, K. D., Adams, J. E. and Hosobuchi, Y., 1975, Phantom limb pain. Treatment with dorsal column stimulation. *J. Neurosurg.* **42**, 301–307.

Nieuwenhuys, R. and Cornelisz, M., 1971, Ascending projections from the spinal cord in the axolotyl (*Ambystoma mexicanum*). *Anat. Rec.* **169**, 388.

Nijensohn, D. E. and Kerr, F. W. L., 1975, The ascending projections of the dorsolateral funiculus of the spinal cord in the primate. *J. Comp. Neurol.* **161**, 459–470.

Nilsson, B. Y., 1969a, Structure and function of the tactile hair receptors on the cat's foreleg. *Acta Physiol. Scand.* **77**, 396–416.

Nilsson, B. Y., 1969b, Hair discs and Pacinian corpuscles functionally associated with the carpal tactile hairs in the cat. *Acta Physiol. Scand.* **77**, 417–428.

Nilsson, B. Y. and Skoglund, C. R., 1965, The tactile hairs on the cat's foreleg. *Acta Physiol. Scand.* **65**, 364–369.

Ninkovic, M. and Hunt, S. P., 1983, α-Bungarotoxin binding sites on sensory neurones and their axonal transport in sensory afferents. *Brain Res.* **272**, 57–69.

Ninkovic, M. and Hunt, S. P., 1985, Opiate and histamine H1 receptors are present on some substance P-containing dorsal root ganglion cells. *Neurosci. Lett.* **53**, 133–137.

Ninkovic, M., Hunt, S. P. and Kelly, J. S., 1981, Effect of dorsal rhizotomy on the autoradiographic distribution of opiate and neurotensin receptors and neurotensin-like immunoreactivity within the rat spinal cord. *Brain Res.* **230**, 111–119.

Ninkovic, M., Hunt, S. P. and Cleave, J. R. W., 1982, Localization of opiate and histamine H1-receptors in the primate sensory ganglia and spinal cord. *Brain Res.* **241**, 197–206.

Ninkovic, M., Beaujouan, J. C., Torrens, Y., Saffroy, M., Hall, M. D. and Glowinski, J., 1985, Differential localization of tachykinin receptors in rat spinal cord. *Eur. J. Pharmacol.* **106**, 463–464.

Nishikawa, N., Bennett, G. J., Ruda, M. A., Lu, G. W. and Dubner, R., 1983, Immunocytochemical evidence for a serotoninergic innervation of dorsal column postsynaptic neurons in cat and monkey: light- and electron-microscopic observations. *Neuroscience* **10**, 1333–1340.

Noback, C. R., 1951, Morphology and phylogeny of hair. *Ann. N.Y. Acad. Sci.* **53**, 476–491.

Noble, R. and Riddell, J. S., 1988, Cutaneous excitatory and inhibitory input to neurones of the postsynaptic dorsal column system in the cat. *J. Physiol.* **396**, 497–513.

Noble, R. and Riddell, J. S., 1989, Descending influences on the cutaneous receptive fields of postsynaptic dorsal column neurones in the cat. *J. Physiol.* **408**, 167–183.

Noguchi, K., Morita, Y., Kiyama, H., Ono, K. and Tohyama, M., 1988, A noxious stimulus induces the preprotachykinin-A gene expression in the rat dorsal root ganglion: a quantitative study using in situ hybridization histochemistry. *Mol. Brain Res.* **4**, 31–35.

Noguchi, K., Senba, E., Morita, Y., Sato, M. and Tohyama, M., 1990a, Co-expression of alpha-CGRP and β-CGRP mRNAs in the rat dorsal root ganglion cells. *Neurosci. Lett.* **100**, 1–5.

Noguchi, K., Senba, E., Morita, Y., Sato, M. and Tohyama, M., 1990b, α-CGRP and β-CGRP mRNAs are differentially regulated in the rat spinal cord and dorsal root ganglion. *Mol. Brain Res.* **7**, 299–304.

Noordenbos, W., 1959, *Pain*. Elsevier, Amsterdam.

Noordenbos, W. and Wall, P. D., 1976, Diverse sensory functions with an almost totally divided spinal cord. A case of spinal cord transection with preservation of part of one anterolateral quadrant. *Pain* **2**, 185–195.

Norcio, R. and DeSantis, M., 1976, The organization of neuronal somata in the 1st sacral spinal ganglion of the cat. *Exp. Neurol.* **50**, 246–258.

Nord, S. G., 1967, Somatotopic organization in the spinal trigeminal nucleus, the dorsal column nuclei and related structures in the rat. *J. Comp. Neurol.* **130**, 343–356.

Nordin, M., 1990, Low-threshold mechanoreceptive and nociceptive units with unmyelinated (C) fibres in the human supraorbital nerve. *J. Physiol.* **426**, 229–240.

Norrsell, U., 1966a, An evoked potential study of spinal pathways projecting to the cerebral somatosensory areas in the dog. *Exp. Brain Res.* **2**, 261–268.

Norrsell, U., 1966b, The spinal afferent pathways of conditioned reflexes to cutaneous stimuli in the dog. *Exp. Brain Res.* **2**, 269–282.

Norrsell, U., 1979, Thermosensory defects after cervical spinal cord lesions in the cat. *Exp. Brain Res.* **35**, 479–494.

Norrsell, U., 1983, Unilateral behavioural thermosensitivity after transection of one lateral funiculus in the cervical spinal cord of the cat. *Exp. Brain Res.* **53**, 71–80.

Norrsell, U., 1989a, Behavioural thermosensitivity after unilateral, partial lesions of the lateral funiculus in the cervical spinal cord of the cat. *Exp. Brain Res.* **78**, 369–373.

Norrsell, U., 1989b, Behavioural thermosensitivity after bilateral lesions of the lateral funiculi in the cervical spinal cord of the cat. *Exp. Brain Res.* **78**, 374–379.

Norrsell, U. and Ullman, M., 1978, Note on the conduction velocity of warm afferent fibres from the skin of the human leg. *Acta Physiol. Scand.* **103**, 337–339.

Norrsell, U. and Voorhoeve, P., 1962, Tactile pathways from the hindlimb to the cerebral cortex in cat. *Acta Physiol. Scand.* **54**, 9–17.

Norrsell, U. and Wolpow, E. R., 1966, An evoked potential study of different pathways from the hindlimb to the somatosensory areas in the cat. *Acta Physiol. Scand.* **66**, 19–33.

North, R. A. and Yoshimura, M., 1984, The actions of noradrenaline on neurones of the rat substantia gelatinosa in vitro. *J. Physiol.* **349**, 43–55.

Northcutt, R. G. and Ebbesson, S. O. E., 1980, Ascending spinal pathways in the sea lamprey. *Neurosci. Abstr.* **6**, 628.

Novikoff, A. B., 1967a, Lysosomes in nerve cells. In: *The Neuron*, pp. 319–376. H. Hyden (ed.). Elsevier, Amsterdam.

Novikoff, A. B., 1967b, Enzyme localization and ultrastructure of neurons. In: *The Neuron*, pp. 255–318. H. Hyden (ed.). Elsevier, Amsterdam.

Novikoff, A. B., Quintana, N., Villanerde, H. and Forschirm, R., 1966, Nucleoside phosphatase and cholinesterase activities in dorsal root ganglia and peripheral nerve. *J. Cell Biol.* **29**, 525–545.

Novikoff, P. M., Novikoff, A. B., Quintana, N. and Hauw, J. J., 1971, Golgi apparatus, GERL, and lysosomes of neurons in rat dorsal root ganglia, studied by thick section and thin section cytochemistry. *J. Cell Biol.* **50**, 859–886.

Nowak, L. M. and MacDonald, R. L., 1982, Substance P: ionic basis for depolarizing responses of mouse spinal cord neurons in cell culture. *J. Neurosci.* **2**, 1119–1128.

Nowycky, M. C., Fox, A. P. and Tsien, R. W., 1985, Three types of neuronal calcium channel with different calcium agonist sensitivity. *Nature* **316**, 440–443.

Nyberg, G., 1988, Representation of the forepaw in the feline cuneate nucleus: a transganglionic transport study. *J. Comp. Neurol.* **271**, 143–152.

Nyberg, G. and Blomqvist, A., 1982, The termination of forelimb nerves in the feline cuneate nucleus demonstrated by the transganglionic transport method. *Brain Res.* **248**, 209–222.

Nyberg, G. and Blomqvist, A., 1984, The central projections of muscle afferent fibers to the lower medulla and upper spinal cord: an anatomical study in the cat with the transganglionic transport method. *J. Comp. Neurol.* **230**, 99–109.

Nyberg, G. and Blomqvist, A., 1985, The somatotopic organization of forelimb cutaneous nerves in the brachial dorsal horn: An anatomical study in the cat. *J. Comp. Neurol.* **242**, 28–39.

Nygren, L. G. and Olson, L., 1977, A new major projection from the locus coeruleus: the main source of noradrenergic nerve terminals in the ventral and dorsal columns of the spinal cord. *Brain Res.* **131**, 85–93.

O'Brien, C., Woolf, C. J., Fitzgerald, M., Lindsay, R. M. and Molander, C., 1989, Differences in the chemical expression of rat primary afferent neurons which innervate skin, muscle or joint. *Neuroscience* **32**, 493–502.

Obrocki, J. and Borroni, E., 1988, Immunocytochemical evaluation of a cholinergic-specific ganglioside antigen (Chol-1) in the central nervous system of the rat. *Exp. Brain Res.* **72**, 71–82.

Ochoa, J. and Torebjörk, E., 1983, Sensations evoked by intraneural microstimulation of single mechanoreceptor units innervating the human hand. *J. Physiol.* **342**, 633–654.

Ochoa, J. and Torebjörk, E., 1989, Sensations evoked by intraneural microstimulation of C nociceptor fibres in human skin nerves. *J. Physiol.* **415**, 583–599.

O'Conner, B. L. and McConnaughey, J. S., 1978, The structure and innervation of cat knee menisci and their relation to a "sensory hypothesis" of meniscal function. *Amer. J. Anat.* **153**, 431–442.

O'Conner, B. L. and Woodbury, P., 1982, The primary articular nerves to the dog knee. *J. Anat.* **134**, 563–572.

O'Donohue, T. L., Massari, V. J., Pazoles, C. J., Chronwall, B. M., Shults, C. W., Quirion, R., Chase, T. N. and Moody, T. W., 1984, A role for bombesin in sensory processing in the spinal cord. *J. Neurosci.* **4**, 2956–2962.

Odutola, A. B., 1972, The organization of cholinesterase-containing systems of the monkey spinal cord. *Brain Res.* **39**, 353–368.

Odutola, A. B., 1977a, On the location of reticular neurons projecting to the cuneo-gracile nuclei in the rat. *Exp. Neurol.* **54**, 54–59.

Odutola, A. B., 1977b, Patterns and fields of dorsal column fiber terminals in the cuneo-gracile nuclei of the rat. *Exp. Neurol.* **57**, 112–120.

Ogawa, T., Kanazawa, I. and Kimura, S., 1985, Regional distribution of substance P, neurokinin alpha and neurokinin β in rat spinal cord, nerve roots and dorsal root ganglion, and the effects of dorsal root section or spinal transection. *Brain Res.* **359**, 152–157.

Oh, U. T., Kim, K. J., Baik-Han, E. J. and Chung, J. M., 1989, Electrophysiological evidence for an increase in the number of ventral root afferent fibers after neonatal peripheral neurectomy in the rat. *Brain Res.* **501**, 90–99.

Ohnishi, A. and Dyck, P. J., 1974, Loss of small peripheral sensory neurons in Fabray's disease. *Arch. Neurol.* **31**, 120–127.

Ohnishi, A. and Ikeda, M., 1980, Morphometric evaluation of primary sensory neurons in experimental p-bromophenylacetylurea intoxication. *Acta Neuropath. (Berl.)* **52**, 111–118.

Ohnishi, A. and Ogawa, M., 1986, Preferential loss of large lumbar primary sensory neurons in carcinomatous sensory neuropathy. *Ann. Neurol.* **20**, 102–104.
Ohta, M., Offord, K. and Dyck, P. J., 1974, Morphometric evaluation of first sacral ganglion of man. *J. Neurol. Sci.* **22**, 73–82.
O'Keefe, J. and Gaffan, D., 1971, Response properties of units in the dorsal column nuclei of the freely moving rat: changes as a function of behaviour. *Brain Res.* **31**, 374–375.
Oku, R., Satoh, M., Fujii, N., Otaka, A., Yajima, H. and Takagi, H., 1987, Calcitonin gene-related peptide promotes mechanical nociception by potentiating release of substance P from the spinal dorsal horn in rats. *Brain Res.* **403**, 350–354.
Oku, R., Nanayama, T. and Satoh, M., 1988, Calcitonin gene-related peptide modulates calcium mobilization in synaptosomes of rat spinal cord horn. *Brain Res.* **475**, 356–360.
O'Leary, J. L., Heinbecker, P. and Bishop, G. H., 1932, Dorsal root fibers which contribute to the tract of Lissauer. *Proc. Soc. Exp. Biol.* **30**, 302–303.
Oliveras, J. L., Besson, J. M., Guilbaud, G. and Liebeskind, J. C., 1974, Behavioral and electrophysiological evidence of pain inhibition from midbrain stimulation in the cat. *Exp. Brain Res.* **20**, 32–44.
Oliveras, J. L., Bourgoin, S., Hery, F., Besson, J. M. and Hamon, M., 1977, The topographical distribution of serotoninergic terminals in the spinal cord of the cat: biochemical mapping by the combined use of microdissection and microassay procedures. *Brain Res.* **138**, 393–406.
Olschowka, J. A., O'Donohue, T. L., Mueller, G. P. and Jacobowitz, D. M., 1982, The distribution of corticotrophin releasing factor-like immunoreactive neurons in rat brain. *Peptides* **3**, 995–1015.
Olson, G. A., Olson, R. D. and Kastin, A. J., 1989, Endogenous opiates: 1988. *Peptides* **10**, 1253–1280.
Olszewski, J., 1952, *The Thalamus of Macaca mulatta*. Karger, New York.
Olszewski, J., 1954, The cytoarchitecture of the human reticular formation. In: *Brain Mechanisms and Consciousness*, pp. 54–76. E. D. Adrian, F. Bremer and H. H. Jasper (eds.). Blackwell, Oxford.
Olszewski, J. and Baxter, D., 1954, *Cytoarchitecture of the Human Brain Stem*. Karger, New York.
Orr, D. and Rows, R. G., 1901, The nerve cells of the human posterior root ganglia and their changes in general paralysis of the insane. *Brain* **24**, 286–309.
Oscarsson, O., 1973, Functional organization of spinocerebellar paths. In: *Handbook of Sensory Physiology*, Vol. II, *Somatosensory System*, pp. 339–380. A. Iggo (ed.). Springer-Verlag, New York.
Oscarsson, O. and Rosén, I., 1963, Projection to cerebral cortex of large muscle-spindle afferents in forelimb nerves of the cat. *J. Physiol.* **169**, 924–945.
Oscarsson, O. and Rosén, I., 1966, Short-latency projections to the cat's cerebral cortex from skin and muscle afferents in the contralateral forelimb. *J. Physiol.* **182**, 164–184.
O'Shaughnessy, D. J., McGregory, G. P., Ghatei, M. A., Blank, M. A., Springall, D. R., Gu, J., Polak, J. M. and Bloom, S. R., 1983, Distribution of bombesin, somatostatin, substance P and vasoactive intestinal polypeptide in feline and porcine skin. *Life Sci.* **32**, 2827–2836.
Ositelu, D. O., Morris, R. and Vaillant, V., 1987, Innervation of facial skin but not masticatory muscles or the tongue by trigeminal primary afferents containing somatostatin in the rat. *Neurosci. Lett.* **78**, 271–276.
Ostapoff, E. M., Johnson, J. I. and Albright, B. C., 1983, Mechanosensory projections to dorsal column nuclei in a tree squirrel (fox squirrel, *Sciurus niger*). *Neuroscience* **6**, 107–127.
Oswaldo-Cruz, E. and Kidd, E., 1964, Functional properties of neurons in the lateral cervical nucleus of cat. *J. Neurophysiol.* **27**, 1–14.
Otsuka, M. and Konishi, S., 1976, Release of substance P-like immunoreactivity from isolated spinal cord of newborn rat. *Nature* **264**, 83–84.
Otsuka, M. and Takahashi, T., 1977, Putative peptide neurotransmitters. *Ann. Rev. Pharmacol. Toxicol.* **17**, 425–439.
Otten, U. and Lorez, H. P., 1983, Nerve growth factor increases substance P, cholecystokinin and vasoactive intestinal polypeptide immunoreactivities in primary sensory neurones of newborn rats. *Neurosci. Lett.* **34**, 153–158.
Ovelmen-Levitt, J., Johnson, B., Bedenbaugh, P. and Nashold, B. S., 1984, Dorsal root rhizotomy and avulsion in the cat: A comparison of long term effects on dorsal horn neuronal activity. *Neurosurgery* **15**, 921–927.
Owman, C. and Santini, M., 1966, Adrenergic nerves in spinal ganglia of the cat. *Acta Physiol. Scand.* **68**, 127–128.
Pacini, F., 1840, *Nuovi Organi Scoperti nel Corpo Umano*. Ciro, Pistoja. Cited in: Pease and Quilliam, 1957.
Pagni, C. A., 1989, Central pain due to spinal cord and brain stem damage. In: *Textbook of Pain*, 2nd ed., pp. 634–655. P. D. Wall and R. Melzack (eds.). Churchill Livingstone, London.
Paik, K. S., Nam, S. C. and Chung, J. M., 1988, Differential inhibition produced by peripheral conditioning stimulation on noxious mechanical and thermal responses of different classes of spinal neurons in the cat. *Exp. Neurol.* **99**, 498–511.
Paintal, A. S., 1957, Responses from mucosal mechanoreceptors in the small intestine of the cat. *J. Physiol.* **139**, 353–368.
Paintal, A. S., 1959, Intramuscular propagation of sensory impulses. *J. Physiol.* **148**, 240–251.
Paintal, A. S., 1960, Functional analysis of group III afferent fibres of mammalian muscles. *J. Physiol.* **152**, 250–270.

Palacios, J. M., Wamsley, J. K. and Kuhar, M. J., 1981, High affinity GABA receptors—autoradiographic localization. *Brain Res.* **222**, 285–307.
Palmer, C. I. and Gardner, E. P., 1990, Simulation of motion on the skin. IV. Responses of Pacinian corpuscle afferents innervating the primate hand to stripe patterns on the OPTACON. *J. Neurophysiol.*. **64**, 236–247.
Pang, I. H. and Vasko, M. R., 1986, Morphine and norepinephrine but not 5-hydroxytryptamine and γM-amino-butyric acid inhibit the potassium-stimulated release of substance P from rat spinal cord slices. *Brain Res.* **376**, 268–279.
Panneton, W. M. and Burton, H., 1985, Projections from the paratrigeminal nucleus and the medullary and spinal dorsal horns to the peribrachial area in the cat. *Neuroscience* **15**, 779–797.
Panula, P., 1986, Histochemistry and function of bombesin-like peptides. *Med. Biol.* **64**, 177–192.
Panula, P., Yang, H-Y.T. and Costa, E., 1982, Neuronal location of the bombesin-like immunoreactivity in the central nervous system of the rat. *Reg. Peptides* **4**, 275–283.
Panula, P., Hadjiconstantinou, M., Yang, H-Y.T. and Costa, E., 1983, Immunohistochemical localization of bombesin/gastrin-releasing peptide and substance P in primary sensory neurons. *J. Neurosci.* **3**, 2021–2029.
Panula, P., Kivipelto, L., Nieminen, O., Majane, E. A. and Yang, H-T.T., 1987, Neuroanatomy of morphine-modulating peptides. *Med. Biol.* **65**, 127–135.
Panula, P., Nieminen, O., Falkenberg, M. and Auvinen, S., 1989a, Localization and development of bombesin/GRP-like immunoreactivity in the rat central nervous system. *Ann. N.Y. Acad. Sci.* **547**, 54–69.
Panula, P., Flugge, G., Fuchs, E., Pirvola, U., Auvinen, S. and Airaksinen, M. S., 1989b, Histamine-immunoreactive nerve fibers in the mammalian spinal cord. *Brain Res.* **484**, 234–239.
Papadopoulos, G. C., Karamanlidis, A. N., Antonopoulos, J. and Dinopoulos, A., 1986a, Neurotensinlike immunoreactive neurons in the hedgehog (*Erinaceus europaeus*) and the sheep (*Ovis aries*) central nervous system. *J. Comp. Neurol.* **244**, 193–203.
Papadopoulos, G. C., Karamanlidis, A. N., Dinopoulos, A. and Antonopoulos, J., 1986b, Somatostatinlike immunoreactive neurons in the hedgehog (*Erinaceus europaeus*) and the sheep (*Ovis aries*) central nervous system. *J. Comp. Neurol.* **244**, 174–192.
Parthe, V., 1981, Histochemical localization of carbonic anhydrase in vertebrate nervous tissue. *J. Neurosci. Res.* **6**, 119–131.
Patrick, H. T., 1896, On the course and destination of Gowers' tract. *J. Nerv. Ment. Dis.* **21**, 85–107.
Patterson, J. T., Head, P. A., McNeill, D. L., Chung, K. and Coggeshall, R. E., 1989, Ascending unmyelinated primary afferent fibers in the dorsal funiculus. *J. Comp. Neurol.* **290**, 384–390.
Patterson, J. T., Coggeshall, R. E., Lee, W. T. and Chung, K., 1990, Long ascending unmyelinated primary afferent axons in the rat dorsal column: immunohistochemical localizations. *Neurosci. Lett.* **108**, 6–10.
Pattle, R. E. and Weddell, G., 1948, Observations on electrical stimulation of pain fibres in an exposed human sensory nerve. *J. Neurophysiol.* **11**, 93–98.
Peach, R., 1972, Acid phosphatase distribution in the trigeminal ganglion of the rat. *Anat. Rec.* **174**, 236–250.
Pearson, A. A., 1952, Role of gelatinous substance of spinal cord in conduction of pain. *Arch. Neurol. Psychiat.* **68**, 515–529.
Pearson, J. C. and Goldfinger, M. D., 1987, The morphology and distribution of serotonin-like immunoreactive fibers in the cat dorsal column nuclei. *Neuroscience Lett.* **74**, 125–131.
Pearson, J. C. and Haines, D. E., 1980, Somatosensory thalamus of a prosimian primate (*Galago senegalensis*). I. Configuration of nuclei and termination of spinothalamic fibers. *J. Comp. Neurol.* **190**, 533–558.
Pease, D. C. and Quilliam, T. A., 1957, Electron microscopy of the Pacinian corpuscle. *J. Biophys. Biochem. Cytol.* **3**, 331–357.
Pechura, C. and Liu, R., 1986, Spinal neurons which project to the periaqueductal gray and the medullary reticular formation via axon collaterals: a double-label fluorescence study in the rat. *Brain Res.* **374**, 357–361.
Peet, M. J., Leah, J. D. and Curtis, D. R., 1983, Antagonists of synaptic and amino acid excitation of neurones in the cat spinal cord. *Brain Res.* **266**, 83–95.
Pelletier, G., LeClerc, R. and Dupont, A., 1977, Electron microscope immunohistochemical localization of substance P in the central nervous system of the rat. *J. Histochem. Cytochem.* **25**, 1373–1380.
Perez de la Mora, M., Possani, L. D., Tapia, R., Teran, L., Palacios, R., Fuxe, K., Hökfelt, T. and Ljungdahl, Å., 1981, Demonstration of central γA-aminobutyrate-containing nerve terminals by means of antibodies against glutamate decarboxylase. *Neuroscience* **6**, 875–895.
Perl, E. R., 1968, Myelinated afferent fibres innervating the primate skin and their response to noxious stimuli. *J. Physiol.* **197**, 593–615.
Perl, E. R., 1971, Is pain a specific sensation? *J. Psychiat. Res.* **8**, 273–287.
Perl, E. R. and Whitlock, D. G., 1961, Somatic stimuli exciting spinothalamic projections to thalamic neurons in cat and monkey. *Exp. Neurol.* **3**, 256–296.
Perl, E. R., Whitlock, D. G. and Gentry, J. R., 1962, Cutaneous projection to second order neurons of the dorsal column system. *J. Neurophysiol.* **25**, 337–358.
Pertovaara, A., Huopaniemi, T. and Tukeva, T., 1986, Liminal and supraliminal response characteristics of mechanoreceptive neurons in the cuneate nucleus of cat. *Exp. Brain Res.* **62**, 486–494.
Peschanski, M. and Besson, J. M., 1984, A spino-reticulo-thalamic pathway in the rat: an anatomical study with reference to pain transmission. *Neuroscience* **12**, 165–178.

Peschanski, M., Mantyh, P. W. and Besson, J. M., 1983, Spinal afferents to the ventrobasal thalamic complex in the rat: an anatomical study using wheat-germ agglutinin conjugated to horseradish peroxidase. *Brain Res.* **278**, 240–244.

Peschanski, M., Roudier, F., Ralston, H. J. and Besson, J. M., 1985, Ultrastructural analysis of the terminals of various somatosensory pathways in the ventrobasal complex of the rat thalamus: an electron-microscopic study using wheatgerm agglutinin conjugated to horseradish peroxidase as an axonal tracer. *Somatosensory Res.* **3**, 75–87.

Peschanski, M., Kayser, V. and Besson, J. M., 1986, Behavioral evidence for a crossed ascending pathway for pain transmission in the anterolateral quadrant of the rat spinal cord. *Brain Res.* **376**, 164–168.

Peterson, D. F. and Brown, A. M., 1973, Functional afferent innervation of testis. *J. Neurophysiol.* **36**, 425–433.

Petit, D., 1972, Postsynaptic fibres in the dorsal columns and their relay in the nucleus gracilis. *Brain Res.* **48**, 380–384.

Petit, D. and Burgess, P. R., 1968, Dorsal column projection of receptors in cat hairy skin supplied by myelinated fibers. *J. Neurophysiol.* **31**, 849–855.

Peto, T. E. A., 1980, Corticofugal actions on the lateral cervical nucleus of the cat. *Exp. Neurol.* **68**, 531–547.

Petras, J. M., 1977, Spinocerebellar neurons in the rhesus monkey. *Brain Res.* **130**, 146–151.

Petren, K., 1902, Ein Beitrag zur Frage vom Verlaufe der Bahnen der Hautsinne im Rueckenmarke. *Skand. Arch. Physiol.* **13**, 9–98.

Petrusz, P., Merchenthaler, I. and Maderdrut, J. L., 1985, Distribution of enkephalin-containing neurons in the central nervous system. In: *Handbook of Chemical Neuroanatomy*, pp. 273–323. A. Björklund and T. Hökfelt (eds.). Elsevier, Amsterdam.

Peyronnard, J., Charron, L., Lavoie, J. and Messier, J., 1986, Motor, sympathetic sensory innervation of rat skeletal muscles. *Brain Res.* **373**, 288–302.

Peyronnard, J., Charron, L., Messier, J. and Lavoie, J., 1988a, Differential effects of distal and proximal nerve lesions on carbonic anhydrase activity in rat primary sensory neurons, ventral and dorsal root axons. *Exp. Brain Res.* **70**, 550–560.

Peyronnard, J. M., Charron, L., Lavoie, J., Messier, J. P. and Dubreuil, M., 1988b, Carbonic anhydrase and horseradish peroxidase: double labeling of rat dorsal root ganglion neurons innervating motor and sensory peripheral nerves. *Anat. Embryol.* **177**, 353–359.

Peyronnard, J-M., Charron, L., Messier, J-P., Lavoie, J., Leger, C. and Faraco-Cantin, F., 1989, Changes in lectin binding of lumbar dorsal root ganglia neurons and peripheral axons after sciatic and spinal nerve injury in the rat. *Cell Tiss. Res.* **257**, 379–388.

Phelps, P. E., Barber, R. P., Houser, C. R., Crawford, G. D., Salvaterra, P. M. and Vaughn, J. E., 1984, Postnatal development of neurons containing choline acetyltransferase in rat spinal cord: an immunocytochemical study. *J. Comp. Neurol.* **229**, 347–361.

Phelps, P. E., Barber, R. P. and Vaughn, J. E., 1988, Generation patterns of four groups of cholinergic neurons in rat cervical spinal cord: a combined tritiated thymidine autoradiographic and choline acetyltransferase immunocytochemical study. *J. Comp. Neurol.* **273**, 459–472.

Phillips, J. R. and Johnson, K. O., 1981a, Tactile spatial resolution. II. Neural representation of bars, edges, and gratings in monkey primary afferents. *J. Neurophysiol.* **46**, 1192–1203.

Phillips, J. R. and Johnson, K. O., 1981b, Tactile spatial resolution. III. A continuum mechanics model of skin predicting mechanoreceptor responses to bars, edges, and gratings. *J. Neurophysiol.* **46**, 1204–1225.

Pickel, V. M., Reis, D. J. and Leeman, S. E., 1977, Ultrastructural localization of substance P in neurons of rat spinal cord. *Brain Res.* **122**, 534–540.

Piepmeier, J. M., Kauer, J. S. and Greer, C. A., 1983, Laminar distributions of 2-deoxyglucose uptake in the rat spinal cord following electrical stimulation of the sciatic nerve. *Brain Res.* **259**, 167–171.

Pierau, F. K., Torrey, P. and Carpenter, D. O., 1975, Afferent nerve fiber activity responding to temperature changes of scrotal skin of the rat. *J. Neurophysiol.* **38**, 601–612.

Pierau, F.-K., Taylor, D. C. M., Abel, W. and Friedrich, B., 1982, Dichotomizing peripheral fibres revealed by intracellular recording from rat sensory neurones. *Neurosci. Lett.* **31**, 123–128.

Pierau, F. K., Fellner, G. and Taylor, D. C. M., 1984, Somatovisceral convergence in cat dorsal root ganglia neurones demonstrated by double labeling with fluorescent tracers. *Brain Res.* **3221**, 63–70.

Pierce, J. P., Weinberg, R. J. and Rustioni, A., 1990, Single fiber studies of ascending input to the cuneate nucleus of cats: II. Postsynaptic afferents. *J. Comp. Neurol.* **300**, 134–152.

Pincus, F., 1902, Ueber einen bisher unbekannten Nebenapparat am Haarsystem des Menschen. *Haarscheiben. Derm. Z.* **9**, 465–469.

Pinkus, H., 1964, Pinkus's Haarscheibe and tactile receptors in cats. *Science* **144**, 891.

Plenderleith, M. B., Cameron, A., Key, B. and Snow, P. J., 1988, Soybean agglutinin binds to a subpopulation of primary sensory neurones in the cat. *Neurosci. Lett.* **86**, 257–262.

Plenderleith, M. B., Cameron, A. A., Key, B. and Snow, P. J., 1989, The plant lectin soybean agglutinin binds to the soma, axon and central terminals of a subpopulation of small-diameter primary sensory neurons in the rat and cat. *Neuroscience* **31**, 683–695.

Plenderleith, M. B., Haller, C. J. and Snow, P. J., 1990, Peptide coexistence in axon terminals within the superficial dorsal horn of the rat spinal cord. *Synapse* **6**, 344–350.

Poggio, G. F. and Mountcastle, V. B., 1960, A study of the functional contributions of the lemniscal and spinothalamic systems to somatic sensibility. *Bull. Johns Hopkins Hosp.* **106**, 266–316.

Poggio, G. F. and Mountcastle, V. B., 1963, The functional properties of ventrobasal thalamic neurons studied in unanesthetized monkeys. *J. Neurophysiol.* **26**, 775–806.

Poirier, L. J. and Bertrand, C., 1955, Experimental and anatomical investigation of the lateral spino-thalamic and spino-tectal tracts. *J. Comp. Neurol.* **102**, 745–757.

Polacek, P., 1961, Differences in the structure and variability of encapsulated nerve endings in the joints of some species of mammals. *Acta Anat.* **47**, 112–124.

Polistina, D. C., Murray, M. and Goldberger, M. E., 1990, Plasticity of dorsal root and descending serotoninergic projections after partial deafferentation of the adult rat spinal cord. *J. Comp. Neurol.* **299**, 349–363.

Pomeranz, B., 1973, Specific nociceptive fibers projecting from spinal cord neurons to the brain: A possible pathway for pain. *Brain Res.* **50**, 447–451.

Pomeranz, B. and Cheng, R., 1979, Suppression of noxious responses in single neurons of cat spinal cord by electroacupuncture and its reversal by the opiate antagonist naloxone. *Exp. Neurol.* **64**, 327–341.

Pomeranz, B., Wall, P. D. and Weber, W. V., 1968, Cord cells responding to fine myelinated afferents from viscera, muscle and skin. *J. Physiol.* **199**, 511–532.

Pompeiano, O. and Brodal, A., 1957, Spino-vestibular fibers in the cat. An experimental study. *J. Comp. Neurol.* **108**, 353–382.

Potanos, J. N., Wolf, A. and Cowen, D., 1959, Cytochemical localization of oxidative enzymes in human nerve cells and neuroglia. *J. Neuropath. Exp. Neurol.* **18**, 627–635.

Poulos, D. A., 1971, Temperature related changes in discharge patterns of squirrel monkey thermoreceptors. In: *Research in Physiology*, pp. 441–455. F. F. Kao, K. Koizumi and M. Vassalle (eds.). A. Gaggi, Bologna.

Poulos, D. A., 1975, Central processing of peripheral temperature information. In: *The Somatosensory System*, pp. 78–93. H. H. Kornhuber (ed.). Thieme, Stuttgart.

Poulos, D. A. and Benjamin, R. M., 1968, Response of thalamic neurons to thermal stimulation of the tongue. *J. Neurophysiol.* **31**, 28–43.

Poulos, D. A. and Lende, R. A., 1970a, Response of trigeminal ganglion neurons to thermal stimulation of oral-facial regions. I. Steady-state response. *J. Neurophysiol.* **33**, 508–517.

Poulos, D. A. and Lende, R. A., 1970b, Response of trigeminal ganglion neurons to thermal stimulation of oral-facial regions. II. Temperature change response. *J. Neurophysiol.* **33**, 518–526.

Presley, R. W., Hammond, D. L., Gogas, K. R., Levine, J. D. and Basbaum, A. I., 1989, Morphine and U50488H suppress noxious visceral stimulation-evoked fos protein immunoreactivity in the spinal cord and nucleus of the solitary tract (NTS) of the rat. *Neurosci. Abstr.* **15**, 155.

Pretel, S., Guinan, M. J. and Carstens, E., 1988, Inhibition of the responses of cat dorsal horn neurons to noxious skin heating by stimulation in medial or lateral medullary reticular formation. *Exp. Brain Res.* **72**, 51–62.

Price, D. D., 1972, Characteristics of second pain and flexion reflexes indicative of prolonged central summation. *Exp. Neurol.* **37**, 371–387.

Price, D. D. and Browe, A. C., 1973, Responses of spinal cord neurons to graded noxious and non-noxious stimuli. *Brain Res.* **64**, 425–429.

Price, D. D. and Dubner, R., 1977, Neurons that subserve the sensory-discriminative aspects of pain. *Pain* **3**, 307–338.

Price, D. D. and Mayer, D. J., 1974, Physiological laminar organization of the dorsal horn of *M. mulatta*. *Brain Res.* **79**, 321–325.

Price, D. D. and Mayer, D. J., 1975, Neurophysiological characterization of the anterolateral quadrant neurons subserving pain in *M. mulatta*. *Pain* **1**, 59–72.

Price, D. D. and Wagman, I. H., 1970, Physiological roles of A and C fiber inputs to spinal dorsal horn of *Macaca mulatta*. *Exp. Neurol.* **29**, 383–399.

Price, D. D., Hull, C. D. and Buchwald, N. A., 1971, Intracellular responses of dorsal horn cells to cutaneous and sural nerve A and C fiber stimuli. *Exp. Neurol.* **33**, 291–309.

Price, D. D., Hu, J. W., Dubner, R. and Gracely, R., 1977, Peripheral suppression of first pain and central summation of second pain evoked by noxious heat pulses. *Pain* **3**, 57–68.

Price, D. D., Hayes, R. L., Ruda, M. and Dubner, R., 1978, Spatial and temporal transformations of input to spinothalamic tract neurons and their relation to somatic sensations. *J. Neurophysiol.* **41**, 933–947.

Price, D. D., Hayashi, H., Dubner, R. and Ruda, M. A., 1979, Functional relationships between neurons of marginal and substantia gelatinosa layers of primate dorsal horn. *J. Neurophysiol.* **42**, 1590–1608.

Price, D. D., Bushnell, M. C. and Iadarola, M. J., 1981, Primary afferent and sacral dorsal horn neuron responses to vaginal probing in the cat. *Neurosci. Lett.* **26**, 67–72.

Price, G. W., Wilkin, G. P., Turnbull, M. J. and Bowery, N. G., 1984, Are baclofen-sensitive GABAb receptors present on primary afferent terminals of the spinal cord? *Nature* **307**, 71–74.

Price, G. W., Kelly, J. S. and Bowery, N. G., 1987, The location of GABAb receptor binding sites in mammalian spinal cord. *Synapse* **1**, 530–538.

Price, J., 1985, An immunohistochemical and quantitative examination of dorsal root ganglion neuronal subpopulations. *J. Neurosci.* **5**, 2051–2059.

Price, J. and Mudge, A. W., 1983, A subpopulation of rat dorsal root ganglion neurones is catecholaminergic. *Nature* **301**, 241–245.

Priestley, J. V., Bramwell, S., Butcher, L. L. and Cuello, A. C., 1982a, Effect of capsaicin on neuropeptides in areas of termination of primary sensory neurones. *Neurochem. Intl.* **4**, 57–65.

Priestley, J. V., Somogyi, P. and Cuello, A. C., 1982b, Immunocytochemical localization of substance P in the spinal trigeminal nucleus of the rat: a light and electron microscopic study. *J. Comp. Neurol.* **211**, 31–49.

Pritz, M. B., 1983, The dorsal column nucleus in a reptile, *Caiman crocodilus. Neurosci. Lett.* **39**, 119–123.

Pritz, M. B. and Northcutt, R. G., 1980, Anatomical evidence for an ascending somatosensory pathway to the telencephalon in crocodiles, *Caiman crocodilus. Exp. Brain Res.* **40**, 342–345.

Pritz, M. B. and Stritzel, M. E., 1986, Dorsal funicular projections to the dorsal column nucleus in a reptile, *Caiman crocodilus. J. Comp. Neurol.* **249**, 1–12.

Pritz, M. B. and Stritzel, M. E., 1989, Reptilian somatosensory midbrain: identification based on input from the spinal cord and dorsal column nucleus. *Brain Behav. Evol.* **33**, 1–14.

Probert, L. and Hanley, M. R., 1987, The immunocytochemical localisation of "substance-P-degrading enzyme" within the rat spinal cord. *Neurosci. Lett.* **78**, 132–137.

Probst, A., Cortés, R. and Palacios, J. M., 1986, The distribution of glycine receptors in the human brain. A light microscopic autoradiographic study using [^{3}H]strychnine. *Neuroscience* **17**, 11–35.

Probst, M., 1902, Zur Kenntnis der Schleifenschicht und ueber centripetale Rueckenmarksfasern zum Deiters'schen Kern, zum Sehhuegel und zur Substantia reticularis. *Mschr. Psychiat. Neurol.* **11**, 3–12.

Proshansky, E. and Egger, M. D., 1977, Staining of the dorsal root projection to the cat's dorsal horn by anterograde movement of horseradish peroxidase. *Neurosci. Lett.* **5**, 103–110.

Proshansky, E., Kauer, J. S., Stewart, W. B. and Egger, M. D., 1980, 2-deoxyglucose uptake in the cat spinal cord during sustained and habituated activity in the plantar cushion reflex pathway. *J. Comp. Neurol.* **194**, 505–517.

Proske, U., Schaible, H. G. and Schmidt, R. F., 1988, Joint receptors and kinaesthesia. *Exp. Brain Res.* **72**, 219–224.

Proudfit, H. K. and Anderson, E. G., 1974, New long latency bulbospinal evoked potentials blocked by 5HT antagonists. *Brain Res.* **65**, 542–546.

Proudfit, H. K., Larson, A. A. and Anderson, E. G., 1980, The role of GABA and serotonin in the mediation of raphe-evoked spinal cord dorsal root potentials. *Brain Res.* **195**, 149–165.

Provins, K. A., 1958, The effect of peripheral nerve block on the appreciation and execution of finger movements. *J. Physiol.* **143**, 55–67.

Pubols, B. H., 1990, Slowly adapting Type I mechanoreceptor discharge as a function of dynamic force versus dynamic displacement of glabrous skin of raccoon and squirrel monkey hand. *Neurosci. Lett.* **110**, 86–90.

Pubols, B. H. and Benkich, M. E., 1986, Relations between stimulus force, skin displacement, and discharge characteristics of slowly adapting type I cutaneous mechanoreceptors in glabrous skin of squirrel monkey hand. *Somatosensory Res.* **4**, 111–125.

Pubols, B. H. and Pubols, L. M., 1976, Coding of mechanical stimulus velocity and indentation depth by squirrel monkey and raccoon glabrous skin mechanoreceptors. *J. Neurophysiol.* **39**, 773–787.

Pubols, B. H., Welker, W. I. and Johnson, J. I., 1965, Somatic sensory representation of forelimb in dorsal root fibers of raccoon, coatimundi, and cat. *J. Neurophysiol.* **28**, 312–341.

Pubols, B. H., Haring, J. H. and Rowinski, M. J., 1989, Patterns of resting discharge in neurons of the raccoon main cuneate nucleus. *J. Neurophysiol.* **61**, 1131–1141.

Pubols, L. M., 1984, The boundary of proximal hindlimb representation in the dorsal horn following peripheral nerve lesions in cats: a reevaluation of plasticity in the somatotopic map. *Somatosensory Res.* **2**, 19–32.

Pubols, L. M. and Bowen, D. C., 1988, Lack of central sprouting of primary afferent fibers after ricin deafferentation. *J. Comp. Neurol.* **275**, 282–287.

Pubols, L. M. and Brenowitz, G. L., 1981, Maintenance of dorsal horn somatotopic organization and increased high-threshold response after single root or spared root deafferentation in cats. *J. Neurophysiol.* **47**, 103–112.

Pubols, L. M. and Goldberger, M. E., 1980, Recovery of function in dorsal horn following partial deafferentation. *J. Neurophysiol.* **43**, 102–117.

Pubols, L. M. and Pubols, B. H., 1973, Modality composition and functional characteristics of dorsal column mechanoreceptive afferent fibers innervating the raccoon's forepaw. *J. Neurophysiol.* **36**, 1023–1037.

Pubols, L. M., Pubols, B. H. and Munger, B. L., 1971, Functional properties of mechanoreceptors in glabrous skin of the raccoon's forepaw. *Exp. Neurol.* **31**, 165–182.

Pubols, L. M., Foglesong, M. E. and Vahle-Hinz, C., 1986, Electrical stimulation reveals relatively ineffective sural nerve projections to dorsal horn neurons in the cat. *Brain Res.* **371**, 109–122.

Puil, E., 1981, S-glutamate: its interactions with spinal neurons. *Brain Res. Rev.* **3**, 299–322.

Puletti, F. and Blomqvist, A. J., 1967, Single neuron activity in posterior columns of the human spinal cord. *J. Neurosurg.* **27**, 255–259.

Putnam, J. E. and Whitehorn, D., 1973, Polarization changes in the terminals of identified primary afferent fibers at the gracile nucleus. *Exp. Neurol.* **41**, 246–259.

Quensel, F., 1898, Ein Fall von Sarcom der Dura spinalis. Beitrag zur Kenntnis der secundaeren Degenerationen nach Rueckenmarkscompression. *Neurol. Zbl.* **17**, 482–493.

Quilliam, T. A., 1975, Neuro-cutaneous relationships in fingerprint skin. In: *The Somatosensory System*, pp. 193–199. H. H. Kornhuber (ed.). Geprg Thieme Verlag, Stuttgart.

Quilliam, T. A. and Sato, M., 1955, The distribution of myelin on nerve fibres from Pacinian corpuscles. *J. Physiol.* **129**, 167–176.

Rabiner, A. M. and Browder, J., 1948, Concerning the conduction of touch and deep sensibilities through the spinal cord. *Trans. Amer. Neurol. Assoc.* **73**, 137–142.

Rakic, P., 1975, Local circuit neurons. *Neurosci. Res. Prog. Bull.* **13**, 297–446.

Ralston, H. J., 1965, The organization of the substantia gelatinosa Rolandi in the cat lumbosacral cord. *Z. Zellforsch.* **67**, 1–23.

Ralston, H. J., 1968a, The fine structure of neurons in the dorsal horn of the cat spinal cord. *J. Comp. Neurol.* **132**, 275–302.

Ralston, H. J., 1968b, Dorsal root projections to dorsal horn neurons in the cat spinal cord. *J. Comp. Neurol.* **132**, 303–330.

Ralston, H. J., 1969, The synaptic organization of lemniscal projections to the ventrobasal thalamus of the cat. *Brain Res.* **14**, 99–115.

Ralston, H. J., 1971, The synaptic organization in the dorsal horn of the spinal cord and in the ventrobasal thalamus in the cat. In: *Oral-Facial Sensory and Motor Mechanisms*, pp. 229–250. R. Dubner and Y. Kawamura (eds.). Appleton-Century-Crofts, New York.

Ralston, H. J., 1979, The fine structure of laminae I, II and III of the macaque spinal cord. *J. Comp. Neurol.* **184**, 619–642.

Ralston, H. J., 1982, The fine structure of laminae IV, V and VI of the macaque spinal cord. *J. Comp. Neurol.* **212**, 425–434.

Ralston, H. J., 1983, The synaptic organization of the ventrobasal thalamus in the rat, cat and monkey. In: *Somatosensory Integration in the Thalamus*, pp. 241–250. G. Macchi, A. Rustioni and R. Spreafico, (eds.). Elsevier, Amsterdam.

Ralston, H. J., 1984, Synaptic organization of spinothalamic tract projections to the thalamus, with special reference to pain. *Adv. Pain Res. Ther.* **6**, 183–195.

Ralston, H. J. and Ralston, D. D., 1979, The distribution of dorsal root axons in laminae I, II and III of the macaque spinal cord: A quantitative electron microscope study. *J. Comp. Neurol.* **184**, 643–684.

Ralston, H. J. and Ralston, D. D., 1982, The distribution of dorsal root axons to laminae IV, V, and VI of the macaque spinal cord: A quantitative electron microscopic study. *J. Comp. Neurol.* **212**, 435–448.

Ralston, H. J., Light, A. R., Ralston, D. D. and Perl, E. R., 1984, Morphology and synaptic relationships of physiologically identified low-threshold dorsal root axons stained with intra-axonal horseradish in the cat and monkey. *J. Neurophysiol.* **51**, 777–792.

Rambourg, A., Clermont, Y. and Beaudet, A., 1983, Ultrastructural features of six types of neurons in rat dorsal root ganglia. *J. Neurocytol.* **12**, 47–66.

Ramón-Moliner, E. and Nauta, W. J. H., 1966, The isodendritic core of the brain stem. *J. Comp. Neurol.* **126**, 311–335.

Ramon y Cajal, S., 1909, Histologie du systeme nerveux de l'homme et des vertebres. Vol. I, pp. 908–911. Inst. Cajal, Madrid, reprinted in 1952.

Randić, M. and Miletić, V., 1977, Effects of substance P in cat dorsal horn neurones activated by noxious stimuli. *Brain Res.* **128**, 164–169.

Randić, M. and Miletić, V., 1978, Depressant actions of methionine-enkephalin and somatostatin in cat dorsal horn neurones activated by noxious stimuli. *Brain Res.* **152**, 196–202.

Randić, M. and Yu, H. Y., 1976, Effects of 5-hydroxytryptamine and bradykinin in cat dorsal horn neurones activated by noxious stimuli. *Brain Res.* **111**, 197–203.

Randić, M., Carstens, E., Zimmermann, M. and Klumpp, D., 1982, Dual effects of substance P on the excitability of single cutaneous primary afferent C- and A-fibers in the cat spinal cord. *Brain Res.* **233**, 389– 393.

Randić, M., Ryu, P. D. and Urban, L., 1986, Effects of polyclonal and monoclonal antibodies to substance P on slow excitatory transmission in rat spinal dorsal horn. *Brain Res.* **383**, 15–27.

Randić, M., Hećimović, H. and Ryu, P. D., 1990, Substance P modulates glutamate-induced currents in acutely isolated rat spinal dorsal horn neurones. *Neurosci. Lett.* **117**, 74–80.

Ranieri, F., Mei, N. and Crousillat, J., 1973, Les afferences splanchniques provenant des mecanorecepteurs gastro-intestinaux et peritoneaux. *Exp. Brain Res.* **16**, 276–290.

Ranson, S. W., 1912, The structure of the spinal ganglia and of the spinal nerves. *J. Comp. Neurol.* **22**, 159–175.

Ranson, S. W., 1913a, The fasciculus cerebro-spinalis in the albino rat. *Amer. J. Anat.* **14**, 411–424.

Ranson, S. W., 1913b, The course within the spinal cord of the non-medullated fibers of the dorsal roots: a study of Lissauer's tract in cat. *J. Comp. Neurol.* **23**, 259–281.

Ranson, S. W., 1914, An experimental study of Lissauer's tract and the dorsal roots. *J. Comp. Neurol.* **24**, 531–545.

Ranson, S. W. and Billingsley, P. R., 1916, The conduction of painful afferent impulses in the spinal nerves. *Amer. J. Anat.* **40**, 571–584.

Ranson, S. W. and Hess, C. L. von, 1915, The conduction within the spinal cord of the afferent impulses producing pain and the vasomotor reflexes. *Amer. J. Physiol.* **38**, 128–152.

Ranson, S. W. and Ingram, W. R., 1932, The diencephalic course and termination of the medial lemniscus and the brachium conjunctivum. *J. Comp. Neurol.* **56**, 257–275.

Ranson, S. W., Davenport, H. K. and Doles, E. A., 1932, Intramedullary course of the dorsal root fibers of the first three cervical nerves. *J. Comp. Neurol.* **54**, 1–12.

Rao, G. S., Breazile, J. E. and Kitchell, R. L., 1969, Distribution and termination of spinoreticular afferents in the brain stem of sheep. *J. Comp. Neurol.* **137**, 185–196.

Rasmussen, A. T. and Peyton, W. T., 1948, The course and termination of the medial lemniscus in man. *J. Comp. Neurol.* **88**, 411–424.
Rasmusson, D. D., 1988, Projections of digit afferents to the cuneate nucleus in the raccoon before and after partial deafferentation. *J. Comp. Neurol.* **277**, 549–556.
Rasmusson, D. D., 1989, The projection pattern of forepaw nerves to the cuneate nucleus of the raccoon. *Neurosci. Lett.* **98**, 129–134.
Rastad, J., 1981a, Morphology of synaptic vesicles in axodendritic and axosomatic collateral terminals of two feline spinocervical tract cells stained intracellularly with horseradish peroxidase. *Exp. Brain Res.* **41**, 390–398.
Rastad, J., 1981b, Quantitative analysis of axodendritic and axosomatic collateral terminals of two feline spinocervical tract cells. *J. Neurocytol.* **10**, 475–496.
Rastad, J., Jankowska, E. and Westman, J., 1977, Arborization of initial axon collaterals of spinocervical tract cells stained intracellularly with horseradish peroxidase. *Brain Res.* **135**, 1–10.
Ray, R. H., Mallach, L. E. and Kruger, L., 1985, The response of single guard and down hair mechanoreceptors to moving air-jet stimulation. *Brain Res.* **346**, 333–347.
Raymond, S. A., Thalhammer, J. G., Pipitz-Bergez, F. and Strichartz, G. R., 1990, Changes in axonal impulse conduction correlate with sensory modality in primary afferent fibers in the rat. *Brain Res.* **526**, 318–321.
Reeh, P. W., Bayer, J., Kocher, L. and Handwerker, H. O., 1987, Sensitization of nociceptive cutaneous nerve fibers from the rat's tail by noxious mechanical stimulation. *Exp. Brain Res.* **65**, 505–512.
Rees, H. and Roberts, M. H. T., 1987, Anterior pretectal stimulation alters the responses of spinal dorsal horn neurones to cutaneous stimulation in the rat. *J. Physiol.* **385**, 415–436.
Regan, L. J., Dodd, J., Barondes, S. H. and Jessell, T. M., 1986, Selective expression of endogenous lactose-binding lectins and lactoseries glycoconjugates in subsets of rat sensory neurons. *Proc. Natl. Acad. Sci. USA* **83**, 2248–2252.
Reichling, D. B. and Basbaum, A. I., 1990, Contribution of brainstem GABAergic circuitry to descending antinociceptive controls: I. GABA-immunoreactive projection neurons in the periaqueductal gray and nucleus raphe magnus. *J. Comp. Neurol.* **302**, 370–377.
Rein, H., 1925, Ueber die Topographie der Warmempfindung. *Z. Biol.* **82**, 513–535.
Repkin, A. H., Wolf, P. and Anderson, E. G., 1976, Non-GABA mediated primary afferent depolarization. *Brain Res.* **117**, 147–152.
Réthelyi, M., 1977, Preterminal and terminal axon arborizations in the substantia gelatinosa of cat's spinal cord. *J. Comp. Neurol.* **172**, 511–528.
Réthelyi, M., 1984a, Synaptic connectivity in the spinal dorsal horn. In: *Handbook of the Spinal Cord*, pp. 137–175. R. A. Davidoff (ed.). Dekker, New York.
Réthelyi, M., 1984b, Types of synaptic connections in the core of the spinal grey matter. In: Handbook of the Spinal Cord, pp. 179–197. R. A. Davidoff (ed.). Dekker, New York.
Réthelyi, M. and Capowski, J. J., 1977, The terminal arborization pattern of primary afferent fibers in the substantia gelatinosa of the spinal cord in the cat. *J. Physiol.* **73**, 269–277.
Réthelyi, M. and Szentágothai, J., 1969, The large synaptic complexes of the substantia gelatinosa. *Exp. Brain Res.* **7**, 258–274.
Réthelyi, M. and Szentágothai, J., 1973, Distribution and connections of afferent fibres in the spinal cord. In: *Handbook of Sensory Physiology*, Vol. II. *Somatosensory System*, pp. 207–252. A. Iggo (ed.). Springer-Verlag, New York.
Réthelyi, M., Trevino, D. L. and Perl, E. R., 1979, Distribution of primary afferent fibers within the sacrococcygeal dorsal horn: an autoradiographic study. *J. Comp. Neurol.* **185**, 603–622.
Réthelyi, M., Light, A. R. and Perl, E. R., 1982, Synaptic complexes formed by functionally defined primary afferent units with fine myelinated fibers. *J. Comp. Neurol.* **207**, 381–393.
Réthelyi, M., Light, A. R. and Perl, E. R., 1989, Synaptic ultrastructure of functionally and morphologically characterized neurons of the superficial spinal dorsal horn of cat. *J. Neurosci.* **9**, 1846–1863.
Réthelyi, M., Metz, C. B. and Lund, P. K., 1989, Distribution of neurons expressing calcitonin gene-related peptide mRNAs in the brain stem, spinal cord and dorsal root ganglia of rat and guinea-pig. *Neuroscience* **29**, 225–239.
Reubi, J. C. and Maurer, R., 1985, Autoradiographic mapping of somatostatin receptors in the rat central nervous system and pituitary. *Neuroscience* **15**, 1183–1193.
Rexed, B., 1951, The nucleus cervicalis lateralis, a spinocerebellar relay nucleus. *Acta Physiol. Scand.* Suppl. **89**, 67–68.
Rexed, B., 1952, The cytoarchitectonic organization of the spinal cord in the rat. *J. Comp. Neurol.* **96**, 415–466.
Rexed, B., 1954, A cytoarchitectonic atlas of the spinal cord in the cat. *J. Comp. Neurol.* **100**, 297–380.
Rexed, B. and Brodal, A., 1951, The nucleus cervicalis lateralis. A spino-cerebellar relay nucleus. *J. Neurophysiol.* **14**, 399–407.
Rexed, B. and Ström, G., 1952, Afferent nervous connexions of the lateral cervical nucleus. *Acta Physiol. Scand.* **25**, 219–229.
Rexed, B. and Therman, P., 1948, Calibre spectra of motor and sensory nerve fibres to flexor and extensor muscles. *J. Neurophysiol.* **11**, 133–139.

Reynolds, P. J., Talbott, R. E. and Brookhart, J. M., 1972, Control of postural reactions in the dog: the role of the dorsal column feedback pathway. *Brain Res.* **40**, 159–164.
Ribeiro-da-Silva, A. and Coimbra, A., 1980, Neuronal uptake of [^{3}H]GABA and [^{3}H]glycine in laminae I-III (substantia gelatinosa Rolandi) of the rat spinal cord. An autoradiographic study. *Brain Res.* **188**, 449–464.
Ribeiro-DaSilva, A. and Coimbra, A., 1982, Two types of synaptic glomeruli and their distribution in laminae I-III of the rat spinal cord. *J. Comp. Neurol.* **209**, 176–186.
Ribeiro-DaSilva, A. and Coimbra, A., 1984, Capsaicin causes selective damage to type I synaptic glomeruli in rats substantia gelatinosa. *Brain Res.* **290**, 380–383.
Ribeiro-DaSilva, A., Pignatelli, D. and Coimbra, A., 1985, Synaptic architecture of glomeruli in superficial dorsal horn of rat spinal cord, as shown in serial reconstructions. *J. Neurocytol.* **14**, 203–220.
Ribeiro-DaSilva, A., Castro-Lopes, J. M. and Coimbra, A., 1986, Distribution of glomeruli with fluoride-resistant acid phosphatase (FRAP)-containing terminals in the substantia gelatinosa of the rat. *Brain Res.* **377**, 323–329.
Ribeiro-DaSilva, A., Tagari, P. and Cuello, A. C., 1989, Morphological characterization of substance P-like immunoreactive glomeruli in the superficial dorsal horn of the rat spinal cord and trigeminal subnucleus caudalis: a quantitative study. *J. Comp. Neurol.* **281**, 497–515.
Richardson, P. M. and Riopelle, R. J., 1984, Uptake of nerve growth factor along peripheral and spinal axons of primary sensory neurons. *J. Neurosci.* **4**, 1683–1689.
Richardson, P. M., Verge Issa, V. M. K. and Riopelle, R. J., 1986, Distribution of neuronal receptors for nerve growth factor in the rat. *J. Neurosci.* **6**, 2312–2321.
Riedel, W., 1976, Warm receptors in the dorsal abdominal wall of the rabbit. *Pfluegers Arch.* **361**, 205–206.
Rigamonti, D. and DeMichelle, D., 1977, Visceral afferent projection to the lateral cervical nucleus. In: *Nerves and the Gut*, pp. 327–333. F. P. Brooks and P. W. Evers (eds.). Slack, Thorofare, N.J.
Rigamonti, D. D. and Hancock, M. B., 1974, Analysis of field potentials elicited in the dorsal column nuclei by splanchnic nerve Aβ afferents. *Brain Res.* **77**, 326–329.
Rigamonti, D. D. and Hancock, M. B., 1978, Viscerosomatic convergence in the dorsal column nuclei of the cat. *Exp. Neurol.* **61**, 337–348.
Riley, D. A., Ellis, S. and Bain, J. L. W., 1984, Ultrastructural cytochemical localization of carbonic anhydrase activity in rat peripheral sensory and motor nerves, dorsal root ganglia and dorsal column nuclei. *Neuroscience* **13**, 189–206.
Rinvik, E. and Walberg, F., 1975, Studies on the cerebellar projections from the main and external cuneate nuclei in the cat by means of retrograde axonal transport of horseradish peroxidase. *Brain Res.* **95**, 371–381.
Risling, M. and Hildebrand, C., 1982, Occurrence of unmyelinated axon profiles at distal, middle and proximal levels in the ventral root L7 of cats and kittens. *J. Neurol. Sci.* **56**, 219–231.
Risling, M., Hildebrand, C. and Aldskogius, H., 1981, Postnatal increase of unmyelinated axon profiles in the feline ventral root L7. *J. Comp. Neurol.* **201**, 343–351.
Risling, M., Dalsgaard, C. J. and Cuello, A. C., 1984a, Invasion of lumbosacral ventral roots and spinal pia mater by substance P-immunoreactive axons after sciatic nerve lesion in kittens. *Brain Res.* **307**, 351–354.
Risling, M., Dalsgaard, C. J., Cukieman, A. and Cuello, A. C., 1984b, Electron microscopic and immunohistochemical evidence that unmyelinated ventral root axons make U-turns or enter the spinal pia mater. *J. Comp. Neurol.* **225**, 53–63.
Risling, M., Hildebrand, C. and Cullheim, S., 1984c, Invasion of the L7 ventral root and spinal pia mater by new axons after sciatic nerve division in kittens. *Exp. Neurol.* **83**, 84–97.
Riss, W., Pedersen, R. A., Jakway, J. S. and Ware, C. B., 1972, Levels of function and their representation in the vertebrate thalamus. *Brain Behav. Evol.* **6**, 26–41.
Ritz, L. A. and Greenspan, J. D., 1985, Morphological features of lamina V neurons receiving nociceptive input in cat sacrocaudal spinal cord. *J. Comp. Neurol.* **238**, 440–452.
Rivot, J. P., Chaouch, A. and Besson, J. M., 1980, Nucleus raphe magnus modulation of response of rat dorsal horn neurons to unmyelinated fiber inputs: partial involvement of serotonergic pathways. *J. Neurophysiol.* **44**, 1039–1057.
Rivot, J. P., Chiang, C. Y. and Besson, J. M., 1982, Increase of serotonin metabolism within the dorsal horn of the spinal cord during nucleus raphe magnus stimulation, as revealed by in vivo electrochemical detection. *Brain Res.* **238**, 117–126.
Rivot, J. P., Calvino, B. and Besson, J. M., 1987, Is there a serotonergic tonic descending inhibition on the responses of dorsal horn convergent neurons to C-fibre inputs? *Brain Res.* **403**, 142–146.
Rizzoli, A. A., 1968, Distribution of glutamic acid, aspartic acid, γ-aminobutyric acid and glycine in six areas of cat spinal cord before and after transection. *Brain Res.* **11**, 11–18.
Robain, O. and Jardin, L., 1972, Histoenzymologie du ganglion spinal du lapin. *J. Neurol. Sci.* **17**, 419–433.
Robards, M. J., Watkins, D. W. and Masterson, R. B., 1976, An anatomical study of some somesthetic afferents to the intercollicular terminal zone of the midbrain of the opossum. *J. Comp. Neurol.* **170**, 499–524.
Roberts, F. and Hill, R. G., 1978, The effects of dorsal column lesions on amino acid levels and glutamate uptake in rat dorsal column nuclei. *J. Neurochem.* **31**, 1549–1551.
Roberts, F., Taberner, P. V. and Hill, R. G., 1978, The effect of 3-mercaptoproprionate, an inhibitor of glutamate

decarboxylase, on the levels of GABA and other amino acids, and on presynaptic inhibition in the rat cuneate nucleus. *Neuropharmacology* **17**, 715–720.

Roberts, P. J., 1974a, The release of amino acids with proposed neurotransmitter function from the cuneate and gracile nuclei of the rat in vivo. *Brain Res.* **67**, 419–428.

Roberts, P. J., 1974b, Amino acid release from isolated rat dorsal root ganglia. *Brain Res.* **74**, 327–332.

Roberts, P. J. and Keen, P., 1974, Effect of dorsal root section on amino acids of rat spinal cord. *Brain Res.* **74**, 333–337.

Roberts, P. J., Keen, P. and Mitchell, J. F., 1973, The distribution and axonal transport of free amino acids and related compounds in the dorsal sensory neuron of the rat, as determined by the dansyl reaction. *J. Neurochem.* **21**, 199–209.

Robertson, B. and Grant, G., 1985, A comparison between wheat germ agglutinin- and choleragenoid-horseradish peroxidase as anterogradely transported markers in central branches of primary sensory neurones in the rat with some observations in the cat. *Neuroscience* **14**, 895–905.

Robertson, B. and Grant, G., 1989, Immunocytochemical evidence for the localization of the GM1 ganglioside in carbonic anhydrase-containing and RT 97–immunoreactive rat primary sensory neurons. *J. Neurocytol.* **18**, 77–86.

Robertson, B. and Taylor, W. R., 1986, Effects of γ--aminobutyric acid and (-)-baclofen on calcium and potassium currents in cat dorsal root ganglion neurones in vitro. *Brit. J. Pharmacol.* **89**, 661–672.

Roby-Brami, A., Bussel, B., Willer, J. C. and LeBars, D., 1987, An electrophysiological investigation into the pain-relieving effects of heterotopic nociceptive stimuli: probable involvement of a supraspinal loop. *Brain* **110**, 1497–1508.

Rockel, A. J., Heath, C. J. and Jones, E. G., 1972, Afferent connections to the diencephalon in the marsupial phalanger and the question of sensory convergence in the "posterior group" of the thalamus. *J. Comp. Neurol.* **145**, 105–130.

Rodin, B. E., Sampogna, S. and Kruger, L., 1983, An examination of intraspinal sprouting in dorsal root axons with the tracer horseradish peroxidase. *J. Comp. Neurol.* **215**, 187–198.

Rodriguez, R. E., Hill, R. G. and Hughes, J., 1987, Cholecystokinin releases [^{3}H]GABA from the perfused subarachnoid space of the anaesthetized rat spinal cord. *Neurosci. Lett.* **83**, 173–178.

Roessmann, U. and Friede, R. L., 1967, The segmental distribution of acetyl cholinesterase in the cat spinal cord. *J. Anat.* **101**, 27–32.

Roettger, V. R., Pearson, J. C. and Goldfinger, M. D., 1989, Identification of γ-aminobutyric acid-like immunoreactive neurons in the rat cuneate nucleus. *Neurosci. Lett.* **97**, 46–50.

Rökaeus, A., Melander, T., Hökfelt, T., Lundberg, J. M., Tatemoto, K., Carlquist, M. and Mutt, V., 1984, A galanin-like peptide in the central nervous system and intestine of the rat. *Neurosci. Lett.* **47**, 161–166.

Roland, P. E. and Ladegaard-Pedersen, H., 1977, A quantitative analysis of sensations of tendon and of kinaesthesia in man. *Brain* **100**, 671–692.

Ronan, M. C. and Northcutt, R. G., 1981, Ascending and descending spinal pathways in the Pacific hagfish. *Neurosci. Abstr.* **7**, 84.

Ronan, M. and Northcutt, R. G., 1990, Projections ascending from the spinal cord to the brain in petromyzontid and myxinoid agnathans. *J. Comp. Neurol.* **291**, 491–508.

Rose, R. D., Koerber, H. R., Sedivec, M. J. and Mendell, L. M., 1986, Somal action potential duration differs in identified primary afferents. *Neurosci. Lett.* **63**, 259–264.

Rosén, I., 1967, Functional organization of group I activated neurones in the cuneate nucleus of the cat. *Brain Res.* **6**, 770–772.

Rosén, I., 1969, Afferent connections to group I activated cells in the main cuneate nucleus of the cat. *J. Physiol.* **205**, 209–236.

Rosén, I. and Sjölund, B., 1973a, Organization of group I activated cells in the main and external cuneate nuclei of the cat: identification of muscle receptors. *Exp. Brain Res.* **16**, 221–237.

Rosén, I. and Sjölund, B., 1973b, Organization of group I activated cells in the main and external cuneate nuclei of the cat: convergence patterns demonstrated by natural stimulation. *Exp. Brain Res.* **16**, 238–246.

Rosenfeld, M. G., Mermod, J. J., Amara, S. G., Swanson, L. W., Sawchenko, P. E., Rivier, J., Vale, W. W. and Evans, R. M., 1983, Production of a novel neuropeptide encoded by the calcitonin gene via tissue-specific RNA processing. *Nature* **304**, 129–135.

Rosenthal, B. M. and Ho, R. H., 1989, An electron microscopic study of somatostatin immunoreactive structures in lamina II of the rat spinal cord. *Brain Res. Bull.* **22**, 439–451.

Rosomoff, H. L., Carroll, F., Brown, J. and Sheptak, P., 1965, Percutaneous radiofrequency cervical cordotomy: technique. *J. Neurosurg.* **23**, 639–644.

Rosomoff, H. L., Sheptak, P. and Carroll, F., 1966, Modern pain relief: percutaneous chordotomy. *J.A.M.A.* **196**, 482–486.

Ross, C. A., Armstrong, D. M., Ruggiero, D. A., Pickel, V. M., Joh, T. H. and Reis, D. J., 1981, Adrenaline neurons in the rostral ventrolateral medulla innervate thoracic spinal cord: a combined immunocytochemical and retrograde transport demonstration. *Neurosci. Lett.* **25**, 257–262.

Rossi, A. and Grigg, P., 1982, Characteristics of hip joint mechanoreceptors in the cat. *J. Neurophysiol.* **47**, 1029–1042.

Rossi, G. F. and Brodal, A., 1957, Terminal distribution of spinoreticular fibers in the cat. *Arch. Neurol. Psychiat.* **78**, 439–453.

Rossi, G. F. and Zanchetti, A., 1957, The brainstem reticular formation. Anatomy and physiology. *Arch. Ital. Biol.* **95**, 199–435.

Rougon, G., Hirsch, M. R., Hirn, M., Guenet, J. L. and Goridis, C., 1983, Monoclonal antibody to neural cell surface protein: identification of a glycoprotein family of restricted cellular localization. *Neuroscience* **10**, 511–520.

Rowinski, M. J., Haring, J. H. and Pubols, B. H., 1981, Correlation of peripheral receptive field area and rostrocaudal locus of neurons within the raccoon cuneate nucleus. *Brain Res.* **211**, 463–467.

Rowinski, M. J., Haring, J. H. and Pubols, B. H., 1985, Response properties of raccoon cuneothalamic neurons. *Somatosensory Res.* **2**, 263–280.

Rozsos, I., 1958, The synapses of Burdach's nucleus. *Act Morphol. Acad. Sci. Hung.* **8**, 105–109.

Ruch, T. C., 1946, Visceral sensation and referred pain. In: *Fulton, Howell's Textbook of Physiology*, 15th ed., pp. 385–401. Saunders, Philadelphia.

Ruch, T. C., Patton, H. D. and Amassian, V. E., 1952, Topographical and functional determinants of cortical localization patterns. *Assoc. Res. Nervous Ment. Dis.* **30**, 403–429.

Rucker, H. K. and Holloway, J. A., 1982, Viscerosomatic convergence onto spinothalamic tract neurons in the cat. *Brain Res.* **243**, 155–157.

Rucker, H. K., Holloway, J. A. and Keyser, G. F., 1984, Response characteristics of cat spinothalamic tract neurons to splanchnic nerve stimulation. *Brain Res.* **291**, 383–387.

Ruda, M. A., 1982, Opiates and pain pathways: demonstration of enkephalin synapses on dorsal horn projection neurons. *Science* **215**, 1523–1525.

Ruda, M. A., 1988, Spinal dorsal horn circuitry involved in the brain stem control of nociception. *Prog. Brain Res.* **77**, 129–140.

Ruda, M. A. and Gobel, S., 1980, Ultrastructural characterization of axonal endings in the substantia gelatinosa which take up [^{3}H]serotonin. *Brain Res.* **184**, 57–83.

Ruda, M. A., Coffield, J. and Steinbusch, H. W. M., 1982, Immunohistochemical analysis of serotonergic axons in lamina I and II of the lumbar spinal cord of the cat. *J. Neurosci.* **2**, 1660–1671.

Ruda, M. A., Coffield, J. and Dubner, R., 1984, Demonstration of postsynaptic opioid modulation of thalamic projection neurons by the combined techniques of retrograde horseradish peroxidase and enkephalin immunocytochemistry. *J. Neurosci.* **4**, 2117–2132.

Ruda, M. A., Bennett, G. J. and Dubner, R., 1986, Neurochemistry and neural circuitry in the dorsal horn. In: *Progress in Brain Research*, Vol. 66, pp. 219–268. P. C. Emson, M. N. Rossor and M. Tonyama (eds.). Elsevier, Amsterdam.

Ruda, M. A., Iadarola, M. J., Cohen, L. V. and Young, W. S., 1988, In situ hybridization histochemistry and immunocytochemistry reveal an increase in spinal dynorphin biosynthesis in a rat model of peripheral inflammation and hyperalgesia. *Proc. Natl. Acad. Sci. USA* **85**, 622–626.

Rudolph, G. and Klein, H. J., 1964, Histochemische darstellung und verteilung der glukose-6-phosphat-dehydrogenase in normalen rattenorganen. *Histochemie* **4**, 238–251.

Rudomín, P., Engberg, I. and Jiménez, I., 1981, Mechanisms involved in presynaptic depolarization of group I and rubrospinal fibers in cat spinal cord. *J. Neurophysiol.* **46**, 532–548.

Rudomín, P., Solodkin, M. and Jiménez, I., 1987, I. Synaptic potentials of primary afferent fibers and motoneurons evoked by single intermediate nucleus interneurons in the cat spinal cord. *J. Neurophysiol.* **57**, 1288–1313.

Rudomín, P., Jiménez, I., Quevedo, J. and Solodkin, M., 1990, Pharmacologic analysis of inhibition produced by last-order intermediate nucleus interneurons mediating nonreciprocal inhibition of motoneurons in cat spinal cord. *J. Neurophysiol.* **63**, 147–160.

Ruffini, A., 1894, Sur un nouvel organe nerveux terminal et sur la presence des corpuscles Golgi-Mazzoni dans le conjunctif sous-cutane de la pulpe des doigts de l'homme. *Arch. Ital. Biol.* **21**, 249–265.

Rustioni, A., 1973, Non-primary afferents to the nucleus gracilis from the lumbar cord of the cat. *Brain Res.* **51**, 81–95.

Rustioni, A., 1974, Non-primary afferents to the cuneate nucleus in the brachial dorsal funiculus of the cat. *Brain Res.* **75**, 247–259.

Rustioni, A. and Cuénod, M., 1982, Selective retrograde transport of d-aspartate in spinal interneurons and cortical neurons of rats. *Brain Res.* **236**, 143–155.

Rustioni, A. and Kaufman, A. B., 1977, Identification of cells of origin of non-primary afferents to the dorsal column nuclei of the cat. *Exp. Brain Res.* **27**, 1–14.

Rustioni, A. and Macchi, G., 1968, Distribution of dorsal root fibers in the medulla oblongata of the cat. *J. Comp. Neurol.* **134**, 113–126.

Rustioni, A. and Molenaar, I., 1975, Dorsal column nuclei afferents in the lateral funiculus of the cat: distribution, pattern and absence of sprouting after chronic deafferentation. *Exp. Brain Res.* **23**, 1–12.

Rustioni, A. and Sotelo, C., 1974, Synaptic organization of the nucleus gracilis of the cat. Experimental identification of dorsal root fibers and cortical afferents. *J. Comp. Neurol.* **155**, 441–468.

Rustioni, A. and Weinberg, R. J., 1989, Chapter III. The somatosensory system. In: *Handbook of Chemical*

Neuroanatomy, Vol. 7. *Integrated Systems of the CNS*, Part II. A. Bjorklund, T. Hokfelt and L. W. Swanson (eds.). Elsevier, Amsterdam.
Rustioni, A., Hayes, N. L. and O'Neill, S., 1979, Dorsal column nuclei and ascending spinal afferents in macaques. *Brain* **102**, 95–125.
Rustioni, A., Schmechel, D. E., Cheema, S. and Fitzpatrick, D., 1984, Glutamic acid decarboxylase containing neurons in the dorsal column of the cat. *Somatosensory Res.* **1**, 329–357.
Ryall, R. W. and Piercey, M. F., 1970, Visceral afferent and efferent fibers in sacral ventral roots in cats. *Brain Res.* **97**, 57–65.
Rymer, W. Z. and D'Almeida, A., 1980, Joint position sense. The effects of muscle contraction. *Brain* **103**, 1–22.
Ryu, P. D., Gerber, G., Murase, K. and Randić, M., 1988a, Actions of calcitonin gene-related peptide on rat spinal dorsal horn neurons. *Brain Res.* **441**, 357–361.
Ryu, P. D., Gerber, G., Murase, K. and Randić, M., 1988b, Calcitonin gene-related peptide enhances calcium current of rat dorsal root ganglion neurons and spinal excitatory synaptic transmission. *Neurosci. Lett.* **89**, 305–312.
Sadjapour, K. and Brodal, A., 1968, The vestibular nuclei in man. A morphological study in the light of experimental findings in the cat. *J. Hirnforsch.* **10**, 299–319.
Sakanaka, M., Shibasaki, T. and Lederis, K., 1987, Corticotropin releasing factor-like immunoreactivity in the rat brain as revealed by a modified cobalt-glucose oxidase-diaminobenzidine method. *J. Comp. Neurol.* **260**, 256–298.
Sales, N., Charnay, Y., Zajac, J. M., Dubois, P. M. and Roques, B. P., 1989, Ontogeny of mu and delta opioid receptors and of neural endopeptidase in human spinal cord: an autoradiographic study. *J. Chem. Neuroanat.* **2**, 179–188.
Salt, T. E. and Hill, R. G., 1983, Neurotransmitter candidates of somatosensory primary afferent fibres. *Neuroscience* **10**, 1083–1103.
Salter, M. W. and Henry, J. L., 1985, Effects of adenosine 5′-monophosphate and adenosine 5′-triphosphate on functionally identified units in the cat spinal dorsal horn. Evidence for a differential effect of adenosine 5′-triphosphate on nociceptive vs. non-nociceptive units. *Neuroscience* **15**, 815–825.
Salter, M. W. and Henry, J. L., 1987, Evidence that adenosine mediates the depression of spinal dorsal horn neurones induced by peripheral vibration in the cat. *Neuroscience* **22**, 631–650.
Sambucetti, L. C. and Curran, T., 1986, The fos protein complex is associated with DNA in isolated nuclei and binds to DNA cellulose. *Science* **234**, 1417–1419.
Samorajski, T., 1960, The application of diphosphopyridine nucleotide diaphorase methods in a study of dorsal root ganglia. *J. Neurochem.* **5**, 349–353.
Samuel, E. P., 1952, The autonomic and somatic innervation of the articular capsule. *Anat. Rec.* **113**, 53–70.
Sandkühler, J. and Zimmermann, M., 1987, Spinal pathways mediating tonic or stimulation-produced descending inhibition from the periaqueductal gray or nucleus raphe magnus are separate in the cat. *J. Neurophysiol.* **58**, 1–15.
Sandkühler, J., Helmchen, C., Fu, Q. G. and Zimmermann, M., 1988, Inhibition of spinal nociceptive neurons by excitation of cell bodies or fibers of passage at various brainstem sites in the cat. *Neurosci. Lett.* **93**, 67–72.
Sanyal, S. and Rustioni, A., 1974, Phosphatases in the substantia gelatinosa and motoneurones: a comparative histochemical study. *Brain Res.* **76**, 161–166.
Saper, C. B., Hurley, K. M., Moga, M. M., Holmes, H. R., Adams, S. A., Leahy, K. M. and Needleman, P., 1989, Brain natriuretic peptides: differential localization of a new family of neuropeptides. *Neurosci. Lett.* **96**, 29–34.
Saria, A., Gamse, R., Petermann, J., Fischer, J. A., Theodorsson-Norheim, E. and Lundberg, J.M., 1986, Simultaneous release of several tachykinins and calcitonin gene-related peptide from rat spinal cord slices. *Neurosci. Lett.* **63**, 310–314.
Sarrat, R., 1970, Zur Chemodifferenzierung des Ruckenmarks und der Spinalganglien der Ratte. *Histochemie* **24**, 202–213.
Sasek, C. A. and Elde, R. P., 1985, Distribution of neuropeptide Y-like immunoreactivity and its relationship to FMRF-amide-like immunoreactivity in the sixth lumbar and first sacral spinal cord segments of the rat. *J. Neurosci.* **7**, 1729–1739.
Sastry, B. R., 1978, Morphine and met-enkephalin effects on sural A-delta afferent terminal excitability. *Europ. J. Pharmacol.* **50**, 269–273.
Sastry, B. R., 1979a, Presynaptic effects of morphine and methionine-enkephalin in feline spinal cord. *Neuropharmacology* **18**, 367–375.
Sastry, B. R., 1979b, Substance P effects on spinal nociceptive neurones. *Life Sci.* **24**, 2169–2178.
Sastry, B. R. and Goh, J. W., 1983, Actions of morphine and met-enkephaline-amide on nociceptor driven neurones in substantia gelatinosa and deeper dorsal horn. *Neuropharmacology* **22**, 119–122.
Sato, J., Mizumura, K. and Kumazawa, T., 1989, Effect of ionic calcium on the responses of canine testicular polymodal receptors to algesic substances. *J. Neurophysiol.* **62**, 119–125.
Satterfield, J. H., 1962, Effect of sensorimotor cortical stimulation upon cuneate nuclear output through medial lemniscus in cat. *J. Nerv. Ment. Dis.* **135**, 507–512.
Sauer, M. E., 1954, Acetylcholinesterase in spinal ganglia of the dog. *Anat. Rec.* **118**, 437.

Sawynok, J. and Sweeney, M. I., 1989, The role of purines in nociception. *Neuroscience* **32**, 557–569.
Scatton, B., Dubois, A., Javoy-Agid, F. and Camus, A., 1984, Autoradiographic localization of muscarinic cholinergic receptors at various segmental levels of the human spinal cord. *Neurosci. Lett.* **49**, 239–245.
Schady, W. and Torebjörk, H. E., 1983, Projected and receptive fields. A comparison of projected areas of sensations evoked by intraneural stimulation of mechanoreceptive units and their innervation territories. *Acta Physiol. Scand.* **119**, 267–275.
Schady, W. J. L., Torebjörk, H. E. and Ochoa, J. L., 1983, Peripheral projections of nerve fibres in the human median nerve. *Brain Res.* **277**, 249–261.
Schaible, H. G. and Schmidt, R. F., 1983a, Activation of groups III and IV sensory units in medial articular nerve by local mechanical stimulation of knee joint. *J. Neurophysiol.* **49**, 35–44.
Schaible, H. G. and Schmidt, R. F., 1983b, Responses of fine medial articular nerve afferents to passive movements of knee joint. *J. Neurophysiol.* **49**, 1118–1126.
Schaible, H. G. and Schmidt, R. F., 1985, Effects of an experimental arthritis on the sensory properties of fine articular afferent units. *J. Neurophysiol.* **54**, 1109–1126.
Schaible, H. G. and Schmidt, R. F., 1988, Excitation and sensitization of fine articular afferents from cat's knee joint by prostaglandin E2. *J. Physiol.* **403**, 91–104.
Schaible, H. G., Schmidt, R. F. and Willis, W. D., 1986, Responses of spinal cord neurones to stimulation of articular afferent fibres in the cat. *J. Physiol.* **372**, 575–593.
Schaible, H. G., Schmidt, R. F. and Willis, W. D., 1987a, Convergent inputs from articular, cutaneous and muscle receptors onto ascending tract cells in the cat spinal cord. *Exp. Brain Res.* **66**, 479–488.
Schaible, H. G., Schmidt, R. F. and Willis, W. D., 1987b, Enhancement of the responses of ascending spinal tract cells in the cat spinal cord by acute inflammation of the knee joint. *Exp. Brain Res.* **66**, 489–499.
Schaible, H. G., Jarrott, B., Hope, P. J. and Duggan, A. W., 1990, Release of immunoreactive substance P in the spinal cord during development of acute arthritis in the knee joint of the cat: a study with antibody microprobes. *Brain Res.* **529**, 214–223.
Scharf, J. H., 1958, Sensible Ganglien. Nervensystem. In: *Handbuck der Mikroskopischen Anatomie des Menschen*, Bd. IV/1. W. v. Mollendorff and W. Bargmann (eds.). Springer-Verlag, Berlin.
Scharf, J. H. and Rowe, C. P., 1957, Zur verteilung der kohlenhydrate und einiger fermente im ganglion semilunare des Rindes. *Acta Histochem.* Bd. **5**, 129–145.
Scheibel, M. E. and Scheibel, A. B., 1968, Terminal axon patterns in cat spinal cord. II: The dorsal horn. *Brain Res.* **9**, 32–58.
Schipper, J., Steinbuch, W. M., Vermes, I. and Tildes, F. J. H., 1983, Mapping of CRF-immunoreactive nerve fibers in the medulla oblongata and spinal cord of the rat. *Brain Res.* **267**, 145–150.
Schlaepfer, W., 1968, Acetylcholinesterase activity of motor and sensory nerve fibers in the spinal nerve roots of the rat. *Zeits. fur Zellforsch.* **88**, 441–456.
Schmalbruch, H., 1987, The number of neurons in dorsal root ganglia L4-L6 of the rat. *Anat. Rec.* **219**, 315–322.
Schmidt, R. F., 1963, Pharmacological studies on the primary afferent depolarization of the toad spinal cord. *Pfluegers Arch.* **277**, 325–346.
Schmidt, R. F., 1971, Presynaptic inhibition in the vertebrate central nervous system. *Ergebn. Physiol.* **63**, 20–101.
Schmidt, R. F. and Weller, E., 1970, Reflex activity in the cervical and lumbar sympathetic trunk induced by unmyelinated somatic afferents. *Brain Res.* **24**, 207–218.
Schneider, A. and Necker, R., 1989, Spinothalamic projections in the pigeon. *Brain Res.* **484**, 139–149.
Schneider, S. P. and Perl, E. R., 1985, Selective excitation of neurons in the mammalian spinal dorsal horn by aspartate and glutamate in vitro: correlation with location and synaptic input. *Brain Res.* **360**, 339–343.
Schneider, S. P. and Perl, E. R., 1988, Comparison of primary afferent and glutamate excitation of neurons in the mammalian spinal dorsal horn. *J. Neurosci.* **8**, 2062–2073.
Schoenen, J., 1978, Histoenzymology of the developing rat spinal cord. *Neuropath. Appl. Neurobiol.* **4**, 37–46.
Schoenen, J., 1982, The dendritic organization of the human spinal cord: the dorsal horn. *Neuroscience* **7**, 2057–2087.
Schoenen, J., Budo, C., Poncelet, G. and Gerebtzoff, M. A., 1968, Effet de la section du sciatique surl'activite de l'isoenzyme fluororesistant de la phosphatase acide dans la moelle epiniere du rat. *C. R. Soc. Biol.* **32**, 2035–2037.
Schoenen, J., Lotstra, F., Vierendeels, G., Reznik, M. and Vanderhaeghen, J. J., 1985a, Substance P, enkephalins, somatostatin, cholecystokinin, oxytocin, and vasopressin in human spinal cord. *Neurology* **35**, 881–890.
Schoenen, J., Van Hees, J., Gybels, J., de Castro Costa, M., and Vanderhaeghen, J. J., 1985b, Histochemical changes of substance P, FRAP, serotonin and succinic dehydrogenase in the spinal cord of rats with adjuvant arthritis. *Life Sci.* **36**, 1247–1254.
Schoenen, J., Lotstra, F., Liston, D., Rossier, J. and Vanderhaeghen, J. J., 1986, Synenkephalin in bovine and human spinal cord. *Cell Tiss. Res.* **246**, 641–645.
Schouenborg, J., 1984, Functional and topographical properties of field potentials evoked in rat dorsal horn by cutaneous C-fibre stimulation. *J. Physiol.* **356**, 169–192.
Schouenborg, J. and Sjölund, B. H., 1986, First-order nociceptive synapses in rat dorsal horn are blocked by an amino acid antagonist. *Brain Res.* **379**, 394–398.

Schoultz, T. W. and Swett, J. E., 1972, The fine structure of the Golgi tendon organ. *J. Neurocytol.* **1**, 1–26.
Schroder, H. D., 1983, Localization of cholecystokinin-like immunoreactivity in the rat spinal cord, with particular reference to the autonomic innervation of the pelvic organs. *J. Comp. Neurol.* **217**, 176–186.
Schroder, H. D., 1984, Somatostatin in the caudal spinal cord: an immunohistochemical study of the spinal centers involved in the innervation of pelvic organs. *J. Comp. Neurol.* **223**, 400–414.
Schroder, H. D. and Skagerberg, G., 1985, Catecholamine innervation of the caudal spinal cord in the rat. *J. Comp. Neurol.* **242**, 358–368.
Schroeder, D. M. and Jane, J. A., 1971, Projection of dorsal column nuclei and spinal cord to brainstem and thalamus in the tree shrew, *Tupaia glis*. *J. Comp. Neurol.* **142**, 309–350.
Schultz, R. A., Miller, D. G., Kerr, C. S. and Micheli, L., 1984, Mechano-receptors in human cruciate ligaments. A histological study. *J. Bone Joint Surg.* **66A**, 1072–1076.
Schultzberg, M., Dockray, G. J. and Williams, R. G., 1982, Capsaicin depletes CCK-like immunoreactivity detected by immunohistochemistry, but not that measured by radioimmunoassay in rat dorsal spinal cord. *Brain Res.* **235**, 198–204.
Schwartz, A. S., Eidelberg, E., Marchok, P. and Azulay, A., 1972, Tactile discrimination in the monkey after section of the dorsal funiculus and lateral lemniscus. *Exp. Neurol.* **37**, 582–596.
Schwartz, H. G. and O'Leary, J. L., 1941, Section of the spinothalamic tract in the medulla with observations on the pathway for pain. *Surgery* **9**, 183–193.
Schwartzkroin, P. A., Duijn, H. van and Prince, D. A., 1974, Effects of projected cortical epileptiform discharges on unit activity in the cat cuneate nucleus. *Exp. Neurol.* **43**, 106–123.
Schwartzman, R. J. and Bogdonoff, M. D., 1968, Behavioral and anatomical analysis of vibration sensibility. *Exp. Neurol.* **20**, 43–51.
Schwartzman, R. J. and Bogdonoff, M. D., 1969, Proprioception and vibration sensibility discrimination in the absence of the posterior columns. *Arch. Neurol.* **20**, 349–353.
Sedivec, M. J., Ovelmen-Levitt, J., Karp, R. and Mendell, L. M., 1983, Increase in nociceptive input to spinocervical tract neurons following chronic partial deafferentation. *J. Neurosci.* **3**, 1511–1519.
Sedivec, M. J., Capowski, J. J. and Mendell, L. M., 1986, Morphology of HRP-injected spinocervical tract neurons: effect of dorsal rhizotomy. *J. Neurosci.* **6**, 661–672.
Segu, L. and Calas, A., 1978, The topographical distribution of serotoninergic terminals in the spinal cord of the cat: quantitative radioautographic studies. *Brain Res.* **153**, 449–464.
Seifert, H., Chesnut, J., DeSouza, E., Rivier, J. and Vale, W., 1985, Binding sites for calcitonin gene-related peptide in distinct areas of rat brain. *Brain Res.* **346**, 195–198.
Seki, Y., 1962, Some aspects of comparative anatomy of the spinal cord. *Recent Adv. Nerv. System* **6**, 908–924. (In Japanese; quoted by Mizuno et al., 1967).
Seltzer, Z. and Devor, M., 1984, Effect of nerve section on the spinal distribution of neighboring nerves. *Brain Res.* **306**, 31–37.
Selzer, M. and Spencer, W. A., 1969, Convergence of visceral and cutaneous afferent pathways in the lumbar spinal cord. *Brain Res.* **14**, 331–348.
Semba, K., Masarachia, P., Malamed, S., Jacquin, M. and Harris, S., 1983, An electron microscopic study of primary afferent terminals from slowly adapting type I receptors in the cat. *J. Comp. Neurol.* **221**, 466–481.
Semba, K., Masarachia, P., Malamed, S., Jacquin, M., Harris, S. and Egger, M. D., 1984, Ultrastructure of pacinian corpuscle primary afferent terminals in the cat spinal cord. *Brain Res.* **302**, 135–150.
Semba, K., Masaracha, P., Malamed, S., Jacquin, M., Harris, S., Yang, G. and Egger, M. D., 1985, An electron microscopic study of terminals of rapidly adapting mechanoreceptive afferent fibers in the cat spinal cord. *J. Comp. Neurol.* **232**, 229–240.
Senba, E. and Tohyama, M., 1988, Calcitonin gene-related peptide containing autonomic efferent pathways to the pelvic ganglia of the rat. *Brain Res.* **449**, 386–390.
Senba, E., Shiosaka, S., Hara, Y., Inagaki, S., Sakanaka, M., Takatsuki, K., Kawai, Y. and Tohyama, M., 1982, Ontogeny of the peptidergic system in the rat spinal cord: immunohistochemical analysis. *J. Comp. Neurol.* **208**, 54–66.
Senba, E., Yanaihara, C., Yanaihara, N. and Tohyama, M., 1988, Co-localization of substance P and met-enkephalin-Arg^6-Gly^7-Leu^8 in the intraspinal neurons of the rat, with special reference to the neurons in the substantia gelatinosa. *Brain Res.* **453**, 110–116.
Senba, E., Yanaihara, C., Yanaihara, N. and Tohyama, M., 1989, Proenkephalin opioid peptides product in the sensory ganglia of the rat: a developmental immunohistochemical study. *Devel. Brain Res.* **48**, 263–271.
Sengupta, J. N., Saha, J. K. and Goyal, R. K., 1990, Stimulus-response function studies of esophageal mechanosensitivie nociceptors in sympathetic afferents of opossum. *J. Neurophysiol.* **64**, 796–812.
Sethi, J. S. and Tewari, H. B., 1971, Histochemical studies on the distribution of some hydrolytic and oxidative enzymes in the neurons of spinal and trigeminal ganglia and in the trigeminal of squirrel (*Funambulus palmarum*). *Acta Morph. Neerl.-Scand.* **9**, 101–115.
Sethi, J. S., Tewari, H. B. and Sood, P. P., 1969, On the distributive patterns of alkaline phosphatase activity and their functional significance amongst the spinal ganglion cells of squirrel. *Acta Neurol.* **69**, 51–57.
Seybold, V. S., 1985, Distribution of histaminergic, muscarinic and serotonergic binding sites in cat spinal cord with emphasis on the region surrounding the central canal. *Brain Res.* **342**, 291–296.

Seybold, V. S. and Elde, R. P., 1980, Immunohistochemical studies of peptidergic neurons in the dorsal horn of the spinal cord. *J. Histochem. Cytochem.* **28**, 367–370.

Seybold, V. S. and Elde, R. P., 1982, Neurotensin immunoreactivity in the superficial laminae of the dorsal horn of the rat. I. Light microscope studies of cell bodies and proximal dendrites. *J. Comp. Neurol.* **205**, 89–100.

Seybold, V. S. and Maley, B., 1984, Ultrastructural study of neurotensin immunoreactivity in the superficial laminae of the dorsal horn of the rat. *Peptides* **5**, 1179–1189.

Shantha, T. R., Manocha, C. L. and Bourne, G. H., 1967, Enzyme histochemistry of the mesenteric and dorsal root ganglion cells of cat and squirrel monkey. *Histochemie* **10**, 234–245.

Shanthaveerappa, T. R. and Bourne, G. H., 1965, The thiamine pyrophosphatase technique as an indicator of the morphology of the Golgi apparatus in the neurons. *Acta Histochem.* **22**, 155–178.

Sharif, N. A. and Burt, D. R., 1985, Limbic, hypothalamic, cortical and spinal regions are enriched in receptors for thyrotropin-releasing hormone: evidence from [^{3}H]ultrofilm autoradiography and correlation with central effects of the tripeptide in rat brain. *Neurosci. Lett.* **60**, 337–342.

Sharkey, K. A., Williams, R. G. and Dockray, G. J., 1984, Sensory substance P innervation of the stomach and pancreas. *Gastroenterology* **87**, 914–921.

Sharkey, K. A., Sobrino, J. A. and Cervero, F., 1987, Evidence for a visceral afferent origin of substance P-like immunoreactivity in lamina V of the rat thoracic spinal cord. *Neuroscience* **22**, 1077–1083.

Shealy, C. N., Mortimer, J. T. and Reswick, J. B., 1967, Electrical inhibition of pain by stimulation of the dorsal columns. *Anesth. Anal.* **46**, 489–491.

Shealy, C. N., Mortimer, J. T. and Hagfors, N. R., 1970, Dorsal column electroanalgesia. *J. Neurosurg.* **32**, 560–564.

Sheehan, D., 1932, The afferent nerve supply of the mesentery and its significance in the causation of abdominal pain. *J. Anat.* **67**, 233–249.

Shehab, S. A. S. and Atkinson, M. E., 1986a, Vasoactive intestinal polypeptide increases in areas of the dorsal horn of the spinal cord from which other neuropeptides are depleted following peripheral axotomy. *Exp. Brain Res.* **62**, 422–430.

Shehab, S. A. S. and Atkinson, M. E., 1986b, Vasoactive intestinal polypeptide (VIP) increases in the spinal cord after peripheral axotomy of the sciatic nerve originate from primary afferent neurons. *Brain Res.* **372**, 37–44.

Shehab, S. A. S., Atkinson, M. E. and Payne, J. N., 1986c, The origins of the sciatic nerve and changes in neuropeptides after axotomy: a double labelling study using retrograde transport of True Blue and vasoactive intestinal polypeptide immunohistochemistry. *Brain Res.* **376**, 180–185.

Sherman, I. C. and Arieff, A. J., 1948, Dissociation between pain and temperature in spinal cord lesions. *J. Nerv. Ment. Dis.* **108**, 285–292.

Sherrington, C. S., 1893, Experiments in examination of the peripheral distribution of the fibres of the posterior roots of some spinal nerves. *Phil. Trans. Roy. Soc. B* **184**, 641–765.

Sherrington, C. S., 1894, On the anatomical constitution of nerves of skeletal muscles; with remarks on recurrent fibers in the ventral spinal nerve root. *J. Physiol.* **17**, 211–258.

Sherrington, C. S., 1898, Experiments in examination of the peripheral distribution of the fibres of the posterior roots of some spinal nerves. Part II. *Phil. Trans. B* **190**, 45–186.

Sherrington, C. S., 1900, The muscular sense. In: *Text-book of Physiology*, pp. 1002–1025. E. A. Schaefer (ed.). Pentland, Edinburgh. Cited in Goodwin et al., 1972b.

Sherrington, C. S., 1906, *The Integrative Action of the Nervous System.* Yale Univ. Press, New Haven, 2nd ed., 1947.

Shimizu, N., 1950, Histochemical studies on the phosphatase of the nervous system. *J. Comp. Neurol.* **93**, 201–213.

Shimoji, K., Matsuki, M. and Shimizu, H., 1977, Wave-form characteristics and spatial distribution of evoked spinal electrogram in man. *J. Neurosurg.* **46**, 304–313.

Shimosegawa, T., Koizumi, M., Toyota, T., Goto, Y., Yanaihara, C. and Yanaihara, N., 1986, An immunohistochemical study of methionine-enkephalin-Arg-Gly-Leu-like immunoreactivity-containing liquor-contacting neurones (LCNs) in the rat spinal cord. *Brain Res.* **379**, 1–9.

Shin, H. K., Kim, J. and Chung, J. M., 1985, Flexion reflex elicited by ventral root afferents in the cat. *Neurosci. Lett.* **62**, 353–358.

Short, A. D., Brown, A. G. and Maxwell, D. J., 1990, Afferent inhibition and facilitation of transmission through the spinocervical tract in the anaesthetized cat. *J. Physiol.* **429**, 511–528.

Shortland, P., Woolf, C. J. and Fitzgerald, M., 1989, Morphology and somatotopic organization of the central terminals of hindlimb hair follicle afferents in the rat lumbar spinal cord. *J. Comp. Neurol.* **289**, 416–433.

Shriver, J. E. and Noback, C. R., 1967, Cortical projections to the lower brain stem and spinal cord in the tree shrew (*Tupaia glis*). *J. Comp. Neurol.* **130**, 25–54.

Shriver, J. E., Stein, B. M. and Carpenter, M. B., 1968, Central projections of spinal dorsal roots in the monkey. I. Cervical and upper thoracic dorsal roots. *Amer. J. Anat.* **123**, 27–74.

Shults, C. W., Quirion, R., Chronwall, B., Chase, T. N. and O'Donohue, T. L., 1984, A comparison of the anatomical distribution of substance P and substance P receptors in the rat central nervous system. *Peptides* **5**, 1097–1128.

Silfvenius, H., 1970, Projections to the cerebral cortex from afferents of the interosseous nerves of the cat. *Acta Physiol. Scand.* **80**, 196–214.

Sillar, K. T. and Roberts, A., 1988, Unmyelinated cutaneous afferent neurons activate two types of excitatory amino acid receptors in the spinal cord of *Xenopus laevis* embryos. *J. Neurosci.* **8**, 1350–1360.

Silver, A. and Wolstencroft, J. H., 1971, The distribution of cholinesterases in relation to the structure of the spinal cord in the cat. *Brain Res.* **34**, 205–227.

Silverman, J. D. and Kruger, L., 1988a, Lectin and neuropeptide labeling of separate populations of dorsal root ganglion neurons and associated "nociceptor" thin axons in rat testis and cornea whole-mount preparations. *Somatosensory Res.* **5**, 259–267.

Silverman, J. D. and Kruger, L., 1988b, Acid phosphatase as a selective marker for a class of small sensory ganglion cells in several mammals: spinal cord distribution, histochemical properties, and relation to fluoride-resistant acid phosphatase (FRAP) of rodents, *Somatosensory Res.* **5**, 219–246.

Silvey, G. E., Gulley, R. L. and Davidoff, R. A., 1974, The frog dorsal column nucleus. *Brain Res.* **73**, 421–437.

Siminoff, R., 1968, Quantitative properties of slowly adapting mechanoreceptors in alligator skin. *Exp. Neurol.* **21**, 290–306.

Siminoff, R. and Kruger, L., 1968, Properties of reptilian cutaneous mechanoreceptors. *Exp. Neurol.* **20**, 403–414.

Simmonds, M. A., 1978, Presynaptic actions of γ-aminobutyric acid and some antagonists in a slice preparation of cuneate nucleus. *Br. J. Pharmacol.* **63**, 495–502.

Simmonds, M. A., 1980, Evidence that bicuculline and picrotoxin act at separate sites to antagonize γ-aminobutyric acid in rat cuneate nucleus. *Neuropharmacology* **19**, 39–45.

Simmonds, M. A., 1982, Classification of some GABA antagonists with regard to site of action and potency in slices of rat cuneate nucleus. *Eur. J. Pharmacol.* **80**, 347–358.

Simone, D. A., Ngeow, J. Y. F., Putterman, G. J. and LaMotte, R. H., 1987, Hyperalgesia to heat after intradermal injection of capsaicin. *Brain Res.* **418**, 201–203.

Simone, D. A., Baumann, T. K. and LaMotte, R. H., 1989a, Dose-dependent pain and mechanical hyperalgesia in humans after intradermal injection of capsaicin. *Pain* **38**, 99–107.

Simone, D. A., Baumann, T. K., Collins, J. G. and LaMotte, R. H., 1989b, Sensitization of cat dorsal horn neurons to innocuous mechanical stimulation after intradermal injection of capsaicin. *Brain Res.* **486**, 185–189.

Simone, D. A., Sorkin, L. S., Oh, U. T., Chung, J. M., Owens, C., LaMotte, R. H. and Willis, W. D., 1991, Neurogenic hyperalgesia: Central neural correlates in responses of spinothalamic tract neurons. *J. Neurophysiol.*, in press.

Sims, T. J. and Vaughn, J. E., 1979, The generation of neurons involved in an early reflex pathway of embryonic mouse spinal cord. *J. Comp. Neurol.* **183**, 707–720.

Sinclair, D. C., 1955, Cutaneous sensation and the doctrine of specific energy. *Brain* **78**, 584–614.

Sinclair, D., 1981, *Mechanisms of Cutaneous Sensation*. Oxford Univ. Press, Oxford.

Sinclair, D. C. and Stokes, B. A. R., 1964, The production and characteristics of "second pain." *Brain* **87**, 609–618.

Sinclair, D. C., Weddell, G. and Feindel, W. H., 1948, Referred pain and associated phenomena. *Brain* **71**, 184–211.

Sindou, M., Quoex, C. and Baleydier, C., 1974, Fiber organization at the posterior spinal cord-rootlet junction in man. *J. Comp. Neurol.* **153**, 15–26.

Sindou, M., Mifsud, J. J., Rosaiti, C. and Boisson, D., 1987, Microsurgical selective posterior rhizotomy in the dorsal root entry zone for treatment of limb spasticity. *Acta Neurochiurgica* **39**, 99–102.

Sinn, R., 1913, Beitrag zur Kenntnis der Medulla oblongata der Voegel. *Monatschr. f. Psych. u. Neurol.* **33**, 1–39.

Sjölander, P., Johansson, H., Sojka, P. and Rehnholm, A., 1989, Sensory nerve endings in the cat cruciate ligaments: a morphological investigation. *Neurosci. Lett.* **102**, 33–38.

Skagerberg, G. and Björklund, A., 1985, Topographic principles in the spinal projections of serotonergic and non-serotonergic brainstem neurons in the rat. *Neuroscience* **15**, 445–480.

Skagerberg, G. and Lindvall, O., 1985, Organization of diencephalic dopamine neurones projecting to the spinal cord in the rat. *Brain Res.* **342**, 340–351.

Skagerberg, G., Björklund, A., Lindvall, O. and Schmidt, R. H., 1982, Origin and termination of the diencephalo-spinal dopamine system in the rat. *Brain Res. Bull.* **9**, 237–244.

Skofitsch, G. and Jacobowitz, D. M., 1985a, Autoradiographic distribution of ^{125}I-calcitonin gene-related peptide binding sites in the rat central nervous system. *Peptides* **4**, 975–986.

Skofitsch, G. and Jacobowitz, D. M., 1985b, Calcitonin gene-related peptide: Detailed immunohistochemical distribution in the central nervous system. *Peptides* **6**, 721–745.

Skofitsch, G. and Jacobowitz, D. M., 1985c, Calcitonin gene-related peptide coexists with substance P in capsaicin sensitive neurons and sensory ganglia of the rat. *Peptides* **6**, 747–754.

Skofitsch, G. and Jacobowitz, D. M., 1985d, Galanin-like immunoreactivity in capsaicin sensitive sensory neurons and ganglia. *Brain Res. Bull.* **15**, 1–195.

Skofitsch, G., Zamir, N., Helke, C., Savitt, J. and Jacobowitz, D., 1985a, Corticotropin releasing factor-like immunoreactivity in sensory ganglia and capsaicin sensitive neurons of the rat central nervous system: colocalization with other neuropeptides. *Peptides* **6**, 307–318.

Skofitsch, G., Insel, T. R. and Jacobowitz, D. M., 1985b, Binding sites for corticotropin releasing factor in sensory areas of the rat hindbrain and spinal cord. *Brain Res. Bull.* **15**, 519–522.

Skoglund, S., 1956, Anatomical and physiological studies of knee joint innervation in the cat. *Acta Physiol. Scand.* **36** (Suppl. 124), 1–101.
Skoglund, S., 1973, Joint receptors and kinesthesia. In: *Handbook of Sensory Physiology*. Vol. II. *Somatosensory System*, pp. 111–136. A. Iggo (ed.). Springer, Berlin.
Smith, C. L., 1983, Development and postnatal organization of primary afferent projections to the rat thoracic spinal cord. *J. Comp. Neurol.* **220**, 29–43.
Smith, K. R., 1968, The structure and function of the Haarscheibe. *J. Comp. Neurol.* **131**, 459–474.
Smith, K. R., 1970, The ultrastructure of the human Haarscheibe and Merkel cell. *J. Invest. Dermatol.* **54**, 150–159.
Snow, P. J. and Meyers, D. E. R., 1981, Observations on the synaptology of intracellularly injected spinothalamic tract neurons in the cat. *Brain Res.* **229**, 491–495.
Snow, P. J. and Meyers, D. E. R., 1983, A hypothesis on the functional significance of the dendritic projections of spinothalamic tract cells to the white matter of the cat spinal cord. *Neurosci. Lett.* **35**, 105–110.
Snyder, R., 1977, The organization of the dorsal root entry zone in cats and monkeys. *J. Comp. Neurol.* **174**, 47–70.
Snyder, R. L., 1982, Light and electron microscopic autoradiography study of the dorsal root projections to the cat dorsal horn. *Neuroscience* **7**, 1417–1437.
Snyder, R. L., Faull, R. L. M. and Mehler, W. R., 1978, Comparative study of the neurons of origin of the spinocerebellar afferents in the rat, cat and squirrel monkey based on the retrograde transport of horseradish peroxidase. *J. Comp. Neurol.* **181**, 833–852.
Sobue, G., Yasuda, T., Mitsuma, T. and Pleasure, D., 1989, Nerve growth factor receptor immunoreactivity in the neuronal perikarya of human sensory and sympathetic nerve ganglia. *Neurology* **39**, 937–941.
Soja, P. J. and Sinclair, J. G., 1980, Evidence against a serotonin involvement in the tonic descending inhibition of nociceptor-driven neurons in the cat spinal cord. *Brain Res.* **199**, 225–230.
Solomon, R. E., Aimone, L. D., Yaksh, T. L. and Gebhart, G. F., 1989, Release of calcitonin gene-related peptide (CGRP) from rat spinal cord *in vitro*. *Neurosci. Abstr.* **15**, 549.
Somjen, G. G. and Lothman, E. W., 1974, Potassium, sustained focal potential shifts, and dorsal root potentials of the mammalian spinal cord. *Brain Res.* **69**, 153–157.
Sommer, E. W., Kazimierczak, J. and Droz, B., 1985, Neuronal subpopulations in the dorsal root ganglion of the mouse as characterized by combination of ultrastructural and cytochemical features. *Brain Res.* **346**, 310–326.
Sorkin, L. S., Ferrington, D. G. and Willis, W. D., 1986, Somatotopic organization and response characteristics of dorsal horn neurons in the cervical spinal cord of the cat. *Somatosensory Res.* **3**, 323–338.
Sorkin, L. S., Steinman, J. L., Hughes, M. G., Willis, W. D. and McAdoo, D. J., 1988a, Microdialysis recovery of serotonin released in spinal cord dorsal horn. *J. Neurosci. Meth.* **23**, 131–138.
Sorkin, L. S., Morrow, T. J. and Casey, K. L., 1988b, Physiological identification of afferent fibers and postsynaptic sensory neurons in the spinal cord of the intact, awake cat. *Exp. Neurol.* **99**, 412–427.
Sorkin, L. S., McAdoo, D. J. and Willis, W. D., 1990, Release of serotonin following brainstem stimulation: correlation with inhibition of nociceptive dorsal horn neurons. In: *Serotonin and Pain*, pp. 105–115. J. M. Besson (ed.). Elsevier, Amsterdam.
Sotgiu, M. L. and Cesa-Bianchi, M. G., 1970, Primary afferent depolarization in the cuneate nucleus induced by stimulation of cerebellar and thalamic non-specific nuclei. *EEG Clin. Neurophysiol.* **29**, 156–165.
Sotgiu, M. L. and Cesa-Bianchi, M. G., 1972, Thalamic and cerebellar influence on single units of the cat cuneate nucleus. *Exp. Neurol.* **34**, 394–408.
Sotgiu, M. L. and Margnelli, M., 1976, Electrophysiological identification of pontomedullary reticular neurons directly projecting into the dorsal column nuclei. *Brain Res.* **103**, 443–453.
Sotgiu, M. L. and Marini, G., 1977, Reticulo-cuneate projections as revealed by horseradish peroxidase axonal transport. *Brain Res.* **128**, 341–345.
Spiller, W. G., 1905, The occasional clinical resemblance between caries of the vertebrae and lumbothoracic syringomyelia, and the location within the spinal cord of the fibres for the sensations of pain and temperature. *Univ. Pennsylvania Med. Bull.* **18**, 147–154.
Spiller, W. G. and Martin, E., 1912, The treatment of persistent pain of organic origin in the lower part of the body by division of the anterolateral column of the spinal cord. *J.A.M.A.* **58**, 1489–1490.
Sprague, J. M. and Ha, H., 1964, The terminal fields of dorsal root fibers in the lumbosacral spinal cord of the cat and the dendritic organization of the motor nuclei. In: *Organization of the Spinal Cord. Progress in Brain Research*, Vol. 11, pp. 120–152. J. C. Eccles and J. P. Schade (eds.). Elsevier, New York.
Srinivasan, M. A., Whitehouse, J. M. and LaMotte, R. H., 1990, Tactile detection of slip: surface microgeometry and peripheral neural codes. *J. Neurophysiol.*, in press.
Stacey, M. J., 1969, Free nerve endings in skeletal muscle of the cat. *J. Anat.* **105**, 231–254.
Standaert, D. G., Watson, S. J., Houghten, R. A. and Saper, C. B., 1986, Opioid peptide immunoreactivity in spinal and trigeminal dorsal horn neurons projecting to the parabrachial nucleus in the rat. *J. Neurosci.* **6**, 1220–1226.
Stanfield, P. R., Nakajima, Y. and Yamaguchi, K., 1985, Substance P raises neuronal membrane excitability by reducing inward rectification. *Nature* **315**, 498–501.
Stanzione, P. and Zieglgänsberger, W., 1983, Action of neurotensin on spinal cord neurons in the rat. *Brain Res.* **268**, 111–118.

Starr, A., McKeon, B., Skuse, N. and Burke, D., 1981, Cerebral potentials evoked by muscle stretch in man. *Brain* **104**, 149–166.

Steedman, W. M., Molony, V. and Iggo, A., 1985, Nociceptive neurons in the superficial dorsal horn of cat lumbar spinal cord and their primary afferent input. *Exp. Brain Res.* **58**, 171–182.

Steenbergh, P. H., Hoppener, J. W. M., Zandberg, J., Lips, C. J. M. and Jansz, H. S., 1985, A second human calcitonin/CGRP gene. *FEBS Lett.* **183**, 403–407.

Steinbusch, H. W. M., 1981, Distribution of serotonin-immunoreactivity in the central nervous system of the rat-cell bodies and terminals. *Neuroscience* **6**, 557–618.

Steiner, F. A. and Meyer, M., 1966, Actions of L-glutamate, acetylcholine and dopamine on single neurones in the nuclei cuneatus and gracilis of the cat. *Experientia* **22**, 58–59

Sterling, P. and Kuypers, H. G. J. M., 1967, Anatomical organization of the brachial spinal cord of the cat. I. The distribution of dorsal root fibers. *Brain Res.* **4**, 1–15.

Sternbach, R. A., Ignelzi, R. J., Deems, L. M. and Timmermans, G., 1976, Transcutaneous electrical analgesia: a follow-up analysis. *Pain* **2**, 35–41.

Stevens, R. T., Hodge, C. J. and Apkarian, A. V., 1982, Koelliker-Fuse nucleus: the principal source of pontine catecholaminergic cells projecting to the lumbar spinal cord of cat. *Brain Res.* **239**, 589–594.

Stevens, R. T., Hodge, C. J. and Apkarian, A. V., 1983, Catecholine varicosities in cat dorsal root ganglion and spinal ventral roots. *Brain Res.* **261**, 151–154.

Stevens, R. T., Hodge, C. J. and Apkarian, A. V., 1989, Medial, intralaminar and lateral terminations of lumbar spinothalamic tract neurons: a fluorescent double label study. *Somatosensory and Motor Res.* **6**, 285–308.

Stevens, S. S., 1970, Neural events and the psychophysical law. *Science* **170**, 1043–1050.

Stoeckel, K., Schwab, M. and Thoenen, H., 1975, Specificity of retrograde transport of nerve growth factor (NGF) in sensory neurons: a biochemical and morphological study. *Brain Res.* **89**, 1–14.

Stolwijk, J. A. J. and Wexler, I., 1971, Peripheral nerve activity in response to heating the cat's skin. *J. Physiol.* **214**, 377–392.

Stookey, B., 1929, Further light on the transmission of pain and temperature within the spinal cord: human cordotomy to abolish pain sense without destroying temperature sense. *J. Nerv. Ment. Dis.* **69**, 552–557.

Straile, W. E., 1960, Sensory hair follicles in mammalian skin: the tylotrich follicle. *Amer. J. Anat.* **106**, 133–147.

Straile, W. E., 1961, The morphology of tylotrich follicles in the skin of the rabbit. *Amer. J. Anat.* **109**, 1–13.

Straile, W. E., 1969, Encapsulated nerve end-organs in the rabbit, mouse, sheep and man. *J. Comp. Neurol.* **136**, 317–336.

Straughan, D. W., Neal, M. J., Simmonds, M. A., Collins, G. G. S. and Hill, R. G., 1971, Evaluation of bicuculline as a GABA antagonist. *Nature* **233**, 352–354.

Streit, W. J., Schulte, B. A., Balentine, J. D. and Spicer, S. S., 1985, Histochemical localization of galactose-containing glycoconjugates in sensory neurons and their processes in the central and peripheral nervous system of the rat. *J. Histochem. Cytochem.* **33**, 1042–1052.

Streit, W. J., Schulte, B. A., Balentine, J. D. and Spicer, S. S., 1986, Evidence for glycoconjugate in nociceptive primary sensory neurons and its origin from the Golgi complex. *Brain Res.* **377**, 1–17.

Strughold, H. and Porz, R., 1931, Die Dichte der Kaltpunkte auf der Haut des menschlichen Koerpers. *Z. Biol.* **91**, 563–571.

Su, H. C., Bishop, A. E., Power, R. F., Hamada, Y. and Polak, J. M., 1987, Dual intrinsic and extrinsic origins of CGRP- and NPY-immunoreactive nerves of rat gut and pancreas. *J. Neurosci.* **7**, 2674–2687.

Sugimoto, T. and Gobel, S., 1984, Dendritic changes in the dorsal horn following transection of a peripheral nerve. *Brain Res.* **321**, 199–208.

Sugimoto, T., Takemura, M. and Wakisaka, S., 1988, Cell size analysis of primary neurons innervating the cornea and tooth pulp of the rat. *Pain* **32**, 375–381.

Sugiura, Y., 1975, Three dimensional analysis of the neurons in the substantia gelatinosa Rolandi. *Proc. Jap. Acad.* **51**, 336–341.

Sugiura, Y., Lee, C. L. and Perl, E. R., 1986, Central projections of identified, unmyelinated (C) afferent fibers innervating mammalian skin. *Science* **234**, 358–361.

Sugiura, Y., Hosoya, Y., Ito, R. and Kohno, K., 1988, Ultrastructural features of functionally identified primary afferent neurons with C (unmyelinated) fibers of the guinea pig: classification of dorsal root ganglion cell type with reference to sensory modality. *J. Comp. Neurol.* **276**, 265–278.

Sugiura, Y., Terui, N. and Hosoya, Y., 1989, Difference in distribution of central terminals between visceral and somatic unmyelinated (C) primary afferent fibers. *J. Neurophysiol.* **62**, 834–840.

Sumal, K. K., Pickel, V. M., Miller, R. J. and Reis, D. J., 1982, Enkephalin-containing neurons in substantia gelatinosa of spinal trigeminal complex: ultrastructure and synaptic interaction with primary sensory afferents. *Brain Res.* **248**, 223–236.

Sumino, R. and Dubner, R., 1981, Response characteristics of specific thermoreceptive afferents innervating monkey facial skin and their relationship to human thermal sensitivity. *Brain Res. Rev.* **3**, 105–122.

Surmeier, D. J. and Towe, A. L., 1987a, Properties of proprioceptive neurons in the cuneate nucleus of the cat. *J. Neurophysiol.* **57**, 938–961.

Surmeier, D. J. and Towe, A. L., 1987b, Intrinsic features contributing to spike train patterning in proprioceptive cuneate neurons. *J. Neurophysiol.* **57**, 962–976.

Surmeier, D. J., Honda, C. N. and Willis, W. D., 1986a, Responses of primate spinothalamic neurons to noxious thermal stimulation of glabrous and hairy skin. *J. Neurophysiol.* **56**, 328–350.

Surmeier, D. J., Honda, C. N. and Willis, W. D., 1986b, Temporal features of the responses of primate spinothalamic neurons to noxious thermal stimulation of hairy and glabrous skin. *J. Neurophysiol.* **56**, 351–369.

Surmeier, D. J., Honda, C. N. and Willis, W. D., 1988, Natural grouping of primate spinothalamic neurons based on cutaneous stimulation. Physiological and anatomical features. *J. Neurophysiol.* **59**, 833–860.

Surmeier, D. J., Honda, C. N. and Willis, W. D., 1989, Patterns of spontaneous discharge in primate spinothalamic neurons. *J. Neurophysiol.* **61**, 106–115.

Suzue, T. and Jessell, T., 1980, Opiate analgesics and endorphins inhibit rat dorsal root potential in vitro. *Neurosci. Lett.* **16**, 161–166.

Svensson, B. A., Rastad, J., Westman, J. and Wiberg, M., 1985a, Somatotopic termination of spinal afferents to the feline lateral cervical nucleus. *Exp. Brain Res.* **57**, 576–584.

Svensson, B. A., Westman, J. and Rastad, J., 1985b, Light and electron microscopic study of neurones in the feline lateral cervical nucleus with a descending projection. *Brain Res.* **361**, 114–124.

Svensson, B. A., Griph, S., Rastad, J. and Westman, J., 1987, Quantitative ultrastructural study of boutons of ascending afferents to the feline lateral cervical nucleus. *Brain Res.* **423**, 229–236.

Svoboda, J., Motin, V., Hajek, I. and Sykova, E., 1988, Increase in extracellular potassium level in rat spinal dorsal horn induced by noxious stimulation and peripheral injury. *Brain Res.* **458**, 97–105.

Sweeney, M., White, T. and Sawynok, J., 1988, 5-hydroxytryptamine releases adenosine from primary afferent nerve terminals in the spinal cord. *Brain Res.* **462**, 346–349.

Sweet, W. H. and Wepsic, J. G., 1968, Treatment of chronic pain by stimulation of fibers of primary afferent neuron. *Trans. Amer. Neurol. Assoc.* **93**, 103–105.

Sweet, W. H., White, J. C., Selverstone, B. and Nilges, R., 1950, Sensory responses from anterior roots and from surface and interior of spinal cord in man. *Trans. Amer. Neurol. Assoc.* pp. 165–169.

Swett, J. E. and Woolf, C. J., 1985, The somatotopic organization of primary afferent terminals in the superficial laminae of the dorsal horn of the rat spinal cord. *J. Comp. Neurol.* **231**, 66–77.

Swett, J. E., McMahon, S. B. and Wall, P. D., 1985, Long ascending projections to the midbrain from cells of lamina I and nucleus of the dorsolateral funiculus of the rat spinal cord. *J. Comp. Neurol.* **238**, 401–416.

Syková, E. and Vyklický, L., 1977, Changes of extracellular potassium activity in isolated cord of frog under high Mg^{++} concentration. *Neurosci. Lett.* **4**, 161–165,

Syková, E. and Vyklický, L., 1978, Effects of picrotoxin on potassium accumulation and dorsal root potentials in the frog spinal cord. *Neuroscience* **3**, 1061–1067.

Syková, E., Shirayev, B., Kříž, N. and Vyklický, L., 1976, Accumulation of extracellular potassium in the spinal cord of frog. *Brain Res.* **106**, 413–417.

Syková, E., Czeh, G. and Kříž, N., 1980, Potassium accumulation in the frog spinal cord induced by nociceptive stimulation of the skin. *Neurosci. Lett.* **17**, 253–258.

Szarijanni, N. and Réthelyi, M., 1979, Differential distribution of small and large neurons in the sacrococcygeal dorsal root ganglia of the cat. *Acta Morph. Acad. Sci. Hung.* **27**, 25–35.

Szentágothai, J., 1964, Neuronal and synaptic arrangement in the substantia gelatinosa Rolandi. *J. Comp. Neurol.* **122**, 219–239.

Szentágothai, J. and Réthelyi, M., 1973, Cyto- and neuropil architecture of the spinal cord. In: *New Development in Electromyography and Clinical Neurophysiology*, pp. 20–37. J. E. Desmedt (ed.). Karger, Basel.

Szolcsanyi, J., 1988, Antidromic vasodilatation and neurogenic inflammation. *Agents and Actions* **23**, 4–11.

Szonyi, G., Knyihár, E. and Csillik, B., 1979, Extra-lysosomal fluoride-resistant acid phosphatase-active neuronal system. Surviving nociception in rat cornea. *Zeits. Mikros. Anat. Forsch.* **93**, 974–981.

Taber, E., 1961, The cytoarchitecture of the brain stem of the cat. I. Brain stem nuclei of cat. *J. Comp. Neurol.* **116**, 27–70.

Takahashi, M. and Yokota, T., 1983, Convergence of cardiac and cutaneous afferents onto neurons in the dorsal horn of the spinal cord in the cat. *Neurosci. Lett.* **38**, 251–256.

Takahashi, O., Traub, R. J. and Ruda, M. A., 1988, Demonstration of calcitonin gene-related peptide immunoreactive axons contacting dynorphin A 1-8 immunoreactive spinal neurons in a rat model of peripheral inflammation and hyperalgesia. *Brain Res.* **475**, 168–172.

Takahashi, T. and Otsuka, M., 1975, Regional distribution of substance P in the spinal cord and nerve roots of the cat and the effect of dorsal root section. *Brain Res.* **87**, 1–11.

Takano, Y., Martin, J. E., Leeman, S. E. and Loewy, A. D., 1984, Substance P immunoreactivity released from rat spinal cord after kainic acid excitation of the ventral medulla oblongata: a correlation with increases in blood pressure. *Brain Res.* **291**, 168–172.

Takeuchi, T., 1958, Histochemical demonstration of branching enzyme (amylo-1,4-1,6-transglucosidase) in animal tissues. *J. Histochem. Cytochem.* **6**, 208–216.

Talaat, M., 1937, Afferent impulses in the nerves supplying the bladder. *J. Physiol.* **89**, 1–13.

Talbot, J. D., Duncan, G. H., Bushnell, M. C. and Boyer, M., 1987, Diffuse noxious inhibitory controls (DNICs): psychophysical evidence in man for intersegmental suppression of noxious heat perception by cold pressor pain. *Pain* **30**, 221–232.

Talbot, J. D., Duncan, G. H. and Bushnell, M. C., 1989, Effects of diffuse noxious inhibitory controls (DNICs) on the sensory-discriminative dimension of pain perception. *Pain* **36**, 231–238.

Talbot, W. H., Darian-Smith, I., Kornhuber, H. H. and Mountcastle, V. B., 1968, The sense of flutter-vibration: comparison of the human capacity with response patterns of mechanoreceptive afferents from the monkey hand. *J. Neurophysiol.* **31**, 301–334.

Tamatani, M., Senba, E. and Tohyama, M., 1989, Calcitonin gene-related peptide- and substance P-containing primary afferent fibers in the dorsal column of the rat. *Brain Res.* **495**, 122–130.

Tan, C. K. and Lieberman, A. R., 1974, The glomerular synaptic complexes of the rat cuneate nucleus: some ultrastructural observations. *J. Anat.* **118**, 344–345.

Tan, C. K. and Lieberman, A. R., 1978, Identification of thalamic projection cells in the rat cuneate nucleus: a light and electron microscopic study using horseradish peroxidase. *Neurosci. Lett.* **10**, 19–22.

Tan, J. and Holstege, G., 1986, Anatomical evidence that the pontine lateral tegmental field projects to lamina I of the caudal spinal trigeminal nucleus and spinal cord and to the Edinger-Westphal nucleus in the cat. *Neurosci. Lett.* **64**, 317–322.

Tapper, D. N., 1965, Stimulus-response relationships in the cutaneous slowly-adapting mechanoreceptor in hairy skin of the cat. *Exp. Neurol.* **13**, 364–385.

Tapper, D. N., 1970, Behavioral evaluation of the tactile pad receptor system in hairy skin of the cat. *Exp. Neurol.* **26**, 447–459.

Tapper, D. N. and Mann, M. D., 1968, Single presynaptic impulse evokes postsynaptic discharge. *Brain Res.* **11**, 688–690.

Tapper, D. N. and Wiesenfeld, Z., 1980, A dorsal spinal neural network in cat. I. Responses to single impulses in single type I cutaneous fibers. *J. Neurophysiol.* **44**, 1190–1213.

Tapper, D. N., Brown, P. B. and Moraff, H., 1973, Functional organization of the cat's dorsal horn: connectivity of myelinated fiber systems of hairy skin. *J. Neurophysiol.* **36**, 817–826.

Tapper, D. N., Wiesenfeld, Z. and Craig, A. D., 1983, A dorsal spinal neural network in cat. II. Changes in responsiveness initiated by single conditioning impulses in single type I cutaneous input fibers. *J. Neurophysiol.* **49**, 534–547.

Tashiro, T. and Ruda, M. A., 1988, Immunocytochemical identification of axons containing coexistent serotonin and substance P in the cat lumbar spinal cord. *Peptides* **9**, 383–391.

Tashiro, T., Takahashi, O., Satoda, T., Matsushima, R. and Mizuno, N., 1987, Immunohistochemical demonstration of coexistence of enkephalin- and substance P-like immunoreactivities in axonal components in the lumbar segments of cat spinal cord. *Brain Res.* **424**, 391–395.

Tashiro, T., Satoda, T., Takahashi, O., Matsushima, R. and Mizuno, N., 1988, Distribution of axons exhibiting both enkephalin- and serotonin-like immunoreactivities in the lumbar cord segments: and immunohistochemical study in the cat. *Brain Res.* **440**, 357–362.

Tasker, R. R. and Dostrovsky, J. O., 1989, Deafferentation and central pain. In: *Textbook of Pain*, pp. 154–180. P. D. Wall and R. Melzack (eds.), 2nd ed. Churchill Livingstone, London.

Tattersall, J. E. H., Cervero, F. and Lumb, B. M., 1986a, Viscerosomatic neurons in the lower thoracic spinal cord of the cat: Excitations and inhibitions evoked by splanchnic and somatic nerve volleys and by stimulation of brain stem nuclei. *J. Neurophysiol.* **56**, 1411–1423.

Tattersall, J. E. H., Cervero, F. and Lumb, B. M., 1986b, Effects of reversible spinalization on the visceral input to viscerosomatic neurons in the lower thoracic spinal cord of the cat. *J. Neurophysiol.* **56**, 785–796.

Taub, A., 1964, Local, segmental and supraspinal interaction with a dorsolateral spinal cutaneous afferent system. *Exp. Neurol.* **10**, 357–374.

Taub, A. and Bishop, P. O., 1965, The spinocervical tract: dorsal column linkage, conduction velocity, primary afferent spectrum. *Exp. Neurol.* **13**, 1–21.

Taub, E. and Berman, A. J., 1968, Movement and learning in the absence of sensory feedback. In: *The Neuropsychology of Spatially Oriented Behavior*, pp. 173–191. S. J. Freedman (ed.). Dorsey, Homewood, Ill.

Taylor, D. C. M. and Pierau, F. K., 1982, Double fluorescent labeling supports electrophysiological evidence for dichotomizing peripheral sensory nerve fibers in rats. *Neurosci. Lett.* **33**, 1–6.

Taylor, D. C. M., Pierau, F. K. and Schmid, H., 1983, The use of fluorescent tracers in the peripheral sensory nervous system. *J. Neurosci. Meth.* **8**, 211–224.

Tennyson, V. M., 1964, Electron microscopic study of the developing neuroblast of the dorsal root ganglion of the rabbit embryo. *J. Comp. Neurol.* **124**, 267–318.

Tennyson, V. and Brzin, M., 1970, The appearance of acetylcholinesterase in the neuroblast of the developing rat embryo. A study by electron microscope cytochemistry and microgasometric analysis with the magnetic diver. *J. Cell Biol.* **46**, 64–80.

Tenser, R. B., 1985, Sequential changes of sensory neuron (fluoride-resistant) acid phosphatase in dorsal root ganglion neurons following neurectomy and rhizotomy. *Brain Res.* **332**, 386–389.

Terenghi, G., Polak, J., Ghatei, M., Mulderry, P., Butler, J., Unger, W. and Bloom, S., 1985, Distribution and origin of calcitonin gene-related peptide (CGRP) immunoreactivity in the sensory innervation of the mammalian eye. *J. Comp. Neurol.* **233**, 506–516.

Terenghi, G., Polak, J., Rodrigo, J., Mulderry, P. and Bloom, S., 1986, Calcitonin gene-related-immunoreactive nerves in the tongue, epiglottis and pharynx of the rat: occurrence, distribution and origin. *Brain Res.* **365**, 1–14.

Tervo, K., Tervo, T., Eränkö, L., Eränkö, O. and Cuello, A. C., 1981, Immunoreactivity for substance P in the Gasserian ganglion, ophthalmic nerve and anterior segment of the rabbit eye. *Histochem. J.* **13**, 435–443.

Tessler, A., Gazer, E., Artymyshyn, R., Murray, M. and Goldberger, M. E., 1980, Recovery of substance P in the cat spinal cord after unilateral lumbosacral deafferentation. *Brain Res.* **191**, 459–470.

Tessler, A., Himes, B. T., Artymyshyn, R., Murray, M. and Goldberger, M. E., 1981, Spinal neurons mediate return of substance P following deafferentation of cat spinal cord. *Brain Res.* **230**, 263–281.

Tessler, A., Himes, B. T., Soper, K., Murray, M., Goldberger, M. E. and Reichlin, S., 1984, Recovery of substance P but not somatostatin in the cat spinal cord after unilateral lumbosacral dorsal rhizotomy: a quantitative study. *Brain Res.* **305**, 95–102.

Teuber, H. L., 1960, Perception. In: *Handbook of Physiology*, Section I. *Neurophysiology*, Vol. III, pp. 1595–1668. J. Field (ed.). American Physiological Society, Washington, D.C.

Tewari, H. B. and Bourne, G. H., 1962a, The histochemistry of the nucleus and nucleolus with reference to nucleo-cytoplasmic relations in the spinal ganglion neuron of the rat. *Acta Histochem.* **13**, 323–350.

Tewari, H. B. and Bourne, G. H., 1962b, Histochemical evidence of metabolic cycles in rat spinal ganglion cells. *J. Histochem. Cytochem.* **10**, 42–64.

Tewari, H. B. and Bourne, G. H., 1962c, Histochemical studies on the distribution of β-gluminidase and succinic dehydrogenase in young and old spinal ganglion cells of the rat. *Zeit. f. Zellforsch.* **58**, 70–75.

Tewari, H. B. and Bourne, G. H., 1963a, On the topographical differences in the localization of certain enzymes in trigeminal ganglion cells of rat. *Experientia* **15**, 238–240.

Tewari, H. B. and Bourne, G. H., 1963b, Histochemical studies on the distribution of adenosine triphosphatase in the trigeminal ganglion cells of the rat. *J. Histochem. Cytochem.* **11**, 511–519.

Tewari, N., Chaturvedi, R. P. and Ujwal, Z.S., 1970, A comparative histoenzymological study on the distribution of alkaline phosphatase amongst the neurons of spinal ganglia of rat and guinea pig. *Anat. Anz.* **126**, 411–417.

Thakar, D. S. and Tewari, H. B., 1967, Histochemical studies on the distribution of alkaline and acid phosphatases amongst the neurons of the cerebellum, spinal and trigeminal ganglia of bat. *Acta Histochem.* **28**, 359–367.

Therman, P. O., 1941, Transmission of impulses through the Burdach nucleus. *J. Neurophysiol.* **4**, 153–166.

Thiele, F. H. and Horsley, V., 1901, A study of the degenerations observed in the central nervous system in a case of fracture dislocation of the spine. *Brain* **24**, 519–531.

Thies, R., 1985, Activation of lumbar spinoreticular neurons by stimulation of muscle, cutaneous and sympathetic afferents. *Brain Res.* **333**, 151–155.

Thies, R. and Foreman, R. D., 1983, Inhibition and excitation of thoracic spinoreticular neurons by electrical stimulation of vagal afferent nerves. *Exp. Neurol.* **82**, 1–16.

Thomas, E., 1963, Dehydrogenasen und Esterasen in unveranderten und geschadigten Spinalganglienzellen vom Menschen. *Acta Neuropath.* **2**, 231–245.

Thomas, E., 1972, Contribution of enzyme histochemistry as to the nature and differences among mantle cells (satellite cells), amphicytes, and Schwann cells in Remak fiber bundles and of myelinated fibers in dorsal root ganglia, sympathetic ganglia and in peripheral nerves of various animals. *Neuropath. Pol.* **10**, 275–283.

Thomas, E., 1977, *Histochemie der Enzyme im Peripheren Nervensystem*. Gustav Fisher Verlag, Stuttgart, 310 pp.

Thomas, E. and Pearse, A. G. E., 1961, The fine localization of dehydrogenases in the nervous system. *Histochemie* **2**, 266–282.

Todd, A. J., 1988, Electron microscope study of Golgi-stained cells in lamina II of the rat spinal dorsal horn. *J. Comp. Neurol.* **275**, 145–157.

Todd, A. J., 1989, Cells in laminae III and IV of rat spinal dorsal horn receive monosynaptic primary afferent input in lamina II. *J. Comp. Neurol.* **289**, 676–686.

Todd, A. J., 1990, An electron microscope study of glycine-like immunoreactivity in laminae I–III of the spinal dorsal horn of the rat. *Neuroscience* **39**, 387–394.

Todd, A. J. and Lewis, S. G., 1986, The morphology of Golgi-stained neurons in lamina II of the rat spinal cord. *J. Anat.* **19**, 113–119.

Todd, A. J. and Lochhead, V., 1990, GABA-like immunoreactivity in type I glomeruli of rat substantia gelatinosa. *Brain Res.* **514**, 171–174.

Todd, A. J. and McKenzie, J., 1989, GABA-immunoreactive neurons in the dorsal horn of the rat spinal cord. *Neuroscience* **31**, 799–806.

Todd, A. J. and Millar, J., 1983, Receptive fields and responses to iontophoretically applied noradrenaline and 5-hydroxytryptamine of units recorded in laminae I-III of cat dorsal horn. *Brain Res.* **288**, 159–167.

Todd, J. K., 1964, Afferent impulses in the pudendal nerves of the cat. *Quart. J. Physiol.* **49**, 258–267.

Toennies, J. F., 1938, Reflex discharge from the spinal cord over the dorsal roots. *J. Neurophysiol.* **1**, 378–390.

Tohyama, M. and Shiotani, Y., 1986, Neuropeptides in spinal cord. In: *Progress in Brain Research*, Vol. 66, pp. 177–218. P. C. Emson, M. N. Rossor and M. Tohyama (eds.). Elsevier, Amsterdam.

Tohyama, M., Sakai, K., Salvert, D., Touret, M. and Jouvet, M., 1979a, Spinal projections from the lower brain stem in the cat as demonstrated by the horseradish peroxidase technique. I. Origins of the reticulospinal tracts and their funicular projections. *Brain Res.* **173**, 383–403.

Tohyama, M., Sakai, T., Touret, M., Salvert, D. and Jouvet, M., 1979b, Spinal projections from the lower brain stem in the cat as demonstrated by the horseradish peroxidase technique. II. Projections from the dorsolateral pontine tegmentum and raphe nuclei. *Brain Res.* **176**, 215–231.

Tomasulo, K. C. and Emmers, R., 1972, Activation of neurons in the gracile nucleus by two afferent pathways in the rat. *Exp. Neurol.* **36**, 197–206.
Tomlinson, R. W. W., Gray, B. G. and Dostrovsky, J. O., 1983, Inhibition of rat spinal cord dorsal horn neurons by non-segmental, noxious cutaneous stimuli. *Brain Res.* **279**, 291–294.
Torebjörk, H. E., 1974, Afferent C units responding to mechanical, thermal and chemical stimuli in human non-glabrous skin. *Acta Physiol. Scand.* **92**, 374–390.
Torebjörk, H. E. and Hallin, R. G., 1973, Perceptual changes accompanying controlled preferential blocking of A and C fibre responses in intact human skin nerves. *Exp. Brain Res.* **16**, 321–332.
Torebjörk, H. E. and Hallin, R. G., 1974, Identification of afferent C units in intact human skin nerves. *Brain Res.* **67**, 387–403.
Torebjörk, H. E. and Hallin, R. G., 1976, Skin receptors supplied by unmyelinated (C) fibers in man. In: *Sensory Functions of the Skin in Primates, with Special Reference to Man*, pp. 475–485. Y. Zotterman, (ed.). Pergamon Press, Oxford.
Torebjörk, H. E. and Ochoa, J. L., 1980, Specific sensations evoked by activity in single identified sensory units in man. *Acta Physiol. Scand.* **110**, 445–447.
Torebjörk, H. E. and Ochoa, J. L., 1981, Pain and itch from C fiber stimulation. *Neurosci. Abstr.* **7**, 228.
Torebjörk, H. E. and Ochoa, J. L., 1983, Selective stimulation of sensory units in man. *Adv. Pain Res. Ther.* **5**, 99–104.
Torebjörk, H. E. and Ochoa, J. L., 1990, New method to identify nociceptor units innervating glabrous skin of the human hand. *Exp. Brain Res.* **81**, 509–514.
Torebjörk, H. E., Ochoa, J. L. and Schady, W., 1984a, Referred pain from intraneural stimulation of muscle fascicles in the median nerve. *Pain* **18**, 145–156.
Torebjörk, H. E., Schady, W. and Ochoa, J., 1984b, Sensory correlates of somatic afferent fibre activation. *Human Neurobiol.* **3**, 15–20.
Torebjörk, H. E., Vallbo, A. B. and Ochoa, J. L., 1987, Intraneural microstimulation in man: its relation to specificity of tactile sensations. *Brain* **110**, 1509–1529.
Towe, A. L., 1973, Somatosensory cortex: descending influences on ascending systems. In: *Handbook of Sensory Physiology*, II. *The Somatosensory System*, pp. 701–718. A. Iggo (ed.). Springer, New York.
Towe, A. L. and Jabbur, S. J., 1961, Cortical inhibition of neurons in dorsal column nuclei of cat. *J. Neurophysiol.* **24**, 488–498.
Tracey, D. J., 1979, Characteristics of wrist joint receptors in cat. *Exp. Brain Res.* **34**, 165–176.
Tracey, D. J., 1980, The projection of joint receptors to the cuneate nucleus in the cat. *J. Physiol.* **305**, 433–449.
Traub, R. J., Iadarola, M. J. and Ruda, M. A., 1989a, Effect of multiple dorsal rhizotomies on calcitonin gene-related peptide-like immunoreactivity in the lumbosacral dorsal spinal cord of the cat: a radioimmunoassay analysis. *Peptides* **10**, 979–983.
Traub, R. J., Solodkin, A. and Ruda, M. A., 1989b, Calcitonin gene-related peptide immunoreactivity in the cat lumbosacral spinal cord and the effects of multiple dorsal rhizotomies. *J. Comp. Neurol.* **287**, 225–237.
Tredici, G., Tarelli, L. T., Cavaletti, G. and Marmiroli, P., 1985, Ultrastructural organization of lamina VI of the spinal cord of the cat. *Prog. Neurobiol.* **24**, 293–311.
Treede, R. D., Meyer, R. A. and Campbell, J. N., 1990, Comparison of heat and mechanical receptive fields of cutaneous C-fiber nociceptors in monkey. *J. Neurophysiol.* **64**, 1502–1513.
Trevino, D. L., 1976, The origin and projections of a spinal nociceptive and thermoreceptive pathway. In: *Sensory Functions of the Skin in Primates, with Special Reference to Man*, pp. 367–376. Y. Zotterman (ed.). Pergamon Press, New York.
Trevino, D. L. and Carstens, E., 1975, Confirmation of the location of spinothalamic neurons in the cat and monkey by the retrograde transport of horseradish peroxidase. *Brain Res.* **98**, 177–182.
Trevino, D. L., Maunz, R. A., Bryan, R. N. and Willis, W. D., 1972, Location of cells of origin of the spinothalamic tract in the lumbar enlargement of cat. *Exp. Neurol.* **34**, 64–77.
Trevino, D. L., Coulter, J. D. and Willis, W. D., 1973, Location of cells of origin of spinothalamic tract in lumbar enlargement of the monkey. *J. Neurophysiol.* **36**, 750–761.
Triepel, J., Metz, J., Monroe, D., London, S., Sweriduk, S. and Forssmann, W. G., 1987, Vasoactive intestinal polypeptide immunoreactivity in the spinal cord of the guinea pig. *Cell Tiss. Res.* **249**, 145–150.
Trotter, W. and Davies, H. M., 1909, Experimental studies in the innervation of the skin. *J. Physiol.* **38**, 134–246.
Truex, R. C., Taylor, M. J., Smythe, M. Q. and Gildenberg, P. L., 1965, The lateral cervical nucleus of the cat, dog, and man. *J. Comp. Neurol.* **139**, 93–104.
Tsai, S., Tew, J., McLean, J. and Shipley, M., 1988, Cerebral arterial innervation by nerve fibers containing calcitonin gene-related peptide (CGRP): I. Distribution and origin of CGRP perivascular innervation in the rat. *J. Comp. Neurol.* **271**, 435–444.
Tschopp, R., Henke, H., Petermann, J., Tobler, P., Janzer, R., Hökfelt, T., Lundberg, J. and Cuello, C., 1985, Calcitonin gene-related peptide and its binding sites in the human central nervous system and pituitary. *Proc. Natl. Acad. Sci.* **82**, 248–252.
Tsuruo, Y., Hokfelt, T. and Visser, T., 1987, Thyrotropin releasing hormone (TRH)-immunoreactive cell groups in the rat central nervous system. *Exp. Brain Res.* **68**, 213–217.
Tsuruoka, M., Li, Q. J., Matsui, A. and Matsui, Y., 1990, Inhibition of nociceptive responses of wide-dynamic-range neurons by peripheral nerve stimulation. *Brain Res. Bull.* **25**, 387–392.

Tuchscherer, M. M. and Seybold, V. S., 1985, Immunohistochemical studies of substance P, cholecystokinin-octapeptide and somatostatin in dorsal root ganglia of the rat. *Neuroscience* **14**, 593–605.

Tuchscherer, M. M. and Seybold, V. S., 1989, A quantitative study of the coexistence of peptides in varicosities within the superficial laminae of the dorsal horn of the rat spinal cord. *J. Neurosci.* **9**, 195–205.

Tuchscherer, M. M., Knox, C. and Seybold, V. S., 1987, Substance P and cholecystokinin-like immunoreactive varicosities in somatosensory and autonomic regions of the rat spinal cord: a quantitative study of coexistence. *J. Neurosci.* **7**, 3984–3995.

Tuckett, R. P., 1982, Innervation pattern of cutaneous hair receptors in cat. *Brain Res.* **249**, 255–263.

Tuckett, R. P. and Wei, J. Y., 1987a, Response to an itch-producing substance in cat. I. Cutaneous receptor populations with myelinated axons. *Brain Res.* **413**, 87–94.

Tuckett, R. P. and Wei, J. Y., 1987b, Response to an itch-producing substance in cat. II. Cutaneous receptor populations with unmyelinated axons. *Brain Res.* **413**, 95–103.

Tuckett, R. P., Horch, K. W. and Burgess, P. R., 1978, Response of cutaneous hair and field mechanoreceptors in cat to threshold stimuli. *J. Neurophysiol.* **41**, 138–149.

Tyce, G. M. and Yaksh, T. L., 1981, Monoamine release from cat spinal cord by somatic stimuli: an intrinsic modulatory system. *J. Physiol.* **314**, 513–529.

Uchida, Y. and Murao, S., 1974, Excitation of afferent cardiac sympathetic nerve fibers during coronary occlusion. *Amer. J. Physiol.* **226**, 1094–1099.

Uddenberg, N., 1968a, Differential localization in dorsal funiculus of fibres originating from different receptors. *Exp. Brain Res.* **4**, 367–376.

Uddenberg, N., 1968b, Functional organization of long, second-order afferents in the dorsal funiculus. *Exp. Brain Res.* **4**, 377–382.

Uddman, R., Edvinsson, L., Ekman, R., Kingman, T. and McCulloch, J., 1985a, Innervation of the feline cerebral vasculature by nerve fibers containing calcitonin gene-related peptide: trigeminal origin and co-existence with substance P. *Neurosci. Lett.* **62**, 134–136.

Uddman, R., Luts, A. and Sundler, F., 1985b, Occurrence and distribution of calcitonin gene-related peptide in the mammalian respiratory tract and middle ear. *Cell Tiss. Res.* **241**, 551–555.

Uddman, R., Edvinsson, L., Ekblad, E., Hokanson, R. and Sundler, F., 1986, Calcitonin gene-related peptide (CGRP): perivascular distribution and vasodilator effects. *Reg. Peptides* **15**, 1–23.

Ueyama, T., Arakawa, H. and Mizuno, N., 1985, Contralateral termination of pudendal nerve fibers in the gracile nucleus of the rat. *Neurosci. Lett.* **62**, 113–117.

Uhl, G. R., Goodman, M. J., Kuhar, M. J., Childers, S. R. and Snyder, S. H., 1979a, Immunohistochemical mapping of enkephalin containing cell bodies, fibers and nerve terminals in the brain stem of the rat. *Brain Res.* **166**, 75–94.

Uhl, G. R., Goodman, R. R. and Snyder, S. H., 1979b, Neurotensin-containing cell bodies, fibers and nerve terminals in the brain stem of the rat: immunohistochemical mapping. *Brain Res.* **167**, 77–91.

Ulfhake, B., Arvidsson, U., Cullheim, S., Hökfelt, T. and Visser, T. J., 1987, Thyrotropin-releasing hormone (TRH)-immunoreactive boutons and nerve cell bodies in the dorsal horn of the cat L7 spinal cord. *Neurosci. Lett.* **73**, 3–8.

Urban, L. and Randić, M., 1984, Slow excitatory transmission in rat dorsal horn: possible mediation by peptides. *Brain Res.* **290**, 336–341.

Urban, L., Aitken, P. G. and Somjen, G. G., 1985, Interstitial potassium concentration, slow depolarization and focal potential responses in the dorsal horn of the rat spinal slice. *Brain Res.* **331**, 168–171.

Urban, L., Willetts, J., Murase, K. and Randić, M., 1989, Cholinergic effects on spinal dorsal horn neurons in vitro: an intracellular study. *Brain Res.* **500**, 12–20.

Vallbo, A. B., 1971, Muscle spindle response at the onset of isometric voluntary contractions in man. Time difference between fusimotor and skeletomotor effects. *J. Physiol.* **218**, 405–431.

Vallbo, A. B., 1974a, Afferent discharge from human muscle spindles in non-contracting muscles. Steady state impulse frequency as a function of joint angle. *Acta Physiol. Scand.* **90**, 303–318.

Vallbo, A. B., 1974b, Human muscle spindle discharge during isometric voluntary contractions. Amplitude relations between spindle frequency and torque. *Acta Physiol. Scand.* **90**, 319–336.

Vallbo, A. B., 1981, Sensations evoked from the glabrous skin of the human hand by electrical stimulation of unitary mechanosensitive afferents. *Brain Res.* **215**, 359–363.

Vallbo, A. B. and Hagbarth, K. E., 1968, Activity from skin mechanoreceptors recorded percutaneously in awake human subjects. *Exp. Neurol.* **21**, 270–289.

Vallbo, A. B., Hagbarth, K. E., Torebjörk, H. E. and Wallin, B. G., 1979, Somatosensory, proprioceptive and sympathetic activity in human peripheral nerves. *Physiol. Rev.* **59**, 919–957.

Vallbo, A. B., Hulliger, M. and Nordh, E., 1981, Do spindle afferents monitor joint position in man? A study with active position holding. *Brain Res.* **204**, 209–213.

Vallbo, A. B., Olsson, K. A., Westberg, K. G. and Clark, F. J., 1984, Microstimulation of single tactile afferents from the human hand. *Brain* **107**, 727–749.

Valverde, F., 1966, The pyramidal tract in rodents. A study of its relations with the posterior column nuclei, dorsolateral reticular formation of the medulla oblongata, and cervical spinal cord (Golgi and electron microscopic observations). *Z. Zellforsch.* **71**, 297–363.

van Beusekom, G. T., 1955, *Fiber Analysis of the Anterior and Lateral Funiculi of the Cord in the Cat*. Eduard Ijdo, Leiden, N.V.
van Beusekom, G. T., 1955, *Fiber Analysis of the Anterior and Lateral Funiculi of the Cord in the Cat*. Eduard Ijdo, Leiden, N.V.
van den Pol, A. N. and Gorcs, T., 1988, Glycine and glycine receptor immunoreactivity in brain and spinal cord. *J. Neurosci.* **8**, 472–492.

Vanderhaeghen, J. J., Lotstra, F., De Mey, J. and Gilles, C., 1980, Immunohistochemical localization of cholecystokinin- and gastrin-like peptides in the brain and hypophysis of the rat. *Proc. Natl. Acad. Sci.* **77**, 1190–1194.

Van Hees, J., 1976a, Human C fibre input during painful and nonpainful skin stimulation with radiant heat. In: *Advances in Pain Research and Therapy*, pp. 35–40. J. J. Bonica and D. Albe-Fessard (eds.). Raven Press, New York.

Van Hees, J., 1976b, Single afferent C fiber activity in the human nerve during painful and nonpainful skin stimulation with radiant heat. In: *Sensory Function of the Skin in Primates*, pp. 503–505. Y. Zotterman (ed.). Oxford, Pergamon.

Van Hees, J. and Gybels, J. M., 1972, Pain related to single afferent C fibers from human skin. *Brain Res.* **48**, 397–400.

Van Hees, J. and Gybels, J., 1981, C nociceptor activity in human nerve during painful and non painful skin stimulation. *J. Neurol. Neurosurg. Psychiat.* **44**, 600–607.

Vater, A., 1741, Dissertatio de consensu partium corporis humani. In: *Haller, Disputationum Anatomicarum Selectarum*, Vol. II. *Gottingae*, pp. 953–972. Cited in Cauna and Mannan (1958).

Vega, J. A., Rodriguez, C., Medina, M., Terreria, A. M., Bengoechea, M. E. and Perez-Casa, A., 1989, Acetylcholinesterase and fluoride-resistant acid phosphatase activities in dorsal root ganglia. *Cell. Mol. Biol.* **35**, 39–46.

Vergara, I., Oberpaur, B. and Alvarez, J., 1986, Ventral root nonmedulated fibers: proportion, calibers and microtubular content. *J. Comp. Neurol.* **248**, 550–554.

Vierck, C. J., 1966, Spinal pathways mediating limb position sense. *Anat. Rec.* **154**, 437.

Vierck, C. J., 1973, Alterations of spatio-tactile discrimination after lesions of primate spinal cord. *Brain Res.* **58**, 69–79.

Vierck, C. J., 1974, Tactile movement detection and discrimination following dorsal column lesions in monkeys. *Exp. Brain Res.* **20**, 331–346.

Vierck, C. J., 1982, Comparison of the effects of dorsal rhizotomy or dorsal column transection on motor performance of monkeys. *Exp. Neurol.* **75**, 566–575.

Vierck, C. J. and Luck, M. M., 1979, Loss and recovery of reactivity to noxious stimuli in monkeys with primary spinothalamic cordotomies, followed by secondary and tertiary lesions of other cord sectors. *Brain* **102**, 233–248.

Vierck, C. J., Hamilton, D. M. and Thornby, J. I., 1971, Pain reactivity of monkeys after lesions to the dorsal and lateral columns of the spinal cord. *Exp. Brain Res.* **13**, 140–158.

Villanueva, L., Cadden, S. W. and Le Bars, D., 1984, Evidence that diffuse noxious inhibitory controls (DNIC) are mediated by a final post-synaptic inhibitory mechanism. *Brain Res.* **298**, 67–74.

Villanueva, L., Chitour, D. and Le Bars, D., 1986a, Involvement of the dorsolateral funiculus in the descending spinal projections responsible for diffuse noxious inhibitory controls in the rat. *J. Neurophysiol.* **56**, 1185–1195.

Villanueva, L., Peschanski, M., Calvino, B. and Le Bars, D., 1986b, Ascending pathways in the spinal cord involved in triggering of diffuse noxious inhibitory controls in the rat. *J. Neurophysiol.* **55**, 35–55.

Villanueva, L., Bouhassira, D., Bing, Z. and Le Bars, D., 1988, Convergence of heterotopic and nociceptive information onto subnucleus reticularis dorsalis neurons in the rat medulla. *J. Neurophysiol.* **60**, 980–1009.

Villanueva, L., Cliffer, K. D., Sorkin, L. S., Le Bars, D. and Willis, W. D., 1990, Convergence of heterotopic nociceptive information onto neurons of caudal medullary reticular formation in monkey (*Macaca fascicularis*). *J. Neurophysiol.* **63**, 1118–1127.

Villar, M. J., Cortés, R., Theodorsson, E., Wiesenfeld-Hallin, Z., Schalling, M., Fahrenkrug, J., Emson, P. C. and Hokfelt, T., 1989, Neuropeptide expression in rat dorsal root ganglion cells and spinal cord after peripheral nerve injury with special reference to galanin. *Neuroscience* **33**, 587–604.

Villiger, J. W. and Faull, R. L. M., 1985, Muscarinic cholinergic receptors in the human spinal cord: differential localization of [^{3}H]pirenzepine and [^{3}H]quinuclidinylbenzilate binding sites. *Brain Res.* **345**, 196–199.

Vincent, S. B., 1913, The tactile hair of the white rat. *J. Comp. Neurol.* **23**, 1–34.

Vincent, S. R., Hökfelt, T., Christensson, I. and Terenius, L., 1982, Dynorphin-immunoreactive neurons in the central nervous system of the rat. *Neurosci. Lett.* **33**, 185–190.

Vincent, S. R., McIntosh, C. H. S., Bueham, A. M. J. and Brown, J. C., 1985, Central somatostatin systems revealed with monoclonal antibodies. *J. Comp. Neurol.* **238**, 169–186.

Vlachová, V., Vyklický, L., Vyklichý, L., Jr. and Vyskocil, F., 1987, The action of excitatory amino acids on chick spinal cord neurones in culture. *J. Physiol.* **386**, 425–438.

von Frey, M., 1896, Treatise on the sensory functions of the human skin. Transl. from German by Biederman-Thorson in Handwerker, H.O. and Brunke, K. (eds.), 1987, pp. 69–131. *Classical German Contributions to Pain Research, Vth World Congress on Pain, Hamburg*. Original German: Untersuchungen ueber die Sinnesfunctionen der Menschlichen Haut. vol. 23, Koenigl. Saech. Ges. Wissensch., Leipsig, S. Hirzel.

von Frey, M., 1906, The distribution of afferent nerves in the skin. *J.A.M.A.* **47**, 645–648.
von Frey, M., 1910, Physiologie der Sinnesorgane der Menschlichen Haut. *Ergebnisse Physiol.* **9**, 351–368.
von Holst, E., 1954, Relations between the central nervoous system and the peripheral organs. *Brit. J. Animal Behav.* **2**, 89–94.
Voorhoeve, P. E. and Zwaagstra, B., 1984, Central effects by ventral root nociceptive afferents. *Exp. Brain Res.* Suppl. 9, 116–125.
Voris, H. C., 1951, Ipsilateral sensory loss following chordotomy: report of a case. *Arch. Neurol. Psychiat.* **65**, 95–96.
Voris, H. C., 1957, Variations in the spinothalamic tract in man. *J. Neurosurg.* **14**, 55–60.
Vyklický, L., Syková, E., Kříž, N. and Ujec, E., 1972, Post-stimulation changes of extracellular potassium concentration in the spinal cord of the rat. *Brain Res.* **45**, 608–611.
Vyklický, L., Syková, E. and Kříž, N., 1975, Slow potentials induced by changes of extracellular potassium in the spinal cord of the cat. *Brain Res.* **87**, 77–80.
Vyklický, L., Syková, E. and Mellerová, B., 1976, Depolarization of primary afferents in the frog spinal cord under high Mg++ concentrations. *Brain Res.* **117**, 153–156.
Wagman, I. H. and Price, D. D., 1969, Responses of dorsal horn cells of *M. mulatta* to cutaneous and sural A and C fiber stimuli. *J. Neurophysiol.* **32**, 803–817.
Walberg, F., 1957, Corticofugal fibers to the nuclei of the dorsal columns. An experimental study in the cat. *Brain* **80**, 273–287.
Walberg, F., 1965, Axoaxonic contacts in the cuneate nucleus, probable basis for presynaptic depolarization. *Exp. Neurol.* **13**, 218–231.
Walberg, F., 1966, The fine structure of the cuneate nucleus in normal cats and following interruption of afferent fibres. An electron microscopic study with particular reference to findings made in Glees and Nauta sections. *Exp. Brain Res.* **2**, 107–128.
Walberg, F., Bowsher, D. and Brodal, A., 1958, The termination of primary vestibular fibers in the vestibular nuclei in the cat. *J. Comp. Neurol.* **110**, 391–419.
Waldeyer, H., 1888, *Das Gorilla-Rueckenmark*. Akad. Wissensch. Berlin, 147 pp.
Waldron, H. A., 1969, The morphology of the lateral cervical nucleus in the hedgehog. *Brain Res.* **16**, 301–306.
Walker, A. E., 1938, The thalamus of the chimpanzee. I. Terminations of the somatic afferent systems. *Confinia Neurol.* **1**, 99–127.
Walker, A. E., 1940, The spinothalamic tract in man. *Arch. Neurol. Psychiat.* **43**, 284–298.
Walker, A. E., 1942a, Somatotopic localization of spinothalamic and sensory trigeminal tracts in mesencephalon. *Arch. Neurol. Psychiat.* **48**, 884–889.
Walker, A. E., 1942b, Relief of pain by mesencephalic tractotomy. *Arch. Neurol. Psychiat.* **48**, 865–880.
Walker, A. E. and Weaver, T. A., 1942, The topical organization and termination of the fibers of the posterior columns in *Macaca mulatta*. *J. Comp. Neurol.* **76**, 145–158.
Wall, P. D., 1958, Excitability changes in afferent fibre terminations and their relation to slow potentials. *J. Physiol.* **142**, 1–21.
Wall, P. D., 1959, Repetitive discharge of neurons. *J. Neurophysiol.* **22**, 305–320.
Wall, P. D., 1960, Cord cells responding to touch, damage, and temperature of skin. *J. Neurophysiol.* **23**, 197–210.
Wall, P. D., 1962, The origin of a spinal cord slow potential. *J. Physiol.* **164**, 508–526.
Wall, P. D., 1964, Presynaptic control of impulses at the first central synapse in the cutaneous pathway. In: *Physiology of Spinal Neurons* J. C. Eccles and J. P. Schade (eds.), *Prog. Brain Res.* **12**, 92–115.
Wall, P. D., 1965, Impulses originating in the region of dendrites. *J. Physiol.* **180**, 116–133.
Wall, P. D., 1967, The laminar organization of dorsal horn and effects of descending impulses. *J. Physiol.* **188**, 403–423.
Wall, P. D., 1970, The sensory and motor role of impulses traveling in the dorsal columns toward cerebral cortex. *Brain* **93**, 505–524.
Wall, P. D., 1977, The presence of ineffective synapses and the circumstances which unmask them. *Phil. Trans. Roy. Soc. B* **278**, 361–372.
Wall, P. D., 1978, The gate control theory of pain mechanisms. A re-examination and re-statement. *Brain* **101**, 1–18.
Wall, P. D., 1982, The effect of peripheral nerve lesions and of neonatal capsaicin in the rat on primary afferent depolarization. *J. Physiol.* **329**, 21–35.
Wall, P. D. and Cronly-Dillon, J. R., 1960, Pain, itch, and vibration. *Arch. Neurol.* **2**, 365–375.
Wall, P. D. and Devor, M., 1981, The effect of peripheral nerve injury on dorsal root potentials and on transmission of afferent signals into the spinal cord. *Brain Res.* **209**, 95–111.
Wall, P. D. and Dubner, R., 1972, Somatosensory pathways. *Ann. Rev. Physiol.* **34**, 315–336.
Wall, P. and Fitzgerald, M., 1982, If substance P fails to fulfill the criteria as a neurotransmitter in somatosensory afferents, what might be its function? In: *Substance P in the Nervous System*, pp. 249–266. R. Porter and M. O'Connor (eds.). Pitman, London.
Wall, P. D. and McMahon, S. B., 1985, Microneurography and its relation to perceived sensation. A critical review. *Pain* **21**, 209–229.
Wall, P. D. and Noordenbos, W., 1977, Sensory functions which remain in man after complete transection of dorsal columns. *Brain* **100**, 641–653.

Wall, P. D. and Sweet, W. H., 1967, Temporary abolition of pain in man. *Science* **155**, 108–109.
Wall, P. D. and Werman, R., 1976, The physiology and anatomy of long ranging afferent fibres within the spinal cord. *J. Physiol.* **255**, 321–334.
Wall, P. D., Freeman, J. and Major, D., 1967, Dorsal horn cells in spinal and in freely moving rats. *Exp. Neurol.* **19**, 519–529.
Wall, P. D., Merrill, E. G. and Yaksh, T. L., 1979a, Responses of single units in laminae 2 and 3 of cat spinal cord. *Brain Res.* **160**, 245–260.
Wall, P. D., Devor, M., Inbal, R., Scadding, J. W., Schonfeld, D., Seltzer, Z. and Tomkiewicz, M. M., 1979b, Autotomy following peripheral nerve lesions: experimental anaesthesia dolorosa. *Pain* **7**, 103–115.
Wallenberg, A., 1900, Secundaere sensible Bahnen im Gehirnstamme des Kaninchens, ihre gegenseitige Lage und ihre Bedeutung fuer den Aufbau des Thalamus. *Anat. Anz.* **18**, 81–105.
Walsh, T. M. and Ebner, F. F., 1973, Distribution of cerebellar and somatic lemniscal projections in the ventral nuclear complex of the Virginia opossum. *J. Comp. Neurol.* **147**, 427–446.
Walshe, F. M. R., 1942, The anatomy and physiology of cutaneous sensibility: a critical review. *Brain* **65**, 48–112.
Wamsley, J. K., Zarbin, M. A. and Kuhar, M. J., 1981a, Muscarinic cholinergic receptors flow in the sciatic nerve. *Brain Res.* **217**, 155–161.
Wamsley, J. K., Lewis, M. S., Young, W. S. and Kuhar, M. J., 1981b, Autoradiographic localization of muscarinic cholinergic receptors in rat brainstem. *J. Neurosci.* **1**, 176–191.
Wanaka, A., Matsuyama, T., Yoneda, S., Kimura, K., Kamada, T., Girgis, S., MacIntyre, I., Emson, P. C. and Tohyama, M., 1986, Origins and distribution of calcitonin gene-related peptide-containing nerves in the wall of the cerebral arteries of the guinea pig with special reference to the coexistence with substance P. *Brain Res.* **369**, 185–192.
Wanaka, A., Shiotani, Y., Kuyama, H., Matsuyama, T., Shiosaka, S. and Tohyama, M., 1987, Glutamate-like immunoreactive structures in primary sensory neurons in the rat detected by a specific antiserum against glutamate. *Exp. Brain Res.* **65**, 691–694.
Warden, M. K. and Young, W. S., 1988, Distribution of cells containing mRNAs encoding substance P and neurokinin B in the rat central nervous system. *J. Comp. Neurol.* **272**, 90–113.
Warrington, W. G. and Griffith, F., 1904, On the cells of the spinal ganglia and on the relationship of their histological structure to the axonal distribution. *Brain* **27**, 297–325.
Watkins, J. C. and Evans, R. H., 1981, Excitatory amino acid transmitters. *Ann. Rev. Pharmacol. Toxicol.* **21**, 165–204.
Weaver, T. A. and Walker, A. E., 1941, Topical arrangement within the spinothalamic tract of the monkey. *Arch. Neurol. Psychiat.* **46**, 877–883.
Weber, E. H., 1846, Der Tastsinn und das Gemeingefuehl. In: *Wagner's Handwoerterbuch der Physiologie*, Vol. III/2, pp. 481–588. Vieweg, Braunschweig. Quoted in Hensel *et al.*, 1960.
Weddell, G., 1955, Somesthesis and the chemical senses. *Ann. Rev. Psychol.* **6**, 19–136.
Weddell, G. and Miller, S., 1962, Cutaneous sensibility. *Ann. Rev. Physiol.* **24**, 199–222.
Weddell, G. and Sinclair, D. C., 1953, The anatomy of pain sensibility. *Acta Neuroveg.* **7**, 135–146.
Weddell, G., Sinclair, D. C. and Feindel, W. H., 1948, An anatomical basis for alterations in quality of pain sensibility. *J. Neurophysiol.* **11**, 99–109.
Wee, B. E. F., Emery, D. G. and Blanchard, J. L., 1985, Unmyelinated fibers in the cervical and lumbar ventral roots of the cat. *Amer. J. Anat.* **172**, 307–316.
Weight, F. F. and Salmoirhagi, G. C., 1966, Responses of spinal cord interneurons to acetylcholine, norepinephrine and serotonin administered by microelectrophoresis. *J. Pharmacol. Exp. Ther.* **153**, 420–427.
Weihe, E., Leibold, A., Nohr, D., Fink, T. and Gauweiler, B., 1986, Co-existence of prodynorphin-opioid peptides and substance P in primary sensory afferents of guinea pigs. *Natl. Inst. Drug Abuse* **75**, 295–298.
Weihe, E., Nohr, D., Millan, M. J., Stein, C., Muller, S., Gramsch, C. and Herz, A., 1988a, Peptide neuroanatomy of adjuvant-induced arthritic inflammation in rat. *Agents and Actions* **25**, 255–259.
Weihe, E., Millan, M. J., Leibold, A., Nohr, D. and Herz, A., 1988b, Co-localization of proenkephalin- and prodynorphin-derived opioid peptides in laminae IV/V spinal neurons revealed in arthritic rats. *Neurosci. Lett.* **85**, 187–192.
Weihe, E., Millan, M. J., Hollt, V., Nohr, D. and Herz, A., 1989, Induction of the gene encoding pro-dynorphin by experimentally induced arthritis enhances staining for dynorphin the spinal cord of rats. *Neuroscience* **31**, 77–95.
Weinberg, R. and Rustioni, A., 1985, Brainstem projections to and from the cuneate nucleus of rats and cats. *Neurosci. Abstr.* **11**, 561.
Weinberg, R. J. and Rustioni, A., 1989, Brainstem projections to the rat cuneate nucleus. *J. Comp. Neurol.* **282**, 142–156.
Weinberg, R. J., Conti, F., Van Eyck, S. L., Petrusz, P. and Rustioni, A., 1987, Glutamate immunoreactivity in superficial laminae of rat dorsal horn and spinal trigeminal nucleus. In: *Excitatory Amino Acid Transmission*. T. P. Hicks, D. Lodge and H. McLennan (eds.). Alan R. Liss, New York.
Weinberg, R. J., Pierce, J. P. and Rustioni, A., 1990, Single fiber studies of ascending input to the cuneate nuclei of cuts: I. Morphometry of primary afferent fibers. *J. Comp. Neurol.* **300**, 113–133.
Weinstein, E. A. and Bender, M. B., 1940, Dissociation of deep sensibility at different levels of the central nervous system. *Arch. Neurol. Psychiat.* **43**, 488–497.

Weisberg, J. A. and Rustioni, A., 1976, Cortical cells projecting to the dorsal column nuclei of cats. *J. Comp. Neurol.* **168**, 625–638.

Weisberg, J. A. and Rustioni, A., 1977, Cortical cells projecting to the dorsal column nuclei of Rhesus monkey. *Exp. Brain Res.* **28**, 521–528.

Wen, C. Y., Tan, C. K. and Wang, W. C., 1977, Presynaptic dendrites in the cuneate nucleus of the monkey (*Macaca fascicularis*). *Neurosci. Lett.* **5**, 129–132.

Werman, R., 1966, Criteria for identification of a central nervous system transmitter. *Comp. Biochem. Physiol.* **18**, 745–766.

Werman, R., Davidoff, R. A. and Aprison, M. H., 1968, Inhibitory action of glycine on spinal neurons in the cat. *J. Neurophysiol.* **31**, 81–95.

Werner, G. and Mountcastle, V. B., 1965, Neural activity in mechanoreceptive cutaneous afferents: stimulus-response relations, Weber functions, and information transmission. *J. Neurophysiol.* **28**, 359–397.

Werner, G. and Whitsel, B. L., 1967, The topology of dermatomal projection in the medial lemniscal system. *J. Physiol.* **192**, 123–144.

Wessendorf, M. W. and Elde, R., 1987, The coexistence of serotonin-and substance P-like immunoreactivity in the spinal cord of the rat as shown by immunofluorescent double labeling. *J. Neurosci.* **7**, 2352–2363.

Westbrook, G. L. and Mayer, M. L., 1984, Glutamate currents in mammalian spinal neurons: resolution of a paradox. *Brain Res.* **301**, 375–379.

Westling, G. and Johansson, R. S., 1984, Factors influencing the force control during precision grip. *Exp. Brain Res.* **53**, 277–284.

Westling, G. and Johansson, R. S., 1987, Responses in glabrous skin mechanoreceptors during precision grip in humans. *Exp. Brain Res.* **66**, 128–140.

Westlund, K. N. and Coulter, J. D., 1980, Descending projections of the locus coeruleus and subcoeruleus/medial parabrachial nuclei in monkey: axonal transport studies and dopamine-β-hydroxylase immunocytochemistry. *Brain Res. Rev.* **2**, 235–264.

Westlund, K. N., Bowker, R. M., Ziegler, M. G. and Coulter, J. D., 1981, Origins of spinal noradrenergic pathways demonstrated by retrograde transport of antibody to dopamine-β-hydroxylase. *Neurosci. Lett.* **25**, 243–249.

Westlund, K. N., Bowker, R. M., Ziegler, M. G. and Coulter, J. D., 1982, Descending noradrenergic projections and their spinal terminations. *Prog. Brain Res.* **57**, 219–238.

Westlund, K. N., Bowker, R. M., Ziegler, M. G. and Coulter, J. D., 1983, Noradrenergic projections to the spinal cord of the rat. *Brain Res.* **263**, 15–31.

Westlund, K. N., McNeill, D. L. and Coggeshall, R. E., 1989a, Glutamate immunoreactivity in rat dorsal roots. *Neurosci. Lett.* **96**, 13–17.

Westlund, K. N., McNeill, D. L., Patterson, J. T. and Coggeshall, R. E., 1989b, Aspartate immunoreactive axons in normal rat L4 dorsal roots. *Brain Res.* **489**, 347–351.

Westlund, K. N., Carlton, S. M., Zhang, D. and Willis, W. D., 1990, Direct catecholaminergic innervation of primate spinothalamic tract neurons. *J. Comp. Neurol.*, in press.

Westman, J., 1968a, The lateral cervical nucleus in the cat. I. A Golgi study. *Brain Res.* **10**, 352–368.

Westman, J., 1968b, The lateral cervical nucleus in the cat. II. An electron microscopic study of the normal structure. *Brain Res.* **11**, 107–123.

Westman, J., 1969, The lateral cervical nucleus in the cat. III. An electron microscopic study after transection of spinal afferents. *Exp. Brain Res.* **7**, 32–50.

Westman, J., 1971, The lateral cervical nucleus in the cat. V. A quantitative evaluation on the bouton- and glia-covered surface area of different LCN-neurons. *Z. Zellforsch.* **115**, 377–387.

Westman, J. and Bowsher, D., 1971, Ultrastructural observations on the degeneration of spinal afferents to the nucleus medullae oblongatae centralis (pars caudalis) of the cat. *Brain Res.* **26**, 395–398.

Westman, J., Blomqvist, A., Koehler, C. and Wu, J. Y., 1984, Light and electron microscopic localization of glutamic acid decarboxylase and substance P in the dorsal column nuclei of the cat. *Neurosci. Lett.* **51**, 347–352.

Westrum, L. E., Johnson, L. R. and Canfield, R. C., 1984, Ultrastructure of transganglionic degeneration in brain stem trigeminal nuclei during normal primary tooth exfoliation and permanent tooth eruption in the cat. *J. Comp. Neurol.* **230**, 198–206.

White, J. C., 1941, Spinothalamic tractotomy in the medulla oblongata. An operation for the relief of intractable neuralgias of the occiput, neck and shoulder. *Arch. Surg.* **43**, 113–127.

White, J. C., 1954, Conduction of pain in man. *Arch. Neurol. Psychiatr.* **71**, 1/23.

White, J. C. and Sweet, W. H., 1955, *Pain, Its Mechanisms and Neurosurgical Control*. Thomas, Springfield.

White, J. C. and Sweet, W. H., 1969, *Pain and the Neurosurgeon*. Thomas, Springfield.

White, J. C., Sweet, W. H., Hawkins, R. and Nilges, R. G., 1950, Anterolateral cordotomy: results, complications and causes of failure. *Brain* **73**, 346–367.

White, J. C., Richardson, E. P. and Sweet, W. H., 1956, Upper thoracic cordotomy for relief of pain. Postmortem correlation of spinal incision with analgesic levels in 18 cases. *Ann. Surg.* **144**, 407–420.

Whitehorn, D. and Burgess, P. R., 1973, Changes in polarization of central branches of myelinated mechanoreceptor and nociceptor fibers during noxious and innocuous stimulation of the skin. *J. Neurophysiol.* **36**, 226–237.

Whitehorn, D., Morse, R. W. and Towe, A. L., 1969, Role of the spinocervical tract in production of the primary cortical response evoked by forepaw stimulation. *Exp. Neurol.* **25**, 349–364.

Whitehorn, D., Bromberg, M. B., Howe, J. F., Putnam, J. E. and Burgess, P. R., 1972, Activation of gracile nucleus: time distribution of activity in presynaptic and postsynaptic elements. *Exp. Neurol.* **37**, 312–321.

Whitehorn, D., Howe, J. F., Lessler, M. J. and Burgess, P. R., 1974, Cutaneous receptors supplied by myelinated fibers in the cat. I. Number of receptors innervated by a single nerve. *J. Neurophysiol.* **37**, 1361–1372.

Whitehouse, P. J., Wamsley, J. K., Zarbin, M. A., Price, D. L., Tourtellotte, W. W. and Kuhar, M. J., 1983, Amyotrophic lateral sclerosis: alterations in neurotransmitter receptors. *Ann. Neurol.* **14**, 8–16.

Whitlock, D. G. and Perl, E. R., 1959, Afferent projections through ventrolateral funiculi to thalamus of cat. *J. Neurophysiol.* **22**, 133–148.

Whitlock, D. G. and Perl, E. R., 1961, Thalamic projections of spinothalamic pathways in monkey. *Exp. Neurol.* **3**, 240–255.

Whitsel, B. L., Petrucelli, L. M. and Sapiro, G., 1969, Modality representation in the lumbar and cervical fasciculus gracilis of squirrel monkeys. *Brain Res.* **15**, 67–78.

Whitsel, B. L., Petrucelli, L. M., Sapiro, G. and Ha, H., 1970, Fiber sorting in the fasciculus gracilis of squirrel monkeys. *Exp. Neurol.* **29**, 227–242.

Wiberg, M. and Blomqvist, A., 1984, The spinomesencephalic tract in the cat: its cells of origin and termination pattern as demonstrated by the intraaxonal transport method. *Brain Res.* **291**, 1–18.

Wiberg, M., Westman, J. and Blomqvist, A., 1987, Somatosensory projection to the mesencephalon: an anatomical study in the monkey. *J. Comp. Neurol.* **264**, 92–117.

Wiesenfeld, Z. and Lindblom, U., 1980, Behavioral and electrophysiological effects of various types of peripheral nerve lesions in the rat: a comparison of possible models for chronic pain. *Pain* **8**, 285–299.

Wiesenfeld-Hallin, Z., Hökfelt, T., Lundberg, J. M., Firssmann, W. G., Reuneche, M., Tschopp, F. A. and Fischer, J. A., 1984, Immunoreactive calcitonin gene-related peptide and substance P coexist in sensory neurons to the spinal cord and interact in spinal behavioral responses of the rat. *Neurosci. Lett.* **52**, 199–204.

Wild, J. M., 1985, The avian somatosensory systems. I. Primary spinal afferent input to the spinal cord and brainstem of the pigeon (*Columba livia*). *J. Comp. Neurol.* **240**, 377–395.

Wild, J. M., 1989, Avian somatosensory system: II. Ascending projections of the dorsal column and external cuneate nuclei in the pigeon. *J. Comp. Neurol.* **287**, 1–18.

Willcockson, H. H., Carlton, S. M. and Willis, W. D., 1987, Mapping study of serotoninergic input to diencephalic-projecting dorsal column neurons in the rat. *J. Comp. Neurol.* **261**, 467–480.

Willcockson, W. S., Chung, J. M., Hori, Y., Lee, K. H. and Willis, W. D., 1984a, Effects of iontophoretically released amino acids and amines on primate spinothalamic tract cells. *J. Neurosci.* **4**, 732–740.

Willcockson, W. S., Chung, J. M., Hori, Y., Lee, K. H. and Willis, W. D., 1984b, Effects of iontophoretically released peptides on primate spinothalamic tract cells. *J. Neurosci.* **4**, 741–750.

Willcockson, W. S., Kim, J., Shin, H. K., Chung, J. M. and Willis, W. D., 1986, Actions of opioids on primate spinothalamic tract neurons. *J. Neurosci.* **6**, 2509–2520.

Willer, J. C. and Albe-Fessard, D., 1983, Further studies on the role of afferent input from relatively large diameter fibers in transmission of nociceptive messages in humans. *Brain Res.* **278**, 318–321.

Willer, J. C., Roby, A. and LeBars, D., 1984, Psychophysical and electrophysiological approaches to the pain-relieving effects of heterotopic nociceptive stimuli. *Brain* **107**, 1095–1112.

Willer, J. C., De Brouker, T. and LeBars, D., 1989, Encoding of nociceptive thermal stimuli by diffuse noxious inhibitory controls in humans. *J. Neurophysiol.* **62**, 1028–1038.

Williams, S., Wisden, W. and Hunt, S. P., 1989, Gene expression in spinal cord following brief sensory stimulation. *Neurosci. Abstr.* **15**, 468.

Williams, W. J., BeMent, S. L., Yin, T. C. T. and McCall, W. D., 1973, Nucleus gracilis responses to knee joint motion: a frequency response study. *Brain Res.* **64**, 123–140.

Willis, W. D., 1982, Control of Nociceptive Transmission in the Spinal Cord. In: *Progress in Sensory Physiology 3*. D. Ottoson (ed.). Springer-Verlag, Berlin.

Willis, W. D., 1984, The raphe-spinal system. In: *Brainstem Control of Spinal Cord Function*, pp. 141–214. C. D. Barnes (ed.). Academic Press, New York.

Willis, W. D., 1985, *The Pain System*. Karger, Basel.

Willis, W. D., 1988, Anatomy and physiology of descending control of nociceptive responses of dorsal horn neurons: comprehensive review. In: *Progress in Brain Research*, pp. 1–29. H. L. Fields and J. M. Besson (eds.). Elsevier, Amsterdam.

Willis, W. D., 1989, Neural mechanisms of pain discrimination. In: *Sensory Processing in the Mammalian Brain*, pp. 130–143. J. S. Lund, (ed.). Oxford University Press, New York.

Willis, W. D. and Grossman, R. G., 1981, *Medical Neurobiology*, 3rd edition. C. V. Mosby Co., St. Louis.

Willis, W. D., Weir, M. A., Skinner, R. D. and Bryan, R. N., 1973, Differential distribution of spinal cord field potentials. *Exp. Brain Res.* **17**, 169–176.

Willis, W. D., Trevino, D. L., Coulter, J. D. and Maunz, R. A., 1974, Responses of primate spinothalamic tract neurons to natural stimulation of hindlimb. *J. Neurophysiol.* **37**, 358–372.

Willis, W. D., Maunz, R. A., Foreman, R. D. and Coulter, J. D., 1975, Static and dynamic responses of spinothalamic tract neurons to mechanical stimuli. *J. Neurophysiol.* **38**, 587–600.

Willis, W. D., Haber, L. H. and Martin, R. F., 1977, Inhibition of spinothalamic tract cells and interneurons by brain stem stimulation in the monkey. *J. Neurophysiol.* **40**, 968–981.

Willis, W. D., Leonard, R. B. and Kenshalo, D. R., Jr., 1978, Spinothalamic tract neurons in the substantia gelatinosa. *Science* **202**, 986–988.

Willis, W. D., Kenshalo, D. R., Jr. and Leonard, R. B., 1979, The cells of origin of the primate spinothalamic tract. *J. Comp. Neurol.* **188**, 543–574.

Wilson, G. and Fay, T., 1929, Two cases of chordotomy, indicating the distinct separation of pain and temperature fibers in the anterolateral aspect of the spinal cord, as well as the relative position of these fibers supplying the trunk and lower extremity. *Arch. Neurol. Psychiat.* **22**, 638–641.

Wilson, P. and Snow, P. J., 1988, Somatotopic organization of the dorsal horn in the lumbosacral enlargement of the spinal cord in the neonatal cat. *Exp. Neurol.* **101**, 428–444.

Wilson, P., Meyers, D. E. R. and Snow, P. J., 1986, The detailed somatotopic organization of the dorsal horn in the lumbosacral enlargement of the cat spinal cord. *J. Neurophysiol.* **55**, 604–617.

Windle, W. F., 1931, Neurons of the sensory type in the ventral roots of man and of other mammals. *Arch. Neurol. Psychiat.* **26**, 791–800.

Winokur, A., Manaker, S. and Kreider, M. S., 1989, TRH and TRH receptors in the spinal cord. *Ann. N.Y. Acad. Sci.* **553**, 314–324.

Winter, D. L., 1965, N. gracilis of cat. Functional organization and corticofugal effects. *J. Neurophysiol.* **28**, 48–70.

Winter, D. L., 1971, Receptor characteristics and conduction velocities in bladder afferents. *J. Psychiat. Res.* **8**, 225–235.

Witt, I. and Hensel, H., 1959, Afferente impulse aud der Exremitaetenhaut der Katze bei thermischer und mechanischer Reizung. *Pfluegers Arch.* **268**, 582–596.

Wolf, S. and Hardy, J. D., 1941, Studies on pain. Observations on pain due to local cooling and on factors involved in the "cold pressor" effect. *J. Clin. Invest.* **20**, 521–533.

Womack, M. D., MacDermott, A. B. and Jessell, T. M., 1988, Sensory transmitters regulate intracellular calcium in dorsal horn neurons. *Nature* **334**, 351–353.

Wong, V., Barrett, C. P., Donati, E. J., Eng, L. F. and Guth, L., 1983, Carbonic anhydrase activity in first-order sensory neurons in the rat. *J. Histochem. Cytochem.* **31**, 293–300.

Wong-Riley, M. T. T. and Kageyama, G. H., 1986, Localization of cytochrome oxidase in the mammalian spinal cord and dorsal root ganglia, with quantitative analysis of ventral horn cells in monkeys. *J. Comp. Neurol.* **245**, 41–61.

Woodbury, J. W. and Patton, H. D., 1952, Electrical activity of single spinal cord elements. *Cold Spring Harbor Symp. Quant. Biol.* **17**, 185–188.

Woolf, C. J., 1983, C-primary afferent fibre mediated inhibitions in the dorsal horn of the decerebrate-spinal rat. *Exp. Brain Res.* **51**, 283–290.

Woolf, C. J., 1984, A selective effect of naloxone on heterosynaptic C-fibre-mediated inhibitions in the rat dorsal horn. *Neurosci. Lett.* **45**, 169–174.

Woolf, C. J., 1987a, Central terminations of cutaneous mechanoreceptive afferents in the rat lumbar spinal cord. *J. Comp. Neurol.* **261**, 105–119.

Woolf, C. J., 1987b, Excitatory amino acids increase glycogen phosphorylase activity in the rat spinal cord. *Neurosci. Lett.* **73**, 209–214.

Woolf, C. J. and Fitzgerald, M., 1982, Do opioid peptides mediate a presynaptic control of C-fibre transmission in the rat spinal cord? *Neurosci. Lett.* **29**, 67–72.

Woolf, C. J. and Fitzgerald, M., 1983, The properties of neurones recorded in the superficial dorsal horn of the rat spinal cord. *J. Comp. Neurol.* **221**, 313–328.

Woolf, C. J. and Fitzgerald, M., 1986, Somatotopic organization of cutaneous afferent terminals and dorsal horn receptive fields in the superficial and deep laminae of the rat lumbar spinal cord. *J. Comp. Neurol.* **251**, 517–531.

Woolf, C. J. and King, A. E., 1987, Physiology and morphology of multireceptive neurons with C-afferent inputs in the deep dorsal horn of the rat lumbar spinal cord. *J. Neurophysiol.* **58**, 460–479.

Woolf, C. J. and King, A. E., 1988, Subliminal fringes and the plasticity of dorsal horn neurones receptive field properties. *Neurosci. Abstr.* **14**, 696.

Woolf, C. J. and King, A. E., 1989, Subthreshold components of the cutaneous mechanoreceptive fields of dorsal horn neurons in the rat lumbar spinal cord. *J. Neurophysiol.* **62**, 907–916.

Woolf, C. J. and Wall, P. D., 1982, Chronic peripheral nerve section diminishes the primary afferent A-fibre mediated inhibition of rat dorsal horn neurones. *Brain Res.* **242**, 77–85.

Woolf, C. J., Chong, M. S. and Rashdi, T. A., 1985, Mapping increased glycogen phosphorylase activity in dorsal root ganglia and in the spinal cord following peripheral stimuli. *J. Comp. Neurol.* **234**, 60–76.

Woudenberg, R. A., 1970, Projections of mechanoreceptive fields to cuneate-gracile and spinal trigeminal nuclear regions in sheep. *Brain Res.* **17**, 417–437.

Wright, D. M. and Roberts, M. H. T., 1978, Supersensitivity to a substance P analogue following dorsal root section. *Life Sci.* **22**, 19–24.

Wynn Parry, C. B., 1980, Pain in avulsion lesions of the brachial plexus. *Pain* **9**, 41–53.

Xian-Min, Y. and Mense, S., 1990, Somatotopical arrangement of rat spinal dorsal horn cells processing input from deep tissues. *Neurosci. Lett.* **108**, 43–47.

Xu, Q. and Grant, G., 1988, Collateral projections of neurons from the lower part of the spinal cord to anterior and posterior cerebellar termination areas. A retrograde fluorescent double labeling study in the cat. *Exp. Brain Res.* **72**, 562–576.

Yaksh, T. L. and Elde, R. P., 1980, Release of methionine-enkephalin immunoreactivity from the rat spinal cord in vivo. *Europ. J. Pharmacol.* **63**, 359–362.

Yaksh, T. L. and Elde, R. P., 1981, Factors governing release of methionine enkephalin-like immunoreactivity from mesencephalon and spinal cord of the cat in vivo. *J. Neurophysiol.* **46**, 1056–1075.

Yaksh, T. L. and Hammond, D. L., 1982, Peripheral and central substrates in the rostral transmission of nociceptive information. *Pain* **13**, 1–85.

Yaksh, T. L. and Noueihed, R., 1985, The physiology and pharmacology of spinal opiates. *Ann. Rev. Pharmcol. Toxicol.* **25**, 433–462.

Yaksh, T. L., Jessell, T. M., Gamse, R., Mudge, A. W. and Leeman, S. E., 1980, Intrathecal morphine inhibits substance P release from mammalian spinal cord in vivo. *Nature* **286**, 155–156.

Yaksh, T. L., Schmauss, C., Micevych, P. E., Abay, E. O. and Go, V. L. W., 1982, Pharmacological studies on the application, disposition, and release of neurotensin in the spinal cord. *Ann. N.Y. Acad. Sci.* **400**, 228–243.

Yamamoto, S. and Miyajima, M., 1961, Unit discharges recorded from dorsal portion of medulla responding to adequate exteroceptive and proprioceptive stimulation in cats. *Jap. J. Physiol.* **11**, 619–626.

Yamamoto, S. and Sugihara, S., 1956, Microelectrode studies on sensory afferents in the posterior funiculus of cat. *Jap. J. Physiol.* **6**, 68–85.

Yamamoto, T., Takahashi, K., Satomi, H. and Ise, H., 1977, Origins of primary afferent fibers in the spinal ventral roots in the cat as demonstrated by the horseradish peroxidase method. *Brain Res.* **16**, 350–354.

Yamamoto, T., Geiger, J. D., DaDonna, P. E. and Nagy, J. I., 1987, Subcellular, regional and immunohistochemical localization of adenosine deaminase in various species. *Brain Res. Bull.* **19**, 473–484.

Yashpal, K., Dam, T. V. and Quirion, R., 1990, Quantitative autoradiographic distribution of multiple neurokinin binding sites in rat spinal cord. *Brain Res.* **506**, 259–266.

Yates, B. J. and Thompson, F. J., 1985, Properties of spinal cord processing of femoral venous afferent input revealed by analysis of evoked potentials. *J. Auton. Nerv. System* **14**, 201–207.

Yezierski, R. P., 1988, Spinomesencephalic tract: projections from the lumbosacral spinal cord of the rat, cat, and monkey. *J. Comp. Neurol.* **267**, 131–146.

Yezierski, R. P., 1990, The effects of midbrain and medullary stimulation on spinomesencephalic tract cells in the cat. *J. Neurophysiol.* **63**, 240–255.

Yezierski, R. P. and Schwartz, R. H., 1986, Response and receptive-field properties of spinomesencephalic tract cells in the cat. *J. Neurophysiol.* **55**, 76–96.

Yezierski, R. P., Wilcox, T. K. and Willis, W. D., 1982, The effects of serotonin antagonists on the inhibition of primate spinothalamic tract cells produced by stimulation in nucleus raphe magnus or periaqueductal gray. *J. Pharm. Exp. Therap.* **220**, 266–277.

Yezierski, R. P., Gerhart, K. D., Schrock, B. J. and Willis, W. D., 1983, A further examination of effects of cortical stimulation on primate spinothalamic tract cells. *J. Neurophysiol.* **49**, 424–441.

Yezierski, R. P., Sorkin, L. S. and Willis, W. D., 1987, Response properties of spinal neurons projecting to midbrain or midbrain-thalamus in the monkey. *Brain Res.* **437**, 165–170.

Yezierski, R. P., Mendez, C. M. and Black, D. J., 1991, Cells of origin of the rat spinomesencephalic tract: spinal distribution and collateral projections. *Neuroscience*, in press.

Ygge, J. and Grant, G., 1983, The organization of the thoracic spinal nerve projection in the rat dorsal horn demonstrated with transganglionic transport of horseradish peroxidase. *J. Comp. Neurol.* **216**, 1–9.

Yip, H. K. and Johnson, E. M., 1984, Developing dorsal root ganglion neurons require trophic support from their central processes: evidence for a role of retrogradely transported nerve growth factor from the central nervous system to the periphery. *Proc. Natl. Acad. Sci. USA* **81**, 6245–6249.

Yip, H. K. and Johnson, E. M., 1987, Nerve growth factor receptors in rat spinal cord: an autoradiographic and immunohistochemical study. *Neuroscience* **22**, 267–279.

Yokokawa, K., Tohyama, M., Shiosaka, S., Shiotani, Y., Sonoda, T., Emson, P. C., Hillyard, C. V., Girgis, S. and MacIntyre, I., 1986, Distribution of calcitonin gene-related peptide-containing fibers in the urinary bladder of the rat and their origin. *Cell Tiss. Res.* **244**, 271–278.

Yoshida, S. and Matsuda, Y., 1979, Studies on sensory neurons of the mouse with intracellular recording and horseradish peroxidase injection techniques. *J. Neurophysiol.* **42**, 1134–1146.

Yoshida, S., Matsuda, Y. and Samejima, A., 1978, Tetrodotoxin-resistant sodium and calcium components of action potentials in dorsal root ganglion cells of the adult mouse. *J. Neurophysiol.* **41**, 1096–1106.

Yoshimura, M. and Jessell, T. M., 1989a, Primary afferent-evoked synaptic responses and slow potential generation in rat substantia gelatinosa neurons in vitro. *J. Neurophysiol.* **62**, 96–108.

Yoshimura, M. and Jessell, T. M., 1989b, Membrane properties of rat substantia gelatinosa neurons in vitro. *J. Neurophysiol.* **62**, 109–118.

Yoshimura, M. and Jessell, T., 1990, Amino acid-mediated EPSPs at primary afferent synapses with substantia gelatinosa neurones in the rat spinal cord. *J. Physiol.* **430**, 315–335.

Yoshimura, M. and North, R. A., 1983, Substantia gelatinosa neurons in vitro hyperpolarized by enkephalin. *Nature* **305**, 529–530.

Yoss, R. E., 1953, Studies of the spinal cord. Part 3. Pathways for deep pain within the spinal cord and brain. *Neurology* **3**, 163–175.

Young, W. S. and Kuhar, M. J., 1981, Neurotensin receptor localization by light microscopic autoradiography in rat brain. *Brain Res.* **206**, 273–285.

Zajac, J.-M., Lombard, M.-C., Peschanski, M., Besson, J.-M. and Roques, B. P., 1989, Autoradiographic study of μ and δ opioid binding sites and neutral endopeptidase-24.11 in rat after dorsal root rhizotomy. *Brain Res.* **477**, 400–403.

Zarbin, M. A., Wamsley, J. K. and Kuhar, M. J., 1981, Glycine receptor: light microscopic autoradiographic localization with [^{3}H]strychnine. *J. Neurosci.* **1**, 532–547.

Zarbin, M. A., Kuhar, M. J., O'Donohue, T. L., Wolf, S. S. and Moody, T. W., 1985, Autoradiographic localization of (^{125}I-TYR4) bombesin-binding sites in rat brain. *J. Neurosci.* **5**, 429–437.

Zarr, G. P., Werling, L. L., Brown, S. R. and Cox, B. M., 1986, Opioid ligand binding sites in the spinal cord of the guinea-pig. *Neuropharmacology* **25**, 471–480.

Zeehandelaar, I., 1921, Ontogenese und Phylogenese der Hinterstrangkerne in Verband mit der Sensibilitaet. *Fol. Neurobiol.* **12**, 1–133.

Zemlan, F. P., Leonard, C. M., Kow, L. M. and Pfaff, D. W., 1978, Ascending tracts of the lateral columns of the rat spinal cord: a study using the silver impregnation and horseradish peroxidase techniques. *Exp. Neurol.* **62**, 298–334.

Zemlan, F. P., Behbehani, M. M. and Beckstead, R. M., 1984, Ascending and descending projections from nucleus reticularis magnocellularis and nucleus reticularis gigantocellularis: an autoradiographic and horseradish peroxidase study in the rat. *Brain Res.* **292**, 207–220.

Zhang, D., Owens, C. M. and Willis, W. D., 1988, The mechanism of inhibition of spinothalamic tract (STT) neurons by electrical stimulation of periaqueductal gray (PAG). *Neurosci. Abstr.* **14**, 857.

Zhang, D., Carlton, S. M., Sorkin, L. S. and Willis, W. D., 1990, Collaterals of primate spinothalamic tract neurons to the periaqueductal gray. *J. Comp. Neurol.* **296**, 277–290.

Zhao, Z. Q. and Duggan, A. W., 1988, Idazoxan blocks the action of noradrenaline but not spinal inhibition from electrical stimulation of the locus coeruleus and nucleus Kolliker-Fuse of the cat. *Neuroscience* **25**, 997–1005.

Zhu, C. G., Sandri, C. and Akert, K., 1981, Morphological identification of axo-axonic and dendro-dendritic synapses in the rat substantia gelatinosa. *Brain Res.* **230**, 25–40.

Zhuo, M. and Gebhart, G. F., 1990, Spinal cholinergic and monoaminergic receptors mediate descending inhibition from the nuclei reticularis gigantocellularis and gigantocellularis pars alpha in the rat. *Brain Res.* **535**, 67–78.

Zieglgänsberger, W. and Herz, A., 1971, Changes of cutaneous receptive fields of spino-cervical-tract neurones and other dorsal column neurones by microelectrophoretically administered amino acids. *Exp. Brain Res.* **13**, 111–126.

Zieglgänsberger, W. and Puil, E. A., 1973, Actions of glutamic acid on spinal neurones. *Exp. Brain Res.* **17**, 35–49.

Zieglgänsberger, W. and Sutor, B., 1983, Responses of substantia gelatinosa neurons to putative neurotransmitters in an in vitro preparation of the adult rat spinal cord. *Brain Res.* **279**, 316–320.

Zieglgänsberger, W. and Tulloch, I. F., 1979a, Effects of substance P on neurones in the dorsal horn of the spinal cord of the cat. *Brain Res.* **166**, 273–282.

Zieglgänsberger, W. and Tulloch, I. F., 1979b, The effects of methionine- and leucine-enkephalin on spinal neurones of the cat. *Brain Res.* **167**, 53–64.

Zimmermann, M., 1968, Dorsal root potentials after C-fiber stimulation. *Science* **160**, 896–898.

Zimmermann, M., 1976, Neurophysiology of nociception. *Int. Rev. Physiol. Neurophysiol.* II, **10**, 179–221.

Zimmermann, R. J. and Stevens, J. C., 1982, Temperature-touch interactions: is there a reverse Weber phenomenon? *Bull. Psychonom. Soc.* **19**, 269–270.

Zimny, M. L., Schutte, M. and Dabezies, E., 1986, Mechanoreceptors in the human anterior cruciate ligament. *Anat. Rec.* **214**, 204–209.

Zotterman, Y., 1939, Touch, pain and tickling: an electrophysiological investigation on cutaneous sensory nerves. *J. Physiol.* **95**, 1–28.

Index

DATE DUE

NOV 1 2 1993
MAR 2 3 1994

DEMCO, INC. 38-2971